Problem Books in Mathematics

Series Editor
Peter Winkler, Department of Mathematics, Dartmouth College, Hanover, NH,
USA

Books in this series are devoted exclusively to problems - challenging, difficult, but accessible problems. They are intended to help at all levels - in college, in graduate school, and in the profession. Arthur Engels "Problem-Solving Strategies" is good for elementary students and Richard Guys "Unsolved Problems in Number Theory" is the classical advanced prototype. The series also features a number of successful titles that prepare students for problem-solving competitions.

Cornel Ioan Vălean

More (Almost) Impossible Integrals, Sums, and Series

A New Collection of Fiendish Problems and Surprising Solutions

 Springer

Cornel Ioan Vălean
Timiş, Romania

ISSN 0941-3502 ISSN 2197-8506 (electronic)
Problem Books in Mathematics
ISBN 978-3-031-21264-2 ISBN 978-3-031-21262-8 (eBook)
https://doi.org/10.1007/978-3-031-21262-8

Mathematics Subject Classification: 40-01, 11M06, 33B99, 26-01

This Springer imprint is published by the registered company Springer Nature Switzerland AG
The registered company address is: Gewerbestrasse 11, 6330 Cham, Switzerland

To all those that have encouraged me to keep writing and publishing my mathematics

Preface

As many of you may easily guess after reading the title of the book, the present work is the sequel to my first book, *(Almost) Impossible Integrals, Sums, and Series*, published by Springer in 2019, or to put it simply, we may view it as the second volume of it. The title *More (Almost) Impossible Integrals, Sums, and Series* comes from an old discussion with Paul Nahin, the author of the famous book *Inside Interesting Integrals*, at the end of 2015, when he suggested me to also consider the possibility of writing a second title (at that time, I didn't know how the publisher would react to my first book proposal, but we both talked about these matters with a positive thinking and expecting a good outcome). Fortunately, the course of my first book project was a good one and eventually got published.

The thing that has played a major motivational part and has given me all the necessary stamina for the continuation of the work for another book has been *Many of You, Dear Readers!*, your positive reactions I have received after the publication day of my first book, May 11, 2019. A few days after that day, I received more messages highlighting more or less directly that it would be nice if I continued writing *such books*, which, as time passed by, made all crystal clear to me that it is a good idea to go on and write a second book.

Writing a book is one thing, but *giving the proper soul* to your book, in order to conquer the hearts of your readers with the beauty of the mathematical results and calculations in it, is another thing, and so important and challenging, a point which I have always tried to carefully consider.

As my first book, *(Almost) Impossible Integrals, Sums, and Series*, the present book is dominated by a strong influence of the harmonic number world. If in my first book I focused on the harmonic series with a classical structure, in this work, special attention and treatment will be given to the atypical harmonic series, especially the ones involving harmonic numbers of the type $(H_{2n}^{(m)})^p$, $(\overline{H}_n^{(m)})^p$, where the last one is known, in simple terms, as a skew-harmonic number.

Let me now use the power of the examples, and I'll start from the famous quadratic series of Au-Yeung, $\sum_{n=1}^{\infty} \left(\dfrac{H_n}{n} \right)^2 = \dfrac{17}{4}\zeta(4)$, which I calculated in my first book title. What if I now used $H_n H_{2n}$ instead of H_n^2? And what if I also modified the denominator and used $(2n + 1)^2$ instead of n^2? You might find that such modifications almost bring us in a different world of calculations with different challenges, often difficult or very difficult challenges!

Here are the versions in closed form I just talked about:

$$\sum_{n=1}^{\infty} \frac{H_n H_{2n}}{n^2}$$

$$= \frac{13}{8}\zeta(4) + \frac{7}{2}\log(2)\zeta(3) - \log^2(2)\zeta(2) + \frac{1}{6}\log^4(2) + 4\operatorname{Li}_4\left(\frac{1}{2}\right)$$

and

$$\sum_{n=1}^{\infty} \frac{H_n H_{2n}}{(2n + 1)^2}$$

$$= \frac{1}{12}\log^4(2) - \frac{1}{2}\log^2(2)\zeta(2) + \frac{7}{8}\log(2)\zeta(3) - \frac{1}{4}\zeta(4) + 2\operatorname{Li}_4\left(\frac{1}{2}\right).$$

Very different closed forms when compared to the one of Au-Yeung series given earlier! And with some courage, we may also take a look at the more advanced versions of the series above I recently obtained in my research and that were published in **JCA** (*Journal of Classical Analysis*), this time with a weight 5 structure:

$$\sum_{n=1}^{\infty} \frac{H_n H_{2n}}{n^3}$$

$$= \frac{307}{16}\zeta(5) - \frac{1}{2}\zeta(2)\zeta(3) - 7\log^2(2)\zeta(3) + \frac{8}{3}\log^3(2)\zeta(2) - \frac{8}{15}\log^5(2)$$

$$- 16\log(2)\operatorname{Li}_4\left(\frac{1}{2}\right) - 16\operatorname{Li}_5\left(\frac{1}{2}\right)$$

and

$$\sum_{n=1}^{\infty} \frac{H_n H_{2n}}{(2n + 1)^3}$$

$$= \frac{1}{12}\log^5(2) - \frac{1}{2}\log^3(2)\zeta(2) + \frac{7}{4}\log^2(2)\zeta(3) - \frac{17}{8}\log(2)\zeta(4) + \frac{31}{128}\zeta(5)$$

$$+ 2\log(2)\operatorname{Li}_4\left(\frac{1}{2}\right).$$

And then, since we are talking about harmonic series with a weight 5 structure, it is also worth mentioning the ones with summands involving the generalized skew-harmonic numbers! Here are two splendid examples:

$$\sum_{n=1}^{\infty} \frac{H_n \overline{H}_n}{n^3}$$

$$= \frac{1}{6} \log^3(2)\zeta(2) - \frac{7}{8}\log^2(2)\zeta(3) + 4\log(2)\zeta(4) - \frac{193}{64}\zeta(5) + \frac{3}{8}\zeta(2)\zeta(3)$$

$$- \frac{1}{60}\log^5(2) + 2\operatorname{Li}_5\left(\frac{1}{2}\right)$$

and then

$$\sum_{n=1}^{\infty} \frac{H_n \overline{H}_n^{(2)}}{n^2}$$

$$= \frac{29}{64}\zeta(5) - \frac{5}{4}\zeta(2)\zeta(3) + \frac{7}{4}\log^2(2)\zeta(3) - \frac{2}{3}\log^3(2)\zeta(2) + \frac{2}{15}\log^5(2)$$

$$+ 4\log(2)\operatorname{Li}_4\left(\frac{1}{2}\right) + 4\operatorname{Li}_5\left(\frac{1}{2}\right).$$

Observe that so far the atypical harmonic series presented above as examples have been all non-alternating! Let's jump now to the atypical alternating ones! Have you ever encountered the alternating harmonic series of weight 4,

$$\sum_{n=1}^{\infty} (-1)^{n-1} \frac{H_n}{n^3}$$

$$= \frac{11}{4}\zeta(4) - \frac{7}{4}\log(2)\zeta(3) + \frac{1}{2}\log^2(2)\zeta(2) - \frac{1}{12}\log^4(2) - 2\operatorname{Li}_4\left(\frac{1}{2}\right),$$

which is also presented in my first book title? People interested in the world of harmonic series will (inevitably) meet it one day, and they might find it challenging to some extent! And if someone intends to find an elegant way to calculate it, then the difficulty will increase! An elegant solution to this alternating harmonic series, one that exploits two beta function representations, may be found in the present book. Having said that, let's imagine now we replace H_n by H_{2n} in the previous alternating harmonic series. What closed form would we obtain? Well, as we'll see later in the book, we'll (surprisingly) find that

$$\sum_{n=1}^{\infty}(-1)^{n-1}\frac{H_{2n}}{n^3}$$

$$= \frac{195}{32}\zeta(4) - \frac{35}{8}\log(2)\zeta(3) + \frac{5}{4}\log^2(2)\zeta(2) - \frac{5}{24}\log^4(2) - 5\operatorname{Li}_4\left(\frac{1}{2}\right).$$

Why *surprisingly*? Because during the calculations, one might first arrive at a closed form with the real part of a tetralogarithmic value involving a complex argument and think that there is no way to get a simpler form! However, we can get a simpler form, which is a matter discussed later in the book! My way, as you'll see, circumvents the appearance of the tetralogarithmic value with a complex argument.

Let me emphasize now that this is one of the most difficult harmonic series I have ever met and successfully calculated, counting the harmonic series of small weights (≤ 7)! In the opening of its solution section, I didn't start at random with the words, *From Agony to Ecstasy* ...! Happily, the solution I have discovered and presented in the book is mainly based on real methods, with no use of contour integration! Definitely one of the most wonderful gems of the book, which the reader should not miss, together with the related advanced series that make use of it!

And since I just said in the last sentence above ... *the related advanced series* ..., please take a look (again) at the first series above I gave as an example from the present book. *What if we considered now the alternating version of it?* That is,

$$\sum_{n=1}^{\infty}(-1)^{n-1}\frac{H_n H_{2n}}{n^2}$$

$$= 2G^2 - 2\log(2)\pi G - \frac{1}{8}\log^4(2) + \frac{3}{2}\log^2(2)\zeta(2) - \frac{21}{8}\log(2)\zeta(3) + \frac{773}{64}\zeta(4)$$

$$- 4\pi\Im\left\{\operatorname{Li}_3\left(\frac{1+i}{2}\right)\right\} - 3\operatorname{Li}_4\left(\frac{1}{2}\right).$$

A definitely (very) challenging alternating harmonic series where the previous series plays a key part, but finding a whole elegant solution is again a moment of *the art of solving mathematical problems*!

And not to be unfair at this point, let me also state one of the harmonic series alike involving the skew-harmonic number:

$$\sum_{n=1}^{\infty}(-1)^{n-1}\frac{\overline{H}_n H_{2n}}{n^2}$$

$$= \frac{5}{48}\log^4(2) - \frac{5}{8}\log^2(2)\zeta(2) + \frac{7}{2}\log(2)\zeta(3) - \frac{77}{32}\zeta(4) + \log(2)\pi G$$

$$- 2G^2 + \frac{5}{2}\operatorname{Li}_4\left(\frac{1}{2}\right).$$

If you think that the alternating version of a certain non-alternating harmonic series is *always* more difficult to calculate (and you count this fact as a rule), then wait to see later in the book how difficult to compute is the non-alternating version of the alternating harmonic series given above!

Additionally, besides the types of harmonic series that have been presented above, there are the *binoharmonic* series, which are those series whose summands contain both the binomial coefficient (often the central binomial coefficient) and various types of harmonic numbers. Here is an example from the fourth chapter, involving a weight 5 structure:

$$\sum_{n=1}^{\infty} \frac{1}{n^3 2^{2n}} \binom{2n}{n} \left(\zeta(2) - H_n^{(2)} \right)$$

$$= \frac{8}{15} \log^5(2) - \frac{4}{3} \log^3(2)\zeta(2) + 4\log^2(2)\zeta(3) + \log(2)\zeta(4) - \frac{123}{8}\zeta(5)$$

$$+ 16\log(2)\operatorname{Li}_4\left(\frac{1}{2}\right) + 16\operatorname{Li}_5\left(\frac{1}{2}\right).$$

In some places, the reader might meet sections involving the harmonic series with a classical structure because there is something special to share! How interesting does it sound to prove the following classical result:

$$\sum_{k=1}^{\infty}(-1)^{k-1}\frac{H_k}{k^{2m}} = \left(m + \frac{1}{2}\right)\eta(2m+1) - \frac{1}{2}\zeta(2m+1) - \sum_{i=1}^{m-1}\eta(2i)\zeta(2m-2i+1),$$

by mainly using series manipulations and no use of integrals? In this book, you'll find how to obtain such an approach! The generalization above is proving to be very useful in various places of the book.

In the following, I'll mention some examples of integrals appearing in the book that are strongly connected to the harmonic series, like

$$\int_0^{1/2} \frac{(\operatorname{Li}_2(x))^2}{x}\,dx$$

$$= \frac{1}{2}\log^3(2)\zeta(2) - \frac{7}{8}\log^2(2)\zeta(3) - \frac{5}{8}\log(2)\zeta(4) + \frac{27}{32}\zeta(5) + \frac{7}{8}\zeta(2)\zeta(3)$$

$$- \frac{7}{60}\log^5(2) - 2\log(2)\operatorname{Li}_4\left(\frac{1}{2}\right) - 2\operatorname{Li}_5\left(\frac{1}{2}\right),$$

which is a (very) difficult integral, and in its dedicated solution section, I'll ingeniously make a fast reduction to its corresponding alternating harmonic series of weight 5 using a Landen-type identity in the form of a series.

Another example, an exceptional one, is

$$\int_0^1 \frac{\log(1+x)\log^2(x)\operatorname{Li}_3(-x)}{x}dx = \frac{115}{64}\zeta(7) - \frac{19}{16}\zeta(2)\zeta(5),$$

where the corresponding alternating harmonic series behind the integral are from the weight 7 class.

Or we may count the fascinating integrals with parameter as

$$\int_0^1 \frac{\log^3(x)\operatorname{Li}_4(x)}{1-ax}dx = 3\frac{(\operatorname{Li}_4(a))^2}{a} - 12\zeta(4)\frac{\operatorname{Li}_4(a)}{a} - 120\zeta(2)\frac{\operatorname{Li}_6(a)}{a} + 210\frac{\operatorname{Li}_8(a)}{a},$$

amazingly beautifully related to the world of harmonic series.

Also, the following trio of integrals with parameter might deserve to be stated as remarkable examples:

$$\int_0^1 \frac{\log^m(x)\log\left(\dfrac{1+x}{2}\right)}{1-x}dx$$

$$= \frac{1}{2}(-1)^{m-1}m!\left((m+1)\zeta(m+2) - \sum_{k=0}^m \eta(k+1)\eta(m-k+1)\right)$$

then

$$\int_0^1 \frac{\log^m(x)\log\left(\dfrac{1+x^2}{2}\right)}{1-x}dx$$

$$= (-1)^{m-1}m!\left((m+1)\zeta(m+2) - \frac{1}{2^{m+2}}\sum_{k=0}^m \eta(k+1)\eta(m-k+1)\right.$$

$$\left. - \sum_{k=0}^m \beta(k+1)\beta(m-k+1)\right)$$

next

$$\int_0^1 \frac{\log^m(x)\log\left(\dfrac{1+x^2}{2}\right)}{1+x}dx$$

$$= (-1)^{m-1}m!\left((m+1)\eta(m+2) + \frac{1}{2^{m+2}}\sum_{k=0}^m \eta(k+1)\eta(m-k+1)\right)$$

$$-\sum_{k=0}^{m}\beta(k+1)\beta(m-k+1)\Bigg).$$

And the list of fascinating integrals involving/related to harmonic series continues, and you'll find such integrals stated in the first chapter of the book!

I've talked so far about the influences of the harmonic series and given various examples, but other problems in the book are out of the reach of the harmonic series (at least at first sight).

Remaining on the realm of integrals, I'll show you in the following a beautiful result involving Catalan's constant containing ideas that orbit around the properties of the inverse tangent integral, a result that according to my best knowledge seems to be new in the mathematical literature, like many of the main results in the book:

$$\int_0^{\tan(5\pi/24)}\frac{\log(1-x)}{1+x^2}dx+\int_0^{\cot(11\pi/24)}\frac{\log(1-x)}{1+x^2}dx=\log(2)\frac{\pi}{8}-\frac{2}{3}G.$$

Another new fascinating example, despite the daunting halo surrounding it, especially when we first meet it, is

$$\int_0^1\arctan\left(\sqrt{\frac{1+x^2}{x(1-x)}}\right)\frac{\log(1+x)}{x\sqrt{1+x^2}}dx=\frac{1}{2}\log^2(\sqrt{2}-1)\pi-\frac{\pi^3}{8}-3\pi\operatorname{Li}_2(1-\sqrt{2}).$$

The details of a solution will be found in the third chapter of the book where we'll see that at the heart of the solution, there lies a result that is proved both with and without harmonic series means.

The book also contains examples with those integrals that must be understood as a *Cauchy principal value*, which in this case are proving to be astounding and counterintuitive. Here is an example:

$$\text{P. V.}\int_0^1\frac{1}{x(1-x)}\log\left(\frac{2(3-\sqrt{5})x^2+(3\sqrt{5}-5)x+2}{(3-\sqrt{5})x^2+(\sqrt{5}-3)x+\sqrt{5}+1}\right)dx=\log^2(\varphi),$$

where $\varphi=\dfrac{1+\sqrt{5}}{2}$ is the golden ratio.

Then, there are those generalizations where the integrand may appear (very) sophisticated, but when we take a look at the closed form, a moment of wonderment shows up! One such example may be seen below

$$\int_0^1\frac{\operatorname{arctanh}^n(x)}{x}\Re\left\{\operatorname{Li}_{n+1}\left(\frac{1-x^2}{1+x^2}+i\frac{2x}{1+x^2}\right)\right\}dx$$

$$= \int_0^{\pi/4} \csc(x)\sec(x)\operatorname{arctanh}^n(\tan(x))\Re\big\{\operatorname{Li}_{n+1}(\cos(2x)+i\sin(2x))\big\}\mathrm{d}x$$

$$= \frac{1}{2^{3n+1}}(2^{n+1}-1)n!\zeta^2(n+1).$$

Further, the fans of multiple integrals (who I didn't forget) might enjoy double integrals with special functions like

$$\int_0^\infty \int_0^\infty \frac{(\tanh^2(x)+\tanh^2(y))\log(\tanh(x))\log(\tanh(y))\chi_2(\tanh(x)\tanh(y))}{\tanh(x)\tanh(y)}\mathrm{d}x\mathrm{d}y$$

$$= \frac{7}{11520}\pi^6.$$

Wow, how can this even be true?, just one possible reaction at the sight of it, since the closed form looks so simple, neat!

Like my first book, the present title comprises six chapters, a set of three chapters (problems, hints, and solutions) dedicated to each of the two major areas, *Integrals* and *Sums and Series*. A tiny idea on the image of the content can be formed based on the examples already given above. However, the book offers a broad panel of results, of various nature, which are not caught in the examples above: a large range of integrals with parameter (some of them in a trigonometric form), often leading to closed forms containing a mix of elementary and special functions; integrals with parameter (again, some of them in a trigonometric form) expressible in terms of diverse types of harmonic numbers; generalized integrals that can be immediately turned into generalized harmonic series; curious Cauchy principal value integrals; fascinating integrals related to the exponential integrals; integrals with a structure of the integrand involving the Fresnel integrals; special sums of integrals with logarithms and the inverse tangent function leading to surprising closed forms; out-of-order integrals with radicals, logarithms, and the inverse tangent function; integrals with logarithms and polylogarithms strongly related to the harmonic series; intriguing double integrals involving special functions (e.g., the Legendre chi function of order 2, the inverse tangent integral), and last but not least, we won't miss a few good results by Ramanujan in one of the sections, or a good (and awkward) integral involving the fractional part, together with other results besides them, which are all part of the *Integrals* area. Further, if we refer to the *Sums and Series* area, then I can also enumerate interesting and useful finite sums involving the central binomial coefficient; series with the central binomial coefficient; lots of Cauchy products, some of them less known or (possibly) not known in the mathematical literature; many generating functions involving the generalized harmonic numbers and the generalized skew-harmonic numbers; atypical harmonic series of weights 4 and 5; generalized atypical harmonic series; two (very) nice sets of atypical harmonic series involving the tail of the Riemann zeta function, one set involving the weight 4 type and the other one with the weight 5 type; the practical Fourier series expansions of the Bernoulli polynomials derived by simple, elegant means; sets of powerful good-to-know Fourier series of functions involving logarithms, polylogarithms, and trigonometric functions, where again most of them you might find less known or not known (in which case, possibly because they didn't appear before in the mathematical literature); fantastic binoharmonic series; special and very useful polylogarithmic values involving a complex argument where *the icing on the cake* is represented by the real part of some tetralogarithmic values, and (very) elegant derivations of their forms to simpler constants are presented; and, of course, other results belonging to the *Sums and Series* area you'll discover on your own.

From the perspective of a solver, there are always at least two major ways to look at the results (problems) in the book: one way is to find a solution, no matter how complicated the steps are, how beautiful it is, or how long it is, and then there are those ways that look like a *magical* moment which makes you think that they are too beautiful to be true, emanating much amazement around! My focus and struggle have permanently been on discovering ways of the latter type, all the time a hard task to accomplish, which could be viewed as part of *the art of solving mathematical problems*, but how much I have succeeded in my efforts I leave to you to decide.

Most of the selected problems are the fruit of my personal research in the area of integrals, sums, and series, arriving at them naturally during the course of my work, and as regards the solutions, a great focus has been put to provide original, unique ways, always when this was possible, when I managed to discover such ways. So, let me be confident and make a guess that during the present work, you'll find lots of wonderful problems and ingenious solutions that likely you have never met before. I strongly hope you'll benefit from them and you'll enjoy them!

One fact we often observe when working with the harmonic series is that in the process of calculating them, we usually want to first reduce them to simpler, easier harmonic series, with a classical structure, and this is the case for the major part of the harmonic series presented in the book, which also means that sometimes it is possible that we need results with harmonic series that are not found with solutions in the present title, but in *(Almost) Impossible Integrals, Sums, and Series*. In such cases, to preserve a good comfort for the readers, I usually proceeded as follows: if a certain result is also reducing to some series that do not appear with solutions in the present book, but only in my first book title, at the needed places, I considered writing the harmonic series together with their closed forms in order to have a clear picture of the auxiliary result(s) and, of course, together with the needed reference(s). Therefore, the reader can follow the whole solution to that problem, having in front all the needed results, and return at a later time at the given references for more information, if necessary. This is an unavoidable part I've wanted to highlight and make you aware of it since many of the problems in the book are related to harmonic series, or it is simply about calculating some harmonic series. The same strategy I've also considered for other results that are needed at some point and that only appeared in my previous title.

From the above, it is clear that the reader will benefit from having both book titles at hand, but at the same time, I would like to strongly make it clear that this necessity, as previously explained, has come about in a very natural way.

Regarding my brave approaches, as in my first book title, which are not meant to be fit in the confines of *rigor*, as the professional mathematicians would expect, some comfort might be found in the words of Bruce C. Berndt, more precisely in the final part of the book *Ramanujan's Notebooks, Part I*:

It is rather remarkable that Ramanujan's formulas are almost invariably correct, even though his methods were generally without a sound theoretical foundation. His amazing insights enabled him to determine when his formal arguments led to bona fide formulas and when they did not. Perhaps Ramanujan's work contains a message for contemporary mathematicians. We might allow our untamed, formal arguments more freedom to roam without worrying about how to return home in order to find new paths to the other side of the mountain.

In the following, I'll mention some of the contemporary sources I'm familiar with and that would help to go through before being exposed to the *orcs* under the form of integrals, sums, and series presented both in this book and in *(Almost) Impossible Integrals, Sums, and Series*. Now, let me begin in a chronological order, and I'll first mention the great book *Series Associated with the Zeta and Related Functions* by Hari Mohan Srivastava and Junesang Choi, published in 2001. Then, I'll bring up the already famous book *Irresistible Integrals* by Victor Moll and George Boros, which appeared in 2004. Almost a decade later, in 2013, another interesting book was published, called *Limits, Series, and Fractional Part Integrals* by Ovidiu Furdui. In 2014, the fantastic title *Inside Interesting Integrals* by Paul Nahin made an entry in the mathematical community, and in 2020, Paul finalized and published a second edition of the mentioned book. Next, we return to Victor Moll who published another two precious books, *Special Integrals of Gradshteyn and Ryzhik: The Proofs—Vol. I* in 2014 and *Special Integrals of Gradshteyn and Ryzhik: The Proofs—Vol. II* in 2015. All the mathematical work written by Victor Moll I have read so far has been enjoyable and instructive! Then, we have another title, released in 2018, *How to Integrate It: A Practical Guide to Finding Elementary Integrals* by Seán M. Stewart, which is a great book for newcomers in the realm of integration, further we find the 2021 enjoyable title, *An Introduction To The Harmonic Series And Logarithmic Integrals: For High School Students Up To Researchers* by Ali Shadhar Olaikhan who is showing much interest and passion to the world of harmonic series, and last but not least, we return to Paul Nahin, and I'll mention his excellent title, *In Pursuit of Zeta-3: The World's Most Mysterious Unsolved Math Problem*, published at the end of 2021. Apart from these book titles, I would emphasize the interesting large article *On logarithmic integrals, harmonic sums and variations* by Ming Hao Zhao with a great focus on the world of harmonic series. Surely, going over the whole reference table, the curious reader will find more helpful resources to consider (it's good not to skip them!).

The book is ready to *embrace* people with a good knowledge of calculus (from self-taught people to researchers), and it might serve as a resource for various purposes: entertaining with challenging mathematical problems involving integrals, sums, and series, practice for those who are trying to become better at solving problems like the ones featured in various journals such as *The American Mathematical Monthly*, as part of the preparation for exams, or for the explorers of infinity who are always looking for new perspectives, new lighthouses to use in their odyssey to the unknown of the mathematical universe.

At last, I would underline that *one of the strongest points of the book*, maybe the strongest, to put it in a few words, is that the present title shows you simple, (almost) elementary techniques for solving advanced, sometimes highly non-trivial problems, which, if they existed in the mathematical literature, you would probably find in specialized papers featuring sophisticated methods. The book, by the techniques presented, offers *accessibility* to such fascinating results for anyone that shares a passion for these calculations and wants to dare for more, without being necessary to consider (very) advanced techniques of solving the problems. At the same time, it is a marvellous moment to witness how powerful various simple techniques can be!

I would like to thank all people I have met so far and encouraged me to continue writing and publishing my mathematics, and a very special place will ever be occupied by Paul Nahin who cared about my work, trusted me, and put me in contact with Springer! Many special thanks to all of you! I'm very grateful for every supportive word from you!

I'm very thankful to Springer, New York; to my editor Dr. Sam Harrison, who decided to say *Yes* again to my second book project and to whom I had all the time a perfect communication and understanding on all the matters to be discussed; to my project coordinator Raghavendra Mohan and project manager Vinesh Velayudham for being tremendously helpful and great communicators; and to all their colleagues for the magnificent work they have done during the publishing process of the book. I feel blessed that I worked with you!

Also, I'm grateful for meeting in the recent years skilled people with an interest in the calculations of integrals and series. One of them is Khalaf Al-Ruhemi (Jerash, Jordan), who, like me, shows much interest in finding atypical, non-obvious ways of solving problems, and he does it very well. Then, I'll mention Moti Levy (Rehovot, Israel), who contributed numerous ingenious problems and solutions to various journals. We also published together a nice paper on some special Euler sums, which you'll find out more about later in the book. Next, another name who has displayed much talent at evaluating integrals and series is Félix Próspero Marin (Venezuela). During the period of writing the book, I also had a chance to meet electronically the renowned Larry Glasser (Potsdam, New York) when I sent him a solution to a wonderful problem he proposed in *La Gaceta de la RSME* (which you'll also find in the present book), and I want thank him for his kindness and the fruitful exchange of information we had. Further, I mention Kam Cheong Au (Hong Kong) who has managed to develop a powerful package for *Mathematica* (periodically updated) which will allow you to promptly find the closed forms of lots of difficult integrals. Good to be aware of it and use it, in particular when checking your work! At the moment of writing the book, the package can be found for free on his ResearchGate account as he told me. In closing, I'll mention Shivam Sharma (India), who also submitted more problems to *The American Mathematical Monthly*, and Sujeethan Balendran (Sri Lanka), who proposed more challenging problems in the *Romanian Mathematical Magazine*.

Many thanks go to Daniel Sitaru, the editor-in-chief of the *Romanian Mathematical Magazine*, who has always promoted and published my work in the form of problems and solutions. Some of my beautiful proposals there, which have never been solved by other solvers, you'll find in the present title.

I express much appreciation to Şodincă family, Adrian and Aurora, for the great real friendship they have always shown to me! (Adrian Şodincă and I were work colleagues for some years.)

A wave of greetings and thanks also goes to my friends in England, the Putineanu family, Andrea and Ionatan, with whom I often had a great time.

Finally, I thank my parents, Ileana Ursachi and Ionel Vălean, for always being on my side, supporting my options, and fully trusting me. My father didn't have a chance to see any of the two titles since he died of cancer at the end of 2017, but he knew I would succeed with my plans of publishing my mathematics.

I finalized the book at the end of August 2022 when I also submitted the full manuscript to Springer.

I'll put an end to the present Preface by inviting you to continue the journey started in my previous book and explore further the stunning world of integrals, sums, and series and have a great time!

Timiş County, Romania   Cornel Ioan Vălean
September 2022

Contents

3   Solutions...   89

Chapter 1
Integrals

Do everything in love.—1 Corinthians 16:14

1.1 A Beautiful Integral by the English Mathematician James Joseph Sylvester

Prove that

$$\int_0^{\pi/2} \frac{\log(\sin(x))}{\cos^2(x) + y^2 \sin^2(x)}dx = \int_0^{\pi/2} \frac{\log(\cos(x))}{\sin^2(x) + y^2 \cos^2(x)}dx$$

$$= -\frac{\pi}{2}\frac{\log(1+y)}{y}, \ y > 0. \tag{1.1}$$

A first form based on the main integral:

$$\int_0^{\pi/2} \frac{\log(\sin(x))}{1 + y^2 \cos^2(x)}dx = \int_0^{\pi/2} \frac{\log(\cos(x))}{1 + y^2 \sin^2(x)}dx$$

$$= -\frac{\pi}{2}\frac{1}{\sqrt{1+y^2}} \log\left(\frac{1 + \sqrt{1+y^2}}{\sqrt{1+y^2}}\right), \ y \in \mathbb{R}. \tag{1.2}$$

A second form based on the main integral:

$$\int_0^{\pi/2} \frac{\log(\sin(x))}{1 - y^2 \sin^2(x)}dx = \int_0^{\pi/2} \frac{\log(\cos(x))}{1 - y^2 \cos^2(x)}dx$$

(continued)

© The Author(s), under exclusive license to Springer Nature Switzerland AG 2023
C. I. Vălean, *More (Almost) Impossible Integrals, Sums, and Series*, Problem
Books in Mathematics, https://doi.org/10.1007/978-3-031-21262-8_1

$$= -\frac{\pi}{2} \frac{\log\left(1 + \sqrt{1 - y^2}\right)}{\sqrt{1 - y^2}}, \quad |y| < 1. \tag{1.3}$$

A challenging question: How would we prove the form in (1.3) by using results involving harmonic series?

1.2 Strange Limits with Trigonometric Integrals

Let $a > 0$ be a real number. Calculate

$$(i) \lim_{a \to 0} \int_a^\infty \frac{\sin(b_1 x) \sin(b_2 x) \cdots \sin(b_n x)}{x} (\log(a + x) - \log(x))^d \mathrm{d}x,$$

$$\tag{1.4}$$

by using high school knowledge.
   Exploiting (i), calculate

$$(ii) \lim_{a \to 0} \int_a^\infty \frac{\cos(b_1 x) \cos(b_2 x) \cdots \cos(b_n x)}{x} (\log(a + x) - \log(x)) \mathrm{d}x,$$

$$\tag{1.5}$$

where $d > 0$, $b_1, b_2, ..., b_n, d \in \mathbb{R}$.

1.3 Two Curious Logarithmic Integrals with Parameter

Let $a > 0$ be a real number. Show that

$$(i) \int_0^1 \frac{x \log(1 - x)}{1 + ax^2} \mathrm{d}x = -\frac{1}{2a} \arcsin^2\left(\sqrt{\frac{a}{a + 1}}\right) - \frac{1}{4a} \mathrm{Li}_2\left(\frac{a}{a + 1}\right)$$

$$= -\frac{1}{2a} \arctan^2(\sqrt{a}) - \frac{1}{4a} \mathrm{Li}_2\left(\frac{a}{a + 1}\right); \tag{1.6}$$

(continued)

(ii) $\displaystyle\int_0^1 \frac{x\log(1+x)}{1+ax^2}dx = \frac{1}{2a}\arcsin^2\left(\sqrt{\frac{a}{a+1}}\right) - \frac{1}{4a}\operatorname{Li}_2\left(\frac{a}{a+1}\right)$

$$= \frac{1}{2a}\arctan^2(\sqrt{a}) - \frac{1}{4a}\operatorname{Li}_2\left(\frac{a}{a+1}\right), \qquad (1.7)$$

where Li_2 denotes the dilogarithm function.

An appealing relation with the main integrals:
Show that

$$(iii)\ \frac{1}{a}\int_0^1 \frac{x\log(1+x)}{a+x^2}dx + \int_0^1 \frac{x\log(1-x)}{1+ax^2}dx$$

$$= \frac{\pi^2}{12}\frac{1}{a} - \frac{\pi}{2}\frac{\arctan(\sqrt{a})}{a} - \frac{1}{4}\frac{\log(a)\log(1+a)}{a} + \frac{1}{4}\frac{\log^2(1+a)}{a}. \qquad (1.8)$$

A challenging question: Prove the results at the points (i) and (ii) without using differentiation under the integral sign or complex numbers.

1.4   Exploring More Appealing Logarithmic Integrals with Parameter: The First Part

Prove that

$$(i)\ \int_0^1 \frac{x\log(1-ax)}{1+a^2x^2}dx$$

$$= \frac{1}{4a^2}\left(\frac{1}{2}\log^2\left(1+a^2\right) - 2\operatorname{Li}_2(a) + \frac{1}{2}\operatorname{Li}_2\left(-a^2\right) + \operatorname{Li}_2\left(\frac{2a}{1+a^2}\right)\right),$$
$$(1.9)$$

$a \le 1,\ a \ne 0,\ a \in \mathbb{R}$.

The point (ii) is an immediate consequence of the result from the point (i):

$$(ii)\ \int_0^1 \frac{x\log(1+ax)}{1+a^2x^2}dx$$

(continued)

$$= \frac{1}{4a^2}\left(\frac{1}{2}\log^2\left(1+a^2\right) - 2\operatorname{Li}_2(-a) + \frac{1}{2}\operatorname{Li}_2\left(-a^2\right) + \operatorname{Li}_2\left(\frac{-2a}{1+a^2}\right)\right),$$

(1.10)

$-1 \le a,\ a \ne 0,\ a \in \mathbb{R};$

$$(iii)\ \int_0^1 \frac{x\log(1-ax)}{1+ax^2}dx$$

$$= \frac{1}{2a}\left(\frac{1}{4}\log^2(1+a) + \arctan^2(\sqrt{a}) - \operatorname{Li}_2(a)\right),\ 0 < a \le 1,\ a \in \mathbb{R};$$

(1.11)

$$(iv)\ \int_0^1 \frac{x\log(1+ax)}{1+ax^2}dx$$

$$= \frac{1}{2a}\left(\frac{1}{4}\log^2(1+a) - \arctan^2(\sqrt{a}) - \operatorname{Li}_2(-a)\right),\ 0 < a,\ a \in \mathbb{R}.$$

(1.12)

A special identity with three dilogarithms:
Prove that

$$(v)\ \operatorname{Li}_2\left(2\sqrt{2}-3\right) + 2\operatorname{Li}_2\left(-\frac{1}{\sqrt{2}}\right) - 4\operatorname{Li}_2\left(1-\sqrt{2}\right)$$

$$= \frac{\pi^2}{24} - \log^2\left(2-\sqrt{2}\right),$$

(1.13)

where Li_2 represents the dilogarithm function.

1.5 Exploring More Appealing Logarithmic Integrals with Parameter: The Second Part

Prove that

$$(i)\ \int_0^1 \frac{x\log(1-ax)}{1+x^2}dx$$

$$= \frac{\pi}{4}\arctan(a) + \frac{1}{4}\log(2)\log\left(1+a^2\right) - \frac{1}{8}\log^2\left(1+a^2\right) - \frac{1}{2}\operatorname{Li}_2(a)$$

(continued)

$$-\frac{1}{8}\operatorname{Li}_2\left(-a^2\right)-\frac{1}{4}\operatorname{Li}_2\left(\frac{2a}{1+a^2}\right),\ a\le 1,\ a\in\mathbb{R}.\qquad(1.14)$$

The point (ii) *is an immediate consequence of the result from the point* (i):

$$(ii)\ \int_0^1\frac{x\log(1+ax)}{1+x^2}dx$$

$$=\frac{1}{4}\log(2)\log\left(1+a^2\right)-\frac{\pi}{4}\arctan(a)-\frac{1}{8}\log^2\left(1+a^2\right)-\frac{1}{2}\operatorname{Li}_2(-a)$$

$$-\frac{1}{8}\operatorname{Li}_2\left(-a^2\right)-\frac{1}{4}\operatorname{Li}_2\left(-\frac{2a}{1+a^2}\right),\ -1\le a,\ a\in\mathbb{R}.\qquad(1.15)$$

$$(iii)\ \int_0^a\frac{x\log(1-ax)}{1+x^2}dx$$

$$=\frac{1}{2}\left(\frac{1}{4}\log^2\left(1+a^2\right)+\arctan^2(a)-\operatorname{Li}_2\left(a^2\right)\right),\ |a|\le 1,\ a\in\mathbb{R};$$

$$\qquad(1.16)$$

$$(iv)\ \int_0^a\frac{x\log(1+ax)}{1+x^2}dx$$

$$=\frac{1}{2}\left(\frac{1}{4}\log^2\left(1+a^2\right)-\arctan^2(a)-\operatorname{Li}_2\left(-a^2\right)\right),\ a\in\mathbb{R},\qquad(1.17)$$

where Li_2 represents the dilogarithm function.

A challenging question: Show that if $x\in(0,1]$, then $f(x)=0$ has a single real solution

$$f(x)=\pi\arctan(x)+2\frac{\arctan^2(\sqrt{x})}{x}+\log(2)\log\left(1+x^2\right)-\frac{1}{2}\log^2\left(1+x^2\right)$$

$$-2\operatorname{Li}_2(x)-\frac{1}{2}\operatorname{Li}_2\left(-x^2\right)+\frac{1}{x}\operatorname{Li}_2\left(\frac{x}{x+1}\right)-\operatorname{Li}_2\left(\frac{2x}{1+x^2}\right).\qquad(1.18)$$

1.6 More Good-Looking Logarithmic Integrals: The First Part

Prove that

(i) $\displaystyle\int_0^1 \frac{\log(1-x)}{1+ax^2}dx = \frac{1}{2\sqrt{a}}\log(1+a)\arctan\left(\sqrt{a}\right) - \frac{1}{\sqrt{a}}\mathrm{Ti}_2\left(\sqrt{a}\right),$

$$(1.19)$$

$0 < a,\ a \in \mathbb{R};$

(ii) $\displaystyle\int_0^1 \frac{\log(1+x)}{1+ax^2}dx = \frac{\log(2)}{\sqrt{a}}\arctan\left(\sqrt{a}\right) - \frac{1}{\sqrt{a}}\mathrm{Ti}_2\left(\sqrt{a}\right) + \frac{1}{\sqrt{a}}\Im\left\{\mathrm{Li}_2\left(\frac{1}{2}+i\frac{\sqrt{a}}{2}\right)\right\},$

$$(1.20)$$

where $0 < a,\ a \in \mathbb{R}$, Li_2 represents the dilogarithm function, and $\mathrm{Ti}_2(x) = \displaystyle\int_0^x \frac{\arctan(t)}{t}dt$ denotes the inverse tangent integral.

A special value derived without well-known polylogarithmic identities:
Show that

$$(iii)\ \Im\left\{\mathrm{Li}_2\left(\frac{1+i}{2}\right)\right\} = G - \frac{1}{8}\log(2)\pi, \qquad (1.21)$$

where G is the Catalan's constant.

1.7 More Good-Looking Logarithmic Integrals: The Second Part

Show that

$$(i)\ \int_0^a \frac{\log(1-x)}{1+x^2}dx$$

$$= \arctan(a)\log(1-a) + \frac{\pi}{4}\operatorname{arctanh}(a) - \frac{1}{2}G - \frac{1}{2}\mathrm{Ti}_2(a)$$

$$+ \frac{1}{2}\mathrm{Ti}_2\left(\frac{1-a}{1+a}\right) + \frac{1}{4}\mathrm{Ti}_2\left(\frac{2a}{1-a^2}\right),\quad |a| < 1,\ a \in \mathbb{R}; \qquad (1.22)$$

(continued)

$$(ii) \quad \int_0^a \frac{\log(1+x)}{1+x^2}dx$$

$$= \arctan(a)\log(1+a) - \frac{\pi}{4}\operatorname{arctanh}(a) + \frac{1}{2}G - \frac{1}{2}\operatorname{Ti}_2(a)$$

$$- \frac{1}{2}\operatorname{Ti}_2\left(\frac{1-a}{1+a}\right) + \frac{1}{4}\operatorname{Ti}_2\left(\frac{2a}{1-a^2}\right), \ |a| < 1, \ a \in \mathbb{R}; \qquad (1.23)$$

$$(iii) \quad \int_0^1 \frac{\log(1-ax)}{1+x^2}dx$$

$$= \arctan(a)\log(1-a) - \frac{1}{2}\log(2)\arctan(a) + \frac{\pi}{8}\log(1+a^2) + \frac{\pi}{4}\operatorname{arctanh}(a)$$

$$- \frac{1}{2}G - \frac{1}{2}\operatorname{Ti}_2(a) + \frac{1}{2}\operatorname{Ti}_2\left(\frac{1-a}{1+a}\right) + \frac{1}{4}\operatorname{Ti}_2\left(\frac{2a}{1-a^2}\right), \ |a| < 1, \ a \in \mathbb{R};$$
$$(1.24)$$

$$(iv) \quad \int_0^1 \frac{\log(1+ax)}{1+x^2}dx$$

$$= \frac{1}{2}\log(2)\arctan(a) - \arctan(a)\log(1+a) + \frac{\pi}{8}\log(1+a^2) + \frac{\pi}{4}\operatorname{arctanh}(a)$$

$$- \frac{1}{2}G + \frac{1}{2}\operatorname{Ti}_2(a) + \frac{1}{2}\operatorname{Ti}_2\left(\frac{1-a}{1+a}\right) - \frac{1}{4}\operatorname{Ti}_2\left(\frac{2a}{1-a^2}\right), \ |a| < 1, \ a \in \mathbb{R},$$
$$(1.25)$$

where G represents the Catalan's constant and $\operatorname{Ti}_2(x) = \int_0^x \frac{\arctan(t)}{t}dt$ denotes the inverse tangent integral.

A *challenging question:* Prove the following curious result, involving a sum of integrals, by exploiting the integral at the point (i):

$$\int_0^{\tan(5\pi/24)} \frac{\log(1-x)}{1+x^2}dx + \int_0^{\cot(11\pi/24)} \frac{\log(1-x)}{1+x^2}dx = \log(2)\frac{\pi}{8} - \frac{2}{3}G.$$
$$(1.26)$$

1.8   More Good-Looking Logarithmic Integrals: The Third Part

Show that

$$(i) \int_0^a \frac{\log(1 - ax)}{1 + x^2} dx$$

$$= \arctan(a) \log(a) + \arctan(a) \log\left(1 - a^2\right) + \frac{\pi}{2} \log\left(1 - a^2\right)$$

$$- \frac{\pi}{4} \log\left(1 + a^2\right) - \text{Ti}_2(a) + \Im\left\{ \text{Li}_2\left(\frac{1}{1 - a^2} + i\frac{a}{1 - a^2}\right) \right\}, 0 < a < 1, \ a \in \mathbb{R}.$$

$$(1.27)$$

A special generalized logarithmic integral:

$$(ii) \int_0^a \frac{\log(1 + ax)}{1 + x^2} dx = \frac{1}{2} \arctan(a) \log\left(1 + a^2\right), \ a \in \mathbb{R}. \qquad (1.28)$$

$$(iii) \int_0^1 \frac{\log(1 - ax)}{1 + ax^2} dx$$

$$= \frac{1}{\sqrt{a}} \left(\frac{1}{2} \arctan(\sqrt{a}) \log(a) + \arctan(\sqrt{a}) \log(1 - a) + \frac{\pi}{2} \log(1 - a) \right.$$

$$- \frac{\pi}{4} \log(1 + a) - \text{Ti}_2(\sqrt{a}) + \Im\left\{ \text{Li}_2\left(\frac{1}{1 - a} + i\frac{\sqrt{a}}{1 - a}\right) \right\} \right), \ 0 < a < 1, \ a \in \mathbb{R};$$

$$(1.29)$$

$$(iv) \int_0^1 \frac{\log(1 + ax)}{1 + ax^2} dx = \frac{1}{2\sqrt{a}} \arctan(\sqrt{a}) \log(1 + a), \ 0 < a, \ a \in \mathbb{R}, $$

$$(1.30)$$

where Li_2 represents the dilogarithm function and $\text{Ti}_2(x) = \int_0^x \frac{\arctan(t)}{t} dt$ denotes the inverse tangent integral.

1.9   Special and Challenging Integrals with Parameter Involving the Inverse Hyperbolic Tangent: The First Part

Prove that

$$(i) \int_0^1 \frac{x \operatorname{arctanh}(x)}{1 + a^2 x^2} dx = \frac{1}{2} \frac{\arctan^2(a)}{a^2}, \ a \in \mathbb{R} \setminus \{0\}; \tag{1.31}$$

$$(ii) \int_0^1 \frac{x \operatorname{arctanh}(x)}{1 - a^2 x^2} dx = \frac{1}{2} \frac{\operatorname{arctanh}^2(a)}{a^2}, \ |a| < 1, \ a \in \mathbb{R} \setminus \{0\}; \tag{1.32}$$

$$(iii) \int_0^1 \frac{\operatorname{arctanh}(x)}{1 + a^2 x^2} dx$$

$$= \frac{1}{2} \log(2) \frac{\arctan(a)}{a} - \frac{1}{4} \frac{\arctan(a) \log(1 + a^2)}{a} + \frac{1}{2a} \Im \left\{ \operatorname{Li}_2 \left(\frac{1}{2} + i \frac{a}{2} \right) \right\}, \ a \in \mathbb{R} \setminus \{0\}; \tag{1.33}$$

$$(iv) \int_0^1 \frac{\operatorname{arctanh}(x)}{1 - a^2 x^2} dx$$

$$= \frac{1}{4a} \left(\log^2(2) - \frac{\pi^2}{6} - 2 \log(2) \log(1 - a) + \log^2(1 - a) \right.$$

$$\left. - 2 \operatorname{arctanh}^2(a) + 2 \operatorname{Li}_2 \left(\frac{1 + a}{2} \right) \right), \ |a| < 1, \ a \in \mathbb{R} \setminus \{0\}. \tag{1.34}$$

A marvellous integral relation with $\arctan(x)$, $\operatorname{arctanh}(x)$, *and* $\operatorname{Ti}_2(x)$:

$$(v) \int_0^1 \arctan(x) \operatorname{arctanh}(x) \log \left(1 + x^2 \right) dx$$

$$= 2 \int_0^1 \operatorname{arctanh}(x) \operatorname{Ti}_2(x) dx + \int_0^1 \frac{\arctan^2(x) \log(1 - x)}{x^2} dx, \tag{1.35}$$

where Li_2 represents the dilogarithm function and $\operatorname{Ti}_2(x) = \int_0^x \frac{\arctan(t)}{t} dt$ denotes the inverse tangent integral.

A challenging question: Prove the result at the point (i) by real methods.

1.10   Special and Challenging Integrals with Parameter Involving the Inverse Hyperbolic Tangent: The Second Part

Let n be a natural number. Prove that

$$(i) \int_0^1 \frac{x \operatorname{arctanh}^n(x)}{1 + a^2 x^2} dx$$

$$= \frac{n!}{a^2 2^n} \left((1 - 2^{-n})\zeta(n+1) + \Re \left\{ \operatorname{Li}_{n+1} \left(\frac{a^2 - 1}{a^2 + 1} + i \frac{2a}{1 + a^2} \right) \right\} \right)$$

$$= \frac{n!}{\tan^2(\theta/2)2^n} \left((1 - 2^{-n})\zeta(n+1) + \Re \left\{ \operatorname{Li}_{n+1} \left(-\cos(\theta) + i \sin(\theta) \right) \right\} \right),$$
$$(1.36)$$

where $a = \tan(\theta/2)$, $\theta \in (-\pi, \pi) \setminus \{0\}$;

$$(ii) \int_0^1 \frac{x \operatorname{arctanh}^n(x)}{1 - a^2 x^2} dx$$

$$= \frac{n!}{a^2} \left((1 - 2^n)4^{-n}\zeta(n+1) - 2^{-n-1} \operatorname{Li}_{n+1} \left(\frac{a-1}{a+1} \right) - 2^{-n-1} \operatorname{Li}_{n+1} \left(\frac{a+1}{a-1} \right) \right),$$
$$(1.37)$$

where $|a| < 1$, $a \in \mathbb{R} \setminus \{0\}$;

$$(iii) \int_0^1 \frac{\operatorname{arctanh}^n(x)}{1 + a^2 x^2} dx$$

$$= \frac{n!}{a2^n} \Im \left\{ \operatorname{Li}_{n+1} \left(\frac{a^2 - 1}{a^2 + 1} + i \frac{2a}{1 + a^2} \right) \right\}$$

$$= \frac{n!}{\tan(\theta/2)2^n} \Im \left\{ \operatorname{Li}_{n+1} \left(-\cos(\theta) + i \sin(\theta) \right) \right\},$$
$$(1.38)$$

where $a = \tan(\theta/2)$, $\theta \in (-\pi, \pi) \setminus \{0\}$;

$$(iv) \int_0^1 \frac{\operatorname{arctanh}^n(x)}{1 - a^2 x^2} dx$$

(continued)

$$= \frac{n!}{a2^{n+1}} \left(\mathrm{Li}_{n+1} \left(\frac{a-1}{a+1} \right) - \mathrm{Li}_{n+1} \left(\frac{a+1}{a-1} \right) \right), \tag{1.39}$$

where $|a| < 1$, $a \in \mathbb{R} \setminus \{0\}$.

Note that ζ is the Riemann zeta function and Li_n denotes the polylogarithm function.

1.11   Some Startling Generalizations Involving Logarithmic Integrals with Trigonometric Parameters

Let n be a natural number and $a = \tan(\theta)$, $\theta \in \left(-\frac{\pi}{2}, \frac{\pi}{2} \right)$. Show that

$$(i) \quad \int_0^1 \frac{x \log^n(1-x)}{a^2 + x^2} dx$$

$$= (-1)^n n! \Re \left\{ \mathrm{Li}_{n+1} \left(\frac{1}{1 \pm ia} \right) \right\} = (-1)^n n! \sum_{k=1}^{\infty} \frac{\cos(k\theta) \cos^k(\theta)}{k^{n+1}}; \tag{1.40}$$

$$(ii) \quad \int_0^1 \frac{\log^n(1-x)}{a^2 + x^2} dx$$

$$= (-1)^n n! \frac{1}{a} \Im \left\{ \mathrm{Li}_{n+1} \left(\frac{1}{1 - ia} \right) \right\} = (-1)^n n! \csc(\theta) \sum_{k=1}^{\infty} \frac{\sin(k\theta) \cos^{k+1}(\theta)}{k^{n+1}}, \quad a, \theta \neq 0; \tag{1.41}$$

$$(iii) \quad \int_0^1 \frac{x \log^n(1+x)}{a^2 + x^2} dx$$

$$= \sum_{k=1}^{n} (-1)^{k-1} \frac{n!}{(n-k)!} \log^{n-k}(2) \Re \left\{ \mathrm{Li}_{k+1} \left(\frac{2}{1 - ia} \right) \right\}$$

$$+ (-1)^n n! \Re \left\{ \mathrm{Li}_{n+1} \left(\frac{1}{1 - ia} \right) \right\}$$

(continued)

$$= \sum_{k=1}^{n} (-1)^{k-1} \frac{n!}{(n-k)!} \log^{n-k}(2) \Re \left\{ \mathrm{Li}_{k+1} \left(1 + e^{i2\theta} \right) \right\}$$

$$+ (-1)^n n! \sum_{k=1}^{\infty} \frac{\cos(k\theta) \cos^k(\theta)}{k^{n+1}}; \qquad (1.42)$$

$$(iv) \int_0^1 \frac{\log^n(1+x)}{a^2+x^2} dx$$

$$= 2\frac{1}{a} \arctan \left(\frac{1}{a} \right) \log^n(2) - \frac{1}{a} \sum_{k=1}^{n} (-1)^{k-1} \frac{n!}{(n-k)!} \log^{n-k}(2) \Im \left\{ \mathrm{Li}_{k+1} \left(\frac{2}{1-ia} \right) \right\}$$

$$+ (-1)^{n-1} n! \frac{1}{a} \Im \left\{ \mathrm{Li}_{n+1} \left(\frac{1}{1-ia} \right) \right\}$$

$$= (\pi \operatorname{sgn}(\theta) - 2\theta) \cot(\theta) \log^n(2)$$

$$- \cot(\theta) \sum_{k=1}^{n} (-1)^{k-1} \frac{n!}{(n-k)!} \log^{n-k}(2) \Im \left\{ \mathrm{Li}_{k+1} \left(1 + e^{i2\theta} \right) \right\}$$

$$+ (-1)^{n-1} n! \cot(\theta) \sum_{k=1}^{\infty} \frac{\sin(k\theta) \cos^k(\theta)}{k^{n+1}}, \ a, \theta \neq 0, \qquad (1.43)$$

where Li_n represents the polylogarithm function.

1.12 More Startling and Enjoyable Logarithmic Integrals Involving Trigonometric Parameters

Let $a = \tan(\theta), \theta \in \left(-\frac{\pi}{2}, \frac{\pi}{2} \right)$. Show that

$$(i) \int_0^1 \frac{x \log^2(1+x)}{a^2+x^2} dx$$

(continued)

$$= \frac{3}{4}\zeta(3) + \frac{1}{2}\operatorname{Li}_3\left(\frac{1}{1+a^2}\right) - 2\Re\left\{\operatorname{Li}_3\left(\frac{1}{1\pm ia}\right)\right\} + \Re\left\{\operatorname{Li}_3\left(\frac{1-a^2}{1+a^2} + i\frac{2a}{1+a^2}\right)\right\}$$

$$= \frac{3}{4}\zeta(3) + \frac{1}{2}\operatorname{Li}_3\left(\cos^2(\theta)\right) + \sum_{n=1}^{\infty}\frac{\cos(2n\theta)}{n^3} - 2\sum_{n=1}^{\infty}\frac{\cos(n\theta)\cos^n(\theta)}{n^3};$$

$$(1.44)$$

$$(ii) \int_0^1 \frac{x\log(1-x)\log(1+x)}{a^2+x^2}dx$$

$$= -\frac{3}{8}\zeta(3) + \frac{1}{4}\operatorname{Li}_3\left(\frac{1}{1+a^2}\right) - \frac{1}{2}\Re\left\{\operatorname{Li}_3\left(\frac{1-a^2}{1+a^2} + i\frac{2a}{1+a^2}\right)\right\}$$

$$= -\frac{3}{8}\zeta(3) + \frac{1}{4}\operatorname{Li}_3\left(\cos^2(\theta)\right) - \frac{1}{2}\sum_{n=1}^{\infty}\frac{\cos(2n\theta)}{n^3}; \qquad (1.45)$$

$$(iii) \int_0^1 \frac{x\log(1-x)\log^2(1+x)}{a^2+x^2}dx$$

$$= \frac{1}{3}\arctan^4(a) - \frac{2}{3}\pi\arctan^3(|a|) + 2\zeta(2)\arctan^2(a) - \frac{15}{8}\zeta(4)$$

$$- \frac{1}{2}\operatorname{Li}_4\left(\frac{1}{1+a^2}\right) + 2\sum_{n=1}^{\infty}\frac{1}{n^4}\left(1+a^2\right)^{-n/2}\cos(n\arctan(a))$$

$$= \frac{1}{3}\theta^4 - \frac{2}{3}\pi|\theta|^3 + 2\zeta(2)\theta^2 - \frac{15}{8}\zeta(4) - \frac{1}{2}\operatorname{Li}_4\left(\cos^2(\theta)\right) + 2\sum_{n=1}^{\infty}\frac{\cos(n\theta)\cos^n(\theta)}{n^4},$$

$$(1.46)$$

where ζ represents the Riemann zeta function and Li_n denotes the polylogarithm function.

Enchanting real parts of dilogarithmic and trilogarithmic values involving the complex argument $1 + e^{i2\theta}$:

Provided that $\theta \in \left[-\dfrac{\pi}{2}, \dfrac{\pi}{2}\right]$, show

$$(iv) \ \Re\left\{\operatorname{Li}_2\left(1+e^{i2\theta}\right)\right\} = \left(\frac{\pi}{2} - |\theta|\right)^2 = \frac{3}{2}\zeta(2) - \pi|\theta| + \theta^2; \qquad (1.47)$$

$$(v) \ \Re\left\{\operatorname{Li}_3\left(1+e^{i2\theta}\right)\right\}$$

(continued)

$$= \frac{3}{2}\log(2)\zeta(2) - \frac{3}{8}\zeta(3) - \log(2)\pi|\theta| + \log(2)\theta^2 - \frac{1}{4}\operatorname{Li}_3(\cos^2(\theta))$$

$$-\frac{1}{2}\sum_{n=1}^{\infty}\frac{\cos(2n\theta)}{n^3} + 2\sum_{n=1}^{\infty}\frac{\cos(n\theta)\cos^n(\theta)}{n^3}. \tag{1.48}$$

1.13 Surprisingly Bewitching Trigonometric Integrals with Appealing Closed Forms: The First Act

Let n be a positive integer. Find

$$(i)\ I_n = \int_0^{\pi/2} \sin(n\theta)\sin^n(\theta)\mathrm{d}\theta; \tag{1.49}$$

$$(ii)\ J_n = \int_0^{\pi/2} \sin(n\theta)\sin^{n-1}(\theta)\mathrm{d}\theta; \tag{1.50}$$

$$(iii)\ K_n = \int_0^{\pi/2} \cos(n\theta)\sin^n(\theta)\mathrm{d}\theta; \tag{1.51}$$

$$(iv)\ L_n = \int_0^{\pi/2} \cos(n\theta)\sin^{n-1}(\theta)\mathrm{d}\theta. \tag{1.52}$$

Four more integrals with a slightly atypical look:

$$(v)\ P_n = \int_0^{\pi/2} \tan\left(\frac{\theta}{2}\right)\sin(n\theta)\cos^n(\theta)\mathrm{d}\theta; \tag{1.53}$$

$$(vi)\ Q_n = \int_0^{\pi/2} \cot\left(\frac{\theta}{2}\right)\sin(n\theta)\cos^n(\theta)\mathrm{d}\theta; \tag{1.54}$$

$$(vii)\ R_n = \int_0^{\pi/2} \frac{\sin(n\theta)}{\sin(\theta)}\cos^{n+1}(\theta)\mathrm{d}\theta; \tag{1.55}$$

(continued)

$$(viii) \ S_n = \int_0^{\pi/2} \frac{\sin(n\theta)}{\sin(\theta)} \cos^n(\theta) d\theta. \tag{1.56}$$

A challenging question: May we design solutions by only exploiting high school knowledge?

1.14 Surprisingly Bewitching Trigonometric Integrals with Appealing Closed Forms: The Second Act

Show that

$$(i) \ \int_0^{\pi/2} \cos\left(\left(\frac{\pi}{2} - \theta\right) n\right) \sin^n(\theta) d\theta$$

$$= \int_0^{\pi/2} \cos(n\theta) \cos^n(\theta) d\theta = \frac{\pi}{2^{n+1}}; \tag{1.57}$$

$$(ii) \ \int_0^{\pi/2} \cos\left(\left(\frac{\pi}{2} + \theta\right) n\right) \sin^n(\theta) d\theta = (-1)^n \frac{\pi}{2^{n+1}}; \tag{1.58}$$

$$(iii) \ \int_0^{\pi/2} \cos\left(\left(\frac{\pi}{2} - \theta\right) n\right) \sin^{n-1}(\theta) d\theta$$

$$= \int_0^{\pi/2} \cos(n\theta) \cos^{n-1}(\theta) d\theta = \frac{2^n}{n\binom{2n}{n}}; \tag{1.59}$$

$$(iv) \ \int_0^{\pi/2} \cos\left(\left(\frac{\pi}{2} + \theta\right) n\right) \sin^{n-1}(\theta) d\theta = (-1)^n \frac{2^n}{n\binom{2n}{n}}; \tag{1.60}$$

$$(v) \ \int_0^{\pi/2} \sin\left(\left(\frac{\pi}{2} - \theta\right) n\right) \sin^n(\theta) d\theta$$

$$= \int_0^{\pi/2} \sin(n\theta) \cos^n(\theta) d\theta = \frac{1}{2^{n+1}} \sum_{k=1}^n \frac{2^k}{k}; \tag{1.61}$$

(continued)

(vi) $\displaystyle\int_0^{\pi/2} \sin\left(\left(\frac{\pi}{2}+\theta\right)n\right)\sin^n(\theta)\mathrm{d}\theta = (-1)^{n-1}\frac{1}{2^{n+1}}\sum_{k=1}^{n}\frac{2^k}{k};$ $\quad$ (1.62)

(vii) $\displaystyle\int_0^{\pi/2} \sin\left(\left(\frac{\pi}{2}-\theta\right)n\right)\sin^{n-1}(\theta)\mathrm{d}\theta$

$$= \int_0^{\pi/2} \sin(n\theta)\cos^{n-1}(\theta)\mathrm{d}\theta$$

$$= \frac{2^n}{n\binom{2n}{n}}\sum_{k=0}^{n-1}\frac{1}{2^k}\binom{2k}{k} = \frac{1}{2^{n-1}}\sum_{k=1}^{n}\frac{1}{2k-1}\binom{n-1}{k-1};$$ $\quad$ (1.63)

$(viii)$ $\displaystyle\int_0^{\pi/2} \sin\left(\left(\frac{\pi}{2}+\theta\right)n\right)\sin^{n-1}(\theta)\mathrm{d}\theta$

$$= (-1)^{n-1}\frac{2^n}{n\binom{2n}{n}}\sum_{k=0}^{n-1}\frac{1}{2^k}\binom{2k}{k} = (-1)^{n-1}\frac{1}{2^{n-1}}\sum_{k=1}^{n}\frac{1}{2k-1}\binom{n-1}{k-1}.$$

$$(1.64)$$

1.15 Surprisingly Bewitching Trigonometric Integrals with Appealing Closed Forms: The Third Act

Show that

(i) $\displaystyle I_n = \int_0^{\pi/2} \theta\sin(n\theta)\cos^n(\theta)\mathrm{d}\theta = \frac{\pi}{4}\frac{H_n}{2^n};$ $\quad$ (1.65)

(ii) $\displaystyle J_n = \int_0^{\pi/2} \theta^2\cos(n\theta)\cos^n(\theta)\mathrm{d}\theta = \frac{\pi^3}{24}\frac{1}{2^n} - \frac{\pi}{8}\frac{1}{2^n}\left(H_n^2 + H_n^{(2)}\right);$

$$(1.66)$$

(continued)

$$(iii) \ K_n = \int_0^{\pi/2} \theta^3 \sin(n\theta) \cos^n(\theta) d\theta$$

$$= \frac{\pi^3}{16} \frac{H_n}{2^n} - \frac{\pi}{16} \frac{1}{2^n} \left(H_n^3 + 3 H_n H_n^{(2)} + 2 H_n^{(3)} \right); \tag{1.67}$$

$$(iv) \ L_n = \int_0^{\pi/2} \theta^4 \cos(n\theta) \cos^n(\theta) d\theta$$

$$= \frac{\pi^5}{160} \frac{1}{2^n} - \frac{\pi^3}{16} \frac{1}{2^n} \left(H_n^2 + H_n^{(2)} \right)$$

$$+ \frac{\pi}{32} \frac{1}{2^n} \left(H_n^4 + 6 H_n^2 H_n^{(2)} + 8 H_n H_n^{(3)} + 3 \left(H_n^{(2)} \right)^2 + 6 H_n^{(4)} \right). \tag{1.68}$$

Similar appealing integrals which involve a logarithm:

$$(v) \int_0^{\pi/2} \cos(n\theta) \cos^n(\theta) \log(\cos(\theta)) d\theta = \frac{\pi}{4} \frac{H_n}{2^n} - \log(2) \frac{\pi}{2} \frac{1}{2^n}; \tag{1.69}$$

$$(vi) \int_0^{\pi/2} \theta \sin(n\theta) \cos^n(\theta) \log(\cos(\theta)) d\theta$$

$$= -\log(2) \frac{\pi}{4} \frac{H_n}{2^n} - \frac{\pi}{4} \frac{H_n^{(2)}}{2^n} + \frac{\pi}{8} \frac{1}{2^n} \left(H_n^2 + H_n^{(2)} \right); \tag{1.70}$$

$$(vii) \int_0^{\pi/2} \theta^2 \cos(n\theta) \cos^n(\theta) \log(\cos(\theta)) d\theta$$

$$= -\log(2) \frac{\pi^3}{24} \frac{1}{2^n} - \frac{\pi}{4} \zeta(3) \frac{1}{2^n} + \frac{\pi^3}{48} \frac{H_n}{2^n} + \frac{\pi}{4} \frac{H_n H_n^{(2)}}{2^n} + \frac{\pi}{4} \frac{H_n^{(3)}}{2^n}$$

$$+ \log(2) \frac{\pi}{8} \frac{1}{2^n} \left(H_n^2 + H_n^{(2)} \right) - \frac{\pi}{16} \frac{1}{2^n} \left(H_n^3 + 3 H_n H_n^{(2)} + 2 H_n^{(3)} \right); \tag{1.71}$$

$$(viii) \int_0^{\pi/2} \theta^3 \sin(n\theta) \cos^n(\theta) \log(\cos(\theta)) d\theta$$

$$= - \left(\frac{3}{8} \pi \zeta(3) + \log(2) \frac{\pi^3}{16} \right) \frac{H_n}{2^n} + \frac{3}{8} \pi \frac{H_n H_n^{(3)}}{2^n}$$

(continued)

$$
+ \frac{3}{16}\pi \frac{H_n^2 H_n^{(2)}}{2^n} - \frac{2}{32}\pi^3 \frac{H_n^{(2)}}{2^n} + \frac{3}{16}\pi \frac{\left(H_n^{(2)}\right)^2}{2^n} + \frac{3}{8}\pi \frac{H_n^{(4)}}{2^n}
$$

$$
+ \frac{\pi^3}{32}\frac{1}{2^n}\left(H_n^2 + H_n^{(2)}\right) + \log(2)\frac{\pi}{16}\frac{1}{2^n}\left(H_n^3 + 3H_n H_n^{(2)} + 2H_n^{(3)}\right)
$$

$$
- \frac{\pi}{32}\frac{1}{2^n}\left(H_n^4 + 6H_n^2 H_n^{(2)} + 8H_n H_n^{(3)} + 3\left(H_n^{(2)}\right)^2 + 6H_n^{(4)}\right). \tag{1.72}
$$

Similar appealing integrals which involve a squared logarithm:

$$
(ix) \int_0^{\pi/2} \cos(n\theta)\cos^n(\theta)\log^2(\cos(\theta))\mathrm{d}\theta
$$

$$
= \frac{\pi}{24}(\pi^2 + 12\log^2(2))\frac{1}{2^n} - \log(2)\frac{\pi}{2}\frac{H_n}{2^n} - \frac{\pi}{2}\frac{H_n^{(2)}}{2^n} + \frac{\pi}{8}\frac{1}{2^n}(H_n^2 + H_n^{(2)});
$$
$$\tag{1.73}$$

$$
(x) \int_0^{\pi/2} \theta\sin(n\theta)\cos^n(\theta)\log^2(\cos(\theta))\mathrm{d}\theta
$$

$$
= \frac{\pi}{48}(\pi^2 + 12\log^2(2))\frac{H_n}{2^n} - \frac{\pi}{2}\frac{H_n H_n^{(2)}}{2^n} + \log(2)\frac{\pi}{2}\frac{H_n^{(2)}}{2^n}
$$

$$
- \log(2)\frac{\pi}{4}\frac{1}{2^n}(H_n^2 + H_n^{(2)}) + \frac{\pi}{16}\frac{1}{2^n}(H_n^3 + 3H_n H_n^{(2)} + 2H_n^{(3)}); \tag{1.74}
$$

$$
(xi) \int_0^{\pi/2} \theta^2\cos(n\theta)\cos^n(\theta)\log^2(\cos(\theta))\mathrm{d}\theta
$$

$$
= \frac{\pi}{1440}(11\pi^4 + 60\log^2(2)\pi^2 + 720\log(2)\zeta(3))\frac{1}{2^n}
$$

$$
- \frac{\pi}{24}(\log(2)\pi^2 + 6\zeta(3))\frac{H_n}{2^n} - \frac{\pi}{16}\frac{H_n^4}{2^n} - \log(2)\frac{\pi}{2}\frac{H_n H_n^{(2)}}{2^n}
$$

$$
- \frac{\pi}{4}\frac{H_n H_n^{(3)}}{2^n} - \frac{\pi^3}{24}\frac{H_n^{(2)}}{2^n} - \frac{\pi}{16}\frac{(H_n^{(2)})^2}{2^n} - \log(2)\frac{\pi}{2}\frac{H_n^{(3)}}{2^n} - \frac{3}{8}\pi\frac{H_n^{(4)}}{2^n}
$$

$$
- \log^2(2)\frac{\pi}{8}\frac{1}{2^n}(H_n^2 + H_n^{(2)}) + \log(2)\frac{\pi}{8}\frac{1}{2^n}(H_n^3 + 3H_n H_n^{(2)} + 2H_n^{(3)})
$$

(continued)

$$+ \frac{\pi}{32} \frac{1}{2^n} (H_n^4 + 6H_n^2 H_n^{(2)} + 8H_n H_n^{(3)} + 3(H_n^{(2)})^2 + 6H_n^{(4)}), \qquad (1.75)$$

where $H_n^{(m)} = 1 + \frac{1}{2^m} + \cdots + \frac{1}{n^m}$, $m \geq 1$, is the nth generalized harmonic number of order m and ζ represents the Riemann zeta function.

1.16 Surprisingly Bewitching Trigonometric Integrals with Appealing Closed Forms: The Fourth Act

Show that

$$(i) \int_0^{\pi/2} \theta \cos(n\theta) \cos^n(\theta) d\theta$$

$$= \frac{3}{4} \zeta(2) \frac{1}{2^n} - \frac{1}{8} \frac{1}{2^n} (H_n^2 + H_n^{(2)}) - \frac{1}{4} \frac{1}{2^n} \int_0^1 \frac{(1 - (1+x)^n) \log(x)}{x} dx;$$

$$(1.76)$$

$$(ii) \int_0^{\pi/2} \theta^2 \sin(n\theta) \cos^n(\theta) d\theta$$

$$= \frac{3}{4} \zeta(2) \frac{H_n}{2^n} - \frac{1}{24} \frac{1}{2^n} (H_n^3 + 3H_n H_n^{(2)} + 2H_n^{(3)}) + \frac{1}{8} \frac{1}{2^n} \int_0^1 \frac{(1 - (1+x)^n) \log^2(x)}{x} dx;$$

$$(1.77)$$

$$(iii) \int_0^{\pi/2} \theta^3 \cos(n\theta) \cos^n(\theta) \, d\theta$$

$$= \frac{45}{32} \zeta(4) \frac{1}{2^n} - \frac{9}{16} \zeta(2) \frac{1}{2^n} (H_n^2 + H_n^{(2)})$$

$$+ \frac{1}{64} \frac{1}{2^n} (H_n^4 + 6H_n^2 H_n^{(2)} + 8H_n H_n^{(3)} + 3(H_n^{(2)})^2 + 6H_n^{(4)})$$

$$+ \frac{1}{16} \frac{1}{2^n} \int_0^1 \frac{(1 - (1+x)^n) \log^3(x)}{x} dx. \qquad (1.78)$$

(continued)

Similar examples of captivating integrals which involve a logarithm:

$$(iv) \int_0^{\pi/2} \theta \cos(n\theta) \cos^n(\theta) \log(\cos(\theta)) d\theta$$

$$= -\frac{1}{4}(3\log(2)\zeta(2) + \zeta(3))\frac{1}{2^n} + \frac{1}{2}\zeta(2)\frac{H_n}{2^n} + \frac{1}{4}\frac{H_n H_n^{(2)}}{2^n} + \frac{1}{4}\frac{H_n^{(3)}}{2^n}$$

$$+ \frac{1}{8}\log(2)\frac{1}{2^n}(H_n^2 + H_n^{(2)}) - \frac{1}{24}\frac{1}{2^n}(H_n^3 + 3H_n H_n^{(2)} + 2H_n^{(3)})$$

$$+ \frac{1}{4}\log(2)\frac{1}{2^n}\int_0^1 \frac{(1-(1+x)^n)\log(x)}{x}dx + \frac{1}{8}\frac{1}{2^n}\int_0^1 \frac{(1-(1+x)^n)\log^2(x)}{x}dx$$

$$+ \frac{1}{4}\frac{1}{2^n}\int_0^1 \frac{(1+x)^n \log(x)\log(1+x)}{x}dx; \qquad (1.79)$$

$$(v) \int_0^{\pi/2} \theta^2 \sin(n\theta) \cos^n(\theta) \log(\cos(\theta)) d\theta$$

$$= \frac{7}{32}\zeta(4)\frac{1}{2^n} - \frac{1}{4}(3\log(2)\zeta(2) + \zeta(3))\frac{H_n}{2^n} + \frac{1}{4}\frac{H_n H_n^{(3)}}{2^n} + \frac{1}{8}\frac{H_n^2 H_n^{(2)}}{2^n} - \frac{3}{4}\zeta(2)\frac{H_n^{(2)}}{2^n}$$

$$+ \frac{1}{8}\frac{(H_n^{(2)})^2}{2^n} + \frac{1}{4}\frac{H_n^{(4)}}{2^n} + \frac{7}{16}\zeta(2)\frac{1}{2^n}(H_n^2 + H_n^{(2)}) + \frac{1}{24}\log(2)\frac{1}{2^n}(H_n^3 + 3H_n H_n^{(2)} + 2H_n^{(3)})$$

$$- \frac{1}{64}\frac{1}{2^n}(H_n^4 + 6H_n^2 H_n^{(2)} + 8H_n H_n^{(3)} + 3(H_n^{(2)})^2 + 6H_n^{(4)})$$

$$- \frac{1}{8}\log(2)\frac{1}{2^n}\int_0^1 \frac{(1-(1+x)^n)\log^2(x)}{x}dx - \frac{1}{16}\frac{1}{2^n}\int_0^1 \frac{(1-(1+x)^n)\log^3(x)}{x}dx$$

$$- \frac{1}{8}\frac{1}{2^n}\int_0^1 \frac{(1+x)^n \log^2(x)\log(1+x)}{x}dx. \qquad (1.80)$$

A similar captivating integral which involves a squared logarithm:

$$(vi) \int_0^{\pi/2} \theta \cos(n\theta) \cos^n(\theta) \log^2(\cos(\theta)) d\theta$$

$$= \frac{1}{320}(630\zeta(4) + 160\log(2)\zeta(3) + 240\log^2(2)\zeta(2))\frac{1}{2^n} - \log(2)\zeta(2)\frac{H_n}{2^n}$$

(continued)

$$-\frac{1}{2}\log(2)\frac{H_n H_n^{(2)}}{2^n}+\frac{1}{4}\frac{H_n^2 H_n^{(2)}}{2^n}-\zeta(2)\frac{H_n^{(2)}}{2^n}-\frac{1}{2}\log(2)\frac{H_n^{(3)}}{2^n}-\frac{1}{4}\frac{H_n^{(4)}}{2^n}$$

$$-\frac{1}{96}(12\log^2(2)-30\zeta(2))\frac{1}{2^n}(H_n^2+H_n^{(2)})+\frac{1}{12}\log(2)\frac{1}{2^n}(H_n^3+3H_n H_n^{(2)}+2H_n^{(3)})$$

$$-\frac{1}{64}\frac{1}{2^n}(H_n^4+6H_n^2 H_n^{(2)}+8H_n H_n^{(3)}+3(H_n^{(2)})^2+6H_n^{(4)})$$

$$-\frac{1}{4}\log^2(2)\frac{1}{2^n}\int_0^1\frac{(1-(1+x)^n)\log(x)}{x}dx-\frac{1}{4}\log(2)\frac{1}{2^n}\int_0^1\frac{(1-(1+x)^n)\log^2(x)}{x}dx$$

$$-\frac{1}{16}\frac{1}{2^n}\int_0^1\frac{(1-(1+x)^n)\log^3(x)}{x}dx-\frac{1}{2}\log(2)\frac{1}{2^n}\int_0^1\frac{(1+x)^n\log(x)\log(1+x)}{x}dx$$

$$-\frac{1}{4}\frac{1}{2^n}\int_0^1\frac{(1+x)^n\log^2(x)\log(1+x)}{x}dx+\frac{1}{4}\frac{1}{2^n}\int_0^1\frac{(1+x)^n\log(x)\log^2(1+x)}{x}dx,$$

$$(1.81)$$

where $H_n^{(m)}=1+\frac{1}{2^m}+\cdots+\frac{1}{n^m}$, $m\geq 1$, is the nth generalized harmonic number of order m and ζ represents the Riemann zeta function.

1.17 Playing with a Hand of Fabulous Integrals: The First Part

Let $m\geq 0$ be an integer. Prove that

$$(i)\quad\int_0^{\pi/2}\sin^{2m}(x)\tan(x)\log\left(\tan\left(\frac{x}{2}\right)\right)dx$$

$$=-\frac{\pi^2}{8}+\frac{1}{2}\sum_{k=1}^m\frac{(2k-2)!!}{k\cdot(2k-1)!!}=-\frac{\pi^2}{8}+\sum_{k=1}^m\frac{2^{2k-2}}{k^2\binom{2k}{k}};\qquad(1.82)$$

$$(ii)\quad\int_0^{\pi/2}\sin^{2m+1}(x)\tan(x)\log\left(\tan\left(\frac{x}{2}\right)\right)dx$$

(continued)

$$= -\frac{\pi^2}{4} + \frac{\pi}{2} \sum_{k=0}^{m} \frac{(2k-1)!!}{(2k+1)\cdot(2k)!!} = -\frac{\pi^2}{4} + \frac{\pi}{2} \sum_{k=0}^{m} \frac{1}{(2k+1)2^{2k}} \binom{2k}{k};$$

$$(1.83)$$

$$(iii) \int_0^{\pi/2} \sin^{2m}(x) \tan(x) \log(1+\cos(x)) \, dx$$

$$= \frac{\pi^2}{12} + \frac{1}{4} H_m^{(2)} - \frac{1}{2} \sum_{k=1}^{m} \frac{(2k-2)!!}{k\cdot(2k-1)!!} = \frac{\pi^2}{12} + \frac{1}{4} H_m^{(2)} - \sum_{k=1}^{m} \frac{2^{2k-2}}{k^2 \binom{2k}{k}};$$

$$(1.84)$$

$$(iv) \int_0^{\pi/2} \sin^{2m+1}(x) \tan(x) \log(1+\cos(x)) \, dx$$

$$= \frac{\pi^2}{8} + H_{2m+1}^{(2)} - \frac{1}{4} H_m^{(2)} - \frac{\pi}{2} \sum_{k=0}^{m} \frac{(2k-1)!!}{(2k+1)\cdot(2k)!!}$$

$$= \frac{\pi^2}{8} + H_{2m+1}^{(2)} - \frac{1}{4} H_m^{(2)} - \frac{\pi}{2} \sum_{k=0}^{m} \frac{1}{(2k+1)2^{2k}} \binom{2k}{k}; \qquad (1.85)$$

$$(v) \int_0^{\pi/2} \sin^{2m}(x) \tan(x) \log(1-\cos(x)) \, dx$$

$$= -\frac{\pi^2}{6} + \frac{1}{4} H_m^{(2)} + \frac{1}{2} \sum_{k=1}^{m} \frac{(2k-2)!!}{k\cdot(2k-1)!!} = -\frac{\pi^2}{6} + \frac{1}{4} H_m^{(2)} + \sum_{k=1}^{m} \frac{2^{2k-2}}{k^2 \binom{2k}{k}};$$

$$(1.86)$$

$$(vi) \int_0^{\pi/2} \sin^{2m+1}(x) \tan(x) \log(1-\cos(x)) \, dx$$

$$= -\frac{3}{8}\pi^2 + H_{2m+1}^{(2)} - \frac{1}{4} H_m^{(2)} + \frac{\pi}{2} \sum_{k=0}^{m} \frac{(2k-1)!!}{(2k+1)\cdot(2k)!!}$$

$$= -\frac{3}{8}\pi^2 + H_{2m+1}^{(2)} - \frac{1}{4} H_m^{(2)} + \frac{\pi}{2} \sum_{k=0}^{m} \frac{1}{(2k+1)2^{2k}} \binom{2k}{k}, \qquad (1.87)$$

(continued)

$$\text{where } n!! = \begin{cases} n \cdot (n-2) \cdots 5 \cdot 3 \cdot 1, \ n > 0 \text{ odd}; \\ n \cdot (n-2) \cdots 6 \cdot 4 \cdot 2, \ n > 0 \text{ even}; \\ 1, \qquad\qquad\qquad n = -1, 0, \end{cases} \text{ is the double factorial}$$

and $H_n^{(m)} = 1 + \frac{1}{2^m} + \cdots + \frac{1}{n^m}$, $m \geq 1$, is the nth generalized harmonic number of order m.

A challenging question: Is it possible to do the calculations without using the beta function?

1.18 Playing with a Hand of Fabulous Integrals: The Second Part

Let $|y| < 1$, $y \neq 0$, be a real number. Then, prove that

(i)
$$\int_0^{\pi/2} \frac{\sin^2(x) \tan(x) \log\left(\tan\left(\frac{x}{2}\right)\right)}{1 - y^2 \sin^2(x)} dx = \frac{1}{2} \frac{\arcsin^2(y)}{y^2(1 - y^2)} - \frac{\pi^2}{8} \frac{1}{1 - y^2};$$
(1.88)

(ii)
$$\int_0^{\pi/2} \frac{\sin(x) \tan(x) \log\left(\tan\left(\frac{x}{2}\right)\right)}{1 - y^2 \sin^2(x)} dx = \frac{\pi}{2} \frac{\arcsin(y)}{y(1 - y^2)} - \frac{\pi^2}{4} \frac{1}{1 - y^2};$$
(1.89)

(iii)
$$\int_0^{\pi/2} \frac{\sin^2(x) \tan(x) \log(1 + \cos(x))}{1 - y^2 \sin^2(x)} dx$$

$$= \frac{\pi^2}{12} \frac{1}{1 - y^2} - \frac{1}{2} \frac{\arcsin^2(y)}{y^2(1 - y^2)} + \frac{1}{4} \frac{\text{Li}_2(y^2)}{y^2(1 - y^2)};$$
(1.90)

(iv)
$$\int_0^{\pi/2} \frac{\sin(x) \tan(x) \log(1 + \cos(x))}{1 - y^2 \sin^2(x)} dx$$

$$= \frac{\pi^2}{8} \frac{1}{1-y^2} - \frac{\pi}{2} \frac{\arcsin(y)}{y(1-y^2)} + \frac{1}{2} \frac{y \text{Li}_2(y)}{1-y^2} - \frac{1}{2} \frac{y \text{Li}_2(-y)}{1-y^2} + \frac{1}{2} \frac{\text{Li}_2(y)}{y} - \frac{1}{2} \frac{\text{Li}_2(-y)}{y};$$
(1.91)

(v)
$$\int_0^{\pi/2} \frac{\sin^2(x) \tan(x) \log(1 - \cos(x))}{1 - y^2 \sin^2(x)} dx$$

(continued)

$$= \frac{1}{2} \frac{\arcsin^2(y)}{y^2(1-y^2)} + \frac{1}{4} \frac{\text{Li}_2(y^2)}{y^2(1-y^2)} - \frac{\pi^2}{6} \frac{1}{1-y^2}; \tag{1.92}$$

$$(vi) \int_0^{\pi/2} \frac{\sin(x)\tan(x)\log(1-\cos(x))}{1-y^2\sin^2(x)}dx$$

$$= \frac{\pi}{2} \frac{\arcsin(y)}{y(1-y^2)} + \frac{1}{2} \frac{y\,\text{Li}_2(y)}{1-y^2} - \frac{1}{2} \frac{y\,\text{Li}_2(-y)}{1-y^2} - \frac{3}{8}\pi^2 \frac{1}{1-y^2} + \frac{1}{2} \frac{\text{Li}_2(y)}{y} - \frac{1}{2} \frac{\text{Li}_2(-y)}{y}, \tag{1.93}$$

where Li_2 represents the dilogarithm function.

1.19 Another Fabulous Integral Related to a Curious Binoharmonic Series Representation of $\zeta(3)$

Let $m \geq 0$ be an integer. Prove that

$$(i) \int_0^{\pi/2} x \sin^{2m+1}(x) \sec(x) \log(\sin(x))dx$$

$$= \frac{\pi}{4} \sum_{k=1}^m \frac{1}{k2^{2k}} \binom{2k}{k} \left(H_{2k} - H_k - \frac{1}{2k} - \log(2) \right) + \frac{\pi}{8} H_m^{(2)} + \log^2(2)\frac{\pi}{4} - \frac{\pi^3}{48}, \tag{1.94}$$

where $H_n^{(m)} = 1 + \frac{1}{2^m} + \cdots + \frac{1}{n^m}$, $m \geq 1$, is the nth generalized harmonic number of order m.

A *straightforward binoharmonic series result based on the point* (*i*): Show that

$$(ii) \log^2(2) = \sum_{n=1}^{\infty} \frac{1}{n2^{2n}} \binom{2n}{n} \left(H_n - H_{2n} + \frac{1}{2n} + \log(2) \right). \tag{1.95}$$

A *binoharmonic series representation of* $\zeta(3)$:

(continued)

Prove that

$$(iii)\ \zeta(3) = 2\sum_{n=1}^{\infty} \frac{H_n}{n}\left(\frac{1}{2^{2n}}\binom{2n}{n}\left(H_{2n} - H_n - \frac{1}{2n} - \log(2)\right) + \frac{1}{2n}\right),$$

(1.96)

by using the integral at the point (i), where ζ represents the Riemann zeta function.

1.20 Pairs of Appealing Integrals with the Logarithm Function, the Dilogarithm Function, and Trigonometric Functions

Show that

$$(i)\ I = \int_0^{\pi/2} \log(\sin(x))\,\mathrm{Li}_2(\sin^2(x))\mathrm{d}x = \int_0^{\pi/2} \log(\cos(x))\,\mathrm{Li}_2(\cos^2(x))\mathrm{d}x$$

$$= \log^3(2)\pi - \log(2)\frac{\pi^3}{6} + \frac{5}{8}\pi\zeta(3);$$

(1.97)

$$(ii)\ J = \int_0^{\pi/2} \log(\sin(x))\,\mathrm{Li}_2(\cos^2(x))\mathrm{d}x = \int_0^{\pi/2} \log(\cos(x))\,\mathrm{Li}_2(\sin^2(x))\mathrm{d}x$$

$$= \log^3(2)\pi + \log(2)\frac{\pi^3}{12} - \frac{9}{8}\pi\zeta(3),$$

(1.98)

where ζ represents the Riemann zeta function and Li_2 denotes the dilogarithm function.

1.21 Valuable Logarithmic Integrals Involving Skew-Harmonic Numbers: A First Partition

Let n be a positive integer. Prove that

$$(i)\ \int_0^1 x^{n-1}\log(1+x)\mathrm{d}x = \log(2)\frac{1}{n} + \log(2)(-1)^{n-1}\frac{1}{n} - (-1)^{n-1}\frac{\overline{H}_n}{n}.$$

(1.99)

(continued)

Two special cases of (i) based on parity:

$$(ii) \int_0^1 x^{2n-1} \log(1+x)\mathrm{d}x = \frac{H_{2n} - H_n}{2n};$$
(1.100)

$$(iii) \int_0^1 x^{2n} \log(1+x)\mathrm{d}x = 2\log(2)\frac{1}{2n+1} - \frac{1}{(2n+1)^2} + \frac{H_n}{2n+1} - \frac{H_{2n}}{2n+1}.$$
(1.101)

$$(iv) \int_0^1 x^{n-1} \log^2(1+x)\mathrm{d}x$$

$$= \log^2(2)\frac{1}{n} + \log^2(2)(-1)^{n-1}\frac{1}{n} - 2\log(2)(-1)^{n-1}\frac{H_n}{n} - 2\log(2)(-1)^{n-1}\frac{\overline{H}_n}{n}$$

$$+ 2(-1)^{n-1}\frac{H_n\overline{H}_n}{n} + 2(-1)^{n-1}\frac{\overline{H}_n^{(2)}}{n} - 2(-1)^{n-1}\frac{1}{n}\sum_{k=1}^{n}(-1)^{k-1}\frac{H_k}{k}$$

$$= \log^2(2)\frac{1}{n} + \log^2(2)(-1)^{n-1}\frac{1}{n} - 2\log(2)(-1)^{n-1}\frac{H_n}{n} - 2\log(2)(-1)^{n-1}\frac{\overline{H}_n}{n}$$

$$+ 2(-1)^{n-1}\frac{1}{n}\sum_{k=1}^{n}\frac{\overline{H}_k}{k}.$$
(1.102)

Two special cases of (iv) based on parity:

$$(v) \int_0^1 x^{2n-1} \log^2(1+x)\mathrm{d}x$$

$$= 2\log(2)\frac{H_{2n}}{n} - \log(2)\frac{H_n}{n} - \frac{H_{2n}^2}{2n} + \frac{H_n^2}{4n} - \frac{H_{2n}^{(2)}}{2n} + \frac{H_n^{(2)}}{4n} + \frac{1}{n}\sum_{k=1}^{n-1}\frac{H_k}{2k+1};$$
(1.103)

$$(vi) \int_0^1 x^{2n} \log^2(1+x)\mathrm{d}x$$

$$= 2\log^2(2)\frac{1}{2n+1} - 4\log(2)\frac{1}{(2n+1)^2} + 2\frac{1}{(2n+1)^3} + 2\log(2)\frac{H_n}{2n+1}$$

$$- 4\log(2)\frac{H_{2n}}{2n+1} + 2\frac{H_{2n}}{(2n+1)^2} - \frac{1}{2}\frac{H_n^2}{2n+1} + \frac{H_{2n}^2}{2n+1} - \frac{1}{2}\frac{H_n^{(2)}}{2n+1} + \frac{H_{2n}^{(2)}}{2n+1}$$

(continued)

$$-2\frac{1}{2n+1}\sum_{k=1}^{n}\frac{H_k}{2k+1}. \tag{1.104}$$

$$(vii) \int_0^1 x^{n-1}\log^3(1+x)dx$$

$$= \log^3(2)\frac{1+(-1)^{n-1}}{n} - 3\log^2(2)(-1)^{n-1}\frac{H_n+\overline{H}_n}{n} + 3\log(2)(-1)^{n-1}\frac{H_n^2+H_n^{(2)}}{n}$$

$$+6(-1)^{n-1}\left(\log(2)\frac{1}{n}-\frac{H_n}{n}\right)\sum_{k=1}^{n}\frac{\overline{H}_k}{k} - 6(-1)^{n-1}\frac{1}{n}\sum_{k=1}^{n}\frac{\overline{H}_k}{k^2} + 6(-1)^{n-1}\frac{1}{n}\sum_{k=1}^{n}\frac{H_k\overline{H}_k}{k}. \tag{1.105}$$

A somewhat twisted logarithmic integral:

$$(viii) \int_0^1 x^{n-1}\frac{\log(1-x)}{1+x}dx$$

$$= \frac{1}{2}\left(\log^2(2)-\frac{\pi^2}{6}\right)(-1)^{n-1} + (-1)^{n-1}\sum_{k=1}^{n-1}(-1)^{k-1}\frac{H_k}{k}$$

$$= \frac{1}{2}\left(\log^2(2)-\frac{\pi^2}{6}\right)(-1)^{n-1} - \frac{H_n}{n} + (-1)^{n-1}H_n\overline{H}_n + (-1)^{n-1}\overline{H}_n^{(2)}$$

$$-(-1)^{n-1}\sum_{k=1}^{n}\frac{\overline{H}_k}{k}. \tag{1.106}$$

Two special cases of (viii) based on parity:

$$(ix) \int_0^1 x^{2n-1}\frac{\log(1-x)}{1+x}dx$$

$$= \frac{1}{2}\left(\frac{\pi^2}{6}-\log^2(2)\right) - \frac{1}{2}\frac{H_{2n}}{n} - \frac{1}{4}H_n^2 + H_nH_{2n} - \frac{1}{2}H_{2n}^2 + \frac{1}{4}H_n^{(2)} - \frac{1}{2}H_{2n}^{(2)}$$

$$-\sum_{k=1}^{n-1}\frac{H_k}{2k+1}; \tag{1.107}$$

$$(x) \int_0^1 x^{2n}\frac{\log(1-x)}{1+x}dx$$

(continued)

$$= \frac{1}{2}\left(\log^2(2) - \frac{\pi^2}{6}\right) + \frac{1}{4}H_n^2 + \frac{1}{2}H_{2n}^2 - H_n H_{2n} - \frac{1}{4}H_n^{(2)} + \frac{1}{2}H_{2n}^{(2)} + \sum_{k=1}^{n-1}\frac{H_k}{2k+1},$$

$$(1.108)$$

where $H_n^{(m)} = 1 + \frac{1}{2^m} + \cdots + \frac{1}{n^m}$, $m \geq 1$, is the nth generalized harmonic number of order m and $\overline{H}_n^{(m)} = 1 - \frac{1}{2^m} + \cdots + (-1)^{n-1}\frac{1}{n^m}$, $m \geq 1$, denotes the nth generalized skew-harmonic number of order m.

1.22 Valuable Logarithmic Integrals Involving Skew-Harmonic Numbers: A Second Partition

Let n be a positive integer. Prove that

$$(i) \int_0^1 x^{n-1} \operatorname{arctanh}(x)dx$$

$$= \frac{1}{2}\log(2)\frac{1}{n} + \frac{1}{2}\log(2)(-1)^{n-1}\frac{1}{n} + \frac{1}{2}\frac{H_n}{n} - \frac{1}{2}(-1)^{n-1}\frac{\overline{H}_n}{n}. \qquad (1.109)$$

Two special cases of (i) based on parity:

$$(ii) \int_0^1 x^{2n-1} \operatorname{arctanh}(x)dx = \frac{1}{2}\frac{H_{2n}}{n} - \frac{1}{4}\frac{H_n}{n}; \qquad (1.110)$$

$$(iii) \int_0^1 x^{2n} \operatorname{arctanh}(x)dx = \log(2)\frac{1}{2n+1} + \frac{1}{2}\frac{H_n}{2n+1}. \qquad (1.111)$$

A version with the squared inverse hyperbolic tangent:

$$(iv) \int_0^1 x^{n-1} \operatorname{arctanh}^2(x)dx$$

$$= \frac{\pi^2}{24}\frac{1}{n} + \frac{\pi^2}{24}(-1)^{n-1}\frac{1}{n} + \frac{1}{2}\log(2)\frac{H_n}{n} - \frac{1}{2}\log(2)(-1)^{n-1}\frac{H_n}{n} + \frac{1}{4}\frac{H_n^2}{n}$$

(continued)

$$+\frac{1}{2}(-1)^{n-1}\frac{H_n\overline{H}_n}{n}+\frac{1}{2}\log(2)\frac{\overline{H}_n}{n}-\frac{1}{2}\log(2)(-1)^{n-1}\frac{\overline{H}_n}{n}-\frac{1}{4}\frac{\overline{H}_n^2}{n}+\frac{1}{2}(-1)^{n-1}\frac{\overline{H}_n^{(2)}}{n}$$

$$-(-1)^{n-1}\frac{1}{n}\sum_{k=1}^{n}(-1)^{k-1}\frac{H_k}{k}$$

$$=\frac{\pi^2}{24}\frac{1}{n}+\frac{\pi^2}{24}(-1)^{n-1}\frac{1}{n}+\frac{1}{2}\log(2)\frac{H_n}{n}-\frac{1}{2}\log(2)(-1)^{n-1}\frac{H_n}{n}+\frac{1}{4}\frac{H_n^2}{n}$$

$$-\frac{1}{2}(-1)^{n-1}\frac{H_n\overline{H}_n}{n}+\frac{1}{2}\log(2)\frac{\overline{H}_n}{n}-\frac{1}{2}\log(2)(-1)^{n-1}\frac{\overline{H}_n}{n}-\frac{1}{4}\frac{\overline{H}_n^2}{n}-\frac{1}{2}(-1)^{n-1}\frac{\overline{H}_n^{(2)}}{n}$$

$$+(-1)^{n-1}\frac{1}{n}\sum_{k=1}^{n}\frac{\overline{H}_k}{k}. \tag{1.112}$$

Two special cases of (iv) based on parity:

$$(v)\ \int_0^1 x^{2n-1}\operatorname{arctanh}^2(x)\mathrm{d}x$$

$$=\log(2)\frac{H_{2n}}{n}-\frac{1}{2}\log(2)\frac{H_n}{n}+\frac{1}{2}\frac{1}{n}\sum_{k=1}^{n-1}\frac{H_k}{2k+1}; \tag{1.113}$$

$$(vi)\ \int_0^1 x^{2n}\operatorname{arctanh}^2(x)\mathrm{d}x$$

$$=\frac{\pi^2}{12}\frac{1}{2n+1}-\frac{1}{2}\frac{H_n^2}{2n+1}+\frac{H_nH_{2n}}{2n+1}-\frac{1}{2n+1}\sum_{k=1}^{n-1}\frac{H_k}{2k+1}. \tag{1.114}$$

Three special tricky logarithmic integrals:

$$(vii)\ \int_0^1 x^{n/2-1}\log(1-x)\mathrm{d}x$$

$$=2\log(2)\frac{1}{n}+2\log(2)(-1)^{n-1}\frac{1}{n}-2\frac{H_n}{n}-2(-1)^{n-1}\frac{\overline{H}_n}{n}; \tag{1.115}$$

$$(viii)\ \int_0^1 x^{n/2-1}\log^2(1-x)\mathrm{d}x$$

$$=\left(4\log^2(2)-\frac{\pi^2}{3}\right)\frac{1}{n}+\left(4\log^2(2)-\frac{\pi^2}{3}\right)(-1)^{n-1}\frac{1}{n}-4\log(2)\frac{H_n}{n}$$

(continued)

$$-4\log(2)(-1)^{n-1}\frac{H_n}{n}+2\frac{H_n^2}{n}+4\frac{H_n^{(2)}}{n}+4(-1)^{n-1}\frac{H_n\overline{H}_n}{n}$$

$$-4\log(2)\frac{\overline{H}_n}{n}-4\log(2)(-1)^{n-1}\frac{\overline{H}_n}{n}+2\frac{\overline{H}_n^2}{n}+4(-1)^{n-1}\frac{\overline{H}_n^{(2)}}{n};\qquad(1.116)$$

$$(ix)\ \int_0^1 x^{n-1}\log(1-x)\log(1+x)\mathrm{d}x$$

$$=\frac{1}{2}\left(\log^2(2)-\frac{\pi^2}{6}\right)\frac{1}{n}+\frac{1}{2}\left(\log^2(2)-\frac{\pi^2}{6}\right)(-1)^{n-1}\frac{1}{n}-\log(2)\frac{H_n}{n}$$

$$+\frac{1}{2}\frac{H_n^{(2)}}{n}-\log(2)\frac{\overline{H}_n}{n}+\frac{1}{2}\frac{\overline{H}_n^2}{n}+(-1)^{n-1}\frac{1}{n}\sum_{k=1}^n(-1)^{k-1}\frac{H_k}{k}$$

$$=\frac{1}{2}\left(\log^2(2)-\frac{\pi^2}{6}\right)\frac{1}{n}+\frac{1}{2}\left(\log^2(2)-\frac{\pi^2}{6}\right)(-1)^{n-1}\frac{1}{n}-\log(2)\frac{H_n}{n}$$

$$+\frac{1}{2}\frac{H_n^{(2)}}{n}+(-1)^{n-1}\frac{H_n\overline{H}_n}{n}-\log(2)\frac{\overline{H}_n}{n}+\frac{1}{2}\frac{\overline{H}_n^2}{n}+(-1)^{n-1}\frac{\overline{H}_n^{(2)}}{n}-(-1)^{n-1}\frac{1}{n}\sum_{k=1}^n\frac{\overline{H}_k}{k}.$$
$$(1.117)$$

Two special cases of (ix) based on parity:

$$(x)\ \int_0^1 x^{2n-1}\log(1-x)\log(1+x)\mathrm{d}x$$

$$=\frac{1}{2}\log(2)\frac{H_n}{n}-\log(2)\frac{H_{2n}}{n}+\frac{1}{8}\frac{H_n^2}{n}+\frac{1}{8}\frac{H_n^{(2)}}{n}-\frac{1}{2}\frac{1}{n}\sum_{k=1}^{n-1}\frac{H_k}{2k+1}.\qquad(1.118)$$

$$(xi)\ \int_0^1 x^{2n}\log(1-x)\log(1+x)\mathrm{d}x$$

$$=\left(\log^2(2)-\frac{\pi^2}{6}\right)\frac{1}{2n+1}-2\log(2)\frac{1}{(2n+1)^2}+2\frac{1}{(2n+1)^3}+\log(2)\frac{H_n}{2n+1}$$

$$-2\log(2)\frac{H_{2n}}{2n+1}-\frac{H_n}{(2n+1)^2}+2\frac{H_{2n}}{(2n+1)^2}+\frac{3}{4}\frac{H_n^2}{2n+1}+\frac{H_{2n}^2}{2n+1}$$

(continued)

$$-2\frac{H_n H_{2n}}{2n+1} - \frac{1}{4}\frac{H_n^{(2)}}{2n+1} + \frac{H_{2n}^{(2)}}{2n+1} + \frac{1}{2n+1}\sum_{k=1}^{n-1}\frac{H_k}{2k+1}, \tag{1.119}$$

where $H_n^{(m)} = 1 + \frac{1}{2^m} + \cdots + \frac{1}{n^m}$, $m \geq 1$, is the nth generalized harmonic number of order m and $\overline{H}_n^{(m)} = 1 - \frac{1}{2^m} + \cdots + (-1)^{n-1}\frac{1}{n^m}$, $m \geq 1$, denotes the nth generalized skew-harmonic number of order m.

1.23 Valuable Logarithmic Integrals Involving Skew-Harmonic Numbers: A Third Partition

Let n, p be positive integers. Show and use

$$(i) \ \sum_{k=1}^{n}(-1)^{k-1}\frac{\overline{H}_k^{(p)}}{k^p} = \frac{1}{2}\left(\left(\overline{H}_n^{(p)}\right)^2 + H_n^{(2p)}\right) \tag{1.120}$$

in order to prove that

$$(ii) \ \int_0^1 \frac{1-x^n}{1-x}\log(1+x)\mathrm{d}x = \log(2)H_n + \log(2)\overline{H}_n - \frac{1}{2}\left(\overline{H}_n^2 + H_n^{(2)}\right). \tag{1.121}$$

An alternative way of presenting the result at the point (ii):

$$(iii) \ \int_0^1 x^n \frac{\log(1-x)}{1+x}\mathrm{d}x + n\int_0^1 x^{n-1}\log(1-x)\log(1+x)\mathrm{d}x$$

$$= \frac{1}{2}\left(\log^2(2) - \frac{\pi^2}{6}\right) - \log(2)H_n - \log(2)\overline{H}_n + \frac{1}{2}\left(\overline{H}_n^2 + H_n^{(2)}\right). \tag{1.122}$$

(continued)

Another (special) sum involving skew-harmonic numbers:

$$(iv) \sum_{k=1}^{n}(-1)^{k-1}\frac{\overline{H}_k^2 + H_k^{(2)}}{k} = \frac{1}{3}\left(\overline{H}_n^3 + 3\overline{H}_n H_n^{(2)} + 2\overline{H}_n^{(3)}\right).$$ (1.123)

Prove that

$$(v) \int_0^1 \sum_{k=1}^{n}(-1)^{k-1}\frac{1}{k}\frac{1-x^k}{1-x}\log(1+x)\mathrm{d}x$$

$$= \frac{1}{2}\log(2)\left(\overline{H}_n^2 + H_n^{(2)}\right) - \frac{1}{6}\left(\overline{H}_n^3 + 3\overline{H}_n H_n^{(2)} + 2\overline{H}_n^{(3)}\right) + \log(2)\sum_{k=1}^{n}(-1)^{k-1}\frac{H_k}{k}$$

$$= \log(2)H_n\overline{H}_n + \log(2)\overline{H}_n^{(2)} + \frac{1}{2}\log(2)\left(\overline{H}_n^2 + H_n^{(2)}\right) - \frac{1}{6}\left(\overline{H}_n^3 + 3\overline{H}_n H_n^{(2)} + 2\overline{H}_n^{(3)}\right)$$

$$- \log(2)\sum_{k=1}^{n}\frac{\overline{H}_k}{k},$$ (1.124)

where $H_n^{(m)} = 1 + \frac{1}{2^m} + \cdots + \frac{1}{n^m}$, $m \geq 1$, is the nth generalized harmonic number of order m and $\overline{H}_n^{(m)} = 1 - \frac{1}{2^m} + \cdots + (-1)^{n-1}\frac{1}{n^m}$, $m \geq 1$, denotes the nth generalized skew-harmonic number of order m.

1.24 A Pair of Precious Generalized Logarithmic Integrals

Let m, n be positive integers. Prove that

$$(i) \int_0^1 x^{n-1}\log^m(x)\log(1-x)\mathrm{d}x$$

$$= (-1)^{m-1}m!\frac{H_n}{n^{m+1}} - (-1)^{m-1}m!\sum_{k=1}^{m}\frac{1}{n^{m-k+1}}\left(\zeta(k+1) - H_n^{(k+1)}\right);$$ (1.125)

$$(ii) \int_0^1 x^{n-1}\log^m(x)\log^2(1-x)\mathrm{d}x$$

(continued)

$$= (-1)^{m-1} m! \sum_{i=1}^{m} \frac{1}{n^i} \left(\zeta(m-i+2) - H_n^{(m-i+2)} \right) \times \left(H_n - \sum_{j=1}^{i-1} n^j \left(\zeta(j+1) - H_n^{(j+1)} \right) \right)$$

$$+ (-1)^{m-1} m! \left(\sum_{i=1}^{m} \frac{H_n}{n^{m-i+1}} \left(\zeta(i+1) - H_n^{(i+1)} \right) + \sum_{i=1}^{m} \frac{i+1}{n^{m-i+1}} \left(\zeta(i+2) - H_n^{(i+2)} \right) \right)$$

$$- (-1)^{m-1} m! \frac{1}{n^{m+1}} (H_n^2 + H_n^{(2)}), \tag{1.126}$$

where $H_n^{(m)} = 1 + \frac{1}{2^m} + \cdots + \frac{1}{n^m}$, $m \geq 1$, is the nth generalized harmonic number of order m and ζ represents the Riemann zeta function.

Examples at the point (i):
For $m = 1$,

$$\int_0^1 x^{n-1} \log(x) \log(1-x) dx = \frac{H_n}{n^2} + \frac{H_n^{(2)}}{n} - \frac{\pi^2}{6} \frac{1}{n}.$$

For $m = 2$,

$$\int_0^1 x^{n-1} \log^2(x) \log(1-x) dx = 2\zeta(3) \frac{1}{n} + \frac{\pi^2}{3} \frac{1}{n^2} - 2 \frac{H_n}{n^3} - 2 \frac{H_n^{(2)}}{n^2} - 2 \frac{H_n^{(3)}}{n}.$$

For $m = 3$,

$$\int_0^1 x^{n-1} \log^3(x) \log(1-x) dx$$

$$= 6 \frac{H_n}{n^4} + 6 \frac{H_n^{(2)}}{n^3} + 6 \frac{H_n^{(3)}}{n^2} + 6 \frac{H_n^{(4)}}{n} - \frac{\pi^4}{15} \frac{1}{n} - 6\zeta(3) \frac{1}{n^2} - 6\zeta(2) \frac{1}{n^3}.$$

Examples at the point (ii):
For $m = 1$,

$$\int_0^1 x^{n-1} \log(x) \log^2(1-x) dx$$

$$= 2\zeta(3) \frac{1}{n} + \frac{\pi^2}{3} \frac{H_n}{n} - \frac{H_n^2}{n^2} - \frac{H_n^{(2)}}{n^2} - 2 \frac{H_n^{(3)}}{n} - 2 \frac{H_n H_n^{(2)}}{n}.$$

(continued)

For $m = 2$,

$$\int_0^1 x^{n-1} \log^2(x) \log^2(1-x) \mathrm{d}x$$

$$= 2\frac{H_n^2}{n^3} + 2\frac{H_n^{(2)}}{n^3} + 2\frac{(H_n^{(2)})^2}{n} + 4\frac{H_n^{(3)}}{n^2} + 6\frac{H_n^{(4)}}{n} + 4\frac{H_n H_n^{(2)}}{n^2} + 4\frac{H_n H_n^{(3)}}{n}$$

$$- \frac{2}{3}\pi^2 \frac{H_n^{(2)}}{n} - \frac{2}{3}\pi^2 \frac{H_n}{n^2} - 4\zeta(3)\frac{H_n}{n} - 4\zeta(3)\frac{1}{n^2} - \frac{\pi^4}{90}\frac{1}{n}.$$

1.25 Two Neat and Useful Generalizations with Logarithmic Integrals Involving Skew-Harmonic Numbers

Let m, n be positive integers. Show that

$$(i) \int_0^1 x^{n-1} \log^m(x) \log(1+x) \mathrm{d}x$$

$$= (-1)^{m+n} m! \frac{\overline{H}_n}{n^{m+1}} - \log(2)(-1)^{m-1} m! \frac{1}{n^{m+1}} - \log(2)(-1)^{m+n} m! \frac{1}{n^{m+1}}$$

$$+ (-1)^{m+n} m! \sum_{k=1}^{m} \frac{\overline{H}_n^{(k+1)}}{n^{m-k+1}} - (-1)^{m+n} m! \sum_{k=2}^{m+1} \frac{1}{n^{m-k+2}} (1 - 2^{1-k}) \zeta(k). \tag{1.127}$$

Another generalization in polygamma values, reducible to forms with (skew-)harmonic numbers as shown in the examples below:

$$(ii) \int_0^1 \frac{x^n \log^m(x) \log(1+x)}{1-x} \mathrm{d}x$$

$$= -\frac{1}{2}\psi^{(m+1)}(n+1) - 2\log(2)\psi^{(m)}(n+1) + \psi(n+1)\psi^{(m)}(n+1)$$

(continued)

$$-\psi\left(\frac{n}{2}+1\right)\psi^{(m)}(n+1)+\log(2)\frac{1}{2^m}\psi^{(m)}\left(\frac{n}{2}+1\right)-\frac{1}{2^m}\psi(n+1)\psi^{(m)}\left(\frac{n}{2}+1\right)$$

$$+\frac{1}{2^m}\psi\left(\frac{n}{2}+1\right)\psi^{(m)}\left(\frac{n}{2}+1\right)+\sum_{k=1}^{m-1}\binom{m}{k}\left(\frac{1}{2}\psi^{(k)}(n+1)\psi^{m-k}(n+1)\right.$$

$$\left.-\frac{1}{2^k}\psi^{(k)}\left(\frac{n}{2}+1\right)\psi^{(m-k)}(n+1)+\frac{1}{2^{m+1}}\psi^{(k)}\left(\frac{n}{2}+1\right)\psi^{(m-k)}\left(\frac{n}{2}+1\right)\right).$$
$$(1.128)$$

Examples at the point (i):
For $m = 1$,

$$\int_0^1 x^{n-1}\log(x)\log(1+x)\mathrm{d}x$$

$$=-\frac{\pi^2}{12}(-1)^{n-1}\frac{1}{n}-\log(2)\frac{1}{n^2}-\log(2)(-1)^{n-1}\frac{1}{n^2}+(-1)^{n-1}\frac{\overline{H}_n}{n^2}+(-1)^{n-1}\frac{\overline{H}_n^{(2)}}{n}.$$

For $m = 2$,

$$\int_0^1 x^{n-1}\log^2(x)\log(1+x)\mathrm{d}x$$

$$=2\log(2)\frac{1}{n^3}+\frac{3}{2}\zeta(3)(-1)^{n-1}\frac{1}{n}+\frac{\pi^2}{6}(-1)^{n-1}\frac{1}{n^2}+2\log(2)(-1)^{n-1}\frac{1}{n^3}$$

$$-2(-1)^{n-1}\frac{\overline{H}_n}{n^3}-2(-1)^{n-1}\frac{\overline{H}_n^{(2)}}{n^2}-2(-1)^{n-1}\frac{\overline{H}_n^{(3)}}{n}.$$

For $m = 3$,

$$\int_0^1 x^{n-1}\log^3(x)\log(1+x)\mathrm{d}x$$

$$=-6\log(2)\frac{1}{n^4}-\frac{7}{120}\pi^4(-1)^{n-1}\frac{1}{n}-\frac{9}{2}\zeta(3)(-1)^{n-1}\frac{1}{n^2}-\frac{\pi^2}{2}(-1)^{n-1}\frac{1}{n^3}$$

$$-6\log(2)(-1)^{n-1}\frac{1}{n^4}+6(-1)^{n-1}\frac{\overline{H}_n}{n^4}+6(-1)^{n-1}\frac{\overline{H}_n^{(2)}}{n^3}$$

$$+6(-1)^{n-1}\frac{\overline{H}_n^{(3)}}{n^2}+6(-1)^{n-1}\frac{\overline{H}_n^{(4)}}{n}.$$

(continued)

Examples at the point (ii):

For $m = 1$,

$$\int_0^1 \frac{x^n \log(x) \log(1+x)}{1-x} dx$$

$$= \zeta(3) - \log(2)\frac{\pi^2}{4} + \log(2)H_n^{(2)} - H_n^{(3)} + \frac{\pi^2}{12}\overline{H}_n + \log(2)\overline{H}_n^{(2)} - \overline{H}_n\overline{H}_n^{(2)}.$$

For $m = 2$,

$$\int_0^1 \frac{x^n \log^2(x) \log(1+x)}{1-x} dx$$

$$= -\frac{19}{720}\pi^4 + \frac{7}{2}\log(2)\zeta(3) - 2\log(2)H_n^{(3)} + 3H_n^{(4)} - \frac{3}{2}\zeta(3)\overline{H}_n + 2\overline{H}_n\overline{H}_n^{(3)}$$

$$- \frac{\pi^2}{6}\overline{H}_n^{(2)} + \left(\overline{H}_n^{(2)}\right)^2 - 2\log(2)\overline{H}_n^{(3)}.$$

For $m = 3$,

$$\int_0^1 \frac{x^n \log^3(x) \log(1+x)}{1-x} dx$$

$$= 12\zeta(5) - \frac{1}{8}\log(2)\pi^4 - \frac{3}{8}\pi^2\zeta(3) + 6\log(2)H_n^{(4)} - 12H_n^{(5)} + \frac{7}{120}\pi^4\overline{H}_n$$

$$- 6\overline{H}_n\overline{H}_n^{(4)} + \frac{9}{2}\zeta(3)\overline{H}_n^{(2)} - 6\overline{H}_n^{(2)}\overline{H}_n^{(3)} + \frac{\pi^2}{2}\overline{H}_n^{(3)} + 6\log(2)\overline{H}_n^{(4)},$$

where $H_n^{(m)} = 1 + \frac{1}{2^m} + \cdots + \frac{1}{n^m}$, $m \geq 1$, is the nth generalized harmonic number of order m, $\overline{H}_n^{(m)} = 1 - \frac{1}{2^m} + \cdots + (-1)^{n-1}\frac{1}{n^m}$, $m \geq 1$, denotes the nth generalized skew-harmonic number of order m, ζ represents the Riemann zeta function, and $\psi^{(n)}$ designates the polygamma function.

1.26   A Special Logarithmic Integral and a Generalization of It

Show, without using the beta function, polylogarithm function, or Euler sums, that

$$(i) \quad \int_0^1 \frac{\log(x)\log(1-x)}{1+x}dx = \frac{13}{8}\zeta(3) - \frac{3}{2}\log(2)\zeta(2). \qquad (1.129)$$

Let $n \geq 1$ be a positive integer. Prove that

$$(ii) \quad \int_0^1 \frac{\log^{2n-1}(x)\log(1-x)}{1+x}dx$$

$$= \frac{1}{2^{2n+1}}\left((n+1)2^{2n+1} - 2n - 1\right)(2n-1)!\zeta(2n+1) - \log(2)(2n-1)!\zeta(2n)$$

$$- \left(1 - \frac{1}{2^{2n-1}}\right)(2n-1)!\sum_{k=1}^{n-1}\zeta(2k)\zeta(2n-2k+1)$$

$$+ (2n-1)!\sum_{k=1}^{n-1}\eta(2k)\zeta(2n-2k+1) - (2n-1)!\sum_{k=0}^{n-1}\eta(2k+1)\eta(2n-2k),$$

$$\qquad (1.130)$$

where ζ represents the Riemann zeta function and η denotes the Dirichlet eta function.

Examples:
For $n = 2$,

$$\int_0^1 \frac{\log^3(x)\log(1-x)}{1+x}dx$$

$$= \frac{273}{16}\zeta(5) - \frac{9}{2}\zeta(2)\zeta(3) - \frac{45}{4}\log(2)\zeta(4).$$

For $n = 3$,

$$\int_0^1 \frac{\log^5(x)\log(1-x)}{1+x}dx$$

$$= \frac{7575}{16}\zeta(7) - \frac{225}{2}\zeta(2)\zeta(5) - 90\zeta(3)\zeta(4) - \frac{945}{4}\log(2)\zeta(6).$$

(continued)

For $n = 4$,

$$\int_0^1 \frac{\log^7(x)\log(1-x)}{1+x}dx$$

$$= \frac{803565}{32}\zeta(9) - \frac{19845}{4}\zeta(2)\zeta(7) - 3780\zeta(3)\zeta(6) - 4725\zeta(4)\zeta(5)$$

$$- \frac{80325}{8}\log(2)\zeta(8).$$

1.27   Three Useful Generalized Logarithmic Integrals Connected to Useful Generalized Alternating Harmonic Series

Let m be a non-negative integer. Prove that

$$(i) \int_0^1 \frac{\log^m(x)\log\left(\dfrac{1+x}{2}\right)}{1-x}dx$$

$$= \frac{1}{2}(-1)^{m-1}m!\left((m+1)\zeta(m+2) - \sum_{k=0}^m \eta(k+1)\eta(m-k+1)\right); \quad (1.131)$$

$$(ii) \int_0^1 \frac{\log^m(x)\log\left(\dfrac{1+x^2}{2}\right)}{1-x}dx$$

$$= (-1)^{m-1}m!\left((m+1)\zeta(m+2) - \frac{1}{2^{m+2}}\sum_{k=0}^m \eta(k+1)\eta(m-k+1)\right.$$

$$\left. - \sum_{k=0}^m \beta(k+1)\beta(m-k+1)\right); \quad (1.132)$$

$$(iii) \int_0^1 \frac{\log^m(x)\log\left(\dfrac{1+x^2}{2}\right)}{1+x}dx$$

(continued)

$$= (-1)^{m-1} m! \left((m+1)\eta(m+2) + \frac{1}{2^{m+2}} \sum_{k=0}^{m} \eta(k+1)\eta(m-k+1) \right.$$

$$\left. - \sum_{k=0}^{m} \beta(k+1)\beta(m-k+1) \right), \tag{1.133}$$

where ζ represents the Riemann zeta function, η denotes the Dirichlet eta function, and β designates the Dirichlet beta function.

Examples at the point (i):

For $m = 0$,

$$\int_0^1 \frac{\log\left(\dfrac{1+x}{2}\right)}{1-x} dx = \frac{1}{2}\log^2(2) - \frac{1}{2}\zeta(2).$$

For $m = 1$,

$$\int_0^1 \frac{\log(x)\log\left(\dfrac{1+x}{2}\right)}{1-x} dx = \zeta(3) - \frac{1}{2}\log(2)\zeta(2).$$

For $m = 2$,

$$\int_0^1 \frac{\log^2(x)\log\left(\dfrac{1+x}{2}\right)}{1-x} dx = \frac{3}{2}\log(2)\zeta(3) - \frac{19}{8}\zeta(4).$$

For $m = 3$,

$$\int_0^1 \frac{\log^3(x)\log\left(\dfrac{1+x}{2}\right)}{1-x} dx = 12\zeta(5) - \frac{9}{4}\zeta(2)\zeta(3) - \frac{21}{4}\log(2)\zeta(4).$$

Examples at the point (ii):

For $m = 0$,

$$\int_0^1 \frac{\log\left(\dfrac{1+x^2}{2}\right)}{1-x} dx = \frac{1}{4}\log^2(2) - \frac{5}{8}\zeta(2).$$

(continued)

For $m = 1$,

$$\int_0^1 \frac{\log(x) \log\left(\dfrac{1+x^2}{2}\right)}{1-x} dx = 2\zeta(3) - \frac{1}{8}\log(2)\zeta(2) - \frac{\pi}{2}G.$$

For $m = 2$,

$$\int_0^1 \frac{\log^2(x) \log\left(\dfrac{1+x^2}{2}\right)}{1-x} dx = 2G^2 + \frac{3}{16}\log(2)\zeta(3) - \frac{199}{64}\zeta(4).$$

For $m = 3$,

$$\int_0^1 \frac{\log^3(x) \log\left(\dfrac{1+x^2}{2}\right)}{1-x} dx$$

$$= 24\zeta(5) - \frac{3}{128}\pi^2\zeta(3) - \frac{7}{1920}\log(2)\pi^4 + \frac{\pi^5}{32} - \frac{3}{8}\pi^3 G - \frac{\pi}{256}\psi^{(3)}\left(\frac{1}{4}\right).$$

Examples at the point (iii):
For $m = 0$,

$$\int_0^1 \frac{\log\left(\dfrac{1+x^2}{2}\right)}{1+x} dx = -\frac{1}{4}\log^2(2) - \frac{1}{8}\zeta(2).$$

For $m = 1$,

$$\int_0^1 \frac{\log(x) \log\left(\dfrac{1+x^2}{2}\right)}{1+x} dx = \frac{3}{2}\zeta(3) + \frac{1}{8}\log(2)\zeta(2) - \frac{\pi}{2}G.$$

For $m = 2$,

$$\int_0^1 \frac{\log^2(x) \log\left(\dfrac{1+x^2}{2}\right)}{1+x} dx = 2G^2 - \frac{3}{16}\log(2)\zeta(3) - \frac{161}{64}\zeta(4).$$

For $m = 3$,

(continued)

$$\int_0^1 \frac{\log^3(x) \log\left(\frac{1+x^2}{2}\right)}{1+x} dx$$

$$= \frac{45}{2}\zeta(5) + \frac{9}{64}\zeta(2)\zeta(3) + \frac{21}{64}\log(2)\zeta(4) + \frac{\pi^5}{32} - \frac{3}{8}\pi^3 G - \frac{\pi}{256}\psi^{(3)}\left(\frac{1}{4}\right).$$

1.28 Three Atypical Logarithmic Integrals Involving $\log(1+x^3)$, Related to a Special Form of Symmetry in Double Integrals

Show that

$$(i) \ I = \int_0^1 \frac{\log\left(1+x^3\right)}{1+x+x^2} dx$$

$$= \frac{2}{9}\pi^2 + \frac{1}{\sqrt{3}}\log(2)\pi - \frac{1}{3}\psi^{(1)}\left(\frac{1}{3}\right); \tag{1.134}$$

$$(ii) \ J = \int_0^1 \frac{x\log\left(1+x^3\right)}{1+x+x^2} dx$$

$$= \frac{1}{2}\log(2)\log(3) - \frac{5}{36}\pi^2 - \frac{1}{2\sqrt{3}}\log(2)\pi + \frac{1}{6}\psi^{(1)}\left(\frac{1}{3}\right); \tag{1.135}$$

$$(iii) \ K = \int_0^1 \frac{x^2\log\left(1+x^3\right)}{1+x+x^2} dx$$

$$= 2\log(2) - \frac{1}{2}\log(2)\log(3) - 3 + \frac{\pi}{\sqrt{3}} - \frac{\pi^2}{12} - \frac{1}{2\sqrt{3}}\log(2)\pi + \frac{1}{6}\psi^{(1)}\left(\frac{1}{3}\right), \tag{1.136}$$

where $\psi^{(1)}$ denotes the trigamma function.

A challenging question: Prove the results at the points (i) and (ii) by exploiting the symmetry in double integrals, with no use of beta function.

1.29 Four Curious Integrals with Cosine Function Which Lead to Beautiful Closed Forms, Plus Two Exotic Integrals

Show that

$$(i) \int_0^{\pi/2} x \log\left(\left|\frac{1 - 2\cos(2x)}{1 + 2\cos(2x)}\right|\right) dx = \frac{7}{9}\zeta(3); \tag{1.137}$$

$$(ii) \int_0^{\pi/2} x^2 \log\left(\left|\frac{1 - 2\cos(2x)}{1 + 2\cos(2x)}\right|\right) dx = \frac{7}{18}\pi\zeta(3); \tag{1.138}$$

$$(iii) \int_0^{\pi/2} x \log\left(\left|\frac{2\cos(2x) - 2\cos(4x) - 1}{2\cos(2x) + 2\cos(4x) + 1}\right|\right) dx = \frac{21}{25}\zeta(3); \tag{1.139}$$

$$(iv) \int_0^{\pi/2} x^2 \log\left(\left|\frac{2\cos(2x) - 2\cos(4x) - 1}{2\cos(2x) + 2\cos(4x) + 1}\right|\right) dx = \frac{21}{50}\pi\zeta(3). \tag{1.140}$$

Then, prove the exotic results with integrals

$$(v) \int_0^{\pi/2} x \left(3\frac{2\cos(2/3x) - 1}{2\cos(2/3x) + 1}\right) \cdot \left(3\frac{2\cos(2/3^2x) - 1}{2\cos(2/3^2x) + 1}\right) \cdot \left(3\frac{2\cos(2/3^3x) - 1}{2\cos(2/3^3x) + 1}\right) \cdots dx$$

$$= \frac{3}{2}\log(2)\zeta(2) - \frac{7}{8}\zeta(3); \tag{1.141}$$

$$(vi) \int_0^{\pi/2} x^2 \left(3\frac{2\cos(2/3x) - 1}{2\cos(2/3x) + 1}\right) \cdot \left(3\frac{2\cos(2/3^2x) - 1}{2\cos(2/3^2x) + 1}\right) \cdot \left(3\frac{2\cos(2/3^3x) - 1}{2\cos(2/3^3x) + 1}\right) \cdots dx$$

$$= \frac{1}{8}\log(2)\pi^3 - \frac{9}{16}\pi\zeta(3), \tag{1.142}$$

where ζ represents the Riemann zeta function.

1.30   Aesthetic Integrals That Must Be Understood as Cauchy Principal Values, Generated by a Beautiful Generalization

Prove that

(i) $\mathrm{P.V.} \displaystyle\int_0^1 \frac{1}{x(1-x)} \log\left(\frac{2x^2+x+1}{x^2-x+2}\right) dx = \mathrm{P.V.} \int_0^1 \frac{1}{x(1-x)} \log\left(\frac{2x^2-5x+4}{x^2-x+2}\right) dx$

$= \mathrm{P.V.} \displaystyle\int_0^\infty \frac{1}{x} \log\left(\frac{4x^2+3x+1}{2x^2+3x+2}\right) dx = \mathrm{P.V.} \int_0^\infty \frac{1}{x} \log\left(\frac{x^2+3x+4}{2x^2+3x+2}\right) dx$

$$= \log^2(2); \tag{1.143}$$

(ii) $\mathrm{P.V.} \displaystyle\int_0^1 \frac{1}{x(1-x)} \log\left(\frac{2(3-\sqrt{5})x^2+(3\sqrt{5}-5)x+2}{(3-\sqrt{5})x^2+(\sqrt{5}-3)x+\sqrt{5}+1}\right) dx$

$= \mathrm{P.V.} \displaystyle\int_0^1 \frac{1}{x(1-x)} \log\left(\frac{2(\sqrt{5}-3)x^2+(7-\sqrt{5})x-\sqrt{5}-3}{(\sqrt{5}-3)x^2+(3-\sqrt{5})x-\sqrt{5}-1}\right) dx$

$= \mathrm{P.V.} \displaystyle\int_0^\infty \frac{1}{x} \log\left(\frac{(\sqrt{5}+3)x^2+(3\sqrt{5}-1)x+2}{(\sqrt{5}+1)x^2+(3\sqrt{5}-1)x+\sqrt{5}+1}\right) dx$

$= \mathrm{P.V.} \displaystyle\int_0^\infty \frac{1}{x} \log\left(\frac{2x^2+(3\sqrt{5}-1)x+\sqrt{5}+3}{(\sqrt{5}+1)x^2+(3\sqrt{5}-1)x+\sqrt{5}+1}\right) dx$

$$= \log^2(\varphi), \tag{1.144}$$

where $\varphi = \dfrac{1+\sqrt{5}}{2}$ is the golden ratio.

(iii) $\mathrm{P.V.} \displaystyle\int_0^1 \frac{1}{x(1-x)} \arctan\left(\frac{4\sqrt{3}x^2-2\sqrt{3}x^3+5\sqrt{3}x-3\sqrt{3}}{2x^4-3x^3-8x^2+12x+9}\right) dx$

$= \mathrm{P.V.} \displaystyle\int_0^1 \frac{1}{x(1-x)} \arctan\left(\frac{2\sqrt{3}x^3-2\sqrt{3}x^2-7\sqrt{3}x+4\sqrt{3}}{2x^4-5x^3-5x^2+5x+12}\right) dx$

(continued)

$$= \text{P.V.} \int_0^\infty \frac{1}{x} \arctan \left(\frac{4\sqrt{3}x^4 + 9\sqrt{3}x^3 + \sqrt{3}x^2 - 7\sqrt{3}x - 3\sqrt{3}}{12x^4 + 53x^3 + 82x^2 + 48x + 9} \right) dx$$

$$= \text{P.V.} \int_0^\infty \frac{1}{x} \arctan \left(\frac{4\sqrt{3} + 9\sqrt{3}x + \sqrt{3}x^2 - 7\sqrt{3}x^3 - 3\sqrt{3}x^4}{12 + 53x + 82x^2 + 48x^3 + 9x^4} \right) dx$$

$$= \log \left(\frac{4}{3} \right) \frac{\pi}{6}. \tag{1.145}$$

1.31 A Special Integral Generalization, Attacked with Strategies Involving the Cauchy Principal Value Integrals, That Also Leads to Two Famous Results by Ramanujan

Let $0 < s < 3$ and $a, c > 0$ be real numbers. Then, prove that

$$(i) \, \sin \left(\frac{\pi s}{2} \right) \int_0^\infty \frac{x^{s-1} \cos(ax)}{c^2 + x^2} dx - \cos \left(\frac{\pi s}{2} \right) \int_0^\infty \frac{x^{s-1} \sin(ax)}{c^2 + x^2} dx$$

$$= \frac{\pi}{2} c^{s-2} e^{-ac}. \tag{1.146}$$

A famous result by Ramanujan:

$$(ii) \, \frac{\pi}{2} \int_0^\infty \frac{\sin(ax)}{1 + x^2} dx + \int_0^\infty \frac{\cos(ax) \log(x)}{1 + x^2} dx = 0. \tag{1.147}$$

Another result by Ramanujan obtained by setting $c = 1$ in (iv):

$$(iii) \, \pi \int_0^\infty \frac{\sin(ax) \log(x)}{1 + x^2} dx + \int_0^\infty \frac{\cos(ax) \log^2(x)}{1 + x^2} dx = \frac{\pi^3}{8} e^{-a}. \tag{1.148}$$

The generalization of the result by Ramanujan stated at (iii):

(continued)

(iv) $\pi \int_0^\infty \dfrac{\sin(ax)\log(x)}{c^2+x^2}dx + \int_0^\infty \dfrac{\cos(ax)\log^2(x)}{c^2+x^2}dx = \dfrac{\pi\left(4\log^2(c)+\pi^2\right)}{8c}e^{-ac}.$

$$(1.149)$$

One more special result:

(v) $2\pi \int_0^\infty \dfrac{\sin(ax)\log^3(x)}{1+x^2}dx - \pi^2 \int_0^\infty \dfrac{\cos(ax)\log^2(x)}{1+x^2}dx + \int_0^\infty \dfrac{\cos(ax)\log^4(x)}{1+x^2}dx$

$$= \dfrac{\pi^5}{32}e^{-a}. \qquad (1.150)$$

A challenging question: Prove the results without using contour integration.

1.32 Four *Magical* Integrals Beautifully Calculated, Which Generate the Closed Forms of Four Classical Integrals

Let $\theta > 0$. Then, prove that

$$(i) \quad \int_0^\infty \frac{\cos(\theta x) - x\sin(\theta x)}{1+x^2}dx = 0; \qquad (1.151)$$

$$(ii) \quad \int_0^\infty \frac{\cos(\theta x) + x\sin(\theta x)}{1+x^2}dx = \pi e^{-\theta}; \qquad (1.152)$$

$$(iii) \quad \int_0^\infty \frac{x\cos(\theta x) - \sin(\theta x)}{1+x^2}dx = -e^{-\theta}\operatorname{Ei}(\theta); \qquad (1.153)$$

$$(iv) \quad \int_0^\infty \frac{x\cos(\theta x) + \sin(\theta x)}{1+x^2}dx = -e^{\theta}\operatorname{Ei}(-\theta), \qquad (1.154)$$

where $\operatorname{Ei}(x)$ denotes the exponential integral, defined by the representations

$\operatorname{Ei}(x) = -\lim_{\epsilon \to 0^+}\left(\int_{-x}^{-\epsilon}\frac{e^{-t}}{t}dt + \int_\epsilon^\infty \frac{e^{-t}}{t}dt\right)$ and $\operatorname{Ei}(-x) = -\int_x^\infty \frac{e^{-t}}{t}dt, \ x > 0.$

(continued)

Classical integrals generated by the main integrals above $(a, b > 0)$:

$$\int_0^\infty \frac{\cos(ax)}{b^2 + x^2} dx = \frac{\pi}{2b} e^{-ab}; \qquad \int_0^\infty \frac{x \sin(ax)}{b^2 + x^2} dx = \frac{\pi}{2} e^{-ab};$$

$$(1.155) \hspace{6cm} (1.156)$$

$$\int_0^\infty \frac{x \cos(ax)}{b^2 + x^2} dx = -\frac{1}{2} \left(e^{-ab} \operatorname{Ei}(ab) + e^{ab} \operatorname{Ei}(-ab) \right); \qquad (1.157)$$

$$\int_0^\infty \frac{\sin(ax)}{b^2 + x^2} dx = \frac{1}{2b} \left(e^{-ab} \operatorname{Ei}(ab) - e^{ab} \operatorname{Ei}(-ab) \right). \qquad (1.158)$$

1.33   A Bouquet of Captivating Integrals Involving Trigonometric and Hyperbolic Functions

Let $a, b, c > 0$ be real numbers. Then, show that

$$(i) \int_0^\infty \frac{\sin(ax)}{\cosh(bx) + \cos(c)} dx = \frac{2a}{\sin(c)} \sum_{n=1}^\infty (-1)^{n-1} \frac{\sin(cn)}{a^2 + b^2 n^2}; \qquad (1.159)$$

$$(ii) \int_0^\infty \frac{\cos(ax)}{\cosh(bx) + \cos(c)} dx = \frac{\pi}{b} \frac{\sinh\left(\frac{ac}{b}\right)}{\sin(c) \sinh\left(\frac{a\pi}{b}\right)}. \qquad (1.160)$$

Prove that

$$(iii) \text{ P. V.} \int_0^\infty \frac{\cos(ax)}{\cosh(x) - \cosh(b)} dx = -\pi \coth(a\pi) \frac{\sin(ab)}{\sinh(b)}, \qquad (1.161)$$

where the integral is understood as a *Cauchy principal value*.

(continued)

Show that

$$(iv) \int_0^\infty \frac{\cos(ax)}{d\cosh(bx)+c}dx = \begin{cases} \dfrac{\pi \sinh\left(\frac{a}{b}\arccos\left(\frac{c}{d}\right)\right)}{b\sqrt{d^2-c^2}\sinh\left(\frac{a\pi}{b}\right)}, & d>c>0; \\[4mm] \dfrac{\pi \sin\left(\frac{a}{b}\text{arccosh}\left(\frac{c}{d}\right)\right)}{b\sqrt{c^2-d^2}\sinh\left(\frac{a\pi}{b}\right)}, & c>d>0. \end{cases}$$

$$(1.162)$$

1.34   Interesting Integrals to Evaluate, One of Them Coming from Lord Kelvin's Work

Prove that

$$(i) \int_0^\infty \cos(2tx - z\sqrt{x})e^{-z\sqrt{x}}dx = \frac{\sqrt{\pi}}{4}\frac{z}{t^{3/2}}e^{-z^2/(4t)}, \ t,z>0, \ t,z \in \mathbb{R}.$$

$$(1.163)$$

By exploiting (i), show that

$$(ii) \ 2\int_0^\infty \left(\cos(x)C\left(\frac{x}{\sqrt{\pi s}}\right) + \sin(x)S\left(\frac{x}{\sqrt{\pi s}}\right)\right)e^{-x}dx = e^{-s}, \ s>0, \ s \in \mathbb{R},$$

$$(1.164)$$

where $C(u) = \int_0^u \cos(\frac{\pi}{2}x^2)dx$ and $S(u) = \int_0^u \sin(\frac{\pi}{2}x^2)dx$ are the Fresnel integrals.

A neat special case of the generalization in (ii) *for* $s = 1/\pi$:

$$\int_0^\infty (\sin(x)S(x) + \cos(x)C(x))\,e^{-x}dx = \frac{1}{2}e^{-1/\pi}.$$

$$(1.165)$$

1.35   A Surprisingly Awesome Fractional Part Integral with Forms Involving $\sqrt{\tan(x)}$ and $\sqrt{\cot(x)}$

Prove that

$$\int_0^{\pi/2} \left\{\sqrt{\tan(x)}\right\} dx = \int_0^{\pi/2} \left\{\sqrt{\cot(x)}\right\} dx$$

$$= \left(2\sqrt{2} - 1\right)\frac{\pi}{4} + \arctan\left(\frac{\tanh\left(\dfrac{\pi}{\sqrt{2}}\right)}{\tan\left(\dfrac{\pi}{\sqrt{2}}\right)}\right), \qquad (1.166)$$

where $\{x\}$ represents the fractional part of x.

1.36   A Superb Integral with Logarithms and the Inverse Tangent Function and a Wonderful Generalization of It

Show that

$$(i)\ 3 \int_0^1 \frac{\arctan(x)\log(1+x^2)}{x} dx - 2 \int_0^1 \frac{\arctan(x)\log(1+x)}{x} dx = 0.$$

$$(1.167)$$

A wonderful generalization of the point (i):

$$(ii)\ 3 \int_0^x \frac{\arctan(t)\log(1+t^2)}{t} dt - 2 \int_0^x \frac{\arctan(t)\log(1+xt)}{t} dt = 0,\ x \in \mathbb{R}.$$

$$(1.168)$$

A challenging question: Prove the results without calculating the integrals separately.

1.37   More Wonderful Results Involving Integrals with Logarithms and the Inverse Tangent Function

Show that

$$(i) \int_0^x \frac{\arctan(t)\log(1+xt)}{t}dt - 3\int_0^1 \frac{\arctan(xt)\log(1-t)}{t}dt$$

$$= 3\sum_{n=1}^\infty (-1)^{n-1}\frac{x^{2n-1}}{(2n-1)^3}, \ |x| \le 1. \qquad (1.169)$$

An extended version of the generalization at the point (i):

$$(ii) \int_0^x \frac{\arctan(t)\log(1+xt)}{t}dt - 3\int_0^1 \frac{\arctan(xt)\log(1-t)}{t}dt = 3\operatorname{Ti}_3(x), \ x \in \mathbb{R},$$

$$\qquad (1.170)$$

where $\operatorname{Ti}_3(x) = \displaystyle\int_0^x \frac{\operatorname{Ti}_2(y)}{y}dy = \int_0^x \frac{1}{y}\left(\int_0^y \frac{\arctan(z)}{z}dz\right)dy$ is the inverse tangent integral of order 3.

The special case $x = 1$ *of the generalizations in* (i) *and* (ii):

$$(iii) \int_0^1 \frac{\arctan(t)\log(1+t)}{t}dt - 3\int_0^1 \frac{\arctan(t)\log(1-t)}{t}dt = \frac{3}{32}\pi^3.$$

$$\qquad (1.171)$$

A curious sum with four integrals:

$$(iv) \int_0^x \frac{\arctan(t)\log(1+xt)}{t}dt + \int_0^{1/x} \frac{\arctan(t)\log(1+t/x)}{t}dt$$

$$- 3\int_0^1 \frac{\arctan(xt)\log(1-t)}{t}dt - 3\int_0^1 \frac{\arctan(t/x)\log(1-t)}{t}dt$$

$$= \operatorname{sgn}(x)3\left(\frac{\pi^3}{16} + \frac{\pi}{4}\log^2(|x|)\right). \qquad (1.172)$$

(continued)

A curious representation of π involving Euler's number:

$$(v)\ \pi = \frac{4}{3} \int_0^e \frac{\arctan(t) \log(1 + et)}{t} dt + \frac{4}{3} \int_0^{1/e} \frac{\arctan(t) \log(1 + t/e)}{t} dt$$

$$- 4 \int_0^1 \frac{\arctan(et) \log(1 - t)}{t} dt - 4 \int_0^1 \frac{\arctan(t/e) \log(1 - t)}{t} dt$$

$$- \frac{8}{3} \int_0^1 \frac{\arctan(t) \log(1 + t)}{t} dt + 8 \int_0^1 \frac{\arctan(t) \log(1 - t)}{t} dt. \qquad (1.173)$$

1.38   Powerful and Useful Sums of Integrals with Logarithms and the Inverse Tangent Function

Prove, without calculating each integral, that

$$(i)\ \int_0^1 \frac{\arctan(x) \log(1 + x^2)}{x} dx - 2 \int_0^1 \frac{x \arctan(x) \log(1 + x^2)}{1 + x^2} dx$$

$$+ 2 \int_0^1 \frac{x \arctan(x) \log(x)}{1 + x^2} dx = \frac{3}{8} \log^2(2)\pi - \frac{\pi^3}{48}; \qquad (1.174)$$

$$(ii)\ \int_0^1 \frac{\arctan(x) \log(1 - x)}{x} dx + \int_0^1 \frac{\arctan(x) \log(1 + x)}{x} dx$$

$$+ 2 \int_0^1 \frac{x \arctan(x) \log(x)}{1 + x^2} dx = 2 \log(2)G - \frac{\pi^3}{16}; \qquad (1.175)$$

$$(iii)\ 2 \int_0^1 \frac{\arctan(x) \log(1 + x)}{x} dx + \int_0^1 \frac{\arctan(x) \log(1 + x^2)}{x} dx$$

$$+ 4 \int_0^1 \frac{x \arctan(x) \log(x)}{1 + x^2} dx = 4 \log(2)G - \frac{\pi^3}{16}; \qquad (1.176)$$

(continued)

$$(iv) \int_0^1 \frac{\arctan(x)\log(1+x^2)}{x}dx + 2\int_0^1 \frac{x\arctan(x)\log(1+x^2)}{1+x^2}dx$$

$$= 2\log(2)G - \frac{3}{8}\log^2(2)\pi - \frac{\pi^3}{96}, \tag{1.177}$$

where G is the Catalan's constant.

A (super) challenging question: Is it possible to prove all four relations with integrals exclusively by real methods (without using complex numbers)?

1.39   Two Beautiful Sums of Integrals, Each One Involving Three Integrals, Leading to a Possible Unexpected Result

Calculate

$$(i) \int_0^1 \frac{\arctan(x)}{x\sqrt{1+x^2}}dx - \frac{1}{2}\int_0^1 \frac{\arctan(x)}{x\sqrt{1+x}}dx - \int_0^1 \frac{1}{\sqrt{1-x^2}}\operatorname{arctanh}\left(x\sqrt{\frac{1-x^2}{1+x^2}}\right)dx.$$

$$\tag{1.178}$$

A wonderful sum with integrals involving logarithms:
Find, without calculating each integral separately,

$$(ii) \int_0^1 \frac{\log(1+x)}{x\sqrt{1+x^2}}dx - \frac{1}{2}\int_0^1 \frac{\log(1+x)}{x\sqrt{1+x}}dx - \int_0^1 \frac{\operatorname{arcsinh}(x)}{\sqrt{1+x^2}}dx.$$

$$\tag{1.179}$$

A challenging question: Perform the calculations to both points of the problem without using series.

1.40   Tackling Curious Logarithmic Integrals with a Radical in the Denominator

Prove, without using (bino)harmonic series, that

$$(i) \int_0^1 \frac{\log(1+x)}{x\sqrt{1+x}}dx = -4\operatorname{Li}_2\left(1-\sqrt{2}\right) - \frac{\pi^2}{12}. \qquad (1.180)$$

Then, also exploiting (i), show that

$$(ii) \int_0^1 \frac{\log(1+x)}{x\sqrt{1+x^2}}dx = \frac{1}{2}\log^2\left(\sqrt{2}-1\right) - \frac{\pi^2}{24} - 2\operatorname{Li}_2\left(1-\sqrt{2}\right).$$
$$(1.181)$$

Prove, with no use of (bino)harmonic series, that

$$(iii) \int_0^1 \frac{\operatorname{arctanh}(x)}{x\sqrt{1+x^2}}dx = \frac{7}{48}\pi^2 + \operatorname{Li}_2\left(1-\sqrt{2}\right). \qquad (1.182)$$

A result immediately derived based on (ii) and (iii):

$$(iv) \int_0^1 \frac{\log(1-x)}{x\sqrt{1+x^2}}dx = \frac{1}{2}\log^2\left(\sqrt{2}-1\right) - \frac{\pi^2}{3} - 4\operatorname{Li}_2\left(1-\sqrt{2}\right),$$
$$(1.183)$$

where Li_2 represents the dilogarithm function.

A *challenging question*: Combine wisely a form of the beta function and a result from the previous points during the calculations to the point (iii).

1.41   Calculating More Logarithmic Integrals with a Radical in the Denominator

Prove, without using (bino)harmonic series, that

$$(i) \int_0^1 \frac{\log(1-x^2)}{\sqrt{1+x^2}}dx = \frac{11}{24}\pi^2 + 2\log(2)\log\left(\sqrt{2}-1\right) + 10\operatorname{Li}_2\left(1-\sqrt{2}\right).$$
$$(1.184)$$

(continued)

Show the integral transformation

(ii) $\displaystyle\int_0^1 \frac{\operatorname{arctanh}(x)}{\sqrt{1+x^2}}dx = \int_0^1 \frac{\log(1+x)}{x\sqrt{1+x^2}}dx - \frac{1}{4}\int_0^1 \frac{\log(1+x)}{x\sqrt{1+x}}dx,$

$$(1.185)$$

and then exploit it to prove that

(iii) $\displaystyle\int_0^1 \frac{\operatorname{arctanh}(x)}{\sqrt{1+x^2}}dx = -\frac{\pi^2}{48} + \frac{1}{2}\log^2\left(\sqrt{2}-1\right) - \operatorname{Li}_2\left(1-\sqrt{2}\right).$

$$(1.186)$$

(iv) $\displaystyle\int_0^1 \frac{\log(1-x)}{\sqrt{1+x^2}}dx$

$$= \frac{\pi^2}{4} + \log(2)\log\left(\sqrt{2}-1\right) - \frac{1}{2}\log^2\left(\sqrt{2}-1\right) + 6\operatorname{Li}_2\left(1-\sqrt{2}\right);$$

$$(1.187)$$

(v) $\displaystyle\int_0^1 \frac{\log(1+x)}{\sqrt{1+x^2}}dx$

$$= \frac{5}{24}\pi^2 + \log(2)\log\left(\sqrt{2}-1\right) + \frac{1}{2}\log^2\left(\sqrt{2}-1\right) + 4\operatorname{Li}_2\left(1-\sqrt{2}\right).$$

$$(1.188)$$

A useful identity related to the Legendre chi function of order 2:

(vi) $\displaystyle\operatorname{Li}_2\left(-\left(1-\sqrt{2}\right)\right) - \operatorname{Li}_2\left(1-\sqrt{2}\right) = \frac{\pi^2}{8} - \frac{1}{2}\log^2\left(\sqrt{2}-1\right),$

$$(1.189)$$

where Li_2 represents the dilogarithm function.

1.42 Somewhat Atypical Integrals with Curious Closed Forms

Prove that

(i) $\displaystyle\int_0^{\pi/2} \frac{\arctan(\sin(x))\log(1+\sin^2(x))}{\sin(x)}dx = \int_0^{\pi/2} \frac{\arctan(\cos(x))\log(1+\cos^2(x))}{\cos(x)}dx$

$$= \frac{5}{24}\pi^3 - \frac{1}{2}\log^2\left(\sqrt{2}-1\right)\pi + 4\pi\operatorname{Li}_2\left(1-\sqrt{2}\right). \qquad (1.190)$$

(continued)

A curious sum with two integrals:
Show, without calculating each integral separately, that

$$(ii) \ \int_0^{\pi/2} \frac{\arctan(\sin(x)) \log(1 + \sin^2(x))}{\sin(x)} dx + 2 \int_0^1 \arctan\left(\sqrt{\frac{1+x^2}{x(1-x)}}\right) \frac{\log(1+x)}{x\sqrt{1+x^2}} dx$$

$$= \frac{1}{2} \log^2\left(\sqrt{2} - 1\right) \pi - \frac{\pi^3}{24} - 2\pi \operatorname{Li}_2\left(1 - \sqrt{2}\right). \qquad (1.191)$$

Revealing another closed form based on the previous two points:

$$(iii) \ \int_0^1 \arctan\left(\sqrt{\frac{1+x^2}{x(1-x)}}\right) \frac{\log(1+x)}{x\sqrt{1+x^2}} dx$$

$$= \frac{1}{2} \log^2\left(\sqrt{2} - 1\right) \pi - \frac{\pi^3}{8} - 3\pi \operatorname{Li}_2\left(1 - \sqrt{2}\right), \qquad (1.192)$$

where Li_2 represents the dilogarithm function.

1.43 More Atypical Integrals with Curious Closed Forms

Show that

$$(i) \ \int_0^1 \frac{\arctan(x)}{(1+x^2)\sqrt{1-x^2}} dx$$

$$= \frac{7}{96}\sqrt{2}\pi^2 + \frac{1}{4}\sqrt{2} \log(2) \log\left(\sqrt{2} - 1\right) + \frac{1}{2}\sqrt{2} \operatorname{Li}_2\left(1 - \sqrt{2}\right) \qquad (1.193)$$

by also proving and exploiting the result

$$(ii) \int_0^1 \frac{\log(x)}{x} \left(1 - \frac{1}{\sqrt{1+x}}\right) dx$$

$$= -\frac{\pi^2}{2} - 2\log^2(2) - 4\log(2)\log\left(\sqrt{2}-1\right) - 8\,\mathrm{Li}_2\left(1-\sqrt{2}\right), \quad (1.194)$$

where Li_2 represents the dilogarithm function.
Another special identity with three dilogarithms:
Prove that

$$(iii)\; 8\,\mathrm{Li}_2\left(-\left(1-\sqrt{2}\right)\right) - 4\,\mathrm{Li}_2\left(-\left(1-\sqrt{2}\right)^2\right) - 4\,\mathrm{Li}_2\left(-2\left(1-\sqrt{2}\right)\right)$$

$$= \frac{\pi^2}{6} + 8\log(2)\log\left(\sqrt{2}-1\right) + 4\log^2\left(\sqrt{2}-1\right). \quad (1.195)$$

1.44   A Wonderful Trigonometric Integral by Larry Glasser

Let $0 < \phi \le \pi/2$. Evaluate

$$\int_0^\pi \frac{\arctan^2(\cot(\phi)\sin(\theta))}{1+\cos(\theta)} d\theta. \quad (1.196)$$

A challenging question: Find a way of exploiting the auxiliary integral

$$\int_0^1 \frac{x\,\mathrm{arctanh}(x)}{\sqrt{1+\cot^2(\phi)x^2}} dx = \sec(\phi)\tan(\phi)\log(\sin(\phi)) + \tan(\phi)\log\left(\cot\left(\frac{\phi}{2}\right)\right)$$
$$(1.197)$$

in order to extract the value of the main integral.

1.45   Resistant Logarithmic Integrals That Are Good to Know

Prove that

$$(i) \int_0^1 \frac{\log(x) \log^3(1+x)}{x} dx$$

$$= \frac{99}{16}\zeta(5) + 3\zeta(2)\zeta(3) + 2\log^3(2)\zeta(2) - \frac{21}{4}\log^2(2)\zeta(3) - \frac{2}{5}\log^5(2)$$

$$- 12\log(2)\operatorname{Li}_4\left(\frac{1}{2}\right) - 12\operatorname{Li}_5\left(\frac{1}{2}\right); \qquad (1.198)$$

$$(ii) \int_0^1 \frac{\log(1-x)\log^2(x)\log(1+x)}{1+x} dx$$

$$= \frac{213}{16}\zeta(5) - \frac{5}{2}\zeta(2)\zeta(3) - \frac{49}{8}\log(2)\zeta(4) + \frac{7}{4}\log^2(2)\zeta(3) + \frac{1}{3}\log^3(2)\zeta(2)$$

$$- \frac{1}{10}\log^5(2) - 4\log(2)\operatorname{Li}_4\left(\frac{1}{2}\right) - 8\operatorname{Li}_5\left(\frac{1}{2}\right); \qquad (1.199)$$

$$(iii) \int_0^1 \frac{\log^2(1-x)\log^2(x)}{1+x} dx$$

$$= \frac{15}{2}\zeta(5) - \frac{13}{2}\zeta(2)\zeta(3) + \log(2)\zeta(4) + \frac{2}{3}\log^3(2)\zeta(2) - \frac{1}{15}\log^5(2) + 8\operatorname{Li}_5\left(\frac{1}{2}\right); \qquad (1.200)$$

$$(iv) \int_0^1 \frac{\log^2(1+x)\log^2(x)}{1-x} dx$$

$$= \frac{47}{2}\zeta(5) - \frac{13}{4}\zeta(2)\zeta(3) - \frac{1}{5}\log^5(2) + \frac{2}{3}\log^3(2)\zeta(2) + \frac{7}{2}\log^2(2)\zeta(3)$$

$$- \frac{49}{4}\log(2)\zeta(4) - 8\log(2)\operatorname{Li}_4\left(\frac{1}{2}\right) - 16\operatorname{Li}_5\left(\frac{1}{2}\right); \qquad (1.201)$$

$$(v) \int_0^1 \frac{\operatorname{arctanh}^2(x)\log(1+x^2)}{x} dx$$

(continued)

$$= G^2 + \frac{53}{32}\zeta(4) - \frac{7}{4}\log(2)\zeta(3) + \frac{1}{2}\log^2(2)\zeta(2) - \frac{1}{12}\log^4(2) - 2\operatorname{Li}_4\left(\frac{1}{2}\right);$$
$$(1.202)$$

$$(vi) \int_0^1 \frac{\log(1-x)\log(1+x)\log(1+x^2)}{x}dx$$

$$= \frac{3}{32}\log^4(2) - G^2 - \frac{9}{16}\log^2(2)\zeta(2) + \frac{63}{32}\log(2)\zeta(3) - \frac{111}{64}\zeta(4) + \frac{9}{4}\operatorname{Li}_4\left(\frac{1}{2}\right),$$
$$(1.203)$$

where G is the Catalan's constant, ζ represents the Riemann zeta function, and Li_n denotes the polylogarithm function.

1.46 Appealing Parameterized Integrals with Logarithms and the Dilogarithm Function, Related to Harmonic Series

Let $m \geq 1$ be an integer. Show that

$$(i) \int_0^1 \frac{\log(1-x)\log^m(x)\operatorname{Li}_2(x)}{x}dx$$

$$= \frac{1}{2}(-1)^{m-1}m!\left((m+5)\zeta(m+4) - \zeta(2)\zeta(m+2)\right)$$

$$- \frac{3}{2}(-1)^{m-1}m!\sum_{k=1}^m \zeta(k+1)\zeta(m-k+3) + (-1)^{m-1}m!\sum_{k=1}^m\sum_{n=1}^\infty \frac{H_n^{(k+1)}}{n^{m-k+3}};$$
$$(1.204)$$

$$(ii) \int_0^1 \frac{\log(1-x)\log^m(x)\operatorname{Li}_2(-x)}{x}dx$$

$$= (-1)^{m-1}m!\sum_{k=1}^m \zeta(k+1)\eta(m-k+3) - (-1)^{m-1}m!\sum_{n=1}^\infty (-1)^{n-1}\frac{H_n}{n^{m+3}}$$

$$- (-1)^{m-1}m!\sum_{k=1}^m\sum_{n=1}^\infty (-1)^{n-1}\frac{H_n^{(k+1)}}{n^{m-k+3}};$$
$$(1.205)$$

(continued)

$$(iii) \quad \int_0^1 \frac{\log(1+x)\log^m(x)\operatorname{Li}_2(x)}{x}dx$$

$$= -\zeta(2)(-1)^{m-1}m!\eta(m+2)-(-1)^{m-1}m!\sum_{k=1}^m(m-k+1)\zeta(k+1)\eta(m-k+3)$$

$$+ (-1)^{m-1}(m+1)!\sum_{n=1}^\infty(-1)^{n-1}\frac{H_n}{n^{m+3}}$$

$$+ (-1)^{m-1}m!\sum_{k=1}^m(m-k+1)\sum_{n=1}^\infty(-1)^{n-1}\frac{H_n^{(k+1)}}{n^{m-k+3}}; \tag{1.206}$$

$$(iv) \quad \int_0^1 \frac{\log(1+x)\log^m(x)\operatorname{Li}_2(-x)}{x}dx$$

$$= 3(-1)^{m-1}m!\eta(m+4) - 2(-1)^{m-1}m!\sum_{n=1}^\infty(-1)^{n-1}\frac{H_n}{n^{m+3}}$$

$$- (-1)^{m-1}m!\sum_{n=1}^\infty(-1)^{n-1}\frac{H_n^{(2)}}{n^{m+2}}, \tag{1.207}$$

where $H_n^{(m)} = 1 + \frac{1}{2^m} + \cdots + \frac{1}{n^m}$, $m \geq 1$, is the nth generalized harmonic number of order m, ζ represents the Riemann zeta function, η denotes the Dirichlet eta function, and Li_2 designates the dilogarithm function.

A simple, beautiful, and (very) useful relation:
Prove that

$$(v) \quad \mathscr{S}_m = \int_0^1 \frac{\log(1-x)\log^m(x)\operatorname{Li}_2(-x)}{x}dx + \int_0^1 \frac{\log(1+x)\log^m(x)\operatorname{Li}_2(x)}{x}dx$$

$$+ \int_0^1 \frac{\log(1+x)\log^m(x)\operatorname{Li}_2(-x)}{x}dx$$

$$= (-1)^{m-1}m!\left(1 - \frac{1}{2^{m+2}}\right)\left(\frac{1}{2}\left(\zeta(2)\zeta(m+2) - (m+5)\zeta(m+4)\right)\right.$$

$$+ \frac{3}{2}\sum_{k=1}^m\zeta(k+1)\zeta(m-k+3) - \sum_{k=1}^m\sum_{n=1}^\infty\frac{H_n^{(k+1)}}{n^{m-k+3}}\right). \tag{1.208}$$

1.47   An Encounter with Six Useful Integrals Involving Logarithms and the Dilogarithm

Prove that

$$(i) \int_0^1 \frac{\log^2(1-x)\,\mathrm{Li}_2(x)}{x}\,dx = 2\zeta(2)\zeta(3) - \zeta(5); \qquad (1.209)$$

$$(ii) \int_0^1 \frac{\log^2(1-x)\,\mathrm{Li}_2(-x)}{x}\,dx$$

$$= \frac{2}{3}\log^3(2)\zeta(2) - \frac{7}{4}\log^2(2)\zeta(3) + \frac{3}{4}\zeta(2)\zeta(3) + \frac{15}{16}\zeta(5) - \frac{2}{15}\log^5(2)$$

$$- 4\log(2)\,\mathrm{Li}_4\left(\frac{1}{2}\right) - 4\,\mathrm{Li}_5\left(\frac{1}{2}\right); \qquad (1.210)$$

$$(iii) \int_0^1 \frac{\log^2(1-x)\,\mathrm{Li}_2(x^2)}{x}\,dx$$

$$= \frac{4}{3}\log^3(2)\zeta(2) - \frac{7}{2}\log^2(2)\zeta(3) + \frac{11}{2}\zeta(2)\zeta(3) - \frac{1}{8}\zeta(5) - \frac{4}{15}\log^5(2)$$

$$- 8\log(2)\,\mathrm{Li}_4\left(\frac{1}{2}\right) - 8\,\mathrm{Li}_5\left(\frac{1}{2}\right); \qquad (1.211)$$

$$(iv) \int_0^1 \frac{\log^2(1+x)\,\mathrm{Li}_2(x)}{x}\,dx$$

$$= \frac{4}{15}\log^5(2) - \frac{4}{3}\log^3(2)\zeta(2) + \frac{7}{2}\log^2(2)\zeta(3) - \frac{39}{8}\zeta(5) - \frac{3}{2}\zeta(2)\zeta(3)$$

$$+ 8\log(2)\,\mathrm{Li}_4\left(\frac{1}{2}\right) + 8\,\mathrm{Li}_5\left(\frac{1}{2}\right); \qquad (1.212)$$

$$(v) \int_0^1 \frac{\log^2(1+x)\,\mathrm{Li}_2(-x)}{x}\,dx$$

$$= \frac{2}{15}\log^5(2) - \frac{2}{3}\log^3(2)\zeta(2) + \frac{7}{4}\log^2(2)\zeta(3) - \frac{1}{8}\zeta(2)\zeta(3) - \frac{125}{32}\zeta(5)$$

(continued)

$$+ 4\log(2)\operatorname{Li}_4\left(\frac{1}{2}\right) + 4\operatorname{Li}_5\left(\frac{1}{2}\right);\qquad (1.213)$$

$$(vi)\ \int_0^1 \frac{\log^2(1+x)\operatorname{Li}_2(x^2)}{x}dx$$

$$= \frac{4}{5}\log^5(2) - \frac{13}{4}\zeta(2)\zeta(3) - \frac{281}{16}\zeta(5) - 4\log^3(2)\zeta(2) + \frac{21}{2}\log^2(2)\zeta(3)$$

$$+ 24\log(2)\operatorname{Li}_4\left(\frac{1}{2}\right) + 24\operatorname{Li}_5\left(\frac{1}{2}\right),\qquad (1.214)$$

where ζ represents the Riemann zeta function and Li_n denotes the polylogarithm function.

1.48 A *Battle* with Three Challenging Integrals Involving Logarithms and the Dilogarithm

Prove that

$$(i)\ \int_0^1 \frac{\log(1-x)\log(1+x)\operatorname{Li}_2(x)}{x}dx = \frac{29}{64}\zeta(5) - \frac{5}{8}\zeta(2)\zeta(3);\qquad (1.215)$$

$$(ii)\ \int_0^1 \frac{\log(1-x)\log(1+x)\operatorname{Li}_2(-x)}{x}dx$$

$$= \frac{123}{32}\zeta(5) + \frac{5}{16}\zeta(2)\zeta(3) + \frac{2}{3}\log^3(2)\zeta(2) - \frac{7}{4}\log^2(2)\zeta(3) - \frac{2}{15}\log^5(2)$$

$$- 4\log(2)\operatorname{Li}_4\left(\frac{1}{2}\right) - 4\operatorname{Li}_5\left(\frac{1}{2}\right);\qquad (1.216)$$

$$(iii)\ \int_0^1 \frac{\log(1-x)\log(1+x)\operatorname{Li}_2(x^2)}{x}dx$$

$$= \frac{275}{32}\zeta(5) - \frac{5}{8}\zeta(2)\zeta(3) + \frac{4}{3}\log^3(2)\zeta(2) - \frac{7}{2}\log^2(2)\zeta(3) - \frac{4}{15}\log^5(2)$$

(continued)

$$-8\log(2)\operatorname{Li}_4\left(\frac{1}{2}\right)-8\operatorname{Li}_5\left(\frac{1}{2}\right), \qquad (1.217)$$

where ζ represents the Riemann zeta function and Li_n denotes the polylogarithm function.

1.49   Fascinating Polylogarithmic Integrals with Parameter Involving the Cauchy Product of Two Series

Let $|a| \le 1$, $a \ne 0$, be a real number. Show that

(i) $\displaystyle\int_0^1 \frac{\log(x)\operatorname{Li}_2(x)}{1-ax}\,\mathrm{d}x = \frac{1}{2}\frac{(\operatorname{Li}_2(a))^2}{a} - 2\zeta(2)\frac{\operatorname{Li}_2(a)}{a} + 3\frac{\operatorname{Li}_4(a)}{a};$

$$\hspace{11cm}(1.218)$$

(ii) $\displaystyle\int_0^1 \frac{\log^2(x)\operatorname{Li}_3(x)}{1-ax}\,\mathrm{d}x = 20\frac{\operatorname{Li}_6(a)}{a} - 12\zeta(2)\frac{\operatorname{Li}_4(a)}{a} + \frac{(\operatorname{Li}_3(a))^2}{a};$

$$\hspace{11cm}(1.219)$$

(iii) $\displaystyle\int_0^1 \frac{\log^3(x)\operatorname{Li}_4(x)}{1-ax}\,\mathrm{d}x$

$$= 3\frac{(\operatorname{Li}_4(a))^2}{a} - 12\zeta(4)\frac{\operatorname{Li}_4(a)}{a} - 120\zeta(2)\frac{\operatorname{Li}_6(a)}{a} + 210\frac{\operatorname{Li}_8(a)}{a}, \qquad (1.220)$$

where ζ represents the Riemann zeta function and Li_n denotes the polylogarithm function.

Special cases obtained by using $a = 1/2$:

(iv) $\displaystyle\int_0^1 \frac{\log(1-x)\operatorname{Li}_2(1-x)}{1+x}\,\mathrm{d}x$

$$= \frac{3}{4}\log^2(2)\zeta(2) - \frac{35}{16}\zeta(4) + \frac{1}{8}\log^4(2) + 3\operatorname{Li}_4\left(\frac{1}{2}\right); \qquad (1.221)$$

(continued)

$$(v) \int_0^1 \frac{\log^2(1-x)\operatorname{Li}_3(1-x)}{1+x}dx$$

$$= \frac{49}{64}\zeta^2(3) - \frac{7}{8}\log(2)\zeta(2)\zeta(3) - \frac{1}{6}\log^4(2)\zeta(2) + \frac{7}{24}\log^3(2)\zeta(3)$$

$$+ \frac{5}{8}\log^2(2)\zeta(4) + \frac{1}{36}\log^6(2) - 12\zeta(2)\operatorname{Li}_4\left(\frac{1}{2}\right) + 20\operatorname{Li}_6\left(\frac{1}{2}\right);$$

$$(1.222)$$

$$(vi) \int_0^1 \frac{\log^3(1-x)\operatorname{Li}_4(1-x)}{1+x}dx$$

$$= 210\operatorname{Li}_8\left(\frac{1}{2}\right) - 120\zeta(2)\operatorname{Li}_6\left(\frac{1}{2}\right) - 12\zeta(4)\operatorname{Li}_4\left(\frac{1}{2}\right) + 3\left(\operatorname{Li}_4\left(\frac{1}{2}\right)\right)^2.$$

$$(1.223)$$

1.50 A Titan Involving Alternating Harmonic Series of Weight 7

Show that

$$\int_0^1 \frac{\log(1+x)\log^2(x)\operatorname{Li}_3(-x)}{x}dx = \frac{115}{64}\zeta(7) - \frac{19}{16}\zeta(2)\zeta(5), \qquad (1.224)$$

where ζ represents the Riemann zeta function and Li_n denotes the polylogarithm function.

A (super) challenging question: Obtain the desired value without calculating separately each of the alternating harmonic series of weight 7 involved, except for the series, $\displaystyle\sum_{n=1}^{\infty}(-1)^{n-1}\frac{H_n}{n^6}$.

1.51   A Tough Integral Approached by Clever Transformations

Show that

$$\int_0^{1/2} \frac{(\text{Li}_2(x))^2}{x}\,dx$$

$$= \frac{1}{2}\log^3(2)\zeta(2) - \frac{7}{8}\log^2(2)\zeta(3) - \frac{5}{8}\log(2)\zeta(4) + \frac{27}{32}\zeta(5) + \frac{7}{8}\zeta(2)\zeta(3)$$

$$- \frac{7}{60}\log^5(2) - 2\log(2)\,\text{Li}_4\left(\frac{1}{2}\right) - 2\,\text{Li}_5\left(\frac{1}{2}\right), \tag{1.225}$$

where ζ represents the Riemann zeta function and Li_n denotes the polylogarithm function.

A challenging question: May we find a (pretty) simple way of transforming the integral into a sum of alternating harmonic series of weight 5?

1.52   An Unexpected Closed Form, Involving Catalan's Constant, of a Nice Integral with the Dilogarithm

Show that

$$\int_0^{\pi/4} \text{Li}_2\left(\frac{\sec^2(\theta)}{2}\right) d\theta = \int_0^1 \frac{1}{1+x^2}\,\text{Li}_2\left(\frac{1+x^2}{2}\right) dx = \log(2)G,$$
$$\tag{1.226}$$

where G is the Catalan's constant and Li_2 represents the dilogarithm function.

A (super) challenging question: May we reduce the integral to calculations involving beta function and a multiple integral where the symmetry may be successfully exploited?

1.53 A Group of Six Special, Challenging Generalized Integrals Involving Curious Closed Forms

Show that

$$(i)\ \int_0^1 \frac{\log^{2m}(x)}{x} \operatorname{Li}_2\left(\frac{2x}{1+x^2}\right) dx$$

$$= (-1)^m \frac{1}{2m+1} \sum_{k=0}^{2m+1} (2\pi)^k \binom{2m+1}{k} B_k \mathscr{C}_{2m-k+1}; \qquad (1.227)$$

$$(ii)\ \int_0^1 \frac{\log^{2m}(x)}{x} \operatorname{Li}_2\left(-\frac{2x}{1+x^2}\right) dx$$

$$= (-1)^{m-1} \frac{1}{2m+1} \sum_{k=0}^{2m+1} \left(1 - \frac{1}{2^{k-1}}\right) (2\pi)^k \binom{2m+1}{k} B_k \mathscr{C}_{2m-k+1};$$

$$(1.228)$$

$$(iii)\ \int_0^1 \frac{x \log^{2m+1}(x) \log(1-x)}{1+x^2} dx$$

$$= \frac{1}{2}\left(1 - \frac{1}{2^{2m+3}}\right)(2m+1)!\,\zeta(2m+3) + \frac{1}{2^{2m+4}}(2m+2)!\,\eta(2m+3)$$

$$- \frac{1}{2^{2m+3}}(2m+1)! \sum_{k=1}^m \eta(2k)\zeta(2m-2k+3)$$

$$+ \frac{1}{4}(-1)^{m-1} \sum_{k=0}^{2m+1} (2\pi)^k \binom{2m+1}{k} B_k \mathscr{C}_{2m-k+1}; \qquad (1.229)$$

$$(iv)\ \int_0^1 \frac{x \log^{2m+1}(x) \log(1+x)}{1+x^2} dx$$

$$= \frac{1}{2^{2m+3}}\left(1 + m - 2^{2m+2}\right)(2m+1)!\,\eta(2m+3) - \frac{1}{2^{2m+4}}(2m+1)!\,\zeta(2m+3)$$

$$- \frac{1}{2^{2m+3}}(2m+1)! \sum_{k=1}^m \eta(2k)\zeta(2m-2k+3)$$

(continued)

$$-\frac{1}{4}(-1)^{m-1}\sum_{k=0}^{2m+1}\left(1-\frac{1}{2^{k-1}}\right)(2\pi)^k\binom{2m+1}{k}B_k\mathscr{C}_{2m-k+1};\qquad(1.230)$$

$$(v)\ \int_0^1\frac{\log(1-x)\log^{2m}(x)\log(1+x^2)}{x}dx$$

$$=\left(1+2m+\frac{1}{2^{2m+3}}\right)(2m)!\zeta(2m+3)$$

$$-\log(2)(2m)!\zeta(2m+2)-\frac{1}{2^{2m+2}}(m+1)(2m)!\eta(2m+3)$$

$$+\frac{1}{2^{2m+2}}(2m)!\sum_{k=1}^{m}\eta(2k)\zeta(2m-2k+3)-\frac{1}{2^{2m+3}}(2m)!\sum_{k=0}^{2m+1}\eta(k+1)\eta(2m-k+2)$$

$$-(2m)!\sum_{k=0}^{2m+1}\beta(k+1)\beta(2m-k+2)-\frac{1}{2}\frac{(-1)^{m-1}}{2m+1}\sum_{k=0}^{2m+1}(2\pi)^k\binom{2m+1}{k}B_k\mathscr{C}_{2m-k+1};$$

$$(1.231)$$

$$(vi)\ \int_0^1\frac{\log(1+x)\log^{2m}(x)\log(1+x^2)}{x}dx$$

$$=\left(\frac{1}{2^{4m+3}}m+3\frac{1}{2^{4m+4}}-5\frac{1}{2^{2m+3}}-2m-1\right)(2m)!\zeta(2m+3)$$

$$+\log(2)(2m)!\zeta(2m+2)+\frac{1}{2^{2m+2}}(m+1)(2m)!\eta(2m+3)$$

$$+\frac{1}{2^{2m+2}}\left(1-\frac{1}{2^{2m+1}}\right)(2m)!\sum_{k=1}^{2m}\zeta(k+1)\zeta(2m-k+2)$$

$$-3\frac{1}{2^{2m+2}}(2m)!\sum_{k=1}^{m}\eta(2k)\zeta(2m-2k+3)+\frac{1}{2^{2m+3}}(2m)!\sum_{k=0}^{2m+1}\eta(k+1)\eta(2m-k+2)$$

$$+(2m)!\sum_{k=0}^{2m+1}\beta(k+1)\beta(2m-k+2)+\frac{1}{2}\frac{(-1)^{m-1}}{2m+1}\sum_{k=0}^{2m+1}(2\pi)^k\binom{2m+1}{k}B_k\mathscr{C}_{2m-k+1},$$

$$(1.232)$$

(continued)

where ζ is the Riemann zeta function, η represents the Dirichlet eta function, β denotes the Dirichlet beta function, Li_2 signifies the dilogarithm function, B_n designates the nth Bernoulli number, and $\mathscr{C}_n$ is defined by

$$\mathscr{C}_n = \int_0^{\pi/2} x^n \log(\cos(x))dx$$

$$= -\log(2)\frac{1}{n+1}\left(\frac{\pi}{2}\right)^{n+1} - \sin\left(\frac{\pi}{2}n\right)\frac{1}{2^{n+1}}\left(1 - \frac{1}{2^{n+1}}\right)n!\zeta(n+2)$$

$$- \frac{1}{2^{n+1}}n!\sum_{k=0}^{n}\sin\left(\frac{\pi}{2}k\right)\frac{1}{(n-k)!}\pi^{n-k}\zeta(k+2). \qquad (1.233)$$

Examples at the point (i):
For $m = 1$,

$$\int_0^1 \frac{\log^2(x)}{x}\text{Li}_2\left(\frac{2x}{1+x^2}\right)dx = \frac{135}{32}\log(2)\zeta(4) + \frac{5}{8}\zeta(2)\zeta(3) - \frac{31}{128}\zeta(5).$$

For $m = 2$,

$$\int_0^1 \frac{\log^4(x)}{x}\text{Li}_2\left(\frac{2x}{1+x^2}\right)dx$$

$$= \frac{6237}{128}\log(2)\zeta(6) + \frac{267}{32}\zeta(3)\zeta(4) + \frac{39}{16}\zeta(2)\zeta(5) - \frac{381}{512}\zeta(7).$$

For $m = 3$,

$$\int_0^1 \frac{\log^6(x)}{x}\text{Li}_2\left(\frac{2x}{1+x^2}\right)dx$$

$$= \frac{1480275}{1024}\log(2)\zeta(8) + \frac{5085}{64}\zeta(4)\zeta(5) + \frac{67725}{256}\zeta(3)\zeta(6) + \frac{2475}{128}\zeta(2)\zeta(7)$$

$$- \frac{22995}{4096}\zeta(9).$$

Examples at the point (ii):

(continued)

For $m = 1$,

$$\int_0^1 \frac{\log^2(x)}{x} \operatorname{Li}_2\left(-\frac{2x}{1+x^2}\right) dx = -\frac{105}{32}\log(2)\zeta(4) - \frac{1}{2}\zeta(2)\zeta(3) - \frac{31}{128}\zeta(5).$$

For $m = 2$,

$$\int_0^1 \frac{\log^4(x)}{x} \operatorname{Li}_2\left(-\frac{2x}{1+x^2}\right) dx$$

$$= -\frac{5859}{128}\log(2)\zeta(6) - \frac{273}{32}\zeta(3)\zeta(4) - \frac{57}{32}\zeta(2)\zeta(5) - \frac{381}{512}\zeta(7).$$

For $m = 3$,

$$\int_0^1 \frac{\log^6(x)}{x} \operatorname{Li}_2\left(-\frac{2x}{1+x^2}\right) dx$$

$$= -\frac{1457325}{1024}\log(2)\zeta(8) - \frac{315}{4}\zeta(4)\zeta(5) - \frac{68355}{256}\zeta(3)\zeta(6) - \frac{3555}{256}\zeta(2)\zeta(7)$$

$$- \frac{22995}{4096}\zeta(9).$$

Examples at the point (iii):
For $m = 1$,

$$\int_0^1 \frac{x\log^3(x)\log(1-x)}{1+x^2} dx = \frac{1761}{512}\zeta(5) - \frac{9}{16}\zeta(2)\zeta(3) - \frac{405}{128}\log(2)\zeta(4).$$

For $m = 2$,

$$\int_0^1 \frac{x\log^5(x)\log(1-x)}{1+x^2} dx$$

$$= \frac{129495}{2048}\zeta(7) - \frac{225}{64}\zeta(2)\zeta(5) - \frac{45}{4}\zeta(3)\zeta(4) - \frac{31185}{512}\log(2)\zeta(6).$$

For $m = 3$,

$$\int_0^1 \frac{x\log^7(x)\log(1-x)}{1+x^2} dx$$

(continued)

$$= \frac{42010605}{16384}\zeta(9) - \frac{19845}{512}\zeta(2)\zeta(7) - \frac{945}{2}\zeta(3)\zeta(6) - \frac{4725}{32}\zeta(4)\zeta(5)$$

$$- \frac{10361925}{4096}\log(2)\zeta(8).$$

Examples at the point (iv):
For $m = 1$,

$$\int_0^1 \frac{x\log^3(x)\log(1+x)}{1+x^2}dx = \frac{315}{128}\log(2)\zeta(4) + \frac{9}{32}\zeta(2)\zeta(3) - \frac{1215}{512}\zeta(5).$$

For $m = 2$,

$$\int_0^1 \frac{x\log^5(x)\log(1+x)}{1+x^2}dx$$

$$= \frac{29295}{512}\log(2)\zeta(6) + \frac{315}{32}\zeta(3)\zeta(4) + \frac{225}{128}\zeta(2)\zeta(5) - \frac{114345}{2048}\zeta(7).$$

For $m = 3$,

$$\int_0^1 \frac{x\log^7(x)\log(1+x)}{1+x^2}dx$$

$$= \frac{10201275}{4096}\log(2)\zeta(8) + \frac{33075}{256}\zeta(4)\zeta(5) + \frac{29295}{64}\zeta(3)\zeta(6) + \frac{19845}{1024}\zeta(2)\zeta(7)$$

$$- \frac{40403475}{16384}\zeta(9).$$

Examples at the point (v):
For $m = 1$,

$$\int_0^1 \frac{\log(1-x)\log^2(x)\log(1+x^2)}{x}dx$$

$$= \frac{1461}{256}\zeta(5) + \frac{7}{128}\pi^2\zeta(3) + \frac{\pi^5}{96} - \frac{\pi^3}{8}G - \frac{\pi}{768}\psi^{(3)}\left(\frac{1}{4}\right).$$

For $m = 2$,

(continued)

$$\int_0^1 \frac{\log(1-x)\log^4(x)\log(1+x^2)}{x}dx$$

$$= \frac{121557}{1024}\zeta(7) + \frac{105}{512}\pi^2\zeta(5) + \frac{121}{2560}\pi^4\zeta(3) + \frac{9}{320}\pi^7 - \frac{5}{32}\pi^5 G$$

$$- \frac{\pi^3}{512}\psi^{(3)}\left(\frac{1}{4}\right) - \frac{\pi}{20480}\psi^{(5)}\left(\frac{1}{4}\right).$$

For $m = 3$,

$$\int_0^1 \frac{\log(1-x)\log^6(x)\log(1+x^2)}{x}dx$$

$$= \frac{41184405}{8192}\zeta(9) + \frac{6615}{4096}\pi^2\zeta(7) + \frac{1815}{4096}\pi^4\zeta(5) + \frac{2017}{14336}\pi^6\zeta(3) + \frac{479}{3584}\pi^9$$

$$- \frac{61}{128}\pi^7 G - \frac{25}{4096}\pi^5\psi^{(3)}\left(\frac{1}{4}\right) - \frac{3}{16384}\pi^3\psi^{(5)}\left(\frac{1}{4}\right) - \frac{\pi}{458752}\psi^{(7)}\left(\frac{1}{4}\right).$$

Examples at the point (vi):
For $m = 1$,

$$\int_0^1 \frac{\log(1+x)\log^2(x)\log(1+x^2)}{x}dx$$

$$= \frac{\pi}{768}\psi^{(3)}\left(\frac{1}{4}\right) + \frac{\pi^3}{8}G - \frac{\pi^5}{96} - \frac{5}{128}\pi^2\zeta(3) - \frac{1515}{256}\zeta(5);$$

For $m = 2$,

$$\int_0^1 \frac{\log(1+x)\log^4(x)\log(1+x^2)}{x}dx$$

$$= \frac{\pi^3}{512}\psi^{(3)}\left(\frac{1}{4}\right) + \frac{\pi}{20480}\psi^{(5)}\left(\frac{1}{4}\right) + \frac{5}{32}\pi^5 G - \frac{9}{320}\pi^7 - \frac{122283}{1024}\zeta(7)$$

$$- \frac{75}{512}\pi^2\zeta(5) - \frac{119}{2560}\pi^4\zeta(3);$$

For $m = 3$,

(continued)

$$\int_0^1 \frac{\log(1+x)\log^6(x)\log(1+x^2)}{x}dx$$

$$= \frac{\pi}{458752}\psi^{(7)}\left(\frac{1}{4}\right) + \frac{3}{16384}\pi^3\psi^{(5)}\left(\frac{1}{4}\right) + \frac{25}{4096}\pi^5\psi^{(3)}\left(\frac{1}{4}\right) + \frac{61}{128}\pi^7 G$$

$$- \frac{479}{3584}\pi^9 - \frac{2015}{14336}\pi^6\zeta(3) - \frac{1785}{4096}\pi^4\zeta(5) - \frac{4725}{4096}\pi^2\zeta(7) - \frac{41229675}{8192}\zeta(9).$$

1.54   Amazing and (Very) Useful Integral Beasts Involving $\log^2(\sin(x))$, $\log^2(\cos(x))$, $\log^3(\sin(x))$, and $\log^3(\cos(x))$

Prove that

$$(i) \int_0^{\pi/2} x\log^2(\sin(x))dx$$

$$= \frac{1}{24}\log^4(2) + \frac{1}{2}\log^2(2)\zeta(2) - \frac{19}{32}\zeta(4) + \text{Li}_4\left(\frac{1}{2}\right); \qquad (1.234)$$

$$(ii) \int_0^{\pi/2} x\log^2(\cos(x))dx$$

$$= \frac{79}{32}\zeta(4) + \log^2(2)\zeta(2) - \frac{1}{24}\log^4(2) - \text{Li}_4\left(\frac{1}{2}\right); \qquad (1.235)$$

$$(iii) \int_0^{\pi/2} x^2\log^2(\sin(x))dx$$

$$= \frac{1}{24}\log^4(2)\pi + \frac{1}{2}\log(2)\pi\zeta(3) - \frac{3}{320}\pi^5 + \pi\,\text{Li}_4\left(\frac{1}{2}\right); \qquad (1.236)$$

$$(iv) \int_0^{\pi/2} x^2\log^2(\cos(x))dx$$

$$= \frac{11}{1440}\pi^5 + \frac{1}{24}\log^2(2)\pi^3 + \frac{1}{2}\log(2)\pi\zeta(3); \qquad (1.237)$$

(continued)

$$(v) \int_0^{\pi/2} x \log^3(\sin(x))dx$$

$$= \frac{57}{32}\log(2)\zeta(4) - \frac{1}{2}\log^3(2)\zeta(2) - \frac{9}{8}\zeta(2)\zeta(3) - \frac{93}{128}\zeta(5)$$

$$- \frac{1}{40}\log^5(2) + 3\operatorname{Li}_5\left(\frac{1}{2}\right); \qquad (1.238)$$

$$(vi) \int_0^{\pi/2} x \log^3(\cos(x))dx$$

$$= \frac{93}{128}\zeta(5) - \frac{9}{8}\zeta(2)\zeta(3) - \log^3(2)\zeta(2) - \frac{237}{32}\log(2)\zeta(4)$$

$$+ \frac{1}{40}\log^5(2) - 3\operatorname{Li}_5\left(\frac{1}{2}\right); \qquad (1.239)$$

$$(vii) \int_0^{\pi/2} x^2 \log^3(\sin(x))dx$$

$$= \frac{9}{320}\log(2)\pi^5 - \frac{1}{40}\log^5(2)\pi - \frac{1}{8}\pi^3\zeta(3) - \frac{189}{128}\pi\zeta(5)$$

$$- \frac{3}{4}\log^2(2)\pi\zeta(3) + 3\pi\operatorname{Li}_5\left(\frac{1}{2}\right); \qquad (1.240)$$

$$(viii) \int_0^{\pi/2} x^2 \log^3(\cos(x))dx$$

$$= -\frac{11}{480}\log(2)\pi^5 - \frac{1}{24}\log^3(2)\pi^3 - \frac{3}{4}\log^2(2)\pi\zeta(3) - \frac{1}{8}\pi^3\zeta(3) - \frac{3}{4}\pi\zeta(5), \qquad (1.241)$$

where ζ represents the Riemann zeta function and Li_n denotes the polylogarithm function.

A *(super) challenging question:* Find an elegant solution to either point (v) or (vi), which avoids the use of advanced alternating harmonic series of weight 5.

1.55   Four Challenging Integrals with the Logarithm and Trigonometric Functions, Giving Nice Closed Forms

Prove that

$$(i) \int_0^{\pi/2} \theta \log(\sin(\theta)) \log^2(\cos(\theta)) d\theta$$

$$= \frac{1}{120} \log^5(2) - \frac{5}{6} \log^3(2)\zeta(2) - \frac{49}{32} \log(2)\zeta(4) + \frac{13}{32}\zeta(2)\zeta(3)$$

$$+ \frac{155}{128}\zeta(5) - \operatorname{Li}_5\left(\frac{1}{2}\right); \qquad (1.242)$$

$$(ii) \int_0^{\pi/2} \theta \log^2(\sin(\theta)) \log(\cos(\theta)) d\theta$$

$$= \frac{49}{32} \log(2)\zeta(4) - \frac{2}{3} \log^3(2)\zeta(2) - \frac{1}{120} \log^5(2) - \frac{1}{32}\zeta(2)\zeta(3)$$

$$- \frac{155}{128}\zeta(5) + \operatorname{Li}_5\left(\frac{1}{2}\right); \qquad (1.243)$$

$$(iii) \int_0^{\pi/2} \theta^2 \log(\sin(\theta)) \log^2(\cos(\theta)) d\theta$$

$$= \frac{121}{128}\pi\zeta(5) + \frac{3}{64}\pi^3\zeta(3) + \log^5(2)\frac{\pi}{120} - \log^3(2)\frac{\pi^3}{18} - \log(2)\frac{\pi^5}{72}$$

$$+ \log^2(2)\frac{\pi}{8}\zeta(3) - \pi \operatorname{Li}_5\left(\frac{1}{2}\right); \qquad (1.244)$$

$$(iv) \int_0^{\pi/2} \theta^2 \log^2(\sin(\theta)) \log(\cos(\theta)) d\theta$$

$$= \log(2)\frac{\pi^5}{320} - \log^3(2)\frac{\pi^3}{24} + \log^2(2)\frac{\pi}{8}\zeta(3) + \frac{\pi^3}{96}\zeta(3) - \frac{17}{64}\pi\zeta(5), \qquad (1.245)$$

where ζ represents the Riemann zeta function and Li_n denotes the polylogarithm function.

1.56 Advanced Integrals with Trigonometric Functions, Related to Fourier Series and Harmonic Series

Show that

$$(i)\ I_n(y) = \int_0^\pi \frac{\cos(n\theta)}{1 - 2y\cos(\theta) + y^2}\,d\theta = \pi\frac{y^n}{1 - y^2},\ |y| < 1. \qquad (1.246)$$

A special form obtained from the integral result at the point (i):

$$(ii)\ J_n(y) = \int_0^{\pi/2} \frac{\cos(2n\theta)}{\cos^2(\theta) + y^2\sin^2(\theta)}\,d\theta = \frac{\pi}{2}\frac{1}{y}\left(\frac{y-1}{y+1}\right)^n,\ y > 0. \qquad (1.247)$$

Advanced integrals with a Sylvester-type structure:

$$(iii)\ P(y) = \int_0^{\pi/2} \frac{\log^2(\cos(\theta))}{\cos^2(\theta) + y^2\sin^2(\theta)}\,d\theta = \frac{1}{4}\int_0^\infty \frac{\log^2(1 + x^2)}{1 + y^2 x^2}\,dx$$

$$= \frac{\pi^3}{24}\frac{1}{y} + \frac{\pi}{2}\frac{1}{y}\log^2\left(1 + \frac{1}{y}\right) + \frac{\pi}{2}\frac{1}{y}\mathrm{Li}_2\left(\frac{1-y}{1+y}\right),\ y > 0; \qquad (1.248)$$

$$(iv)\ Q(y) = \int_0^{\pi/2} \frac{\log^3(\cos(\theta))}{\cos^2(\theta) + y^2\sin^2(\theta)}\,d\theta = -\frac{1}{8}\int_0^\infty \frac{\log^3(1 + x^2)}{1 + y^2 x^2}\,dx$$

$$= \frac{\pi}{8}\left(6\zeta(3) - \log(2)\pi^2 - 4\log^3(2)\right)\frac{1}{y} + \frac{3}{8}\pi(4\log^2(2) + \pi^2)\frac{1}{y}\log\left(\frac{2y}{1+y}\right)$$

$$-\frac{3}{4}\log(2)\pi\frac{1}{y}\log^2\left(\frac{2y}{1+y}\right) - \frac{\pi}{4}\frac{1}{y}\log^3\left(\frac{2y}{1+y}\right) + \frac{3}{4}\pi\frac{\log(y)}{y}\log^2\left(\frac{2y}{1+y}\right)$$

$$-\frac{3}{4}\pi\frac{\log(1-y)}{y}\log^2\left(\frac{2y}{1+y}\right) - \frac{3}{2}\log(2)\pi\frac{1}{y}\mathrm{Li}_2\left(\frac{1-y}{1+y}\right)$$

$$-\frac{3}{4}\pi\frac{1}{y}\mathrm{Li}_3\left(\frac{1-y}{1+y}\right) - \frac{3}{2}\pi\frac{1}{y}\mathrm{Li}_3\left(\frac{2y}{1+y}\right),\ y > 0; \qquad (1.249)$$

$$(v)\ R(y) = \int_0^{\pi/2} \frac{\log^4(\cos(\theta))}{\cos^2(\theta) + y^2\sin^2(\theta)}\,d\theta = \frac{1}{16}\int_0^\infty \frac{\log^4(1 + x^2)}{1 + y^2 x^2}\,dx$$

$$= \frac{\pi}{96}(7\pi^4 + 24\log^2(2)\pi^2 + 48\log^4(2) - 288\log(2)\zeta(3))\frac{1}{y}$$

(continued)

$$- \frac{1}{2} \log(2)\pi (4 \log^2(2) + 3\pi^2) \frac{1}{y} \log \left(\frac{2y}{1+y} \right) + \frac{\pi^3}{2} \frac{1}{y} \log^2 \left(\frac{2y}{1+y} \right)$$

$$- 3 \log(2)\pi \frac{\log(y)}{y} \log^2 \left(\frac{2y}{1+y} \right) + 3 \log(2)\pi \frac{\log(1-y)}{y} \log^2 \left(\frac{2y}{1+y} \right)$$

$$+ \frac{3}{2} \log(2)\pi \frac{1}{y} \log^3 \left(\frac{2y}{1+y} \right) + \frac{\pi}{2} \frac{\log(y)}{y} \log^3 \left(\frac{2y}{1+y} \right) - \frac{\pi}{2} \frac{\log(1-y)}{y} \log^3 \left(\frac{2y}{1+y} \right)$$

$$- \frac{\pi}{8} \frac{1}{y} \log^4 \left(\frac{2y}{1+y} \right) + \frac{\pi}{4} (12 \log^2(2) + \pi^2) \frac{1}{y} \mathrm{Li}_2 \left(\frac{1-y}{1+y} \right)$$

$$+ 3 \log(2)\pi \frac{1}{y} \mathrm{Li}_3 \left(\frac{1-y}{1+y} \right) + 6 \log(2)\pi \frac{1}{y} \mathrm{Li}_3 \left(\frac{2y}{1+y} \right)$$

$$- \frac{3}{2} \pi \frac{1}{y} \mathrm{Li}_4 \left(\frac{1-y}{1+y} \right) - 3\pi \frac{1}{y} \mathrm{Li}_4 \left(\frac{2y}{1+y} \right) - 3\pi \frac{1}{y} \mathrm{Li}_4 \left(\frac{y-1}{2y} \right), \quad y > 0,$$

$$\tag{1.250}$$

where ζ represents the Riemann zeta function and Li_n denotes the polylogarithm function.

1.57 Gems of Integration Involving Splendid Ideas About Symmetry in Two Dimensions

Show that

$$(i) \int_0^1 \left(\frac{\pi}{4} + \arctan(x) \right)^2 \frac{\log(x)}{x^2 - 1} dx = \int_0^{\pi/4} \left(\frac{\pi}{4} + x \right)^2 \sec(2x) \log(\cot(x)) dx = \frac{\pi^4}{64}.$$

$$\tag{1.251}$$

A stunning generalization following the symmetry ideas at the point (i):

$$(ii) \int_0^1 \frac{\mathrm{arctanh}^n(x)}{x} \Re \left\{ \mathrm{Li}_{n+1} \left(\frac{1-x^2}{1+x^2} + i \frac{2x}{1+x^2} \right) \right\} dx$$

$$= \int_0^{\pi/4} \csc(x) \sec(x) \mathrm{arctanh}^n(\tan(x)) \Re \left\{ \mathrm{Li}_{n+1} (\cos(2x) + i \sin(2x)) \right\} dx$$

(continued)

$$= \frac{1}{2^{3n+1}}(2^{n+1} - 1)n!\zeta^2(n+1), \tag{1.252}$$

where ζ represents the Riemann zeta function and Li_n denotes the polylogarithm function.

1.58   More Gems of Integration, This Time Involving Splendid Ideas About Symmetry in Three Dimensions

Show that

(i) $\displaystyle\int_0^1 \frac{x \arctan(x)}{x^2 - 1} \log\left(\frac{1+x^2}{2}\right) dx = \int_0^{\pi/4} x \log(2\cos^2(x)) \sec(2x) \tan(x)dx = \frac{\pi^3}{192}.$
$$\tag{1.253}$$

Two special double integrals with the Legendre chi function of order 2 and the inverse tangent integral:

(ii) $\displaystyle\int_0^\infty \int_0^\infty \frac{(\tanh^2(x) + \tanh^2(y)) \log(\tanh(x)) \log(\tanh(y)) \chi_2(\tanh(x)\tanh(y))}{\tanh(x)\tanh(y)} dxdy$

$$= \frac{7}{11520}\pi^6; \tag{1.254}$$

(iii) $\displaystyle\int_0^{\pi/4} \int_0^{\pi/4} \frac{(\tan^2(x) + \tan^2(y)) \log(\tan(x)) \log(\tan(y)) \text{Ti}_2(\tan(x)\tan(y))}{\tan(x)\tan(y)} dxdy$

$$= \frac{1}{368640}\psi^{(5)}\left(\frac{1}{4}\right) - \frac{\pi^6}{1440} - \frac{2}{3}G^3, \tag{1.255}$$

where G is the Catalan's constant, $\psi^{(n)}$ designates the polygamma function, $\chi_2(x) = \displaystyle\int_0^x \frac{\text{arctanh}(t)}{t} dt$ denotes the Legendre chi function of order 2, and $\text{Ti}_2(x) = \displaystyle\int_0^x \frac{\arctan(t)}{t} dt$ indicates the inverse tangent integral.

1.59   The Complete Elliptic Integral of the First Kind at Play

Prove that

$$\int_0^1 \frac{K(x)}{1-x} \log\left(\frac{4x}{(1+x)^2}\right) dx = \frac{7}{4}\zeta(3) - \pi G, \tag{1.256}$$

where G is the Catalan's constant, ζ represents the Riemann zeta function,
and $K(k) = \displaystyle\int_0^{\pi/2} \frac{d\theta}{\sqrt{1-k^2\sin^2(\theta)}} = \int_0^1 \frac{dt}{\sqrt{(1-t^2)(1-k^2t^2)}}$ denotes the
complete elliptic integral of the first kind.

   A challenging question: Find a (simple) solution by reducing the calculations to a series result given in the fourth chapter.

1.60   Evaluating An Esoteric-Looking Integral Involving a Triple Series with Factorial Numbers, Leading to $\zeta(4)$

If we consider that $\Lambda(x) = \displaystyle\sum_{i=0}^{\infty}\sum_{j=0}^{\infty}\sum_{k=0}^{\infty} x^{i+j+k} \frac{i!\,j!\,k!}{(i+j+k+2)!}$, then show
that

$$\zeta(4)$$

$$= \frac{8}{9}\int_0^1 \left((4-x)(1-x)^2\Lambda(x-1) + \frac{(5-2x)(1-x)^2}{(2-x)^3}\Lambda\left(\frac{1-x}{2-x}\right)\right)\frac{\log(x)}{x-1}dx, \tag{1.257}$$

where ζ represents the Riemann zeta function.

Chapter 2
Hints

2.1 A Beautiful Integral by the English Mathematician James Joseph Sylvester

One might use that $\log(\sin(x)) = -\dfrac{1}{2}\displaystyle\int_0^{\pi/2}\dfrac{\cos^2(x)\sin(2a)}{1-\cos^2(x)\cos^2(a)}da,\ 0 < x < \pi,$
in order to prove the result at the point (i).

2.2 Strange Limits with Trigonometric Integrals

Consider the use of elementary inequalities with $\sin(x)$ and $\log(1+x)$.

2.3 Two Curious Logarithmic Integrals with Parameter

One way to go is based on the use of the generating function in (4.34), Sect. 4.6.

2.4 Exploring More Appealing Logarithmic Integrals with Parameter: The First Part

Do you know there is a cool closed form involving dilogarithms for the generalized integral $\displaystyle\int_0^a\dfrac{x\log(1\pm x)}{1+x^2}dx$ that you might find useful here? It also appeared in *(Almost) Impossible Integrals, Sums, and Series*.

© The Author(s), under exclusive license to Springer Nature Switzerland AG 2023
C. I. Vălean, *More (Almost) Impossible Integrals, Sums, and Series*, Problem
Books in Mathematics, https://doi.org/10.1007/978-3-031-21262-8_2

2.5 Exploring More Appealing Logarithmic Integrals with Parameter: The Second Part

For the first four points of the problem, one may exploit double integrals. As regards the last point, one may use established results with integrals.

2.6 More Good-Looking Logarithmic Integrals: The First Part

For the point (i) of the problem, one may exploit (3.13), the case $n = 1$, Sect. 3.3. Then, for the point (ii), one may use a generalized integral of the type, $\int_0^x \dfrac{\log(1+bt)}{1+at} dt$.

2.7 More Good-Looking Logarithmic Integrals: The Second Part

One might create two useful relations with the integrals at the points (i) and (ii) by differentiating and integrating back $\text{Ti}_2\left(\dfrac{1-x}{1+x}\right)$ and $\text{Ti}_2\left(\dfrac{2x}{1-x^2}\right)$.

2.8 More Good-Looking Logarithmic Integrals: The Third Part

One may obtain good inspiration from the previous sections.

2.9 Special and Challenging Integrals with Parameter Involving the Inverse Hyperbolic Tangent: The First Part

For a first solution to the point (i), one may start by considering a restriction like $0 < a \leq 1$ and then exploiting geometric series.

2.10 Special and Challenging Integrals with Parameter Involving the Inverse Hyperbolic Tangent: The Second Part

For all points, make the variable change $y = (1-x)/(1+x)$ and then, when needed, force a factorization in the denominator of the integrand in order to exploit complex numbers.

2.11 Some Startling Generalizations Involving Logarithmic Integrals with Trigonometric Parameters

For the last two points, one might find useful to calculate and consider the generalized integral in the form $\int_0^1 \frac{\log^n(1+x)}{a+x}\mathrm{d}x$.

2.12 More Startling and Enjoyable Logarithmic Integrals Involving Trigonometric Parameters

For the points (ii) and (iii) of the problem, combine simple algebraic identities with results connected to the previous sections.

2.13 Surprisingly Bewitching Trigonometric Integrals with Appealing Closed Forms: The First Act

Exploit results related to the next section.

2.14 Surprisingly Bewitching Trigonometric Integrals with Appealing Closed Forms: The Second Act

A possible option is to approach the integrals by constructing and exploiting recurrence relations.

2.15 Surprisingly Bewitching Trigonometric Integrals with Appealing Closed Forms: The Third Act

Again, one may think of exploiting recurrence relations. Also, a good idea is to think of expressing $\cos(n\theta)\cos^n(\theta)$ and $\sin(n\theta)\cos^n(\theta)$ in terms of series.

2.16 Surprisingly Bewitching Trigonometric Integrals with Appealing Closed Forms: The Fourth Act

As in the previous section, one might find useful to express $\cos(n\theta)\cos^n(\theta)$ and $\sin(n\theta)\cos^n(\theta)$ in terms of series.

2.17 Playing with a Hand of Fabulous Integrals: The First Part

For the first two points of the problem, one might establish recurrence relations which involve the use of Wallis' integrals.

2.18 Playing with a Hand of Fabulous Integrals: The Second Part

Exploit the results with integrals from the previous section.

2.19 Another Fabulous Integral Related to a Curious Binoharmonic Series Representation of $\zeta(3)$

For the point (i) of the problem, establish and exploit a recurrence relation.

2.20 Pairs of Appealing Integrals with the Logarithm Function, the Dilogarithm Function, and Trigonometric Functions

For a first solution, use the well-known Landen's dilogarithmic identity, $\operatorname{Li}_2(x) + \operatorname{Li}_2(x/(x-1)) = -1/2\log^2(1-x)$, to reduce the main integral to simpler integrals.

Then, for a second solution, find and exploit a Fourier-like series of the dilogarithm with a squared cosine argument.

2.21   Valuable Logarithmic Integrals Involving Skew-Harmonic Numbers: A First Partition

The logarithmic integrals at the points (i), (iv), and (vii) are interrelated, and for deriving a higher log power integral, we might want to exploit integrals with smaller log powers. For example, to derive the integral at the point (iv), we consider the integral result from the point (i) in variable k instead of n, where we might want to start with multiplying both sides by $(-1)^{k-1}$ and then sum from $k = 1$ to $k = n$.

2.22   Valuable Logarithmic Integrals Involving Skew-Harmonic Numbers: A Second Partition

Results linked to the nearby sections might be found helpful.

2.23   Valuable Logarithmic Integrals Involving Skew-Harmonic Numbers: A Third Partition

For the sum at the point (i), one might exploit the change of summation order in a double sum. Then, for the sum at the point (iv), we might exploit the complete homogeneous symmetric polynomial expressed in terms of power sums.

2.24   A Pair of Precious Generalized Logarithmic Integrals

For the point (i) of the problem, one might develop and exploit a recurrence relation, and for the point (ii), one might combine differentiation and the logarithmic integral, $\int_0^1 x^{n-1} \log^2(1-x)\mathrm{d}x = \dfrac{(\psi(n+1)+\gamma)^2}{n} + \dfrac{\zeta(2) - \psi^{(1)}(n+1)}{n}$.

2.25   Two Neat and Useful Generalizations with Logarithmic Integrals Involving Skew-Harmonic Numbers

To get a solution to the point (i), one might start from $\overline{H}_n^{(m)}$ and derive a useful recurrence relation. Then, to obtain a solution to the point (ii), one might combine a result tied to the previous section together with differentiation.

2.26   A Special Logarithmic Integral and a Generalization of It

One might successfully exploit a result related to the next section.

2.27   Three Useful Generalized Logarithmic Integrals Connected to Useful Generalized Alternating Harmonic Series

Reduce the first two integrals to double integrals where one may usefully exploit the symmetry.

2.28   Three Atypical Logarithmic Integrals Involving $\log(1 + x^3)$, Related to a Special Form of Symmetry in Double Integrals

Relate the integral at the point (i) of the problem to the double integral $\int_0^1 \left(\int_0^1 \frac{t^3}{(1+t^3)(1+tu+(tu)^2)} du \right) dt$, where one may exploit the symmetry.

2.29   Four Curious Integrals with Cosine Function Which Lead to Beautiful Closed Forms, plus Two Exotic Integrals

Find the proper trigonometric transformations for the arguments of the logarithms.

2.30   Aesthetic Integrals That Must Be Understood as Cauchy Principal Values, Generated by a Beautiful Generalization

It may be useful to find a generalized integral with parameter, leading to a simple closed form.

2.31 A Special Integral Generalization, Attacked with Strategies Involving the Cauchy Principal Value Integrals, That Also Leads to Two Famous Results by Ramanujan

At the point (i) of the problem, calculate and exploit the integral P.V. $\int_0^\infty \frac{\cos(ax)}{b^2-x^2}dx$, which we understand as a Cauchy principal value.

2.32 Four *Magical* Integrals Beautifully Calculated, Which Generate the Closed Forms of Four Classical Integrals

For the point (i) of the problem, prove and exploit the elementary integral representation, $e^\theta \int_\theta^\infty \cos(xa)e^{-a}da = \frac{\cos(\theta x) - x\sin(\theta x)}{1 + x^2}$.

2.33 A Bouquet of Captivating Integrals Involving Trigonometric and Hyperbolic Functions

As regards the first point of the problem, one might want to prove and exploit the fact that $\frac{2}{\sin(c)} \sum_{n=1}^\infty (-1)^{n-1} \sin(cn)e^{-bnx} = \frac{1}{\cosh(bx) + \cos(c)}$.

2.34 Interesting Integrals to Evaluate, One of Them Coming from Lord Kelvin's Work

Find $\int_0^\infty xe^{-ax^2-bx}dx$ and exploit it.

2.35 A Surprisingly Awesome Fractional Part Integral with Forms Involving $\sqrt{\tan(x)}$ and $\sqrt{\cot(x)}$

Use that $\{x\} = x - \lfloor x \rfloor$, where $\lfloor x \rfloor$ represents the integer part of x, and reduce the main calculations to the corresponding series.

2.36 A Superb Integral with Logarithms and the Inverse Tangent Function and a Wonderful Generalization of It

One may start with proving and using the integral result

$$\int_0^1 \frac{\arctan(yt)}{1+y^2t}\,dt = \frac{1}{2y^2}\arctan(y)\log(1+y^2).$$

2.37 More Wonderful Results Involving Integrals with Logarithms and the Inverse Tangent Function

The first two points of the problem may be obtained by using results from the problem statement sections corresponding to the preceding ones. For the point (iv) of the problem, we need to show and use that $\text{Ti}_3(x) + \text{Ti}_3\left(\dfrac{1}{x}\right) = \text{sgn}(x)\left(\dfrac{\pi^3}{16} + \dfrac{\pi}{4}\log^2(|x|)\right)$.

2.38 Powerful and Useful Sums of Integrals with Logarithms and the Inverse Tangent Function

Regarding the first point of the problem, one might attack the middle integral with integration by parts and then cleverly exploit algebraic identities.

2.39 Two Beautiful Sums of Integrals, Each One Involving Three Integrals, Leading to a Possible Unexpected Result

For each of the two points of the problem, we might like to cleverly merge the first two integrals and then turn everything into a useful double integral which we may further exploit successfully in order to arrive at the desired result.

2.40 Tackling Curious Logarithmic Integrals with a Radical in the Denominator

For the point (ii) of the problem, exploit the result in (1.179) from Sect. 1.39. Then, for the point (iii) of the problem, one may start with proving and using that

$$\int_0^y \frac{1}{(1 - y^2 x^2)\sqrt{1 + x^2}} dx = \frac{\operatorname{arctanh}(y)}{\sqrt{1 + y^2}}, \quad |y| < 1.$$

2.41 Calculating More Logarithmic Integrals with a Radical in the Denominator

For the point (i) of the problem, one might consider the variable change $1 + 2x^2 - 2x\sqrt{1 + x^2} = t$ that gives $x = \dfrac{1 - t}{2\sqrt{t}}$. Then, for the point (ii), one might get inspiration from the solution section connected to the previous section. As regards the point (vi) of the problem, one might find useful to think of the Legendre chi function of order 2 and its classical identities.

2.42 Somewhat Atypical Integrals with Curious Closed Forms

One way to get a solution to the point (i) of the problem involves the use of the series representation, $\displaystyle\sum_{k=1}^{\infty} (-1)^{k-1} x^{2k+1} \frac{H_{2k}}{2k + 1} = \frac{1}{2} \arctan(x) \log(1 + x^2), \ |x| \leq 1.$
For the point (ii) of the problem, one might exploit the result in (1.28), Sect. 1.8.

2.43 More Atypical Integrals with Curious Closed Forms

At the first point of the problem, we might exploit the result in (1.76), Sect. 1.16.

2.44 A Wonderful Trigonometric Integral by Larry Glasser

Exploit the integral result in (1.31), Sect. 1.9.

2.45 Resistant Logarithmic Integrals That Are Good to Know

At the first point of the problem, one might design a solution by combining ingeniously two forms of the beta function in order to create two useful relations from which to extract the desired result. The integrals at the points (ii)-(iv) could be turned into sums of harmonic series.

2.46 Appealing Parameterized Integrals with Logarithms and the Dilogarithm Function, Related to Harmonic Series

For instance, at the first two points of the problem, one might exploit the integral result in (1.125), Sect. 1.24.

2.47 An Encounter with Six Useful Integrals Involving Logarithms and the Dilogarithm

As a first step, one might want to reduce the integrals at the first two points to their corresponding harmonic series.

2.48 A *Battle* with Three Challenging Integrals Involving Logarithms and the Dilogarithm

One may start with the point (ii) of the problem, where it is useful to employ the Cauchy product of two series of $\log(1 - x)\operatorname{Li}_2(x)$, next use this result to extract the value of the series $\displaystyle\sum_{n=1}^{\infty} \frac{H_n H_{2n}}{n^3}$, and then return to the point (i) of the problem.

2.49 Fascinating Polylogarithmic Integrals with Parameter Involving the Cauchy Product of Two Series

Derive and exploit the Cauchy products of $(\operatorname{Li}_2(x))^2$, $(\operatorname{Li}_3(x))^2$, and $(\operatorname{Li}_4(x))^2$.

2.50 A Titan Involving Alternating Harmonic Series of Weight 7

Exploiting the relations with integrals in Sect. 1.46 leads to a key series result to use here.

2.51 A Tough Integral Approached by Clever Transformations

A Landen-type identity in the form of a series might be useful here.

2.52 An Unexpected Closed Form, Involving Catalan's Constant, of a Nice Integral with the Dilogarithm

For a first solution, exploit the result in (1.253), Sect. 1.58.

2.53 A Group of Six Special, Challenging Generalized Integrals Involving Curious Closed Forms

For the first two points, exploit the Fourier series of the Bernoulli polynomials.

2.54 Amazing and (Very) Useful Integral Beasts Involving $\log^2(\sin(x))$, $\log^2(\cos(x))$, $\log^3(\sin(x))$, and $\log^3(\cos(x))$

An (excellent) idea would be to start from the Fourier series of $\log^2(\sin(x))$ and $\log^3(\sin(x))$.

2.55 Four Challenging Integrals with the Logarithm and Trigonometric Functions, Giving Nice Closed Forms

Some special Fourier series could be useful, and the same as regards exploiting simple algebraic identities.

2.56 Advanced Integrals with Trigonometric Functions, Related to Fourier Series and Harmonic Series

Snake Oil Method is one of the possible options to consider at the point (i). For the last three points of the problem, we may exploit the point (ii) together with Fourier series found in the fourth chapter.

2.57 Gems of Integration Involving Splendid Ideas About Symmetry in Two Dimensions

Prove and exploit the integral result, $\left(\dfrac{\pi}{2} - \arctan(t)\right)^2 = 2\displaystyle\int_0^1 \dfrac{x \operatorname{arctanh}(x)}{t^2 + x^2}\mathrm{d}x,\, t>0.$

2.58 More Gems of Integration, This Time Involving Splendid Ideas About Symmetry in Three Dimensions

For every point, consider a reduction to a triple integral where to exploit the symmetry.

2.59 The Complete Elliptic Integral of the First Kind at Play

Prove and exploit the fact that $\displaystyle\int_y^1 \dfrac{1}{(1+x)(y+x)}\mathrm{d}x = -\dfrac{1}{1-y}\log\left(\dfrac{4y}{(1+y)^2}\right).$

2.60 Evaluating An Esoteric-Looking Integral Involving a Triple Series with Factorial Numbers, Leading to $\zeta(4)$

Establish and exploit a functional equation of $\Lambda(x)$.

Chapter 3
Solutions

3.1 A Beautiful Integral by the English Mathematician James Joseph Sylvester

Solution I have always found it curious and fascinating to learn, when possible, historical facts behind the integrals I meet. There is a great *habit* I noticed in *Inside Interesting Integrals* (in both 2014 and 2020 versions), where Paul Nahin shares such historical information about various interesting integrals he presents, and in the following I will mention, as examples, Dalzell's integral,[1] $\int_0^1 \frac{x^4(1-x)^4}{1+x^2}dx$ (see [66, pp.23–25]), which leads to the neat inequality $\frac{22}{7} - \pi > 0$, and Dini's integral, $\int_0^\pi \log(1 - 2\alpha\cos(x) + \alpha^2)dx$ (see [66, pp.140–142]). They both are well-known in the mathematical literature.

Now, on the pages of the book *Polylogarithms and Associated Functions* by Leonard Lewin, more precisely in [21, Chapter 8, p.275], we find the integral at the point (*i*), called *Sylvester's formula*, which leads to a reference by the English mathematician James Joseph Sylvester (1814–1897), that is, *Note on Certain Definite Integrals*, Quart. J. Math. 3:319–324 (1858). Another integral with a name!

---

[1] By exploiting the polynomial long division and making simple calculations, as shown by Paul in *Inside Interesting Integrals*, we obtain that $\int_0^1 \frac{x^4(1-x)^4}{1+x^2}dx = \frac{22}{7} - \pi > 0$, and the inequality comes from the fact that the integrand is never negative. The rational number $\frac{22}{7}$ might be seen as an approximation of π, one that exceeds π (these details are analyzed in the mentioned book). Also, the integral appeared as a first problem in the Putnam Competition (1968).

© The Author(s), under exclusive license to Springer Nature Switzerland AG 2023   89
C. I. Vălean, *More (Almost) Impossible Integrals, Sums, and Series*, Problem Books in Mathematics, https://doi.org/10.1007/978-3-031-21262-8_3

In order to perform the calculations, we'll want to use two simple results, that is, $\log(\sin(x)) = -\dfrac{1}{2}\displaystyle\int_0^{\pi/2} \dfrac{\cos^2(x)\sin(2a)}{1-\cos^2(x)\cos^2(a)}\,da$, $0 < x < \pi$, and the following trigonometric integral also found and calculated in *(Almost) Impossible Integrals, Sums, and Series*: $\displaystyle\int_0^{\pi/2} \dfrac{dx}{1-a\sin^2(x)} = \int_0^{\pi/2}\dfrac{dx}{1-a\cos^2(x)} = \dfrac{\pi}{2}\dfrac{1}{\sqrt{1-a}}$, $a <$ 1 (see [76, Chapter 3, p.212]). So, let's proceed with the calculations to the point (*i*), and then we write

$$\int_0^{\pi/2} \frac{\log(\sin(x))}{\cos^2(x) + y^2\sin^2(x)}\,dx \overset{\pi/2-x=t}{=} \int_0^{\pi/2} \frac{\log(\cos(t))}{\sin^2(t) + y^2\cos^2(t)}\,dt$$

$$= -\frac{1}{2}\int_0^{\pi/2}\left(\int_0^{\pi/2} \frac{\cos^2(x)\sin(2a)}{(\cos^2(x)+y^2\sin^2(x))(1-\cos^2(x)\cos^2(a))}\,da\right)dx$$

{reverse the order of integration}

$$= -\frac{1}{2y^2}\int_0^{\pi/2}\sin(2a)\left(\int_0^{\pi/2}\frac{\cos^2(x)}{(1-(1-1/y^2)\cos^2(x))(1-\cos^2(a)\cos^2(x))}\,dx\right)da$$

$$= \frac{1}{2}\int_0^{\pi/2}\frac{\sin(2a)}{1-y^2+y^2\cos^2(a)}\left(\int_0^{\pi/2}\frac{1}{1-(1-1/y^2)\cos^2(x)}\,dx\right.$$

$$-\left.\int_0^{\pi/2}\frac{1}{1-\cos^2(a)\cos^2(x)}\,dx\right)da = \frac{1}{2}\int_0^{\pi/2}\frac{\sin(2a)}{1-y^2\sin^2(a)}\cdot\frac{\pi}{2}\left(\frac{y\sin(a)-1}{\sin(a)}\right)da$$

$$= -\frac{\pi}{2}\int_0^{\pi/2}\frac{\cos(a)}{1+y\sin(a)}\,da = -\frac{\pi}{2}\frac{1}{y}\int_0^{\pi/2}\frac{(1+y\sin(a))'}{1+y\sin(a)}\,da$$

$$= -\frac{\pi}{2}\frac{1}{y}\log(1+y\sin(a))\Big|_{a=0}^{a=\pi/2} = -\frac{\pi}{2}\frac{\log(1+y)}{y},$$

and the solution to the point (*i*) is complete.

For a second (wonderful) solution, the curious reader might want to exploit a powerful idea that you'll find in a later section of the book, where I calculate more advanced integrals with a Sylvester-type structure. More precisely, it is about Sect. 3.56. For example, exploiting the strategy there together with the Fourier series, $\log(\sin(x)) = -\log(2) - \displaystyle\sum_{k=1}^{\infty}\frac{\cos(2nx)}{n}$, will elegantly give the desired result.

Further, if we replace y by $1/\sqrt{1+y^2}$ in the main integral, then consider the trigonometric identity $\sin^2(x) + \cos^2(x) = 1$, and rearrange, we get

$$\int_0^{\pi/2} \frac{\log(\sin(x))}{1 + y^2 \cos^2(x)} dx \overset{\pi/2 - x = t}{=} \int_0^{\pi/2} \frac{\log(\cos(t))}{1 + y^2 \sin^2(t)} dt$$

$$= -\frac{\pi}{2} \frac{1}{\sqrt{1 + y^2}} \log\left(\frac{1 + \sqrt{1 + y^2}}{\sqrt{1 + y^2}}\right),$$

which is the first form obtained based on the principal integral.

Also, if we replace y by $\sqrt{1 - y^2}$ in the main integral, we have

$$\int_0^{\pi/2} \frac{\log(\sin(x))}{1 - y^2 \sin^2(x)} dx \overset{\pi/2 - x = t}{=} \int_0^{\pi/2} \frac{\log(\cos(t))}{1 - y^2 \cos^2(t)} dt = -\frac{\pi}{2} \frac{\log\left(1 + \sqrt{1 - y^2}\right)}{\sqrt{1 - y^2}},$$

which is the second form obtained based on the principal integral.

As regards this last integral form, we might also like to consider the *challenging question*, that is, to prove it by using results involving harmonic series.

So, without loss of generality, we assume that $0 < y < 1$, and then we write

$$\int_0^{\pi/2} \frac{\log(\sin(x))}{1 - y^2 \sin^2(x)} dx = \int_0^{\pi/2} \log(\sin(x)) \sum_{n=0}^{\infty} (y \sin(x))^{2n} dx$$

$$= \sum_{n=0}^{\infty} y^{2n} \int_0^{\pi/2} \sin^{2n}(x) \log(\sin(x)) dx$$

{exploit the integral result in (3.126), Sect. 3.19}

$$= \frac{\pi}{2} \sum_{n=0}^{\infty} y^{2n} \frac{1}{2^{2n}} \binom{2n}{n} (H_{2n} - H_n - \log(2))$$

$$\left\{\text{use that } \int_0^1 \frac{x^{2n}}{1 + x} dx = H_n - H_{2n} + \log(2), \text{ which is straightforward if}\right\}$$

{we use the integral result in (1.100), Sect. 1.21, and integrate by parts}

$$= -\frac{\pi}{2} \sum_{n=0}^{\infty} y^{2n} \frac{1}{2^{2n}} \binom{2n}{n} \int_0^1 \frac{x^{2n}}{1 + x} dx = -\frac{\pi}{2} \int_0^1 \frac{1}{1 + x} \sum_{n=0}^{\infty} (xy)^{2n} \frac{1}{2^{2n}} \binom{2n}{n} dx$$

$$= -\frac{\pi}{2} \int_0^1 \frac{1}{(1 + x)\sqrt{1 - y^2 x^2}} dx \overset{\frac{1}{y} \frac{1 - t^2}{1 + t^2} = x}{=} -\frac{\pi}{1 + y} \int_{\sqrt{\frac{1-y}{1+y}}}^1 \frac{1}{1 - (\sqrt{(1 - y)/(1 + y)})^2 t^2} dt$$

$$= -\frac{\pi}{\sqrt{1-y^2}} \operatorname{arctanh}\left(\sqrt{\frac{1-y}{1+y}}\, t\right)\Bigg|_{t=\sqrt{\frac{1-y}{1+y}}}^{t=1} = -\frac{\pi}{2}\frac{\log\left(1+\sqrt{1-y^2}\right)}{\sqrt{1-y^2}},$$

where the last equality is obtained by using that $\operatorname{arctanh}(x) = \frac{1}{2}\log\left(\frac{1+x}{1-x}\right)$,

and hence, by simple algebraic manipulations we have that $\operatorname{arctanh}\left(\sqrt{\frac{1-y}{1+y}}\right) =$

$\frac{1}{2}\log\left(\frac{1+\sqrt{1-y^2}}{y}\right)$ and $\operatorname{arctanh}\left(\frac{1-y}{1+y}\right) = -\frac{1}{2}\log(y)$, and this brings an end

to the solution. In the calculations above I also used that $\displaystyle\sum_{n=0}^{\infty} x^n \binom{2n}{n} = \frac{1}{\sqrt{1-4x}}$,

known as the generating function of the central binomial coefficients, which is

immediately obtained by using the binomial series and the fact that $\binom{2n}{n} =$

$(-1)^n 4^n \binom{-1/2}{n}$, or, to derive it differently, we can simply exploit the following

Wallis' integral case, $\displaystyle\int_0^{\pi/2} \sin^{2n}(x)\mathrm{d}x = \frac{\pi}{2}\cdot\frac{(2n-1)!!}{(2n)!!} = \frac{\pi}{2^{2n+1}}\binom{2n}{n}.$

Finally, I would add two notes. First, do not miss the second solution suggested above to the main integral, which uses Fourier series!

Second, exploiting $\displaystyle\int_0^{\pi/2} \frac{\log(\sin(x))}{\cos^2(x) + y^2\sin^2(x)}\mathrm{d}x = -\frac{\pi}{2}\frac{\log(1+y)}{y}$ and inte-

grating from $y = 0$ to $y = c$, we get another curious representation of the dilogarithm,

$$\int_0^{\pi/2} \csc(2x)\arctan(c\tan(x))\log(\sin(x))\mathrm{d}x = \frac{\pi}{4}\operatorname{Li}_2(-c). \tag{3.1}$$

The work above also answers the *challenging question*. What other ways would you like to consider?

3.2 Strange Limits with Trigonometric Integrals

Solution In this section I have prepared two unexpectedly beautiful limits, an enjoyable moment, and it will be nice to see that despite the fact that the first limit looks daunting, it is easily manageable by cleverly juggling with high school knowledge.

Before reading further, I highly recommend you to take a break and try your best first and have fun with the first limit (at least)!

For the first part of the problem, let's initially split the limit as follows:

$$\lim_{a\to 0}\int_a^\infty \frac{\sin(b_1 x)\sin(b_2 x)\cdots\sin(b_n x)}{x}(\log(a+x)-\log(x))^d dx$$

$$=\lim_{a\to 0}\left(\int_a^c+\int_c^\infty\right)\frac{\sin(b_1 x)\sin(b_2 x)\cdots\sin(b_n x)}{x}(\log(a+x)-\log(x))^d dx$$

$$=\lim_{a\to 0}\int_a^c \frac{\sin(b_1 x)\sin(b_2 x)\cdots\sin(b_n x)}{x}(\log(a+x)-\log(x))^d dx$$

$$+\lim_{a\to 0}\int_c^\infty \frac{\sin(b_1 x)\sin(b_2 x)\cdots\sin(b_n x)}{x}(\log(a+x)-\log(x))^d dx, \qquad (3.2)$$

where we chose a, c with $c > a > 0$.

Then, considering the absolute value of the first resulting integral in (3.2), and assuming $d \neq 1$, we have that

$$\left|\int_a^c \frac{\sin(b_1 x)\sin(b_2 x)\cdots\sin(b_n x)}{x}(\log(a+x)-\log(x))^d dx\right|$$

$$\leq \int_a^c \frac{|\sin(b_1 x)\sin(b_2 x)\cdots\sin(b_n x)|}{x}(\log(a+x)-\log(x))^d dx$$

$$= \int_a^c \frac{|\sin(b_1 x)|\cdot|\sin(b_2 x)|\cdots|\sin(b_n x)|}{x}(\log(a+x)-\log(x))^d dx$$

{use that $|\sin(b_2 x)| \leq 1, |\sin(b_3 x)| \leq 1, \ldots, |\sin(b_n x)| \leq 1$}

$$\leq \int_a^c \frac{|\sin(b_1 x)|}{x}(\log(a+x)-\log(x))^d dx$$

{make use of the inequality $|\sin(x)| \leq |x|, \ x \in \mathbb{R}$}

$$\leq |b_1|\int_a^c \log^d\left(1+\frac{a}{x}\right) dx$$

{employ the elementary inequality $\log(1+x) \leq x, \ x > -1$}

$$\leq |b_1|a^d\int_a^c \frac{1}{x^d} dx = |b_1|a^d c^{1-d}\frac{1-\left(\frac{a}{c}\right)^{1-d}}{1-d}. \qquad (3.3)$$

To treat the case $d = 1$, we first take the limit $d \to 1$ in (3.3), which gives

$$|b_1| \lim_{d \to 1} a^d c^{1-d} \frac{1 - \left(\frac{a}{c}\right)^{1-d}}{1 - d} = |b_1| a (\log(c) - \log(a)), \qquad (3.4)$$

where we recognize the use of the elementary limit, $\displaystyle\lim_{x \to 0} \frac{a^x - 1}{x} = \log(a), \ a > 0$.

Now, if we combine the results in (3.3) and (3.4), and then let $a \to 0$, we get

$$\lim_{a \to 0} \left| \int_a^c \frac{\sin(b_1 x) \sin(b_2 x) \cdots \sin(b_n x)}{x} (\log(a + x) - \log(x)) dx \right|$$

$$\leq |b_1| \lim_{a \to 0} a (\log(c) - \log(a)) = |b_1| \log(c) \underbrace{\lim_{a \to 0} a}_{0} - |b_1| \underbrace{\lim_{a \to 0} a \log(a)}_{0} = 0,$$

where for the last limit I applied L'Hospital's rule, and therefore we have that

$$\lim_{a \to 0} \int_a^c \frac{\sin(b_1 x) \sin(b_2 x) \cdots \sin(b_n x)}{x} (\log(a + x) - \log(x)) dx = 0. \qquad (3.5)$$

If $d > 0$, $d \neq 1$, based on (3.3) we have that

$$\left| \int_a^c \frac{\sin(b_1 x) \sin(b_2 x) \cdots \sin(b_n x)}{x} (\log(a + x) - \log(x))^d dx \right|$$

$$\leq |b_1| a^d c^{1-d} \frac{1 - \left(\frac{a}{c}\right)^{1-d}}{1 - d} = |b_1| \frac{a^d c^{1-d} - a}{1 - d},$$

which promptly leads to

$$\lim_{a \to 0} \int_a^c \frac{\sin(b_1 x) \sin(b_2 x) \cdots \sin(b_n x)}{x} (\log(a + x) - \log(x))^d dx = 0. \qquad (3.6)$$

For the other resulting integral in (3.2), we proceed as follows:

$$\left| \int_c^\infty \frac{\sin(b_1 x) \sin(b_2 x) \cdots \sin(b_n x)}{x} (\log(a + x) - \log(x))^d dx \right|$$

$$\leq \int_c^\infty \frac{|\sin(b_1 x) \sin(b_2 x) \cdots \sin(b_n x)|}{x} \log^d \left(1 + \frac{a}{x}\right) dx$$

$$= \int_c^\infty \frac{|\sin(b_1 x)| \cdot |\sin(b_2 x)| \cdots |\sin(b_n x)|}{x} \log^d \left(1 + \frac{a}{x}\right) dx$$

$$\{\text{use that } |\sin(b_1 x)| \leq 1, |\sin(b_2 x)| \leq 1, \ldots, |\sin(b_n x)| \leq 1\}$$

{together with the fact that $\log(1 + x) \leq x,\ x > -1$}

$$\leq a^d \int_c^\infty \frac{1}{x^{d+1}}\,dy = \frac{a^d}{dc^d}. \tag{3.7}$$

Letting $a \to 0$ in (3.7), we get

$$\lim_{a \to 0} \left| \int_c^\infty \frac{\sin(b_1 x)\sin(b_2 x)\cdots\sin(b_n x)}{x}(\log(a + x) - \log(x))^d\,dx \right| \leq \lim_{a \to 0}\frac{a^d}{dc^d} = 0,$$

whence we obtain that

$$\lim_{a \to 0} \int_c^\infty \frac{\sin(b_1 x)\sin(b_2 x)\cdots\sin(b_n x)}{x}(\log(a + x) - \log(x))^d\,dx = 0. \tag{3.8}$$

Collecting (3.5), (3.6), and (3.8) in (3.2), we conclude that for the generalized case with the given conditions we get

$$\lim_{a \to 0} \int_a^\infty \frac{\sin(b_1 x)\sin(b_2 x)\cdots\sin(b_n x)}{x}(\log(a + x) - \log(x))^d\,dx = 0,$$

and the solution to the point (i) of the problem is finalized.

To calculate the generalized limit at the point (ii), we exploit the previous result, and then we write

$$\lim_{a \to 0} \int_a^\infty \frac{\cos(b_1 x)\cos(b_2 x)\cdots\cos(b_n x)}{x}(\log(a + x) - \log(x))\,dx$$

$\left\{\text{make use of the elementary trigonometric identity } \cos(x) = 1 - 2\sin^2\left(\frac{x}{2}\right)\right\}$

$$= \lim_{a \to 0} \int_a^\infty \frac{\displaystyle\prod_{k=1}^n \left(1 - 2\sin^2\left(\frac{b_k x}{2}\right)\right)}{x}(\log(a + x) - \log(x))\,dx,$$

where expanding the product yields 2^n limits of which $2^n - 1$ are of the type

$$\lim_{a \to 0} \int_a^\infty \frac{\sin(t_1 x)\sin(t_2 x)\cdots\sin(t_m x)}{x}(\log(a + x) - \log(x))\,dx,$$

and one limit of the type

$$\lim_{a \to 0} \int_a^\infty \frac{\log(a + x) - \log(x)}{x}\,dx.$$

Since all $2^n - 1$ limits mentioned above equal 0, our initial limit becomes

$$\lim_{a\to 0} \int_a^\infty \frac{\cos(b_1 x)\cos(b_2 x)\cdots\cos(b_n x)}{x}(\log(a+x) - \log(x))$$

$$= \lim_{a\to 0} \int_a^\infty \frac{\log(a+x) - \log(x)}{x}dx = \lim_{a\to 0} \int_a^\infty \frac{\log\left(1+\dfrac{a}{x}\right)}{x}dx$$

$$\overset{a/x=t}{=} \int_0^1 \frac{\log(1+t)}{t}dt = \int_0^1 \sum_{n=1}^\infty (-1)^{n-1}\frac{t^{n-1}}{n}dt = \sum_{n=1}^\infty (-1)^{n-1}\frac{1}{n^2} = \frac{1}{2}\zeta(2),$$

which is the desired limit, and the solution to the point (ii) of the problem is finalized.

We might agree that looking at the problem statement and then at the values of the limits found at the end of each of the two solutions, there is an impressive antithesis to admire!

3.3 Two Curious Logarithmic Integrals with Parameter

Solution The present section is part of a cascade of sections which contain solutions to various fun and intriguing logarithmic integrals with parameter.

In the following, for a first approach, the work will be based on a restricted value of a, $0 < a \leq 1$, without considering the requirements of the *challenging question*, like not using harmonic series or complex numbers.

So, if we consider the generating function in (4.34), the first equality, Sect. 4.6, where we replace x by $i\sqrt{x}$, divide both sides by x, and take the real part, we get

$$\sum_{n=1}^\infty (-1)^n x^{n-1}\frac{H_{2n}}{2n} = \Re\left\{\frac{1}{2}\frac{\log^2(1-i\sqrt{x})}{x} + \frac{\text{Li}_2(i\sqrt{x})}{x}\right\}. \tag{3.9}$$

So far, so good! This is one of the key subresults needed in this way of approaching the problem. Further, we also need the fact that

$$\int_0^1 x^{n-1}\log(1-x)dx = -\frac{H_n}{n}, \tag{3.10}$$

which is straightforward if we use the Taylor series[2] of $\log(1-x)$. The integral may also be found elementarily calculated in [76, Chapter 3, p.59].

---

[2] We note we may write that $\displaystyle\int_0^1 x^{n-1}\log(1-x)dx = -\int_0^1 \sum_{k=1}^\infty \frac{x^{k+n-1}}{k}dx =$

$\displaystyle -\sum_{k=1}^\infty \int_0^1 \frac{x^{k+n-1}}{k}dx = -\sum_{k=1}^\infty \frac{1}{k(k+n)} = -\frac{1}{n}\sum_{k=1}^\infty \sum_{j=1}^n \frac{1}{(j+k-1)(j+k)} =$

At this point, we return to the integral at the point (i) of the problem, and assuming $0 < a \leq 1$, we write

$$\int_0^1 \frac{x \log(1-x)}{1+ax^2} dx = \int_0^1 \log(1-x) \sum_{n=1}^\infty (-1)^{n-1} a^{n-1} x^{2n-1} dx$$

{reverse the order of summation and integration}

$$= \sum_{n=1}^\infty (-1)^{n-1} a^{n-1} \int_0^1 x^{2n-1} \log(1-x) dx$$

$$\overset{\text{use (3.10)}}{=} \sum_{n=1}^\infty (-1)^n a^{n-1} \frac{H_{2n}}{2n} \overset{\text{use (3.9)}}{=} \Re \left\{ \frac{1}{2} \frac{\log^2(1-i\sqrt{a})}{a} + \frac{\text{Li}_2(i\sqrt{a})}{a} \right\}$$

$$= -\frac{1}{2a} \arctan^2(\sqrt{a}) - \frac{1}{4a} \text{Li}_2 \left(\frac{a}{a+1} \right),$$

which is one of the desired closed forms. In the calculations I used the logarithm with a complex argument, $\log(a+ib) = \log(\sqrt{a^2+b^2}) + i \arctan(b/a)$, $a > 0$, the fact that $\Re\{\text{Li}_2(i\sqrt{a})\} = \Re \left\{ \sum_{n=1}^\infty \frac{(i\sqrt{a})^n}{n^2} \right\} = \sum_{n=1}^\infty \frac{(i\sqrt{a})^{2n}}{(2n)^2} = \frac{1}{4} \sum_{n=1}^\infty (-1)^n \frac{a^n}{n^2} = \frac{1}{4} \text{Li}_2(-a)$, and the Landen's dilogarithmic identity, (see [21, Chapter 1, p.5], [70, Chapter 2, p.107]) $\text{Li}_2(x) + \text{Li}_2(x/(x-1)) = -\frac{1}{2} \log^2(1-x)$, with x replaced by $-x$. To obtain the Landen's dilogarithmic identity, we simply differentiate $\text{Li}_2(x/(x-1))$ and then integrate back.

For a different solution above, we could cleverly exploit the Cauchy product as in the solution flow of (6.441), Sect. 6.57.

To get the other closed form involving the inverse sine function, we use the trigonometric identity, $\arctan(x) = \arcsin(x/\sqrt{1+x^2})$, which readily gives

$$\int_0^1 \frac{x \log(1-x)}{1+ax^2} dx = -\frac{1}{2a} \arcsin^2 \left(\sqrt{\frac{a}{a+1}} \right) - \frac{1}{4a} \text{Li}_2 \left(\frac{a}{a+1} \right).$$

$$-\frac{1}{n} \sum_{j=1}^n \sum_{k=1}^\infty \frac{1}{(j+k-1)(j+k)} = -\frac{1}{n} \sum_{j=1}^n \sum_{k=1}^\infty \left(\frac{1}{j+k-1} - \frac{1}{j+k} \right) = -\frac{1}{n} \sum_{j=1}^n \frac{1}{j} = -\frac{H_n}{n},$$

which is the desired result.

In order to obtain the result at the point (ii), we'll want to exploit the previous result, and then we write

$$\int_0^1 \frac{x \log(1+x)}{1+ax^2} dx = \int_0^1 \frac{x \log(1-x^2)}{1+ax^2} dx - \int_0^1 \frac{x \log(1-x)}{1+ax^2} dx. \quad (3.11)$$

Then, we focus on the first integral in the right-hand side, where we let the variable change $x^2 = y$, and returning to the variable in x, we have

$$\int_0^1 \frac{x \log(1-x^2)}{1+ax^2} dx = \frac{1}{2} \int_0^1 \frac{\log(1-x)}{1+ax} dx = \frac{1}{2} \int_0^1 \sum_{n=1}^{\infty} (-1)^{n-1} (ax)^{n-1} \log(1-x) dx$$

{reverse the order of summation and integration}

$$= \frac{1}{2} \sum_{n=1}^{\infty} (-1)^{n-1} a^{n-1} \int_0^1 x^{n-1} \log(1-x) dx$$

$$\overset{\text{use (3.10)}}{=} -\frac{1}{2} \sum_{n=1}^{\infty} (-1)^{n-1} a^{n-1} \frac{H_n}{n} = -\frac{1}{2a} \text{Li}_2\left(\frac{a}{a+1}\right), \quad (3.12)$$

where the last equality comes from the generating function in (4.34), the second equality, Sect. 4.6. Alternatively, if we make the variable change $1 - x = y$ in the integral after the first equal sign in (3.12), and rearrange, we get $\frac{1}{1+a} \int_0^1 \frac{\log(y)}{1-a/(1+a)y} dy$, and if we use the generalization,

$$\int_0^1 \frac{y \log^n(x)}{1-yx} dx = (-1)^n n! \text{Li}_{n+1}(y), \ y \in (-\infty, 1], \quad (3.13)$$

which may be found in [76, Chapter 1, p.4], we arrive at the value, $-\frac{1}{a} \text{Li}_2\left(\frac{a}{a+1}\right)$.

So, if we combine the result from the point (i) and the results from (3.11) and (3.12), we arrive at the closed form of the integral at the point (ii).

A non-obvious approach

Now, before getting the desired results in a special way, which follows the requirements of the *challenging question*, you might carefully follow Sects. 3.17 and 3.18 and then return to this point and step further. If we go back to the integral at the point (i), denote it by $\mathscr{I}$, and make the variable change $x = \cos(y)$, we get

$$\mathscr{I} = \int_0^1 \frac{x \log(1-x)}{1+ax^2} dx = \int_0^{\pi/2} \frac{\sin(y) \cos(y) \log(1-\cos(y))}{1+a\cos^2(y)} dy$$

$\left\{\text{in the denominator consider that } \sin^2(y) + \cos^2(y) = 1 \text{ and at the same time}\right\}$

$\left\{\text{we rearrange the numerator using that } \sin(y)\cos(y) = \tan(y) - \sin^2(y)\tan(y)\right\}$

$$= \frac{1}{1+a} \int_0^{\pi/2} \frac{(\tan(y) - \sin^2(y)\tan(y))\log(1 - \cos(y))}{1 - a/(a+1)\sin^2(y)} dy$$

$$= \frac{1}{1+a} \left(\int_0^{\pi/2} \frac{\tan(y)\log(1 - \cos(y))}{1 - a/(a+1)\sin^2(y)} dy - \int_0^{\pi/2} \frac{\sin^2(y)\tan(y)\log(1 - \cos(y))}{1 - a/(a+1)\sin^2(y)} dy \right).$$

$$(3.14)$$

What a wonderful moment now! We immediately notice that the second integral in (3.14) may be extracted by using (1.92), Sect. 1.18, and then we arrive at

$$\int_0^{\pi/2} \frac{\sin^2(y)\tan(y)\log(1 - \cos(y))}{1 - a/(a+1)\sin^2(y)} dy$$

$$= \frac{(1+a)^2}{2a} \arcsin^2\left(\sqrt{\frac{a}{a+1}}\right) + \frac{(1+a)^2}{4a}\text{Li}_2\left(\frac{a}{a+1}\right) - \frac{\pi^2}{6}(1+a). \quad (3.15)$$

For the first integral in (3.14), we consider the variable change $\cos(y) = x$, and then we write

$$\int_0^{\pi/2} \frac{\tan(y)\log(1 - \cos(y))}{1 - a/(a+1)\sin^2(y)} dy = (1+a) \int_0^1 \frac{\log(1-x)}{x(1+ax^2)} dx$$

$$= (1+a) \int_0^1 \frac{\log(1-x)}{x} dx - a(1+a) \int_0^1 \frac{x\log(1-x)}{1+ax^2} dx = -\frac{\pi^2}{6}(1+a) - a(1+a)\mathscr{I},$$

$$(3.16)$$

where I used that $\int_0^1 \frac{\log(1-x)}{x} dx = -\text{Li}_2(x)\Big|_{x=0}^{x=1} = -\text{Li}_2(1) = -\zeta(2) = -\frac{\pi^2}{6}$ and at the same time we notice that we have recovered the main integral multiplied by $-a(1+a)$, which allows us to extract it. We also observe that $\text{Li}_2(1) = \sum_{n=1}^{\infty} \frac{1}{n^2} = \frac{\pi^2}{6}$, where the last series is a famous series in the mathematical literature. It is known as the Basel problem, first solved by Leonhard Euler in 1734. A solution to the Basel problem may also be found in [76, Chapter 3, pp.55–57].

Then, if we plug the results from (3.15) and (3.16) in (3.14), we conclude that

$$\int_0^1 \frac{x \log(1-x)}{1+ax^2} dx = -\frac{1}{2a} \arcsin^2\left(\sqrt{\frac{a}{a+1}}\right) - \frac{1}{4a} \text{Li}_2\left(\frac{a}{a+1}\right)$$

$$= -\frac{1}{2a} \arctan^2(\sqrt{a}) - \frac{1}{4a} \text{Li}_2\left(\frac{a}{a+1}\right),$$

where for getting the second closed form we simply use again the trigonometric identity, $\arctan(x) = \arcsin(x/\sqrt{1+x^2})$.

Of course, no matter how we calculate the integral at the point (i), if we combine it with the integral relation in (3.11) and the value of the integral in (3.12), we obtain the value of the integral at the point (ii).

For another solution, the curious reader might extract the values of the desired integrals by creating a system of relations involving (3.11) and (1.31), Sect. 1.9.

Finally, to prove the result at the point (iii) we make use of the results from the points (i) and (ii). Things are pretty clear on how to go further: we need the integral from the point (i) as it is posed and then the integral from the point (ii), with a replaced by $1/a$, and then multiplied by $1/a^2$, and afterwards we add together the two integrals that gives:

$$\frac{1}{a} \int_0^1 \frac{x \log(1+x)}{a+x^2} dx + \int_0^1 \frac{x \log(1-x)}{1+ax^2} dx$$

$$= \frac{1}{2a} \arctan^2\left(\frac{1}{\sqrt{a}}\right) - \frac{1}{2a} \arctan^2(\sqrt{a}) - \frac{1}{4a} \text{Li}_2\left(\frac{1}{1+a}\right) - \frac{1}{4a} \text{Li}_2\left(\frac{a}{1+a}\right)$$

$$= \frac{\pi^2}{12} \frac{1}{a} - \frac{\pi}{2} \frac{\arctan(\sqrt{a})}{a} - \frac{1}{4} \frac{\log(a) \log(1+a)}{a} + \frac{1}{4} \frac{\log^2(1+a)}{a},$$

where to get the last equality I used that for $a > 0$ we have $\arctan(a) + \arctan(1/a) = \pi/2$ and $\text{Li}_2(1/(1+a)) + \text{Li}_2(a/(a+1)) = \log(a) \log(1+a) - \log^2(1+a) + \pi^2/6$, where the last result is obtained by deriving and integrating back the dilogarithmic side of the identity. Note that the last identity can also be deduced by exploiting the Landen's dilogarithmic identity. At this point our calculations are finalized!

The curious reader might like to try to prove the result at the point (iii) more directly, avoiding the use of the results from the points (i) and (ii).

3.4 Exploring More Appealing Logarithmic Integrals with Parameter: The First Part

Solution In this section we continue to approach logarithmic integrals similar to the ones in the previous section, parameterized in different ways.

In my first book, *(Almost) Impossible Integrals, Sums, and Series*, after treating the beautiful logarithmic integrals, $\displaystyle\int_0^1 \frac{x\log(1-x)}{1+x^2}dx = \frac{1}{8}\left(\log^2(2) - \frac{5}{2}\zeta(2)\right)$ and $\displaystyle\int_0^1 \frac{x\log(1+x)}{1+x^2}dx = \frac{1}{8}\left(\log^2(2) + \frac{1}{2}\zeta(2)\right)$, by adding and subtracting them in order to get simpler manageable integral forms, thus creating a system of relations, I also mentioned in passing the use of the *magical* integral result,

$$\int_0^x \frac{t\log(1-t)}{1+t^2}dt = \frac{1}{4}\left(\frac{1}{2}\log^2(1+x^2) - 2\mathrm{Li}_2(x) + \frac{1}{2}\mathrm{Li}_2\left(-x^2\right) + \mathrm{Li}_2\left(\frac{2x}{1+x^2}\right)\right),$$
(3.17)

with $x < 1$, where the particular cases stated above are obtained by letting $x \to 1^-$, and $x \to -1$, respectively (see [76, Chapter 3, p.97]). We need this result here, too!

Proof To get a solution to the result in (3.17), we first differentiate $\mathrm{Li}_2(2t/(1+t^2))$,

$$\left(\mathrm{Li}_2\left(\frac{2t}{1+t^2}\right)\right)' = \frac{\log(1+t^2)}{t} - 2\frac{\log(1-t)}{t} - 2\frac{t\log(1+t^2)}{1+t^2} + 4\frac{t\log(1-t)}{1+t^2},$$

and if we integrate then from $t = 0$ to $t = x$ and rearrange, we obtain

$$\int_0^x \frac{t\log(1-t)}{1+t^2}dt = \frac{1}{4}\left(\frac{1}{2}\log^2(1+x^2) - 2\mathrm{Li}_2(x) + \frac{1}{2}\mathrm{Li}_2\left(-x^2\right) + \mathrm{Li}_2\left(\frac{2x}{1+x^2}\right)\right),$$

which gives us the desired result. ∎

Now, considering the result in (3.17), where we replace x by a, let the variable change $t/a = u$, and then return to the notation in t and rearrange, we get

$$\int_0^1 \frac{t\log(1-at)}{1+a^2t^2}dt = \frac{1}{4a^2}\left(\frac{1}{2}\log^2\left(1+a^2\right) - 2\mathrm{Li}_2(a) + \frac{1}{2}\mathrm{Li}_2\left(-a^2\right) + \mathrm{Li}_2\left(\frac{2a}{1+a^2}\right)\right),$$

and the solution to the point (i) of the problem is finalized.

The result at the point (ii) is immediately obtained based on the result from the point (i), by replacing a by $-a$.

For the next two points, we'll also want to make use of results from the next section, and in order establish such connections we need to consider variable changes. So, by letting the variable change $\sqrt{a}x = t$, we get

$$\int_0^1 \frac{x\log(1-ax)}{1+ax^2}dx = \int_0^1 \frac{x\log(1-\sqrt{a}\sqrt{a}x)}{1+(\sqrt{a}x)^2}dx = \frac{1}{a}\int_0^{\sqrt{a}} \frac{t\log(1-\sqrt{a}t)}{1+t^2}dt$$

{make use of the result in (1.16), Sect. 1.5, where we replace a by $\sqrt{a}$}

$$= \frac{1}{2a}\left(\frac{1}{4}\log^2(1+a) + \arctan^2(\sqrt{a}) - \text{Li}_2(a)\right),$$

and the solution to the point (iii) of the problem is finalized.

Proceeding in a similar style as before, and letting the variable change $\sqrt{a}x = t$, we have

$$\int_0^1 \frac{x\log(1+ax)}{1+ax^2}\,dx = \int_0^1 \frac{x\log(1+\sqrt{a}\sqrt{a}x)}{1+(\sqrt{a}x)^2}\,dx = \frac{1}{a}\int_0^{\sqrt{a}} \frac{t\log(1+\sqrt{a}t)}{1+t^2}\,dt$$

$$\{\text{make use of the result in (1.17), Sect. 1.5, where we replace } a \text{ by } \sqrt{a}\}$$

$$= \frac{1}{2a}\left(\frac{1}{4}\log^2(1+a) - \arctan^2(\sqrt{a}) - \text{Li}_2(-a)\right),$$

and the solution to the point (iv) of the problem is finalized.

Finally, to prove the identity with three dilogarithms from the last point, we proceed as in my paper "Two identities with special dilogarithmic values" (see [90, January 15, 2021]), where I considered the identity in (3.17), and use the version

$$\int_0^x \frac{t\log(1+t)}{1+t^2}\,dt = \frac{1}{4}\left(\frac{1}{2}\log^2(1+x^2) - 2\text{Li}_2(-x) + \frac{1}{2}\text{Li}_2\left(-x^2\right) + \text{Li}_2\left(\frac{-2x}{1+x^2}\right)\right),$$

$$(3.18)$$

with $-1 < x$.

If we set $x = \sqrt{2} - 1$ in (3.18), we get

$$\text{Li}_2\left(2\sqrt{2}-3\right) + 2\text{Li}_2\left(-\frac{1}{\sqrt{2}}\right) - 4\text{Li}_2\left(1-\sqrt{2}\right)$$

$$= -\log^2(2) - 2\log(2)\log\left(2-\sqrt{2}\right) - \log^2\left(2-\sqrt{2}\right) + 8\int_0^{\sqrt{2}-1} \frac{t\log(1+t)}{1+t^2}\,dt.$$

$$(3.19)$$

There is one more step needed to finalize the calculations above, that is, to evaluate the integral in (3.19). If we let the variable change $u = (1-t)/(1+t)$, and then return to the notation in t, we get

$$\int_0^{\sqrt{2}-1} \frac{t\log(1+t)}{1+t^2}\,dt = \log(2)\int_{\sqrt{2}-1}^1 \frac{1}{1+t}\,dt - \log(2)\int_{\sqrt{2}-1}^1 \frac{t}{1+t^2}\,dt$$

$$- \int_{\sqrt{2}-1}^1 \frac{\log(1+t)}{1+t}\,dt + \int_0^1 \frac{t\log(1+t)}{1+t^2}\,dt - \int_0^{\sqrt{2}-1} \frac{t\log(1+t)}{1+t^2}\,dt,$$

whence we obtain that

$$\int_0^{\sqrt{2}-1} \frac{t\log(1+t)}{1+t^2}dt = \frac{1}{16}\log^2(2)+\frac{1}{4}\log(2)\log\left(2-\sqrt{2}\right)+\frac{1}{2}\int_0^1 \frac{t\log(1+t)}{1+t^2}dt$$

$$= \frac{\pi^2}{192} + \frac{1}{8}\log^2(2) + \frac{1}{4}\log(2)\log\left(2-\sqrt{2}\right), \tag{3.20}$$

where in the calculations I also used that $\int_0^1 \dfrac{t\log(1+t)}{1+t^2}dt = \dfrac{\pi^2}{96} + \dfrac{1}{8}\log^2(2)$, which is immediately obtained based on the result in (3.18).

Finally, combining the results in (3.19) and (3.20), we conclude that

$$\text{Li}_2\left(2\sqrt{2}-3\right) + 2\text{Li}_2\left(-\frac{1}{\sqrt{2}}\right) - 4\text{Li}_2\left(1-\sqrt{2}\right) = \frac{\pi^2}{24} - \log^2\left(2-\sqrt{2}\right),$$

and the solution to the point (v) of the problem is finalized.

We'll also meet the integral result in (3.17) in other sections, including the next section where we'll be pretty glad to take advantage of its usefulness.

I would finally add another identity similar to the one presented at the point (v) and published in the same paper mentioned above, that is,

$$\text{Li}_2\left(2\sqrt{2}-3\right) + 2\text{Li}_2\left(2-\sqrt{2}\right) - \text{Li}_2\left(3-2\sqrt{2}\right)$$

$$= \frac{\pi^2}{8} + \frac{1}{4}\log^2(2) - \log^2\left(2-\sqrt{2}\right). \tag{3.21}$$

The result in (3.21) is immediately derived if we consider the powerful relation with four dilogarithms,

$$\text{Li}_2\left(\frac{4x}{(1+x)^2}\right) + 4\text{Li}_2\left(\frac{1+x}{2}\right) + 2\text{Li}_2(-x) - 2\text{Li}_2(x)$$

$$= \frac{\pi^2}{3} - 2\log^2(2) - 2\log^2(1+x) + 4\log(2)\log(1+x), \tag{3.22}$$

where $-1 < x \le 1$, and then plug in $x = (\sqrt{2}-1)^2 = 3 - 2\sqrt{2}$. The result in (3.22) can be proved by differentiating and integrating back the first dilogarithmic part in the left-hand side.

More similar integrals with parameter are waiting for us in the next section!

3.5 Exploring More Appealing Logarithmic Integrals with Parameter: The Second Part

Solution As you may expect from the title of the section, we go on with evaluating integrals similar to the ones in the previous two sections.

For the first two points of the problem, we'll want to return to the *magical* result in (3.17), considered in the previous section. Then, the points (iii) and (iv) will be a wonderful experience with double integrals. And the last point is another enjoyable one, using ideas involving some generalized integrals and simple inequalities.

So, using that $\log(1 - ax) = -\int_0^a \dfrac{x}{1 - xy}dy$, we have

$$\int_0^1 \frac{x \log(1 - ax)}{1 + x^2}dx = -\int_0^1 \left(\int_0^a \frac{x^2}{(1 + x^2)(1 - xy)}dy \right) dx$$

{reverse the order of integration and use the partial fraction expansion}

$$= \int_0^a \left(\int_0^1 \left(\frac{1}{(1 + y^2)(1 + x^2)} + \frac{yx}{(1 + y^2)(1 + x^2)} - \frac{1}{(1 + y^2)(1 - xy)} \right) dx \right) dy$$

$$= \frac{\pi}{4} \int_0^a \frac{1}{1 + y^2}dy + \frac{1}{2} \log(2) \int_0^a \frac{y}{1 + y^2}dy + \int_0^a \frac{\log(1 - y)}{y}dy - \int_0^a \frac{y \log(1 - y)}{1 + y^2}dy$$

{make use of the result in (3.17) from the previous section}

$$= \frac{\pi}{4} \arctan(a) + \frac{1}{4} \log(2) \log \left(1 + a^2 \right) - \frac{1}{8} \log^2 \left(1 + a^2 \right) - \frac{1}{2} \operatorname{Li}_2(a)$$

$$- \frac{1}{8} \operatorname{Li}_2 \left(-a^2 \right) - \frac{1}{4} \operatorname{Li}_2 \left(\frac{2a}{1 + a^2} \right),$$

and the solution to the point (i) of the problem is complete.

As mentioned in the problem statement, the result at the point (ii) is an immediate consequence of the result from the point (i), and we get it by simply replacing a by $-a$ and considering the restriction changed.

In the following, I'll approach the integral at the point (iii) of the problem by considering a double integral representation of the given integral and exploiting the symmetry (one of my favorite ways to build solutions), and then we write

$$\int_0^a \frac{x \log(1 - ax)}{1 + x^2}dx = -\int_0^a \left(\int_0^a \frac{x^2}{(1 + x^2)(1 - xy)}dy \right) dx$$

$$= -\frac{1}{2} \left(\int_0^a \left(\int_0^a \frac{x^2}{(1 + x^2)(1 - xy)}dy \right) dx + \int_0^a \left(\int_0^a \frac{y^2}{(1 + y^2)(1 - xy)}dx \right) dy \right)$$

{reverse the order of integration in the last double integral, then}

{merge the two integrals and write the numerator in a useful way}

$$= -\frac{1}{2} \int_0^a \left(\int_0^a \frac{(1+x^2)(1+y^2) - (1-(xy)^2)}{(1+x^2)(1+y^2)(1-xy)} dy \right) dx$$

$$= -\frac{1}{2} \int_0^a \left(\int_0^a \frac{1}{1-xy} dy \right) dx + \frac{1}{2} \int_0^a \left(\int_0^a \frac{1}{(1+x^2)(1+y^2)} dy \right) dx$$

$$+ \frac{1}{2} \int_0^a \left(\int_0^a \frac{xy}{(1+x^2)(1+y^2)} dy \right) dx$$

$$= \frac{1}{2} \left(\frac{1}{4} \log^2(1+a^2) + \arctan^2(a) - \operatorname{Li}_2(a^2) \right),$$

and the solution to the point (iii) of the problem is complete.

The curious reader interested in approaching the problem in a slightly different way may start with considering the integral $I(a) = \int_0^a \frac{x \log(1-ax)}{1+x^2} dx$ which is then differentiated once with respect to a (we apply *Leibniz's rule for differentiation under the integral sign*).

Passing to the point (iv) of the problem, we get a first solution by employing a similar strategy to the one given in the solution to the point (iii) of the problem. So, we have

$$\int_0^a \frac{x \log(1+ax)}{1+x^2} dx = \int_0^a \left(\int_0^a \frac{x^2}{(1+x^2)(1+xy)} dy \right) dx$$

$$= \frac{1}{2} \left(\int_0^a \left(\int_0^a \frac{x^2}{(1+x^2)(1+xy)} dy \right) dx + \int_0^a \left(\int_0^a \frac{y^2}{(1+y^2)(1+xy)} dx \right) dy \right)$$

{swap the order of integration in the last double integral, and then}

{merge the two integrals and write the numerator in a useful way}

$$= \frac{1}{2} \int_0^a \left(\int_0^a \frac{(1+x^2)(1+y^2) - (1-(xy)^2)}{(1+x^2)(1+y^2)(1+xy)} dy \right) dx$$

$$= \frac{1}{2} \int_0^a \left(\int_0^a \frac{1}{1+xy} dy \right) dx - \frac{1}{2} \int_0^a \left(\int_0^a \frac{1}{(1+x^2)(1+y^2)} dy \right) dx$$

$$+ \frac{1}{2} \int_0^a \left(\int_0^a \frac{xy}{(1+x^2)(1+y^2)} dy \right) dx$$

$$= \frac{1}{2} \left(\frac{1}{4} \log^2(1+a^2) - \arctan^2(a) - \operatorname{Li}_2(-a^2) \right),$$

and the solution to the point (iv) of the problem is complete.

Again, for a slightly different solution, the curious reader might start with considering the integral $J(a) = \int_0^a \frac{x \log(1 + ax)}{1 + x^2} dx$ which is then differentiated once with respect to a (where we use *Leibniz's rule for differentiation under the integral sign*).

Here we are, at the last point of the problem that looks rather daunting! A first key observation is that $f(a)$ may be recovered by the results in (1.6), Sect. 1.3 and (1.14), Sect. 1.5,

$$f(a) = \pi \arctan(a) + 2\frac{\arctan^2(\sqrt{a})}{a} + \log(2)\log(1 + a^2) - \frac{1}{2}\log^2(1 + a^2)$$

$$- 2\mathrm{Li}_2(a) - \frac{1}{2}\mathrm{Li}_2\left(-a^2\right) + \frac{1}{a}\mathrm{Li}_2\left(\frac{a}{a+1}\right) - \mathrm{Li}_2\left(\frac{2a}{1+a^2}\right)$$

$$= 4\int_0^1 \frac{x\log(1 - ax)}{1 + x^2}dx - 4\int_0^1 \frac{x\log(1 - x)}{1 + ax^2}dx$$

$$= 4\int_0^1 \frac{x(1 + ax^2)\log(1 - ax) - x(1 + x^2)\log(1 - x)}{(1 + x^2)(1 + ax^2)}dx$$

$$= 4\int_0^1 \frac{x\left(ax^2\log\left(\frac{1 - ax}{1 - x}\right) + \log\left(\frac{1 - ax}{1 - x}\right) + (a - 1)x^2\log(1 - x)\right)}{(1 + x^2)(1 + ax^2)}dx,$$

where it's easy to see that for $a = 1$ we have $f(1) = 0$, and for $a \in (0, 1)$, we have $f(a) > 0$, and this may be immediately noted by inspecting the integrand of the last integral. Note that since $\frac{1 - ax}{1 - x} > 1, \forall a, x \in (0, 1)$, we have $\log\left(\frac{1 - ax}{1 - x}\right) > 0$, and the solution to the point (v) of the problem is complete.

3.6 More Good-Looking Logarithmic Integrals: The First Part

Solution One of the most popular integrals in the mathematical literature, also known as *Serret's integral*, mentioned both in *Inside Interesting Integrals* (see [76, Chapter 3, p.69]) and *(Almost) Impossible Integrals, Sums, and Series* (see [76, Chapter 3, p.218]), and that also appeared in the 66th Putnam Competition, 2005, is $\int_0^1 \frac{\log(1 + x)}{1 + x^2}dx = \frac{1}{8}\log(2)\pi$. Now, imagine trying to evaluate this integral parameterized in various ways! How does it sound to you? I hope you agree with me that ... *it sounds so interesting, fun, and challenging!*

Compared to the closed forms in the previous sections, this section is mainly populated with closed forms involving the inverse tangent integral (see [21]). However, since we have the result[3] $\text{Li}_2(ix) = \frac{1}{4}\text{Li}_2(-x^2) + i\,\text{Ti}_2(x)$, we obtain the representation $\text{Ti}_2(x) = \Im\{\text{Li}_2(ix)\}$, and then we may also say that in this section we have closed forms involving the imaginary parts of the dilogarithm with a complex argument.

Now, some of the usual representations of the inverse tangent integral we need in this book are the series representation, $\text{Ti}_2(x) = \sum_{n=1}^{\infty}(-1)^{n-1}\dfrac{x^{2n-1}}{(2n-1)^2}$, $|x| \leq 1$, and the integral representations, $\text{Ti}_2(x) = \displaystyle\int_0^x \dfrac{\arctan(t)}{t}dt$ and $\text{Ti}_2(x) = -\displaystyle\int_0^1 \dfrac{x\log(y)}{1+x^2y^2}dy$, where $x \in \mathbb{R}$ (see also [76, Chapter 3, p.141]). For example, exploiting the variable change $xy = t$ in the latter integral representation of the inverse tangent integral, and integrating by parts once, we may show we arrive at the former integral representation of the inverse tangent integral.

First, I emphasize that we could exploit a strategy involving harmonic series, with a very similar solution flow to the first one in Sect. 3.3, when we have the restriction $0 < a \leq 1$, which the curious reader may finalize.

So, to calculate the first main integral given at the point (i) we proceed as follows:

$$\int_0^1 \frac{\log(1-x)}{1+ax^2}dx = \Re\left\{\int_0^1 \frac{\log(1-x)}{1-i\sqrt{a}x}dx\right\}$$

$$\overset{1-x=y}{=} \Re\left\{-\frac{1}{i\sqrt{a}}\int_0^1 \frac{(i\sqrt{a}/(i\sqrt{a}-1))\log(y)}{1-i\sqrt{a}/(i\sqrt{a}-1)y}dy\right\}$$

{use the extended version to complex numbers of (3.13), the case $n = 1$, Sect. 3.3}

$$= \Re\left\{\frac{1}{i\sqrt{a}}\text{Li}_2\left(\frac{i\sqrt{a}}{i\sqrt{a}-1}\right)\right\} = \Im\left\{\frac{1}{\sqrt{a}}\text{Li}_2\left(\frac{i\sqrt{a}}{i\sqrt{a}-1}\right)\right\}$$

$$\left\{\text{use the Landen's identity, } \text{Li}_2(x) + \text{Li}_2\left(\frac{x}{x-1}\right) = -\frac{1}{2}\log^2(1-x)\right\}$$

---

[3] To see this fact is true, we may consider the dilogarithm series representation, $\text{Li}_2(z) = \sum_{n=1}^{\infty}\dfrac{z^n}{n^2}$, $|z| \leq 1$, where if we plug in $z = ix$, we get $\text{Li}_2(ix) = \sum_{n=1}^{\infty}(-1)^n\dfrac{x^{2n}}{(2n)^2} + i\sum_{n=1}^{\infty}(-1)^{n-1}\dfrac{x^{2n-1}}{(2n-1)^2} = \frac{1}{4}\text{Li}_2(-x^2) + i\,\text{Ti}_2(x)$, from which we deduce that $\text{Ti}_2(x) = \Im\{\text{Li}_2(ix)\}$. Also, the values of x in the last result may be extended to $x \in \mathbb{R}$.

$$= \Im\left\{ -\frac{1}{2}\frac{\log^2(1 - i\sqrt{a})}{\sqrt{a}} - \frac{\text{Li}_2(i\sqrt{a})}{\sqrt{a}} \right\} = \frac{1}{2\sqrt{a}}\log(1+a)\arctan(\sqrt{a}) - \frac{\text{Ti}_2(\sqrt{a})}{\sqrt{a}},$$

where to get the last equality I also used the logarithm with a complex argument, $\log(a + ib) = \log(\sqrt{a^2 + b^2}) + i\arctan\left(\frac{b}{a}\right)$, $a > 0$, and the fact that $\text{Ti}_2(x) = \Im\{\text{Li}_2(ix)\}$, and the solution to the point (i) of the problem is complete.

What about the next integral? It appeared in [35], in 2013, and a solution was given about 1 year later, but one leading to a different closed form than the one in my solution. In the following, I'll propose a different solution that will lead directly to one of its simplest closed forms, involving one single dilogarithmic value.

Further, to calculate the result at the point (ii), I'll consider the following generalized logarithmic integral result:

$$\int_0^x \frac{\log(1 + bt)}{1 + at}dt$$

$$= \frac{1}{a}\left(\log(b - a)\log(1 + ax) - \log(b - a)\log(1 + bx) \right.$$

$$+ \log(x)\log(1 + ax) - \log(x)\log(1 + bx) + \frac{1}{2}\log^2(1 + bx)$$

$$\left. + \text{Li}_2(-ax) - \text{Li}_2(-bx) - \frac{\pi^2}{6} + \text{Li}_2\left(\frac{1 + ax}{1 + bx}\right) \right). \tag{3.23}$$

Proof If we consider $\text{Li}_2((1+at)/(1+bt))$ which we differentiate once with respect to t and then integrate back from $t = 0$ to $t = x$, we get

$$\text{Li}_2\left(\frac{1 + ax}{1 + bx}\right) = \frac{\pi^2}{6} + \int_0^x \left(\text{Li}_2\left(\frac{1 + at}{1 + bt}\right) \right)' dt$$

$$\{\text{expand the integral after differentiation}\}$$

$$= \log(b - a)\int_0^x \frac{b}{1 + bt}dt - \log(b - a)\int_0^x \frac{a}{1 + at}dt + b\int_0^x \frac{\log(t)}{1 + bt}dt$$

$$- a\int_0^x \frac{\log(t)}{1 + at}dt - \int_0^x \frac{b\log(1 + bt)}{1 + bt}dt + a\int_0^x \frac{\log(1 + bt)}{1 + at}dt + \frac{\pi^2}{6}$$

$$\left\{\text{use that } \int_0^x \frac{\log(t)}{1 + at}dt = \frac{1}{a}\log(x)\log(1 + ax) + \frac{1}{a}\text{Li}_2(-ax)\right\}$$

$$= \log(b-a)\log(1+bx) - \log(b-a)\log(1+ax)$$

$$+ \log(x)\log(1+bx) - \log(x)\log(1+ax) - \frac{1}{2}\log^2(1+bx)$$

$$+ \operatorname{Li}_2(-bx) - \operatorname{Li}_2(-ax) + \frac{\pi^2}{6} + a\int_0^x \frac{\log(1+bt)}{1+at}dt,$$

whence the desired result in (3.23) follows. For the calculation of $\int_0^x \dfrac{\log(t)}{1+at}dt$ I used a simple integration by parts. ∎

It is worth mentioning that the integral from (3.23), in a slightly more general form, and derived by using different ideas, may be found in [21, Chapter 8, pp.243–244], which leads to a different form of the value of the integral in (3.23).

At this point, we replace a by $i\sqrt{a}$ and set $x = 1$, $b = 1$ in (3.23). Then, considering the real part of both sides, we obtain

$$\Re\left\{\int_0^1 \frac{\log(1+t)}{1+i\sqrt{a}\,t}dt\right\} = \int_0^1 \frac{\log(1+t)}{1+at^2}dt$$

$$= \Re\left\{i\frac{\pi^2}{12\sqrt{a}} - i\frac{\log^2(2)}{2\sqrt{a}} + i\log(2)\frac{\log\left(1-i\sqrt{a}\right)}{\sqrt{a}} - i\frac{\log\left(1-i\sqrt{a}\right)\log\left(1+i\sqrt{a}\right)}{\sqrt{a}}\right.$$

$$\left. - i\frac{1}{\sqrt{a}}\operatorname{Li}_2(-i\sqrt{a}) - \frac{i}{\sqrt{a}}\operatorname{Li}_2\left(\frac{1}{2}+i\frac{\sqrt{a}}{2}\right)\right\}$$

$$= \Im\left\{\frac{\log(1-i\sqrt{a})\log(1+i\sqrt{a})}{\sqrt{a}} - \log(2)\frac{\log(1-i\sqrt{a})}{\sqrt{a}}\right.$$

$$\left. + \frac{1}{\sqrt{a}}\operatorname{Li}_2(-i\sqrt{a}) + \frac{1}{\sqrt{a}}\operatorname{Li}_2\left(\frac{1}{2}+i\frac{\sqrt{a}}{2}\right)\right\}$$

$$= \frac{1}{\sqrt{a}}\Im\left\{\left(\frac{1}{2}\log(1+a) - i\arctan(\sqrt{a})\right)\left(\frac{1}{2}\log(1+a) + i\arctan(\sqrt{a})\right)\right\}$$

$$- \frac{\log(2)}{\sqrt{a}}\Im\left\{\frac{1}{2}\log(1+a) - i\arctan(\sqrt{a})\right\} + \frac{1}{\sqrt{a}}\operatorname{Ti}_2(-\sqrt{a})$$

$$+ \frac{1}{\sqrt{a}}\Im\left\{\operatorname{Li}_2\left(\frac{1}{2}+i\frac{\sqrt{a}}{2}\right)\right\}$$

$$= \frac{\log(2)}{\sqrt{a}}\arctan(\sqrt{a}) - \frac{1}{\sqrt{a}}\operatorname{Ti}_2(\sqrt{a}) + \frac{1}{\sqrt{a}}\Im\left\{\operatorname{Li}_2\left(\frac{1}{2}+i\frac{\sqrt{a}}{2}\right)\right\},$$

where I considered the logarithm with a complex argument and the simple fact that $\text{Ti}_2(x) = -\text{Ti}_2(-x)$, and the solution to the point (ii) is complete.

For the dilogarithmic value at the point (iii), we set $a = 1$ in the result above:

$$\Im\left\{\text{Li}_2\left(\frac{1+i}{2}\right)\right\} = G - \frac{1}{8}\log(2)\pi,$$

where I used $\arctan(1) = \pi/4$, $\text{Ti}_2(1) = G$, and the well-known integral result $\int_0^1 \frac{\log(1+x)}{1+x^2}dx = \frac{1}{8}\log(2)\pi$, previously mentioned and that is easily derived (e.g., by the variable change $(1-x)/(1+x) = y$).

Good to know that the polylogarithmic value above may also be obtained by other means involving identities like the Landen's identity, $\text{Li}_2(x) + \text{Li}_2(x/(x-1)) = -\frac{1}{2}\log^2(1-x)$ (see [21, Chapter 1, p.5], [70, Chapter 2, p.107]), which the curious reader might like to consider.

In some of the next sections, we will discover the power of the integrals at the points (i) and (ii), revealing unexpected connections with other results.

As a final word, the curious reader might like to know that it is also possible to consider the result in (3.23) for deriving the integral result at the point (i).

3.7 More Good-Looking Logarithmic Integrals: The Second Part

Solution *How would the integral version without x in front look like?* is one of the possible questions to think of after seeing the integral form in (3.17), Sect. 3.4.

In fact, from the book *Polylogarithms and Associated Functions* we find that Leonard Lewin treated integrals alike, more precisely in the chapter "The Inverse Tangent Integral" (see [21, Chapter 2, pp.40–42]).

Now, we may take good inspiration from the section with the statement of the problem where the closed forms are given, and we are particularly interested in the values involving the inverse tangent integral, that is, $\text{Ti}_2(2a/(1-a^2))$ and $\text{Ti}_2((1-a)/(1+a))$. *Why?* By differentiating and integrating back both these forms of the inverse tangent integral, we'll be able to arrive at the integrals we are interested in.

So, recollecting the integral representation of the inverse tangent integral, $\text{Ti}_2(x) = \int_0^x \frac{\arctan(t)}{t}dt$, and considering differentiation, we have $\left(\text{Ti}_2(2x/(1-x^2))\right)' = 2\arctan(x)/x + 2\arctan(x)/(1-x) - 2\arctan(x)/(1+x)$, where if we integrate back from $x = 0$ to $x = a$, we get

$$\text{Ti}_2\left(\frac{2a}{1-a^2}\right) = 2\int_0^a \frac{\arctan(x)}{x}dx + 2\int_0^a \frac{\arctan(x)}{1-x}dx - 2\int_0^a \frac{\arctan(x)}{1+x}dx$$

{integrate by parts for the last two integrals}

$$= 2\text{Ti}_2(a) - 2\log(1-a)\arctan(a) - 2\log(1+a)\arctan(a)$$

$$+ 2\int_0^a \frac{\log(1-x)}{1+x^2}dx + 2\int_0^a \frac{\log(1+x)}{1+x^2}dx. \tag{3.24}$$

In the derivation process above I also used the identity $\arctan\left(2x/(1-x^2)\right) = 2\arctan(x)$, $|x| < 1$.

If we focus on the other form of the inverse tangent integral, $\text{Ti}_2((1-x)/(1+x))$, which we differentiate, we obtain that $(\text{Ti}_2((1-x)/(1+x)))' = -\pi/4(1/(1-x)) - \pi/4(1/(1+x)) + \arctan(x)/(1-x) + \arctan(x)/(1+x)$, where if we integrate back from $x = 0$ to $x = a$ and rearrange, considering that $\text{Ti}_2(1) = G$, we get

$$\text{Ti}_2\left(\frac{1-a}{1+a}\right) = G + \frac{\pi}{4}\log(1-a) - \frac{\pi}{4}\log(1+a) + \int_0^a \frac{\arctan(x)}{1-x}dx + \int_0^a \frac{\arctan(x)}{1+x}dx$$

{integrate by parts both remaining integrals}

$$= G - \frac{\pi}{2}\text{arctanh}(a) - \log(1-a)\arctan(a) + \log(1+a)\arctan(a)$$

$$+ \int_0^a \frac{\log(1-x)}{1+x^2}dx - \int_0^a \frac{\log(1+x)}{1+x^2}dx. \tag{3.25}$$

Note that in the derivation process I also used the identity, $\arctan((1-x)/(1+x)) = \pi/4 - \arctan(x)$, $-1 < x$.

Happily, since we have managed to establish in (3.24) and (3.25) two distinct relations with the integrals $\int_0^a \frac{\log(1-x)}{1+x^2}dx$ and $\int_0^a \frac{\log(1+x)}{1+x^2}dx$, we are able to extract their values in closed form, and we obtain

$$\int_0^a \frac{\log(1-x)}{1+x^2}dx = \arctan(a)\log(1-a) + \frac{\pi}{4}\text{arctanh}(a) - \frac{1}{2}G$$

$$-\frac{1}{2}\text{Ti}_2(a) + \frac{1}{2}\text{Ti}_2\left(\frac{1-a}{1+a}\right) + \frac{1}{4}\text{Ti}_2\left(\frac{2a}{1-a^2}\right), \quad |a| < 1, \ a \in \mathbb{R},$$

and

$$\int_0^a \frac{\log(1+x)}{1+x^2}dx = \arctan(a)\log(1+a) - \frac{\pi}{4}\text{arctanh}(a) + \frac{1}{2}G$$

$$-\frac{1}{2}\text{Ti}_2(a) - \frac{1}{2}\text{Ti}_2\left(\frac{1-a}{1+a}\right) + \frac{1}{4}\text{Ti}_2\left(\frac{2a}{1-a^2}\right), \quad |a| < 1, \ a \in \mathbb{R},$$

and the common solution to the points (i) and (ii) of the problem is complete.

The reader might like to observe that knowing any of the results from the points (i) and (ii), we are able to promptly extract the value of the other integral.

Then, for the point (iii) of the problem we start by considering the integral representation, $\log(1 - ax) = -\int_0^a \dfrac{x}{1 - xy}\,dy$, and we write

$$\int_0^1 \frac{\log(1 - ax)}{1 + x^2}\,dx = -\int_0^1 \left(\int_0^a \frac{x}{(1 + x^2)(1 - xy)}\,dy\right) dx$$

{reverse the order of integration and use the partial fraction expansion}

$$= \int_0^a \left(\int_0^1 \frac{y}{(1 + x^2)(1 + y^2)} - \frac{x}{(1 + x^2)(1 + y^2)} - \frac{y}{(1 + y^2)(1 - xy)}\,dx\right) dy$$

$$= \frac{\pi}{4}\int_0^a \frac{y}{1 + y^2}\,dy - \frac{1}{2}\log(2)\int_0^a \frac{1}{1 + y^2}\,dy + \int_0^a \frac{\log(1 - y)}{1 + y^2}\,dy$$

{observe the third integral is previously calculated}

$$= \arctan(a)\log(1 - a) - \frac{1}{2}\log(2)\arctan(a) + \frac{\pi}{8}\log(1 + a^2) + \frac{\pi}{4}\operatorname{arctanh}(a)$$

$$- \frac{1}{2}G - \frac{1}{2}\operatorname{Ti}_2(a) + \frac{1}{2}\operatorname{Ti}_2\left(\frac{1 - a}{1 + a}\right) + \frac{1}{4}\operatorname{Ti}_2\left(\frac{2a}{1 - a^2}\right),$$

and the solution to the point (iii) of the problem is complete.

Finally, proceeding in a similar style as before and considering the integral representation, $\log(1 + ax) = \int_0^a \dfrac{x}{1 + xy}\,dy$, we write

$$\int_0^1 \frac{\log(1 + ax)}{1 + x^2}\,dx = \int_0^1 \left(\int_0^a \frac{x}{(1 + x^2)(1 + xy)}\,dy\right) dx$$

{reverse the order of integration and use the partial fraction expansion}

$$= \int_0^a \left(\int_0^1 \frac{x}{(1 + y^2)(1 + x^2)} + \frac{y}{(1 + y^2)(1 + x^2)} - \frac{y}{(1 + y^2)(1 + xy)}\,dx\right) dy$$

$$= \frac{1}{2}\log(2)\int_0^a \frac{1}{1 + y^2}\,dy + \frac{\pi}{4}\int_0^a \frac{y}{1 + y^2}\,dy - \int_0^a \frac{\log(1 + y)}{1 + y^2}\,dy$$

{notice the third integral is previously calculated}

$$= \frac{1}{2} \log(2) \arctan(a) - \arctan(a) \log(1 + a) + \frac{\pi}{8} \log(1 + a^2) + \frac{\pi}{4} \operatorname{arctanh}(a)$$

$$- \frac{1}{2} G + \frac{1}{2} \operatorname{Ti}_2(a) + \frac{1}{2} \operatorname{Ti}_2\left(\frac{1 - a}{1 + a}\right) - \frac{1}{4} \operatorname{Ti}_2\left(\frac{2a}{1 - a^2}\right),$$

and the solution to the point (iv) of the problem is complete.

Good to observe that knowing any of the results from the points (iii) and (iv), we promptly extract the value of the other integral. For example, if we replace a by $-a$ in the integral at the point (iii) and use the inverse relation,

$$\operatorname{Ti}_2(x) - \operatorname{Ti}_2\left(\frac{1}{x}\right) = \frac{\pi}{2} \operatorname{sgn}(x) \log(|x|), \tag{3.26}$$

for $\operatorname{Ti}_2((1 - a)/(1 + a))$, which appears in the closed form, we arrive at the result given at the point (iv). The derivation of (3.26) is explained in Sect. 3.37 together with a more advanced version of it.

What about the *challenging question*? How would we like to proceed for getting a proof? First, let's observe that $\cot\left(\dfrac{11\pi}{24}\right) = \dfrac{\cos(11\pi/24)}{\sin(11\pi/24)} = $

$\dfrac{\cos(\pi/4 + 5\pi/24)}{\sin(\pi/4 + 5\pi/24)} = \dfrac{\sin(\pi/4 - 5\pi/24)}{\sin(\pi/4 + 5\pi/24)} = \sqrt{\dfrac{\sin^2(\pi/4 - 5\pi/24)}{\sin^2(\pi/4 + 5\pi/24)}} = $

$\sqrt{\dfrac{1 - \sin(5\pi/12)}{1 + \sin(5\pi/12)}} = \dfrac{1 - \tan(5\pi/24)}{1 + \tan(5\pi/24)}$, where for the last equality I used that

$\sqrt{\dfrac{1 - \sin(\theta)}{1 + \sin(\theta)}} = \left|\dfrac{1 - \tan(\theta/2)}{1 + \tan(\theta/2)}\right|$, which is derived by using the tangent half-angle

formula, $\sin(\theta) = \dfrac{2 \tan(\theta/2)}{1 + \tan^2(\theta/2)}$. More directly, we can employ the well-known

formula, $\cot(x + y) = \dfrac{\cot(x) \cot(y) - 1}{\cot(x) + \cot(y)}$, where we set $x = \pi/4$ and $y = 5\pi/24$.

Thus, the result to prove may be put in the form

$$\int_0^{\tan(5\pi/24)} \frac{\log(1 - x)}{1 + x^2} dx + \int_0^{\frac{1 - \tan(5\pi/24)}{1 + \tan(5\pi/24)}} \frac{\log(1 - x)}{1 + x^2} dx = \log(2)\frac{\pi}{8} - \frac{2}{3} G. \tag{3.27}$$

Now, we make a nice observation! If we replace a by $\dfrac{1 - a}{1 + a}$ in (i) and then add the relation obtained to the initial relation in (i), we arrive at

$$\int_0^a \frac{\log(1 - x)}{1 + x^2} dx + \int_0^{\frac{1 - a}{1 + a}} \frac{\log(1 - x)}{1 + x^2} dx$$

$$= \log(2)\frac{\pi}{8} - G - \log\left(\frac{2a}{1 - a^2}\right) \arctan(a) + \frac{1}{2} \operatorname{Ti}_2\left(\frac{2a}{1 - a^2}\right), \tag{3.28}$$

where $0 \le a < 1$.

So, if we set $a = \tan\left(\dfrac{5\pi}{24}\right)$ in (3.28) and then consider that $\tan(2x) = \dfrac{2\tan(x)}{1 - \tan^2(x)}$, we obtain that

$$\int_0^{\tan(5\pi/24)} \frac{\log(1-x)}{1+x^2}dx + \int_0^{\frac{1-\tan(5\pi/24)}{1+\tan(5\pi/24)}} \frac{\log(1-x)}{1+x^2}dx$$

$$= \log(2)\frac{\pi}{8} - G - \log\left(\tan\left(\frac{5\pi}{12}\right)\right)\arctan\left(\tan\left(\frac{5\pi}{24}\right)\right) + \frac{1}{2}\mathrm{Ti}_2\left(\tan\left(\frac{5\pi}{12}\right)\right)$$

$$\left\{\text{let's use that } \tan\left(\frac{5\pi}{12}\right) = \tan\left(\frac{\pi}{2} - \frac{\pi}{12}\right) = \cot\left(\frac{\pi}{12}\right) = 2 + \sqrt{3}\right\}$$

$$= \log(2)\frac{\pi}{8} - G - \frac{5}{24}\log\left(2 + \sqrt{3}\right)\pi + \frac{1}{2}\mathrm{Ti}_2\left(2 + \sqrt{3}\right)$$

$$= \log(2)\frac{\pi}{8} - \frac{2}{3}G,$$

which is the desired result.

Hold on! Any story to share about $\mathrm{Ti}_2(2+\sqrt{3})$?, which is how some of you might possibly react at this point. Good to know that both $\mathrm{Ti}_2(2 - \sqrt{3})$ and $\mathrm{Ti}_2(2 + \sqrt{3})$ are well-known special values of the inverse tangent integral, that is,

$$\mathrm{Ti}_2\left(2 - \sqrt{3}\right) = \frac{2}{3}G + \log\left(2 - \sqrt{3}\right)\frac{\pi}{12}, \tag{3.29}$$

and if we combine (3.29) and the inverse relation given in (3.26), we obtain

$$\mathrm{Ti}_2\left(2 + \sqrt{3}\right) = \frac{2}{3}G - \frac{5}{12}\log\left(2 - \sqrt{3}\right)\pi. \tag{3.30}$$

In *(Almost) Impossible Integrals, Sums, and Series* (see [76, Chapter 3, pp.215–216]), I presented the extraction of the value $\mathrm{Ti}_2(2 - \sqrt{3})$ by using a well-known formula of Ramanujan.[4]

---

[4] By the Fourier series, $\log(\tan(x)) = -2\sum_{n=1}^{\infty} \dfrac{\cos(2(2n-1)x)}{2n-1}$, $0 < x < \pi/2$, together with differentiation, we prove the well-known result by Ramanujan, $\sum_{n=1}^{\infty} \dfrac{\sin(2(2n-1)x)}{(2n-1)^2} = \mathrm{Ti}_2(\tan(x)) - x\log(\tan(x))$, $0 < x < \pi/2$. Setting $x = \pi/12$ in Ramanujan's result and using that $\tan(\pi/12) = 2 - \sqrt{3}$, together with the fact that $\sum_{n=1}^{\infty} \dfrac{1}{(2n-1)^2}\sin\left(\dfrac{(2n-1)\pi}{6}\right) =$

Another wonderful idea to prove that $\mathrm{Ti}_2(2 - \sqrt{3}) = \int_0^{2-\sqrt{3}} \dfrac{\arctan(x)}{x}\,dx =$
$\int_0^{\tan(\pi/12)} \dfrac{\arctan(x)}{x}\,dx = \dfrac{2}{3}G + \log(2 - \sqrt{3})\dfrac{\pi}{12}$ is based on exploiting the identity,
$\tan(3x) = \tan\left(\dfrac{\pi}{3} + x\right)\tan(x)\tan\left(\dfrac{\pi}{3} - x\right)$, that is part of a strategy presented by
David M. Bradley in the paper "Representations of Catalan's Constant", which the
curious reader wouldn't like to miss!

3.8 More Good-Looking Logarithmic Integrals: The Third Part

Solution We'll continue with similar integrals to the ones in the previous sections,
and we note that two of the integrals in this section have wonderful closed forms
when compared to the other two integral results, the ones at the points (ii) and (iv)
(maybe some elegant ways are possible?). In one of the next sections we'll see the
usefulness of the result at the point (ii) in the solution to a beautiful problem.

Returning to the point (i) of the problem, recollecting and using the result in
(3.23), Sect. 3.6, where we set $a = i$, then replace x by a and b by $-a$, and taking
the real part of both sides, we obtain that

$$\Re\left\{\int_0^a \frac{\log(1 - ax)}{1 + ix}\,dx\right\} = \int_0^a \frac{\log(1 - ax)}{1 + x^2}\,dx$$

$$= \Re\left\{i\frac{\pi^2}{6} - i\log(-a - i)\log(1 + ia) + i\log(-a - i)\log(1 - a^2) - i\log(1 + ia)\log(a)\right.$$

$$\left. - i\frac{1}{2}\log^2(1 - a^2) + i\log(a)\log(1 - a^2) - i\mathrm{Li}_2(-ia) + i\mathrm{Li}_2(a^2) - i\mathrm{Li}_2\left(\frac{1 + ia}{1 - a^2}\right)\right\}$$

$$\left\{\text{use } \log(a + ib) = \frac{1}{2}\log(a^2 + b^2) + i\arctan\left(\frac{b}{a}\right), a > 0,\right\}$$

$$\left\{\text{and } \log(a + ib) = \frac{1}{2}\log(a^2 + b^2) + i\left(\arctan\left(\frac{b}{a}\right) - \pi\right), a, b < 0,\right\}$$

$\dfrac{1}{2}\left(1 - \dfrac{1}{3^2} + \dfrac{1}{5^2} - \cdots\right) + \dfrac{1}{6}\left(1 - \dfrac{1}{3^2} + \dfrac{1}{5^2} - \cdots\right) = \dfrac{2}{3}G$, where the terms have been grouped
wisely, that is, for the numbers generated by $\sin((2n - 1)\pi/6)$ I have used $3 \cdot 1/2 - 1/2$
when I have got 1, and $1/2 - 3 \cdot 1/2$ when I have obtained -1, we arrive at $\mathrm{Ti}_2(2 - \sqrt{3}) =$
$2G/3 + \log(2 - \sqrt{3})\pi/12$.

{and at the same time we recollect that $\Re\{i\mathrm{Li}_2(ix)\} = -\mathrm{Ti}_2(x) = \mathrm{Ti}_2(-x)\}$

$\left\{\text{and the inverse relation, } \arctan(x) + \arctan\left(\frac{1}{x}\right) = \mathrm{sgn}(x)\frac{\pi}{2}, \; x \in \mathbb{R} \setminus \{0\}\right\}$

$$= \arctan(a)\log(a) + \arctan(a)\log(1-a^2) + \frac{\pi}{2}\log(1-a^2) - \frac{\pi}{4}\log(1+a^2)$$

$$- \mathrm{Ti}_2(a) + \Im\left\{\mathrm{Li}_2\left(\frac{1}{1-a^2} + i\frac{a}{1-a^2}\right)\right\},$$

and the solution to the point (i) of the problem is finalized.

The integral at the point (ii) I prepare to evaluate is already included as the result **4.291.18** in the well-known *Table of Integrals, Series, and Products* by I.S. Gradshteyn and I.M. Ryzhik (8th Edition). The same thing to point out about the integral at the point (iv) which is included in the previously mentioned book. A treatment of both integrals at the points (ii) and (iv) may also be found in *Special Integrals of Gradshteyn and Ryzhik: The Proofs, Volume II* by Victor H. Moll.

For a first solution, I'll present a strategy involving the use of symmetry in double integrals, similar to one presented in *(Almost) Impossible Integrals, Sums, and Series* (see [76, Chapter 3, pp.162–163]).

Since we have that $\log(1+ax) = \displaystyle\int_0^a \frac{x}{1+xy}\,dy$, we may write

$$\int_0^a \frac{\log(1+ax)}{1+x^2}\,dx = \int_0^a \left(\int_0^a \frac{x}{(1+x^2)(1+xy)}\,dy\right)dx$$

{exploit the symmetry with respect to variables x and y}

$$= \frac{1}{2}\left(\int_0^a \left(\int_0^a \frac{x}{(1+x^2)(1+xy)}\,dy\right)dx + \int_0^a \left(\int_0^a \frac{y}{(1+y^2)(1+xy)}\,dy\right)dx\right)$$

$$= \frac{1}{2}\int_0^a \left(\int_0^a \frac{(x+y)(1+xy)}{(1+x^2)(1+y^2)(1+xy)}\,dy\right)dx = \frac{1}{2}\int_0^a \left(\int_0^a \frac{x+y}{(1+x^2)(1+y^2)}\,dy\right)dx$$

{exploit again the symmetry with respect to variables x and y}

$$= \int_0^a \frac{x}{1+x^2}\,dx \int_0^a \frac{1}{1+y^2}\,dy = \frac{1}{2}\arctan(a)\log(1+a^2),$$

and the first solution to the point (ii) of the problem is finalized.

For a second solution, we may consider the use of a *magical* variable change popularized in [22]. So, if we consider the variable change $(a-x)/(1+ax) = y$, we obtain that

$$\int_0^a \frac{\log(1+ax)}{1+x^2}dx = \int_0^a \frac{1}{1+y^2}\log\left(\frac{1+a^2}{1+ay}\right)dy = \log(1+a^2)\int_0^a \frac{1}{1+y^2}dy$$

$$-\int_0^a \frac{\log(1+ay)}{1+y^2}dy = \arctan(a)\log(1+a^2) - \int_0^a \frac{\log(1+ay)}{1+y^2}dy,$$

and since the last integral obtained after the last equal sign is the initial integral with the opposite sign, we are able the extract the value of the desired integral,

$$\int_0^a \frac{\log(1+ax)}{1+x^2}dx = \frac{1}{2}\arctan(a)\log(1+a^2),$$

and the second solution to the point (ii) of the problem is finalized.

For a third solution, I'll give another fast way also presented by Victor H. Moll in [64, pp.127–128], the author of the famous book, *Irresistible Integrals*. Essentially, we apply *Leibniz's rule for differentiation under the integral sign*, and if we denote the main integral by $I(a)$, and differentiate with respect to a, we get

$$I'(a) = \frac{\log(1+a^2)}{1+a^2} + \int_0^a \frac{x}{(1+x^2)(1+ax)}dx$$

{use the partial fraction decomposition}

$$= \frac{\log(1+a^2)}{1+a^2} + \frac{1}{1+a^2}\int_0^a \frac{x}{1+x^2}dx + \frac{a}{1+a^2}\int_0^a \frac{1}{1+x^2}dx$$

$$-\frac{a}{1+a^2}\int_0^a \frac{1}{1+ax}dx = \frac{1}{2}\frac{\log(1+a^2)}{1+a^2} + \frac{a\arctan(a)}{1+a^2} = \frac{1}{2}(\arctan(a)\log(1+a^2))'.$$

$$(3.31)$$

Finally, integrating with respect to a in (3.31), and considering that the integration constant is 0, we arrive at the desired result

$$I(a) = \int_0^a \frac{\log(1+ax)}{1+x^2}dx = \frac{1}{2}\arctan(a)\log(1+a^2),$$

and the third solution to the point (ii) of the problem is finalized.

Also, a fourth solution is possible, where the curious reader may use the strategy in the solution to the point (i) and exploit the result in (3.23), Sect. 3.6.

For the two remaining integrals we'll proceed as I did for the last two integrals in Sect. 3.4. So, by considering the variable change $\sqrt{a}x = t$, we write

$$\int_0^1 \frac{\log(1-ax)}{1+ax^2}dx = \int_0^1 \frac{\log(1-\sqrt{a}\sqrt{a}x)}{1+(\sqrt{a}x)^2}dx = \frac{1}{\sqrt{a}}\int_0^{\sqrt{a}} \frac{\log(1-\sqrt{a}t)}{1+t^2}dt$$

{employ the result from the point (i) and replace a by $\sqrt{a}$}

$$= \frac{1}{\sqrt{a}} \left(\frac{1}{2} \arctan(\sqrt{a}) \log(a) + \arctan(\sqrt{a}) \log(1-a) + \frac{\pi}{2} \log(1-a) - \frac{\pi}{4} \log(1+a) \right.$$

$$\left. - \operatorname{Ti}_2(\sqrt{a}) + \Im \left\{ \operatorname{Li}_2 \left(\frac{1}{1-a} + i \frac{\sqrt{a}}{1-a} \right) \right\} \right),$$

and the solution to the point (iii) of the problem is finalized.

By proceeding similarly as before for the last point of the problem, and letting the variable change $\sqrt{a}x = t$, we write

$$\int_0^1 \frac{\log(1+ax)}{1+ax^2} dx = \int_0^1 \frac{\log(1+\sqrt{a}\sqrt{a}x)}{1+(\sqrt{a}x)^2} dx = \frac{1}{\sqrt{a}} \int_0^{\sqrt{a}} \frac{\log(1+\sqrt{a}t)}{1+t^2} dt$$

{employ the result from the point (ii) and replace a by $\sqrt{a}$}

$$= \frac{1}{2\sqrt{a}} \arctan(\sqrt{a}) \log(1+a),$$

and the solution to the point (iv) of the problem is finalized.

The last logarithmic integral also appears as the result **4.291.19** in [17].

3.9 Special and Challenging Integrals with Parameter Involving the Inverse Hyperbolic Tangent: The First Part

Solution To put it directly, the existence of the present section is mainly due to the integral at the point (i), which deserves a special attention! For example, we'll see later in the book its power in solving a wonderful integral proposed by the renowned Larry Glasser or a splendid integral found in one of the last sections of this chapter.

So, how would we like to start? The point (i) of the problem is straightforward if we have in hand the first two points of Sect. 1.3, namely, the results in (1.6) and (1.7), which if we combine with the simple fact that $\operatorname{arctanh}(x) = 1/2 \log((1+x)/(1-x)) = 1/2 \log(1+x) - 1/2 \log(1-x)$, we arrive at

$$\int_0^1 \frac{x \operatorname{arctanh}(x)}{1+a^2x^2} dx = \frac{1}{2} \frac{\arctan^2(a)}{a^2}, \quad a \in \mathbb{R} \setminus \{0\},$$

and the first solution to the point (i) of the problem is complete.

How would we go differently, using a more direct way and avoiding the dilogarithm appearance?, you might wonder! In order to get a different way, we first might like to restrict a and consider $0 < a \leq 1$, and then we have

$$\int_0^1 \frac{x \, \text{arctanh}(x)}{1 + a^2 x^2} dx = \int_0^1 \sum_{n=1}^{\infty} (-1)^{n-1} x (ax)^{2n-2} \, \text{arctanh}(x) dx$$

{reverse the order of summation and integration}

$$= \sum_{n=1}^{\infty} (-1)^{n-1} a^{2n-2} \int_0^1 x^{2n-1} \, \text{arctanh}(x) dx$$

{make use of the result in (1.110), Sect. 1.22}

$$= \frac{1}{4} \sum_{n=1}^{\infty} (-1)^{n-1} a^{2n-2} \frac{2H_{2n} - H_n}{n} = \frac{1}{2} \frac{\text{arctan}^2(a)}{a^2}, \tag{3.32}$$

where the last equality is given by (4.17), Sect. 4.5.

Everything worked so smoothly in (3.32)! How about the version $1 < a < \infty$? In fact, to prove this version I'll also exploit the case $0 < a \leq 1$! So, considering $1 < a < \infty$ and using that $\text{arctanh}(x) = \int_0^1 \frac{x}{1 - x^2 y^2} dy$, we write

$$\int_0^1 \frac{x \, \text{arctanh}(x)}{1 + a^2 x^2} dx = \int_0^1 \left(\int_0^1 \frac{x^2}{(1 + a^2 x^2)(1 - x^2 y^2)} dy \right) dx$$

$$= \frac{1}{a^2} \int_0^1 \frac{1}{1 + (1/a^2) x^2} \left(\int_0^1 \frac{1}{1 - x^2 y^2} dy \right) dx - \int_0^1 \frac{1}{1 + a^2 x^2} \left(\int_0^1 \frac{1}{a^2 + y^2} dy \right) dx$$

$$\left\{ \text{use that } \int \frac{1}{a^2 + x^2} dx = \frac{1}{a} \text{arctan}\left(\frac{x}{a} \right), \ a \in \mathbb{R}, \ a \neq 0 \right\}$$

$$= \frac{1}{a^2} \int_0^1 \frac{\text{arctanh}(x)}{x} dx - \frac{1}{a^4} \int_0^1 \frac{x \, \text{arctanh}(x)}{1 + (1/a^2) x^2} dx - \frac{1}{a^2} \text{arctan}(a) \text{arctan}\left(\frac{1}{a} \right). \tag{3.33}$$

As regards the first integral in (3.33), we exploit the Taylor series that gives

$$\int_0^1 \frac{\text{arctanh}(x)}{x} dx = \int_0^1 \sum_{n=1}^{\infty} \frac{x^{2n-2}}{2n-1} dx = \sum_{n=1}^{\infty} \frac{1}{2n-1} \int_0^1 x^{2n-2} dx = \sum_{n=1}^{\infty} \frac{1}{(2n-1)^2}$$

$$= \sum_{n=1}^{\infty} \frac{1}{n^2} - \sum_{n=1}^{\infty} \frac{1}{(2n)^2} = \frac{3}{4} \sum_{n=1}^{\infty} \frac{1}{n^2} = \frac{\pi^2}{8}. \tag{3.34}$$

Now, a critical observation is that the second integral in (3.33) is already obtained in (3.32). I remind you that this is possible since $0 < 1/a^2 < 1$ due to the fact that in this part of the solution we treat the case $1 < a < \infty$, and thus we have

$$\int_0^1 \frac{x \arctanh(x)}{1 + (1/a^2)x^2} dx = \frac{1}{2} a^2 \arctan^2 \left(\frac{1}{a}\right). \tag{3.35}$$

By combining (3.33), (3.34), (3.35), and the inverse relation $\arctan(x) + \arctan\left(\dfrac{1}{x}\right) = \mathrm{sgn}(x)\dfrac{\pi}{2}$, $x \in \mathbb{R} \setminus \{0\}$, we get that

$$\int_0^1 \frac{x \arctanh(x)}{1 + a^2 x^2} dx = \frac{1}{2} \frac{\arctan^2(a)}{a^2}, \quad 1 < a < \infty. \tag{3.36}$$

Finally, based on (3.32) and (3.36), and then also extending the values of a to the negative real numbers, we conclude that

$$\int_0^1 \frac{x \arctanh(x)}{1 + a^2 x^2} dx = \frac{1}{2} \frac{\arctan^2(a)}{a^2}, \quad a \in \mathbb{R} \setminus \{0\},$$

and the second solution to the point (i) of the problem is complete.

A (very) elegant solution may also be obtained by exploiting the differentiation under the integral sign, and upon multiplying both sides of the stated equality by $2a^2$ and then integrating by parts, we get

$$\int_0^1 2a^2 \frac{x \arctanh(x)}{1 + a^2 x^2} dx = \int_0^1 \left(\log\left(\frac{1 + a^2 x^2}{1 + a^2}\right) \right)' \arctanh(x) dx$$

$$= \underbrace{\log\left(\frac{1 + a^2 x^2}{1 + a^2}\right) \arctanh(x) \Big|_{x=0}^{x=1}}_{0} - \int_0^1 \log\left(\frac{1 + a^2 x^2}{1 + a^2}\right) \frac{1}{1 - x^2} dx$$

$$= -\int_0^1 \frac{1}{1 - x^2} \left(\int_0^a \frac{\partial}{\partial t}\left\{ \log\left(\frac{1 + t^2 x^2}{1 + t^2}\right) \right\} dt \right) dx$$

$$= \int_0^1 \left(\int_0^a \frac{2t}{(1 + t^2)(1 + t^2 x^2)} dt \right) dx = \int_0^a \frac{2t}{1 + t^2} \left(\int_0^1 \frac{1}{1 + t^2 x^2} dx \right) dt$$

$$= \int_0^a \frac{2t}{1 + t^2} \left(\frac{\arctan(tx)}{t} \Big|_{x=0}^{x=1} \right) dt = 2 \int_0^a \frac{\arctan(t)}{1 + t^2} dt = \arctan^2(a),$$

whence the desired equality follows, and the third solution to the point (i) of the problem is complete.

To get a fourth solution, you might like to visit the next section where I treat generalizations with such integrals.

The point (ii) of the problem is straightforward if we replace a by ia at the point (i), thus giving us

$$\int_0^1 \frac{x \arctanh(x)}{1 - a^2 x^2} dx = \frac{1}{2} \frac{\arctanh^2(a)}{a^2}, \quad |a| < 1, \ a \in \mathbb{R} \setminus \{0\};$$

and the solution to the point (ii) of the problem is complete.

Next, the result at the point (iii) is again straightforward if we combine the results from the points (1.19) and (1.20), Sect. 1.6, that gives

$$\int_0^1 \frac{\arctanh(x)}{1 + a^2 x^2} dx$$

$$= \frac{1}{2} \log(2) \frac{\arctan(a)}{a} - \frac{1}{4} \frac{\arctan(a) \log(1 + a^2)}{a} + \frac{1}{2a} \Im \left\{ \mathrm{Li}_2 \left(\frac{1}{2} + i\frac{a}{2} \right) \right\},$$

where $a \in \mathbb{R} \setminus \{0\}$, and the solution to the point (iii) of the problem is complete. For example, it is easy to see that by the dilogarithm function reflection formula (see [21, Chapter 1, p.5], [70, Chapter 2, p.107]), $\mathrm{Li}_2(z) + \mathrm{Li}_2(1 - z) = \zeta(2) - \log(z) \log(1 - z)$, we readily find out that $\Im \left\{ \mathrm{Li}_2 \left(1/2 + ia/2 \right) \right\} = \Im \left\{ -\mathrm{Li}_2 \left(1/2 - ia/2 \right) \right\}$, which easily explains why $g(a) = g(-a)$, where $g(a) = \Im \left\{ \mathrm{Li}_2 \left(1/2 + ia/2 \right) \right\}/a$, used for extending the values of a to the negative real numbers in the main formula.

As regards the point (iv), we split the given integral by employing the fact that $\arctanh(x) = \frac{1}{2} \log((1 + x)/(1 - x)) = \frac{1}{2} \log(1 + x) - \frac{1}{2} \log(1 - x)$, and we have

$$\int_0^1 \frac{\arctanh(x)}{1 - a^2 x^2} dx = \frac{1}{4} \int_0^1 \frac{\log(1 + x)}{1 + ax} dx + \frac{1}{4} \int_0^1 \frac{\log(1 + x)}{1 - ax} dx$$

$$- \frac{1}{4} \int_0^1 \frac{\log(1 - x)}{1 + ax} dx - \frac{1}{4} \int_0^1 \frac{\log(1 - x)}{1 - ax} dx. \tag{3.37}$$

So, for the first integral in the right-hand side of (3.37) we start with integrating by parts, and then we write

$$\int_0^1 \frac{\log(1 + x)}{1 + ax} dx = \frac{1}{a} \int_0^1 (\log(1 + ax))' \log(1 + x) dx = \log(2) \frac{\log(1 + a)}{a}$$

$$- \frac{1}{a} \int_0^1 \frac{\log(1 + ax)}{1 + x} dx = \log(2) \frac{\log(1 + a)}{a} - \frac{1}{a} \int_0^1 \left(\int_0^a \frac{x}{(1 + x)(1 + xy)} dy \right) dx. \tag{3.38}$$

For the double integral in (3.38), we reverse the order of integration, and then we have that

$$\int_0^1 \left(\int_0^a \frac{x}{(1+x)(1+xy)} dy \right) dx = \int_0^a \left(\int_0^1 \frac{x}{(1+x)(1+xy)} dx \right) dy$$

{use the partial fraction decomposition and expand the double integral}

$$= \int_0^a \frac{1}{1-y} \left(\int_0^1 \frac{1}{1+yx} dx \right) dy - \int_0^a \frac{1}{1-y} \left(\int_0^1 \frac{1}{1+x} dx \right) dy$$

{calculate the inner integrals and then rearrange the resulting ones}

$$= \int_0^a \frac{\log((1+y)/2)}{1-y} dy + \int_0^a \frac{\log(1+y)}{y} dy = \text{Li}_2 \left(\frac{1-y}{2} \right) \Big|_{y=0}^{y=a} - \text{Li}_2(-y) \Big|_{y=0}^{y=a}$$

$$= \frac{1}{2} \log^2(2) - \frac{\pi^2}{12} - \text{Li}_2(-a) + \text{Li}_2 \left(\frac{1-a}{2} \right), \qquad (3.39)$$

where in the calculations I used that

$$\int_0^a \frac{\log((1+y)/2)}{1-y} dy = - \int_0^a \frac{\log(1-(1-y)/2)}{(1-y)/2} ((1-y)/2)' dy$$

$$\overset{(1-y)/2=t}{=\!=\!=} - \int_{1/2}^{(1-a)/2} \frac{\log(1-t)}{t} dt = \text{Li}_2 \left(\frac{1-a}{2} \right) - \text{Li}_2 \left(\frac{1}{2} \right)$$

$$\left\{ \text{make use of the special value, } \text{Li}_2 \left(\frac{1}{2} \right) = \frac{\pi^2}{12} - \frac{1}{2} \log^2(2) \right\}$$

$$= \text{Li}_2 \left(\frac{1-a}{2} \right) + \frac{1}{2} \log^2(2) - \frac{\pi^2}{12}. \qquad (3.40)$$

Thus, if we plug (3.39) in (3.38), we immediately arrive at

$$\int_0^1 \frac{\log(1+x)}{1+ax} dx$$

$$= \frac{1}{a} \left(\frac{\pi^2}{12} - \frac{1}{2} \log^2(2) + \log(2) \log(1+a) + \text{Li}_2(-a) - \text{Li}_2 \left(\frac{1-a}{2} \right) \right). \qquad (3.41)$$

The second integral in (3.37) is readily derived from (3.41), by replacing a with $-a$, and thus we get

$$\int_0^1 \frac{\log(1+x)}{1-ax}dx$$

$$= \frac{1}{a}\left(\frac{1}{2}\log^2(2) - \frac{\pi^2}{12} - \log(2)\log(1-a) - \text{Li}_2(a) + \text{Li}_2\left(\frac{1+a}{2}\right)\right).$$
(3.42)

For the third integral in (3.37) we may recall that this one is derived in (3.12), Sect. 3.3, and we have

$$\int_0^1 \frac{\log(1-t)}{1+at}dt = -\frac{1}{a}\text{Li}_2\left(\frac{a}{a+1}\right) = \frac{1}{a}\left(\frac{1}{2}\log^2(1+a) + \text{Li}_2(-a)\right),$$
(3.43)

where the last equality is a consequence of the Landen's dilogarithmic identity (see [21, Chapter 1, p.5], [70, Chapter 2, p.107]), $\text{Li}_2(x) + \text{Li}_2\left(\frac{x}{x-1}\right) = -\frac{1}{2}\log^2(1-x)$.

Finally, for the fourth integral in (3.37) we simply replace a by $-a$ in (3.43) that gives

$$\int_0^1 \frac{\log(1-t)}{1-at}dt = \frac{1}{a}\text{Li}_2\left(\frac{a}{a-1}\right) = -\frac{1}{a}\left(\frac{1}{2}\log^2(1-a) + \text{Li}_2(a)\right). \quad (3.44)$$

So, collecting the results from (3.41), (3.42), (3.43), and (3.44) in (3.37), we arrive at

$$\int_0^1 \frac{\text{arctanh}(x)}{1-a^2x^2}dx$$

$$= \frac{1}{4a}\left(\log^2(2) - \frac{\pi^2}{6} - 2\log(2)\log(1-a) + \log^2(1-a)\right.$$

$$\left. - 2\,\text{arctanh}^2(a) + 2\text{Li}_2\left(\frac{1+a}{2}\right)\right),$$

where to get the final result I also used that $\text{Li}_2((1-x)/2) + \text{Li}_2((1+x)/2) = \pi^2/6 - \log((1-x)/2)\log((1+x)/2)$, which is immediately derived from the dilogarithm function reflection formula, $\text{Li}_2(x) + \text{Li}_2(1-x) = \pi^2/6 - \log(x)\log(1-x)$, and the solution to the point (iv) of the problem is complete.

The last point of the problem is immediate if we consider the result from the point (i) where we multiply both sides by $\log(1-a)$, then integrate from $a = 0$ to $a = 1$, and finally employ the result in (1.19), Sect. 1.6,

$$\int_0^1 \arctan(x)\,\text{arctanh}(x)\log(1+x^2)dx$$

$$= 2 \int_0^1 \operatorname{arctanh}(x) \mathrm{Ti}_2(x) \mathrm{d}x + \int_0^1 \frac{\arctan^2(x) \log(1 - x)}{x^2} \mathrm{d}x,$$

and the solution to the point (v) of the problem is complete.

And how would we like to attack this last point differently?, one of your possible thoughts! Well, for a second solution, the curious reader might think of a transformation of the integrals in the relation into series. For example, as regards the integral in the left-hand side we may return to *(Almost) Impossible Integrals, Sums, and Series* and use the Cauchy product, $\sum_{k=1}^{\infty} (-1)^{k-1} x^{2k+1} \dfrac{H_{2k}}{2k + 1} = \dfrac{1}{2} \arctan(x) \log(1 + x^2)$, $|x| \le 1$, found in [76, Chapter 4, p.283]. Further, for the first integral in the right-hand side we may combine the power series, $\mathrm{Ti}_2(x) = \sum_{n=1}^{\infty} (-1)^{n-1} \dfrac{x^{2n-1}}{(2n - 1)^2}$, $|x| \le 1$, and the integral result in (1.110), Sect. 1.22, and for the second integral in the right-hand side we may combine the Cauchy product given in (4.17), Sect. 4.5, and the logarithmic integral in (3.10), Sect. 3.3.

At last, I remind you and strongly encourage you to enter the next section where I'll attack generalizations of the integrals at the four main points!

3.10   Special and Challenging Integrals with Parameter Involving the Inverse Hyperbolic Tangent: The Second Part

Solution   One thing you might easily figure out is that in this section the generalizations of the four main points from the previous section are reflected. And it is even more than that! The strategy of solving the generalizations is built on different ideas than the ones exploited in the previous section.

For example, if one takes a look at the case $n = 1$ of the point (i) and check the resulting closed form, it is not immediately noticeable that everything may be reduced to the nice closed form given at the point (i) from the previous section.

The core of the solutions to all four points of the section is represented by the reduction of the calculations to the powerful integral representation of the Polylogarithm function,

$$\int_0^1 \frac{z \log^n(t)}{1 - zt} \mathrm{d}t = (-1)^n n! \mathrm{Li}_{n+1}(z), \quad z \in \mathbb{C} \setminus (1, \infty), \tag{3.45}$$

which is an extended version to the complex plane of the result previously stated in (3.13), Sect. 3.3.

So, as regards the first point of the problem we start from the fact that $\arctanh(x) = 1/2 \log((1 + x)/(1 - x))$, and then we write

$$\int_0^1 \frac{x \arctanh^n(x)}{1 + a^2 x^2} dx = (-1)^n \frac{1}{2^n} \int_0^1 \frac{x}{1 + a^2 x^2} \log^n \left(\frac{1 - x}{1 + x}\right) dx$$

$$\overset{(1-x)/(1+x)=y}{=} (-1)^n \frac{1}{2^{n-1}} \int_0^1 \frac{(1 - y) \log^n(y)}{(1 + y)((1 + y)^2 + a^2(1 - y)^2)} dy$$

$$= (-1)^n \frac{1}{2^{n-1}} \Re \left\{ \int_0^1 \frac{(1 - y) \log^n(y)}{(1 + y)^2((1 + y) + ia(1 - y))} dy \right\}$$

{use the partial fraction decomposition and then split the integral}

$$= (-1)^n \frac{1}{a^2 2^n} \int_0^1 \frac{\log^n(y)}{1 + y} dy + (-1)^n \frac{1}{a^2 2^n} \Re \left\{ \int_0^1 \frac{((ia - 1)/(ia + 1)) \log^n(y)}{1 - ((ia - 1)/(ia + 1))y} dy \right\}.$$
$$(3.46)$$

For the first integral in (3.46), we have

$$\int_0^1 \frac{\log^n(y)}{1 + y} dy = \int_0^1 \log^n(y) \sum_{k=1}^\infty (-1)^{k-1} y^{k-1} dy = \sum_{k=1}^\infty (-1)^{k-1} \int_0^1 y^{k-1} \log^n(y) dy$$

$$= (-1)^n n! \sum_{k=1}^\infty (-1)^{k-1} \frac{1}{k^{n+1}} = (-1)^n n! \eta(n + 1) = (-1)^n (1 - 2^{-n}) n! \zeta(n + 1),$$
$$(3.47)$$

where I used that the Dirichlet eta function (see [93]), $\eta(s) = \sum_{n=1}^\infty (-1)^{n-1} \frac{1}{n^s}$, can be expressed in terms of the Riemann zeta function by the relation $\eta(s) = (1 - 2^{1-s})\zeta(s)$, and the simple fact that $\int_0^1 x^m \log^n(x) dx = (-1)^n \frac{n!}{(m + 1)^{n+1}}$, $m, n \in \mathbb{N}$, also presented in [76, Chapter 1, p.1].

Then, for the second integral in (3.46), we simply employ the result in (3.45), and then we arrive at

$$\int_0^1 \frac{((ia - 1)/(ia + 1)) \log^n(y)}{1 - ((ia - 1)/(ia + 1))y} dy = (-1)^n n! \mathrm{Li}_{n+1} \left(\frac{ia - 1}{ia + 1}\right)$$

$$= (-1)^n n! \mathrm{Li}_{n+1} \left(\frac{a^2 - 1}{a^2 + 1} + i \frac{2a}{1 + a^2}\right).$$
$$(3.48)$$

Upon collecting the results from (3.47) and (3.48) in (3.46), we conclude that

$$\int_0^1 \frac{x \arctanh^n(x)}{1+a^2x^2}\,dx$$

$$= \frac{n!}{a^2 2^n}\left((1-2^{-n})\zeta(n+1)+\Re\left\{\mathrm{Li}_{n+1}\left(\frac{a^2-1}{a^2+1}+i\,\frac{2a}{1+a^2}\right)\right\}\right)$$

$$= \frac{n!}{\tan^2(\theta/2)2^n}\left((1-2^{-n})\zeta(n+1)+\Re\left\{\mathrm{Li}_{n+1}\left(-\cos(\theta)+i\sin(\theta)\right)\right\}\right),$$

where $a = \tan(\theta/2)$, $\theta \in (-\pi,\pi)\setminus\{0\}$, and the solution to the point (i) of the problem is finalized.

Now, if we take a look at the case $n = 1$ of the generalization above, we get

$$\int_0^1 \frac{x \arctanh(x)}{1+a^2x^2}\,dx = \frac{1}{2a^2}\left(\frac{\pi^2}{12}+\Re\left\{\mathrm{Li}_2\left(\frac{a^2-1}{a^2+1}+i\,\frac{2a}{1+a^2}\right)\right\}\right)$$

$$= \frac{1}{2\tan^2(\theta/2)}\left(\frac{\pi^2}{12}+\Re\left\{\mathrm{Li}_2\left(-\cos(\theta)+i\sin(\theta)\right)\right\}\right). \tag{3.49}$$

Wait! The closed form obtained in (3.49) *looks different from the one at the point* (i) *in the previous section!*, which is how you might immediately react. Indeed, that's true, and in the following I'll show how to obtain that form!

For art's sake, I'll use the trigonometric form in (3.49), which I combine with the celebrated Euler's formula, $e^{i\theta} = \cos(\theta)+i\sin(\theta)$, and the dilogarithm series, $\mathrm{Li}_2(z) = \sum_{n=1}^{\infty}\frac{z^n}{n^2}$, $|z| \le 1$, and then we have

$$\int_0^1 \frac{x \arctanh(x)}{1+a^2x^2}\,dx = \frac{1}{2\tan^2(\theta/2)}\left(\frac{\pi^2}{12}+\Re\left\{\mathrm{Li}_2\left(-\cos(\theta)+i\sin(\theta)\right)\right\}\right)$$

$$= \frac{1}{2\tan^2(\theta/2)}\left(\frac{\pi^2}{12}+\Re\left\{\mathrm{Li}_2\left(-e^{-i\theta}\right)\right\}\right)$$

$$= \frac{1}{2\tan^2(\theta/2)}\left(\frac{\pi^2}{12}-\sum_{n=1}^{\infty}(-1)^{n-1}\frac{\Re\{e^{-in\theta}\}}{n^2}\right)$$

$$= \frac{1}{2\tan^2(\theta/2)}\left(\frac{\pi^2}{12}-\sum_{n=1}^{\infty}(-1)^{n-1}\frac{\cos(n\theta)}{n^2}\right) = \frac{\theta^2}{8\tan^2(\theta/2)}$$

$$\stackrel{\text{use } a=\tan(\theta/2)}{=} \frac{1}{2}\frac{\arctan^2(a)}{a^2}, \tag{3.50}$$

where in the calculations I used the Fourier series, $\sum_{n=1}^{\infty}(-1)^{n-1}\dfrac{\cos(nx)}{n^2} = \dfrac{\pi^2}{12} -$

$\dfrac{x^2}{4}$, $-\pi \le x \le \pi$ (see **1.443.4** in [17]), which is immediately derived by integration

from the Fourier series, $\sum_{n=1}^{\infty}(-1)^{n-1}\dfrac{\sin(nx)}{n} = \dfrac{x}{2}$, $-\pi < x < \pi$ (see **1.441.3** in

[17]), and this last Fourier series may also be obtained from the well-known Fourier

series, $\sum_{n=1}^{\infty}\dfrac{\sin(nx)}{n} = \dfrac{\pi - x}{2}$, $0 < x < 2\pi$ (see **1.441.1** in [17]), if we exploit the

case with x replaced by $2x$. Also, check the second footnote within Sect. 6.47.

Alternatively, the curious reader might like to perform the extraction of the desired closed form based on the first equality in (3.49).

Following the starting step from the previous point, for the point (ii) of the problem we write

$$\int_0^1 \frac{x \operatorname{arctanh}^n(x)}{1 - a^2 x^2}\, dx = (-1)^n \frac{1}{2^n} \int_0^1 \frac{x}{1 - a^2 x^2} \log^n\left(\frac{1-x}{1+x}\right) dx$$

$$\overset{(1-x)/(1+x)=y}{=} (-1)^n \frac{1}{2^{n-1}} \int_0^1 \frac{(1-y)\log^n(y)}{(1+y)((1+y)^2 - a^2(1-y)^2)}\, dy$$

{use the partial fraction decomposition and then split the integral}

$$= (-1)^{n-1}\frac{1}{a^2 2^n}\int_0^1 \frac{\log^n(y)}{1+y}\,dy + (-1)^{n-1}\frac{1}{a^2 2^{n+1}}\int_0^1 \frac{((a-1)/(a+1))\log^n(y)}{1 - ((a-1)/(a+1))y}\,dy$$

$$+ (-1)^{n-1}\frac{1}{a^2 2^{n+1}}\int_0^1 \frac{((a+1)/(a-1))\log^n(y)}{1 - ((a+1)/(a-1))y}\,dy$$

$$= \frac{n!}{a^2}\left((1 - 2^n)4^{-n}\zeta(n+1) - 2^{-n-1}\operatorname{Li}_{n+1}\left(\frac{a-1}{a+1}\right) - 2^{-n-1}\operatorname{Li}_{n+1}\left(\frac{a+1}{a-1}\right)\right),$$

where $|a| < 1$, $a \in \mathbb{R} \setminus \{0\}$, and the solution to the point (ii) of the problem is finalized. Note that during the calculations I also used the result in (3.45).

So, if we consider the case $n = 1$ of the generalization, we have

$$\int_0^1 \frac{x \operatorname{arctanh}(x)}{1 - a^2 x^2}\, dx = -\frac{1}{4a^2}\left(\frac{\pi^2}{6} + \operatorname{Li}_2\left(\frac{a-1}{a+1}\right) + \operatorname{Li}_2\left(\frac{a+1}{a-1}\right)\right). \tag{3.51}$$

Again, as happened with the case presented in (3.49) at the point (i), we easily see that the closed form obtained in (3.51) is different from the one given at the point (ii) from the previous section. However, it is not hard to obtain the closed form given in (1.32), Sect. 1.9.

Let's first observe that for $|a| < 1$, both $\dfrac{a-1}{a+1}$ and $\dfrac{a+1}{a-1}$ are negative. On the other hand, we have a well-known dilogarithm function identity (see [21, Chapter 1, p. 4]) that says that

$$\text{Li}_2(-x) + \text{Li}_2\left(-\frac{1}{x}\right) = -\frac{\pi^2}{6} - \frac{1}{2}\log^2(x), \quad x > 0, \tag{3.52}$$

which is easy to obtain if we differentiate $\text{Li}_2(-1/x)$ and then integrate back.

So, by exploiting (3.52), we write that

$$\text{Li}_2\left(\frac{a-1}{a+1}\right) + \text{Li}_2\left(\frac{a+1}{a-1}\right) = \text{Li}_2\left(-\frac{1-a}{1+a}\right) + \text{Li}_2\left(-\frac{1+a}{1-a}\right)$$

$$= -\frac{\pi^2}{6} - \frac{1}{2}\log^2\left(\frac{1+a}{1-a}\right) = -\frac{\pi^2}{6} - 2\operatorname{arctanh}^2(a). \tag{3.53}$$

Then, if we plug the result from (3.53) in (3.51), we arrive at the desired form,

$$\int_0^1 \frac{x\operatorname{arctanh}(x)}{1 - a^2 x^2}\,dx = \frac{1}{2}\frac{\operatorname{arctanh}^2(a)}{a^2}, \quad |a| < 1,\ a \in \mathbb{R} \setminus \{0\}.$$

Surely, for an alternative approach one might like to check the previous section.

Let's go further and attack the point (iii) with a style similar to the one at the point (i). So, we write

$$\int_0^1 \frac{\operatorname{arctanh}^n(x)}{1 + a^2 x^2}\,dx = (-1)^n \frac{1}{2^n} \int_0^1 \frac{1}{1 + a^2 x^2}\log^n\left(\frac{1-x}{1+x}\right)dx$$

$$\overset{(1-x)/(1+x)=y}{=} (-1)^n \frac{1}{2^{n-1}} \int_0^1 \frac{\log^n(y)}{(1+y)^2 + a^2(1-y)^2}\,dy$$

$$= (-1)^{n-1}\frac{1}{2^{n-1}}\Im\left\{\int_0^1 \frac{\log^n(y)}{a(1-y)((1+y) + ia(1-y))}\,dy\right\}$$

$$= (-1)^n \frac{1}{a2^n}\Im\left\{\int_0^1 \frac{((ia-1)/(ia+1))\log^n(y)}{1 - ((ia-1)/(ia+1))y}\,dy\right\}$$

{make use of the integral result in (3.48)}

$$= \frac{n!}{a2^n}\Im\left\{\text{Li}_{n+1}\left(\frac{a^2-1}{a^2+1} + i\frac{2a}{1+a^2}\right)\right\}$$

$$= \frac{n!}{\tan(\theta/2)2^n} \Im\{\mathrm{Li}_{n+1}(-\cos(\theta) + i\sin(\theta))\},$$

and the solution to the point (iii) of the problem is finalized.

Having arrived at the last point of the section, we act in a way similar to the one found at the point (ii). So, we write

$$\int_0^1 \frac{\mathrm{arctanh}^n(x)}{1 - a^2 x^2} dx = (-1)^n \frac{1}{2^n} \int_0^1 \frac{1}{1 - a^2 x^2} \log^n\left(\frac{1-x}{1+x}\right) dx$$

$$\underset{=}{\overset{(1-x)/(1+x)=y}{=}} (-1)^n \frac{1}{2^{n-1}} \int_0^1 \frac{\log^n(y)}{(1+y)^2 - a^2(1-y)^2} dy$$

{use the partial fraction decomposition and then split the integral}

$$= (-1)^n \frac{1}{a2^{n+1}} \int_0^1 \frac{(a-1)/(a+1)\log^n(y)}{1 - (a-1)/(a+1)y} dy$$

$$- (-1)^n \frac{1}{a2^{n+1}} \int_0^1 \frac{(a+1)/(a-1)\log^n(y)}{1 - (a+1)/(a-1)y} dy$$

$$= \frac{n!}{a2^{n+1}} \left(\mathrm{Li}_{n+1}\left(\frac{a-1}{a+1}\right) - \mathrm{Li}_{n+1}\left(\frac{a+1}{a-1}\right) \right),$$

where to get the last equality I used the result in (3.45), and the solution to the point (iv) of the problem is finalized.

Nice to see that everything flows so smoothly in the calculations above! The same smooth flow to expect in the next section!

3.11 Some Startling Generalizations Involving Logarithmic Integrals with Trigonometric Parameters

Solution The generalizations in this section may also be viewed as preliminary *tools*, together with the ones in the previous section, which are good to consider for obtaining the curious results in Sect. 1.12.

The first two points of the problem do not pose some particular difficulties, especially if strategies like the ones presented in the previous section are considered. The last two points might be found a bit tricky, where a slightly different approach is needed, based on clever applications of integration by parts.

In respect of the first point of the problem, we might exploit the denominator of the integrand to force a factorization and express the integrand by exploiting complex numbers, and then we have

$$\int_0^1 \frac{x \log^n(1-x)}{a^2+x^2}\,dx = \Re\left\{\int_0^1 \frac{\log^n(1-x)}{\pm ia + x}\,dx\right\} \overset{1-x=t}{=} \Re\left\{\int_0^1 \frac{1/(1 \pm ia)\log^n(t)}{1 - 1/(1\pm ia)t}\,dt\right\}$$

{make use of the Polylogarithm integral definition in (3.45), Sect. 3.10}

$$= (-1)^n n! \Re\left\{\operatorname{Li}_{n+1}\left(\frac{1}{1 \pm ia}\right)\right\}$$

$$\left\{\text{consider making the setting } a = \tan(\theta),\ \theta \in \left(-\frac{\pi}{2}, \frac{\pi}{2}\right)\right\}$$

$$= (-1)^n n! \Re\left\{\operatorname{Li}_{n+1}\left(\frac{1}{1 \pm i\tan(\theta)}\right)\right\}$$

$$= (-1)^n n! \Re\left\{\operatorname{Li}_{n+1}\left(\cos(\theta)(\cos(\theta) \mp i\sin(\theta))\right)\right\} = (-1)^n n! \sum_{k=1}^\infty \frac{\cos(k\theta)\cos^k(\theta)}{k^{n+1}},$$

where to get the last equality I combined the dilogarithm series, $\operatorname{Li}_2(z) = \sum_{n=1}^\infty \frac{z^n}{n^2}$, $|z| \le 1$, and Euler's formula, $e^{i\theta} = \cos(\theta) + i\sin(\theta)$, and the solution to the point (i) of the problem is complete.

Further, for the point (ii) of the problem we continue by using a plan of action similar to the one at the previous point, and then we write

$$\int_0^1 \frac{\log^n(1-x)}{a^2+x^2}\,dx = \frac{1}{a}\Im\left\{\int_0^1 \frac{\log^n(1-x)}{-ia+x}\,dx\right\} \overset{1-x=t}{=} \frac{1}{a}\Im\left\{\int_0^1 \frac{1/(1-ia)\log^n(t)}{1-1/(1-ia)t}\,dt\right\}$$

{make use of the Polylogarithm integral definition in (3.45), Sect. 3.10}

$$= (-1)^n n! \frac{1}{a}\Im\left\{\operatorname{Li}_{n+1}\left(\frac{1}{1-ia}\right)\right\}$$

$$\left\{\text{consider making the setting } a = \tan(\theta),\ \theta \in \left(-\frac{\pi}{2}, \frac{\pi}{2}\right) \setminus \{0\}\right\}$$

$$= (-1)^n n! \cot(\theta)\Im\left\{\operatorname{Li}_{n+1}\left(\frac{1}{1-i\tan(\theta)}\right)\right\}$$

$$= (-1)^n n! \cot(\theta)\Im\left\{\operatorname{Li}_{n+1}\left(\cos(\theta)(\cos(\theta)+i\sin(\theta))\right)\right\}$$

$$= (-1)^n n! \csc(\theta)\sum_{k=1}^\infty \frac{\sin(k\theta)\cos^{k+1}(\theta)}{k^{n+1}},$$

and the solution to the point (ii) of the problem is complete.

Regarding the point (iii) of the problem, which looks somewhat more *dangerous* than the previous versions above, we want first to prove the following generalization:

$$\int_0^1 \frac{\log^n(1+x)}{a+x} dx = \log^n(2) \log\left(\frac{a+1}{a-1}\right)$$

$$+ \sum_{k=1}^n (-1)^{k-1} \frac{n!}{(n-k)!} \log^{n-k}(2) \mathrm{Li}_{k+1}\left(\frac{2}{1-a}\right) + (-1)^n n! \mathrm{Li}_{n+1}\left(\frac{1}{1-a}\right).$$

$$(3.54)$$

Proof The solution of the result in (3.54) flows smoothly if we cleverly integrate by parts. So, we write

$$\int_0^1 \frac{\log^n(1+x)}{a+x} dx = \int_0^1 (\log((a+x)/(a-1)))' \log^n(1+x) dx$$

$$= \log^n(2) \log\left(\frac{a+1}{a-1}\right) - n \int_0^1 \frac{\log((a+x)/(a-1))}{1+x} \log^{n-1}(1+x) dx$$

$$\left\{ \text{use} \int \frac{\log(1+(1+x)/(a-1))}{(1+x)/(a-1)} ((1+x)/(a-1))' dx = -\mathrm{Li}_2\left(\frac{1+x}{1-a}\right) \right\}$$

{and then prepare to perform integration by parts by using the mentioned result}

$$= \log^n(2) \log\left(\frac{a+1}{a-1}\right) - n \int_0^1 \left(-\mathrm{Li}_2\left(\frac{1+x}{1-a}\right)\right)' \log^{n-1}(1+x) dx$$

$$= \log^n(2) \log\left(\frac{a+1}{a-1}\right) + n \log^{n-1}(2) \mathrm{Li}_2\left(\frac{2}{1-a}\right)$$

$$- n(n-1) \int_0^1 \mathrm{Li}_2\left(\frac{1+x}{1-a}\right) \frac{\log^{n-2}(1+x)}{1+x} dx$$

$$\left\{ \text{note and use that} \int \frac{1}{1+x} \mathrm{Li}_n\left(\frac{1+x}{1-a}\right) dx = \mathrm{Li}_{n+1}\left(\frac{1+x}{1-a}\right) \right\}$$

{and afterwards prepare again to perform integration by parts}

$$= \log^n(2) \log\left(\frac{a+1}{a-1}\right) + n \log^{n-1}(2) \mathrm{Li}_2\left(\frac{2}{1-a}\right)$$

$$-n(n-1)\int_0^1 \left(\mathrm{Li}_3\left(\frac{1+x}{1-a}\right)\right)' \log^{n-2}(1+x)dx$$

$$= \log^n(2)\log\left(\frac{a+1}{a-1}\right)+n\log^{n-1}(2)\mathrm{Li}_2\left(\frac{2}{1-a}\right)-n(n-1)\log^{n-2}(2)\mathrm{Li}_3\left(\frac{2}{1-a}\right)$$

$$+n(n-1)(n-2)\int_0^1\left(\mathrm{Li}_4\left(\frac{1+x}{1-a}\right)\right)'\log^{n-3}(1+x)dx$$

{integrate by parts for another $n-3$ times}

$$= \log^n(2)\log\left(\frac{a+1}{a-1}\right)+\sum_{k=1}^{n}(-1)^{k-1}\frac{n!}{(n-k)!}\log^{n-k}(2)\mathrm{Li}_{k+1}\left(\frac{2}{1-a}\right)$$

$$+(-1)^n n!\mathrm{Li}_{n+1}\left(\frac{1}{1-a}\right),$$

which brings an end to the solution of the auxiliary result. We might also like to note that for $a \in (-1, 0)$ the result must be understood as a Cauchy principal value. ∎

If we replace a by ia in (3.54) and then take the real part of both sides, we immediately arrive at

$$\int_0^1 \frac{x\log^n(1+x)}{a^2+x^2}dx$$

$$= \sum_{k=1}^{n}(-1)^{k-1}\frac{n!}{(n-k)!}\log^{n-k}(2)\Re\left\{\mathrm{Li}_{k+1}\left(\frac{2}{1-ia}\right)\right\}$$

$$+(-1)^n n!\Re\left\{\mathrm{Li}_{n+1}\left(\frac{1}{1-ia}\right)\right\}$$

$$\left\{\text{consider making the setting } a=\tan(\theta), \theta\in\left(-\frac{\pi}{2},\frac{\pi}{2}\right)\right\}$$

$$= \sum_{k=1}^{n}(-1)^{k-1}\frac{n!}{(n-k)!}\log^{n-k}(2)\Re\left\{\mathrm{Li}_{k+1}\left(1+e^{i2\theta}\right)\right\}$$

$$+(-1)^n n!\Re\left\{\mathrm{Li}_{n+1}\left(\cos(\theta)e^{i\theta}\right)\right\}$$

$$= \sum_{k=1}^{n}(-1)^{k-1}\frac{n!}{(n-k)!}\log^{n-k}(2)\Re\left\{\operatorname{Li}_{k+1}\left(1+e^{i2\theta}\right)\right\}$$

$$+ (-1)^n n! \sum_{k=1}^{\infty}\frac{\cos(k\theta)\cos^k(\theta)}{k^{n+1}},$$

and the solution to the point (iii) of the problem is complete. During the calculations I also used the complex form of the inverse tangent, $\arctan(x) = \frac{1}{2}i\log((1 - ix)/(1 + ix))$, that shows that $\log((ia + 1)/(ia - 1)) = -i2\arctan(1/a)$, which appears when replacing a by ia in (3.54). The particular case $a = 1$ of the generalization above also appeared in [41].

Finally, for the last point of the problem we proceed in a similar style to the one at the point (iii), except that after replacing a by ia in (3.54), we multiply both sides by $-1/a$ and take the imaginary part that leads to

$$\int_0^1 \frac{\log^n(1+x)}{a^2+x^2}dx$$

$$= \frac{2}{a}\arctan\left(\frac{1}{a}\right)\log^n(2) - \frac{1}{a}\sum_{k=1}^{n}(-1)^{k-1}\frac{n!}{(n-k)!}\log^{n-k}(2)\Im\left\{\operatorname{Li}_{k+1}\left(\frac{2}{1-ia}\right)\right\}$$

$$+ (-1)^{n-1}n!\frac{1}{a}\Im\left\{\operatorname{Li}_{n+1}\left(\frac{1}{1-ia}\right)\right\}$$

$$\left\{\text{consider making the setting } a = \tan(\theta), \theta \in \left(-\frac{\pi}{2},\frac{\pi}{2}\right)\setminus\{0\}\right\}$$

$$= (\pi\operatorname{sgn}(\theta) - 2\theta)\cot(\theta)\log^n(2)$$

$$- \cot(\theta)\sum_{k=1}^{n}(-1)^{k-1}\frac{n!}{(n-k)!}\log^{n-k}(2)\Im\left\{\operatorname{Li}_{k+1}\left(1+e^{i2\theta}\right)\right\}$$

$$+ (-1)^{n-1}n!\cot(\theta)\Im\left\{\operatorname{Li}_{n+1}\left(\cos(\theta)e^{i\theta}\right)\right\}$$

$$= (\pi\operatorname{sgn}(\theta) - 2\theta)\cot(\theta)\log^n(2)$$

$$- \cot(\theta)\sum_{k=1}^{n}(-1)^{k-1}\frac{n!}{(n-k)!}\log^{n-k}(2)\Im\left\{\operatorname{Li}_{k+1}\left(1+e^{i2\theta}\right)\right\}$$

$$+ (-1)^{n-1} n! \cot(\theta) \sum_{k=1}^{\infty} \frac{\sin(k\theta) \cos^k(\theta)}{k^{n+1}},$$

and the solution to the point (iv) of the problem is complete. In the calculations I used that based on the identity, $\arctan(x) + \arctan(1/x) = \pi/2 \operatorname{sgn}(x)$, $x \in \mathbb{R} \setminus \{0\}$, we immediately have that $\arctan(1/a)/a = (\pi/2 \operatorname{sgn}(a) - \arctan(a))/a = \cot(\theta)(\pi/2 \operatorname{sgn}(\tan(\theta)) - \theta) = \cot(\theta)(\pi/2 \operatorname{sgn}(\theta) - \theta)$.

Some of the particular cases of the present generalizations will be successfully exploited in the next section!

3.12   More Startling and Enjoyable Logarithmic Integrals Involving Trigonometric Parameters

Solution   Here we are, preparing to enjoy a section with spectacular and (probably) unexpected integral results strongly related to the ones in the previous sections!

Essentially, for getting the values of the desired integrals, we'll want to exploit particular cases of the generalized integrals in Sects. 1.10 and 1.11.

First, let's pick up the result at the point (ii) and prove it. We observe that since we have the algebraic identity $ab = ((a+b)^2 - (a-b)^2)/4$, if we set $a = \log(1+x)$ and $b = \log(1-x)$, and then multiply both sides by $x/(a^2+x^2)$ and integrate from $x = 0$ to $x = 1$, we arrive at

$$\int_0^1 \frac{x \log(1-x) \log(1+x)}{a^2 + x^2} dx = \frac{1}{4} \int_0^1 \frac{x \log^2(1-x^2)}{a^2 + x^2} dx - \int_0^1 \frac{x \operatorname{arctanh}^2(x)}{a^2 + x^2} dx. \tag{3.55}$$

For the first integral in the right-hand side of (3.55), we start with the variable change $1 - x^2 = y$, and then we have

$$\int_0^1 \frac{x \log^2(1-x^2)}{a^2 + x^2} dx = \frac{1}{2} \int_0^1 \frac{(1/(1+a^2)) \log^2(y)}{1 - (1/(1+a^2))y} dy$$

{employ the Polylogarithm integral definition in (3.45), Sect. 3.10}

$$= \operatorname{Li}_3 \left(\frac{1}{1+a^2} \right) \overset{a=\tan(\theta)}{=} \operatorname{Li}_3 \left(\cos^2(\theta) \right). \tag{3.56}$$

On the other hand, to get the value of the second integral in the right-hand side of (3.55), we want to return to the first equality in (1.36), Sect. 1.10, where if we replace a by $1/a$ and then multiply both sides by $1/a^2$, we arrive at

$$\int_0^1 \frac{x \operatorname{arctanh}^n(x)}{a^2 + x^2} dx = \frac{n!}{2^n} \left((1 - 2^{-n})\zeta(n+1) + \Re\left\{ \operatorname{Li}_{n+1}\left(\frac{1-a^2}{1+a^2} + i\frac{2a}{1+a^2} \right) \right\} \right)$$

$$\overset{a = \tan(\theta)}{=} \frac{n!}{2^n} \left((1 - 2^{-n})\zeta(n+1) + \Re\{ \operatorname{Li}_{n+1}\left(\cos(2\theta) + i\sin(2\theta) \right) \} \right)$$

$$= \frac{n!}{2^n} \left((1 - 2^{-n})\zeta(n+1) + \sum_{k=1}^{\infty} \frac{\cos(2k\theta)}{k^{n+1}} \right). \tag{3.57}$$

So, if we set $n = 2$ in (3.57), we immediately get the needed auxiliary result,

$$\int_0^1 \frac{x \operatorname{arctanh}^2(x)}{a^2 + x^2} dx$$

$$= \frac{3}{8}\zeta(3) + \frac{1}{2}\Re\left\{ \operatorname{Li}_3\left(\frac{1-a^2}{1+a^2} + i\frac{2a}{1+a^2} \right) \right\} = \frac{3}{8}\zeta(3) + \frac{1}{2}\sum_{k=1}^{\infty} \frac{\cos(2k\theta)}{k^3}. \tag{3.58}$$

Collecting the results from (3.56) and (3.58) in (3.55), we obtain that

$$\int_0^1 \frac{x \log(1 - x)\log(1 + x)}{a^2 + x^2} dx$$

$$= -\frac{3}{8}\zeta(3) + \frac{1}{4}\operatorname{Li}_3\left(\frac{1}{1+a^2} \right) - \frac{1}{2}\Re\left\{ \operatorname{Li}_3\left(\frac{1-a^2}{1+a^2} + i\frac{2a}{1+a^2} \right) \right\}$$

$$= -\frac{3}{8}\zeta(3) + \frac{1}{4}\operatorname{Li}_3(\cos^2(\theta)) - \frac{1}{2}\sum_{n=1}^{\infty} \frac{\cos(2n\theta)}{n^3},$$

and the solution to the point (ii) of the problem is finalized.

Returning to the result in (3.56), in particular focusing on the opposite sides expressed in terms of parameter a, we are able to write the integral at the point (i) in a useful form, after expanding the left-hand side, and then we have

$$\int_0^1 \frac{x \log^2(1 + x)}{a^2 + x^2} dx$$

$$= \operatorname{Li}_3\left(\frac{1}{1+a^2} \right) - \int_0^1 \frac{x \log^2(1 - x)}{a^2 + x^2} dx - 2\int_0^1 \frac{x \log(1 - x)\log(1 + x)}{a^2 + x^2} dx$$

$$= \frac{3}{4}\zeta(3) + \frac{1}{2}\operatorname{Li}_3\left(\frac{1}{1+a^2} \right) - 2\Re\left\{ \operatorname{Li}_3\left(\frac{1}{1 \pm ia} \right) \right\} + \Re\left\{ \operatorname{Li}_3\left(\frac{1-a^2}{1+a^2} + i\frac{2a}{1+a^2} \right) \right\}$$

$$\overset{a=\tan(\theta)}{=} \frac{3}{4}\zeta(3) + \frac{1}{2}\mathrm{Li}_3(\cos^2(\theta)) + \sum_{n=1}^{\infty}\frac{\cos(2n\theta)}{n^3} - 2\sum_{n=1}^{\infty}\frac{\cos(n\theta)\cos^n(\theta)}{n^3},$$

where in the calculations I used the case $n = 2$ of the generalization in (1.40), Sect. 1.11, and the result from the point (ii) previously calculated, and the solution to the point (i) of the problem is finalized.

Further, for the third point we consider the algebraic identity $ab^2 = \frac{1}{6}((a-b)^3 + (a+b)^3 - 2a^3)$, where if we set $a = \log(1-x)$ and $b = \log(1+x)$ and then multiply both sides by $x/(a^2 + x^2)$ and integrate from $x = 0$ to $x = 1$, we get that

$$\int_0^1 \frac{x\log(1-x)\log^2(1+x)}{a^2+x^2}\,dx$$

$$= -\frac{4}{3}\int_0^1\frac{x\operatorname{arctanh}^3(x)}{a^2+x^2}\,dx + \frac{1}{6}\int_0^1\frac{x\log^3(1-x^2)}{a^2+x^2}\,dx - \frac{1}{3}\int_0^1\frac{x\log^3(1-x)}{a^2+x^2}\,dx.$$
$$(3.59)$$

Now, for the first integral in the right-hand side of (3.59) we use the opposite sides of the result in (3.57), the case $n = 3$, met during the calculations to the point (ii) of the problem, and then we have

$$\int_0^1\frac{x\operatorname{arctanh}^3(x)}{a^2+x^2}\,dx = \frac{7}{960}\pi^4 + \frac{3}{4}\sum_{n=1}^{\infty}\frac{\cos(2n\theta)}{n^4} = \frac{\pi^4}{64} - \frac{\pi^2}{4}\theta^2 + \frac{\pi}{2}|\theta|^3 - \frac{1}{4}\theta^4,$$
$$(3.60)$$

where in the calculations I used the Fourier series $\displaystyle\sum_{n=1}^{\infty}\frac{\cos(nx)}{n^4} = \frac{\pi^4}{90} - \frac{\pi^2}{12}x^2 +$

$\dfrac{\pi}{12}x^3 - \dfrac{x^4}{48}$, $0 \le x \le 2\pi$ (see **1.443.6** in [17]), which is obtained from the well-known Fourier series, $\displaystyle\sum_{n=1}^{\infty}\frac{\sin(nt)}{n} = \frac{\pi-t}{2}$, $0 < t < 2\pi$ (see **1.441.1** in [17]), by integrating three times both sides, from 0 to x. Also, observe that the use of the absolute value in (3.60) is natural to consider when extending the values of x to $-2\pi \le x \le 2\pi$, since cosine in the Fourier series is even.

Then, for the second integral in the right-hand side of (3.59) we let the variable change $1 - x^2 = y$ that gives

$$\int_0^1\frac{x\log^3(1-x^2)}{a^2+x^2}\,dx = \frac{1}{2}\int_0^1\frac{(1/(1+a^2))\log^3(y)}{1-(1/(1+a^2))y}\,dy$$

{employ the Polylogarithm integral definition in (3.45), Sect. 3.10}

$$= -3\mathrm{Li}_4\left(\frac{1}{1+a^2}\right) \overset{a=\tan(\theta)}{=\!=} -3\mathrm{Li}_4\left(\cos^2(\theta)\right). \qquad (3.61)$$

At this point, collecting the case $n = 3$ of the generalization in (1.40), Sect. 1.11, and the results previously derived in (3.60) and (3.61) in (3.59), we get that

$$\int_0^1 \frac{x\log(1-x)\log^2(1+x)}{a^2+x^2}dx$$

$$= \frac{1}{3}\theta^4 - \frac{2}{3}\pi|\theta|^3 + 2\zeta(2)\theta^2 - \frac{15}{8}\zeta(4) - \frac{1}{2}\mathrm{Li}_4(\cos^2(\theta)) + 2\sum_{n=1}^{\infty}\frac{\cos(n\theta)\cos^n(\theta)}{n^4}$$

$$\overset{\theta=\arctan(a)}{=\!=} \frac{1}{3}\arctan^4(a) - \frac{2}{3}\pi\arctan^3(|a|) + 2\zeta(2)\arctan^2(a) - \frac{15}{8}\zeta(4)$$

$$- \frac{1}{2}\mathrm{Li}_4\left(\frac{1}{1+a^2}\right) + 2\sum_{n=1}^{\infty}\frac{1}{n^4}(1+a^2)^{-n/2}\cos(n\arctan(a)),$$

and the solution to the point (iii) of the problem is finalized.

At last, we prepare to derive beautiful forms for the given real parts of dilogarithmic and trilogarithmic values involving the complex argument $1 + e^{i2\theta}$ found at the last two points of the problem.

So, for the point (iv) we assume $-\pi/2 < \theta < \pi/2$ and start with the dilogarithm function reflection formula, $\mathrm{Li}_2(x) + \mathrm{Li}_2(1-x) = \zeta(2) - \log(x)\log(1-x)$, where if we set $x = -e^{i2\theta}$, we obtain

$$\Re\left\{\mathrm{Li}_2\left(1+e^{i2\theta}\right)\right\} = \zeta(2) - \Re\left\{\mathrm{Li}_2\left(-e^{i2\theta}\right)\right\} - \Re\left\{\log\left(-e^{i2\theta}\right)\log\left(1+e^{i2\theta}\right)\right\}$$

$$= \zeta(2) + \sum_{n=1}^{\infty}(-1)^{n-1}\frac{\cos(2n\theta)}{n^2} - \Re\left\{\log\left(-e^{i2\theta}\right)\log\left(1+e^{i2\theta}\right)\right\}$$

$$\left\{\text{use the Fourier series, } \sum_{n=1}^{\infty}(-1)^{n-1}\frac{\cos(n\theta)}{n^2} = \frac{\pi^2}{12} - \frac{\theta^2}{4}, \ -\pi \le \theta \le \pi,\right\}$$

$$\left\{\text{and then that } \log\left(-e^{i2\theta}\right) = i(2\theta - \mathrm{sgn}(\theta)\pi), \ \theta \in \left[-\frac{\pi}{2}, \frac{\pi}{2}\right]\setminus\{0\}\right\}$$

$$\left\{\text{together with the fact that } \Im\left\{\log\left(1+e^{i2\theta}\right)\right\} = \theta, \ \theta \in \left(-\frac{\pi}{2}, \frac{\pi}{2}\right)\right\}$$

$$= \left(\frac{\pi}{2} - |\theta|\right)^2 = \frac{3}{2}\zeta(2) - \pi|\theta| + \theta^2,$$

and the solution to the point (iv) of the problem is finalized. Observe that the Fourier series employed in the calculations above, $\sum_{n=1}^{\infty}(-1)^{n-1}\dfrac{\cos(nx)}{n^2} = \dfrac{\pi^2}{12} - \dfrac{x^2}{4}$, $-\pi \leq x \leq \pi$ (see **1.443.4** in [17]), is easily obtained by integration from the Fourier series, $\sum_{n=1}^{\infty}(-1)^{n-1}\dfrac{\sin(2n\theta)}{n} = \theta$, $\theta \in \left(-\dfrac{\pi}{2}, \dfrac{\pi}{2}\right)$ (see **1.441.3** in [17]). For more details on its derivation, see the last part of the current section.

Finally, for the point (v) we assume again that $-\pi/2 < \theta < \pi/2$ and start by considering the case $n = 2$ of the generalization in (1.42), Sect. 1.11, that leads to

$$\Re\left\{\mathrm{Li}_3\left(1 + e^{i2\theta}\right)\right\}$$

$$= \log(2)\Re\left\{\mathrm{Li}_2\left(1 + e^{i2\theta}\right)\right\} + \sum_{n=1}^{\infty}\frac{\cos(n\theta)\cos^n(\theta)}{n^3} - \frac{1}{2}\int_0^1 \frac{x\log^2(1+x)}{\tan^2(\theta) + x^2}dx$$

$$\{\text{employ the results from the points } (i) \text{ and } (iv)\}$$

$$= \frac{3}{2}\log(2)\zeta(2) - \frac{3}{8}\zeta(3) - \log(2)\pi\,|\theta| + \log(2)\theta^2 - \frac{1}{4}\mathrm{Li}_3(\cos^2(\theta))$$

$$- \frac{1}{2}\sum_{n=1}^{\infty}\frac{\cos(2n\theta)}{n^3} + 2\sum_{n=1}^{\infty}\frac{\cos(n\theta)\cos^n(\theta)}{n^3},$$

and the solution to the point (v) of the problem is finalized.

Observe that in the calculations I exploited the fact that $e^{\pm i\pi} = -1$ which leads to $\log\left(-e^{i2\theta}\right) = \log\left(e^{i(2\theta - \mathrm{sign}(\theta)\pi)}\right) = i(2\theta - \mathrm{sgn}(\theta)\pi)$, $\theta \in \left[-\dfrac{\pi}{2}, \dfrac{\pi}{2}\right] \setminus \{0\}$.

Further, since $\log\left(1 + e^{i2\theta}\right) = \log\left(2\cos^2(\theta) + i\sin(2\theta)\right) = \dfrac{1}{2}\log(4\cos^2(\theta)) + i\arctan(\tan(\theta)) = \log(2\cos(\theta)) + i\theta$, $\theta \in \left(-\dfrac{\pi}{2}, \dfrac{\pi}{2}\right)$, based on the fact that, $\log(a + ib) = \log(\sqrt{a^2 + b^2}) + i\arctan(b/a)$, $a > 0$, we obtain $\Im\left\{\log\left(1 + e^{i2\theta}\right)\right\} = \theta$, $\theta \in \left(-\dfrac{\pi}{2}, \dfrac{\pi}{2}\right)$. From this last result we immediately get the Fourier series, $\sum_{n=1}^{\infty}(-1)^{n-1}\dfrac{\sin(2n\theta)}{n} = \theta$, $\theta \in \left(-\dfrac{\pi}{2}, \dfrac{\pi}{2}\right)$. Useful generalizations with such Fourier series as the ones used in this section may be found in the fourth chapter.

3.13 Surprisingly Bewitching Trigonometric Integrals with Appealing Closed Forms: The First Act

Solution While looking over the statement section with the first four points of the problem, one might wonder why these integrals and the ones in the next section are not put together since they are very similar. Well, it could be the case that they deserve more attention!

There is something simple and special about them that one might not notice at the first sight: unlike the integrals in the next section, you'll find out that this time the desired closed forms are extracted based on parity.

Having said that, let's go straight to the calculations, and for the point (i) we start with splitting the integral according to the parity of n, and then we write that

$$I_{2n-1} = \int_0^{\pi/2} \sin((2n-1)\theta)\sin^{2n-1}(\theta)d\theta$$

$$\overset{\pi/2-\theta=u}{=} (-1)^{n-1}\int_0^{\pi/2}\cos((2n-1)u)\cos^{2n-1}(u)du$$

{consider the integral result in (1.57), Sect. 1.14}

$$= (-1)^{n-1}\frac{\pi}{2^{2n}}. \tag{3.62}$$

On the other hand, we have

$$I_{2n} = \int_0^{\pi/2}\sin(2n\theta)\sin^{2n}(\theta)d\theta \overset{\pi/2-\theta=u}{=} (-1)^{n-1}\int_0^{\pi/2}\sin(2nu)\cos^{2n}(u)du$$

{employ the integral result in (1.61), Sect. 1.14}

$$= (-1)^{n-1}\frac{1}{2^{2n+1}}\sum_{k=1}^{2n}\frac{2^k}{k}. \tag{3.63}$$

So, combining (3.62) and (3.63), we see that everything may be put in the form

$$I_n = \int_0^{\pi/2}\sin(n\theta)\sin^n(\theta)d\theta = \begin{cases} I_{2n-1} = (-1)^{n-1}\dfrac{\pi}{2^{2n}}; \\[2ex] I_{2n} = (-1)^{n-1}\dfrac{1}{2^{2n+1}}\displaystyle\sum_{k=1}^{2n}\dfrac{2^k}{k}, \end{cases} \tag{3.64}$$

and the solution to the point (i) of the problem is finalized.

Passing to the point (ii) of the problem, we proceed in a similar style and perform an analysis based on the parity of n, and then we write

$$J_{2n-1} = \int_0^{\pi/2} \sin((2n-1)\theta) \sin^{2n-2}(\theta)d\theta$$

$$\overset{\pi/2-\theta=u}{=} (-1)^{n-1} \int_0^{\pi/2} \cos((2n-1)u) \cos^{2n-2}(u)du$$

{consider the integral result in (1.59), Sect. 1.14}

$$= (-1)^{n-1} \frac{2^{2n-1}}{(2n-1)\binom{4n-2}{2n-1}} = (-1)^{n-1} \frac{(4n-1)2^{2n-1}}{n(2n-1)\binom{4n}{2n}}, \qquad (3.65)$$

where in the calculations I used that $\binom{4n-2}{2n-1} = \dfrac{n}{4n-1}\binom{4n}{2n}$.

In the second place, we have

$$J_{2n} = \int_0^{\pi/2} \sin(2n\theta) \sin^{2n-1}(\theta)d\theta \overset{\pi/2-\theta=u}{=} (-1)^{n-1} \int_0^{\pi/2} \sin(2nu) \cos^{2n-1}(u)du$$

{employ the integral result in (1.63), Sect. 1.14}

$$= (-1)^{n-1} \frac{2^{2n-1}}{n\binom{4n}{2n}} \sum_{k=0}^{2n-1} \frac{1}{2^k}\binom{2k}{k} = (-1)^{n-1} \frac{1}{2^{2n-1}} \sum_{k=1}^{2n} \frac{1}{2k-1}\binom{2n-1}{k-1}.$$

$$\qquad (3.66)$$

So, combining (3.65) and (3.66), we arrive at

$$J_n = \int_0^{\pi/2} \sin(n\theta) \sin^{n-1}(\theta)d\theta$$

$$= \begin{cases} J_{2n-1} = (-1)^{n-1} \dfrac{2^{2n-1}}{(2n-1)\binom{4n-2}{2n-1}} = (-1)^{n-1} \dfrac{(4n-1)2^{2n-1}}{n(2n-1)\binom{4n}{2n}}; \\[4ex] J_{2n} = (-1)^{n-1} \dfrac{2^{2n-1}}{n\binom{4n}{2n}} \displaystyle\sum_{k=0}^{2n-1} \dfrac{1}{2^k}\binom{2k}{k} = (-1)^{n-1} \dfrac{1}{2^{2n-1}} \displaystyle\sum_{k=1}^{2n} \dfrac{1}{2k-1}\binom{2n-1}{k-1}, \end{cases}$$

$$\qquad (3.67)$$

and the solution to the point (ii) of the problem is finalized.

Moving to the point (iii) of the problem and using again the treatment based on the parity of n, we write

$$K_{2n-1} = \int_0^{\pi/2} \cos((2n-1)\theta) \sin^{2n-1}(\theta)d\theta$$

$$\overset{\pi/2-\theta=u}{=} (-1)^{n-1} \int_0^{\pi/2} \sin((2n-1)u) \cos^{2n-1}(u)du$$

{employ the integral result in (1.61), Sect. 1.14}

$$= (-1)^{n-1} \frac{1}{2^{2n}} \sum_{k=1}^{2n-1} \frac{2^k}{k}. \tag{3.68}$$

Next, we have that

$$K_{2n} = \int_0^{\pi/2} \cos(2n\theta) \sin^{2n}(\theta)d\theta \overset{\pi/2-\theta=u}{=} (-1)^n \int_0^{\pi/2} \cos(2n\theta) \cos^{2n}(\theta)d\theta$$

{consider the integral result in (1.57), Sect. 1.14}

$$= (-1)^n \frac{\pi}{2^{2n+1}}. \tag{3.69}$$

So, if we put together (3.68) and (3.69), we obtain that

$$K_n = \int_0^{\pi/2} \cos(n\theta) \sin^n(\theta)d\theta = \begin{cases} K_{2n-1} = (-1)^{n-1} \dfrac{1}{2^{2n}} \displaystyle\sum_{k=1}^{2n-1} \dfrac{2^k}{k}; \\ K_{2n} = (-1)^n \dfrac{\pi}{2^{2n+1}}, \end{cases} \tag{3.70}$$

and the solution to the point (iii) of the problem is finalized.

Then, by following the routine of the approach based on the parity of n, we write

$$L_{2n-1} = \int_0^{\pi/2} \cos((2n-1)\theta) \sin^{2n-2}(\theta)d\theta$$

$$\overset{\pi/2-\theta=u}{=} (-1)^{n-1} \int_0^{\pi/2} \sin((2n-1)u) \cos^{2n-2}(u)du$$

{bring into play the integral result in (1.63), Sect. 1.14}

$$= (-1)^{n-1} \frac{2^{2n-1}}{(2n-1)\binom{4n-2}{2n-1}} \sum_{k=0}^{2n-2} \frac{1}{2^k} \binom{2k}{k} = (-1)^{n-1} \frac{1}{2^{2n-2}} \sum_{k=1}^{2n-1} \frac{1}{2k-1} \binom{2n-2}{k-1}.$$

$$(3.71)$$

Further, we also have that

$$L_{2n} = \int_0^{\pi/2} \cos(2n\theta) \sin^{2n-1}(\theta) d\theta \overset{\pi/2-\theta=u}{=} (-1)^n \int_0^{\pi/2} \cos(2nu) \cos^{2n-1}(u) du$$

{use the integral result in (1.59), Sect. 1.14}

$$= (-1)^n \frac{2^{2n-1}}{n\binom{4n}{2n}}.$$

$$(3.72)$$

Collecting (3.71) and (3.72), and putting them together, we may get the form

$$L_n = \int_0^{\pi/2} \cos(n\theta) \sin^{n-1}(\theta) d\theta$$

$$= \begin{cases} L_{2n-1} = (-1)^{n-1} \dfrac{2^{2n-1}}{(2n-1)\binom{4n-2}{2n-1}} \displaystyle\sum_{k=0}^{2n-2} \frac{1}{2^k} \binom{2k}{k} \\[2em] \qquad = (-1)^{n-1} \dfrac{1}{2^{2n-2}} \displaystyle\sum_{k=1}^{2n-1} \frac{1}{2k-1} \binom{2n-2}{k-1}; \\[2em] L_{2n} = (-1)^n \dfrac{2^{2n-1}}{n\binom{4n}{2n}}, \end{cases}$$

$$(3.73)$$

and the solution to the point (iv) of the problem is finalized.

As regards the last four points of the problem, we need a slightly different approach that is not based on parity as before.

So, let's start with the point (vii) where we want to employ the angle difference formula $\sin(a-b) = \sin(a)\cos(b) - \sin(b)\cos(a)$, and then we write

$$R_n = \int_0^{\pi/2} \frac{\sin(n\theta)}{\sin(\theta)} \cos^{n+1}(\theta) d\theta = \int_0^{\pi/2} \frac{\sin((n+1)\theta - \theta)}{\sin(\theta)} \cos^{n+1}(\theta) d\theta$$

$$= \int_0^{\pi/2} \frac{\sin((n+1)\theta) \cos(\theta) - \sin(\theta) \cos((n+1)\theta)}{\sin(\theta)} \cos^{n+1}(\theta) d\theta$$

$$= \int_0^{\pi/2} \frac{\sin((n+1)\theta)}{\sin(\theta)} \cos^{n+2}(\theta)d\theta - \underbrace{\int_0^{\pi/2} \cos((n+1)\theta) \cos^{n+1}(\theta)d\theta}_{R_{n+1}}$$

{the last integral is calculated in (1.57), Sect. 1.14}

$$= R_{n+1} - \frac{\pi}{2^{n+2}},$$

or if we consider the notation in k, we get the recurrence relation

$$R_{k+1} - R_k = \frac{\pi}{2^{k+2}}. \tag{3.74}$$

Now, if we consider that $R_1 = \frac{\pi}{4}$ and then make the sum from $k = 1$ to $k = n-1$ in (3.74), we obtain

$$\sum_{k=1}^{n-1}(R_{k+1} - R_k) = R_n - R_1 = R_n - \frac{\pi}{4} = \frac{\pi}{8}\sum_{k=1}^{n-1}\left(\frac{1}{2}\right)^{k-1} = \pi\frac{2^{n-1}-1}{2^{n+1}},$$

or if we arrange, we get that

$$R_n = \int_0^{\pi/2} \frac{\sin(n\theta)}{\sin(\theta)} \cos^{n+1}(\theta)d\theta = \pi\frac{2^n - 1}{2^{n+1}}, \tag{3.75}$$

and the solution to the point (vii) of the problem is finalized.

For the point $(viii)$ we need a similar starting strategy to the one at the point (vii), and then we write

$$S_n = \int_0^{\pi/2} \frac{\sin(n\theta)}{\sin(\theta)} \cos^n(\theta)d\theta = \int_0^{\pi/2} \frac{\sin((n+1)\theta - \theta)}{\sin(\theta)} \cos^n(\theta)d\theta$$

$$= \int_0^{\pi/2} \frac{\sin((n+1)\theta)\cos(\theta) - \sin(\theta)\cos((n+1)\theta)}{\sin(\theta)} \cos^n(\theta)d\theta$$

$$= \underbrace{\int_0^{\pi/2} \frac{\sin((n+1)\theta)}{\sin(\theta)} \cos^{n+1}(\theta)d\theta}_{S_{n+1}} - \int_0^{\pi/2} \cos((n+1)\theta) \cos^n(\theta)d\theta$$

{the last integral is calculated in (1.59), Sect. 1.14}

$$= S_{n+1} - \frac{2^{n+1}}{(n+1)\binom{2n+2}{n+1}},$$

or if we change to the notation in k, we have the recurrence

$$S_{k+1} - S_k = \frac{2^{k+1}}{(k+1)\binom{2k+2}{k+1}}. \tag{3.76}$$

So, if we consider that $S_1 = 1$ and then make the sum from $k = 1$ to $k = n - 1$ in (3.76), we obtain

$$\sum_{k=1}^{n-1}(S_{k+1} - S_k) = S_n - S_1 = \sum_{k=1}^{n-1} \frac{2^{k+1}}{(k+1)\binom{2k+2}{k+1}},$$

or if we arrange, we get that

$$S_n = \int_0^{\pi/2} \frac{\sin(n\theta)}{\sin(\theta)} \cos^n(\theta)\,d\theta = \sum_{k=0}^{n-1} \frac{2^{k+1}}{(k+1)\binom{2k+2}{k+1}} = \sum_{k=1}^{n} \frac{2^k}{k\binom{2k}{k}}, \tag{3.77}$$

and the solution to the point $(viii)$ of the problem is finalized.

At this time, we are ready to return to the points (v) and (vi) where my plan is to prove that the integrals nicely reduce to calculations involving the integrals from the points (vii) and $(viii)$, just calculated above. Therefore, we make up the following system of relations where I also exploit two simple trigonometric identities[5] $\tan(\theta/2) + \cot(\theta/2) = 2\csc(\theta)$ and $\tan(\theta/2) - \cot(\theta/2) = -2\cot(\theta)$ that give

---

[5] Both trigonometric identities are immediately derived by simple means. For the first identity, we write that $\tan\left(\frac{\theta}{2}\right) + \cot\left(\frac{\theta}{2}\right) = \frac{\sin(\theta/2)}{\cos(\theta/2)} + \frac{\cos(\theta/2)}{\sin(\theta/2)} = \frac{\sin^2(\theta/2) + \cos^2(\theta/2)}{\sin(\theta/2)\cos(\theta/2)} = \frac{2}{\sin(\theta)}$, where I used the identities $\sin^2(\theta) + \cos^2(\theta) = 1$ and $\sin(2\theta) = 2\sin(\theta)\cos(\theta)$. For the second identity, we have that $\tan\left(\frac{\theta}{2}\right) - \cot\left(\frac{\theta}{2}\right) = \frac{\sin(\theta/2)}{\cos(\theta/2)} - \frac{\cos(\theta/2)}{\sin(\theta/2)} = \frac{\sin^2(\theta/2) - \cos^2(\theta/2)}{\sin(\theta/2)\cos(\theta/2)} = -2\frac{\cos(\theta)}{\sin(\theta)} = -2\cot(\theta)$. In the last calculations I used that $\cos(a+b) = \cos(a)\cos(b) - \sin(a)\sin(b)$.

$$\begin{cases} P_n + Q_n = 2 \underbrace{\int_0^{\pi/2} \frac{\sin(n\theta)}{\sin(\theta)} \cos^n(\theta) d\theta}_{S_n \text{ is calculated in (3.77)}} = \sum_{k=1}^{n} \frac{2^{k+1}}{k \binom{2k}{k}}; \\[3mm] P_n - Q_n = -2 \underbrace{\int_0^{\pi/2} \frac{\sin(n\theta)}{\sin(\theta)} \cos^{n+1}(\theta) d\theta}_{R_n \text{ is calculated in (3.75)}} = \pi \frac{1 - 2^n}{2^n}. \end{cases} \qquad (3.78)$$

Based on the system of relations in (3.78), we obtain that

$$P_n = \int_0^{\pi/2} \tan\left(\frac{\theta}{2}\right) \sin(n\theta) \cos^n(\theta) d\theta = \pi \frac{1 - 2^n}{2^{n+1}} + \sum_{k=1}^{n} \frac{2^k}{k \binom{2k}{k}} \qquad (3.79)$$

and

$$Q_n = \int_0^{\pi/2} \cot\left(\frac{\theta}{2}\right) \sin(n\theta) \cos^n(\theta) d\theta = \pi \frac{2^n - 1}{2^{n+1}} + \sum_{k=1}^{n} \frac{2^k}{k \binom{2k}{k}}, \qquad (3.80)$$

and the solution to the points (v) and (vi) of the problem is finalized.

As seen, all the given integrals are easily reducible to integrals from the next section which, as shown, may be approached by constructing simple recurrence relations. So, the present solutions also answer the *challenging question*!

Besides, the curious reader might want to observe that the particular cases of the first four main integrals (based on parity) appearing above may also be extracted by using the first forms of the integrals in the next sections.

3.14 Surprisingly Bewitching Trigonometric Integrals with Appealing Closed Forms: The Second Act

Solution The reader familiar with the already famous books, *Ramanujan's Notebooks, Part I* in [4] and *Table of Integrals, Series, and Products* in [17], might immediately recognize a couple of the integrals found in the present section. For instance, in the former mentioned book, more precisely in *Entry* 33, (ii) (see [4, p.290]), it is stated that $\int_0^{\pi/2} \sin(nx) \cos^n(x) dx = \frac{1}{2^{n+1}} \sum_{k=1}^{n} \frac{2^k}{k}$, which is given at the point (v) of the present problem.

Now, this previously mentioned integral is also popularized by the latter book (see [17, **3.631.16**, p.400]). In the same book, one may also find the version

$$\int_0^{\pi/2} \cos(nx)\cos^n(x)dx = \frac{\pi}{2^{n+1}},$$ which is found at the point (i) within the statement of the current problem.

So, a good rule to consider when dealing with such generalizations is to calculate the first few values of the integrals and see if we could figure out a possible pattern.

We note that the first equality at the point (i) easily follows by the variable change $\pi/2 - \theta = u$, and we have

$$\int_0^{\pi/2} \cos\left(\left(\frac{\pi}{2} - \theta\right)n\right)\sin^n(\theta)d\theta = \int_0^{\pi/2} \cos(n\theta)\cos^n(\theta)d\theta.$$

The same way we'll go for the points (iii), (v), and (vii).

Checking the small cases like $n = 1, 2, 3$, we are immediately tempted to assume that the general form of the integral is $\int_0^{\pi/2} \cos(n\theta)\cos^n(\theta)d\theta = \frac{\pi}{2^{n+1}}$.

The first solution is straightforward if we assume that n is a positive integer and start with a simple integration by parts. So, using the notation I_n we write that

$$I_n = \int_0^{\pi/2} \cos(n\theta)\cos^n(\theta)d\theta$$

$$= \int_0^{\pi/2} \frac{1}{n}(\sin(n\theta))' \cos^n(\theta)d\theta = \int_0^{\pi/2} \sin(\theta)\sin(n\theta)\cos^{n-1}(\theta)d\theta,$$

and if we add I_n to the opposite sides and merge the integrals in the right-hand side, we obtain

$$2I_n = \int_0^{\pi/2} (\cos(n\theta)\cos(\theta) + \sin(n\theta)\sin(\theta))\cos^{n-1}(\theta)d\theta$$

{use the angle difference formula, $\cos(a - b) = \cos(a)\cos(b) + \sin(a)\sin(b)$}

$$= \int_0^{\pi/2} \cos((n - 1)\theta)\cos^{n-1}(\theta)d\theta = I_{n-1}. \tag{3.81}$$

So, if we replace n by k in the recurrence relation obtained in (3.81), then multiply both sides by 2^{k-1} and make the summation from $k = 1$ to $k = n$, we arrive at the following telescoping sum:

$$\sum_{k=1}^{n} \left(2^k I_k - 2^{k-1}I_{k-1}\right) = 2^n I_n - I_0 = 2^n I_n - \frac{\pi}{2} = 0,$$

whence we obtain that

$$I_n = \int_0^{\pi/2} \cos(n\theta)\cos^n(\theta)d\theta = \frac{\pi}{2^{n+1}},$$

and the first solution to the point (i) of the problem is complete.

For a second solution, it is natural to consider using Euler's formula, $e^{i\theta} = \cos(\theta) + i\sin(\theta)$, and hence the fact that $\cos(\theta) = \dfrac{e^{i\theta} + e^{-i\theta}}{2}$. We also use an extension of n to the real numbers $n > -1$, and then we write

$$I_n = \int_0^{\pi/2} \cos(n\theta)\cos^n(\theta)d\theta$$

$$= \int_0^{\pi/2} \Re\{e^{in\theta}\}\left(\frac{e^{i\theta} + e^{-i\theta}}{2}\right)^n d\theta = \Re\left\{\int_0^{\pi/2} e^{in\theta}\left(\frac{e^{i\theta} + e^{-i\theta}}{2}\right)^n d\theta\right\}$$

$$= \frac{1}{2^n}\Re\left\{\int_0^{\pi/2}(1 + e^{i2\theta})^n d\theta\right\} = \frac{1}{2^n}\Re\left\{\sum_{k=0}^{\infty}\binom{n}{k}\int_0^{\pi/2} e^{i2k\theta}d\theta\right\}$$

$$= \frac{1}{2^n}\sum_{k=0}^{\infty}\binom{n}{k}\int_0^{\pi/2}\Re\{e^{i2k\theta}\}d\theta = \frac{1}{2^n}\sum_{k=0}^{\infty}\binom{n}{k}\int_0^{\pi/2}\cos(2k\theta)d\theta = \frac{\pi}{2^{n+1}},$$

and the second solution to the point (i) of the problem is complete. Observe that in the calculations I used the simple facts that $\int_0^{\pi/2}\cos(2k\theta)d\theta = \dfrac{\pi}{2}$ when $k = 0$ and $\int_0^{\pi/2}\cos(2k\theta)d\theta = 0, \forall\, k \geq 1,\ k \in \mathbb{N}$. Alternatively, one can use a slightly different way for such a solution as presented in the second solution to the first problem in the next section, where we start with deriving first the value of the key binomial series needed.

Such strategies like the ones above are not new in the literature. For example, the reference found in [4, p.290] sends us to an older book by Grigorii Mikhailovich Fichtenholz (1888–1959), a Russian mathematician, where such a solution by constructing a recurrence relation is suggested, based on a similar integral that is worked out in [13, p.136], also given in the current section.

For a third solution, the curious reader might enjoy to explore a strategy similar to the one found in the second solution to the point (v) of the problem.

The point (ii) of the problem is straightforward if we combine the simple fact that $\cos\left(\left(\frac{\pi}{2} + \theta\right)n\right) = (-1)^n\cos\left(\left(\frac{\pi}{2} - \theta\right)n\right)$ and the point (i) of the problem,

$$\int_0^{\pi/2}\cos\left(\left(\frac{\pi}{2} + \theta\right)n\right)\sin^n(\theta)d\theta = (-1)^n\frac{\pi}{2^{n+1}},$$

and the solution to the point (ii) of the problem is complete.

The result at the point (iii) is somewhat less usual compared to the previous results, and this time we have n in the argument of the first cosine, and a power of $n - 1$ for the second cosine in the product of the integrand. However, maybe we can use a similar style to the one found in the first solution to the point (i) of the problem where a reduction to a recurrence relation has been considered!

The first equality at the point (iii) is obtained by the variable change $\pi/2 - \theta = u$. As regards the second equality, we start with the angle difference formula, $\cos(a - b) = \cos(a)\cos(b) + \sin(a)\sin(b)$, and then, considering the notation J_n, we write

$$J_n = \int_0^{\pi/2} \cos(n\theta)\cos^{n-1}(\theta)d\theta = \int_0^{\pi/2} \cos((n+1)\theta - \theta)\cos^{n-1}(\theta)d\theta$$

$$= \int_0^{\pi/2} (\cos((n+1)\theta)\cos(\theta) + \sin((n+1)\theta)\sin(\theta))\cos^{n-1}(\theta)d\theta$$

$$= \underbrace{\int_0^{\pi/2} \cos((n+1)\theta)\cos^n(\theta)d\theta}_{J_{n+1}} + \int_0^{\pi/2} \sin((n+1)\theta)\sin(\theta)\cos^{n-1}(\theta)d\theta.$$

$$(3.82)$$

For the second resulting integral after the last equal sign in (3.82), we apply integration by parts that leads to

$$\int_0^{\pi/2} \sin((n+1)\theta)\sin(\theta)\cos^{n-1}(\theta)d\theta = -\frac{1}{n}\int_0^{\pi/2} \sin((n+1)\theta)(\cos^n(\theta))'d\theta$$

$$= \underbrace{-\frac{1}{n}\sin((n+1)\theta)\cos^n(\theta)\Big|_{\theta=0}^{\theta=\pi/2}}_{0} + \frac{n+1}{n}\underbrace{\int_0^{\pi/2} \cos((n+1)\theta)\cos^n(\theta)d\theta}_{J_{n+1}}.$$

$$(3.83)$$

By combining (3.82) and (3.83), and rearranging, we arrive at

$$J_n - \frac{2n+1}{n}J_{n+1} = 0. \qquad (3.84)$$

At this point a key step comes into play! We observe the simple fact that $\binom{2n+2}{n+1}/\binom{2n}{n} = 2(2n+1)/(n+1)$ where if we multiply both sides by $(n+1)2^n/(n2^{n+1})$ and rearrange, we arrive at $\frac{n+1}{2^{n+1}}\binom{2n+2}{n+1}/\left(\frac{n}{2^n}\binom{2n}{n}\right) = \frac{2n+1}{n}$, and if we use it in (3.84), then everything *almost magically* turns into the convenient recurrence relation in k,

$$\underbrace{\frac{k}{2^k}\binom{2k}{k}J_k}_{f(k)} - \underbrace{\frac{k+1}{2^{k+1}}\binom{2k+2}{k+1}J_{k+1}}_{f(k+1)} = 0. \qquad (3.85)$$

Based on (3.85) we may develop a telescoping sum in order to extract the value of the desired integral, and then we have

$$0 = \sum_{k=1}^{n-1}(f(k) - f(k+1)) = f(1) - \frac{n}{2^n}\binom{2n}{n}J_n = 1 - \frac{n}{2^n}\binom{2n}{n}J_n$$

from which we obtain that

$$J_n = \int_0^{\pi/2}\cos(n\theta)\cos^{n-1}(\theta)\mathrm{d}\theta = \frac{2^n}{n\binom{2n}{n}},$$

and the first solution to the point (iii) of the problem is complete.

A second solution to the point (iii) of the problem is obtained by following a strategy similar to the one found in the second solution to the point (i).

So, by also using an extension of n to the real numbers $n > 0$, we write

$$J_n = \int_0^{\pi/2}\cos(n\theta)\cos^{n-1}(\theta)\mathrm{d}\theta$$

$$= \int_0^{\pi/2}\Re\{e^{in\theta}\}\left(\frac{e^{i\theta}+e^{-i\theta}}{2}\right)^{n-1}\mathrm{d}\theta = \Re\left\{\int_0^{\pi/2}e^{in\theta}\left(\frac{e^{i\theta}+e^{-i\theta}}{2}\right)^{n-1}\mathrm{d}\theta\right\}$$

$$= \frac{1}{2^{n-1}}\Re\left\{\int_0^{\pi/2}e^{i\theta}(1+e^{i2\theta})^{n-1}\mathrm{d}\theta\right\} = \frac{1}{2^{n-1}}\Re\left\{\sum_{k=0}^{\infty}\binom{n-1}{k}\int_0^{\pi/2}e^{i(2k+1)\theta}\mathrm{d}\theta\right\}$$

$$= \frac{1}{2^{n-1}}\sum_{k=0}^{\infty}\binom{n-1}{k}\Re\left\{\int_0^{\pi/2}e^{i(2k+1)\theta}\mathrm{d}\theta\right\} = \frac{1}{2^{n-1}}\sum_{k=0}^{\infty}(-1)^k\frac{1}{2k+1}\binom{n-1}{k}$$

$$= \frac{1}{2^{n-1}}\sum_{k=0}^{\infty}(-1)^k\int_0^1 x^{2k}\mathrm{d}x\binom{n-1}{k} = \frac{1}{2^{n-1}}\int_0^1\sum_{k=0}^{\infty}(-1)^k x^{2k}\binom{n-1}{k}\mathrm{d}x$$

$$= \frac{1}{2^{n-1}}\int_0^1(1-x^2)^{n-1}\mathrm{d}x = \frac{1}{2^n}\int_{-1}^1(1-x^2)^{n-1}\mathrm{d}x$$

$$\overset{(1-x)/2=t}{=} 2^{n-1}\int_0^1 t^{n-1}(1-t)^{n-1}\mathrm{d}t$$

$$\left\{\text{employ the Beta function, } B(x, y) = \int_0^1 t^{x-1}(1-t)^{y-1}dt, \text{ and}\right\}$$

$$\left\{\text{then exploit the Beta-Gamma identity, } B(x, y) = \frac{\Gamma(x)\Gamma(y)}{\Gamma(x+y)}\right\}$$

$$= 2^{n-1} B(n, n) = 2^{n-1}\frac{\Gamma^2(n)}{\Gamma(2n)} = 2^{n-1}\frac{(n-1)!^2}{(2n-1)!} = \frac{2^n}{n\binom{2n}{n}},$$

and the second solution to the point (iii) of the problem is complete.

As a first remark, observe that the binomial sum appearing in the calculations above is a notorious one:

$$\sum_{k=0}^{\infty}(-1)^k\frac{1}{2k+1}\binom{n}{k} = \sum_{k=0}^{n}(-1)^k\frac{1}{2k+1}\binom{n}{k} = \frac{2^{2n}}{(2n+1)\binom{2n}{n}}. \tag{3.86}$$

For example, (3.86) is usually seen as a particular case of a well-known binomial sum extension in the mathematical literature,

$$\sum_{k=0}^{n}(-1)^k\frac{1}{x+k}\binom{n}{k} = \frac{n!}{x(x+1)\cdots(x+n)}, \tag{3.87}$$

where $x \notin \{0, -1, -2, \ldots, -n\}$, $n \in \mathbb{N}$.

In order to get the result in (3.86), it is enough to set $x = 1/2$ in (3.87) and then multiply both sides by $1/2$. One way to prove the result in (3.87) is by mathematical induction as shown in [61, Chapter 3, pp.239–240].

Then, as a second remark, during the calculations the use of the Beta function could have been avoided if one had developed a recurrence relation by using integration by parts for the integral $\int_0^1 (1-x^2)^{n-1}dx$.

A third creative solution may be constructed by using a strategy similar to the one found in the second solution to the point (v) of the problem.

Based on the same consideration met at the point (ii) of the problem, we combine the simple fact that $\cos\left(\left(\frac{\pi}{2}+\theta\right)n\right) = (-1)^n \cos\left(\left(\frac{\pi}{2}-\theta\right)n\right)$ and the point (iii) of the problem that gives

$$\int_0^{\pi/2} \cos\left(\left(\frac{\pi}{2}+\theta\right)n\right)\sin^{n-1}(\theta)d\theta = (-1)^n\frac{2^n}{n\binom{2n}{n}},$$

and the solution to the point (iv) of the problem is complete.

Stepping further and considering the point (v) of the problem, it is easy to see that the first equality is shown by the variable change $\pi/2 - \theta = u$. Then, for the second equality we'll want to start with integration by parts with the hope of getting a useful recurrence relation. So, by employing the notation K_n, we write that

$$K_n = \int_0^{\pi/2} \sin(n\theta) \cos^n(\theta) d\theta$$

$$= \int_0^{\pi/2} -\frac{1}{n}(\cos(n\theta))' \cos^n(\theta) d\theta = \frac{1}{n} - \int_0^{\pi/2} \sin(\theta) \cos(n\theta) \cos^{n-1}(\theta) d\theta,$$

and if we add up the integral K_n to the opposite sides and merge the integrals in the right-hand side, we arrive at

$$2K_n = \frac{1}{n} + \int_0^{\pi/2} (\sin(n\theta) \cos(\theta) - \sin(\theta) \cos(n\theta)) \cos^{n-1}(\theta) d\theta$$

{use the angle difference formula, $\sin(a - b) = \sin(a) \cos(b) - \sin(b) \cos(a)$}

$$= \frac{1}{n} + \int_0^{\pi/2} \sin((n - 1)\theta) \cos^{n-1}(\theta) d\theta = \frac{1}{n} + K_{n-1}. \qquad (3.88)$$

Now, since $K_0 = 0$, if we replace n by i in the recurrence relation obtained in (3.88), multiply both sides by 2^i, rearrange, and then consider the sum from $i = 1$ to $i = n$, we have

$$\sum_{i=1}^{n}(2^{i+1} K_i - 2^i K_{i-1}) = 2^{n+1} K_n = \sum_{i=1}^{n} \frac{2^i}{i},$$

whence we obtain that

$$K_n = \int_0^{\pi/2} \sin(n\theta) \cos^n(\theta) d\theta = \frac{1}{2^{n+1}} \sum_{k=1}^{n} \frac{2^k}{k},$$

and the first solution to the point (v) of the problem is complete. The solution follows the same lines of the strategy described in [13, p.136].

For a second solution, we might like to consider a powerful strategy known as *Snake Oil Method*, also described in [99, pp.126–138]. Essentially, we calculate $\sum_{n=0}^{\infty} t^n \mathscr{I}_n$ and arrive at $G(t)$, and afterwards we want to identify the coefficients of the generating function $G(t)$ in order to extract the values of $\mathscr{I}_n$.

So, considering the strategy described above, together with the fact that $\Im\{e^{in\theta}\} = \sin(n\theta)$, we write

$$\sum_{n=0}^{\infty} x^n K_n = \sum_{n=0}^{\infty} x^n \int_0^{\pi/2} \sin(n\theta) \cos^n(\theta) d\theta = \sum_{n=0}^{\infty} x^n \int_0^{\pi/2} \Im\{e^{in\theta}\} \cos^n(\theta) d\theta$$

{reverse the order of integration and summation}

$$= \int_0^{\pi/2} \sum_{n=0}^{\infty} x^n \Im\{e^{in\theta}\} \cos^n(\theta) d\theta = \int_0^{\pi/2} \Im\left\{\sum_{n=0}^{\infty} x^n e^{in\theta} \cos^n(\theta)\right\} d\theta$$

$$= \int_0^{\pi/2} \Im\left\{\frac{1}{1 - x\cos(\theta)e^{i\theta}}\right\} d\theta = \frac{1}{2(2-x)} \int_0^{\pi/2} \frac{2x(2-x)\sin(2\theta)}{x^2 - 2x + 2 - x(2-x)\cos(2\theta)} d\theta$$

$$= \frac{1}{2(2-x)} \int_0^{\pi/2} \frac{(x^2 - 2x + 2 - x(2-x)\cos(2\theta))'}{x^2 - 2x + 2 - x(2-x)\cos(2\theta)} d\theta$$

$$= \frac{1}{2(2-x)} \log(x^2 - 2x + 2 - x(2-x)\cos(2\theta))\bigg|_{\theta=0}^{\theta=\pi/2} = -\frac{\log(1-x)}{2-x}. \qquad (3.89)$$

Happily, the generating function obtained in (3.89) is easily manageable if we rearrange it and employ the Cauchy product of two series as shown in (6.15), Sect. 6.5, and then we have

$$-\frac{1}{2}\frac{\log(1-x)}{1-x/2} = \frac{1}{x}\left(\sum_{n=1}^{\infty} \frac{x^n}{n}\right)\left(\sum_{n=1}^{\infty} \left(\frac{x}{2}\right)^n\right) = \sum_{n=1}^{\infty} x^n \left(\frac{1}{2^{n+1}} \sum_{k=1}^{n} \frac{2^k}{k}\right). \qquad (3.90)$$

At this point, by inspecting and equating the corresponding coefficients of the two power series found in (3.89) and (3.90), we conclude that

$$K_n = \int_0^{\pi/2} \sin(n\theta) \cos^n(\theta) d\theta = \frac{1}{2^{n+1}} \sum_{k=1}^{n} \frac{2^k}{k},$$

and the second solution to the point (v) of the problem is complete.

The point (vi) is obtained if we use $\sin\left(\left(\frac{\pi}{2} + \theta\right)n\right) = (-1)^{n-1}\sin\left(\left(\frac{\pi}{2} - \theta\right)n\right)$ and the result from the point (v). So, we have that

$$\int_0^{\pi/2} \sin\left(\left(\frac{\pi}{2} + \theta\right)n\right) \sin^n(\theta) d\theta = (-1)^{n-1}\frac{1}{2^{n+1}} \sum_{k=1}^{n} \frac{2^k}{k},$$

and the solution to the point (vi) of the problem is complete.

As in the cases from the points (i), (iii), and (v), the first equality of the result at the point (vii) follows by the variable change $\pi/2 - \theta = u$. For the second equality, we combine the angle difference formula $\sin(a - b) = \sin(a)\cos(b) - \sin(b)\cos(a)$ and integration by parts, trying to obtain a useful recurrence relation.

So, by considering the notation L_n, we write

$$L_n = \int_0^{\pi/2} \sin(n\theta)\cos^{n-1}(\theta)d\theta = \int_0^{\pi/2} \sin((n+1)\theta - \theta)\cos^{n-1}(\theta)d\theta$$

$$= \int_0^{\pi/2} (\sin((n+1)\theta)\cos(\theta) - \sin(\theta)\cos((n+1)\theta))\cos^{n-1}(\theta)d\theta$$

$$= \underbrace{\int_0^{\pi/2} \sin((n+1)\theta)\cos^n(\theta)d\theta}_{L_{n+1}} - \int_0^{\pi/2} \cos((n+1)\theta)\sin(\theta)\cos^{n-1}(\theta)d\theta.$$

$$(3.91)$$

If we consider the second residual integral after the last equal sign in (3.91), we want to integrate by parts, and we have that

$$\int_0^{\pi/2} \cos((n+1)\theta)\sin(\theta)\cos^{n-1}(\theta)d\theta = -\frac{1}{n}\int_0^{\pi/2} \cos((n+1)\theta)(\cos^n(\theta))'d\theta$$

$$= \underbrace{-\frac{1}{n}\cos((n+1)\theta)\cos^n(\theta)\Big|_{\theta=0}^{\theta=\pi/2}}_{1/n} - \underbrace{\frac{n+1}{n}\int_0^{\pi/2} \sin((n+1)\theta)\cos^n(\theta)d\theta}_{L_{n+1}}$$

$$= \frac{1}{n} - \frac{n+1}{n}L_{n+1}. \qquad (3.92)$$

Combining (3.91) and (3.92), we arrive at the recurrence relation in k,

$$\frac{2k+1}{k}L_{k+1} - L_k = \frac{1}{k}. \qquad (3.93)$$

At this point, we multiply both sides of (3.93) by $(2k - 1)!!/(k - 1)!$, where you might want to observe that I also used the double factorial denoted by $n!!$ (see the end of Sect. 1.17 and [94]). So, we have that

$$\frac{(2k+1)!!}{k!}L_{k+1} - \frac{(2k-1)!!}{(k-1)!}L_k = \frac{(2k-1)!!}{k!}, \qquad (3.94)$$

or if we use $\dfrac{(2k-1)!!}{k!} = \dfrac{(2k)!}{2^k k!^2} = \dfrac{1}{2^k}\dbinom{2k}{k}$ in (3.94), we get

$$\frac{2k+1}{2^k}\binom{2k}{k}L_{k+1} - \frac{2k-1}{2^{k-1}}\binom{2k-2}{k-1}L_k = \frac{1}{2^k}\binom{2k}{k}. \tag{3.95}$$

Since $L_1 = 1$, if we sum up both sides of (3.95) from $k = 1$ to $k = n - 1$, we get the telescoping sum

$$\sum_{k=1}^{n-1}\left(\frac{2k+1}{2^k}\binom{2k}{k}L_{k+1} - \frac{2k-1}{2^{k-1}}\binom{2k-2}{k-1}L_k\right) = \frac{2n-1}{2^{n-1}}\binom{2n-2}{n-1}L_n - 1$$

$$= \frac{n}{2^n}\binom{2n}{n}L_n - 1 = \sum_{k=1}^{n-1}\frac{1}{2^k}\binom{2k}{k},$$

which can be brought to the form

$$L_n = \frac{2^n}{n\binom{2n}{n}}\sum_{k=0}^{n-1}\frac{1}{2^k}\binom{2k}{k},$$

and the first solution with the first given closed form involving the central binomial coefficient to the point (vii) of the problem is complete.

A second solution that leads directly to the other closed form may be constructed by employing a very similar strategy to the one found in the second solutions to the points (i) and (iii) of the problem. Thus, we write

$$L_n = \int_0^{\pi/2}\sin(n\theta)\cos^{n-1}(\theta)d\theta$$

$$= \int_0^{\pi/2}\Im\{e^{in\theta}\}\left(\frac{e^{i\theta}+e^{-i\theta}}{2}\right)^{n-1}d\theta = \Im\left\{\int_0^{\pi/2}e^{in\theta}\left(\frac{e^{i\theta}+e^{-i\theta}}{2}\right)^{n-1}d\theta\right\}$$

$$= \Im\left\{\frac{1}{2^{n-1}}\int_0^{\pi/2}e^{i\theta}(1+e^{i2\theta})^{n-1}d\theta\right\} = \Im\left\{\frac{1}{2^{n-1}}\sum_{k=0}^{n-1}\binom{n-1}{k}\int_0^{\pi/2}e^{i(2k+1)\theta}d\theta\right\}$$

$$= \frac{1}{2^{n-1}}\sum_{k=0}^{n-1}\binom{n-1}{k}\Im\left\{\int_0^{\pi/2}e^{i(2k+1)\theta}d\theta\right\} = \frac{1}{2^{n-1}}\sum_{k=0}^{n-1}\frac{1}{2k+1}\binom{n-1}{k}$$

$$\underset{\text{the sum}}{\overset{\text{reindex}}{=}} \frac{1}{2^{n-1}}\sum_{k=1}^{n}\frac{1}{2k-1}\binom{n-1}{k-1},$$

and the second solution involving the second given closed form to the point (vii) of the problem is complete.

Finally, to calculate the integral at the last point, we combine the simple fact that $\sin\left(\left(\frac{\pi}{2}+\theta\right)n\right) = (-1)^{n-1}\sin\left(\left(\frac{\pi}{2}-\theta\right)n\right)$ and the result from the point (vii) that leads to

$$\int_0^{\pi/2} \sin\left(\left(\frac{\pi}{2}+\theta\right)n\right)\sin^{n-1}(\theta)d\theta = (-1)^{n-1}\int_0^{\pi/2}\sin\left(\left(\frac{\pi}{2}-\theta\right)n\right)\sin^{n-1}(\theta)d\theta$$

$$= (-1)^{n-1}\frac{2^n}{n\binom{2n}{n}}\sum_{k=0}^{n-1}\frac{1}{2^k}\binom{2k}{k} = (-1)^{n-1}\frac{1}{2^{n-1}}\sum_{k=1}^n\frac{1}{2k-1}\binom{n-1}{k-1},$$

and the solution to the point $(viii)$ of the problem is complete.

Hold on! The equality with the binomial sums is just wonderful!, a possible reaction from you at the sight of the relation obtained above (slightly rearranged now):

$$\frac{2^n}{n\binom{2n}{n}}\sum_{k=0}^{n-1}\frac{1}{2^k}\binom{2k}{k} = \frac{1}{2^{n-1}}\sum_{k=0}^{n-1}\frac{1}{2k+1}\binom{n-1}{k}. \tag{3.96}$$

The curious reader might also like to explore other ways of proving (3.96)!

To summarize, obtaining and using recurrence relations is proving again to be an excellent way of approaching such integrals. Then, exploiting the complex exponential forms of sine and cosine may lead to elegant solutions as seen above. Last but not least, considering *Snake Oil Method* is a wonderful and powerful way of elegantly getting the desired results as seen during the calculations.

Let's jump now in the next section and get prepared for another round with similar (special) integrals.

3.15 Surprisingly Bewitching Trigonometric Integrals with Appealing Closed Forms: The Third Act

Solution I started the previous section by referring to *Entry* 33 of *Ramanujan's Notebooks, Part I* in [4, p.290], and a beautiful fact there, particularly for the readers interested in the results with harmonic numbers, is precisely the first integral which in this section is given at the point (i), and to have right now its image, I'll write it in the following: $\int_0^{\pi/2}\theta\sin(n\theta)\cos^n(\theta)d\theta = \frac{\pi}{4}\frac{H_n}{2^n}$. *How would we prove and further exploit such a result? How would we make it useful in the extraction process of other results?* are some of the questions immediately coming to mind.

First, I'll think of a simple strategy like the one involving recurrence relations presented at the first point in the previous section, and then, for the point (i), we write that

$$I_n = \int_0^{\pi/2} \theta \sin(n\theta) \cos^n(\theta) d\theta = \int_0^{\pi/2} -\frac{1}{n}(\cos(n\theta))' \theta \cos^n(\theta) d\theta$$

{integrate by parts and then split the integral}

$$= \frac{1}{n} \int_0^{\pi/2} \cos(n\theta) \cos^n(\theta) d\theta - \int_0^{\pi/2} \theta \sin(\theta) \cos(n\theta) \cos^{n-1}(\theta) d\theta,$$

and if we add I_n to the opposite sides and at the same time consider the result in (1.57), Sect. 1.14, we get

$$2I_n = \frac{\pi}{n2^{n+1}} - \int_0^{\pi/2} \theta \sin(\theta) \cos(n\theta) \cos^{n-1}(\theta) d\theta + \underbrace{\int_0^{\pi/2} \theta \sin(n\theta) \cos^n(\theta) d\theta}_{I_n}$$

{merge the two integrals and use that $\sin(a - b) = \sin(a)\cos(b) - \sin(b)\cos(a)$}

$$2I_n = \frac{\pi}{n2^{n+1}} + \int_0^{\pi/2} \theta \sin((n - 1)\theta) \cos^{n-1}(\theta) d\theta = \frac{\pi}{n2^{n+1}} + I_{n-1}$$

or to put it in a simpler way, we have

$$2I_n - I_{n-1} = \frac{\pi}{n2^{n+1}}. \tag{3.97}$$

If we consider that $I_0 = 0$ and then replace n by k in (3.97), multiply both sides by 2^{k-1}, and make the sum from $k = 1$ to $k = n$, we get

$$\sum_{k=1}^n (2^k I_k - 2^{k-1} I_{k-1}) = 2^n I_n - I_0 = 2^n I_n = \frac{\pi}{4} \sum_{k=1}^n \frac{1}{k} = \frac{\pi}{4} H_n,$$

whence we obtain that

$$I_n = \int_0^{\pi/2} \theta \sin(n\theta) \cos^n(\theta) d\theta = \frac{\pi}{4} \frac{H_n}{2^n},$$

and the first solution to the point (i) of the problem is complete.

For a second solution, we can go with exploiting Euler's formula and considering the complex exponential forms of sine and cosine. In fact, Ramanujan uses such ideas of performing the calculations, and he starts from the beginning with the series representation of the integrand. Such useful series together with their derivations are presented in [4, p.246]. So, we want to combine Euler's formula, $e^{ix} = \cos(x) + i\sin(x)$, and binomial series, and then we write

$$\sum_{k=0}^{\infty} \binom{n}{k} \cos(2k\theta) + i \sum_{k=0}^{\infty} \binom{n}{k} \sin(2k\theta) = \sum_{k=0}^{\infty} \binom{n}{k}(\cos(2k\theta) + i\sin(2k\theta))$$

$$= \sum_{k=0}^{\infty} \binom{n}{k} e^{2k\theta i} = (1 + e^{2\theta i})^n = (1 + \cos(2\theta) + i\sin(2\theta))^n$$

$$= (2\cos^2(\theta) + i2\sin(\theta)\cos(\theta))^n = (2\cos(\theta)(\cos(\theta) + i\sin(\theta)))^n$$

$$= 2^n \cos^n(\theta)(\cos(n\theta) + i\sin(n\theta)) = 2^n \cos(n\theta)\cos^n(\theta) + i2^n \sin(n\theta)\cos^n(\theta),$$

where by equating the real and imaginary parts of the opposite sides above, we arrive at the following results:

$$\frac{1}{2^n} \sum_{k=0}^{\infty} \binom{n}{k} \cos(2k\theta) = \cos(n\theta)\cos^n(\theta) \qquad (3.98)$$

and

$$\frac{1}{2^n} \sum_{k=0}^{\infty} \binom{n}{k} \sin(2k\theta) = \frac{1}{2^n} \sum_{k=1}^{\infty} \binom{n}{k} \sin(2k\theta) = \sin(n\theta)\cos^n(\theta). \qquad (3.99)$$

So, upon returning to the integral I_n and using (3.99), we write

$$I_n = \int_0^{\pi/2} \theta \sin(n\theta)\cos^n(\theta)d\theta = \frac{1}{2^n} \int_0^{\pi/2} \theta \sum_{k=1}^{\infty} \binom{n}{k} \sin(2k\theta)d\theta$$

{reverse the order of summation and integration}

$$= \frac{1}{2^n} \sum_{k=1}^{\infty} \binom{n}{k} \int_0^{\pi/2} \theta \sin(2k\theta)d\theta = \frac{\pi}{2^{n+2}} \sum_{k=1}^{\infty} (-1)^{k-1} \frac{1}{k} \binom{n}{k}$$

$$= \frac{\pi}{2^{n+2}} \sum_{k=1}^{\infty} (-1)^{k-1} \int_0^1 x^{k-1} dx \binom{n}{k} = \frac{\pi}{2^{n+2}} \int_0^1 \sum_{k=1}^{\infty} (-1)^{k-1} x^{k-1} \binom{n}{k} dx$$

$$= \frac{\pi}{2^{n+2}} \int_0^1 \frac{1-(1-x)^n}{x} dx \overset{1-x=y}{=} \frac{\pi}{2^{n+2}} \int_0^1 \frac{1-y^n}{1-y} dy = \frac{\pi}{4} \frac{\gamma + \psi(n+1)}{2^n},$$

$$(3.100)$$

where the last integral is evaluated by expressing the integral in terms of a limit with

the Beta function, $\displaystyle \int_0^1 \frac{1-x^n}{1-x} dx = \lim_{a\to 0} \left(\int_0^1 (1-x)^{a-1} dx - \int_0^1 x^n (1-x)^{a-1} dx \right)$,

leading immediately to the desired closed form. The details of such an approach, in

a more generalized form, that is, $\displaystyle \int_0^1 \frac{x^{p-1} - x^{q-1}}{1-x} dx = \psi(q) - \psi(p)$, are given in

[63, Chapter 10, pp.126–127].

From the last equality of (3.100), we see that when n is a non-negative integer we

have that $\gamma + \psi(n+1) = H_n$. This fact is easily noted if we consider that $\displaystyle \sum_{k=1}^n x^{k-1} =$

$\dfrac{1-x^n}{1-x}$, and thus we have $\displaystyle \int_0^1 \frac{1-x^n}{1-x} dx = \sum_{k=1}^n \int_0^1 x^{k-1} dx = \sum_{k=1}^n \frac{1}{k} = H_n$.

Therefore, when n is a non-negative integer, we get that

$$I_n = \int_0^{\pi/2} \theta \sin(n\theta) \cos^n(\theta) d\theta = \frac{\pi}{4} \frac{H_n}{2^n},$$

and the second solution to the point (i) of the problem is complete.

Well, how would we like to design a third solution? I will use the powerful *Snake Oil Method* described in the previous section. So, we write

$$\sum_{n=0}^\infty x^n I_n = \sum_{n=0}^\infty x^n \int_0^{\pi/2} \theta \sin(n\theta) \cos^n(\theta) d\theta = \sum_{n=0}^\infty x^n \int_0^{\pi/2} \Im\{e^{in\theta}\} \theta \cos^n(\theta) d\theta$$

{reverse the order of integration and summation}

$$= \int_0^{\pi/2} \theta \sum_{n=0}^\infty x^n \Im\{e^{in\theta}\} \cos^n(\theta) d\theta = \int_0^{\pi/2} \theta \Im \left\{ \sum_{n=0}^\infty x^n e^{in\theta} \cos^n(\theta) \right\} d\theta$$

$$= \int_0^{\pi/2} \theta \Im \left\{ \frac{1}{1 - x\cos(\theta)e^{i\theta}} \right\} d\theta = \frac{1}{2(2-x)} \int_0^{\pi/2} \theta \frac{2x(2-x)\sin(2\theta) d\theta}{x^2 - 2x + 2 - x(2-x)\cos(2\theta)}$$

$$\left\{ \text{in the denominator exploit the trigonometric identity, } 1 + \cos(x) = 2\cos^2\left(\frac{x}{2}\right) \right\}$$

$$= \frac{1}{2(2-x)} \int_0^{\pi/2} \theta \frac{(1 - x(2-x)\cos^2(\theta))'}{1 - x(2-x)\cos^2(\theta)} d\theta$$

$$\overset{\text{integrate by}}{\underset{\text{parts}}{=}} -\frac{1}{2(2-x)}\int_0^{\pi/2} \log(1 - x(2-x)\cos^2(\theta))d\theta. \tag{3.101}$$

At this point, we may fix $0 < x < 1$ and then recognize in (3.101) a classical integral,

$$\int_0^{\pi/2} \log(1 - a\cos^2(\theta))d\theta = \pi \log\left(\frac{1+\sqrt{1-a}}{2}\right), \ a < 1. \tag{3.102}$$

which is easily obtained by differentiating under the integral sign.[6]

Returning with the result from (3.102) in (3.101), we get

$$\sum_{n=0}^{\infty} x^n I_n = \sum_{n=0}^{\infty} x^n \int_0^{\pi/2} \theta \sin(n\theta)\cos^n(\theta)d\theta = -\frac{\pi}{4}\frac{\log(1-x/2)}{1-x/2}. \tag{3.103}$$

The generating function obtained in (3.103) may be transformed into a series by the Cauchy product of two series, as shown in (6.15), Sect. 6.5, and then we have

$$-\frac{\pi}{4}\frac{\log(1-x/2)}{1-x/2} = \frac{\pi}{2}\frac{1}{x}\left(\sum_{n=1}^{\infty}\frac{x^n}{n2^n}\right)\left(\sum_{n=1}^{\infty}\frac{x^n}{2^n}\right) = \frac{\pi}{2}\sum_{n=1}^{\infty}x^n\left(\frac{1}{2^{n+1}}\sum_{k=1}^{n}\frac{1}{k}\right)$$

$$= \sum_{n=1}^{\infty} x^n \frac{\pi}{4}\frac{H_n}{2^n}. \tag{3.104}$$

Finally, by inspecting and equating the corresponding coefficients of the two power series given in (3.103) and (3.104), we conclude that

---

[6] If we consider the classical integral $\mathscr{I}(a) = \int_0^{\pi/2} \log(1 - a\cos^2(\theta))d\theta$ and afterwards differentiate once with respect to the parameter a, we obtain that $\mathscr{I}'(a) = -\int_0^{\pi/2}\frac{\cos^2(\theta)}{1-a\cos^2(\theta)}d\theta =$
$\frac{1}{a}\int_0^{\pi/2}\frac{(1-a\cos^2(\theta))-1}{1-a\cos^2(\theta)}d\theta = \frac{1}{a}\int_0^{\pi/2}d\theta - \frac{1}{a}\int_0^{\pi/2}\frac{1}{1-a\cos^2(\theta)}d\theta = \frac{\pi}{2a} - \frac{\pi}{2a\sqrt{1-a}}$,
where in the calculations I used that $\int_0^{\pi/2}\frac{dx}{1-a\sin^2(x)} = \int_0^{\pi/2}\frac{dx}{1-a\cos^2(x)} =$
$\frac{\pi}{2}\frac{1}{\sqrt{1-a}}$, $a < 1$ (for a proof, see [76, Chapter 3, p.212]). Finally, $\mathscr{I}(a) =$
$\int_0^a \mathscr{I}'(x)dx = \frac{\pi}{2}\int_0^a\left(\frac{1}{x} - \frac{1}{x\sqrt{1-x}}\right)dx = \frac{\pi}{2}\left(\log(x) + 2\operatorname{arctanh}(\sqrt{1-x})\right)\Big|_{x=0}^{x=a} =$
$\pi\log(1+\sqrt{1-x})\Big|_{x=0}^{x=a} = \pi\log\left(\frac{1+\sqrt{1-a}}{2}\right)$.

$$I_n = \int_0^{\pi/2} \theta \sin(n\theta) \cos^n(\theta) d\theta = \frac{\pi}{4} \frac{H_n}{2^n},$$

and the third solution to the point (i) of the problem is complete.

Jumping to the point (ii) of the problem, we first want to go with integration by parts as I did at the point (i), and then we write

$$J_n = \int_0^{\pi/2} \theta^2 \cos(n\theta) \cos^n(\theta) d\theta = \int_0^{\pi/2} \frac{1}{n}(\sin(n\theta))' \theta^2 \cos^n(\theta) d\theta$$

{integrate by parts and then split the integral}

$$= -2\frac{1}{n} \underbrace{\int_0^{\pi/2} \theta \sin(n\theta) \cos^n(\theta) d\theta}_{I_n} + \int_0^{\pi/2} \theta^2 \sin(\theta) \sin(n\theta) \cos^{n-1}(\theta) d\theta,$$

and if we add J_n to the opposite sides and at the same time consider the value of I_n given at the point (i), we get

$$2J_n = -\frac{\pi}{2} \frac{H_n}{n2^n} + \int_0^{\pi/2} \theta^2 \sin(\theta) \sin(n\theta) \cos^{n-1}(\theta) d\theta + \underbrace{\int_0^{\pi/2} \theta^2 \cos(n\theta) \cos^n(\theta) d\theta}_{J_n}$$

{merge the two integrals and use that $\cos(a - b) = \cos(a)\cos(b) + \sin(a)\sin(b)$}

$$= -\frac{\pi}{2} \frac{H_n}{n2^n} + \underbrace{\int_0^{\pi/2} \theta^2 \cos((n - 1)\theta) \cos^{n-1} \theta d\theta}_{J_{n-1}},$$

and to put it in a simpler form, we get

$$2J_n - J_{n-1} = -\frac{\pi}{2} \frac{H_n}{n2^n}. \qquad (3.105)$$

Now, if we consider that $J_0 = \frac{\pi^3}{24}$ and then replace n by k in (3.105), multiply both sides by 2^{k-1}, and make the sum from $k = 1$ to $k = n$, we obtain that

$$\sum_{k=1}^n (2^k J_k - 2^{k-1} J_{k-1}) = 2^n J_n - J_0 = 2^n J_n - \frac{\pi^3}{24} = -\frac{\pi}{4} \sum_{k=1}^n \frac{H_k}{k}$$

$$\left\{ \text{use that } \sum_{k=1}^{n} \frac{H_k}{k} = \frac{1}{2}(H_n^2 + H_n^{(2)}) \text{ which is proved in [76, Chapter 3, p.60]} \right\}$$

$$= -\frac{\pi}{8}(H_n^2 + H_n^{(2)}),$$

whence we obtain that

$$J_n = \int_0^{\pi/2} \theta^2 \cos(n\theta) \cos^n(\theta) d\theta = \frac{\pi^3}{24} \frac{1}{2^n} - \frac{\pi}{8} \frac{1}{2^n}(H_n^2 + H_n^{(2)}),$$

and the first solution to the point (ii) of the problem is complete. One might also notice that $\sum_{k=1}^{n} \frac{H_k}{k} = \frac{1}{2}(H_n^2 + H_n^{(2)})$ is easily derived by Abel's summation (see [76, Chapter 2, pp.39–40]). The sum may also be viewed as the case $p = q = 1$ of the result in (6.102), Sect. 6.13.

For another solution to the point (ii), we employ the series result in (3.98), and then we write

$$J_n = \int_0^{\pi/2} \theta^2 \cos(n\theta) \cos^n(\theta) d\theta = \frac{1}{2^n} \int_0^{\pi/2} \theta^2 \sum_{k=0}^{\infty} \binom{n}{k} \cos(2k\theta) d\theta$$

$$\{\text{leave out the term of the series for } k = 0\}$$

$$= \frac{\pi^3}{24} \frac{1}{2^n} + \frac{1}{2^n} \int_0^{\pi/2} \theta^2 \sum_{k=1}^{\infty} \binom{n}{k} \cos(2k\theta) d\theta$$

$$\{\text{reverse the order of summation and integration}\}$$

$$= \frac{\pi^3}{24} \frac{1}{2^n} + \frac{1}{2^n} \sum_{k=1}^{\infty} \binom{n}{k} \int_0^{\pi/2} \theta^2 \cos(2k\theta) d\theta = \frac{\pi^3}{24} \frac{1}{2^n} + \frac{\pi}{2^{n+2}} \sum_{k=1}^{\infty} (-1)^k \frac{1}{k^2} \binom{n}{k}$$

$$= \frac{\pi^3}{24} \frac{1}{2^n} - \frac{\pi}{4} \frac{1}{2^n} \sum_{k=1}^{\infty} (-1)^k \int_0^1 x^{k-1} \log(x) dx \binom{n}{k}$$

$$= \frac{\pi^3}{24} \frac{1}{2^n} + \frac{\pi}{4} \frac{1}{2^n} \int_0^1 \log(x) \sum_{k=1}^{\infty} (-1)^{k-1} x^{k-1} \binom{n}{k} dx$$

$$= \frac{\pi^3}{24} \frac{1}{2^n} + \frac{\pi}{4} \frac{1}{2^n} \int_0^1 \frac{1 - (1-x)^n}{x} \log(x) dx$$

$$\overset{1-x=y}{=} \frac{\pi^3}{24} \frac{1}{2^n} + \frac{\pi}{4} \frac{1}{2^n} \int_0^1 \frac{1-y^n}{1-y} \log(1-y)dy$$

$$\overset{\text{integrate by}}{\underset{\text{parts}}{=}} \frac{\pi^3}{24} \frac{1}{2^n} - \frac{\pi}{8} \frac{n}{2^n} \int_0^1 y^{n-1} \log^2(1-y)dy = \frac{\pi^3}{24} \frac{1}{2^n} - \frac{\pi}{8} \frac{1}{2^n}(H_n^2 + H_n^{(2)}),$$

and the second solution to the point (ii) of the problem is complete. In the calculations, I used the logarithmic integral

$$\int_0^1 x^{n-1} \log^2(1-x)dx = \frac{H_n^2 + H_n^{(2)}}{n}, \tag{3.106}$$

which may also be found in [76, Chapter 1, p.2].

We prepare now to calculate the integral at the point (iii), and I will use a strategy similar to the ones found in the second solution to the point (i) and in the previous solution. Therefore, employing the series in (3.99), we write

$$K_n = \int_0^{\pi/2} \theta^3 \sin(n\theta) \cos^n(\theta)d\theta = \frac{1}{2^n} \int_0^{\pi/2} \theta^3 \sum_{k=1}^{\infty} \binom{n}{k} \sin(2k\theta)d\theta$$

{reverse the order of summation and integration}

$$= \frac{1}{2^n} \sum_{k=1}^{\infty} \binom{n}{k} \int_0^{\pi/2} \theta^3 \sin(2k\theta)d\theta$$

$$= \frac{\pi^3}{16} \frac{1}{2^n} \sum_{k=1}^{\infty} (-1)^{k-1} \frac{1}{k} \binom{n}{k} - \frac{3}{8} \pi \frac{1}{2^n} \sum_{k=1}^{\infty} (-1)^{k-1} \frac{1}{k^3} \binom{n}{k}$$

$$= \frac{\pi^3}{16} \frac{1}{2^n} \sum_{k=1}^{\infty} (-1)^{k-1} \int_0^1 x^{k-1}dx \binom{n}{k} - \frac{3}{16} \pi \frac{1}{2^n} \sum_{k=1}^{\infty} (-1)^{k-1} \int_0^1 x^{k-1} \log^2(x)dx \binom{n}{k}$$

$$= \frac{\pi^3}{16} \frac{1}{2^n} \int_0^1 \sum_{k=1}^{\infty} (-1)^{k-1} x^{k-1} \binom{n}{k} dx - \frac{3}{16} \pi \frac{1}{2^n} \int_0^1 \sum_{k=1}^{\infty} (-1)^{k-1} x^{k-1} \binom{n}{k} \log^2(x)dx$$

$$= \frac{\pi^3}{16} \frac{1}{2^n} \int_0^1 \frac{1-(1-x)^n}{x} dx - \frac{3}{16} \pi \frac{1}{2^n} \int_0^1 \frac{1-(1-x)^n}{x} \log^2(x)dx$$

$$\overset{1-x=y}{=} \frac{\pi^3}{16} \frac{1}{2^n} \int_0^1 \frac{1-y^n}{1-y} dy - \frac{3}{16} \pi \frac{1}{2^n} \int_0^1 \frac{1-y^n}{1-y} \log^2(1-y)dy$$

$$\overset{\text{integrate by}}{\underset{=}{\text{parts}}} \quad -\frac{\pi^3}{16}\frac{n}{2^n}\int_0^1 y^{n-1}\log(1-y)dy + \frac{\pi}{16}\frac{n}{2^n}\int_0^1 y^{n-1}\log^3(1-y)dy$$

$$= \frac{\pi^3}{16}\frac{H_n}{2^n} - \frac{\pi}{16}\frac{1}{2^n}\left(H_n^3 + 3H_n H_n^{(2)} + 2H_n^{(3)}\right),$$

and the solution to the point (iii) of the problem is complete. During the calculations I used that $\int_0^1 x^{n-1}\log(1-x)dx = -\dfrac{H_n}{n}$, which also appears in (3.10), Sect. 3.3, and

$$\int_0^1 x^{n-1}\log^3(1-x)dx = -\frac{H_n^3 + 3H_n H_n^{(2)} + 2H_n^{(3)}}{n}, \tag{3.107}$$

which is considered in [76, Chapter 1, p.2].

Next, for the integral at the point (iv) we follow a strategy similar to the one found in the second solution to the point (ii). Thus, exploiting (3.98) we write

$$L_n = \int_0^{\pi/2} \theta^4 \cos(n\theta)\cos^n(\theta)d\theta = \frac{1}{2^n}\int_0^{\pi/2}\theta^4\sum_{k=0}^{\infty}\binom{n}{k}\cos(2k\theta)d\theta$$

$$\{\text{leave out the term of the series for } k = 0\}$$

$$= \frac{\pi^5}{160}\frac{1}{2^n} + \frac{1}{2^n}\int_0^{\pi/2}\theta^4\sum_{k=1}^{\infty}\binom{n}{k}\cos(2k\theta)d\theta$$

$$= \frac{\pi^5}{160}\frac{1}{2^n} + \frac{1}{2^n}\sum_{k=1}^{\infty}\binom{n}{k}\int_0^{\pi/2}\theta^4\cos(2k\theta)d\theta$$

$$= \frac{\pi^5}{160}\frac{1}{2^n} - \frac{\pi^3}{8}\frac{1}{2^n}\sum_{k=1}^{\infty}(-1)^{k-1}\frac{1}{k^2}\binom{n}{k} + \frac{3}{4}\pi\frac{1}{2^n}\sum_{k=1}^{\infty}(-1)^{k-1}\frac{1}{k^4}\binom{n}{k}$$

$$= \frac{\pi^5}{160}\frac{1}{2^n} + \frac{\pi^3}{8}\frac{1}{2^n}\sum_{k=1}^{\infty}(-1)^{k-1}\int_0^1 x^{k-1}\log(x)dx\binom{n}{k}$$

$$- \frac{\pi}{8}\frac{1}{2^n}\sum_{k=1}^{\infty}(-1)^{k-1}\int_0^1 x^{k-1}\log^3(x)dx\binom{n}{k}$$

$$= \frac{\pi^5}{160}\frac{1}{2^n} + \frac{\pi^3}{8}\frac{1}{2^n}\int_0^1 \log(x)\sum_{k=1}^{\infty}(-1)^{k-1}x^{k-1}\binom{n}{k}dx$$

$$-\frac{\pi}{8}\frac{1}{2^n}\int_0^1 \log^3(x)\sum_{k=1}^{\infty}(-1)^{k-1}x^{k-1}\binom{n}{k}dx$$

$$=\frac{\pi^5}{160}\frac{1}{2^n}+\frac{\pi^3}{8}\frac{1}{2^n}\int_0^1 \frac{1-(1-x)^n}{x}\log(x)dx-\frac{\pi}{8}\frac{1}{2^n}\int_0^1 \frac{1-(1-x)^n}{x}\log^3(x)dx$$

$$\overset{1-x=y}{=}\frac{\pi^5}{160}\frac{1}{2^n}+\frac{\pi^3}{8}\frac{1}{2^n}\int_0^1 \frac{1-y^n}{1-y}\log(1-y)dy-\frac{\pi}{8}\frac{1}{2^n}\int_0^1 \frac{1-y^n}{1-y}\log^3(1-y)dy$$

$$\overset{\underset{\text{parts}}{\text{integrate by}}}{=}\frac{\pi^5}{160}\frac{1}{2^n}-\frac{\pi^3}{16}\frac{n}{2^n}\int_0^1 y^{n-1}\log^2(1-y)dy+\frac{\pi}{32}\frac{n}{2^n}\int_0^1 y^{n-1}\log^4(1-y)dy$$

$$=\frac{\pi^5}{160}\frac{1}{2^n}-\frac{\pi^3}{16}\frac{1}{2^n}(H_n^2+H_n^{(2)})$$

$$+\frac{\pi}{32}\frac{1}{2^n}\left(H_n^4+6H_n^2 H_n^{(2)}+8H_n H_n^{(3)}+3(H_n^{(2)})^2+6H_n^{(4)}\right),$$

and the solution to the point (iv) of the problem is complete. In the calculations I used the logarithmic integral in (3.106), and then the logarithmic integral,

$$\int_0^1 x^{n-1}\log^4(1-x)dx = \frac{H_n^4+6H_n^2 H_n^{(2)}+8H_n H_n^{(3)}+3\left(H_n^{(2)}\right)^2+6H_n^{(4)}}{n},$$
(3.108)

which may also be found in [76, Chapter 1, p.2].

One thing to observe is that the solutions constructed by exploiting the results in (3.98) and (3.99) show the possibility of an extension of n to real numbers through *binomial series*. On the other hand, the values of n from the resulting logarithmic integrals appearing in the solutions where (3.98) and (3.99) are used may be extended to real numbers (all these integrals may also be viewed in terms of the derivatives of the Beta function). I remind you that the argument of the generalized harmonic numbers may be extended to real values through the use of the Polygamma function. I will use these facts during the next derivations!

Coming back to the point (v), I will consider the result from the point (i) in the previous section, which if we differentiate once with respect to n, we get

$$\frac{d}{dn}\left(\int_0^{\pi/2}\cos(n\theta)\cos^n(\theta)d\theta\right)=\int_0^{\pi/2}\cos(n\theta)\cos^n(\theta)\log(\cos(\theta))d\theta$$

$$-\int_0^{\pi/2}\theta\sin(n\theta)\cos^n(\theta)d\theta=\frac{d}{dn}\left(\frac{\pi}{2^{n+1}}\right)=-\log(2)\frac{\pi}{2}\frac{1}{2^n},$$

and since the value of the second emerging integral is given at the point (i), then we obtain that

$$\int_0^{\pi/2} \cos(n\theta)\cos^n(\theta)\log(\cos(\theta))d\theta = \frac{\pi}{4}\frac{H_n}{2^n} - \log(2)\frac{\pi}{2}\frac{1}{2^n},$$

and the solution to the point (v) of the problem is complete.

Further, to derive the result at the point (vi), we start with the integral from the point (i), which if we differentiate once with respect to n, we have

$$\frac{d}{dn}\left(\int_0^{\pi/2}\theta\sin(n\theta)\cos^n(\theta)d\theta\right)$$

$$= \int_0^{\pi/2}\theta\sin(n\theta)\cos^n(\theta)\log(\cos(\theta))d\theta + \int_0^{\pi/2}\theta^2\cos(n\theta)\cos^n(\theta)d\theta$$

$$= \frac{\pi}{4}\frac{d}{dn}\left(\frac{\psi(n+1)+\gamma}{2^n}\right) = \frac{\pi^3}{24}\frac{1}{2^n} - \log(2)\frac{\pi}{4}\frac{H_n}{2^n} - \frac{\pi}{4}\frac{H_n^{(2)}}{2^n},$$

where since the second integral coming to light is given and calculated at the point (ii), we get

$$\int_0^{\pi/2}\theta\sin(n\theta)\cos^n(\theta)\log(\cos(\theta))d\theta$$

$$= -\log(2)\frac{\pi}{4}\frac{H_n}{2^n} - \frac{\pi}{4}\frac{H_n^{(2)}}{2^n} + \frac{\pi}{8}\frac{1}{2^n}\left(H_n^2 + H_n^{(2)}\right),$$

and the solution to the point (vi) of the problem is complete.

As regards the extraction of the integral result at the point (vii), we want to differentiate once with respect to n the integral result from the point (ii), and then we obtain that

$$\frac{d}{dn}\left(\int_0^{\pi/2}\theta^2\cos(n\theta)\cos^n(\theta)d\theta\right)$$

$$= \int_0^{\pi/2}\theta^2\cos(n\theta)\cos^n(\theta)\log(\cos(\theta))d\theta - \int_0^{\pi/2}\theta^3\sin(n\theta)\cos^n(\theta)d\theta$$

$$= \frac{d}{dn}\left(\frac{\pi^3}{24}\frac{1}{2^n} - \frac{\pi}{8}\frac{1}{2^n}\left((\psi(n+1)+\gamma)^2 + \zeta(2) - \psi^{(1)}(n+1)\right)\right)$$

$$= -\log(2)\frac{\pi^3}{24}\frac{1}{2^n} - \frac{\pi}{4}\zeta(3)\frac{1}{2^n} - \frac{\pi^3}{24}\frac{H_n}{2^n} + \frac{\pi}{4}\frac{H_n H_n^{(2)}}{2^n} + \frac{\pi}{4}\frac{H_n^{(3)}}{2^n}$$

$$+ \log(2) \frac{\pi}{8} \frac{1}{2^n} \left(H_n^2 + H_n^{(2)} \right),$$

and based on the fact that the value of the second resulting integral is evaluated at the point (iii), we get

$$\int_0^{\pi/2} \theta^2 \cos(n\theta) \cos^n(\theta) \log(\cos(\theta)) d\theta$$

$$= - \log(2) \frac{\pi^3}{24} \frac{1}{2^n} - \frac{\pi}{4} \zeta(3) \frac{1}{2^n} + \frac{\pi^3}{48} \frac{H_n}{2^n} + \frac{\pi}{4} \frac{H_n H_n^{(2)}}{2^n} + \frac{\pi}{4} \frac{H_n^{(3)}}{2^n}$$

$$+ \log(2) \frac{\pi}{8} \frac{1}{2^n} \left(H_n^2 + H_n^{(2)} \right) - \frac{\pi}{16} \frac{1}{2^n} \left(H_n^3 + 3 H_n H_n^{(2)} + 2 H_n^{(3)} \right),$$

and the solution to the point (vii) of the problem is complete.

Next, for obtaining the integral result at the point $(viii)$, I consider the integral from the point (iii) which if we differentiate once with respect to n, we get that

$$\frac{d}{dn} \left(\int_0^{\pi/2} \theta^3 \sin(n\theta) \cos^n(\theta) d\theta \right)$$

$$= \int_0^{\pi/2} \theta^3 \sin(n\theta) \cos^n(\theta) \log(\cos(\theta)) d\theta + \int_0^{\pi/2} \theta^4 \cos(n\theta) \cos^n(\theta) d\theta$$

$$= \frac{d}{dn} \left(\frac{\pi^3}{16} \frac{H_n}{2^n} - \frac{\pi}{16} \frac{1}{2^n} \left((\psi(n+1) + \gamma)^3 + 3(\psi(n+1) + \gamma)(\zeta(2) - \psi^{(1)}(n+1)) \right. \right.$$

$$\left. \left. + 2 \left(\zeta(3) + \frac{1}{2} \psi^{(2)}(n+1) \right) \right) \right)$$

$$= \frac{\pi^5}{160} \frac{1}{2^n} - \left(\frac{3}{8} \pi \zeta(3) + \log(2) \frac{\pi^3}{16} \right) \frac{H_n}{2^n} + \frac{3}{8} \pi \frac{H_n H_n^{(3)}}{2^n} + \frac{3}{16} \pi \frac{H_n^2 H_n^{(2)}}{2^n}$$

$$- \frac{\pi^3}{16} \frac{H_n^{(2)}}{2^n} + \frac{3}{16} \pi \frac{(H_n^{(2)})^2}{2^n} + \frac{3}{8} \pi \frac{H_n^{(4)}}{2^n} - \frac{\pi^3}{32} \frac{1}{2^n} (H_n^2 + H_n^{(2)})$$

$$+ \log(2) \frac{\pi}{16} \frac{1}{2^n} \left(H_n^3 + 3 H_n H_n^{(2)} + 2 H_n^{(3)} \right),$$

but since the second integral above, taking shape during the calculations, is given at the point (iv), we obtain

$$\int_0^{\pi/2} \theta^3 \sin(n\theta) \cos^n(\theta) \log(\cos(\theta)) d\theta$$

$$= -\left(\frac{3}{8}\pi\zeta(3) + \log(2)\frac{\pi^3}{16}\right)\frac{H_n}{2^n} + \frac{3}{8}\pi\frac{H_n H_n^{(3)}}{2^n}$$

$$+ \frac{3}{16}\pi\frac{H_n^2 H_n^{(2)}}{2^n} - \frac{2}{32}\pi^3\frac{H_n^{(2)}}{2^n} + \frac{3}{16}\pi\frac{(H_n^{(2)})^2}{2^n} + \frac{3}{8}\pi\frac{H_n^{(4)}}{2^n}$$

$$+ \frac{\pi^3}{32}\frac{1}{2^n}\left(H_n^2 + H_n^{(2)}\right) + \log(2)\frac{\pi}{16}\frac{1}{2^n}\left(H_n^3 + 3H_n H_n^{(2)} + 2H_n^{(3)}\right)$$

$$- \frac{\pi}{32}\frac{1}{2^n}\left(H_n^4 + 6H_n^2 H_n^{(2)} + 8H_n H_n^{(3)} + 3(H_n^{(2)})^2 + 6H_n^{(4)}\right),$$

and the solution to the point $(viii)$ of the problem is complete.

For the last group of integrals involving the squared logarithms, we want to differentiate with respect to n the integrals from the points (v)–(vii), involving a logarithm.

So, for deriving the result at the point (ix), we start with the integral from the point (v) which if we differentiate once with respect to n, we have

$$\frac{d}{dn}\left(\int_0^{\pi/2} \cos(n\theta) \cos^n(\theta) \log(\cos(\theta)) d\theta\right)$$

$$= \int_0^{\pi/2} \cos(n\theta) \cos^n(\theta) \log^2(\cos(\theta)) d\theta - \int_0^{\pi/2} \theta \sin(n\theta) \cos^n(\theta) \log(\cos(\theta)) d\theta$$

$$= \frac{d}{dn}\left(\frac{\pi}{4}\frac{\psi(n+1) + \gamma}{2^n} - \frac{1}{2}\log(2)\frac{\pi}{2^n}\right)$$

$$= \frac{\pi}{24}(12\log^2(2) + \pi^2)\frac{1}{2^n} - \log(2)\frac{\pi}{4}\frac{H_n}{2^n} - \frac{\pi}{4}\frac{H_n^{(2)}}{2^n},$$

and due to the fact that the second emerging integral is given at the point (vi), we arrive at

$$\int_0^{\pi/2} \cos(n\theta) \cos^n(\theta) \log^2(\cos(\theta)) d\theta$$

$$= \frac{\pi}{24}(\pi^2 + 12\log^2(2))\frac{1}{2^n} - \log(2)\frac{\pi}{2}\frac{H_n}{2^n} - \frac{\pi}{2}\frac{H_n^{(2)}}{2^n} + \frac{\pi}{8}\frac{1}{2^n}(H_n^2 + H_n^{(2)}),$$

and the solution to the point (ix) of the problem is complete.

For the result at the penultimate point, we want to differentiate the integral result from the point (vi), and then we write

$$\frac{d}{dn}\left(\int_0^{\pi/2}\theta\sin(n\theta)\cos^n(\theta)\log(\cos(\theta))d\theta\right)$$

$$=\int_0^{\pi/2}\theta\sin(n\theta)\cos^n(\theta)\log^2(\cos(\theta))d\theta$$

$$+\int_0^{\pi/2}\theta^2\cos(n\theta)\cos^n(\theta)\log(\cos(\theta))d\theta$$

$$=\frac{d}{dn}\left(-\log(2)\frac{\pi}{4}\frac{\psi(n+1)+\gamma}{2^n}-\frac{\pi}{4}\frac{\zeta(2)-\psi^{(1)}(n+1)}{2^n}\right.$$

$$\left.+\frac{\pi}{8}\frac{1}{2^n}((\psi(n+1)+\gamma)^2+\zeta(2)-\psi^{(1)}(n+1))\right)$$

$$=-\frac{\pi}{24}(\log(2)\pi^2+6\zeta(3))\frac{1}{2^n}+\frac{\pi}{24}(\pi^2+6\log^2(2))\frac{H_n}{2^n}-\frac{\pi}{4}\frac{H_nH_n^{(2)}}{2^n}$$

$$+\log(2)\frac{\pi}{2}\frac{H_n^{(2)}}{2^n}+\frac{\pi}{4}\frac{H_n^{(3)}}{2^n}-\log(2)\frac{\pi}{8}\frac{1}{2^n}(H_n^2+H_n^{(2)}),$$

where since the second resulting integral above is evaluated at the point (vii), we get that

$$\int_0^{\pi/2}\theta\sin(n\theta)\cos^n(\theta)\log^2(\cos(\theta))d\theta$$

$$=\frac{\pi}{48}(\pi^2+12\log^2(2))\frac{H_n}{2^n}-\frac{\pi}{2}\frac{H_nH_n^{(2)}}{2^n}+\log(2)\frac{\pi}{2}\frac{H_n^{(2)}}{2^n}$$

$$-\log(2)\frac{\pi}{4}\frac{1}{2^n}(H_n^2+H_n^{(2)})+\frac{\pi}{16}\frac{1}{2^n}(H_n^3+3H_nH_n^{(2)}+2H_n^{(3)}),$$

and the solution to the point (x) of the problem is complete.

Ultimately, to get the result at the last point, we want to set about deriving with respect to n the result from the point (vii), and then we write

$$\frac{d}{dn}\left(\int_0^{\pi/2}\theta^2\cos(n\theta)\cos^n(\theta)\log(\cos(\theta))d\theta\right)$$

$$=\int_0^{\pi/2}\theta^2\cos(n\theta)\cos^n(\theta)\log^2(\cos(\theta))d\theta$$

$$- \int_0^{\pi/2} \theta^3 \sin(n\theta) \cos^n(\theta) \log(\cos(\theta)) d\theta$$

$$= \frac{d}{dn} \left(- \log(2) \frac{\pi^3}{24} \frac{1}{2^n} - \frac{\pi}{4} \zeta(3) \frac{1}{2^n} + \frac{\pi^3}{48} \frac{\psi(n+1) + \gamma}{2^n} \right.$$

$$+ \frac{\pi}{4} \frac{(\psi(n+1) + \gamma)(\zeta(2) - \psi^{(1)}(n+1))}{2^n} + \frac{\pi}{4} \frac{\zeta(3) + 1/2 \psi^{(2)}(n+1)}{2^n}$$

$$+ \log(2) \frac{\pi}{8} \frac{1}{2^n} ((\psi(n+1) + \gamma)^2 + \zeta(2) - \psi^{(1)}(n+1))$$

$$- \frac{\pi}{16} \frac{1}{2^n} \left((\psi(n+1) + \gamma)^3 + 3(\psi(n+1) + \gamma)(\zeta(2) - \psi^{(1)}(n+1)) \right.$$

$$\left. \left. + 2 \left(\zeta(3) + \frac{1}{2} \psi^{(2)}(n+1) \right) \right) \right)$$

$$= \frac{\pi}{1440} (60 \log^2(2) \pi^2 + 11\pi^4 + 720 \log(2) \zeta(3)) \frac{1}{2^n} + \frac{\pi}{48} (\log(2) \pi^2 + 6\zeta(3)) \frac{H_n}{2^n}$$

$$- \frac{\pi}{32} (4 \log^2(2) + \pi^2) \frac{H_n^2}{2^n} + \log(2) \frac{\pi}{16} \frac{H_n^3}{2^n} - \frac{5}{16} \log(2) \pi \frac{H_n H_n^{(2)}}{2^n} - \frac{\pi}{8} \frac{H_n H_n^{(3)}}{2^n}$$

$$+ \frac{3}{16} \pi \frac{H_n^2 H_n^{(2)}}{2^n} - \frac{\pi}{96} (12 \log^2(2) + \pi^2) \frac{H_n^{(2)}}{2^n} - \frac{\pi}{16} \frac{(H_n^{(2)})^2}{2^n}$$

$$- \frac{3}{8} \log(2) \pi \frac{H_n^{(3)}}{2^n} - \frac{3}{8} \pi \frac{H_n^{(4)}}{2^n},$$

but since the second resulting integral is given at the point $(viii)$, we obtain

$$\int_0^{\pi/2} \theta^2 \cos(n\theta) \cos^n(\theta) \log^2(\cos(\theta)) d\theta$$

$$= \frac{\pi}{1440} (11\pi^4 + 60 \log^2(2) \pi^2 + 720 \log(2) \zeta(3)) \frac{1}{2^n}$$

$$- \frac{\pi}{24} (\log(2) \pi^2 + 6\zeta(3)) \frac{H_n}{2^n} - \frac{\pi}{16} \frac{H_n^4}{2^n} - \log(2) \frac{\pi}{2} \frac{H_n H_n^{(2)}}{2^n}$$

$$- \frac{\pi}{4} \frac{H_n H_n^{(3)}}{2^n} - \frac{\pi^3}{24} \frac{H_n^{(2)}}{2^n} - \frac{\pi}{16} \frac{(H_n^{(2)})^2}{2^n} - \log(2) \frac{\pi}{2} \frac{H_n^{(3)}}{2^n} - \frac{3}{8} \pi \frac{H_n^{(4)}}{2^n}$$

$$- \log^2(2)\frac{\pi}{8}\frac{1}{2^n}(H_n^2 + H_n^{(2)}) + \log(2)\frac{\pi}{8}\frac{1}{2^n}(H_n^3 + 3H_n H_n^{(2)} + 2H_n^{(3)})$$

$$+ \frac{\pi}{32}\frac{1}{2^n}(H_n^4 + 6H_n^2 H_n^{(2)} + 8H_n H_n^{(3)} + 3(H_n^{(2)})^2 + 6H_n^{(4)}),$$

and the solution to the point (xi) of the problem is complete.

A *long section!*, you probably exclaim! The results in this section will be helpful later in the book!

3.16 Surprisingly Bewitching Trigonometric Integrals with Appealing Closed Forms: The Fourth Act

Solution If you happened to wonder why the integrands in the previous section have a certain pattern like $\theta \sin(n\theta)\cos^n(\theta), \theta^2 \cos(n\theta)\cos^n(\theta), \theta^3 \sin(n\theta)\cos^n(\theta)$, and $\theta^4 \cos(n\theta)\cos^n(\theta)$ (e.g., note I skipped a case like $\theta^2 \sin(n\theta)\cos^n(\theta)$), in this section you may easily guess the reason by inspecting the problem statement.

In case we analyze the integrals from the points (i)–(iv) in the preceding section, we may easily note that they are related to the series of the kind $\displaystyle\sum_{k=1}^{\infty}(-1)^{k-1}\frac{1}{k^m}\binom{n}{k}$ which may be connected to an integral of the type $\displaystyle\int_0^1 x^{n-1}\log^m(1-x)dx$, and this is nicely expressible in terms of harmonic numbers. As seen in [76, Chapter 3, p.63], we have $\displaystyle\int_0^1 x^{n-1}\log^m(1-x)dx = \frac{(-1)^m m!}{n}h_m\left(1, \frac{1}{2}, \ldots, \frac{1}{n}\right)$, where $h_k(x_1, x_2, \ldots, x_n) = \sum_{1 \le i_1 \le i_2 \le \ldots \le i_k \le n} x_{i_1} x_{i_2} \ldots x_{i_k}$ is the complete homogeneous symmetric polynomial, and such results may be represented by harmonic numbers.

Now, if we try to calculate the first three points of the current problem, then we may note that the integrals also reduce to the series of the type $\displaystyle\sum_{k=1}^{\infty}\frac{1}{k^m}\binom{n}{k}$ besides the alternating series of the form $\displaystyle\sum_{k=1}^{\infty}(-1)^{k-1}\frac{1}{k^m}\binom{n}{k}$ mentioned above, and for the former series we'll be happy to consider a reduction to a useful integral form.

To get a better picture of the things written above, let's start with the calculation of the first integral, and using the series result in (3.98), which is found right in the previous section, we write

$$\int_0^{\pi/2} \theta \cos(n\theta)\cos^n(\theta)d\theta = \frac{1}{2^n}\int_0^{\pi/2} \theta \sum_{k=0}^{\infty}\binom{n}{k}\cos(2k\theta)d\theta$$

{leave out the term of the series for $k = 0$}

$$= \frac{3}{4}\zeta(2)\frac{1}{2^n} + \frac{1}{2^n}\int_0^{\pi/2}\theta\sum_{k=1}^{\infty}\binom{n}{k}\cos(2k\theta)d\theta$$

{reverse the order of summation and integration}

$$= \frac{3}{4}\zeta(2)\frac{1}{2^n} + \frac{1}{2^n}\sum_{k=1}^{\infty}\binom{n}{k}\int_0^{\pi/2}\theta\cos(2k\theta)d\theta$$

$$= \frac{3}{4}\zeta(2)\frac{1}{2^n} - \frac{1}{4}\frac{1}{2^n}\sum_{k=1}^{\infty}(-1)^{k-1}\frac{1}{k^2}\binom{n}{k} - \frac{1}{4}\frac{1}{2^n}\sum_{k=1}^{\infty}\frac{1}{k^2}\binom{n}{k}$$

$$\left\{\text{use that }\sum_{k=1}^{\infty}(-1)^{k-1}\frac{1}{k^2}\binom{n}{k} = \frac{1}{2}(H_n^2 + H_n^{(2)})\right\}$$

$$= \frac{3}{4}\zeta(2)\frac{1}{2^n} - \frac{1}{8}\frac{1}{2^n}(H_n^2 + H_n^{(2)}) + \frac{1}{4}\frac{1}{2^n}\sum_{k=1}^{\infty}\binom{n}{k}\int_0^1 x^{k-1}\log(x)dx$$

$$= \frac{3}{4}\zeta(2)\frac{1}{2^n} - \frac{1}{8}\frac{1}{2^n}(H_n^2 + H_n^{(2)}) - \frac{1}{4}\frac{1}{2^n}\int_0^1 \frac{(1 - (1+x)^n)\log(x)}{x}dx,$$

where to get the last equality I reversed the order of integration and summation, and the solution to the point (i) of the problem is finalized. The alternating binomial series above, $\sum_{k=1}^{\infty}(-1)^{k-1}\frac{1}{k^2}\binom{n}{k} = \frac{1}{2}(H_n^2 + H_n^{(2)})$, is already found in the previous section (e.g., see the second solution to the point (ii) of the problem).

Could we turn such results into useful tools in the calculations with integrals and series?, you might wonder. Absolutely, and you'll see this later in the book!

Passing to the point (ii) of the problem and using the series result in (3.99), in the previous section, we write

$$\int_0^{\pi/2}\theta^2\sin(n\theta)\cos^n(\theta)d\theta = \frac{1}{2^n}\int_0^{\pi/2}\theta^2\sum_{k=1}^{\infty}\binom{n}{k}\sin(2k\theta)d\theta$$

{reverse the order of summation and integration}

$$= \frac{1}{2^n}\sum_{k=1}^{\infty}\binom{n}{k}\int_0^{\pi/2}\theta^2\sin(2k\theta)d\theta$$

$$= \frac{3}{4}\zeta(2)\frac{1}{2^n}\sum_{k=1}^{\infty}(-1)^{k-1}\frac{1}{k}\binom{n}{k} - \frac{1}{4}\frac{1}{2^n}\sum_{k=1}^{\infty}(-1)^{k-1}\frac{1}{k^3}\binom{n}{k} - \frac{1}{4}\frac{1}{2^n}\sum_{k=1}^{\infty}\frac{1}{k^3}\binom{n}{k}$$

$$\left\{\text{observe and use that } \sum_{k=1}^{\infty}(-1)^{k-1}\frac{1}{k}\binom{n}{k} = H_n, \text{ and at the same time}\right\}$$

$$\left\{\text{we want to consider that } \sum_{k=1}^{\infty}(-1)^{k-1}\frac{1}{k^3}\binom{n}{k} = \frac{1}{6}(H_n^3 + 3H_n H_n^{(2)} + 2H_n^{(3)})\right\}$$

$$= \frac{3}{4}\zeta(2)\frac{H_n}{2^n} - \frac{1}{24}\frac{1}{2^n}(H_n^3 + 3H_n H_n^{(2)} + 2H_n^{(3)}) - \frac{1}{8}\frac{1}{2^n}\sum_{k=1}^{\infty}\binom{n}{k}\int_0^1 x^{k-1}\log^2(x)dx$$

$$= \frac{3}{4}\zeta(2)\frac{H_n}{2^n} - \frac{1}{24}\frac{1}{2^n}(H_n^3 + 3H_n H_n^{(2)} + 2H_n^{(3)}) + \frac{1}{8}\frac{1}{2^n}\int_0^1 \frac{(1-(1+x)^n)\log^2(x)}{x}dx,$$

and the solution to the point (ii) of the problem is finalized. The alternating binomial sums appearing in the calculations above are easily extracted from the second solution to the point (i) and the solution to the point (iii) of the previous section.

Next, for the point (iii) we use a style similar to the one at the point (i), and employing the series result in (3.98), in the previous section, we write

$$\int_0^{\pi/2} \theta^3 \cos(n\theta)\cos^n(\theta)d\theta = \frac{1}{2^n}\int_0^{\pi/2}\theta^3\sum_{k=0}^{\infty}\binom{n}{k}\cos(2k\theta)d\theta$$

$$= \frac{45}{32}\zeta(4)\frac{1}{2^n} + \frac{1}{2^n}\int_0^{\pi/2}\theta^3\sum_{k=1}^{\infty}\binom{n}{k}\cos(2k\theta)d\theta$$

$$\{\text{reverse the order of summation and integration}\}$$

$$= \frac{45}{32}\zeta(4)\frac{1}{2^n} + \frac{1}{2^n}\sum_{k=1}^{\infty}\binom{n}{k}\int_0^{\pi/2}\theta^3\cos(2k\theta)d\theta = \frac{45}{32}\zeta(4)\frac{1}{2^n}$$

$$-\frac{9}{8}\zeta(2)\frac{1}{2^n}\sum_{k=1}^{\infty}(-1)^{k-1}\frac{1}{k^2}\binom{n}{k} + \frac{3}{8}\frac{1}{2^n}\sum_{k=1}^{\infty}(-1)^{k-1}\frac{1}{k^4}\binom{n}{k} + \frac{3}{8}\frac{1}{2^n}\sum_{k=1}^{\infty}\frac{1}{k^4}\binom{n}{k}$$

$$\left\{\text{employ the series result } \sum_{k=1}^{\infty}(-1)^{k-1}\frac{1}{k^2}\binom{n}{k} = \frac{1}{2}(H_n^2 + H_n^{(2)}) \text{ together with}\right\}$$

$$\left\{ \sum_{k=1}^{\infty} (-1)^{k-1} \frac{1}{k^4} \binom{n}{k} = \frac{1}{24}(H_n^4 + 6H_n^2 H_n^{(2)} + 8H_n H_n^{(3)} + 3(H_n^{(2)})^2 + 6H_n^{(4)}) \right\}$$

$$= \frac{45}{32}\zeta(4)\frac{1}{2^n} - \frac{9}{16}\zeta(2)\frac{1}{2^n}(H_n^2 + H_n^{(2)})$$

$$+ \frac{1}{64}\frac{1}{2^n}(H_n^4 + 6H_n^2 H_n^{(2)} + 8H_n H_n^{(3)} + 3(H_n^{(2)})^2 + 6H_n^{(4)})$$

$$- \frac{1}{16}\frac{1}{2^n} \sum_{k=1}^{\infty} \binom{n}{k} \int_0^1 x^{k-1} \log^3(x)dx$$

$$= \frac{45}{32}\zeta(4)\frac{1}{2^n} - \frac{9}{16}\zeta(2)\frac{1}{2^n}(H_n^2 + H_n^{(2)})$$

$$+ \frac{1}{64}\frac{1}{2^n}(H_n^4 + 6H_n^2 H_n^{(2)} + 8H_n H_n^{(3)} + 3(H_n^{(2)})^2 + 6H_n^{(4)})$$

$$+ \frac{1}{16}\frac{1}{2^n} \int_0^1 \frac{(1 - (1+x)^n)\log^3(x)}{x}dx,$$

and the solution to the point (*iii*) of the problem is finalized. The alternating binomial sums arising in the calculations are readily obtained from the second solution to the point (*ii*) and the solution to the point (*iv*) in the previous section.

At the following two points we'll be exposed to similar integrals where this time the integrands also contain a logarithm.

So, as regards the point (*iv*) of the problem, we want to consider the point (*i*), where if we differentiate both sides once with respect to n, we get

$$\frac{d}{dn}\left(\int_0^{\pi/2} \theta \cos(n\theta) \cos^n(\theta)d\theta \right)$$

$$= \int_0^{\pi/2} \theta \cos(n\theta) \cos^n(\theta) \log(\cos(\theta))d\theta - \int_0^{\pi/2} \theta^2 \sin(n\theta) \cos^n(\theta)d\theta$$

$$= \frac{d}{dn}\left(\frac{3}{4}\zeta(2)\frac{1}{2^n} - \frac{1}{8}\frac{1}{2^n}((\psi(n+1) + \gamma)^2 + \zeta(2) - \psi^{(1)}(n+1)) \right.$$

$$\left. - \frac{1}{4}\frac{1}{2^n} \int_0^1 \frac{(1 - (1+x)^n)\log(x)}{x}dx \right)$$

$$= -\frac{1}{8}(6\log(2)\zeta(2) + 2\zeta(3))\frac{1}{2^n} - \frac{1}{4}\zeta(2)\frac{H_n}{2^n} + \frac{1}{4}\frac{H_n H_n^{(2)}}{2^n} + \frac{1}{4}\frac{H_n^{(3)}}{2^n}$$

$$+ \frac{1}{8}\log(2)\frac{1}{2^n}(H_n^2 + H_n^{(2)}) + \frac{1}{4}\log(2)\frac{1}{2^n}\int_0^1 \frac{(1-(1+x)^n)\log(x)}{x}dx$$

$$+ \frac{1}{4}\frac{1}{2^n}\int_0^1 \frac{(1+x)^n\log(x)\log(1+x)}{x}dx,$$

and if we use the value of the integral at the point (ii) and then rearrange, we obtain

$$\int_0^{\pi/2} \theta\cos(n\theta)\cos^n(\theta)\log(\cos(\theta))d\theta$$

$$= -\frac{1}{8}(\log(2)\pi^2 + 2\zeta(3))\frac{1}{2^n} + \frac{\pi^2}{12}\frac{H_n}{2^n} + \frac{1}{4}\frac{H_n H_n^{(2)}}{2^n} + \frac{1}{4}\frac{H_n^{(3)}}{2^n}$$

$$+ \frac{1}{8}\log(2)\frac{1}{2^n}(H_n^2 + H_n^{(2)}) - \frac{1}{24}\frac{1}{2^n}(H_n^3 + 3H_n H_n^{(2)} + 2H_n^{(3)})$$

$$+ \frac{1}{4}\log(2)\frac{1}{2^n}\int_0^1 \frac{(1-(1+x)^n)\log(x)}{x}dx + \frac{1}{8}\frac{1}{2^n}\int_0^1 \frac{(1-(1+x)^n)\log^2(x)}{x}dx$$

$$+ \frac{1}{4}\frac{1}{2^n}\int_0^1 \frac{(1+x)^n\log(x)\log(1+x)}{x}dx,$$

and the solution to the point (iv) of the problem is finalized.

Jumping to the point (v) of the problem, we want to consider the point (ii) where if we differentiate its both sides with respect to n, we have

$$\frac{d}{dn}\left(\int_0^{\pi/2} \theta^2\sin(n\theta)\cos^n(\theta)d\theta\right)$$

$$= \int_0^{\pi/2} \theta^2\sin(n\theta)\cos^n(\theta)\log(\cos(\theta))d\theta + \int_0^{\pi/2} \theta^3\cos(n\theta)\cos^n(\theta)d\theta$$

$$= \frac{d}{dn}\left(\frac{3}{4}\zeta(2)\frac{\psi(n+1)+\gamma}{2^n}\right.$$

$$-\frac{1}{24}\frac{1}{2^n}\Big((\psi(n+1)+\gamma)^3 + 3(\psi(n+1)+\gamma)(\zeta(2)-\psi^{(1)}(n+1))$$

$$+ 2\Big(\zeta(3) + \frac{1}{2}\psi^{(2)}(n+1)\Big)\Big) + \frac{1}{8}\frac{1}{2^n}\int_0^1 \frac{(1-(1+x)^n)\log^2(x)}{x}dx\right)$$

$$= \frac{13}{8}\zeta(4)\frac{1}{2^n} - \frac{1}{8}(6\log(2)\zeta(2) + 2\zeta(3))\frac{H_n}{2^n} + \frac{1}{4}\frac{H_n H_n^{(3)}}{2^n} + \frac{1}{8}\frac{H_n^2 H_n^{(2)}}{2^n}$$

$$- \frac{3}{4}\zeta(2)\frac{H_n^{(2)}}{2^n} + \frac{1}{8}\frac{(H_n^{(2)})^2}{2^n} + \frac{1}{4}\frac{H_n^{(4)}}{2^n} - \frac{1}{8}\zeta(2)\frac{1}{2^n}(H_n^2 + H_n^{(2)})$$

$$+ \frac{1}{24}\log(2)\frac{1}{2^n}(H_n^3 + 3H_n H_n^{(2)} + 2H_n^{(3)}) - \frac{1}{8}\log(2)\frac{1}{2^n}\int_0^1 \frac{(1 - (1+x)^n)\log^2(x)}{x}dx$$

$$- \frac{1}{8}\frac{1}{2^n}\int_0^1 \frac{(1+x)^n \log^2(x)\log(1+x)}{x}dx,$$

where if we employ the result from the point (iii) and then rearrange, we arrive at

$$\int_0^{\pi/2} \theta^2 \sin(n\theta) \cos^n(\theta) \log(\cos(\theta))d\theta$$

$$= \frac{7}{32}\zeta(4)\frac{1}{2^n} - \frac{1}{4}(3\log(2)\zeta(2) + \zeta(3))\frac{H_n}{2^n} + \frac{1}{4}\frac{H_n H_n^{(3)}}{2^n} + \frac{1}{8}\frac{H_n^2 H_n^{(2)}}{2^n} - \frac{3}{4}\zeta(2)\frac{H_n^{(2)}}{2^n}$$

$$+ \frac{1}{8}\frac{(H_n^{(2)})^2}{2^n} + \frac{1}{4}\frac{H_n^{(4)}}{2^n} + \frac{7}{16}\zeta(2)\frac{1}{2^n}(H_n^2 + H_n^{(2)}) + \frac{1}{24}\log(2)\frac{1}{2^n}(H_n^3 + 3H_n H_n^{(2)} + 2H_n^{(3)})$$

$$- \frac{1}{64}\frac{1}{2^n}(H_n^4 + 6H_n^2 H_n^{(2)} + 8H_n H_n^{(3)} + 3(H_n^{(2)})^2 + 6H_n^{(4)})$$

$$- \frac{1}{8}\log(2)\frac{1}{2^n}\int_0^1 \frac{(1 - (1+x)^n)\log^2(x)}{x}dx - \frac{1}{16}\frac{1}{2^n}\int_0^1 \frac{(1 - (1+x)^n)\log^3(x)}{x}dx$$

$$- \frac{1}{8}\frac{1}{2^n}\int_0^1 \frac{(1+x)^n \log^2(x)\log(1+x)}{x}dx,$$

and the solution to the point (v) of the problem is finalized.

Finally, to extract the result at the point (vi), we need to consider the result from the point (iv) where if we differentiate its both sides with respect to n, we obtain

$$\frac{d}{dn}\left(\int_0^{\pi/2} \theta \cos(n\theta) \cos^n(\theta) \log(\cos(\theta))d\theta\right)$$

$$= \int_0^{\pi/2} \theta \cos(n\theta) \cos^n(\theta) \log^2(\cos(\theta))d\theta$$

$$-\int_0^{\pi/2}\theta^2\sin(n\theta)\cos^n(\theta)\log(\cos(\theta))d\theta$$

$$=\frac{d}{dn}\left(-\frac{1}{4}(3\log(2)\zeta(2)+\zeta(3))\frac{1}{2^n}+\frac{1}{2}\zeta(2)\frac{H_n}{2^n}+\frac{1}{4}\frac{H_nH_n^{(2)}}{2^n}+\frac{1}{4}\frac{H_n^{(3)}}{2^n}\right.$$

$$+\frac{1}{8}\log(2)\frac{1}{2^n}(H_n^2+H_n^{(2)})-\frac{1}{24}\frac{1}{2^n}(H_n^3+3H_nH_n^{(2)}+2H_n^{(3)})$$

$$+\frac{1}{4}\log(2)\frac{1}{2^n}\int_0^1\frac{(1-(1+x)^n)\log(x)}{x}dx+\frac{1}{8}\frac{1}{2^n}\int_0^1\frac{(1-(1+x)^n)\log^2(x)}{x}dx$$

$$\left.+\frac{1}{4}\frac{1}{2^n}\int_0^1\frac{(1+x)^n\log(x)\log(1+x)}{x}dx\right)$$

$$=\frac{1}{360}(630\zeta(4)+180\log(2)\zeta(3)+270\log^2(2)\zeta(2))\frac{1}{2^n}-\frac{1}{4}(\log(2)\zeta(2)-\zeta(3))\frac{H_n}{2^n}$$

$$-\frac{1}{2}\log(2)\frac{H_nH_n^{(2)}}{2^n}-\frac{1}{4}\frac{H_nH_n^{(3)}}{2^n}+\frac{1}{8}\frac{H_n^2H_n^{(2)}}{2^n}-\frac{1}{4}\zeta(2)\frac{H_n^{(2)}}{2^n}-\frac{1}{8}\frac{(H_n^{(2)})^2}{2^n}$$

$$-\frac{1}{2}\log(2)\frac{H_n^{(3)}}{2^n}-\frac{1}{2}\frac{H_n^{(4)}}{2^n}-\frac{1}{8}(\log^2(2)+\zeta(2))\frac{1}{2^n}(H_n^2+H_n^{(2)})$$

$$+\frac{1}{24}\log(2)\frac{1}{2^n}(H_n^3+3H_nH_n^{(2)}+2H_n^{(3)})-\frac{1}{4}\log^2(2)\frac{1}{2^n}\int_0^1\frac{(1-(1+x)^n)\log(x)}{x}dx$$

$$-\frac{1}{8}\log(2)\frac{1}{2^n}\int_0^1\frac{(1-(1+x)^n)\log^2(x)}{x}dx-\frac{1}{8}\frac{1}{2^n}\int_0^1\frac{(1+x)^n\log^2(x)\log(1+x)}{x}dx$$

$$-\frac{1}{2}\log(2)\frac{1}{2^n}\int_0^1\frac{(1+x)^n\log(x)\log(1+x)}{x}dx$$

$$+\frac{1}{4}\frac{1}{2^n}\int_0^1\frac{(1+x)^n\log(x)\log^2(1+x)}{x}dx,$$

and if we consider the result from the point (v), we conclude that

$$\int_0^{\pi/2}\theta\cos(n\theta)\cos^n(\theta)\log^2(\cos(\theta))d\theta$$

$$=\frac{1}{320}(630\zeta(4)+160\log(2)\zeta(3)+240\log^2(2)\zeta(2))\frac{1}{2^n}-\log(2)\zeta(2)\frac{H_n}{2^n}$$

$$-\frac{1}{2}\log(2)\frac{H_n H_n^{(2)}}{2^n} + \frac{1}{4}\frac{H_n^2 H_n^{(2)}}{2^n} - \zeta(2)\frac{H_n^{(2)}}{2^n} - \frac{1}{2}\log(2)\frac{H_n^{(3)}}{2^n} - \frac{1}{4}\frac{H_n^{(4)}}{2^n}$$

$$-\frac{1}{96}(12\log^2(2) - 30\zeta(2))\frac{1}{2^n}(H_n^2 + H_n^{(2)}) + \frac{1}{12}\log(2)\frac{1}{2^n}(H_n^3 + 3H_n H_n^{(2)} + 2H_n^{(3)})$$

$$-\frac{1}{64}\frac{1}{2^n}(H_n^4 + 6H_n^2 H_n^{(2)} + 8H_n H_n^{(3)} + 3(H_n^{(2)})^2 + 6H_n^{(4)})$$

$$-\frac{1}{4}\log^2(2)\frac{1}{2^n}\int_0^1 \frac{(1-(1+x)^n)\log(x)}{x}dx - \frac{1}{4}\log(2)\frac{1}{2^n}\int_0^1 \frac{(1-(1+x)^n)\log^2(x)}{x}dx$$

$$-\frac{1}{16}\frac{1}{2^n}\int_0^1 \frac{(1-(1+x)^n)\log^3(x)}{x}dx - \frac{1}{2}\log(2)\frac{1}{2^n}\int_0^1 \frac{(1+x)^n \log(x)\log(1+x)}{x}dx$$

$$-\frac{1}{4}\frac{1}{2^n}\int_0^1 \frac{(1+x)^n \log^2(x)\log(1+x)}{x}dx + \frac{1}{4}\frac{1}{2^n}\int_0^1 \frac{(1+x)^n \log(x)\log^2(1+x)}{x}dx,$$

and the solution to the point (vi) of the problem is finalized.

This section closes a series of four sections involving integrands with similar trigonometric structures! I hope you had a lot of fun with these enjoyable integrals!

3.17 Playing with a Hand of Fabulous Integrals: The First Part

Solution If you read the book from the beginning, page by page, then let me guess you probably arrived here at some point during the reading of a solution in Sect. 3.3, as I suggested. A good idea to better understand the solution provided!

The integrals treated in this section will prepare the ground to extract other results that will beautifully lead to *the tool* we need in the *non-obvious* solution from Sect. 3.3. If we are careful enough not to get into complicated calculations, everything will flow smoothly and beautifully. Now, we assume $n \geq 2$, and then, for the integral at the point (i) of the problem, which we denote by I_n, we write

$$I_n = \int_0^{\pi/2} \sin^n(x)\tan(x)\log\left(\tan\left(\frac{x}{2}\right)\right)dx$$

$$= \int_0^{\pi/2} \sin^2(x)\sin^{n-2}(x)\tan(x)\log\left(\tan\left(\frac{x}{2}\right)\right)dx$$

$$= \int_0^{\pi/2} (1 - \cos^2(x))\sin^{n-2}(x)\tan(x)\log\left(\tan\left(\frac{x}{2}\right)\right)dx$$

$$= \int_0^{\pi/2} \sin^{n-2}(x)\tan(x)\log\left(\tan\left(\frac{x}{2}\right)\right)dx - \int_0^{\pi/2} \sin^{n-1}(x)\cos(x)\log\left(\tan\left(\frac{x}{2}\right)\right)dx,$$

$$\underbrace{\phantom{\int_0^{\pi/2} \sin^{n-2}(x)\tan(x)\log\left(\tan\left(\frac{x}{2}\right)\right)dx}}_{I_{n-2}}$$

whence we obtain that

$$I_n - I_{n-2}$$

$$= -\int_0^{\pi/2} \sin^{n-1}(x)\cos(x)\log\left(\tan\left(\frac{x}{2}\right)\right)dx = -\frac{1}{n}\int_0^{\pi/2} (\sin^n(x))'\log\left(\tan\left(\frac{x}{2}\right)\right)dx$$

$$\{\text{apply integration by parts}\}$$

$$= \frac{1}{n}\int_0^{\pi/2} \sin^{n-1}(x)dx. \tag{3.109}$$

Then, we treat the recurrence relation in (3.109) according to n odd and even, using the celebre cases of Wallis' integral, $\int_0^{\pi/2} \sin^{2k}(x)dx = \frac{\pi}{2}\cdot\frac{(2k-1)!!}{(2k)!!} =$

$\frac{\pi}{2^{2k+1}}\binom{2k}{k}$ and $\int_0^{\pi/2} \sin^{2k-1}(x)dx = \frac{(2k-2)!!}{(2k-1)!!} = \frac{2^{2k-1}}{k\binom{2k}{k}}$. Wallis' integral

cases are straightforward if we use the recurrence relation, $kJ_k = (k-1)J_{k-2}$, $k \geq 2$, obtained with integration by parts, where $J_k = \int_0^{\pi/2} \sin^k(x)dx$, and then split k based on parity.

So, considering n even in (3.109), and then replacing n by $2k$, we get

$$I_{2k} - I_{2k-2} = \frac{1}{2k}\int_0^{\pi/2} \sin^{2k-1}(x)dx,$$

and summing over both sides from $k = 1$ to m, we obtain

$$\sum_{k=1}^m (I_{2k} - I_{2k-2}) = I_{2m} - I_0 = \frac{1}{2}\sum_{k=1}^m \frac{1}{k}\int_0^{\pi/2} \sin^{2k-1}(x)dx$$

$$= \frac{1}{2}\sum_{k=1}^m \frac{(2k-2)!!}{k\cdot(2k-1)!!} = \sum_{k=1}^m \frac{2^{2k-2}}{k^2\binom{2k}{k}}$$

or

$$I_{2m} = I_0 + \frac{1}{2} \sum_{k=1}^{m} \frac{(2k-2)!!}{k \cdot (2k-1)!!} = I_0 + \sum_{k=1}^{m} \frac{2^{2k-2}}{k^2 \binom{2k}{k}}. \qquad (3.110)$$

Next, if we consider n odd in (3.109), and then replace n by $2k+1$, we have

$$I_{2k+1} - I_{2k-1} = \frac{1}{2k+1} \int_0^{\pi/2} \sin^{2k}(x) dx,$$

and if we further sum over both sides from $k=1$ to m, we get

$$\sum_{k=1}^{m} (I_{2k+1} - I_{2k-1}) = I_{2m+1} - I_1 = \sum_{k=1}^{m} \frac{1}{2k+1} \int_0^{\pi/2} \sin^{2k}(x) dx$$

$$= \frac{\pi}{2} \sum_{k=1}^{m} \frac{(2k-1)!!}{(2k+1) \cdot (2k)!!} = \frac{\pi}{2} \sum_{k=1}^{m} \frac{1}{(2k+1)2^{2k}} \binom{2k}{k}$$

or

$$I_{2m+1} = I_1 + \frac{\pi}{2} \sum_{k=1}^{m} \frac{(2k-1)!!}{(2k+1) \cdot (2k)!!} = I_1 + \frac{\pi}{2} \sum_{k=1}^{m} \frac{1}{(2k+1)2^{2k}} \binom{2k}{k}. \qquad (3.111)$$

Now, based upon the calculations so far, using (3.110) and (3.111), we have

$$I_n = \begin{cases} I_{2m} = I_0 + \dfrac{1}{2} \displaystyle\sum_{k=1}^{m} \dfrac{(2k-2)!!}{k \cdot (2k-1)!!} = I_0 + \displaystyle\sum_{k=1}^{m} \dfrac{2^{2k-2}}{k^2 \binom{2k}{k}}; \\[4mm] I_{2m+1} = I_1 + \dfrac{\pi}{2} \displaystyle\sum_{k=1}^{m} \dfrac{(2k-1)!!}{(2k+1) \cdot (2k)!!} = I_1 + \dfrac{\pi}{2} \displaystyle\sum_{k=1}^{m} \dfrac{1}{(2k+1)2^{2k}} \binom{2k}{k}, \end{cases}$$
$$(3.112)$$

where $I_n = \int_0^{\pi/2} \sin^n(x) \tan(x) \log\left(\tan\left(\frac{x}{2}\right)\right) dx$, $m, n \ge 0$.

We need to finalize one more task in (3.112), that is, to compute the integrals I_0 and I_1.

In order to calculate the integral I_0, we proceed as follows:

$$I_0 = \int_0^{\pi/2} \tan(x) \log\left(\tan\left(\frac{x}{2}\right)\right) dx \overset{\tan^2(x/2)=y}{=} \int_0^1 \frac{\log(y)}{1-y^2} dy = \int_0^1 \sum_{k=1}^{\infty} y^{2k-2} \log(y) dy$$

$$= \sum_{k=1}^{\infty} \int_0^1 y^{2k-2} \log(y)dy = -\sum_{k=1}^{\infty} \frac{1}{(2k-1)^2} = -\left(\sum_{k=1}^{\infty} \frac{1}{k^2} - \sum_{k=1}^{\infty} \frac{1}{(2k)^2}\right) = -\frac{\pi^2}{8},$$

$$(3.113)$$

and the calculations to the integral I_0 are finalized since $\sum_{n=1}^{\infty} \frac{1}{n^2} = \frac{\pi^2}{6}$. Note that above I also exploited the identity, $\tan(2\arctan(x)) = 2x/(1-x^2)$.

Then, to calculate the integral I_1 we employ the *Weierstrass substitution* which gives

$$I_1 = \int_0^{\pi/2} \sin(x)\tan(x) \log\left(\tan\left(\frac{x}{2}\right)\right) dx \overset{\tan(x/2)=y}{=} 4\int_0^1 \frac{\log(y)}{1-y^4}dy - 4\int_0^1 \frac{\log(y)}{(1+y^2)^2}dy$$

$$= 4\int_0^1 \sum_{n=1}^{\infty} y^{4n-4} \log(y)dy - 4\int_0^1 \sum_{n=1}^{\infty}(-1)^{n-1}ny^{2n-2}\log(y)dy$$

$$= 4\sum_{n=1}^{\infty}\int_0^1 y^{4n-4}\log(y)dy - 4\sum_{n=1}^{\infty}(-1)^{n-1}n\int_0^1 y^{2n-2}\log(y)dy$$

$$= -4\sum_{n=1}^{\infty}\frac{1}{(4n-3)^2} + 2\sum_{n=1}^{\infty}(-1)^{n-1}\frac{(2n-1)+1}{(2n-1)^2}$$

$$= -2\left(\underbrace{\sum_{n=1}^{\infty}\frac{1}{(4n-3)^2} - \sum_{n=1}^{\infty}\frac{1}{(4n-1)^2}}_{=\sum_{n=1}^{\infty}(-1)^{n-1}1/(2n-1)^2} + \underbrace{\sum_{n=1}^{\infty}\frac{1}{(4n-3)^2} + \sum_{n=1}^{\infty}\frac{1}{(4n-1)^2}}_{=\sum_{n=1}^{\infty}1/(2n-1)^2}\right)$$

$$+ 2\underbrace{\sum_{n=1}^{\infty}(-1)^{n-1}\frac{1}{2n-1}}_{\pi/4} + 2\sum_{n=1}^{\infty}(-1)^{n-1}\frac{1}{(2n-1)^2}$$

$$= \frac{\pi}{2} - 2\sum_{n=1}^{\infty}\frac{1}{(2n-1)^2} = \frac{\pi}{2} - 2\left(\sum_{n=1}^{\infty}\frac{1}{n^2} - \sum_{n=1}^{\infty}\frac{1}{(2n)^2}\right) = \frac{\pi}{2} - \frac{\pi^2}{4}, \quad (3.114)$$

and the calculations to the integral I_1 are finalized. It's easy to see that based on

the Taylor series, $\arctan(x) = \sum_{n=1}^{\infty}(-1)^{n-1}\frac{x^{2n-1}}{2n-1}$, where we set $x = 1$, we obtain

immediately that $\frac{\pi}{4} = \sum_{n=1}^{\infty}(-1)^{n-1}\frac{1}{2n-1}$.

Finally, collecting the results from (3.113) and (3.114) in (3.112), we conclude that

$$I_n = \begin{cases} I_{2m} = -\dfrac{\pi^2}{8} + \dfrac{1}{2}\displaystyle\sum_{k=1}^{m}\dfrac{(2k-2)!!}{k \cdot (2k-1)!!} = -\dfrac{\pi^2}{8} + \displaystyle\sum_{k=1}^{m}\dfrac{2^{2k-2}}{k^2\binom{2k}{k}}; \\[4mm] I_{2m+1} = -\dfrac{\pi^2}{4} + \dfrac{\pi}{2}\displaystyle\sum_{k=0}^{m}\dfrac{(2k-1)!!}{(2k+1)\cdot(2k)!!} \\[4mm] \qquad\quad = -\dfrac{\pi^2}{4} + \dfrac{\pi}{2}\displaystyle\sum_{k=0}^{m}\dfrac{1}{(2k+1)2^{2k}}\binom{2k}{k}, \end{cases}$$

where $I_n = \displaystyle\int_0^{\pi/2}\sin^n(x)\tan(x)\log\left(\tan\left(\frac{x}{2}\right)\right)dx$, $m, n \geq 0$, and the solutions to the first two points of the problem are complete.

Further, for the next two points of the problem we also need the trigonometric identity,[7] $\tan\left(\dfrac{x}{2}\right) = \dfrac{\sin(x)}{1+\cos(x)}$. Then, for the point (iii) of the problem we start with the integral from the point (i), and then we write

$$\int_0^{\pi/2}\sin^{2m}(x)\tan(x)\log\left(\tan\left(\frac{x}{2}\right)\right)dx$$

$$= \int_0^{\pi/2}\sin^{2m}(x)\tan(x)\log(\sin(x))dx - \int_0^{\pi/2}\sin^{2m}(x)\tan(x)\log(1+\cos(x))dx.$$

$$(3.115)$$

Now, by letting the variable change $\sin^2(x) = y$ in the first integral from the right-hand side of (3.115), we get

$$\int_0^{\pi/2}\sin^{2m}(x)\tan(x)\log(\sin(x))dx = \frac{1}{4}\int_0^1\frac{y^m\log(y)}{1-y}dy$$

---

[7] A useful trigonometric identity easy to prove! So, we simply write $\tan\left(\dfrac{x}{2}\right) = \dfrac{\sin(x/2)}{\cos(x/2)} = \dfrac{\sin(x/2)}{\cos(x/2)}\cdot\dfrac{2\cos(x/2)}{2\cos(x/2)} = \dfrac{\sin(x)}{1+\cos(x)}$, where in the last equality I used the trigonometric identities, $\sin(2x) = 2\sin(x)\cos(x)$ and $1 + \cos(x) = 2\cos^2\left(\dfrac{x}{2}\right)$.

$$= \frac{1}{4} \int_0^1 \log(y) \sum_{n=1}^{\infty} y^{m+n-1} dy = \frac{1}{4} \sum_{n=1}^{\infty} \int_0^1 y^{m+n-1} \log(y) dy = -\frac{1}{4} \sum_{n=1}^{\infty} \frac{1}{(n+m)^2}$$

$$= -\frac{1}{4} \sum_{n=m+1}^{\infty} \frac{1}{n^2} = -\frac{1}{4} \left(\sum_{n=1}^{\infty} \frac{1}{n^2} - \sum_{n=1}^{m} \frac{1}{n^2} \right) = -\frac{1}{4} \left(\frac{\pi^2}{6} - H_m^{(2)} \right) = \frac{1}{4} H_m^{(2)} - \frac{\pi^2}{24}.$$

$$(3.116)$$

Therefore, by combining the results in (3.115), (3.116), and the one from the point (i), we obtain

$$\int_0^{\pi/2} \sin^{2m}(x) \tan(x) \log(1 + \cos(x)) \, dx$$

$$= \frac{\pi^2}{12} + \frac{1}{4} H_m^{(2)} - \frac{1}{2} \sum_{k=1}^{m} \frac{(2k-2)!!}{k \cdot (2k-1)!!} = \frac{\pi^2}{12} + \frac{1}{4} H_m^{(2)} - \sum_{k=1}^{m} \frac{2^{2k-2}}{k^2 \binom{2k}{k}},$$

and the solution to the point (iii) of the problem is complete.

For the next point, using a strategy similar to the one at the previous point and employing the result from the point (ii), we write

$$\int_0^{\pi/2} \sin^{2m+1}(x) \tan(x) \log \left(\tan \left(\frac{x}{2} \right) \right) dx$$

$$= \int_0^{\pi/2} \sin^{2m+1}(x) \tan(x) \log(\sin(x)) dx - \int_0^{\pi/2} \sin^{2m+1}(x) \tan(x) \log(1 + \cos(x)) dx.$$

$$(3.117)$$

By making the variable change $\sin(x) = y$ in the first integral from the right-hand side of (3.117), we get

$$\int_0^{\pi/2} \sin^{2m+1}(x) \tan(x) \log(\sin(x)) dx = \int_0^1 \frac{y^{2m+2} \log(y)}{1 - y^2} dy = \int_0^1 \log(y) \sum_{n=1}^{\infty} y^{2m+2n} dy$$

$$= \sum_{n=1}^{\infty} \int_0^1 y^{2m+2n} \log(y) dy = -\sum_{n=1}^{\infty} \frac{1}{(2n+2m+1)^2} = -\sum_{n=1}^{\infty} \frac{1}{(2n+1)^2} + \sum_{n=1}^{m} \frac{1}{(2n+1)^2}$$

$$\underset{\text{the sums}}{\overset{\text{reindex}}{=}} -\sum_{n=2}^{\infty} \frac{1}{(2n-1)^2} + \sum_{n=2}^{m+1} \frac{1}{(2n-1)^2} = -\sum_{n=1}^{\infty} \frac{1}{(2n-1)^2} + \sum_{n=1}^{m+1} \frac{1}{(2n-1)^2}$$

$$= -\left(\sum_{n=1}^{\infty} \frac{1}{n^2} - \sum_{n=1}^{\infty} \frac{1}{(2n)^2}\right) + \sum_{n=1}^{m} \frac{1}{(2n-1)^2} + \frac{1}{(2m+1)^2}$$

$$\left\{\text{make use of the fact that } \sum_{n=1}^{m} \frac{1}{(2n-1)^2} = H_{2m}^{(2)} - \frac{1}{4} H_m^{(2)}\right\}$$

$$= -\frac{\pi^2}{8} + H_{2m+1}^{(2)} - \frac{1}{4} H_m^{(2)}. \tag{3.118}$$

At this point, we combine the results in (3.117), (3.118), and the one from the point (ii), which lead to

$$\int_0^{\pi/2} \sin^{2m+1}(x) \tan(x) \log(1 + \cos(x)) \, dx$$

$$= \frac{\pi^2}{8} + H_{2m+1}^{(2)} - \frac{1}{4} H_m^{(2)} - \frac{\pi}{2} \sum_{k=0}^{m} \frac{(2k-1)!!}{(2k+1) \cdot (2k)!!}$$

$$= \frac{\pi^2}{8} + H_{2m+1}^{(2)} - \frac{1}{4} H_m^{(2)} - \frac{\pi}{2} \sum_{k=0}^{m} \frac{1}{(2k+1)2^{2k}} \binom{2k}{k},$$

and the solution to the point (iv) of the problem is complete.

Further, for the point (v) we'll exploit previous results, and then we write

$$\int_0^{\pi/2} \sin^{2m}(x) \tan(x) \log(1 - \cos(x)) dx$$

$$= \int_0^{\pi/2} \sin^{2m}(x) \tan(x) \log(1 - \cos^2(x)) dx - \int_0^{\pi/2} \sin^{2m}(x) \tan(x) \log(1 + \cos(x)) dx$$

$$= 2 \int_0^{\pi/2} \sin^{2m}(x) \tan(x) \log(\sin(x)) dx - \int_0^{\pi/2} \sin^{2m}(x) \tan(x) \log(1 + \cos(x)) dx$$

{make use of the results from (3.116) and the point (iii)}

$$= -\frac{\pi^2}{6} + \frac{1}{4} H_m^{(2)} + \frac{1}{2} \sum_{k=1}^{m} \frac{(2k-2)!!}{k \cdot (2k-1)!!} = -\frac{\pi^2}{6} + \frac{1}{4} H_m^{(2)} + \sum_{k=1}^{m} \frac{2^{2k-2}}{k^2 \binom{2k}{k}},$$

and the solution to the point (v) of the problem is complete.

Finally, for the point (vi) of the problem we proceed similarly, and then we write

$$\int_0^{\pi/2} \sin^{2m+1}(x) \tan(x) \log(1 - \cos(x)) dx$$

$$= \int_0^{\pi/2} \sin^{2m+1}(x) \tan(x) \log(1 - \cos^2(x)) dx - \int_0^{\pi/2} \sin^{2m+1}(x) \tan(x) \log(1 + \cos(x)) dx$$

$$= 2 \int_0^{\pi/2} \sin^{2m+1}(x) \tan(x) \log(\sin(x)) dx - \int_0^{\pi/2} \sin^{2m+1}(x) \tan(x) \log(1 + \cos(x)) dx$$

{make use of the results from (3.118) and the point (iv)}

$$= -\frac{3}{8}\pi^2 + H_{2m+1}^{(2)} - \frac{1}{4}H_m^{(2)} + \frac{\pi}{2} \sum_{k=0}^{m} \frac{(2k-1)!!}{(2k+1) \cdot (2k)!!}$$

$$= -\frac{3}{8}\pi^2 + H_{2m+1}^{(2)} - \frac{1}{4}H_m^{(2)} + \frac{\pi}{2} \sum_{k=0}^{m} \frac{1}{(2k+1)2^{2k}} \binom{2k}{k},$$

and the solution to the point (vi) of the problem is complete.

Good to observe that if we group the integrals from the last four points, such as in one group we have the integrals at the points (iii) and (v), and in another group the integrals at the points (iv) and (vi), and then exploit the facts that $\dfrac{1 - \cos(x)}{1 + \cos(x)} =$ $\tan^2\left(\dfrac{x}{2}\right)$ and $(1 - \cos(x))(1 + \cos(x)) = 1 - \cos^2(x) = \sin^2(x)$, we may extract these integrals from systems of relations. Well, not way different from the strategy above where the calculations get reduced to the same main integrals to evaluate.

This section emphasizes the importance of the strategy based on the use of the recurrence relations which, in this case, makes everything so easy.

The solutions also answer the *challenging question* which asks us to avoid the use of the Beta function.

3.18 Playing with a Hand of Fabulous Integrals: The Second Part

Solution Looking at the integrals in this section and comparing them with the ones from the previous section, there is a crystal-clear similarity in the numerators. Now, the reader familiar with series may already see how these integrals are connected to the ones from the previous section. *But how do we go further and complete the calculations?*, you might wonder.

So, let's start the *game* and calculate the integral at the point (i)! As a general rule, which is available for all points of the problem, we'll want to explore the results with integrals calculated in the previous section. A first step is to return to the integral result in (1.82), Sect. 1.17, the first equality, and multiplying its sides by y^{2m} and then considering the summation from $m = 1$ to ∞, we have

$$\sum_{m=1}^{\infty} y^{2m} \int_0^{\pi/2} \sin^{2m}(x) \tan(x) \log\left(\tan\left(\frac{x}{2}\right)\right) dx$$

{reverse the order of integration and summation}

$$\int_0^{\pi/2} \sum_{m=1}^{\infty} (y \sin(x))^{2m} \tan(x) \log\left(\tan\left(\frac{x}{2}\right)\right) dx$$

$$= y^2 \int_0^{\pi/2} \frac{\sin^2(x) \tan(x) \log\left(\tan\left(\frac{x}{2}\right)\right)}{1 - y^2 \sin^2(x)} dx$$

$$= \sum_{m=1}^{\infty} y^{2m} \left(-\frac{\pi^2}{8} + \frac{1}{2} \sum_{k=1}^{m} \frac{(2k-2)!!}{k \cdot (2k-1)!!}\right)$$

{make use of the result in (4.4), Sect. 4.2}

$$= \frac{1}{2} \frac{\arcsin^2(y)}{1 - y^2} - \frac{\pi^2}{8} \frac{y^2}{1 - y^2},$$

from which we obtain that

$$\int_0^{\pi/2} \frac{\sin^2(x) \tan(x) \log\left(\tan\left(\frac{x}{2}\right)\right)}{1 - y^2 \sin^2(x)} dx = \frac{1}{2} \frac{\arcsin^2(y)}{y^2(1 - y^2)} - \frac{\pi^2}{8} \frac{1}{1 - y^2},$$

and the solution to the point (i) of the problem is finalized.

Next, passing to the result in (1.83), Sect. 1.17, the first equality, multiplying both sides by y^{2m} and considering the summation from $m = 0$ to ∞, we get

$$\sum_{m=0}^{\infty} \int_0^{\pi/2} y^{2m} \sin^{2m+1}(x) \tan(x) \log\left(\tan\left(\frac{x}{2}\right)\right) dx$$

{reverse the order of integration and summation}

$$\int_0^{\pi/2} \sum_{m=0}^{\infty} (y \sin(x))^{2m} \sin(x) \tan(x) \log\left(\tan\left(\frac{x}{2}\right)\right) dx$$

$$= \int_0^{\pi/2} \frac{\sin(x)\tan(x)\log\left(\tan\left(\frac{x}{2}\right)\right)}{1 - y^2\sin^2(x)} dx$$

$$= \sum_{m=0}^{\infty} y^{2m}\left(-\frac{\pi^2}{4} + \frac{\pi}{2}\sum_{k=0}^{m}\frac{(2k-1)!!}{(2k+1)\cdot(2k)!!}\right)$$

{make use of the result in (4.5), Sect. 4.2}

$$= \frac{\pi}{2}\frac{\arcsin(y)}{y(1 - y^2)} - \frac{\pi^2}{4}\frac{1}{1 - y^2},$$

or to put it simpler

$$\int_0^{\pi/2} \frac{\sin(x)\tan(x)\log\left(\tan\left(\frac{x}{2}\right)\right)}{1 - y^2\sin^2(x)} dx = \frac{\pi}{2}\frac{\arcsin(y)}{y(1 - y^2)} - \frac{\pi^2}{4}\frac{1}{1 - y^2},$$

and the solution to the point (ii) of the problem is finalized.

For the next two points of the problem we also need to make use of a classical result, $\sum_{n=1}^{\infty} x^n H_n^{(m)} = \frac{\text{Li}_m(x)}{1 - x}$, $|x| < 1$, $m \geq 1$, $m \in \mathbb{N}$, which is elementarily shown in [76, Chapter 6, pp.348–349], and it is also proved in Sect. 6.6.

Further, multiplying both sides of the first equality in (1.84), Sect. 1.17, by y^{2m} and then considering the summation from $m = 1$ to ∞, we obtain

$$\sum_{m=1}^{\infty}\int_0^{\pi/2} y^{2m}\sin^{2m}(x)\tan(x)\log\left(1 + \cos(x)\right)dx$$

{reverse the order of integration and summation}

$$= \int_0^{\pi/2}\sum_{m=1}^{\infty}(y\sin(x))^{2m}\tan(x)\log\left(1 + \cos(x)\right)dx$$

$$= y^2\int_0^{\pi/2}\frac{\sin^2(x)\tan(x)\log\left(1 + \cos(x)\right)}{1 - y^2\sin^2(x)}dx$$

$$= \sum_{m=1}^{\infty} y^{2m}\left(\frac{\pi^2}{12} + \frac{1}{4}H_m^{(2)} - \frac{1}{2}\sum_{k=1}^{m}\frac{(2k-2)!!}{k\cdot(2k-1)!!}\right)$$

{rearrange the series and split it}

$$= \sum_{m=1}^{\infty} y^{2m} \left(\frac{\pi^2}{8} - \frac{1}{2} \sum_{k=1}^{m} \frac{(2k-2)!!}{k \cdot (2k-1)!!} \right) - \frac{\pi^2}{24} \sum_{m=1}^{\infty} y^{2m} + \frac{1}{4} \sum_{m=1}^{\infty} y^{2m} H_m^{(2)}$$

{make use of the result in (4.4), Sect. 4.2}

$$= \frac{\pi^2}{12} \frac{y^2}{1-y^2} - \frac{1}{2} \frac{\arcsin^2(y)}{1-y^2} + \frac{1}{4} \frac{\mathrm{Li}_2(y^2)}{1-y^2},$$

and from here we get

$$\int_0^{\pi/2} \frac{\sin^2(x) \tan(x) \log (1 + \cos(x))}{1 - y^2 \sin^2(x)} dx = \frac{\pi^2}{12} \frac{1}{1-y^2} - \frac{1}{2} \frac{\arcsin^2(y)}{y^2(1-y^2)} + \frac{1}{4} \frac{\mathrm{Li}_2(y^2)}{y^2(1-y^2)},$$

and the solution to the point (iii) of the problem is finalized.

For the point (iv) of the problem, we multiply both sides of the result in (1.85), Sect. 1.17, the first equality, by y^{2m} and then consider the summation from $m = 0$ to ∞, which gives

$$\sum_{m=0}^{\infty} \int_0^{\pi/2} y^{2m} \sin^{2m+1}(x) \tan(x) \log (1 + \cos(x)) dx$$

$$= \int_0^{\pi/2} \sum_{m=0}^{\infty} (y \sin(x))^{2m} \sin(x) \tan(x) \log (1 + \cos(x)) dx$$

$$= \int_0^{\pi/2} \frac{\sin(x) \tan(x) \log(1 + \cos(x))}{1 - y^2 \sin^2(x)} dx$$

$$= \sum_{m=0}^{\infty} y^{2m} \left(\frac{\pi^2}{8} + H_{2m+1}^{(2)} - \frac{1}{4} H_m^{(2)} - \frac{\pi}{2} \sum_{k=0}^{m} \frac{(2k-1)!!}{(2k+1) \cdot (2k)!!} \right)$$

{rearrange the series and split it}

$$= \frac{\pi}{2} \sum_{m=0}^{\infty} y^{2m} \left(\frac{\pi}{2} - \sum_{k=0}^{m} \frac{(2k-1)!!}{(2k+1) \cdot (2k)!!} \right) - \frac{\pi^2}{8} \sum_{m=0}^{\infty} y^{2m} - \frac{1}{4} \sum_{m=0}^{\infty} y^{2m} H_m^{(2)}$$

$$+ \sum_{m=0}^{\infty} y^{2m} H_{2m}^{(2)} + \frac{1}{y} \sum_{m=0}^{\infty} \frac{y^{2m+1}}{(2m+1)^2}$$

{make use of the result in (4.5), Sect. 4.2}

$$= \frac{\pi^2}{8} \frac{1}{1-y^2} - \frac{\pi}{2} \frac{\arcsin(y)}{y(1-y^2)} + \frac{1}{2} \frac{y\mathrm{Li}_2(y)}{1-y^2} - \frac{1}{2} \frac{y\mathrm{Li}_2(-y)}{1-y^2} + \frac{1}{2} \frac{\mathrm{Li}_2(y)}{y} - \frac{1}{2} \frac{\mathrm{Li}_2(-y)}{y},$$

and the solution to the point (iv) of the problem is finalized. In the calculations I also used that $\displaystyle\sum_{m=1}^{\infty} y^{2m} H_{2m}^{(2)} = \frac{1}{2}\left(\frac{\mathrm{Li}_2(y)}{1-y} + \frac{\mathrm{Li}_2(-y)}{1+y}\right)$, $|y| < 1$, then

$$\sum_{m=0}^{\infty} \frac{y^{2m+1}}{(2m+1)^2} = \sum_{m=1}^{\infty} \frac{y^m}{m^2} - \sum_{m=1}^{\infty} \frac{y^{2m}}{(2m)^2} = \sum_{m=1}^{\infty} \frac{y^m}{m^2} - \frac{1}{2}\left(\sum_{m=1}^{\infty} \frac{y^m}{m^2} - \sum_{m=1}^{\infty} (-1)^{m-1} \frac{y^m}{m^2}\right)$$

$$= \frac{1}{2}\sum_{m=1}^{\infty} \frac{y^m}{m^2} - \frac{1}{2}\sum_{m=1}^{\infty} \frac{(-y)^m}{m^2} = \frac{1}{2}(\mathrm{Li}_2(y) - \mathrm{Li}_2(-y)), \text{ and finally the dilogarithmic}$$

identity, $\mathrm{Li}_2(y) + \mathrm{Li}_2(-y) = \frac{1}{2}\mathrm{Li}_2(y^2)$. The recommendation to *rearrange the series and split it* found in the solution steps can be skipped, if one wants to, and then we simply expand the summand and calculate each resulting series (which is applicable to the previous solution, too).

Non-trigonometric forms of the integrals at the points (iii) and (iv)

In the following, I'll make a variable change of the type $\cos(x) = t$ and little rearrangements that together lead to

$$\int_0^1 \frac{(1-t^2)\log(1+t)}{t(1-y^2t^2)}dt = \frac{\pi^2}{12} - \frac{1-y^2}{2y^2}\operatorname{arctanh}^2(y) - \frac{1-y^2}{4y^2}\mathrm{Li}_2\left(\frac{y^2}{y^2-1}\right),$$

$$\tag{3.119}$$

for the integral at the point (iii).

Then, proceeding similarly as before for the integral at the point (iv), we have

$$\int_0^1 \frac{\sqrt{1-t^2}\log(1+t)}{t(1-y^2t^2)}dt$$

$$= \frac{\pi^2}{8} - \frac{\pi}{2}\frac{\sqrt{1-y^2}}{y}\operatorname{arcsinh}\left(\frac{y}{\sqrt{1-y^2}}\right) + \frac{\sqrt{1-y^2}}{y}\mathrm{Ti}_2\left(\frac{y}{\sqrt{1-y^2}}\right). \quad (3.120)$$

So, let's go further and finalize the last two points of the problem, where we'll make use of the results from the previous points. Also, remember the simple trigonometric identity, $\dfrac{1-\cos(x)}{1+\cos(x)} = \tan^2\left(\dfrac{x}{2}\right)$, which is going to be useful in our strategy. Thus, for the next point of the problem, we write

$$\int_0^{\pi/2} \frac{\sin^2(x)\tan(x)\log(1-\cos(x))}{1-y^2\sin^2(x)}dx$$

$$= \int_0^{\pi/2} \frac{\sin^2(x)\tan(x)\log\left(\frac{1-\cos(x)}{1+\cos(x)}\right)}{1-y^2\sin^2(x)}dx + \int_0^{\pi/2} \frac{\sin^2(x)\tan(x)\log(1+\cos(x))}{1-y^2\sin^2(x)}dx$$

$$= 2\int_0^{\pi/2} \frac{\sin^2(x)\tan(x)\log\left(\tan\left(\frac{x}{2}\right)\right)}{1-y^2\sin^2(x)}dx + \int_0^{\pi/2} \frac{\sin^2(x)\tan(x)\log(1+\cos(x))}{1-y^2\sin^2(x)}dx$$

{make use of the results from (i) and (iii)}

$$= \frac{1}{2}\frac{\arcsin^2(y)}{y^2(1-y^2)} + \frac{1}{4}\frac{\text{Li}_2(y^2)}{y^2(1-y^2)} - \frac{\pi^2}{6}\frac{1}{1-y^2},$$

and the solution to the point (v) of the problem is finalized.

Finally, for the last point of the problem, we have

$$\int_0^{\pi/2} \frac{\sin(x)\tan(x)\log(1-\cos(x))}{1-y^2\sin^2(x)}dx$$

$$= \int_0^{\pi/2} \frac{\sin(x)\tan(x)\log\left(\frac{1-\cos(x)}{1+\cos(x)}\right)}{1-y^2\sin^2(x)}dx + \int_0^{\pi/2} \frac{\sin(x)\tan(x)\log(1+\cos(x))}{1-y^2\sin^2(x)}dx$$

$$= 2\int_0^{\pi/2} \frac{\sin(x)\tan(x)\log\left(\tan\left(\frac{x}{2}\right)\right)}{1-y^2\sin^2(x)}dx + \int_0^{\pi/2} \frac{\sin(x)\tan(x)\log(1+\cos(x))}{1-y^2\sin^2(x)}dx$$

{make use of the results from (ii) and (iv)}

$$= \frac{\pi}{2}\frac{\arcsin(y)}{y(1-y^2)} + \frac{1}{2}\frac{y\text{Li}_2(y)}{1-y^2} - \frac{1}{2}\frac{y\text{Li}_2(-y)}{1-y^2} - \frac{3}{8}\pi^2\frac{1}{1-y^2} + \frac{1}{2}\frac{\text{Li}_2(y)}{y} - \frac{1}{2}\frac{\text{Li}_2(-y)}{y},$$

and the solution to the point (vi) of the problem is finalized.

Non-trigonometric forms of the integrals at the points (v) *and* (vi)

Again, with the variable change $\cos(x) = t$ and some rearrangements, we have

$$\int_0^1 \frac{(1-t^2)\log(1-t)}{t(1-y^2t^2)}dt$$

$$= \frac{1-y^2}{2y^2}\text{arcsinh}^2\left(\frac{y}{\sqrt{1-y^2}}\right) - \frac{1-y^2}{4y^2}\text{Li}_2\left(\frac{y^2}{y^2-1}\right) - \frac{\pi^2}{6}, \qquad (3.121)$$

for the integral at the point (v).

Then, as regards the integral at the point (vi), we may obtain the form

$$\int_0^1 \frac{\sqrt{1-t^2}\log(1-t)}{t(1-y^2t^2)}dt$$

$$= \frac{\pi}{2}\frac{\sqrt{1-y^2}}{y}\operatorname{arcsinh}\left(\frac{y}{\sqrt{1-y^2}}\right) + \frac{\sqrt{1-y^2}}{y}\operatorname{Ti}_2\left(\frac{y}{\sqrt{1-y^2}}\right) - \frac{3}{8}\pi^2.$$

$$\tag{3.122}$$

Perhaps some of you might wonder why the calculations of the resulting series aren't continued in this section. Well, these main resulting series I consider valuable to be treated separately, in more ways, where you can pay a careful attention to them, which will also be a good opportunity for practice, and therefore you may find them with full solutions in the sixth chapter.

3.19   Another Fabulous Integral Related to a Curious Binoharmonic Series Representation of $\zeta(3)$

Solution Did you manage to go through Sect. 3.17? I'll *borrow* the recurrence relation idea I used in the mentioned section and start by employing a similar strategy.

After completing the solution to the point (i) of the problem, I'll make use of this result in order to get solutions to the points (ii) and (iii), where for the latter point I'll also want to employ some specific *tools* particularly useful in the work with the series manipulations and related resulting limits.

To proceed with the calculations, we assume that $n \geq 2$, and exploiting the fact that $\sin^2(x) + \cos^2(x) = 1$, we write

$$I_n = \int_0^{\pi/2} x\sin^n(x)\sec(x)\log(\sin(x))dx$$

$$= \int_0^{\pi/2} x(1 - \cos^2(x))\sin^{n-2}(x)\sec(x)\log(\sin(x))dx$$

{expand the integral, and for the first resulting integral use the notation I_{n-2}}

$$= I_{n-2} - \int_0^{\pi/2} x\cos(x)\sin^{n-2}(x)\log(\sin(x))dx,$$

whence we obtain that

$$I_n - I_{n-2}$$

$$= -\int_0^{\pi/2} x \cos(x) \sin^{n-2}(x) \log(\sin(x))dx = \frac{1}{1-n} \int_0^{\pi/2} (\sin^{n-1}(x))' x \log(\sin(x))dx$$

{integrate by parts and expand the integral}

$$= \frac{1}{n-1} \int_0^{\pi/2} x \cos(x) \sin^{n-2}(x)dx + \frac{1}{n-1} \int_0^{\pi/2} \log(\sin(x)) \sin^{n-1}(x)dx$$

{use integration by parts for the first integral, and observe that the second}

{integral may be expressed by using the derivative of a Wallis-like integral}

$$= \frac{\pi}{2(n-1)^2} - \frac{1}{(n-1)^2} \int_0^{\pi/2} \sin^{n-1}(x)dx + \frac{1}{n-1} \frac{\partial}{\partial s} \left(\int_0^{\pi/2} \sin^{s-1}(x)dx \right) \Bigg|_{s=n}.$$
$$(3.123)$$

Now, if we set $n = 2m + 1$ in (3.123), then the first integral in the right-hand side turns into the classical Wallis' integral:

$$\int_0^{\pi/2} \sin^{2m}(x)dx = \frac{\pi}{2^{2m+1}} \binom{2m}{m}. \tag{3.124}$$

As regards the derivative with respect to s of the second integral in the right-hand side of (3.123), we consider the following trigonometric form of the Beta function:

$$\int_0^{\pi/2} \sin^x(t) \cos^y(t)dt = \frac{1}{2} B\left(\frac{1}{2}(x+1), \frac{1}{2}(y+1)\right) = \frac{1}{2} \frac{\Gamma\left(\frac{1}{2}(x+1)\right)\Gamma\left(\frac{1}{2}(y+1)\right)}{\Gamma\left(1+\frac{1}{2}(x+y)\right)},$$
$$(3.125)$$

where if we set $y = 0$, then differentiate once with respect to x, and replace x by $2m$, we obtain

$$\int_0^{\pi/2} \log(\sin(x)) \sin^{2m}(x)dx = \frac{\sqrt{\pi}}{4} \frac{\Gamma\left(m+\frac{1}{2}\right)}{\Gamma(m+1)} \left(\psi\left(m+\frac{1}{2}\right) - \psi(m+1)\right)$$

$$= \frac{\pi}{2} \frac{1}{2^{2m}} \binom{2m}{m} (H_{2m} - H_m - \log(2)), \tag{3.126}$$

where in the calculations I used the Legendre duplication formula,

$$\Gamma(2x) = \frac{1}{\sqrt{\pi}} 2^{2x-1} \Gamma(x)\Gamma\left(x+\frac{1}{2}\right), \tag{3.127}$$

together with the fact that

$$\psi\left(m+\frac{1}{2}\right) - \psi(m+1) = 2(H_{2m} - H_m - \log(2)). \tag{3.128}$$

The result from (3.128) comes immediately by combining the fact that

$$\psi(2x) = \frac{1}{2}\psi(x) + \frac{1}{2}\psi\left(x+\frac{1}{2}\right) + \log(2), \tag{3.129}$$

and the relation between the harmonic number and the Digamma function, that is, $H_n = \psi(n+1) + \gamma$. The identity in (3.129) may be extracted immediately from the Legendre duplication formula in (3.127), if we take the log of both sides and then differentiate once. For a different solution to the formula in (3.129), see [76, Chapter 3, pp.68–69].

Returning in (3.123) with the results from (3.124) and (3.126), we obtain

$$I_{2m+1} - I_{2m-1} = \frac{\pi}{8}\frac{1}{m^2} + \frac{\pi}{4}\frac{1}{m2^{2m}}\binom{2m}{m}\left(H_{2m} - H_m - \frac{1}{2m} - \log(2)\right). \tag{3.130}$$

If we replace m by k in (3.130) and consider the summation from $k = 1$ to m, we have that

$$\sum_{k=1}^{m}(I_{2k+1} - I_{2k-1}) = I_{2m+1} - I_1$$

$$= \frac{\pi}{8}\sum_{k=1}^{m}\frac{1}{k^2} + \frac{\pi}{4}\sum_{k=1}^{m}\frac{1}{k2^{2k}}\binom{2k}{k}\left(H_{2k} - H_k - \frac{1}{2k} - \log(2)\right),$$

and if we take into account that $I_1 = \log^2(2)\dfrac{\pi}{4} - \dfrac{\pi^3}{48}$ and $H_m^{(2)} = \sum_{k=1}^{m}\dfrac{1}{k^2}$, we get

$$I_{2m+1} = \frac{\pi}{4}\sum_{k=1}^{m}\frac{1}{k2^{2k}}\binom{2k}{k}\left(H_{2k} - H_k - \frac{1}{2k} - \log(2)\right) + \frac{\pi}{8}H_m^{(2)} + \log^2(2)\frac{\pi}{4} - \frac{\pi^3}{48},$$

which is the desired result to obtain at the point (i).

Could we have some more details on the integral $I_1 = \displaystyle\int_0^{\pi/2} x\tan(x)\log(\sin(x))$ dx?, *which is one of the reactions I would expect from you.* This integral can be approached in more ways.

For example, we may recall and successfully use the following powerful Fourier series presented in *(Almost) Impossible Integrals, Sums, and Series*:

$$\sum_{n=1}^{\infty} \left(\psi \left(\frac{n+1}{2} \right) - \psi \left(\frac{n}{2} \right) - \frac{1}{n} \right) \sin(2nx) = -\tan(x) \log(\sin(x)), \ 0 < x < \frac{\pi}{2},$$

$$(3.131)$$

proceeding as in the solutions presented in [76, Chapter 3, pp.242–252]. The Fourier series in (3.131) is also discussed in Sect. 6.49.

For another solution to the integral I_1, we start with a clever integration by parts (which might not be easy to observe at first sight), and then we have that

$$I_1 = \int_0^{\pi/2} x \tan(x) \log(\sin(x))dx = \frac{1}{4} \int_0^{\pi/2} x \left(\mathrm{Li}_2(\cos^2(x)) \right)' dx$$

$$= \underbrace{\frac{1}{4} x \mathrm{Li}_2(\cos^2(x)) \Big|_{x=0}^{x=\pi/2}}_{0} - \frac{1}{4} \int_0^{\pi/2} \mathrm{Li}_2(\cos^2(x))dx = -\frac{1}{4} \int_0^{\pi/2} \mathrm{Li}_2(\cos^2(x))dx.$$

$$(3.132)$$

Now, for the remaining integral in (3.132), if we consider the dilogarithm function reflection formula, $\mathrm{Li}_2(x) + \mathrm{Li}_2(1 - x) = \pi^2/6 - \log(x)\log(1 - x)$, where we replace x by $\cos^2(x)$, then integrate both sides from $x = 0$ to $x = \pi/2$, and use the fact that $\int_0^{\pi/2} \mathrm{Li}_2(\sin^2(x))dx \overset{\pi/2-x=y}{=} \int_0^{\pi/2} \mathrm{Li}_2(\cos^2(y))dy$, we arrive at

$$\int_0^{\pi/2} \mathrm{Li}_2(\cos^2(x))dx = \frac{\pi^3}{24} - 2\int_0^{\pi/2} \log(\sin(x)) \log(\cos(x))dx$$

$$\{\text{exploit the result in (3.125)}\}$$

$$= \frac{\pi^3}{24} - \lim_{\substack{x \to 0 \\ y \to 0}} \frac{\partial^2}{\partial x \partial y} \mathrm{B} \left(\frac{1}{2}(x + 1), \frac{1}{2}(y + 1) \right) = \frac{\pi^3}{12} - \log^2(2)\pi, \qquad (3.133)$$

where the final result may be obtained either by using *Mathematica* or manually, by simple manipulations and results with the Digamma function $\Big($e.g., use that

$\psi \left(\frac{1}{2} \right) = -\gamma - 2\log(2)$, which is derived in Sect. 6.19$\Big)$.

Finally, upon plugging (3.133) in (3.132), we get the value of I_1,

$$I_1 = \int_0^{\pi/2} x \tan(x) \log(\sin(x))dx = \log^2(2)\frac{\pi}{4} - \frac{\pi^3}{48}, \qquad (3.134)$$

which puts an end to the additional explanations to the point (i) of the problem.

For the points (ii) and (iii) of the problem, we want to start with looking at the behavior of the integral at the point (i), when m tends to ∞. We note that $g(x) =$

$x \sec(x) \log(\sin(x))$ is a well-behaved function when x in $(0, \pi/2)$. Usually, our main concern would be to see what happens when x approaches to 0 and $\pi/2$, and in our cases[8] $g(x)$ approaches 0. Then, we have that

$$\left| \underbrace{\int_0^{\pi/2} x \sin^{2m+1}(x) \sec(x) \log(\sin(x)) dx}_{I_{2m+1}} \right| \leq \int_0^{\pi/2} \underbrace{| x \sec(x) \log(\sin(x)) |}_{g(x)} || \sin^{2m+1}(x)| dx$$

$$\leq M \int_0^{\pi/2} \sin^{2m+1}(x) dx = M \frac{2^{2m}}{(2m+1) \binom{2m}{m}}, \qquad (3.135)$$

where M is the maximum of $|g(x)|_{0 < x < \pi/2}$.

If we let $m \to \infty$ in (3.135) and consider the asymptotic expansion behavior of the central binomial coefficient, $\binom{2n}{n} \approx \frac{4^n}{\sqrt{\pi n}}$, as $n \to \infty$, which is obtained via Stirling's formula, $n! \approx \sqrt{2\pi n} \left(\frac{n}{e} \right)^n$ (see [70, Chapter 1, p.8], [97]), we get that

$$\lim_{m \to \infty} I_{2m+1} = 0. \qquad (3.136)$$

As regards the previous calculations, a nice exercise for the curious reader is to also approach differently a limit of the type, $\lim_{n \to \infty} \int_0^{\pi/2} \sin^n(x) dx$.

So, by letting $m \to \infty$ in the main result from the previous point, considering (3.135), then using the limit $\lim_{m \to \infty} \left(\frac{\pi^2}{6} - H_m^{(2)} \right) = 0$, and finally replacing the summation variable letter k by n, we arrive at

$$\log^2(2) = \sum_{n=1}^{\infty} \frac{1}{n 2^{2n}} \binom{2n}{n} \left(H_n - H_{2n} + \frac{1}{2n} + \log(2) \right),$$

and the point (ii) of the problem is finalized.

---

[8] These details are easily clarified if we compute two elementary limits. So, $\lim_{x \to 0+} g(x) = \lim_{x \to 0+} x \sec(x) \log(\sin(x)) = \lim_{x \to 0+} x \csc(x) \sec(x) \sin(x) \log(\sin(x)) = \lim_{x \to 0+} x \csc(x) \cdot \lim_{x \to 0+} \sec(x) \cdot \lim_{x \to 0+} \sin(x) \log(\sin(x)) = 1 \cdot 1 \cdot 0 = 0$, where I exploited the classical elementary limits, $\lim_{x \to 0} x \csc(x) = 1$ and $\lim_{x \to 0+} x \log(x) = 0$. For the other main limit, we have $\lim_{x \to \pi/2} g(x) = \lim_{x \to \pi/2} x \sec(x) \log(\sin(x)) = \lim_{x \to 0} (\pi/2 - x) \csc(x) \log(\cos(x)) = 1/2 \lim_{x \to 0} (\pi/2 - x) \sin(x) \csc^2(x) \log(1 - \sin^2(x)) = 1/2 \lim_{x \to 0} (\pi/2 - x) \sin(x) \cdot \lim_{x \to 0} \csc^2(x) \log(1 - \sin^2(x)) = 0 \cdot (-1) = 0$, where in the calculations I used the elementary limit, $\lim_{x \to 0} \log(1 - x)/x = -1$.

To go further, another *tool* we need in our subsequent calculations is *Stolz–Cesàro theorem* (see [15, Appendix, pp.263–266]), and we need the variant $(0/0)$, which says that if we have two sequences of real numbers, $(a_n)_{n\geq 1}$ and $(b_n)_{n\geq 1}$, such that $\lim_{n\to\infty} a_n = \lim_{n\to\infty} b_n = 0$, b_n is strictly decreasing, and $\lim_{n\to\infty} \dfrac{a_{n+1} - a_n}{b_{n+1} - b_n} = l$, then we have that

$$\lim_{n\to\infty} \frac{a_n}{b_n} = \lim_{n\to\infty} \frac{a_{n+1} - a_n}{b_{n+1} - b_n} = l. \tag{3.137}$$

Now, it's time to return to the main calculations of the last point of the problem! If we replace m by $n - 1$ in (1.94), multiply both sides by $1/n$, and consider the summation from $n = 1$ to ∞, we get that

$$\frac{\pi}{4} \sum_{n=1}^{\infty} \underbrace{\frac{1}{n}}_{a_n} \underbrace{\left(\sum_{k=1}^{n-1} \frac{1}{k 2^{2k}} \binom{2k}{k} \left(H_{2k} - H_k - \frac{1}{2k} - \log(2) \right) + \frac{1}{2} H_{n-1}^{(2)} + \log^2(2) - \frac{\pi^2}{12} \right)}_{b_n},$$

$$\tag{3.138}$$

and if we apply Abel's summation, the series version in (6.7), Sect. 6.2, in (3.138), we arrive at

$$-\frac{\pi}{4} \sum_{n=1}^{\infty} \frac{H_n}{n} \left(\frac{1}{2^{2n}} \binom{2n}{n} \right) \left(H_{2n} - H_n - \frac{1}{2n} - \log(2) \right) + \frac{1}{2n} \right)$$

$$= \int_0^{\pi/2} x \sec(x) \log(\sin(x)) \sum_{n=1}^{\infty} \frac{\sin^{2n-1}(x)}{n} dx = -2 \underbrace{\int_0^{\pi/2} x \frac{\log(\sin(x)) \log(\cos(x))}{\sin(x) \cos(x)} dx}_{J}.$$

$$\tag{3.139}$$

Regarding the remaining integral in (3.139), we let the variable change $\pi/2 - x = y$ and then return to the notation in x, and we write

$$J = \int_0^{\pi/2} x \frac{\log(\sin(x)) \log(\cos(x))}{\sin(x) \cos(x)} dx = \int_0^{\pi/2} \left(\frac{\pi}{2} - x \right) \frac{\log(\sin(x)) \log(\cos(x))}{\sin(x) \cos(x)} dx$$

$$= \frac{\pi}{2} \int_0^{\pi/2} \frac{\log(\sin(x)) \log(\cos(x))}{\sin(x) \cos(x)} dx - J,$$

from which we obtain that

$$J = \frac{\pi}{4} \int_0^{\pi/2} \frac{\log(\sin(x)) \log(\cos(x))}{\sin(x) \cos(x)} dx$$

$$= \frac{\pi}{4} \int_0^{\pi/2} \log(\sin(x)) \log(\cos(x))(\tan(x) + \cot(x)) dx$$

$$= \frac{\pi}{4} \int_0^{\pi/2} \tan(x) \log(\sin(x)) \log(\cos(x)) dx + \frac{\pi}{4} \int_0^{\pi/2} \cot(x) \log(\sin(x)) \log(\cos(x)) dx$$

{in the second integral, make the variable change $\pi/2 - x = y$ and}

{afterwards return to the notation in x and add it to the first integral}

$$= \frac{\pi}{2} \int_0^{\pi/2} \tan(x) \log(\sin(x)) \log(\cos(x)) dx = \frac{\pi}{8} \int_0^{\pi/2} (\text{Li}_2(\cos^2(x)))' \log(\cos(x)) dx$$

$$\overset{\substack{\text{integrate by}\\ \text{parts}}}{=} \frac{\pi}{8} \int_0^{\pi/2} \text{Li}_2(\cos^2(x)) \tan(x) dx = -\frac{\pi}{16} \text{Li}_3(\cos^2(x)) \Big|_{x=0}^{x=\pi/2} = \frac{\pi}{16} \zeta(3).$$

$$(3.140)$$

Finally, if we combine (3.139) and (3.140), we conclude that

$$\zeta(3) = 2 \sum_{n=1}^{\infty} \frac{H_n}{n} \left(\frac{1}{2^{2n}} \binom{2n}{n} \right) \left(H_{2n} - H_n - \frac{1}{2n} - \log(2) \right) + \frac{1}{2n} \right),$$

and the point (iii) of the problem is finalized.

As regards the limit that appears during the application of Abel's summation in (3.138), it's easy to see why it vanishes. So, if we consider the limit

$$\lim_{N \to \infty} \frac{1}{1/H_N} \left(\sum_{k=1}^{N} \frac{1}{k2^{2k}} \binom{2k}{k} \left(H_{2k} - H_k - \frac{1}{2k} - \log(2) \right) + \frac{1}{2} H_N^{(2)} + \log^2(2) - \frac{\pi^2}{12} \right),$$

and apply the *Stolz–Cesàro theorem* in (3.137), which we combine with the asymptotic expansion behavior of the central binomial coefficient, $\binom{2n}{n} \approx \frac{4^n}{\sqrt{\pi n}}$, as $n \to \infty$, and the fact that $H_n = \gamma + \log(n) + O\left(\frac{1}{n}\right)$, as $n \to \infty$, we immediately get that the limit is 0.

We prepare now to enter the next section that is related in some ways to the binoharmonic series result from the point (iii)!

3.20 Pairs of Appealing Integrals with the Logarithm Function, the Dilogarithm function, and Trigonometric Functions

Solution As I mentioned in the very last part of the previous section, the main results in this section are related to the binoharmonic series representation of $\zeta(3)$ from the preceding section. A first fact to observe for each of the two points of the problem, when inspecting the integrands, is that knowing the value of any of the two integrals, we easily derive the other one based on the simple substitution $\pi/2 - x = y$.

Now, my proposal with the binoharmonic series representation of $\zeta(3)$, calculated in the previous section, was also considered in [23] where, in another solution, an integral like the ones at the point (i) arose, which in this section I'll consider to calculate in two different ways.

In order to get a first solution, we start with recalling and using the Landen's dilogarithmic identity (see [21, Chapter 1, p.5], [70, Chapter 2, p.107]), $\text{Li}_2(x) + \text{Li}_2(x/(x-1)) = -1/2 \log^2(1-x)$, where if we replace x by $\sin^2(x)$, we get $\text{Li}_2(\sin^2(x)) = -2\log^2(\cos(x)) - \text{Li}_2(-\tan^2(x))$. Then, by exploiting this last relation in the main sine integral at the point (i), we get

$$I = \int_0^{\pi/2} \log(\cos(x))\text{Li}_2(\cos^2(x))dx \overset{\pi/2-x=y}{=} \int_0^{\pi/2} \log(\sin(y))\text{Li}_2(\sin^2(y))dy$$

$$\overset{y=x}{=} \underbrace{-2\int_0^{\pi/2} \log(\sin(x))\log^2(\cos(x))dx}_{U} - \underbrace{\int_0^{\pi/2} \log(\sin(x))\text{Li}_2(-\tan^2(x))dx}_{V}.$$

$$(3.141)$$

The integral U from (3.141) is immediately extracted by considering the limits with the derivatives of the trigonometric form of the Beta function found in the previous section, more precisely in (3.125):

$$U = \int_0^{\pi/2} \log(\sin(x))\log^2(\cos(x))dx = \frac{1}{2}\lim_{\substack{x\to 0 \\ y\to 0}} \frac{\partial^3}{\partial x\partial y^2} B\left(\frac{1}{2}(x+1), \frac{1}{2}(y+1)\right)$$

$$= \frac{\pi}{8}\zeta(3) - \frac{1}{2}\log^3(2)\pi, \qquad (3.142)$$

and the calculation of the limit can be done either with *Mathematica* or manually. It is also easy to extract this information based on (4.172), Sect. 4.51.

As regards the integral V in (3.141), we need the following integral representation of the dilogarithm, $-\int_0^1 \dfrac{x\log(y)}{1-xy}dy = \text{Li}_2(x)$, found stated as a generalization in (3.13), Sect. 3.3, where if we replace x by $-\tan^2(x)$, we get

$$\mathrm{Li}_2(-\tan^2(x)) = \int_0^1 \frac{\tan^2(x)\log(y)}{1+\tan^2(x)y}\mathrm{d}y. \qquad (3.143)$$

Returning to the integral V in (3.141) and using the result in (3.143), we have

$$V = \int_0^{\pi/2} \log(\sin(x))\mathrm{Li}_2(-\tan^2(x))\mathrm{d}x$$

$$= \int_0^{\pi/2} \log(\sin(x))\left(\int_0^1 \frac{\tan^2(x)\log(y)}{1+\tan^2(x)y}\mathrm{d}y\right)\mathrm{d}x$$

$$\overset{\substack{\text{reverse the order}\\\text{of integration}}}{=} \int_0^1 \log(y)\left(\int_0^{\pi/2} \frac{\tan^2(x)\log(\sin(x))}{1+y\tan^2(x)}\mathrm{d}x\right)\mathrm{d}y$$

{make the variable change $\tan(x) = t$ and expand the inner integral}

$$= \int_0^1 \frac{\log(y)}{1-y}\left(\frac{1}{2}\int_0^\infty \frac{\log(1+t^2)}{1+t^2}\mathrm{d}t - \int_0^\infty \frac{\log(t)}{1+t^2}\mathrm{d}t + \int_0^\infty \frac{\log(t)}{1+yt^2}\mathrm{d}t\right.$$

$$\left. - \frac{1}{2}\int_0^\infty \frac{\log(1+t^2)}{1+yt^2}\mathrm{d}t\right)\mathrm{d}y$$

$$= \frac{\pi}{2}\int_0^1 \frac{\log(y)}{1-y}\left(\log(2) - \frac{\log(1+\sqrt{y})}{\sqrt{y}}\right)\mathrm{d}y$$

{make the variable change $y = u^2$, expand, and rearrange}

$$= -\log(2)\pi\int_0^1 \frac{\log(u)}{1+u}\mathrm{d}u - \pi\int_0^1 \frac{\log(u)\log\left(\frac{1+u}{2}\right)}{1-u}\mathrm{d}u - \pi\int_0^1 \frac{\log(u)\log(1+u)}{1+u}\mathrm{d}u$$

$$= \frac{1}{6}\log(2)\pi^3 - \frac{7}{8}\pi\zeta(3). \qquad (3.144)$$

In the following, I'll provide more details on the resulting integrals that appeared in (3.144). So, we have that $\int_0^\infty \frac{\log(1+t^2)}{1+t^2}\mathrm{d}t \overset{t=\cot(x)}{=} -2\int_0^{\pi/2} \log(\sin(x))\mathrm{d}x = \log(2)\pi$, where the last sine integral[9] is well-known. Then, with the variable change

---

[9] Since $\int_0^{\pi/2} \log(\sin(x))\mathrm{d}x = \int_0^{\pi/2} \log(\cos(x))\mathrm{d}x$, and using the symmetry of $\log(\sin(x))$ when we have $x \in (0, \pi)$, we have that $\int_0^{\pi/2} \log(\sin(x))\mathrm{d}x = \frac{1}{2}\int_0^\pi \log(\sin(x))\mathrm{d}x \overset{x=2y}{=}$ $\int_0^{\pi/2} \log(\sin(2y))\mathrm{d}y = \log(2)\int_0^{\pi/2} \mathrm{d}y + \int_0^{\pi/2} \log(\sin(y))\mathrm{d}y + \int_0^{\pi/2} \log(\cos(y))\mathrm{d}y =$ $\frac{\pi}{2}\log(2) + 2\int_0^{\pi/2} \log(\sin(x))\mathrm{d}x$, from which we obtain that $\int_0^{\pi/2} \log(\sin(x))\mathrm{d}x = -\frac{\pi}{2}\log(2)$.

$t = 1/x$, we get that $\displaystyle\int_0^\infty \frac{\log(t)}{1+t^2}\,dt = 0$, and here I also used the fact that

the integral exists.[10] Next, we have $\displaystyle\int_0^\infty \frac{\log(t)}{1+yt^2}\,dt \overset{\sqrt{y}t=u}{=} \frac{1}{\sqrt{y}}\int_0^\infty \frac{\log(u)}{1+u^2}\,du -$

$\dfrac{\log(y)}{2\sqrt{y}}\displaystyle\int_0^\infty \frac{1}{1+u^2}\,du = -\frac{\pi}{4}\frac{\log(y)}{\sqrt{y}}$. Further, we obtain that $\displaystyle\int_0^\infty \frac{\log(1+t^2)}{1+yt^2}\,dt =$

$\dfrac{\pi}{\sqrt{y}}\log\left(1+\dfrac{1}{\sqrt{y}}\right)$, where in the calculations I used the more (well-known) general

result

$$\int_0^\infty \frac{\log(1+a^2x^2)}{1+b^2x^2}\,dx = \frac{\pi}{b}\log\left(1+\frac{a}{b}\right),\quad a,b>0. \tag{3.145}$$

Proof To briefly prove the result in (3.145), we start with the integral representation

$\log(1+a^2x^2) = \displaystyle\int_0^a \frac{2x^2y}{1+x^2y^2}\,dy$, and then we write

$$\int_0^\infty \frac{\log(1+a^2x^2)}{1+b^2x^2}\,dx = \int_0^\infty \frac{1}{1+b^2x^2}\left(\int_0^a \frac{2x^2y}{1+x^2y^2}\,dy\right)dx$$

{reverse the order of integration and expand the inner integral}

$$= \int_0^a \frac{2y}{b^2-y^2}\left(\int_0^\infty \frac{1}{1+y^2x^2}\,dx - \int_0^\infty \frac{1}{1+b^2x^2}\,dx\right)dy = \frac{\pi}{2}\int_0^a \frac{2y}{b^2-y^2}\cdot\frac{b-y}{by}\,dy$$

$$= \frac{\pi}{b}\int_0^a \frac{1}{b+y}\,dy = \frac{\pi}{b}\log\left(1+\frac{a}{b}\right),$$

and the proof to the auxiliary result is complete.   ∎

Finally, to explain the last equality in (3.144), for the first integral we integrate

by parts and write $\displaystyle\int_0^1 \frac{\log(u)}{1+u}\,du = \int_0^1 (\log(1+u))'\log(u)\,du = \log(u)\log(1+u)$

$u)\Big|_{u=0}^{u=1} - \displaystyle\int_0^1 \frac{\log(1+u)}{u}\,du = \text{Li}_2(-u)\Big|_{u=0}^{u=1} = -\frac{\pi^2}{12}$. Then, the second integral is

---

[10] The existence of the integral is easy to see if we observe that near the possible problematic point $t = 0$ the integrand behaves like $\log(t)$. On the other hand, we may easily note that $\displaystyle\int_1^\infty \frac{\log(t)}{1+t^2}\,dt \le \int_1^\infty \frac{\log(t)}{t^2}\,dt = -\frac{1}{t} - \frac{\log(t)}{t}\Big|_{t=1}^{t=\infty} = 1$. Therefore, we may safely conclude that our integral converges.

$$\int_0^1 \frac{\log(u)\log\left(\frac{1+u}{2}\right)}{1-u}du = \zeta(3) - \frac{1}{2}\log(2)\zeta(2), \text{ which is the particular case } m =$$

1 of the generalization in (1.131), Sect. 1.27. At last, for the third integral we could

use that $\sum_{n=1}^{\infty} x^n H_n = -\frac{\log(1-x)}{1-x}$, which is the case $m = 1$ in (4.32), Sect. 4.6, and

then we obtain that $\int_0^1 \frac{\log(u)\log(1+u)}{1+u}du = \int_0^1 \log(u)\sum_{n=1}^{\infty}(-1)^{n-1}u^n H_n du =$

$$\sum_{n=1}^{\infty}(-1)^{n-1}H_n\int_0^1 u^n\log(u)du \quad = \quad \sum_{n=1}^{\infty}(-1)^n\frac{H_{n+1}-1/(n+1)}{(n+1)^2} \quad \overset{\text{reindex and}}{\underset{\text{expand}}{=}}$$

$$\sum_{n=1}^{\infty}(-1)^{n-1}\frac{H_n}{n^2} - \sum_{n=1}^{\infty}(-1)^{n-1}\frac{1}{n^3} = -\frac{1}{8}\zeta(3), \text{ where I used that } \sum_{n=1}^{\infty}(-1)^{n-1}\frac{H_n}{n^2} =$$

$\frac{5}{8}\zeta(3)$, which is the particular case $m = 1$ of the alternating harmonic series
generalization in (4.105), Sect. 4.21.

Ultimately, returning with the values from (3.142) and (3.144) in (3.141), we
conclude that

$$I = \int_0^{\pi/2} \log(\sin(x))\text{Li}_2(\sin^2(x))dx = \int_0^{\pi/2} \log(\cos(x))\text{Li}_2(\cos^2(x))dx$$

$$= \log^3(2)\pi - \log(2)\frac{\pi^3}{6} + \frac{5}{8}\pi\zeta(3),$$

and the first solution to the point (i) of the problem is finalized.

My solution above was first presented, in large steps, in [24, November 7, 2019]
where the sine integral from the point (i) appears.

A second solution will be constructed by exploiting a (very) useful Fourier series
that appeared both in *(Almost) Impossible Integrals, Sums, and Series* (see [76,
Chapter 3, p.243]) and in the previous section, in (3.131).

A key starting point for a second solution is the simple fact that $(\text{Li}_2(\cos^2(t)))' =$
$4\tan(t)\log(\sin(t))$, which if we combine with (3.131), we get

$$(\text{Li}_2(\cos^2(t)))' = 4\sum_{n=1}^{\infty}\left(\frac{1}{n} + \psi\left(\frac{n}{2}\right) - \psi\left(\frac{n+1}{2}\right)\right)\sin(2nt). \qquad (3.146)$$

Now, if we integrate both sides of (3.146) from $t = 0$ to $t = x$, we obtain that

$$\text{Li}_2(\cos^2(x)) = \frac{\pi^2}{6} + 4\sum_{n=1}^{\infty}\frac{1}{n}\left(\frac{1}{n} + \psi\left(\frac{n}{2}\right) - \psi\left(\frac{n+1}{2}\right)\right)\sin^2(nx). \qquad (3.147)$$

The Fourier-like series in (3.147) looks so cool!, a reaction I wouldn't be surprised to receive from you! *But wait!* The Fourier series of $\text{Li}_2(\cos^2(x))$, in a different form, may be found in (4.161), Sect. 4.49, and a form with the coefficients expressed in terms of integrals may be extracted based on the related solution in Sect. 6.49.

Further, another result we need in our calculations is

$$A_n = \int_0^{\pi/2} \sin^2(nx) \log(\cos(x))dx = -\frac{\pi}{4}\log(2) - \frac{\pi}{8}(-1)^{n-1}\frac{1}{n}, \qquad (3.148)$$

and it is clear to see why if we look at the cosine integral at the point (i) and combine it with (3.147).

Proof To prove the result in (3.148), we start with considering the trigonometric identity $1-\cos(x) = 2\sin^2(x/2)$ that further leads to $1/2(1-\cos(2nx)) = \sin^2(nx)$, and then we have

$$A_n = \frac{1}{2}\int_0^{\pi/2}(1-\cos(2nx))\log(\cos(x))dx = \frac{1}{2}\int_0^{\pi/2}\log(\cos(x))dx$$

$$-\frac{1}{2}\int_0^{\pi/2}\cos(2nx)\log(\cos(x))dx = -\frac{\pi}{4}\log(2) - \frac{1}{2}\int_0^{\pi/2}\cos(2nx)\log(\cos(x))dx$$

$$\overset{\substack{\text{integrate by}\\\text{parts}}}{=} -\frac{\pi}{4}\log(2) - \frac{1}{4n}\int_0^{\pi/2}\sin(2nx)\tan(x)dx$$

$$\overset{\pi/2-x=y}{=} -\frac{\pi}{4}\log(2) - (-1)^{n-1}\frac{1}{4n}\underbrace{\int_0^{\pi/2}\sin(2nx)\cot(x)dx}_{B_n}. \qquad (3.149)$$

Now, for the last integral in (3.149) we exploit the difference $B_k - B_{k-1}$, $k \geq 2$, using the fact that $\sin(a) - \sin(b) = 2\sin\left(\dfrac{a-b}{2}\right)\cos\left(\dfrac{a+b}{2}\right)$, and then we get

$$B_k - B_{k-1} = \int_0^{\pi/2}(\sin(2kx) - \sin(2(k-1)x))\cot(x)dx$$

$$= 2\int_0^{\pi/2}\cos((2k-1)x)\cos(x)dx$$

{use the trigonometric identity, $\cos(a)\cos(b) = 1/2\,(\cos(a+b) + \cos(a-b))$}

$$= \int_0^{\pi/2}(\cos(2kx) + \cos(2(k-1)x))dx = \frac{\sin(2kx)}{2k} + \frac{\sin((2k-2)x)}{2k-2}\bigg|_{x=0}^{x=\pi/2} = 0, \qquad (3.150)$$

and since $B_1 = 2 \int_0^{\pi/2} \cos^2(x)\mathrm{d}x = \int_0^{\pi/2} (1 + \cos(2x))\mathrm{d}x = x +$
$\dfrac{1}{2}\sin(2x)\Big|_{x=0}^{x=\pi/2} = \dfrac{\pi}{2}$, if we sum the opposite sides of (3.150), from $k = 2$ to $k = n$, we have

$$\sum_{k=2}^{n}(B_k - B_{k-1}) = B_n - B_1 = B_n - \frac{\pi}{2} = 0$$

that gives

$$B_n = \int_0^{\pi/2} \sin(2nx)\cot(x)\mathrm{d}x = \frac{\pi}{2}. \tag{3.151}$$

So, if we plug the result from (3.151) in (3.149), we immediately get that

$$A_n = \int_0^{\pi/2} \sin^2(nx)\log(\cos(x))\mathrm{d}x = -\frac{\pi}{4}\log(2) - \frac{\pi}{8}(-1)^{n-1}\frac{1}{n},$$

which brings an end to the auxiliary result in (3.148). $\blacksquare$

Before passing to the main calculations we need one more result,

$$\int_0^1 t^{n-1}\frac{1-t}{1+t}\mathrm{d}t = \psi\left(\frac{n+1}{2}\right) - \psi\left(\frac{n}{2}\right) - \frac{1}{n}, \tag{3.152}$$

that is easily derived by using that $\int_0^1 \dfrac{x^{s-1}}{1+x}\mathrm{d}x = \dfrac{1}{2}\left(\psi\left(\dfrac{1+s}{2}\right) - \psi\left(\dfrac{s}{2}\right)\right)$, which may be found in [76, Chapter 1, p.3], and the recurrence relation of Digamma function, $\psi(x+1) = \psi(x) + \dfrac{1}{x}$.

Returning to the main integral at the point (i) and exploiting the results from (3.147) and (3.152), we write

$$I = \int_0^{\pi/2} \log(\sin(x))\mathrm{Li}_2(\sin^2(x))\mathrm{d}x = \int_0^{\pi/2} \log(\cos(x))\mathrm{Li}_2(\cos^2(x))\mathrm{d}x$$

$$= \frac{\pi^2}{6}\int_0^{\pi/2} \log(\cos(x))\mathrm{d}x - 4\int_0^{\pi/2} \log(\cos(x))\sum_{n=1}^{\infty}\frac{1}{n}\left(\int_0^1 t^{n-1}\frac{1-t}{1+t}\mathrm{d}t\right)\sin^2(nx)\mathrm{d}x$$

{reverse the order of integration and summation}

$$= -\log(2)\frac{\pi^3}{12} - 4\int_0^1 \frac{1-t}{1+t}\sum_{n=1}^{\infty} t^{n-1}\frac{1}{n}\left(\int_0^{\pi/2} \sin^2(nx)\log(\cos(x))dx\right) dt$$

{make use of the result in (3.148)}

$$= -\log(2)\frac{\pi^3}{12} - \frac{\pi}{2}\int_0^1 \frac{(2\log(2)\log(1-t) + \mathrm{Li}_2(-t))(1-t)}{t(1+t)}dt$$

$$= -\log(2)\frac{\pi^3}{12} - \log(2)\pi\int_0^1 \frac{\log(1-t)}{t}dt + 2\log(2)\pi\int_0^1 \frac{\log(1-t)}{1+t}dt$$

$$- \frac{\pi}{2}\int_0^1 \frac{\mathrm{Li}_2(-t)}{t}dt + \pi\int_0^1 \frac{\mathrm{Li}_2(-t)}{1+t}dt$$

$$= \log^3(2)\pi - \log(2)\frac{\pi^3}{6} + \frac{5}{8}\pi\zeta(3),$$

where in the calculations I used that $\displaystyle\int_0^1 \frac{\log(1-t)}{t}dt = -\mathrm{Li}_2(t)\Big|_{t=0}^{t=1} = -\frac{\pi^2}{6}$,

then $\displaystyle\int_0^1 \frac{\log(1-t)}{1+t}dt \overset{(1-t)/(1+t)=u}{=} \log(2)\int_0^1 \frac{1}{1+u}du + \int_0^1 \frac{\log(u)}{1+u}du -$

$\displaystyle\int_0^1 \frac{\log(1+u)}{1+u}du = \frac{1}{2}\log^2(2) + (\log(u)\log(1+u) + \mathrm{Li}_2(-u))\Big|_{u=0}^{u=1} = \frac{1}{2}\log^2(2) -$

$\frac{\pi^2}{12}$, next $\displaystyle\int_0^1 \frac{\mathrm{Li}_2(-t)}{t}dt = \mathrm{Li}_3(-t)\Big|_{t=0}^{t=1} = -\frac{3}{4}\zeta(3)$, and finally, integrating by parts

twice, we have $\displaystyle\int_0^1 \frac{\mathrm{Li}_2(-t)}{1+t}dt = \log(1+t)\mathrm{Li}_2(-t)\Big|_{t=0}^{t=1} + \int_0^1 \frac{\log^2(1+t)}{t}dt =$

$-\log(2)\frac{\pi^2}{12} + \frac{1}{4}\zeta(3)$, where the last integral is treated in [76, Chapter 1, p.4], and
the second solution to the point (i) of the problem is finalized. Observe that the last
integral can also be attacked with the use of harmonic series as I did for the integral
$\displaystyle\int_0^1 \frac{\log(t)\log(1+t)}{1+t}dt$, which is found calculated during the first solution.

How about a third solution? The curious reader might want to combine the
integral representation of the dilogarithm based on the generalization (3.13) in
Sect. 3.3, together with a form of the Sylvester's integral presented in (1.3), Sect. 1.1.
Accordingly, using these results, we immediately arrive at

$$I = \frac{\pi}{2}\int_0^1 \frac{\log(y)}{y}\left(\frac{\log(1+\sqrt{1-y})}{\sqrt{1-y}} - \log(2)\right) dy,$$

where the curious reader might like to continue the calculations and get a third solution to the point (i) of the problem.

And how about a fourth solution? Another valuable solution may be created by exploiting the identities in (1.6) and (1.7), Sect. 1.3, which the curious reader might like to consider.

Regarding the second point of the problem, we have a similar story as before in the sense that letting the variable change $\pi/2 - x = y$ in any of the two integrals, we get the other one.

So, it is enough to consider the dilogarithm function reflection formula (see [21, Chapter 1, p.5], [70, Chapter 2, p.107]), $\text{Li}_2(x) + \text{Li}_2(1-x) = \pi^2/6 - \log(x)\log(1-x)$, and then we have

$$J = \int_0^{\pi/2} \log(\sin(x))\text{Li}_2(\cos^2(x))dx = \int_0^{\pi/2} \log(\cos(x))\text{Li}_2(\sin^2(x))dx$$

$$= \frac{\pi^2}{6} \int_0^{\pi/2} \log(\cos(x))dx - 4 \int_0^{\pi/2} \log(\sin(x))\log^2(\cos(x))dx$$

$$- \int_0^{\pi/2} \log(\cos(x))\text{Li}_2(\cos^2(x))dx$$

$$= \log^3(2)\pi + \log(2)\frac{\pi^3}{12} - \frac{9}{8}\pi\zeta(3),$$

where I used that $\int_0^{\pi/2} \log(\cos(x))dx = \int_0^{\pi/2} \log(\sin(x))dx = -\frac{1}{2}\log(2)\pi$, the second remaining integral was calculated in (3.142), and the last remaining integral is the main integral from the point (i) of the problem.

It's so nice to have more perspectives on such beautiful integrals and therefore more ways of evaluating them!

3.21 Valuable Logarithmic Integrals Involving Skew-Harmonic Numbers: A First Partition

Solution No doubt that very likely the readers of *(Almost) Impossible Integrals, Sums, and Series* will find in this section much similarity to [76, Chapter 1, Sect. 1.3, p.2], and this time instead of $\log(1-x)$, $\log^2(1-x)$, and $\log^3(1-x)$ we have $\log(1+x)$, $\log^2(1+x)$, and $\log^3(1+x)$. That sign modification in the log argument changes things and we are also brought to the realm of skew-harmonic numbers.

Let's recollect that $\overline{H}_n^{(m)} = 1 - \dfrac{1}{2^m} + \cdots + (-1)^{n-1}\dfrac{1}{n^m}$, $m \geq 1$, denotes the nth generalized skew-harmonic number of order m, and two versions of small orders we'll meet in the solutions below.

For the first point, we start with the simple integral $\displaystyle\int_0^1 x^{k-1}dx = \dfrac{1}{k}$ where if we multiply both sides by $(-1)^{k-1}$ and then consider the sum from $k = 1$ to $k = n$, we have that

$$\sum_{k=1}^n (-1)^{k-1}\frac{1}{k} = \overline{H}_n = \sum_{k=1}^n (-1)^{k-1}\int_0^1 x^{k-1}dx = \int_0^1 \sum_{k=1}^n (-x)^{k-1}dx$$

$$= \int_0^1 \frac{1 - (-x)^n}{1+x}dx = \int_0^1 (1 - (-x)^n)(\log(1 + x))'dx$$

$$\overset{\substack{\text{integrate by}\\ \text{parts}}}{=} \underbrace{(1 - (-x)^n)\log(1 + x)\Big|_{x=0}^{x=1}}_{(1 + (-1)^{n-1})\log(2)} - (-1)^{n-1}n\int_0^1 x^{n-1}\log(1 + x)dx,$$

whence we obtain

$$\int_0^1 x^{n-1}\log(1 + x)dx = \log(2)\frac{1}{n} + \log(2)(-1)^{n-1}\frac{1}{n} - (-1)^{n-1}\frac{\overline{H}_n}{n},$$

and the point (i) of the problem is complete.

The results from the points (ii) and (iii) are straightforward if in the previous result we replace n by $2n$, and n by $2n + 1$, respectively, where we also use that $\overline{H}_{2n} = H_{2n} - H_n$ and $\overline{H}_{2n+1} = H_{2n+1} - H_n = H_{2n} - H_n + 1/(2n + 1)$, and thus we get

$$\int_0^1 x^{2n-1}\log(1 + x)dx = \frac{H_{2n} - H_n}{2n}$$

and

$$\int_0^1 x^{2n}\log(1 + x)dx = 2\log(2)\frac{1}{2n + 1} - \frac{1}{(2n + 1)^2} + \frac{H_n}{2n + 1} - \frac{H_{2n}}{2n + 1}.$$

In 2020 I created and uploaded a small paper treating these two particular cases which you may find in [86]. It is worth mentioning that they also appear in

4.293.4 and **4.293.5** from [17]. The curious reader may consider more ways[11] of approaching these integrals.

At this point, I would like to emphasize that the integrals at the points (i)–(iii) could be generalized with the help of Digamma function. From *(Almost) Impossible Integrals, Sums, and Series* (see [76, Chapter 1, p.3]), we have that

$$\int_0^1 \frac{x^{s-1}}{1+x} dx = \frac{1}{2}\left(\psi\left(\frac{1+s}{2}\right) - \psi\left(\frac{s}{2}\right)\right) = \psi(s) - \psi\left(\frac{s}{2}\right) - \log(2), \ s > 0.$$
(3.153)

Now, integrating by parts in (3.153) and rearranging, we have

$$\int_0^1 x^{s-2}\log(1+x)dx = \frac{1}{s-1}\left(\log(2) - \frac{1}{2}\psi\left(\frac{1+s}{2}\right) + \frac{1}{2}\psi\left(\frac{s}{2}\right)\right)$$

$$= \frac{1}{s-1}\left(2\log(2) - \psi(s) + \psi\left(\frac{s}{2}\right)\right) \ s > 0 \wedge s \neq 1,$$
(3.154)

where the first equality is also given in **4.293.1** from [17].

For example, the cases from the points (ii) and (iii) can be easily extracted by exploiting (3.154) if we use the Digamma recurrence relation, $\psi(1+x) = \psi(x) + \frac{1}{x}$, then $\psi(n) + \gamma = H_{n-1}$ and the identity $\psi(2x) = \frac{1}{2}\psi(x) + \frac{1}{2}\psi\left(x + \frac{1}{2}\right) + \log(2)$ (see [76, Chapter 3, p.68]) that together lead to

$$\psi\left(n + \frac{1}{2}\right) = 2\psi(2n) - \psi(n) - 2\log(2)$$

$$\{\text{employ the relation } \psi(n) + \gamma = H_{n-1}\}$$

$$= 2H_{2n} - H_n - \gamma - 2\log(2).$$
(3.155)

---

[11] If we denote $I_n = \int_0^1 x^{2n-1}\log(1+x)dx$, one might derive and exploit the recurrence relation $2(n+1)I_{n+1} - 2nI_n = \frac{1}{2n+1} - \frac{1}{2}\frac{1}{n+1}$. For the same integral, we might exploit the identity $\log(1+x) = \log((1-x^2)/(1-x))$ together with the use of the integral $\int_0^1 x^{n-1}\log(1-x)dx = -\frac{H_n}{n}$ you may find in (3.10), Sect. 3.3. For the integral $J_n = \int_0^1 x^{2n}\log(1+x)dx$, observe that by exploiting integration by parts we obtain $J_n = \frac{1}{2}\log(2)\frac{1}{n} - \frac{1}{2}\frac{1}{n}\int_0^1 \frac{1+x^{2n+1}}{1+x}dx + \frac{1}{2}\frac{1}{n}\int_0^1 \frac{1}{1+x}dx - \frac{1}{2}\frac{1}{n}J_n$.

Let's pass now to the point (iv) where I'll want to exploit the result from the point (i), and if we consider the notation in k, multiply both sides by $(-1)^{k-1}$, and then sum from $k = 1$ to $k = n$, we have

$$\sum_{k=1}^{n}(-1)^{k-1}\int_{0}^{1}x^{k-1}\log(1+x)dx = \int_{0}^{1}\sum_{k=1}^{n}(-x)^{k-1}\log(1+x)dx$$

$$= \int_{0}^{1}\frac{1-(-x)^n}{1+x}\log(1+x)dx = \frac{1}{2}\int_{0}^{1}(1-(-x)^n)(\log^2(1+x))'dx$$

$$\underset{\text{parts}}{\overset{\text{integrate by}}{=}}\ \underbrace{\frac{1}{2}(1-(-x)^n)\log^2(1+x)\Big|_{x=0}^{x=1}}_{(1+(-1)^{n-1})\log^2(2)/2} - \frac{1}{2}(-1)^{n-1}n\int_{0}^{1}x^{n-1}\log^2(1+x)dx$$

$$= \log(2)H_n + \log(2)\overline{H}_n - \sum_{k=1}^{n}\frac{\overline{H}_k}{k},$$

whence we obtain that

$$\int_{0}^{1}x^{n-1}\log^2(1+x)dx$$

$$= \log^2(2)\frac{1}{n} + \log^2(2)(-1)^{n-1}\frac{1}{n} - 2\log(2)(-1)^{n-1}\frac{H_n}{n} - 2\log(2)(-1)^{n-1}\frac{\overline{H}_n}{n}$$

$$+ 2(-1)^{n-1}\frac{1}{n}\sum_{k=1}^{n}\frac{\overline{H}_k}{k}, \tag{3.156}$$

which is the desired second closed form.

To get the first given closed form we also need based on Abel's summation (see [76, Chapter 2, pp.39–40]), where we set $a_k = 1/k$ and $b_k = \overline{H}_k$, that

$$\sum_{k=1}^{n}\frac{\overline{H}_k}{k} = H_n\overline{H}_{n+1} + \sum_{k=1}^{n}(-1)^{k-1}\frac{H_{k+1}-1/(k+1)}{k+1}$$

{reindex the sum and start from $k = 1$}

$$= H_n\left(\overline{H}_n + \frac{(-1)^n}{n+1}\right) - \sum_{k=1}^{n+1}(-1)^{k-1}\frac{H_k-1/k}{k} = H_n\overline{H}_n + \overline{H}_n^{(2)} - \sum_{k=1}^{n}(-1)^{k-1}\frac{H_k}{k}.$$

$$\tag{3.157}$$

The result in (3.157) can also be obtained if we turn the initial sum into a double sum and then use the change of summation order.

So, if we plug (3.157) in (3.156), we obtain that

$$\int_0^1 x^{n-1} \log^2(1+x)dx$$

$$= \log^2(2)\frac{1}{n} + \log^2(2)(-1)^{n-1}\frac{1}{n} - 2\log(2)(-1)^{n-1}\frac{H_n}{n} - 2\log(2)(-1)^{n-1}\frac{\overline{H}_n}{n}$$

$$+ 2(-1)^{n-1}\frac{H_n\overline{H}_n}{n} + 2(-1)^{n-1}\frac{\overline{H}_n^{(2)}}{n} - 2(-1)^{n-1}\frac{1}{n}\sum_{k=1}^{n}(-1)^{k-1}\frac{H_k}{k},$$

which is the first desired closed form, and the point (iv) of the problem is complete.

The results at the points (v) and (vi) are derived from the previous point by replacing, in one case, n by $2n$, and in the other case, n by $2n+1$. When needed, we use the relations between the skew-harmonic numbers and harmonic numbers, based on parity, mentioned in the solutions to the points (ii) and (iii), and it may also be useful to consider $\overline{H}_{2n}^{(2)} = H_{2n}^{(2)} - \frac{1}{2}H_n^{(2)}$ and/or $\overline{H}_{2n+1}^{(2)} = H_{2n+1}^{(2)} - \frac{1}{2}H_n^{(2)} = H_{2n}^{(2)} - \frac{1}{2}H_n^{(2)} + \frac{1}{(2n+1)^2}$. Depending on which of the two closed forms we start from, in the calculations we might also need (3.157) and the fact that

$$\sum_{k=1}^{2n}(-1)^{k-1}\frac{H_k}{k} = \sum_{k=1}^{2n}\frac{H_k}{k} - 2\sum_{k=1}^{n}\frac{H_{2k}}{2k} = \frac{1}{2}(H_{2n}^2 + H_{2n}^{(2)}) - \sum_{k=1}^{n}\frac{H_{2k}}{k}$$

{use Abel's summation (see [76, Chapter 2, pp.39–40]) with $a_k = k$ and $b_k = H_{2k}$}

$$= \frac{1}{2}(H_{2n}^2 + H_{2n}^{(2)}) - H_n H_{2n+2} + \frac{1}{2}\sum_{k=1}^{n}\frac{H_{k+1} - 1/(k+1)}{k+1} + \sum_{k=1}^{n}\frac{H_k}{2k+1}$$

$$= \frac{1}{4}\left(\left(H_n + \frac{1}{n+1}\right)^2 - \left(H_n^{(2)} + \frac{1}{(n+1)^2}\right)\right) + \frac{1}{2}(H_{2n}^2 + H_{2n}^{(2)})$$

$$- H_n\left(H_{2n} + \frac{1}{2n+1} + \frac{1}{2n+2}\right) + \sum_{k=1}^{n}\frac{H_k}{2k+1}$$

$$= \frac{1}{4}H_n^2 + \frac{1}{2}H_{2n}^2 - H_n H_{2n} - \frac{1}{4}H_n^{(2)} + \frac{1}{2}H_{2n}^{(2)} + \sum_{k=1}^{n-1}\frac{H_k}{2k+1}. \qquad (3.158)$$

Note that above we used that $\sum_{k=1}^{n}\frac{H_k}{k} = \frac{1}{2}(H_n^2 + H_n^{(2)})$, which is the case $p = q = 1$ in (6.102), Sect. 6.13, and we'll also consider in the calculations below.

Next, for the result at the point (vii) we want to consider the result from the point (iv), using the notation in k instead of n, where if we multiply both sides by $(-1)^{k-1}$ and then sum from $k = 1$ to $k = n$, we get

$$\sum_{k=1}^{n}(-1)^{k-1}\int_{0}^{1}x^{k-1}\log^2(1+x)dx = \int_{0}^{1}\sum_{k=1}^{n}(-x)^{k-1}\log^2(1+x)dx$$

$$= \int_{0}^{1}\frac{1-(-x)^n}{1+x}\log^2(1+x)dx = \frac{1}{3}\int_{0}^{1}(1-(-x)^n)(\log^3(1+x))'dx$$

$$\overset{\substack{\text{integrate by}\\ \text{parts}}}{=} \underbrace{\frac{1}{3}(1-(-x)^n)\log^3(1+x)\Big|_{x=0}^{x=1}}_{(1+(-1)^{n-1})\log^3(2)/3} - \frac{1}{3}(-1)^{n-1}n\int_{0}^{1}x^{n-1}\log^3(1+x)dx$$

$$= \log^2(2)H_n + \log^2(2)\overline{H}_n - 2\log(2)\sum_{k=1}^{n}\frac{H_k}{k} - 2\log(2)\sum_{k=1}^{n}\frac{\overline{H}_k}{k} + 2\sum_{k=1}^{n}\frac{1}{k}\sum_{i=1}^{k}\frac{\overline{H}_i}{i}$$

$$= \log^2(2)H_n + \log^2(2)\overline{H}_n - \log(2)(H_n^2 + H_n^{(2)}) - 2\log(2)\sum_{k=1}^{n}\frac{\overline{H}_k}{k} + 2\sum_{k=1}^{n}\frac{1}{k}\sum_{i=1}^{k}\frac{\overline{H}_i}{i}.$$
$$(3.159)$$

For the double sum in (3.159), we change the summation order, and then we write

$$\sum_{k=1}^{n}\frac{1}{k}\sum_{i=1}^{k}\frac{\overline{H}_i}{i} = \sum_{i=1}^{n}\frac{\overline{H}_i}{i}\sum_{k=i}^{n}\frac{1}{k} = \sum_{i=1}^{n}\frac{\overline{H}_i}{i}\left(H_n - H_i + \frac{1}{i}\right) = \sum_{k=1}^{n}\frac{\overline{H}_k}{k}\left(H_n - H_k + \frac{1}{k}\right)$$

$$= H_n\sum_{k=1}^{n}\frac{\overline{H}_k}{k} + \sum_{k=1}^{n}\frac{\overline{H}_k}{k^2} - \sum_{k=1}^{n}\frac{H_k\overline{H}_k}{k}.$$
$$(3.160)$$

So, if we plug the result from (3.160) in (3.159) and rearrange, we arrive at

$$\int_{0}^{1}x^{n-1}\log^3(1+x)dx$$

$$= \log^3(2)\frac{1+(-1)^{n-1}}{n} - 3\log^2(2)(-1)^{n-1}\frac{H_n + \overline{H}_n}{n} + 3\log(2)(-1)^{n-1}\frac{H_n^2 + H_n^{(2)}}{n}$$

$$+ 6(-1)^{n-1}\left(\log(2)\frac{1}{n} - \frac{H_n}{n}\right)\sum_{k=1}^{n}\frac{\overline{H}_k}{k} - 6(-1)^{n-1}\frac{1}{n}\sum_{k=1}^{n}\frac{\overline{H}_k}{k^2} + 6(-1)^{n-1}\frac{1}{n}\sum_{k=1}^{n}\frac{H_k\overline{H}_k}{k},$$

and the point (*vii*) of the problem is complete.

For simplicity, I might denote the integral at the point (*viii*) by I_n, and then we observe that

$$I_{n+1} + I_n = \int_0^1 (x^n + x^{n-1}) \frac{\log(1-x)}{1+x} dx = \int_0^1 x^{n-1} \log(1-x) dx = -\frac{H_n}{n},$$

$$(3.161)$$

where the last resulting integral may be found in (3.10), Sect. 3.3.

If we consider the resulting recurrence in (3.161), $I_{k+1} + I_k = -\dfrac{H_k}{k}$, and then multiply both sides by $(-1)^k$ and sum from $k = 1$ to $k = n - 1$, we get that

$$\sum_{k=1}^{n-1} ((-1)^k I_{k+1} - (-1)^{k-1} I_k) = (-1)^{n-1} I_n - I_1 = \sum_{k=1}^{n-1} (-1)^{k-1} \frac{H_k}{k},$$

which further can be written as

$$I_n = \int_0^1 x^{n-1} \frac{\log(1-x)}{1+x} dx = (-1)^{n-1} \underbrace{\int_0^1 \frac{\log(1-x)}{1+x} dx}_{I_1} + (-1)^{n-1} \sum_{k=1}^{n-1} (-1)^{k-1} \frac{H_k}{k}.$$

$$(3.162)$$

Since we have that

$$I_1 = \int_0^1 \frac{\log(1-x)}{1+x} dx \overset{(1-x)/(1+x)=t}{=} \int_0^1 \frac{\log(t)}{1+t} dt + \underbrace{\int_0^1 \frac{\log(2)}{1+t} dt}_{\log^2(2)} - \underbrace{\int_0^1 \frac{\log(1+t)}{1+t} dt}_{\log^2(2)/2}$$

$$= \frac{1}{2} \log^2(2) + \int_0^1 \frac{\log(t)}{1+t} dt = \frac{1}{2} \log^2(2) + \int_0^1 \sum_{n=1}^{\infty} (-1)^{n-1} t^{n-1} \log(t) dt$$

$$= \frac{1}{2} \log^2(2) + \sum_{n=1}^{\infty} (-1)^{n-1} \int_0^1 t^{n-1} \log(t) dt = \frac{1}{2} \log^2(2) - \sum_{n=1}^{\infty} (-1)^{n-1} \frac{1}{n^2}$$

$$= \frac{1}{2} \left(\log^2(2) - \frac{\pi^2}{6} \right),$$

$$(3.163)$$

then by plugging (3.163) in (3.162), we obtain that

$$I_n = \int_0^1 x^{n-1} \frac{\log(1-x)}{1+x} dx$$

$$= \frac{1}{2} \left(\log^2(2) - \frac{\pi^2}{6} \right) (-1)^{n-1} + (-1)^{n-1} \sum_{k=1}^{n-1} (-1)^{k-1} \frac{H_k}{k}, \qquad (3.164)$$

which is the first desired closed form.

It is straightforward to get the second closed form if we combine (3.164) and (3.157), and therefore we conclude that

$$\int_0^1 x^{n-1} \frac{\log(1-x)}{1+x} dx$$

$$= \frac{1}{2} \left(\log^2(2) - \frac{\pi^2}{6} \right) (-1)^{n-1} - \frac{H_n}{n} + (-1)^{n-1} H_n \overline{H}_n + (-1)^{n-1} \overline{H}_n^{(2)}$$

$$- (-1)^{n-1} \sum_{k=1}^n \frac{\overline{H}_k}{k},$$

and the point $(viii)$ of the problem is complete.

Exploiting the parity in (3.164), with the first closed form, together with the use of (3.158), we arrive at

$$\int_0^1 x^{2n-1} \frac{\log(1-x)}{1+x} dx$$

$$= \frac{1}{2} \left(\frac{\pi^2}{6} - \log^2(2) \right) - \frac{1}{2} \frac{H_{2n}}{n} - \frac{1}{4} H_n^2 + H_n H_{2n} - \frac{1}{2} H_{2n}^2 + \frac{1}{4} H_n^{(2)} - \frac{1}{2} H_{2n}^{(2)}$$

$$- \sum_{k=1}^{n-1} \frac{H_k}{2k+1}$$

and

$$\int_0^1 x^{2n} \frac{\log(1-x)}{1+x} dx$$

$$= \frac{1}{2}\left(\log^2(2) - \frac{\pi^2}{6}\right) + \frac{1}{4}H_n^2 + \frac{1}{2}H_{2n}^2 - H_n H_{2n} - \frac{1}{4}H_n^{(2)} + \frac{1}{2}H_{2n}^{(2)} + \sum_{k=1}^{n-1}\frac{H_k}{2k+1},$$

$$(3.165)$$

and the solution to the points (ix) and (x) of the problem is complete.

We'll find such integrals as good-to-know auxiliary results in the work with more advanced integrals and series. For example, in the last chapter we'll meet special Fourier series involving coefficients with skew-harmonic numbers and that will be derived with the help of the integrals from the points (i) and (iv).

Some results in the present section will also be found useful in the next section!

3.22   Valuable Logarithmic Integrals Involving Skew-Harmonic Numbers: A Second Partition

Solution If you have found interesting to calculate the logarithmic integrals in the previous section, be ready to go further and approach other exciting versions too, some of them involving forms with arctanh(x).

While some of the points will be easily calculated based on the previous results, as you'll see, other points might be perceived as being more challenging and there we'll need more creativity.

Since we know that $\operatorname{arctanh}(x) = \frac{1}{2}\log\left(\frac{1+x}{1-x}\right)$, by using the logarithmic integral $\int_0^1 x^{n-1}\log(1-x)\mathrm{d}x = -\frac{H_n}{n}$ (see (3.10), Sect. 3.3) and the one at the first point, in the previous section, we have

$$\int_0^1 x^{n-1}\operatorname{arctanh}(x)\mathrm{d}x = \frac{1}{2}\int_0^1 x^{n-1}\log(1+x)\mathrm{d}x - \frac{1}{2}\int_0^1 x^{n-1}\log(1-x)\mathrm{d}x$$

$$= \frac{1}{2}\log(2)\frac{1}{n} + \frac{1}{2}\log(2)(-1)^{n-1}\frac{1}{n} + \frac{1}{2}\frac{H_n}{n} - \frac{1}{2}(-1)^{n-1}\frac{\overline{H}_n}{n},$$

and the solution to the point (i) is finalized.

The results from the points (ii) and (iii) are promptly derived from the previous main result if we replace n by $2n$, respectively n by $2n+1$, and then also consider that $\overline{H}_{2n} = H_{2n} - H_n$ and $\overline{H}_{2n+1} = H_{2n+1} - H_n = H_{2n} - H_n + 1/(2n+1)$ which gives

$$\int_0^1 x^{2n-1}\operatorname{arctanh}(x)\mathrm{d}x = \frac{1}{2}\frac{H_{2n}}{n} - \frac{1}{4}\frac{H_n}{n}$$

and

$$\int_0^1 x^{2n} \operatorname{arctanh}(x)\mathrm{d}x = \log(2)\frac{1}{2n+1} + \frac{1}{2}\frac{H_n}{2n+1},$$

and the solution to the points (ii) and (iii) is finalized.

Now, the integrals given at the points (i)–(iii) could be brought to a generalized form by using Digamma function. So, if we exploit the extended form in Digamma function of the result in (3.10), Sect. 3.3, that is, $\int_0^1 x^{s-2}\log(1-x)\mathrm{d}x = -\dfrac{\psi(s)+\gamma}{s-1}$, and then combine it with the result in (3.154), in the previous section, we get

$$\int_0^1 x^{s-2} \operatorname{arctanh}(x)\mathrm{d}x = \frac{1}{2}\int_0^1 x^{s-2}\log(1+x)\mathrm{d}x - \frac{1}{2}\int_0^1 x^{s-2}\log(1-x)\mathrm{d}x$$

$$= \frac{1}{2}\frac{1}{s-1}\left(\log(2)+\gamma+\psi(s)+\frac{1}{2}\psi\left(\frac{s}{2}\right)-\frac{1}{2}\psi\left(\frac{1+s}{2}\right)\right)$$

$$= \frac{1}{2}\frac{1}{s-1}\left(2\log(2)+\gamma+\psi\left(\frac{s}{2}\right)\right), \quad s>0 \wedge s\neq 1. \tag{3.166}$$

How exactly would we take the bull (the squared inverse hyperbolic tangent version) by the horns?, which you might think about after seeing the integral result at the point (iv).

We might start with integrating by parts first, and then we write

$$\int_0^1 x^{n-1}\operatorname{arctanh}^2(x)\mathrm{d}x = \frac{1}{n}\int_0^1 (x^n-1)'\operatorname{arctanh}^2(x)\mathrm{d}x = \underbrace{\frac{1}{n}(x^n-1)\operatorname{arctanh}^2(x)\Big|_{x=0}^{x=1}}_{0}$$

$$+2\frac{1}{n}\int_0^1 \frac{1-x^n}{1-x^2}\operatorname{arctanh}(x)\mathrm{d}x = \frac{1}{n}\int_0^1 \frac{1-x^n}{1-x}\operatorname{arctanh}(x)\mathrm{d}x$$

$$+\frac{1}{2}\frac{1}{n}\int_0^1 (1-x^n)\frac{\log(1+x)}{1+x}\mathrm{d}x - \frac{1}{2}\frac{1}{n}\int_0^1 \frac{\log(1-x)}{1+x}\mathrm{d}x + \frac{1}{2}\frac{1}{n}\int_0^1 \frac{x^n\log(1-x)}{1+x}\mathrm{d}x. \tag{3.167}$$

For the first resulting integral in (3.167), we need to know and use the value of the integral $\int_0^1 x^{k-1}\operatorname{arctanh}(x)\mathrm{d}x$, which is given at the point (i), together with the

sums $\displaystyle\sum_{k=1}^n \frac{H_k}{k} = \frac{1}{2}(H_n^2 + H_n^{(2)})$, which is the case $p=q=1$ in (6.102), Sect.

6.13, and $\sum_{k=1}^{n}(-1)^{k-1}\dfrac{\overline{H}_k}{k} = \dfrac{1}{2}\left(\overline{H}_n^2 + H_n^{(2)}\right)$, given in (1.120), the case $p = 1$, Sect. 1.23, and then we write

$$\int_0^1 \frac{1-x^n}{1-x}\operatorname{arctanh}(x)dx = \int_0^1 \sum_{k=1}^{n} x^{k-1}\operatorname{arctanh}(x)dx = \sum_{k=1}^{n}\int_0^1 x^{k-1}\operatorname{arctanh}(x)dx$$

$$= \frac{1}{2}\log(2)H_n + \frac{1}{2}\log(2)\overline{H}_n + \frac{1}{4}H_n^2 - \frac{1}{4}\overline{H}_n^2. \tag{3.168}$$

For the second resulting integral in (3.167), we integrate by parts in order to relate it to an integral from the previous section:

$$\int_0^1 (1-x^n)\frac{\log(1+x)}{1+x}dx = \frac{1}{2}\int_0^1 (1-x^n)(\log^2(1+x))'dx$$

$$= \underbrace{\frac{1}{2}(1-x^n)\log^2(1+x)\Big|_{x=0}^{x=1}}_{0} + \frac{1}{2}n\int_0^1 x^{n-1}\log^2(1+x)dx. \tag{3.169}$$

Now, if we combine the results in (3.167), (3.168), and (3.169) and the results in (1.102) and (1.106), Sect. 1.21, where we consider both closed forms, we arrive at

$$\int_0^1 x^{n-1}\operatorname{arctanh}^2(x)dx$$

$$= \frac{\pi^2}{24}\frac{1}{n} + \frac{\pi^2}{24}(-1)^{n-1}\frac{1}{n} + \frac{1}{2}\log(2)\frac{H_n}{n} - \frac{1}{2}\log(2)(-1)^{n-1}\frac{H_n}{n} + \frac{1}{4}\frac{H_n^2}{n}$$

$$+\frac{1}{2}(-1)^{n-1}\frac{H_n\overline{H}_n}{n} + \frac{1}{2}\log(2)\frac{\overline{H}_n}{n} - \frac{1}{2}\log(2)(-1)^{n-1}\frac{\overline{H}_n}{n} - \frac{1}{4}\frac{\overline{H}_n^2}{n} + \frac{1}{2}(-1)^{n-1}\frac{\overline{H}_n^{(2)}}{n}$$

$$- (-1)^{n-1}\frac{1}{n}\sum_{k=1}^{n}(-1)^{k-1}\frac{H_k}{k}$$

$$= \frac{\pi^2}{24}\frac{1}{n} + \frac{\pi^2}{24}(-1)^{n-1}\frac{1}{n} + \frac{1}{2}\log(2)\frac{H_n}{n} - \frac{1}{2}\log(2)(-1)^{n-1}\frac{H_n}{n} + \frac{1}{4}\frac{H_n^2}{n}$$

$$-\frac{1}{2}(-1)^{n-1}\frac{H_n\overline{H}_n}{n} + \frac{1}{2}\log(2)\frac{\overline{H}_n}{n} - \frac{1}{2}\log(2)(-1)^{n-1}\frac{\overline{H}_n}{n} - \frac{1}{4}\frac{\overline{H}_n^2}{n} - \frac{1}{2}(-1)^{n-1}\frac{\overline{H}_n^{(2)}}{n}$$

$$+ (-1)^{n-1} \frac{1}{n} \sum_{k=1}^{n} \frac{\overline{H}_k}{k},$$

and the solution to the point (iv) is finalized.

The results given at the points (v) and (vi) are easily derived based on the result from the point (iv), where we need to use relations between the skew-harmonic numbers and harmonic numbers as presented in the solutions to the points (ii) and (iii), the current section, and the solution to the points (v) and (vi), in the previous section. We also need (3.158), and depending on which of the two closed forms we start from, we might also want to employ (3.157), the previous section. Therefore, we obtain that

$$\int_0^1 x^{2n-1} \operatorname{arctanh}^2(x)\mathrm{d}x$$

$$= \log(2)\frac{H_{2n}}{n} - \frac{1}{2}\log(2)\frac{H_n}{n} + \frac{1}{2}\frac{1}{n}\sum_{k=1}^{n-1}\frac{H_k}{2k+1},$$

and

$$\int_0^1 x^{2n} \operatorname{arctanh}^2(x)\mathrm{d}x$$

$$= \frac{\pi^2}{12}\frac{1}{2n+1} - \frac{1}{2}\frac{H_n^2}{2n+1} + \frac{H_n H_{2n}}{2n+1} - \frac{1}{2n+1}\sum_{k=1}^{n-1}\frac{H_k}{2k+1},$$

which brings an end to the solution to the points (v) and (vi).

Further, for the next point we might consider two short solutions. If we let the variable change $x = t^2$, we have

$$\int_0^1 x^{n/2-1} \log(1-x)\mathrm{d}x = 2\int_0^1 t^{n-1}\log(1-t^2)\mathrm{d}t = 2\int_0^1 t^{n-1}\log(1-t)\mathrm{d}t$$

$$+2\int_0^1 t^{n-1}\log(1+t)\mathrm{d}t = 2\log(2)\frac{1}{n} + 2\log(2)(-1)^{n-1}\frac{1}{n} - 2\frac{H_n}{n} - 2(-1)^{n-1}\frac{\overline{H}_n}{n},$$

and the first solution to the point (vii) is finalized.

For a second solution, we start with the extended version of the integral in (3.10), Sect. 3.3, that is, $\int_0^1 x^{n-1}\log(1-x)\mathrm{d}x = -\dfrac{\psi(n+1)+\gamma}{n}$, $n > 0$, $n \in \mathbb{R}$, and then we write

$$\int_0^1 x^{n/2-1} \log(1-x)\mathrm{d}x = -\frac{2}{n}\left(\psi\left(\frac{n}{2}+1\right)+\gamma\right)$$

$$= 2\log(2)\frac{1}{n} + 2\log(2)(-1)^{n-1}\frac{1}{n} - 2\frac{H_n}{n} - 2(-1)^{n-1}\frac{\overline{H}_n}{n},$$

where to get the last equality I used the result in (4.97), Sect. 4.18, and the second solution to the point (vii) is finalized.

For the result at the point $(viii)$ we might use again the strategy in the second solution to the previous point, and then starting with the extended version of the integral $\int_0^1 x^{n-1} \log^2(1-x)\mathrm{d}x = \dfrac{(\psi(n+1)+\gamma)^2 + \zeta(2) - \psi^{(1)}(n+1)}{n}$, $n > 0$, $n \in \mathbb{R}$, in [76, Chapter 1, p.2], we write

$$\int_0^1 x^{n/2-1} \log^2(1-x)\mathrm{d}x = \frac{2}{n}\left(\left(\psi\left(\frac{n}{2}+1\right)+\gamma\right)^2 + \zeta(2) - \psi^{(1)}\left(\frac{n}{2}+1\right)\right)$$

$$= \left(4\log^2(2) - \frac{\pi^2}{3}\right)\frac{1}{n} + \left(4\log^2(2) - \frac{\pi^2}{3}\right)(-1)^{n-1}\frac{1}{n} - 4\log(2)\frac{H_n}{n}$$

$$- 4\log(2)(-1)^{n-1}\frac{H_n}{n} + 2\frac{H_n^2}{n} + 4\frac{H_n^{(2)}}{n} + 4(-1)^{n-1}\frac{H_n\overline{H}_n}{n}$$

$$- 4\log(2)\frac{\overline{H}_n}{n} - 4\log(2)(-1)^{n-1}\frac{\overline{H}_n}{n} + 2\frac{\overline{H}_n^2}{n} + 4(-1)^{n-1}\frac{\overline{H}_n^{(2)}}{n},$$

where to get the last equality I used the results in (4.97) and (4.99), the case $m = 1$, Sect. 4.18, and the solution to the point $(viii)$ is finalized.

Of course, the curious reader might also want to exploit the variable change $x = t^2$ together with the algebraic identity, $\log^2(1-x^2) = \log^2(1-x) + 2\log(1-x)\log(1+x) + \log^2(1+x)$.

What about the result at the point (ix)? Well, we will obtain a *lightning strike* solution if we combine the result from the point (1.106), Sect. 1.21, and the one given in (1.122), Sect. 1.23, leading to

$$\int_0^1 x^{n-1} \log(1-x)\log(1+x)\mathrm{d}x$$

$$= \frac{1}{2}\left(\log^2(2) - \frac{\pi^2}{6}\right)\frac{1}{n} + \frac{1}{2}\left(\log^2(2) - \frac{\pi^2}{6}\right)(-1)^{n-1}\frac{1}{n} - \log(2)\frac{H_n}{n}$$

$$+ \frac{1}{2}\frac{H_n^{(2)}}{n} - \log(2)\frac{\overline{H}_n}{n} + \frac{1}{2}\frac{\overline{H}_n^2}{n} + (-1)^{n-1}\frac{1}{n}\sum_{k=1}^n(-1)^{k-1}\frac{H_k}{k}$$

$$= \frac{1}{2}\left(\log^2(2) - \frac{\pi^2}{6}\right)\frac{1}{n} + \frac{1}{2}\left(\log^2(2) - \frac{\pi^2}{6}\right)(-1)^{n-1}\frac{1}{n} - \log(2)\frac{H_n}{n}$$

$$+ \frac{1}{2}\frac{H_n^{(2)}}{n} + (-1)^{n-1}\frac{H_n\overline{H}_n}{n} - \log(2)\frac{\overline{H}_n}{n} + \frac{1}{2}\frac{\overline{H}_n^2}{n} + (-1)^{n-1}\frac{\overline{H}_n^{(2)}}{n} - (-1)^{n-1}\frac{1}{n}\sum_{k=1}^{n}\frac{\overline{H}_k}{k},$$

and the solution to the point (ix) is finalized.

Again, the curious reader might want to look at things from another perspective and attack the problem differently (e.g., by using algebraic identities as suggested at the end of the previous solution).

At last, the results at the points (x) and (xi) are immediately extracted by exploiting the parity of n, in a way similar to the strategy I presented in the beginning of the solution to the points (v) and (vi) above, and then we get

$$\int_0^1 x^{2n-1}\log(1-x)\log(1+x)dx$$

$$= \frac{1}{2}\log(2)\frac{H_n}{n} - \log(2)\frac{H_{2n}}{n} + \frac{1}{8}\frac{H_n^2}{n} + \frac{1}{8}\frac{H_n^{(2)}}{n} - \frac{1}{2}\frac{1}{n}\sum_{k=1}^{n-1}\frac{H_k}{2k+1}$$

and

$$\int_0^1 x^{2n}\log(1-x)\log(1+x)dx$$

$$= \frac{1}{2n+1}\left(\log^2(2) - \frac{\pi^2}{6}\right) - 2\log(2)\frac{1}{(2n+1)^2} + 2\frac{1}{(2n+1)^3} + \log(2)\frac{H_n}{2n+1}$$

$$- 2\log(2)\frac{H_{2n}}{2n+1} - \frac{H_n}{(2n+1)^2} + 2\frac{H_{2n}}{(2n+1)^2} + \frac{3}{4}\frac{H_n^2}{2n+1} + \frac{H_{2n}^2}{2n+1}$$

$$- 2\frac{H_n H_{2n}}{2n+1} - \frac{1}{4}\frac{H_n^{(2)}}{2n+1} + \frac{H_{2n}^{(2)}}{2n+1} + \frac{1}{2n+1}\sum_{k=1}^{n-1}\frac{H_k}{2k+1},$$

which finalizes the solution to the last points of the section.

For instance, the integral at the point (ix) plays an important part in my derivation strategy of another interesting Fourier series, as you'll see in the sixth chapter.

Also, the curious reader might want to consider other strategies for deriving every point above, by employing different means, which could be an excellent exercise.

3.23   Valuable Logarithmic Integrals Involving Skew-Harmonic Numbers: A Third Partition

Solution As mentioned in the title, this is the third section from a series of (three) sections where we treat integrals involving skew-harmonic numbers, and as in the first section of this series, I'll start by highlighting the similarities of some results to the ones in *(Almost) Impossible Integrals, Sums, and Series*. More precisely, I refer now to the auxiliary sums needed (for deriving the integrals), which to a certain extent look like the ones in [76, Chapter 1, Sect. 1.3, p.2]. For example, if at the indicated reference we had $\displaystyle\sum_{k=1}^{n}\frac{H_k}{k} = \frac{1}{2}\left(H_n^2 + H_n^{(2)}\right)$, now, we need the

version $\displaystyle\sum_{k=1}^{n}(-1)^{k-1}\frac{\overline{H}_k}{k} = \frac{1}{2}\left(\overline{H}_n^2 + H_n^{(2)}\right)$. Or if at the mentioned reference we

had $\displaystyle\sum_{k=1}^{n}\frac{H_k^2 + H_k^{(2)}}{k} = \frac{1}{3}(H_n^3 + 3H_n H_n^{(2)} + 2H_n^{(3)})$, later we'll want to prove the

following version involving skew-harmonic numbers, $\displaystyle\sum_{k=1}^{n}(-1)^{k-1}\frac{\overline{H}_k^2 + H_k^{(2)}}{k} =$

$\frac{1}{3}\left(\overline{H}_n^3 + 3\overline{H}_n H_n^{(2)} + 2\overline{H}_n^{(3)}\right)$.

   Now, the sum at the point (i), the case $p = 1$, is also found in [8] where the author proves the result by induction. In the following, I'll first prove a more general case,

$$\sum_{k=1}^{n}(-1)^{k-1}\frac{\overline{H}_k^{(p)}}{k^q} + \sum_{k=1}^{n}(-1)^{k-1}\frac{\overline{H}_k^{(q)}}{k^p} = H_n^{(p+q)} + \overline{H}_n^{(p)}\overline{H}_n^{(q)}. \qquad (3.170)$$

Proof The result is again straightforward, and we have

$$\sum_{k=1}^{n}(-1)^{k-1}\frac{\overline{H}_k^{(p)}}{k^q} = \sum_{k=1}^{n}\sum_{i=1}^{k}(-1)^{k+i}\frac{1}{k^q i^p} \overset{\substack{\text{reverse the order}\\\text{of summation}}}{=} \sum_{i=1}^{n}\sum_{k=i}^{n}(-1)^{k+i}\frac{1}{k^q i^p}$$

$$= \sum_{i=1}^{n}(-1)^{i-1}\frac{1}{i^p}\left((-1)^{i-1}\frac{1}{i^q} + \overline{H}_n^{(q)} - \overline{H}_i^{(q)}\right) = \sum_{i=1}^{n}\frac{1}{i^{p+q}}$$

$$+ \sum_{i=1}^{n}(-1)^{i-1}\frac{\overline{H}_n^{(q)}}{i^p} - \sum_{i=1}^{n}(-1)^{i-1}\frac{\overline{H}_i^{(q)}}{i^p} = H_n^{(p+q)} + \overline{H}_n^{(p)}\overline{H}_n^{(q)} - \sum_{k=1}^{n}(-1)^{k-1}\frac{\overline{H}_k^{(q)}}{k^p},$$

and the proof of the auxiliary result is done.   ∎

Alternatively, the result may be extracted by using Abel's summation (see [76, Chapter 2, pp.39–40]).

Setting $p = q$ in (3.170), we arrive at the desired result,

$$\sum_{k=1}^{n}(-1)^{k-1}\frac{\overline{H}_k^{(p)}}{k^p} = \frac{1}{2}\left(\left(\overline{H}_n^{(p)}\right)^2 + H_n^{(2p)}\right),$$

and the solution to the point (i) is finalized.

The value of the integral at the point (ii) is easily extracted if we combine (1.99), Sect. 1.21, in variable k, the use of $\sum_{k=1}^{n} x^{k-1} = \dfrac{1-x^n}{1-x}$, and the result from the previous point, the case $p = 1$. Therefore, we have that

$$\int_0^1 \frac{1-x^n}{1-x}\log(1+x)\mathrm{d}x = \int_0^1 \sum_{k=1}^{n} x^{k-1}\log(1+x)\mathrm{d}x = \sum_{k=1}^{n}\int_0^1 x^{k-1}\log(1+x)\mathrm{d}x$$

$$= \log(2)\sum_{k=1}^{n}\frac{1}{k} + \log(2)\sum_{k=1}^{n}(-1)^{k-1}\frac{1}{k} - \sum_{k=1}^{n}(-1)^{k-1}\frac{\overline{H}_k}{k}$$

$$= \log(2)H_n + \log(2)\overline{H}_n - \frac{1}{2}\left(\overline{H}_n^2 + H_n^{(2)}\right),$$

or to put the essential part, we have

$$\int_0^1 \frac{1-x^n}{1-x}\log(1+x)\mathrm{d}x = \log(2)H_n + \log(2)\overline{H}_n - \frac{1}{2}\left(\overline{H}_n^2 + H_n^{(2)}\right),$$

and the solution to the point (ii) is finalized.

The point (iii) is straightforward if we consider the result from the previous point and integrate by parts. Therefore, we write

$$\int_0^1 \frac{1-x^n}{1-x}\log(1+x)\mathrm{d}x = -\int_0^1 (\log(1-x))'(1-x^n)\log(1+x)\mathrm{d}x$$

$$= \int_0^1 \frac{\log(1-x)}{1+x}\mathrm{d}x - \int_0^1 x^n\frac{\log(1-x)}{1+x}\mathrm{d}x - n\int_0^1 x^{n-1}\log(1-x)\log(1+x)\mathrm{d}x$$

$$= \log(2)H_n + \log(2)\overline{H}_n - \frac{1}{2}\left(\overline{H}_n^2 + H_n^{(2)}\right),$$

whence we obtain that

$$\int_0^1 x^n\frac{\log(1-x)}{1+x}\mathrm{d}x + n\int_0^1 x^{n-1}\log(1-x)\log(1+x)\mathrm{d}x$$

$$= \frac{1}{2}\left(\log^2(2) - \frac{\pi^2}{6}\right) - \log(2)H_n - \log(2)\overline{H}_n + \frac{1}{2}\left(\overline{H}_n^2 + H_n^{(2)}\right),$$

where in the calculations I used that $\int_0^1 \frac{\log(1-x)}{1+x}dx = \frac{1}{2}\left(\log^2(2) - \frac{\pi^2}{6}\right)$, which is given (3.163), Sect. 3.21, and the solution to the point (iii) is finalized.

Regarding the sum at the point (iv), at the beginning of the section I already emphasized the similarity to the sum $\sum_{k=1}^{n} \frac{H_k^2 + H_k^{(2)}}{k} = \frac{1}{3}(H_n^3 + 3H_n H_n^{(2)} + 2H_n^{(3)})$.

So, for a first approach we may follow closely the solution to the previously mentioned sum, which may be found in [76, Chapter 6, pp.61–62]. Thus, one might want to start by applying Abel's summation (see [76, Chapter 2, pp.39–40]) with $a_k = (-1)^{k-1}/k$ and $b_k = \overline{H}_k^2 + H_k^{(2)}$, and hence the curious reader may continue the calculations.

For a second solution, we could choose to use a powerful result involving the complete homogeneous symmetric polynomial expressed in terms of power sums, which as at the preceding suggested solution brings us back to *(Almost) Impossible Integrals, Sums, and Series*, more precisely in [76, Chapter 6, p.359],

$$\sum_{k=1}^{n}\sum_{l=1}^{k}\sum_{m=1}^{l} x_k x_l x_m = \frac{1}{6}\left(\left(\sum_{k=1}^{n} x_k\right)^3 + 3\left(\sum_{k=1}^{n} x_k\right)\left(\sum_{k=1}^{n} x_k^2\right) + 2\sum_{k=1}^{n} x_k^3\right).$$
(3.171)

So, if we set $x_k = (-1)^{k-1}1/k$ in (3.171) and count that $\sum_{k=1}^{n}(-1)^{k-1}\frac{1}{k^p} = \overline{H}_n^{(p)}$, we obtain that

$$\sum_{k=1}^{n}(-1)^{k-1}\frac{1}{k}\sum_{l=1}^{k}(-1)^{l-1}\frac{1}{l}\sum_{m=1}^{l}(-1)^{m-1}\frac{1}{m} = \sum_{k=1}^{n}(-1)^{k-1}\frac{1}{k}\sum_{l=1}^{k}(-1)^{l-1}\frac{\overline{H}_l}{l}$$

{make use of the sum result found at the point (i), the case $p = 1$}

$$= \frac{1}{2}\sum_{k=1}^{n}(-1)^{k-1}\frac{\overline{H}_k^2 + H_k^{(2)}}{k} = \frac{1}{6}\left(\overline{H}_n^3 + 3\overline{H}_n H_n^{(2)} + 2\overline{H}_n^{(3)}\right),$$

where if we count the last equality and simplify, we get

$$\sum_{k=1}^{n}(-1)^{k-1}\frac{\overline{H}_k^2 + H_k^{(2)}}{k} = \frac{1}{3}\left(\overline{H}_n^3 + 3\overline{H}_n H_n^{(2)} + 2\overline{H}_n^{(3)}\right),$$

which brings an end to the solution above.

What about the result at the point (v)? Well, again, using results from the previous points we'll manage to finalize it promptly.

So, I will consider the result from the point (ii), in variable k, and if we multiply both sides by $(-1)^{k-1}/k$ and then sum from $k = 1$ to $k = n$, we get

$$\int_0^1 \sum_{k=1}^n (-1)^{k-1} \frac{1}{k} \frac{1-x^k}{1-x} \log(1+x)\mathrm{d}x$$

$$= \log(2) \sum_{k=1}^n (-1)^{k-1} \frac{H_k}{k} + \log(2) \sum_{k=1}^n (-1)^{k-1} \frac{\overline{H}_k}{k} - \frac{1}{2} \sum_{k=1}^n (-1)^{k-1} \frac{\overline{H}_k^2 + H_k^{(2)}}{k}$$

$$= \frac{1}{2} \log(2) \left(\overline{H}_n^2 + H_n^{(2)} \right) - \frac{1}{6} \left(\overline{H}_n^3 + 3\overline{H}_n H_n^{(2)} + 2\overline{H}_n^{(3)} \right) + \log(2) \sum_{k=1}^n (-1)^{k-1} \frac{H_k}{k},$$

where in the calculations I used the main results from the point (i), the case $p = 1$, and point (iv), in the current section, and the first equality of the point (v) is proved.

To get the second equality of (v), consider above the sum transformation in (3.157), Sect. 3.21, and then we have

$$\int_0^1 \sum_{k=1}^n (-1)^{k-1} \frac{1}{k} \frac{1-x^k}{1-x} \log(1+x)\mathrm{d}x$$

$$= \log(2) H_n \overline{H}_n + \log(2) \overline{H}_n^{(2)} + \frac{1}{2} \log(2) \left(\overline{H}_n^2 + H_n^{(2)} \right) - \frac{1}{6} \left(\overline{H}_n^3 + 3\overline{H}_n H_n^{(2)} + 2\overline{H}_n^{(3)} \right)$$

$$- \log(2) \sum_{k=1}^n \frac{\overline{H}_k}{k},$$

and the solution to the point (v) is finalized.

Before moving further, keep in mind that being equipped with one of the main results in this section is of great help in one of the upcoming sections!

3.24 A Pair of Precious Generalized Logarithmic Integrals

Solution The reader of *(Almost) Impossible Integrals, Sums, and Series* will probably find the two proposed generalized logarithmic integrals familiar to some extent since integrals of the type $\int_0^1 x^{n-1} \log^m(1-x)\mathrm{d}x$, $m, n \in \mathbb{N}$, are given as problems in [76, Chapter 1, Sect. 1.3, p.2]. For example, I remind you the main result from the point (i), with the cases $m = 1, 2$, I needed during the calculations in [76, Chapter 3, Sect. 3.55, pp.258–260], and, as seen there, one way to go is by combining $\displaystyle\int_0^1 x^{n-1}\log(1-x)\mathrm{d}x = -\frac{\psi(n+1)+\gamma}{n}$ and differentiation. So, we have an idea on how to attack the point (i) of the problem, and the same idea works for the other point too, if we want to extract the generalizations!

For a first solution to the point (i) I will focus on a strategy that uses recurrence relations and avoids the use of differentiation. In order to do that, I'll cleverly integrate by parts (as in [76, Chapter 3, pp.59–62]), and then we write

$$\int_0^1 x^{n-1}\log^m(x)\log(1-x)\mathrm{d}x = \frac{1}{n}\int_0^1 (x^n-1)'\log^m(x)\log(1-x)\mathrm{d}x$$

$$= -\frac{m}{n}\int_0^1 x^{n-1}\log^{m-1}(x)\log(1-x)\mathrm{d}x - \frac{1}{n}\int_0^1 \frac{1-x^n}{1-x}\log^m(x)\mathrm{d}x$$

$$+ \frac{m}{n}\int_0^1 \frac{\log(1-x)\log^{m-1}(x)}{x}\mathrm{d}x. \tag{3.172}$$

Now, for the second integral in the right-hand side of (3.172), we have

$$\int_0^1 \frac{1-x^n}{1-x}\log^m(x)\mathrm{d}x = \sum_{k=1}^n \int_0^1 x^{k-1}\log^m(x)\mathrm{d}x = -(-1)^{m-1}m!\sum_{k=1}^n \frac{1}{k^{m+1}}$$

$$= -(-1)^{m-1}m!H_n^{(m+1)}, \tag{3.173}$$

where the resulting logarithmic integral is known (see [76, Chapter 1, p.1]) and it can be easily derived by using recurrence relations.

Then, for the third integral in (3.172) we exploit the Taylor series $\log(1-x) = -\sum_{n=1}^\infty \frac{x^n}{n}$ and the fact that $\displaystyle\int_0^1 x^{n-1}\log^{m-1}(x)\mathrm{d}x = (-1)^{m-1}(m-1)!\frac{1}{n^m}$ that lead to

$$\int_0^1 \frac{\log(1-x)\log^{m-1}(x)}{x}\mathrm{d}x = -\int_0^1 \sum_{n=1}^\infty \frac{x^{n-1}}{n}\log^{m-1}(x)\mathrm{d}x$$

$$= -\sum_{n=1}^\infty \frac{1}{n}\int_0^1 x^{n-1}\log^{m-1}(x)\mathrm{d}x = -(-1)^{m-1}(m-1)!\zeta(m+1). \tag{3.174}$$

If we denote $I_{m,n} = \displaystyle\int_0^1 x^{n-1}\log^m(x)\log(1-x)\mathrm{d}x$ and return to (3.172) together with the results in (3.173) and (3.174), and then rearrange, we have

$$m!(\zeta(m+1) - H_n^{(m+1)}) = n(-1)^m I_{m,n} - (-1)^{m-1} m I_{m-1,n}. \qquad (3.175)$$

Further, if we replace m by k in (3.175), then multiply both sides of (3.175) by $n^{k-1}/k!$, and sum from $k=1$ to $k=m$, we get

$$\sum_{k=1}^{m} \left((-1)^k \frac{1}{k!} n^k I_{k,n} - (-1)^{k-1} \frac{1}{(k-1)!} n^{k-1} I_{k-1,n} \right) = -(-1)^{m-1} \frac{1}{m!} n^m I_{m,n} - I_{0,n}$$

$$= \sum_{k=1}^{m} n^{k-1} (\zeta(k+1) - H_n^{(k+1)}),$$

and if we take the last equality and rearrange it, we arrive at

$$I_{m,n} = \int_0^1 x^{n-1} \log^m(x) \log(1-x) dx$$

$$= -(-1)^{m-1} m! \frac{1}{n^m} I_{0,n} - (-1)^{m-1} m! \sum_{k=1}^{m} \frac{1}{n^{m-k+1}} (\zeta(k+1) - H_n^{(k+1)})$$

$$= (-1)^{m-1} m! \frac{H_n}{n^{m+1}} - (-1)^{m-1} m! \sum_{k=1}^{m} \frac{1}{n^{m-k+1}} (\zeta(k+1) - H_n^{(k+1)}),$$

where in the calculations I used that $I_{0,n} = \int_0^1 x^{n-1} \log(1-x) dx = -\frac{H_n}{n}$, which is given in (3.10), Sect. 3.3, and the first solution to the point (i) is complete.

In order to construct a second solution, we might want to consider an extended form of the last mentioned integral in the previous solution, that is,

$$\int_0^1 x^{n-1} \log(1-x) dx = -\frac{\psi(n+1) + \gamma}{n}, \qquad (3.176)$$

which is achieved by noting and using that $H_n = \psi(n+1) + \gamma$.

We observe that if we differentiate m times with respect to n both sides of (3.176), we immediately get that

$$\int_0^1 x^{n-1} \log^m(x) \log(1-x) dx = -\frac{d^m}{dn^m} \left\{ \frac{\psi(n+1) + \gamma}{n} \right\}. \qquad (3.177)$$

Considering that $\dfrac{d^m}{dn^m}\left(\dfrac{1}{n}\right) = (-1)^m m!\dfrac{1}{n^{m+1}}$, $\dfrac{d^m}{dn^m}(\psi(n+1)+\gamma) = \psi^{(m)}(n+$

1), and using that for n a positive integer, $\psi^{(m)}(n+1) = (-1)^{m-1}m!\displaystyle\sum_{k=n+1}^{\infty}\dfrac{1}{k^{m+1}} =$

$(-1)^{m-1}m!\left(\zeta(m+1) - \displaystyle\sum_{k=1}^{n}\dfrac{1}{k^{m+1}}\right) = (-1)^{m-1}m!\left(\zeta(m+1) - H_n^{(m+1)}\right)$, if

we apply the *general Leibniz rule*, $(fg)^{(m)}(x) = \displaystyle\sum_{k=0}^{m}\binom{m}{k}f^{(m-k)}(x)g^{(k)}(x)$,

or if we let the term for $k = 0$ out of the sum, $(fg)^{(m)}(x) = f^{(m)}g +$

$\displaystyle\sum_{k=1}^{m}\binom{m}{k}f^{(m-k)}(x)g^{(k)}(x)$, where we set $f = \dfrac{1}{n}$ and $g = \psi(n+1)+\gamma$, we

obtain that

$$\frac{d^m}{dn^m}\left\{\frac{\psi(n+1)+\gamma}{n}\right\}$$

$$= \left(\frac{1}{n}\right)^{(m)}(\psi(n+1)+\gamma) + \sum_{k=1}^{m}\binom{m}{k}\left(\frac{1}{n}\right)^{(m-k)}(\psi(n+1)+\gamma)^{(k)}$$

$$= -(-1)^{m-1}m!\frac{H_n}{n^{m+1}} + (-1)^{m-1}\sum_{k=1}^{m}\binom{m}{k}\frac{(m-k)!k!}{n^{m-k+1}}\left(\zeta(k+1) - H_n^{(k+1)}\right)$$

$$= -(-1)^{m-1}m!\frac{H_n}{n^{m+1}} + (-1)^{m-1}m!\sum_{k=1}^{m}\frac{1}{n^{m-k+1}}\left(\zeta(k+1) - H_n^{(k+1)}\right),$$

$$(3.178)$$

where I have counted that after differentiation we set n a positive integer.

Finally, if we plug (3.178) in (3.177), we obtain that

$$\int_0^1 x^{n-1}\log^m(x)\log(1-x)dx$$

$$= (-1)^{m-1}m!\frac{H_n}{n^{m+1}} - (-1)^{m-1}m!\sum_{k=1}^{m}\frac{1}{n^{m-k+1}}(\zeta(k+1) - H_n^{(k+1)}),$$

and the second solution to the point (i) is complete.

Regarding the point (ii) of the problem, again we'll want to construct a solution involving a strategy similar to the one found in the second solution to the point (i), which uses differentiation. Before proceeding with the main calculations, we want to prepare some key auxiliary results.

Using that $\dfrac{d^m}{dn^m}\{\psi(n+1)+\gamma\} = \psi^{(m)}(n+1)$, where we consider that

when n is a positive integer we get $\psi^{(m)}(n+1) = (-1)^{m-1}m!\displaystyle\sum_{k=n+1}^{\infty}\frac{1}{k^{m+1}} =$

$(-1)^{m-1}m!\left(\zeta(m+1)-\displaystyle\sum_{k=1}^{n}\frac{1}{k^{m+1}}\right) = (-1)^{m-1}m!\left(\zeta(m+1)-H_n^{(m+1)}\right)$, together

with the result in (3.178), and then applying *general Leibniz rule*, where we set

$f = \dfrac{\psi(n+1)+\gamma}{n}$ and $g = \psi(n+1)+\gamma$, we obtain that

$$\frac{d^m}{dn^m}\left\{\frac{(\psi(n+1)+\gamma)^2}{n}\right\} = \left(\frac{\psi(n+1)+\gamma}{n}\right)^{(m)}(\psi(n+1)+\gamma)$$

$$+\sum_{k=1}^{m}\binom{m}{k}\left(\frac{\psi(n+1)+\gamma}{n}\right)^{(m-k)}(\psi(n+1)+\gamma)^{(k)}$$

$$= -(-1)^{m-1}m!\frac{H_n}{n^{m+1}}\left(H_n - \sum_{i=1}^{m}n^i\left(\zeta(i+1)-H_n^{(i+1)}\right)\right)$$

$$+(-1)^{m-1}m!\sum_{i=1}^{m}\frac{1}{n^{m-i+1}}\left(\zeta(i+1)-H_n^{(i+1)}\right)\left(H_n-\sum_{j=1}^{m-i}n^j\left(\zeta(j+1)-H_n^{(j+1)}\right)\right)$$

{in the second sum let the variable change $m-i+1 = l$, and then replace l by i}

$$= -(-1)^{m-1}m!\frac{H_n}{n^{m+1}}\left(H_n-\sum_{j=1}^{m}n^j\left(\zeta(j+1)-H_n^{(j+1)}\right)\right)$$

$$+(-1)^{m-1}m!\sum_{i=1}^{m}\frac{1}{n^i}\left(\zeta(m-i+2)-H_n^{(m-i+2)}\right)\left(H_n-\sum_{j=1}^{i-1}n^j\left(\zeta(j+1)-H_n^{(j+1)}\right)\right),$$

$$(3.179)$$

where I have considered that after differentiation we set n a positive integer.

Further, since we have that $\dfrac{d^m}{dn^m}\{\zeta(2)-\psi^{(1)}(n+1)\} = -\psi^{(m+1)}(n+1)$, next

$-\psi^{(m+1)}(n+1) = (-1)^{m-1}(m+1)!\displaystyle\sum_{k=n+1}^{\infty}\frac{1}{k^{m+2}} = (-1)^{m-1}(m+1)!\Big(\zeta(m+2)-$

$\displaystyle\sum_{k=1}^{n}\frac{1}{k^{m+2}}\Big) = (-1)^{m-1}(m+1)!\left(\zeta(m+2)-H_n^{(m+2)}\right)$, for n a positive integer, and $\dfrac{d^m}{dn^m}\left(\dfrac{1}{n}\right) = (-1)^m m! \dfrac{1}{n^{m+1}}$, if we apply *general Leibniz rule*, where we set $f = \dfrac{1}{n}$ and $g = \zeta(2) - \psi^{(1)}(n+1)$, we get

$$\dfrac{d^m}{dn^m}\left\{\dfrac{\zeta(2) - \psi^{(1)}(n+1)}{n}\right\}$$

$$= \left(\dfrac{1}{n}\right)^{(m)}(\zeta(2) - \psi^{(1)}(n+1)) + \sum_{i=1}^{m}\binom{m}{i}\left(\dfrac{1}{n}\right)^{(m-i)}(\zeta(2) - \psi^{(1)}(n+1))^{(i)}$$

$$= -(-1)^{m-1}m!\dfrac{H_n^{(2)}}{n^{m+1}} + (-1)^{m-1}\sum_{i=1}^{m}\binom{m}{i}\dfrac{(m-i)!(i+1)!}{n^{m-i+1}}\left(\zeta(i+2) - H_n^{(i+2)}\right)$$

$$= -(-1)^{m-1}m!\dfrac{H_n^{(2)}}{n^{m+1}} + (-1)^{m-1}m!\sum_{i=1}^{m}\dfrac{i+1}{n^{m-i+1}}\left(\zeta(i+2) - H_n^{(i+2)}\right),$$

(3.180)

where, as at other previous results, we assume that after differentiation we set n a positive integer.

Now, we are ready to proceed with the main calculations, and then we start with the following integral:

$$\int_0^1 x^{n-1}\log^2(1-x)dx = \dfrac{(\psi(n+1) + \gamma)^2}{n} + \dfrac{\zeta(2) - \psi^{(1)}(n+1)}{n}, \qquad (3.181)$$

which is an extended version of the integral $\displaystyle\int_0^1 x^{n-1}\log^2(1-x)dx = \dfrac{H_n^2 + H_n^{(2)}}{n}$, appearing in [76, Chapter 1, Section 1.3, p.2].

At this point, we differentiate m times with respect to n both sides of (3.181), and by plugging in the results from (3.179) and (3.180), we obtain that

$$\int_0^1 x^{n-1}\log^m(x)\log^2(1-x)dx$$

$$= \dfrac{d^m}{dn^m}\left\{\dfrac{(\psi(n+1) + \gamma)^2}{n}\right\} + \dfrac{d^m}{dn^m}\left\{\dfrac{\zeta(2) - \psi^{(1)}(n+1)}{n}\right\}$$

$$= (-1)^{m-1}m!\sum_{i=1}^{m}\dfrac{1}{n^i}\left(\zeta(m-i+2) - H_n^{(m-i+2)}\right)\left(H_n - \sum_{j=1}^{i-1}n^j\left(\zeta(j+1) - H_n^{(j+1)}\right)\right)$$

$$+ (-1)^{m-1} m! \left(\sum_{i=1}^{m} \frac{H_n}{n^{m-i+1}} \left(\zeta(i+1) - H_n^{(i+1)} \right) + \sum_{i=1}^{m} \frac{i+1}{n^{m-i+1}} \left(\zeta(i+2) - H_n^{(i+2)} \right) \right)$$

$$- (-1)^{m-1} m! \frac{1}{n^{m+1}} (H_n^2 + H_n^{(2)}),$$

and the solution to the point (ii) is complete. One might also get a slightly different version of the closed form by observing that $\displaystyle\sum_{i=1}^{m} \frac{H_n}{n^{m-i+1}} \left(\zeta(i+1) - H_n^{(i+1)} \right)$ can be extracted two times from above, one time with a reversed summation of the terms.

In the next section, we'll continue to calculate similar logarithmic integrals to the ones above, but with $\log(1+x)$ instead of $\log(1-x)$, and some of the ideas exploited in the present section might be found (very) useful.

3.25 Two Neat and Useful Generalizations with Logarithmic Integrals Involving Skew-Harmonic Numbers

Solution From a certain perspective, both given results from the present section can be seen as more complex generalizations with logarithmic integrals than the one given in (1.99), Sect. 1.21. And we'll find useful such results! For example, a particular case of the generalized integral at the point (i) arises in the extraction process of the coefficient (expressed in terms of skew-harmonic numbers) of a certain Fourier series given in the fourth chapter.

In my solution below I'll exploit a strategy that relies on the use of recurrence relations.

So, using the fact that $(-1)^{m-1} \dfrac{1}{(m-1)!} \displaystyle\int_0^1 x^{k-1} \log^{m-1}(x)\mathrm{d}x = \dfrac{1}{k^m}$, and then returning to the nth generalized skew-harmonic number of order m, $m \geq 2$, we write

$$\overline{H}_n^{(m)} = \sum_{k=1}^{n} (-1)^{k-1} \frac{1}{k^m} = (-1)^{m-1} \frac{1}{(m-1)!} \sum_{k=1}^{n} (-1)^{k-1} \int_0^1 x^{k-1} \log^{m-1}(x)\mathrm{d}x$$

$$= (-1)^{m-1} \frac{1}{(m-1)!} \int_0^1 \log^{m-1}(x) \sum_{k=1}^{n} (-1)^{k-1} x^{k-1} \mathrm{d}x$$

$$= (-1)^{m-1} \frac{1}{(m-1)!} \int_0^1 \frac{1 - (-1)^n x^n}{1+x} \log^{m-1}(x)\mathrm{d}x$$

$$\stackrel{\substack{\text{integrate by}\\\text{parts}}}{=} -(-1)^{m-1}\frac{1}{(m-2)!}\int_0^1 \frac{\log(1+x)\log^{m-2}(x)}{x}dx$$

$$+ (-1)^{(m-1)+n}\frac{1}{(m-2)!}\int_0^1 x^{n-1}\log^{m-2}(x)\log(1+x)dx$$

$$- (-1)^{m+n}\frac{1}{(m-1)!}\int_0^1 nx^{n-1}\log^{m-1}(x)\log(1+x)dx. \tag{3.182}$$

Now, for the first remaining integral in (3.182), we use that $\log(1+x) = \sum_{n=1}^{\infty}(-1)^{n-1}\frac{x^n}{n}$, and then we have

$$\int_0^1 \frac{\log(1+x)\log^{m-2}(x)}{x}dx = \int_0^1 \sum_{n=1}^{\infty}(-1)^{n-1}\frac{x^{n-1}}{n}\log^{m-2}(x)dx$$

$$= \sum_{n=1}^{\infty}(-1)^{n-1}\frac{1}{n}\int_0^1 x^{n-1}\log^{m-2}(x)dx = (-1)^m(m-2)!\sum_{n=1}^{\infty}(-1)^{n-1}\frac{1}{n^m}$$

$$= (-1)^m(m-2)!\eta(m) = (-1)^m(m-2)!(1-2^{1-m})\zeta(m), \quad m \geq 2, \tag{3.183}$$

where I used the fact that the Dirichlet eta function (see [93]), $\eta(s) = \sum_{n=1}^{\infty}(-1)^{n-1}\frac{1}{n^s}$, can be expressed in terms of the Riemann zeta function by the relation $\eta(s) = (1-2^{1-s})\zeta(s)$.

If we denote $I_{m,n} = (-1)^{m-1}\frac{1}{(m-2)!}\int_0^1 x^{n-1}\log^{m-2}(x)\log(1+x)dx$ and then return to (3.182) together with (3.183), and rearrange, we get

$$(-1)^{n-1}\overline{H}_n^{(m)} - (-1)^{n-1}(1-2^{1-m})\zeta(m) = nI_{m+1,n} - I_{m,n}. \tag{3.184}$$

At this point, we replace m by k in (3.184), then multiply both sides by n^k, and sum from $k = 2$ to $k = m+1$ (note that this time we consider $m \geq 1$) that gives

$$\sum_{k=2}^{m+1}(n^{k+1}I_{k+1,n} - n^k I_{k,n}) = n^{m+2}I_{m+2,n} - n^2I_{2,n}$$

$$= (-1)^{n-1}\sum_{k=2}^{m+1}n^k\overline{H}_n^{(k)} - (-1)^{n-1}\sum_{k=2}^{m+1}n^k(1-2^{1-k})\zeta(k),$$

whence after multiplying both sides of the last equality by $(-1)^{m-1}m!\frac{1}{n^{m+2}}$, reindexing the first sum, and rearranging, we obtain that

$$(-1)^{m-1}m!I_{m+2,n} = \int_0^1 x^{n-1}\log^m(x)\log(1+x)dx = (-1)^{m-1}m!\frac{I_{2,n}}{n^m}$$

$$+ (-1)^{m+n}m!\sum_{k=1}^{m}\frac{\overline{H}_n^{(k+1)}}{n^{m-k+1}} - (-1)^{m+n}m!\sum_{k=2}^{m+1}\frac{1}{n^{m-k+2}}(1-2^{1-k})\zeta(k)$$

$$= (-1)^{m+n}m!\frac{\overline{H}_n}{n^{m+1}} - \log(2)(-1)^{m-1}m!\frac{1}{n^{m+1}} - \log(2)(-1)^{m+n}m!\frac{1}{n^{m+1}}$$

$$+ (-1)^{m+n}m!\sum_{k=1}^{m}\frac{\overline{H}_n^{(k+1)}}{n^{m-k+1}} - (-1)^{m+n}m!\sum_{k=2}^{m+1}\frac{1}{n^{m-k+2}}(1-2^{1-k})\zeta(k),$$

where in the calculations I used $I_{2,n}$ which is extracted based on (1.99), Sect. 1.21, and the solution to the point (i) is finalized.

For a second solution, the curious reader may start from $\int_0^1 x^{n-1}\log(1+x)dx$ and then exploit differentiation. A strategy exploiting such ideas is given in the solution to the point (ii) you may find below.

At the second point of the section another beautiful generalization lies in front of us! *How to (elegantly) prove it?*, you might wonder.

A first key result needed in the main derivation is obtained by combining the results in (1.121), Sect. 1.23, and (4.97), Sect. 4.18, that is,

$$\int_0^1 \frac{1-x^n}{1-x}\log(1+x)dx$$

$$= \log(2)\gamma + 2\log(2)\psi(n+1) - \log(2)\psi\left(\frac{n}{2}+1\right) - \frac{1}{2}(\psi(n+1))^2$$

$$+ \psi(n+1)\psi\left(\frac{n}{2}+1\right) - \frac{1}{2}\left(\psi\left(\frac{n}{2}+1\right)\right)^2 - \frac{1}{2}(\zeta(2) - \psi^{(1)}(n+1)). \quad (3.185)$$

Why a closed form in Polygamma function (in particular, in Digamma and Trigamma functions)? We need a form in Polygamma function since we want to exploit differentiation (this step assures the extension to real numbers, which we need in order to differentiate).

At this point, we gather some results we need during the main calculations. For example, we need a slightly modified result of the one found in (6.143), Sect. 6.19,

$$\frac{d^m}{dn^m}\{(\psi(n+1))^2\}$$

$$= 2\psi(n+1)\psi^{(m)}(n+1) + \sum_{k=1}^{m-1}\binom{m}{k}\psi^{(m-k)}(n+1)\psi^{(k)}(n+1), \quad m \geq 1. \quad (3.186)$$

By using again *general Leibniz rule* and following a similar way to go as explained at mentioned reference above, we obtain that

$$\frac{d^m}{dn^m}\left\{\left(\psi\left(\frac{n}{2}+1\right)\right)^2\right\}$$

$$= \frac{1}{2^{m-1}}\psi\left(\frac{n}{2}+1\right)\psi^{(m)}\left(\frac{n}{2}+1\right) + \frac{1}{2^m}\sum_{k=1}^{m-1}\binom{m}{k}\psi^{(m-k)}\left(\frac{n}{2}+1\right)\psi^{(k)}\left(\frac{n}{2}+1\right)$$

(3.187)

and

$$\frac{d^m}{dn^m}\left\{\psi(n+1)\psi\left(\frac{n}{2}+1\right)\right\}$$

$$= \psi^{(m)}(n+1)\psi\left(\frac{n}{2}+1\right) + \frac{1}{2^m}\psi(n+1)\psi^{(m)}\left(\frac{n}{2}+1\right)$$

$$+ \sum_{k=1}^{m-1}\frac{1}{2^k}\binom{m}{k}\psi^{(m-k)}(n+1)\psi^{(k)}\left(\frac{n}{2}+1\right),\ m\geq 1.$$

(3.188)

If we differentiate m times with respect to n both sides of (3.185) and then plug in the results from (3.186), (3.187), and (3.188), where we also consider that $\frac{d^m}{dn^m}\{\psi(n+1)\} = \psi^{(m)}(n+1)$, $\frac{d^m}{dn^m}\{\psi^{(1)}(n+1)\} = \psi^{(m+1)}(n+1)$, and $\frac{d^m}{dn^m}\left\{\psi\left(\frac{n}{2}+1\right)\right\} = \frac{1}{2^m}\psi^{(m)}\left(\frac{n}{2}+1\right)$, $m\geq 1$, we arrive at the desired result

$$\int_0^1 \frac{x^n\log^m(x)\log(1+x)}{1-x}dx$$

$$= -\frac{1}{2}\psi^{(m+1)}(n+1) - 2\log(2)\psi^{(m)}(n+1) + \psi(n+1)\psi^{(m)}(n+1)$$

$$- \psi\left(\frac{n}{2}+1\right)\psi^{(m)}(n+1) + \log(2)\frac{1}{2^m}\psi^{(m)}\left(\frac{n}{2}+1\right) - \frac{1}{2^m}\psi(n+1)\psi^{(m)}\left(\frac{n}{2}+1\right)$$

$$+ \frac{1}{2^m}\psi\left(\frac{n}{2}+1\right)\psi^{(m)}\left(\frac{n}{2}+1\right) + \sum_{k=1}^{m-1}\binom{m}{k}\left(\frac{1}{2}\psi^{(k)}(n+1)\psi^{m-k}(n+1)\right.$$

$$\left. - \frac{1}{2^k}\psi^{(k)}\left(\frac{n}{2}+1\right)\psi^{(m-k)}(n+1) + \frac{1}{2^{m+1}}\psi^{(k)}\left(\frac{n}{2}+1\right)\psi^{(m-k)}\left(\frac{n}{2}+1\right)\right),$$

and the solution to the point (ii) is finalized.

A *wonderful fact* to observe is that the closed form of the last result above is expressible in terms of the generalized harmonic numbers and the generalized skew-harmonic numbers through results like the ones given in Sect. 4.18.

One may also notice that the generalized integral can also be reduced directly to a sum of integrals involving forms of the derivatives of the Beta function if we exploit simple algebraic identities.

What other options would we have? A beautiful idea to exploit is based on simple observations that consist of relating the two main integrals from the points (i) and (ii). Note that if we replace n by $n + p$ in the result at the point (i) and make the summation from $p = 1$ to ∞, we obtain that

$$\sum_{p=1}^{\infty} \int_0^1 x^{n+p-1} \log^m(x) \log(1+x)dx = \int_0^1 \sum_{p=1}^{\infty} x^{n+p-1} \log^m(x) \log(1+x)dx$$

$$= \int_0^1 \frac{x^n \log^m(x) \log(1+x)}{1-x}dx. \qquad (3.189)$$

To easily see my point, we can take, say, $m = 1$ in (3.189), and then we have

$$\int_0^1 \frac{x^n \log(x) \log(1+x)}{1-x}dx$$

$$= \sum_{p=1}^{\infty} \left(-\frac{\pi^2}{12}(-1)^{n+p-1}\frac{1}{n+p} - \log(2)\frac{1}{(n+p)^2} - \log(2)(-1)^{n+p-1}\frac{1}{(n+p)^2} \right.$$

$$\left. +(-1)^{n+p-1}\frac{\overline{H}_{n+p}}{(n+p)^2} + (-1)^{n+p-1}\frac{\overline{H}_{n+p}^{(2)}}{n+p} \right)$$

$$\stackrel{\substack{\text{reindex} \\ \text{the series}}}{=} \sum_{p=n+1}^{\infty} \left(-\frac{\pi^2}{12}(-1)^{p-1}\frac{1}{p} - \log(2)\frac{1}{p^2} - \log(2)(-1)^{p-1}\frac{1}{p^2} \right.$$

$$\left. +(-1)^{p-1}\frac{\overline{H}_p}{p^2} + (-1)^{p-1}\frac{\overline{H}_p^{(2)}}{p} \right),$$

where the curious reader might like to use that $\displaystyle\sum_{p=n+1}^{\infty} f(p) = \sum_{p=1}^{\infty} f(p) - \sum_{p=1}^{n} f(p)$, where $f(p)$ is the summand above, together with simple manipulations with sums and series, and we also exploit that $\displaystyle\sum_{k=1}^{n}(-1)^{k-1}\frac{\overline{H}_k}{k^2} + \sum_{k=1}^{n}(-1)^{k-1}\frac{\overline{H}_k^{(2)}}{k} =$

$\overline{H}_n \overline{H}_n^{(2)} + H_n^{(3)}$, based on the result in (3.170), the case $p = 1, q = 2$ (or $p = 2, q = 1$), Sect. 3.23. Thus, we immediately extract the desired result.

3.26 A Special Logarithmic Integral and a Generalization of It

Solution There is a good chance, in case you read *(Almost) Impossible Integrals, Sums, and Series*, you might recall the integral from the first point appearing as an auxiliary result during the extraction of two integrals[12] (see [76, Chapter 3, p.98]). There I also used Beta function during the calculations, but this time we are given a series of restrictions: *no use of Beta function, Polylogarithm, or Euler sums!* They look like tough conditions, right? However, we have an ace up our sleeve!

So, I'll consider writing the integral at the point (i) by exploiting the case $m = 1$ of the generalized integral in (1.131), Sect. 1.27, and then, integrating by parts, rearranging, and splitting the resulting integral, we write

$$\int_0^1 \frac{\log(x)\log(1-x)}{1+x}dx = \int_0^1 (\log(1+x))' \log(x)\log(1-x)dx$$

$$= \underbrace{\int_0^1 \frac{\log(x)\log\left(\frac{1+x}{2}\right)}{1-x}dx}_{\zeta(3) - \log(2)\zeta(2)/2} - \int_0^1 \frac{\log(1-x)\log(1+x)}{x}dx + \log(2)\int_0^1 \frac{\log(x)}{1-x}dx.$$

(3.190)

To make it clear, the first integral obtained in (3.190) is found and calculated, in a generalized form, in the next section, by exploiting the symmetry in double integrals and avoiding the use of Beta function, Polylogarithm, and Euler sums.

Next, the second derived integral in (3.190) is elementarily calculated if we exploit algebraic identities, and then we write

$$\int_0^1 \frac{\log(1-x)\log(1+x)}{x}dx = \frac{1}{4}\int_0^1 \frac{\log^2(1-x^2)}{x}dx - \frac{1}{4}\int_0^1 \frac{1}{x}\log^2\left(\frac{1-x}{1+x}\right)dx$$

$$\left\{\text{in the first integral let } x^2 = y \text{ and in the second integral let } \frac{1-x}{1-x} = y\right\}$$

---

[12] It is about the integrals $\int_0^1 \frac{x\log(x)\log(1-x)}{1+x^2}dx$ and $\int_0^1 \frac{x\log(x)\log(1+x)}{1+x^2}dx$ that are attacked by using a system of relations. When considering their sum, after making the variable change $x^2 = y$, we arrive at the integral from the point (i).

$$= \frac{1}{8} \int_0^1 \frac{\log^2(1-y)}{y} dy - \frac{1}{2} \int_0^1 \frac{\log^2(y)}{1-y^2} dy = -\frac{5}{8}\zeta(3), \tag{3.191}$$

where in the calculations I used the simple facts that $\int_0^1 \frac{\log^2(1-y)}{y} dy \overset{1-y=z}{=}$

$$\int_0^1 \frac{\log^2(z)}{1-z} dz = \int_0^1 \sum_{n=1}^{\infty} z^{n-1} \log^2(z) dz = \sum_{n=1}^{\infty} \int_0^1 z^{n-1} \log^2(z) dz = 2 \sum_{n=1}^{\infty} \frac{1}{n^3} =$$

$$2\zeta(3) \text{ and } \int_0^1 \frac{\log^2(y)}{1-y^2} dy = \int_0^1 \sum_{n=1}^{\infty} y^{2n-2} \log^2(y) dy = \sum_{n=1}^{\infty} \int_0^1 y^{2n-2} \log^2(y) dy =$$

$$\sum_{n=1}^{\infty} \frac{2}{(2n-1)^3} = 2\left(\sum_{n=1}^{\infty} \frac{1}{n^3} - \sum_{n=1}^{\infty} \frac{1}{(2n)^3}\right) = \frac{7}{4}\zeta(3).$$

Further, for the last resulting integral in (3.190) we exploit geometric series, and then we have that

$$\int_0^1 \frac{\log(x)}{1-x} dx = \int_0^1 \sum_{n=1}^{\infty} x^{n-1} \log(x) dx = \sum_{n=1}^{\infty} \int_0^1 x^{n-1} \log(x) dx = -\sum_{n=1}^{\infty} \frac{1}{n^2} = -\zeta(2).$$
$$\tag{3.192}$$

At last, plugging (3.191) and (3.192) in (3.190), we arrive at

$$\int_0^1 \frac{\log(x) \log(1-x)}{1+x} dx = \frac{13}{8}\zeta(3) - \frac{3}{2}\log(2)\zeta(2),$$

and the solution to the point (i) is complete.

I largely presented in the paper "A note presenting the generalization of a special logarithmic integral" (see [78, August 13, 2019]) the strategy above together with the way to go for the generalized integral at the point (ii) (but with slight modifications as regards the closed form), which I'll consider below.

Further, as regards the second point of the problem, integrating by parts we have

$$\int_0^1 \frac{\log^{2n-1}(x) \log(1-x)}{1+x} dx = \int_0^1 (\log(1+x))' \log^{2n-1}(x) \log(1-x) dx$$

$$= \int_0^1 \frac{\log^{2n-1}(x) \log\left(\frac{1+x}{2}\right)}{1-x} dx$$

$$- (2n-1) \int_0^1 \frac{\log(1-x) \log^{2n-2}(x) \log(1+x)}{x} dx + \log(2) \int_0^1 \frac{\log^{2n-1}(x)}{1-x} dx.$$
$$\tag{3.193}$$

If we pick up the first integral in (3.193) and consider the result in (1.131), Sect. 1.27, we get

$$\int_0^1 \frac{\log^{2n-1}(x) \log\left(\frac{1+x}{2}\right)}{1-x} dx$$

$$= \frac{1}{2}(2n-1)! \left(2n\zeta(2n+1) - \sum_{k=0}^{2n-1} \eta(k+1)\eta(2n-k) \right). \tag{3.194}$$

Regarding the finite sum in (3.194), we exploit the parity that gives

$$\sum_{k=0}^{2n-1} \eta(k+1)\eta(2n-k) = \sum_{k=0}^{n-1} \eta(2k+1)\eta(2n-2k) + \sum_{k=0}^{n-1} \eta(2k+2)\eta(2n-2k-1)$$

$$= 2\sum_{k=0}^{n-1} \eta(2k+1)\eta(2n-2k), \tag{3.195}$$

where in the first line of (3.195), I reversed the order of summing the terms for the second sum after the equal sign in order to arrive at the last equal sign.

Further, if we plug (3.195) in (3.194), we have

$$\int_0^1 \frac{\log^{2n-1}(x) \log\left(\frac{1+x}{2}\right)}{1-x} dx = (2n-1)! \left(n\zeta(2n+1) - \sum_{k=0}^{n-1} \eta(2k+1)\eta(2n-2k) \right). \tag{3.196}$$

As regards the second integral in (3.193), we recall the following generalized integral presented in *(Almost) Impossible Integrals, Sums, and Series*:

$$\int_0^1 \frac{\log(1-x) \log^{2n}(x) \log(1+x)}{x} dx = \frac{1}{2^{2n+3}}(2n+3 - 2^{2n+3})(2n)!\zeta(2n+3)$$

$$+ \frac{1}{2}\left(1 - \frac{1}{2^{2n+1}}\right)(2n)! \sum_{k=1}^{2n} \zeta(k+1)\zeta(2n-k+2) - (2n)! \sum_{k=1}^{n} \eta(2k)\zeta(2n-2k+3), \tag{3.197}$$

where compared to the form in [76, Chapter 1, p.6], I also used the Dirichlet eta function.

Now, based on parity reasons, we have

$$\sum_{k=1}^{2n-2} \zeta(k+1)\zeta(2n-k) = \sum_{k=1}^{n-1} \zeta(2k)\zeta(2n-2k+1) + \sum_{k=1}^{n-1} \zeta(2k+1)\zeta(2n-2k)$$

$$= 2\sum_{k=1}^{n-1} \zeta(2k)\zeta(2n-2k+1), \tag{3.198}$$

where in the first line of (3.198), I reversed the order of summing the terms for the second sum after the equal sign in order to arrive at the last equal sign.

Then, if we replace n by $n-1$ in (3.197), where we also plug in (3.198), we get

$$\int_0^1 \frac{\log(1-x)\log^{2n-2}(x)\log(1+x)}{x}\,dx = \frac{1}{2^{2n+1}}(2n+1-2^{2n+1})(2n-2)!\zeta(2n+1)$$

$$+\left(1-\frac{1}{2^{2n-1}}\right)(2n-2)!\sum_{k=1}^{n-1}\zeta(2k)\zeta(2n-2k+1)-(2n-2)!\sum_{k=1}^{n-1}\eta(2k)\zeta(2n-2k+1).$$

$$(3.199)$$

Next, if for the last integral in (3.193) we use the fact that $\displaystyle\int_0^1 x^k \log^n(x)dx =$
$(-1)^n\dfrac{n!}{(k+1)^{n+1}}$, $k, n \in \mathbb{N}$, also presented in [76, Chapter 1, p.1], we have

$$\int_0^1 \frac{\log^{2n-1}(x)}{1-x}\,dx = \int_0^1 \log^{2n-1}(x)\sum_{k=1}^\infty x^{k-1}dx = \sum_{k=1}^\infty \int_0^1 x^{k-1}\log^{2n-1}(x)dx$$

$$= -(2n-1)!\sum_{k=1}^\infty \frac{1}{k^{2n}} = -(2n-1)!\zeta(2n). \qquad (3.200)$$

Collecting (3.196), (3.199), and (3.200) in (3.193), we conclude that

$$\int_0^1 \frac{\log^{2n-1}(x)\log(1-x)}{1+x}\,dx$$

$$= \frac{1}{2^{2n+1}}((n+1)2^{2n+1}-2n-1)(2n-1)!\zeta(2n+1)-\log(2)(2n-1)!\zeta(2n)$$

$$-\left(1-\frac{1}{2^{2n-1}}\right)(2n-1)!\sum_{k=1}^{n-1}\zeta(2k)\zeta(2n-2k+1)$$

$$+ (2n-1)!\sum_{k=1}^{n-1}\eta(2k)\zeta(2n-2k+1)-(2n-1)!\sum_{k=0}^{n-1}\eta(2k+1)\eta(2n-2k),$$

and the solution to the point (ii) is complete.

The next section will also reveal how to deal with a key generalized integral needed in this section! So, let's check it!

3.27   Three Useful Generalized Logarithmic Integrals Connected to Useful Generalized Alternating Harmonic Series

Solution The generalized integrals in the current section play a crucial part in the derivation of some atypical alternating harmonic series found in the fourth chapter. The solutions below follow the lines of the strategy I presented in the paper "A simple strategy of calculating two alternating harmonic series generalizations" ([80, May 24, 2019]). The first two ones also appeared in [46, May 23, 2019].

Elegant evaluations here are possible if we exploit the symmetry in double integrals! Relevant, similar examples are found in *(Almost) Impossible Integrals, Sums, and Series* in at least three sections (see [76, Chapter 1, Sect. 1.28–1.30, pp.18–21]).

So, passing to the point (i) of the problem, we note and use that $\log\left(\dfrac{1+x}{2}\right) = -\displaystyle\int_x^1 \frac{1}{1+y}\mathrm{d}y$, and then we write

$$\int_0^1 \frac{\log^m(x)\log\left(\dfrac{1+x}{2}\right)}{1-x}\mathrm{d}x$$

$$= -\int_0^1\left(\int_x^1 \frac{\log^m(x)}{(1+y)(1-x)}\mathrm{d}y\right)\mathrm{d}x \overset{\substack{\text{reverse the order}\\\text{of integration}}}{=} -\int_0^1\left(\int_0^y \frac{\log^m(x)}{(1+y)(1-x)}\mathrm{d}x\right)\mathrm{d}y$$

$$\overset{x=yz}{=} -\int_0^1\left(\int_0^1 \frac{y\log^m(yz)}{(1+y)(1-yz)}\mathrm{d}z\right)\mathrm{d}y \overset{z=x}{=} -\underbrace{\int_0^1\left(\int_0^1 \frac{y\log^m(xy)}{(1+y)(1-xy)}\mathrm{d}x\right)\mathrm{d}y}_{D}.$$

$$\text{(3.201)}$$

By symmetry reasons, we note we have that $D=\displaystyle\int_0^1\left(\int_0^1 \frac{x\log^m(xy)}{(1+x)(1-xy)}\mathrm{d}y\right)\mathrm{d}x$

$\overset{\substack{\text{reverse the order}\\\text{of integration}}}{=}\displaystyle\int_0^1\left(\int_0^1 \frac{x\log^m(xy)}{(1+x)(1-xy)}\mathrm{d}x\right)\mathrm{d}y$, and then, continuing the calculations in (3.201), we get

$$\int_0^1 \frac{\log^m(x)\log\left(\dfrac{1+x}{2}\right)}{1-x}\mathrm{d}x$$

$$= -\frac{1}{2}\left(\int_0^1\left(\int_0^1 \frac{y\log^m(xy)}{(1+y)(1-xy)}\mathrm{d}x\right)\mathrm{d}y + \int_0^1\left(\int_0^1 \frac{x\log^m(xy)}{(1+x)(1-xy)}\mathrm{d}x\right)\mathrm{d}y\right)$$

$$= -\frac{1}{2}\int_0^1 \left(\int_0^1 \frac{x+y+2xy}{(1+x)(1+y)(1-xy)} \log^m(xy)dx \right) dy$$

$$= -\frac{1}{2}\int_0^1 \left(\int_0^1 \frac{(1+x)(1+y)-(1-xy)}{(1+x)(1+y)(1-xy)} \log^m(xy)dx \right) dy$$

$$= \underbrace{-\frac{1}{2}\int_0^1 \left(\int_0^1 \frac{\log^m(xy)}{1-xy}dx \right) dy}_{X} + \underbrace{\frac{1}{2}\int_0^1 \left(\int_0^1 \frac{\log^m(xy)}{(1+x)(1+y)}dx \right) dy}_{Y}.$$

$$(3.202)$$

For the integral X, we make the variable change $yx = t$, and then we have

$$X = \int_0^1 \left(\int_0^1 \frac{\log^m(xy)}{1-xy}dx \right) dy = \int_0^1 \frac{1}{y} \left(\int_0^y \frac{\log^m(t)}{1-t}dt \right) dy$$

$$\overset{\substack{\text{reverse the order} \\ \text{of integration}}}{=} \int_0^1 \frac{\log^m(t)}{1-t} \left(\int_t^1 \frac{1}{y}dy \right) dt = -\int_0^1 \frac{\log^{m+1}(t)}{1-t}dt$$

$$= -\int_0^1 \log^{m+1}(t) \sum_{n=1}^{\infty} t^{n-1}dt = -\sum_{n=1}^{\infty} \int_0^1 t^{n-1} \log^{m+1}(t)dt$$

$$= (-1)^m(m+1)! \sum_{n=1}^{\infty} \frac{1}{n^{m+2}} = (-1)^m(m+1)!\zeta(m+2), \qquad (3.203)$$

where in the calculations I used that $\int_0^1 x^n \log^m(x)dx = (-1)^m \dfrac{m!}{(n+1)^{m+1}}$, given with swapped letters in [76, Chapter 1, p.1].

Next, for the integral Y, we use the simple fact that $\log^m(xy) = (\log(x) + \log(y))^m = \sum_{k=0}^m \binom{m}{k} \log^k(x) \log^{m-k}(y)$, and then we write

$$Y = \int_0^1 \left(\int_0^1 \frac{\log^m(xy)}{(1+x)(1+y)}dx \right) dy = \sum_{k=0}^m \binom{m}{k} \int_0^1 \frac{\log^{m-k}(y)}{1+y} \left(\int_0^1 \frac{\log^k(x)}{1+x}dx \right) dy$$

$$= (-1)^m m! \sum_{k=0}^m \eta(k+1)\eta(m-k+1), \qquad (3.204)$$

where in the calculations I used that $\displaystyle\int_0^1 \frac{\log^k(x)}{1+x}dx = \int_0^1 \sum_{n=1}^{\infty}(-1)^{n-1}x^{n-1}\log^k(x)dx$

$$= \sum_{n=1}^{\infty}(-1)^{n-1}\int_0^1 x^{n-1}\log^k(x)dx = (-1)^k k! \sum_{n=1}^{\infty}(-1)^{n-1}\frac{1}{n^{k+1}} = (-1)^k k!\eta(k+1).$$

Plugging the results from (3.203) and (3.204) in (3.202), we arrive at

$$\int_0^1 \frac{\log^m(x)\log\left(\dfrac{1+x}{2}\right)}{1-x}dx$$

$$= \frac{1}{2}(-1)^{m-1}m!\left((m+1)\zeta(m+2) - \sum_{k=0}^{m}\eta(k+1)\eta(m-k+1)\right),$$

and the solution to the point (i) of the problem is finalized.

As regards the point (ii) of the problem, we observe and use that $\log\left(\dfrac{1+x^2}{2}\right) = -2\int_x^1 \dfrac{y}{1+y^2}dy$, and then we write

$$\int_0^1 \frac{\log^m(x)\log\left(\dfrac{1+x^2}{2}\right)}{1-x}dx = -2\int_0^1 \left(\int_x^1 \frac{y\log^m(x)}{(1+y^2)(1-x)}dy\right)dx$$

$$\underset{\text{of integration}}{\overset{\text{swap the order}}{=}} -2\int_0^1 \left(\int_0^y \frac{y\log^m(x)}{(1+y^2)(1-x)}dx\right)dy$$

$$\overset{x=yz}{=} -2\int_0^1 \left(\int_0^1 \frac{y^2\log^m(yz)}{(1+y^2)(1-yz)}dz\right)dy \overset{z=x}{=} -2\int_0^1 \left(\underbrace{\int_0^1 \frac{y^2\log^m(xy)}{(1+y^2)(1-xy)}dx}_{S}\right)dy.$$

$$\tag{3.205}$$

Due to the symmetry, we have $S = \displaystyle\int_0^1 \left(\int_0^1 \frac{x^2\log^m(xy)}{(1+x^2)(1-xy)}dy\right)dx \overset{\text{swap the order}}{\underset{\text{of integration}}{=}}$

$\displaystyle\int_0^1 \left(\int_0^1 \frac{x^2\log^m(xy)}{(1+x^2)(1-xy)}dx\right)dy$, and proceeding with the calculations in (3.205), we obtain

$$\int_0^1 \frac{\log^m(x)\log\left(\dfrac{1+x^2}{2}\right)}{1-x}dx$$

$$= -\int_0^1 \left(\int_0^1 \frac{y^2 \log^m(xy)}{(1+y^2)(1-xy)} dx \right) dy - \int_0^1 \left(\int_0^1 \frac{x^2 \log^m(xy)}{(1+x^2)(1-xy)} dx \right) dy$$

$$= -\int_0^1 \left(\int_0^1 \frac{x^2 + y^2 + 2x^2 y^2}{(1+x^2)(1+y^2)(1-xy)} \log^m(xy) dx \right) dy$$

$$= -\int_0^1 \left(\int_0^1 \frac{(1+x^2)(1+y^2) - (1-(xy)^2)}{(1+x^2)(1+y^2)(1-xy)} \log^m(xy) dx \right) dy$$

$$= \underbrace{-\int_0^1 \left(\int_0^1 \frac{\log^m(xy)}{1-xy} dx \right) dy}_{X} + \underbrace{\int_0^1 \left(\int_0^1 \frac{\log^m(xy)}{(1+x^2)(1+y^2)} dx \right) dy}_{W}$$

$$+ \underbrace{\int_0^1 \left(\int_0^1 \frac{xy \log^m(xy)}{(1+x^2)(1+y^2)} dx \right) dy}_{Z}. \qquad (3.206)$$

For the double integral W in (3.206), we proceed as in the case of the double integral Y in (3.204), and then we have

$$W = \int_0^1 \left(\int_0^1 \frac{\log^m(xy)}{(1+x^2)(1+y^2)} dx \right) dy = \sum_{k=0}^m \binom{m}{k} \int_0^1 \frac{\log^{m-k}(y)}{1+y^2} \left(\int_0^1 \frac{\log^k(x)}{1+x^2} dx \right) dy$$

$$= (-1)^m m! \sum_{k=0}^m \beta(k+1)\beta(m-k+1), \qquad (3.207)$$

where in the calculations I used that $\displaystyle \int_0^1 \frac{\log^k(x)}{1+x^2} dx = \int_0^1 \sum_{n=1}^\infty (-1)^{n-1} x^{2n-2} \log^k(x) dx$

$$= \sum_{n=1}^\infty (-1)^{n-1} \int_0^1 x^{2n-2} \log^k(x) dx = (-1)^k k! \sum_{n=1}^\infty \frac{(-1)^{n-1}}{(2n-1)^{k+1}} = (-1)^k k! \beta(k+1).$$

Further, for the double integral Z in (3.206), we let the variable changes $x^2 = u$ and $y^2 = v$, and use the result in (3.204) that together give

$$Z = \int_0^1 \left(\int_0^1 \frac{xy \log^m(xy)}{(1+x^2)(1+y^2)} dx \right) dy = \frac{1}{2^{m+2}} \int_0^1 \left(\int_0^1 \frac{\log^m(uv)}{(1+u)(1+v)} du \right) dv$$

$$= \frac{1}{2^{m+2}} Y = (-1)^m \frac{1}{2^{m+2}} m! \sum_{k=0}^m \eta(k+1)\eta(m-k+1). \qquad (3.208)$$

The double integral X in (3.206) has already appeared during the calculations to the point (i), more precisely in (3.203), and if we plug it in (3.206), together with the results in (3.207) and (3.208), we arrive at

$$\int_0^1 \frac{\log^m(x) \log\left(\dfrac{1+x^2}{2}\right)}{1-x} dx$$

$$= (-1)^{m-1} m! \left((m+1)\zeta(m+2) - \frac{1}{2^{m+2}} \sum_{k=0}^m \eta(k+1)\eta(m-k+1) \right.$$

$$\left. - \sum_{k=0}^m \beta(k+1)\beta(m-k+1) \right),$$

and the solution to the point (ii) of the problem is finalized.

The result at the third point is straightforward if we exploit the generalized integrals at the previous points, and then we write

$$\int_0^1 \frac{\log^m(x) \log\left(\dfrac{1+x^2}{2}\right)}{1+x} dx$$

$$= \int_0^1 \frac{\log^m(x) \log\left(\dfrac{1+x^2}{2}\right)}{1-x} dx - 2 \int_0^1 \frac{x \log^m(x) \log\left(\dfrac{1+x^2}{2}\right)}{1-x^2} dx$$

{let $x^2 = y$ in the second integral, and then return to the notation in x}

$$= \int_0^1 \frac{\log^m(x) \log\left(\dfrac{1+x^2}{2}\right)}{1-x} dx - \frac{1}{2^m} \int_0^1 \frac{\log^m(x) \log\left(\dfrac{1+x}{2}\right)}{1-x} dx$$

$$= (-1)^{m-1} m! \left((m+1)\eta(m+2) + \frac{1}{2^{m+2}} \sum_{k=0}^m \eta(k+1)\eta(m-k+1) \right.$$

$$\left. - \sum_{k=0}^m \beta(k+1)\beta(m-k+1) \right),$$

and the solution to the point (iii) of the problem is finalized.

There are *two important notes* to emphasize at the end of the section. First, observe that by using a similar strategy to the ones above we might attack the more general case

$$\int_0^1 \frac{\log^m(x)\log\left(\dfrac{1+x^k}{2}\right)}{1-x}\,dx, \tag{3.209}$$

where $k \geq 1$, $m \geq 0$ are integers {e.g., if $k = 3$, in the arranged resulting double integral we exploit that $x^3(1+y^3)+y^3(1+x^3) = x^3+y^3+2x^3y^3 = (1+x^3)(1+y^3) - (1-(xy)^3)$}.

Second, the integrals of this type, like the more generalized version in (3.209), can be brought to forms involving (limits with) Beta function. For instance, if we take the generalized integral from the point (i), $m \geq 1$, we may write

$$\int_0^1 \frac{\log^m(x)\log\left(\dfrac{1+x}{2}\right)}{1-x}\,dx = \int_0^1 \frac{\log^m(x)\log\left(\dfrac{1-x^2}{2(1-x)}\right)}{1-x}\,dx$$

$$= \int_0^1 \frac{(1+x)\log^m(x)\log(1-x^2)}{1-x^2}\,dx - \int_0^1 \frac{\log^m(x)\log(1-x)}{1-x}\,dx$$

$$- \log(2)\int_0^1 \frac{\log^m(x)}{1-x}\,dx.$$

Reductions of such integrals to forms with Beta function may also be found in [32].

So, let's wrap *this story* up and pass now to the next section where we may also find curious integrals involving special forms of symmetry!

3.28 Three Atypical Logarithmic Integrals Involving $\log(1+x^3)$, Related to a Special Form of Symmetry in Double Integrals

Solution Exploiting the symmetry in a multiple integral is often a *wonderful moment*, a powerful way of attacking an integral, a thing I also stated in my book, *(Almost) Impossible Integrals, Sums, and Series* (see [76, Chapter 3, pp.159–160]).

In general, the challenging part with respect to the use of the symmetry in a multiple integral is that such ways are not *that visible*, and *the eye* of the solver may easily miss opportunities of this kind to construct a solution, particularly when additional results and rearrangements are needed. Some beautiful, relevant examples may be found in my first book, previously mentioned, and you may check [76, Chapter 3, Sect.1.27–1.30, pp.17–21].

In this section, I'll use a form of symmetry you could possibly perceive it more subtle when compared to what I previously suggested at the given reference.

So, first we want to observe a simple relation to extract between the three integrals I, J, and K, and then we write

$$I + J + K = \int_0^1 \log(1 + x^3)dx = \int_0^1 x' \log(1 + x^3)dx$$

$$= x \log(1 + x^3)\Big|_{x=0}^{x=1} - 3\int_0^1 \frac{(1 + x^3) - 1}{1 + x^3}dx = \log(2) - 3 + 3\int_0^1 \frac{1}{1 + x^3}dx$$

$$= \log(2) - 3 + \int_0^1 \frac{1}{1 + x}dx - \frac{1}{2}\int_0^1 \frac{2x - 1}{1 - x + x^2}dx + \frac{3}{2}\int_0^1 \frac{1}{(x - 1/2)^2 + (\sqrt{3}/2)^2}dx$$

$$= 2\log(2) - 3 - \frac{1}{2}\log(1 - x + x^2)\Big|_{x=0}^{x=1} + \sqrt{3}\arctan\left(\frac{2x - 1}{\sqrt{3}}\right)\Big|_{x=0}^{x=1}$$

$$= 2\log(2) - 3 + \frac{\pi}{\sqrt{3}}. \tag{3.210}$$

Returning to the first point of the problem, we want to exploit the integral representation $\log\left(\dfrac{1 + x^3}{2}\right) = -3\displaystyle\int_x^1 \frac{t^2}{1 + t^3}dt$, and then we get

$$I = \int_0^1 \frac{\log(1 + x^3)}{1 + x + x^2}dx = \log(2)\int_0^1 \frac{1}{1 + x + x^2}dx + \int_0^1 \frac{1}{1 + x + x^2}\log\left(\frac{1 + x^3}{2}\right)dx$$

$$= \log(2)\int_0^1 \frac{1}{(x + 1/2)^2 + (\sqrt{3}/2)^2}dx - 3\int_0^1 \left(\int_x^1 \frac{t^2}{(1 + t^3)(1 + x + x^2)}dt\right)dx$$

$$\underset{\substack{\text{reverse the order} \\ \text{of integration}}}{=} \frac{2}{\sqrt{3}}\log(2)\arctan\left(\frac{2x + 1}{\sqrt{3}}\right)\Big|_{x=0}^{x=1} - 3\int_0^1 \left(\int_0^t \frac{t^2}{(1 + t^3)(1 + x + x^2)}dx\right)dt$$

$$\underset{x/t=u}{=} \frac{1}{3\sqrt{3}}\log(2)\pi - 3\int_0^1 \left(\int_0^1 \frac{t^3}{(1 + t^3)(1 + tu + (tu)^2)}du\right)dt. \tag{3.211}$$

For the remaining double integral in (3.211), we consider the notation $g(t, u) = \dfrac{t^3}{(1 + t^3)(1 + tu + (tu)^2)}$, exploit the symmetry, and use the change of integration order in a double integral, and then we write

$$\int_0^1 \left(\int_0^1 \frac{t^3}{(1+t^3)(1+tu+(tu)^2)} du \right) dt$$

$$= \frac{1}{2} \left(\int_0^1 \left(\int_0^1 g(t,u) du \right) dt + \int_0^1 \left(\int_0^1 g(u,t) du \right) dt \right)$$

$$= \frac{1}{2} \int_0^1 \left(\int_0^1 (g(t,u) + g(u,t)) du \right) dt$$

$$= \frac{1}{2} \int_0^1 \left(\int_0^1 \frac{t^3 + u^3 + 2t^3 u^3}{(1+t^3)(1+u^3)(1+tu+(tu)^2)} du \right) dt$$

$$= \frac{1}{2} \int_0^1 \left(\int_0^1 \frac{(1+t^3)(1+u^3) - (1-(tu)^3)}{(1+t^3)(1+u^3)(1+tu+(tu)^2)} du \right) dt$$

$$= \frac{1}{2} \int_0^1 \left(\int_0^1 \frac{1}{1+tu+(tu)^2} du \right) dt - \frac{1}{2} \int_0^1 \left(\int_0^1 \frac{1}{(1+t^3)(1+u^3)} du \right) dt$$

$$+ \frac{1}{2} \int_0^1 \left(\int_0^1 \frac{tu}{(1+t^3)(1+u^3)} du \right) dt. \qquad (3.212)$$

Now, for the first double integral in (3.212), we make the change of variable $tu = v$, and we have

$$\int_0^1 \left(\int_0^1 \frac{1}{1+tu+(tu)^2} du \right) dt = \int_0^1 \left(\int_0^t \frac{1}{(1+v+v^2)t} dv \right) dt$$

$$\underset{\text{of integration}}{\overset{\text{reverse the order}}{=}} \int_0^1 \left(\int_v^1 \frac{1}{(1+v+v^2)t} dt \right) dv = - \int_0^1 \frac{(1-v)\log(v)}{1-v^3} dv$$

$$= \int_0^1 \frac{v \log(v)}{1-v^3} dv - \int_0^1 \frac{\log(v)}{1-v^3} dv = \int_0^1 \sum_{n=1}^{\infty} v^{3n-2} \log(v) dv - \int_0^1 \sum_{n=1}^{\infty} v^{3n-3} \log(v) dv$$

{reverse the order of summation and integration}

$$= \sum_{n=1}^{\infty} \int_0^1 v^{3n-2} \log(v) dv - \sum_{n=1}^{\infty} \int_0^1 v^{3n-3} \log(v) dv$$

$$= \frac{1}{9} \sum_{n=1}^{\infty} \frac{1}{(n-2/3)^2} - \frac{1}{9} \sum_{n=1}^{\infty} \frac{1}{(n-1/3)^2} = \frac{1}{9} \psi^{(1)} \left(\frac{1}{3} \right) - \frac{1}{9} \psi^{(1)} \left(\frac{2}{3} \right)$$

$$\left\{\text{based on the reflection formula } (-1)^m \psi^{(m)}(1-z) - \psi^{(m)}(z) = \pi \frac{d^m}{dz^m} \cot(\pi z)\right\}$$

$$\left\{\text{where we set } m = 1 \text{ and } z = 1/3, \text{ we obtain that } \psi^{(1)}\left(\frac{1}{3}\right) + \psi^{(1)}\left(\frac{2}{3}\right) = \frac{4}{3}\pi^2\right\}$$

$$= \frac{2}{9}\psi^{(1)}\left(\frac{1}{3}\right) - \frac{4}{27}\pi^2. \tag{3.213}$$

Next, the second double integral in (3.212) is straightforward since based on the calculations in (3.210) we have $\int_0^1 \frac{1}{1+t^3} dt = \frac{\pi}{3\sqrt{3}} + \frac{1}{3}\log(2)$, and then we obtain

$$\int_0^1 \left(\int_0^1 \frac{1}{(1+t^3)(1+u^3)} du\right) dt = \frac{\pi^2}{27} + \frac{1}{9}\log^2(2) + \frac{2}{9\sqrt{3}}\log(2)\pi. \tag{3.214}$$

Finally, for the third double integral in (3.212) we use that

$$\int_0^1 \frac{t}{1+t^3} dt = \frac{1}{3}\int_0^1 \frac{t+1}{t^2-t+1} dt - \frac{1}{3}\int_0^1 \frac{1}{1+t} dt = \frac{1}{6}\int_0^1 \frac{2t+2}{t^2-t+1} dt - \frac{1}{3}\log(2)$$

$$= \frac{1}{6}\int_0^1 \frac{2t-1}{t^2-t+1} dt + \frac{1}{2}\int_0^1 \frac{1}{t^2-t+1} dt - \frac{1}{3}\log(2) = \frac{1}{6}\log(t^2-t+1)\Big|_{t=0}^{t=1}$$

$$+ \frac{1}{\sqrt{3}}\arctan\left(\frac{2t-1}{\sqrt{3}}\right)\Big|_{t=0}^{t=1} - \frac{1}{3}\log(2) = \frac{\pi}{3\sqrt{3}} - \frac{1}{3}\log(2). \tag{3.215}$$

Thereafter, based on the result in (3.215), we get

$$\int_0^1 \left(\int_0^1 \frac{tu}{(1+t^3)(1+u^3)} du\right) dt = \frac{\pi^2}{27} + \frac{1}{9}\log^2(2) - \frac{2}{9\sqrt{3}}\log(2)\pi. \tag{3.216}$$

Collecting the results from (3.213), (3.214), and (3.216) in (3.212), we obtain

$$\int_0^1 \left(\int_0^1 \frac{t^3}{(1+t^3)(1+tu+(tu)^2)} du\right) dt = \frac{1}{9}\psi^{(1)}\left(\frac{1}{3}\right) - \frac{2}{27}\pi^2 - \frac{2}{9\sqrt{3}}\log(2)\pi. \tag{3.217}$$

At last, if we plug the result from (3.217) in (3.211), we conclude that

$$I = \int_0^1 \frac{\log(1+x^3)}{1+x+x^2} dx = \frac{2}{9}\pi^2 + \frac{1}{\sqrt{3}}\log(2)\pi - \frac{1}{3}\psi^{(1)}\left(\frac{1}{3}\right),$$

and the solution to the point (i) of the problem is finalized.

For a second approach, the curious reader might like to start with writing that
$$\int_0^1 \frac{\log(1+x^3)}{1+x+x^2}dx = \int_0^\infty \frac{\log(1+x^3)}{1+x+x^2}dx - \int_1^\infty \frac{\log(1+x^3)}{1+x+x^2}dx,$$ and if we let the
variable change $1/x = y$ in the second integral from the right-hand side and then
rearrange, we get

$$I = \int_0^1 \frac{\log(1+x^3)}{1+x+x^2}dx = \frac{1}{2}\underbrace{\int_0^\infty \frac{\log(1+x^3)}{1+x+x^2}dx}_{2\log(2)\pi/\sqrt{3}} + \frac{3}{2}\underbrace{\int_0^1 \frac{\log(x)}{1+x+x^2}dx}_{4\pi^2/27 - 2\psi^{(1)}(1/3)/9},$$

$$(3.218)$$

where both integrals in the right-hand side of (3.218) are easily manageable (e.g.,
the first integral works with differentiation under the integral sign).

Then, for a third approach, the curious reader might exploit algebraic identities
to reduce the main integral to Beta function forms and limits, which is essentially
possible to do for all three main integrals of the problem.

Passing to the point (ii) of the problem, we might proceed in a similar style as
before and use the symmetry in double integrals. Using similar steps as in (3.211),
we may write that

$$J = \int_0^1 \frac{x\log(1+x^3)}{1+x+x^2}dx = \log(2)\int_0^1 \frac{x}{1+x+x^2}dx + \int_0^1 \frac{x}{1+x+x^2}\log\left(\frac{1+x^3}{2}\right)dx$$

$$= \frac{1}{2}\log(2)\log(3) - \frac{1}{6\sqrt{3}}\log(2)\pi - 3\int_0^1 \left(\int_x^1 \frac{t^2 x}{(1+t^3)(1+x+x^2)}dt\right)dx$$

$$\overset{\substack{\text{reverse the order}\\\text{of integration}}}{=} \frac{1}{2}\log(2)\log(3) - \frac{1}{6\sqrt{3}}\log(2)\pi - 3\int_0^1 \left(\int_0^t \frac{t^2 x}{(1+t^3)(1+x+x^2)}dx\right)dt$$

$$\overset{x/t=u}{=} \frac{1}{2}\log(2)\log(3) - \frac{1}{6\sqrt{3}}\log(2)\pi - 3\int_0^1 \left(\int_0^1 \frac{t^4 u}{(1+t^3)(1+tu+(tu)^2)}du\right)dt,$$

$$(3.219)$$

where in the calculations above I also used the simple fact that $\int_0^1 \frac{x}{1+x+x^2}dx =$
$\frac{1}{2}\int_0^1 \frac{2x+1-1}{1+x+x^2}dx = \frac{1}{2}\int_0^1 \frac{2x+1}{1+x+x^2}dx - \frac{1}{2}\int_0^1 \frac{1}{1+x+x^2}dx = \frac{1}{2}\log(1+$
$x+x^2)\Big|_{x=0}^{x=1} - \frac{1}{\sqrt{3}}\arctan\left(\frac{2x+1}{\sqrt{3}}\right)\Big|_{x=0}^{x=1} = \frac{1}{2}\log(3) - \frac{\pi}{6\sqrt{3}}.$

Then, for the remaining double integral in (3.219) we consider the notation
$$h(t,u) = \frac{t^4 u}{(1+t^3)(1+tu+(tu)^2)},$$ exploit the symmetry, and use the change of
integration order in a double integral, and therefore we write

$$\int_0^1 \left(\int_0^1 \frac{t^4 u}{(1+t^3)(1+tu+(tu)^2)} du \right) dt$$

$$= \frac{1}{2} \left(\int_0^1 \left(\int_0^1 h(t,u) du \right) dt + \int_0^1 \left(\int_0^1 h(u,t) du \right) dt \right)$$

$$= \frac{1}{2} \int_0^1 \left(\int_0^1 (h(t,u) + h(u,t)) du \right) dt$$

$$= \frac{1}{2} \int_0^1 \left(\int_0^1 \frac{tu(t^3 + u^3 + 2t^3 u^3)}{(1+t^3)(1+u^3)(1+tu+(tu)^2)} du \right) dt$$

$$= \frac{1}{2} \int_0^1 \left(\int_0^1 \frac{tu((1+t^3)(1+u^3) - (1-(tu)^3))}{(1+t^3)(1+u^3)(1+tu+(tu)^2)} du \right) dt$$

$$= \frac{1}{2} \int_0^1 \left(\int_0^1 \frac{tu}{1+tu+(tu)^2} du \right) dt - \frac{1}{2} \int_0^1 \left(\int_0^1 \frac{tu}{(1+t^3)(1+u^3)} du \right) dt$$

$$+ \frac{1}{2} \int_0^1 \left(\int_0^1 \frac{t^2 u^2}{(1+t^3)(1+u^3)} du \right) dt. \tag{3.220}$$

Further, for the first double integral in (3.220), we make the change of variable $tu = v$, and then we get

$$\int_0^1 \left(\int_0^1 \frac{tu}{1+tu+(tu)^2} du \right) dt = \int_0^1 \left(\int_0^t \frac{v}{(1+v+v^2)t} dv \right) dt$$

$$\overset{\substack{\text{reverse the order} \\ \text{of integration}}}{=\!=\!=} \int_0^1 \left(\int_v^1 \frac{v}{(1+v+v^2)t} dt \right) dv = -\int_0^1 \frac{(v-v^2)\log(v)}{1-v^3} dv$$

$$= \int_0^1 \frac{v^2 \log(v)}{1-v^3} dv - \int_0^1 \frac{v\log(v)}{1-v^3} dv = \int_0^1 \sum_{n=1}^\infty v^{3n-1} \log(v) dv - \int_0^1 \sum_{n=1}^\infty v^{3n-2} \log(v) dv$$

{reverse the order of summation and integration}

$$= \sum_{n=1}^\infty \int_0^1 v^{3n-1} \log(v) dv - \sum_{n=1}^\infty \int_0^1 v^{3n-2} \log(v) dv$$

$$= \frac{1}{9} \sum_{n=1}^\infty \frac{1}{(n-1/3)^2} - \frac{1}{9} \sum_{n=1}^\infty \frac{1}{n^2} = \frac{1}{9} \psi^{(1)} \left(\frac{2}{3} \right) - \frac{\pi^2}{54}$$

$$\left\{ \text{use that } \psi^{(1)}\left(\frac{1}{3}\right) + \psi^{(1)}\left(\frac{2}{3}\right) = \frac{4}{3}\pi^2, \text{ which is explained in (3.213)} \right\}$$

$$= \frac{7}{54}\pi^2 - \frac{1}{9}\psi^{(1)}\left(\frac{1}{3}\right). \tag{3.221}$$

Next, observe the second double integral in (3.220) is already calculated in (3.216) during the calculations to the point (i) of the problem.

Ultimately, the third double integral in (3.220) is straightforward, and then we have that

$$\int_0^1 \left(\int_0^1 \frac{t^2 u^2}{(1+t^3)(1+u^3)} du \right) dt = \left(\int_0^1 \frac{t^2}{1+t^3} dt \right)^2 = \frac{1}{9}\log^2(2). \tag{3.222}$$

Collecting the values of the integrals from (3.221), (3.216), and (3.222) in (3.220), we arrive at

$$\int_0^1 \left(\int_0^1 \frac{t^4 u}{(1+t^3)(1+tu+(tu)^2)} du \right) dt = \frac{5}{108}\pi^2 + \frac{1}{9\sqrt{3}}\log(2)\pi - \frac{1}{18}\psi^{(1)}\left(\frac{1}{3}\right). \tag{3.223}$$

Finally, if we plug the result from (3.223) in (3.219), we conclude that

$$J = \int_0^1 \frac{x \log\left(1+x^3\right)}{1+x+x^2} dx$$

$$= \frac{1}{2}\log(2)\log(3) - \frac{5}{36}\pi^2 - \frac{1}{2\sqrt{3}}\log(2)\pi + \frac{1}{6}\psi^{(1)}\left(\frac{1}{3}\right),$$

and the solution to the point (ii) of the problem is finalized.

Having arrived at this point, it is enough to recollect the result we got in (3.210) involving the three main integrals. Since we already calculated the integrals at the points (i) and (ii), based on the result in (3.210), we conclude that

$$K = \int_0^1 \frac{x^2 \log\left(1+x^3\right)}{1+x+x^2} dx$$

$$= 2\log(2) - \frac{1}{2}\log(2)\log(3) - 3 + \frac{\pi}{\sqrt{3}} - \frac{\pi^2}{12} - \frac{1}{2\sqrt{3}}\log(2)\pi + \frac{1}{6}\psi^{(1)}\left(\frac{1}{3}\right),$$

and the solution to the point (iii) of the problem is finalized. Observe that during the calculations I also used the Polygamma series representation, $\psi^{(m)}(z) = (-1)^{m-1}m! \sum_{k=0}^{\infty} \frac{1}{(z+k)^{m+1}}$.

Surely, the integral K, like the integrals I and J, may also be calculated by exploiting the symmetry in double integrals as previously shown.

The solutions in this section also answer the appealing *challenging question*.

As a final note, observe that another nice relation may be established between the integrals I and J, and the curious reader might like to find an elegant way to calculate the sum $\frac{1}{2}I + J$, without exploiting the symmetry in double integrals.

3.29 Four Curious Integrals with Cosine Function Which Lead to Beautiful Closed Forms, Plus Two Exotic Integrals

Solution Back on March 26, 1949, during the Putnam contest sessions the following beautiful and interesting product was given to the contestants:

Prove that for every real or complex x

$$\prod_{n=1}^{\infty} \frac{1 + 2\cos\left(2x/3^n\right)}{3} = \frac{\sin(x)}{x}.$$

As shown in [16], if we use the identity $\sin(3x) - \sin(x) = 2\sin(x)\cos(2x)$, we immediately arrive at $\dfrac{\sin(3x)}{\sin(x)} = 2\cos(2x) + 1$, where if we replace x by $x/3^n$, divide both sides by 3, and then consider the partial product from $n = 1$ to m, we get

$$\prod_{n=1}^{m} \frac{1 + 2\cos\left(2x/3^n\right)}{3} = \prod_{n=1}^{m} \frac{\sin(x/3^{n-1})}{3\sin(x/3^n)} = \frac{x/3^m}{\sin(x/3^m)} \cdot \frac{\sin(x)}{x},$$

and if we let $m \to \infty$, we immediately obtain the desired result,

$$\prod_{n=1}^{\infty} \frac{1 + 2\cos\left(2x/3^n\right)}{3} = \frac{\sin(x)}{x} \cdot \lim_{m\to\infty} \frac{x/3^m}{\sin(x/3^m)} = \frac{\sin(x)}{x}.$$

To get the last equality, we use the limit, $\lim\limits_{x\to0} \dfrac{\sin(x)}{x} = 1$.

A nice infinite product from the Putnam contest!, you might say! Indeed, that's true! In general, the solutions involving telescoping sums and products are very elegant.

Now, if we return to the main results to prove, and take a close look at the log argument, we notice that the key trigonometric identity $\dfrac{\sin(3x)}{\sin(x)} = 2\cos(2x) + 1$ employed in the infinite product could be useful in our calculations! Further, if we replace x by $\pi/2 - x$ in this identity, we arrive at $\dfrac{\cos(3x)}{\cos(x)} = 2\cos(2x) - 1$.

Thus, based on the two trigonometric identities above, we have that

$$\frac{\tan(x)}{\tan(3x)} = \frac{\cos(3x)/\cos(x)}{\sin(3x)/\sin(x)} = \frac{2\cos(2x) - 1}{2\cos(2x) + 1}, \tag{3.224}$$

which looks like what we need in the next calculations.

Alternatively, we might directly employ a slightly different trigonometric identity,

$$\tan\left(\frac{\pi}{6} + x\right)\tan\left(\frac{\pi}{6} - x\right) = \frac{2\cos(2x) - 1}{2\cos(2x) + 1}, \tag{3.225}$$

which provides another route to get a solution (and we may see it at work in the solution to the second point of the problem).

With the result (3.224) in hand, we are ready to start the main calculations, and then, for the point (*i*) of the problem we write

$$\int_0^{\pi/2} x \log\left(\left|\frac{1 - 2\cos(2x)}{1 + 2\cos(2x)}\right|\right) dx = \int_0^{\pi/2} x \log\left(\left|\frac{\tan(x)}{\tan(3x)}\right|\right) dx$$

{split the integral taking into account the sign of $\tan(3x)$}

$$= \int_0^{\pi/2} x \log(\tan(x)) dx - \int_0^{\pi/6} x \log(\tan(3x)) dx - \int_{\pi/6}^{\pi/3} x \log(-\tan(3x)) dx$$

$$- \int_{\pi/3}^{\pi/2} x \log(\tan(3x)) dx$$

{make the change of variable $\pi/3 - x = y$ in the third integral and}

{for the fourth integral make the change of variable $\pi/2 - x = y$}

$$= \int_0^{\pi/2} x \log(\tan(x)) dx - \int_0^{\pi/6} x \log(\tan(3x)) dx - \int_0^{\pi/6} \left(\frac{\pi}{3} - x\right) \log(\tan(3x)) dx$$

$$+ \int_0^{\pi/6} \left(\frac{\pi}{2} - x\right) \log(\tan(3x)) dx$$

$$= \int_0^{\pi/2} x \log(\tan(x)) dx - \int_0^{\pi/6} x \log(\tan(3x)) dx + \frac{\pi}{6} \int_0^{\pi/6} \log(\tan(3x)) dx$$

{in the last two integrals make the change of variable $3x = y$}

$$= \frac{8}{9} \int_0^{\pi/2} x \log(\tan(x)) dx + \frac{\pi}{18} \underbrace{\int_0^{\pi/2} \log(\tan(x)) dx}_{0} = \frac{8}{9} \int_0^{\pi/2} x \log(\tan(x)) dx$$

$$\left\{ \text{use the Fourier series,} \ \log(\tan(x)) = -2 \sum_{k=1}^{\infty} \frac{\cos(2(2k-1)x)}{2k-1}, \ 0 < x < \frac{\pi}{2} \right\}$$

$$= -\frac{16}{9} \int_0^{\pi/2} x \sum_{k=1}^{\infty} \frac{\cos(2(2k-1)x)}{2k-1} dx = -\frac{16}{9} \sum_{k=1}^{\infty} \int_0^{\pi/2} x \frac{\cos(2(2k-1)x)}{2k-1} dx$$

$$= \frac{8}{9} \sum_{k=1}^{\infty} \frac{1}{(2k-1)^3} = \frac{7}{9} \zeta(3),$$

and the point (i) of the problem is finalized. One thing to note is that during the calculations we could have used the Fourier series earlier, without showing the beautiful reduction of the main integral to $\dfrac{8}{9} \displaystyle\int_0^{\pi/2} x \log(\tan(x)) dx$.

For the point (ii) of the problem we may use again the trigonometric identity in (3.224), but this time we might go further with the trigonometric identity in (3.225), and then we write

$$\int_0^{\pi/2} x^2 \log\left(\left| \frac{1 - 2\cos(2x)}{1 + 2\cos(2x)} \right| \right) dx = \int_0^{\pi/2} x^2 \log\left(\left| \tan\left(\frac{\pi}{6} + x\right) \tan\left(\frac{\pi}{6} - x\right) \right| \right) dx$$

$$= \int_0^{\pi/2} x^2 \log\left(\left| \tan\left(\frac{\pi}{6} + x\right) \right| \right) dx + \int_0^{\pi/2} x^2 \log\left(\left| \tan\left(\frac{\pi}{6} - x\right) \right| \right) dx$$

{in the second integral make the change of variable $x = -y$}

$$= \int_{-\pi/2}^{\pi/2} y^2 \log\left(\left| \tan\left(\frac{\pi}{6} + y\right) \right| \right) dy \stackrel{\pi/6 + y = z}{=} \int_{-\pi/3}^{2\pi/3} \left(z - \frac{\pi}{6}\right)^2 \log(|\tan(z)|) dz$$

$$\left\{ \text{we use the Fourier series,} \ \log(|\tan(x)|) = -2 \sum_{k=1}^{\infty} \frac{\cos(2(2k-1)x)}{2k-1}, \ x \neq \frac{n\pi}{2} \right\}$$

$$= -2 \int_{-\pi/3}^{2\pi/3} \left(z - \frac{\pi}{6}\right)^2 \sum_{k=1}^{\infty} \frac{\cos(2(2k-1)z)}{2k-1} dz$$

$$= -2 \sum_{k=1}^{\infty} \frac{1}{2k-1} \int_{-\pi/3}^{2\pi/3} \left(z - \frac{\pi}{6}\right)^2 \cos(2(2k-1)z) dz = \pi \sum_{k=1}^{\infty} \frac{\sin\left(\frac{\pi}{6}(4k+1)\right)}{(2k-1)^3}$$

$$\left\{ \text{inspect the pattern generated by the numerator:} \ \frac{1}{2}, -1, \frac{1}{2}, \frac{1}{2}, -1, \frac{1}{2}, \dots \right\}$$

{and notice that the resulting series can be written as a sum of two series}

$$= \frac{\pi}{2} \sum_{k=1}^{\infty} \frac{1}{(2k-1)^3} - \frac{\pi}{18} \sum_{k=1}^{\infty} \frac{1}{(2k-1)^3} = \pi \sum_{k=1}^{\infty} \frac{1}{(2k-1)^3} = \frac{7}{18} \pi \zeta(3),$$

and the point (ii) of the problem is finalized.

Next, to attack the integral at the point (iii), we first need another key trigonometric identity, $\dfrac{\sin(5x)}{\sin(x)} = 1 + 2\cos(2x) + 2\cos(4x)$. *But how do we get it?* Since we have that $\sin(5x) - \sin(x) = 2\sin(2x)\cos(3x) = 4\sin(x)\cos(x)\cos(3x)$, then $\sin(5x) = \sin(x) + 4\sin(x)\cos(x)\cos(3x) = \sin(x)(1 + 4\cos(x)\cos(3x)) = \sin(x)(1 + 2\cos(2x) + 2\cos(4x))$ which leads to the stated identity. On the other hand, if we replace x by $\pi/2 - x$ in the key trigonometric identity above, we obtain another useful one, $\dfrac{\cos(5x)}{\cos(x)} = 1 - 2\cos(2x) + 2\cos(4x)$.

Now, based on the trigonometric identities above, we get

$$\frac{\tan(x)}{\tan(5x)} = \frac{1 - 2\cos(2x) + 2\cos(4x)}{1 + 2\cos(2x) + 2\cos(4x)}. \tag{3.226}$$

Returning to the main calculations with the result in (3.226), we write

$$\int_0^{\pi/2} x \log\left(\left|\frac{2\cos(2x) - 2\cos(4x) - 1}{2\cos(2x) + 2\cos(4x) + 1}\right|\right) dx = \int_0^{\pi/2} x \log\left(\left|\frac{\tan(x)}{\tan(5x)}\right|\right) dx$$

{after splitting the logarithmic integral, we observe that from}

$$\left\{\text{the calculations to the point } (i) \text{ we have } \int_0^{\pi/2} x \log(\tan(x)) dx = \frac{7}{8}\zeta(3); \text{ and for}\right\}$$

$$\left\{\text{the other one use the Fourier series, } \log(|\tan(x)|) = -2\sum_{k=1}^{\infty} \frac{\cos(2(2k-1)x)}{2k-1}\right\}$$

$$= \frac{7}{8}\zeta(3) - \int_0^{\pi/2} x \log(|\tan(5x)|) dx = \frac{7}{8}\zeta(3) + 2\int_0^{\pi/2} x \sum_{k=1}^{\infty} \frac{\cos(10(2k-1)x)}{2k-1} dx$$

$$= \frac{7}{8}\zeta(3) + 2\sum_{k=1}^{\infty} \frac{1}{2k-1} \int_0^{\pi/2} x\cos(10(2k-1)x) dx = \frac{7}{8}\zeta(3) - \frac{1}{25}\sum_{k=1}^{\infty} \frac{1}{(2k-1)^3}$$

$$= \frac{7}{8}\zeta(3) - \frac{7}{200}\zeta(3) = \frac{21}{25}\zeta(3),$$

and the point (iii) of the problem is finalized.

For the integral at the point (iv), we use the strategy in the previous case, and then we have

$$\int_0^{\pi/2} x^2 \log\left(\left|\frac{2\cos(2x) - 2\cos(4x) - 1}{2\cos(2x) + 2\cos(4x) + 1}\right|\right) dx = \int_0^{\pi/2} x^2 \log\left(\left|\frac{\tan(x)}{\tan(5x)}\right|\right) dx$$

$$= \int_0^{\pi/2} x^2 \log(\tan(x)) dx - \int_0^{\pi/2} x^2 \log(|\tan(5x)|) dx$$

$$\left\{ \text{use the Fourier series, } \log(|\tan(x)|) = -2\sum_{k=1}^{\infty} \frac{\cos(2(2k-1)x)}{2k-1} \right\}$$

$$= -2\int_0^{\pi/2} x^2 \sum_{k=1}^{\infty} \frac{\cos(2(2k-1)x)}{2k-1} dx + 2\int_0^{\pi/2} x^2 \sum_{k=1}^{\infty} \frac{\cos(10(2k-1)x)}{2k-1} dx$$

{reverse the order of summation and integration}

$$= -2\sum_{k=1}^{\infty} \frac{1}{2k-1} \int_0^{\pi/2} x^2 \cos(2(2k-1)x) dx + 2\sum_{k=1}^{\infty} \frac{1}{2k-1} \int_0^{\pi/2} x^2 \cos(10(2k-1)x) dx$$

$$= \frac{\pi}{2}\sum_{k=1}^{\infty} \frac{1}{(2k-1)^3} - \frac{\pi}{50}\sum_{k=1}^{\infty} \frac{1}{(2k-1)^3} = \frac{12}{25}\pi\sum_{k=1}^{\infty} \frac{1}{(2k-1)^3} = \frac{21}{50}\pi\zeta(3),$$

and the point (iv) of the problem is finalized.

Further, for the last two points of the section we might like to notice that if we denote the cosine product of both integrands by $P(x)$, then the integrand from the point (v) is $xP(x)$, and the one from the point (vi) is $x^2P(x)$. We will focus on $P(x)$ and use the trigonometric identity in (3.224), $\dfrac{\tan(x)}{\tan(3x)} = \dfrac{2\cos(2x) - 1}{2\cos(2x) + 1}$, and then we have

$$P(x) = \left(3\frac{2\cos(2/3x) - 1}{2\cos(2/3x) + 1}\right) \cdot \left(3\frac{2\cos(2/3^2x) - 1}{2\cos(2/3^2x) + 1}\right) \cdot \left(3\frac{2\cos(2/3^3x) - 1}{2\cos(2/3^3x) + 1}\right) \cdots$$

$$= \lim_{N\to\infty} \prod_{n=1}^{N} \left(3\frac{2\cos(2/3^n x) - 1}{2\cos(2/3^n x) + 1}\right) = \lim_{N\to\infty} \prod_{n=1}^{N} 3\frac{\tan(x/3^n)}{\tan(x/3^{n-1})}$$

$$= \lim_{N\to\infty} 3\frac{\tan(x/3^1)}{\tan(x/3^0)} \cdot 3\frac{\tan(x/3^2)}{\tan(x/3^1)} \cdot 3\frac{\tan(x/3^3)}{\tan(x/3^2)} \cdots 3\frac{\tan(x/3^{N-1})}{\tan(x/3^{N-2})} \cdot 3\frac{\tan(x/3^N)}{\tan(x/3^{N-1})}$$

$$= \lim_{N\to\infty} 3^N \frac{\tan(x/3^N)}{\tan(x)} = \frac{x}{\tan(x)} \cdot \underbrace{\lim_{N\to\infty} \frac{\tan(x/3^N)}{x/3^N}}_{1} = \frac{x}{\tan(x)}. \tag{3.227}$$

Now, with the result in (3.227), we return to the point (v) of the problem, and then we have

$$\int_0^{\pi/2} x \left(3 \frac{2\cos(2/3x) - 1}{2\cos(2/3x) + 1}\right) \cdot \left(3 \frac{2\cos(2/3^2x) - 1}{2\cos(2/3^2x) + 1}\right) \cdot \left(3 \frac{2\cos(2/3^3x) - 1}{2\cos(2/3^3x) + 1}\right) \cdots dx$$

$$= \int_0^{\pi/2} \frac{x^2}{\tan(x)} dx = -2 \int_0^{\pi/2} x \log(\sin(x)) dx$$

$$\left\{ \text{use the Fourier series, } \log(\sin(x)) = -\log(2) - \sum_{k=1}^{\infty} \frac{\cos(2kx)}{k} \right\}$$

$$= 2\log(2) \int_0^{\pi/2} x dx + 2 \int_0^{\pi/2} x \sum_{k=1}^{\infty} \frac{\cos(2kx)}{k} dx$$

{reverse the order of summation and integration}

$$= \frac{1}{4} \log(2)\pi^2 + 2 \sum_{k=1}^{\infty} \frac{1}{k} \int_0^{\pi/2} x \cos(2kx) dx$$

$$= \frac{1}{4} \log(2)\pi^2 - \frac{1}{2} \sum_{k=1}^{\infty} \frac{1}{k^3} - \frac{1}{2} \sum_{k=1}^{\infty} \frac{(-1)^{k-1}}{k^3} = \frac{3}{2} \log(2)\zeta(2) - \frac{7}{8}\zeta(3),$$

and the point (v) of the problem is finalized.

To deal with the last integral, the one at the point (vi), we proceed as in the solution to the previous point, and then we write

$$\int_0^{\pi/2} x^2 \left(3 \frac{2\cos(2/3x) - 1}{2\cos(2/3x) + 1}\right) \cdot \left(3 \frac{2\cos(2/3^2x) - 1}{2\cos(2/3^2x) + 1}\right) \cdot \left(3 \frac{2\cos(2/3^3x) - 1}{2\cos(2/3^3x) + 1}\right) \cdots dx$$

$$= \int_0^{\pi/2} \frac{x^3}{\tan(x)} dx = -3 \int_0^{\pi/2} x^2 \log(\sin(x)) dx$$

$$\left\{ \text{use the Fourier series, } \log(\sin(x)) = -\log(2) - \sum_{k=1}^{\infty} \frac{\cos(2kx)}{k} \right\}$$

$$= 3\log(2) \int_0^{\pi/2} x^2 dx + 3 \int_0^{\pi/2} x^2 \sum_{k=1}^{\infty} \frac{\cos(2kx)}{k} dx$$

{reverse the order of summation and integration}

$$= \frac{1}{8}\log(2)\pi^3 + 3\sum_{k=1}^{\infty}\frac{1}{k}\int_0^{\pi/2} x^2\cos(2kx)dx = \frac{1}{8}\log(2)\pi^3 - \frac{3}{4}\pi\sum_{k=1}^{\infty}\frac{(-1)^{k-1}}{k^3}$$

$$= \frac{1}{8}\log(2)\pi^3 - \frac{9}{16}\pi\zeta(3),$$

and the point (vi) of the problem is finalized.

In conclusion, we see that a key step in the solutions to all points of the problem is represented by the use of the proper trigonometric transformations.

3.30 Aesthetic Integrals That Must Be Understood as Cauchy Principal Values, Generated by a Beautiful Generalization

Solution The kind of integrals we have to deal with in the present section falls in the category of the *Cauchy principal value* integrals. In simple words, in such cases we may imagine a *clash* between quantities tending to infinity (e.g., one tending to ∞ and the other one going to $-\infty$, leading to an indeterminate case, $\infty - \infty$).

The strategy I'll present relies largely on my paper "A short presentation of a parameterized logarithmic integral with a Cauchy principal value meaning" which appeared in [89, January 31, 2021].

Let's prepare for a *non-obvious* way, one that relies on the following different forms of a parameterized logarithmic integral:

$$I(a) = \text{P. V.}\int_0^1 \frac{1}{x(1-x)}\log\left(\frac{2a^2x^2 + (2-a)ax + 1}{a^2x^2 - a^2x + a + 1}\right)dx$$

$$= \text{P. V.}\int_0^1 \frac{1}{x(1-x)}\log\left(\frac{2a^2x^2 - (3a+2)ax + (a+1)^2}{a^2x^2 - a^2x + a + 1}\right)dx$$

$$= \text{P. V.}\int_0^\infty \frac{1}{x}\log\left(\frac{(a^2 - 2a - 2)x - (a+1)^2x^2 - 1}{(a^2 - 2a - 2)x - (a+1)x^2 - a - 1}\right)dx$$

$$= \text{P. V.}\int_0^\infty \frac{1}{x}\log\left(\frac{(a^2 - 2a - 2)x - x^2 - (a+1)^2}{(a^2 - 2a - 2)x - (a+1)x^2 - a - 1}\right)dx$$

$$= \log^2(1+a), \tag{3.228}$$

where to keep things simple, we pick up $0 \le a \le 1$. However, if we focus, say, on the first integral form, it is not hard to note that we may extend the initial interval to $a \in (2(1 - \sqrt{2}), 2(1 + \sqrt{2}))$, and we do it by forcing $\Delta < 0$ for both quadratic equations in x in the fraction of the log argument.

Proof We start with the first integral version, where if we denote its integrand by $f(a, x)$, consider that $I(a) = \text{P. V.} \int_0^1 f(a, x)dx = \lim_{\epsilon \to 0^+} \int_\epsilon^{1-\epsilon} f(a, x)dx$, then differentiate with respect to the parameter a, rearrange, and cleverly expand, we get

$$I'(a) = \frac{1}{1 + a} \underbrace{\lim_{\epsilon \to 0^+} \int_\epsilon^{1-\epsilon} \left(\frac{1}{1 - x} - \frac{1}{x} \right) dx}_{0} + \frac{1}{1 + a} \int_0^1 \frac{(2a^2x^2 + (2a - a^2)x + 1)'}{2a^2x^2 + (2a - a^2)x + 1} dx$$

$$+ \frac{2a + a^2}{1 + a} \underbrace{\int_0^1 \left(\frac{1}{a^2x^2 - a^2x + a + 1} - \frac{1}{2a^2x^2 + (2 - a)ax + 1} \right) dx}_{0} = 2\frac{\log(1 + a)}{1 + a},$$

$$(3.229)$$

where we have $\lim_{\epsilon \to 0^+} \int_\epsilon^{1-\epsilon} \left(\frac{1}{1 - x} - \frac{1}{x} \right) dx = - \lim_{\epsilon \to 0^+} \left(\log(x(1 - x)) \Big|_{x=\epsilon}^{x=1-\epsilon} \right) = $ 0, and here the curious reader may also observe the integration could have been avoided. *And how about the third integral in* (3.229)? This may be also be viewed as a separate little challenging question for the curious reader. We observe that if we split it and then make any of the *magical* variable changes, $(1 + a)y/(1 + ay) = x$, or $(1 - y)/(1 + ay) = x$ as in the paper mentioned above, we readily get that

$$\int_0^1 \frac{1}{a^2x^2 - a^2x + a + 1} dx = \int_0^1 \frac{1}{2a^2y^2 + (2 - a)ay + 1} dy, \qquad (3.230)$$

which explains why the third integral in (3.229) equals 0.

Now, integrating back in (3.229) and noting that by setting $a = 0$ we get that the integration constant is 0, we arrive at the desired result

$$I(a) = \text{P. V.} \int_0^1 \frac{1}{x(1 - x)} \log \left(\frac{2a^2x^2 + (2 - a)ax + 1}{a^2x^2 - a^2x + a + 1} \right) dx = \log^2(1 + a),$$

which is the first wanted form in (3.228). The second desired form in (3.228) is obtained from the first form by the variable change $1 - y = x$, and the last two integral forms may be derived from the first two ones by the variable change $y/(1 + y) = x$, and the proof to the auxiliary result is complete. ∎

Since we have just finished the core generalized integral, we are ready to extract the desired special cases! For example, if we set $a = 1$ in (3.228), we immediately obtain the result at the first point,

$$I(1) = \text{P. V.} \int_0^1 \frac{1}{x(1-x)} \log\left(\frac{2x^2+x+1}{x^2-x+2}\right) dx$$

$$= \text{P. V.} \int_0^1 \frac{1}{x(1-x)} \log\left(\frac{2x^2-5x+4}{x^2-x+2}\right) dx$$

$$= \text{P. V.} \int_0^\infty \frac{1}{x} \log\left(\frac{4x^2+3x+1}{2x^2+3x+2}\right) dx = \text{P. V.} \int_0^\infty \frac{1}{x} \log\left(\frac{x^2+3x+4}{2x^2+3x+2}\right) dx$$

$$= \log^2(2),$$

and the solution to the point (i) of the problem is complete.

Similarly, for the second point of the problem we may easily guess the needed value by inspecting the given closed form in the statement question, and if we consider $1 + a = \varphi = (1+\sqrt{5})/2$, then we obtain $a = (\sqrt{5}-1)/2$. Thus, we have

$$I\left(\frac{\sqrt{5}-1}{2}\right) = \text{P. V.} \int_0^1 \frac{1}{x(1-x)} \log\left(\frac{2(3-\sqrt{5})x^2+(3\sqrt{5}-5)x+2}{(3-\sqrt{5})x^2+(\sqrt{5}-3)x+\sqrt{5}+1}\right) dx$$

$$= \text{P. V.} \int_0^1 \frac{1}{x(1-x)} \log\left(\frac{2(\sqrt{5}-3)x^2+(7-\sqrt{5})x-\sqrt{5}-3}{(\sqrt{5}-3)x^2+(3-\sqrt{5})x-\sqrt{5}-1}\right) dx$$

$$= \text{P. V.} \int_0^\infty \frac{1}{x} \log\left(\frac{(\sqrt{5}+3)x^2+(3\sqrt{5}-1)x+2}{(\sqrt{5}+1)x^2+(3\sqrt{5}-1)x+\sqrt{5}+1}\right) dx$$

$$= \text{P. V.} \int_0^\infty \frac{1}{x} \log\left(\frac{2x^2+(3\sqrt{5}-1)x+\sqrt{5}+3}{(\sqrt{5}+1)x^2+(3\sqrt{5}-1)x+\sqrt{5}+1}\right) dx$$

$$= \log^2(\varphi),$$

and the solution to the point (ii) of the problem is complete.

For the last point of the problem, we assume the possibility of an extension of (3.228) to complex numbers. We start from the simple fact that $\log(a + ib) = \log(\sqrt{a^2+b^2}) + i \arctan(b/a)$, $a > 0$, where if we square both sides and then take the imaginary part, we get $\Im\{\log^2(a+ib)\} = \log(a^2+b^2)\arctan(b/a)$. If we inspect the closed form given in the problem statement, it is not hard to see that we need $a = 1$ and $b = 1/\sqrt{3}$ to obtain that $\log(4/3)\pi/6 = \Im\{\log^2(1+i/\sqrt{3})\}$, and with this result in (3.228), where we also take the imaginary part of all sides, we get

$$\Im\left\{I\left(\frac{i}{\sqrt{3}}\right)\right\} = \text{P. V.} \int_0^1 \frac{1}{x(1-x)} \arctan\left(\frac{4\sqrt{3}x^2 - 2\sqrt{3}x^3 + 5\sqrt{3}x - 3\sqrt{3}}{2x^4 - 3x^3 - 8x^2 + 12x + 9}\right) dx$$

$$= \text{P. V.} \int_0^1 \frac{1}{x(1-x)} \arctan\left(\frac{2\sqrt{3}x^3 - 2\sqrt{3}x^2 - 7\sqrt{3}x + 4\sqrt{3}}{2x^4 - 5x^3 - 5x^2 + 5x + 12}\right) dx$$

$$= \text{P. V.} \int_0^\infty \frac{1}{x} \arctan\left(\frac{4\sqrt{3}x^4 + 9\sqrt{3}x^3 + \sqrt{3}x^2 - 7\sqrt{3}x - 3\sqrt{3}}{12 + 53x + 82x^2 + 48x^3 + 9x^4}\right) dx$$

$$= \text{P. V.} \int_0^\infty \frac{1}{x} \arctan\left(\frac{4\sqrt{3} + 9\sqrt{3}x + \sqrt{3}x^2 - 7\sqrt{3}x^3 - 3\sqrt{3}x^4}{12 + 53x + 82x^2 + 48x^3 + 9x^4}\right) dx$$

$$= \log\left(\frac{4}{3}\right)\frac{\pi}{6},$$

and the solution to the point (iii) of the problem is complete.

Finding different ways to go is very tempting now that we have arrived at the end of the section, and the curious reader might enjoy such a ride!

3.31 A Special Integral Generalization, Attacked with Strategies Involving the Cauchy Principal Value Integrals, That Also Leads to Two Famous Results by Ramanujan

Solution In the paper called "Certain Integrals Arising from Ramanujan's Notebooks" (2015) by Bruce C. Berndt and Armin Straub, one may find contour integration solutions to two fascinating results by Ramanujan, and at the same time the authors offer and prove more general theorems starting from these results.

In this section, I'll extract the two fascinating results differently, based on a generalization I'll establish by using integrals which we'll understand in the sense of Cauchy principal values. How does it sound? It's going to be fast and fun!

In the first step of the solution, we prove the auxiliary result

$$\text{P. V.} \int_0^\infty \frac{\cos(ax)}{b^2 - x^2} dx = \frac{\pi}{2b} \sin(ab), \ a, b > 0. \tag{3.231}$$

Proof For a first fast solution, we consider the classical result,

$$\int_0^\infty \frac{\cos(ax)}{b^2 + x^2}\,dx = \frac{\pi}{2b}e^{-ab}, \tag{3.232}$$

where upon replacing b by ib and considering the real part of both sides, the conclusion follows. The present result may also be found in Paul's book, *Inside Interesting Integrals*, where the author reduces the calculations to the Dirichlet's integral (see [65, pp.375–376]). In the following, I'll present this (natural) way to go as Paul did:

$$\text{P. V.} \int_0^\infty \frac{\cos(ax)}{b^2 - x^2}\,dx = \frac{1}{2}\,\text{P. V.} \int_{-\infty}^\infty \frac{\cos(ax)}{b^2 - x^2}\,dx$$

$$= \text{P. V.}\frac{1}{4b} \int_{-\infty}^\infty \frac{\cos(ax)}{b + x}\,dx + \text{P. V.}\frac{1}{4b} \int_{-\infty}^\infty \frac{\cos(ax)}{b - x}\,dx$$

{make the variable change $b + x = y$ in the first integral}

{and let the variable change $b - x = y$ in the second one}

$$= \text{P. V.}\frac{1}{4b} \int_{-\infty}^\infty \frac{\cos(ay - ab)}{y}\,dy + \text{P. V.}\frac{1}{4b} \int_{-\infty}^\infty \frac{\cos(ab - ay)}{y}\,dy$$

{in the second integral we use that $\cos(x) = \cos(-x)$}

$$= \text{P. V.}\frac{1}{2b} \int_{-\infty}^\infty \frac{\cos(ay - ab)}{y}\,dy$$

$$= \frac{\cos(ab)}{2b}\,\text{P. V.} \int_{-\infty}^\infty \frac{\cos(ay)}{y}\,dy + \frac{\sin(ab)}{2b} \int_{-\infty}^\infty \frac{\sin(ay)}{y}\,dy = \frac{\pi}{2b}\sin(ab),$$

where the Cauchy principal value of the first integral is 0 since the integrand is odd, and for the second integral we find useful to employ the Dirichlet's integral, $\int_0^\infty \frac{\sin(ax)}{x}\,dx = \frac{\pi}{2}\,\text{sgn}(a)$. ∎

Another auxiliary result we need in our calculations is

$$\int_0^\infty \frac{x^{s-1}}{x + c}\,dx = \begin{cases} \pi\csc(\pi s)c^{s-1}, & c > 0; \\ -\pi\cot(\pi s)(-c)^{s-1}, & c < 0, \end{cases} \tag{3.233}$$

where $0 < s < 1$ is a real number.

Proof This is a special case of a classical form of the Beta function. The case $c < 0$ must be understood as a Cauchy principal value. The result is straightforward and a proof of it may be found in [63, pp.57–58]. We can also go as follows:

for the case $c > 0$, we can make the variable change $x = cy$ followed by the variable change $y/(1 + y) = z$, and then we get $c^{s-1} \int_0^1 z^{s-1}(1 - z)^{1-s-1} dz =$ $c^{s-1} B(s, 1 - s) = c^{s-1} \Gamma(s)\Gamma(1 - s) = \pi \csc(\pi s)c^{s-1}$, where I used the relation between Beta function and Gamma function, $B(x, y) = \dfrac{\Gamma(x)\Gamma(y)}{\Gamma(x + y)}$, together with Euler's reflection formula, $\Gamma(x)\Gamma(1 - x) = \dfrac{\pi}{\sin(\pi x)}$. The case $c < 0$ is obtained by replacing c by $-c$ in the first case of (3.233) and taking the real part of the right-hand side (the extraction is also straightforward by using the limit form of the Cauchy principal value, which the curious reader might like to cleverly exploit). ∎

Now, if we make the variable change $x = y^2$, return to the notation in x, and replace s by $(s + 1)/2$ and c by c^2, we get

$$\int_0^\infty \frac{x^s}{x^2 + c^2} dx = \frac{\pi}{2} \sec\left(\frac{\pi}{2}s\right) c^{s-1}, \quad -1 < s < 1, \ c > 0 \qquad (3.234)$$

and

$$\mathrm{P.\,V.} \int_0^\infty \frac{x^s}{x^2 - c^2} dx = \frac{\pi}{2} \tan\left(\frac{\pi}{2}s\right) c^{s-1}, \qquad (3.235)$$

in accordance to the two cases in (3.233). We need these last two integrals in the main calculations.

Let's start out and prove first the generalization at the point (*i*)! If we multiply both sides of (3.231) by $\dfrac{b^s}{b^2 + c^2}$ and integrate from $b = 0$ to ∞, we get

$$\frac{\pi}{2} \int_0^\infty \frac{b^{s-1} \sin(ab)}{c^2 + b^2} db = \int_0^\infty \frac{b^s}{c^2 + b^2} \left(\mathrm{P.\,V.} \int_0^\infty \frac{\cos(ax)}{b^2 - x^2} dx\right) db$$

{reverse the order of integration in the double integral}

$$= \int_0^\infty \left(\mathrm{P.\,V.} \int_0^\infty \frac{b^s \cos(ax)}{(b^2 - x^2)(c^2 + b^2)} db\right) dx$$

$$= \int_0^\infty \frac{\cos(ax)}{x^2 + c^2} \left(\mathrm{P.\,V.} \int_0^\infty \frac{b^s}{b^2 - x^2} db\right) dx - \int_0^\infty \frac{\cos(ax)}{c^2 + x^2} \left(\int_0^\infty \frac{b^s}{c^2 + b^2} db\right) dx$$

{use the results in (3.232), (3.234), and (3.235)}

$$= \frac{\pi}{2} \tan\left(\frac{\pi s}{2}\right) \int_0^\infty \frac{x^{s-1} \cos(ax)}{x^2 + c^2} dx - \frac{\pi^2}{4} \sec\left(\frac{\pi s}{2}\right) c^{s-2} e^{-ac},$$

which can be immediately turned into our desired form,

$$\sin\left(\frac{\pi s}{2}\right)\int_0^\infty \frac{x^{s-1}\cos(ax)}{c^2+x^2}dx - \cos\left(\frac{\pi s}{2}\right)\int \frac{x^{s-1}\sin(ax)}{c^2+x^2}dx = \frac{\pi}{2}c^{s-2}e^{-ac},$$

and the point (i) of the problem is finalized.

For the point (ii) of the problem, we consider the previous result with $c = 1$, and then differentiate both sides once with respect to s and set $s = 1$, which immediately leads to the first result by Ramanujan we wanted to prove,

$$\frac{\pi}{2}\int_0^\infty \frac{\sin(ax)}{1+x^2}dx + \int_0^\infty \frac{\cos(ax)\log(x)}{1+x^2}dx = 0,$$

and the point (ii) of the problem is finalized.

As regards the point (iv), we consider the result from the point (i), where we differentiate twice both sides with respect to s, and then set $s = 1$, which gives

$$\pi\int_0^\infty \frac{\sin(ax)\log(x)}{c^2+x^2}dx + \int_0^\infty \frac{\cos(ax)\log^2(x)}{c^2+x^2}dx - \frac{\pi^2}{4}\int_0^\infty \frac{\cos(ax)}{c^2+x^2}dx = \frac{\pi\log^2(c)}{2c}e^{-ac},$$

and if we use the classical result in (3.232), we arrive at

$$\pi\int_0^\infty \frac{\sin(ax)\log(x)}{c^2+x^2}dx + \int_0^\infty \frac{\cos(ax)\log^2(x)}{c^2+x^2}dx = \frac{\pi\left(4\log^2(c)+\pi^2\right)}{8c}e^{-ac},$$

and the point (iv) of the problem is finalized.

At this point, if we set $c = 1$ in the previous integral result, we get the second result obtained by Ramanujan, as stated in [5],

$$\pi\int_0^\infty \frac{\sin(ax)\log(x)}{1+x^2}dx + \int_0^\infty \frac{\cos(ax)\log^2(x)}{1+x^2}dx = \frac{\pi^3}{8}e^{-a},$$

and the point (iii) of the problem is finalized.

If we use the result in (i), with $c = 1$, where we differentiate four times its both sides with respect to s and set $s = 1$, we get

$$\frac{\pi^4}{16}\int_0^\infty \frac{\cos(ax)}{1+x^2}dx - \frac{3}{2}\pi^2\int_0^\infty \frac{\cos(ax)\log^2(x)}{1+x^2}dx + \int_0^\infty \frac{\cos(ax)\log^4(x)}{1+x^2}dx$$

$$- \frac{\pi^3}{2}\int_0^\infty \frac{\sin(ax)\log(x)}{1+x^2}dx + 2\pi\int_0^\infty \frac{\sin(ax)\log^3(x)}{1+x^2}dx = 0,$$

which if we combine with the classical result stated in (3.232) and Ramanujan's result in (iii), we obtain another desired result

$$2\pi\int_0^\infty \frac{\sin(ax)\log^3(x)}{1+x^2}dx - \pi^2\int_0^\infty \frac{\cos(ax)\log^2(x)}{1+x^2}dx + \int_0^\infty \frac{\cos(ax)\log^4(x)}{1+x^2}dx$$

$$= \frac{\pi^5}{32} e^{-a},$$

and the point (v) of the problem is finalized.

As seen in the first part of the problem, the solution to the key generalization that leads to the extraction of the subsequent points relies on the use of integrals we understand as a Cauchy principal value.

Also, the solutions answer the *challenging question* that asks to avoid the use of contour integration.

3.32 Four *Magical* Integrals Beautifully Calculated, Which Generate the Closed Forms of Four Classical Integrals

Solution Words like *magical integrals* in the title of the section almost give a fairy tale nuance to the present section. I characterize them so for two reasons. One of them is obvious from the statement of the problem, and it is about the possibility of extracting four classical challenging integrals based on the four main integrals, where at least two of the classical integrals can be definitely counted as famous integrals, and I refer to these integrals as $\int_0^\infty \frac{\cos(ax)}{b^2 + x^2} dx = \frac{\pi}{2b} e^{-ab}$ and $\int_0^\infty \frac{x \sin(ax)}{b^2 + x^2} dx = \frac{\pi}{2} e^{-ab}$, which are both always a great experience for any lover of integrals. Now, the second reason for which I call the main integrals *magical* is given by the unexpectedly simple and beautiful ways that one could use here.

So, let's do the calculations in a cool fashion! For the point (i) of the problem, we start with using the elementary representation

$$e^\theta \int_\theta^\infty \cos(xa) e^{-a} da = \frac{\cos(\theta x) - x \sin(\theta x)}{1 + x^2}, \tag{3.236}$$

which is a key step of the solution (and similar key steps we'll consider for the solutions to the other points of the problem). Observe the integral is elementary and it can be calculated by exploiting Euler's formula, $e^{ix} = \cos(x) + i \sin(x)$, since for the real part we have $\Re\{e^{ixa}\} = \cos(xa)$.

Multiplying both sides of (3.236) by e^{-sx} and integrating from $x = 0$ to ∞, we obtain

$$\int_0^\infty \frac{\cos(\theta x) - x \sin(\theta x)}{1 + x^2} e^{-sx} dx = e^\theta \int_0^\infty \left(\int_\theta^\infty \cos(xa) e^{-a-sx} da \right) dx$$

{reverse the order of integration}

$$= e^\theta \int_\theta^\infty e^{-a} \left(\int_0^\infty \cos(ax)e^{-sx}dx \right) da = se^\theta \underbrace{\int_\theta^\infty \frac{e^{-a}}{s^2+a^2}da}_{<\infty},$$

where we notice the last integral converges independently of the value of $s > 0$, and by letting $s \to 0^+$ in the opposite sides of the result above, we obtain that

$$\int_0^\infty \frac{\cos(\theta x) - x\sin(\theta x)}{1+x^2}dx = 0,$$

and the point (i) of the problem is complete.

To perform the calculations at the point (ii), we use a similar strategy to the one found at the previous point, and consider the version of the result in (3.236) where θ is replaced by $-\theta$,

$$e^{-\theta} \int_{-\theta}^\infty \cos(xa)e^{-a}da = \frac{\cos(\theta x) + x\sin(\theta x)}{1+x^2}. \tag{3.237}$$

Then, by multiplying both sides of (3.237) by e^{-sx} and integrating from $x = 0$ to ∞, we have

$$\int_0^\infty \frac{\cos(\theta x) + x\sin(\theta x)}{1+x^2}e^{-sx}dx = e^{-\theta} \int_0^\infty \left(\int_{-\theta}^\infty \cos(xa)e^{-a-sx}da \right) dx$$

$$\{\text{reverse the order of integration}\}$$

$$= e^{-\theta} \int_{-\theta}^\infty \left(\int_0^\infty \cos(ax)e^{-a-sx}dx \right) da = e^{-\theta} \int_{-\theta}^\infty \frac{e^{-a}s}{s^2+a^2}da,$$

and by letting $s \to 0^+$ in the opposite sides of the result above, we get that

$$\int_0^\infty \frac{\cos(\theta x) + x\sin(\theta x)}{1+x^2}dx = \pi e^{-\theta},$$

and the point (ii) of the problem is complete. The question also appeared in [54].

Wait! How did you get that limit leading to π?, might be your kind of reaction at this point, and I agree with it. Calculating the limit, $\lim\limits_{s\to0^+} \int_{-\theta}^\infty \frac{e^{-a}s}{s^2+a^2}da$, is a good problem itself you might like to try.

Let's prove that

$$\lim_{s\to0^+} \int_{-\theta}^\infty \frac{e^{-a}s}{s^2+a^2}da = \lim_{s\to0^+} \int_{-\sqrt{s}}^{\sqrt{s}} \frac{e^{-a}s}{s^2+a^2}da = \pi, \quad \sqrt{s} < \theta. \tag{3.238}$$

Proof We observe we can write the limit as follows:

$$\lim_{s \to 0^+} \int_{-\theta}^{\infty} \frac{e^{-as}}{s^2 + a^2} da$$

$$= \underbrace{\lim_{s \to 0^+} \int_{-\theta}^{-\sqrt{s}} \frac{e^{-as}}{s^2 + a^2} da}_{L_1} + \underbrace{\lim_{s \to 0^+} \int_{-\sqrt{s}}^{\sqrt{s}} \frac{e^{-as}}{s^2 + a^2} da}_{L_2} + \underbrace{\lim_{s \to 0^+} \int_{\sqrt{s}}^{\infty} \frac{e^{-as}}{s^2 + a^2} da}_{L_3}.$$

$$(3.239)$$

We'll easily note that $L_1 = 0$ by a similar reasoning to the one given for the third limit you'll find in the following. So, to see that $L_3 = 0$, we write that

$$0 \le \int_{\sqrt{s}}^{\infty} \frac{e^{-as}}{s^2 + a^2} da \le s e^{-\sqrt{s}} \int_{\sqrt{s}}^{\infty} \frac{1}{s^2 + a^2} da = e^{-\sqrt{s}} \left(\frac{\pi}{2} - \arctan \left(\frac{1}{\sqrt{s}} \right) \right),$$

and if we let $s \to 0^+$, we get $L_3 = 0$ as announced. Finally, applying the mean value theorem for the integral under the remaining limit, that is, the limit L_2, we

have $\int_{-\sqrt{s}}^{\sqrt{s}} \frac{e^{-as}}{s^2 + a^2} da = s e^{-t(s)} \int_{-\sqrt{s}}^{\sqrt{s}} \frac{1}{s^2 + a^2} da = 2 \arctan \left(\frac{1}{\sqrt{s}} \right) e^{-t(s)}$ with

$-\sqrt{s} < t(s) < \sqrt{s}$, and by letting $s \to 0^+$, we immediately get $L_2 = \pi$. The limit L_2 can also be calculated by using the *squeeze theorem*. Thus, returning with all these subresults in (3.239), we obtain that

$$\lim_{s \to 0^+} \int_{-\theta}^{\infty} \frac{e^{-as}}{s^2 + a^2} da = \pi,$$

and the proof to the auxiliary limit result is finalized. ∎

Further, to calculate the integral at the point (iii), we consider the following integral representation:

$$e^{-\theta} \int_{-\theta}^{\infty} \sin(xa) e^{-a} da = \frac{x \cos(\theta x) - \sin(\theta x)}{1 + x^2}. \tag{3.240}$$

Thus, in view of (3.240), where we multiply both sides by e^{-sx} and integrate from $x = 0$ to ∞, we obtain

$$\int_0^{\infty} \frac{x \cos(\theta x) - \sin(\theta x)}{1 + x^2} e^{-sx} dx = e^{-\theta} \int_0^{\infty} \left(\int_{-\theta}^{\infty} \sin(xa) e^{-a-sx} da \right) dx$$

{reverse the order of integration}

$$= e^{-\theta} \int_{-\theta}^{\infty} \left(\int_0^{\infty} \sin(ax) e^{-a-sx} dx \right) da = e^{-\theta} \int_{-\theta}^{\infty} \frac{a e^{-a}}{s^2 + a^2} da,$$

and if we let $s \to 0^+$ in the opposite sides, we get

$$\int_0^\infty \frac{x\cos(\theta x) - \sin(\theta x)}{1 + x^2}dx = e^{-\theta}\, \text{P. V.} \int_{-\theta}^\infty \frac{e^{-a}}{a}da = -e^{-\theta}\,\text{Ei}(\theta),$$

where I used the following definition of the exponential integral (we understand in the sense of the Cauchy principal value), $\text{Ei}(x) = -\lim\limits_{\epsilon \to 0^+} \left(\int_{-x}^{-\epsilon} \frac{e^{-t}}{t}dt + \int_\epsilon^\infty \frac{e^{-t}}{t}dt \right)$, with $x > 0$, and the point (iii) of the problem is complete.

Finally, for the last part of the problem we consider the integral representation

$$e^\theta \int_\theta^\infty \sin(xa)e^{-a}da = \frac{x\cos(\theta x) + \sin(\theta x)}{1 + x^2}. \tag{3.241}$$

Further, using the result in (3.241) where we multiply both sides by e^{-sx} and integrate from $x = 0$ to ∞, we have

$$\int_0^\infty \frac{x\cos(\theta x) + \sin(\theta x)}{1 + x^2}e^{-sx}dx = e^\theta \int_0^\infty \left(\int_\theta^\infty \sin(xa)e^{-a-sx}da \right) dx$$

{reverse the order of integration}

$$= e^\theta \int_\theta^\infty \left(\int_0^\infty \sin(xa)e^{-a-sx}dx \right) da = e^\theta \int_\theta^\infty \frac{ae^{-a}}{s^2 + a^2}da,$$

and if we let $s \to 0^+$ in the opposite sides, we have

$$\int_0^\infty \frac{x\cos(\theta x) + \sin(\theta x)}{1 + x^2}dx = e^\theta \int_\theta^\infty \frac{e^{-a}}{a}da = -e^\theta\,\text{Ei}(-\theta),$$

where I used the exponential integral representation, $\text{Ei}(-x) = -\int_x^\infty \frac{e^{-t}}{t}dt$, $x > 0$, and the point (iv) of the problem is complete.

All four classical integrals listed at the end of the section with the problem statement are obtained from the previous integrals if we set $\theta = ab$, $a, b > 0$, and then make the variable change $bx = y$.

If you've had a chance to read my first book, *(Almost) Impossible Integrals, Sums, and Series*, you've probably noticed that I managed to obtain the value of the integral, $\int_0^\infty \frac{\cos(ax)}{b^2 + x^2}dx$, by exploiting the Cauchy-Schlömilch transformation, and that way is so awesome! At that time, to me it looked (almost) impossible to come up with something simpler, and still, as seen from this section, the technique I provided here could definitely challenge my first solution, at least in terms of simplicity. Which one of the two solutions do you like most?

3.33   A Bouquet of Captivating Integrals Involving Trigonometric and Hyperbolic Functions

Solution If you are a connoisseur of *Table of Integrals, Series, and Products*, then you probably noticed that the integrals at the points (ii)–(iv) are part of it and may be found in [17, **3.983.1–3.983.3**, p.515].

As regards the first point of the problem, we use the following key result:

$$\frac{2}{\sin(c)} \sum_{n=1}^{\infty} (-1)^{n-1} \sin(cn) e^{-bnx} = \frac{1}{\cosh(bx) + \cos(c)}. \tag{3.242}$$

Proof Using the simple fact that $\displaystyle\sum_{n=1}^{N} z^n = \frac{z - z^{N+1}}{1 - z}$ and then replacing z by $-e^{-bx+ci}$ and letting $N \to \infty$, we get that $\displaystyle\sum_{n=1}^{\infty} (-1)^{n-1} (e^{-bx+ci})^n = \frac{1}{1 + e^{bx-ci}}$.

Finally, taking the imaginary parts and multiplying both sides of the last result by $2/\sin(c)$, we get the desired result. Also, we may recall the well-known entry

1.447.1 from [17], $\displaystyle\sum_{n=1}^{\infty} p^n \sin(nx) = \frac{p \sin(x)}{1 - 2p \cos(x) + p^2}$, which immediately leads

to the series form in (3.242). ∎∎ ■

Returning to the point (i) of the problem and using the result in (3.242), we have

$$\int_0^{\infty} \frac{\sin(ax)}{\cosh(bx) + \cos(c)} dx = \frac{2}{\sin(c)} \int_0^{\infty} \sin(ax) \sum_{n=1}^{\infty} (-1)^{n-1} \sin(cn) e^{-bnx} dx$$

{reverse the order of summation and integration}

$$= \frac{2}{\sin(c)} \sum_{n=1}^{\infty} (-1)^{n-1} \sin(cn) \int_0^{\infty} \sin(ax) e^{-bnx} dx$$

$$= \frac{2a}{\sin(c)} \sum_{n=1}^{\infty} (-1)^{n-1} \frac{\sin(cn)}{a^2 + b^2 n^2},$$

and the point (i) of the problem is complete.

Now, to calculate the integral at the point (ii) we need one more result besides the key result in (3.242) from the previous part of the problem. If we expand $f(x) = \sinh(ax), -\pi < x < \pi$, in a Fourier series,[13] we observe immediately that since $f(x)$ is an odd function, the Fourier coefficients are given by $b_n = \dfrac{1}{\pi}\displaystyle\int_{-\pi}^{\pi} f(x)\sin(nx)\mathrm{d}x = \dfrac{1}{\pi}\displaystyle\int_{-\pi}^{\pi} \sinh(ax)\sin(nx)\mathrm{d}x = \dfrac{2}{\pi}\sinh(a\pi)(-1)^{n-1}\dfrac{n}{a^2+n^2}$ (observe the coefficients a_n vanish). Therefore, we have the following Fourier series:

$$\frac{\pi}{2}\frac{\sinh(ax)}{\sinh(a\pi)} = \sum_{n=1}^{\infty}(-1)^{n-1}\frac{n\sin(nx)}{a^2+n^2}, \tag{3.243}$$

which is the second result we need during the main calculations.

At this point, we are ready to prove the main result. So, with the result from (3.242) in mind, we write

$$\int_0^{\infty}\frac{\cos(ax)}{\cosh(bx)+\cos(c)}\mathrm{d}x = \frac{2}{\sin(c)}\int_0^{\infty}\cos(ax)\sum_{n=1}^{\infty}(-1)^{n-1}\sin(cn)e^{-bnx}\mathrm{d}x$$

{reverse the order of summation and integration}

$$= \frac{2}{\sin(c)}\sum_{n=1}^{\infty}(-1)^{n-1}\sin(cn)\int_0^{\infty}\cos(ax)e^{-bnx}\mathrm{d}x = \frac{2b}{\sin(c)}\sum_{n=1}^{\infty}(-1)^{n-1}\frac{n\sin(cn)}{a^2+b^2n^2}$$

{make use of the result in (3.243)}

$$= \frac{\pi}{b}\frac{\sinh\left(\dfrac{ac}{b}\right)}{\sin(c)\sinh\left(\dfrac{a\pi}{b}\right)},$$

and the point (ii) of the problem is complete.

---

[13] The Fourier expansion of a function $f(x)$ of period 2π is usually given by $f(x) = \dfrac{a_0}{2} + \displaystyle\sum_{n=1}^{\infty}(a_n\cos(nx) + b_n\sin(nx))$, where the coefficients of the expansion are $a_0 = \dfrac{1}{\pi}\displaystyle\int_{-\pi}^{\pi} f(x)\mathrm{d}x$, $a_n = \dfrac{1}{\pi}\displaystyle\int_{-\pi}^{\pi} f(x)\cos(nx)\mathrm{d}x$, and $b_n = \dfrac{1}{\pi}\displaystyle\int_{-\pi}^{\pi} f(x)\sin(nx)\mathrm{d}x$. Based on the parity of $f(x)$, we have two special forms of the Fourier series. If $f(x)$ is *even*, we have the expansion, $f(x) = \dfrac{a_0}{2} + \displaystyle\sum_{n=1}^{\infty} a_n\cos(nx)$, and this short form is given by the fact that b_n coefficients vanish. Otherwise, if $f(x)$ is *odd*, we get the expansion, $f(x) = \displaystyle\sum_{n=1}^{\infty} b_n\sin(nx)$, and note that a_n coefficients vanish.

The integral at the point (iii) can be extracted based on the generalization from the point (ii), where if we assume the validity of an extension of the parameter c to complex numbers, and then set $b = 1$ and $c = bi + \pi$, we immediately arrive at the desired result,

$$\text{P. V.} \int_0^\infty \frac{\cos(ax)}{\cosh(x) - \cosh(b)} dx = -\pi \coth(a\pi) \frac{\sin(ab)}{\sinh(b)},$$

and the point (iii) of the problem is complete. In the calculations I used the formula, $\sin(x - iy) = \sin(x)\cosh(y) - i\cos(x)\sinh(y)$. The curious reader might also be interested to arrive at this result by considering other approaches of solving the problem (e.g., to make the extraction differently, one might promptly employ contour integration).

Finally, for the last part of the problem we use again the result from the point (ii) where if we replace c by $\arccos(c/d)$, $d > c$, divide both sides by d, and consider the identity, $\sin(\arccos(x)) = \sqrt{1 - x^2}$, we get the first case of the integral. The other case, $c > d$, is proved in a similar way, by also taking into account that $\arccos(x) = i \operatorname{arccosh}(x)$, $x > 1$, $\sinh(ix) = i\sin(x)$. Thus, we obtain that

$$\int_0^\infty \frac{\cos(ax)}{d\cosh(bx) + c} dx = \begin{cases} \dfrac{\pi \sinh\left(\frac{a}{b}\arccos\left(\frac{c}{d}\right)\right)}{b\sqrt{d^2 - c^2}\sinh\left(\frac{a\pi}{b}\right)}, & d > c > 0; \\[4mm] \dfrac{\pi \sin\left(\frac{a}{b}\operatorname{arccosh}\left(\frac{c}{d}\right)\right)}{b\sqrt{c^2 - d^2}\sinh\left(\frac{a\pi}{b}\right)}, & c > d > 0, \end{cases}$$

and the point (iv) of the problem is complete.

The curious reader might exploit the power of the generalizations proposed and solved in this section in order to extract other exotic integrals.

3.34 Interesting Integrals to Evaluate, One of Them Coming from Lord Kelvin's Work

Solution In the summer of 2017 (at the end of June) I received an email from Paul Nahin, which contained an attachment with the integral at the point (i) accompanied by words like *a little puzzle*, suggesting it could be a good entry for my book. Paul told me that he came across the integral while reading a paper by William Thomson (later Lord Kelvin), who considered this integral, and where, without further analysis, he just wrote the value of the integral. So, we may count this fact as a puzzle and try to find a simple way of obtaining the result, but how simple the way I'll present is will remain up to you to decide.

Let me remind you first that error function is defined as $\operatorname{erf}(x) = \dfrac{2}{\sqrt{\pi}} \int_0^x e^{-t^2} dt$, and although in our calculations forms of this integral will appear, I won't use $\operatorname{erf}(x)$ dedicated notation to preserve a more elementary image of the solution.

I'll begin by stating and proving that

$$J(a, b) = \int_0^\infty e^{-ax^2 - bx} \mathrm{d}x = \frac{\sqrt{\pi}}{2} \frac{1}{\sqrt{a}} e^{b^2/(4a)} \left(1 - \frac{2}{\sqrt{\pi}} \int_0^{b/(2\sqrt{a})} e^{-t^2} \mathrm{d}t\right),$$

$$(3.244)$$

where $a > 0$, $a, b \in \mathbb{R}$.

Proof Completing the square and using the Gaussian integral, $\int_0^\infty e^{-x^2} \mathrm{d}x = \frac{\sqrt{\pi}}{2}$, which can be found calculated in [95], we have that

$$J(a, b) = \int_0^\infty e^{-ax^2 - bx} \mathrm{d}x = e^{b^2/(4a)} \int_0^\infty e^{-((\sqrt{a}x)^2 + 2b/2x + (b/(2\sqrt{a}))^2)} \mathrm{d}x$$

$$= e^{b^2/(4a)} \int_0^\infty e^{-(\sqrt{a}x + b/(2\sqrt{a}))^2} \mathrm{d}x$$

$$\{\text{consider making the variable change } \sqrt{a}x + b/(2\sqrt{a}) = t\}$$

$$= \frac{1}{\sqrt{a}} e^{b^2/(4a)} \int_{b/(2\sqrt{a})}^\infty e^{-t^2} \mathrm{d}t = \frac{1}{\sqrt{a}} e^{b^2/(4a)} \left(\int_0^\infty e^{-t^2} \mathrm{d}t - \int_0^{b/(2\sqrt{a})} e^{-t^2} \mathrm{d}t\right)$$

$$= \frac{\sqrt{\pi}}{2} \frac{1}{\sqrt{a}} e^{b^2/(4a)} \left(1 - \frac{2}{\sqrt{\pi}} \int_0^{b/(2\sqrt{a})} e^{-t^2} \mathrm{d}t\right),$$

and the auxiliary result is proved.   ∎

Differentiating (3.244) once with respect to b, we get

$$\frac{\partial}{\partial b}\{J(a, b)\} = -\int_0^\infty x e^{-ax^2 - bx} \mathrm{d}x$$

$$= -\frac{1}{2}\frac{1}{a} + \frac{\sqrt{\pi}}{4}\frac{b}{a^{3/2}} e^{b^2/(4a)} - \frac{1}{2}\frac{b}{a^{3/2}} e^{b^2/(4a)} \int_0^{b/(2\sqrt{a})} e^{-t^2} \mathrm{d}t. \quad (3.245)$$

Considering replacing a by ia and b by $1 - i$ in (3.245), where we also use Euler's formula $e^{i\theta} = \cos(\theta) + i \sin(\theta)$ to obtain and use the values $i^{1/2} = e^{i\pi/4} = 1/\sqrt{2}(1 + i)$, $i^{3/2} = e^{i3\pi/4} = -1/\sqrt{2}(1 - i)$, and rearranging, we obtain

$$\int_0^\infty x e^{-x - ix(ax-1)} \mathrm{d}x$$

$$= -i\frac{1}{2}\frac{1}{a} + \frac{1}{2}\sqrt{\frac{\pi}{2}}\frac{1}{a^{3/2}} e^{-1/(2a)} - \frac{1}{\sqrt{2}}\frac{1}{a^{3/2}} e^{-1/(2a)} \underbrace{\int_0^{-i/\sqrt{2a}} e^{-t^2} \mathrm{d}t}_{it = u}$$

$$= \frac{1}{2}\sqrt{\frac{\pi}{2}}\frac{1}{a^{3/2}}e^{-1/(2a)} - i\left(\frac{1}{2}\frac{1}{a} - \frac{1}{\sqrt{2}}\frac{1}{a^{3/2}}e^{-1/(2a)}\int_0^{1/\sqrt{2a}} e^{u^2}\,du\right). \quad (3.246)$$

Taking the real part of (3.246), we get that

$$\Re\left\{\int_0^\infty xe^{-x-ix(ax-1)}\,dx\right\} = \int_0^\infty x\cos(ax^2 - x)e^{-x}\,dx = \frac{1}{2}\sqrt{\frac{\pi}{2}}\frac{1}{a^{3/2}}e^{-1/(2a)}. \quad (3.247)$$

Returning to our integral found at the first point of the problem, where we let the change of variable $x = y^2/z^2$, we get

$$\int_0^\infty \cos(2tx - z\sqrt{x})e^{-z\sqrt{x}}\,dx = \frac{2}{z^2}\int_0^\infty y\cos\left(\frac{2t}{z^2}y^2 - y\right)e^{-y}\,dy$$

$$\overset{\text{use (3.247)}}{=} \frac{\sqrt{\pi}}{4}\frac{z}{t^{3/2}}e^{-z^2/(4t)},$$

and the solution to the point (i) is finalized.

Regarding the point (ii) of the problem, if we consider the following forms of the Fresnel integrals, $C(u) = \int_0^u \cos\left(\frac{\pi}{2}x^2\right)dx \overset{x/u=t}{=} u\int_0^1 \cos\left(\frac{\pi}{2}u^2t^2\right)dt$ and $S(u) = \int_0^u \sin\left(\frac{\pi}{2}x^2\right)dx \overset{x/u=t}{=} u\int_0^1 \sin\left(\frac{\pi}{2}u^2t^2\right)dt$, and then return to the previous result, where we replace t by t^2, then z by $2\sqrt{s}$, and integrate both sides from $t = 0$ to $t = 1$, we arrive at

$$\int_0^1\left(\int_0^\infty \cos(2t^2x - 2\sqrt{sx})e^{-2\sqrt{sx}}\,dx\right)dt = \int_0^\infty\left(\int_0^1 \cos(2xt^2 - 2\sqrt{sx})e^{-2\sqrt{sx}}\,dt\right)dx$$

{use that $\cos(a - b) = \cos(a)\cos(b) + \sin(a)\sin(b)$ and expand the inner integral}

$$= \int_0^\infty\left(\cos(2\sqrt{sx})\int_0^1 \cos(2xt^2)\,dt + \sin(2\sqrt{sx})\int_0^1 \sin(2xt^2)\,dt\right)e^{-2\sqrt{sx}}\,dx$$

$$= \frac{\sqrt{\pi}}{2}\int_0^\infty \frac{\cos(2\sqrt{sx})C\left(\frac{2\sqrt{x}}{\sqrt{\pi}}\right) + \sin(2\sqrt{sx})S\left(\frac{2\sqrt{x}}{\sqrt{\pi}}\right)}{\sqrt{x}}e^{-2\sqrt{sx}}\,dx$$

$$\overset{2\sqrt{sx}=y}{=} \frac{\sqrt{\pi}}{2}\frac{1}{\sqrt{s}}\int_0^\infty\left(\cos(y)C\left(\frac{y}{\sqrt{\pi s}}\right) + \sin(y)S\left(\frac{y}{\sqrt{\pi s}}\right)\right)e^{-y}\,dy$$

$$= \frac{\sqrt{\pi}}{2}\sqrt{s}\int_0^1 \frac{1}{t^3}e^{-s/t^2}\,dt = \frac{\sqrt{\pi}}{4}\frac{e^{-s}}{\sqrt{s}},$$

whence we conclude that

$$2 \int_0^\infty \left(\cos(x) C \left(\frac{x}{\sqrt{\pi s}} \right) + \sin(x) S \left(\frac{x}{\sqrt{\pi s}} \right) \right) e^{-x} dx = e^{-s},$$

and the solution to the point (ii) is finalized.

Another solution to my proposed integral at the point (ii), one that exploits contour integration, may be found in *La Gaceta de la RSME*, Vol. 22, No. 3 (2019).

3.35 A Surprisingly Awesome Fractional Part Integral with Forms Involving $\sqrt{\tan(x)}$, $\sqrt{\cot(x)}$

Solution Let me tell first that at the sight of the proposed integral, many (if not all) of the integration enthusiasts will immediately recollect another two integrals, at least in the indefinite forms,[14] $\int \sqrt{\tan(x)} dx$ and $\int \sqrt{\cot(x)} dx$, which are classical.

Then we have other known variants, when x goes from 0 to $\pi/2$, and we'll meet such a case below.

At this point, you might remember another two fractional part integrals met in *(Almost) Impossible Integrals, Sums, and Series*, that is, $\int_0^{\pi/2} \frac{\{\cot(x)\}}{\cot(x)} dx$ and $\int_0^{\pi/2} \{\cot(x)\} dx$ (see [76, Chapter 1, p.32]). *Maybe you could get some inspiration from those calculations there?* I think so—at least as regards the beginning!

So, using the fact that $\{x\} = x - \lfloor x \rfloor$, where $\lfloor x \rfloor$ denotes the integer part of x, we write

$$\int_0^{\pi/2} \left\{ \sqrt{\tan(x)} \right\} dx = \int_0^{\pi/2} \left(\sqrt{\tan(x)} - \lfloor \sqrt{\tan(x)} \rfloor \right) dx$$

$$= \underbrace{\int_0^{\pi/2} \sqrt{\tan(x)} dx}_{\pi/\sqrt{2}} - \sum_{k=0}^\infty \int_{\arctan(k^2)}^{\arctan((k+1)^2)} \lfloor \sqrt{\tan(x)} \rfloor dx = \frac{\pi}{\sqrt{2}} - \sum_{k=0}^\infty k \int_{\arctan(k^2)}^{\arctan((k+1)^2)} dx$$

---

[14] Both $\int \sqrt{\tan(x)} dx$ and $\int \sqrt{\cot(x)} dx$ may also be found as an exercise in *How to Integrate It: A Practical Guide to Finding Elementary Integrals* by Seán M. Stewart, Cambridge University Press (2017), more precisely on page 202, where the author indicates to first consider adding and subtracting the two integrals to form a system of two relations and calculate them this way, and then extract each integral. Following this way, we obtain that $\int \sqrt{\tan(x)} dx = \frac{1}{\sqrt{2}} \arcsin(\sin(x) - \cos(x)) - \frac{1}{\sqrt{2}} \log|\sqrt{\sin(2x)} + \sin(x) + \cos(x)| + C$ and $\int \sqrt{\cot(x)} dx = \frac{1}{\sqrt{2}} \arcsin(\sin(x) - \cos(x)) + \frac{1}{\sqrt{2}} \log|\sqrt{\sin(2x)} + \sin(x) + \cos(x)| + C$.

$$\left\{ \text{start from } k = 1 \text{ and use the relation } \arctan(x) + \arctan\left(\frac{1}{x}\right) = \frac{\pi}{2}, \, x > 0 \right\}$$

$$= \frac{\pi}{\sqrt{2}} - \sum_{k=1}^{\infty} k \left(\arctan\left(\frac{1}{k^2}\right) - \arctan\left(\frac{1}{(k+1)^2}\right) \right)$$

$$= \frac{\pi}{\sqrt{2}} - \lim_{N \to \infty} \sum_{k=1}^{N} \left((k-1) \arctan\left(\frac{1}{k^2}\right) - k \arctan\left(\frac{1}{(k+1)^2}\right) \right) - \sum_{k=1}^{\infty} \arctan\left(\frac{1}{k^2}\right)$$

$$= \frac{\pi}{\sqrt{2}} + \lim_{N \to \infty} N \arctan\left(\frac{1}{(N+1)^2}\right) - \sum_{k=1}^{\infty} \arctan\left(\frac{1}{k^2}\right) = \frac{\pi}{\sqrt{2}} - \sum_{k=1}^{\infty} \arctan\left(\frac{1}{k^2}\right),$$

$$(3.248)$$

where we see that $\lim_{N \to \infty} N \arctan\left(1/(N+1)^2\right) = \lim_{N \to \infty} \frac{N}{(N+1)^2} \frac{\arctan(1/(N+1)^2)}{1/(N+1)^2}$

$$= \underbrace{\lim_{N \to \infty} \frac{N}{(N+1)^2}}_{0} \cdot \underbrace{\lim_{N \to \infty} \frac{\arctan(1/(N+1)^2)}{1/(N+1)^2}}_{1} = 0 \text{ by using that } \lim_{x \to 0} \frac{\arctan(x)}{x} =$$

1.

In order to get that $\int_0^{\pi/2} \sqrt{\tan(x)} dx = \frac{\pi}{\sqrt{2}}$, as seen in (3.248), we might choose either to use the form of the indefinite integral or to start with the variable change $\sqrt{\tan(x)} = t$ and continue as follows:

$$\int_0^{\pi/2} \sqrt{\tan(x)} dx = \int_0^{\infty} \frac{2t^2}{1+t^4} dt = \int_0^{\infty} \frac{(1+1/t^2) + (1 - 1/t^2)}{t^2 + 1/t^2} dt$$

$$= \underbrace{\int_0^{\infty} \frac{(t - 1/t)'}{2 + (t - 1/t)^2} dt}_{t - 1/t = u} + \underbrace{\int_0^{\infty} \frac{t^2 - 1}{1 + t^4} dt}_{0} = \int_{-\infty}^{\infty} \frac{1}{(\sqrt{2})^2 + u^2} du$$

$$= \frac{1}{\sqrt{2}} \arctan\left(\frac{u}{\sqrt{2}}\right) \Big|_{u=-\infty}^{u=\infty} = \frac{\pi}{\sqrt{2}},$$

where in the calculations I also used that $\int_0^{\infty} \frac{1}{1+t^4} dt = \int_0^{\infty} \frac{t^2}{1+t^4} dt$, easily seen by letting the variable change $1/t = u$ in any of the two integrals.

But how would we approach the remaining arctan series in (3.248)? Well, we may start with the Taylor series, $\arctan(x) = \sum_{n=1}^{\infty} (-1)^{n-1} \frac{x^{2n-1}}{2n-1}$, and then we write

$$\sum_{k=1}^{\infty} \arctan\left(\frac{1}{k^2}\right) = \sum_{k=1}^{\infty}\left(\sum_{n=1}^{\infty}(-1)^{n-1}\frac{1}{(2n-1)k^{4n-2}}\right)$$

$$\underset{\text{of summation}}{\overset{\text{reverse the order}}{=}} \sum_{n=1}^{\infty}(-1)^{n-1}\frac{1}{2n-1}\left(\sum_{k=1}^{\infty}\frac{1}{k^{4n-2}}\right) = \sum_{n=1}^{\infty}(-1)^{n-1}\frac{\zeta(4n-2)}{2n-1}.$$

$$(3.249)$$

From (6.316), Sect. 6.47, we have $\pi x \cot(\pi x) = 1 - 2\sum_{n=1}^{\infty}x^{2n}\zeta(2n)$, and if we

replace x by ix, we get $\pi x \coth(\pi x) = 1 + 2\sum_{n=1}^{\infty}(-1)^{n-1}x^{2n}\zeta(2n)$. Now, combining

the two series results and rearranging, we get

$$\frac{\pi}{4}(\coth(\pi x) - \cot(\pi x)) = \sum_{n=1}^{\infty}x^{4n-3}\zeta(4n-2).\qquad(3.250)$$

Replacing x by $\sqrt{x}$ in (3.250) and then dividing both sides by $\sqrt{x}$, we arrive at

$$\frac{\pi}{4}\frac{\coth(\pi\sqrt{x}) - \cot(\pi\sqrt{x})}{\sqrt{x}} = \sum_{n=1}^{\infty}x^{2n-2}\zeta(4n-2).\qquad(3.251)$$

Considering (3.251) in t, and then integrating from $t = 0$ to $t = x$, we have

$$\frac{1}{2}\log\left(\frac{\sinh(\pi\sqrt{x})}{\sin(\pi\sqrt{x})}\right) = \sum_{n=1}^{\infty}x^{2n-1}\frac{\zeta(4n-2)}{2n-1}.\qquad(3.252)$$

Upon setting $x = i$ in (3.252), and rearranging, we get

$$\sum_{n=1}^{\infty}(-1)^{n-1}\frac{\zeta(4n-2)}{2n-1} = \frac{1}{2i}\log\left(\frac{\sinh(\pi\sqrt{i})}{\sin(\pi\sqrt{i})}\right).\qquad(3.253)$$

Since we have that $\sqrt{i} = \pm\frac{1}{\sqrt{2}}(1+i)$ {e.g., given that $z = x + iy$, we note that $z^2 = (x^2 - y^2) + i2xy = i$, and then solve for the resulting conditions, $x^2 - y^2 = 0$ and $2xy = 1$}, we may write that

$$\frac{\sinh(\pi\sqrt{i})}{\sin(\pi\sqrt{i})} = \frac{\sinh(\pi/\sqrt{2} + i\pi/\sqrt{2})}{\sin(\pi/\sqrt{2} + i\pi/\sqrt{2})}$$

$$= \frac{\overbrace{\cos(\pi/\sqrt{2})\sinh(\pi/\sqrt{2})}^{v} + i\,\overbrace{\sin(\pi/\sqrt{2})\cosh(\pi/\sqrt{2})}^{u}}{\underbrace{\sin(\pi/\sqrt{2})\cosh(\pi/\sqrt{2})}_{u} + i\,\underbrace{\cos(\pi/\sqrt{2})\sinh(\pi/\sqrt{2})}_{v}} = \frac{2uv}{u^2+v^2} + i\frac{u^2-v^2}{u^2+v^2},$$

$$\tag{3.254}$$

and we briefly note that since $\dfrac{\pi}{2} < \dfrac{\pi}{\sqrt{2}} < \dfrac{3}{4}\pi$, in this part of the second quadrant we have $uv < 0$, since $\cos(\pi/\sqrt{2}) < 0$, $\sin(\pi/\sqrt{2}) > 0$, and then $u^2 - v^2 > 0$, which is easily noted if we count that $\cosh^2(\pi/\sqrt{2}) > \sinh^2(\pi/\sqrt{2})$ (e.g., using the identity $\cosh^2(x) - \sinh^2(x) = 1$) and $\sin^2(\pi/\sqrt{2}) > \cos^2(\pi/\sqrt{2})$.

So, if we take log of the opposite sides of (3.254) and consider that $\log(a+ib) = \frac{1}{2}\log(a^2+b^2) + i\left(\arctan\left(\dfrac{b}{a}\right) + \pi\right)$, $a < 0$, $b \geq 0$, we write

$$\log\left(\frac{\sinh(\pi\sqrt{i})}{\sin(\pi\sqrt{i})}\right)$$

$$= \frac{1}{2}\log\left(\underbrace{\left(\frac{2uv}{u^2+v^2}\right)^2 + \left(\frac{u^2-v^2}{u^2+v^2}\right)^2}_{1}\right) + i\left(\arctan\left(\frac{u^2-v^2}{2uv}\right) + \pi\right)$$

$$= i\left(\arctan\left(\frac{u}{v}\right) - \arctan\left(\frac{v}{u}\right) + \pi\right)$$

$$\left\{\text{use the relation } \arctan(x) + \arctan\left(\frac{1}{x}\right) = -\frac{\pi}{2},\ x < 0\right\}$$

$$= i\left(\frac{\pi}{2} - 2\arctan\left(\frac{v}{u}\right)\right) = i\left(\frac{\pi}{2} - 2\arctan\left(\frac{\tanh(\pi/\sqrt{2})}{\tan(\pi/\sqrt{2})}\right)\right). \tag{3.255}$$

Combining (3.255), (3.253), and (3.249), we get

$$\sum_{k=1}^{\infty}\arctan\left(\frac{1}{k^2}\right) = \sum_{k=1}^{\infty}(-1)^{k-1}\frac{\zeta(4k-2)}{2k-1} = \frac{\pi}{4} - \arctan\left(\frac{\tanh(\pi/\sqrt{2})}{\tan(\pi/\sqrt{2})}\right). \tag{3.256}$$

Finally, plugging (3.256) in (3.248), we conclude that

$$\int_0^{\pi/2}\{\sqrt{\tan(x)}\}dx = \int_0^{\pi/2}\{\sqrt{\cot(x)}\}dx = (2\sqrt{2}-1)\frac{\pi}{4} + \arctan\left(\frac{\tanh\left(\dfrac{\pi}{\sqrt{2}}\right)}{\tan\left(\dfrac{\pi}{\sqrt{2}}\right)}\right),$$

where the first equality is provided by the variable change $x = \pi/2 - y$, and the solution is complete.

Good to know that for a different approach of extracting the value of the arctan series in (3.256), one may consider the strategy in the paper "Sums of arctangents and some formulas of Ramanujan" by George Boros and Victor H. Moll (see [7]) or the similar approach in [33]. As regards the latter reference, exploiting that $\log(1 + ix) = \frac{1}{2}\log(1 + x^2) + i\arctan(x)$, we have

$$\sum_{k=1}^{\infty}\arctan\left(\frac{1}{k^2}\right) = \Im\left\{\log\left(\prod_{k=1}^{\infty}\left(1 - \frac{(1/\sqrt{2} - i/\sqrt{2})^2}{k^2}\right)\right)\right\},$$

and if we further combine the previous result with Euler's infinite product for the sine, that is, $\sin(x) = x\prod_{n=1}^{\infty}\left(1 - \frac{x^2}{n^2\pi^2}\right)$ (see [10, pp.251–252]), and rearrange, we arrive at the desired result as seen at the given reference.

3.36   A Superb Integral with Logarithms and the Inverse Tangent Function, and a Wonderful Generalization of It

Solution Allow me to start this section a bit differently and tell you first that one of my proposed problems that have been (very) well received by the integral solvers is the problem **12054** I submitted to *The American Mathematical Monthly* (see [74]).

Not long after the publishing moment of the problem I just mentioned above, a similar relation has been presented in [29] with the following title at the moment of writing these lines in the book, *An AMM-like integral* $\int_0^1 \frac{\arctan x}{x}\ln\frac{(1 + x^2)^3}{(1 + x)^2}dx$, where you can see that for accuracy I also kept the notation of the natural logarithm as it appears in the title since for it I prefer to use log instead of ln. And, *again*, it is a *wonderful* problem with integrals one wouldn't like to miss!

Before proceeding with the solution to the main integrals, I'll consider a small digression and state the relation with integrals from the problem **12054**,

$$\int_0^1 \frac{\arctan(t)\log(1 + t^2)}{t}dt - 2\int_0^1 \frac{\arctan(t)\log(1 - t)}{t}dt = \frac{\pi^3}{16}, \tag{3.257}$$

and then its generalization in the form

$$\int_0^x \frac{\arctan(t)\log(1 + t^2)}{t}dt - 2\int_0^1 \frac{\arctan(xt)\log(1 - t)}{t}dt$$

$$= 2\sum_{n=1}^{\infty}(-1)^{n-1}\frac{x^{2n-1}}{(2n - 1)^3}, \quad |x| \leq 1, \tag{3.258}$$

or its generalization in a more extended form

$$\int_0^x \frac{\arctan(t)\log(1+t^2)}{t}dt - 2\int_0^1 \frac{\arctan(xt)\log(1-t)}{t}dt = 2\mathrm{Ti}_3(x), \ x \in \mathbb{R},$$
(3.259)

where $\mathrm{Ti}_3(x) = \int_0^x \frac{\mathrm{Ti}_2(y)}{y}dy = \int_0^x \frac{1}{y}\left(\int_0^y \frac{\arctan(z)}{z}dz\right)dy$ is the inverse tangent integral of order 3.

Both results in (3.257) and (3.258) are treated in *(Almost) Impossible Integrals, Sums, and Series* (see [76, Chapter 3, pp.150–154]), and here I emphasize that also a solution by real methods exclusively is given. Essentially, for showing the result in (3.259) it is enough to consider the second solution found at the reference above, where the only difference is that in the solution we won't use the reduction to a power series as it appears in the right-hand side of (3.258), but instead we employ the inverse tangent integral of order 3 to keep x in $\mathbb{R}$.

Hold on! Did you manage to carefully look over the several opening sections of the book where I presented various parameterized logarithmic integrals?

If we consider the result in (1.19), Sect. 1.6, we immediately get another solution to the result in (3.259). So, if we replace in (1.19) the integration variable x by t, then replace a by y^2, and finally integrate its both sides from $y = 0$ to $y = x$, we get

$$\int_0^x \left(\int_0^1 \frac{\log(1-t)}{1+y^2t^2}dt\right)dy \overset{\substack{\text{reverse the order} \\ \text{of integration}}}{=} \int_0^1 \left(\int_0^x \frac{\log(1-t)}{1+y^2t^2}dy\right)dt$$

$$= \int_0^1 \frac{\arctan(xt)\log(1-t)}{t}dt$$

$$= \frac{1}{2}\int_0^x \frac{\arctan(y)\log(1+y^2)}{y}dy - \underbrace{\int_0^x \frac{\mathrm{Ti}_2(y)}{y}dy}_{\mathrm{Ti}_3(x)},$$

whence, by properly changing the letters in the first integral from the rightmost-hand side, we obtain that

$$\int_0^x \frac{\arctan(t)\log(1+t^2)}{t}dt - 2\int_0^1 \frac{\arctan(xt)\log(1-t)}{t}dt = 2\mathrm{Ti}_3(x),$$

and the new solution to the result in (3.259) is complete, and hence both (3.257) and (3.258) are true. At this point, I'll also put an end to the digression started above.

Let's return now to the main problem, and we'll focus on the generalization at the point (*ii*). In order to get a solution, I'll proceed by following the same strategy I initially used (in 2018) for the particular case in [29], where I started from **4.535.1** in *Table of Integrals, Series, and Products* by I.S. Gradshteyn and I.M. Ryzhik (8th edition):

$$\int_0^1 \frac{\arctan(yt)}{1 + y^2 t} dt = \frac{1}{2y^2} \arctan(y) \log(1 + y^2). \tag{3.260}$$

Proof We start with the variable change $yt = u$ that gives $\dfrac{1}{y} \displaystyle\int_0^y \frac{\arctan(u)}{1 + yu} du$, and

if we integrate by parts, using that $\dfrac{d}{du}(\log(1 + yu)) = \dfrac{y}{1 + yu}$, we arrive at

$$\int_0^1 \frac{\arctan(yt)}{1 + y^2 t} dt = \frac{1}{y^2} \int_0^y \arctan(u)(\log(1 + yu))' du$$

$$= \frac{1}{y^2} \arctan(y) \log(1 + y^2) - \frac{1}{y^2} \int_0^y \frac{\log(1 + yu)}{1 + u^2} du$$

{make use of the result in (1.28), Sect. 1.8}

$$= \frac{1}{2y^2} \arctan(y) \log(1 + y^2),$$

and the solution to the result in (3.260) is complete.   ∎

So, if we multiply both sides of (3.260) by y and then integrate from $y = 0$ to $y = x$, we get

$$\frac{1}{2} \int_0^x \frac{\arctan(y) \log(1 + y^2)}{y} dy = \int_0^x \left(\int_0^1 \frac{y \arctan(yt)}{1 + y^2 t} dt \right) dy$$

$$\overset{yt=u}{=} \int_0^x \left(\int_0^y \frac{\arctan(u)}{1 + yu} du \right) dy \overset{\substack{\text{reverse the order} \\ \text{of integration}}}{=} \int_0^x \left(\int_u^x \frac{\arctan(u)}{1 + yu} dy \right) du$$

$$= \int_0^x \frac{\arctan(u)}{u} \log \left(\frac{1 + xu}{1 + u^2} \right) du$$

$$= \int_0^x \frac{\arctan(u) \log(1 + xu)}{u} du - \int_0^x \frac{\arctan(u) \log(1 + u^2)}{u} du,$$

where upon changing the letters of the integration variables, we conclude that

$$3 \int_0^x \frac{\arctan(t) \log(1 + t^2)}{t} dt - 2 \int_0^x \frac{\arctan(t) \log(1 + xt)}{t} dt = 0,$$

and the solution to the point (ii) of the problem is finalized.

A good point to note is that the integral $\int_0^y \dfrac{\log(1 + yu)}{1 + u^2}\,du$ may also be viewed as a transformed form of $\int_0^1 \dfrac{\arctan(yt)}{1 + y^2 t}\,dt$ and vice versa, which is easily seen from the solution to (3.260), and one may also get a solution to the point (ii) starting from the former integral.

I also treated the generalization at the point (ii) slightly differently in my paper "A symmetry-related treatment of two fascinating sums of integrals" (see [87, April 28, 2020]).

Surely, at last we observe the result from the point (i) is the case $x = 1$ of the generalization found at the point (ii).

I'm ready now to announce to you that in the next section I'll continue treating similar results that will generate two spectacular results!

3.37 More Wonderful Results Involving Integrals with Logarithms and the Inverse Tangent Function

Solution Although at some point there was the temptation to consider these results together with other results in the same section, like the ones in the previous section, I have finally concluded they deserve a separate section (for getting more attention).

After completing the solutions to the first three points of the problem, we'll want to use the result from the point (ii) in order to prove the result at the point (iv), which in turn will produce a wonderful representation of π, found at the point (v).

We immediately get a solution to the point (i) of the problem if we combine the results in (1.168), Sect. 1.36, and (3.258), Sect. 3.36, that together lead to

$$\int_0^x \frac{\arctan(t)\log(1 + xt)}{t}\,dt - 3\int_0^1 \frac{\arctan(xt)\log(1 - t)}{t}\,dt$$

$$= 3\sum_{n=1}^{\infty}(-1)^{n-1}\frac{x^{2n-1}}{(2n - 1)^3},$$

and the solution to the point (i) of the problem is complete.

Here, I emphasize that the result in (3.258), in the previous section, is treated by real methods exclusively in the first solution given in *(Almost) Impossible Integrals, Sums, and Series* (see [76, Chapter 3, pp.150–152]). Further, as seen in the previous section, also the other result needed above may be extracted by real methods only, and therefore we may safely assert that the generalization at the point (i) of the current section may be extracted by real methods exclusively (and this fact may matter for some of you that are possibly *hunters* of real methods).

Essentially, the point (ii) of the problem is an extended version of the generalization from the point (i), and this is built in a similar style as I proceeded to the point (i) except that now we need the result in (3.259), in the previous section, which uses the inverse tangent integral of order 3 instead of (3.258), we combine with the result in (1.168), Sect. 1.36, leading to

$$\int_0^x \frac{\arctan(t)\log(1+xt)}{t}dt - 3\int_0^1 \frac{\arctan(xt)\log(1-t)}{t}dt = 3\mathrm{Ti}_3(x),$$

and the solution to the point (ii) of the problem is complete.

It is clear from both points of the main problem that for the point (iii) we have

$$\int_0^1 \frac{\arctan(t)\log(1+t)}{t}dt - 3\int_0^1 \frac{\arctan(t)\log(1-t)}{t}dt = 3\mathrm{Ti}_3(1)$$

$$= 3\sum_{n=1}^\infty \frac{(-1)^{n-1}}{(2n-1)^3} = \frac{3}{32}\pi^3,$$

where the resulting alternating series is calculated both in Sect. 3.38, during the solution to the point (i), and also in *(Almost) Impossible Integrals, Sums, and Series* (see [76, Chapter 3, p.141]).

Next, to attack the point (iv) of the problem, we want to focus on the result from the point (ii)! In the following, we'll note the beautiful fact that the result at the point (iv) may be obtained by considering the results generated by the point (ii), once in the form from the problem statement and once in the form with x replaced by $1/x$, which we then rearrange and add together, leading to

$$\int_0^x \frac{\arctan(t)\log(1+xt)}{t}dt + \int_0^{1/x} \frac{\arctan(t)\log(1+t/x)}{t}dt$$

$$- 3\int_0^1 \frac{\arctan(xt)\log(1-t)}{t}dt - 3\int_0^1 \frac{\arctan(t/x)\log(1-t)}{t}dt$$

$$= 3\left(\mathrm{Ti}_3(x) + \mathrm{Ti}_3\left(\frac{1}{x}\right)\right) = \mathrm{sgn}(x)3\left(\frac{\pi^3}{16} + \frac{\pi}{4}\log^2(|x|)\right),$$

and the solution to the point (iv) of the problem is complete.

Wait a second! How do we obtain the last equality above?, you might immediately ask. Now, a (very) nice fact about the last equality is that there we recognize the inverse relation of the inverse tangent integral of order 3,

$$\text{Ti}_3(x) + \text{Ti}_3\left(\frac{1}{x}\right) = \text{sgn}(x)\left(\frac{\pi^3}{16} + \frac{\pi}{4}\log^2(|x|)\right). \tag{3.261}$$

Proof To get a proof, we start from two well-known identities, that is, $\arctan(x) + \arctan\left(\frac{1}{x}\right) = \text{sgn}(x)\frac{\pi}{2}$, $x \in \mathbb{R} \setminus \{0\}$, and

$$\text{Ti}_2(x) - \text{Ti}_2\left(\frac{1}{x}\right) = \frac{\pi}{2}\text{sgn}(x)\log(|x|). \tag{3.262}$$

To obtain a solution to (3.262), we first consider the restriction $x > 0$, and then differentiate and integrate the left-hand side of (3.262) that gives

$$\text{Ti}_2(x) - \text{Ti}_2\left(\frac{1}{x}\right) = \int_1^x \left(\text{Ti}_2(y) - \text{Ti}_2\left(\frac{1}{y}\right)\right)' dy = \int_1^x \frac{\arctan(y) + \arctan(1/y)}{y} dy$$

$$\left\{\text{for } 0 < x \text{ we use that } \arctan(x) + \arctan\left(\frac{1}{x}\right) = \frac{\pi}{2}\right\}$$

$$= \frac{\pi}{2}\int_1^x \frac{1}{y}dy = \frac{\pi}{2}\log(x). \tag{3.263}$$

As regards the restriction $x < 0$, we proceed in a similar style as before, and differentiating and integrating back the left-hand side of (3.262), we have

$$\text{Ti}_2(x) - \text{Ti}_2\left(\frac{1}{x}\right) = \int_{-1}^x \left(\text{Ti}_2(y) - \text{Ti}_2\left(\frac{1}{y}\right)\right)' dy = \int_{-1}^x \frac{\arctan(y) + \arctan(1/y)}{y} dy$$

$$\left\{\text{for } x < 0 \text{ we use that } \arctan(x) + \arctan\left(\frac{1}{x}\right) = -\frac{\pi}{2}\right\}$$

$$= -\frac{\pi}{2}\int_{-1}^x \frac{1}{y}dy = -\frac{\pi}{2}\log(-x). \tag{3.264}$$

Combining the results in (3.263) and (3.264), we obtain the result in (3.262).

Finally, to get the result in (3.261) I'll use a similar way to the one previously given for derivation of (3.262). And in the calculations, the result in (3.262) is a key part in the derivation process!

So, let's begin again with the restriction $0 < x$, and then we differentiate and integrate the left-hand side of (3.261), taking into account that $\text{Ti}_3(1) = \frac{\pi^3}{32}$, and then we obtain

$$\text{Ti}_3(x) + \text{Ti}_3\left(\frac{1}{x}\right) = \frac{\pi^3}{16} + \int_1^x \left(\text{Ti}_3(y) + \text{Ti}_3\left(\frac{1}{y}\right)\right)' dy$$

$$= \frac{\pi^3}{16} + \int_1^x \frac{\text{Ti}_2(y) - \text{Ti}_2(1/y)}{y} dy$$

$$\left\{\text{based on (3.262), for } 0 < x \text{ we have that } \text{Ti}_2(x) - \text{Ti}_2\left(\frac{1}{x}\right) = \frac{\pi}{2}\log(x)\right\}$$

$$= \frac{\pi^3}{16} + \frac{\pi}{2}\int_1^x \frac{\log(y)}{y} dy = \frac{\pi^3}{16} + \frac{\pi}{4}\log^2(x). \tag{3.265}$$

For the other case, where the restriction is $x < 0$, we differentiate and integrate the left-hand side of (3.261), taking into account that $\text{Ti}_3(-1) = -\text{Ti}_3(1) = -\frac{\pi^3}{32}$, and then we get

$$\text{Ti}_3(x) + \text{Ti}_3\left(\frac{1}{x}\right) = -\frac{\pi^3}{16} + \int_{-1}^x \left(\text{Ti}_3(y) + \text{Ti}_3\left(\frac{1}{y}\right)\right)' dy$$

$$= -\frac{\pi^3}{16} + \int_{-1}^x \frac{\text{Ti}_2(y) - \text{Ti}_2(1/y)}{y} dy$$

$$\left\{\text{based on (3.262), for } x < 0 \text{ we have that } \text{Ti}_2(x) - \text{Ti}_2\left(\frac{1}{x}\right) = -\frac{\pi}{2}\log(-x)\right\}$$

$$= -\frac{\pi^3}{16} - \frac{\pi}{2}\int_{-1}^x \frac{\log(-y)}{y} dy = -\frac{\pi^3}{16} - \frac{\pi}{4}\log^2(-x). \tag{3.266}$$

At last, by combining (3.265) and (3.266), we finalize the solution to the auxiliary result in (3.261). ∎

In order to get the π representation from the last point of the problem, we consider the result from the previous point where we set $x = e$ which we combine with the result from the point (iii), and then we have

$$\pi = \frac{4}{3}\int_0^e \frac{\arctan(t)\log(1+et)}{t}dt + \frac{4}{3}\int_0^{1/e}\frac{\arctan(t)\log(1+t/e)}{t}dt$$

$$- 4\int_0^1 \frac{\arctan(et)\log(1-t)}{t}dt - 4\int_0^1 \frac{\arctan(t/e)\log(1-t)}{t}dt$$

$$- \frac{8}{3}\int_0^1 \frac{\arctan(t)\log(1+t)}{t}dt + 8\int_0^1 \frac{\arctan(t)\log(1-t)}{t}dt,$$

and the solution to the point (v) of the problem is complete.

The curious reader might wonder if other such results as the one from the point (v) are possible. Well, the answer is simple: *Yes!* Another example may be found below:

$$\pi = 2\int_0^e \frac{\arctan(t)\log(1+t^2)}{t}dt + 2\int_0^{1/e}\frac{\arctan(t)\log(1+t^2)}{t}dt$$

$$-4\int_0^1 \frac{\arctan(et)\log(1-t)}{t}dt - 4\int_0^1 \frac{\arctan(t/e)\log(1-t)}{t}dt$$

$$-4\int_0^1 \frac{\arctan(t)\log(1+t^2)}{t}dt + 8\int_0^1 \frac{\arctan(t)\log(1-t)}{t}dt, \qquad (3.267)$$

which appears in my paper "A symmetry-related treatment of two fascinating sums of integrals" (see [87, April 28, 2020]).

3.38 Powerful and Useful Sums of Integrals with Logarithms and the Inverse Tangent Function

Solution As it appears mentioned in the title of the section, the relations with integrals involving logarithms and the inverse tangent function I prepare to prove are indeed powerful & useful. *Why so?* With them in hand, we are able to avoid complicated calculations in potentially many problems, which in some cases may seem endless calculations while trying to reduce some sums of integrals to simpler, manageable forms, but then we realize this doesn't happen (easily).

A relevant example may be represented by the second solution I provided for the challenging integral in Sect. 1.52 (a good problem to understand the power of such relations with integrals).

In order to get a solution to the point (i), we might like to take a careful look at the second integral, which if we integrate by parts, we obtain that

$$\int_0^1 \frac{x\arctan(x)\log(1+x^2)}{1+x^2}dx = \frac{1}{4}\int_0^1 (\log^2(1+x^2))'\arctan(x)dx$$

$$= \frac{1}{16}\log^2(2)\pi - \frac{1}{4}\int_0^1 \frac{\log^2(1+x^2)}{1+x^2}dx. \qquad (3.268)$$

At this point, we may consider the algebraic identity, $\log^2\left(\dfrac{2x}{1+x^2}\right) = (\log(2)+$

$\log(x)-\log(1+x^2))^2 = \log^2(2)+2\log(2)\log(x)+\log^2(x)-2\log(2)\log(1+x^2)-$

$2\log(x)\log(1+x^2)+\log^2(1+x^2)$, where if we multiply both sides by $1/(1+x^2)$, rearrange, and then integrate from $x = 0$ to $x = 1$, we get

$$\int_0^1 \frac{\log^2(1+x^2)}{1+x^2}dx$$

$$= -\log^2(2)\underbrace{\int_0^1 \frac{1}{1+x^2}dx}_{\pi/4} -2\log(2)\underbrace{\int_0^1 \frac{\log(x)}{1+x^2}dx}_{-G} - \int_0^1 \frac{\log^2(x)}{1+x^2}dx$$

$$+ \int_0^1 \log^2\left(\frac{2x}{1+x^2}\right)\frac{1}{1+x^2}dx + 2\log(2)\underbrace{\int_0^1 \frac{\log(1+x^2)}{1+x^2}dx}_{\log(2)\pi/2 - G}$$

$$+ 2\int_0^1 \frac{\log(x)\log(1+x^2)}{1+x^2}dx, \tag{3.269}$$

where the second integral in the right-hand side of (3.269) equals $-G$ as shown in Sect. 3.52, then for the third integral we have

$$\int_0^1 \frac{\log^2(x)}{1+x^2}dx = \int_0^1 \sum_{n=1}^{\infty}(-1)^{n-1}x^{2n-2}\log^2(x)dx$$

$$= \sum_{n=1}^{\infty}(-1)^{n-1}\int_0^1 x^{2n-2}\log^2(x)dx$$

$$= 2\sum_{n=1}^{\infty}(-1)^{n-1}\frac{1}{(2n-1)^3} = 2\cdot\frac{\pi^3}{32} = \frac{\pi^3}{16}, \tag{3.270}$$

and this last series may be viewed as a particular case of the Dirichlet beta function, $\beta(s) = \sum_{n=0}^{\infty}\frac{(-1)^n}{(2n+1)^s}$, i.e., $\beta(3) = \frac{\pi^3}{32}$, also met with a solution in *(Almost) Impossible Integrals, Sums, and Series* (for details, see [76, Chapter 3, p.141]).

To get a different way of calculating $\sum_{n=1}^{\infty}(-1)^{n-1}\frac{1}{(2n-1)^3}$ than the one suggested at the reference above, we may start from the well-known Fourier series $\sum_{n=1}^{\infty}\frac{\sin(nx)}{n} = \frac{\pi - x}{2}$, $0 < x < 2\pi$, where if we multiply both sides by x and integrate from $x = 0$ to $x = \pi/2$, we obtain that

$$\frac{1}{2}\int_0^{\pi/2} x(\pi-x)dx = \frac{\pi^3}{24} = \int_0^{\pi/2} x\sum_{n=1}^{\infty}\frac{\sin(nx)}{n}dx = \sum_{n=1}^{\infty}\frac{1}{n}\int_0^{\pi/2} x\sin(nx)dx =$$

$$\sum_{n=1}^{\infty}\left(\frac{\sin(n\pi/2)}{n^3} - \frac{\pi\cos(n\pi/2)}{2n^2}\right) = \sum_{n=1}^{\infty}(-1)^{n-1}\frac{1}{(2n-1)^3} + \frac{\pi}{8}\sum_{n=1}^{\infty}(-1)^{n-1}\frac{1}{n^2},$$

and since the last series is the particular case $\eta(2) = \dfrac{\pi^2}{12}$ of the Dirichlet eta

function, $\eta(s) = \sum_{n=1}^{\infty}(-1)^{n-1}\dfrac{1}{n^s}$, we may immediately extract the desired result,

$$\sum_{n=1}^{\infty}(-1)^{n-1}\frac{1}{(2n-1)^3} = \frac{\pi^3}{32}.$$

Next, for the fourth integral in the right-hand side of (3.269), we let the variable change $x = \tan\left(\dfrac{y}{2}\right)$, and then we have

$$\int_0^1 \log^2\left(\frac{2x}{1+x^2}\right)\frac{1}{1+x^2}dx = \frac{1}{2}\int_0^{\pi/2}\log^2(\sin(x))dx$$

$$= \frac{1}{4}\lim_{x\to 0}\frac{\partial^2}{\partial x^2}B\left(\frac{1}{2}(x+1),\frac{1}{2}\right) = \frac{\pi^3}{48} + \frac{1}{4}\log^2(2)\pi, \tag{3.271}$$

and the calculation of the limit can be done either with *Mathematica* or manually.

Further, the fifth integral in the right-hand side of (3.269) equals $\dfrac{1}{2}\log(2)\pi - G$ as seen in (3.347), Sect. 3.52.

Finally, for the last integral in (3.269) it is enough to only rearrange it by integrating by parts, and then we have

$$\int_0^1 \frac{\log(x)\log(1+x^2)}{1+x^2}dx = \int_0^1 (\arctan(x))'\log(x)\log(1+x^2)dx$$

$$= -\int_0^1 \frac{\arctan(x)\log(1+x^2)}{x}dx - 2\int_0^1 \frac{x\arctan(x)\log(x)}{1+x^2}dx. \tag{3.272}$$

Combining (3.268), (3.269), (3.270), (3.271), and (3.272), we conclude that

$$\int_0^1 \frac{\arctan(x)\log(1+x^2)}{x}dx - 2\int_0^1 \frac{x\arctan(x)\log(1+x^2)}{1+x^2}dx$$

$$+ 2\int_0^1 \frac{x\arctan(x)\log(x)}{1+x^2}dx = \frac{3}{8}\log^2(2)\pi - \frac{\pi^3}{48},$$

and the solution to the point (*i*) of the problem is finalized.

The curious reader might also want to invest efforts to find other ways of solving the present point of the problem, which is an excellent opportunity to get more practice in the work with such integral relations.

After spending a good amount of time with the previous point of the problem, one might conclude not after long that the point (ii) of the problem is more challenging.

So, let's start with the third integral which we'll turn into a triple integral by using that $\log(x) = -\int_x^1 \frac{1}{y} dy$ and $\arctan(x) = \int_0^1 \frac{x}{1 + x^2 z^2} dz$, and then we may write

$$\int_0^1 \frac{x \arctan(x) \log(x)}{1 + x^2} dx = -\int_0^1 \left(\int_0^1 \left(\int_x^1 \frac{x^2}{y(1 + x^2)(1 + x^2 z^2)} dy \right) dz \right) dx$$

{reverse the order of integration in the outer double integral}

$$= -\int_0^1 \left(\int_0^1 \left(\int_x^1 \frac{x^2}{y(1 + x^2)(1 + x^2 z^2)} dy \right) dx \right) dz$$

{reverse the order of integration in the inner double integral}

$$= -\int_0^1 \left(\int_0^1 \left(\int_0^y \frac{x^2}{y(1 + x^2)(1 + x^2 z^2)} dx \right) dy \right) dz$$

$$= \int_0^1 \left(\int_0^1 \frac{z \arctan(y) - \arctan(yz)}{yz(1 - z^2)} dy \right) dz$$

{reverse the order of integration and prepare it for integration by parts}

$$= \int_0^1 \left(\int_0^1 \left(\log(z) - \frac{1}{2} \log(1 - z^2) \right)' \frac{z \arctan(y) - \arctan(yz)}{y} dz \right) dy$$

$$= \int_0^1 \left(\int_0^1 \frac{\log(z)}{1 + y^2 z^2} dz \right) dy - \int_0^1 \frac{\arctan(y)}{y} \left(\int_0^1 \log(z) dz \right) dy$$

$$+ \frac{1}{2} \int_0^1 \frac{\arctan(y)}{y} \left(\int_0^1 \log(1 - z^2) dz \right) dy - \frac{1}{2} \int_0^1 \left(\int_0^1 \frac{\log(1 - z^2)}{1 + y^2 z^2} dz \right) dy$$

{reverse the order of integration in the last double integral}

$$= \log(2)G - \frac{\pi^3}{32} - \frac{1}{2} \int_0^1 \left(\int_0^1 \frac{\log(1 - z^2)}{1 + y^2 z^2} dy \right) dz$$

$$= \log(2)G - \frac{\pi^3}{32} - \frac{1}{2} \int_0^1 \frac{\arctan(z) \log(1 - z^2)}{z} dz,$$

whence we obtain that

$$\int_0^1 \frac{\arctan(x) \log(1 - x)}{x} dx$$

$$+ \int_0^1 \frac{\arctan(x) \log(1 + x)}{x} dx + 2 \int_0^1 \frac{x \arctan(x) \log(x)}{1 + x^2} dx$$

$$= 2 \log(2)G - \frac{\pi^3}{16},$$

and the solution to the point (ii) of the problem is finalized. In the calculations, I also used that $\displaystyle\int_0^1 \left(\int_0^1 \frac{\log(z)}{1 + y^2 z^2} dz \right) dy = \sum_{n=1}^{\infty}(-1)^{n-1} \int_0^1 y^{2n-2} \left(\int_0^1 z^{2n-2} \log(z) dz \right)$

$dy = -\displaystyle\sum_{n=1}^{\infty} \frac{(-1)^{n-1}}{(2n-1)^3} = -\frac{\pi^3}{32}$, where the last series already appeared at the previous point.

Passing to the point (iii) of the problem, we first observe the similarity between the present relation and the one from the point (ii). *Then, could we exploit the result from the point (ii)?* Yes, *we could* as you'll see in the following.

The first solution to the point (iii) of the problem is straightforward, and all we need is to combine the relation from the point (ii) and the one in (3.257), Sect. 3.36, and we conclude that

$$2 \int_0^1 \frac{\arctan(x) \log(1 + x)}{x} dx + \int_0^1 \frac{\arctan(x) \log(1 + x^2)}{x} dx$$

$$+ 4 \int_0^1 \frac{x \arctan(x) \log(x)}{1 + x^2} dx = 4 \log(2)G - \frac{\pi^3}{16},$$

and the first solution to the point (iii) of the problem is finalized.

How about getting a more direct way? If in the previous solution I used two distinct relations with integrals, for a second solution I'll use a more direct way by exploiting a special generalized logarithmic integral given in (1.20), Sect. 1.6, where if we replace a by a^2 and then integrate its both sides from $a = 0$ to $a = 1$, we have

$$\int_0^1 \left(\int_0^1 \frac{\log(1 + x)}{1 + a^2 x^2} dx \right) da = \int_0^1 \left(\int_0^1 \frac{\log(1 + x)}{1 + a^2 x^2} da \right) dx$$

$$= \int_0^1 \frac{\arctan(x) \log(1 + x)}{x} dx.$$

$$= \log(2) \underbrace{\int_0^1 \frac{\arctan(a)}{a}da}_{G} - \underbrace{\int_0^1 \frac{\text{Ti}_2(a)}{a}da}_{\pi^3/32} + \int_0^1 (\log(a))' \Im\left\{\text{Li}_2\left(\frac{1}{2} + i\frac{a}{2}\right)\right\}da$$

{use integration by parts in the last integral}

$$= \log(2)G - \frac{\pi^3}{32} - \log(2) \underbrace{\int_0^1 \frac{\log(a)}{1+a^2}da}_{-G} - \int_0^1 \frac{a\arctan(a)\log(a)}{1+a^2}da$$

$$+ \frac{1}{2}\int_0^1 \frac{\log(a)\log(1+a^2)}{1+a^2}da$$

{again, use integration by parts in the last integral}

$$= 2\log(2)G - \frac{\pi^3}{32} - 2\int_0^1 \frac{a\arctan(a)\log(a)}{1+a^2}da - \frac{1}{2}\int_0^1 \frac{\arctan(a)\log(1+a^2)}{a}da,$$

whence we obtain that

$$2\int_0^1 \frac{\arctan(x)\log(1+x)}{x}dx + \int_0^1 \frac{\arctan(x)\log(1+x^2)}{x}dx$$

$$+ 4\int_0^1 \frac{x\arctan(x)\log(x)}{1+x^2}dx = 4\log(2)G - \frac{\pi^3}{16},$$

and the second solution to the point (iii) of the problem is finalized.

In the calculations above I used two well-known integral representations of the Catalan's constant as follows: $\displaystyle\int_0^1 \frac{\arctan(x)}{x}dx = \int_0^1 \sum_{n=1}^{\infty}(-1)^{n-1}\frac{x^{2n-2}}{2n-1}dx =$

$$\sum_{n=1}^{\infty}(-1)^{n-1}\int_0^1 \frac{x^{2n-2}}{2n-1}dx = \sum_{n=1}^{\infty}(-1)^{n-1}\frac{1}{(2n-1)^2} = G \text{ and } \int_0^1 \frac{\log(a)}{1+a^2}da =$$

$-G$, where the latter one may be derived with integration by parts from the previous integral. Further, I also used that $\displaystyle\int_0^1 \frac{\text{Ti}_2(x)}{x}dx = \frac{\pi^3}{32}$, and to see this is true, we start from the following series representation of the inverse tangent integral, $\text{Ti}_2(x) = \displaystyle\sum_{n=1}^{\infty}(-1)^{n-1}\frac{x^{2n-1}}{(2n-1)^2}$, and then we get $\displaystyle\int_0^1 \frac{\text{Ti}_2(x)}{x}dx =$

$$\int_0^1 \sum_{n=1}^{\infty}(-1)^{n-1}\frac{x^{2n-2}}{(2n-1)^2}dx = \sum_{n=1}^{\infty}(-1)^{n-1}\frac{1}{(2n-1)^2}\int_0^1 x^{2n-2}dx = \sum_{n=1}^{\infty}(-1)^{n-1}$$

$\dfrac{1}{(2n-1)^3} = \dfrac{\pi^3}{32}$, where the last series also appears in (3.270), during the solution to the point (i) of the problem.

Finally, one might like to know how to deal with the resulting limits after integrating by parts for $\int_0^1 (\log(a))' \Im\left\{ \mathrm{Li}_2\left(\frac{1}{2} + i\frac{a}{2}\right)\right\} da$. In fact, the less obvious limit is represented by the case $\lim_{a\to 0^+} \log(a) \Im\left\{\mathrm{Li}_2\left(\frac{1}{2} + i\frac{a}{2}\right)\right\}$.

One way to analyze this limit is to start from the dilogarithm representation, $\mathrm{Li}_2(z) = -\int_0^1 \frac{\log(1-zt)}{t} dt$, where if we consider the form of the logarithm with a complex argument, $\log(a + ib) = \log(\sqrt{a^2 + b^2}) + i \arctan\left(\frac{b}{a}\right)$, $a > 0$, we immediately arrive at $\Im\left\{\mathrm{Li}_2\left(\frac{1}{2} + i\frac{a}{2}\right)\right\} = \int_0^1 \frac{1}{t} \arctan\left(\frac{at}{2-t}\right) dt$. To easily see what happens at this point, it is enough to use that $\lim_{a\to 0} \frac{\arctan(a)}{a} = 1$, and this fact tells us that when a is very small, approaching to 0, we have that $\arctan(a) \approx a$, and then in our specific case above we'll have that $\Im\left\{\mathrm{Li}_2\left(\frac{1}{2} + i\frac{a}{2}\right)\right\} \approx a \int_0^1 \frac{1}{2-t} dt = \log(2)a$. Thus, returning to the main limit, we get $\lim_{a\to 0^+} \log(a) \Im\left\{\mathrm{Li}_2\left(\frac{1}{2} + i\frac{a}{2}\right)\right\} = \log(2) \lim_{a\to 0^+} a\log(a) = 0$, where the last limit is well-known, immediately derived from the L'Hospital's rule. To make the whole process above more rigorous, we might consider the Taylor series of $\arctan(a)$ and use Big-O notation.

Now, it is worth mentioning that by combining the relation with integrals at this point and the result in (3.257), Sect. 3.36, we arrive at the relation with integrals from the previous point of the problem.

Finally, the last relation is obtained immediately if we combine the relations with integrals from the points (i) and (iii) together with the one in (1.167), Sect. 1.36,

$$\int_0^1 \frac{\arctan(x)\log(1 + x^2)}{x} dx + 2\int_0^1 \frac{x\arctan(x)\log(1 + x^2)}{1 + x^2} dx$$

$$= 2\log(2)G - \frac{3}{8}\log^2(2)\pi - \frac{\pi^3}{96},$$

and the solution to the point (iv) of the problem is finalized.

Looking over the solutions above, we may answer the *challenging question* and say that, *yes*, it is possible to prove all four relations with integrals exclusively by real methods (without using complex numbers). Note that in the first solution proposed to the point (iii) of the problem I also used the relation in (3.257), Sect. 3.36, and this one, in a generalized form, appears in *(Almost) Impossible Integrals, Sums, and Series* and is also proved by real methods (see [76, Chapter 3, pp. 150–152]).

3.39   Two Beautiful Sums of Integrals, Each One Involving Three Integrals, Leading to a Possible Unexpected Result

Solution At the end of 2018, I proposed in R.M.M. (*Romanian Mathematical Magazine*, the e-version) the tricky sum of integrals you may find at the point (i). Somewhat curiously, till the end of 2019 when I revealed my solution, no other solution had been submitted to the mentioned magazine! So, I concluded it could be a useful problem from which one might like to learn more about *the art of solving problems*.

Before reading further, I *highly recommend* you to ponder over the possibility of investing more efforts for finding a solution! If you are inspired to attack the problem in a proper way, everything becomes easy and fast to finalize!

For both points of the problem we'll want to focus on the middle integral where in a first key step we use the variable change $x = y^2$, and in a second key step we want to merge the first two integrals and then turn them into a useful double integral.

According to the strategy stated above, for the point (i) of the problem we consider the variable change $x = y^2$ in the middle integral, and then we have $\int_0^1 \frac{\arctan(x)}{x\sqrt{1+x}} dx = 2 \int_0^1 \frac{\arctan(y^2)}{y\sqrt{1+y^2}} dy$. So, returning to the main sum of integrals and using the previous result, where we replace y by x and then merge the first two integrals, we have

$$\int_0^1 \frac{\arctan(x)}{x\sqrt{1+x^2}} dx - \frac{1}{2} \int_0^1 \frac{\arctan(x)}{x\sqrt{1+x}} dx - \int_0^1 \frac{1}{\sqrt{1-x^2}} \operatorname{arctanh}\left(x\sqrt{\frac{1-x^2}{1+x^2}}\right) dx$$

$$= \int_0^1 \frac{\arctan(x) - \arctan(x^2)}{x\sqrt{1+x^2}} dx - \int_0^1 \frac{1}{\sqrt{1-x^2}} \operatorname{arctanh}\left(x\sqrt{\frac{1-x^2}{1+x^2}}\right) dx.$$

$$(3.273)$$

Now, in the right-hand side of (3.273), we observe that the numerator in the integrand of the first integral may be written as $\arctan(x) - \arctan(x^2) = \int_x^1 \frac{x}{1+x^2y^2} dy$, and then we have that

$$\int_0^1 \frac{\arctan(x) - \arctan(x^2)}{x\sqrt{1+x^2}} dx = \int_0^1 \frac{1}{\sqrt{1+x^2}} \left(\int_x^1 \frac{1}{1+x^2y^2} dy\right) dx$$

$$\overset{\substack{\text{reverse the order} \\ \text{of integration}}}{=\!=\!=} \int_0^1 \left(\int_0^y \frac{1}{\sqrt{1+x^2}(1+x^2y^2)} dx\right) dy$$

$$\overset{x=\frac{t}{\sqrt{1-t^2}}}{=}\int_0^1 \frac{1}{1-y^2}\left(\int_0^{y/\sqrt{1+y^2}} \frac{1}{\left(1/\sqrt{1-y^2}\right)^2 - t^2}dt\right)dy$$

$$\left\{\text{use that } \int_0^x \frac{1}{a^2 - t^2}dt = \frac{1}{a}\operatorname{arctanh}\left(\frac{x}{a}\right), \ |x| < a\right\}$$

$$= \int_0^1 \frac{1}{\sqrt{1-y^2}}\operatorname{arctanh}\left(y\sqrt{\frac{1-y^2}{1+y^2}}\right)dy \overset{y=x}{=} \int_0^1 \frac{1}{\sqrt{1-x^2}}\operatorname{arctanh}\left(x\sqrt{\frac{1-x^2}{1+x^2}}\right)dx.$$

$$(3.274)$$

By plugging (3.274) in (3.273), we conclude that

$$\int_0^1 \frac{\arctan(x)}{x\sqrt{1+x^2}}dx - \frac{1}{2}\int_0^1 \frac{\arctan(x)}{x\sqrt{1+x}}dx - \int_0^1 \frac{1}{\sqrt{1-x^2}}\operatorname{arctanh}\left(x\sqrt{\frac{1-x^2}{1+x^2}}\right)dx = 0,$$

and the solution to the point (i) of the problem is complete.

It's so exciting to see how beautifully and elegantly everything plays out in the calculations above!

Before passing to the second sum of integrals at the point (ii), we need to prove the following auxiliary result:

$$\int_0^y \frac{1}{(1+yx)\sqrt{1+x^2}}dx = \frac{\operatorname{arcsinh}(y)}{\sqrt{1+y^2}}, \ y \geq 0, \tag{3.275}$$

where, for simplicity, we want to slightly rearrange it, before providing a solution.

So, if we let the variable change $x = \sinh(t)$ in (3.275) and at the same time we replace y by $\sinh(a)$, then all reduces to proving that

$$\int_0^a \frac{1}{1+\sinh(a)\sinh(t)}dt = \frac{a}{\cosh(a)}. \tag{3.276}$$

Proof To get a solution to (3.276) (some of you could possibly count it as *non-obvious*), I'll exploit similar simple ideas to the ones presented in [19]. On one hand, we have the key hyperbolic identity, $1 = \cosh^2(u) - \sinh^2(u)$, and on the other hand since $(1+\sinh(a)\sinh(t))^2 = \cosh^2(a)\cosh^2(t) - (\sinh(a) - \sinh(t))^2$, after dividing both sides by $(1+\sinh(a)\sinh(t))^2$, we get $1 = \left(\dfrac{\cosh(a)\cosh(t)}{1+\sinh(a)\sinh(t)}\right)^2 - \left(\dfrac{\sinh(a) - \sinh(t)}{1+\sinh(a)\sinh(t)}\right)^2$, which suggests that we may pick up $\cosh(u)$ and $\sinh(u)$ such that $\cosh(u) = \dfrac{\cosh(a)\cosh(t)}{1+\sinh(a)\sinh(t)}$ and $\sinh(u) = \dfrac{\sinh(a) - \sinh(t)}{1+\sinh(a)\sinh(t)}$. So, based on the last substitution, we have $\sinh(t) = \dfrac{\sinh(a) - \sinh(u)}{1 + \sinh(a)\sinh(u)}$ that further gives

$$\int_0^a \frac{1}{1 + \sinh(a)\sinh(t)}\, dt = \frac{1}{\cosh(a)} \int_0^a du = \frac{a}{\cosh(a)},$$

which is the desired result in (3.276), and hence a first solution to the auxiliary result in (3.275) is finalized. ∎

Next, if we multiply both sides of (3.276) by $\cosh(a)$, and denote the integral by $I(a)$, we arrive at

$$I(a) = \int_0^a \frac{\cosh(a)}{1 + \sinh(a)\sinh(t)}\, dt = a, \qquad (3.277)$$

which we want to prove in order to get a second solution to (3.275).

Proof A fast and interesting proof to the result in (3.277) is obtained by using *Leibniz's rule for differentiation under the integral sign* that gives

$$I'(a) = \frac{\cosh(a)}{1 + \sinh^2(a)} + \int_0^a \frac{\sinh(a) - \sinh(t)}{(1 + \sinh(a)\sinh(t))^2}\, dt$$

$$\left\{ \text{exploit the identity } \cosh^2(t) - \sinh^2(t) = 1 \right\}$$

$$= \frac{1}{\cosh(a)} + \int_0^a \frac{\sinh(a)(\cosh^2(t) - \sinh^2(t)) - \sinh(t)}{(1 + \sinh(a)\sinh(t))^2}\, dt$$

$$= \frac{1}{\cosh(a)} - \int_0^a \frac{\sinh(t)(1 + \sinh(a)\sinh(t)) - \sinh(a)\cosh^2(t)}{(1 + \sinh(a)\sinh(t))^2}\, dt$$

$$= \frac{1}{\cosh(a)} - \int_0^a \frac{(\cosh(t))'(1 + \sinh(a)\sinh(t)) - \cosh(t)(1 + \sinh(a)\sinh(t))'}{(1 + \sinh(a)\sinh(t))^2}\, dt$$

$$\left\{ \text{use the quotient rule, } \left(\frac{f}{g}\right)' = \frac{f'g - fg'}{g^2} \right\}$$

$$= \frac{1}{\cosh(a)} - \int_0^a \left(\frac{\cosh(t)}{1 + \sinh(a)\sinh(t)}\right)' dt = \frac{1}{\cosh(a)} - \left(\frac{1}{\cosh(a)} - 1\right) = 1.$$

$$(3.278)$$

Integrating back with respect to a in (3.278), we have that $I(a) = a + C$, and upon setting $a = 0$, and using that $I(0) = 0$, we get that the integration constant is 0 (i.e., $C = 0$). Hence, we obtain that

$$I(a) = \int_0^a \frac{\cosh(a)}{1 + \sinh(a)\sinh(t)} dt = a,$$

which is the desired result in (3.277), and hence a second solution to the result in (3.276) is finalized, and therefore also to (3.275). ∎

For another approach, the curious reader might like to get inspiration from the strategy described in [76, Chapter 3, pp.55–56].

It's time to return to the point (ii) of the problem, where we want to adopt a similar strategy to the one from the point (i)!

If we consider the change of variable $x = y^2$ in the middle integral, we get
$$\int_0^1 \frac{\log(1 + x)}{x\sqrt{1 + x}} dx = 2 \int_0^1 \frac{\log(1 + y^2)}{y\sqrt{1 + y^2}} dy.$$ Then, with the preceding result in hand
we return to the main sum of integrals, and we write

$$\int_0^1 \frac{\log(1 + x)}{x\sqrt{1 + x^2}} dx - \frac{1}{2} \int_0^1 \frac{\log(1 + x)}{x\sqrt{1 + x}} dx - \int_0^1 \frac{\operatorname{arcsinh}(x)}{\sqrt{1 + x^2}} dx$$

$$= \int_0^1 \frac{\log(1 + x) - \log(1 + x^2)}{x\sqrt{1 + x^2}} dx - \int_0^1 \frac{\operatorname{arcsinh}(x)}{\sqrt{1 + x^2}} dx. \qquad (3.279)$$

As in the previous solution, we observe that the numerator in the integrand of the first integral in the right-hand side of (3.279) may be written in a useful way, that is,
$$\log(1 + x) - \log(1 + x^2) = \int_x^1 \frac{x}{1 + xy} dy,$$ and this further gives

$$\int_0^1 \frac{\log(1 + x) - \log(1 + x^2)}{x\sqrt{1 + x^2}} dx = \int_0^1 \frac{1}{\sqrt{1 + x^2}} \left(\int_x^1 \frac{1}{1 + xy} dy \right) dx$$

$$\underset{\equiv}{\overset{\text{reverse the order}}{\underset{\text{of integration}}{}}} \int_0^1 \left(\int_0^y \frac{1}{(1 + yx)\sqrt{1 + x^2}} dx \right) dy$$

{make use of the auxiliary result in (3.275)}

$$= \int_0^1 \frac{\operatorname{arcsinh}(y)}{\sqrt{1 + y^2}} dy = \int_0^1 \frac{\operatorname{arcsinh}(x)}{\sqrt{1 + x^2}} dx. \qquad (3.280)$$

Finally, if we plug the result from (3.280) in (3.279), we conclude that

$$\int_0^1 \frac{\log(1 + x)}{x\sqrt{1 + x^2}} dx - \frac{1}{2} \int_0^1 \frac{\log(1 + x)}{x\sqrt{1 + x}} dx - \int_0^1 \frac{\operatorname{arcsinh}(x)}{\sqrt{1 + x^2}} dx = 0,$$

and the point (ii) of the problem is complete. Surely, it is easy to observe that the third integral above is elementary.

In this section we have learned again that sometimes it is wiser to attack a group
of integrals together rather than trying to calculate them separately (e.g., when taken
one by one, the integrals from the first point may be perceived as a tough challenge).

In the next section, we make ready to meet some similar integrals, where you
might find one of the results presented in this section (very) useful to consider!

3.40   Tackling Curious Logarithmic Integrals with a Radical in the Denominator

Solution One of the things that might please someone at most in this section is to
see how curiously and beautifully the extraction process of the desired values of
each integral takes place (and I particularly refer to the second and third points).

As mentioned in the last part of the previous section, one of the points presented
there will be useful in our present section (and it is not hard to make a guess)!

Another thing that some of you might find interesting and challenging is the
restriction about the use of bino(harmonic) series. No use of (bino)harmonic series
is allowed! Therefore, the panel of the possible approaches here is tighter!

To begin with, we attack the integral at the point (i) by the variable change
$\sqrt{1+x} - 1 = y$ leading to $x = 2y + y^2$ that further gives

$$\int_0^1 \frac{\log(1+x)}{x\sqrt{1+x}}dx = 4\int_0^{\sqrt{2}-1} \frac{\log(1+y)}{y(2+y)}dy$$

$$= 2\int_0^{\sqrt{2}-1} \frac{\log(1+y)}{y}dy - 2\int_0^{\sqrt{2}-1} \frac{\log(1+y)}{2+y}dy$$

{for the second integral we integrate by parts}

$$= -2\operatorname{Li}_2\left(1-\sqrt{2}\right) + \log(2)\log\left(\sqrt{2}-1\right) + 2\int_0^{\sqrt{2}-1} \frac{\log(1+(1+y))}{1+y}dy$$

$$\left\{\text{use that } (-\operatorname{Li}_2(-(1+x)))' = \frac{\log(1+(1+x))}{1+x}\right\}$$

$$= -2\operatorname{Li}_2\left(-\sqrt{2}\right) - 2\operatorname{Li}_2\left(1-\sqrt{2}\right) + \log(2)\log\left(\sqrt{2}-1\right) - \frac{\pi^2}{6}. \qquad (3.281)$$

To get a more convenient form of (3.281), we consider the following identity:

$$\text{Li}_2(-x) - \text{Li}_2(1-x) + \frac{1}{2}\text{Li}_2\left(\frac{1}{x^2}\right) = \log(x)\log(x-1) - \log^2(x), \ x > 1,$$

$$(3.282)$$

which is immediately obtained by deriving and integrating back the left-hand side of (3.282). So, if we set $x = \sqrt{2}$ in (3.282), we get that

$$\text{Li}_2\left(-\sqrt{2}\right) = \text{Li}_2\left(1 - \sqrt{2}\right) - \frac{1}{2}\text{Li}_2\left(\frac{1}{2}\right) + \frac{1}{2}\log(2)\log\left(\sqrt{2} - 1\right) - \frac{1}{4}\log^2(2)$$

$$\left\{ \text{use the special value of the dilogarithm, } \text{Li}_2\left(\frac{1}{2}\right) = \frac{1}{2}\left(\frac{\pi^2}{6} - \log^2(2)\right) \right\}$$

$$= \text{Li}_2\left(1 - \sqrt{2}\right) + \frac{1}{2}\log(2)\log\left(\sqrt{2} - 1\right) - \frac{\pi^2}{24}. \qquad (3.283)$$

At last, if we plug the result from (3.283) in (3.281), we arrive at

$$\int_0^1 \frac{\log(1+x)}{x\sqrt{1+x}}dx = -4\text{Li}_2\left(1 - \sqrt{2}\right) - \frac{\pi^2}{12},$$

and the point (i) of the problem is complete.

To calculate the integral at the point (ii), we might exploit the sum of integrals from the point (1.179), Sect. 1.39, where our integral appears!

We observe the second integral from (1.179), Sect. 1.39, is calculated at the previous point, and the third integral is straightforward since it has a simple, elementary primitive as follows:

$$\int_0^1 \frac{\text{arcsinh}(x)}{\sqrt{1+x^2}}dx = \frac{1}{2}\int_0^1 (\text{arcsinh}^2(x))'dx = \frac{1}{2}\text{arcsinh}^2(1)$$

$$\left\{ \text{consider the definition, } \text{arcsinh}(x) = \log\left(x + \sqrt{1+x^2}\right) \right\}$$

$$= \frac{1}{2}\log^2\left(\sqrt{2} + 1\right) = \frac{1}{2}\log^2\left(\sqrt{2} - 1\right). \qquad (3.284)$$

Therefore, by combining the results from the point (i), (3.284), and (1.179), Sect. 1.39, we are able to extract the desired integral,

$$\int_0^1 \frac{\log(1+x)}{x\sqrt{1+x^2}}dx = \frac{1}{2}\log^2\left(\sqrt{2} - 1\right) - \frac{\pi^2}{24} - 2\text{Li}_2\left(1 - \sqrt{2}\right),$$

and the point (ii) of the problem is complete.

I expect you'll find the solution to the point (iii) as an exciting moment of the section with unexpected connections and calculations. One of the key results to prove and use is the following:

$$\int_0^y \frac{1}{(1-y^2x^2)\sqrt{1+x^2}}dx = \frac{\operatorname{arctanh}(y)}{\sqrt{1+y^2}}, \quad |y| < 1. \tag{3.285}$$

Proof If we consider the variable change $x = \sinh(t)$, we write

$$\int_0^y \frac{1}{(1-y^2x^2)\sqrt{1+x^2}}dx = \int_0^{\operatorname{arcsinh}(y)} \frac{1}{1-y^2\sinh^2(t)}dt$$

$\left\{\text{use the hyperbolic identity } 1 = \cosh^2(x) - \sinh^2(x) \text{ and then rearrange}\right\}$

$$= \int_0^{\operatorname{arcsinh}(y)} \frac{\operatorname{sech}^2(t)}{1-(\sqrt{1+y^2})^2\tanh^2(t)}dt$$

$$\stackrel{\tanh(t)=u}{=\!=} \int_0^{\tanh(\operatorname{arcsinh}(y))} \frac{1}{1-(\sqrt{1+y^2})^2u^2}du$$

$\left\{\text{employ the fact that } \int_0^x \frac{1}{1-a^2t^2}dt = \frac{1}{a}\operatorname{arctanh}(ax),\ |ax| < 1\right\}$

$$= \frac{\operatorname{arctanh}\left(\sqrt{1+y^2}\tanh(\operatorname{arcsinh}(y))\right)}{\sqrt{1+y^2}} = \frac{\operatorname{arctanh}(y)}{\sqrt{1+y^2}},$$

where to get the last equality I used that $\tanh(\operatorname{arcsinh}(x)) = \dfrac{x}{\sqrt{1+x^2}}$, and the solution to the auxiliary result in (3.285) is finalized. ∎

What now? Well, we'll return to the integral we want to calculate and exploit the result in (3.285) I just proved, and then we obtain the following transformation:

$$\int_0^1 \frac{\operatorname{arctanh}(x)}{x\sqrt{1+x^2}}dx = \int_0^1 \frac{1}{x}\left(\int_0^x \frac{1}{(1-x^2y^2)\sqrt{1+y^2}}dy\right)dx$$

$$\stackrel{\substack{\text{reverse the order}\\\text{of integration}}}{=\!=} \int_0^1 \frac{1}{\sqrt{1+y^2}}\left(\int_y^1 \frac{1}{x(1-y^2x^2)}dx\right)dy$$

$\left\{\text{for the inner integral we can write } \dfrac{1}{x(1-y^2x^2)} = \dfrac{1}{x} + \dfrac{2y^2x}{2(1-y^2x^2)} \text{ that gives}\right\}$

$$\left\{ \int \frac{1}{x(1-y^2x^2)}dx = \int \left(\frac{1}{x} + \frac{2y^2x}{2(1-y^2x^2)} \right) dx = \log(x) - \frac{1}{2}\log(1-y^2x^2) + C \right\}$$

$$= \int_0^1 \frac{1/2\log(1+y^2) - \log(y)}{\sqrt{1+y^2}}dy. \tag{3.286}$$

Let's take a break at this stage! To make it very clear, we may continue the calculations in (3.286) by trying to evaluate the resulting integral in a more direct manner, either by using a variable change or by first splitting the integral and then considering variable changes.

What about if we take into account the *challenging question* and consider another transformation? You might find (highly) *non-obvious* the next step I'll do!

If we consider the following *special form of the Beta function*, also met with a derivation in [76, Chapter 3, p.72], $\int_0^1 \frac{t^{a-1} + t^{b-1}}{(1+t)^{a+b}} dt = B(a, b)$, which we differentiate once with respect to a, then let $a \to 1/2$ and $b \to 0$ and wisely group, we obtain that

$$-\int_0^1 \frac{\log(1+t) - \log(t)}{\sqrt{t}\sqrt{1+t}}dt - \int_0^1 \frac{\log(1+t)}{t\sqrt{1+t}}dt$$

$$\left\{ \text{in the first integral make the variable change } t = y^2 \right\}$$

$$= -4\int_0^1 \frac{1/2\log(1+y^2) - \log(y)}{\sqrt{1+y^2}}dy - \int_0^1 \frac{\log(1+t)}{t\sqrt{1+t}}dt$$

$$= \lim_{\substack{a\to 1/2 \\ b\to 0}} \frac{\partial}{\partial a} B(a, b)$$

$$\left\{ \text{use the definition of the Beta function, } B(a, b) = \int_0^1 t^{a-1}(1-t)^{b-1}dt \right\}$$

$$= \int_0^1 \frac{\log(t)}{\sqrt{t}(1-t)}dt \overset{t=u^2}{=} 4\int_0^1 \frac{\log(t)}{1-t^2}dt = 4\int_0^1 \log(t) \sum_{n=1}^{\infty} t^{2n-2}dt$$

$$= 4\sum_{n=1}^{\infty} \int_0^1 t^{2n-2}\log(t)dt = -4\sum_{n=1}^{\infty} \frac{1}{(2n-1)^2} = -4\left(\sum_{n=1}^{\infty} \frac{1}{n^2} - \sum_{n=1}^{\infty} \frac{1}{(2n)^2} \right)$$

$$= -3\sum_{n=1}^{\infty} \frac{1}{n^2} = -\frac{\pi^2}{2}, \tag{3.287}$$

whence, if we replace y by t, we arrive at

$$\int_0^1 \frac{1/2\log(1+t^2) - \log(t)}{\sqrt{1+t^2}} dt = \frac{\pi^2}{8} - \frac{1}{4}\int_0^1 \frac{\log(1+t)}{t\sqrt{1+t}} dt. \qquad (3.288)$$

The transformation in (3.288) is just wonderful and useful! If we plug the result from (i) in (3.288), we get that

$$\int_0^1 \frac{1/2\log(1+t^2) - \log(t)}{\sqrt{1+t^2}} dt = \frac{7}{48}\pi^2 + \mathrm{Li}_2(1 - \sqrt{2}). \qquad (3.289)$$

Finally, by plugging the result from (3.289) in (3.286), we conclude that

$$\int_0^1 \frac{\operatorname{arctanh}(x)}{x\sqrt{1+x^2}} dx = \frac{7}{48}\pi^2 + \mathrm{Li}_2\left(1 - \sqrt{2}\right),$$

and the point (iii) of the problem is complete.

It is straightforward to obtain that

$$\int_0^1 \frac{\log(1-x)}{x\sqrt{1+x^2}} dx = \frac{1}{2}\log^2(\sqrt{2} - 1) - \frac{\pi^2}{3} - 4\mathrm{Li}_2\left(1 - \sqrt{2}\right),$$

if we combine the results from the points (ii) and (iii), and the point (iv) of the problem is complete.

A *wonderful section for the lovers of integral transformations!*, could be one of your possible reactions here!

3.41   Calculating More Logarithmic Integrals with a Radical in the Denominator

Solution In this section, we meet integrals with a radical in the denominator, similar to the ones in the previous section! And like in the preceding section, we'll want to perform the calculations without using (bino)harmonic series.

In contrast to other sections, we'll start with proving the identity from the last point that, as suggested in the subtitle of the problem statement, is related to the Legendre's chi function of order 2 [21, Chapter 1, p.18 –p.21], which is defined by the following integral representation: $\chi_2(x) = \int_0^x \frac{\operatorname{arctanh}(t)}{t} dt$.

More precisely, we need the special value $\chi_2(\sqrt{2}-1)$, and it can be derived from the classical identity (see [21, Chapter 1, p.19], [70, Chapter 2, p.108]),

$$\chi_2\left(\frac{1-x}{1+x}\right) + \chi_2(x) = \frac{\pi^2}{8} + \frac{1}{2}\log(x)\log\left(\frac{1+x}{1-x}\right), \qquad (3.290)$$

also obtained by Euler, Legendre, and Landen, where if we set $x = \sqrt{2} - 1$, we immediately get that

$$\chi_2(\sqrt{2} - 1) = \frac{\pi^2}{16} - \frac{1}{4}\log^2\left(\sqrt{2} - 1\right). \tag{3.291}$$

The identity in (3.290) can be easily obtained if we differentiate $\chi_2\left(\dfrac{1-x}{1+x}\right)$ and then integrate back. The necessity of employing the special value $\chi_2(\sqrt{2} - 1)$ also appears in *(Almost) Impossible Integrals, Sums, and Series* during the calculation of a wonderful double integral (see [76, Chapter 3, p.217]). Did you see it?

Now, the Legendre's chi function of order 2 is defined in terms of dilogarithm function, $\chi_2(x) = \displaystyle\sum_{n=1}^{\infty} \frac{x^{2n-1}}{(2n-1)^2} = \frac{1}{2}(\operatorname{Li}_2(x) - \operatorname{Li}_2(-x))$. Do you see at this point the continuation of the story?

Based on the series definition above and the identity in (3.291), we arrive at the desired result

$$\operatorname{Li}_2\left(-\left(1 - \sqrt{2}\right)\right) - \operatorname{Li}_2\left(1 - \sqrt{2}\right) = \frac{\pi^2}{8} - \frac{1}{2}\log^2\left(\sqrt{2} - 1\right),$$

and the point (vi) of the problem is complete.

As regards the integral at the point (i), we consider the variable change $1 + 2x^2 - 2x\sqrt{1 + x^2} = t$ that gives $x = \dfrac{1-t}{2\sqrt{t}}$, which is inspired by the well-known Euler substitution $\sqrt{1 + x^2} - x = t$. Thus, we get

$$\int_0^1 \frac{\log(1 - x^2)}{\sqrt{1 + x^2}}dx \overset{x=(1-t)/(2\sqrt{t})}{=} \frac{1}{2}\int_{3-2\sqrt{2}}^1 \frac{\log((-1 + 6t - t^2)/(4t))}{t}dt$$

$$= \frac{1}{2}\int_{3-2\sqrt{2}}^1 \frac{\log((t - (3 - 2\sqrt{2}))(1 - (3 - 2\sqrt{2})t)/(4(3 - 2\sqrt{2})t))}{t}dt$$

$$= \frac{1}{2}\int_{3-2\sqrt{2}}^1 \frac{\log(t - (3 - 2\sqrt{2}))}{t}dt + \frac{1}{2}\int_{3-2\sqrt{2}}^1 \frac{\log(1 - (3 - 2\sqrt{2})t)}{t}dt$$

$$- \log(2)\int_{3-2\sqrt{2}}^1 \frac{1}{t}dt - \frac{1}{2}\log(3 - 2\sqrt{2})\int_{3-2\sqrt{2}}^1 \frac{1}{t}dt - \frac{1}{2}\int_{3-2\sqrt{2}}^1 \frac{\log(t)}{t}dt$$

$$= \frac{1}{2}\underbrace{\int_{3-2\sqrt{2}}^1 \frac{\log(t - (3 - 2\sqrt{2}))}{t}dt}_{X} + \frac{1}{2}\underbrace{\int_{3-2\sqrt{2}}^1 \frac{\log(1 - (3 - 2\sqrt{2})t)}{t}dt}_{Y}$$

$$+ 2\log(2)\log(\sqrt{2} - 1) + 3\log^2(\sqrt{2} - 1). \tag{3.292}$$

To calculate the first integral in (3.292), we consider the variable change $t = (3 - 2\sqrt{2})u$, and then we have

$$X = \int_{3-2\sqrt{2}}^{1} \frac{\log(t - (3 - 2\sqrt{2}))}{t} dt = \int_{1}^{1/(3-2\sqrt{2})} \frac{\log((3 - 2\sqrt{2})u - (3 - 2\sqrt{2}))}{u} du$$

$$= \log(3 - 2\sqrt{2}) \int_{1}^{1/(3-2\sqrt{2})} \frac{1}{u} du + \underbrace{\int_{1}^{1/(3-2\sqrt{2})} \frac{\log(u - 1)}{u} du}_{u = 1/v}$$

$$= -\log^2(3 - 2\sqrt{2}) + \int_{3-2\sqrt{2}}^{1} \frac{\log(1 - v)}{v} dv - \int_{3-2\sqrt{2}}^{1} \frac{\log(v)}{v} dv$$

$$= -\frac{1}{2}\log^2(3 - 2\sqrt{2}) + \int_{3-2\sqrt{2}}^{1} \frac{\log(1 - v)}{v} dv$$

$$= -\frac{1}{2}\log^2(3 - 2\sqrt{2}) - \text{Li}_2(v)\Big|_{v=3-2\sqrt{2}}^{v=1}$$

$$= -\frac{\pi^2}{6} - \frac{1}{2}\log^2(3 - 2\sqrt{2}) + \text{Li}_2((1 - \sqrt{2})^2)$$

$$\left\{ \text{employ the dilogarithmic identity, } \text{Li}_2(x) + \text{Li}_2(-x) = \frac{1}{2}\text{Li}_2(x^2) \right\}$$

$$= -\frac{\pi^2}{6} - 2\log^2(\sqrt{2} - 1) + 2\text{Li}_2(1 - \sqrt{2}) + 2\text{Li}_2(-(1 - \sqrt{2})). \qquad (3.293)$$

Further, to calculate the second integral in (3.292), we make the variable change $(3 - 2\sqrt{2})t = u$, and then we get

$$Y = \int_{3-2\sqrt{2}}^{1} \frac{\log(1 - (3 - 2\sqrt{2})t)}{t} dt = \int_{(3-2\sqrt{2})^2}^{3-2\sqrt{2}} \frac{\log(1 - u)}{u} du$$

$$\overset{3-2\sqrt{2}=(1-\sqrt{2})^2}{\underset{\text{use that}}{=}} -\text{Li}_2(u)\Big|_{u=(1-\sqrt{2})^4}^{u=(1-\sqrt{2})^2} = \text{Li}_2((1 - \sqrt{2})^4) - \text{Li}_2((1 - \sqrt{2})^2)$$

$$\left\{ \text{use the dilogarithmic identity, } \text{Li}_2(x) + \text{Li}_2(-x) = \frac{1}{2}\text{Li}_2(x^2) \right\}$$

$$= 2\text{Li}_2(1 - \sqrt{2}) + 2\text{Li}_2(-(1 - \sqrt{2})) + 2\text{Li}_2(-(1 - \sqrt{2})^2). \qquad (3.294)$$

If we plug the results from (3.293) and (3.294) in (3.292), we arrive at

$$\int_0^1 \frac{\log(1 - x^2)}{\sqrt{1 + x^2}}\,dx$$

$$= 2\text{Li}_2(1 - \sqrt{2}) + 2\text{Li}_2(-(1 - \sqrt{2})) + \text{Li}_2(-(1 - \sqrt{2})^2)$$

$$- \frac{\pi^2}{12} + 2\log(2)\log(\sqrt{2} - 1) + 2\log^2(\sqrt{2} - 1). \tag{3.295}$$

In order to get a form reduction in (3.295), we start from the simple dilogarithmic identity, also appearing in (3.52), Sect. 3.10, $\text{Li}_2(-x) + \text{Li}_2(-1/x) = -\pi^2/6 - 1/2\log^2(x)$, $x > 0$, where if we set $x = \sqrt{2}$, we obtain that

$$\text{Li}_2\left(-\frac{1}{\sqrt{2}}\right) = -\frac{\pi^2}{6} - \frac{1}{8}\log^2(2) - \text{Li}_2(-\sqrt{2}). \tag{3.296}$$

Now, if we combine (3.296) and the identity in (3.283), Sect. 3.40, we arrive at

$$\text{Li}_2\left(-\frac{1}{\sqrt{2}}\right) = -\frac{\pi^2}{8} - \frac{1}{8}\log^2(2) - \frac{1}{2}\log(2)\log(\sqrt{2}-1) - \text{Li}_2(1-\sqrt{2}). \tag{3.297}$$

Further, if we consider $2\sqrt{2} - 3 = -(1 - \sqrt{2})^2$ in the special identity with three dilogarithms from (1.13), Sect. 1.4, we get

$$\text{Li}_2\left(-(1 - \sqrt{2})^2\right) + 2\text{Li}_2\left(-\frac{1}{\sqrt{2}}\right) - 4\text{Li}_2\left(1 - \sqrt{2}\right)$$

$$= \frac{\pi^2}{24} - \log^2\left(2 - \sqrt{2}\right). \tag{3.298}$$

At this point, if we combine (3.297) and (3.298), we manage to extract a key dilogarithmic value needed in the simplification of the closed form in (3.295),

$$\text{Li}_2\left(-(1 - \sqrt{2})^2\right) = \frac{7}{24}\pi^2 - \log^2(\sqrt{2} - 1) + 6\text{Li}_2(1 - \sqrt{2}). \tag{3.299}$$

Lastly, by plugging both the identity from the point (vi), previously derived, and the result from (3.299) in (3.295), we conclude that

$$\int_0^1 \frac{\log(1 - x^2)}{\sqrt{1 + x^2}}\,dx = \frac{11}{24}\pi^2 + 2\log(2)\log(\sqrt{2} - 1) + 10\text{Li}_2(1 - \sqrt{2}),$$

and the point (i) of the problem is complete.

Further, to prove the transformation at the point (ii) we need the result in (3.285), Sect. 3.40, where if we integrate both sides from $y = 0$ to $y = 1$, we obtain that

$$\int_0^1 \frac{\text{arctanh}(y)}{\sqrt{1+y^2}}dy \overset{y=x}{=} \int_0^1 \frac{\text{arctanh}(x)}{\sqrt{1+x^2}}dx = \int_0^1 \left(\int_0^x \frac{1}{(1-x^2y^2)\sqrt{1+y^2}}dy \right) dx$$

{reverse the order of integration and then split the inner integral}

$$= \int_0^1 \frac{1}{\sqrt{1+y^2}} \left(\left(\int_0^1 - \int_0^y \right) \frac{1}{1-y^2x^2}dx \right) dy$$

$$\left\{ \text{employ the fact that } \int_0^x \frac{1}{1-a^2t^2}dt = \frac{1}{a}\text{arctanh}(ax), \ |ax| < 1 \right\}$$

$$= \int_0^1 \frac{\text{arctanh}(y) - \text{arctanh}(y^2)}{y\sqrt{1+y^2}}dy$$

$$\left\{ \text{use that } \text{arctanh}(x) - \text{arctanh}(x^2) = \log(1+x) - 1/2\log(1+x^2) \right\}$$

{and afterwards consider splitting the resulting logarithmic integral}

$$= \int_0^1 \frac{\log(1+y)}{y\sqrt{1+y^2}}dy - \frac{1}{2}\int_0^1 \frac{\log(1+y^2)}{y\sqrt{1+y^2}}dy$$

$$\left\{ \text{in the second integral let the variable change } y^2 = t \right\}$$

{and then return to the notation in x in both integrals}

$$= \int_0^1 \frac{\log(1+x)}{x\sqrt{1+x^2}}dx - \frac{1}{4}\int_0^1 \frac{\log(1+x)}{x\sqrt{1+x}}dx,$$

and the point (ii) of the problem is complete.

Next, the point (iii) is straightforward if we combine the result from the point (ii) and the ones from (1.180) and (1.181) Sect. 1.40,

$$\int_0^1 \frac{\text{arctanh}(x)}{\sqrt{1+x^2}}dx = -\frac{\pi^2}{48} + \frac{1}{2}\log^2(\sqrt{2}-1) - \text{Li}_2(1-\sqrt{2}),$$

and the point (iii) of the problem is complete.

Finally, the results from the points (iv) and (v) are straightforward if we combine the points (i) and (iii) that together form a system of relations from which we may extract the desired values as follows:

$$\int_0^1 \frac{\log(1-x)}{\sqrt{1+x^2}}dx$$

$$= \frac{\pi^2}{4} + \log(2)\log(\sqrt{2}-1) - \frac{1}{2}\log^2(\sqrt{2}-1) + 6\mathrm{Li}_2(1-\sqrt{2})$$

and

$$\int_0^1 \frac{\log(1+x)}{\sqrt{1+x^2}}dx$$

$$= \frac{5}{24}\pi^2 + \log(2)\log(\sqrt{2}-1) + \frac{1}{2}\log^2(\sqrt{2}-1) + 4\mathrm{Li}_2(1-\sqrt{2}),$$

and the points (iv) and (v) of the problem are complete.

The curious reader might also consider treating the integrals from the points (iv) and (v) separately, without making use of a system of relations as presented above. Moreover, finding different ways to go for every point of the problem could be an enjoyable exercise for the curious reader.

3.42 Somewhat Atypical Integrals with Curious Closed Forms

Solution In order to get solutions in the present section, we'll also want to exploit the power of some results found in the first sections with generalizations of the present book.

As regards the point (i) of the problem, we observe that after replacing $\sin(x)$ by x in the numerator of the integrand we get a form which is known, that is, $\arctan(x)\log(1+x^2)$. Does this ring any bell?

Let's recall a useful series representation

$$\sum_{k=1}^{\infty}(-1)^{k-1}x^{2k+1}\frac{H_{2k}}{2k+1} = \frac{1}{2}\arctan(x)\log(1+x^2)\ |x| \le 1, \qquad (3.300)$$

that is also met and derived in two different ways in *(Almost) Impossible Integrals, Sums, and Series* (see [76, Chapter 6, pp.346–347]).

So, if we replace x by $\sin(x)$ in (3.300), divide both sides by $\sin(x)$, and then rearrange and integrate from $x = 0$ to $x = \pi/2$, we arrive at

$$\int_0^{\pi/2} \frac{\arctan(\sin(x))\log(1+\sin^2(x))}{\sin(x)}dx = 2\int_0^{\pi/2}\sum_{k=1}^{\infty}(-1)^{k-1}\sin^{2k}(x)\frac{H_{2k}}{2k+1}dx$$

{rearrange and wisely split the integral}

$$= 2\int_0^{\pi/2}\sum_{k=1}^{\infty}(-1)^{k-1}\sin^{2k}(x)\frac{H_{2k+1}}{2k+1}dx - 2\int_0^{\pi/2}\sum_{k=1}^{\infty}(-1)^{k-1}\frac{\sin^{2k}(x)}{(2k+1)^2}dx.$$

$$(3.301)$$

For the first remaining integral in (3.301), we write

$$\int_0^{\pi/2}\sum_{k=1}^{\infty}(-1)^{k-1}\sin^{2k}(x)\frac{H_{2k+1}}{2k+1}dx$$

$$\left\{ \text{use the logarithmic integral in (3.10), Sect. 3.3, } \int_0^1 t^{n-1}\log(1-t)dt = -\frac{H_n}{n}\right\}$$

$$= -\int_0^{\pi/2}\sum_{k=1}^{\infty}(-1)^{k-1}\sin^{2k}(x)\left(\int_0^1 t^{2k}\log(1-t)dt\right)dx$$

{reverse the order of integration and summation}

$$= -\int_0^1 \log(1-t)\left(\int_0^{\pi/2}\sum_{k=1}^{\infty}(-1)^{k-1}(t\sin(x))^{2k}dx\right)dt$$

$$= -\int_0^1 \log(1-t)\left(\int_0^{\pi/2}\left(1-\frac{1}{1+t^2\sin^2(x)}\right)dx\right)dt$$

$$= \int_0^1 \log(1-t)\left(\int_0^{\pi/2}\frac{1}{1+t^2\sin^2(x)}dx\right)dt - \int_0^1 \log(1-t)\left(\int_0^{\pi/2}dx\right)dt$$

$$\left\{ \text{employ the result (see [76, Chapter 3, p.212]), } \int_0^{\pi/2}\frac{1}{1-a\sin^2(x)}dx = \frac{\pi}{2}\frac{1}{\sqrt{1-a}}\right\}$$

$$= \frac{\pi}{2}\int_0^1 \frac{\log(1-t)}{\sqrt{1+t^2}}dt + \frac{\pi}{2}$$

$$= \frac{\pi}{2} + \frac{\pi^3}{8} + \frac{1}{2}\log(2)\log(\sqrt{2}-1)\pi - \frac{1}{4}\log^2(\sqrt{2}-1)\pi + 3\pi\operatorname{Li}_2(1-\sqrt{2}),$$

$$(3.302)$$

where we might like to observe the last integral is derived in the previous section.

Then, for the second remaining integral in (3.301), we write

$$\int_0^{\pi/2} \sum_{k=1}^{\infty} (-1)^{k-1} \frac{\sin^{2k}(x)}{(2k+1)^2} dx$$

$$\left\{ \text{use the logarithmic integral, } \int_0^1 t^{2k} \log(t) dt = -\frac{1}{(2k+1)^2} \right\}$$

$$= -\int_0^{\pi/2} \sum_{k=1}^{\infty} (-1)^{k-1} \sin^{2k}(x) \left(\int_0^1 t^{2k} \log(t) dt \right) dx$$

{reverse the order of integration and summation}

$$= -\int_0^1 \log(t) \left(\int_0^{\pi/2} \sum_{k=1}^{\infty} (-1)^{k-1} (t \sin(x))^{2k} dx \right) dt$$

$$= -\int_0^1 \log(t) \left(\int_0^{\pi/2} \left(1 - \frac{1}{1 + t^2 \sin^2(x)} \right) dx \right) dt$$

$$= \int_0^1 \log(t) \left(\int_0^{\pi/2} \frac{1}{1 + t^2 \sin^2(x)} dx \right) dt - \int_0^1 \log(t) \left(\int_0^{\pi/2} dx \right) dt$$

$$= \frac{\pi}{2} \int_0^1 \frac{\log(t)}{\sqrt{1+t^2}} dt + \frac{\pi}{2}. \tag{3.303}$$

Next, for the remaining integral in (3.303) we consider the substitution $\sqrt{1+t^2} - t = u$ that gives $t = \dfrac{1-u^2}{2u}$, and then we have

$$\int_0^1 \frac{\log(t)}{\sqrt{1+t^2}} dt = \int_{\sqrt{2}-1}^1 \frac{1}{u} \log\left(\frac{1-u^2}{2u} \right) du = \int_{\sqrt{2}-1}^1 \frac{\log(1-u)}{u} du$$

$$+ \int_{\sqrt{2}-1}^1 \frac{\log(1+u)}{u} du - \log(2) \int_{\sqrt{2}-1}^1 \frac{1}{u} du - \int_{\sqrt{2}-1}^1 \frac{\log(u)}{u} du$$

$$= -\frac{\pi^2}{12} + \log(2) \log(\sqrt{2}-1) + \frac{1}{2} \log^2(\sqrt{2}-1)$$

$$+ \text{Li}_2(1 - \sqrt{2}) + \text{Li}_2(-(1-\sqrt{2}))$$

{employ the identity in (1.189), Sect. 1.41}

$$= \frac{\pi^2}{24} + \log(2)\log(\sqrt{2} - 1) + 2\text{Li}_2(1 - \sqrt{2}). \tag{3.304}$$

So, by plugging the result from (3.304) in (3.303), we arrive at

$$\int_0^{\pi/2} \sum_{k=1}^{\infty} (-1)^{k-1} \frac{\sin^{2k}(x)}{(2k+1)^2} dx$$

$$= \frac{\pi}{2} + \frac{\pi^3}{48} + \frac{1}{2}\log(2)\log(\sqrt{2} - 1)\pi + \pi\text{Li}_2(1 - \sqrt{2}). \tag{3.305}$$

Collecting the results from (3.302) and (3.305) in (3.301), we conclude that

$$\int_0^{\pi/2} \frac{\arctan(\sin(x))\log(1+\sin^2(x))}{\sin(x)} dx = \int_0^{\pi/2} \frac{\arctan(\cos(x))\log(1+\cos^2(x))}{\cos(x)} dx$$

$$= \frac{5}{24}\pi^3 - \frac{1}{2}\log^2(\sqrt{2} - 1)\pi + 4\pi\text{Li}_2(1 - \sqrt{2}),$$

where it is easy to see the first equality is obtained by the variable change $\pi/2 - x = y$, and the first solution to the point (i) of the problem is finalized.

How about getting a way that doesn't exploit harmonic series?, you might wonder. In this case, we might think of the identity in (1.19), Sect. 1.6, where the second solution provided uses no harmonic series. Thus, exploiting the mentioned identity where we replace x by y and then a by $\sin^2(x)$, rearrange, and integrate from $x = 0$ to $x = \pi/2$, we obtain that

$$\int_0^{\pi/2} \frac{\arctan(\sin(x))\log(1 + \sin^2(x))}{\sin(x)} dx$$

$$= 2\int_0^{\pi/2} \left(\int_0^1 \frac{\log(1-y)}{1 + \sin^2(x)y^2} dy \right) dx + 2\int_0^{\pi/2} \frac{\text{Ti}_2(\sin(x))}{\sin(x)} dx$$

{switch the order of integration in the first double integral and in}

{the second integral use the series representation of the inverse}

$$\left\{ \text{tangent integral defined by } \text{Ti}_2(x) = \sum_{n=1}^{\infty} (-1)^{n-1} \frac{x^{2n-1}}{(2n-1)^2}, \ |x| \le 1 \right\}$$

$$= 2 \int_0^1 \log(1-y) \left(\int_0^{\pi/2} \frac{1}{1+y^2 \sin^2(x)} dx \right) dy + 2 \int_0^{\pi/2} \sum_{n=1}^{\infty} (-1)^{n-1} \frac{\sin^{2n-2}(x)}{(2n-1)^2} dx$$

{the first double integral is met in the previous solution, and for the second series}

{under the second integral we reindex the series and leave out the term for $n = 0$}

$$= \pi \int_0^1 \frac{\log(1-y)}{\sqrt{1+y^2}} dy + \pi - 2 \int_0^{\pi/2} \sum_{n=1}^{\infty} (-1)^{n-1} \frac{\sin^{2n}(x)}{(2n+1)^2} dx$$

$$= \frac{5}{24}\pi^3 - \frac{1}{2} \log^2(\sqrt{2}-1)\pi + 4\pi \operatorname{Li}_2(1-\sqrt{2}),$$

where the last equality is obtained easily since the needed results already appeared in the previous solution, more exactly in (3.302) and (3.305), and the second solution to the point (i) of the problem is finalized.

Let's go further and attack the exotic relation with integrals at the point (ii)! So, if we consider the result from (1.28), Sect. 1.8, where we replace a by $\sin(a)$, rearrange, divide both sides by $\sin(a)$, and integrate from $a = 0$ to $a = \pi/2$, we get

$$\int_0^{\pi/2} \frac{\arctan(\sin(a))\log(1+\sin^2(a))}{\sin(a)} da \overset{a=x}{=} \int_0^{\pi/2} \frac{\arctan(\sin(x))\log(1+\sin^2(x))}{\sin(x)} dx$$

$$= 2 \int_0^{\pi/2} \left(\int_0^{\sin(a)} \frac{\log(1+\sin(a)x)}{\sin(a)(1+x^2)} dx \right) da$$

$$\overset{\sin(a)x=y}{=} 2 \int_0^{\pi/2} \left(\int_0^{\sin^2(a)} \frac{\log(1+y)}{\sin^2(a)+y^2} dy \right) da$$

$$\overset{\substack{\text{reverse the order} \\ \text{of integration}}}{=} 2 \int_0^1 \log(1+y) \left(\int_{\arcsin(\sqrt{y})}^{\pi/2} \frac{1}{y^2+\sin^2(a)} da \right) dy$$

{in the denominator of the inner integral exploit that $y^2(\sin^2(a) + \cos^2(a)) = y^2$}

$$= 2 \int_0^1 \frac{\log(1+y)}{1+y^2} \left(\int_{\arcsin(\sqrt{y})}^{\pi/2} \frac{(\tan(a))'}{(y/\sqrt{1+y^2})^2 + \tan^2(a)} da \right) dy$$

$$= 2 \int_0^1 \frac{\log(1+y)}{y\sqrt{1+y^2}} \left(\arctan\left(\frac{\sqrt{1+y^2}}{y} \tan(a) \right) \Big|_{a=\arcsin(\sqrt{y})}^{a=\pi/2} \right) dy$$

{after performing the calculations, return to the variable x}

$$= \pi \int_0^1 \frac{\log(1+x)}{x\sqrt{1+x^2}}dx - 2\int_0^1 \arctan\left(\sqrt{\frac{1+x^2}{x(1-x)}}\right)\frac{\log(1+x)}{x\sqrt{1+x^2}}dx,$$

whence we obtain that

$$\int_0^{\pi/2}\frac{\arctan(\sin(x))\log(1+\sin^2(x))}{\sin(x)}dx + 2\int_0^1 \arctan\left(\sqrt{\frac{1+x^2}{x(1-x)}}\right)\frac{\log(1+x)}{x\sqrt{1+x^2}}dx$$

$$= \pi\int_0^1 \frac{\log(1+x)}{x\sqrt{1+x^2}}dx$$

{make use of the result in (1.181), Sect. 1.40}

$$= \frac{1}{2}\log^2(\sqrt{2}-1)\pi - \frac{\pi^3}{24} - 2\pi\operatorname{Li}_2(1-\sqrt{2}),$$

and the solution to the point (ii) of the problem is finalized. Note that in the calculations I also used that $\tan(\arcsin(x)) = \dfrac{x}{\sqrt{1-x^2}}$.

Finally, combining the previous two points we immediately arrive at

$$\int_0^1 \arctan\left(\sqrt{\frac{1+x^2}{x(1-x)}}\right)\frac{\log(1+x)}{x\sqrt{1+x^2}}dx$$

$$= \frac{1}{2}\log^2(\sqrt{2}-1)\pi - \frac{\pi^3}{8} - 3\pi\operatorname{Li}_2(1-\sqrt{2}),$$

and the solution to the point (iii) of the problem is finalized.

A moment of (possible) wonderment (to some extent) if we carefully take a look at the last integral result! In (1.181), Sect. 1.40, the integral result we have to calculate is $\int_0^1 \dfrac{\log(1+x)}{x\sqrt{1+x^2}}dx$. Now, if we take the integrand of the previously mentioned integral and multiply it by $g(x) = \arctan\left(\sqrt{\dfrac{1+x^2}{x(1-x)}}\right)$, we get the integrand of the integral of this final point of the problem.

Of course, it is tempting to think about other curious, non-trivial cases involving functions like $g(x)$ that would lead to integrals with *nice* closed forms!

3.43   More Atypical Integrals with Curious Closed Forms

Solution There are moments when we look over some results and immediately wonder about their usefulness in deriving other results. Well, if you have already passed through Sect. 1.16, it is possible you *embraced* such a moment, and later we'll want to use the first result presented there.

For now, let's pass directly to the integral at the point (ii), where if we make the variable change $(\sqrt{1+x} - 1)/2 = t$, we have

$$
\int_0^1 \frac{\log(x)}{x} \left(1 - \frac{1}{\sqrt{1+x}} \right) dx = 2 \int_0^{(\sqrt{2}-1)/2} \frac{\log(4t(1+t))}{1+t} dt
$$

$$
= 4 \log(2) \int_0^{(\sqrt{2}-1)/2} \frac{1}{1+t} dt + 2 \int_0^{(\sqrt{2}-1)/2} \frac{\log(1+t)}{1+t} dt + 2 \int_0^{(\sqrt{2}-1)/2} \frac{\log(t)}{1+t} dt
$$

$$
= -\log^2(2) - 2\log(2)\log(\sqrt{2} - 1) - \log^2(\sqrt{2} - 1) + 2\text{Li}_2 \left(\frac{1}{2} \left(1 - \sqrt{2} \right) \right),
$$
$$
(3.306)
$$

where the first two integrals from the second line of calculations above are straightforward, and the third one is promptly evaluated by using integration by parts,
$$
\int_0^x \frac{\log(t)}{1+t} dt = \log(x) \log(1+x) - \int_0^x \frac{\log(1+t)}{t} dt = \log(x) \log(1+x) + \text{Li}_2(-x).
$$
For the dilogarithmic part in (3.306), we employ the Landen's dilogarithmic identity (see [21, Chapter 1, p.5], [70, Chapter 2, p.107]), $\text{Li}_2(x) + \text{Li}_2 \left(\dfrac{x}{x-1} \right) = -\dfrac{1}{2} \log^2(1-x)$, with $x = (1 - \sqrt{2})/2$, and rearranging, we obtain that

$$
\text{Li}_2 \left(\frac{1}{2}(1 - \sqrt{2}) \right) = -\frac{1}{2} \log^2(2(\sqrt{2} - 1)) - \text{Li}_2((1 - \sqrt{2})^2)
$$

$$
\left\{ \text{employ the dilogarithmic identity, } \text{Li}_2(x) + \text{Li}_2(-x) = \frac{1}{2} \text{Li}_2(x^2) \right\}
$$

$$
= -\frac{1}{2} \log^2(2(\sqrt{2} - 1)) - 2\text{Li}_2(1 - \sqrt{2}) - 2\text{Li}_2(-(1 - \sqrt{2}))
$$

$$
= -\frac{\pi^2}{4} - \frac{1}{2} \log^2(2) - \log(2) \log(\sqrt{2} - 1) + \frac{1}{2} \log^2(\sqrt{2} - 1) - 4\text{Li}_2(1 - \sqrt{2}),
$$
$$
(3.307)
$$

where for getting the last equality I also used the identity in (1.189), Sect. 1.41.

Then, if we plug (3.307) in (3.306), we obtain that

$$\int_0^1 \frac{\log(x)}{x}\left(1 - \frac{1}{\sqrt{1+x}}\right)dx$$

$$= -\frac{\pi^2}{2} - 2\log^2(2) - 4\log(2)\log(\sqrt{2}-1) - 8\mathrm{Li}_2(1-\sqrt{2}),$$

and the solution to the point (ii) of the problem is complete.

At this point, we return to the main integral at the point (i), and letting the variable change $x = \tan(t/2)$, we get

$$\int_0^1 \frac{\arctan(x)}{(1+x^2)\sqrt{1-x^2}}dx = \frac{1}{4}\int_0^{\pi/2} \frac{t\cos(t/2)}{\sqrt{\cos(t)}}dt = \frac{1}{4}\int_0^{\pi/2} t\cos(nt)\cos^n(t)dt\bigg|_{n=-1/2}$$

$\{$exploit the integral result in (1.76), Sect. 1.16, the extended version to reals$\}$

$$= \frac{3}{16}\zeta(2)\frac{1}{2^n} - \frac{1}{32}\frac{1}{2^n}((\psi(n+1)+\gamma)^2 + \zeta(2) - \psi^{(1)}(n+1))$$

$$- \frac{1}{16}\frac{1}{2^n}\int_0^1 \frac{(1-(1+x)^n)\log(x)}{x}dx\bigg|_{n=-1/2}$$

$$= \frac{7}{96}\sqrt{2}\pi^2 + \frac{1}{4}\sqrt{2}\log(2)\log(\sqrt{2}-1) + \frac{1}{2}\sqrt{2}\mathrm{Li}_2(1-\sqrt{2}),$$

where the last equality follows by employing the facts that $\psi\left(\dfrac{1}{2}\right) = -\gamma - 2\log(2)$, $\psi^{(1)}\left(\dfrac{1}{2}\right) = 3\zeta(2)$, which are given in (6.146) and (6.147), the case $k = 1$, Sect. 6.19, together with the result from the point (ii), in the current section, and the solution to the point (i) of the problem is complete. Also, observe that since I considered using the extended version to reals of (1.76), Sect. 1.16, I switched from the use of cases of the generalized harmonic numbers to the Polygamma function.

For the last point of the problem, we start from the fact that $\pi^2/12 = -\mathrm{Li}_2(-1) = \displaystyle\int_0^1 \frac{\log(1+x)}{x}dx$, and then, letting the variable change $\sqrt{x(1+x)} = x+t$, we have

$$\frac{\pi^2}{12} = \int_0^1 \frac{\log(1+x)}{x}dx = 2\int_0^{\sqrt{2}-1} \frac{1-t}{t(1-2t)}\log\left(\frac{(1-t)^2}{1-2t}\right)dt$$

$$= 2 \underbrace{\int_0^{\sqrt{2}-1} -\frac{\log(1-2t)}{1-2t}dt}_{\log^2(\sqrt{2}-1)} + 4 \underbrace{\int_0^{\sqrt{2}-1} \frac{\log(1-t)}{t}dt}_{-\mathrm{Li}_2(-(1-\sqrt{2}))} - 2 \underbrace{\int_0^{\sqrt{2}-1} \frac{\log(1-2t)}{t}dt}_{-\mathrm{Li}_2(-2(1-\sqrt{2}))}$$

$$+ 4 \underbrace{\int_0^{\sqrt{2}-1} \frac{\log(1-t)}{1-2t}dt}_{\pi^2/24 + \log(2)\log(\sqrt{2}-1) + \mathrm{Li}_2(-(1-\sqrt{2})^2)/2}$$

$$= \frac{\pi^2}{6} + 4\log(2)\log(\sqrt{2}-1) + 2\log^2(\sqrt{2}-1)$$

$$- 4\mathrm{Li}_2(-(1-\sqrt{2})) + 2\mathrm{Li}_2(-(1-\sqrt{2})^2) + 2\mathrm{Li}_2(-2(1-\sqrt{2})),$$

whence we obtain that

$$8\mathrm{Li}_2(-(1-\sqrt{2})) - 4\mathrm{Li}_2\left(-(1-\sqrt{2})^2\right) - 4\mathrm{Li}_2(-2(1-\sqrt{2}))$$

$$= \frac{\pi^2}{6} + 8\log(2)\log(\sqrt{2}-1) + 4\log^2(\sqrt{2}-1),$$

and the solution to the point (iii) of the problem is complete. In the calculations I also used that $\int_0^x \frac{\log(1-t)}{1-2t}dt = \int_0^x \frac{-\log(2) + \log(2-2t)}{1-2t}dt = -\log(2)\int_0^x \frac{dt}{1-2t} + \int_0^x \frac{\log(1+1-2t)}{1-2t}dt = \frac{\pi^2}{24} + \frac{1}{2}\log(2)\log(1-2x) + \frac{1}{2}\mathrm{Li}_2(2x-1),\ x < \frac{1}{2}.$

The curious reader might also enjoy considering other creative ways to go before passing to another section! A final thought to share is that exploiting results like (1.76), Sect. 1.16, as seen above, is another powerful way of solving and creating mathematical problems.

3.44 A Wonderful Trigonometric Integral by Larry Glasser

Solution In March 2021 I prepared and sent an email with my solution, in large steps, to the given integral to one of the well-known names in the mathematical community, Larry Glasser, the proposer of the problem (the author of numerous wonderful proposals in the mathematical journals during the time). It is about the problem **398** that appeared in *La Gaceta de la RSME*, published in Vol. 23, No. 2

(2020). *After some work on it, I realized that the problem is perfect for the present book—so I assigned a separate statement section to it!*

In the following, I'll exploit an auxiliary result presented in one of the previous sections, but before doing that my plan is to bring the integral to a convenient form. We first want to split the integral at $x = \pi/2$, and then we write

$$\int_0^\pi \frac{\arctan^2(\cot(\phi)\sin(\theta))}{1 + \cos(\theta)} d\theta$$

$$= \int_0^{\pi/2} \frac{\arctan^2(\cot(\phi)\sin(\theta))}{1 + \cos(\theta)} d\theta + \int_{\pi/2}^\pi \frac{\arctan^2(\cot(\phi)\sin(\theta))}{1 + \cos(\theta)} d\theta$$

{let $\pi - \theta = \alpha$ in the second integral and then merge the two integrals}

$$= 2 \int_0^{\pi/2} \frac{\arctan^2(\cot(\phi)\sin(\theta))}{\sin^2(\theta)} d\theta. \tag{3.308}$$

Does it ring any bells? In fact, we may exploit the integral result in (1.31), Sect. 1.9, which leads to

$$\int_0^{\pi/2} \frac{\arctan^2(\cot(\phi)\sin(\theta))}{\sin^2(\theta)} d\theta$$

$$= 2 \cot^2(\phi) \int_0^{\pi/2} \left(\int_0^1 \frac{x \operatorname{arctanh}(x)}{1 + (\cot(\phi)\sin(\theta))^2 x^2} dx \right) d\theta$$

$$\overset{\substack{\text{reverse the order} \\ \text{of integration}}}{=} 2 \cot^2(\phi) \int_0^1 x \operatorname{arctanh}(x) \left(\int_0^{\pi/2} \frac{1}{1 + (\cot(\phi)x)^2 \sin^2(\theta)} d\theta \right) dx$$

$$\left\{ \text{use} \int_0^{\pi/2} \frac{dx}{1 - a\sin^2(x)} = \frac{\pi}{2} \frac{1}{\sqrt{1-a}}, \ a < 1, \ \text{given in [76, Chapter 3, p.212]} \right\}$$

$$= \pi \cot^2(\phi) \int_0^1 \frac{x \operatorname{arctanh}(x)}{\sqrt{1 + \cot^2(\phi)x^2}} dx. \tag{3.309}$$

One interesting thing with respect to the last integral in (3.309) is that we have an elementary antiderivative, and I'll exploit this fact! So, we prepare to integrate by parts, and then we have

$$\int \frac{x \operatorname{arctanh}(x)}{\sqrt{1 + \cot^2(\phi)x^2}} dx = \tan^2(\phi) \int \left(\sqrt{1 + \cot^2(\phi)x^2} \right)' \operatorname{arctanh}(x) dx$$

$$= \tan^2(\phi)\sqrt{1 + \cot^2(\phi)x^2}\,\text{arctanh}(x) - \tan^2(\phi)\int \frac{\sqrt{1 + \cot^2(\phi)x^2}}{1 - x^2}dx.$$

$$(3.310)$$

Then, for the remaining integral in (3.310) we write

$$\int \frac{\sqrt{1 + \cot^2(\phi)x^2}}{1 - x^2}dx = \int \frac{1 + \cot^2(\phi) - \cot^2(\phi)(1 - x^2)}{(1 - x^2)\sqrt{1 + \cot^2(\phi)x^2}}dx$$

$$= (1 + \cot^2(\phi))\int \frac{1}{(1 - x^2)\sqrt{1 + \cot^2(\phi)x^2}}dx - \cot(\phi)\int \frac{(\cot(\phi)x)'}{\sqrt{1 + \cot^2(\phi)x^2}}dx$$

$$= \csc^2(\phi)\int \frac{1}{(1 - x^2)\sqrt{1 + \cot^2(\phi)x^2}}dx - \cot(\phi)\,\text{arcsinh}(\cot(\phi)x). \quad (3.311)$$

Further, if we let $\cot(\phi)x = \tan(y)$ in the last integral of (3.311), and rearrange, we get that

$$\int \frac{1}{(1 - x^2)\sqrt{1 + \cot^2(\phi)x^2}}dx = \sin(\phi)\int \frac{(\sec(\phi)\sin(y))'}{1 - \sec^2(\phi)\sin^2(y)}dy$$

$$= \sin(\phi)\,\text{arctanh}(\sec(\phi)\sin(y)) + C = \sin(\phi)\,\text{arctanh}\left(\frac{\csc(\phi)x}{\sqrt{1 + \cot^2(\phi)x^2}}\right) + C,$$

$$(3.312)$$

where to get the last equality I used the identity, $\sin(\arctan(x)) = \dfrac{x}{\sqrt{1 + x^2}}$.

Combining (3.311) and (3.312), we arrive at

$$\int \frac{\sqrt{1 + \cot^2(\phi)x^2}}{1 - x^2}dx$$

$$= \csc(\phi)\,\text{arctanh}\left(\frac{\csc(\phi)x}{\sqrt{1 + \cot^2(\phi)x^2}}\right) - \cot(\phi)\,\text{arcsinh}(\cot(\phi)x) + C. \quad (3.313)$$

By plugging (3.313) in (3.310), we have

$$\int \frac{x\,\text{arctanh}(x)}{\sqrt{1 + \cot^2(\phi)x^2}}dx$$

$$= \tan^2(\phi)\sqrt{1 + \cot^2(\phi)x^2}\,\text{arctanh}(x) - \sec(\phi)\tan(\phi)\,\text{arctanh}\left(\frac{\csc(\phi)x}{\sqrt{1 + \cot^2(\phi)x^2}}\right)$$

$$+ \tan(\phi)\,\text{arcsinh}(\cot(\phi)x) + C = f(x, \phi). \quad (3.314)$$

So, if we integrate in (3.314) from $x = 0$ to $x = 1$, we obtain that

$$\int_0^1 \frac{x \operatorname{arctanh}(x)}{\sqrt{1 + \cot^2(\phi) x^2}} dx = \underbrace{\lim_{\substack{x \to 1 \\ x < 1}} f(x, \phi) - \lim_{\substack{x \to 0 \\ x > 0}} f(x, \phi)}_{0} = \lim_{\substack{x \to 1 \\ x < 1}} f(x, \phi)$$

{use an artifice of calculation and cleverly distribute the limit}

$$= \sec(\phi) \tan(\phi) \underbrace{\lim_{\substack{x \to 1 \\ x < 1}} \left(\sin(\phi) \sqrt{1 + \cot^2(\phi) x^2} - 1 \right) \operatorname{arctanh}(x)}_{0}$$

$$+ \sec(\phi) \tan(\phi) \underbrace{\lim_{\substack{x \to 1 \\ x < 1}} \left(\operatorname{arctanh}(x) - \operatorname{arctanh}\left(\frac{\csc(\phi) x}{\sqrt{1 + \cot^2(\phi) x^2}} \right) \right)}_{\log(\sin(\phi))}$$

$$+ \tan(\phi) \operatorname{arcsinh}(\cot(\phi))$$

$$= \sec(\phi) \tan(\phi) \log(\sin(\phi)) + \tan(\phi) \operatorname{arcsinh}(\cot(\phi))$$

$$= \sec(\phi) \tan(\phi) \log(\sin(\phi)) + \tan(\phi) \log\left(\cot\left(\frac{\phi}{2} \right) \right), \qquad (3.315)$$

where the last equality comes from the fact that $\operatorname{arcsinh}(x) = \log(x + \sqrt{1 + x^2})$.
Finally, combining (3.308), (3.309), and (3.315), we conclude that

$$\int_0^\pi \frac{\arctan^2(\cot(\phi) \sin(\theta))}{1 + \cos(\theta)} d\theta$$

$$= 2\pi \left(\csc(\phi) \log(\sin(\phi)) + \cot(\phi) \log\left(\cot\left(\frac{\phi}{2} \right) \right) \right),$$

and the solution is finalized.
As regards the limits in (3.315), we observe that

$$\lim_{\substack{x \to 1 \\ x < 1}} \left(\sin(\phi) \sqrt{1 + \cot^2(\phi) x^2} - 1 \right) \operatorname{arctanh}(x)$$

$$= \cos^2(\phi) \lim_{\substack{x \to 1 \\ x < 1}} \frac{x^2 - 1}{\sin(\phi) \sqrt{1 + \cot^2(\phi) x^2} + 1} \operatorname{arctanh}(x)$$

$$= \cos^2(\phi) \underbrace{\lim_{\substack{x \to 1 \\ x<1}} \frac{1+x}{\sin(\phi)\sqrt{1+\cot^2(\phi)x^2}+1}}_{1} \cdot \underbrace{\lim_{\substack{x \to 1 \\ x<1}} (x-1)\operatorname{arctanh}(x) = 0}_{0},$$

where the last limit is derived by applying *L'Hospital's rule* once. Further, since $\operatorname{arctanh}(x) \pm \operatorname{arctanh}(y) = \operatorname{arctanh}\left(\dfrac{x \pm y}{1 \pm xy}\right)$, we get

$$\lim_{\substack{x \to 1 \\ x<1}} \left(\operatorname{arctanh}(x) - \operatorname{arctanh}\left(\frac{\csc(\phi)x}{\sqrt{1+\cot^2(\phi)x^2}}\right) \right)$$

$$= \lim_{\substack{x \to 1 \\ x<1}} \operatorname{arctanh} \left(\frac{x - x\sin(\phi)\sqrt{1+\cot^2(\phi)x^2}}{x^2 - \sin(\phi)\sqrt{1+\cot^2(\phi)x^2}} \right)$$

$$= \lim_{\substack{x \to 1 \\ x<1}} \operatorname{arctanh} \left(-\frac{x^2\cos^2(\phi)}{x^2+\sin^2(\phi)} \cdot \frac{x^2+\sin(\phi)\sqrt{1+\cot^2(\phi)x^2}}{x+x\sin(\phi)\sqrt{1+\cot^2(\phi)x^2}} \right)$$

$$= \log(\sin(\phi)).$$

Another solution, which involves nice ideas, may be found in [20]. If we consider the problem with restrictions on ϕ, then we might attack it by exploiting the Taylor series of $\arctan^2(x)$.

3.45   Resistant Logarithmic Integrals That Are Good to Know

Solution   A collection of special and useful logarithmic integrals is waiting for us in the present section! In fact, we need to use almost all of them in various places of the book, and therefore we want to know how to evaluate them. Moreover, some of the evaluations might be definitely viewed as a part of *the art of integration*, and we'll see this right from the first point of the problem!

For the first integral of the section I also wrote a paper called "A new perspective on the evaluation of the logarithmic integral", $\int_0^1 \frac{\log(x)\log^3(1+x)}{x}dx$ (see [85, February 4, 2020]), as soon as I discovered a (magically) beautiful, non-obvious way of attacking the integral, and in the following I'll consider the strategy in the paper.

There are two key results I want to prove first:

$$\lim_{\substack{a \to 0 \\ b \to 0}} \frac{\partial^4}{\partial a^3 \partial b} B(a, b) = 18\zeta(5) - 6\zeta(2)\zeta(3)$$

$$= 2\int_0^1 \frac{\log^4(1+x)}{x}dx - 4\int_0^1 \frac{\log(x)\log^3(1+x)}{x}dx + 3\int_0^1 \frac{\log^2(x)\log^2(1+x)}{x}dx$$

$$-\int_0^1 \frac{\log^3(x)\log(1+x)}{x}dx \qquad (3.316)$$

and

$$\lim_{\substack{a \to 0 \\ b \to 0}} \frac{\partial^4}{\partial a^2 \partial b^2} B(a, b) = 16\zeta(5) - 8\zeta(2)\zeta(3)$$

$$= 2\int_0^1 \frac{\log^4(1+x)}{x}dx - 4\int_0^1 \frac{\log(x)\log^3(1+x)}{x}dx + 2\int_0^1 \frac{\log^2(x)\log^2(1+x)}{x}dx,$$

$$(3.317)$$

where $B(a, b)$ represents the Beta function.

Proof In order to prove the first result above, we use the Beta function defined by $B(x, y) = \int_0^1 t^{x-1}(1-t)^{y-1}dt$, $\Re(x), \Re(y) > 0$, and reduce the particular Beta function limits to particular cases of the Euler sum generalization presented in (6.149), Sect. 6.19.

Let's observe the simple fact that

$$\lim_{\substack{a \to 0 \\ b \to 0}} \frac{\partial^4}{\partial a^3 \partial b} B(a, b) = \int_0^1 \frac{\log^3(x)\log(1-x)}{x(1-x)}dx = -\int_0^1 \log^3(x)\sum_{n=1}^{\infty} x^{n-1}H_n dx$$

$$= -\sum_{n=1}^{\infty} H_n \int_0^1 x^{n-1}\log^3(x)dx = 6\sum_{n=1}^{\infty} \frac{H_n}{n^4} = 18\zeta(5) - 6\zeta(2)\zeta(3). \qquad (3.318)$$

On the other hand, if we use the following different Beta function definition, $\int_0^1 \frac{x^{a-1} + x^{b-1}}{(1+x)^{a+b}}dx = B(a, b)$, we get that

$$\lim_{\substack{a \to 0 \\ b \to 0}} \frac{\partial^4}{\partial a^3 \partial b} B(a, b) = 2\int_0^1 \frac{\log^4(1+x)}{x}dx - 4\int_0^1 \frac{\log(x)\log^3(1+x)}{x}dx$$

$$+ 3\int_0^1 \frac{\log^2(x)\log^2(1+x)}{x}dx - \int_0^1 \frac{\log^3(x)\log(1+x)}{x}dx. \qquad (3.319)$$

Combining (3.318) and (3.319), the result in (3.316) follows immediately. You might check Sect. 6.20 where I use a similar strategy, which involves the use of the two different definitions of the Beta function.

Following the same strategy as before, we have

$$\lim_{\substack{a \to 0 \\ b \to 0}} \frac{\partial^4}{\partial a^2 \partial b^2} B(a, b) = \int_0^1 \frac{\log^2(x) \log^2(1 - x)}{x(1 - x)} dx = \int_0^1 \frac{\log^2(x) \log^2(1 - x)}{x} dx$$

$$+ \underbrace{\int_0^1 \frac{\log^2(x) \log^2(1 - x)}{1 - x} dx}_{1 - x = y} = 2 \int_0^1 \frac{\log^2(x) \log^2(1 - x)}{x} dx$$

$$= 4 \int_0^1 \log^2(x) \sum_{n=1}^{\infty} x^{n-1} \frac{H_{n-1}}{n} dx = 4 \sum_{n=1}^{\infty} \frac{H_{n-1}}{n} \int_0^1 x^{n-1} \log^2(x) dx = 8 \sum_{n=1}^{\infty} \frac{H_{n-1}}{n^4}$$

$$= 8 \sum_{n=1}^{\infty} \frac{H_n}{n^4} - 8 \sum_{n=1}^{\infty} \frac{1}{n^5} = 16\zeta(5) - 8\zeta(2)\zeta(3), \tag{3.320}$$

where the resulting Euler sum is the particular case $n = 4$ of the Euler sum generalization in (6.149), Sect. 6.19. I also exploited, based on (4.33), Sect. 4.6, that

$$\frac{1}{2} \log^2(1 - x) = \sum_{n=1}^{\infty} x^{n+1} \frac{H_n}{n+1} = \sum_{n=2}^{\infty} x^n \frac{H_{n-1}}{n} = \sum_{n=1}^{\infty} x^n \frac{H_{n-1}}{n}.$$

So, if we consider the other Beta function definition, $\int_0^1 \frac{x^{a-1} + x^{b-1}}{(1 + x)^{a+b}} dx = $ $B(a, b)$, as at the previous point to prove, we have

$$\lim_{\substack{a \to 0 \\ b \to 0}} \frac{\partial^4}{\partial a^2 \partial b^2} B(a, b)$$

$$= 2 \int_0^1 \frac{\log^4(1 + x)}{x} dx - 4 \int_0^1 \frac{\log(x) \log^3(1 + x)}{x} dx + 2 \int_0^1 \frac{\log^2(x) \log^2(1 + x)}{x} dx. \tag{3.321}$$

Combining (3.320) and (3.321), the result in (3.317) follows.   ∎

Based on (3.316) and (3.317), we obtain the desired result

$$\int_0^1 \frac{\log(x) \log^3(1 + x)}{x} dx = \frac{1}{2} \lim_{\substack{a \to 0 \\ b \to 0}} \frac{\partial^4}{\partial a^3 \partial b} B(a, b) - \frac{3}{4} \lim_{\substack{a \to 0 \\ b \to 0}} \frac{\partial^4}{\partial a^2 \partial b^2} B(a, b)$$

$$+ \frac{1}{2} \int_0^1 \frac{\log^3(x) \log(1+x)}{x} dx + \frac{1}{2} \int_0^1 \frac{\log^4(1+x)}{x} dx$$

$$= 3\zeta(2)\zeta(3) - 3\zeta(5) + \frac{1}{2} \int_0^1 \frac{\log^3(x) \log(1+x)}{x} dx + \frac{1}{2} \int_0^1 \frac{\log^4(1+x)}{x} dx$$

$$= \frac{99}{16}\zeta(5) + 3\zeta(2)\zeta(3) + 2\log^3(2)\zeta(2) - \frac{21}{4}\log^2(2)\zeta(3) - \frac{2}{5}\log^5(2)$$

$$- 12\log(2)\text{Li}_4\left(\frac{1}{2}\right) - 12\text{Li}_5\left(\frac{1}{2}\right),$$

where in the calculations I used that $\int_0^1 \frac{\log^3(x) \log(1+x)}{x} dx = -\frac{45}{8}\zeta(5)$, which also appears in (6.164), Sect. 6.20, and then

$$\int_0^1 \frac{\log^4(1+x)}{x} dx$$

$$= 4\log^3(2)\zeta(2) - \frac{21}{2}\log^2(2)\zeta(3) + 24\zeta(5) - \frac{4}{5}\log^5(2) - 24\log(2)\text{Li}_4\left(\frac{1}{2}\right)$$

$$- 24\text{Li}_5\left(\frac{1}{2}\right),$$

where the last integral appears in [76, Chapter 3, pp.78–79], and it is immediately evaluated if we make the variable change $1/(1+x) = y$, exploit geometric series, and finally employ well-known special dilogarithmic and trilogarithmic values, and the solution to the point (i) is finalized.

It is wonderful to see how beautifully we can evaluate such integrals! The strategy above can be adapted to also attack other more difficult results with integrals and series. The main integral also appeared in 2014 in [50].

The integrals at the points (ii)–(iv) will be *somewhat straightforward* if we reduce them to their series representations, but, of course, it is important to have at hand the values of all those series!

Now, since we have based on (4.32), Sect. 4.6, that $\sum_{n=1}^{\infty}(-1)^{n-1}x^n H_n = \frac{\log(1+x)}{1+x}$, we write

$$\int_0^1 \frac{\log(1-x)\log^2(x)\log(1+x)}{1+x} dx = \int_0^1 \log^2(x)\log(1-x)\sum_{n=1}^{\infty}(-1)^{n-1}x^n H_n dx$$

{reverse the order of summation and integration}

$$= \sum_{n=1}^{\infty} (-1)^{n-1} H_n \int_0^1 x^n \log^2(x) \log(1-x) dx$$

{reindex the series and start from $n = 1$}

$$= -\sum_{n=1}^{\infty} (-1)^{n-1} \left(H_n - \frac{1}{n} \right) \int_0^1 x^{n-1} \log^2(x) \log(1-x) dx$$

{make use of the result in (1.125), the case $m = 2$, Sect. 1.24}

$$= 2\zeta(3) \sum_{n=1}^{\infty} (-1)^{n-1} \frac{1}{n^2} + 2\zeta(2) \sum_{n=1}^{\infty} (-1)^{n-1} \frac{1}{n^3} - 2\zeta(3) \sum_{n=1}^{\infty} (-1)^{n-1} \frac{H_n}{n}$$

$$-2\zeta(2) \sum_{n=1}^{\infty} (-1)^{n-1} \frac{H_n}{n^2} - 2 \sum_{n=1}^{\infty} (-1)^{n-1} \frac{H_n}{n^4} + 2 \sum_{n=1}^{\infty} (-1)^{n-1} \frac{H_n^2}{n^3} - 2 \sum_{n=1}^{\infty} (-1)^{n-1} \frac{H_n^{(2)}}{n^3}$$

$$- 2 \sum_{n=1}^{\infty} (-1)^{n-1} \frac{H_n^{(3)}}{n^2} + 2 \sum_{n=1}^{\infty} (-1)^{n-1} \frac{H_n H_n^{(2)}}{n^2} + 2 \sum_{n=1}^{\infty} (-1)^{n-1} \frac{H_n H_n^{(3)}}{n}$$

$$= \frac{213}{16} \zeta(5) - \frac{5}{2} \zeta(2)\zeta(3) - \frac{49}{8} \log(2)\zeta(4) + \frac{7}{4} \log^2(2)\zeta(3) + \frac{1}{3} \log^3(2)\zeta(2)$$

$$- \frac{1}{10} \log^5(2) - 4 \log(2)\mathrm{Li}_4 \left(\frac{1}{2} \right) - 8 \mathrm{Li}_5 \left(\frac{1}{2} \right),$$

where the third series is the particular case $m = 1$ of the special alternating harmonic series generalization in (4.106), Sect. 4.22, the fourth and fifth harmonic series are the particular cases $m = 1, 2$ of the alternating Euler sum generalization in (4.105), Sect. 4.21, then the sixth harmonic series is also found in (3.331), Sect. 3.48, the seventh and eighth harmonic series have

$$\sum_{n=1}^{\infty} (-1)^{n-1} \frac{H_n^{(2)}}{n^3} = \frac{5}{8} \zeta(2)\zeta(3) - \frac{11}{32} \zeta(5) \text{ (see [76, Chapter 4, p.311]) and}$$

$$\sum_{n=1}^{\infty} (-1)^{n-1} \frac{H_n^{(3)}}{n^2} = \frac{3}{4} \zeta(2)\zeta(3) - \frac{21}{32} \zeta(5) \text{ [76, Chapter 4, p.311], the ninth}$$

harmonic series is also given in (3.332), Sect. 3.48, and the value of the tenth series is $\sum_{n=1}^{\infty} (-1)^{n-1} \frac{H_n H_n^{(3)}}{n} = \frac{1}{6} \log^3(2)\zeta(2) + \frac{3}{8} \log^2(2)\zeta(3) - \frac{49}{16} \log(2)\zeta(4) +$

$\frac{167}{32}\zeta(5) - \frac{1}{16}\zeta(2)\zeta(3) - \frac{1}{20}\log^5(2) - 2\log(2)\mathrm{Li}_4\left(\frac{1}{2}\right) - 4\mathrm{Li}_5\left(\frac{1}{2}\right)$, which is found in [76, Chapter 6, Section 6.58, pp.523–529], and the solution to the point (ii) is finalized.

How would we try to go differently? Of course, having at our disposal the values of the advanced alternating harmonic series is comfortable, as seen above, but one idea to attack the integral differently could be based on exploiting algebraic identities as in [76, Chapter 6, Section 6.58, pp.526–529].

The integral at the third point of the problem was also posed in 2014 in [51], and a full solution has never been given since then. With the values of some advanced alternating harmonic series of weight 5 in hand, we immediately get the desired result. In fact, we'll proceed in a similar style as above. So, we write that

$$\int_0^1 \frac{\log^2(1-x)\log^2(x)}{1+x}dx = \int_0^1 \log^2(x)\log^2(1-x)\sum_{n=1}^{\infty}(-1)^{n-1}x^{n-1}dx$$

{reverse the order of summation and integration}

$$= \sum_{n=1}^{\infty}(-1)^{n-1}\int_0^1 x^{n-1}\log^2(x)\log^2(1-x)dx$$

{make use of the result in (1.126), the case $m=2$, Sect. 1.24}

$$= -\zeta(4)\sum_{n=1}^{\infty}(-1)^{n-1}\frac{1}{n} - 4\zeta(3)\sum_{n=1}^{\infty}(-1)^{n-1}\frac{1}{n^2} - 4\zeta(3)\sum_{n=1}^{\infty}(-1)^{n-1}\frac{H_n}{n}$$

$$- 4\zeta(2)\sum_{n=1}^{\infty}(-1)^{n-1}\frac{H_n}{n^2} - 4\zeta(2)\sum_{n=1}^{\infty}(-1)^{n-1}\frac{H_n^{(2)}}{n} + 6\sum_{n=1}^{\infty}(-1)^{n-1}\frac{H_n^{(4)}}{n}$$

$$+ 2\sum_{n=1}^{\infty}(-1)^{n-1}\frac{H_n^2}{n^3} + 2\sum_{n=1}^{\infty}(-1)^{n-1}\frac{H_n^{(2)}}{n^3} + 4\sum_{n=1}^{\infty}(-1)^{n-1}\frac{H_n^{(3)}}{n^2}$$

$$+ 4\sum_{n=1}^{\infty}(-1)^{n-1}\frac{H_n H_n^{(2)}}{n^2} + 4\sum_{n=1}^{\infty}(-1)^{n-1}\frac{H_n H_n^{(3)}}{n} + 2\sum_{n=1}^{\infty}(-1)^{n-1}\frac{(H_n^{(2)})^2}{n}$$

$$= \frac{15}{2}\zeta(5) - \frac{13}{2}\zeta(2)\zeta(3) + \log(2)\zeta(4) + \frac{2}{3}\log^3(2)\zeta(2) - \frac{1}{15}\log^5(2) + 8\mathrm{Li}_5\left(\frac{1}{2}\right),$$

where most of the series above have already been met at the end of the previous solution, but we also need the values of the harmonic series given by (4.106), the

cases $m = 2, 4$, Sect. 4.22, and $\displaystyle\sum_{n=1}^{\infty}(-1)^{n-1}\frac{(H_n^{(2)})^2}{n} = \frac{1}{5}\log^5(2) - \frac{2}{3}\log^3(2)\zeta(2) +$
$\dfrac{29}{4}\log(2)\zeta(4) - \dfrac{259}{16}\zeta(5) + \dfrac{5}{8}\zeta(2)\zeta(3) + 8\log(2)\mathrm{Li}_4\left(\dfrac{1}{2}\right) + 16\mathrm{Li}_5\left(\dfrac{1}{2}\right)$, found and
calculated in [76, Chapter 6, Section 6.58, pp.523–529], and the solution to the point
(iii) is finalized.

The integral at the fourth point remained without a solution in [52] for more than
4 years. So, we prepare an integration by parts, and then we write

$$\int_0^1 \frac{\log^2(1+x)\log^2(x)}{1-x}dx = -\int_0^1 (\log(1-x))'\log^2(1+x)\log^2(x)dx$$

$$= 2\int_0^1 \frac{\log(1-x)\log^2(x)\log(1+x)}{1+x}dx + 2\int_0^1 \frac{\log(1-x)\log(x)\log^2(1+x)}{x}dx.$$
$$\tag{3.322}$$

Let's observe that the first remaining logarithmic integral in (3.322) is
found right at the point (ii), and the second remaining one is a popular
one, usually calculated elegantly by exploiting the algebraic identity, $ab^2 = 1/6((a+b)^3 + (a-b)^3 - 2a^3)$, as seen in [53]. However, I'll go differently
this time, by exploiting the advanced alternating harmonic series of weight 5, and
using that $\dfrac{1}{2}\dfrac{\log^2(1+x)}{x} = \displaystyle\sum_{n=1}^{\infty}(-1)^{n-1}x^n\frac{H_n}{n+1} = -\sum_{n=2}^{\infty}(-1)^{n-1}x^{n-1}\frac{H_{n-1}}{n} =$
$\displaystyle\sum_{n=1}^{\infty}(-1)^{n-1}x^{n-1}\left(\frac{1}{n^2} - \frac{H_n}{n}\right)$, based on (4.33), Sect. 4.6, we write

$$\int_0^1 \frac{\log(1-x)\log(x)\log^2(1+x)}{x}dx$$

$$= 2\int_0^1 \log(1-x)\log(x)\sum_{n=1}^{\infty}(-1)^{n-1}x^{n-1}\left(\frac{1}{n^2} - \frac{H_n}{n}\right)dx$$

{reverse the order of summation and integration}

$$= 2\sum_{n=1}^{\infty}(-1)^{n-1}\left(\frac{1}{n^2} - \frac{H_n}{n}\right)\int_0^1 x^{n-1}\log(x)\log(1-x)dx$$

{use the result in (1.125), case $m = 1$, Sect. 1.24}

$$= -2\zeta(2)\sum_{n=1}^{\infty}(-1)^{n-1}\frac{1}{n^3} + 2\zeta(2)\sum_{n=1}^{\infty}(-1)^{n-1}\frac{H_n}{n^2} + 2\sum_{n=1}^{\infty}(-1)^{n-1}\frac{H_n}{n^4}$$

$$- 2\sum_{n=1}^{\infty}(-1)^{n-1}\frac{H_n^2}{n^3} + 2\sum_{n=1}^{\infty}(-1)^{n-1}\frac{H_n^{(2)}}{n^3} - 2\sum_{n=1}^{\infty}(-1)^{n-1}\frac{H_n H_n^{(2)}}{n^2}$$

$$= \frac{7}{8}\zeta(2)\zeta(3) - \frac{25}{16}\zeta(5), \tag{3.323}$$

where all the needed harmonic series have already been met at the previous points.

Returning with the results from (3.323) and the one at the point (ii) in (3.322), we obtain that

$$\int_0^1 \frac{\log^2(1+x)\log^2(x)}{1-x}dx$$

$$= \frac{47}{2}\zeta(5) - \frac{13}{4}\zeta(2)\zeta(3) - \frac{1}{5}\log^5(2) + \frac{2}{3}\log^3(2)\zeta(2) + \frac{7}{2}\log^2(2)\zeta(3)$$

$$- \frac{49}{4}\log(2)\zeta(4) - 8\log(2)\mathrm{Li}_4\left(\frac{1}{2}\right) - 16\mathrm{Li}_5\left(\frac{1}{2}\right),$$

and the solution to the point (iv) is finalized.

Next, I'll start with the variable change $(1-x)/(1+x) = t$ to attack the integral at the point (v), and then we have

$$\int_0^1 \frac{\mathrm{arctanh}^2(x)\log(1+x^2)}{x}dx = \frac{1}{2}\int_0^1 \frac{\log^2(t)}{1-t^2}\log\left(\frac{2(1+t^2)}{(1+t)^2}\right)dt$$

{expand the integral cleverly}

$$= \frac{1}{2}\log(2)\int_0^1 \frac{\log^2(t)}{1+t}dt - \frac{1}{2}\int_0^1 \frac{\log^2(t)\log(1+t)}{1+t}dt$$

$$- \frac{1}{2}\int_0^1 \frac{\log^2(t)\log\left(\frac{1+t}{2}\right)}{1-t}dt + \frac{1}{4}\int_0^1 \frac{\log^2(t)\log\left(\frac{1+t^2}{2}\right)}{1-t}dt$$

$$+ \frac{1}{4}\int_0^1 \frac{\log^2(t)\log\left(\frac{1+t^2}{2}\right)}{1+t}dt$$

$$= G^2 + \frac{53}{32}\zeta(4) - \frac{7}{4}\log(2)\zeta(3) + \frac{1}{2}\log^2(2)\zeta(2) - \frac{1}{12}\log^4(2) - 2\mathrm{Li}_4\left(\frac{1}{2}\right),$$

where for the first resulting integral above we immediately see that $\displaystyle\int_0^1 \frac{\log^2(t)}{1+t}dt =$

$$\sum_{n=1}^{\infty}(-1)^{n-1}\int_0^1 t^{n-1}\log^2(t)dt = 2\sum_{n=1}^{\infty}(-1)^{n-1}\frac{1}{n^3} = \frac{3}{2}\zeta(3), \text{ then for the}$$

second remaining logarithmic integral we write $\displaystyle\int_0^1 \frac{\log^2(t)\log(1+t)}{1+t}dt =$

$$\sum_{n=1}^{\infty}(-1)^{n-1}H_n\int_0^1 t^n\log^2(t)dt = 2\sum_{n=1}^{\infty}(-1)^{n-1}\frac{H_n}{(n+1)^3} = 2\sum_{n=1}^{\infty}\frac{(-1)^{n-1}}{n^4} -$$

$$2\sum_{n=1}^{\infty}(-1)^{n-1}\frac{H_n}{n^3} = \frac{1}{6}\log^4(2) - \log^2(2)\zeta(2) + \frac{7}{2}\log(2)\zeta(3) - \frac{15}{4}\zeta(4) + 4\operatorname{Li}_4\left(\frac{1}{2}\right),$$

next, the third integral is the case $m = 2$ of the generalization in (1.131), after that the fourth integral is the case $m = 2$ of the generalization in (1.132), and finally, the last integral is the case $m = 2$ of the generalization in (1.133), Sect. 1.27, and the solution to the point (v) is finalized.

To evaluate the last integral of the section, I'll also exploit the integral at the previous point. To do that, I'll consider the algebraic identity, $ab = ((a+b)^2 - (a-b)^2)/4$, where if we set $a = \log(1+x)$ and $b = \log(1-x)$, then multiply both sides by $\log(1+x^2)/x$ and integrate from $x = 0$ to $x = 1$, we get

$$\int_0^1 \frac{\log(1-x)\log(1+x)\log(1+x^2)}{x}dx$$

$$= \frac{1}{4}\underbrace{\int_0^1 \frac{\log^2(1-x^2)\log(1+x^2)}{x}dx}_{\text{let } x^2 = t} - \int_0^1 \frac{\operatorname{arctanh}^2(x)\log(1+x^2)}{x}dx$$

$$= \frac{1}{8}\int_0^1 \frac{\log^2(1-x)\log(1+x)}{x}dx - \int_0^1 \frac{\operatorname{arctanh}^2(x)\log(1+x^2)}{x}dx. \qquad (3.324)$$

As regards the first integral in (3.324), we may use the algebraic identity $ab^2 = \frac{1}{6}((a-b)^3 + (a+b)^3 - 2a^3)$, where if we set $a = \log(1+x)$ and $b = \log(1-x)$, multiply both sides by $1/x$, and then integrate from $x = 0$ to $x = 1$, we obtain that

$$\int_0^1 \frac{\log^2(1-x)\log(1+x)}{x}dx$$

$$= \frac{1}{6}\underbrace{\int_0^1 \frac{\log^3(1-x^2)}{x}dx}_{\text{let } x^2 = t} + \frac{4}{3}\underbrace{\int_0^1 \frac{\operatorname{arctanh}^3(x)}{x}dx}_{\text{let } (1-x)/(1+x) = t} - \frac{1}{3}\int_0^1 \frac{\log^3(1+x)}{x}dx$$

$$= \frac{1}{12} \int_0^1 \frac{\log^3(1-x)}{x} dx - \frac{1}{3} \int_0^1 \frac{\log^3(x)}{1-x^2} dx - \frac{1}{3} \int_0^1 \frac{\log^3(1+x)}{x} dx$$

$$= \frac{1}{12} \log^4(2) - \frac{1}{2} \log^2(2)\zeta(2) + \frac{7}{4} \log(2)\zeta(3) - \frac{5}{8}\zeta(4) + 2\mathrm{Li}_4\left(\frac{1}{2}\right), \quad (3.325)$$

where the first resulting integral is given in (6.182), Sect. 6.23, then for the second one we have $\int_0^1 \frac{\log^3(x)}{1-x^2} dx = \sum_{n=1}^{\infty} \int_0^1 x^{2n-2} \log^3(x)dx = -6 \sum_{n=1}^{\infty} \frac{1}{(2n-1)^4} =$

$-6\sum_{n=1}^{\infty} \frac{1}{n^4} + \frac{3}{8} \sum_{n=1}^{\infty} \frac{1}{n^4} = -\frac{45}{8}\zeta(4)$, and the value of the last integral is found in (6.155), Sect. 6.20.

If we collect the result from (3.325) together with the one at the previous point of the problem in (3.324), we conclude that

$$\int_0^1 \frac{\log(1-x)\log(1+x)\log(1+x^2)}{x} dx$$

$$= \frac{3}{32} \log^4(2) - G^2 - \frac{9}{16} \log^2(2)\zeta(2) + \frac{63}{32} \log(2)\zeta(3) - \frac{111}{64}\zeta(4) + \frac{9}{4}\mathrm{Li}_4\left(\frac{1}{2}\right),$$
$$(3.326)$$

and the solution to the point (vi) is finalized.

Finding viable approaches to such integrals is a fascinating challenge. There is always that enticing thought coming back to mind periodically and making you wonder if there are also other special (maybe *magical?*) ways to evaluate them.

3.46   Appealing Parameterized Integrals with Logarithms and the Dilogarithm Function, Related to Harmonic Series

Solution We go further with another round of integrals involving logarithms and the dilogarithm, and at last a simple, beautiful integral relation is given, which we'll find helpful in the work with harmonic series.

Besides, the curious reader might want to consider, as an enjoyable exercise, the calculation in closed form of the integrals from all first four points when $m = 1$.

Let's start with the result at the point (i). Thus, we write

$$\int_0^1 \frac{\log(1-x)\log^m(x)\mathrm{Li}_2(x)}{x} dx = \int_0^1 \log(1-x)\log^m(x) \sum_{n=1}^{\infty} \frac{x^{n-1}}{n^2} dx$$

{reverse the order of summation and integration}

$$= \sum_{n=1}^{\infty} \frac{1}{n^2} \int_0^1 x^{n-1} \log^m(x) \log(1-x) dx$$

{make use of the result in (1.125), Sect. 1.24, and expand}

$$= (-1)^{m-1} m! \sum_{n=1}^{\infty} \frac{H_n}{n^{m+3}} - (-1)^{m-1} m! \sum_{k=1}^{m} \zeta(k+1) \zeta(m-k+3)$$

$$+ (-1)^{m-1} m! \sum_{k=1}^{m} \sum_{n=1}^{\infty} \frac{H_n^{(k+1)}}{n^{m-k+3}}$$

$$= \frac{1}{2} (-1)^{m-1} m! ((m+5) \zeta(m+4) - \zeta(2) \zeta(m+2))$$

$$- \frac{3}{2} (-1)^{m-1} m! \sum_{k=1}^{m} \zeta(k+1) \zeta(m-k+3) + (-1)^{m-1} m! \sum_{k=1}^{m} \sum_{n=1}^{\infty} \frac{H_n^{(k+1)}}{n^{m-k+3}},$$

where to get the last equality I used the Euler sum generalization in (6.149), Sect. 6.19, and the solution to the point (i) is complete.

Next, for the point (ii) of the problem we proceed in a similar style as before, and then we have

$$\int_0^1 \frac{\log(1-x) \log^m(x) \text{Li}_2(-x)}{x} dx = \int_0^1 \log(1-x) \log^m(x) \sum_{n=1}^{\infty} (-1)^n \frac{x^{n-1}}{n^2} dx$$

{reverse the order of summation and integration}

$$= \sum_{n=1}^{\infty} (-1)^n \frac{1}{n^2} \int_0^1 x^{n-1} \log^m(x) \log(1-x) dx$$

{employ the result in (1.125), Sect. 1.24, and expand}

$$= (-1)^{m-1} m! \sum_{k=1}^{m} \zeta(k+1) \eta(m-k+3) - (-1)^{m-1} m! \sum_{n=1}^{\infty} (-1)^{n-1} \frac{H_n}{n^{m+3}}$$

$$- (-1)^{m-1} m! \sum_{k=1}^{m} \sum_{n=1}^{\infty} (-1)^{n-1} \frac{H_n^{(k+1)}}{n^{m-k+3}},$$

and the solution to the point (ii) is complete.

Passing to the point (iii) of the problem, we first prepare some auxiliary results. Having said that, a first auxiliary result we need is

$$\int_0^1 x^{n-1}\operatorname{Li}_2(x)\mathrm{d}x = \underbrace{\frac{1}{n}x^n\operatorname{Li}_2(x)\Big|_{x=0}^{x=1}}_{\zeta(2)/n} + \frac{1}{n}\int_0^1 x^{n-1}\log(1-x)\mathrm{d}x = \zeta(2)\frac{1}{n} - \frac{H_n}{n^2},$$

(3.327)

where the last integral may be found elementarily calculated in [76, Chapter 3, p.59]. As regards (3.327), more generally, we have that

$$J_{m,n} = \int_0^1 x^{n-1}\operatorname{Li}_m(x)\mathrm{d}x = (-1)^{m-1}\frac{H_n}{n^m} + \sum_{k=1}^{m-1}(-1)^{m+k-1}\frac{1}{n^{m-k}}\zeta(k+1).$$

(3.328)

Proof Integrating by parts once, we get

$$J_{m,n} = \int_0^1 x^n(\operatorname{Li}_{m+1}(x))'\mathrm{d}x = \underbrace{\operatorname{Li}_{m+1}(1)}_{\zeta(m+1)} - n\underbrace{\int_0^1 x^{n-1}\operatorname{Li}_{m+1}(x)\mathrm{d}x}_{J_{m+1,n}},$$

or to put it simply

$$J_{m,n} + nJ_{m+1,n} = \zeta(m+1),$$

where if we replace m by k, then multiply both sides by $(-1)^k n^k$, and consider the summation from $k=1$ to $k=m-1$, we obtain

$$\sum_{k=1}^{m-1}\left((-1)^k n^k J_{k,n} - (-1)^{k+1}n^{k+1}J_{k+1,n}\right)$$

$$= -n\underbrace{J_{1,n}}_{H_n/n} + (-1)^{m-1}n^m J_{m,n} = \sum_{k=1}^{m-1}(-1)^k n^k \zeta(k+1),$$

whence we arrive at

$$J_{m,n} = \int_0^1 x^{n-1}\operatorname{Li}_m(x)\mathrm{d}x = (-1)^{m-1}\frac{H_n}{n^m} + \sum_{k=1}^{m-1}(-1)^{m+k-1}\frac{1}{n^{m-k}}\zeta(k+1),$$

which has been the result to prove. I also used that $J_{1,n} = \int_0^1 x^{n-1} \text{Li}_1(x)dx =$
$- \int_0^1 x^{n-1} \log(1-x)dx = \dfrac{H_n}{n}$, where the last integral is elementarily evaluated in
[76, Chapter 3, p.59].   ∎

Further, since $\dfrac{d^m}{dn^m}\left(\dfrac{1}{n^2}\right) = (-1)^m(m+1)!\dfrac{1}{n^{m+2}}$, $\dfrac{d^m}{dn^m}(\psi(n+1)+$
$\gamma) = \psi^{(m)}(n+1)$, and using that for n a positive integer, $\psi^{(m)}(n+$
$1) = (-1)^{m-1}m!\displaystyle\sum_{k=n+1}^{\infty}\dfrac{1}{k^{m+1}} = (-1)^{m-1}m!\left(\zeta(m+1) - \displaystyle\sum_{k=1}^{n}\dfrac{1}{k^{m+1}}\right) =$
$(-1)^{m-1}m!\left(\zeta(m+1) - H_n^{(m+1)}\right)$, if we apply the *general Leibniz rule*, the form
$(fg)^{(m)}(x) = f^{(m)}g + \displaystyle\sum_{k=1}^{m}\binom{m}{k}f^{(m-k)}(x)g^{(k)}(x)$, where we set $f = \dfrac{1}{n^2}$ and
$g = \psi(n+1)+\gamma$, and following the flow steps in (3.178), Sect. 3.24, we have

$$\frac{d^m}{dn^m}\left\{\frac{\psi(n+1)+\gamma}{n^2}\right\}$$

$$= \left(\frac{1}{n^2}\right)^{(m)}(\psi(n+1)+\gamma) + \sum_{k=1}^{m}\binom{m}{k}\left(\frac{1}{n^2}\right)^{(m-k)}(\psi(n+1)+\gamma)^{(k)}$$

$$= -(-1)^{m-1}(m+1)!\frac{H_n}{n^{m+2}} + (-1)^{m-1}\sum_{k=1}^{m}\binom{m}{k}\frac{(m-k+1)!k!}{n^{m-k+2}}\left(\zeta(k+1) - H_n^{(k+1)}\right)$$

$$= -(-1)^{m-1}(m+1)!\frac{H_n}{n^{m+2}} + (-1)^{m-1}m!\sum_{k=1}^{m}\frac{m-k+1}{n^{m-k+2}}\left(\zeta(k+1) - H_n^{(k+1)}\right).$$

$$(3.329)$$

Returning to (3.327), where we differentiate its opposite sides m times with respect to n and then use that $\dfrac{d^m}{dn^m}\left(\dfrac{1}{n}\right) = (-1)^m m!\dfrac{1}{n^{m+1}}$, together with the result in (3.329), we get that

$$\int_0^1 x^{n-1}\log^m(x)\text{Li}_2(x)dx = \zeta(2)\frac{d^m}{dn^m}\left\{\frac{1}{n}\right\} - \frac{d^m}{dn^m}\left\{\frac{\psi(n+1)+\gamma}{n^2}\right\}$$

$$= -\zeta(2)(-1)^{m-1}m!\frac{1}{n^{m+1}} + (-1)^{m-1}(m+1)!\frac{H_n}{n^{m+2}}$$

$$- (-1)^{m-1}m!\sum_{k=1}^{m}\frac{m-k+1}{n^{m-k+2}}\left(\zeta(k+1) - H_n^{(k+1)}\right).$$

$$(3.330)$$

We are ready now to start with the main calculations, and then we have

$$\int_0^1 \frac{\log(1+x)\log^m(x)\text{Li}_2(x)}{x}dx = \int_0^1 \text{Li}_2(x)\log^m(x)\sum_{n=1}^{\infty}(-1)^{n-1}\frac{x^{n-1}}{n}dx$$

{reverse the order of summation and integration}

$$= \sum_{n=1}^{\infty}(-1)^{n-1}\frac{1}{n}\int_0^1 x^{n-1}\log^m(x)\text{Li}_2(x)dx$$

{employ the result in (3.330) and then expand}

$$= -\zeta(2)(-1)^{m-1}m!\eta(m+2) - (-1)^{m-1}m!\sum_{k=1}^{m}(m-k+1)\zeta(k+1)\eta(m-k+3)$$

$$+ (-1)^{m-1}(m+1)!\sum_{n=1}^{\infty}(-1)^{n-1}\frac{H_n}{n^{m+3}}$$

$$+ (-1)^{m-1}m!\sum_{k=1}^{m}(m-k+1)\sum_{n=1}^{\infty}(-1)^{n-1}\frac{H_n^{(k+1)}}{n^{m-k+3}},$$

and the solution to the point (iii) is complete.

For dealing with the integral at the point (iv), we employ the Cauchy product of two series in (4.23), Sect. 4.5, and then we write

$$\int_0^1 \frac{\log(1+x)\log^m(x)\text{Li}_2(-x)}{x}dx$$

$$= \int_0^1 \left(3\sum_{n=1}^{\infty}(-1)^n\frac{x^{n-1}}{n^3} - 2\sum_{n=1}^{\infty}(-1)^n x^{n-1}\frac{H_n}{n^2} - \sum_{n=1}^{\infty}(-1)^n x^{n-1}\frac{H_n^{(2)}}{n}\right)\log^m(x)dx$$

{expand and reverse the order of summation and integration}

$$= 3\sum_{n=1}^{\infty}(-1)^n\frac{1}{n^3}\int_0^1 x^{n-1}\log^m(x)dx - 2\sum_{n=1}^{\infty}(-1)^n\frac{H_n}{n^2}\int_0^1 x^{n-1}\log^m(x)dx$$

$$- \sum_{n=1}^{\infty}(-1)^n\frac{H_n^{(2)}}{n}\int_0^1 x^{n-1}\log^m(x)dx$$

$$= 3(-1)^{m-1}m!\eta(m+4) - 2(-1)^{m-1}m! \sum_{n=1}^{\infty}(-1)^{n-1}\frac{H_n}{n^{m+3}}$$

$$- (-1)^{m-1}m! \sum_{n=1}^{\infty}(-1)^{n-1}\frac{H_n^{(2)}}{n^{m+2}},$$

where in the calculations I used that $\int_0^1 x^n \log^m(x)dx = (-1)^m \frac{m!}{(n+1)^{m+1}}$, given with swapped letters in [76, Chapter 1, p.1], and the solution to the point (iv) is complete.

Finally, in order to prove the relation at the point (v), we'll want to exploit the dilogarithmic identity, $\text{Li}_2(x) + \text{Li}_2(-x) = \frac{1}{2}\text{Li}_2(x^2)$, and then we write

$$\mathscr{S}_m = \int_0^1 \frac{\log(1-x)\log^m(x)\text{Li}_2(-x)}{x}dx + \int_0^1 \frac{\log(1+x)\log^m(x)\text{Li}_2(x)}{x}dx$$

$$+ \int_0^1 \frac{\log(1+x)\log^m(x)\text{Li}_2(-x)}{x}dx$$

$$= \underbrace{\mathscr{S}_m + \int_0^1 \frac{\log(1-x)\log^m(x)\text{Li}_2(x)}{x}dx - \int_0^1 \frac{\log(1-x)\log^m(x)\text{Li}_2(x)}{x}dx}_{\text{Sum up all integrals}}$$

$$= \frac{1}{2}\underbrace{\int_0^1 \frac{\log(1-x^2)\log^m(x)\text{Li}_2(x^2)}{x}dx}_{\text{let } x^2 = y} - \int_0^1 \frac{\log(1-x)\log^m(x)\text{Li}_2(x)}{x}dx$$

$$= \left(\frac{1}{2^{m+2}}-1\right)\int_0^1 \frac{\log(1-x)\log^m(x)\text{Li}_2(x)}{x}dx$$

$$= (-1)^{m-1}m!\left(1-\frac{1}{2^{m+2}}\right)\left(\frac{1}{2}(\zeta(2)\zeta(m+2) - (m+5)\zeta(m+4))\right.$$

$$\left. + \frac{3}{2}\sum_{k=1}^{m}\zeta(k+1)\zeta(m-k+3) - \sum_{k=1}^{m}\sum_{n=1}^{\infty}\frac{H_n^{(k+1)}}{n^{m-k+3}}\right),$$

and the solution to the point (v) is complete.

Later in the book we'll see the power of the relation above I've just finalized!

3.47 An Encounter with Six Useful Integrals Involving Logarithms and the Dilogarithm

Solution One of the places where we might see the usefulness of such integrals is the area of the atypical harmonic series of weight 5, involving harmonic numbers of the type $H_{2n}^{(p)}$. As relevant examples, I could mention two of my papers uploaded on ResearchGate, "The evaluation of a special harmonic series with a weight 5 structure, involving harmonic numbers of the type H_{2n}" [84, October 1, 2019], whose central element is a wonderful and surprising series given in the fourth chapter, and "The calculation of a harmonic series with a weight 5 structure, involving the product of harmonic numbers, $H_n H_{2n}^{(2)}$" [82, October 10, 2019].

Let's start with the evaluation of the integral at the first point! We need to use that $\text{Li}_2(x) = \sum_{n=1}^{\infty} \frac{x^n}{n^2}$ together with the fact that $\int_0^1 x^{n-1} \log^2(1-x)\mathrm{d}x = \frac{H_n^2 + H_n^{(2)}}{n}$ (see [76, Chapter 1, Section 1.3, p.2]), and then we write

$$\int_0^1 \frac{\log^2(1-x)\text{Li}_2(x)}{x}\mathrm{d}x = \int_0^1 \log^2(1-x)\sum_{n=1}^{\infty} \frac{x^{n-1}}{n^2}\mathrm{d}x$$

$$= \sum_{n=1}^{\infty} \frac{1}{n^2}\int_0^1 x^{n-1}\log^2(1-x)\mathrm{d}x = \sum_{n=1}^{\infty} \frac{H_n^2}{n^3} + \sum_{n=1}^{\infty} \frac{H_n^{(2)}}{n^3} = 2\zeta(2)\zeta(3) - \zeta(5),$$

where the last equality is obtained by using $\sum_{n=1}^{\infty} \frac{H_n^2}{n^3} = \frac{7}{2}\zeta(5) - \zeta(2)\zeta(3)$ (see [76, Chapter 4, p.293]) and $\sum_{n=1}^{\infty} \frac{H_n^{(2)}}{n^3} = 3\zeta(2)\zeta(3) - \frac{9}{2}\zeta(5)$ (see [76, Chapter 6, p.386]), and the solution to the point (i) is finalized.

Further, for the integral at the point (ii) we proceed similarly and consider the same starting results as before. Thus, we have

$$\int_0^1 \frac{\log^2(1-x)\text{Li}_2(-x)}{x}\mathrm{d}x = \int_0^1 \log^2(1-x)\sum_{n=1}^{\infty}(-1)^n\frac{x^{n-1}}{n^2}\mathrm{d}x$$

$$= \sum_{n=1}^{\infty}(-1)^n\frac{1}{n^2}\int_0^1 x^{n-1}\log^2(1-x)\mathrm{d}x = -\sum_{n=1}^{\infty}(-1)^{n-1}\frac{H_n^2}{n^3} - \sum_{n=1}^{\infty}(-1)^{n-1}\frac{H_n^{(2)}}{n^3}$$

$$= \frac{2}{3}\log^3(2)\zeta(2) - \frac{7}{4}\log^2(2)\zeta(3) + \frac{3}{4}\zeta(2)\zeta(3) + \frac{15}{16}\zeta(5) - \frac{2}{15}\log^5(2)$$

$$- 4\log(2)\text{Li}_4\left(\frac{1}{2}\right) - 4\text{Li}_5\left(\frac{1}{2}\right),$$

where the last equality is obtained by employing the alternating harmonic series of weight 5, $\displaystyle\sum_{n=1}^{\infty}(-1)^{n-1}\frac{H_n^2}{n^3} = \frac{2}{15}\log^5(2) - \frac{11}{8}\zeta(2)\zeta(3) - \frac{19}{32}\zeta(5) + \frac{7}{4}\log^2(2)\zeta(3) -$

$\displaystyle\frac{2}{3}\log^3(2)\zeta(2) + 4\log(2)\text{Li}_4\left(\frac{1}{2}\right) + 4\text{Li}_5\left(\frac{1}{2}\right)$ and $\displaystyle\sum_{n=1}^{\infty}(-1)^{n-1}\frac{H_n^{(2)}}{n^3} =$

$\displaystyle\frac{5}{8}\zeta(2)\zeta(3) - \frac{11}{32}\zeta(5)$ (see [76, Chapter 4, p.311]), and the solution to the point (ii) is finalized. The integral also appeared in [57], and for about 5 years no full solution has been provided.

The integral result at the point (iii) is straightforward if we combine the dilogarithmic identity, $\text{Li}_2(x) + \text{Li}_2(-x) = \frac{1}{2}\text{Li}_2(x^2)$, and the results from the points (i) and (ii), and then we have

$$\int_0^1 \frac{\log^2(1-x)\text{Li}_2(x^2)}{x}dx = 2\int_0^1\frac{\log^2(1-x)\text{Li}_2(x)}{x}dx + 2\int_0^1\frac{\log^2(1-x)\text{Li}_2(-x)}{x}dx$$

$$= \frac{4}{3}\log^3(2)\zeta(2) - \frac{7}{2}\log^2(2)\zeta(3) + \frac{11}{2}\zeta(2)\zeta(3) - \frac{1}{8}\zeta(5) - \frac{4}{15}\log^5(2)$$

$$- 8\log(2)\text{Li}_4\left(\frac{1}{2}\right) - 8\text{Li}_5\left(\frac{1}{2}\right),$$

and the solution to the point (iii) is finalized.

Moving to the point (iv) of the problem where we exploit the generating function in (4.33), Sect. 4.6, we have the fact that $\displaystyle\frac{\log^2(1+x)}{x} = 2\sum_{n=1}^{\infty}(-1)^{n-1}x^n\frac{H_n}{n+1} =$

$\displaystyle 2\sum_{n=1}^{\infty}(-1)^{n-1}x^n\frac{H_{n+1} - 1/(n+1)}{n+1} = 2\sum_{n=1}^{\infty}(-1)^{n-1}x^{n-1}\left(\frac{1}{n^2} - \frac{H_n}{n}\right)$, and returning to the main result, where we also use (3.327), Sect. 3.46, we write

$$\int_0^1 \frac{\log^2(1+x)\text{Li}_2(x)}{x}dx = 2\int_0^1\text{Li}_2(x)\sum_{n=1}^{\infty}(-1)^{n-1}x^{n-1}\left(\frac{1}{n^2} - \frac{H_n}{n}\right)dx$$

$$= 2\sum_{n=1}^{\infty}(-1)^{n-1}\left(\frac{1}{n^2} - \frac{H_n}{n}\right)\int_0^1 x^{n-1}\text{Li}_2(x)dx$$

$$= 2\sum_{n=1}^{\infty}(-1)^{n-1}\left(\frac{1}{n^2} - \frac{H_n}{n}\right)\left(\zeta(2)\frac{1}{n} - \frac{H_n}{n^2}\right) = 2\zeta(2)\sum_{n=1}^{\infty}(-1)^{n-1}\frac{1}{n^3}$$

$$- 2\zeta(2) \sum_{n=1}^{\infty} (-1)^{n-1} \frac{H_n}{n^2} - 2 \sum_{n=1}^{\infty} (-1)^{n-1} \frac{H_n}{n^4} + 2 \sum_{n=1}^{\infty} (-1)^{n-1} \frac{H_n^2}{n^3}$$

$$= \frac{4}{15} \log^5(2) - \frac{4}{3} \log^3(2)\zeta(2) + \frac{7}{2} \log^2(2)\zeta(3) - \frac{39}{8}\zeta(5) - \frac{3}{2}\zeta(2)\zeta(3)$$

$$+ 8 \log(2)\mathrm{Li}_4\left(\frac{1}{2}\right) + 8\mathrm{Li}_5\left(\frac{1}{2}\right),$$

where in the calculations I used the alternating Euler sum generalization in (4.105), Sect. 4.21, the cases $m = 2, 4$, and the value of the series $\sum_{n=1}^{\infty} (-1)^{n-1} \frac{H_n^2}{n^3}$, which is also given and used in the solution to the point (ii), and the solution to the point (iv) is finalized.

As regards the point (v) of the problem, we may start with the integral from the point (i), and then we write

$$2\zeta(2)\zeta(3) - \zeta(5) = \int_0^1 \frac{\log^2(1-x)\mathrm{Li}_2(x)}{x} dx \overset{x=t^2}{=} 2 \int_0^1 \frac{\log^2(1-t^2)\mathrm{Li}_2(t^2)}{t} dt$$

$$\left\{ \text{employ the dilogarithmic identity, } \mathrm{Li}_2(t) + \mathrm{Li}_2(-t) = \frac{1}{2}\mathrm{Li}_2(t^2), \text{ and expand} \right\}$$

$$\overset{t=x}{=} 4 \int_0^1 \frac{\log(1-x)\log(1+x)\mathrm{Li}_2(x^2)}{x} dx + 2 \int_0^1 \frac{\log^2(1-x)\mathrm{Li}_2(x^2)}{x} dx$$

$$+ 4 \int_0^1 \frac{\log^2(1+x)\mathrm{Li}_2(x)}{x} dx + 4 \int_0^1 \frac{\log^2(1+x)\mathrm{Li}_2(-x)}{x} dx,$$

and since the first integral is found in (1.217), Sect. 1.48, and the second and third integrals are given at the points (iii) and (iv), in the current section, we are able to extract the value of the desired integral,

$$\int_0^1 \frac{\log^2(1+x)\mathrm{Li}_2(-x)}{x} dx$$

$$= \frac{2}{15} \log^5(2) - \frac{2}{3} \log^3(2)\zeta(2) + \frac{7}{4} \log^2(2)\zeta(3) - \frac{1}{8}\zeta(2)\zeta(3) - \frac{125}{32}\zeta(5)$$

$$+ 4 \log(2)\mathrm{Li}_4\left(\frac{1}{2}\right) + 4\mathrm{Li}_5\left(\frac{1}{2}\right),$$

and the solution to the point (v) is finalized.

The integral result at the point (vi) is immediately obtained if we combine the dilogarithmic identity $\text{Li}_2(x) + \text{Li}_2(-x) = \frac{1}{2}\text{Li}_2(x^2)$ and the results from the points (iv) and (v), and then we obtain

$$\int_0^1 \frac{\log^2(1+x)\text{Li}_2(x^2)}{x}\,dx$$

$$= \frac{4}{5}\log^5(2) - \frac{13}{4}\zeta(2)\zeta(3) - \frac{281}{16}\zeta(5) - 4\log^3(2)\zeta(2) + \frac{21}{2}\log^2(2)\zeta(3)$$

$$+ 24\log(2)\text{Li}_4\left(\frac{1}{2}\right) + 24\text{Li}_5\left(\frac{1}{2}\right),$$

and the solution to the point (vi) is finalized.

In the next section, we'll find how to derive one of the key integrals used in the evaluation of the penultimate main integral!

3.48 A *Battle* with Three Challenging Integrals Involving Logarithms and the Dilogarithm

Solution Besides the fact that the given integrals are quite challenging, you might also find pretty unexpected the way to go in this section and prefer to better start with the integral at the point (ii), which plays an important part in the calculation of the harmonic series with a weight 5 structure presented in Sect. 4.33. After finishing the calculation of the second integral, we'll want to prove the series result in Sect. 4.33 and then return to the present section to continue with the evaluation of the first integral. A bit of going *back and forth* between sections!

In the calculation of the second integral, I'll use the strategy presented in my article "On the calculation of two essential harmonic series with a weight 5 structure, involving harmonic numbers of the type H_{2n}" that was published in *Journal of Classical Analysis*, Vol. 16, No. 1, 2020, and it is a critical point in the evaluation process of the series previously mentioned.

So, for the integral at the point (ii) we consider the Cauchy product in (4.23), Sect. 4.5, with x replaced by $-x$, and then we write

$$\int_0^1 \frac{\log(1-x)\log(1+x)\text{Li}_2(-x)}{x}\,dx$$

$$= \int_0^1 \frac{\log(1-x)}{x}\left(2\sum_{n=1}^\infty(-1)^{n-1}x^n\frac{H_n}{n^2} + \sum_{n=1}^\infty(-1)^{n-1}x^n\frac{H_n^{(2)}}{n} - 3\sum_{n=1}^\infty(-1)^{n-1}\frac{x^n}{n^3}\right)dx$$

{reverse the order of summation and integration}

$$= 2\sum_{n=1}^{\infty}(-1)^{n-1}\frac{H_n}{n^2}\int_0^1 x^{n-1}\log(1-x)dx+\sum_{n=1}^{\infty}(-1)^{n-1}\frac{H_n^{(2)}}{n}\int_0^1 x^{n-1}\log(1-x)dx$$

$$- 3\sum_{n=1}^{\infty}(-1)^{n-1}\frac{1}{n^3}\int_0^1 x^{n-1}\log(1-x)dx$$

$$\left\{\text{use that } \int_0^1 x^{n-1}\log(1-x)dx = -\frac{H_n}{n} \text{ (see [76, Chapter 1, p.2])}\right\}$$

$$= 3\sum_{n=1}^{\infty}(-1)^{n-1}\frac{H_n}{n^4} - 2\sum_{n=1}^{\infty}(-1)^{n-1}\frac{H_n^2}{n^3} - \sum_{n=1}^{\infty}(-1)^{n-1}\frac{H_n H_n^{(2)}}{n^2}$$

$$= \frac{123}{32}\zeta(5) + \frac{5}{16}\zeta(2)\zeta(3) + \frac{2}{3}\log^3(2)\zeta(2) - \frac{7}{4}\log^2(2)\zeta(3) - \frac{2}{15}\log^5(2)$$

$$- 4\log(2)\text{Li}_4\left(\frac{1}{2}\right) - 4\text{Li}_5\left(\frac{1}{2}\right),$$

where in the calculations I used that the value of the first series is found in (4.104), Sect. 4.20, and the values of the last two series are given in the book *(Almost) Impossible Integrals, Sums, and Series* as follows:

$$\sum_{n=1}^{\infty}(-1)^{n-1}\frac{H_n^2}{n^3}$$

$$= \frac{2}{15}\log^5(2) - \frac{11}{8}\zeta(2)\zeta(3) - \frac{19}{32}\zeta(5) + \frac{7}{4}\log^2(2)\zeta(3) - \frac{2}{3}\log^3(2)\zeta(2)$$

$$+ 4\log(2)\text{Li}_4\left(\frac{1}{2}\right) + 4\text{Li}_5\left(\frac{1}{2}\right), \tag{3.331}$$

which is given in [76, Chapter 4, p.311], and then

$$\sum_{n=1}^{\infty}(-1)^{n-1}\frac{H_n H_n^{(2)}}{n^2}$$

$$= \frac{23}{8}\zeta(5) - \frac{7}{4}\log^2(2)\zeta(3) + \frac{2}{3}\log^3(2)\zeta(2) + \frac{15}{16}\zeta(2)\zeta(3) - \frac{2}{15}\log^5(2)$$

$$- 4\log(2)\mathrm{Li}_4\left(\frac{1}{2}\right) - 4\mathrm{Li}_5\left(\frac{1}{2}\right), \tag{3.332}$$

which may be found in [76, Chapter 4, p.312], and the solution to the point (ii) of the problem is finalized.

To calculate the integral at the point (i), we exploit the results in (6.175), Sect. 6.23, and (3.327), Sect. 3.46, and then we have

$$\int_0^1 \frac{\log(1-x)\log(1+x)\mathrm{Li}_2(x)}{x}dx$$

$$= -\int_0^1 \sum_{n=1}^\infty x^{2n-1}\left(\frac{H_{2n}-H_n}{n}+\frac{1}{2n^2}\right)\mathrm{Li}_2(x)dx$$

{reverse the order of summation and integration}

$$= -\sum_{n=1}^\infty \left(\frac{H_{2n}-H_n}{n}+\frac{1}{2n^2}\right)\int_0^1 x^{2n-1}\mathrm{Li}_2(x)dx$$

$$= -\sum_{n=1}^\infty \left(\frac{H_{2n}-H_n}{n}+\frac{1}{2n^2}\right)\left(\zeta(2)\frac{1}{2n}-\frac{H_{2n}}{(2n)^2}\right)$$

$$= -2\zeta(2)\sum_{n=1}^\infty \frac{H_{2n}}{(2n)^2}+2\sum_{n=1}^\infty \frac{H_{2n}}{(2n)^4}+2\sum_{n=1}^\infty \frac{H_{2n}^2}{(2n)^3}+\frac{1}{2}\zeta(2)\sum_{n=1}^\infty \frac{H_n}{n^2}$$

$$-\frac{1}{4}\sum_{n=1}^\infty \frac{H_n H_{2n}}{n^3}-\frac{1}{4}\zeta(2)\sum_{n=1}^\infty \frac{1}{n^3}$$

$$\left\{\text{for the first three series use that } \sum_{n=1}^\infty a_{2n}=\frac{1}{2}\left(\sum_{n=1}^\infty a_n-\sum_{n=1}^\infty(-1)^{n-1}a_n\right)\right\}$$

$$= -\frac{1}{2}\zeta(2)\sum_{n=1}^\infty \frac{H_n}{n^2}+\sum_{n=1}^\infty \frac{H_n}{n^4}+\sum_{n=1}^\infty \frac{H_n^2}{n^3}+\zeta(2)\sum_{n=1}^\infty(-1)^{n-1}\frac{H_n}{n^2}-\sum_{n=1}^\infty(-1)^{n-1}\frac{H_n}{n^4}$$

$$-\sum_{n=1}^\infty(-1)^{n-1}\frac{H_n^2}{n^3}-\frac{1}{4}\sum_{n=1}^\infty \frac{H_n H_{2n}}{n^3}-\frac{1}{4}\zeta(2)\zeta(3)$$

$$= \frac{29}{64}\zeta(5)-\frac{5}{8}\zeta(2)\zeta(3),$$

where in the calculations I used that the values of the first two series are particular cases of the Euler sum generalization in (6.149), Sect. 6.19, then the third series may be found both in [76, Chapter 4, p.293] and in my article "A new proof for a classical quadratic harmonic series" that was published in *Journal of Classical Analysis*, Vol. 8, No. 2, 2016 (see [72]), next the fourth and fifth series are particular cases of the Euler sum generalization in (4.105), Sect. 4.21, after that the value of the sixth series is found in (3.331), and finally, the value of the last series is found in (4.124), Sect. 4.33, and the solution to the point (i) of the problem is finalized.

At last, it's easy to observe that the last integral is straightforward to calculate if we consider the use of the dilogarithm function identity, $\text{Li}_2(x) + \text{Li}_2(-x) = \frac{1}{2}\text{Li}_2(x^2)$, together with the results from the points (i) and (ii) of the problem, and then we obtain

$$\int_0^1 \frac{\log(1-x)\log(1+x)\text{Li}_2(x^2)}{x}dx$$

$$= \frac{275}{32}\zeta(5) - \frac{5}{8}\zeta(2)\zeta(3) + \frac{4}{3}\log^3(2)\zeta(2) - \frac{7}{2}\log^2(2)\zeta(3) - \frac{4}{15}\log^5(2)$$

$$- 8\log(2)\text{Li}_4\left(\frac{1}{2}\right) - 8\text{Li}_5\left(\frac{1}{2}\right),$$

which is the desired result, and the point (iii) of the problem is finalized.

Pretty interesting the solution flow of the problem, right? I started with the integral from the point (ii) of the problem, then I jumped to Sect. 6.33 to calculate the series $\sum_{n=1}^{\infty} \frac{H_n H_{2n}}{n^3}$, next I returned to prove the integral result from the point (i) where the previously mentioned series was needed, and finally, the point (iii) was easily derived with the help of a dilogarithmic identity and the integral results from the points (i) and (ii).

I wouldn't be surprised if at first sight you were inclined to call this section *a truly bumpy ride*! Not even mentioning now the advanced alternating harmonic series of weight 5 needed in the extraction process of the desired values!

3.49 Fascinating Polylogarithmic Integrals with Parameter Involving the Cauchy Product of Two Series

Solution Some of the results we discover sometimes are not only *too beautiful to be true*, but they can also be proved by means that make you think exactly the same way, *That's too beautiful to be true!*, and such examples we'll find right in the present section (I strongly assume you will agree with me in the following).

The first two generalized integrals are part of my paper "A simple idea to calculate a class of polylogarithmic integrals by using the Cauchy product of squared Polylogarithm function" in [79, December 4, 2019].

All solutions are based on a very simple idea, but at the same time a counterintuitive one, and very surprising. At the first point, I'll expand the integral in series and then I'll exploit the Cauchy product,

$$(\text{Li}_2(x))^2 = 4 \sum_{n=1}^{\infty} x^n \frac{H_n}{n^3} + 2 \sum_{n=1}^{\infty} x^n \frac{H_n^{(2)}}{n^2} - 6 \sum_{n=1}^{\infty} \frac{x^n}{n^4}, \qquad (3.333)$$

which is already stated in *(Almost) Impossible Integrals, Sums, and Series* (see [76, Chapter 3, p.181]) and we can derive it either by applying the Cauchy product of two series for $(\text{Li}_2(x))^2 = \left(\sum_{n=1}^{\infty} \frac{x^n}{n^2} \right) \left(\sum_{n=1}^{\infty} \frac{x^n}{n^2} \right)$ or by considering the Cauchy product in (4.23), Sect. 4.5, where we replace x by t, multiply both sides by $-2/t$, and then integrate from $t = 0$ to $t = x$.

Now, exploiting geometric series and then the result $\int_0^1 x^{n-1} \text{Li}_2(x) \text{d}x = \zeta(2)\frac{1}{n} - \frac{H_n}{n^2}$, as seen in (3.327), Sect. 3.46, we write

$$\int_0^1 \frac{\log(x)\text{Li}_2(x)}{1 - ax} \text{d}x = \int_0^1 \log(x)\text{Li}_2(x) \sum_{n=1}^{\infty} (ax)^{n-1} \text{d}x$$

{reverse the order of summation and integration}

$$= \sum_{n=1}^{\infty} a^{n-1} \int_0^1 x^{n-1} \log(x)\text{Li}_2(x)\text{d}x = \sum_{n=1}^{\infty} a^{n-1} \frac{d}{dn} \left(\int_0^1 x^{n-1} \text{Li}_2(x)\text{d}x \right)$$

$$= \sum_{n=1}^{\infty} a^{n-1} \frac{d}{dn} \left(\zeta(2)\frac{1}{n} - \frac{\psi(n+1) + \gamma}{n^2} \right)$$

$$= \sum_{n=1}^{\infty} a^{n-1} \left(2\frac{H_n}{n^3} + \frac{H_n^{(2)}}{n^2} - 2\zeta(2)\frac{1}{n^2} \right)$$

{the *magic* happens at this point since we observe we can exploit (3.333)}

$$= \frac{1}{2}\frac{1}{a} \left(\underbrace{4 \sum_{n=1}^{\infty} a^n \frac{H_n}{n^3} + 2 \sum_{n=1}^{\infty} a^n \frac{H_n^{(2)}}{n^2} - 6 \sum_{n=1}^{\infty} \frac{a^n}{n^4}}_{(\text{Li}_2(a))^2} \right) - 2\zeta(2)\frac{1}{a} \sum_{n=1}^{\infty} \frac{a^n}{n^2} + 3\frac{1}{a} \sum_{n=1}^{\infty} \frac{a^n}{n^4}$$

$$= \frac{1}{2} \frac{(\text{Li}_2(a))^2}{a} - 2\zeta(2) \frac{\text{Li}_2(a)}{a} + 3 \frac{\text{Li}_4(a)}{a},$$

and the solution to the point (i) of the problem is complete.

Passing to the point (ii) of the problem we want to proceed similarly as before, and this time we need that

$$(\text{Li}_3(x))^2 = 12 \sum_{n=1}^{\infty} x^n \frac{H_n}{n^5} + 6 \sum_{n=1}^{\infty} x^n \frac{H_n^{(2)}}{n^4} + 2 \sum_{n=1}^{\infty} x^n \frac{H_n^{(3)}}{n^3} - 20 \sum_{n=1}^{\infty} \frac{x^n}{n^6}, \quad (3.334)$$

which may be obtained either by applying the Cauchy product of two series for

$$(\text{Li}_3(x))^2 = \left(\sum_{n=1}^{\infty} \frac{x^n}{n^3} \right) \left(\sum_{n=1}^{\infty} \frac{x^n}{n^3} \right) \text{ or by considering the following Cauchy product}$$

that already appeared in *(Almost) Impossible Integrals, Sums, and Series* (see [76, Chapter 6, p.515]):

$$\text{Li}_2(x)\text{Li}_3(x) = 6 \sum_{n=1}^{\infty} x^n \frac{H_n}{n^4} + 3 \sum_{n=1}^{\infty} x^n \frac{H_n^{(2)}}{n^3} + \sum_{n=1}^{\infty} x^n \frac{H_n^{(3)}}{n^2} - 10 \sum_{n=1}^{\infty} \frac{x^n}{n^5}, \quad (3.335)$$

where if we replace x by t, then multiply both sides by $2/t$ and integrate from $t = 0$ to $t = x$, we get the desired result in (3.334).

Returning to the main result, using again geometric series combined with the fact that $\int_0^1 x^{n-1} \text{Li}_3(x) dx = \zeta(3) \frac{1}{n} - \zeta(2) \frac{1}{n^2} + \frac{H_n}{n^3}$, which is the case $m = 3$ of the result in (3.328), Sect. 3.46, we write

$$\int_0^1 \frac{\log^2(x)\text{Li}_3(x)}{1 - ax} dx = \int_0^1 \log^2(x)\text{Li}_3(x) \sum_{n=1}^{\infty} (ax)^{n-1} dx$$

{reverse the order of summation and integration}

$$= \sum_{n=1}^{\infty} a^{n-1} \int_0^1 x^{n-1} \log^2(x)\text{Li}_3(x) dx = \sum_{n=1}^{\infty} a^{n-1} \frac{d^2}{dn^2} \left(\int_0^1 x^{n-1}\text{Li}_3(x) dx \right)$$

$$= \sum_{n=1}^{\infty} a^{n-1} \frac{d^2}{dn^2} \left(\zeta(3)\frac{1}{n} - \zeta(2)\frac{1}{n^2} + \frac{\psi(n+1) + \gamma}{n^3} \right)$$

$$= \sum_{n=1}^{\infty} a^{n-1} \left(2\frac{H_n^{(3)}}{n^3} + 6\frac{H_n^{(2)}}{n^4} + 12\frac{H_n}{n^5} - 12\zeta(2)\frac{1}{n^4} \right)$$

{rearrange everything in order to exploit the Cauchy product in (3.334)}

$$= \frac{1}{a} \left(\underbrace{12\sum_{n=1}^{\infty} a^n \frac{H_n}{n^5} + 6\sum_{n=1}^{\infty} a^n \frac{H_n^{(2)}}{n^4} + 2\sum_{n=1}^{\infty} a^n \frac{H_n^{(3)}}{n^3} - 20\sum_{n=1}^{\infty} \frac{a^n}{n^6}}_{(\mathrm{Li}_3(a))^2} \right.$$

$$\left. - 12\zeta(2)\frac{1}{a}\sum_{n=1}^{\infty} \frac{a^n}{n^4} + 20\frac{1}{a}\sum_{n=1}^{\infty} \frac{a^n}{n^6} = 20\frac{\mathrm{Li}_6(a)}{a} - 12\zeta(2)\frac{\mathrm{Li}_4(a)}{a} + \frac{(\mathrm{Li}_3(a))^2}{a}, \right.$$

and the solution to the point (ii) of the problem is complete.

Finally, for the last point of the problem we want to act similarly as at the previous two points, and then we first want to prove and use that

$$(\mathrm{Li}_4(x))^2 = 40\sum_{n=1}^{\infty} x^n \frac{H_n}{n^7} + 20\sum_{n=1}^{\infty} x^n \frac{H_n^{(2)}}{n^6} + 8\sum_{n=1}^{\infty} x^n \frac{H_n^{(3)}}{n^5} + 2\sum_{n=1}^{\infty} x^n \frac{H_n^{(4)}}{n^4} - 70\sum_{n=1}^{\infty} \frac{x^n}{n^8}.$$
$$(3.336)$$

Proof Employing the Cauchy product of two series, as seen in (6.15), Sect. 6.5, using that $\mathrm{Li}_4(x) = \sum_{n=1}^{\infty} \frac{x^n}{n^4}$, we have

$$(\mathrm{Li}_4(x))^2 = \left(\sum_{n=1}^{\infty} \frac{x^n}{n^4}\right)\left(\sum_{n=1}^{\infty} \frac{x^n}{n^4}\right) = \sum_{n=1}^{\infty} x^{n+1}\left(\sum_{k=1}^{n} \frac{1}{k^4(n-k+1)^4}\right)$$

$$= \sum_{n=1}^{\infty} x^{n+1}\left(20\sum_{k=1}^{n} \frac{1}{k(n+1)^7} + 20\sum_{k=1}^{n} \frac{1}{(n-k+1)(n+1)^7} + 10\sum_{k=1}^{n} \frac{1}{k^2(n+1)^6}\right.$$

$$+ 10\sum_{k=1}^{n} \frac{1}{(n-k+1)^2(n+1)^6} + 4\sum_{k=1}^{n} \frac{1}{k^3(n+1)^5} + 4\sum_{k=1}^{n} \frac{1}{(n-k+1)^3(n+1)^5}$$

$$\left. + \sum_{k=1}^{n} \frac{1}{k^4(n+1)^4} + \sum_{k=1}^{n} \frac{1}{(n-k+1)^4(n+1)^4}\right)$$

$$= 40\sum_{n=1}^{\infty} x^{n+1}\frac{H_{n+1} - \frac{1}{n+1}}{(n+1)^7} + 20\sum_{n=1}^{\infty} x^{n+1}\frac{H_{n+1}^{(2)} - \frac{1}{(n+1)^2}}{(n+1)^6} + 8\sum_{n=1}^{\infty} x^{n+1}\frac{H_{n+1}^{(3)} - \frac{1}{(n+1)^3}}{(n+1)^5}$$

$$+ 2\sum_{n=1}^{\infty} x^{n+1}\frac{H_{n+1}^{(4)} - \frac{1}{(n+1)^4}}{(n+1)^4}$$

{reindex the series and expand them}

$$= 40 \sum_{n=1}^{\infty} x^n \frac{H_n}{n^7} + 20 \sum_{n=1}^{\infty} x^n \frac{H_n^{(2)}}{n^6} + 8 \sum_{n=1}^{\infty} x^n \frac{H_n^{(3)}}{n^5} + 2 \sum_{n=1}^{\infty} x^n \frac{H_n^{(4)}}{n^4} - 70 \sum_{n=1}^{\infty} \frac{x^n}{n^8},$$

which is the desired auxiliary result. ∎

Coming back to the result at the point (iii) and using the *routine procedure*, as seen above, together with the fact that $\int_0^1 x^{n-1} \text{Li}_4(x) dx = \zeta(2) \frac{1}{n^3} - \zeta(3) \frac{1}{n^2} + \zeta(4) \frac{1}{n} - \frac{H_n}{n^4}$, which is the case $m = 4$ of the result in (3.328), Sect. 3.46, we write

$$\int_0^1 \frac{\log^3(x) \text{Li}_4(x)}{1 - ax} dx = \int_0^1 \log^3(x) \text{Li}_4(x) \sum_{n=1}^{\infty} (ax)^{n-1} dx$$

$$= \sum_{n=1}^{\infty} a^{n-1} \int_0^1 x^{n-1} \log^3(x) \text{Li}_4(x) dx = \sum_{n=1}^{\infty} a^{n-1} \frac{d^3}{dn^3} \left(\int_0^1 x^{n-1} \text{Li}_4(x) dx \right)$$

$$= \sum_{n=1}^{\infty} a^{n-1} \frac{d^3}{dn^3} \left(\zeta(2) \frac{1}{n^3} - \zeta(3) \frac{1}{n^2} + \zeta(4) \frac{1}{n} - \frac{\psi(n+1) + \gamma}{n^4} \right)$$

$$= \sum_{n=1}^{\infty} a^{n-1} \left(120 \frac{H_n}{n^7} + 60 \frac{H_n^{(2)}}{n^6} + 24 \frac{H_n^{(3)}}{n^5} + 6 \frac{H_n^{(4)}}{n^4} - 12\zeta(4) \frac{1}{n^4} - 120\zeta(2) \frac{1}{n^6} \right)$$

{expand and wisely group the series to use (3.336)}

$$= 3\frac{1}{a} \underbrace{\left(40 \sum_{n=1}^{\infty} a^n \frac{H_n}{n^7} + 20 \sum_{n=1}^{\infty} a^n \frac{H_n^{(2)}}{n^6} + 8 \sum_{n=1}^{\infty} a^n \frac{H_n^{(3)}}{n^5} + 2 \sum_{n=1}^{\infty} a^n \frac{H_n^{(4)}}{n^4} - 70 \sum_{n=1}^{\infty} \frac{a^n}{n^8} \right)}_{(\text{Li}_4(a))^2}$$

$$- 12\zeta(4) \frac{1}{a} \sum_{n=1}^{\infty} \frac{a^n}{n^4} - 120\zeta(2) \frac{1}{a} \sum_{n=1}^{\infty} \frac{a^n}{n^6} + 210 \frac{1}{a} \sum_{n=1}^{\infty} \frac{a^n}{n^8}$$

$$= 3 \frac{(\text{Li}_4(a))^2}{a} - 12\zeta(4) \frac{\text{Li}_4(a)}{a} - 120\zeta(2) \frac{\text{Li}_6(a)}{a} + 210 \frac{\text{Li}_8(a)}{a},$$

and the solution to the point (iii) of the problem is complete.

As mentioned in the problem statement, the special cases at the points (iv)–(vi) can be derived by setting $a = 1/2$ in the three generalizations, where we also need the special dilogarithmic and trilogarithmic values, $\text{Li}_2(1/2) = \pi^2/12 - 1/2\log^2(2)$ and $\text{Li}_3(1/2) = 7/8\zeta(3) + 1/6\log^3(2) - 1/2\log(2)\zeta(2)$.

Looking back at how smoothly and beautifully everything has flowed, the whole picture looks like a dream, and happily it is a real dream!

There are two more things to add at the final of this section. One thing is that the curious reader might expect to be able to employ such a strategy for any case of the generalization, $\displaystyle\int_0^1 \frac{\log^n(x)\text{Li}_{n+1}(x)}{1 - ax}dx$, $n \geq 1, n \in \mathbb{N}$.

Numerical experiments suggest that these results can be extended to the whole complex plane except for $a \in (1, \infty)$. However, when $a \in (1, \infty)$, the integrals should be viewed as Cauchy principal values, and there we expect that we need to take the real part of the closed form. The curious reader might also try to attack these integrals by combining the result in (3.45), Sect. 3.10, together with the inverse relation of the Polylogarithm, which is well-known and may be found in [21, Chapter 7, p.192]). To be more specific, to attack differently the integral at the point (i), the curious reader might consider using that $\text{Li}_2(z) = -\displaystyle\int_0^1 \frac{z\log(t)}{1 - zt}dt$, $z \in \mathbb{C} \setminus (1, \infty)$ and $\text{Li}_2(z) + \text{Li}_2(1/z) = -\zeta(2) - \frac{1}{2}\log^2(-z)$, $z \in \mathbb{C} \setminus [0, 1)$.

3.50   A Titan Involving Alternating Harmonic Series of Weight 7

Solution In this section, we continue with another integral, this time involving a combination of logarithms and the Trilogarithm.

As we'll see, the key of the solution I'll present below relies on a beautiful relation with alternating harmonic series of weight 7, which I'll extract by also exploiting results from a previous section. Essentially, to be more precise, we'll want to exploit integral relations given in Sect. 1.46!

So, if we set $m = 3$ in (1.205), Sect. 1.46, we get

$$\int_0^1 \frac{\log(1 - x)\log^3(x)\text{Li}_2(-x)}{x}dx = \frac{45}{8}\zeta(2)\zeta(5) + \frac{39}{4}\zeta(3)\zeta(4)$$

$$-6\sum_{n=1}^{\infty}(-1)^{n-1}\frac{H_n}{n^6} - 6\sum_{n=1}^{\infty}(-1)^{n-1}\frac{H_n^{(2)}}{n^5} - 6\sum_{n=1}^{\infty}(-1)^{n-1}\frac{H_n^{(3)}}{n^4} - 6\sum_{n=1}^{\infty}(-1)^{n-1}\frac{H_n^{(4)}}{n^3}$$

{use the alternating Euler sum in (4.105), the case $m = 3$, Sect. 4.21}

$$= 15\zeta(3)\zeta(4) + \frac{69}{8}\zeta(2)\zeta(5) - \frac{1131}{64}\zeta(7)$$

$$- 6\sum_{n=1}^{\infty}(-1)^{n-1}\frac{H_n^{(2)}}{n^5} - 6\sum_{n=1}^{\infty}(-1)^{n-1}\frac{H_n^{(3)}}{n^4} - 6\sum_{n=1}^{\infty}(-1)^{n-1}\frac{H_n^{(4)}}{n^3}. \qquad (3.337)$$

Next, if we set $m = 3$ in (1.206), Sect. 1.46, we have

$$\int_0^1 \frac{\log(1+x)\log^3(x)\mathrm{Li}_2(x)}{x}dx = -\frac{45}{2}\zeta(2)\zeta(5)-15\zeta(3)\zeta(4)+24\sum_{n=1}^{\infty}(-1)^{n-1}\frac{H_n}{n^6}$$

$$+ 18\sum_{n=1}^{\infty}(-1)^{n-1}\frac{H_n^{(2)}}{n^5} + 12\sum_{n=1}^{\infty}(-1)^{n-1}\frac{H_n^{(3)}}{n^4} + 6\sum_{n=1}^{\infty}(-1)^{n-1}\frac{H_n^{(4)}}{n^3}$$

{use the alternating Euler sum in (4.105), the case $m = 3$, Sect. 4.21}

$$= \frac{1131}{16}\zeta(7) - 36\zeta(3)\zeta(4) - \frac{69}{2}\zeta(2)\zeta(5)$$

$$+ 18\sum_{n=1}^{\infty}(-1)^{n-1}\frac{H_n^{(2)}}{n^5} + 12\sum_{n=1}^{\infty}(-1)^{n-1}\frac{H_n^{(3)}}{n^4} + 6\sum_{n=1}^{\infty}(-1)^{n-1}\frac{H_n^{(4)}}{n^3}. \qquad (3.338)$$

Further, if we set $m = 3$ in (1.207), Sect. 1.46, we arrive at

$$\int_0^1 \frac{\log(1+x)\log^3(x)\mathrm{Li}_2(-x)}{x}dx$$

$$= \frac{567}{32}\zeta(7) - 12\sum_{n=1}^{\infty}(-1)^{n-1}\frac{H_n}{n^6} - 6\sum_{n=1}^{\infty}(-1)^{n-1}\frac{H_n^{(2)}}{n^5}$$

{use the alternating Euler sum in (4.105), the case $m = 3$, Sect. 4.21}

$$= 6\zeta(2)\zeta(5) + \frac{21}{2}\zeta(3)\zeta(4) - \frac{141}{8}\zeta(7) - 6\sum_{n=1}^{\infty}(-1)^{n-1}\frac{H_n^{(2)}}{n^5}. \qquad (3.339)$$

If we plug (3.337), (3.338), and (3.339) in (1.208), with the case $m = 3$, Sect. 1.46, we obtain that

$$\frac{2265}{64}\zeta(7) - \frac{21}{2}\zeta(3)\zeta(4) - \frac{159}{8}\zeta(2)\zeta(5)+6\sum_{n=1}^{\infty}(-1)^{n-1}\frac{H_n^{(2)}}{n^5} +6\sum_{n=1}^{\infty}(-1)^{n-1}\frac{H_n^{(3)}}{n^4}$$

$$= \frac{93}{8}\zeta(2)\zeta(5) + \frac{279}{16}\zeta(3)\zeta(4) - \frac{93}{4}\zeta(7) - \frac{93}{16}\sum_{n=1}^{\infty}\frac{H_n^{(2)}}{n^5}$$

$$-\frac{93}{16}\left(\sum_{n=1}^{\infty}\frac{H_n^{(3)}}{n^4}+\sum_{n=1}^{\infty}\frac{H_n^{(4)}}{n^3}\right)=\frac{465}{16}\zeta(7)-\frac{279}{16}\zeta(2)\zeta(5),$$

whence we get that

$$\sum_{n=1}^{\infty}(-1)^{n-1}\frac{H_n^{(2)}}{n^5}+\sum_{n=1}^{\infty}(-1)^{n-1}\frac{H_n^{(3)}}{n^4}=\frac{13}{32}\zeta(2)\zeta(5)+\frac{7}{4}\zeta(3)\zeta(4)-\frac{135}{128}\zeta(7),$$

$$(3.340)$$

where during the calculations above I used that $\displaystyle\sum_{n=1}^{\infty}\frac{H_n^{(2)}}{n^5}=5\zeta(2)\zeta(5)+$

$2\zeta(3)\zeta(4)-10\zeta(7)$, which is found in [76, Chapter 6, p.389], and $\displaystyle\sum_{n=1}^{\infty}\frac{H_n^{(3)}}{n^4}+$

$\displaystyle\sum_{n=1}^{\infty}\frac{H_n^{(4)}}{n^3}=\zeta(7)+\zeta(3)\zeta(4)$, which is obtained by using (6.102), with $p=3$,
$q=4$ (or $p=4$, $q=3$), and letting $n\to\infty$, Sect. 6.13.

Our next step is to consider the Cauchy product in (4.24), Sect. 4.5, the case with
x replaced by $-x$, that gives

$$\int_0^1\frac{\log(1+x)\log^2(x)\mathrm{Li}_3(-x)}{x}\mathrm{d}x$$

$$=\int_0^1\frac{\log^2(x)}{x}\left(2\sum_{n=1}^{\infty}(-1)^{n-1}x^n\frac{H_n}{n^3}+\sum_{n=1}^{\infty}(-1)^{n-1}x^n\frac{H_n^{(2)}}{n^2}+\sum_{n=1}^{\infty}(-1)^{n-1}x^n\frac{H_n^{(3)}}{n}\right.$$

$$\left.-4\sum_{n=1}^{\infty}(-1)^{n-1}\frac{x^n}{n^4}\right)\mathrm{d}x$$

$$=2\sum_{n=1}^{\infty}(-1)^{n-1}\frac{H_n}{n^3}\int_0^1 x^{n-1}\log^2(x)\mathrm{d}x+\sum_{n=1}^{\infty}(-1)^{n-1}\frac{H_n^{(2)}}{n^2}\int_0^1 x^{n-1}\log^2(x)\mathrm{d}x$$

$$+\sum_{n=1}^{\infty}(-1)^{n-1}\frac{H_n^{(3)}}{n}\int_0^1 x^{n-1}\log^2(x)\mathrm{d}x-4\sum_{n=1}^{\infty}\frac{(-1)^{n-1}}{n^4}\int_0^1 x^{n-1}\log^2(x)\mathrm{d}x$$

$$=4\sum_{n=1}^{\infty}(-1)^{n-1}\frac{H_n}{n^6}+2\left(\sum_{n=1}^{\infty}(-1)^{n-1}\frac{H_n^{(2)}}{n^5}+\sum_{n=1}^{\infty}(-1)^{n-1}\frac{H_n^{(3)}}{n^4}\right)-8\eta(7)$$

$$=\frac{115}{64}\zeta(7)-\frac{19}{16}\zeta(2)\zeta(5),$$

where the last equality is obtained by using that $\eta(s) = (1 - 2^{1-s})\zeta(s)$, the alternating Euler sum in (4.105), the case $m = 3$, Sect. 4.21, and the result in (3.340), and the solution is complete.

Indeed, it is a *non-obvious* way to go, but at the same time it looks so *magical*! Besides, the present solution also answers the challenging question! The curious reader with experience in the calculations with integrals and series might find *trying to get other ways to go here* pretty seductive (or maybe simply irresistible)!

3.51 A Tough Integral Approached by Clever Transformations

Solution One of the most beautiful and challenging integrals I naturally found during the calculations of the advanced harmonic series of weight 5 with integer powers of 2 in the denominator is the present one (although they are not treated in this book, having in hand this integral result and some generating functions found in the fourth chapter is of great help in deriving them all). Depending on the way to go, the calculations may prove to be a hard task. In the following, I'll consider a fast reduction to advanced alternating harmonic series of weight 5 which will be obtained based on the use of a Landen-type identity in the form of a series found in *(Almost) Impossible Integrals, Sums, and Series* (see [76, Chapter 4, p.285]), that is,

$$-\frac{1}{6}\sum_{n=1}^{\infty}\frac{t^{n-1}}{n}(H_n^3 + 3H_n H_n^{(2)} + 2H_n^{(3)}) = \frac{1}{t}\text{Li}_4\left(\frac{t}{t-1}\right). \tag{3.341}$$

If we replace t by $-t$ in (3.341), multiply both sides by -1, and integrate from $t = 0$ to $t = 1$, then for the right-hand side of (3.341) we obtain that

$$\int_0^1 \frac{1}{t}\text{Li}_4\left(\frac{t}{t+1}\right)dt \overset{t/(t+1)=x}{=} \int_0^{1/2}\frac{\text{Li}_4(x)}{x(1-x)}dx = \int_0^{1/2}\frac{\text{Li}_4(x)}{x}dx + \int_0^{1/2}\frac{\text{Li}_4(x)}{1-x}dx$$

$$= \text{Li}_5(x)\Big|_{x=0}^{x=1/2} + \int_0^{1/2}\frac{\text{Li}_4(x)}{1-x}dx = \text{Li}_5\left(\frac{1}{2}\right) + \int_0^{1/2}\frac{\text{Li}_4(x)}{1-x}dx. \tag{3.342}$$

As regards the last integral in (3.342), we want to integrate by parts two times, and then we write

$$\int_0^{1/2}\frac{\text{Li}_4(x)}{1-x}dx = -\int_0^{1/2}(\log(1-x))'\text{Li}_4(x)dx = \underbrace{-\log(1-x)\text{Li}_4(x)\Big|_{x=0}^{x=1/2}}_{\log(2)\text{Li}_4(1/2)}$$

$$+ \int_0^{1/2} \frac{\log(1-x)\mathrm{Li}_3(x)}{x} dx = \log(2)\mathrm{Li}_4\left(\frac{1}{2}\right) - \int_0^{1/2} (\mathrm{Li}_2(x))' \mathrm{Li}_3(x) dx$$

$$= \log(2)\mathrm{Li}_4\left(\frac{1}{2}\right) - \mathrm{Li}_2(x)\mathrm{Li}_3(x)\Big|_{x=0}^{x=1/2} + \int_0^{1/2} \frac{(\mathrm{Li}_2(x))^2}{x} dx$$

$$= \frac{5}{8}\log(2)\zeta(4) + \frac{7}{16}\log^2(2)\zeta(3) - \frac{1}{3}\log^3(2)\zeta(2) - \frac{7}{16}\zeta(2)\zeta(3)$$

$$+ \frac{1}{12}\log^5(2) + \log(2)\mathrm{Li}_4\left(\frac{1}{2}\right) + \int_0^{1/2} \frac{(\mathrm{Li}_2(x))^2}{x} dx, \tag{3.343}$$

where I also considered the special values of the Dilogarithm and Trilogarithm, that is $\mathrm{Li}_2\left(\frac{1}{2}\right) = \frac{1}{2}\left(\frac{\pi^2}{6} - \log^2(2)\right)$ and $\mathrm{Li}_3\left(\frac{1}{2}\right) = \frac{7}{8}\zeta(3) + \frac{1}{6}\log^3(2) - \frac{1}{2}\log(2)\zeta(2)$.

Putting (3.342) and (3.343) together, we get that

$$\int_0^1 \frac{1}{x}\mathrm{Li}_4\left(\frac{x}{x+1}\right) dx$$

$$= \frac{5}{8}\log(2)\zeta(4) + \frac{7}{16}\log^2(2)\zeta(3) - \frac{1}{3}\log^3(2)\zeta(2) - \frac{7}{16}\zeta(2)\zeta(3)$$

$$+ \frac{1}{12}\log^5(2) + \log(2)\mathrm{Li}_4\left(\frac{1}{2}\right) + \mathrm{Li}_5\left(\frac{1}{2}\right) + \int_0^{1/2} \frac{(\mathrm{Li}_2(x))^2}{x} dx. \tag{3.344}$$

Returning to the left-hand side of the result in (3.341), and proceeding similarly as before, we have

$$\frac{1}{6}\int_0^1 \sum_{n=1}^\infty (-1)^{n-1} \frac{t^{n-1}}{n}(H_n^3 + 3H_n H_n^{(2)} + 2H_n^{(3)}) dt$$

$$= \frac{1}{6}\sum_{n=1}^\infty (-1)^{n-1}\frac{1}{n}(H_n^3 + 3H_n H_n^{(2)} + 2H_n^{(3)}) \int_0^1 t^{n-1} dt$$

{integrate and expand the series}

$$= \frac{1}{6}\sum_{n=1}^\infty (-1)^{n-1}\frac{H_n^3}{n^2} + \frac{1}{2}\sum_{n=1}^\infty (-1)^{n-1}\frac{H_n H_n^{(2)}}{n^2} + \frac{1}{3}\sum_{n=1}^\infty (-1)^{n-1}\frac{H_n^{(3)}}{n^2}$$

$$= \frac{1}{6}\log^3(2)\zeta(2) - \frac{7}{16}\log^2(2)\zeta(3) + \frac{27}{32}\zeta(5) + \frac{7}{16}\zeta(2)\zeta(3) - \frac{1}{30}\log^5(2)$$

$$- \log(2)\text{Li}_4\left(\frac{1}{2}\right) - \text{Li}_5\left(\frac{1}{2}\right), \tag{3.345}$$

where I used that for the first alternating harmonic series of weight 5 we have

$$\sum_{n=1}^{\infty}(-1)^{n-1}\frac{H_n^3}{n^2} = \frac{1}{5}\log^5(2) - \log^3(2)\zeta(2) + \frac{21}{8}\log^2(2)\zeta(3) - \frac{27}{16}\zeta(2)\zeta(3) -$$

$$\frac{9}{4}\zeta(5) + 6\log(2)\text{Li}_4\left(\frac{1}{2}\right) + 6\text{Li}_5\left(\frac{1}{2}\right) \text{ (see [76, Chapter 4, p.312]), then the second}$$

series is given in (3.332), Sect. 3.48, and the third one is $\displaystyle\sum_{n=1}^{\infty}(-1)^{n-1}\frac{H_n^{(3)}}{n^2} =$

$\frac{3}{4}\zeta(2)\zeta(3) - \frac{21}{32}\zeta(5)$ ([76, Chapter 4, p.311]).

Combining (3.344) and (3.345) in view of (3.341), we arrive at

$$\int_0^{1/2} \frac{(\text{Li}_2(x))^2}{x}dx$$

$$= \frac{1}{2}\log^3(2)\zeta(2) - \frac{7}{8}\log^2(2)\zeta(3) - \frac{5}{8}\log(2)\zeta(4) + \frac{27}{32}\zeta(5) + \frac{7}{8}\zeta(2)\zeta(3)$$

$$- \frac{7}{60}\log^5(2) - 2\log(2)\text{Li}_4\left(\frac{1}{2}\right) - 2\text{Li}_5\left(\frac{1}{2}\right),$$

and the solution is complete.

Another interesting solution may be found in [38], based on the clever use of the algebraic identities which allow a reduction to simpler integrals. Finding simpler solutions to such integral problems remains an appealing open point!

3.52 An Unexpected Closed Form, Involving Catalan's Constant, of a Nice Integral with the Dilogarithm

Solution The following problem was submitted in April 2021 to R.M.M. (*Romanian Mathematical Magazine*) by Sujeethan Balendran (University of Moratuwa, Sri Lanka). It is worth mentioning that with integration by parts the main integral reduces to the calculation of a known integral in the mathematical literature, that is, $\int_0^1 \frac{x\arctan(x)\log(1-x^2)}{1+x^2}dx$, which appeared in [28]. Although known, it is always a good problem to consider given the difficulties one meets during the calculations. *Well, all fine so far, but how would we actually go here?*

I'll consider a *magical* solution, which is based on a wonderful integral that can be calculated by exploiting the symmetry in three dimensions! So, integrating by parts two times, and rearranging, we get

$$\int_0^1 \frac{1}{1+x^2} \operatorname{Li}_2\left(\frac{1+x^2}{2}\right) dx = \int_0^1 (\arctan(x))' \operatorname{Li}_2\left(\frac{1+x^2}{2}\right) dx$$

$$= \underbrace{\arctan(x)\operatorname{Li}_2\left(\frac{1+x^2}{2}\right)\Big|_{x=0}^{x=1}}_{\pi^3/24} + 2\int_0^1 \frac{x\arctan(x)}{1+x^2}\log\left(\frac{1-x^2}{2}\right) dx = \frac{\pi^3}{24}$$

$$+ \int_0^1 \left(\log\left(\frac{1+x^2}{2}\right)\right)' \arctan(x)\log\left(\frac{1-x^2}{2}\right) dx = \frac{\pi^3}{24}$$

$$+ \underbrace{\log\left(\frac{1+x^2}{2}\right)\arctan(x)\log\left(\frac{1-x^2}{2}\right)\Big|_{x=0}^{x=1}}_{0} - \log^2(2)\underbrace{\int_0^1 \frac{1}{1+x^2}dx}_{\pi/4}$$

$$+ \log(2)\underbrace{\int_0^1 \frac{\log\left(\frac{1-x}{1+x}\right)}{1+x^2}dx}_{-G} + 2\log(2)\underbrace{\int_0^1 \frac{\log(1+x)}{1+x^2}dx}_{\log(2)\pi/8} + \log(2)\int_0^1 \frac{\log(1+x^2)}{1+x^2}dx$$

$$- \int_0^1 \frac{\log(1-x^2)\log(1+x^2)}{1+x^2}dx - 2\int_0^1 \frac{x\arctan(x)}{x^2-1}\log\left(\frac{1+x^2}{2}\right) dx,$$

$$(3.346)$$

where we readily observe that $\int_0^1 \frac{1}{1+x^2}\log\left(\frac{1-x}{1+x}\right) dx \overset{(1-x)/(1+x)=t}{=}$

$$\int_0^1 \frac{\log(t)}{1+t^2}dt = \int_0^1 \log(t)\sum_{n=1}^{\infty}(-1)^{n-1}t^{2n-2}dt = -\sum_{n=1}^{\infty}(-1)^{n-1}\frac{1}{(2n-1)^2} = -G,$$

and $\int_0^1 \frac{\log(1+x)}{1+x^2}dx = \log(2)\frac{\pi}{8}$ is again straightforward with the variable change $(1-x)/(1+x) = t$.

As regards the fourth integral in (3.346), we proceed as follows:

$$\int_0^1 \frac{\log(1+x^2)}{1+x^2}dx = 2\underbrace{\int_0^1 \frac{\log(x)}{1+x^2}dx}_{-G} + \underbrace{\int_0^1 \frac{\log(1+1/x^2)}{1+x^2}dx}_{1/x^2 = y}$$

$$= -2G + \int_1^{\infty} \frac{\log(1+y^2)}{1+y^2}dy = -2G + \int_0^{\infty} \frac{\log(1+y^2)}{1+y^2}dy - \int_0^1 \frac{\log(1+y^2)}{1+y^2}dy,$$

whence the desired integral equals

$$\int_0^1 \frac{\log(1+x^2)}{1+x^2}dx = -G + \frac{1}{2}\int_0^\infty \frac{\log(1+x^2)}{1+x^2}dx \overset{\cot(t)=x}{=} -G - \int_0^{\pi/2}\log(\sin(t))dt$$

$$= \frac{1}{2}\log(2)\pi - G, \tag{3.347}$$

and the last integral from the last equality is evaluated in a footnote in Sect. 3.20.

The fifth integral in (3.346) could also be viewed as a nice separate problem[15] the curious reader might take first before continuing reading the solution I propose. So, using the algebraic identity $ab = 1/4((a+b)^2 - (a-b)^2)$, where we consider $a = \log(1-x^2)$ and $b = \log(1+x^2)$, and rearranging, we write

$$\int_0^1 \frac{\log(1-x^2)\log(1+x^2)}{1+x^2}dx$$

$$= \frac{1}{4}\int_0^1 \left(\frac{\log^2(1-x^4)}{1-x^4} - \frac{x^2\log^2(1-x^4)}{1-x^4}\right)dx - \frac{1}{4}\int_0^1 \frac{1}{1+x^2}\log^2\left(\frac{1-x^2}{1+x^2}\right)dx$$

{in the first integral let $x^4 = t$ and in the second one use $(1-x^2)/(1+x^2) = t^{1/2}$}

$$= \frac{1}{16}\int_0^1 \left(\frac{t^{-3/4}\log^2(1-t)}{1-t} - \frac{t^{-1/4}\log^2(1-t)}{1-t}\right)dt - \frac{1}{64}\int_0^1 \frac{\log^2(t)}{\sqrt{t(1-t)}}dt$$

$$= \frac{1}{16}\lim_{b\to 0^+}\frac{d^2}{db^2}\int_0^1 \left(t^{1/4-1}(1-t)^{b-1} - t^{3/4-1}(1-t)^{b-1}\right)dt$$

$$- \frac{1}{64}\lim_{a\to 1/2}\frac{d^2}{da^2}\int_0^1 t^{a-1}(1-t)^{1/2-1}dt$$

$$\left\{\text{expand first integral and use } B(a,b) = \int_0^1 t^{a-1}(1-t)^{b-1}dt, \ \Re(a), \Re(b) > 0\right\}$$

$$= \frac{1}{16}\lim_{b\to 0^+}\frac{d^2}{db^2}\left(B\left(\frac{1}{4},b\right) - B\left(\frac{3}{4},b\right)\right) - \frac{1}{64}\lim_{a\to 1/2}\frac{d^2}{da^2}B\left(a,\frac{1}{2}\right)$$

$$= \frac{\pi^3}{32} + \log^2(2)\frac{\pi}{2} - 3\log(2)G, \tag{3.348}$$

and the calculation of the limits can be done either with *Mathematica* or manually.

Collecting the results from (3.347), (3.348), and (1.253), Sect. 1.58, in (3.346), we conclude that

---

[15] I first met the integral in 2017, in the form of a problem proposal by Srinivasa Raghava, India, one of the R.M.M. (*Romanian Mathematical Magazine*) proposers. The solution follows the ideas I exploited at that time to put the integral into a nice (and surprising) form involving Beta function.

$$\int_0^{\pi/4} \text{Li}_2\left(\frac{\sec^2(\theta)}{2}\right) d\theta \overset{\tan(\theta)=x}{=} \int_0^1 \frac{1}{1+x^2}\text{Li}_2\left(\frac{1+x^2}{2}\right) dx = \log(2)G,$$

and the first solution is complete.

To get a second solution, I'll exploit an identity I recently found and included in the paper "Two identities with special dilogarithmic values" (see [90, January 15, 2021]), which is also stated in (3.22), Sect. 3.4. So, if we replace x by x^2 in the mentioned identity, divide both sides by $1 + x^2$, and then rearrange and integrate from $x = 0$ to $x = 1$, we get

$$\int_0^1 \frac{1}{1+x^2}\text{Li}_2\left(\frac{1+x^2}{2}\right) dx$$

$$= \frac{1}{4}\int_0^1 \frac{1}{1+x^2}\text{Li}_2\left(\frac{4x^2}{(1+x^2)^2}\right) dx - \frac{1}{2}\int_0^1 \frac{\text{Li}_2(-x^2) - \text{Li}_2(x^2) + \log^2(1+x^2)}{1+x^2} dx$$

$$+ \frac{\pi^3}{48} - \frac{1}{8}\log^2(2)\pi + \log(2)\int_0^1 \frac{\log(1+x^2)}{1+x^2} dx. \tag{3.349}$$

For the first integral in the right-hand side of (3.349), we let the variable change $x = \tan(y/2)$ that gives

$$\int_0^1 \frac{1}{1+x^2}\text{Li}_2\left(\frac{4x^2}{(1+x^2)^2}\right) dx$$

$$= \frac{1}{2}\int_0^{\pi/2} \text{Li}_2(\sin^2(y))dy \overset{\pi/2-y=t}{=} \frac{1}{2}\int_0^{\pi/2} \text{Li}_2(\cos^2(t))dt = \frac{\pi^3}{24} - \frac{1}{2}\log^2(2)\pi, \tag{3.350}$$

where the last integral is calculated in (3.133), Sect. 3.19.

Next, for the second integral in the right-hand side of (3.349) we integrate by parts, and then expand and cleverly rearrange, taking into account some special integral relations with logarithms and the inverse tangent function,

$$\int_0^1 \frac{\text{Li}_2(-x^2) - \text{Li}_2(x^2) + \log^2(1+x^2)}{1+x^2} dx$$

$$= \int_0^1 (\arctan(x))'(\text{Li}_2(-x^2) - \text{Li}_2(x^2) + \log^2(1+x^2))dx = \underbrace{\frac{1}{4}\log^2(2)\pi - \frac{\pi^3}{16}}_{}$$

$$+ \underbrace{\int_0^1 \frac{\arctan(x)\log(1+x^2)}{x}dx - 2\int_0^1 \frac{\arctan(x)\log(1-x)}{x}dx}_{R_1 = \pi^3/16}$$

$$+ 3 \int_0^1 \frac{\arctan(x) \log(1 + x^2)}{x} dx \underbrace{- 2 \int_0^1 \frac{\arctan(x) \log(1 + x)}{x} dx}_{R_2 = 0}$$

$$\underbrace{- 2 \left(\int_0^1 \frac{\arctan(x) \log(1 + x^2)}{x} dx + 2 \int_0^1 \frac{x \arctan(x) \log(1 + x^2)}{1 + x^2} dx \right)}_{R_3 = 2 \log(2)G - 3/8 \log^2(2)\pi - \pi^3/96}$$

$$= \frac{\pi^3}{48} + \log^2(2)\pi - 4 \log(2)G, \tag{3.351}$$

where the last equality is obtained based on the values of R_1, R_2, and R_3 which are beautiful, special relations with integrals given in (3.257), Sect. 3.36, (1.167), Sect. 1.36, and (1.177), Sect. 1.38.

Collecting the results from (3.350), (3.351), and (3.347) in (3.349), we arrive at

$$\int_0^{\pi/4} \text{Li}_2 \left(\frac{\sec^2(\theta)}{2} \right) d\theta \overset{\tan(\theta) = x}{=} \int_0^1 \frac{1}{1 + x^2} \text{Li}_2 \left(\frac{1 + x^2}{2} \right) dx = \log(2)G,$$

and the second solution is complete.

Observe that the first solution also answers the *challenging question*! Besides, we might remark that the panel of results involved in both solutions and how creatively they need to be combined in order to get the desired value is at least fascinating!

3.53 A Group of Six Special, Challenging Generalized Integrals Involving Curious Closed Forms

Solution If you enjoy *discovering mathematics*,[16] then there are those mesmerizing moments we are (very) happy with when we have possibly discovered a problem key step that went unnoticed for us for a long while, and then we finally may obtain a solution. It could be an integral like the ones presented and calculated below!

I first treated the integrals in this section, in the form of particular cases, in the paper "The derivation of eighteen special challenging logarithmic integrals" [83, July 21, 2019]. Later, they were generalized by Ming Hao Zhao in [100], by continuing the strategy on the particular cases presented in the mentioned paper and exploiting the Fourier series of the Bernoulli polynomials. For the generalizations, I'll proceed in a similar style, except that I'll avoid the use of contour integration,

---

[16] This introductory wording might immediately make you think of the celebre quote by the famous French mathematician Siméon Denis Poisson (1781–1840), who allegedly said "Life is good for only two things: *discovering mathematics* and teaching mathematics."

and all auxiliary results involved are presented with full solutions (e.g., the derivation of the Fourier series of the Bernoulli polynomials).

Regarding the first two points of the problem, we consider that $\text{Li}_2\left(\pm\dfrac{2x}{1+x^2}\right) =$

$$\mp 2\int_0^1 \frac{x\log(y)}{1\mp 2xy+x^2}dy, \text{ derived based on } \int_0^1 \frac{y\log^n(x)}{1-yx}dx = (-1)^n n!\text{Li}_{n+1}(y),$$
$y \in (-\infty, 1]$ (see [76, Chapter 1, p.4]), and the result in (6.326), Sect. 6.48, which give

$$\int_0^1 \frac{\log^{2m}(x)}{x}\text{Li}_2\left(\pm\frac{2x}{1+x^2}\right)dx = \mp 2\int_0^1 \log^{2m}(x)\left(\int_0^1 \frac{\log(y)}{1\mp 2xy+x^2}dy\right)dx$$

$$\overset{y=\cos(t)}{=} \mp 2\int_0^1 \log^{2m}(x)\left(\int_0^{\pi/2} \frac{\sin(t)\log(\cos(t))}{1\mp 2x\cos(t)+x^2}dt\right)dx$$

$$= \mp 2\int_0^1 \log^{2m}(x)\left(\int_0^{\pi/2} \log(\cos(t))\sum_{n=1}^{\infty}(\pm 1)^{n-1}x^{n-1}\sin(nt)dt\right)dx$$

{swap the order of summation and integration}

$$= \mp 2\int_0^{\pi/2} \log(\cos(t))\sum_{n=1}^{\infty}(\pm 1)^{n-1}\left(\int_0^1 x^{n-1}\log^{2m}(x)dx\right)\sin(nt)dt$$

$$= \mp 2(2m)!\int_0^{\pi/2} \log(\cos(t))\sum_{n=1}^{\infty}(\pm 1)^{n-1}\frac{\sin(nt)}{n^{2m+1}}dt, \qquad (3.352)$$

where in the calculations I also used that $\displaystyle\int_0^1 x^n\log^m(x)dx = (-1)^m \frac{m!}{(n+1)^{m+1}}$, $m, n \in \mathbb{N}$, found in [76, Chapter 1, p.1], too.

Before going further, we want to prove the following auxiliary result:

$$\mathscr{C}_n = \int_0^{\pi/2} x^n\log(\cos(x))dx$$

$$= -\log(2)\frac{1}{n+1}\left(\frac{\pi}{2}\right)^{n+1} - \sin\left(\frac{\pi}{2}n\right)\frac{1}{2^{n+1}}\left(1-\frac{1}{2^{n+1}}\right)n!\zeta(n+2)$$

$$- \frac{1}{2^{n+1}}n!\sum_{k=0}^{n}\sin\left(\frac{\pi}{2}k\right)\frac{1}{(n-k)!}\pi^{n-k}\zeta(k+2). \qquad (3.353)$$

Proof Since we have $\cos(z) = \dfrac{e^{iz}+e^{-iz}}{2}$, we write that

$$\mathscr{C}_n = \int_0^{\pi/2} x^n \log\left(\frac{1 + e^{i2x}}{2e^{ix}}\right) dx \overset{e^{ix}=y}{=} -(-i)^{n+1} \int_i^1 \frac{\log^n(y)}{y} \log\left(\frac{1 + y^2}{2y}\right) dy$$

{expand the integral, calculate the simple ones, using that $\log(i) = i\pi/2$}

$$= \Re\left\{-(-i)^{n+1} \int_i^1 \frac{\log^n(y) \log(1 + y^2)}{y} dy\right\} - \log(2)\frac{1}{n+1}\left(\frac{\pi}{2}\right)^{n+1}$$

$$\left\{\text{make the variable change } y^2 = t \text{ and use } f_n(t) = \frac{\log^n(t) \log(1 + t)}{t}\right\}$$

$$= \Re\left\{-\frac{(-i)^{n+1}}{2^{n+1}} \lim_{\epsilon \to 0^+} \left(\int_{-1}^{-\epsilon} f_n(t)dt + \int_\epsilon^1 f_n(t)dt\right)\right\} - \log(2)\frac{1}{n+1}\left(\frac{\pi}{2}\right)^{n+1}$$

$$= \Re\left\{-\frac{(-i)^{n+1}}{2^{n+1}} \int_{-1}^0 \frac{\log^n(t) \log(1 + t)}{t}dt - \frac{(-i)^{n+1}}{2^{n+1}} \int_0^1 \frac{\log^n(t) \log(1 + t)}{t}dt\right\}$$

$$- \log(2)\frac{1}{n+1}\left(\frac{\pi}{2}\right)^{n+1}. \tag{3.354}$$

For the first resulting integral in (3.354), we let the variable change $-t = u$ and consider that $\log(-1) = i\pi$, leading to

$$\int_{-1}^0 \frac{\log^n(t) \log(1 + t)}{t}dt = -\int_0^1 \frac{\log^n(-t) \log(1 - t)}{t}dt$$

$$= -\int_0^1 \frac{(i\pi + \log(t))^n \log(1 - t)}{t}dt = \sum_{k=0}^n \binom{n}{k}(i\pi)^{n-k} \int_0^1 \log^k(t) \sum_{j=1}^\infty \frac{t^{j-1}}{j}dt$$

$$= \sum_{k=0}^n \binom{n}{k}(i\pi)^{n-k} \sum_{j=1}^\infty \frac{1}{j} \int_0^1 t^{j-1} \log^k(t)dt = n! \sum_{k=0}^n (-1)^k \frac{1}{(n-k)!}(i\pi)^{n-k} \sum_{j=1}^\infty \frac{1}{j^{k+2}}$$

$$= n! \sum_{k=0}^n (-1)^k \frac{1}{(n-k)!}(i\pi)^{n-k}\zeta(k + 2)$$

$$= n! \sum_{k=0}^n (-1)^k \left(\cos\left(\frac{\pi}{2}(n - k)\right) + i \sin\left(\frac{\pi}{2}(n - k)\right)\right) \frac{1}{(n-k)!}\pi^{n-k}\zeta(k + 2),$$

$$\tag{3.355}$$

where for the last equality I used Euler's formula $e^{i\theta} = \cos(\theta) + i \sin(\theta)$.

Then, for the second resulting integral in (3.354), we have

$$\int_0^1 \frac{\log^n(t)\log(1+t)}{t}dt = \int_0^1 \log^n(t) \sum_{j=1}^{\infty}(-1)^{j-1}\frac{t^{j-1}}{j}dt$$

$$= \sum_{j=1}^{\infty}(-1)^{j-1}\frac{1}{j}\int_0^1 t^{j-1}\log^n(t)dt = (-1)^n n! \sum_{j=1}^{\infty}(-1)^{j-1}\frac{1}{j^{n+2}}$$

$$= (-1)^n n! \left(1 - \frac{1}{2^{n+1}}\right)\zeta(n+2). \qquad (3.356)$$

Upon plugging (3.355) and (3.356) in (3.354), and using Euler's formula previously mentioned, the auxiliary result in (3.353) follows. ∎

Now, we focus on the point (i) of the problem, and returning to (3.352), where we consider the use of the Fourier series in (4.150), the second equality, Sect. 4.47, and combine it with the generalized integral in (3.353), we get

$$\int_0^1 \frac{\log^{2m}(x)}{x}\mathrm{Li}_2\left(\frac{2x}{1+x^2}\right)dx$$

$$= (-1)^m \frac{1}{2m+1} \sum_{k=0}^{2m+1}(2\pi)^k \binom{2m+1}{k}B_k \int_0^{\pi/2} t^{2m-k+1}\log(\cos(t))dt$$

$$= (-1)^m \frac{1}{2m+1} \sum_{k=0}^{2m+1}(2\pi)^k \binom{2m+1}{k}B_k \mathscr{C}_{2m-k+1},$$

and the solution to the point (i) of the problem is complete. To keep the closed form simple I proceeded with the notation in (3.353), and I'll continue to use this style at the next points.

Further, for the point (ii) of the problem we return again to (3.352), where we consider the use of the Fourier series in (4.152), the second equality, Sect. 4.47, and combine it with the generalized integral in (3.353) that gives

$$\int_0^1 \frac{\log^{2m}(x)}{x}\mathrm{Li}_2\left(-\frac{2x}{1+x^2}\right)dx$$

$$= (-1)^{m-1} \frac{1}{2m+1} \sum_{k=0}^{2m+1}\left(1 - \frac{1}{2^{k-1}}\right)(2\pi)^k \binom{2m+1}{k}B_k \int_0^{\pi/2} t^{2m-k+1}\log(\cos(t))dt$$

$$= (-1)^{m-1} \frac{1}{2m+1} \sum_{k=0}^{2m+1}\left(1 - \frac{1}{2^{k-1}}\right)(2\pi)^k \binom{2m+1}{k}B_k \mathscr{C}_{2m-k+1},$$

and the solution to the point (ii) of the problem is complete.

Did you have a chance to pass through Sect. 3.4? We need here the result in (3.17) as well, also employed in *(Almost) Impossible Integrals, Sums, and Series* (see [76, Chapter 3, p.97])! The result can also be written as

$$\int_0^x \frac{t \log(1 \mp t)}{1 + t^2} dt$$

$$= \frac{1}{4} \left(\frac{1}{2} \log^2(1 + x^2) - 2\mathrm{Li}_2(\pm x) + \frac{1}{2}\mathrm{Li}_2\left(-x^2\right) + \mathrm{Li}_2\left(\pm\frac{2x}{1 + x^2}\right) \right).$$

(3.357)

Upon multiplying both sides of (3.357) by $\log^{2m}(x)/x$ and integrating with respect to x, from $x = 0$ to $x = 1$, we get

$$\int_0^1 \frac{\log^{2m}(x)}{x} \left(\int_0^x \frac{t \log(1 \mp t)}{1 + t^2} dt \right) dx = \int_0^1 \frac{t \log(1 \mp t)}{1 + t^2} \left(\int_t^1 \frac{\log^{2m}(x)}{x} dx \right) dt$$

$$\overset{t=x}{=} -\frac{1}{2m + 1} \int_0^1 \frac{x \log^{2m+1}(x) \log(1 \mp x)}{1 + x^2} dx$$

$$= \frac{1}{8} \int_0^1 \frac{\log^2(1 + x^2) \log^{2m}(x)}{x} dx - \frac{1}{2} \int_0^1 \frac{\mathrm{Li}_2(\pm x) \log^{2m}(x)}{x} dx$$

$$+ \frac{1}{8} \int_0^1 \frac{\mathrm{Li}_2\left(-x^2\right) \log^{2m}(x)}{x} dx + \frac{1}{4} \int_0^1 \frac{\log^{2m}(x)}{x} \mathrm{Li}_2\left(\pm\frac{2x}{1 + x^2}\right) dx. \quad (3.358)$$

We need to deal further with the resulting integrals in the rightmost-hand side of (3.358), and then, since upon exploiting the generating function in (4.33), Sect. 4.6, we have the simple fact that $\dfrac{\log^2(1 + x)}{x} = 2\displaystyle\sum_{n=1}^{\infty}(-1)^{n-1}x^n\frac{H_n}{n + 1} =$

$2\displaystyle\sum_{n=1}^{\infty}(-1)^{n-1}x^n\frac{H_{n+1} - 1/(n + 1)}{n + 1} = 2\displaystyle\sum_{n=1}^{\infty}(-1)^{n-1}x^{n-1}\left(\frac{1}{n^2} - \frac{H_n}{n}\right)$, we write

$$\int_0^1 \frac{\log^2(1 + x^2) \log^{2m}(x)}{x} dx = 2 \int_0^1 \log^{2m}(x) \sum_{n=1}^{\infty}(-1)^{n-1}x^{2n-1}\left(\frac{1}{n^2} - \frac{H_n}{n}\right) dx$$

{reverse the order of summation and integration}

$$= 2 \sum_{n=1}^{\infty}(-1)^{n-1}\left(\frac{1}{n^2} - \frac{H_n}{n}\right) \int_0^1 x^{2n-1} \log^{2m}(x) dx$$

$$\left\{ \text{based on [76, Chapter 1, p.1], we have } \int_0^1 x^{2n-1} \log^{2m}(x)dx = \frac{(2m)!}{(2n)^{2m+1}} \right\}$$

$$= \frac{1}{2^{2m}}(2m)! \sum_{n=1}^{\infty}(-1)^{n-1}\frac{1}{n^{2m+3}} - \frac{1}{2^{2m}}(2m)! \sum_{n=1}^{\infty}(-1)^{n-1}\frac{H_n}{n^{2m+2}}$$

$$= \frac{1}{2^{2m+1}}(2m)!\zeta(2m+3) - \frac{1}{2^{2m+1}}(2m+1)!\eta(2m+3)$$

$$+ \frac{1}{2^{2m}}(2m)! \sum_{k=1}^{m} \eta(2k)\zeta(2m-2k+3), \qquad (3.359)$$

where in the calculations I used that $\displaystyle\sum_{n=1}^{\infty}(-1)^{n-1}\frac{1}{n^{2m+3}} = \eta(2m+3)$ and the alternating Euler sum generalization in (4.105), Sect. 4.21.

Next, by similar means as before, we have that

$$\int_0^1 \frac{\text{Li}_2(\pm x) \log^{2m}(x)}{x}dx = \int_0^1 \sum_{n=1}^{\infty}(\pm 1)^n \frac{x^{n-1}}{n^2} \log^{2m}(x)dx$$

$$= \sum_{n=1}^{\infty}(\pm 1)^n \frac{1}{n^2} \int_0^1 x^{n-1} \log^{2m}(x)dx = (2m)! \sum_{n=1}^{\infty}(\pm 1)^n \frac{1}{n^{2m+3}},$$

whence we obtain

$$\int_0^1 \frac{\text{Li}_2(x) \log^{2m}(x)}{x}dx = (2m)!\zeta(2m+3) \qquad (3.360)$$

and

$$\int_0^1 \frac{\text{Li}_2(-x) \log^{2m}(x)}{x}dx = -(2m)!\eta(2m+3). \qquad (3.361)$$

Again, proceeding similarly, we get

$$\int_0^1 \frac{\text{Li}_2\left(-x^2\right) \log^{2m}(x)}{x}dx = \int_0^1 \left(\sum_{n=1}^{\infty}(-1)^n \frac{x^{2n-1}}{n^2}\right) \log^{2m}(x)dx$$

$$= \sum_{n=1}^{\infty}(-1)^n \frac{1}{n^2} \int_0^1 x^{2n-1} \log^{2m}(x)dx = -\frac{1}{2^{2m+1}}(2m)! \sum_{n=1}^{\infty}(-1)^{n-1}\frac{1}{n^{2m+3}}$$

$$= -\frac{1}{2^{2m+1}}(2m)!\eta(2m+3). \tag{3.362}$$

At this point, if we combine (3.358), (3.359), (3.360), (3.362), and the result from the point (i), we obtain that

$$\int_0^1 \frac{x \log^{2m+1}(x) \log(1-x)}{1+x^2} dx$$

$$= \frac{1}{2}\left(1 - \frac{1}{2^{2m+3}}\right)(2m+1)!\zeta(2m+3) + \frac{1}{2^{2m+4}}(2m+2)!\eta(2m+3)$$

$$- \frac{1}{2^{2m+3}}(2m+1)! \sum_{k=1}^m \eta(2k)\zeta(2m-2k+3)$$

$$+ \frac{1}{4}(-1)^{m-1} \sum_{k=0}^{2m+1} (2\pi)^k \binom{2m+1}{k} B_k \mathscr{C}_{2m-k+1},$$

and the solution to the point (iii) of the problem is complete.

Following a similar way to go as before, if we combine (3.358), (3.359), (3.361), (3.362), and the result from the point (ii), we get

$$\int_0^1 \frac{x \log^{2m+1}(x) \log(1+x)}{1+x^2} dx$$

$$= \frac{1}{2^{2m+3}}\left(1 + m - 2^{2m+2}\right)(2m+1)!\eta(2m+3) - \frac{1}{2^{2m+4}}(2m+1)!\zeta(2m+3)$$

$$- \frac{1}{2^{2m+3}}(2m+1)! \sum_{k=1}^m \eta(2k)\zeta(2m-2k+3)$$

$$- \frac{1}{4}(-1)^{m-1} \sum_{k=0}^{2m+1} \left(1 - \frac{1}{2^{k-1}}\right)(2\pi)^k \binom{2m+1}{k} B_k \mathscr{C}_{2m-k+1},$$

and the solution to the point (iv) of the problem is complete.

For the integral at the point (v), we integrate by parts, using the fact that $1/(m+1)(\log^{m+1}(x))' = \log^m(x)/x$, and then we have that

$$\int_0^1 \frac{\log(1-x) \log^{2m}(x) \log(1+x^2)}{x} dx$$

$$= \frac{1}{2m+1} \int_0^1 \frac{\log^{2m+1}(x) \log\left(\frac{1+x^2}{2}\right)}{1-x} dx - \frac{2}{2m+1} \int_0^1 \frac{x \log^{2m+1}(x) \log(1-x)}{1+x^2} dx$$

$$+ \log(2) \frac{1}{2m+1} \int_0^1 \frac{\log^{2m+1}(x)}{1-x} dx$$

$$= \left(1 + 2m + \frac{1}{2^{2m+3}}\right)(2m)! \zeta(2m+3)$$

$$- \log(2)(2m)! \zeta(2m+2) - \frac{1}{2^{2m+2}}(m+1)(2m)! \eta(2m+3)$$

$$+ \frac{1}{2^{2m+2}}(2m)! \sum_{k=1}^{m} \eta(2k)\zeta(2m-2k+3) - \frac{1}{2^{2m+3}}(2m)! \sum_{k=0}^{2m+1} \eta(k+1)\eta(2m-k+2)$$

$$-(2m)! \sum_{k=0}^{2m+1} \beta(k+1)\beta(2m-k+2) - \frac{1}{2}\frac{(-1)^{m-1}}{2m+1} \sum_{k=0}^{2m+1} (2\pi)^k \binom{2m+1}{k} B_k \mathscr{C}_{2m-k+1},$$

where in the calculations I also used the result in (1.132), Sect. 1.27, the one from the point (*iii*), in the current section, and $\int_0^1 \frac{\log^{2m+1}(x)}{1-x} dx =$

$$\int_0^1 \log^{2m+1}(x) \sum_{n=1}^{\infty} x^{n-1} dx = \sum_{n=1}^{\infty} \int_0^1 x^{n-1} \log^{2m+1}(x) dx = -(2m+1)! \sum_{n=1}^{\infty} \frac{1}{n^{2m+2}}$$

$= -(2m+1)! \zeta(2m+2)$, and the solution to the point (*v*) of the problem is complete.

What about the last point of the problem? Well, we prepare to use the result in (3.197), Sect. 3.26. So, we observe the simple fact that

$$\int_0^1 \frac{\log(1+x) \log^{2m}(x) \log(1+x^2)}{x} dx$$

$$= \int_0^1 \frac{\log(1-x^2) \log^{2m}(x) \log(1+x^2)}{x} dx - \int_0^1 \frac{\log(1-x) \log^{2m}(x) \log(1+x^2)}{x} dx$$

$$\left\{\text{in the first integral let } x^2 = y, \text{ and then return to the notation in } x\right\}$$

$$= \frac{1}{2^{2m+1}} \int_0^1 \frac{\log(1-x) \log^{2m}(x) \log(1+x)}{x} dx - \int_0^1 \frac{\log(1-x) \log^{2m}(x) \log(1+x^2)}{x} dx$$

$$= \left(\frac{1}{2^{4m+3}}m + 3\frac{1}{2^{4m+4}} - 5\frac{1}{2^{2m+3}} - 2m - 1\right)(2m)! \zeta(2m+3)$$

$$+ \log(2)(2m)!\zeta(2m+2) + \frac{1}{2^{2m+2}}(m+1)(2m)!\eta(2m+3)$$

$$+ \frac{1}{2^{2m+2}}\left(1 - \frac{1}{2^{2m+1}}\right)(2m)! \sum_{k=1}^{2m} \zeta(k+1)\zeta(2m-k+2)$$

$$-3\frac{1}{2^{2m+2}}(2m)! \sum_{k=1}^{m} \eta(2k)\zeta(2m-2k+3) + \frac{1}{2^{2m+3}}(2m)! \sum_{k=0}^{2m+1} \eta(k+1)\eta(2m-k+2)$$

$$+(2m)! \sum_{k=0}^{2m+1} \beta(k+1)\beta(2m-k+2) + \frac{1}{2}\frac{(-1)^{m-1}}{2m+1} \sum_{k=0}^{2m+1} (2\pi)^k \binom{2m+1}{k} B_k \mathscr{C}_{2m-k+1},$$

where in the calculations I also used the result from the previous point, and the solution to the point (vi) of the problem is complete. Both here and at the previous point slightly different closed forms can be obtained if for some of the finite sums we exploit the symmetry that halves their terms.

At the end of the *voyage* in this section, thinking about obtaining other ways of approaching the problem might be (very) tempting for the curious reader! Such integrals are related to atypical harmonic series, and as an example, note that

$$\int_0^1 \frac{\log(1-x)\log^2(x)\log(1+x^2)}{x}dx$$

$$= \frac{7}{8}\zeta(2)\zeta(3) - \frac{1}{4}\sum_{n=1}^{\infty}(-1)^{n-1}\frac{H_{2n}}{n^4} - \frac{1}{2}\sum_{n=1}^{\infty}(-1)^{n-1}\frac{H_{2n}^{(2)}}{n^3} - \sum_{n=1}^{\infty}(-1)^{n-1}\frac{H_{2n}^{(3)}}{n^2},$$

which may be easily obtained by exploiting (1.125), Sect. 1.24.

3.54 Amazing and (Very) Useful Integral Beasts Involving $\log^2(\sin(x))$, $\log^2(\cos(x))$, $\log^3(\sin(x))$, and $\log^3(\cos(x))$

Solution We prepare to confront now more challenging integrals involving logarithms and trigonometric functions. The integral at the point (i) is a relevant example that also appeared in [25] where it took some years until a first solution emerged. *How about the possibility of getting strategies that simplify everything a lot?*

To pass right to the matter, I'll first exploit the powerful Fourier series presented in Sect. 4.48, and since we'll start with the point (i) of the problem, we need the result in (4.154), Sect. 4.48, where if we multiply both sides by x, integrate from $x = 0$ to $x = \pi/2$, and change the order of summation and integration, we get

$$\int_0^{\pi/2} x\log^2(\sin(x))dx$$

$$= \int_0^{\pi/2} x\left(\log^2(2) + \frac{\pi^2}{12}\right)dx + 2\sum_{n=1}^{\infty}\left(\frac{H_n}{n} - \frac{1}{2n^2} + \log(2)\frac{1}{n}\right)\int_0^{\pi/2} x\cos(2nx)dx$$

$$\left\{\text{use that } \int_0^{\pi/2} x\cos(2nx)dx = -\frac{(-1)^{n-1}+1}{4n^2} \text{ and expand the series}\right\}$$

$$= \frac{3}{4}\log^2(2)\zeta(2) + \frac{15}{16}\zeta(4) - \frac{1}{2}\log(2)\sum_{n=1}^{\infty}\frac{1}{n^3} - \frac{1}{2}\log(2)\sum_{n=1}^{\infty}(-1)^{n-1}\frac{1}{n^3}$$

$$+ \frac{1}{4}\sum_{n=1}^{\infty}\frac{1}{n^4} + \frac{1}{4}\sum_{n=1}^{\infty}(-1)^{n-1}\frac{1}{n^4} - \frac{1}{2}\sum_{n=1}^{\infty}\frac{H_n}{n^3} - \frac{1}{2}\sum_{n=1}^{\infty}(-1)^{n-1}\frac{H_n}{n^3}$$

$$= \frac{1}{24}\log^4(2) + \frac{1}{2}\log^2(2)\zeta(2) - \frac{19}{32}\zeta(4) + \text{Li}_4\left(\frac{1}{2}\right),$$

where in the calculations I also used the Euler sum generalization in (6.149), the case $n = 3$, Sect. 6.19, and the value of the alternating Euler sum in (4.103), Sect. 4.20, and the solution to the point (i) of the problem is complete. For a second solution, a non-obvious one, check and exploit the second solution to the upcoming point.

Next, for the point (ii) of the problem we exploit the previous result involving sine, and if we let the variable change $x = \pi/2 - y$ and expand the integral, we have

$$\int_0^{\pi/2} x\log^2(\sin(x))dx = \frac{\pi}{2}\int_0^{\pi/2}\log^2(\cos(x))dx - \int_0^{\pi/2} x\log^2(\cos(x))dx,$$

from where we get

$$\int_0^{\pi/2} x\log^2(\cos(x))dx = \frac{\pi}{2}\int_0^{\pi/2}\log^2(\cos(x))dx - \int_0^{\pi/2} x\log^2(\sin(x))dx$$

$$= \frac{79}{32}\zeta(4) + \log^2(2)\zeta(2) - \frac{1}{24}\log^4(2) - \text{Li}_4\left(\frac{1}{2}\right),$$

where I used that $\int_0^{\pi/2}\log^2(\cos(x))dx = \frac{1}{2}\log^2(2)\pi + \frac{\pi^3}{24}$, which can also be

extracted from the Fourier series in (4.155), Sect. 4.48, since $\int_0^{\pi/2}\cos(2nx)dx = 0$, $n \geq 1$, or by exploiting (1.57), Sect. 1.14, where we differentiate its both sides twice and let $n \to 0$, and then the value of the integral at the previous point, and the first solution to the point (ii) of the problem is complete.

How about a magical, non-obvious solution? We want to exploit (1.76), Sect. 1.16, the extended version to reals, as in Sect. 3.43, and then we observe that

$$\lim_{n \to 0} \frac{d^2}{dn^2} \left(\int_0^{\pi/2} \theta \cos(n\theta) \cos^n(\theta) d\theta \right) = \int_0^{\pi/2} (-\theta^3 + \theta \log^2(\cos(\theta))) d\theta$$

$$= -\frac{45}{32}\zeta(4) + \int_0^{\pi/2} \theta \log^2(\cos(\theta)) d\theta$$

$$= \lim_{n \to 0} \frac{d^2}{dn^2} \left(\frac{3}{4}\zeta(2)\frac{1}{2^n} - \frac{1}{8}\frac{1}{2^n}((\psi(n+1) + \gamma)^2 + \zeta(2) - \psi^{(1)}(n+1)) \right.$$

$$\left. - \frac{1}{4}\frac{1}{2^n} \int_0^1 \frac{(1 - (1+x)^n) \log(x)}{x} dx \right)$$

$$= \frac{1}{8}\zeta(4) + \frac{3}{4}\log^2(2)\zeta(2) + \frac{1}{2}\log(2)\zeta(3) - \frac{1}{2}\log(2) \int_0^1 \frac{\log(x) \log(1+x)}{x} dx$$

$$+ \frac{1}{4} \int_0^1 \frac{\log(x) \log^2(1+x)}{x} dx,$$

where if we also employ the following integral result, $\displaystyle\int_0^1 \frac{\log(x) \log(1+x)}{x} dx =$

$$\int_0^1 \log(x) \sum_{n=1}^{\infty} (-1)^{n-1} \frac{x^{n-1}}{n} dx = \sum_{n=1}^{\infty} \frac{(-1)^{n-1}}{n} \int_0^1 x^{n-1} \log(x) dx = -\sum_{n=1}^{\infty} (-1)^{n-1}$$

$\displaystyle\frac{1}{n^3} = -\frac{3}{4}\zeta(3)$ together with the fact that $\displaystyle\int_0^1 \frac{\log(t) \log^2(1+t)}{t} dt = \frac{15}{4}\zeta(4) -$

$\displaystyle\frac{7}{2}\log(2)\zeta(3) + \log^2(2)\zeta(2) - \frac{1}{6}\log^4(2) - 4\mathrm{Li}_4\left(\frac{1}{2}\right)$, which is obtained from combining (6.152), (6.153), (6.154), and (6.155) in Sect. 6.20, we arrive at

$$\int_0^{\pi/2} x \log^2(\cos(x)) dx = \frac{79}{32}\zeta(4) + \log^2(2)\zeta(2) - \frac{1}{24}\log^4(2) - \mathrm{Li}_4\left(\frac{1}{2}\right),$$

and the second solution to the point (ii) of the problem is complete.

Also, the calculation of the limit involving the Polygamma function can be done either with *Mathematica* or manually. It is easy to observe now that we may use this integral to derive the one from the previous point, by letting the variable change $\pi/2 - x = y$.

Further, for the point (iii) of the problem we proceed as I did at the first point, and considering the Fourier series in (4.154), Sect. 4.48, we write

$$\int_0^{\pi/2} x^2 \log^2(\sin(x))dx$$

$$= \int_0^{\pi/2} x^2 \left(\log^2(2) + \frac{\pi^2}{12}\right) dx + 2 \sum_{n=1}^{\infty} \left(\frac{H_n}{n} - \frac{1}{2n^2} + \log(2)\frac{1}{n}\right) \int_0^{\pi/2} x^2 \cos(2nx)dx$$

$$\left\{ \text{use that } \int_0^{\pi/2} x^2 \cos(2nx)dx = -\frac{\pi}{4}\frac{(-1)^{n-1}}{n^2} \text{ and expand the series} \right\}$$

$$= \frac{\pi^5}{288} + \frac{1}{24}\log^2(2)\pi^3 - \frac{1}{2}\log(2)\pi \sum_{n=1}^{\infty}(-1)^{n-1}\frac{1}{n^3} + \frac{\pi}{4}\sum_{n=1}^{\infty}(-1)^{n-1}\frac{1}{n^4}$$

$$- \frac{\pi}{2}\sum_{n=1}^{\infty}(-1)^{n-1}\frac{H_n}{n^3} = \frac{1}{24}\log^4(2)\pi + \frac{1}{2}\log(2)\pi\zeta(3) - \frac{3}{320}\pi^5 + \pi \text{Li}_4\left(\frac{1}{2}\right),$$

where in the calculations I also used the value of the alternating Euler sum in (4.103), Sect. 4.20, and the solution to the point (iii) of the problem is complete.

Then, for the point (iv) of the problem we employ the integral result from the previous point, and letting the variable change $x = \pi/2 - y$ and expanding, we get

$$\int_0^{\pi/2} x^2 \log^2(\sin(x))dx$$

$$= \frac{\pi^2}{4} \int_0^{\pi/2} \log^2(\cos(x))dx - \pi \int_0^{\pi/2} x \log^2(\cos(x))dx + \int_0^{\pi/2} x^2 \log^2(\cos(x))dx,$$

from which we obtain that

$$\int_0^{\pi/2} x^2 \log^2(\cos(x))dx$$

$$= \int_0^{\pi/2} x^2 \log^2(\sin(x))dx + \pi \int_0^{\pi/2} x \log^2(\cos(x))dx - \frac{\pi^2}{4} \int_0^{\pi/2} \log^2(\cos(x))dx$$

$$= \frac{11}{1440}\pi^5 + \frac{1}{24}\log^2(2)\pi^3 + \frac{1}{2}\log(2)\pi\zeta(3),$$

where in the calculations I used the value of the integral from the previous point, then the value of the integral from the point (ii), and finally observe that the last integral is calculated in the first solution to the point (ii), and the solution to the point (iv) of the problem is complete. For a second solution, consider (1.65), Sect. 1.15, where if we differentiate three times its both sides and then let $n \to 0$, we arrive at the desired value of the integral. And a first set of integrals with a squared log is finalized!

One of the difficulties in the evaluation process of the next set of integrals is represented by the appearance of an advanced alternating harmonic series of weight 5 during the calculations, but happily it was already treated in my first book, *(Almost) Impossible Integrals, Sums, and Series.*

So, let's exploit again the powerful Fourier series presented in Sect. 4.48, and since we'll start with the point (v) of the problem, we need the result in (4.156) where if we multiply both sides by x, integrate from $x = 0$ to $x = \pi/2$, and change the order of summation and integration, we obtain

$$\int_0^{\pi/2} x \log^3(\sin(x))dx$$

$$= -\int_0^{\pi/2} x \left(\log^3(2) + \frac{1}{4}\log(2)\pi^2 + \frac{3}{2}\zeta(3) \right) dx - \sum_{n=1}^{\infty} \left(3\log^2(2)\frac{1}{n} + \frac{\pi^2}{4}\frac{1}{n} \right.$$

$$\left. - 3\log(2)\frac{1}{n^2} + \frac{3}{2}\frac{1}{n^3} + 6\log(2)\frac{H_n}{n} - 3\frac{H_n}{n^2} + 3\frac{H_n^2}{n} \right) \int_0^{\pi/2} x\cos(2nx)dx$$

$$\left\{ \text{use that } \int_0^{\pi/2} x\cos(2nx)dx = -\frac{(-1)^{n-1}+1}{4n^2} \text{ and expand the series} \right\}$$

$$= -\frac{45}{16}\log(2)\zeta(4) - \frac{3}{4}\log^3(2)\zeta(2) - \frac{9}{8}\zeta(2)\zeta(3)$$

$$+ \frac{3}{4}\left(\log^2(2) + \frac{1}{2}\zeta(2) \right)\sum_{n=1}^{\infty}\frac{1}{n^3} + \frac{3}{4}\left(\log^2(2) + \frac{1}{2}\zeta(2) \right)\sum_{n=1}^{\infty}(-1)^{n-1}\frac{1}{n^3}$$

$$- \frac{3}{4}\log(2)\sum_{n=1}^{\infty}\frac{1}{n^4} - \frac{3}{4}\log(2)\sum_{n=1}^{\infty}(-1)^{n-1}\frac{1}{n^4} + \frac{3}{8}\sum_{n=1}^{\infty}\frac{1}{n^5} + \frac{3}{8}\sum_{n=1}^{\infty}(-1)^{n-1}\frac{1}{n^5}$$

$$+ \frac{3}{2}\log(2)\sum_{n=1}^{\infty}\frac{H_n}{n^3} - \frac{3}{4}\sum_{n=1}^{\infty}\frac{H_n}{n^4} + \frac{3}{4}\sum_{n=1}^{\infty}\frac{H_n^2}{n^3} + \frac{3}{2}\log(2)\sum_{n=1}^{\infty}(-1)^{n-1}\frac{H_n}{n^3}$$

$$- \frac{3}{4}\sum_{n=1}^{\infty}(-1)^{n-1}\frac{H_n}{n^4} + \frac{3}{4}\sum_{n=1}^{\infty}(-1)^{n-1}\frac{H_n^2}{n^3}$$

$$= \frac{57}{32}\log(2)\zeta(4) - \frac{1}{2}\log^3(2)\zeta(2) - \frac{9}{8}\zeta(2)\zeta(3) - \frac{93}{128}\zeta(5)$$

$$- \frac{1}{40}\log^5(2) + 3\text{Li}_5\left(\frac{1}{2}\right),$$

where in the calculations I used the Euler sum generalization in (6.149), the cases $n = 3, 4$, Sect. 6.19, then $\sum_{n=1}^{\infty} \dfrac{H_n^2}{n^3} = \dfrac{7}{2}\zeta(5) - \zeta(2)\zeta(3)$ is given in [76, Chapter 4, p.293] and in my article "A new proof for a classical quadratic harmonic series that was published in *Journal of Classical Analysis*", Vol. 8, No. 2, 2016 (see [72]), next, $\sum_{n=1}^{\infty}(-1)^{n-1}\dfrac{H_n}{n^3}$ and $\sum_{n=1}^{\infty}(-1)^{n-1}\dfrac{H_n}{n^4}$ are both given in (4.103) and (4.104), Sect. 4.20, and finally, $\sum_{n=1}^{\infty}(-1)^{n-1}\dfrac{H_n^2}{n^3} = \dfrac{2}{15}\log^5(2) - \dfrac{11}{8}\zeta(2)\zeta(3) - \dfrac{19}{32}\zeta(5) +$

$\dfrac{7}{4}\log^2(2)\zeta(3) - \dfrac{2}{3}\log^3(2)\zeta(2) + 4\log(2)\mathrm{Li}_4\left(\dfrac{1}{2}\right) + 4\mathrm{Li}_5\left(\dfrac{1}{2}\right)$ is given in (3.331),
Sect. 3.48, and the solution to the point (v) of the problem is complete. If interested in a second solution, a *non-obvious one*, consider the second solution at the next point.

Then, for the point (vi) of the problem we exploit the previous result involving sine, and if we let the variable change $x = \pi/2 - y$ and expand the integral, we get

$$\int_0^{\pi/2} x \log^3(\sin(x))\mathrm{d}x = \frac{\pi}{2}\int_0^{\pi/2}\log^3(\cos(x))\mathrm{d}x - \int_0^{\pi/2} x \log^3(\cos(x))\mathrm{d}x,$$

from which we obtain that

$$\int_0^{\pi/2} x \log^3(\cos(x))\mathrm{d}x = \frac{\pi}{2}\int_0^{\pi/2}\log^3(\cos(x))\mathrm{d}x - \int_0^{\pi/2} x \log^3(\sin(x))\mathrm{d}x$$

$$= \frac{93}{128}\zeta(5) - \frac{9}{8}\zeta(2)\zeta(3) - \log^3(2)\zeta(2) - \frac{237}{32}\log(2)\zeta(4)$$

$$+ \frac{1}{40}\log^5(2) - 3\mathrm{Li}_5\left(\frac{1}{2}\right),$$

where in the calculations I also used that $\int_0^{\pi/2}\log^3(\cos(x))\mathrm{d}x = -\dfrac{1}{2}\log^3(2)\pi -$
$\dfrac{1}{8}\log(2)\pi^3 - \dfrac{3}{4}\pi\zeta(3)$, which is beautifully extracted based on the Fourier series in (4.157), Sect. 4.48, since $\int_0^{\pi/2}\cos(2nx)\mathrm{d}x = 0$, $n \geq 1$, or by exploiting (1.57), Sect. 1.14, where we differentiate its both sides thrice and let $n \to 0$, together with (1.65), Sect. 1.15, where we differentiate twice its both sides and let $n \to 0$, and then the value of the integral at the previous point, and the first solution to the point (vi) of the problem is complete.

Again, how about a magical, non-obvious solution? We can proceed by using the strategy presented in the second solution to the point (ii) of the problem, with some adjustments, and then we write that

$$\lim_{n \to 0} \frac{d^3}{dn^3} \left(\int_0^{\pi/2} \theta \cos(n\theta) \cos^n(\theta) d\theta \right)$$

$$= \int_0^{\pi/2} (-3\theta^3 \log(\cos(\theta)) + \theta \log^3(\cos(\theta))) d\theta$$

$$= -3 \int_0^{\pi/2} \theta^3 \log(\cos(\theta)) d\theta + \int_0^{\pi/2} \theta \log^3(\cos(\theta)) d\theta$$

$$= \lim_{n \to 0} \frac{d^3}{dn^3} \left(\frac{3}{4} \zeta(2) \frac{1}{2^n} - \frac{1}{8} \frac{1}{2^n} ((\psi(n+1) + \gamma)^2 + \zeta(2) - \psi^{(1)}(n+1)) \right.$$

$$\left. - \frac{1}{4} \frac{1}{2^n} \int_0^1 \frac{(1 - (1+x)^n) \log(x)}{x} dx \right)$$

$$= \frac{3}{2} \zeta(2) \zeta(3) - 3\zeta(5) - \frac{3}{8} \log(2) \zeta(4) - \frac{3}{4} \log^2(2) \zeta(3) - \frac{3}{4} \log^3(2) \zeta(2)$$

$$+ \frac{3}{4} \log^2(2) \int_0^1 \frac{\log(x) \log(1+x)}{x} dx - \frac{3}{4} \log(2) \int_0^1 \frac{\log(x) \log^2(1+x)}{x} dx$$

$$+ \frac{1}{4} \int_0^1 \frac{\log(x) \log^3(1+x)}{x} dx,$$

where if we also use that by the known Fourier series, $\log(\cos(\theta)) = -\log(2) +$
$\sum_{n=1}^{\infty} (-1)^{n-1} \frac{\cos(2n\theta)}{n}$, $-\frac{\pi}{2} < \theta < \frac{\pi}{2}$, we obtain that $\int_0^{\pi/2} \theta^3 \log(\cos(\theta)) d\theta =$

$$-\log(2) \int_0^{\pi/2} \theta^3 d\theta + \sum_{n=1}^{\infty} (-1)^{n-1} \frac{1}{n} \int_0^{\pi/2} \theta^3 \cos(2n\theta) d\theta = -\frac{45}{32} \log(2) \zeta(4) -$$

$$\frac{9}{8} \zeta(2) \sum_{n=1}^{\infty} \frac{1}{n^3} + \frac{3}{8} \sum_{n=1}^{\infty} \frac{1}{n^5} + \frac{3}{8} \sum_{n=1}^{\infty} (-1)^{n-1} \frac{1}{n^5} = -\frac{45}{32} \log(2) \zeta(4) - \frac{9}{8} \zeta(2) \zeta(3) +$$

$\frac{93}{128} \zeta(5)$, and then count that the first two integrals after the last equal sign above are given at the end of the second solution to the point (ii), and the third integral is found in (1.198), Sect. 1.45, upon rearranging we obtain the desired result

$$\int_0^{\pi/2} x \log^3(\cos(x)) dx = \frac{93}{128} \zeta(5) - \frac{9}{8} \zeta(2) \zeta(3) - \frac{237}{32} \log(2) \zeta(4) - \log^3(2) \zeta(2)$$

$$+ \frac{1}{40} \log^5(2) - 3\mathrm{Li}_5\left(\frac{1}{2}\right),$$

and the second solution to the point (vi) of the problem is complete, which also answers the *challenging question*. So, now we may exploit this integral to derive the previous one by using the variable change $\pi/2 - x = y$.

Regarding the point (vii) of the problem, we consider the Fourier series in (4.156), Sect. 4.48, as in the case of the point (v), and then we write

$$\int_0^{\pi/2} x^2 \log^3(\sin(x))dx$$

$$= -\int_0^{\pi/2} x^2 \left(\log^3(2) + \frac{1}{4}\log(2)\pi^2 + \frac{3}{2}\zeta(3)\right)dx - \sum_{n=1}^{\infty}\left(3\log^2(2)\frac{1}{n} + \frac{\pi^2}{4}\frac{1}{n}\right.$$

$$\left. - 3\log(2)\frac{1}{n^2} + \frac{3}{2}\frac{1}{n^3} + 6\log(2)\frac{H_n}{n} - 3\frac{H_n}{n^2} + 3\frac{H_n^2}{n}\right)\int_0^{\pi/2} x^2\cos(2nx)dx$$

$$\left\{\text{use that } \int_0^{\pi/2} x^2\cos(2nx)dx = -\frac{\pi}{4}\frac{(-1)^{n-1}}{n^2} \text{ and expand the series}\right\}$$

$$= -\frac{1}{24}\log^3(2)\pi^3 - \frac{1}{96}\log(2)\pi^5 - \frac{1}{16}\pi^3\zeta(3) + \left(\frac{\pi^3}{16} + \frac{3}{4}\log^2(2)\pi\right)\sum_{n=1}^{\infty}(-1)^{n-1}\frac{1}{n^3}$$

$$- \frac{3}{4}\log(2)\pi\sum_{n=1}^{\infty}(-1)^{n-1}\frac{1}{n^4} + \frac{3}{8}\pi\sum_{n=1}^{\infty}(-1)^{n-1}\frac{1}{n^5} + \frac{3}{2}\log(2)\pi\sum_{n=1}^{\infty}(-1)^{n-1}\frac{H_n}{n^3}$$

$$- \frac{3}{4}\pi\sum_{n=1}^{\infty}(-1)^{n-1}\frac{H_n}{n^4} + \frac{3}{4}\pi\sum_{n=1}^{\infty}(-1)^{n-1}\frac{H_n^2}{n^3}$$

$$= \frac{9}{320}\log(2)\pi^5 - \frac{1}{40}\log^5(2)\pi - \frac{1}{8}\pi^3\zeta(3) - \frac{189}{128}\pi\zeta(5)$$

$$- \frac{3}{4}\log^2(2)\pi\zeta(3) + 3\pi\mathrm{Li}_5\left(\frac{1}{2}\right),$$

where $\sum_{n=1}^{\infty}(-1)^{n-1}\dfrac{H_n}{n^3}$ and $\sum_{n=1}^{\infty}(-1)^{n-1}\dfrac{H_n}{n^4}$ are both given in (4.103) and (4.104),

Sect. 4.20, and $\sum_{n=1}^{\infty}(-1)^{n-1}\dfrac{H_n^2}{n^3}$ is given in (3.331), Sect. 3.48, and the point (vii) of the problem is complete.

Finally, for the point $(viii)$ of the problem we employ the integral result from the previous point, and letting the variable change $x = \pi/2 - y$ and expanding, we have

$$\int_0^{\pi/2} x^2 \log^3(\sin(x))dx$$

$$= \frac{\pi^2}{4}\int_0^{\pi/2} \log^3(\cos(x))dx - \pi\int_0^{\pi/2} x\log^3(\cos(x))dx + \int_0^{\pi/2} x^2\log^3(\cos(x))dx,$$

from which we obtain that

$$\int_0^{\pi/2} x^2 \log^3(\cos(x))dx$$

$$= \int_0^{\pi/2} x^2 \log^3(\sin(x))dx + \pi\int_0^{\pi/2} x\log^3(\cos(x))dx - \frac{\pi^2}{4}\int_0^{\pi/2} \log^3(\cos(x))dx$$

$$= -\frac{11}{480}\log(2)\pi^5 - \frac{1}{24}\log^3(2)\pi^3 - \frac{3}{4}\log^2(2)\pi\zeta(3) - \frac{1}{8}\pi^3\zeta(3) - \frac{3}{4}\pi\zeta(5),$$

where in the calculations I used the value of the integral from the previous point, then the value of the integral from the point (vi), and lastly observe that the last integral is calculated in the first solution to the point (vi), and the point $(viii)$ of the problem is complete. For a second solution, we might consider (1.74), Sect. 1.15, where we differentiate two times its both sides and then let $n \to 0$, or the result in (1.75), Sect. 1.15, where we differentiate one time its both sides and then let $n \to 0$.

So pleasant to see how creative we can be when working on such integral problems (and have in hand the proper *tools*). The curious reader might also attack similar integrals to these ones, when the logarithms are replaced by the Polylogarithm function, and such a wonderful example may be found in the penultimate section of the last chapter.

3.55 Four Challenging Integrals with the Logarithm and Trigonometric Functions, Giving Nice Closed Forms

Solution If you had a chance to take a look at Sect. 4.51, in particular at the last two spectacular Fourier series, you probably asked yourself where we would like to employ such results. *Right here, for the proposed integrals!*, I would add.

All four challenging integrals found in this section are *beasts* hard to beat. For example, the integral at the fourth point also appeared in [39], and since 2014 a solution has never been provided.

So, let's start with the first integral! For a first solution, we want to employ the special Fourier series in (4.172), which for simplicity I'll initially denote by

$$\log(\sin(\theta))\log^2(\cos(\theta)) = \frac{1}{4}\zeta(3) - \log^3(2) - \sum_{n=1}^{\infty} f(n)\cos(2n\theta), \text{ and then we have}$$

$$\int_0^{\pi/2} \theta \log(\sin(\theta)) \log^2(\cos(\theta)) d\theta$$

$$= \int_0^{\pi/2} \left(\frac{1}{4}\zeta(3) - \log^3(2)\right)\theta d\theta - \int_0^{\pi/2} \sum_{n=1}^{\infty} f(n)\theta \cos(2n\theta)d\theta$$

$$= \frac{3}{16}\zeta(2)\zeta(3) - \frac{3}{4}\log^3(2)\zeta(2) - \sum_{n=1}^{\infty} f(n)\int_0^{\pi/2} \theta\cos(2n\theta)d\theta$$

$$= \frac{3}{16}\zeta(2)\zeta(3) - \frac{3}{4}\log^3(2)\zeta(2) + \frac{1}{4}\sum_{n=1}^{\infty} \frac{1+(-1)^{n-1}}{n^2}f(n)$$

{take the series and then expand the integral of $f(n)$}

$$= \frac{3}{16}\zeta(2)\zeta(3) - \frac{3}{4}\log^3(2)\zeta(2)$$

$$+ \frac{1}{4}\log^2(2)\underbrace{\int_0^1 \frac{\text{Li}_2(-t)}{t}dt}_{-3/4\zeta(3)} - \frac{1}{4}\log^2(2)\underbrace{\int_0^1 \frac{\text{Li}_2(t)}{t}dt}_{\zeta(3)}$$

$$+ \frac{1}{4}\log(2)\underbrace{\int_0^1 \frac{\log(t)\text{Li}_2(-t)}{t}dt}_{7/8\zeta(4)} - \frac{1}{4}\log(2)\underbrace{\int_0^1 \frac{\log(t)\text{Li}_2(t)}{t}dt}_{-\zeta(4)}$$

$$+ \frac{1}{16}\underbrace{\int_0^1 \frac{\log^2(t)\text{Li}_2(-t)}{t}dt}_{-15/8\zeta(5)} - \frac{1}{16}\underbrace{\int_0^1 \frac{\log^2(t)\text{Li}_2(t)}{t}dt}_{2\zeta(5)}$$

$$+ \frac{1}{2}\log(2)\int_0^1 \frac{\log(1-t)\text{Li}_2(t)}{t}dt - \frac{1}{2}\log(2)\int_0^1 \frac{\log(1-t)\text{Li}_2(-t)}{t}dt$$

$$- \frac{1}{4}\int_0^1 \frac{\log(1-t)\log(t)\text{Li}_2(-t)}{t}dt + \frac{1}{4}\int_0^1 \frac{\log(1-t)\log(t)\text{Li}_2(t)}{t}dt$$

$$+ \frac{1}{2} \int_0^1 \frac{\log(1-t)\log(1+t)\mathrm{Li}_2(-t)}{t}\mathrm{d}t - \frac{1}{2}\int_0^1 \frac{\log(1-t)\log(1+t)\mathrm{Li}_2(t)}{t}\mathrm{d}t$$

$$- \frac{1}{4}\int_0^1 \frac{\log^2(1+t)\mathrm{Li}_2(-t)}{t}\mathrm{d}t + \frac{1}{4}\int_0^1 \frac{\log^2(1+t)\mathrm{Li}_2(t)}{t}\mathrm{d}t, \qquad (3.363)$$

where in the calculations I used the first six integrals are straightforward: the first two by direct integration, the next two derived with one integration by parts, and the fifth and sixth with two integrations by parts.

Then, for the seventh integral we immediately observe that

$$\int_0^1 \frac{\log(1-t)\mathrm{Li}_2(t)}{t}\mathrm{d}t = -\frac{1}{2}(\mathrm{Li}_2(t))^2\Big|_{t=0}^{t=1} = -\frac{5}{4}\zeta(4). \qquad (3.364)$$

Next, for the eighth integral in (3.363), we get

$$\int_0^1 \frac{\log(1-t)\mathrm{Li}_2(-t)}{t}\mathrm{d}t = \int_0^1 \log(1-t)\sum_{n=1}^\infty (-1)^n \frac{t^{n-1}}{n^2}\mathrm{d}t$$

$$= \sum_{n=1}^\infty (-1)^n \frac{1}{n^2}\int_0^1 t^{n-1}\log(1-t)\mathrm{d}t \overset{\text{use (3.10), Sect. 3.3}}{=} \sum_{n=1}^\infty (-1)^{n-1}\frac{H_n}{n^3}$$

$$= \frac{11}{4}\zeta(4) - \frac{7}{4}\log(2)\zeta(3) + \frac{1}{2}\log^2(2)\zeta(2) - \frac{1}{12}\log^4(2) - 2\mathrm{Li}_4\left(\frac{1}{2}\right), \qquad (3.365)$$

where the resulting alternating harmonic series is given in (4.103), Sect. 4.20.

For the ninth integral in (3.363), we consider (1.205), $m = 1$, Sect. 1.46,

$$\int_0^1 \frac{\log(1-t)\log(t)\mathrm{Li}_2(-t)}{t}\mathrm{d}t = \frac{\pi^2}{8}\zeta(3) - \sum_{n=1}^\infty (-1)^{n-1}\frac{H_n}{n^4} - \sum_{n=1}^\infty (-1)^{n-1}\frac{H_n^{(2)}}{n^3}$$

$$= \frac{5}{8}\zeta(2)\zeta(3) - \frac{3}{2}\zeta(5), \qquad (3.366)$$

where the first resulting harmonic series is the case $m = 2$ of the alternating Euler sum generalization in (4.105), Sect. 4.21, and the second one is $\sum_{n=1}^\infty (-1)^{n-1}\frac{H_n^{(2)}}{n^3} = \frac{5}{8}\zeta(2)\zeta(3) - \frac{11}{32}\zeta(5)$ (see [76, Chapter 4, p.311]).

Further, as regards the tenth integral in (3.363), take into account the case $m = 1$ of (1.204), Sect. 1.46, that gives

$$\int_0^1 \frac{\log(1-t)\log(t)\mathrm{Li}_2(t)}{t}dt = 3\zeta(5)-2\zeta(2)\zeta(3)+\sum_{k=1}^{\infty}\frac{H_n^{(2)}}{n^3} = \zeta(2)\zeta(3)-\frac{3}{2}\zeta(5),$$
(3.367)

where in the calculations I used $\displaystyle\sum_{n=1}^{\infty}\frac{H_n^{(2)}}{n^3} = 3\zeta(2)\zeta(3) - \frac{9}{2}\zeta(5)$ (see [76, Chapter 6, p.386])

Finally, if we plug in (3.363) the results from (3.364), (3.365), (3.366), (3.367), (1.216), and (1.215) of Sect. 1.48 and finally (1.213) and (1.212) of Sect. 1.47, we conclude that

$$\int_0^{\pi/2} \theta \log(\sin(\theta))\log^2(\cos(\theta))d\theta$$

$$= \frac{1}{120}\log^5(2) - \frac{5}{6}\log^3(2)\zeta(2) - \frac{49}{32}\log(2)\zeta(4) + \frac{13}{32}\zeta(2)\zeta(3)$$

$$+ \frac{155}{128}\zeta(5) - \mathrm{Li}_5\left(\frac{1}{2}\right),$$

and the first solution to the point (i) is complete.

How does it sound to try to get a second solution? Given the difficulty level of this integral it should be really interesting! As shown in [27], an answer from 2020 presenting my solution, one of the key identities used there was $\log(\sin(\theta))\log(\cos(\theta)) = 1/4\log^2(1/2\sin(2\theta)) - 1/4\log^2(\tan(\theta))$, and this can also be seen by inspecting (6.376) and (6.377), Sect. 6.52. In a similar way, we have the useful identity

$$\log(\sin(\theta))\log^2(\cos(\theta)) = \frac{1}{6}\log^3\left(\frac{1}{2}\sin(2\theta)\right)+\frac{1}{6}\log^3(\tan(\theta))-\frac{1}{3}\log^3(\sin(\theta)).$$
(3.368)

Multiplying both sides of (3.368) by θ and integrating from $\theta = 0$ to $\theta = \pi/2$, we obtain that

$$\int_0^{\pi/2} \theta \log(\sin(\theta))\log^2(\cos(\theta))d\theta = \frac{1}{6}\int_0^{\pi/2} \theta \log^3\left(\frac{1}{2}\sin(2\theta)\right)d\theta$$

$$+ \frac{1}{6}\int_0^{\pi/2} \theta \log^3(\tan(\theta))d\theta - \frac{1}{3}\int_0^{\pi/2} \theta \log^3(\sin(\theta))d\theta.$$
(3.369)

As regards the first integral in the right-hand side of (3.369), we denote by I, we write

$$I = \int_0^{\pi/2} \theta \log^3\left(\frac{1}{2}\sin(2\theta)\right)d\theta \stackrel{\pi/2-\theta=u}{=} \int_0^{\pi/2}\left(\frac{\pi}{2} - u\right)\log^3\left(\frac{1}{2}\sin(2u)\right)du$$

$$= \frac{\pi}{2} \int_0^{\pi/2} \log^3\left(\frac{1}{2}\sin(2u)\right) du - \underbrace{\int_0^{\pi/2} u \log^3\left(\frac{1}{2}\sin(2u)\right) du}_{I},$$

whence we obtain that

$$I = \int_0^{\pi/2} \theta \log^3\left(\frac{1}{2}\sin(2\theta)\right) d\theta = \frac{\pi}{4} \int_0^{\pi/2} \log^3\left(\frac{1}{2}\sin(2\theta)\right) d\theta$$

$$\overset{2\theta=u}{=} \frac{\pi}{8} \int_0^{\pi} \log^3\left(\frac{1}{2}\sin(u)\right) du \overset{\text{exploit the symmetry}}{\underset{\text{over the interval}}{=}} \frac{\pi}{4} \int_0^{\pi/2} \log^3\left(\frac{1}{2}\sin(u)\right) du$$

$$= -\frac{3}{4}\log^3(2)\zeta(2) + \frac{3}{4}\log^2(2)\pi \int_0^{\pi/2} \log(\sin(u)) du$$

$$- \frac{3}{4}\log(2)\pi \int_0^{\pi/2} \log^2(\sin(u)) du + \frac{\pi}{4} \int_0^{\pi/2} \log^3(\sin(u)) du$$

$$= -6\log^3(2)\zeta(2) - \frac{45}{8}\log(2)\zeta(4) - \frac{9}{8}\zeta(2)\zeta(3), \qquad (3.370)$$

where the first remaining log-sine integral is evaluated in a footnote in Sect. 3.20, then the second and third logarithmic integrals are found as cosine versions in the previous section (obtained with the simple variable change, $\pi/2 - u = t$).

Then, for the second integral in the right-hand side of (3.369), we have

$$\int_0^{\pi/2} \theta \log^3(\tan(\theta)) d\theta \overset{\tan(\theta)=\sqrt{u}}{=} \frac{1}{16} \int_0^{\infty} \frac{\arctan(\sqrt{u})}{\sqrt{u}(1+u)} \log^3(u) du$$

$$= \frac{1}{16} \int_0^{\infty} \left(\int_0^1 \frac{\log^3(u)}{(1+u)(1+uv^2)} dv \right) du = \frac{1}{16} \int_0^1 \left(\int_0^{\infty} \frac{\log^3(u)}{(1+u)(1+v^2u)} du \right) dv$$

$$= \frac{1}{16} \int_0^1 \left(\lim_{s\to 0} \frac{d^3}{ds^3} \int_0^{\infty} \frac{u^s}{(1+u)(1+v^2u)} du \right) dv$$

$$= \frac{1}{16} \int_0^1 \frac{1}{1-v^2} \lim_{s\to 0} \frac{d^3}{ds^3} \underbrace{\left(\int_0^{\infty} \frac{u^s}{1+u} du - \int_0^{\infty} \frac{v^2 u^s}{1+v^2 u} du \right)}_{\text{let } v^2 u = t} dv$$

$$= \frac{1}{16} \int_0^1 \frac{1}{1-v^2} \lim_{s\to 0} \frac{d^3}{ds^3} \left((1 - v^{-2s}) \int_0^{\infty} \frac{u^s}{1+u} du \right) dv$$

{consider the integral result in (3.233), Sect. 3.31}

$$= -\frac{\pi}{16} \int_0^1 \frac{1}{1-v^2} \lim_{s \to 0} \frac{d^3}{ds^3} \left(\frac{1-v^{-2s}}{\sin(\pi s)} \right) dv = \frac{\pi^2}{8} \int_0^1 \frac{\log^2(v)}{1-v^2} dv + \frac{1}{4} \int_0^1 \frac{\log^4(v)}{1-v^2} dv$$

$$= \frac{\pi^2}{8} \int_0^1 \log^2(v) \sum_{n=1}^{\infty} v^{2n-2} dv + \frac{1}{4} \int_0^1 \log^4(v) \sum_{n=1}^{\infty} v^{2n-2} dv$$

{reverse the order of summation and integration}

$$= \frac{\pi^2}{8} \sum_{n=1}^{\infty} \int_0^1 v^{2n-2} \log^2(v) dv + \frac{1}{4} \sum_{n=1}^{\infty} \int_0^1 v^{2n-2} \log^4(v) dv = \frac{\pi^2}{4} \sum_{n=1}^{\infty} \frac{1}{(2n-1)^3}$$

$$+ 6 \sum_{n=1}^{\infty} \frac{1}{(2n-1)^5} = \frac{\pi^2}{4} \left(\sum_{n=1}^{\infty} \frac{1}{n^3} - \sum_{n=1}^{\infty} \frac{1}{(2n)^3} \right) + 6 \left(\sum_{n=1}^{\infty} \frac{1}{n^5} - \sum_{n=1}^{\infty} \frac{1}{(2n)^5} \right)$$

$$= \frac{21}{16}\zeta(2) \sum_{n=1}^{\infty} \frac{1}{n^3} + \frac{93}{16} \sum_{n=1}^{\infty} \frac{1}{n^5} = \frac{21}{16}\zeta(2)\zeta(3) + \frac{93}{16}\zeta(5). \qquad (3.371)$$

Collecting (3.370), (3.371), and (1.238), Sect. 1.54, in (3.369), we arrive at

$$\int_0^{\pi/2} \theta \log(\sin(\theta)) \log^2(\cos(\theta)) d\theta$$

$$= \frac{1}{120} \log^5(2) - \frac{5}{6} \log^3(2)\zeta(2) - \frac{49}{32} \log(2)\zeta(4) + \frac{13}{32}\zeta(2)\zeta(3)$$

$$+ \frac{155}{128}\zeta(5) - \operatorname{Li}_5\left(\frac{1}{2}\right),$$

and the second solution to the point (i) is complete. A similar solution was also presented in 2021 in [40].

As regards the integral at the point (ii), we want to let the variable change $\pi/2 - \theta = t$ and exploit the result at the previous point, and then we have

$$\int_0^{\pi/2} \theta \log^2(\sin(\theta)) \log(\cos(\theta)) d\theta = \int_0^{\pi/2} \left(\frac{\pi}{2} - \theta \right) \log^2(\cos(\theta)) \log(\sin(\theta)) d\theta$$

$$= \frac{\pi}{2} \int_0^{\pi/2} \log(\sin(\theta)) \log^2(\cos(\theta)) d\theta - \int_0^{\pi/2} \theta \log(\sin(\theta)) \log^2(\cos(\theta)) d\theta$$

$$= \frac{49}{32} \log(2)\zeta(4) - \frac{2}{3} \log^3(2)\zeta(2) - \frac{1}{120} \log^5(2) - \frac{1}{32}\zeta(2)\zeta(3)$$

$$- \frac{155}{128}\zeta(5) + \text{Li}_5\left(\frac{1}{2}\right),$$

and the solution to the point (ii) is complete. The information related to the fact that $\int_0^{\pi/2} \log(\sin(\theta)) \log^2(\cos(\theta))d\theta = \frac{\pi}{8}\zeta(3) - \log^3(2)\frac{\pi}{2}$ can be extracted from the Fourier series in (4.172), Sect. 4.51, since we have $\int_0^{\pi/2} \cos(2nx)dx = 0$, $n \geq 1$.

Now, I'll jump right to the point (iv) and afterwards return to the point (iii), where I'll use the result at the point (iv). So, for a first solution we could proceed as in the first solution to the point (i) and use the Fourier series in (4.171), Sect. 4.51, where the curious reader might want to continue the calculations.

In the following, I'll present a solution using the strategy in the second solution to the point (i), and then, by exploiting (3.368), where we replace θ by $\pi/2 - \theta$, multiply its both sides by θ^2, and integrate from $\theta = 0$ to $\theta = \pi/2$, we get that

$$\int_0^{\pi/2} \theta^2 \log^2(\sin(\theta)) \log(\cos(\theta))d\theta = \frac{1}{6} \int_0^{\pi/2} \theta^2 \log^3\left(\frac{1}{2}\sin(2\theta)\right)d\theta$$

$$- \frac{1}{6}\int_0^{\pi/2} \theta^2 \log^3(\tan(\theta))d\theta - \frac{1}{3}\int_0^{\pi/2} \theta^2 \log^3(\cos(\theta))d\theta. \qquad (3.372)$$

Regarding the first integral in the right-hand side of (3.372), we write

$$\int_0^{\pi/2} \theta^2 \log^3\left(\frac{1}{2}\sin(2\theta)\right)d\theta \overset{2\theta=t}{=} \frac{1}{8}\int_0^{\pi} t^2 \log^3\left(\frac{1}{2}\sin(t)\right)dt$$

$$= \frac{1}{8}\int_0^{\pi/2} t^2 \log^3\left(\frac{1}{2}\sin(t)\right)dt + \underbrace{\frac{1}{8}\int_{\pi/2}^{\pi} t^2 \log^3\left(\frac{1}{2}\sin(t)\right)dt}_{\text{let } \pi - t = u}$$

$$= \frac{\pi^2}{8}\int_0^{\pi/2} \log^3\left(\frac{1}{2}\sin(t)\right)dt - \frac{\pi}{4}\int_0^{\pi/2} t \log^3\left(\frac{1}{2}\sin(t)\right)dt$$

$$+ \frac{1}{4}\int_0^{\pi/2} t^2 \log^3\left(\frac{1}{2}\sin(t)\right)dt$$

$$= - \log^3(2)\frac{\pi^2}{8}\underbrace{\int_0^{\pi/2} dt}_{\pi/2} + \log^3(2)\frac{\pi}{4}\underbrace{\int_0^{\pi/2} t dt}_{\pi^2/8} - \frac{1}{4}\log^3(2)\underbrace{\int_0^{\pi/2} t^2 dt}_{\pi^3/24}$$

$$+ \frac{3}{8} \log^2(2)\pi^2 \int_0^{\pi/2} \log(\sin(t))dt - \frac{3}{8} \log(2)\pi^2 \int_0^{\pi/2} \log^2(\sin(t))dt$$

$$+ \frac{\pi^2}{8} \int_0^{\pi/2} \log^3(\sin(t))dt - \frac{3}{4} \log^2(2)\pi \int_0^{\pi/2} t \log(\sin(t))dt$$

$$+ \frac{3}{4} \log(2)\pi \int_0^{\pi/2} t \log^2(\sin(t))dt - \frac{\pi}{4} \int_0^{\pi/2} t \log^3(\sin(t))dt$$

$$+ \frac{3}{4} \log^2(2) \int_0^{\pi/2} t^2 \log(\sin(t))dt - \frac{3}{4} \log(2) \int_0^{\pi/2} t^2 \log^2(\sin(t))dt$$

$$+ \frac{1}{4} \int_0^{\pi/2} t^2 \log^3(\sin(t))dt$$

$$= -\frac{13}{480} \log(2)\pi^5 - \frac{1}{3} \log^3(2)\pi^3 - \frac{3}{4}\pi \log^2(2)\zeta(3) - \frac{5}{64}\pi^3\zeta(3) - \frac{3}{16}\pi\zeta(5),$$
$$\tag{3.373}$$

where in the calculations I used that the fourth integral is found evaluated in a footnote of Sect. 3.20, then the fifth and sixth integrals are found as cosine versions in the previous section (so, just use the variable change, $\pi/2 - t = u$), the seventh integral is found in Sect. 3.29, during the solution to the point (v), the eighth and ninth integrals are given in (1.234) and (1.238), Sect. 1.54, the value of the tenth integral may be taken from Sect. 3.29, the calculations to the point (vi), and the values of the last two integrals are found in (1.236) and (1.240), Sect. 1.54.

Further, for the second integral in the right-hand side of (3.372), we might denote by J, we make the variable change $\pi/2 - \theta = t$ that gives

$$J = \int_0^{\pi/2} \theta^2 \log^3(\tan(\theta))d\theta = \int_0^{\pi/2} \left(\frac{\pi}{2} - t\right)^2 \log^3(\cot(t))dt$$

$$= -\frac{\pi^2}{4} \underbrace{\int_0^{\pi/2} \log^3(\tan(t))dt}_{0} + \pi \int_0^{\pi/2} t \log^3(\tan(t))dt - \underbrace{\int_0^{\pi/2} t^2 \log^3(\tan(t))dt}_{J},$$

whence we obtain that

$$J = \int_0^{\pi/2} \theta^2 \log^3(\tan(\theta))d\theta = \frac{\pi}{2} \int_0^{\pi/2} t \log^3(\tan(t))dt$$

$$\overset{\text{use (3.371)}}{=} \frac{7}{64}\pi^3\zeta(3) + \frac{93}{32}\pi\zeta(5), \tag{3.374}$$

where in the calculations I also used that $\int_0^{\pi/2} \log^3(\tan(t))dt = 0$ which is easily seen by using the variable change $\pi/2 - t = u$ together with the fact that the integral has a finite value.[17]

At last, collecting the results from (3.373), (3.374), and (1.241), Sect. 1.54, in (3.372), we arrive at

$$\int_0^{\pi/2} \theta^2 \log^2(\sin(\theta)) \log(\cos(\theta))d\theta$$

$$= \log(2)\frac{\pi^5}{320} - \log^3(2)\frac{\pi^3}{24} + \log^2(2)\frac{\pi}{8}\zeta(3) + \frac{\pi^3}{96}\zeta(3) - \frac{17}{64}\pi\zeta(5),$$

and the solution to the point (iv) is complete.

Returning to the point (iii), we'll want to exploit the previous results, and then, by letting the variable change $\pi/2 - \theta = t$, we have

$$\int_0^{\pi/2} \theta^2 \log(\sin(\theta)) \log^2(\cos(\theta))d\theta = \int_0^{\pi/2} \left(\frac{\pi}{2} - \theta\right)^2 \log^2(\sin(\theta)) \log(\cos(\theta))d\theta$$

$$= \frac{\pi^2}{4}\int_0^{\pi/2} \log^2(\sin(\theta)) \log(\cos(\theta))d\theta - \pi\int_0^{\pi/2} \theta \log^2(\sin(\theta)) \log(\cos(\theta))d\theta$$

$$+ \int_0^{\pi/2} \theta^2 \log^2(\sin(\theta)) \log(\cos(\theta))d\theta$$

$$= \frac{121}{128}\pi\zeta(5) + \frac{3}{64}\pi^3\zeta(3) + \log^5(2)\frac{\pi}{120} - \log^3(2)\frac{\pi^3}{18} - \log(2)\frac{\pi^5}{72}$$

$$+ \log^2(2)\frac{\pi}{8}\zeta(3) - \pi\operatorname{Li}_5\left(\frac{1}{2}\right),$$

where the first resulting integral, when using the variable change $\pi/2 - \theta = t$, leads to $\int_0^{\pi/2} \log^2(\sin(\theta)) \log(\cos(\theta))d\theta = \int_0^{\pi/2} \log(\sin(\theta)) \log^2(\cos(\theta))d\theta = \frac{\pi}{8}\zeta(3) - \log^3(2)\frac{\pi}{2}$, and the latter integral also appears during the solution to the point (ii) above, then the second integral is given at the point (ii), and the last integral has been calculated previously, and the solution to the point (iii) is complete.

---

[17] It is not hard to see that the integral leads to a finite value if we inspect the possible problematic points found at the end of the integration interval and note that the integrand behaves like $\log^3(t)$ near $t = 0$, $(0 < t)$, and like $-\log^3(\pi/2 - t)$ near $t = \pi/2$, $(t < \pi/2)$.

Essentially, each of the points can be attacked by at least two ways (similar to the two ones given at the first point of the problem), which is so pleasant to know given the difficulty of the present integrals!

3.56 Advanced Integrals with Trigonometric Functions, Related to Fourier Series and Harmonic Series

Solution Let me begin by telling you that a beautiful and powerful way of attacking (very) challenging integrals will be outlined in the present section.

The first two points of the problem, which involve classical integrals, are part of the framework needed in the construction of the solutions to the integral results from the other points, besides special Fourier series found in the fourth chapter.

As regards the first point of the problem, I would point out that the given integral is a well-known one in the mathematical literature (see **3.613.2** in [17]). In the following, I'll try to get a somewhat different solution than the popular ones (you'll find the suggested alternative way at the end of the solution below) by forcing an approach involving the use of *Snake Oil Method*, also outlined in [99, pp.126–138].

So, to begin we prepare an auxiliary result, and if we add 1 to both sides of (6.327), Sect. 6.48, and then start the series from $n = 0$, we have

$$\sum_{n=0}^{\infty} x^n \cos(n\theta) = \frac{1 - x\cos(\theta)}{1 - 2x\cos(\theta) + x^2}. \tag{3.375}$$

Next, with the result (3.375) in hand we consider the use of *Snake Oil Method*, and multiplying the main integral by x^n and considering the summation from $n = 0$ to ∞, we obtain that

$$\sum_{n=0}^{\infty} x^n I_n(y) = \sum_{n=0}^{\infty} x^n \left(\int_0^{\pi} \frac{\cos(n\theta)}{1 - 2y\cos(\theta) + y^2} d\theta \right)$$

$$= \int_0^{\pi} \frac{1}{1 - 2y\cos(\theta) + y^2} \sum_{n=0}^{\infty} x^n \cos(n\theta) d\theta$$

$$\overset{\text{use (3.375)}}{=} \int_0^{\pi} \frac{1 - x\cos(\theta)}{(1 - 2x\cos(\theta) + x^2)(1 - 2y\cos(\theta) + y^2)} d\theta$$

$$\overset{\theta/2=t}{=} 2\int_0^{\pi/2} \frac{1 - x\cos(2t)}{(1 - 2x\cos(2t) + x^2)(1 - 2y\cos(2t) + y^2)} dt$$

$$\left\{ \text{consider the trigonometric identity, } 1 + \cos(2t) = 2\cos^2(t) \right\}$$

$$= \int_0^{\pi/2} \frac{((1+x)^2 - 4x\cos^2(t)) + 1 - x^2}{((1+x)^2 - 4x\cos^2(t))((1+y)^2 - 4y\cos^2(t))} dt$$

$$= \frac{1}{(1+y)^2} \int_0^{\pi/2} \frac{1}{1 - 4y/(1+y)^2 \cos^2(t)} dt$$

$$+ \frac{1-x}{(1+x)(1+y)^2} \int_0^{\pi/2} \frac{1}{(1 - 4x/(1+x)^2 \cos^2(t))(1 - 4y/(1+y)^2 \cos^2(t))} dt.$$
$$\tag{3.376}$$

For the resulting integrals in (3.376), recall the elementary trigonometric integral,
$$\int_0^{\pi/2} \frac{dx}{1 - a\sin^2(x)} = \int_0^{\pi/2} \frac{dx}{1 - a\cos^2(x)} = \frac{\pi}{2} \frac{1}{\sqrt{1-a}}, \ a < 1 \text{ (see [76, Chapter}$$
3, p.212]). Thus, for the first remaining integral in (3.376), we have

$$\int_0^{\pi/2} \frac{1}{1 - 4y/(1+y)^2 \cos^2(t)} dt = \frac{\pi}{2} \frac{1+y}{1-y}. \tag{3.377}$$

Further, for the second remaining integral in (3.376), we may assume that $x \in [-a, a]$, where $0 < a < |y| < 1$, and using for simplicity that $f(x) = 4x/(1+x)^2$, we write

$$\int_0^{\pi/2} \frac{1}{(1 - f(x)\cos^2(t))(1 - f(y)\cos^2(t))} dt$$

$$= \int_0^{\pi/2} \left(\frac{f(x)}{f(x) - f(y)} \frac{1}{1 - f(x)\cos^2(t)} + \frac{f(y)}{f(y) - f(x)} \frac{1}{1 - f(y)\cos^2(t)} \right) dt$$

$$= \frac{f(x)}{f(x) - f(y)} \int_0^{\pi/2} \frac{1}{1 - f(x)\cos^2(t)} dt + \frac{f(y)}{f(y) - f(x)} \int_0^{\pi/2} \frac{1}{1 - f(y)\cos^2(t)} dt$$

{exploit the integral result in (3.377)}

$$= \frac{\pi}{2} \frac{(1+x)(1+y)(1+xy)}{(1-x)(1-y)(1-xy)}. \tag{3.378}$$

Returning with the results from (3.377) and (3.378) in (3.376), we get that

$$\sum_{n=0}^{\infty} x^n I_n(y) = \sum_{n=0}^{\infty} x^n \left(\int_0^{\pi} \frac{\cos(n\theta)}{1 - 2y\cos(\theta) + y^2} d\theta \right) = \pi \frac{1}{1-y^2} \cdot \frac{1}{1-xy}$$

$$= \sum_{n=0}^{\infty} x^n \left(\pi \frac{y^n}{1-y^2} \right). \tag{3.379}$$

Inspecting and equating the coefficients of the power series found at the opposite sides of (3.379), we conclude that

$$I_n(y) = \int_0^\pi \frac{\cos(n\theta)}{1 - 2y\cos(\theta) + y^2}d\theta = \pi\frac{y^n}{1 - y^2},$$

and the solution to the point (i) of the problem is complete.

Omran Kouba (Damascus, Syria) is proposing in [31] a more direct solution to a more general version of the integral, that is,

$$\int_0^\pi \frac{\cos(n\theta)}{a^2 - 2ab\cos(\theta) + b^2}d\theta = \frac{\pi}{a^2 - b^2}\left(\frac{b}{a}\right)^n,$$

where n is a positive integer value and it is assumed that $|b| < a$. On the other hand, when $n = 0$ the value of the integral is $\frac{\pi}{a^2 - b^2}$.

The point (ii) is easily solved if we consider the point (i) where we let $\theta/2 = t$, then replace y by $(y-1)/(y+1)$, with the new restriction $y > 0$, and rearrange

$$\int_0^{\pi/2} \frac{(1+y)^2\cos(2n\theta)}{y^2 + 1 - (y^2 - 1)\cos(2\theta)}d\theta = \int_0^{\pi/2} \frac{(1+y)^2\cos(2n\theta)}{1 + \cos(2\theta) + y^2(1 - \cos(2\theta))}d\theta$$

$$\left\{\text{use that } 1 + \cos(2\theta) = 2\cos^2(\theta) \text{ and } 1 - \cos(2\theta) = 2\sin^2(\theta)\right\}$$

$$= \frac{1}{2}(1+y)^2\int_0^{\pi/2} \frac{\cos(2n\theta)}{\cos^2(\theta) + y^2\sin^2(\theta)}d\theta = \frac{\pi}{4}\frac{1}{y}(1+y)^2\left(\frac{y-1}{y+1}\right)^n,$$

whence, based on the last equality above, we arrive at

$$J_n(y) = \int_0^{\pi/2} \frac{\cos(2n\theta)}{\cos^2(\theta) + y^2\sin^2(\theta)}d\theta = \frac{\pi}{2}\frac{1}{y}\left(\frac{y-1}{y+1}\right)^n,$$

and the solution to the point (ii) of the problem is complete.

It is good to know that the first equality of the results from the points (iii)–(v) is immediately obtained by the substitution $\tan(\theta) = x$. We are ready now to pass to the evaluation of the first proposed integral with a Sylvester-type structure! So, if we use the Fourier series in (4.155), Sect. 4.48, that is, $\log^2(\cos(\theta)) = \log^2(2) + \frac{\pi^2}{12} - 2\sum_{n=1}^\infty f(n)\cos(2n\theta)$, where for simplicity I denoted the Fourier series coefficient by $f(n)$, we have that

$$P(y) = \int_0^{\pi/2} \frac{\log^2(\cos(\theta))}{\cos^2(\theta) + y^2\sin^2(\theta)}d\theta$$

$$= \left(\log^2(2) + \frac{\pi^2}{12} \right) \int_0^{\pi/2} \frac{1}{\cos^2(\theta) + y^2 \sin^2(\theta)} d\theta$$

$$- 2 \int_0^{\pi/2} \frac{1}{\cos^2(\theta) + y^2 \sin^2(\theta)} \sum_{n=1}^{\infty} f(n) \cos(2n\theta) d\theta$$

$$\left\{ \text{use that } \int_0^{\pi/2} \frac{1}{\cos^2(\theta) + y^2 \sin^2(\theta)} d\theta = \left. \frac{\arctan(y \tan(\theta))}{y} \right|_{\theta=0}^{\theta=\pi/2} = \frac{\pi}{2} \frac{1}{y} \right\}$$

$$= \left(\frac{1}{2} \log^2(2)\pi + \frac{\pi^3}{24} \right) \frac{1}{y} - 2 \sum_{n=1}^{\infty} f(n) \int_0^{\pi/2} \frac{\cos(2n\theta)}{\cos^2(\theta) + y^2 \sin^2(\theta)} d\theta$$

$$\overset{\text{use } J_n(y)}{=} \left(\frac{1}{2} \log^2(2)\pi + \frac{\pi^3}{24} \right) \frac{1}{y} - \pi \frac{1}{y} \sum_{n=1}^{\infty} \left(\frac{y-1}{y+1} \right)^n f(n)$$

$$= \left(\frac{1}{2} \log^2(2)\pi + \frac{\pi^3}{24} \right) \frac{1}{y} + \log(2)\pi \frac{1}{y} \sum_{n=1}^{\infty} \left(\frac{1-y}{1+y} \right)^n \frac{1}{n} - \frac{\pi}{2} \frac{1}{y} \sum_{n=1}^{\infty} \left(\frac{1-y}{1+y} \right)^n \frac{1}{n^2}$$

$$+ \pi \frac{1}{y} \sum_{n=1}^{\infty} \left(\frac{1-y}{1+y} \right)^n \frac{H_n}{n}$$

$$= \frac{\pi^3}{24} \frac{1}{y} + \frac{\pi}{2} \frac{1}{y} \log^2 \left(1 + \frac{1}{y} \right) + \frac{\pi}{2} \frac{1}{y} \text{Li}_2 \left(\frac{1-y}{1+y} \right),$$

where in the calculations I employed the generating function in (4.34), the first equality, Sect. 4.6, and the solution to the point (iii) of the problem is complete.

Further, we want to continue with similar ideas to the ones presented in the previous solution, and then, using the Fourier series in (4.157), Sect. 4.48, that is,

$$\log^3(\cos(\theta)) = -\log^3(2) - \frac{1}{4} \log(2)\pi^2 - \frac{3}{2}\zeta(3) + \sum_{n=1}^{\infty} g(n) \cos(2n\theta), \text{ where } g(n)$$

is the coefficient of the Fourier series, we write

$$Q(y) = \int_0^{\pi/2} \frac{\log^3(\cos(\theta))}{\cos^2(\theta) + y^2 \sin^2(\theta)} d\theta$$

$$= \left(-\log^3(2) - \frac{1}{4} \log(2)\pi^2 - \frac{3}{2}\zeta(3) \right) \int_0^{\pi/2} \frac{1}{\cos^2(\theta) + y^2 \sin^2(\theta)} d\theta$$

$$+ \int_0^{\pi/2} \frac{1}{\cos^2(\theta) + y^2 \sin^2(\theta)} \sum_{n=1}^{\infty} g(n) \cos(2n\theta) d\theta$$

$$= \left(-\frac{1}{2} \log^3(2)\pi - \frac{1}{8} \log(2)\pi^3 - \frac{3}{4}\pi\zeta(3) \right) \frac{1}{y}$$

$$+ \sum_{n=1}^{\infty} g(n) \int_0^{\pi/2} \frac{\cos(2n\theta)}{\cos^2(\theta) + y^2 \sin^2(\theta)} d\theta$$

$$\overset{\text{use } J_n(y)}{=} \left(-\frac{1}{2} \log^3(2)\pi - \frac{1}{8} \log(2)\pi^3 - \frac{3}{4}\pi\zeta(3) \right) \frac{1}{y}$$

$$- \left(\frac{3}{2} \log^2(2)\pi + \frac{\pi^3}{8} \right) \frac{1}{y} \sum_{n=1}^{\infty} \left(\frac{1-y}{1+y} \right)^n \frac{1}{n} + \frac{3}{2} \log(2)\pi \frac{1}{y} \sum_{n=1}^{\infty} \left(\frac{1-y}{1+y} \right)^n \frac{1}{n^2}$$

$$- \frac{3}{4}\pi \frac{1}{y} \sum_{n=1}^{\infty} \left(\frac{1-y}{1+y} \right)^n \frac{1}{n^3} - 3 \log(2)\pi \frac{1}{y} \sum_{n=1}^{\infty} \left(\frac{1-y}{1+y} \right)^n \frac{H_n}{n}$$

$$+ \frac{3}{2}\pi \frac{1}{y} \sum_{n=1}^{\infty} \left(\frac{1-y}{1+y} \right)^n \frac{H_n}{n^2} - \frac{3}{2}\pi \frac{1}{y} \sum_{n=1}^{\infty} \left(\frac{1-y}{1+y} \right)^n \frac{H_n^2}{n}$$

$$= \frac{\pi}{8} \left(6\zeta(3) - \log(2)\pi^2 - 4 \log^3(2) \right) \frac{1}{y} + \frac{3}{8}\pi (4 \log^2(2) + \pi^2) \frac{1}{y} \log \left(\frac{2y}{1+y} \right)$$

$$- \frac{3}{4} \log(2)\pi \frac{1}{y} \log^2 \left(\frac{2y}{1+y} \right) - \frac{\pi}{4} \frac{1}{y} \log^3 \left(\frac{2y}{1+y} \right) + \frac{3}{4}\pi \frac{\log(y)}{y} \log^2 \left(\frac{2y}{1+y} \right)$$

$$- \frac{3}{4}\pi \frac{\log(1-y)}{y} \log^2 \left(\frac{2y}{1+y} \right) - \frac{3}{2} \log(2)\pi \frac{1}{y} \mathrm{Li}_2 \left(\frac{1-y}{1+y} \right) - \frac{3}{4}\pi \frac{1}{y} \mathrm{Li}_3 \left(\frac{1-y}{1+y} \right)$$

$$- \frac{3}{2}\pi \frac{1}{y} \mathrm{Li}_3 \left(\frac{2y}{1+y} \right),$$

where during the calculations I used the generating functions in (4.34), the first equality, (4.36), the first equality, Sect. 4.6, and (4.42), Sect. 4.7, and the solution to the point (*iv*) of the problem is complete. A particular case of this integral appeared in [58, February 4, 2022], and for about a year no solution to explain *Mathematica* result in the post was given.

For the last point, we proceed similarly as before, and then we consider the Fourier series in (4.159), Sect. 4.48, where to keep things simple we may write it in the form $\log^4(\cos(\theta)) = \log^4(2) + \frac{1}{2} \log^2(2)\pi^2 + 6 \log(2)\zeta(3) + \frac{19}{240}\pi^4 -$

$\sum_{n=1}^{\infty} h(n) \cos(2n\theta)$, where $h(n)$ is the coefficient of the Fourier series. So, we have

$$R(y) = \int_0^{\pi/2} \frac{\log^4(\cos(\theta))}{\cos^2(\theta) + y^2 \sin^2(\theta)} d\theta$$

$$= \left(\log^4(2) + \frac{1}{2}\log^2(2)\pi^2 + 6\log(2)\zeta(3) + \frac{19}{240}\pi^4\right) \int_0^{\pi/2} \frac{d\theta}{\cos^2(\theta) + y^2 \sin^2(\theta)}$$

$$- \int_0^{\pi/2} \frac{1}{\cos^2(\theta) + y^2 \sin^2(\theta)} \sum_{n=1}^{\infty} h(n) \cos(2n\theta) d\theta$$

$$= \left(\frac{1}{2}\log^4(2)\pi + \frac{1}{4}\log^2(2)\pi^3 + 3\log(2)\pi\zeta(3) + \frac{19}{480}\pi^5\right) \frac{1}{y}$$

$$- \sum_{n=1}^{\infty} h(n) \int_0^{\pi/2} \frac{\cos(2n\theta)}{\cos^2(\theta) + y^2 \sin^2(\theta)} d\theta$$

$$\stackrel{\text{use } J_n(y)}{=} \left(\frac{1}{2}\log^4(2)\pi + \frac{1}{4}\log^2(2)\pi^3 + 3\log(2)\pi\zeta(3) + \frac{19}{480}\pi^5\right) \frac{1}{y}$$

$$+ \left(2\log^3(2)\pi + \frac{1}{2}\log(2)\pi^3 + 3\pi\zeta(3)\right) \frac{1}{y} \sum_{n=1}^{\infty} \left(\frac{1-y}{1+y}\right)^n \frac{1}{n}$$

$$- \left(3\log^2(2)\pi + \frac{\pi^3}{4}\right) \frac{1}{y} \sum_{n=1}^{\infty} \left(\frac{1-y}{1+y}\right)^n \frac{1}{n^2} + 3\log(2)\pi \frac{1}{y} \sum_{n=1}^{\infty} \left(\frac{1-y}{1+y}\right)^n \frac{1}{n^3}$$

$$- \frac{3}{2}\pi \frac{1}{y} \sum_{n=1}^{\infty} \left(\frac{1-y}{1+y}\right)^n \frac{1}{n^4} + \left(6\log^2(2)\pi + \frac{\pi^3}{2}\right) \frac{1}{y} \sum_{n=1}^{\infty} \left(\frac{1-y}{1+y}\right)^n \frac{H_n}{n}$$

$$- 6\log(2)\pi \frac{1}{y} \sum_{n=1}^{\infty} \left(\frac{1-y}{1+y}\right)^n \frac{H_n}{n^2} + 3\pi \frac{1}{y} \sum_{n=1}^{\infty} \left(\frac{1-y}{1+y}\right)^n \frac{H_n}{n^3}$$

$$+ 6\log(2)\pi \frac{1}{y} \sum_{n=1}^{\infty} \left(\frac{1-y}{1+y}\right)^n \frac{H_n^2}{n} - 3\pi \frac{1}{y} \sum_{n=1}^{\infty} \left(\frac{1-y}{1+y}\right)^n \frac{H_n^2}{n^2}$$

$$+ 2\pi \frac{1}{y} \sum_{n=1}^{\infty} \left(\frac{1-y}{1+y}\right)^n \frac{H_n^3}{n} + \pi \frac{1}{y} \sum_{n=1}^{\infty} \left(\frac{1-y}{1+y}\right)^n \frac{H_n^{(3)}}{n}$$

$$= \frac{\pi}{96}(7\pi^4 + 24\log^2(2)\pi^2 + 48\log^4(2) - 288\log(2)\zeta(3))\frac{1}{y}$$

$$- \frac{1}{2}\log(2)\pi(4\log^2(2) + 3\pi^2)\frac{1}{y}\log\left(\frac{2y}{1+y}\right) + \frac{\pi^3}{2}\frac{1}{y}\log^2\left(\frac{2y}{1+y}\right)$$

$$- 3\log(2)\pi\frac{\log(y)}{y}\log^2\left(\frac{2y}{1+y}\right) + 3\log(2)\pi\frac{\log(1-y)}{y}\log^2\left(\frac{2y}{1+y}\right)$$

$$+ \frac{3}{2}\log(2)\pi\frac{1}{y}\log^3\left(\frac{2y}{1+y}\right) + \frac{\pi}{2}\frac{\log(y)}{y}\log^3\left(\frac{2y}{1+y}\right) - \frac{\pi}{2}\frac{\log(1-y)}{y}\log^3\left(\frac{2y}{1+y}\right)$$

$$- \frac{\pi}{8}\frac{1}{y}\log^4\left(\frac{2y}{1+y}\right) + \frac{\pi}{4}(12\log^2(2) + \pi^2)\frac{1}{y}\mathrm{Li}_2\left(\frac{1-y}{1+y}\right) + 3\log(2)\pi\frac{1}{y}\mathrm{Li}_3\left(\frac{1-y}{1+y}\right)$$

$$+ 6\log(2)\pi\frac{1}{y}\mathrm{Li}_3\left(\frac{2y}{1+y}\right) - \frac{3}{2}\pi\frac{1}{y}\mathrm{Li}_4\left(\frac{1-y}{1+y}\right) - 3\pi\frac{1}{y}\mathrm{Li}_4\left(\frac{2y}{1+y}\right)$$

$$- 3\pi\frac{1}{y}\mathrm{Li}_4\left(\frac{y-1}{2y}\right),$$

where in the calculations I used the generating functions in (4.34), the first equality, (4.36), the first equality, (4.38), Sect. 4.6, then (4.42), (4.44), and (4.50), Sect. 4.7, and finally (4.56), Sect. 4.8, and the solution to the point (v) of the problem is complete.

Important to know: The way presented at the last three points of the problem can be used for evaluating many exotic, challenging integrals. For example, one can also consider the use of the Fourier series found in Sects. 4.49, 4.50, and 4.51 in order to derive more results. Below I'll provide another integral, as an example, which the curious reader might like to check:

$$T(y) = \int_0^{\pi/2} \frac{\mathrm{Li}_2(\cos^2(\theta))}{\cos^2(\theta) + y^2\sin^2(\theta)}d\theta = \int_0^\infty \frac{1}{1+y^2x^2}\mathrm{Li}_2\left(\frac{1}{1+x^2}\right)dx$$

$$= -\frac{\pi^3}{12}\frac{1}{y} + 2\pi\frac{1}{y}\mathrm{Li}_2\left(\frac{y}{y+1}\right) - \pi\frac{1}{y}\mathrm{Li}_2\left(\frac{y-1}{y+1}\right), \tag{3.380}$$

where the closed form is obtained if we exploit the Fourier series in (4.161), Sect. 4.49, together with the generating function in (4.64), Sect. 4.10.

Are you keen on discovering on your own such curious results? If *yes*, then from the work above, you know how to proceed! Surely, for the previously mentioned result, and the ones from the last three points of the problem, one might also consider exploiting different ways like the use of differentiation under the integral sign.

3.57   Gems of Integration Involving Splendid Ideas About Symmetry in Two Dimensions

Solution The integral at the first point appeared in [34], where a real solution (a possible simple one) to explain the *clean result* has been expected. As we'll see later, the result plays a crucial part in obtaining a wonderful solution to an atypical alternating harmonic series from the weight 4 class given in the fourth chapter.

In the following, I'll present a strategy involving the use of the symmetry in double integrals, but before doing this I'll first let the variable change $(1 - t)/(1 + t) = u$ in the main integral, which we denote by I, and then returning to the variable letter t, we write

$$I = \int_0^1 \left(\frac{\pi}{4} + \arctan(t)\right)^2 \frac{\log(t)}{t^2 - 1} dt = \int_0^1 \frac{(\pi/2 - \arctan(t))^2 \arctanh(t)}{t} dt,$$

(3.381)

where I also used the identity, $\arctan((1 - x)/(1 + x)) = \pi/4 - \arctan(x)$, $-1 < x$.

Further, I'll use the integral result in (1.31), Sect. 1.9, where if we consider $a > 0$, then replace a by $1/a$, simplify, use that $\arctan(a) + \arctan(1/a) = \pi/2$, $a > 0$, and finally replace a by t, and rearrange, we have

$$\left(\frac{\pi}{2} - \arctan(t)\right)^2 = 2 \int_0^1 \frac{x \arctanh(x)}{t^2 + x^2} dx, t > 0.$$

(3.382)

Returning with (3.382) in (3.381), we get

$$I = 2 \int_0^1 \left(\int_0^1 \frac{x \arctanh(t) \arctanh(x)}{t(t^2 + x^2)} dx\right) dt$$

{swap the variable letters and then reverse the integration order}

$$= 2 \int_0^1 \left(\int_0^1 \frac{t \arctanh(x) \arctanh(t)}{x(x^2 + t^2)} dx\right) dt,$$

and if we add up the two versions, divide both sides of the equality by 2, and merge the integrals, we get

$$I = \int_0^1 \left(\int_0^1 \underbrace{\left(\frac{x}{t} + \frac{t}{x}\right)}_{(t^2 + x^2)/(tx)} \frac{\arctanh(t) \arctanh(x)}{t^2 + x^2} dx\right) dt$$

$$= \int_0^1 \frac{\operatorname{arctanh}(t)}{t} \left(\int_0^1 \frac{\operatorname{arctanh}(x)}{x} dx \right) dt = \left(\underbrace{\int_0^1 \frac{\operatorname{arctanh}(x)}{x} dx}_{\pi^2/8} \right)^2 = \frac{\pi^4}{64},$$

(3.383)

where the last integral is calculated in (3.34), Sect. 3.9.

So, according to (3.383), and using the variable change $\tan(y) = x$ to get the second closed form in the problem statement section, we conclude that

$$\int_0^1 \left(\frac{\pi}{4} + \arctan(x) \right)^2 \frac{\log(x)}{x^2 - 1} dx = \int_0^{\pi/4} \left(\frac{\pi}{4} + x \right)^2 \sec(2x) \log(\cot(x)) dx = \frac{\pi^4}{64},$$

and the solution to the point (i) is complete.

What's next? By inspecting the double integral form of the integral at the point (i), we might think of considering the following generalized double integral:

$$I_n = 2 \int_0^1 \left(\int_0^1 \frac{x \operatorname{arctanh}^n(t) \operatorname{arctanh}^n(x)}{t(t^2 + x^2)} dx \right) dt.$$

(3.384)

Further, by swapping the variable letters and reversing the integration order in (3.384), we get

$$I_n = 2 \int_0^1 \left(\int_0^1 \frac{t \operatorname{arctanh}^n(x) \operatorname{arctanh}^n(t)}{x(x^2 + t^2)} dx \right) dt.$$

(3.385)

Then, if we add up the two versions in (3.384) and (3.385), divide both sides of the resulting equality by 2, and merge the integrals, we obtain

$$I_n = \int_0^1 \left(\int_0^1 \left(\underbrace{\frac{x}{t} + \frac{t}{x}}_{(t^2 + x^2)/(tx)} \right) \frac{\operatorname{arctanh}^n(t) \operatorname{arctanh}^n(x)}{t^2 + x^2} dx \right) dt$$

$$= \int_0^1 \frac{\operatorname{arctanh}^n(t)}{t} \left(\int_0^1 \frac{\operatorname{arctanh}^n(x)}{x} dx \right) dt = \left(\int_0^1 \frac{\operatorname{arctanh}^n(x)}{x} dx \right)^2$$

$$= \frac{1}{2^{2n-2}} \left(1 - \frac{1}{2^{n+1}} \right)^2 n!^2 \zeta^2(n+1),$$

(3.386)

where the arctanh integral can be easily calculated if we start with the variable change $(1 - x)/(1 + x) = t$, and then we write

$$\int_0^1 \frac{\operatorname{arctanh}^n(x)}{x}dx = \frac{(-1)^n}{2^{n-1}}\int_0^1 \frac{\log^n(t)}{1-t^2}dt = (-1)^n\frac{1}{2^{n-1}}\int_0^1 \sum_{k=1}^\infty t^{2k-2}\log^n(t)dt$$

$$= (-1)^n\frac{1}{2^{n-1}}\sum_{k=1}^\infty \int_0^1 t^{2k-2}\log^n(t)dt = \frac{1}{2^{n-1}}n!\sum_{k=1}^\infty \frac{1}{(2k-1)^{n+1}}$$

$$= \frac{1}{2^{n-1}}n!\left(\sum_{k=1}^\infty \frac{1}{k^{n+1}} - \sum_{k=1}^\infty \frac{1}{(2k)^{n+1}}\right) = \frac{1}{2^{n-1}}\left(1 - \frac{1}{2^{n+1}}\right)n!\zeta(n+1),$$

$$(3.387)$$

and in the calculations I used the integral result, $\displaystyle\int_0^1 t^k\log^n(t)dt = (-1)^n\frac{n!}{(k+1)^{n+1}}$,
$k, n \in \mathbb{N}$, which is also given in [76, Chapter 1, p.1]).

On the other hand, if we return to (3.384), use the result in (1.36), Sect. 1.10, with a replaced by $1/t$, then simplify, and lastly expand the outer integral and apply (3.387), we have

$$I_n = 2\int_0^1 \frac{\operatorname{arctanh}^n(t)}{t}\left(\int_0^1 \frac{x\operatorname{arctanh}^n(x)}{t^2+x^2}dx\right)dt$$

$$= \frac{1}{2^{2n-2}}\left(1 - \frac{1}{2^n}\right)\left(1 - \frac{1}{2^{n+1}}\right)n!^2\zeta^2(n+1)$$

$$+ \frac{1}{2^{n-1}}n!\int_0^1 \frac{\operatorname{arctanh}^n(t)}{t}\Re\left\{\operatorname{Li}_{n+1}\left(\frac{1-t^2}{1+t^2} + i\frac{2t}{1+t^2}\right)\right\}dt. \qquad (3.388)$$

Combining (3.386) and (3.388), and rearranging, we conclude that

$$\int_0^1 \frac{\operatorname{arctanh}^n(x)}{x}\Re\left\{\operatorname{Li}_{n+1}\left(\frac{1-x^2}{1+x^2} + i\frac{2x}{1+x^2}\right)\right\}dx$$

$$= \int_0^{\pi/4} \csc(x)\sec(x)\operatorname{arctanh}^n(\tan(x))\Re\left\{\operatorname{Li}_{n+1}(\cos(2x) + i\sin(2x))\right\}dx$$

$$= \frac{1}{2^{3n+1}}(2^{n+1} - 1)n!\zeta^2(n+1),$$

where we note the first equality is straightforward by the variable change $\tan(y) = x$, and the solution to the point (ii) is complete.

Finally, as regards the key double integral involved, we may also think of possible generalized forms, like $\displaystyle\int_a^b\left(\int_a^b \frac{xf(x)f(y)}{y(y^2+x^2)}dx\right)dy = \frac{1}{2}\left(\int_a^b \frac{f(x)}{x}dx\right)^2$, or the more complex form,

$$\int_a^b \left(\int_a^b \frac{g(x)f(x)f(y)}{g(y)(g^2(y)+g^2(x))} dx \right) dy = \frac{1}{2} \left(\int_a^b \frac{f(x)}{g(x)} dx \right)^2, \qquad (3.389)$$

obtained by exploiting the symmetry.

Here is another example the curious reader might like to check. If we consider $f(x) = \log(x)/\sqrt{1+x^2}$ and $g(x) = \sqrt{1+x^2}$ together with the needed integration limits in (3.389), we have

$$\int_0^\infty \frac{\log(x) \log(2+x^2)}{(1+x^2)\sqrt{2+x^2}} dx = 0. \qquad (3.390)$$

Letting our imagination run free, we may discover many other curious results like the one in (3.390)!

Unexpected connections of (3.389) *to the famous Ahmed's integral:* One of the most well-known integrals in the mathematical community, which has received a lot of attention, also appearing in *The American Mathematical Monthly* (Problem 10884, **108** 566, 2001), is $\int_0^1 \frac{\arctan(\sqrt{2+x^2})}{(1+x^2)\sqrt{2+x^2}} dx$. Author's original solution relies on a reduction to the double integral $\int_0^1 \left(\int_0^1 \frac{1}{(1+x^2)(2+x^2+y^2)} dx \right) dy$, as seen in [1]. In general, the solutions exploiting the symmetry are very neat and (almost) impossible to beat.

Returning in (3.389), and setting $f(x) = 1/\sqrt{1+x^2}$, $g(x) = \sqrt{1+x^2}$, we are led to the key double integral in Ahmed's solution! So, we immediately obtain that

$$\int_0^1 \left(\int_0^1 \frac{1}{(1+y^2)(2+x^2+y^2)} dx \right) dy = \frac{1}{2} \left(\int_0^1 \frac{1}{1+x^2} dx \right)^2 = \frac{\pi^2}{32}.$$

Finally, for the curious reader I mention that a beautiful solution to Ahmed's integral may also be found in *Inside Interesting Integrals* by Paul Nahin (see [66, pp.230–233]).

3.58   More Gems of Integration, This Time Involving Splendid Ideas About Symmetry in Three Dimensions

Solution As regards the use of the symmetry in double integrals, we have already witnessed its power both in sections of the present book such as Sects. 3.5, 3.8, 3.27, and 3.28 and the previous one, and in *(Almost) Impossible Integrals, Sums, and Series*, for integrals like the ones in [76, Chapter 1, Sections 1.27–1.30, pp.17–21, Section 1.58, p.36].

The question for the first point, in a slightly different form, was asked in 2013 in [36], and only a solution involving contour integration was available for about 9 years at the given reference. Sometimes, the ways to go by real methods are (highly) *non-obvious*. More recently, the question was also asked in [37].

How about exploiting the symmetry in three dimensions? It is exactly what we are going to do in the following! The symmetrical structure I'll build upon is

$$f(x, y, z) = \frac{x}{(1+x)(1+z)(1+xyz)} + \frac{y}{(1+y)(1+x)(1+xyz)}$$

$$+ \frac{z}{(1+z)(1+y)(1+xyz)} = \frac{(1+x)(1+y)(1+z) - (1+xyz)}{(1+x)(1+y)(1+z)(1+xyz)}$$

$$= \frac{1}{1+xyz} - \frac{1}{(1+x)(1+y)(1+z)}. \tag{3.391}$$

Focusing on the first equality in (3.391), with the needed settings, and integrating with respect to x, y, z, from 0 to 1, we easily note that due to the symmetry we get three times the same integral (and we easily see that by swapping the variable letters and reversing the order of integration). Thus, we write

$$\int_0^1 \left(\int_0^1 \left(\int_0^1 f(x^2, y^2, z^2) \mathrm{d}x \right) \mathrm{d}y \right) \mathrm{d}z$$

$$= 3 \int_0^1 \left(\int_0^1 \left(\int_0^1 \frac{x^2}{(1+x^2)(1+z^2)(1+x^2y^2z^2)} \mathrm{d}x \right) \mathrm{d}y \right) \mathrm{d}z$$

$$\overset{\substack{\text{reverse the order} \\ \text{of integration}}}{=} 3 \int_0^1 \left(\int_0^1 \left(\int_0^1 \frac{x^2}{(1+x^2)(1+z^2)(1+x^2y^2z^2)} \mathrm{d}z \right) \mathrm{d}y \right) \mathrm{d}x$$

$$\left\{ \text{use that } \frac{1}{(1+z^2)(1+x^2y^2z^2)} = \frac{1}{(1-x^2y^2)(1+z^2)} - \frac{x^2y^2}{(1-x^2y^2)(1+x^2y^2z^2)} \right\}$$

$$= 3 \int_0^1 \left(\int_0^1 \frac{x^2(\pi/4 - xy\arctan(xy))}{(1+x^2)(1-x^2y^2)} \mathrm{d}y \right) \mathrm{d}x$$

$$\overset{xy=t}{=} 3 \int_0^1 \left(\int_0^x \frac{x(\pi/4 - t\arctan(t))}{(1+x^2)(1-t^2)} \mathrm{d}t \right) \mathrm{d}x$$

$$\overset{\substack{\text{reverse the order} \\ \text{of integration}}}{=} 3 \int_0^1 \left(\int_t^1 \frac{x(\pi/4 - t\arctan(t))}{(1+x^2)(1-t^2)} \mathrm{d}x \right) \mathrm{d}t$$

$$= \frac{3}{2} \int_0^1 \frac{(\pi/4 - t\arctan(t))}{t^2 - 1} \log\left(\frac{1+t^2}{2}\right) dt$$

$$= \frac{3}{8}\pi \underbrace{\int_0^1 \frac{1}{t^2-1} \log\left(\frac{1+t^2}{2}\right) dt}_{\pi^2/16} - \frac{3}{2}\int_0^1 \frac{t\arctan(t)}{t^2-1} \log\left(\frac{1+t^2}{2}\right) dt$$

$$= \frac{3}{128}\pi^3 - \frac{3}{2}\int_0^1 \frac{t\arctan(t)}{t^2-1} \log\left(\frac{1+t^2}{2}\right) dt$$

{exploit the rightmost-hand side of (3.391)}

$$= \underbrace{\int_0^1 \left(\int_0^1 \left(\int_0^1 \frac{1}{1+x^2y^2z^2} dx\right) dy\right) dz}_{\pi^3/32} - \underbrace{\left(\int_0^1 \frac{1}{1+x^2} dx\right)^3}_{\pi^3/64} = \frac{\pi^3}{64},$$

whence we obtain that

$$\int_0^1 \frac{t\arctan(t)}{t^2-1} \log\left(\frac{1+t^2}{2}\right) dt$$

$$\overset{\tan(u)=t}{=} \int_0^{\pi/4} u \log(2\cos^2(u)) \sec(2u)\tan(u) du = \frac{\pi^3}{192},$$

and the solution to the point (i) is complete. In the calculations I used the simple results, $\int_0^1 \frac{1}{t^2-1} \log\left(\frac{1+t^2}{2}\right) dt$ $\overset{(1-t)/(1+t)=u}{=}$ $\int_0^1 \frac{\log(1+u)}{u} du -$

$$\underbrace{\frac{1}{2}\int_0^1 \frac{\log(1+u^2)}{u} du}_{u^2 = v} = \frac{3}{4}\int_0^1 \frac{\log(1+u)}{u} du = \frac{3}{4}\int_0^1 \sum_{n=1}^{\infty}(-1)^{n-1}\frac{u^{n-1}}{n} du =$$

$$\frac{3}{4}\sum_{n=1}^{\infty}(-1)^{n-1}\frac{1}{n}\int_0^1 u^{n-1} du = \frac{3}{4}\sum_{n=1}^{\infty}(-1)^{n-1}\frac{1}{n^2} = \frac{3}{4}\eta(2) = \frac{\pi^2}{16}, \text{ and}$$

then $\int_0^1 \left(\int_0^1 \left(\int_0^1 \frac{1}{1+x^2y^2z^2} dx\right) dy\right) dz = \sum_{n=1}^{\infty}(-1)^{n-1}\int_0^1 \left(\int_0^1 \left(\int_0^1\right.\right.$

$\left.\left.(xyz)^{2n-2} dx\right) dy\right) dz = \sum_{n=1}^{\infty}(-1)^{n-1}\frac{1}{(2n-1)^3} = \frac{\pi^3}{32}$, and the last series is obtained by (4.150) Sect. 4.47, where we set $n = 2$ and $x = 1/4$.

To solve the point (ii) of the problem, we first recall that the Legendre's chi function of order 2 is defined as $\chi_2(a) = \int_0^a \dfrac{\operatorname{arctanh}(t)}{t} dt = \dfrac{1}{2}(\operatorname{Li}_2(a) - \operatorname{Li}_2(-a))$,

but also as $\chi_2(a) = \displaystyle\sum_{n=1}^{\infty} \dfrac{a^{2n-1}}{(2n-1)^2}$.

Now, using the starting ideas met at the previous point, based on (3.391), we have

$$\int_0^1 \left(\int_0^1 \left(\int_0^1 \log(x) \log(y) \log(z) f(-x^2, -y^2, -z^2) dx \right) dy \right) dz$$

$$= -3 \int_0^1 \left(\int_0^1 \left(\int_0^1 \frac{x^2 \log(x) \log(y) \log(z)}{(1-x^2)(1-z^2)(1-x^2 y^2 z^2)} dx \right) dy \right) dz$$

$$\underset{=}{\overset{\text{reverse the order}}{\underset{\text{of integration}}{}}} -3 \int_0^1 \left(\int_0^1 \frac{x^2 \log(x) \log(z)}{(1-x^2)(1-z^2)} \left(\int_0^1 \frac{\log(y)}{1-x^2 y^2 z^2} dy \right) dx \right) dz$$

$$\left\{ \text{by partial fraction expansion we have } \frac{\log(y)}{1 - x^2 y^2 z^2} = \frac{1}{2}\left(\frac{\log(y)}{1 - xyz} + \frac{\log(y)}{1 + xyz} \right), \right\}$$

$\{$and at the same time we want to consider the integral result in (3.13), Sect. 3.3$\}$

$$= 3 \int_0^1 \left(\int_0^1 \frac{x \log(x) \log(z)(\operatorname{Li}_2(xz) - \operatorname{Li}_2(-xz))/2}{z(1-x^2)(1-z^2)} dx \right) dz$$

$$= 3 \int_0^1 \left(\int_0^1 \frac{x \log(x) \log(z) \chi_2(xz)}{z(1-x^2)(1-z^2)} dx \right) dz$$

$\{$make use of the variable changes $x = \tanh(u)$ and $z = \tanh(v)\}$

$$= 3 \int_0^\infty \left(\int_0^\infty \underbrace{\frac{\tanh(u) \log(\tanh(u)) \log(\tanh(v)) \chi_2(\tanh(u) \tanh(v))}{\tanh(v)}}_{g(u,\,v)} du \right) dv$$

$\{$exploit the symmetry of the integrand to put it as below$\}$

$$= \frac{3}{2} \int_0^\infty \left(\int_0^\infty (g(u,v) + g(v,u)) du \right) dv$$

$\{$exploit the rightmost-hand side of (3.391)$\}$

$$= \underbrace{\int_0^1 \left(\int_0^1 \left(\int_0^1 \frac{\log(x) \log(y) \log(z)}{1 - x^2 y^2 z^2} dx \right) dy \right) dz}_{-\pi^6/960} - \underbrace{\left(\int_0^1 \frac{\log(x)}{1 - x^2} dx \right)^3}_{-\pi^6/512} = \frac{7}{7680}\pi^6,$$

whence we obtain that

$$\int_0^\infty \int_0^\infty \frac{(\tanh^2(x) + \tanh^2(y)) \log(\tanh(x)) \log(\tanh(y)) \chi_2(\tanh(x) \tanh(y))}{\tanh(x) \tanh(y)} dx\, dy$$

$$= \frac{7}{11520} \pi^6,$$

and the solution to the point (ii) is complete. During the calculations I used that

$$\int_0^1 \left(\int_0^1 \left(\int_0^1 \frac{\log(x) \log(y) \log(z)}{1 - x^2 y^2 z^2} dx \right) dy \right) dz$$

$$= \sum_{n=1}^\infty \int_0^1 \left(\int_0^1 \left(\int_0^1 (xyz)^{2n-2} \log(x) \log(y) \log(z) dx \right) dy \right) dz = -\sum_{n=1}^\infty \frac{1}{(2n-1)^6}$$

$$= \sum_{n=1}^\infty \frac{1}{(2n)^6} - \sum_{n=1}^\infty \frac{1}{n^6} = -\frac{63}{64}\zeta(6) = -\frac{\pi^6}{960}, \tag{3.392}$$

where the last equality in (3.392) may be obtained by (6.315), Sect. 6.47, and then the fact, $\int_0^1 \frac{\log(x)}{1 - x^2} dx = -\frac{\pi^2}{8}$, obtained from the case $n = 1$ in (3.387), Sect. 3.57. I could have started the solution directly from the given integral, aiming to reach the place where to exploit the symmetry. Such a flow is illustrated at the next point.

I submitted the integral from the last point to the *Romanian Mathematical Magazine (R.M.M.)*, as seen in [68]. From then until the moment of finalizing the present book it has remained without a solution. Undoubtedly, it looks daunting (and similar to the previous integral)!

Let's go differently from the previous solution flow, and we start directly from the given integral, but first we might want to (re)visit how the inverse tangent integral is defined and take a look at Sect. 3.6. In particular, in our calculations we'll need the following useful integral representation of the inverse tangent integral, $\text{Ti}_2(x) = -\int_0^1 \frac{x \log(t)}{1 + x^2 t^2} dt$.

Returning to the main integral, exploiting the symmetry, and making the variable changes $\tan(x) = u$ and $\tan(y) = v$, we write

$$\int_0^{\pi/4} \left(\int_0^{\pi/4} \frac{(\tan^2(x) + \tan^2(y)) \log(\tan(x)) \log(\tan(y)) \text{Ti}_2(\tan(x) \tan(y))}{\tan(x) \tan(y)} dx \right) dy$$

$$= 2 \int_0^{\pi/4} \left(\int_0^{\pi/4} \frac{\tan(x) \log(\tan(x)) \log(\tan(y)) \text{Ti}_2(\tan(x) \tan(y))}{\tan(y)} dx \right) dy$$

$$= -2 \int_0^1 \left(\int_0^1 \left(\int_0^1 \frac{u^2 \log(u) \log(v) \log(t)}{(1 + u^2)(1 + v^2)(1 + u^2 v^2 t^2)} dt \right) du \right) dv$$

{exploit the symmetrical structure in (3.391)}

$$= -\frac{2}{3} \int_0^1 \left(\int_0^1 \left(\int_0^1 \log(u)\log(v)\log(t) f(u^2, v^2, t^2) dt \right) du \right) dv$$

$$= \frac{2}{3} \left(\underbrace{\left(\int_0^1 \frac{\log(t)}{1+t^2} dt \right)^3}_{-G^3} - \underbrace{\int_0^1 \left(\int_0^1 \left(\int_0^1 \frac{\log(u)\log(v)\log(t)}{1+u^2v^2t^2} dt \right) du \right) dv}_{\pi^6/960 - \psi^{(5)}(1/4)/245760} \right)$$

$$= \frac{1}{368640} \psi^{(5)}\left(\frac{1}{4}\right) - \frac{\pi^6}{1440} - \frac{2}{3} G^3,$$

and the solution to the point (iii) is complete. During the calculations I used that with integration by parts we have $\int_0^1 \frac{\log(t)}{1+t^2} dt = -\int_0^1 \frac{\arctan(t)}{t} dt = -G$, and for the triple integral we get that

$$\int_0^1 \left(\int_0^1 \left(\int_0^1 \frac{\log(u)\log(v)\log(t)}{1+u^2v^2t^2} dt \right) du \right) dv$$

$$= \sum_{n=1}^{\infty} (-1)^{n-1} \int_0^1 \left(\int_0^1 \left(\int_0^1 (uvt)^{2n-2} \log(u)\log(v)\log(t) dt \right) du \right) dv$$

$$\overset{\text{exploit}}{\underset{\text{the parity}}{=}} \sum_{n=1}^{\infty} (-1)^n \frac{1}{(2n-1)^6} = \frac{1}{4096} \left(\sum_{n=1}^{\infty} \frac{1}{(n-1/4)^6} - \sum_{n=1}^{\infty} \frac{1}{(n-3/4)^6} \right)$$

$$\left\{ \text{use Polygamma series representation, } \psi^{(m)}(z) = (-1)^{m-1} m! \sum_{k=0}^{\infty} \frac{1}{(z+k)^{m+1}} \right\}$$

$$= \frac{1}{491520} \psi^{(5)}\left(\frac{3}{4}\right) - \frac{1}{491520} \psi^{(5)}\left(\frac{1}{4}\right) = \frac{\pi^6}{960} - \frac{1}{245760} \psi^{(5)}\left(\frac{1}{4}\right), \quad (3.393)$$

where to obtain the last equality in (3.393) I used the Polygamma function reflection formula, $(-1)^m \psi^{(m)}(1-x) - \psi^{(m)}(x) = \pi \frac{d^m}{dx^m} \cot(\pi x)$, with $m = 5$ and $x = \frac{1}{4}$, to derive and use the identity $\psi^{(5)}\left(\frac{1}{4}\right) + \psi^{(5)}\left(\frac{3}{4}\right) = 512\pi^6$.

Exploiting the symmetry in multiple integrals is proving again to be a powerful way to go, often a very beautiful one, which is properly appreciated when we also try to get different solutions to such problems!

3.59 The Complete Elliptic Integral of the First Kind at Play

Solution At the sight of this result, some of you might want to readily check (at least) that the numerical values in both sides of the equality match, and thus run *Mathematica*. I stress that if we employ *Mathematica* command *EllipticK[x]*, then we might want to consider using *EllipticK[x²]* for matching the definition of the complete elliptic integral of the first kind given in the problem statement.

First, we want to observe and use the simple fact that $\displaystyle\int_y^1 \frac{1}{(1+x)(y+x)}dx =$

$$\frac{1}{1-y}\int_y^1 \frac{1}{y+x}dx - \frac{1}{1-y}\int_y^1 \frac{1}{1+x}dx = \frac{1}{1-y}\log(y+x)\Big|_{x=y}^{x=1} - \frac{1}{1-y}\log(1+$$

$$x)\Big|_{x=y}^{x=1} = -\frac{1}{1-y}\log\left(\frac{4y}{(1+y)^2}\right), \text{ and then we write}$$

$$\int_0^1 \frac{K(x)}{1-x}\log\left(\frac{4x}{(1+x)^2}\right)dx = -\int_0^1\left(\int_x^1 \frac{K(x)}{(1+y)(x+y)}dy\right)dx$$

$$\underset{=}{\overset{\text{reverse the order}}{\text{of integration}}} -\int_0^1\left(\int_0^y \frac{K(x)}{(1+y)(x+y)}dx\right)dy \overset{x/y=t}{=} -\int_0^1\left(\int_0^1 \frac{K(yt)}{(1+y)(1+t)}dt\right)dy$$

$$\left\{\text{use the power series, } K(x) = \frac{\pi}{2}\sum_{n=0}^\infty\left(\frac{(2n-1)!!}{(2n)!!}\right)^2 x^{2n} = \frac{\pi}{2}\sum_{n=0}^\infty \frac{1}{16^n}\binom{2n}{n}^2 x^{2n}\right\}$$

$$= -\frac{\pi}{2}\int_0^1\left(\int_0^1 \frac{1}{(1+y)(1+t)}\sum_{n=0}^\infty \frac{1}{16^n}\binom{2n}{n}^2 (yt)^{2n}dt\right)dy$$

$$\{\text{reverse the order of summation and integration}\}$$

$$= -\frac{\pi}{2}\sum_{n=0}^\infty \frac{1}{16^n}\binom{2n}{n}^2\left(\int_0^1 \frac{y^{2n}}{1+y}dy\right)\left(\int_0^1 \frac{t^{2n}}{1+t}dt\right)$$

$$\left\{\text{exploit that } \int_0^1 \frac{x^{2n}}{1+x}dx = H_n - H_{2n} + \log(2), \text{ which is straightforward if}\right\}$$

$$\{\text{we employ the integral result in (1.100), Sect. 1.21, and integrate by parts}\}$$

$$= -\frac{\pi}{2}\sum_{n=0}^\infty \frac{1}{16^n}\binom{2n}{n}^2 (H_{2n} - H_n - \log(2))^2 = \frac{7}{4}\zeta(3) - \pi G,$$

where the last equality is obtained by considering (4.15), Sect. 4.4, and the solution is complete. The power series of $K(x)$ used above is elegantly derived in *(Almost) Impossible Integrals, Sums, and Series* (see [76, Chapter 3, pp.212–213]).

Everything has worked smoothly and fast since in this section I've only invoked the value of the key binoharmonic series needed to get the desired result. The curious reader interested in developing more complex integrals alike might exploit similar ideas combined with integrals like the ones with a Sylvester-type structure found in Sect. 1.56, and this may be better understood after seeing the solution to the last resulting series above. Similar integrals, attacked by different strategies, may be found in [9].

3.60   Evaluating an Esoteric-Looking Integral Involving a Triple Series with Factorial Numbers, Leading to $\zeta(4)$

Solution While meeting the proposed integral in this section, the first immediate thing to remark is its *wow* look! The closed form of the integral allows us to rearrange the result and put it into the form of a beautiful representation of $\zeta(4)$, which is exactly how it appears in the problem statement.

The solution flow is going to be very fast since it will be built upon a result presented in the fourth chapter, *Sums and Series*. Again, it is one of those solutions where the way to go is (highly) *non-obvious*.

Let's start the calculations! We let the variable change $1 - x = y$ and exploit the special functional equation in (4.197), Sect. 4.60, and we have

$$\int_0^1 \left((4-x)(1-x)^2 \Lambda(x-1) + \frac{(5-2x)(1-x)^2}{(2-x)^3} \Lambda\left(\frac{1-x}{2-x}\right)\right) \frac{\log(x)}{x-1} dx$$

$$= -\int_0^1 \left((3+x)x^2 \Lambda(-x) + \frac{(3+2x)x^2}{(1+x)^3} \Lambda\left(\frac{x}{1+x}\right)\right) \frac{\log(1-x)}{x} dx$$

$$= -3\int_0^1 \frac{\log(1-x)\log^2(1+x)}{x} dx = \frac{9}{8}\zeta(4),$$

whence, upon rearranging, we obtain that

$$\zeta(4)$$

$$= \frac{8}{9}\int_0^1 \left((4-x)(1-x)^2 \Lambda(x-1) + \frac{(5-2x)(1-x)^2}{(2-x)^3} \Lambda\left(\frac{1-x}{2-x}\right)\right) \frac{\log(x)}{x-1} dx,$$

and the solution is complete. In the calculations I also used the integral result in (6.183), Sect. 6.23. It is worth mentioning that one can create a lot of interesting integrals by exploiting the functional equation of $\Lambda(x)$ used above.

All in all, we've just reached the end of the first series of three chapters dedicated to the *Integrals* area! I hope it was a very enjoyable one, and the proposed integrals often surprised you pleasantly! The given problems and the strategies presented open the gates to the curious reader to create and solve many other wonderful problems!

Are you ready for the other half of the journey, focused on sums and series? If *yes*, then let's pass to the next chapter and prepare for more fun!

Chapter 4
Sums and Series

Whatever you do, work at it with all your heart, as working for the Lord, not for human masters . . .
– Colossians 3:23

4.1 A Remarkable IMC Limit Problem Involving a Curious Sum with the Reciprocal of a Product with Two Logarithms

Find

$$\lim_{n \to \infty} \frac{\log^2(n)}{n} \sum_{k=2}^{n-2} \frac{1}{\log(k) \log(n-k)}. \tag{4.1}$$

A (little) challenging question: Calculate the limit without using integrals.

4.2 Two Series with Tail Involving the Double Factorial, Their Generalizations, and a $\zeta(2)$ Representation

Show that

$$(i) \sum_{n=1}^{\infty} (-1)^{n-1} \left(\frac{\pi^2}{4} - \sum_{k=1}^{n} \frac{(2k-2)!!}{k \cdot (2k-1)!!} \right) = \sum_{n=1}^{\infty} (-1)^{n-1} \left(\frac{\pi^2}{4} - \sum_{k=1}^{n} \frac{2^{2k-1}}{k^2 \binom{2k}{k}} \right)$$

$$= \frac{1}{8} \left(\pi^2 - 4\log^2(\sqrt{2} - 1) \right); \tag{4.2}$$

(continued)

$$(ii) \sum_{n=0}^{\infty} (-1)^n \left(\frac{\pi}{2} - \sum_{k=0}^{n} \frac{(2k-1)!!}{(2k+1) \cdot (2k)!!} \right) = \sum_{n=0}^{\infty} (-1)^n \left(\frac{\pi}{2} - \sum_{k=0}^{n} \frac{1}{(2k+1)2^{2k}} \binom{2k}{k} \right)$$

$$= \frac{\pi}{4} - \frac{1}{2} \log(\sqrt{2}+1), \tag{4.3}$$

where $n!! = \begin{cases} n \cdot (n-2) \cdots 5 \cdot 3 \cdot 1, & n > 0 \text{ odd}; \\ n \cdot (n-2) \cdots 6 \cdot 4 \cdot 2, & n > 0 \text{ even}; \\ 1, & n = -1, 0, \end{cases}$ is the double factorial.

A special generalization with the double factorial:
Prove that

$$(iii) \sum_{n=1}^{\infty} x^{2n} \left(\frac{\pi^2}{4} - \sum_{k=1}^{n} \frac{(2k-2)!!}{k \cdot (2k-1)!!} \right) = \sum_{n=1}^{\infty} x^{2n} \left(\frac{\pi^2}{4} - \sum_{k=1}^{n} \frac{2^{2k-1}}{k^2 \binom{2k}{k}} \right)$$

$$= \frac{\pi^2}{4} \frac{x^2}{1-x^2} - \frac{\arcsin^2(x)}{1-x^2}, \quad |x| < 1. \tag{4.4}$$

Another special generalization with the double factorial:
Prove that

$$(iv) \sum_{n=0}^{\infty} x^{2n} \left(\frac{\pi}{2} - \sum_{k=0}^{n} \frac{(2k-1)!!}{(2k+1) \cdot (2k)!!} \right) = \sum_{n=0}^{\infty} x^{2n} \left(\frac{\pi}{2} - \sum_{k=0}^{n} \frac{1}{(2k+1)2^{2k}} \binom{2k}{k} \right)$$

$$= \frac{\pi}{2} \frac{1}{1-x^2} - \frac{\arcsin(x)}{x(1-x^2)}, \quad |x| < 1. \tag{4.5}$$

A curious representation of $\zeta(2)$:
Show that

$$(v) \, \zeta(2) = \sum_{n=1}^{\infty} \frac{(2n-1)!}{n(n+1)((2n-2)!!)^2} \left(\frac{\pi^2}{4} - \sum_{k=1}^{n} \frac{(2k-2)!!}{k \cdot (2k-1)!!} \right), \tag{4.6}$$

where ζ represents the Riemann zeta function.

4.3 Six Enjoyable Sums Involving the Reciprocal of the Central Binomial Coefficient and Two Series Derived from Them

Prove that

$$(i) \sum_{k=1}^{n} \frac{2^{2k}}{k\binom{2k}{k}} = \frac{2^{2n+1}}{\binom{2n}{n}} - 2; \qquad (4.7)$$

$$(ii) \sum_{k=1}^{n} \frac{2^{2k}}{k(2k+1)\binom{2k}{k}} = 2 - \frac{2^{2n+1}}{(2n+1)\binom{2n}{n}}. \qquad (4.8)$$

A little challenge for the next two sums is to construct proofs based on the sums from the points (i) and (ii). How would we proceed?

$$(iii) \sum_{k=1}^{n} \frac{2^{4k}(4k-1)}{k^2\binom{2k}{k}^2} = \frac{2^{4n+2}}{\binom{2n}{n}^2} - 4; \qquad (4.9)$$

$$(iv) \sum_{k=1}^{n} \frac{2^{4k}(4k+1)}{k^2(2k+1)^2\binom{2k}{k}^2} = 4 - \frac{2^{4n+2}}{(2n+1)^2\binom{2n}{n}^2}; \qquad (4.10)$$

$$(v) \sum_{k=1}^{n} \frac{2^{8k}(4k-1)(8k^2-4k+1)}{k^4\binom{2k}{k}^4} = \frac{2^{8n+4}}{\binom{2n}{n}^4} - 16; \qquad (4.11)$$

$$(vi) \sum_{k=1}^{n} \frac{2^{8k}(4k+1)(8k^2+4k+1)}{k^4(2k+1)^4\binom{2k}{k}^4} = 16 - \frac{2^{8n+4}}{(2n+1)^4\binom{2n}{n}^4}. \qquad (4.12)$$

Two series derived from the previous finite sums:

$$(vii) \sum_{k=1}^{\infty} \frac{2^{4k}(4k+1)}{k^2(2k+1)^2\binom{2k}{k}^2} = 4; \qquad (4.13)$$

(continued)

$$(viii) \sum_{k=1}^{\infty} \frac{2^{8k}(4k+1)(8k^2+4k+1)}{k^4(2k+1)^4\binom{2k}{k}^4} = 16. \qquad (4.14)$$

4.4 A Great Time with a Special Binoharmonic Series

Show that

$$\sum_{n=0}^{\infty} \frac{1}{16^n}\binom{2n}{n}^2 (H_{2n} - H_n - \log(2))^2 = 2G - \frac{7}{2}\frac{\zeta(3)}{\pi}, \qquad (4.15)$$

where $H_n = \sum_{k=1}^{n} \frac{1}{k}$ is the nth harmonic number, G represents the Catalan's constant, and ζ denotes the Riemann zeta function.

   A challenging question: Perform the calculations by exploiting Sylvester's integral found in the first section of the book.

4.5 A Panel of (Very) Useful Cauchy Products of Two Series: From Known Cauchy Products to Less Known Ones

Prove that

$$(i) \ \frac{\arctan(x)}{1+x^2} = \frac{1}{2}\sum_{n=1}^{\infty}(-1)^{n-1}x^{2n-1}(2H_{2n} - H_n), \ |x| < 1; \qquad (4.16)$$

$$(ii) \ \arctan^2(x) = \frac{1}{2}\sum_{n=1}^{\infty}(-1)^{n-1}x^{2n}\frac{2H_{2n} - H_n}{n}, \ |x| \le 1; \qquad (4.17)$$

$$(iii) \ \frac{\operatorname{arctanh}(x)}{1-x^2} = \frac{1}{2}\sum_{n=1}^{\infty}x^{2n-1}(2H_{2n} - H_n), \ |x| < 1; \qquad (4.18)$$

$$(iv) \ \operatorname{arctanh}^2(x) = \frac{1}{2}\sum_{n=1}^{\infty}x^{2n}\frac{2H_{2n} - H_n}{n}, \ |x| < 1; \qquad (4.19)$$

(continued)

$$(v) \ \arctan(x)\operatorname{arctanh}(x) = \sum_{n=1}^{\infty} \frac{x^{4n-2}}{2n-1} \sum_{k=1}^{2n-1} (-1)^{k-1} \frac{1}{2k-1}, \ |x| < 1;$$

$$(4.20)$$

$$(vi) \ \arctan(x)\operatorname{Li}_2(-x^2) = \sum_{n=1}^{\infty} (-1)^n x^{2n+1} \left(4\frac{H_{2n}}{(2n+1)^2} + \frac{H_n^{(2)}}{2n+1} \right), \ |x| \le 1;$$

$$(4.21)$$

$$(vii) \ \operatorname{arctanh}(x)\operatorname{Li}_2(x^2) = \sum_{n=1}^{\infty} x^{2n+1} \left(4\frac{H_{2n}}{(2n+1)^2} + \frac{H_n^{(2)}}{2n+1} \right), \ |x| < 1;$$

$$(4.22)$$

$$(viii) \ \log(1-x)\operatorname{Li}_2(x) = \sum_{n=1}^{\infty} x^n \left(3\frac{1}{n^3} - 2\frac{H_n}{n^2} - \frac{H_n^{(2)}}{n} \right), \ |x| \le 1 \wedge x \ne 1;$$

$$(4.23)$$

$$(ix) \ \log(1-x)\operatorname{Li}_3(x) = \sum_{n=1}^{\infty} x^n \left(4\frac{1}{n^4} - 2\frac{H_n}{n^3} - \frac{H_n^{(2)}}{n^2} - \frac{H_n^{(3)}}{n} \right), \ |x| \le 1 \wedge x \ne 1;$$

$$(4.24)$$

$$(x) \ \frac{\log(1-x)\log(1+x)}{1-x} = \sum_{n=1}^{\infty} x^n \left(\frac{1}{2}\overline{H}_n^2 + \overline{H}_n^{(2)} - \frac{1}{2}H_n^{(2)} - \sum_{k=1}^{n} \frac{\overline{H}_k}{k} \right), \ |x| < 1;$$

$$(4.25)$$

$$(xi) \ \frac{\log(1-x)\operatorname{Li}_2(x)}{1-x} = \sum_{n=1}^{\infty} x^n \left(2H_n^{(3)} - H_n H_n^{(2)} - \sum_{k=1}^{n} \frac{H_k}{k^2} \right), \ |x| < 1;$$

$$(4.26)$$

$$(xii) \ \frac{\log(1-x)\operatorname{Li}_2(-x)}{1-x}$$

$$= \sum_{n=1}^{\infty} x^n \left(H_n^{(3)} - 2\overline{H}_n^{(3)} + \sum_{k=1}^{n} \frac{\overline{H}_k}{k^2} - \sum_{k=1}^{n} (-1)^{k-1}\frac{\overline{H}_k}{k^2} + \sum_{k=1}^{n} \frac{\overline{H}_k^{(2)}}{k} \right), \ |x| < 1;$$

$$(4.27)$$

$$(xiii) \ \frac{\log(1+x)\operatorname{Li}_2(x)}{1-x} = \sum_{n=1}^{\infty} x^n \left(H_n^{(3)} - \overline{H}_n^{(3)} - \overline{H}_n \overline{H}_n^{(2)} + \sum_{k=1}^{n} \frac{\overline{H}_k}{k^2} \right), \ |x| < 1;$$

$$(4.28)$$

(continued)

$$(xiv) \ \frac{(\text{Li}_2(x))^2}{1-x} = \sum_{n=1}^{\infty} x^n \left((H_n^{(2)})^2 - 5H_n^{(4)} + 4 \sum_{k=1}^{n} \frac{H_k}{k^3} \right), \ |x| < 1;$$

$$(4.29)$$

$$(xv) \ \frac{\text{Li}_2(x) \, \text{Li}_2(-x)}{1-x}$$

$$= \sum_{n=1}^{\infty} x^n \left(3\overline{H}_n^{(4)} + \frac{1}{2} \left(\overline{H}_n^{(2)} \right)^2 - \frac{5}{2} H_n^{(4)} - 2 \sum_{k=1}^{n} \frac{\overline{H}_k}{k^3} + 2 \sum_{k=1}^{n} (-1)^{k-1} \frac{\overline{H}_k}{k^3} - \sum_{k=1}^{n} \frac{\overline{H}_k^{(2)}}{k^2} \right),$$

$$|x| < 1;$$

$$(4.30)$$

$$(xvi) \ - \frac{\log(1-x^2)}{\sqrt{1-x^2}} = \sum_{n=0}^{\infty} x^{2n} \frac{1}{4^n} \binom{2n}{n} (2H_{2n} - H_n), \ |x| < 1, \quad (4.31)$$

where $H_n^{(m)} = 1 + \frac{1}{2^m} + \cdots + \frac{1}{n^m}$, $m \geq 1$, is the nth generalized harmonic number of order m, $\overline{H}_n^{(m)} = 1 - \frac{1}{2^m} + \cdots + (-1)^{n-1} \frac{1}{n^m}$, $m \geq 1$, represents the nth generalized skew-harmonic number of order m, and Li_n denotes the polylogarithm function.

4.6 Good-to-Know Generating Functions: The First Part

Prove that

$$(i) \ \sum_{n=1}^{\infty} x^n H_n^{(m)} = \frac{\text{Li}_m(x)}{1-x}, \ |x| < 1, \ m \in \mathbb{N}, \quad (4.32)$$

where $\text{Li}_1(x) = -\log(1-x)$;

$$(ii) \ \sum_{n=1}^{\infty} x^n \frac{H_n}{n+1} = \frac{1}{2} \frac{\log^2(1-x)}{x}, \ |x| \leq 1 \wedge x \neq 0, 1; \quad (4.33)$$

$$(iii) \ \sum_{n=1}^{\infty} x^n \frac{H_n}{n} = \frac{1}{2} \log^2(1-x) + \text{Li}_2(x) = -\text{Li}_2\left(\frac{x}{x-1} \right), \ |x| \leq 1 \wedge x \neq 1;$$

$$(4.34)$$

(continued)

$$(iv) \sum_{n=1}^{\infty} x^n \frac{H_n}{(n+1)^2}$$

$$= \zeta(2)\frac{\log(1-x)}{x} - \frac{1}{2}\frac{\log(x)\log^2(1-x)}{x} - \frac{\log(1-x)\operatorname{Li}_2(x)}{x} - \frac{\operatorname{Li}_3(1-x)}{x} + \zeta(3)\frac{1}{x}$$

$$= 2\zeta(2)\frac{\log(1-x)}{x} + \frac{1}{6}\frac{\log^3(1-x)}{x} - \frac{\log(x)\log^2(1-x)}{x} - \frac{\log(1-x)\operatorname{Li}_2(x)}{x}$$

$$- \frac{\operatorname{Li}_3(x)}{x} - 2\frac{\operatorname{Li}_3(1-x)}{x} - \frac{1}{x}\operatorname{Li}_3\left(\frac{x}{x-1}\right) + 2\zeta(3)\frac{1}{x}$$

$$= \frac{\operatorname{Li}_3(x)}{x} + \frac{1}{x}\operatorname{Li}_3\left(\frac{x}{x-1}\right) - \frac{\log(1-x)\operatorname{Li}_2(x)}{x} - \frac{1}{6}\frac{\log^3(1-x)}{x}, \ |x| \le 1 \wedge x \ne 0, 1;$$

$$(4.35)$$

$$(v) \sum_{n=1}^{\infty} x^n \frac{H_n}{n^2}$$

$$= \zeta(2)\log(1-x) - \frac{1}{2}\log(x)\log^2(1-x) - \log(1-x)\operatorname{Li}_2(x)$$

$$+ \operatorname{Li}_3(x) - \operatorname{Li}_3(1-x) + \zeta(3)$$

$$= 2\zeta(2)\log(1-x) + \frac{1}{6}\log^3(1-x) - \log(x)\log^2(1-x) - \log(1-x)\operatorname{Li}_2(x)$$

$$- 2\operatorname{Li}_3(1-x) - \operatorname{Li}_3\left(\frac{x}{x-1}\right) + 2\zeta(3)$$

$$= 2\operatorname{Li}_3(x) + \operatorname{Li}_3\left(\frac{x}{x-1}\right) - \log(1-x)\operatorname{Li}_2(x) - \frac{1}{6}\log^3(1-x), \ |x| \le 1 \wedge x \ne 0, 1;$$

$$(4.36)$$

$$(vi) \sum_{n=1}^{\infty} x^n \frac{H_n}{(n+1)^3}$$

$$= \zeta(4)\frac{1}{x} + \zeta(3)\frac{\log(1-x)}{x} + \frac{1}{2}\zeta(2)\frac{\log^2(1-x)}{x} + \frac{1}{24}\frac{\log^4(1-x)}{x}$$

$$- \frac{1}{6}\frac{\log(x)\log^3(1-x)}{x} - \frac{\log(1-x)\operatorname{Li}_3(x)}{x} + \frac{\operatorname{Li}_4(x)}{x} - \frac{\operatorname{Li}_4(1-x)}{x} + \frac{1}{x}\operatorname{Li}_4\left(\frac{x}{x-1}\right),$$

(continued)

$$|x| \leq 1 \wedge x \neq 0, 1; \tag{4.37}$$

$$(vii) \sum_{n=1}^{\infty} x^n \frac{H_n}{n^3}$$

$$= \zeta(4) + \zeta(3) \log(1-x) + \frac{1}{2}\zeta(2) \log^2(1-x) + \frac{1}{24} \log^4(1-x)$$

$$- \frac{1}{6} \log(x) \log^3(1-x) - \log(1-x) \operatorname{Li}_3(x) + 2\operatorname{Li}_4(x) - \operatorname{Li}_4(1-x) + \operatorname{Li}_4\left(\frac{x}{x-1}\right),$$

$$|x| \leq 1 \wedge x \neq 0, 1,$$

$$\tag{4.38}$$

where $H_n^{(m)} = 1 + \frac{1}{2^m} + \cdots + \frac{1}{n^m}$, $m \geq 1$, is the nth generalized harmonic number of order m, ζ represents the Riemann zeta function, and Li_n denotes the polylogarithm function.

A *challenging question*: Could we possibly derive the Landen trilogarithmic identity, the version with $|x| \leq 1$, $x \neq 0, 1$,

$$\operatorname{Li}_3(x) + \operatorname{Li}_3(1-x) + \operatorname{Li}_3\left(\frac{x}{x-1}\right)$$

$$= \zeta(3) + \zeta(2) \log(1-x) - \frac{1}{2} \log(x) \log^2(1-x) + \frac{1}{6} \log^3(1-x), \tag{4.39}$$

by evaluating $\sum_{n=1}^{\infty} x^n \frac{H_n}{n^2}$ in two simple ways?

4.7   Good-to-Know Generating Functions: The Second Part

Prove that

$$(i) \sum_{n=1}^{\infty} x^n H_n^2 = \frac{1}{1-x} \left(\log^2(1-x) + \operatorname{Li}_2(x) \right), \quad |x| < 1; \tag{4.40}$$

$$(ii) \sum_{n=1}^{\infty} x^n \frac{H_n^2}{n+1}$$

(continued)

$$= 2\frac{\mathrm{Li}_3(1-x)}{x} - \frac{\log(1-x)\,\mathrm{Li}_2(1-x)}{x} - \frac{1}{3}\frac{\log^3(1-x)}{x} - \zeta(2)\frac{\log(1-x)}{x} - 2\zeta(3)\frac{1}{x},$$

$$|x| \leq 1 \wedge\ x \neq 0, 1;$$

$$(4.41)$$

$$(iii)\ \sum_{n=1}^{\infty} x^n \frac{H_n^2}{n}$$

$$= \mathrm{Li}_3(x) - \log(1-x)\,\mathrm{Li}_2(x) - \frac{1}{3}\log^3(1-x),\ |x| \leq 1 \wedge\ x \neq 1; \qquad (4.42)$$

$$(iv)\ \sum_{n=1}^{\infty} x^n \frac{H_n^2}{(n+1)^2}$$

$$= -2\zeta(3)\frac{\log(1-x)}{x} - \zeta(2)\frac{\log^2(1-x)}{x} - \frac{1}{12}\frac{\log^4(1-x)}{x} - \frac{\log^2(1-x)\,\mathrm{Li}_2(1-x)}{x}$$

$$+ \frac{1}{2}\frac{(\mathrm{Li}_2(x))^2}{x} + 2\frac{\log(1-x)\,\mathrm{Li}_3(x)}{x} + 2\frac{\log(1-x)\,\mathrm{Li}_3(1-x)}{x}$$

$$- 2\frac{\mathrm{Li}_4(x)}{x} - 2\frac{1}{x}\,\mathrm{Li}_4\left(\frac{x}{x-1}\right),\ |x| \leq 1 \wedge\ x \neq 0, 1; \qquad (4.43)$$

$$(v)\ \sum_{n=1}^{\infty} x^n \frac{H_n^2}{n^2}$$

$$= 2\zeta(4) - \frac{1}{3}\log(x)\log^3(1-x) - \log^2(1-x)\,\mathrm{Li}_2(1-x) + \frac{1}{2}(\mathrm{Li}_2(x))^2$$

$$+ 2\log(1-x)\,\mathrm{Li}_3(1-x) + \mathrm{Li}_4(x) - 2\,\mathrm{Li}_4(1-x),\ |x| \leq 1 \wedge x \neq 0, 1; \qquad (4.44)$$

$$(vi)\ \sum_{n=1}^{\infty} x^n \frac{H_n^{(2)}}{n+1}$$

$$= 2\frac{\mathrm{Li}_3(1-x)}{x} - \frac{\log(1-x)\,\mathrm{Li}_2(1-x)}{x} - \zeta(2)\frac{\log(1-x)}{x} - 2\zeta(3)\frac{1}{x}$$

$$= 2\frac{\mathrm{Li}_3(1-x)}{x} + \frac{\log(1-x)\,\mathrm{Li}_2(x)}{x} + \frac{\log(x)\log^2(1-x)}{x}$$

$$- 2\zeta(2)\frac{\log(1-x)}{x} - 2\zeta(3)\frac{1}{x},\ |x| \leq 1 \wedge\ x \neq 0, 1; \qquad (4.45)$$

(continued)

$$(vii) \sum_{n=1}^{\infty} x^n \frac{H_n^{(2)}}{n}$$

$$= \text{Li}_3(x) + 2\text{Li}_3(1-x) - \log(1-x)\text{Li}_2(1-x) - \zeta(2)\log(1-x) - 2\zeta(3)$$

$$= \text{Li}_3(x) + 2\text{Li}_3(1-x) + \log(1-x)\text{Li}_2(x) + \log(x)\log^2(1-x)$$

$$- 2\zeta(2)\log(1-x) - 2\zeta(3), \quad |x| \le 1 \wedge x \ne 0, 1; \tag{4.46}$$

$$(viii) \sum_{n=1}^{\infty} x^n \frac{H_n^{(2)}}{(n+1)^2}$$

$$= -2\zeta(4)\frac{1}{x} - 2\zeta(3)\frac{\log(1-x)}{x} - \zeta(2)\frac{\log^2(1-x)}{x} + \frac{1}{3}\frac{\log(x)\log^3(1-x)}{x}$$

$$- \frac{1}{12}\frac{\log^4(1-x)}{x} + \frac{1}{2}\frac{(\text{Li}_2(x))^2}{x} + 2\frac{\log(1-x)\text{Li}_3(x)}{x} - 2\frac{\text{Li}_4(x)}{x}$$

$$+ 2\frac{\text{Li}_4(1-x)}{x} - 2\frac{1}{x}\text{Li}_4\left(\frac{x}{x-1}\right), \quad |x| \le 1 \wedge x \ne 0, 1; \tag{4.47}$$

$$(ix) \sum_{n=1}^{\infty} x^n \frac{H_n^{(2)}}{n^2}$$

$$= -2\zeta(4) - 2\zeta(3)\log(1-x) - \zeta(2)\log^2(1-x) - \frac{1}{12}\log^4(1-x)$$

$$+ \frac{1}{3}\log(x)\log^3(1-x) + \frac{1}{2}(\text{Li}_2(x))^2 + 2\log(1-x)\text{Li}_3(x) - \text{Li}_4(x)$$

$$+ 2\text{Li}_4(1-x) - 2\text{Li}_4\left(\frac{x}{x-1}\right), \quad |x| \le 1 \wedge x \ne 0, 1; \tag{4.48}$$

$$(x) \sum_{n=1}^{\infty} x^n \frac{H_n^{(3)}}{n+1} = -\frac{\log(1-x)\text{Li}_3(x)}{x} - \frac{1}{2}\frac{(\text{Li}_2(x))^2}{x}, \quad |x| \le 1 \wedge x \ne 0, 1; \tag{4.49}$$

$$(xi) \sum_{n=1}^{\infty} x^n \frac{H_n^{(3)}}{n} = \text{Li}_4(x) - \log(1-x)\text{Li}_3(x) - \frac{1}{2}(\text{Li}_2(x))^2, \quad |x| \le 1 \wedge x \ne 1, \tag{4.50}$$

(continued)

where $H_n^{(m)} = 1 + \frac{1}{2^m} + \cdots + \frac{1}{n^m}$, $m \geq 1$, is the nth generalized harmonic number of order m, ζ represents the Riemann zeta function, and Li_n denotes the polylogarithm function.

4.8 Good-to-Know Generating Functions: The Third Part

Prove that

$$(i) \sum_{n=1}^{\infty} x^n H_n^4$$

$$= \frac{1}{1-x}\left(4\zeta(4) + 4\zeta(3)\log(1-x) + 2\zeta(2)\log^2(1-x) + \frac{2}{3}\log^4(1-x)\right.$$

$$-\frac{2}{3}\log(x)\log^3(1-x) + (\text{Li}_2(x))^2 + 4\log(1-x)\text{Li}_3(x) - 7\text{Li}_4(x)$$

$$\left.-4\text{Li}_4(1-x) - 8\text{Li}_4\left(\frac{x}{x-1}\right)\right); \tag{4.51}$$

$$(ii) \sum_{n=1}^{\infty} x^n (H_n^{(2)})^2$$

$$= \frac{1}{1-x}\left(-4\zeta(4) - 4\zeta(3)\log(1-x) - 2\zeta(2)\log^2(1-x) - \frac{1}{6}\log^4(1-x)\right.$$

$$+\frac{2}{3}\log(x)\log^3(1-x) + (\text{Li}_2(x))^2 + 4\log(1-x)\text{Li}_3(x) - 3\text{Li}_4(x)$$

$$\left.+4\text{Li}_4(1-x) - 4\text{Li}_4\left(\frac{x}{x-1}\right)\right); \tag{4.52}$$

$$(iii) \sum_{n=1}^{\infty} x^n H_n H_n^{(3)}$$

$$= \frac{1}{1-x}\left(\zeta(4) + \zeta(3)\log(1-x) + \frac{1}{2}\zeta(2)\log^2(1-x) + \frac{1}{24}\log^4(1-x)\right.$$

(continued)

$$-\frac{1}{6}\log(x)\log^3(1-x)-\frac{1}{2}(\mathrm{Li}_2(x))^2-2\log(1-x)\,\mathrm{Li}_3(x)+2\,\mathrm{Li}_4(x)$$

$$-\mathrm{Li}_4(1-x)+\mathrm{Li}_4\left(\frac{x}{x-1}\right)\bigg); \tag{4.53}$$

$$(iv)\ \sum_{n=1}^{\infty}x^n H_n^2 H_n^{(2)}=\frac{1}{1-x}\left(-\frac{1}{12}\log^4(1-x)-\mathrm{Li}_4(x)-2\,\mathrm{Li}_4\left(\frac{x}{x-1}\right)\right),\tag{4.54}$$

where $|x|<1\wedge x\neq 0$, $H_n^{(m)}=1+\frac{1}{2^m}+\cdots+\frac{1}{n^m}$, $m\geq 1$, is the nth generalized harmonic number of order m, ζ denotes the Riemann zeta function, and Li_n represents the polylogarithm function.

More useful related ordinary generating functions:

$$(v)\ \sum_{n=1}^{\infty}x^n\frac{H_n^3}{n+1}$$

$$=3\zeta(3)\frac{\log(1-x)}{x}+\frac{3}{2}\zeta(2)\frac{\log^2(1-x)}{x}+\frac{1}{4}\frac{\log^4(1-x)}{x}+\frac{3}{2}\frac{\log^2(1-x)\,\mathrm{Li}_2(1-x)}{x}$$

$$-\frac{1}{2}\frac{(\mathrm{Li}_2(x))^2}{x}-\frac{\log(1-x)\,\mathrm{Li}_3(x)}{x}-3\frac{\log(1-x)\,\mathrm{Li}_3(1-x)}{x},\ |x|\leq 1\wedge x\neq 0,1;\tag{4.55}$$

$$(vi)\ \sum_{n=1}^{\infty}x^n\frac{H_n^3}{n}$$

$$=3\zeta(4)-\frac{1}{2}\log(x)\log^3(1-x)+\frac{1}{8}\log^4(1-x)-\frac{3}{2}\log^2(1-x)\,\mathrm{Li}_2(1-x)$$

$$+(\mathrm{Li}_2(x))^2+2\log(1-x)\,\mathrm{Li}_3(x)+3\log(1-x)\,\mathrm{Li}_3(1-x)-2\,\mathrm{Li}_4(x)$$

$$-3\,\mathrm{Li}_4(1-x)-3\,\mathrm{Li}_4\left(\frac{x}{x-1}\right),\ |x|\leq 1\wedge x\neq 0,1;\tag{4.56}$$

$$(vii)\ \sum_{n=1}^{\infty}x^n\frac{H_n H_n^{(2)}}{n+1}$$

$$=\zeta(3)\frac{\log(1-x)}{x}+\frac{1}{2}\zeta(2)\frac{\log^2(1-x)}{x}+\frac{1}{2}\frac{\log^2(1-x)\,\mathrm{Li}_2(1-x)}{x}$$

(continued)

$$-\frac{1}{2}\frac{(\text{Li}_2(x))^2}{x}-\frac{\log(1-x)\,\text{Li}_3(x)}{x}-\frac{\log(1-x)\,\text{Li}_3(1-x)}{x},\ |x|\le 1 \wedge x \ne 0,1;$$

$$(4.57)$$

$$(viii)\ \sum_{n=1}^{\infty}x^n\frac{H_n H_n^{(2)}}{n}$$

$$=-\zeta(4)+\frac{1}{6}\log(x)\log^3(1-x)-\frac{1}{24}\log^4(1-x)+\frac{1}{2}\log^2(1-x)\,\text{Li}_2(1-x)$$

$$-\log(1-x)\,\text{Li}_3(1-x)+\text{Li}_4(1-x)-\text{Li}_4\left(\frac{x}{x-1}\right),\ |x|\le 1 \wedge x \ne 0,1.$$

$$(4.58)$$

4.9 Good-to-Know Generating Functions: The Fourth Part

Prove that

$$(i)\ \sum_{n=1}^{\infty}x^n H_n \sum_{k=n+1}^{\infty}\frac{H_k}{k^2}$$

$$=\frac{1}{1-x}\left(\zeta(4)-\zeta(3)\log(1-x)+\frac{1}{3}\log(x)\log^3(1-x)+\frac{1}{2}\log^2(1-x)\,\text{Li}_2(1-x)\right.$$

$$\left.+\log(1-x)\,\text{Li}_3(x)-\text{Li}_4(1-x)-\text{Li}_4(x)\right).$$

$$(4.59)$$

An interesting generalization of the point (i):

$$(ii)\ \sum_{n=1}^{\infty}x^n H_n \sum_{k=n+1}^{\infty}\frac{H_k}{k^s}$$

$$=-\frac{\log(1-x)}{1-x}\sum_{k=1}^{\infty}\frac{H_k}{k^s}-\frac{1}{1-x}\sum_{k=1}^{\infty}x^k\frac{H_k^2}{k^s}-\frac{1}{1-x}\int_0^x\frac{1}{1-t}\sum_{k=1}^{\infty}t^k\frac{H_k}{k^s}dt,\ s>1,\ s\in\mathbb{R}.$$

$$(4.60)$$

(continued)

By also exploiting (i), show that

$$(iii) \sum_{n=1}^{\infty} x^n H_n \sum_{k=n+1}^{\infty} \frac{1}{k}\left(\zeta(2) - H_k^{(2)}\right)$$

$$= \frac{1}{1-x}\left(2\zeta(4) + \zeta(3)\log(1-x) - \frac{1}{24}\log^4(1-x) - \frac{1}{3}\log(x)\log^3(1-x)\right.$$

$$- \frac{1}{2}\log^2(1-x)\operatorname{Li}_2(x) - \zeta(2)\operatorname{Li}_2(x) - \frac{1}{2}(\operatorname{Li}_2(x))^2 - \log(1-x)\operatorname{Li}_3(x)$$

$$\left. - 2\operatorname{Li}_4(1-x) - \operatorname{Li}_4\left(\frac{x}{x-1}\right)\right), \tag{4.61}$$

where $|x| < 1 \wedge x \neq 0$, $H_n^{(m)} = 1 + \frac{1}{2^m} + \cdots + \frac{1}{n^m}$, $m \geq 1$, is the nth generalized harmonic number of order m, ζ denotes the Riemann zeta function, and Li_n represents the polylogarithm function.

4.10   Good-to-Know Generating Functions: The Fifth Part

Prove that

$$(i) \sum_{n=1}^{\infty} x^n \overline{H}_n^{(m)} = -\frac{\operatorname{Li}_m(-x)}{1-x}, \quad |x| < 1,\ m \in \mathbb{N}, \tag{4.62}$$

where $\operatorname{Li}_1(-x) = -\log(1+x)$;

$$(ii) \sum_{n=1}^{\infty} x^n \frac{\overline{H}_n}{n+1}$$

$$= \frac{1}{x}\left(\frac{1}{2}\log^2(2) - \frac{1}{2}\zeta(2) - \log(2)\log(1-x) + \operatorname{Li}_2\left(\frac{1-x}{2}\right)\right), |x| \leq 1 \wedge x \neq 0, 1;$$

$$\tag{4.63}$$

$$(iii) \sum_{n=1}^{\infty} x^n \frac{\overline{H}_n}{n}$$

(continued)

$$= \frac{1}{2}\log^2(2) - \frac{1}{2}\zeta(2) - \log(2)\log(1-x) - \operatorname{Li}_2(-x) + \operatorname{Li}_2\left(\frac{1-x}{2}\right), \ |x| \le 1 \wedge \ x \ne 1;$$

$$(4.64)$$

$$(iv) \ \sum_{n=1}^{\infty} x^n \frac{\overline{H}_n}{(n+1)^2}$$

$$= \frac{1}{x}\left(\frac{1}{6}\log^3(2) - \frac{1}{2}\log(2)\zeta(2) + \frac{7}{8}\zeta(3) - \frac{1}{2}(\log^2(2) - \zeta(2))\log(1+x)\right.$$

$$+ \frac{1}{2}\log(2)\log^2(1+x) + \log(1+x)\operatorname{Li}_2(x) - \operatorname{Li}_3(x)$$

$$\left. - \operatorname{Li}_3\left(\frac{1+x}{2}\right) - \operatorname{Li}_3\left(\frac{x}{1+x}\right) + \operatorname{Li}_3\left(\frac{2x}{1+x}\right)\right), \ |x| \le 1 \wedge \ x \ne -1, 0;$$

$$(4.65)$$

$$(v) \ \sum_{n=1}^{\infty} x^n \frac{\overline{H}_n}{n^2}$$

$$= \frac{1}{6}\log^3(2) - \frac{1}{2}\log(2)\zeta(2) + \frac{7}{8}\zeta(3) - \frac{1}{2}(\log^2(2) - \zeta(2))\log(1+x)$$

$$+ \frac{1}{2}\log(2)\log^2(1+x) + \log(1+x)\operatorname{Li}_2(x) - \operatorname{Li}_3(x) - \operatorname{Li}_3(-x)$$

$$- \operatorname{Li}_3\left(\frac{1+x}{2}\right) - \operatorname{Li}_3\left(\frac{x}{1+x}\right) + \operatorname{Li}_3\left(\frac{2x}{1+x}\right), \ |x| \le 1 \wedge \ x \ne -1, \quad (4.66)$$

where $\overline{H}_n^{(m)} = 1 - \frac{1}{2^m} + \cdots + (-1)^{n-1}\frac{1}{n^m}$, $m \ge 1$, is the nth generalized skew-harmonic number of order m, ζ represents the Riemann zeta function, and Li_n denotes the polylogarithm function.

4.11 Good-to-Know Generating Functions: The Sixth Part

Prove that

$$(i) \ \sum_{n=1}^{\infty} x^n \overline{H}_n^2$$

$$= \frac{1}{1-x}\left(\zeta(2) - \log^2(2) + 2\log(2)\log(1+x) + \operatorname{Li}_2(x) - 2\operatorname{Li}_2\left(\frac{1+x}{2}\right)\right), |x| < 1;$$

$$(4.67)$$

(continued)

$$(ii) \sum_{n=1}^{\infty} x^n \frac{\overline{H}_n^2}{n+1}$$

$$= \frac{1}{x}\left(\frac{2}{3}\log^3(2) - 2\log(2)\zeta(2) + \frac{3}{2}\zeta(3) - \log^2(2)\log(1-x) - \log(1-x)\operatorname{Li}_2(1-x)\right.$$

$$\left. + 2\log(1-x)\operatorname{Li}_2\left(\frac{1-x}{2}\right) + 2\operatorname{Li}_3(1-x) - 4\operatorname{Li}_3\left(\frac{1-x}{2}\right)\right), \ |x| \le 1 \wedge x \ne 0,1;$$

$$(4.68)$$

$$(iii) \sum_{n=1}^{\infty} x^n \frac{\overline{H}_n^2}{n}$$

$$= \frac{1}{3}\log^3(2) - \log(2)\zeta(2) + \frac{7}{4}\zeta(3) - \log(2)\log^2(1-x) + \frac{1}{3}\log^3(1-x)$$

$$+ \log(1-x)\operatorname{Li}_2(x) - 2\log(1-x)\operatorname{Li}_2(-x) + 2\log(1-x)\operatorname{Li}_2\left(\frac{1-x}{2}\right)$$

$$- \operatorname{Li}_3(x) + 2\operatorname{Li}_3(-x) - 2\operatorname{Li}_3\left(\frac{1-x}{2}\right) - 2\operatorname{Li}_3\left(\frac{2x}{x-1}\right), \ |x| \le 1 \wedge x \ne 1;$$

$$(4.69)$$

$$(iv) \sum_{n=1}^{\infty} x^n \frac{\overline{H}_n^{(2)}}{n+1}$$

$$= \frac{1}{x}\left(\frac{1}{6}\log^3(1+x) - \zeta(2)\log(1-x) + \frac{1}{2}\zeta(2)\log(1-x^2) + \frac{1}{2}\log(x)\log^2(1-x)\right.$$

$$- \frac{1}{4}\log(x^2)\log^2(1-x^2) - \log(1+x)\operatorname{Li}_2(x) - \operatorname{Li}_3(-x) + \operatorname{Li}_3(1-x) - \frac{1}{2}\operatorname{Li}_3(1-x^2)$$

$$\left. - \operatorname{Li}_3\left(\frac{x}{1+x}\right) - \frac{1}{2}\zeta(3)\right), \ |x| \le 1 \wedge x \ne -1, 0, 1; \qquad (4.70)$$

$$(v) \sum_{n=1}^{\infty} x^n \frac{\overline{H}_n^{(2)}}{n}$$

$$= \frac{1}{6}\log^3(1+x) - \zeta(2)\log(1-x) + \frac{1}{2}\zeta(2)\log(1-x^2) + \frac{1}{2}\log(x)\log^2(1-x)$$

$$- \frac{1}{4}\log(x^2)\log^2(1-x^2) - \log(1+x)\operatorname{Li}_2(x) - 2\operatorname{Li}_3(-x) + \operatorname{Li}_3(1-x) - \frac{1}{2}\operatorname{Li}_3(1-x^2)$$

(continued)

$$-\operatorname{Li}_3\left(\frac{x}{1+x}\right)-\frac{1}{2}\zeta(3),\ |x|\le 1 \wedge x\ne -1,0,1, \tag{4.71}$$

where $\overline{H}_n^{(m)}=1-\frac{1}{2^m}+\cdots+(-1)^{n-1}\frac{1}{n^m}$, $m\ge 1$, is the nth generalized skew-harmonic number of order m, ζ represents the Riemann zeta function, and Li_n denotes the polylogarithm function.

4.12 Good-to-Know Generating Functions: The Seventh Part

Prove that

$$(i)\ \sum_{n=1}^{\infty}x^n H_n \overline{H}_n$$

$$=\frac{1}{1-x}\left(\frac{1}{2}\log^2(2)-\frac{1}{2}\zeta(2)-\log(2)\log(1-x)-\frac{1}{2}\log^2(1+x)\right.$$

$$\left.-\operatorname{Li}_2(-x)+\operatorname{Li}_2\left(\frac{1-x}{2}\right)\right),\ |x|<1; \tag{4.72}$$

$$(ii)\ \sum_{n=1}^{\infty}x^n\frac{H_n\overline{H}_n}{n+1}$$

$$=\frac{1}{x}\left(\frac{1}{3}\log^3(2)-\log(2)\zeta(2)+\frac{5}{4}\zeta(3)-\frac{1}{2}\log^2(2)\log(1-x)+\frac{1}{2}\log(2)\log^2(1-x)\right.$$

$$+\frac{1}{2}\zeta(2)\log(1+x)-\frac{1}{2}\log(2)\log^2(1+x)+\frac{1}{6}\log^3(1+x)+\frac{1}{2}\log(x)\log^2(1-x)$$

$$-\frac{1}{4}\log(x^2)\log^2(1-x^2)+\frac{1}{2}\log(1-x)\log^2(1+x)-\log(1+x)\operatorname{Li}_2(x)$$

$$+\log(1+x)\operatorname{Li}_2\left(\frac{1+x}{2}\right)-\operatorname{Li}_3(-x)+\operatorname{Li}_3(1-x)-\operatorname{Li}_3\left(\frac{1-x}{2}\right)$$

$$\left.-\frac{1}{2}\operatorname{Li}_3(1-x^2)-\operatorname{Li}_3\left(\frac{1+x}{2}\right)-\operatorname{Li}_3\left(\frac{x}{1+x}\right)\right),\ |x|\le 1 \wedge x\ne -1,0,1; \tag{4.73}$$

(continued)

$$(iii) \sum_{n=1}^{\infty} x^n \frac{H_n \overline{H}_n}{n}$$

$$= \frac{1}{2} \log^3(2) - \frac{3}{2} \log(2)\zeta(2) + \frac{17}{8}\zeta(3) - \frac{1}{2} \log^2(2) \log(1-x)$$

$$+ \frac{1}{2} \log(2) \log^2(1-x) - \frac{1}{2}(\log^2(2) - 2\zeta(2)) \log(1+x) + \frac{1}{3} \log^3(1+x)$$

$$+ \frac{1}{2} \log(x) \log^2(1-x) - \frac{1}{4} \log(x^2) \log^2(1-x^2) + \frac{1}{2} \log(1-x) \log^2(1+x)$$

$$+\log(1+x)\operatorname{Li}_2(-x)+\log(1+x)\operatorname{Li}_2\left(\frac{1+x}{2}\right)-\operatorname{Li}_3(x)-3\operatorname{Li}_3(-x)+\operatorname{Li}_3(1-x)$$

$$-\operatorname{Li}_3\left(\frac{1-x}{2}\right) - \frac{1}{2}\operatorname{Li}_3(1-x^2) - 2\operatorname{Li}_3\left(\frac{1+x}{2}\right) - 3\operatorname{Li}_3\left(\frac{x}{1+x}\right) + \operatorname{Li}_3\left(\frac{2x}{1+x}\right),$$

$$|x| \leq 1 \wedge x \neq -1, 0, 1. \tag{4.74}$$

More impressive generating functions involving skew-harmonic numbers:

$$(iv) \sum_{n=1}^{\infty} x^n \overline{H}_n H_n^{(2)}$$

$$= \frac{1}{1-x}\left(\frac{1}{6} \log^3(2) - \frac{1}{2} \log(2)\zeta(2) + \frac{7}{8}\zeta(3) - \frac{1}{2}(\log^2(2) - \zeta(2)) \log(1+x)\right.$$

$$+ \frac{1}{2} \log(2) \log^2(1+x) - \frac{1}{3} \log^3(1+x)+\log(1+x)\operatorname{Li}_2(x)-\log(1+x)\operatorname{Li}_2(-x)$$

$$\left.-\operatorname{Li}_3(x)+\operatorname{Li}_3(-x)-\operatorname{Li}_3\left(\frac{1+x}{2}\right)+\operatorname{Li}_3\left(\frac{x}{1+x}\right)+\operatorname{Li}_3\left(\frac{2x}{1+x}\right)\right), \quad |x| < 1;$$

$$\tag{4.75}$$

$$(v) \sum_{n=1}^{\infty} x^n H_n \overline{H}_n^{(2)}$$

$$= \frac{1}{1-x}\left(-\frac{1}{2}\zeta(3) - \zeta(2)\log(1-x) + \frac{1}{2}\zeta(2)\log(1-x^2) + \frac{1}{3}\log^3(1+x)\right.$$

$$+ \frac{1}{2}\log(x)\log^2(1-x) - \frac{1}{4}\log(x^2)\log^2(1-x^2) - \log(1+x)\operatorname{Li}_2(x)$$

(continued)

$$+\log(1+x)\operatorname{Li}_2(-x)-3\operatorname{Li}_3(-x)+\operatorname{Li}_3(1-x)-\frac{1}{2}\operatorname{Li}_3(1-x^2)-2\operatorname{Li}_3\left(\frac{x}{1+x}\right)\Bigg),$$

$$|x|<1\wedge x\neq 0;\qquad(4.76)$$

$$(vi)\ \sum_{n=1}^{\infty}x^n\overline{H}_n\overline{H}_n^{(2)}$$

$$=\frac{1}{1-x}\Bigg(\frac{5}{8}\zeta(3)+\frac{1}{2}\log(2)\zeta(2)-\frac{1}{6}\log^3(2)+\frac{1}{2}(\log^2(2)+2\zeta(2))\log(1-x)$$

$$-\frac{1}{2}\log(2)\log^2(1-x)+\frac{1}{6}\log^3(1-x)-\frac{1}{2}\zeta(2)\log(1+x)-\frac{1}{6}\log^3(1+x)$$

$$-\log(x)\log^2(1-x)+\frac{1}{4}\log(x^2)\log^2(1-x^2)+2\operatorname{Li}_3(-x)-2\operatorname{Li}_3(1-x)$$

$$+\frac{1}{2}\operatorname{Li}_3(1-x^2)+\operatorname{Li}_3\left(\frac{1-x}{2}\right)+\operatorname{Li}_3\left(\frac{x}{1+x}\right)-\operatorname{Li}_3\left(\frac{2x}{x-1}\right)\Bigg),\ |x|<1\wedge x\neq 0;$$

$$(4.77)$$

$$(vii)\ \sum_{n=1}^{\infty}x^n\overline{H}_n^3$$

$$=\frac{1}{1-x}\Bigg(-\frac{3}{2}\log^3(2)+\frac{9}{2}\log(2)\zeta(2)-\frac{63}{8}\zeta(3)+\frac{3}{2}(\log^2(2)-\zeta(2))\log(1+x)$$

$$+\frac{3}{2}\log(2)\log^2(1+x)-\log^3(1+x)+3\log(1+x)\operatorname{Li}_2(x)-3\log(1+x)\operatorname{Li}_2(-x)$$

$$-6\log(1+x)\operatorname{Li}_2\left(\frac{1+x}{2}\right)-3\operatorname{Li}_3(x)+5\operatorname{Li}_3(-x)+9\operatorname{Li}_3\left(\frac{1+x}{2}\right)$$

$$+3\operatorname{Li}_3\left(\frac{x}{1+x}\right)+3\operatorname{Li}_3\left(\frac{2x}{1+x}\right)\Bigg),\ |x|<1;\qquad(4.78)$$

$$(viii)\ \sum_{n=1}^{\infty}x^n H_n\overline{H}_n^2$$

$$=\frac{1}{1-x}\Bigg(\frac{1}{4}\zeta(3)+\log(2)\zeta(2)-\frac{1}{3}\log^3(2)+2\zeta(2)\log(1-x)+\frac{1}{3}\log^3(1-x)$$

$$-(2\zeta(2)-\log^2(2))\log(1+x)-\log(2)\log^2(1+x)-\frac{1}{3}\log^3(1+x)$$

(continued)

$$-\frac{3}{2}\log(x)\log^2(1-x) + \frac{1}{2}\log(x^2)\log^2(1-x^2) - \log^2(1-x)\log(1+x)$$

$$-\operatorname{Li}_3(x) + 4\operatorname{Li}_3(-x) - 3\operatorname{Li}_3(1-x) + \operatorname{Li}_3(1-x^2) + 2\operatorname{Li}_3\left(\frac{1+x}{2}\right)$$

$$+ 2\operatorname{Li}_3\left(\frac{x}{1+x}\right) - 2\operatorname{Li}_3\left(\frac{2x}{x-1}\right)\right), \quad |x| < 1 \wedge x \neq 0; \qquad (4.79)$$

$$(ix) \sum_{n=1}^{\infty} x^n H_n^2 \overline{H}_n$$

$$= \frac{1}{1-x}\left(\frac{5}{6}\log^3(2) - \frac{5}{2}\log(2)\zeta(2) + \frac{27}{8}\zeta(3) - \log^2(2)\log(1-x)\right.$$

$$+ \log(2)\log^2(1-x) - \frac{1}{2}(\log^2(2) - 3\zeta(2))\log(1+x) - \frac{1}{2}\log(2)\log^2(1+x)$$

$$+ \frac{2}{3}\log^3(1+x) + \log(x)\log^2(1-x) - \frac{1}{2}\log(x^2)\log^2(1-x^2) + \log(1-x)\log^2(1+x)$$

$$- \log(1+x)\operatorname{Li}_2(x) + \log(1+x)\operatorname{Li}_2(-x) + 2\log(1+x)\operatorname{Li}_2\left(\frac{1+x}{2}\right)$$

$$- \operatorname{Li}_3(x) - 3\operatorname{Li}_3(-x) + 2\operatorname{Li}_3(1-x) - \operatorname{Li}_3(1-x^2)$$

$$- 2\operatorname{Li}_3\left(\frac{1-x}{2}\right) - 3\operatorname{Li}_3\left(\frac{1+x}{2}\right) - 3\operatorname{Li}_3\left(\frac{x}{1+x}\right) + \operatorname{Li}_3\left(\frac{2x}{1+x}\right)\right), \quad |x| < 1 \wedge x \neq 0, \tag{4.80}$$

where $H_n^{(m)} = 1 + \frac{1}{2^m} + \cdots + \frac{1}{n^m}$, $m \geq 1$, is the nth generalized harmonic number of order m, $\overline{H}_n^{(m)} = 1 - \frac{1}{2^m} + \cdots + (-1)^{n-1}\frac{1}{n^m}$, $m \geq 1$, denotes the nth generalized skew-harmonic number of order m, ζ represents the Riemann zeta function, and Li_n designates the polylogarithm function.

4.13 Two Nice Sums Related to the Generalized Harmonic Numbers, an Asymptotic Expansion Extraction, a Neat Representation of $\log^2(2)$, and a Curious Power Series

Show that

$$(i) \quad \sum_{k=1}^{n} \frac{H_{k+n}}{k} = \frac{1}{2}(2H_n^2 + H_n^{(2)});$$
(4.81)

$$(ii) \quad \sum_{k=1}^{n} \frac{H_k}{k+n} = H_n(H_{2n} - H_n) - \frac{2}{n}(H_{2n} - H_n) + \frac{H_{2n}}{n} - \frac{1}{2}H_n^{(2)},$$
(4.82)

where $H_n^{(m)} = 1 + \frac{1}{2^m} + \cdots + \frac{1}{n^m}$, $m \geq 1$, is the nth generalized harmonic number of order m.

A *challenging question:* Could we easily prove the asymptotic behavior

$$(iii) \quad \sum_{k=1}^{n} \frac{H_k}{k+n} = \gamma \log(2) - \frac{\pi^2}{12} + \log(2)\log(n), \text{ when } n \to \infty,$$
(4.83)

without calculating the precise value of $\sum_{k=1}^{n} \frac{H_k}{k+n}$ given at the point (ii)?

A *neat limit based on the previous result:*

$$(iv) \quad \log^2(2) = \lim_{n\to\infty} \sum_{k=1}^{n} \left(\frac{H_{k+n}}{k+3n} - \frac{nH_k}{(k+n)(k+2n)} \right).$$
(4.84)

Find the coefficient c_n of the power series of $f(x)$:

$$(v) \quad f(x) = \int_0^1 \frac{\log(1 - xy^2)}{y(xy-1)(1-xy^2)} dy = \sum_{n=1}^{\infty} c_n x^n, \; |x| < 1.$$
(4.85)

4.14   Opening the World of Harmonic Series with Beautiful Series That Require Athletic Movements During Their Calculations: The First (Enjoyable) Part

Prove with *no use of integrals* that

$$(i) \sum_{k=1}^{\infty} \frac{H_k}{2k(2k+1)} = \log^2(2). \qquad (4.86)$$

Employ the result from the point (i) to prove with *no use of integrals* that

$$(ii) \sum_{k=1}^{\infty} (-1)^{k-1} \frac{H_k}{k} = \frac{1}{2} \left(\zeta(2) - \log^2(2) \right), \qquad (4.87)$$

where $H_n = \sum_{k=1}^{n} \frac{1}{k}$ is the nth harmonic number and ζ represents the Riemann zeta function.

4.15   Opening the World of Harmonic Series with Beautiful Series That Require Athletic Movements During Their Calculations: The Second (Enjoyable) Part

Prove with *no use of integrals* that

$$(i) \sum_{k=1}^{\infty} \frac{H_k^{(2)}}{2k(2k+1)} = \frac{5}{4} \zeta(3) - \log(2)\zeta(2); \qquad (4.88)$$

$$(ii) \sum_{k=1}^{\infty} (-1)^{k-1} \frac{H_k}{k^2} = \frac{5}{8} \zeta(3), \qquad (4.89)$$

where $H_n^{(m)} = 1 + \frac{1}{2^m} + \cdots + \frac{1}{n^m}$, $m \geq 1$, is the nth generalized harmonic number of order m and ζ represents the Riemann zeta function.

A *(super) challenging question*: Show with no use of integrals and no use of the Cauchy product of two series that

$$(iii) \, 2 \sum_{k=1}^{\infty} (-1)^{k-1} \frac{H_k}{k^3} + \sum_{k=1}^{\infty} (-1)^{k-1} \frac{H_k^{(2)}}{k^2} = \frac{37}{16} \zeta(4). \qquad (4.90)$$

4.16   A Special Harmonic Series in Disguise Involving Nice Tricks

Prove that

$$\sum_{n=1}^{\infty} \frac{1}{n(2n+1)} \left(\frac{1}{n+1} + \frac{1}{n+2} + \cdots + \frac{1}{2n+1} \right)^2$$

$$= \zeta(3) - \frac{1}{2}\zeta(2) - 3\log(2)\zeta(2) - 2\log(2) - 2\log^2(2) - \frac{2}{3}\log^3(2) + 6,$$

(4.91)

where ζ represents the Riemann zeta function.

   A *(fascinating) challenging question*: Is it possible to make all calculations by mainly using series manipulations, with no use of integrals?

4.17   A Few Nice Generalized Series: Most of Them May Be Seen as Applications of *The Master Theorem of Series*

Show that

$$(i) \sum_{k=1}^{\infty} \frac{H_{2k} - 1/2 H_k}{(k+1)(k+n+1)}$$

$$= 2\log(2)\frac{1}{n} \left(H_{2n} - \frac{1}{2}H_n \right) - 4\log(2)\frac{1}{2n+1} + \frac{1}{n}\sum_{i=1}^{n} \frac{H_i}{2i+1};$$

(4.92)

$$(ii) \sum_{k=1}^{\infty} \frac{H_{2k}}{(k+1)(k+n+1)}$$

$$= \frac{1}{4}\frac{1}{n}(H_n^2 + H_n^{(2)}) + 2\log(2)\frac{1}{n} \left(H_{2n} - \frac{1}{2}H_n \right) - 4\log(2)\frac{1}{2n+1} + \frac{1}{n}\sum_{i=1}^{n} \frac{H_i}{2i+1};$$

(4.93)

$$(iii) \sum_{k=1}^{\infty} \frac{(-1)^{k-1}H_k}{(k+1)(k+n+1)}$$

(continued)

$$= \frac{1}{2}\log^2(2)\frac{1}{n} + \frac{1}{2}\log^2(2)(-1)^{n-1}\frac{1}{n} - \log(2)(-1)^{n-1}\frac{H_n}{n} - \log(2)(-1)^{n-1}\frac{\overline{H}_n}{n}$$

$$+ (-1)^{n-1}\frac{H_n\overline{H}_n}{n} + (-1)^{n-1}\frac{\overline{H}_n^{(2)}}{n} - (-1)^{n-1}\frac{1}{n}\sum_{k=1}^{n}(-1)^{k-1}\frac{H_k}{k}$$

$$= \frac{1}{2}\log^2(2)\frac{1}{n} + \frac{1}{2}\log^2(2)(-1)^{n-1}\frac{1}{n} - \log(2)(-1)^{n-1}\frac{H_n}{n} - \log(2)(-1)^{n-1}\frac{\overline{H}_n}{n}$$

$$+ (-1)^{n-1}\frac{1}{n}\sum_{k=1}^{n}\frac{\overline{H}_k}{k}; \tag{4.94}$$

$$(iv)\ \sum_{k=1}^{\infty}\frac{\overline{H}_k^{(m)}}{(k+1)(k+n+1)} = \begin{cases} \log(2)\dfrac{H_n}{n} - \dfrac{1}{2}\dfrac{H_n^{(2)}}{n} + \log(2)\dfrac{\overline{H}_n}{n} - \dfrac{1}{2}\dfrac{\overline{H}_n^2}{n}, & m=1; \\[2mm] \log(2)(-1)^{m-1}\dfrac{\overline{H}_n^{(m)}}{n} \\[2mm] + \dfrac{(-1)^{m-1}}{n}\displaystyle\sum_{i=1}^{m}(-1)^{i-1}\eta(i)H_n^{(m-i+1)} \\[2mm] - \dfrac{(-1)^{m-1}}{n}\displaystyle\sum_{i=1}^{n}(-1)^{i-1}\dfrac{\overline{H}_i}{i^m}, & m \geq 2, \end{cases}$$

$$\tag{4.95}$$

where $H_n^{(m)} = 1 + \frac{1}{2^m} + \cdots + \frac{1}{n^m}$, $m \geq 1$, is the nth generalized harmonic number of order m, $\overline{H}_n^{(m)} = 1 - \frac{1}{2^m} + \cdots + (-1)^{n-1}\frac{1}{n^m}$, $m \geq 1$, represents the nth generalized skew-harmonic number of order m, and η denotes the Dirichlet eta function.

Examples at the point (iv):
For $m = 2$,

$$\sum_{k=1}^{\infty}\frac{\overline{H}_k^{(2)}}{(k+1)(k+n+1)}$$

$$= \frac{1}{2}\zeta(2)\frac{H_n}{n} - \log(2)\frac{H_n^{(2)}}{n} - \log(2)\frac{\overline{H}_n^{(2)}}{n} + \frac{1}{n}\sum_{i=1}^{n}(-1)^{i-1}\frac{\overline{H}_i}{i^2}.$$

For $m = 3$,

$$\sum_{k=1}^{\infty}\frac{\overline{H}_k^{(3)}}{(k+1)(k+n+1)}$$

(continued)

$$= \frac{3}{4}\zeta(3)\frac{H_n}{n} - \frac{1}{2}\zeta(2)\frac{H_n^{(2)}}{n} + \log(2)\frac{H_n^{(3)}}{n} + \log(2)\frac{\overline{H}_n^{(3)}}{n} - \frac{1}{n}\sum_{i=1}^{n}(-1)^{i-1}\frac{\overline{H}_i}{i^3}.$$

For $m = 4$,

$$\sum_{k=1}^{\infty}\frac{\overline{H}_k^{(4)}}{(k+1)(k+n+1)}$$

$$= \frac{7}{8}\zeta(4)\frac{H_n}{n} - \frac{3}{4}\zeta(3)\frac{H_n^{(2)}}{n} + \frac{1}{2}\zeta(2)\frac{H_n^{(3)}}{n} - \log(2)\frac{H_n^{(4)}}{n} - \log(2)\frac{\overline{H}_n^{(4)}}{n}$$

$$+ \frac{1}{n}\sum_{i=1}^{n}(-1)^{i-1}\frac{\overline{H}_i}{i^4}.$$

4.18   Useful Relations Involving Polygamma with the Argument $n/2$ and the Generalized Skew-Harmonic Numbers

Let m, n be natural numbers. Show that

$$(i) \ \psi\left(\frac{n}{2}\right)$$

$$= -\log(2) - \gamma - \log(2)(-1)^{n-1} - 2\frac{1}{n} + H_n + (-1)^{n-1}\overline{H}_n; \qquad (4.96)$$

$$(ii) \ \psi\left(\frac{n}{2}+1\right)$$

$$= -\log(2) - \gamma - \log(2)(-1)^{n-1} + H_n + (-1)^{n-1}\overline{H}_n; \qquad (4.97)$$

$$(iii) \ \psi^{(m)}\left(\frac{n}{2}\right)$$

$$= (-1)^{m-1}2^{m+1}m!\frac{1}{n^{m+1}} + (-1)^{m+n}(2^m-1)m!\zeta(m+1)$$

$$+(-1)^{m-1}2^m m!\left(\zeta(m+1) - H_n^{(m+1)}\right) - (-1)^{m+n}2^m m!\overline{H}_n^{(m+1)}; \qquad (4.98)$$

$$(iv) \ \psi^{(m)}\left(\frac{n}{2}+1\right)$$

(continued)

$$= (-1)^{m+n}(2^m - 1)m!\zeta(m + 1) + (-1)^{m-1}2^m m! \left(\zeta(m + 1) - H_n^{(m+1)}\right)$$

$$- (-1)^{m+n}2^m m! \overline{H}_n^{(m+1)}, \tag{4.99}$$

where $H_n^{(m)} = 1 + \frac{1}{2^m} + \cdots + \frac{1}{n^m}$, $m \geq 1$, is the nth generalized harmonic number of order m, $\overline{H}_n^{(m)} = 1 - \frac{1}{2^m} + \cdots + (-1)^{n-1}\frac{1}{n^m}$, $m \geq 1$, represents the nth generalized skew-harmonic number of order m, ζ denotes the Riemann zeta function, and $\psi^{(n)}$ designates the polygamma function.

Examples at the point (iii):
For $m = 1$,

$$\psi^{(1)}\left(\frac{n}{2}\right)$$

$$= 2\zeta(2) + \zeta(2)(-1)^{n-1} + 4\frac{1}{n^2} - 2H_n^{(2)} - 2(-1)^{n-1}\overline{H}_n^{(2)}.$$

For $m = 2$,

$$\psi^{(2)}\left(\frac{n}{2}\right)$$

$$= -8\zeta(3) - 6\zeta(3)(-1)^{n-1} - 16\frac{1}{n^3} + 8H_n^{(3)} + 8(-1)^{n-1}\overline{H}_n^{(3)}.$$

For $m = 3$,

$$\psi^{(3)}\left(\frac{n}{2}\right)$$

$$= 48\zeta(4) + 42\zeta(4)(-1)^{n-1} + 96\frac{1}{n^4} - 48H_n^{(4)} - 48(-1)^{n-1}\overline{H}_n^{(4)}.$$

Examples at the point (iv):
For $m = 1$,

$$\psi^{(1)}\left(\frac{n}{2} + 1\right)$$

$$= 2\zeta(2) + \zeta(2)(-1)^{n-1} - 2H_n^{(2)} - 2(-1)^{n-1}\overline{H}_n^{(2)}.$$

For $m = 2$,

$$\psi^{(2)}\left(\frac{n}{2} + 1\right)$$

(continued)

$$= -8\zeta(3) - 6\zeta(3)(-1)^{n-1} + 8H_n^{(3)} + 8(-1)^{n-1}\overline{H}_n^{(3)}.$$

For $m = 3$,

$$\psi^{(3)}\left(\frac{n}{2}+1\right)$$

$$= 48\zeta(4) + 42\zeta(4)(-1)^{n-1} - 48H_n^{(4)} - 48(-1)^{n-1}\overline{H}_n^{(4)}.$$

4.19 A Key Classical Generalized Harmonic Series

Let $m \geq 1$ be a natural number. Prove that

$$(i) \sum_{k=1}^{\infty} \frac{H_k}{(2k+1)^{2m}}$$

$$= 2m\left(1 - \frac{1}{2^{2m+1}}\right)\zeta(2m+1) - 2\log(2)\left(1 - \frac{1}{2^{2m}}\right)\zeta(2m)$$

$$- \frac{1}{2^{2m}}\sum_{k=1}^{m-1}(1 - 2^{k+1})(1 - 2^{2m-k})\zeta(k+1)\zeta(2m-k); \tag{4.100}$$

$$(ii) \sum_{k=1}^{\infty} \frac{H_k}{(2k+1)^{2m+1}}$$

$$= \left(1 - \frac{1}{2^{2m+2}}\right)(2m+1)\zeta(2m+2)$$

$$- \left(1 - \frac{1}{2^{m+1}}\right)^2 \zeta^2(m+1) - 2\log(2)\left(1 - \frac{1}{2^{2m+1}}\right)\zeta(2m+1)$$

$$- \frac{1}{2^{2m+1}}\sum_{k=1}^{m-1}(1 - 2^{k+1})(1 - 2^{2m-k+1})\zeta(k+1)\zeta(2m-k+1), \tag{4.101}$$

where $H_n = \sum_{k=1}^{n}\frac{1}{k}$ is the nth harmonic number and ζ represents the Riemann zeta function.

(continued)

A larger generalization:
Let $p \geq 2$ be a natural number. Show that

$$(iii) \sum_{k=1}^{\infty} \frac{H_k}{(2k+1)^p}$$

$$= p \left(1 - \frac{1}{2^{p+1}}\right) \zeta(p+1) - 2 \log(2) \left(1 - \frac{1}{2^p}\right) \zeta(p)$$

$$- \frac{1}{2^{p+1}} \sum_{k=1}^{p-2} (1 - 2^{k+1})(1 - 2^{p-k}) \zeta(k+1) \zeta(p-k). \tag{4.102}$$

Examples:
For $p = 2$,

$$\sum_{k=1}^{\infty} \frac{H_k}{(2k+1)^2} = \frac{7}{4} \zeta(3) - \frac{3}{2} \log(2) \zeta(2).$$

For $p = 3$,

$$\sum_{k=1}^{\infty} \frac{H_k}{(2k+1)^3} = \frac{45}{32} \zeta(4) - \frac{7}{4} \log(2) \zeta(3).$$

For $p = 4$,

$$\sum_{k=1}^{\infty} \frac{H_k}{(2k+1)^4} = \frac{31}{8} \zeta(5) - \frac{21}{16} \zeta(2) \zeta(3) - \frac{15}{8} \log(2) \zeta(4).$$

For $p = 5$,

$$\sum_{k=1}^{\infty} \frac{H_k}{(2k+1)^5} = \frac{315}{128} \zeta(6) - \frac{49}{64} \zeta^2(3) - \frac{31}{16} \log(2) \zeta(5).$$

For $p = 6$,

$$\sum_{k=1}^{\infty} \frac{H_k}{(2k+1)^6} = \frac{381}{64} \zeta(7) - \frac{93}{64} \zeta(2) \zeta(5) - \frac{105}{64} \zeta(3) \zeta(4) - \frac{63}{32} \log(2) \zeta(6).$$

4.20   Revisiting Two Classical Challenging Alternating Harmonic Series, Calculated by Exploiting a Beta Function Form

Show that

$$(i) \sum_{n=1}^{\infty}(-1)^{n-1}\frac{H_n}{n^3}$$

$$= \frac{11}{4}\zeta(4) - \frac{7}{4}\log(2)\zeta(3) + \frac{1}{2}\log^2(2)\zeta(2) - \frac{1}{12}\log^4(2) - 2\operatorname{Li}_4\left(\frac{1}{2}\right);$$
$$(4.103)$$

$$(ii) \sum_{n=1}^{\infty}(-1)^{n-1}\frac{H_n}{n^4} = \frac{59}{32}\zeta(5) - \frac{1}{2}\zeta(2)\zeta(3), \qquad (4.104)$$

where $H_n = \sum_{k=1}^{n}\frac{1}{k}$ is the nth harmonic number, ζ represents the Riemann zeta function, and Li_n denotes the polylogarithm function.

4.21   A Famous Classical Generalization with Alternating Harmonic Series, Derived by a New Special Way

Let $m \geq 1$ be a natural number. Prove that

$$\sum_{k=1}^{\infty}(-1)^{k-1}\frac{H_k}{k^{2m}} = \left(m + \frac{1}{2}\right)\eta(2m+1) - \frac{1}{2}\zeta(2m+1) - \sum_{i=1}^{m-1}\eta(2i)\zeta(2m-2i+1),$$
$$(4.105)$$

where $H_n = \sum_{k=1}^{n}\frac{1}{k}$ is the nth harmonic number, ζ represents the Riemann zeta function, and η denotes the Dirichlet eta function.

Examples:
For $m = 1$,

$$\sum_{k=1}^{\infty}(-1)^{k-1}\frac{H_k}{k^2} = \frac{5}{8}\zeta(3).$$

For $m = 2$,

(continued)

$$\sum_{k=1}^{\infty}(-1)^{k-1}\frac{H_k}{k^4} = \frac{59}{32}\zeta(5) - \frac{1}{2}\zeta(2)\zeta(3).$$

For $m = 3$,

$$\sum_{k=1}^{\infty}(-1)^{k-1}\frac{H_k}{k^6} = \frac{377}{128}\zeta(7) - \frac{7}{8}\zeta(3)\zeta(4) - \frac{1}{2}\zeta(2)\zeta(5).$$

For $m = 4$,

$$\sum_{k=1}^{\infty}(-1)^{k-1}\frac{H_k}{k^8} = \frac{2039}{512}\zeta(9) - \frac{7}{8}\zeta(4)\zeta(5) - \frac{31}{32}\zeta(3)\zeta(6) - \frac{1}{2}\zeta(2)\zeta(7).$$

For $m = 5$,

$$\sum_{k=1}^{\infty}(-1)^{k-1}\frac{H_k}{k^{10}}$$

$$= \frac{10229}{2048}\zeta(11) - \frac{31}{32}\zeta(5)\zeta(6) - \frac{7}{8}\zeta(4)\zeta(7) - \frac{127}{128}\zeta(3)\zeta(8) - \frac{1}{2}\zeta(2)\zeta(9).$$

4.22   Seven Useful Generalized Harmonic Series

Let m be a natural number. Prove that

$$(i)\ \sum_{n=1}^{\infty}(-1)^{n-1}\frac{H_n^{(m)}}{n} = \frac{1}{2}\left(m\zeta(m+1) - \sum_{k=1}^{m}\eta(k)\eta(m-k+1)\right); \qquad (4.106)$$

$$(ii)\ \sum_{n=1}^{\infty}(-1)^{n-1}\frac{H_{2n}^{(m)}}{n}$$

$$= m\zeta(m+1) - \frac{1}{2^{m+1}}\sum_{k=1}^{m}\eta(k)\eta(m-k+1) - \sum_{k=1}^{m}\beta(k)\beta(m-k+1);$$

$$(4.107)$$

(continued)

$$(iii) \sum_{n=1}^{\infty} \frac{\overline{H}_n}{n^m}$$

$$= \log(2)\zeta(m) - \frac{1}{2}m\zeta(m+1) + \eta(m+1) + \frac{1}{2}\sum_{k=1}^{m} \eta(k)\eta(m-k+1), \; m \neq 1;$$

$$(4.108)$$

$$(iv) \sum_{n=1}^{\infty} \frac{\overline{H}_n}{(2n+1)^m}$$

$$= \log(2)(1 - 2^{-m})\zeta(m) - m(1 - 2^{-m-1})\zeta(m+1) + \sum_{k=1}^{m} \beta(k)\beta(m-k+1), \; m \neq 1;$$

$$(4.109)$$

$$(v) \sum_{n=1}^{\infty}(-1)^{n-1}\frac{\overline{H}_{2n}^{(m)}}{n}$$

$$= m\eta(m+1) + \frac{1}{2^{m+1}}\sum_{k=1}^{m} \eta(k)\eta(m-k+1) - \sum_{k=1}^{m} \beta(k)\beta(m-k+1);$$

$$(4.110)$$

$$(vi) \sum_{n=1}^{\infty}(-1)^{n-1}\frac{\overline{H}_n}{n^{2m}}$$

$$= \log(2)\zeta(2m) - \frac{1}{2^{2m+1}}(m2^{2m+1} - 2m - 1)\zeta(2m+1)$$

$$+ \left(1 - \frac{1}{2^{2m-1}}\right)\sum_{k=1}^{m-1} \zeta(2k)\zeta(2m-2k+1)$$

$$- \sum_{k=1}^{m-1} \eta(2k)\zeta(2m-2k+1) + \sum_{k=0}^{m-1} \eta(2k+1)\eta(2m-2k); \quad (4.111)$$

$$(vii) \sum_{n=1}^{\infty}(-1)^{n-1}\frac{\overline{H}_n^{(2m)}}{n}$$

$$= \log(2)\eta(2m) - \log(2)\zeta(2m) + \frac{1}{2^{2m+1}}((m+1)2^{2m+1} - 2m - 1)\zeta(2m+1)$$

(continued)

$$-\left(1 - \frac{1}{2^{2m-1}}\right) \sum_{k=1}^{m-1} \zeta(2k)\zeta(2m - 2k + 1)$$

$$+ \sum_{k=1}^{m-1} \eta(2k)\zeta(2m - 2k + 1) - \sum_{k=0}^{m-1} \eta(2k + 1)\eta(2m - 2k), \qquad (4.112)$$

where $H_n^{(m)} = 1 + \frac{1}{2^m} + \cdots + \frac{1}{n^m}$, $m \geq 1$, is the nth generalized harmonic number of order m, $\overline{H}_n^{(m)} = 1 - \frac{1}{2^m} + \cdots + (-1)^{n-1}\frac{1}{n^m}$, $m \geq 1$, denotes the nth generalized skew-harmonic number of order m, ζ represents the Riemann zeta function, η signifies the Dirichlet eta function, and β designates the Dirichlet beta function.

Examples at the point (i):
For $m = 1$,

$$\sum_{n=1}^{\infty} (-1)^{n-1} \frac{H_n}{n} = \frac{1}{2}\zeta(2) - \frac{1}{2}\log^2(2).$$

For $m = 2$,

$$\sum_{n=1}^{\infty} (-1)^{n-1} \frac{H_n^{(2)}}{n} = \zeta(3) - \frac{1}{2}\log(2)\zeta(2).$$

For $m = 3$,

$$\sum_{n=1}^{\infty} (-1)^{n-1} \frac{H_n^{(3)}}{n} = \frac{19}{16}\zeta(4) - \frac{3}{4}\log(2)\zeta(3).$$

For $m = 4$,

$$\sum_{n=1}^{\infty} (-1)^{n-1} \frac{H_n^{(4)}}{n} = 2\zeta(5) - \frac{3}{8}\zeta(2)\zeta(3) - \frac{7}{8}\log(2)\zeta(4).$$

Examples at the point (ii):
For $m = 1$,

$$\sum_{n=1}^{\infty} (-1)^{n-1} \frac{H_{2n}}{n} = \frac{5}{8}\zeta(2) - \frac{1}{4}\log^2(2).$$

(continued)

For $m = 2$,

$$\sum_{n=1}^{\infty} (-1)^{n-1} \frac{H_{2n}^{(2)}}{n} = 2\zeta(3) - \frac{1}{8} \log(2)\zeta(2) - \frac{\pi}{2} G.$$

For $m = 3$,

$$\sum_{n=1}^{\infty} (-1)^{n-1} \frac{H_{2n}^{(3)}}{n} = \frac{199}{128} \zeta(4) - G^2 - \frac{3}{32} \log(2)\zeta(3).$$

For $m = 4$,

$$\sum_{n=1}^{\infty} (-1)^{n-1} \frac{H_{2n}^{(4)}}{n}$$

$$= 4\zeta(5) - \frac{\pi^2}{256} \zeta(3) - \frac{7}{11520} \log(2)\pi^4 + \frac{\pi^5}{192} - \frac{\pi^3}{16} G - \frac{\pi}{1536} \psi^{(3)}\left(\frac{1}{4}\right).$$

Examples at the point (iii):
For $m = 2$,

$$\sum_{n=1}^{\infty} \frac{\overline{H}_n}{n^2} = \frac{3}{2} \log(2)\zeta(2) - \frac{1}{4}\zeta(3).$$

For $m = 3$,

$$\sum_{n=1}^{\infty} \frac{\overline{H}_n}{n^3} = \frac{7}{4} \log(2)\zeta(3) - \frac{5}{16}\zeta(4).$$

For $m = 4$,

$$\sum_{n=1}^{\infty} \frac{\overline{H}_n}{n^4} = \frac{15}{8} \log(2)\zeta(4) + \frac{3}{8}\zeta(2)\zeta(3) - \frac{17}{16}\zeta(5).$$

Examples at the point (iv):
For $m = 2$,

$$\sum_{n=1}^{\infty} \frac{\overline{H}_n}{(2n+1)^2} = \frac{\pi}{2} G + \frac{3}{4} \log(2)\zeta(2) - \frac{7}{4}\zeta(3).$$

(continued)

For $m = 3$,

$$\sum_{n=1}^{\infty} \frac{\overline{H}_n}{(2n+1)^3} = G^2 - \frac{45}{32}\zeta(4) + \frac{7}{8}\log(2)\zeta(3).$$

For $m = 4$,

$$\sum_{n=1}^{\infty} \frac{\overline{H}_n}{(2n+1)^4} = \frac{\pi^3}{16}G + \log(2)\frac{\pi^4}{96} - \frac{\pi^5}{192} - \frac{31}{8}\zeta(5) + \frac{\pi}{1536}\psi^{(3)}\left(\frac{1}{4}\right).$$

Examples at the point (v):
For $m = 1$,

$$\sum_{n=1}^{\infty} (-1)^{n-1} \frac{\overline{H}_{2n}}{n} = \frac{1}{8}\zeta(2) + \frac{1}{4}\log^2(2).$$

For $m = 2$,

$$\sum_{n=1}^{\infty} (-1)^{n-1} \frac{\overline{H}_{2n}^{(2)}}{n} = \frac{3}{2}\zeta(3) + \frac{1}{8}\log(2)\zeta(2) - \frac{\pi}{2}G.$$

For $m = 3$,

$$\sum_{n=1}^{\infty} (-1)^{n-1} \frac{\overline{H}_{2n}^{(3)}}{n} = \frac{3}{32}\log(2)\zeta(3) + \frac{161}{128}\zeta(4) - G^2.$$

For $m = 4$,

$$\sum_{n=1}^{\infty} (-1)^{n-1} \frac{\overline{H}_{2n}^{(4)}}{n}$$

$$= \frac{15}{4}\zeta(5) + \frac{\pi^2}{256}\zeta(3) + \frac{7}{11520}\log(2)\pi^4 + \frac{\pi^5}{192} - \frac{\pi^3}{16}G - \frac{\pi}{1536}\psi^{(3)}\left(\frac{1}{4}\right).$$

Examples at the point (vi):
For $m = 1$,

$$\sum_{n=1}^{\infty} (-1)^{n-1} \frac{\overline{H}_n}{n^2} = \frac{3}{2}\log(2)\zeta(2) - \frac{5}{8}\zeta(3).$$

For $m = 2$,

(continued)

$$\sum_{n=1}^{\infty}(-1)^{n-1}\frac{\overline{H}_n}{n^4}=\frac{15}{8}\log(2)\zeta(4)+\frac{3}{4}\zeta(2)\zeta(3)-\frac{59}{32}\zeta(5).$$

For $m=3$,

$$\sum_{n=1}^{\infty}(-1)^{n-1}\frac{\overline{H}_n}{n^6}=\frac{63}{32}\log(2)\zeta(6)+\frac{3}{4}\zeta(3)\zeta(4)+\frac{15}{16}\zeta(2)\zeta(5)-\frac{377}{128}\zeta(7).$$

For $m=4$,

$$\sum_{n=1}^{\infty}(-1)^{n-1}\frac{\overline{H}_n}{n^8}$$

$$=\frac{255}{128}\log(2)\zeta(8)-\frac{2039}{512}\zeta(9)+\frac{63}{64}\zeta(2)\zeta(7)+\frac{3}{4}\zeta(3)\zeta(6)+\frac{15}{16}\zeta(4)\zeta(5).$$

Examples at the point (vii):
For $m=1$,

$$\sum_{n=1}^{\infty}(-1)^{n-1}\frac{\overline{H}_n^{(2)}}{n}=\frac{13}{8}\zeta(3)-\log(2)\zeta(2).$$

For $m=2$,

$$\sum_{n=1}^{\infty}(-1)^{n-1}\frac{\overline{H}_n^{(4)}}{n}=\frac{91}{32}\zeta(5)-\frac{3}{4}\zeta(2)\zeta(3)-\log(2)\zeta(4).$$

For $m=3$,

$$\sum_{n=1}^{\infty}(-1)^{n-1}\frac{\overline{H}_n^{(6)}}{n}=\frac{505}{128}\zeta(7)-\frac{15}{16}\zeta(2)\zeta(5)-\frac{3}{4}\zeta(3)\zeta(4)-\log(2)\zeta(6).$$

For $m=4$,

$$\sum_{n=1}^{\infty}(-1)^{n-1}\frac{\overline{H}_n^{(8)}}{n}$$

$$=\frac{2551}{512}\zeta(9)-\frac{63}{64}\zeta(2)\zeta(7)-\frac{3}{4}\zeta(3)\zeta(6)-\frac{15}{16}\zeta(4)\zeta(5)-\log(2)\zeta(8).$$

4.23 A Special Challenging Harmonic Series of Weight 4, Involving Harmonic Numbers of the Type H_{2n}

Prove that

$$\sum_{n=1}^{\infty} \frac{H_n H_{2n}}{n^2}$$

$$= \frac{13}{8}\zeta(4) + \frac{7}{2}\log(2)\zeta(3) - \log^2(2)\zeta(2) + \frac{1}{6}\log^4(2) + 4\operatorname{Li}_4\left(\frac{1}{2}\right),$$
(4.113)

where $H_n = \sum_{k=1}^{n} \frac{1}{k}$ is the nth harmonic number, ζ represents the Riemann zeta function, and Li_n denotes the polylogarithm function.

4.24 Two Useful Atypical Harmonic Series of Weight 4 with Denominators of the Type $(2n+1)^2$

Prove that

$$(i) \sum_{n=1}^{\infty} \frac{H_n^{(2)}}{(2n+1)^2}$$

$$= \frac{1}{3}\log^4(2) - 2\log^2(2)\zeta(2) + 7\log(2)\zeta(3) - \frac{121}{16}\zeta(4) + 8\operatorname{Li}_4\left(\frac{1}{2}\right);$$
(4.114)

$$(ii) \sum_{n=1}^{\infty} \frac{H_n^2}{(2n+1)^2}$$

$$= \frac{1}{3}\log^4(2) + \log^2(2)\zeta(2) - \frac{61}{16}\zeta(4) + 8\operatorname{Li}_4\left(\frac{1}{2}\right),$$
(4.115)

where $H_n^{(m)} = 1 + \frac{1}{2^m} + \cdots + \frac{1}{n^m}$, $m \geq 1$, is the nth generalized harmonic number of order m, ζ represents the Riemann zeta function, and Li_n denotes the polylogarithm function.

4.25   Another Special Challenging Harmonic Series of Weight 4, Involving Harmonic Numbers of the Type H_{2n}

Prove that

$$\sum_{n=1}^{\infty} \frac{H_n H_{2n}}{(2n+1)^2}$$

$$= \frac{1}{12} \log^4(2) - \frac{1}{2} \log^2(2)\zeta(2) + \frac{7}{8} \log(2)\zeta(3) - \frac{1}{4}\zeta(4) + 2\operatorname{Li}_4\left(\frac{1}{2}\right),$$
$$(4.116)$$

where $H_n = \sum_{k=1}^{n} \frac{1}{k}$ is the nth harmonic number, ζ represents the Riemann zeta function, and Li_n denotes the polylogarithm function.

4.26   A First Uncommon Series with the Tail of the Riemann Zeta Function $\zeta(2) - H_{2n}^{(2)}$, Related to Weight 4 Harmonic Series

Prove that

$$\sum_{n=1}^{\infty} \frac{H_n}{n}\left(\zeta(2) - H_{2n}^{(2)}\right)$$

$$= \frac{1}{3} \log^4(2) - \frac{1}{2} \log^2(2)\zeta(2) + \frac{7}{2} \log(2)\zeta(3) - \frac{21}{4}\zeta(4) + 8\operatorname{Li}_4\left(\frac{1}{2}\right),$$
$$(4.117)$$

where $H_n^{(m)} = 1 + \frac{1}{2^m} + \cdots + \frac{1}{n^m}$, $m \ge 1$, is the nth generalized harmonic number of order m, ζ denotes the Riemann zeta function, and Li_n represents the polylogarithm function.

4.27 A Second Uncommon Series with the Tail of the Riemann Zeta Function $\zeta(2) - H_n^{(2)}$, Related to Weight 4 Harmonic Series

Prove that

$$\sum_{n=1}^{\infty} \frac{H_{2n}}{n} \left(\zeta(2) - H_n^{(2)} \right)$$

$$= \frac{15}{2}\zeta(4) - \frac{7}{2}\log(2)\zeta(3) + 2\log^2(2)\zeta(2) - \frac{1}{3}\log^4(2) - 8\operatorname{Li}_4\left(\frac{1}{2}\right),$$
$$(4.118)$$

where $H_n^{(m)} = 1 + \frac{1}{2^m} + \cdots + \frac{1}{n^m}$, $m \geq 1$, is the nth generalized harmonic number of order m, ζ denotes the Riemann zeta function, and Li_n represents the polylogarithm function.

4.28 A Third Uncommon Series with the Tail of the Riemann Zeta Function $\zeta(2) - H_{2n}^{(2)}$, Related to Weight 4 Harmonic Series

Prove that

$$\sum_{n=1}^{\infty} \frac{H_{2n}}{n} \left(\zeta(2) - H_{2n}^{(2)} \right)$$

$$= \frac{1}{12}\log^4(2) + \frac{1}{4}\log^2(2)\zeta(2) + \frac{7}{8}\log(2)\zeta(3) - \frac{1}{2}\zeta(4) + 2\operatorname{Li}_4\left(\frac{1}{2}\right),$$
$$(4.119)$$

where $H_n^{(m)} = 1 + \frac{1}{2^m} + \cdots + \frac{1}{n^m}$, $m \geq 1$, is the nth generalized harmonic number of order m, ζ denotes the Riemann zeta function, and Li_n represents the polylogarithm function.

4.29 A Fourth Uncommon Series with the Tail of the Riemann Zeta Function $\zeta(2) - H_n^{(2)}$, Related to Weight 4 Harmonic Series

Prove that

$$\sum_{n=1}^{\infty} \frac{H_n}{2n+1} \left(\zeta(2) - H_n^{(2)} \right)$$

$$= \frac{1}{6} \log^4(2) - \log^2(2)\zeta(2) - \frac{1}{2}\zeta(4) + 4\operatorname{Li}_4\left(\frac{1}{2}\right), \qquad (4.120)$$

where $H_n^{(m)} = 1 + \frac{1}{2^m} + \cdots + \frac{1}{n^m}$, $m \geq 1$, is the nth generalized harmonic number of order m, ζ denotes the Riemann zeta function, and Li_n represents the polylogarithm function.

A (very nice) challenging question: Calculate the series in a simple, elegant way without using any of the harmonic series $\displaystyle\sum_{n=1}^{\infty} \frac{H_n H_{2n}}{n^2}$, $\displaystyle\sum_{n=1}^{\infty} (-1)^{n-1} \frac{H_n}{n^3}$, $\displaystyle\sum_{n=1}^{\infty} (-1)^{n-1} \frac{H_n^2}{n^2}$, or $\displaystyle\sum_{n=1}^{\infty} (-1)^{n-1} \frac{H_n^{(2)}}{n^2}$.

4.30 A Fifth Uncommon Series with the Tail of the Riemann Zeta Function $\zeta(2) - H_{2n}^{(2)}$, Related to Weight 4 Harmonic Series

Prove that

$$\sum_{n=1}^{\infty} \frac{H_n}{2n+1} \left(\zeta(2) - H_{2n}^{(2)} \right)$$

$$= \frac{79}{32}\zeta(4) - \frac{7}{8}\log(2)\zeta(3) - \frac{5}{4}\log^2(2)\zeta(2) - \frac{1}{24}\log^4(2) - \operatorname{Li}_4\left(\frac{1}{2}\right),$$
$$(4.121)$$

where $H_n^{(m)} = 1 + \frac{1}{2^m} + \cdots + \frac{1}{n^m}$, $m \geq 1$, is the nth generalized harmonic number of order m, ζ denotes the Riemann zeta function, and Li_n represents the polylogarithm function.

(continued)

A (super) challenging question: Show the series result by also using the series from the previous section, $\sum_{n=1}^{\infty} \dfrac{H_n}{2n+1}\left(\zeta(2) - H_n^{(2)}\right)$, together with a series identity previously met, generated by *The Master Theorem of Series*.

4.31   A Sixth Uncommon Series with the Tail of the Riemann Zeta Function $\zeta(2) - H_n^{(2)}$, Related to Weight 4 Harmonic Series

Prove that

$$\sum_{n=1}^{\infty} \frac{H_{2n}}{2n+1}\left(\zeta(2) - H_n^{(2)}\right)$$

$$= \frac{1}{4}\log^4(2) - \frac{57}{16}\zeta(4) + \frac{7}{2}\log(2)\zeta(3) - \frac{3}{2}\log^2(2)\zeta(2) + 6\operatorname{Li}_4\left(\frac{1}{2}\right),$$
(4.122)

where $H_n^{(m)} = 1 + \frac{1}{2^m} + \cdots + \frac{1}{n^m}$, $m \geq 1$, is the nth generalized harmonic number of order m, ζ denotes the Riemann zeta function, and Li_n represents the polylogarithm function.

A challenging question: Exploit the series $\sum_{n=1}^{\infty} \dfrac{H_{2n}}{n}\left(\zeta(2) - H_n^{(2)}\right)$ given in Sect. 4.27 to get the evaluation of the main series.

4.32   A Seventh Uncommon Series with the Tail of the Riemann Zeta Function $\zeta(2) - H_{2n}^{(2)}$, Related to Weight 4 Harmonic Series

Prove that

$$\sum_{n=1}^{\infty} \frac{H_{2n}}{2n+1}\left(\zeta(2) - H_{2n}^{(2)}\right)$$

$$= \frac{15}{32}\zeta(4) + \frac{7}{16}\log(2)\zeta(3) - \frac{3}{8}\log^2(2)\zeta(2),$$
(4.123)

(continued)

where $H_n^{(m)} = 1 + \frac{1}{2^m} + \cdots + \frac{1}{n^m}$, $m \geq 1$, is the nth generalized harmonic number of order m and ζ denotes the Riemann zeta function.

A *challenging question:* (i) May we get a fast evaluation if we employ one of the six previous uncommon harmonic series with the tail of the Riemann zeta function? (ii) Is it possible to get the desired result without using work involving the polylogarithm function?

4.33   On the Calculation of an Essential Harmonic Series of Weight 5, Involving Harmonic Numbers of the Type H_{2n}

Show that

$$\sum_{n=1}^{\infty} \frac{H_n H_{2n}}{n^3}$$

$$= \frac{307}{16} \zeta(5) - \frac{1}{2} \zeta(2)\zeta(3) - 7 \log^2(2)\zeta(3) + \frac{8}{3} \log^3(2)\zeta(2) - \frac{8}{15} \log^5(2)$$

$$- 16 \log(2) \operatorname{Li}_4\left(\frac{1}{2}\right) - 16 \operatorname{Li}_5\left(\frac{1}{2}\right), \tag{4.124}$$

where $H_n = \sum_{k=1}^{n} \frac{1}{k}$ is the nth harmonic number, ζ represents the Riemann zeta function, and Li_n denotes the polylogarithm function.

4.34   More Helpful Atypical Harmonic Series of Weight 5 with Denominators of the Type $(2n+1)^2$ and $(2n+1)^3$

Prove that

$$(i) \sum_{n=1}^{\infty} \frac{H_n^{(2)}}{(2n+1)^3} = \frac{49}{8} \zeta(2)\zeta(3) - \frac{93}{8} \zeta(5); \tag{4.125}$$

$$(ii) \sum_{n=1}^{\infty} \frac{H_n^{(3)}}{(2n+1)^2} = \frac{31}{2} \zeta(5) - 8\zeta(2)\zeta(3); \tag{4.126}$$

(continued)

$$(iii) \sum_{n=1}^{\infty} \frac{H_n^2}{(2n+1)^3} = \frac{7}{2} \log^2(2)\zeta(3) - \frac{45}{8} \log(2)\zeta(4) + \frac{31}{8}\zeta(5) - \frac{7}{8}\zeta(2)\zeta(3).$$

$$(4.127)$$

Another pair of wonderful harmonic series alike:

$$(iv) \sum_{n=1}^{\infty} \frac{H_n H_n^{(2)}}{(2n+1)^2}$$

$$= \frac{7}{4}\zeta(2)\zeta(3) - \frac{589}{32}\zeta(5) + \frac{121}{8} \log(2)\zeta(4) - 7\log^2(2)\zeta(3) + \frac{4}{3}\log^3(2)\zeta(2)$$

$$- \frac{2}{15}\log^5(2) + 16\operatorname{Li}_5\left(\frac{1}{2}\right);$$

$$(4.128)$$

$$(v) \sum_{n=1}^{\infty} \frac{H_n^3}{(2n+1)^2}$$

$$= \frac{7}{4}\zeta(2)\zeta(3) - \frac{1271}{32}\zeta(5) + \frac{183}{8} \log(2)\zeta(4) - 2\log^3(2)\zeta(2) - \frac{2}{5}\log^5(2)$$

$$+ 48\operatorname{Li}_5\left(\frac{1}{2}\right),$$

$$(4.129)$$

where $H_n^{(m)} = 1 + \frac{1}{2^m} + \cdots + \frac{1}{n^m}$, $m \geq 1$, is the nth generalized harmonic number of order m, ζ represents the Riemann zeta function, and Li_n denotes the polylogarithm function.

4.35   On the Calculation of Another Essential Harmonic Series of Weight 5, Involving Harmonic Numbers of the Type H_{2n}

Show that

$$\sum_{n=1}^{\infty} \frac{H_n H_{2n}}{(2n+1)^3}$$

(continued)

$$= \frac{1}{12} \log^5(2) - \frac{1}{2} \log^3(2)\zeta(2) + \frac{7}{4}\log^2(2)\zeta(3) - \frac{17}{8}\log(2)\zeta(4) + \frac{31}{128}\zeta(5)$$

$$+ 2\log(2)\operatorname{Li}_4\left(\frac{1}{2}\right), \tag{4.130}$$

where $H_n = \sum_{k=1}^n \frac{1}{k}$ is the nth harmonic number, ζ represents the Riemann zeta function, and Li_n denotes the polylogarithm function.

4.36   A First Unusual Series with the Tail of the Riemann Zeta Function $\zeta(3) - H_{2n}^{(3)}$, Related to Weight 5 Harmonic Series

Prove that

$$\sum_{n=1}^\infty \frac{H_n}{n}\left(\zeta(3) - H_{2n}^{(3)}\right)$$

$$= \frac{23}{8}\zeta(2)\zeta(3) - \frac{69}{16}\zeta(5) + \frac{7}{4}\log^2(2)\zeta(3) - \frac{45}{16}\log(2)\zeta(4), \tag{4.131}$$

where $H_n^{(m)} = 1 + \frac{1}{2^m} + \cdots + \frac{1}{n^m}$, $m \geq 1$, is the nth generalized harmonic number of order m and ζ represents the Riemann zeta function.

4.37   A Second Unusual Series with the Tail of the Riemann Zeta Function $\zeta(3) - H_n^{(3)}$, Related to Weight 5 Harmonic Series

Prove that

$$\sum_{n=1}^\infty \frac{H_{2n}}{n}\left(\zeta(3) - H_n^{(3)}\right)$$

$$= \frac{203}{8}\zeta(5) + \frac{11}{4}\zeta(2)\zeta(3) - \frac{53}{4}\log(2)\zeta(4) + \frac{4}{3}\log^3(2)\zeta(2) - \frac{2}{5}\log^5(2)$$

(continued)

$$-16 \log(2) \operatorname{Li}_4\left(\frac{1}{2}\right) - 32 \operatorname{Li}_5\left(\frac{1}{2}\right), \tag{4.132}$$

where $H_n^{(m)} = 1 + \frac{1}{2^m} + \cdots + \frac{1}{n^m}$, $m \geq 1$, is the nth generalized harmonic number of order m, ζ represents the Riemann zeta function, and Li_n denotes the polylogarithm function.

4.38   A Third Unusual Series with the Tail of the Riemann Zeta Function $\zeta(3) - H_{2n}^{(3)}$, Related to Weight 5 Harmonic Series

Prove that

$$\sum_{n=1}^{\infty} \frac{H_{2n}}{n}\left(\zeta(3) - H_{2n}^{(3)}\right)$$

$$= \frac{1}{6}\log^3(2)\zeta(2) + \frac{7}{8}\log^2(2)\zeta(3) - \frac{49}{16}\log(2)\zeta(4) + \frac{55}{32}\zeta(5) + \frac{23}{16}\zeta(2)\zeta(3)$$

$$- \frac{1}{20}\log^5(2) - 2\log(2)\operatorname{Li}_4\left(\frac{1}{2}\right) - 4\operatorname{Li}_5\left(\frac{1}{2}\right), \tag{4.133}$$

where $H_n^{(m)} = 1 + \frac{1}{2^m} + \cdots + \frac{1}{n^m}$, $m \geq 1$, is the nth generalized harmonic number of order m, ζ represents the Riemann zeta function, and Li_n denotes the polylogarithm function.

4.39   A Fourth Unusual Series with the Tail of the Riemann Zeta Function $\zeta(3) - H_n^{(3)}$, Related to Weight 5 Harmonic Series

Prove that

$$\sum_{n=1}^{\infty} \frac{H_n}{2n+1}\left(\zeta(3) - H_n^{(3)}\right)$$

(continued)

$$= \frac{7}{4}\zeta(2)\zeta(3) - \frac{279}{16}\zeta(5) + \frac{4}{3}\log^3(2)\zeta(2) - 7\log^2(2)\zeta(3) + \frac{53}{4}\log(2)\zeta(4)$$

$$- \frac{2}{15}\log^5(2) + 16\operatorname{Li}_5\left(\frac{1}{2}\right), \tag{4.134}$$

where $H_n^{(m)} = 1 + \frac{1}{2^m} + \cdots + \frac{1}{n^m}$, $m \geq 1$, is the nth generalized harmonic number of order m, ζ denotes the Riemann zeta function, and Li_n represents the polylogarithm function.

4.40   A Fifth Unusual Series with the Tail of the Riemann Zeta Function $\zeta(3) - H_{2n}^{(3)}$, Related to Weight 5 Harmonic Series

Prove that

$$\sum_{n=1}^{\infty} \frac{H_n}{2n+1}\left(\zeta(3) - H_{2n}^{(3)}\right)$$

$$= \frac{1}{15}\log^5(2) - \frac{1}{3}\log^3(2)\zeta(2) - \frac{7}{8}\log^2(2)\zeta(3) + \frac{15}{32}\log(2)\zeta(4) - \frac{31}{64}\zeta(5)$$

$$- \frac{7}{16}\zeta(2)\zeta(3) + 2\log(2)\operatorname{Li}_4\left(\frac{1}{2}\right) + 2\operatorname{Li}_5\left(\frac{1}{2}\right), \tag{4.135}$$

where $H_n^{(m)} = 1 + \frac{1}{2^m} + \cdots + \frac{1}{n^m}$, $m \geq 1$, is the nth generalized harmonic number of order m, ζ denotes the Riemann zeta function, and Li_n represents the polylogarithm function.

A (super) challenging question: Show the series result by also using the series from the previous section, $\displaystyle\sum_{n=1}^{\infty} \frac{H_n}{2n+1}\left(\zeta(3) - H_n^{(3)}\right)$, together with a series identity previously met, generated by *The Master Theorem of Series*.

4.41　A Sixth Unusual Series with the Tail of the Riemann Zeta Function $\zeta(3) - H_n^{(3)}$, Related to Weight 5 Harmonic Series

Prove that

$$\sum_{n=1}^{\infty} \frac{H_{2n}}{2n+1} \left(\zeta(3) - H_n^{(3)} \right)$$

$$= \frac{2}{3} \log^3(2)\zeta(2) - \frac{7}{2} \log^2(2)\zeta(3) + \frac{53}{8} \log(2)\zeta(4) - \frac{1}{15} \log^5(2) - \frac{31}{16}\zeta(5)$$

$$- \frac{21}{8}\zeta(2)\zeta(3) + 8\operatorname{Li}_5\left(\frac{1}{2}\right), \tag{4.136}$$

where $H_n^{(m)} = 1 + \frac{1}{2^m} + \cdots + \frac{1}{n^m}$, $m \geq 1$, is the nth generalized harmonic number of order m, ζ denotes the Riemann zeta function, and Li_n represents the polylogarithm function.

A challenging question: Exploit the series $\sum_{n=1}^{\infty} \frac{H_{2n}}{n} \left(\zeta(3) - H_n^{(3)} \right)$ given in Sect. 4.37 to get the evaluation of the main series.

4.42　A Seventh Unusual Series with the Tail of the Riemann Zeta Function $\zeta(3) - H_{2n}^{(3)}$, Related to Weight 5 Harmonic Series

Prove that

$$\sum_{n=1}^{\infty} \frac{H_{2n}}{2n+1} \left(\zeta(3) - H_{2n}^{(3)} \right)$$

$$= \frac{1}{40} \log^5(2) - \frac{1}{12} \log^3(2)\zeta(2) - \frac{7}{16} \log^2(2)\zeta(3) + \frac{49}{32} \log(2)\zeta(4) - \frac{31}{64}\zeta(5)$$

$$- \frac{27}{32}\zeta(2)\zeta(3) + \log(2)\operatorname{Li}_4\left(\frac{1}{2}\right) + 2\operatorname{Li}_5\left(\frac{1}{2}\right), \tag{4.137}$$

where $H_n^{(m)} = 1 + \frac{1}{2^m} + \cdots + \frac{1}{n^m}$, $m \geq 1$, is the nth generalized harmonic number of order m, ζ denotes the Riemann zeta function, and Li_n represents the polylogarithm function.

4.43 Three More Spectacular Harmonic Series of Weight 5, Involving Harmonic Numbers of the Type H_{2n} and $H_{2n}^{(2)}$

Prove that

$$(i) \sum_{n=1}^{\infty} \frac{H_{2n} H_n^{(2)}}{n^2} = \frac{101}{16} \zeta(5) - \frac{5}{4} \zeta(2)\zeta(3);$$ (4.138)

$$(ii) \sum_{n=1}^{\infty} \frac{H_n H_{2n}^{(2)}}{n^2}$$

$$= \frac{23}{8} \zeta(2)\zeta(3) - \frac{581}{32} \zeta(5) - \frac{8}{3} \log^3(2)\zeta(2) + 7\log^2(2)\zeta(3) + \frac{8}{15} \log^5(2)$$

$$+ 16\log(2)\operatorname{Li}_4\left(\frac{1}{2}\right) + 16\operatorname{Li}_5\left(\frac{1}{2}\right);$$ (4.139)

$$(iii) \sum_{n=1}^{\infty} \frac{H_{2n} H_n^{(2)}}{(2n+1)^2}$$

$$= \frac{713}{64} \zeta(5) - \frac{21}{16} \zeta(2)\zeta(3) - \frac{7}{2} \log^2(2)\zeta(3) + \frac{4}{3} \log^3(2)\zeta(2) - \frac{4}{15} \log^5(2)$$

$$- 8\log(2)\operatorname{Li}_4\left(\frac{1}{2}\right) - 8\operatorname{Li}_5\left(\frac{1}{2}\right),$$ (4.140)

where $H_n^{(m)} = 1 + \frac{1}{2^m} + \cdots + \frac{1}{n^m}$, $m \geq 1$, is the nth generalized harmonic number of order m, ζ represents the Riemann zeta function, and Li_n denotes the polylogarithm function.

4.44 Two Atypical Sums of Series, One of Them Involving the Product of the Generalized Harmonic Numbers $H_n^{(3)} H_n^{(6)}$

Show that

$$(i) \, 5 \sum_{n=1}^{\infty} \frac{H_n^{(2)}}{n^6} + 2 \sum_{n=1}^{\infty} \frac{H_n^{(3)}}{n^5} = 10\zeta(3)\zeta(5) - \frac{21}{4} \zeta(8).$$ (4.141)

(continued)

By also exploiting (i), prove that

$$(ii)\ 2\sum_{n=1}^{\infty}\left(\zeta(3)\zeta(6)-H_n^{(3)}H_n^{(6)}\right)+7\sum_{n=1}^{\infty}\frac{H_n^{(2)}}{n^6}$$

$$=10\zeta(3)\zeta(5)-2\zeta(3)\zeta(6)-\frac{23}{12}\zeta(8),\qquad(4.142)$$

where $H_n^{(m)}=1+\frac{1}{2^m}+\cdots+\frac{1}{n^m}$, $m\geq1$, is the nth generalized harmonic number of order m and ζ represents the Riemann zeta function.

4.45   Amazing, Unexpected Relations with Alternating and Non-alternating Harmonic Series of Weights 5 and 7

Show, without calculating the series separately, that

$$(i)\ 8\sum_{n=1}^{\infty}(-1)^{n-1}\frac{H_n}{n^4}-\sum_{n=1}^{\infty}\frac{H_nH_n^{(2)}}{n^2}=\frac{55}{4}\zeta(5)-5\zeta(2)\zeta(3);\qquad(4.143)$$

$$(ii)\ 32\sum_{n=1}^{\infty}(-1)^{n-1}\frac{H_n}{n^6}+\sum_{n=1}^{\infty}\frac{H_nH_n^{(3)}}{n^3}-\sum_{n=1}^{\infty}\frac{H_nH_n^{(4)}}{n^2}-\sum_{n=1}^{\infty}\frac{H_nH_n^{(2)}}{n^4}$$

$$=111\zeta(7)-28\zeta(2)\zeta(5)-27\zeta(3)\zeta(4),\qquad(4.144)$$

where $H_n^{(m)}=1+\frac{1}{2^m}+\cdots+\frac{1}{n^m}$, $m\geq1$, is the nth generalized harmonic number of order m and ζ denotes the Riemann zeta function.

4.46   A Quintet of Advanced Harmonic Series of Weight 5 Involving Skew-Harmonic Numbers

Show that

$$(i)\ \sum_{n=1}^{\infty}\frac{H_n\overline{H}_n}{n^3}$$

(continued)

$$= \frac{1}{6}\log^3(2)\zeta(2) - \frac{7}{8}\log^2(2)\zeta(3) + 4\log(2)\zeta(4) - \frac{193}{64}\zeta(5) + \frac{3}{8}\zeta(2)\zeta(3)$$

$$- \frac{1}{60}\log^5(2) + 2\operatorname{Li}_5\left(\frac{1}{2}\right); \tag{4.145}$$

$$(ii)\ \sum_{n=1}^{\infty} \frac{(\overline{H}_n)^2}{n^3}$$

$$= \frac{1}{3}\log^3(2)\zeta(2) + \frac{7}{4}\log^2(2)\zeta(3) + \frac{19}{8}\log(2)\zeta(4) - \frac{167}{32}\zeta(5) - \frac{1}{30}\log^5(2)$$

$$+ \frac{3}{4}\zeta(2)\zeta(3) + 4\operatorname{Li}_5\left(\frac{1}{2}\right); \tag{4.146}$$

$$(iii)\ \sum_{n=1}^{\infty} \frac{H_n \overline{H}_n^{(2)}}{n^2}$$

$$= \frac{29}{64}\zeta(5) - \frac{5}{4}\zeta(2)\zeta(3) + \frac{7}{4}\log^2(2)\zeta(3) - \frac{2}{3}\log^3(2)\zeta(2) + \frac{2}{15}\log^5(2)$$

$$+ 4\log(2)\operatorname{Li}_4\left(\frac{1}{2}\right) + 4\operatorname{Li}_5\left(\frac{1}{2}\right); \tag{4.147}$$

$$(iv)\ \sum_{n=1}^{\infty} \frac{\overline{H}_n H_n^{(2)}}{n^2}$$

$$= \frac{27}{4}\zeta(5) + \frac{1}{2}\zeta(2)\zeta(3) + \frac{1}{3}\log^3(2)\zeta(2) - \frac{23}{16}\log(2)\zeta(4) - \frac{1}{10}\log^5(2)$$

$$- 4\log(2)\operatorname{Li}_4\left(\frac{1}{2}\right) - 8\operatorname{Li}_5\left(\frac{1}{2}\right); \tag{4.148}$$

$$(v)\ \sum_{n=1}^{\infty} \frac{\overline{H}_n \overline{H}_n^{(2)}}{n^2}$$

$$= \frac{1}{3}\log^3(2)\zeta(2) - \frac{7}{4}\log^2(2)\zeta(3) + \frac{49}{8}\log(2)\zeta(4) - \frac{23}{8}\zeta(5) - \frac{3}{4}\zeta(2)\zeta(3)$$

(continued)

$$-\frac{1}{30}\log^5(2) + 4\,\mathrm{Li}_5\left(\frac{1}{2}\right),\tag{4.149}$$

where $H_n^{(m)} = 1 + \frac{1}{2^m} + \cdots + \frac{1}{n^m}$, $m \geq 1$, is the nth generalized harmonic number of order m, $\overline{H}_n^{(m)} = 1 - \frac{1}{2^m} + \cdots + (-1)^{n-1}\frac{1}{n^m}$, $m \geq 1$, denotes the nth generalized skew-harmonic number of order m, ζ represents the Riemann zeta function, and Li_n designates the polylogarithm function.

4.47   Fourier Series Expansions of the Bernoulli Polynomials

Show that

$$(i)\ \sum_{k=1}^{\infty}\frac{\sin(2k\pi x)}{k^{2n-1}}$$

$$= (-1)^n\frac{1}{2}\frac{(2\pi)^{2n-1}}{(2n-1)!}B_{2n-1}(x) = (-1)^n\frac{1}{2}\frac{(2\pi)^{2n-1}}{(2n-1)!}\sum_{k=0}^{2n-1}\binom{2n-1}{k}B_k x^{2n-k-1},\tag{4.150}$$

which holds for $0 < x < 1$ if $n = 1$ and for $0 \leq x \leq 1$ if $n \in \mathbb{N}\setminus\{1\}$;

$$(ii)\ \sum_{k=1}^{\infty}\frac{\cos(2k\pi x)}{k^{2n}}$$

$$= (-1)^{n-1}\frac{1}{2}\frac{(2\pi)^{2n}}{(2n)!}B_{2n}(x) = (-1)^{n-1}\frac{1}{2}\frac{(2\pi)^{2n}}{(2n)!}\sum_{k=0}^{2n}\binom{2n}{k}B_k x^{2n-k},\tag{4.151}$$

which holds for $0 \leq x \leq 1$ if $n \in \mathbb{N}$;

$$(iii)\ \sum_{k=1}^{\infty}(-1)^{k-1}\frac{\sin(2k\pi x)}{k^{2n-1}}$$

$$= (-1)^{n-1}\frac{\pi^{2n-1}}{(2n-1)!}\Big(B_{2n-1}(2x) - 2^{2(n-1)}B_{2n-1}(x)\Big)$$

(continued)

$$= (-1)^{n-1} \frac{1}{2} \frac{(2\pi)^{2n-1}}{(2n-1)!} \sum_{k=0}^{2n-1} \left(\frac{1}{2^{k-1}} - 1\right) \binom{2n-1}{k} B_k x^{2n-k-1},$$

(4.152)

which holds for $-\frac{1}{2} < x < \frac{1}{2}$ if $n = 1$ and for $-\frac{1}{2} \le x \le \frac{1}{2}$ if $n \in \mathbb{N} \setminus \{1\}$;

$$(iv) \ \sum_{k=1}^{\infty} (-1)^{k-1} \frac{\cos(2k\pi x)}{k^{2n}}$$

$$= (-1)^{n-1} \frac{\pi^{2n}}{(2n)!} \left(2^{2n-1} B_{2n}(x) - B_{2n}(2x)\right)$$

$$= (-1)^{n-1} \frac{1}{2} \frac{(2\pi)^{2n}}{(2n)!} \sum_{k=0}^{2n} \left(1 - \frac{1}{2^{k-1}}\right) \binom{2n}{k} B_k x^{2n-k}, \qquad (4.153)$$

which holds for $-\frac{1}{2} \le x \le \frac{1}{2}$ if $n \in \mathbb{N}$. B_n represents the nth Bernoulli number, and $B_n(x)$ denotes the nth Bernoulli polynomial.

4.48 Stunning Fourier Series with $\log(\sin(x))$ and $\log(\cos(x))$ Raised to Positive Integer Powers, Related to Harmonic Numbers

Prove that

$$(i) \ \log^2(\sin(x))$$

$$= \log^2(2) + \frac{\pi^2}{12} + 2 \sum_{n=1}^{\infty} \left(\log(2)\frac{1}{n} - \frac{1}{2n^2} + \frac{H_n}{n}\right) \cos(2nx), \ 0 < x < \pi;$$

(4.154)

$$(ii) \ \log^2(\cos(x))$$

$$= \log^2(2) + \frac{\pi^2}{12} - 2 \sum_{n=1}^{\infty} (-1)^{n-1} \left(\log(2)\frac{1}{n} - \frac{1}{2n^2} + \frac{H_n}{n}\right) \cos(2nx), \ -\frac{\pi}{2} < x < \frac{\pi}{2};$$

(4.155)

(continued)

$$(iii)\ \log^3(\sin(x))$$

$$= -\log^3(2) - \frac{1}{4}\log(2)\pi^2 - \frac{3}{2}\zeta(3) + \sum_{n=1}^{\infty}\left(\left(-3\log^2(2) - \frac{\pi^2}{4}\right)\frac{1}{n} + 3\log(2)\frac{1}{n^2}\right.$$

$$\left. - \frac{3}{2}\frac{1}{n^3} - 6\log(2)\frac{H_n}{n} + 3\frac{H_n}{n^2} - 3\frac{H_n^2}{n}\right)\cos(2nx),\ 0 < x < \pi; \qquad (4.156)$$

$$(iv)\ \log^3(\cos(x))$$

$$= -\log^3(2) - \frac{1}{4}\log(2)\pi^2 - \frac{3}{2}\zeta(3) + \sum_{n=1}^{\infty}(-1)^{n-1}\left(\left(3\log^2(2) + \frac{\pi^2}{4}\right)\frac{1}{n}\right.$$

$$\left. -3\log(2)\frac{1}{n^2} + \frac{3}{2}\frac{1}{n^3} + 6\log(2)\frac{H_n}{n} - 3\frac{H_n}{n^2} + 3\frac{H_n^2}{n}\right)\cos(2nx),\ -\frac{\pi}{2} < x < \frac{\pi}{2};$$
$$(4.157)$$

$$(v)\ \log^4(\sin(x))$$

$$= \log^4(2) + \frac{1}{2}\log^2(2)\pi^2 + 6\log(2)\zeta(3) + \frac{19}{240}\pi^4$$

$$+ \sum_{n=1}^{\infty}\left((4\log^3(2) + \log(2)\pi^2 + 6\zeta(3))\frac{1}{n} - \left(6\log^2(2) + \frac{\pi^2}{2}\right)\frac{1}{n^2} + 6\log(2)\frac{1}{n^3}\right.$$

$$- 3\frac{1}{n^4} + (12\log^2(2) + \pi^2)\frac{H_n}{n} - 12\log(2)\frac{H_n}{n^2} + 6\frac{H_n}{n^3} + 12\log(2)\frac{H_n^2}{n}$$

$$\left. - 6\frac{H_n^2}{n^2} + 4\frac{H_n^3}{n} + 2\frac{H_n^{(3)}}{n}\right)\cos(2nx),\ 0 < x < \pi; \qquad (4.158)$$

$$(vi)\ \log^4(\cos(x))$$

$$= \log^4(2) + \frac{1}{2}\log^2(2)\pi^2 + 6\log(2)\zeta(3) + \frac{19}{240}\pi^4$$

$$- \sum_{n=1}^{\infty}(-1)^{n-1}\left((4\log^3(2) + \log(2)\pi^2 + 6\zeta(3))\frac{1}{n} - \left(6\log^2(2) + \frac{\pi^2}{2}\right)\frac{1}{n^2}\right.$$

$$+ 6\log(2)\frac{1}{n^3} - 3\frac{1}{n^4} + (12\log^2(2) + \pi^2)\frac{H_n}{n} - 12\log(2)\frac{H_n}{n^2} + 6\frac{H_n}{n^3} + 12\log(2)\frac{H_n^2}{n}$$

(continued)

$$-6\frac{H_n^2}{n^2}+4\frac{H_n^3}{n}+2\frac{H_n^{(3)}}{n}\Bigg)\cos(2nx),\quad -\frac{\pi}{2}<x<\frac{\pi}{2}, \qquad (4.159)$$

where $H_n^{(m)} = 1 + \frac{1}{2^m} + \cdots + \frac{1}{n^m}$, $m \geq 1$, is the nth generalized harmonic number of order m and ζ denotes the Riemann zeta function.

4.49 More Stunning Fourier Series, Related to Atypical Harmonic Numbers (Skew-Harmonic Numbers)

Prove that

$$(i)\ \mathrm{Li}_2\left(\sin^2(x)\right)$$

$$=\frac{\pi^2}{6}-2\log^2(2)-2\sum_{n=1}^{\infty}\left(2\log(2)\frac{1}{n}+(-1)^{n-1}\frac{1}{n^2}-2\frac{\overline{H}_n}{n}\right)\cos(2nx),\ x\in\mathbb{R};$$

$$(4.160)$$

$$(ii)\ \mathrm{Li}_2\left(\cos^2(x)\right)$$

$$=\frac{\pi^2}{6}-2\log^2(2)+2\sum_{n=1}^{\infty}(-1)^{n-1}\left(2\log(2)\frac{1}{n}+(-1)^{n-1}\frac{1}{n^2}-2\frac{\overline{H}_n}{n}\right)\cos(2nx),\ x\in\mathbb{R};$$

$$(4.161)$$

$$(iii)\ \mathrm{Li}_3\left(\sin^2(x)\right)$$

$$=\frac{4}{3}\log^3(2)-\log(2)\frac{\pi^2}{3}+2\zeta(3)+\sum_{n=1}^{\infty}\left(\left(4\log^2(2)-\frac{\pi^2}{3}\right)\frac{1}{n}-4\log(2)\frac{1}{n^2}\right.$$

$$\left.-2\frac{(-1)^{n-1}}{n^3}+8\log(2)\frac{H_n}{n}+4\frac{\overline{H}_n}{n^2}-4\frac{\overline{H}_n^{(2)}}{n}-8\frac{H_n\overline{H}_n}{n}+8\frac{1}{n}\sum_{k=1}^{n}(-1)^{k-1}\frac{H_k}{k}\right)\cos(2nx)$$

$$=\frac{4}{3}\log^3(2)-\log(2)\frac{\pi^2}{3}+2\zeta(3)+\sum_{n=1}^{\infty}\left(\left(4\log^2(2)-\frac{\pi^2}{3}\right)\frac{1}{n}-4\log(2)\frac{1}{n^2}\right.$$

(continued)

$$-2\frac{(-1)^{n-1}}{n^3}+8\log(2)\frac{H_n}{n}+4\frac{\overline{H}_n}{n^2}+4\frac{\overline{H}_n^{(2)}}{n}-8\frac{1}{n}\sum_{k=1}^n\frac{\overline{H}_k}{k}\Bigg)\cos(2nx),\ x\in\mathbb{R};$$

$$(4.162)$$

$$(iv)\ \mathrm{Li}_3\left(\cos^2(x)\right)$$

$$=\frac{4}{3}\log^3(2)-\log(2)\frac{\pi^2}{3}+2\zeta(3)-\sum_{n=1}^{\infty}(-1)^{n-1}\Bigg(\Bigg(4\log^2(2)-\frac{\pi^2}{3}\Bigg)\frac{1}{n}$$

$$-4\log(2)\frac{1}{n^2}-2\frac{(-1)^{n-1}}{n^3}+8\log(2)\frac{H_n}{n}+4\frac{\overline{H}_n}{n^2}-4\frac{\overline{H}_n^{(2)}}{n}-8\frac{H_n\overline{H}_n}{n}$$

$$+8\frac{1}{n}\sum_{k=1}^n(-1)^{k-1}\frac{H_k}{k}\Bigg)\cos(2nx)$$

$$=\frac{4}{3}\log^3(2)-\log(2)\frac{\pi^2}{3}+2\zeta(3)-\sum_{n=1}^{\infty}(-1)^{n-1}\Bigg(\Bigg(4\log^2(2)-\frac{\pi^2}{3}\Bigg)\frac{1}{n}$$

$$-4\log(2)\frac{1}{n^2}-2\frac{(-1)^{n-1}}{n^3}+8\log(2)\frac{H_n}{n}+4\frac{\overline{H}_n}{n^2}+4\frac{\overline{H}_n^{(2)}}{n}$$

$$-8\frac{1}{n}\sum_{k=1}^n\frac{\overline{H}_k}{k}\Bigg)\cos(2nx),\ x\in\mathbb{R},\qquad(4.163)$$

where $H_n^{(m)}=1+\frac{1}{2^m}+\cdots+\frac{1}{n^m}$, $m\ge1$, is the nth generalized harmonic number of order m, $\overline{H}_n^{(m)}=1-\frac{1}{2^m}+\cdots+(-1)^{n-1}\frac{1}{n^m}$, $m\ge1$, denotes the nth generalized skew-harmonic number of order m, ζ represents the Riemann zeta function, and Li_n designates the polylogarithm function.

4.50 And More Stunning Fourier Series, Related to Atypical Harmonic Numbers (Skew-Harmonic Numbers)

Prove that

$$(i)\ \mathrm{Li}_2\left(-\tan^2(x)\right)$$

(continued)

$$= -\frac{\pi^2}{3} + 4\sum_{n=1}^{\infty}\left(\log(2)\frac{1}{n} + \log(2)\frac{(-1)^{n-1}}{n} + (-1)^{n-1}\frac{H_n}{n} - \frac{\overline{H}_n}{n}\right)\cos(2nx),$$

$$x \in \mathbb{R}, \ x \neq \frac{\pi}{2} + k\pi, \ k \in \mathbb{Z}; \qquad (4.164)$$

$$(ii) \ \text{Li}_2\left(-\cot^2(x)\right)$$

$$= -\frac{\pi^2}{3} - 4\sum_{n=1}^{\infty}(-1)^{n-1}\left(\log(2)\frac{1}{n} + \log(2)\frac{(-1)^{n-1}}{n} + (-1)^{n-1}\frac{H_n}{n} - \frac{\overline{H}_n}{n}\right)\cos(2nx),$$

$$x \in \mathbb{R}, \ x \neq k\pi, \ k \in \mathbb{Z}; \qquad (4.165)$$

$$(iii) \ \text{Li}_3\left(-\tan^2(x)\right)$$

$$= -6\zeta(3) + \sum_{n=1}^{\infty}\left(\frac{2}{3}\pi^2\frac{1}{n} + \frac{2}{3}\pi^2(-1)^{n-1}\frac{1}{n} - 8\log(2)\frac{H_n}{n} + 8\log(2)(-1)^{n-1}\frac{H_n}{n}\right.$$

$$+ 4(-1)^{n-1}\frac{H_n^2}{n} - 8\log(2)\frac{\overline{H}_n}{n} + 8\log(2)(-1)^{n-1}\frac{\overline{H}_n}{n} - 4(-1)^{n-1}\frac{\overline{H}_n^2}{n}$$

$$\left. + 8\frac{\overline{H}_n^{(2)}}{n} + 8\frac{H_n\overline{H}_n}{n} - 16\frac{1}{n}\sum_{k=1}^{n}(-1)^{k-1}\frac{H_k}{k}\right)\cos(2nx)$$

$$= -6\zeta(3) + \sum_{n=1}^{\infty}\left(\frac{2}{3}\pi^2\frac{1}{n} + \frac{2}{3}\pi^2(-1)^{n-1}\frac{1}{n} - 8\log(2)\frac{H_n}{n} + 8\log(2)(-1)^{n-1}\frac{H_n}{n}\right.$$

$$+ 4(-1)^{n-1}\frac{H_n^2}{n} - 8\log(2)\frac{\overline{H}_n}{n} + 8\log(2)(-1)^{n-1}\frac{\overline{H}_n}{n} - 4(-1)^{n-1}\frac{\overline{H}_n^2}{n}$$

$$\left. - 8\frac{\overline{H}_n^{(2)}}{n} - 8\frac{H_n\overline{H}_n}{n} + 16\frac{1}{n}\sum_{k=1}^{n}\frac{\overline{H}_k}{k}\right)\cos(2nx), \ x \in \mathbb{R}, \ x \neq \frac{\pi}{2} + k\pi, \ k \in \mathbb{Z};$$

$$(4.166)$$

$$(iv) \ \text{Li}_3\left(-\cot^2(x)\right)$$

$$= -6\zeta(3) - \sum_{n=1}^{\infty}(-1)^{n-1}\left(\frac{2}{3}\pi^2\frac{1}{n} + \frac{2}{3}\pi^2(-1)^{n-1}\frac{1}{n} - 8\log(2)\frac{H_n}{n}\right.$$

$$+ 8\log(2)(-1)^{n-1}\frac{H_n}{n} + 4(-1)^{n-1}\frac{H_n^2}{n} - 8\log(2)\frac{\overline{H}_n}{n} + 8\log(2)(-1)^{n-1}\frac{\overline{H}_n}{n}$$

(continued)

$$- 4(-1)^{n-1}\frac{\overline{H}_n^2}{n} + 8\frac{\overline{H}_n^{(2)}}{n} + 8\frac{H_n\overline{H}_n}{n} - 16\frac{1}{n}\sum_{k=1}^{n}(-1)^{k-1}\frac{H_k}{k}\Bigg)\cos(2nx)$$

$$= -6\zeta(3) - \sum_{n=1}^{\infty}(-1)^{n-1}\Bigg(\frac{2}{3}\pi^2\frac{1}{n} + \frac{2}{3}\pi^2(-1)^{n-1}\frac{1}{n} - 8\log(2)\frac{H_n}{n}$$

$$+ 8\log(2)(-1)^{n-1}\frac{H_n}{n} + 4(-1)^{n-1}\frac{H_n^2}{n} - 8\log(2)\frac{\overline{H}_n}{n} + 8\log(2)(-1)^{n-1}\frac{\overline{H}_n}{n}$$

$$- 4(-1)^{n-1}\frac{\overline{H}_n^2}{n} - 8\frac{\overline{H}_n^{(2)}}{n} - 8\frac{H_n\overline{H}_n}{n} + 16\frac{1}{n}\sum_{k=1}^{n}\frac{\overline{H}_k}{k}\Bigg)\cos(2nx),$$

$$x \in \mathbb{R},\ x \neq k\pi,\ k \in \mathbb{Z}, \qquad (4.167)$$

where $H_n^{(m)} = 1 + \frac{1}{2^m} + \cdots + \frac{1}{n^m}$, $m \geq 1$, is the nth generalized harmonic number of order m, $\overline{H}_n^{(m)} = 1 - \frac{1}{2^m} + \cdots + (-1)^{n-1}\frac{1}{n^m}$, $m \geq 1$, denotes the nth generalized skew-harmonic number of order m, ζ represents the Riemann zeta function, and Li_n designates the polylogarithm function.

4.51   Yet Other Stunning Fourier Series, This Time with the Coefficients Mainly Kept in an Integral Form

Prove that

$$(i)\ \log^2(\tan(x)) = \log^2(\cot(x))$$

$$= \frac{\pi^2}{4} + \sum_{n=1}^{\infty}\Bigg((1-(-1)^{n-1})\int_0^1 t^{n-1}\left(\frac{1}{n}\frac{1-t}{1+t} - 2\log\left(\frac{1-t}{2\sqrt{t}}\right)\right)\mathrm{d}t\Bigg)\cos(2nx)$$

$$= \frac{\pi^2}{4} + 2\sum_{n=1}^{\infty}(1-(-1)^{n-1})\left(\frac{H_n}{n} + \frac{\overline{H}_n}{n}\right)\cos(2nx),\ 0 < x < \frac{\pi}{2};$$

$$(4.168)$$

$$(ii)\ \log^3(\tan(x))$$

$$= 12\sum_{n=1}^{\infty}\Bigg((1+(-1)^{n-1})\int_0^1 t^{n-1}\left(\frac{\pi^2}{48} - \operatorname{arctanh}^2(t)\right)\mathrm{d}t\Bigg)\cos(2nx),\ 0 < x < \frac{\pi}{2};$$

$$(4.169)$$

(continued)

$$(iii) \ \log^3(\cot(x))$$

$$= -12 \sum_{n=1}^{\infty} \left((1 + (-1)^{n-1}) \int_0^1 t^{n-1} \left(\frac{\pi^2}{48} - \operatorname{arctanh}^2(t) \right) dt \right) \cos(2nx), \ 0 < x < \frac{\pi}{2};$$

$$(4.170)$$

$$(iv) \ \log(\cos(x)) \log^2(\sin(x))$$

$$= \frac{1}{4} \zeta(3) - \log^3(2) + \sum_{n=1}^{\infty} \left(\int_0^1 t^{n-1} \left(4 \operatorname{arctanh}^2(t) - \log^2 \left(\frac{1-t}{2\sqrt{t}} \right) \right. \right.$$

$$\left. \left. - \left(1 - (-1)^{n-1} \right) \log^2 \left(\frac{1+t}{2\sqrt{t}} \right) \right) dt \right) \cos(2nx), \ 0 < x < \frac{\pi}{2}; \quad (4.171)$$

$$(v) \ \log(\sin(x)) \log^2(\cos(x))$$

$$= \frac{1}{4} \zeta(3) - \log^3(2) - \sum_{n=1}^{\infty} (-1)^{n-1} \left(\int_0^1 t^{n-1} \left(4 \operatorname{arctanh}^2(t) - \log^2 \left(\frac{1-t}{2\sqrt{t}} \right) \right. \right.$$

$$\left. \left. - \left(1 - (-1)^{n-1} \right) \log^2 \left(\frac{1+t}{2\sqrt{t}} \right) \right) dt \right) \cos(2nx), \ 0 < x < \frac{\pi}{2}, \quad (4.172)$$

where $H_n = \sum_{k=1}^{n} \frac{1}{k}$ is the nth harmonic number, $\overline{H}_n = \sum_{k=1}^{n} (-1)^{k-1} \frac{1}{k}$ denotes the nth skew-harmonic number, and ζ designates the Riemann zeta function.

4.52 A Pair of (Very) Challenging Alternating Harmonic Series with a Weight 4 Structure, Involving Harmonic Numbers of the Type H_{2n}

Prove that

$$(i) \ \sum_{n=1}^{\infty} (-1)^{n-1} \frac{H_{2n}}{n^3}$$

(continued)

$$= \frac{195}{32}\zeta(4) - \frac{35}{8}\log(2)\zeta(3) + \frac{5}{4}\log^2(2)\zeta(2) - \frac{5}{24}\log^4(2) - 5\,\mathrm{Li}_4\left(\frac{1}{2}\right);$$

(4.173)

$$(ii) \sum_{n=1}^{\infty}(-1)^{n-1}\frac{H_{2n}^{(2)}}{n^2}$$

$$= 2G^2 - \frac{353}{64}\zeta(4) + \frac{35}{8}\log(2)\zeta(3) - \frac{5}{4}\log^2(2)\zeta(2) + \frac{5}{24}\log^4(2) + 5\,\mathrm{Li}_4\left(\frac{1}{2}\right),$$

(4.174)

where $H_n^{(m)} = 1 + \frac{1}{2^m} + \cdots + \frac{1}{n^m}$, $m \geq 1$, is the nth generalized harmonic number of order m, G represents the Catalan's constant, ζ denotes the Riemann zeta function, and Li_n designates the polylogarithm function.

    A (super) challenging question: Prove both harmonic series results by real methods exclusively.

4.53   Important Tetralogarithmic Values and More (Curious) Challenging Alternating Harmonic Series with a Weight 4 Structure, Involving Harmonic Numbers H_{2n}

Show that

$$(i) \; \Re\left\{\mathrm{Li}_4\left(\frac{1+i}{2}\right)\right\} = \Re\left\{\mathrm{Li}_4\left(\frac{1-i}{2}\right)\right\};$$

(4.175)

$$(ii) \; \Re\left\{\mathrm{Li}_4\left(1+i\right)\right\} = \Re\left\{\mathrm{Li}_4\left(1-i\right)\right\};$$

(4.176)

$$(iii) \; \Re\left\{\mathrm{Li}_4\left(\frac{1\pm i}{2}\right)\right\} = \frac{343}{1024}\zeta(4) - \frac{5}{128}\log^2(2)\zeta(2) + \frac{1}{96}\log^4(2) + \frac{5}{16}\mathrm{Li}_4\left(\frac{1}{2}\right);$$

(4.177)

$$(iv) \; \Re\{\mathrm{Li}_4(1\pm i)\} = \frac{485}{512}\zeta(4) + \frac{1}{8}\log^2(2)\zeta(2) - \frac{5}{384}\log^4(2) - \frac{5}{16}\mathrm{Li}_4\left(\frac{1}{2}\right).$$

(4.178)

Three strong alternating harmonic series involving the previous values:

$$(v) \sum_{n=1}^{\infty}(-1)^{n-1}\frac{H_{2n}^2}{n^2}$$

(continued)

$$= 2G^2 - \log(2)\pi G + \frac{231}{32}\zeta(4) - \frac{35}{16}\log(2)\zeta(3) + \log^2(2)\zeta(2) - \frac{5}{48}\log^4(2)$$

$$- 2\pi\Im\left\{\operatorname{Li}_3\left(\frac{1+i}{2}\right)\right\} - \frac{5}{2}\operatorname{Li}_4\left(\frac{1}{2}\right); \tag{4.179}$$

$$(vi) \sum_{n=1}^{\infty}(-1)^{n-1}\frac{H_{2n}^3}{n}$$

$$= 2G^2 - \frac{3}{4}\log(2)\pi G + \frac{1055}{256}\zeta(4) - \frac{93}{64}\log(2)\zeta(3) + \frac{21}{32}\log^2(2)\zeta(2) - \frac{1}{32}\log^4(2)$$

$$- \frac{3}{2}\pi\Im\left\{\operatorname{Li}_3\left(\frac{1+i}{2}\right)\right\}; \tag{4.180}$$

$$(vii) \sum_{n=1}^{\infty}(-1)^{n-1}\frac{H_{2n}H_{2n}^{(2)}}{n}$$

$$= \frac{1}{4}\log(2)\pi G - \frac{137}{128}\zeta(4) + \frac{35}{64}\log(2)\zeta(3) - \frac{3}{8}\log^2(2)\zeta(2) + \frac{5}{96}\log^4(2)$$

$$+ \frac{\pi}{2}\Im\left\{\operatorname{Li}_3\left(\frac{1+i}{2}\right)\right\} + \frac{5}{4}\operatorname{Li}_4\left(\frac{1}{2}\right), \tag{4.181}$$

where $H_n^{(m)} = 1 + \frac{1}{2^m} + \cdots + \frac{1}{n^m}$, $m \geq 1$, is the nth generalized harmonic number of order m, G represents the Catalan's constant, ζ denotes the Riemann zeta function, and Li_n designates the polylogarithm function.

4.54 Two Alternating Euler Sums Involving Special Tails, a Joint Work with Moti Levy, Plus Two Newer Ones

Prove that

$$(i) \sum_{n=1}^{\infty}(-1)^{n-1}H_n \sum_{k=n+1}^{\infty}\frac{H_k}{k^2}$$

(continued)

$$= \frac{1}{16}\zeta(4) + \frac{7}{8}\log(2)\zeta(3) + \frac{1}{8}\log^2(2)\zeta(2) - \frac{1}{48}\log^4(2) - \frac{1}{2}\operatorname{Li}_4\left(\frac{1}{2}\right);$$

(4.182)

$$(ii) \sum_{n=1}^{\infty}(-1)^{n-1}H_n \sum_{k=n+1}^{\infty}\frac{H_k}{k^3}$$

$$= \frac{5}{12}\log^3(2)\zeta(2) + 2\log(2)\zeta(4) + \frac{7}{8}\zeta(2)\zeta(3) - \frac{21}{16}\log^2(2)\zeta(3)$$

$$- \frac{155}{128}\zeta(5) - \frac{3}{40}\log^5(2) - 2\log(2)\operatorname{Li}_4\left(\frac{1}{2}\right) - \operatorname{Li}_5\left(\frac{1}{2}\right).$$

(4.183)

Another two (challenging) atypical alternating harmonic series involving special tetralogarithmic values:

$$(iii) \sum_{n=1}^{\infty}(-1)^{n-1}H_{2n}\sum_{k=2n+1}^{\infty}\frac{H_k}{k^2}$$

$$= \frac{\pi^2}{64}G + \frac{1}{16}\log(2)\pi G - \frac{1}{16}\log^2(2)G + \frac{55}{18432}\pi^4 + \frac{7}{512}\log(2)\pi^3 + \frac{5}{768}\log^2(2)\pi^2$$

$$+ \log^3(2)\frac{\pi}{384} + \frac{35}{256}\pi\zeta(3) + \frac{35}{128}\log(2)\zeta(3) - \frac{5}{768}\log^4(2) - \frac{1}{1536}\psi^{(3)}\left(\frac{1}{4}\right)$$

$$+ \frac{1}{2}\Im\left\{\operatorname{Li}_4\left(\frac{1+i}{2}\right)\right\} - \frac{5}{32}\operatorname{Li}_4\left(\frac{1}{2}\right);$$

(4.184)

$$(iv) \sum_{n=1}^{\infty}(-1)^{n-1}H_{2n-1}\sum_{k=2n}^{\infty}\frac{H_k}{k^2}$$

$$= \frac{\pi^2}{64}G - \frac{1}{16}\log(2)\pi G - \frac{1}{16}\log^2(2)G + \frac{137}{18432}\pi^4 + \frac{7}{512}\log(2)\pi^3 - \frac{5}{768}\log^2(2)\pi^2$$

$$+ \log^3(2)\frac{\pi}{384} + \frac{35}{256}\pi\zeta(3) - \frac{35}{128}\log(2)\zeta(3) + \frac{5}{768}\log^4(2) - \frac{1}{1536}\psi^{(3)}\left(\frac{1}{4}\right)$$

$$+ \frac{1}{2}\Im\left\{\operatorname{Li}_4\left(\frac{1+i}{2}\right)\right\} + \frac{5}{32}\operatorname{Li}_4\left(\frac{1}{2}\right),$$

(4.185)

(continued)

where $H_n = \sum_{k=1}^{n} \frac{1}{k}$ is the nth harmonic number, G denotes the Catalan's constant, ζ represents the Riemann zeta function, $\psi^{(n)}$ designates the polygamma function, and Li_n signifies the polylogarithm function.

4.55   A (Very) Hard Nut to Crack (An Alternating Harmonic Series with a Weight 4 Structure, Involving Harmonic Numbers of the Type H_{2n})

Evaluating and using $\displaystyle\int_0^1 \frac{\arctan^2(x)\,\mathrm{arctanh}(x)}{x}\,\mathrm{d}x$, prove that

$$\sum_{n=1}^{\infty}(-1)^{n-1}\frac{H_n H_{2n}}{n^2}$$

$$= 2G^2 - 2\log(2)\pi G - \frac{1}{8}\log^4(2) + \frac{3}{2}\log^2(2)\zeta(2) - \frac{21}{8}\log(2)\zeta(3) + \frac{773}{64}\zeta(4)$$

$$- 4\pi\Im\left\{\mathrm{Li}_3\left(\frac{1+i}{2}\right)\right\} - 3\,\mathrm{Li}_4\left(\frac{1}{2}\right), \tag{4.186}$$

where $H_n = \sum_{k=1}^{n} \frac{1}{k}$ is the nth harmonic number, G represents the Catalan's constant, ζ denotes the Riemann zeta function, and Li_n designates the polylogarithm function.

4.56   Another (Very) Hard Nut to Crack (An Alternating Harmonic Series with a Weight 4 Structure, Involving Harmonic Numbers of the Type H_{2n})

Evaluating and using the sum of integrals, $\displaystyle 2\int_0^1 \frac{x\,\mathrm{arctanh}(x)\log^2(x)}{1+x^2}\,\mathrm{d}x$
$+ \displaystyle\int_0^1 \frac{\mathrm{arctanh}(x)\,\mathrm{Li}_2(-x^2)}{x}\,\mathrm{d}x$, show that

$$\sum_{n=1}^{\infty}(-1)^{n-1}\frac{H_{2n}H_n^{(2)}}{n}$$

(continued)

$$= \frac{1}{2}\log^2(2)\zeta(2) - \frac{21}{16}\log(2)\zeta(3) - \frac{15}{32}\zeta(4) - \frac{1}{8}\log^4(2)$$

$$+ 2\pi\Im\left\{\operatorname{Li}_3\left(\frac{1+i}{2}\right)\right\} - 3\operatorname{Li}_4\left(\frac{1}{2}\right), \tag{4.187}$$

where $H_n^{(m)} = 1 + \frac{1}{2^m} + \cdots + \frac{1}{n^m}$, $m \geq 1$, is the nth generalized harmonic number of order m, ζ represents the Riemann zeta function, and Li_n denotes the polylogarithm function.

4.57 Two Harmonic Series with a Wicked Look, Involving Skew-Harmonic Numbers and Harmonic Numbers H_{2n}

Prove that

$$(i) \sum_{n=1}^{\infty} \frac{\overline{H}_n H_{2n}}{n^2}$$

$$= \frac{507}{64}\zeta(4) - \frac{7}{4}\log(2)\zeta(3) + \frac{5}{4}\log^2(2)\zeta(2) - \frac{7}{48}\log^4(2)$$

$$- 2\pi\Im\left\{\operatorname{Li}_3\left(\frac{1+i}{2}\right)\right\} - \frac{7}{2}\operatorname{Li}_4\left(\frac{1}{2}\right); \tag{4.188}$$

$$(ii) \sum_{n=1}^{\infty} (-1)^{n-1}\frac{\overline{H}_n H_{2n}}{n^2}$$

$$= \frac{5}{48}\log^4(2) - \frac{5}{8}\log^2(2)\zeta(2) + \frac{7}{2}\log(2)\zeta(3) - \frac{77}{32}\zeta(4) + \log(2)\pi G$$

$$- 2G^2 + \frac{5}{2}\operatorname{Li}_4\left(\frac{1}{2}\right), \tag{4.189}$$

where $H_n = \sum_{k=1}^{n}\frac{1}{k}$ is the nth harmonic number, $\overline{H}_n = \sum_{k=1}^{n}(-1)^{k-1}\frac{1}{k}$ represents the nth skew-harmonic number, G denotes the Catalan's constant, ζ designates the Riemann zeta function, and Li_n signifies the polylogarithm function.

4.58 Nice Series with the Reciprocal of the Central Binomial Coefficient and the Generalized Harmonic Number

Prove that

$$(i) \sum_{n=1}^{\infty} \frac{2^{2n}}{n^4 \binom{2n}{n}}$$

$$= \frac{1}{3}\log^4(2) + 4\log^2(2)\zeta(2) - \frac{19}{4}\zeta(4) + 8\,\mathrm{Li}_4\left(\frac{1}{2}\right); \qquad (4.190)$$

$$(ii) \sum_{n=1}^{\infty} \frac{2^{2n}}{n^5 \binom{2n}{n}}$$

$$= \frac{2}{15}\log^5(2) + \frac{8}{3}\log^3(2)\zeta(2) - \frac{19}{2}\log(2)\zeta(4) + 6\zeta(2)\zeta(3)$$

$$+ \frac{31}{8}\zeta(5) - 16\,\mathrm{Li}_5\left(\frac{1}{2}\right). \qquad (4.191)$$

Two curious series transformations:
Find

$$(iii) \sum_{n=1}^{\infty} \frac{2^{2n} H_n^{(3)}}{n(2n+1)\binom{2n}{n}}; \qquad (4.192)$$

$$(iv) \sum_{n=1}^{\infty} \frac{2^{2n} H_n^{(4)}}{n(2n+1)\binom{2n}{n}}, \qquad (4.193)$$

where $H_n^{(m)} = 1 + \frac{1}{2^m} + \cdots + \frac{1}{n^m}$, $m \geq 1$, is the nth generalized harmonic number of order m, ζ represents the Riemann zeta function, and Li_n denotes the polylogarithm function.

4.59   Marvellous Binoharmonic Series Forged with Nice Ideas

Prove that

$$(i) \sum_{n=1}^{\infty} \frac{1}{n^3 2^{2n}} \binom{2n}{n} \left(\zeta(2) - H_n^{(2)} \right)$$

$$= \frac{8}{15} \log^5(2) - \frac{4}{3} \log^3(2)\zeta(2) + 4\log^2(2)\zeta(3) + \log(2)\zeta(4) - \frac{123}{8}\zeta(5)$$

$$+ 16\log(2)\operatorname{Li}_4\left(\frac{1}{2}\right) + 16\operatorname{Li}_5\left(\frac{1}{2}\right). \qquad (4.194)$$

A special series with a surprisingly beautiful closed form:

$$(ii) \sum_{n=1}^{\infty} \frac{H_{n-1}}{n 2^{2n}} \binom{2n}{n} \left(\left(\zeta(3) - H_n^{(3)} \right) - (H_{2n} - H_n - \log(2)) \left(\zeta(2) - H_n^{(2)} \right) \right)$$

$$= \frac{17}{8}\zeta(5) - \frac{1}{2}\zeta(2)\zeta(3) + 2\log^3(2)\zeta(2) - \log^2(2)\zeta(3) - \frac{9}{4}\log(2)\zeta(4), \qquad (4.195)$$

where $H_n^{(m)} = 1 + \frac{1}{2^m} + \cdots + \frac{1}{n^m}$, $m \geq 1$, is the nth generalized harmonic number of order m, ζ represents the Riemann zeta function, and Li_n denotes the polylogarithm function.

4.60   Presenting an Appealing Triple Infinite Series Together with an Esoteric-Looking Functional Equation

Let's consider $|x| < 1$. Find

$$(i) \; \Lambda(x) = \sum_{i=0}^{\infty} \sum_{j=0}^{\infty} \sum_{k=0}^{\infty} x^{i+j+k} \frac{i! \, j! \, k!}{(i + j + k + 2)!}. \qquad (4.196)$$

Then, prove the functional equation

$$(ii) \; (3+\theta)\theta^2 \Lambda(-\theta) + \frac{(3 + 2\theta)\theta^2}{(1 + \theta)^3} \Lambda\left(\frac{\theta}{\theta + 1} \right) = 3\log^2(1 + \theta). \qquad (4.197)$$

Chapter 5
Hints

5.1 A Remarkable IMC Limit Problem Involving a Curious Sum with the Reciprocal of a Product with Two Logarithms

How about finding a way of using *squeeze theorem*?

5.2 Two Series with Tail Involving the Double Factorial, Their Generalizations, and a $\zeta(2)$ Representation

In order to get the generalizations at the points (iii) and (iv), one might exploit at least three ways: (1) the change of the summation order, (2) the Cauchy product, and (3) Abel's summation.

5.3 Six Enjoyable Sums Involving the Reciprocal of the Central Binomial Coefficient and Two Series Derived from Them

A good option is to reduce the sums to telescoping sums.

5.4 A Great Time with a Special Binoharmonic Series

Consider the integral result in (3.126), Sect. 3.19.

© The Author(s), under exclusive license to Springer Nature Switzerland AG 2023   457
C. I. Vălean, *More (Almost) Impossible Integrals, Sums, and Series*, Problem
Books in Mathematics, https://doi.org/10.1007/978-3-031-21262-8_5

5.5   A Panel of (Very) Useful Cauchy Products of Two Series: From Known Cauchy Products to Less Known Ones

For some of the Cauchy products, one might find useful to calculate and exploit

particular cases of finite sums like $\displaystyle\sum_{k=1}^{n-1} \frac{H_k^{(m)}}{(n-k)^p}$ and $\displaystyle\sum_{k=1}^{n-1} \frac{\overline{H}_k^{(m)}}{(n-k)^p}$.

5.6   Good-to-Know Generating Functions: The First Part

One might find useful to wisely employ and combine the Landen dilogarithmic identity, $\mathrm{Li}_2(x) + \mathrm{Li}_2\left(\dfrac{x}{x-1}\right) = -\dfrac{1}{2}\log^2(1-x)$, Landen trilogarithmic identity

$\mathrm{Li}_3(x) + \mathrm{Li}_3(1-x) + \mathrm{Li}_3\left(\dfrac{x}{x-1}\right) = \zeta(3) + \zeta(2)\log(1-x) - \dfrac{1}{2}\log(x)\log^2(1-x) +$

$\dfrac{1}{6}\log^3(1-x)$, and dilogarithm function reflection formula, $\mathrm{Li}_2(x) + \mathrm{Li}_2(1-x) = \pi^2/6 - \log(x)\log(1-x)$.

5.7   Good-to-Know Generating Functions: The Second Part

Consider the hint in the previous section.

5.8   Good-to-Know Generating Functions: The Third Part

One might think about creative ways of exploiting telescoping sums.

5.9   Good-to-Know Generating Functions: The Fourth Part

To get a first solution to the point (i), we might want to prove and exploit that

$\displaystyle\sum_{n=1}^{k-1} x^n H_n^{(p)} = \frac{1}{1-x}\left(-x^k H_k^{(p)} + \sum_{m=1}^{k} \frac{x^m}{m^p}\right)$, and for a second solution, we could use telescoping sums.

5.10 Good-to-Know Generating Functions: The Fifth Part

For the generating function at the first point, we might exploit at least three ways: (1) one involving the Cauchy product of two series; (2) another one involving a clever change of summation order; and (3) an approach that is based on the use of the telescoping sums. For the next two points, we may consider the result from the point (i) in order to construct solutions.

5.11 Good-to-Know Generating Functions: The Sixth Part

Get inspired by the work related to the generating functions in the solution sections corresponding to the previous sections.

5.12 Good-to-Know Generating Functions: The Seventh Part

The same recommendation as in the previous section.

5.13 Two Nice Sums Related to the Generalized Harmonic Numbers, an Asymptotic Expansion Extraction, a Neat Representation of $\log^2(2)$, and a Curious Power Series

For the first two points of the problem, one might exploit the difference of a sum when one of its variables takes different values.

5.14 Opening the World of Harmonic Series with Beautiful Series That Require Athletic Movements During Their Calculations: The First (Enjoyable) Part

Start with using the identity $\displaystyle\sum_{n=1}^{\infty} \frac{1}{n(n+k)} = \frac{H_k}{k}$ for the point (i) of the problem.

5.15 Opening the World of Harmonic Series with Beautiful Series That Require Athletic Movements During Their Calculations: The Second (Enjoyable) Part

Check the solution section related to the previous section and approach things in a similar style.

5.16 A Special Harmonic Series in Disguise Involving Nice Tricks

Apply Abel's summation and then cleverly rearrange the resulting series.

5.17 A Few Nice Generalized Series: Most of Them May Be Seen as Applications of *The Master Theorem of Series*

For the generalization at the point (iv), it might also be useful to consider using *The Master Theorem of Series* in [73].

5.18 Useful Relations Involving Polygamma with the Argument $n/2$ and the Generalized Skew-Harmonic Numbers

Prove and exploit the result

$$\sum_{k=1}^{\infty}(-1)^{k-1}\frac{1}{k+x} = \log(2) + 2\frac{1}{x} - \psi(1+x) + \psi\left(\frac{x}{2}\right), \quad x > 0.$$

5.19 A Key Classical Generalized Harmonic Series

One way to go is by proving and using that

$$\sum_{k=1}^{\infty}\frac{H_k}{(k+1)(k+n+1)} = \frac{H_n^2 + H_n^{(2)}}{2n} = \frac{(\gamma + \psi(n+1))^2 + \zeta(2) - \psi^{(1)}(n+1)}{2n},$$

where we further properly rearrange the opposite sides and then differentiate with respect to n.

5.20 Revisiting Two Classical Challenging Alternating Harmonic Series, Calculated by Exploiting a Beta Function Form

For both points of the problem, exploit *a special form of the beta function*,

$$\int_0^1 \frac{x^{a-1} + x^{b-1}}{(1+x)^{a+b}} dx = B(a, b).$$

5.21 A Famous Classical Generalization with Alternating Harmonic Series, Derived by a New Special Way

Prove and exploit the identity

$$\sum_{k=1}^{\infty} \frac{1}{2k(2k + 2n - 1)} = \frac{1}{2n - 1}\left(H_{2n} - \frac{1}{2}H_n - \log(2)\right).$$

5.22 Seven Useful Generalized Harmonic Series

Relating some of the given series (e.g., the first two ones) to integrals from the first chapter could immediately lead to obtaining solutions.

5.23 A Special Challenging Harmonic Series of Weight 4, Involving Harmonic Numbers of the Type H_{2n}

To get a first solution, prove and exploit the fact that

$$-\log(1 + x)\log(1 - x) = \sum_{k=1}^{\infty} x^{2k}\frac{H_{2k} - H_k}{k} + \frac{1}{2}\sum_{k=1}^{\infty}\frac{x^{2k}}{k^2}, \quad |x| < 1,$$

together with the result in (1.100), Sect. 1.21.

In order to obtain a second solution, employ the result in (4.82), Sect. 4.13.

5.24 Two Useful Atypical Harmonic Series of Weight 4 with Denominators of the Type $(2n+1)^2$

For the point (i), use Abel's summation, and for the point (ii), exploit the identity

$$\sum_{k=1}^{\infty} \frac{H_k^2 - H_k^{(2)}}{(k+1)(k+n+1)} = \frac{H_n^3 + 3H_n H_n^{(2)} + 2H_n^{(3)}}{3n},$$

with n extended to real numbers.

5.25 Another Special Challenging Harmonic Series of Weight 4, Involving Harmonic Numbers of the Type H_{2n}

Did you have a chance to read my first book, *(Almost) Impossible Integrals, Sums, and Series*, more precisely the penultimate section in the fourth chapter where I give a special representation of $\zeta(4)$ in terms of a sum of seven series?

5.26 A First Uncommon Series with the Tail of the Riemann Zeta Function $\zeta(2) - H_{2n}^{(2)}$, Related to Weight 4 Harmonic Series

One may start with the use of Abel's summation.

5.27 A Second Uncommon Series with the Tail of the Riemann Zeta Function $\zeta(2) - H_n^{(2)}$, Related to Weight 4 Harmonic Series

One way to go for building a solution is by exploiting the integral representation of the tail of the Riemann zeta function

$$\zeta(2) - 1 - \frac{1}{2^2} - \cdots - \frac{1}{n^2} = -\sum_{k=1}^{\infty} \int_0^1 x^{k+n-1} \log(x) dx$$

$$= -\int_0^1 \frac{x^n \log(x)}{1-x} dx \overset{x=y^2}{=} -4 \int_0^1 \frac{y^{2n+1} \log(y)}{1-y^2} dy.$$

5.28   A Third Uncommon Series with the Tail of the Riemann Zeta Function $\zeta(2) - H_{2n}^{(2)}$, Related to Weight 4 Harmonic Series

A possible way to go may be designed if we start with the use of the identity $\zeta(2) - H_{2n}^{(2)} = -\int_0^1 \frac{x^{2n} \log(x)}{1-x} dx$.

5.29   A Fourth Uncommon Series with the Tail of the Riemann Zeta Function $\zeta(2) - H_n^{(2)}$, Related to Weight 4 Harmonic Series

For a first solution, we might exploit the identity in (4.93), Sect. 4.17. For a second solution, cleverly exploit the result in (1.103), Sect. 1.21.

5.30   A Fifth Uncommon Series with the Tail of the Riemann Zeta Function $\zeta(2) - H_{2n}^{(2)}$, Related to Weight 4 Harmonic Series

Prove and exploit that

$$\sum_{k=n}^{\infty} \frac{1}{(2k+1)^2} = \zeta(2) - H_{2n}^{(2)} - \frac{1}{4}\left(\zeta(2) - H_n^{(2)}\right).$$

5.31   A Sixth Uncommon Series with the Tail of the Riemann Zeta Function $\zeta(2) - H_n^{(2)}$, Related to Weight 4 Harmonic Series

One way is to combine the main series and the series suggested to use in the *challenging question* and then cleverly apply Abel's summation.

5.32   A Seventh Uncommon Series with the Tail of the Riemann Zeta Function $\zeta(2) - H_{2n}^{(2)}$, Related to Weight 4 Harmonic Series

Get inspiration from the hint provided in Sect. 5.27.

5.33   On the Calculation of an Essential Harmonic Series of Weight 5, Involving Harmonic Numbers of the Type H_{2n}

Start with the series representation of $\log(1 + x) \log(1 - x)$,

$$-\log(1+x)\log(1-x) = \sum_{n=1}^{\infty} x^{2n} \left(\frac{H_{2n} - H_n}{n} + \frac{1}{2n^2} \right), \ |x| < 1.$$

5.34   More Helpful Atypical Harmonic Series of Weight 5 with Denominators of the Type $(2n + 1)^2$ and $(2n + 1)^3$

For the first three points of the problem, get inspiration from Sect. 6.24.

5.35   On the Calculation of Another Essential Harmonic Series of Weight 5, Involving Harmonic Numbers of the Type H_{2n}

Prove and exploit the identity, $\displaystyle\sum_{k=1}^{\infty} \frac{H_k}{(k+1)(k+n+1)} = \frac{H_n^2 + H_n^{(2)}}{2n}.$

5.36   A First Unusual Series with the Tail of the Riemann Zeta Function $\zeta(3) - H_{2n}^{(3)}$, Related to Weight 5 Harmonic Series

One may consider using Abel's summation.

5.37 A Second Unusual Series with the Tail of the Riemann Zeta Function $\zeta(3) - H_n^{(3)}$, Related to Weight 5 Harmonic Series

Exploit that $\zeta(3) - H_n^{(3)} = 4 \int_0^1 \frac{x^{2n+1} \log^2(x)}{1 - x^2} dx$.

5.38 A Third Unusual Series with the Tail of the Riemann Zeta Function $\zeta(3) - H_{2n}^{(3)}$, Related to Weight 5 Harmonic Series

Use that $\zeta(3) - H_{2n}^{(3)} = \frac{1}{2} \int_0^1 \frac{x^{2n} \log^2(x)}{1 - x} dx$.

5.39 A Fourth Unusual Series with the Tail of the Riemann Zeta Function $\zeta(3) - H_n^{(3)}$, Related to Weight 5 Harmonic Series

Have you seen the ideas and strategies in Sect. 6.29?

5.40 A Fifth Unusual Series with the Tail of the Riemann Zeta Function $\zeta(3) - H_{2n}^{(3)}$, Related to Weight 5 Harmonic Series

Did you have a chance to go through Sect. 6.30? We may act in a similar style.

5.41 A Sixth Unusual Series with the Tail of the Riemann Zeta Function $\zeta(3) - H_n^{(3)}$, Related to Weight 5 Harmonic Series

One might exploit the series suggested in the *challenging question*.

5.42   A Seventh Unusual Series with the Tail of the Riemann Zeta Function $\zeta(3) - H_{2n}^{(3)}$, Related to Weight 5 Harmonic Series

Relate the series to the one in (4.133), Sect. 4.38.

5.43   Three More Spectacular Harmonic Series of Weight 5, Involving Harmonic Numbers of the Type H_{2n} and $H_{2n}^{(2)}$

One might exploit (4.23), Sect. 4.5, for the point (i) of the problem.

5.44   Two Atypical Sums of Series, One of Them Involving the Product of the Generalized Harmonic Numbers $H_n^{(3)} H_n^{(6)}$

For the sum of harmonic series at the point (i), exploit the double series

$$\sum_{k=1}^{\infty} \left(\sum_{n=1}^{\infty} \frac{1}{k^4 (k+n)^4} \right).$$

5.45   Amazing, Unexpected Relations with Alternating and Non-alternating Harmonic Series of Weights 5 and 7

Have you seen the second solution in Sect. 6.23?

5.46   A Quintet of Advanced Harmonic Series of Weight 5 Involving Skew-Harmonic Numbers

A solution to the point (i) may be developed by exploiting the Cauchy product in (4.29), Sect. 4.5. Alternatively, to get another solution, we may exploit the parity, which may also be a good option to consider for the harmonic series at the point (ii).

5.47 Fourier Series Expansions of the Bernoulli Polynomials

What do you think about using induction?

5.48 Stunning Fourier Series with log(sin(x)) and log(cos(x)) Raised to Positive Integer Powers, Related to Harmonic Numbers

One might think of exploiting the integral $\displaystyle\int_0^\infty \log^p(\sinh(x))e^{-2nx}dx$, $n, p \in \mathbb{N}$.

5.49 More Stunning Fourier Series, Related to Atypical Harmonic Numbers (Skew-Harmonic Numbers)

A similar idea as in the previous section, but now consider the integral version with $\cosh(x)$.

5.50 And More Stunning Fourier Series, Related to Atypical Harmonic Numbers (Skew-Harmonic Numbers)

Get inspiration from the derivations of the Fourier series treated in the solution sections connected to the previous sections.

5.51 Yet Other Stunning Fourier Series, This Time with the Coefficients Mainly Kept in an Integral Form

In order to obtain a first solution to the point (i) of the problem, we may start by exploiting the identity $\log^2(\tan(x)) = \log^2(\sin(x)) - 2\log(\sin(x))\log(\cos(x)) + \log^2(\cos(x))$, $0 < x < \dfrac{\pi}{2}$. Then, for a second solution, we may use that $\log(\tan(x)) = -2\operatorname{arctanh}(e^{i2x}) + i\dfrac{\pi}{2}$, $0 < x < \dfrac{\pi}{2}$, together with the Cauchy product of two series.

5.52   A Pair of (Very) Challenging Alternating Harmonic Series with a Weight 4 Structure, Involving Harmonic Numbers of the Type H_{2n}

A solution may be constructed by proving and exploiting the Fourier-like series,

$$\sum_{n=1}^{\infty}\left(2H_{2n} - 2H_n + \frac{1}{2n} - 2\log(2)\right)\frac{\sin^2(2nx)}{n}$$

$$= \frac{1}{4}\log^2\left(\frac{1}{2}\sin(2x)\right) - \frac{1}{4}\log^2(\tan(x)), \ 0 < x < \frac{\pi}{2}.$$

5.53   Important Tetralogarithmic Values and More (Curious) Challenging Alternating Harmonic Series with a Weight 4 Structure, Involving Harmonic Numbers H_{2n}

The tetralogarithmic values may be extracted by exploiting, say, the series given at the point (i) in the problem statement related to the previous section.

5.54   Two Alternating Euler Sums Involving Special Tails, a Joint Work with Moti Levy, plus Two Newer Ones

At the point (i) of the problem, we might exploit a key generating function given in the fourth chapter.

5.55   A (Very) Hard Nut to Crack (An Alternating Harmonic Series with a Weight 4 Structure, Involving Harmonic Numbers of the Type H_{2n})

Approaching the integral $\Re\left\{\int_0^{\infty}\frac{(\pi/2 - \arctan(t))^2\arctanh(t)}{t}\mathrm{d}t\right\}$ in two different ways might be considered a way to go for building a solution.

5.56 Another (Very) Hard Nut to Crack (An Alternating Harmonic Series with a Weight 4 Structure, Involving Harmonic Numbers of the Type H_{2n})

Approach $\Re\left\{\int_0^\infty \frac{\operatorname{arctanh}(t)\operatorname{Li}_2(-t^2)}{t(1+t^2)}\,\mathrm{d}t\right\}$ in two different ways, one leading to the value of the integral and the other one giving a useful connection to the desired series.

5.57 Two Harmonic Series with a Wicked Look, Involving Skew-Harmonic Numbers and Harmonic Numbers H_{2n}

For both points of the problem, start with (1.99), Sect. 1.21.

5.58 Nice Series with the Reciprocal of the Central Binomial Coefficient and the Generalized Harmonic Number

A good start for the first two points of the problem may be represented by the use of the power series $\dfrac{\arcsin(x)}{\sqrt{1-x^2}} = \sum_{n=1}^\infty \dfrac{(2x)^{2n-1}}{n\dbinom{2n}{n}}$. The last two points of the problem may be related to the first two points by a clever use of a telescoping sum given in the fourth chapter.

5.59 Marvellous Binoharmonic Series Forged with Nice Ideas

Start with evaluating and exploiting the integral $\displaystyle\int_0^{\pi/2} x^2 \cos^{2n}(x)\,\mathrm{d}x$.

5.60 Presenting an Appealing Triple Infinite Series Together with an Esoteric-Looking Functional Equation

Show and exploit the relation $\dfrac{\Gamma(a_1)\Gamma(a_2)\Gamma(a_3)}{\Gamma(a_1+a_2+a_3)} = \mathrm{B}(a_1, a_2) \cdot \mathrm{B}(a_1+a_2, a_3)$.

Chapter 6
Solutions

6.1 A Remarkable IMC Limit Problem Involving a Curious Sum with the Reciprocal of a Product with Two Logarithms

Solution In the opening of this last chapter of the book, I prepare a solution to a wonderful **IMC** (International Mathematics Competition for University Students) problem, involving the limit of a sum with the reciprocal of a product with two logarithms. In fact, this limit problem is the sixth problem (the last one) given during the second day of the mentioned contest that took place in Plovdiv, Bulgaria, 1994.

Before going further, I would like to remind the lovers of such limits about the possibility of putting more efforts into finding a solution before continuing to read the way to go I present below.

Let's start the calculations! Assuming that $n \geq 4$ and then using that $\dfrac{1}{\log^2(n)} \leq \dfrac{1}{\log(k)\log(n-k)}$, $\forall k \in \{2, \ldots, n-2\}$, we readily arrive at

$$1 - \frac{3}{n} \leq \frac{\log^2(n)}{n} \sum_{k=2}^{n-2} \frac{1}{\log(k)\log(n-k)}. \tag{6.1}$$

In order to get another useful lower bound, we may use that $\dfrac{1}{\log(k)\log(n)} \leq \dfrac{1}{\log(k)\log(n-k)}$, $\forall k \in \{2, \ldots, n-2\}$, which gives

© The Author(s), under exclusive license to Springer Nature Switzerland AG 2023
C. I. Vălean, *More (Almost) Impossible Integrals, Sums, and Series*, Problem
Books in Mathematics, https://doi.org/10.1007/978-3-031-21262-8_6

$$\frac{\log(n)}{n} \sum_{k=2}^{n-2} \frac{1}{\log(k)} \le \frac{\log^2(n)}{n} \sum_{k=2}^{n-2} \frac{1}{\log(k)\log(n-k)}. \tag{6.2}$$

What's the reasoning here? In view of (6.3) that provides a useful upper bound, we see that also getting the lower bound in (6.2) might be a natural way to go since at last we have to deal with $\lim_{n\to\infty} \dfrac{\log(n)}{n} \displaystyle\sum_{k=2}^{n-2} \dfrac{1}{\log(k)}$, and that's because we'll want to apply the *squeeze theorem* to the main limit.

To get an upper bound, we might exploit, say, either rearrangement inequality or Chebyshev's sum inequality (and I'll consider the latter one, as seen in [98]). Since we have $a_n : \dfrac{1}{\log(2)} \ge \dfrac{1}{\log(3)} \ge \cdots \ge \dfrac{1}{\log(n-2)}$ and $b_n : \dfrac{1}{\log(n-2)} \le \dfrac{1}{\log(n-3)} \le \cdots \le \dfrac{1}{\log(2)}$, we readily obtain that

$$\frac{\log^2(n)}{n} \sum_{k=2}^{n-2} \frac{1}{\log(k)\log(n-k)} \le \frac{\log^2(n)}{n(n-3)} \left(\sum_{k=2}^{n-2} \frac{1}{\log(k)} \right)^2$$

$$= \frac{n}{n-3} \left(\frac{\log(n)}{n} \sum_{k=2}^{n-2} \frac{1}{\log(k)} \right)^2. \tag{6.3}$$

At this point, we want to prove the following auxiliary result:

$$\lim_{n\to\infty} \frac{\log(n)}{n} \sum_{k=2}^{n-2} \frac{1}{\log(k)} = 1. \tag{6.4}$$

Proof If we consider the *Stolz–Cesàro theorem* (see [15, Appendix, pp. 263–266]), we immediately obtain that

$$\lim_{n\to\infty} \frac{\log(n)}{n} \sum_{k=2}^{n-2} \frac{1}{\log(k)} = \lim_{n\to\infty} \frac{1/\log(n-1)}{(n+1)/\log(n+1) - n/\log(n)}$$

$$= \lim_{n\to\infty} \frac{\log(n+1)}{\log(n-1)(1 - (\log(1+1/n)^n)/\log(n))}$$

$$= 1,$$

where I have just used the celebre limit, $\lim_{n\to\infty}(1+1/n)^n = e$, which shows that $\lim_{n\to\infty} \log((1+1/n)^n) = 1$.

Or we may proceed a bit differently as follows:

$$\lim_{n\to\infty} \frac{\log(n)}{n} \sum_{k=2}^{n-2} \frac{1}{\log(k)} = \lim_{n\to\infty} \frac{1/\log(n-1)}{(n+1)/\log(n+1) - n/\log(n)}$$

$$= \lim_{n\to\infty} \frac{\log^2(c_n)}{\log(n-1)(\log(c_n) - 1)}$$

$$= 1,$$

where in the calculations I also used the *mean value theorem* applied to the function $f(x) = x/\log(x)$ on the interval $[n, n+1]$, $\forall n \geq 2$, $n \in \mathbb{N}$, that gives $(n+1)/\log(n+1) - n/\log(n) = (\log(c_n) - 1)/\log^2(c_n)$, $c_n \in (n, n+1)$, and at the same time we also note that when n is large, we have $\log(n-1) \approx \log(c_n) \approx \log(n)$.   ■

Finally, based on the *squeeze theorem*, if we consider either the lower bound in (6.1), or the one in (6.2), then the upper bound in (6.3), and let $n \to \infty$ and use the limit in (6.4), we conclude that

$$\lim_{n\to\infty} \frac{\log^2(n)}{n} \sum_{k=2}^{n-2} \frac{1}{\log(k)\log(n-k)} = 1,$$

and the solution is complete.

The solution presented above also answers the challenging question since the use of the integrals has been avoided as requested.

6.2   Two Series with Tail Involving the Double Factorial, Their Generalizations, and a $\zeta(2)$ Representation

Solution If you didn't skip Sect. 1.18 and its related-solutions section, you probably noticed that generalizations in this section are required to finalize the calculations there. I will provide multiple solutions to the two generalizations which contain strategies that may be adapted to also solve a lot of similar problems. An important prerequisite in the course of the solutions is the knowledge of two famous series in the literature, $\displaystyle \frac{1}{2}\sum_{n=1}^{\infty} \frac{(2x)^{2n}}{n^2\binom{2n}{n}} = \arcsin^2(x)$ and $\displaystyle \sum_{n=0}^{\infty} \frac{x^{2n}}{(2n+1)2^{2n}}\binom{2n}{n} = \frac{\arcsin(x)}{x}$.

For example, the first mentioned inverse sine series may be found in [76, Chapter 4, p.279], where it appears as a problem, and a solution to it may be found in the same book. Then, the second inverse sine series may be easily obtained if we start from the generating function of the central binomial coefficients,[1]

$$\sum_{n=0}^{\infty} \binom{2n}{n} t^n = \frac{1}{\sqrt{1-4t}}, \quad |t| \le \frac{1}{4} \wedge t \ne \frac{1}{4}, \text{ replace t by } t^2/4, \text{ integrate from}$$

$t = 0$ to $t = x$, and then divide both sides by x to arrive at the second desired series,

$$\sum_{n=0}^{\infty} \frac{x^{2n}}{(2n+1)2^{2n}} \binom{2n}{n} = \frac{\arcsin(x)}{x}.$$

For the moment, we let apart the results at the points (i) and (ii) and focus on the generalizations at the points (iii) and (iv). Now, for the point (iii) of the problem,

we use the fact that $\displaystyle\sum_{k=1}^{\infty} \frac{2^{2k-1}}{k^2 \binom{2k}{k}} = \frac{\pi^2}{4}$, which is derived immediately from the

squared inverse sine series representation above, and then we write

$$\sum_{n=1}^{\infty} x^{2n} \left(\frac{\pi^2}{4} - \sum_{k=1}^{n} \frac{(2k-2)!!}{k \cdot (2k-1)!!} \right) = \sum_{n=1}^{\infty} x^{2n} \left(\frac{\pi^2}{4} - \sum_{k=1}^{n} \frac{2^{2k-1}}{k^2 \binom{2k}{k}} \right)$$

$$= \sum_{n=1}^{\infty} x^{2n} \left(\sum_{k=n+1}^{\infty} \frac{2^{2k-1}}{k^2 \binom{2k}{k}} \right)$$

{reverse the order of summation}

$$= \sum_{k=1}^{\infty} \frac{2^{2k-1}}{k^2 \binom{2k}{k}} \sum_{n=1}^{k-1} x^{2n} = \sum_{k=1}^{\infty} \frac{2^{2k-1}}{k^2 \binom{2k}{k}} \frac{x^2 - x^{2k}}{1-x^2} = \frac{\pi^2}{4} \frac{x^2}{1-x^2}$$

---

[1] Usually, the generating function of the central binomial coefficients is obtained by using binomial series. How about a different perspective using a Wallis' integral? Since we know

that $\displaystyle\int_0^{\pi/2} \sin^{2k}(x)\,dx = \frac{\pi}{2^{2k+1}} \binom{2k}{k}$, if we multiply both sides by $(2^{2k+1}/\pi)t^k$ and consider

the summation from $k = 0$ to ∞, we get $\displaystyle\sum_{k=0}^{\infty} \binom{2k}{k} t^k = \frac{2}{\pi} \sum_{k=0}^{\infty} 2^{2k} t^k \int_0^{\pi/2} \sin^{2k}(x)\,dx =$

$\displaystyle\frac{2}{\pi} \int_0^{\pi/2} \sum_{k=0}^{\infty} 2^{2k} t^k \sin^{2k}(x)\,dx = \frac{2}{\pi} \int_0^{\pi/2} \frac{1}{1 - 4t\sin^2(x)}\,dx = \frac{1}{\sqrt{1-4t}}$, where I used that

$\displaystyle\int_0^{\pi/2} \frac{dx}{1 - a\sin^2(x)} = \frac{\pi}{2} \frac{1}{\sqrt{1-a}}$, $a < 1$ (for a proof, see [76, Chapter 3, p.212]).

$$-\frac{1}{2}\frac{1}{1-x^2}\sum_{k=1}^{\infty}\frac{(2x)^{2k}}{k^2\binom{2k}{k}} = \frac{\pi^2}{4}\frac{x^2}{1-x^2} - \frac{\arcsin^2(x)}{1-x^2},$$

and the first solution to the point (iii) is complete.

Further, from a different perspective, we can build a solution based on the Cauchy product of two series, the version which says that if $\sum_{n=1}^{\infty}a_n$ and $\sum_{n=1}^{\infty}b_n$ are absolutely convergent, then we have

$$\left(\sum_{n=1}^{\infty}a_n\right)\left(\sum_{n=1}^{\infty}b_n\right) = \sum_{n=1}^{\infty}\left(\sum_{k=1}^{n}a_k b_{n-k+1}\right). \qquad (6.5)$$

Upon considering the series $\arcsin^2(x) = \frac{1}{2}\sum_{n=1}^{\infty}\frac{(2x)^{2n}}{n^2\binom{2n}{n}}$ and $\frac{1}{1-x^2} = \sum_{n=1}^{\infty}x^{2n-2}$,

and applying the Cauchy product formula in (6.5), we arrive at

$$\frac{\arcsin^2(x)}{1-x^2} = \sum_{n=1}^{\infty}x^{2n}\left(\sum_{k=1}^{n}\frac{2^{2k-1}}{k^2\binom{2k}{k}}\right). \qquad (6.6)$$

Returning to the main series and considering the result in (6.6), we write

$$\sum_{n=1}^{\infty}x^{2n}\left(\frac{\pi^2}{4} - \sum_{k=1}^{n}\frac{(2k-2)!!}{k\cdot(2k-1)!!}\right) = \sum_{n=1}^{\infty}x^{2n}\left(\frac{\pi^2}{4} - \sum_{k=1}^{n}\frac{2^{2k-1}}{k^2\binom{2k}{k}}\right)$$

{expand the summand and split the series}

$$= \frac{\pi^2}{4}\sum_{n=1}^{\infty}x^{2n} - \sum_{n=1}^{\infty}x^{2n}\left(\sum_{k=1}^{n}\frac{2^{2k-1}}{k^2\binom{2k}{k}}\right) = \frac{\pi^2}{4}\frac{x^2}{1-x^2} - \frac{\arcsin^2(x)}{1-x^2},$$

and the second solution to the point (iii) is complete.

A last thought here would be the use of Abel's summation for generating another solution. Now, we consider the following series version of Abel's summation, which states that if $(a_n)_{n\geq 1}$, $(b_n)_{n\geq 1}$ are two sequences of real numbers with $A_n = \sum_{k=1}^{n}a_k$, then we have

$$\sum_{k=1}^{\infty} a_k b_k = \lim_{n \to \infty} (A_n b_{n+1}) + \sum_{k=1}^{\infty} A_k (b_k - b_{k+1}). \qquad (6.7)$$

First, we rearrange the series, and we write

$$\sum_{n=1}^{\infty} x^{2n} \left(\frac{\pi^2}{4} - \sum_{k=1}^{n} \frac{(2k-2)!!}{k \cdot (2k-1)!!} \right) = \sum_{n=1}^{\infty} x^{2n} \left(\frac{\pi^2}{4} - \sum_{k=1}^{n} \frac{2^{2k-1}}{k^2 \binom{2k}{k}} \right)$$

{reindex the series and start from $n = 2$ to ∞}

$$= \sum_{n=2}^{\infty} x^{2n-2} \left(\frac{\pi^2}{4} - \sum_{k=1}^{n-1} \frac{2^{2k-1}}{k^2 \binom{2k}{k}} \right) = -\frac{\pi^2}{4} + \sum_{n=1}^{\infty} x^{2n-2} \left(\frac{\pi^2}{4} - \sum_{k=1}^{n-1} \frac{2^{2k-1}}{k^2 \binom{2k}{k}} \right)$$

{apply Abel's summation, the series version in (6.7) stated above}

$$= -\frac{\pi^2}{4} + \sum_{n=1}^{\infty} \frac{1 - x^{2n}}{1 - x^2} \frac{2^{2n-1}}{n^2 \binom{2n}{n}} = -\frac{\pi^2}{4} + \frac{1}{1 - x^2} \left(\sum_{n=1}^{\infty} \frac{2^{2n-1}}{n^2 \binom{2n}{n}} - \frac{1}{2} \sum_{n=1}^{\infty} \frac{(2x)^{2n}}{n^2 \binom{2n}{n}} \right)$$

$$= \frac{\pi^2}{4} \frac{x^2}{1 - x^2} - \frac{\arcsin^2(x)}{1 - x^2},$$

and the third solution to the point (iii) is complete.

For the solutions to the point (iv), we will use the same strategies we applied in the calculations to the point (iii). So, for a first solution based on reversing the order of summation, we recall the second inverse sine related series representation stated at the beginning of the section which shows that $\sum_{k=0}^{\infty} \frac{1}{(2k+1)2^{2k}} \binom{2k}{k} = \frac{\pi}{2}$, and then we write

$$\sum_{n=0}^{\infty} x^{2n} \left(\frac{\pi}{2} - \sum_{k=0}^{n} \frac{(2k-1)!!}{(2k+1) \cdot (2k)!!} \right) = \sum_{n=0}^{\infty} x^{2n} \left(\frac{\pi}{2} - \sum_{k=0}^{n} \frac{1}{(2k+1)2^{2k}} \binom{2k}{k} \right)$$

$$= \sum_{n=0}^{\infty} x^{2n} \left(\sum_{k=n+1}^{\infty} \frac{1}{(2k+1)2^{2k}} \binom{2k}{k} \right)$$

{reverse the order of summation}

$$= \sum_{k=1}^{\infty} \frac{1}{(2k+1)2^{2k}} \binom{2k}{k} \left(\sum_{n=0}^{k-1} x^{2n} \right) = \sum_{k=1}^{\infty} \frac{1}{(2k+1)2^{2k}} \binom{2k}{k} \frac{1-x^{2k}}{1-x^2}$$

$$= \frac{\pi}{2} \frac{1}{1-x^2} - \frac{\arcsin(x)}{x(1-x^2)},$$

and the first solution to the point (iv) is complete.

How about a second solution based on the Cauchy product of two series? One may use the version

$$\left(\sum_{n=0}^{\infty} a_n \right) \left(\sum_{n=0}^{\infty} b_n \right) = \sum_{n=0}^{\infty} \left(\sum_{k=0}^{n} a_k b_{n-k} \right), \tag{6.8}$$

where $\displaystyle\sum_{n=0}^{\infty} a_n$ and $\displaystyle\sum_{n=0}^{\infty} b_n$ are absolutely convergent.

If we consider the series $\arcsin(x) = \displaystyle\sum_{n=0}^{\infty} \frac{x^{2n+1}}{(2n+1)2^{2n}} \binom{2n}{n}$ and $\dfrac{1}{1-x^2} =$

$\displaystyle\sum_{n=0}^{\infty} x^{2n}$ and apply the Cauchy product formula I just mentioned in (6.8), we obtain

$$\frac{\arcsin(x)}{x(1-x^2)} = \sum_{n=0}^{\infty} x^{2n} \left(\sum_{k=0}^{n} \frac{1}{(2k+1)2^{2k}} \binom{2k}{k} \right). \tag{6.9}$$

Returning to the main series with the result in (6.9), we write

$$\sum_{n=0}^{\infty} x^{2n} \left(\frac{\pi}{2} - \sum_{k=0}^{n} \frac{(2k-1)!!}{(2k+1)\cdot(2k)!!} \right) = \sum_{n=0}^{\infty} x^{2n} \left(\frac{\pi}{2} - \sum_{k=0}^{n} \frac{1}{(2k+1)2^{2k}} \binom{2k}{k} \right)$$

{expand the summand and split the series}

$$= \frac{\pi}{2} \sum_{n=0}^{\infty} x^{2n} - \sum_{n=0}^{\infty} x^{2n} \left(\sum_{k=0}^{n} \frac{1}{(2k+1)2^{2k}} \binom{2k}{k} \right)$$

$$= \frac{\pi}{2} \frac{1}{1-x^2} - \frac{\arcsin(x)}{x(1-x^2)},$$

and the second solution to the point (iv) is complete.

Finally, for an approach involving Abel's summation, we first rearrange the series and write

$$\sum_{n=0}^{\infty} x^{2n}\left(\frac{\pi}{2} - \sum_{k=0}^{n} \frac{(2k-1)!!}{(2k+1)\cdot(2k)!!}\right) = \sum_{n=0}^{\infty} x^{2n}\left(\frac{\pi}{2} - \sum_{k=0}^{n} \frac{1}{(2k+1)2^{2k}}\binom{2k}{k}\right)$$

{reindex the series and start from $n = 1$ to ∞}

$$= \sum_{n=1}^{\infty} x^{2n-2}\left(\frac{\pi}{2} - \sum_{k=0}^{n-1} \frac{1}{(2k+1)2^{2k}}\binom{2k}{k}\right)$$

{apply Abel's summation, the series version in (6.7)}

$$= \sum_{n=1}^{\infty} \frac{1}{(2n+1)2^{2n}}\binom{2n}{n}\sum_{k=1}^{n} x^{2k-2} = \sum_{n=1}^{\infty} \frac{1}{(2n+1)2^{2n}}\binom{2n}{n}\frac{1-x^{2n}}{1-x^2}$$

$\left\{\text{due to } 1 - x^{2n} \text{ in the summand, we can start from } n = 0\right\}$

$$= \sum_{n=0}^{\infty} \frac{1}{(2n+1)2^{2n}}\binom{2n}{n}\frac{1-x^{2n}}{1-x^2} = \frac{\pi}{2}\frac{1}{1-x^2} - \frac{\arcsin(x)}{x(1-x^2)},$$

and the third solution to the point (iv) is complete.

What's next? Remember we left unfinished the first two points of the problem; now it's time to bring them to an end. With the generalizations from the points (iii) and (iv) in our pocket, everything will turn into an easy task to finish. So, if we set $x = i$ in the generalization from the point (iii) and multiply both sides by -1, we get

$$\sum_{n=1}^{\infty}(-1)^{n-1}\left(\frac{\pi^2}{4} - \sum_{k=1}^{n} \frac{(2k-2)!!}{k\cdot(2k-1)!!}\right) = \sum_{n=1}^{\infty}(-1)^{n-1}\left(\frac{\pi^2}{4} - \sum_{k=1}^{n} \frac{2^{2k-1}}{k^2\binom{2k}{k}}\right)$$

$$= \frac{\arcsin^2(i)}{1-i^2} - \frac{\pi^2}{4}\frac{i^2}{1-i^2} = \frac{1}{8}\left(\pi^2 - 4\log^2(\sqrt{2}-1)\right),$$

and the solution to the point (i) is complete. To get the last equality, I used that $\arcsin(ix) = i\arcsinh(x)$ together with the fact that $\arcsinh(x) = \log(x+\sqrt{1+x^2})$.

Further, if we set $x = i$ in the generalization from the point (iv), we obtain

$$\sum_{n=0}^{\infty}(-1)^n\left(\frac{\pi}{2} - \sum_{k=0}^{n} \frac{(2k-1)!!}{(2k+1)\cdot(2k)!!}\right) = \sum_{n=0}^{\infty}(-1)^n\left(\frac{\pi}{2} - \sum_{k=0}^{n} \frac{1}{(2k+1)2^{2k}}\binom{2k}{k}\right)$$

$$= \frac{\pi}{2}\frac{1}{1-i^2} - \frac{\arcsin(i)}{i(1-i^2)} = \frac{\pi}{4} - \frac{1}{2}\log(\sqrt{2}+1),$$

and the solution to the point (ii) is complete.

What is left? To prove a nice series representation of $\zeta(2)$, which we will want to build based upon the generalization from the point (iii). So, considering the result from the point (iii), multiplying the opposite sides by $\sqrt{1-x^2}$ and integrating from $x = 0$ to $x = 1$, we have

$$\frac{\pi^2}{4} \int_0^1 \frac{x^2}{\sqrt{1-x^2}} dx - \int_0^1 \frac{\arcsin^2(x)}{\sqrt{1-x^2}} dx$$

$$= \frac{\pi^2}{8} \left(\arcsin(x) - x\sqrt{1-x^2} \right) \Bigg|_{x=0}^{x=1} - \frac{1}{3} \arcsin^3(x) \Bigg|_{x=0}^{x=1} = \frac{\pi^3}{48}$$

$$= \int_0^1 \sum_{n=1}^{\infty} x^{2n} \sqrt{1-x^2} \left(\frac{\pi^2}{4} - \sum_{k=1}^{n} \frac{(2k-2)!!}{k \cdot (2k-1)!!} \right) dx$$

{reverse the order of the main summation and integration}

$$= \sum_{n=1}^{\infty} \int_0^1 x^{2n} \sqrt{1-x^2} \left(\frac{\pi^2}{4} - \sum_{k=1}^{n} \frac{(2k-2)!!}{k \cdot (2k-1)!!} \right) dx$$

{make the change of variable $x = \sin(t)$ and use that $\sin^2(t) + \cos^2(t) = 1$}

$$= \sum_{n=1}^{\infty} \int_0^{\pi/2} \sin^{2n}(t)(1 - \sin^2(t)) \left(\frac{\pi^2}{4} - \sum_{k=1}^{n} \frac{(2k-2)!!}{k \cdot (2k-1)!!} \right) dt$$

$$\left\{ \text{use Wallis' integral,} \int_0^{\pi/2} \sin^{2n}(t) dt = \frac{\pi}{2} \cdot \frac{(2n-1)!!}{(2n)!!} \right\}$$

$$= \frac{\pi}{8} \sum_{n=1}^{\infty} \frac{(2n-1)!}{n(n+1)((2n-2)!!)^2} \left(\frac{\pi^2}{4} - \sum_{k=1}^{n} \frac{(2k-2)!!}{k \cdot (2k-1)!!} \right),$$

from which we obtain that

$$\zeta(2) = \sum_{n=1}^{\infty} \frac{(2n-1)!}{n(n+1)((2n-2)!!)^2} \left(\frac{\pi^2}{4} - \sum_{k=1}^{n} \frac{(2k-2)!!}{k \cdot (2k-1)!!} \right),$$

and the solution to the point (v) is complete.

What other different ways would you like to consider in order to prove the beautiful series representation of $\zeta(2)$ at the last point of the problem? One option would be to cleverly consider the use of telescoping sums.

6.3   Six Enjoyable Sums Involving the Reciprocal of the Central Binomial Coefficient and Two Series Derived from Them

Solution *The sums of this type can be calculated elegantly, and it's good to know that they may easily turn into a difficult task to do if we are not careful with respect to the strategy to adopt. How would you prefer to go?* This is how I initially started this section (when I only had the first two points of the problem) and then continued with the solution below, which involves the use of Wallis' integral! After that I suddenly noticed the telescoping sums may play a wonderful part here!

In the following, I'll show how I initially derived the sum at the point (i), and then I'll present the beautiful way involving the use of a telescoping sum.

The keywords for the first way to go I propose: *Wallis' integral*! More precisely, we need to use the version, $\displaystyle\int_0^{\pi/2} \sin^{2k-1}(x)\mathrm{d}x = \frac{(2k-2)!!}{(2k-1)!!} = \frac{2^{2k-1}}{k\binom{2k}{k}}$. I remind you that we also meet Wallis' integrals in the solution presented in Sect. 3.17.

So, we write that

$$\sum_{k=1}^n \frac{2^{2k}}{k\binom{2k}{k}} = 2\sum_{k=1}^n \int_0^{\pi/2} \sin^{2k-1}(x)\mathrm{d}x = 2\int_0^{\pi/2} \sum_{k=1}^n \sin^{2k-1}(x)\mathrm{d}x$$

$$= 2\int_0^{\pi/2} \frac{\sin(x)(1-\sin^{2n}(x))}{1-\sin^2(x)}\mathrm{d}x = 2\int_0^{\pi/2} \frac{\sin(x)(1-\sin^{2n}(x))}{\cos^2(x)}\mathrm{d}x$$

$$\left\{\text{integrate by parts, using that } (1/\cos(x))' = \sin(x)/\cos^2(x)\right\}$$

$$= -2 + 4n \int_0^{\pi/2} \sin^{2n-1}(x)\mathrm{d}x = \frac{2^{2n+1}}{\binom{2n}{n}} - 2,$$

and the first solution to the first point of the problem is complete.

For a second solution, we want to take into account the idea of using a telescoping sum. First, we want to observe a very beautiful fact, that is, the summand may be written as follows:

$$\frac{2^{2k}}{k\binom{2k}{k}} = \frac{2^{2k+1}k - 2^{2k}(2k-1)}{k\binom{2k}{k}} = \frac{2^{2k+1}}{\binom{2k}{k}} - \frac{2^{2k}(2k-1)}{k\binom{2k}{k}} = \frac{2^{2k+1}}{\binom{2k}{k}} - \frac{2^{2k-1}}{\binom{2k-2}{k-1}},$$

$$\tag{6.10}$$

where I also used the fact that $\binom{2k}{k} = \frac{2(2k-1)}{k}\binom{2k-2}{k-1}$.

Considering the summation over the opposite sides of (6.10), we have

$$\sum_{k=1}^{n} \frac{2^{2k}}{k\binom{2k}{k}} = \sum_{k=1}^{n}\left(\underbrace{\frac{2^{2k+1}}{\binom{2k}{k}}}_{f(k)} - \underbrace{\frac{2^{2k-1}}{\binom{2k-2}{k-1}}}_{f(k-1)}\right) = \frac{2^{2n+1}}{\binom{2n}{n}} - 2,$$

and the second solution to the first point of the problem is complete. Looks like things have worked awesomely since we've got such a short and elegant solution! This sum is a classical one, and it also appears in [69] where it is calculated by a different method involving generating functions (which the curious readers might want to check, as well).

Next, for the sum at the point (ii), we consider a simple partial fraction decomposition, and we write

$$\sum_{k=1}^{n} \frac{2^{2k}}{k(2k+1)\binom{2k}{k}} = \sum_{k=1}^{n}\left(\frac{2^{2k}}{k\binom{2k}{k}} - \frac{2^{2k+1}}{(2k+1)\binom{2k}{k}}\right)$$

$$\left\{\text{in the first fraction from the right-hand side use that } \binom{2k}{k} = \frac{2(2k-1)}{k}\binom{2k-2}{k-1}\right\}$$

$$= \sum_{k=1}^{n}\left(\underbrace{\frac{2^{2k-1}}{(2k-1)\binom{2k-2}{k-1}}}_{g(k-1)} - \underbrace{\frac{2^{2k+1}}{(2k+1)\binom{2k}{k}}}_{g(k)}\right)$$

$$= 2 - \frac{2^{2n+1}}{(2n+1)\binom{2n}{n}},$$

and the solution to the point (ii) of the problem is complete. The sum, together with a different solution, also appears in [69].

For the next two points, I'll provide two different solutions for each of them. Now, I'll start with the challenging question and prove the result at the point (iii) by using the result from the point (i).

Multiplying both sides of (i) by $\dfrac{2^{2n}}{n\dbinom{2n}{n}}$ and then summing from $n = 1$ to m, we have

$$S = \sum_{n=1}^{m} \frac{2^{2n}}{n\dbinom{2n}{n}} \sum_{k=1}^{n} \frac{2^{2k}}{k\dbinom{2k}{k}} = 2\sum_{n=1}^{m} \frac{2^{4n}}{n\dbinom{2n}{n}^2} - 2\sum_{n=1}^{m} \frac{2^{2n}}{n\dbinom{2n}{n}}$$

$$= 2\sum_{n=1}^{m} \frac{2^{4n}}{n\dbinom{2n}{n}^2} - \frac{2^{2m+2}}{\dbinom{2m}{m}} + 4, \qquad (6.11)$$

where to get the second equality in (6.11), we used again the result from the point (i).

Then, to treat differently the double sum in (6.11), we observe that by simple manipulations, we have

$$S = \sum_{n=1}^{m} \frac{2^{2n}}{n\dbinom{2n}{n}} \sum_{k=1}^{n} \frac{2^{2k}}{k\dbinom{2k}{k}} = \sum_{n=1}^{m} \frac{2^{2n}}{n\dbinom{2n}{n}} \left(\sum_{k=1}^{m} - \sum_{k=n}^{m} + \sum_{k=n}^{m} \right) \frac{2^{2k}}{k\dbinom{2k}{k}}$$

$$= \sum_{n=1}^{m} \frac{2^{2n}}{n\dbinom{2n}{n}} \sum_{k=1}^{m} \frac{2^{2k}}{k\dbinom{2k}{k}} - \sum_{n=1}^{m} \frac{2^{2n}}{n\dbinom{2n}{n}} \sum_{k=n}^{m} \frac{2^{2k}}{k\dbinom{2k}{k}} + \sum_{n=1}^{m} \frac{2^{2n}}{n\dbinom{2n}{n}} \sum_{k=n}^{n} \frac{2^{2k}}{k\dbinom{2k}{k}}$$

{change the summation order in the second double sum}

$$= \left(\sum_{n=1}^{m} \frac{2^{2n}}{n\dbinom{2n}{n}} \right)^2 - \sum_{k=1}^{m} \frac{2^{2k}}{k\dbinom{2k}{k}} \sum_{n=1}^{k} \frac{2^{2n}}{n\dbinom{2n}{n}} + \sum_{n=1}^{m} \frac{2^{4n}}{n^2\dbinom{2n}{n}^2}$$

{for the first sum use the result from the point (i), and in the second sum swap}

{the variables k and n to make it clear there we have the double sum S}

$$= \left(\frac{2^{2m+1}}{\binom{2m}{m}} - 2 \right)^2 - \underbrace{\sum_{n=1}^{m} \frac{2^{2n}}{n\binom{2n}{n}} \sum_{k=1}^{n} \frac{2^{2k}}{k\binom{2k}{k}} + \sum_{n=1}^{m} \frac{2^{4n}}{n^2 \binom{2n}{n}^2}}_{S},$$

from which we obtain that

$$S = \frac{1}{2} \left(\frac{2^{2m+1}}{\binom{2m}{m}} - 2 \right)^2 + \frac{1}{2} \sum_{n=1}^{m} \frac{2^{4n}}{n^2 \binom{2n}{n}^2}. \tag{6.12}$$

We note that in our calculations, we used a (very) useful change of summation order.[2] For example, this particular change of summation order is also presented in the popular books, *Concrete Mathematics* (see [18, p.36]) and *Irresistible Integrals* (see [6, p.22]).

At this point, by combining the results in (6.11) and (6.12), rearranging, and passing to the same letters as in the statement of the result, we get

$$\sum_{k=1}^{n} \frac{2^{4k}(4k-1)}{k^2 \binom{2k}{k}^2} = \frac{2^{4n+2}}{\binom{2n}{n}^2} - 4,$$

and the first solution to the point (*iii*) of the problem is complete, which also answers the *challenging question*.

To get a second solution to the point (*iii*) of the problem, we use the idea of telescoping sums. It's good to be careful on how we perform the splitting to make everything easy and beautiful. So, we observe and write that

---

[2] In general, for a sum of the type $\sum_{j=1}^{n} \sum_{i=1}^{j} a_{(i,j)}$, we have the following result, $\sum_{j=1}^{n} \sum_{i=1}^{j} a_{(i,j)} = \sum_{i=1}^{n} \sum_{j=i}^{n} a_{(i,j)}$. It's easy to understand why the result is true if we place $a_{(i,j)}$s over the triangle $1 \le i \le j \le n$ of a grid and observe that we simply add up the same terms in two different ways. In our calculations in the book, during some work with series, we'll also find quite useful the infinite version, $\sum_{j=1}^{\infty} \left(\sum_{i=1}^{j} a_{(i,j)} \right) = \sum_{i=1}^{\infty} \left(\sum_{j=i}^{\infty} a_{(i,j)} \right)$.

$$\sum_{k=1}^{n} \frac{2^{4k}(4k-1)}{k^2 \binom{2k}{k}^2} = \sum_{k=1}^{n} \frac{2^{4k}((2k)^2 - (2k-1)^2)}{k^2 \binom{2k}{k}^2} = \sum_{k=1}^{n} \left(\frac{2^{4k+2}}{\binom{2k}{k}^2} - \frac{2^{4k}(2k-1)^2}{k^2 \binom{2k}{k}^2} \right)$$

$$= \sum_{k=1}^{n} \left(\underbrace{\frac{2^{4k+2}}{\binom{2k}{k}^2}}_{h(k)} - \underbrace{\frac{2^{4k-2}}{\binom{2k-2}{k-1}^2}}_{h(k-1)} \right) = \frac{2^{4n+2}}{\binom{2n}{n}^2} - 4,$$

where everything is finalized beautifully again, and the second solution to the point (iii) of the problem is complete. During the calculations, we also used the fact that $\binom{2k}{k} = \frac{2(2k-1)}{k} \binom{2k-2}{k-1}$.

Passing to the point (iv) of the problem, we start with the *challenging question* and try to use the result from the point (ii) in order to prove the present result. Multiplying both sides of (ii) by $\dfrac{2^{2n}}{n(2n+1)\binom{2n}{n}}$ and then summing from $n = 1$ to m, we get

$$T = \sum_{n=1}^{m} \frac{2^{2n}}{n(2n+1)\binom{2n}{n}} \sum_{k=1}^{n} \frac{2^{2k}}{k(2k+1)\binom{2k}{k}} = 2 \sum_{n=1}^{m} \frac{2^{2n}}{n(2n+1)\binom{2n}{n}}$$

$$- \sum_{n=1}^{m} \frac{2^{4n+1}}{n(2n+1)^2 \binom{2n}{n}^2} = 4 - \frac{2^{2m+2}}{(2m+1)\binom{2m}{m}} - \sum_{n=1}^{m} \frac{2^{4n+1}}{n(2n+1)^2 \binom{2n}{n}^2},$$

$$(6.13)$$

where to get the second equality, we used one more time the result from the point (ii).

Then, we treat differently the double sum in (6.13), and we write that

$$T = \sum_{n=1}^{m} \frac{2^{2n}}{n(2n+1)\binom{2n}{n}} \sum_{k=1}^{n} \frac{2^{2k}}{k(2k+1)\binom{2k}{k}}$$

$$= \sum_{n=1}^{m} \frac{2^{2n}}{n(2n+1)\binom{2n}{n}} \left(\sum_{k=1}^{m} - \sum_{k=n}^{m} + \sum_{k=n}^{n} \right) \frac{2^{2k}}{k(2k+1)\binom{2k}{k}}$$

$$= \sum_{n=1}^{m} \frac{2^{2n}}{n(2n+1)\binom{2n}{n}} \sum_{k=1}^{m} \frac{2^{2k}}{k(2k+1)\binom{2k}{k}} - \sum_{n=1}^{m} \frac{2^{2n}}{n(2n+1)\binom{2n}{n}} \sum_{k=n}^{m} \frac{2^{2k}}{k(2k+1)\binom{2k}{k}}$$

$$+ \sum_{n=1}^{m} \frac{2^{2n}}{n(2n+1)\binom{2n}{n}} \sum_{k=n}^{n} \frac{2^{2k}}{k(2k+1)\binom{2k}{k}}$$

{change the summation order in the second double sum}

$$= \left(\sum_{n=1}^{m} \frac{2^{2n}}{n(2n+1)\binom{2n}{n}} \right)^2 - \sum_{k=1}^{m} \frac{2^{2k}}{k(2k+1)\binom{2k}{k}} \sum_{n=1}^{k} \frac{2^{2n}}{n(2n+1)\binom{2n}{n}}$$

$$+ \sum_{n=1}^{m} \frac{2^{4n}}{n^2(2n+1)^2\binom{2n}{n}^2}$$

{for the first sum use the result from the point (ii), and in the second sum}

{swap the variables k and n to see clearly there we have the double sum T}

$$= \left(2 - \frac{2^{2m+1}}{(2m+1)\binom{2m}{m}} \right)^2 - \underbrace{\sum_{n=1}^{m} \frac{2^{2n}}{n(2n+1)\binom{2n}{n}} \sum_{k=1}^{n} \frac{2^{2k}}{k(2k+1)\binom{2k}{k}}}_{T}$$

$$+ \sum_{n=1}^{m} \frac{2^{4n}}{n^2(2n+1)^2\binom{2n}{n}^2},$$

from which we obtain that

$$T = \frac{1}{2} \left(2 - \frac{2^{2m+1}}{(2m+1)\binom{2m}{m}} \right)^2 + \sum_{n=1}^{m} \frac{2^{4n-1}}{n^2(2n+1)^2\binom{2n}{n}^2}. \tag{6.14}$$

Now, by combining the results in (6.13) and (6.14), rearranging, and passing to the same letters as in the statement of the result, we obtain

$$\sum_{k=1}^{n} \frac{2^{4k}(4k+1)}{k^2(2k+1)^2\binom{2k}{k}^2} = 4 - \frac{2^{4n+2}}{(2n+1)^2\binom{2n}{n}^2},$$

and the first solution to the point (iv) of the problem is complete, which also answers the *challenging question*.

For a second solution to the point (iv) of the problem, we'll be happy to carefully turn all into a telescoping sum, and then we write

$$\sum_{k=1}^{n} \frac{2^{4k}(4k+1)}{k^2(2k+1)^2\binom{2k}{k}^2} = \sum_{k=1}^{n} \frac{2^{4k}((2k+1)^2 - (2k)^2)}{k^2(2k+1)^2\binom{2k}{k}^2}$$

$$= \sum_{k=1}^{n} \left(\frac{2^{4k}}{k^2\binom{2k}{k}^2} - \frac{2^{4k+2}}{(2k+1)^2\binom{2k}{k}^2} \right)$$

$$= \sum_{k=1}^{n} \left(\underbrace{\frac{2^{4k-2}}{(2k-1)^2\binom{2k-2}{k-1}^2}}_{p(k-1)} - \underbrace{\frac{2^{4k+2}}{(2k+1)^2\binom{2k}{k}^2}}_{p(k)} \right)$$

$$= 4 - \frac{2^{4n+2}}{(2n+1)^2\binom{2n}{n}^2},$$

where, as seen, the desired result is obtained in a very beautiful manner, and the second solution to the point (iv) of the problem is complete. In the calculations, we also used the fact that $\binom{2k}{k} = \frac{2(2k-1)}{k}\binom{2k-2}{k-1}$.

For the points (v) and (vi), we may proceed exactly as before and use the idea of telescoping sums. Indeed, the next two sums look *dangerous*, but happily they can be easily calculated when using the proper ways to go. So, for the point (v) of the problem, we write

$$\sum_{k=1}^{n} \frac{2^{8k}(4k-1)(8k^2-4k+1)}{k^4\binom{2k}{k}^4} = \sum_{k=1}^{n} \frac{2^{8k}((2k)^4-(2k-1)^4)}{k^4\binom{2k}{k}^4}$$

$$= \sum_{k=1}^{n}\left(\frac{2^{8k+4}}{\binom{2k}{k}^4} - \frac{(2k-1)^4 2^{8k}}{k^4\binom{2k}{k}^4}\right) = \sum_{k=1}^{n}\left(\underbrace{\frac{2^{8k+4}}{\binom{2k}{k}^4}}_{q(k)} - \underbrace{\frac{2^{8k-4}}{\binom{2k-2}{k-1}^4}}_{q(k-1)}\right)$$

$$= \frac{2^{8n+4}}{\binom{2n}{n}^4} - 16,$$

and the solution to the point (v) of the problem is complete.

Finally, for the last finite sum of the problem, we have

$$\sum_{k=1}^{n} \frac{2^{8k}(4k+1)(8k^2+4k+1)}{k^4(2k+1)^4\binom{2k}{k}^4} = \sum_{k=1}^{n} \frac{2^{8k}((2k+1)^4-(2k)^4)}{k^4(2k+1)^4\binom{2k}{k}^4}$$

$$= \sum_{k=1}^{n}\left(\frac{2^{8k}}{k^4\binom{2k}{k}^4} - \frac{2^{8k+4}}{(2k+1)^4\binom{2k}{k}^4}\right)$$

$$\left\{\text{for the first term inside the sum use that } \binom{2k}{k} = \frac{2(2k-1)}{k}\binom{2k-2}{k-1}\right\}$$

$$= \sum_{k=1}^{n}\left(\underbrace{\frac{2^{8k-4}}{(2k-1)^4\binom{2k-2}{k-1}^4}}_{r(k-1)} - \underbrace{\frac{2^{8k+4}}{(2k+1)^4\binom{2k}{k}^4}}_{r(k)}\right)$$

$$= 16 - \frac{2^{8n+4}}{(2n+1)^4 \binom{2n}{n}^4},$$

and the solution to the point (vi) of the problem is complete.

At last, it is not hard to observe that the series at the end of the problem statement section are immediately derived by using the sums from the points (iv) and (vi).

So, we write

$$\sum_{k=1}^{\infty} \frac{2^{4k}(4k+1)}{k^2(2k+1)^2 \binom{2k}{k}^2} = \lim_{n \to \infty} \sum_{k=1}^{n} \frac{2^{4k}(4k+1)}{k^2(2k+1)^2 \binom{2k}{k}^2}$$

$$= 4 - \lim_{n \to \infty} \frac{2^{4n+2}}{(2n+1)^2 \binom{2n}{n}^2} = 4,$$

where the desired result follows immediately by considering the asymptotical behavior of the central binomial coefficient, $\binom{2n}{n} \approx \frac{4^n}{\sqrt{\pi n}}$, which is obtained via Stirling's formula, $n! \approx \sqrt{2\pi n} \left(\frac{n}{e}\right)^n$ (see [70, Chapter 1, p.8], [97]), and the solution to the point (vii) of the problem is complete.

In a similar style as before, to calculate the last series, we use the sum from the point (vi), and then we write

$$\sum_{k=1}^{\infty} \frac{2^{8k}(4k+1)(8k^2+4k+1)}{k^4(2k+1)^4 \binom{2k}{k}^4} = \lim_{n \to \infty} \sum_{k=1}^{n} \frac{2^{8k}(4k+1)(8k^2+4k+1)}{k^4(2k+1)^4 \binom{2k}{k}^4}$$

$$= 16 - \lim_{n \to \infty} \frac{2^{8n+4}}{(2n+1)^4 \binom{2n}{n}^4} = 16,$$

and the solution to the point $(viii)$ of the problem is complete.

While the first two sums were known to me when I first considered writing this section, the sums from the points (iii)–(vi) were unknown, and I created them immediately by using the ideas in the solutions to the *challenging question*.

Using the ideas from the solutions to the *challenging question*, one may go on and derive more such sums with the central binomial coefficient.

6.4   A Great Time with a Special Binoharmonic Series

Solution There is no doubt that jumping to this section while reading Sect. 3.59 is irresistible, since the result to prove here is a key series in the finalization process of the solution in the mentioned section. *So, how would we like to go?*

Before continuing, I would like to remind you that also in *(Almost) Impossible Integrals, Sums, and Series*, we may find a fascinating series involving the squared central binomial coefficient (see [76, Chapter 4, p.281]).

At first, I'll combine two results already met in previous sections! More exactly, we begin with observing the wonderful fact that we may express the summand by using the integral result in (3.126), Sect. 3.19, and then we write

$$\sum_{n=0}^{\infty} \frac{1}{16^n} \binom{2n}{n}^2 (H_{2n} - H_n - \log(2))^2$$

$$= \frac{4}{\pi^2} \sum_{n=0}^{\infty} \left(\int_0^{\pi/2} \log(\sin(x)) \sin^{2n}(x)dx \right) \left(\int_0^{\pi/2} \log(\sin(y)) \sin^{2n}(y)dy \right)$$

{reverse the order of integration and summation}

$$= \frac{4}{\pi^2} \int_0^{\pi/2} \log(\sin(y)) \left(\int_0^{\pi/2} \log(\sin(x)) \sum_{n=0}^{\infty} (\sin(x)\sin(y))^{2n}dx \right) dy$$

$$= \frac{4}{\pi^2} \int_0^{\pi/2} \log(\sin(y)) \left(\int_0^{\pi/2} \frac{\log(\sin(x))}{1 - \sin^2(y)\sin^2(x)}dx \right) dy$$

{exploit the Sylvester's integral, the form in (1.3), Sect. 1.1}

$$= -\frac{2}{\pi} \int_0^{\pi/2} \frac{\log(\sin(y))\log(1+\cos(y))}{\cos(y)}dy$$

$$\left\{ \text{employ the variable change } \frac{1 - \tan(y/2)}{1 + \tan(y/2)} = t \text{ together with the following} \right\}$$

$$\left\{ \text{simple identities, } \sin(2\arctan(x)) = \frac{2x}{1 + x^2} \text{ and } \cos(2\arctan(x)) = \frac{1 - x^2}{1 + x^2} \right\}$$

$$= -\frac{2}{\pi} \int_0^1 \frac{\log\left(\dfrac{1 - t^2}{1 + t^2}\right) \log\left(\dfrac{(1+t)^2}{1 + t^2}\right)}{t}dt$$

$$= -\frac{4}{\pi}\int_0^1 \frac{\log^2(1+t)}{t}dt - \frac{2}{\pi}\underbrace{\int_0^1 \frac{\log^2(1+t^2)}{t}dt}_{t^2=u} - \frac{4}{\pi}\int_0^1 \frac{\log(1-t)\log(1+t)}{t}dt$$

$$+ \frac{2}{\pi}\underbrace{\int_0^1 \frac{\log(1-t^2)\log(1+t^2)}{t}dt}_{t^2=u} + \frac{4}{\pi}\int_0^1 \frac{\log(1+t)\log(1+t^2)}{t}dt$$

$$= -\frac{5}{\pi}\int_0^1 \frac{\log^2(1+t)}{t}dt - \frac{3}{\pi}\int_0^1 \frac{\log(1-t)\log(1+t)}{t}dt$$

$$+ \frac{4}{\pi}\int_0^1 \frac{\log(1+t)\log(1+t^2)}{t}dt = 2G - \frac{7}{2}\frac{\zeta(3)}{\pi},$$

where I used that $\int_0^1 \frac{\log^2(1+t)}{t}dt = \frac{1}{4}\zeta(3)$ ([76, Chapter 1, p.4]); afterward, $\int_0^1 \frac{\log(1-t)\log(1+t)}{t}dt = -\frac{5}{8}\zeta(3)$ (see [76, Chapter 1, p.4]); and then, using integration by parts, $\int_0^1 \frac{\log(1+t)\log(1+t^2)}{t}dt = \int_0^1 -(\text{Li}_2(-t))' \log(1+t^2)dt = -\text{Li}_2(-t)\log(1+t^2)\Big|_{t=0}^{t=1} + 2\int_0^1 \frac{t\,\text{Li}_2(-t)}{1+t^2}dt = \frac{1}{2}\pi G - \frac{33}{32}\zeta(3)$, where the last integral is already evaluated in [76, Chapter 3, pp. 125–127], $\int_0^1 \frac{t\,\text{Li}_2(-t)}{1+t^2}dt = \frac{\pi}{4}G - \log(2)\frac{\pi^2}{24} - \frac{33}{64}\zeta(3)$, and the solution is finalized.

Alternatively, we may use the simple fact that $\log(1+t) = \log(1-t^2) - \log(1-t)$ that promptly leads to $\int_0^1 \frac{\log(1+t)\log(1+t^2)}{t}dt = \int_0^1 \frac{\log(1-t^2)\log(1+t^2)}{t}dt - \int_0^1 \frac{\log(1-t)\log(1+t^2)}{t}dt$, where for the second resulting integral, we may exploit the logarithmic integral in (3.10), Sect. 3.3, and the generating function in (4.36), Sect. 4.6, and from here, the curious reader may continue the calculations.

Finally, note that the present solution also answers the *challenging question*!

6.5   A Panel of (Very) Useful Cauchy Products of Two Series: From Known Cauchy Products to Less Known Ones

Solution As in *(Almost) Impossible Integrals, Sums, and Series* (e.g., see [76, Chapter 3, Sect. 3.10, pp. 87–89, Chapter 4, Sect. 4.9, p. 283]), in the present book, there are various places where it is of great help to employ certain series as the ones given in the current section, which are usually derived with the help of the Cauchy product of two series. In the following, I'll derive a bouquet of such results, from known ones to some less known or possibly not known in the mathematical literature.

Recall that according to the Cauchy product of two series, if $\sum\limits_{n=1}^{\infty} a_n$ and $\sum\limits_{n=1}^{\infty} b_n$ are absolutely convergent, then we have

$$\left(\sum_{n=1}^{\infty} a_n\right)\left(\sum_{n=1}^{\infty} b_n\right) = \sum_{n=1}^{\infty}\left(\sum_{k=1}^{n} a_k b_{n-k+1}\right). \tag{6.15}$$

Let's get started! Since we have the power series $\arctan(x) = \sum\limits_{n=1}^{\infty}(-1)^{n-1}\dfrac{x^{2n-1}}{2n-1}$

and $\dfrac{1}{1+x^2} = \sum\limits_{n=1}^{\infty}(-1)^{n-1}x^{2n-2}$, by applying (6.15), we get

$$\begin{aligned}
\frac{\arctan(x)}{1+x^2} &= \left(\sum_{n=1}^{\infty}(-1)^{n-1}\frac{x^{2n-1}}{2n-1}\right)\left(\sum_{n=1}^{\infty}(-1)^{n-1}x^{2n-2}\right) \\
&= \sum_{n=1}^{\infty}(-1)^{n-1}x^{2n-1}\left(\sum_{k=1}^{n}\frac{1}{2k-1}\right) \\
&= \sum_{n=1}^{\infty}(-1)^{n-1}x^{2n-1}\left(\sum_{k=1}^{2n}\frac{1}{k} - \sum_{k=1}^{n}\frac{1}{2k}\right) \\
&= \frac{1}{2}\sum_{n=1}^{\infty}(-1)^{n-1}x^{2n-1}(2H_{2n} - H_n),
\end{aligned}$$

and the solution to the point (i) is complete.

Considering the result from the previous point in variable t, multiplying both sides by 2, and then integrating from $t = 0$ to $t = x$, we have

$$\int_0^x \frac{2\arctan(t)}{1+t^2}\,dt = \arctan^2(x) = \frac{1}{2}\sum_{n=1}^{\infty}(-1)^{n-1}x^{2n}\frac{2H_{2n} - H_n}{n},$$

and the solution to the point (ii) is complete.

Replacing x by ix in the previous two results and then simplifying, we arrive at

$$\frac{\text{arctanh}(x)}{1 - x^2} = \frac{1}{2} \sum_{n=1}^{\infty} x^{2n-1}(2H_{2n} - H_n)$$

and

$$\text{arctanh}^2(x) = \frac{1}{2} \sum_{n=1}^{\infty} x^{2n} \frac{2H_{2n} - H_n}{n},$$

and the solutions to the points (iii) and (iv) of the problem are complete.

As regards the points (v), we consider the following power series, $\arctan(x) = \sum_{n=1}^{\infty} (-1)^{n-1} \frac{x^{2n-1}}{2n-1}$ and $\text{arctanh}(x) = \sum_{n=1}^{\infty} \frac{x^{2n-1}}{2n-1}$, and upon applying (6.15), we have

$$\arctan(x)\,\text{arctanh}(x) = \left(\sum_{n=1}^{\infty} (-1)^{n-1} \frac{x^{2n-1}}{2n-1} \right) \left(\sum_{n=1}^{\infty} \frac{x^{2n-1}}{2n-1} \right)$$

$$= \sum_{n=1}^{\infty} x^{2n} \left(\sum_{k=1}^{n} \frac{(-1)^{k-1}}{(2k-1)(2n-2k+1)} \right)$$

{split the series according to the parity of n}

$$= \sum_{n=1}^{\infty} x^{4n-2} \left(\sum_{k=1}^{2n-1} \frac{(-1)^{k-1}}{(2k-1)(4n-2k-1)} \right) + \underbrace{\sum_{n=1}^{\infty} x^{4n} \left(\sum_{k=1}^{2n} \frac{(-1)^{k-1}}{(2k-1)(4n-2k+1)} \right)}_{0}$$

$$= \sum_{n=1}^{\infty} \frac{x^{4n-2}}{2n-1} \frac{1}{2} \left(\sum_{k=1}^{2n-1} \frac{(-1)^{k-1}}{2k-1} + \underbrace{\sum_{k=1}^{2n-1} \frac{(-1)^{k-1}}{4n-2k-1}}_{\text{let } 2n-k=m} \right)$$

$$= \sum_{n=1}^{\infty} \frac{x^{4n-2}}{2n-1} \left(\sum_{k=1}^{2n-1} (-1)^{k-1} \frac{1}{2k-1} \right),$$

and the solution to the point (v) is complete. Observe that above I also used the simple fact that $\displaystyle\sum_{k=1}^{2n}\frac{(-1)^{k-1}}{(2k-1)(4n-2k+1)} = \frac{1}{4n}\sum_{k=1}^{2n}\frac{(-1)^{k-1}}{2k-1} +$

$\displaystyle\frac{1}{4n}\sum_{k=1}^{2n}\frac{(-1)^{k-1}}{4n-2k+1} = 0$, and this is because $\displaystyle\sum_{k=1}^{2n}\frac{(-1)^{k-1}}{4n-2k+1} = -\sum_{k=1}^{2n}\frac{(-1)^{k-1}}{2k-1}$,

easily seen if we change the order of summing the terms in any of the two sums. The present Cauchy product also appears in [17, **1.517.2**, p.54].

Further, for the point (vi) recall the power series $\arctan(x) = \displaystyle\sum_{n=1}^{\infty}(-1)^{n-1}\frac{x^{2n-1}}{2n-1}$

and $\text{Li}_2(-x^2) = \displaystyle\sum_{n=1}^{\infty}(-1)^n\frac{x^{2n}}{n^2}$, and applying (6.15), we get

$$\arctan(x)\,\text{Li}_2(-x^2) = \left(\sum_{n=1}^{\infty}(-1)^{n-1}\frac{x^{2n-1}}{2n-1}\right)\left(\sum_{n=1}^{\infty}(-1)^n\frac{x^{2n}}{n^2}\right)$$

$$= \sum_{n=1}^{\infty}(-1)^n x^{2n+1}\sum_{k=1}^{n}\frac{1}{(2k-1)(n-k+1)^2} = \sum_{n=1}^{\infty}(-1)^n x^{2n+1}$$

$$\times\left(\frac{4}{(2n+1)^2}\sum_{k=1}^{n}\frac{1}{2k-1} + \frac{2}{(2n+1)^2}\sum_{k=1}^{n}\frac{1}{n-k+1} + \frac{1}{2n+1}\sum_{k=1}^{n}\frac{1}{(n-k+1)^2}\right)$$

$$= \sum_{n=1}^{\infty}(-1)^n x^{2n+1}\left(4\frac{H_{2n}}{(2n+1)^2} + \frac{H_n^{(2)}}{2n+1}\right),$$

and the solution to the point (vi) is complete.

Upon replacing x by ix in the previous result and then simplifying, we arrive at

$$\operatorname{arctanh}(x)\,\text{Li}_2(x^2) = \sum_{n=1}^{\infty}x^{2n+1}\left(4\frac{H_{2n}}{(2n+1)^2} + \frac{H_n^{(2)}}{2n+1}\right),$$

and the solution to the point (vii) is complete.

Based on the Cauchy product of two series, the version in the *Mertens' theorem* (see [11, pp. 82–83]), with both series converging, and at least one converging absolutely, if we consider $\text{Li}_2(x) = \displaystyle\sum_{n=1}^{\infty}\frac{x^n}{n^2}$ and $-\log(1-x) = \displaystyle\sum_{n=1}^{\infty}\frac{x^n}{n}$, we have

$$-\text{Li}_2(x)\log(1-x) = \left(\sum_{n=1}^{\infty}\frac{x^n}{n^2}\right)\left(\sum_{n=1}^{\infty}\frac{x^n}{n}\right) = \sum_{n=1}^{\infty}x^{n+1}\left(\sum_{k=1}^{n}\frac{1}{k^2(n-k+1)}\right)$$

$$= \sum_{n=1}^{\infty} x^{n+1} \left(\sum_{k=1}^{n} \frac{1}{k(n+1)^2} + \sum_{k=1}^{n} \frac{1}{(n-k+1)(n+1)^2} + \sum_{k=1}^{n} \frac{1}{k^2(n+1)} \right)$$

$$\left\{ \text{observe and use that } \sum_{k=1}^{n} \frac{1}{k} = \sum_{k=1}^{n} \frac{1}{n-k+1} = H_n \right\}$$

$$= 2 \sum_{n=1}^{\infty} x^{n+1} \frac{H_{n+1} - \frac{1}{n+1}}{(n+1)^2} + \sum_{n=1}^{\infty} x^{n+1} \frac{H_{n+1}^{(2)} - \frac{1}{(n+1)^2}}{n+1}$$

{reindex the series and expand them}

$$= 2 \sum_{n=1}^{\infty} x^n \frac{H_n}{n^2} + \sum_{n=1}^{\infty} x^n \frac{H_n^{(2)}}{n} - 3 \sum_{n=1}^{\infty} \frac{x^n}{n^3} = \sum_{n=1}^{\infty} x^n \left(2\frac{H_n}{n^2} + \frac{H_n^{(2)}}{n} - 3\frac{1}{n^3} \right),$$

and the result follows after multiplying the opposite sides by -1, and the solution to the point $(viii)$ is complete.

By a similar reasoning as before, if we consider the Cauchy product of two series, the version in the *Mertens' theorem* (see [11, pp. 82–83]), with $\text{Li}_3(x) = \sum_{n=1}^{\infty} \frac{x^n}{n^3}$

and $-\log(1-x) = \sum_{n=1}^{\infty} \frac{x^n}{n}$, we have

$$-\text{Li}_3(x)\log(1-x) = \left(\sum_{n=1}^{\infty} \frac{x^n}{n^3} \right) \left(\sum_{n=1}^{\infty} \frac{x^n}{n} \right) = \sum_{n=1}^{\infty} x^{n+1} \left(\sum_{k=1}^{n} \frac{1}{k^3(n-k+1)} \right)$$

$$= \sum_{n=1}^{\infty} x^{n+1} \left(\sum_{k=1}^{n} \frac{1}{k(n+1)^3} + \sum_{k=1}^{n} \frac{1}{(n-k+1)(n+1)^3} + \sum_{k=1}^{n} \frac{1}{k^2(n+1)^2} \right.$$

$$\left. + \sum_{k=1}^{n} \frac{1}{k^3(n+1)} \right)$$

$$= 2 \sum_{n=1}^{\infty} x^{n+1} \frac{H_{n+1} - \frac{1}{n+1}}{(n+1)^3} + \sum_{n=1}^{\infty} x^{n+1} \frac{H_{n+1}^{(2)} - \frac{1}{(n+1)^2}}{(n+1)^2} + \sum_{n=1}^{\infty} x^{n+1} \frac{H_{n+1}^{(3)} - \frac{1}{(n+1)^3}}{n+1}$$

{reindex the series and merge them}

$$= \sum_{n=1}^{\infty} x^n \left(\frac{H_n^{(3)}}{n} + \frac{H_n^{(2)}}{n^2} + 2\frac{H_n}{n^3} - 4\frac{1}{n^4} \right),$$

and the result follows after multiplying the opposite sides by -1, and the solution to the point (ix) is complete.

I'll provide two solutions to the tenth point of the problem! Observe that since $\overline{H}_{2n} = H_{2n} - H_n$, then the Cauchy product in (6.175), Sect. 6.23, can be put in the form

$$- \log(1+x)\log(1-x) = 2\sum_{n=1}^{\infty} x^{2n}\left(\frac{\overline{H}_{2n}}{2n} + \frac{1}{(2n)^2}\right)$$

$$= \sum_{n=1}^{\infty} x^n (1 - (-1)^{n-1})\left(\frac{\overline{H}_n}{n} + \frac{1}{n^2}\right). \tag{6.16}$$

Now, we are ready to apply (6.15), and using (6.16), we write that

$$\frac{\log(1-x)\log(1+x)}{1-x} = -\left(\sum_{n=1}^{\infty} x^n (1 - (-1)^{n-1})\left(\frac{\overline{H}_n}{n} + \frac{1}{n^2}\right)\right)\left(\sum_{n=1}^{\infty} x^{n-1}\right)$$

$$= \sum_{n=1}^{\infty} x^n \left(\sum_{k=1}^{n}(-1)^{k-1}\frac{\overline{H}_k}{k} + \sum_{k=1}^{n}(-1)^{k-1}\frac{1}{k^2} - \sum_{k=1}^{n}\frac{1}{k^2} - \sum_{k=1}^{n}\frac{\overline{H}_k}{k}\right)$$

$$= \sum_{n=1}^{\infty} x^n \left(\frac{1}{2}\overline{H}_n^2 + \overline{H}_n^{(2)} - \frac{1}{2}H_n^{(2)} - \sum_{k=1}^{n}\frac{\overline{H}_k}{k}\right),$$

where in the calculations I also used that $\sum_{k=1}^{n}(-1)^{k-1}\dfrac{\overline{H}_k}{k} = \dfrac{1}{2}\left(\overline{H}_n^2 + H_n^{(2)}\right)$, which appears in (1.120), the case $p=1$, Sect. 1.23, and the first solution to the point (x) is complete.

In order to obtain a second solution, we denote $S_{m,p}(n) = \sum_{k=1}^{n-1}\dfrac{\overline{H}_k^{(m)}}{(n-k)^p} = \sum_{k=1}^{n-1}\dfrac{\overline{H}_{n-k}^{(m)}}{k^p}$ and observe that the last sum is obtained by reversing the order of summing the terms of the initial sum. Then, if we consider the difference $S_{m,p}(n) - S_{m,p}(n-1)$, we obtain that

$$S_{m,p}(n) - S_{m,p}(n-1) = \sum_{k=1}^{n-1}\frac{\overline{H}_{n-k}^{(m)}}{k^p} - \sum_{k=1}^{n-2}\frac{\overline{H}_{n-k-1}^{(m)}}{k^p} = \sum_{k=1}^{n-1}\frac{\overline{H}_{n-k}^{(m)} - \overline{H}_{n-k-1}^{(m)}}{k^p}$$

$$= \sum_{k=1}^{n-1}\frac{(-1)^{n-k-1}}{k^p(n-k)^m} = \sum_{k=1}^{n-1}\frac{(-1)^{k-1}}{k^m(n-k)^p}. \tag{6.17}$$

Setting $m = 1$, $p = 1$ in (6.17), we get

$$S_{1,1}(n) - S_{1,1}(n-1) = \sum_{k=1}^{n-1} \frac{(-1)^{k-1}}{k(n-k)} = \frac{1}{n}\sum_{k=1}^{n-1}(-1)^{k-1}\frac{1}{k} + \frac{1}{n}\sum_{k=1}^{n-1}(-1)^{k-1}\frac{1}{n-k}$$

$$= \frac{1}{n}\sum_{k=1}^{n-1}(-1)^{k-1}\frac{1}{k} - (-1)^{n-1}\frac{1}{n}\sum_{k=1}^{n-1}(-1)^{k-1}\frac{1}{k} = \frac{\overline{H}_{n-1}}{n} - (-1)^{n-1}\frac{\overline{H}_{n-1}}{n}$$

$$= \frac{1}{n^2} - (-1)^{n-1}\frac{1}{n^2} - (-1)^{n-1}\frac{\overline{H}_n}{n} + \frac{\overline{H}_n}{n}. \tag{6.18}$$

Replacing n by k in the opposite sides of (6.18) and then summing both sides from $k = 1$ to n, where $S_{1,1}(0) = 0$ and $S_{1,1}(1) = 0$, we get

$$\sum_{k=1}^{n}(S_{1,1}(k) - S_{1,1}(k-1)) = S_{1,1}(n) = \sum_{k=1}^{n-1}\frac{\overline{H}_k}{n-k} = \sum_{k=1}^{n}\frac{1}{k^2} - \sum_{k=1}^{n}(-1)^{k-1}\frac{1}{k^2}$$

$$- \sum_{k=1}^{n}(-1)^{k-1}\frac{\overline{H}_k}{k} + \sum_{k=1}^{n}\frac{\overline{H}_k}{k} = \frac{1}{2}H_n^{(2)} - \overline{H}_n^{(2)} - \frac{1}{2}\overline{H}_n^2 + \sum_{k=1}^{n}\frac{\overline{H}_k}{k}, \tag{6.19}$$

where in the calculations I also used that $\sum_{k=1}^{n}(-1)^{k-1}\frac{\overline{H}_k}{k} = \frac{1}{2}\left(\overline{H}_n^2 + H_n^{(2)}\right)$, which appears in (1.120), the case $p = 1$, Sect. 1.23. Careful reader might observe that since I wrote $\sum_{k=1}^{n}(S_{1,1}(k) - S_{1,1}(k-1))$, using the letter k and not another letter like i instead of k, one might automatically count that $S_{1,1}(k)$ takes a form like $S_{1,1}(k) = \sum_{j=1}^{k-1}\frac{\overline{H}_j}{k-j}$, and later, after considering the telescoping sum, we may change the letters and write, say, the sum from $k = 1$ to $n - 1$.

Returning to the main result, using that $\dfrac{\log(1+x)}{1-x} = \sum_{n=1}^{\infty}x^n\overline{H}_n$, as seen in (4.62), Sect. 4.10, and $\log(1 - x) = -\sum_{n=1}^{\infty}\dfrac{x^n}{n}$, and applying (6.15), we get

$$\frac{\log(1-x)\log(1+x)}{1-x} = \left(\sum_{n=1}^{\infty}x^n\overline{H}_n\right)\left(-\sum_{n=1}^{\infty}\frac{x^n}{n}\right) = -\sum_{n=1}^{\infty}x^{n+1}\sum_{k=1}^{n}\frac{\overline{H}_k}{n-k+1}$$

$$= -\sum_{n=2}^{\infty} x^n \sum_{k=1}^{n-1} \frac{\overline{H}_k}{n-k} = -\sum_{n=1}^{\infty} x^n \sum_{k=1}^{n-1} \frac{\overline{H}_k}{n-k}$$

$$= \sum_{n=1}^{\infty} x^n \left(\frac{1}{2}\overline{H}_n^2 + \overline{H}_n^{(2)} - \frac{1}{2}H_n^{(2)} - \sum_{k=1}^{n} \frac{\overline{H}_k}{k} \right),$$

and the second solution to the point (x) is complete. Note that in order to get the last equality I used the result in (6.19).

Continuing and considering the result at the point (xi) together with the Cauchy product of two series in (6.15), and then rearranging everything, we write

$$\frac{\log(1-x)\operatorname{Li}_2(x)}{1-x} = \left(\sum_{n=1}^{\infty} x^n \left(3\frac{1}{n^3} - 2\frac{H_n}{n^2} - \frac{H_n^{(2)}}{n} \right) \right) \left(\sum_{n=1}^{\infty} x^{n-1} \right)$$

$$= \sum_{n=1}^{\infty} x^n \left(3\sum_{k=1}^{n} \frac{1}{k^3} - \sum_{k=1}^{n} \frac{H_k}{k^2} - \sum_{k=1}^{n} \left(\frac{H_k}{k^2} + \frac{H_k^{(2)}}{k} \right) \right)$$

$$= \sum_{n=1}^{\infty} x^n \left(2H_n^{(3)} - H_n H_n^{(2)} - \sum_{k=1}^{n} \frac{H_k}{k^2} \right),$$

and the solution to the point (xi) is complete. In the calculations I also used the Cauchy product given at the point $viii)$ and the generalized result in (6.102), Sect. 6.13, with $p = 1$ and $q = 2$ (or $p = 2$ and $q = 1$).

We'll need again the result in (6.17), this time the version with $m = 2$, $p = 1$, to prepare an extraction of the sum needed to the point (xii). So, we have

$$S_{2,1}(n) - S_{2,1}(n-1) = \sum_{k=1}^{n-1} \frac{(-1)^{k-1}}{k^2(n-k)} = \frac{1}{n^2}\sum_{k=1}^{n-1}(-1)^{k-1}\frac{1}{k} + \frac{1}{n^2}\sum_{k=1}^{n-1}\frac{(-1)^{k-1}}{n-k}$$

$$+ \frac{1}{n}\sum_{k=1}^{n-1}(-1)^{k-1}\frac{1}{k^2} = \frac{1}{n^2}\sum_{k=1}^{n-1}(-1)^{k-1}\frac{1}{k} - \frac{(-1)^{n-1}}{n^2}\sum_{k=1}^{n-1}(-1)^{k-1}\frac{1}{k}$$

$$+ \frac{1}{n}\sum_{k=1}^{n-1}(-1)^{k-1}\frac{1}{k^2} = \frac{1}{n^3} - 2(-1)^{n-1}\frac{1}{n^3} + \frac{\overline{H}_n}{n^2} - (-1)^{n-1}\frac{\overline{H}_n}{n^2} + \frac{\overline{H}_n^{(2)}}{n}. \qquad (6.20)$$

Upon replacing n by k in the opposite sides of (6.20), and then summing both sides from $k = 1$ to n, where $S_{2,1}(0) = 0$ and $S_{2,1}(1) = 0$, we have

$$\sum_{k=1}^{n}(S_{2,1}(k) - S_{2,1}(k-1)) = S_{2,1}(n) = \sum_{k=1}^{n-1} \frac{\overline{H}_k^{(2)}}{n-k} = \sum_{k=1}^{n}\frac{1}{k^3} - 2\sum_{k=1}^{n}(-1)^{k-1}\frac{1}{k^3}$$

$$+ \sum_{k=1}^{n} \frac{\overline{H}_k}{k^2} - \sum_{k=1}^{n}(-1)^{k-1}\frac{\overline{H}_k}{k^2} + \sum_{k=1}^{n} \frac{\overline{H}_k^{(2)}}{k}$$

$$= H_n^{(3)} - 2\overline{H}_n^{(3)} + \sum_{k=1}^{n} \frac{\overline{H}_k}{k^2} - \sum_{k=1}^{n}(-1)^{k-1}\frac{\overline{H}_k}{k^2} + \sum_{k=1}^{n} \frac{\overline{H}_k^{(2)}}{k}. \tag{6.21}$$

Now, since, based on (4.62), the case $m = 2$, Sect. 4.10, we have that $-\dfrac{\operatorname{Li}_2(-x)}{1-x} = \sum_{n=1}^{\infty} x^n \overline{H}_n^{(2)}$, and $\log(1-x) = -\sum_{n=1}^{\infty} \dfrac{x^n}{n}$, if we apply (6.15) and use (6.21), we get

$$\frac{\log(1-x)\operatorname{Li}_2(-x)}{1-x} = \left(\sum_{n=1}^{\infty} x^n \overline{H}_n^{(2)}\right)\left(\sum_{n=1}^{\infty} \frac{x^n}{n}\right) = \sum_{n=1}^{\infty} x^{n+1} \sum_{k=1}^{n} \frac{\overline{H}_k^{(2)}}{n-k+1}$$

$$= \sum_{n=2}^{\infty} x^n \sum_{k=1}^{n-1} \frac{\overline{H}_k^{(2)}}{n-k} = \sum_{n=1}^{\infty} x^n \sum_{k=1}^{n-1} \frac{\overline{H}_k^{(2)}}{n-k}$$

$$= \sum_{n=1}^{\infty} x^n \left(H_n^{(3)} - 2\overline{H}_n^{(3)} + \sum_{k=1}^{n} \frac{\overline{H}_k}{k^2} - \sum_{k=1}^{n}(-1)^{k-1}\frac{\overline{H}_k}{k^2} + \sum_{k=1}^{n} \frac{\overline{H}_k^{(2)}}{k}\right),$$

and the solution to the point (xii) is complete.

In the following, I'll prepare another finite sum needed for the extraction of the next Cauchy product, and setting $m = 1$, $p = 2$ in (6.17), we obtain

$$S_{1,2}(n) - S_{1,2}(n-1) = \sum_{k=1}^{n-1} \frac{(-1)^{k-1}}{k(n-k)^2} = \frac{1}{n^2}\sum_{k=1}^{n-1}(-1)^{k-1}\frac{1}{k} + \frac{1}{n^2}\sum_{k=1}^{n-1} \frac{(-1)^{k-1}}{n-k}$$

$$+\frac{1}{n}\sum_{k=1}^{n-1} \frac{(-1)^{k-1}}{(n-k)^2} = \frac{\overline{H}_{n-1}}{n^2} - (-1)^{n-1}\frac{1}{n^2}\sum_{k=1}^{n-1}(-1)^{k-1}\frac{1}{k} - (-1)^{n-1}\frac{1}{n}\sum_{k=1}^{n-1}(-1)^{k-1}\frac{1}{k^2}$$

$$= 2\frac{1}{n^3} - (-1)^{n-1}\frac{1}{n^3} + \frac{\overline{H}_n}{n^2} - (-1)^{n-1}\frac{\overline{H}_n}{n^2} - (-1)^{n-1}\frac{\overline{H}_n^{(2)}}{n}. \tag{6.22}$$

Observe that the result above can also be extracted from the calculations at the previous point, more precisely by using the result in (6.20). Now, if we replace n by k in the opposite sides of (6.22) and then sum both sides from $k = 1$ to n, where $S_{1,2}(0) = 0$ and $S_{1,2}(1) = 0$, we get

$$\sum_{k=1}^{n}(S_{1,2}(k) - S_{1,2}(k-1)) = S_{1,2}(n) = \sum_{k=1}^{n-1}\frac{\overline{H}_k}{(n-k)^2} = 2\sum_{k=1}^{n}\frac{1}{k^3} - \sum_{k=1}^{n}(-1)^{k-1}\frac{1}{k^3}$$

$$+ \sum_{k=1}^{n}\frac{\overline{H}_k}{k^2} - \left(\sum_{k=1}^{n}(-1)^{k-1}\frac{\overline{H}_k}{k^2} + \sum_{k=1}^{n}(-1)^{k-1}\frac{\overline{H}_k^{(2)}}{k}\right)$$

$$= H_n^{(3)} - \overline{H}_n^{(3)} - \overline{H}_n\overline{H}_n^{(2)} + \sum_{k=1}^{n}\frac{\overline{H}_k}{k^2}, \tag{6.23}$$

where in the calculations I used that $\sum_{k=1}^{n}(-1)^{k-1}\frac{\overline{H}_k}{k^2} + \sum_{k=1}^{n}(-1)^{k-1}\frac{\overline{H}_k^{(2)}}{k} = H_n^{(3)} +$

$\overline{H}_n\overline{H}_n^{(2)}$, which is the case $p = 1, q = 2$ (or $p = 2, q = 1$) in (3.170), Sect. 3.23.

So, since we have that $\dfrac{\log(1+x)}{1-x} = \sum_{n=1}^{\infty}x^n\overline{H}_n$ and $\text{Li}_2(x) = \sum_{n=1}^{\infty}\dfrac{x^n}{n^2}$, if we

apply (6.15) and use (6.23), we arrive at

$$\frac{\log(1+x)\,\text{Li}_2(x)}{1-x} = \left(\sum_{n=1}^{\infty}x^n\overline{H}_n\right)\left(\sum_{n=1}^{\infty}\frac{x^n}{n^2}\right) = \sum_{n=1}^{\infty}x^{n+1}\sum_{k=1}^{n}\frac{\overline{H}_k}{(n-k+1)^2}$$

$$= \sum_{n=2}^{\infty}x^n\sum_{k=1}^{n-1}\frac{\overline{H}_k}{(n-k)^2} = \sum_{n=1}^{\infty}x^n\sum_{k=1}^{n-1}\frac{\overline{H}_k}{(n-k)^2}$$

$$= \sum_{n=1}^{\infty}x^n\left(H_n^{(3)} - \overline{H}_n^{(3)} - \overline{H}_n\overline{H}_n^{(2)} + \sum_{k=1}^{n}\frac{\overline{H}_k}{k^2}\right),$$

and the solution to the point $(xiii)$ is complete.

Next, for the point (xiv) we start with the power series, $\sum_{n=1}^{\infty}x^n H_n^{(2)} = \dfrac{\text{Li}_2(x)}{1-x}$,

which is the case $m = 2$ of (4.32), Sect. 4.6, and $\text{Li}_2(x) = \sum_{n=1}^{\infty}\dfrac{x^n}{n^2}$, and then we have

$$\frac{(\text{Li}_2(x))^2}{1-x} = \left(\sum_{n=1}^{\infty}x^n H_n^{(2)}\right)\left(\sum_{n=1}^{\infty}\frac{x^n}{n^2}\right) = \sum_{n=1}^{\infty}x^{n+1}\sum_{k=1}^{n}\frac{H_k^{(2)}}{(n-k+1)^2}$$

{reindex the series and start from $n = 2$}

$$= \sum_{n=2}^{\infty} x^n \sum_{k=1}^{n-1} \frac{H_k^{(2)}}{(n-k)^2} = \sum_{n=1}^{\infty} x^n \sum_{k=1}^{n-1} \frac{H_k^{(2)}}{(n-k)^2} = \sum_{n=1}^{\infty} x^n \left(\left(H_n^{(2)} \right)^2 - 5 H_n^{(4)} + 4 \sum_{k=1}^{n} \frac{H_k}{k^3} \right),$$

where in the calculations I used the special finite sum, $\sum_{k=1}^{n-1} \dfrac{H_k^{(2)}}{(n-k)^2} = \left(H_n^{(2)} \right)^2 -$

$5 H_n^{(4)} + 4 \sum_{k=1}^{n} \dfrac{H_k}{k^3}$ in [76, Chapter 4, p.288], and the solution to the point (xiv) is
complete.

Like in other solutions before, including the previous solution, for the penultimate point, we need an auxiliary finite sum! We go back to (6.17) and set $m = 2$, $p = 2$, which gives

$$S_{2,2}(n) - S_{2,2}(n-1) = \sum_{k=1}^{n-1} \frac{(-1)^{k-1}}{k^2 (n-k)^2} = \frac{1}{n} \sum_{k=1}^{n-1} \frac{(-1)^{k-1}}{k^2 (n-k)} + \frac{1}{n} \sum_{k=1}^{n-1} \frac{(-1)^{k-1}}{k(n-k)^2}$$

{make use of the results in (6.20) and (6.22)}

$$= 3 \frac{1}{n^4} - 3(-1)^{n-1} \frac{1}{n^4} + 2 \frac{\overline{H}_n}{n^3} - 2(-1)^{n-1} \frac{\overline{H}_n}{n^3} + \frac{\overline{H}_n^{(2)}}{n^2} - (-1)^{n-1} \frac{\overline{H}_n^{(2)}}{n^2}. \quad (6.24)$$

If we consider the opposite sides of the identity in (6.24) in variable letter k, and then sum both sides from $k = 1$ to n, where $S_{2,2}(0) = 0$ and $S_{2,2}(1) = 0$, we have

$$\sum_{k=1}^{n} (S_{2,2}(k) - S_{2,2}(k-1)) = S_{2,2}(n) = \sum_{k=1}^{n-1} \frac{\overline{H}_k^{(2)}}{(n-k)^2} = 3 \sum_{k=1}^{n} \frac{1}{k^4} - 3 \sum_{k=1}^{n} (-1)^{k-1} \frac{1}{k^4}$$

$$+ 2 \sum_{k=1}^{n} \frac{\overline{H}_k}{k^3} - 2 \sum_{k=1}^{n} (-1)^{k-1} \frac{\overline{H}_k}{k^3} + \sum_{k=1}^{n} \frac{\overline{H}_k^{(2)}}{k^2} - \sum_{k=1}^{n} (-1)^{k-1} \frac{\overline{H}_k^{(2)}}{k^2}$$

$$= \frac{5}{2} H_n^{(4)} - 3 \overline{H}_n^{(4)} - \frac{1}{2} (\overline{H}_n^{(2)})^2 + 2 \sum_{k=1}^{n} \frac{\overline{H}_k}{k^3} - 2 \sum_{k=1}^{n} (-1)^{k-1} \frac{\overline{H}_k}{k^3} + \sum_{k=1}^{n} \frac{\overline{H}_k^{(2)}}{k^2},$$
$$(6.25)$$

where during the calculations I also used that $\sum_{k=1}^{n} (-1)^{k-1} \dfrac{\overline{H}_k^{(2)}}{k^2} = \dfrac{1}{2} \Big(H_n^{(4)} +$

$(\overline{H}_n^{(2)})^2 \Big)$, which is the case $p = 2, q = 2$ in (3.170), Sect. 3.23.

With the result from (6.25) in hand, and using that $-\dfrac{\mathrm{Li}_2(-x)}{1-x} = \displaystyle\sum_{n=1}^{\infty} x^n \overline{H}_n^{(2)}$ and

$\mathrm{Li}_2(x) = \displaystyle\sum_{n=1}^{\infty} \dfrac{x^n}{n^2}$, and then applying (6.15), we get

$$\frac{\mathrm{Li}_2(x)\,\mathrm{Li}_2(-x)}{1-x} = -\left(\sum_{n=1}^{\infty} x^n \overline{H}_n^{(2)}\right)\left(\sum_{n=1}^{\infty} \frac{x^n}{n^2}\right) = -\sum_{n=1}^{\infty} x^{n+1} \sum_{k=1}^{n} \frac{\overline{H}_k^{(2)}}{(n-k+1)^2}$$

$$\{\text{reindex the series and start from } n = 2\}$$

$$= -\sum_{n=2}^{\infty} x^n \sum_{k=1}^{n-1} \frac{\overline{H}_k^{(2)}}{(n-k)^2} = -\sum_{n=1}^{\infty} x^n \sum_{k=1}^{n-1} \frac{\overline{H}_k^{(2)}}{(n-k)^2}$$

$$= \sum_{n=1}^{\infty} x^n \left(3\overline{H}_n^{(4)} + \frac{1}{2}(\overline{H}_n^{(2)})^2 - \frac{5}{2}H_n^{(4)} - 2\sum_{k=1}^{n} \frac{\overline{H}_k}{k^3} + 2\sum_{k=1}^{n}(-1)^{k-1}\frac{\overline{H}_k}{k^3} - \sum_{k=1}^{n} \frac{\overline{H}_k^{(2)}}{k^2}\right),$$

and the solution to the point $(x\,v)$ is complete.

And we have arrived at the last point of the problem where I won't consider applying (6.15), but I'll go another route! Based on (1.110), Sect. 1.22, we have that $4 \displaystyle\int_0^1 nt^{2n-1} \operatorname{arctanh}(t)\mathrm{d}t = 2H_{2n} - H_n$, and if we multiply both sides by $x^{2n} \dfrac{1}{4^n}\dbinom{2n}{n}$ and consider the summation from $n = 0$ to ∞, we get

$$\sum_{n=0}^{\infty} x^{2n} \frac{1}{4^n}\binom{2n}{n}(2H_{2n} - H_n) = 4\sum_{n=0}^{\infty} x^{2n} \frac{1}{4^n}\binom{2n}{n}\int_0^1 nt^{2n-1}\operatorname{arctanh}(t)\mathrm{d}t$$

$$\{\text{reverse the order of integration and summation}\}$$

$$= 2\int_0^1 \frac{\operatorname{arctanh}(t)}{t} \sum_{n=0}^{\infty} 2n(xt)^{2n}\frac{1}{4^n}\binom{2n}{n}\mathrm{d}t$$

$$= 2\int_0^1 \frac{x^2 t\,\operatorname{arctanh}(t)}{(1 - x^2 t^2)^{3/2}}\mathrm{d}t, \tag{6.26}$$

where I exploited the generating function of the central binomial coefficients.

Now, I'll denote $I(a) = \displaystyle\int_0^a \frac{x^2 t\,\operatorname{arctanh}(t)}{(1 - x^2 t^2)^{3/2}}\mathrm{d}t$, $0 < a < 1$, and upon integrating by parts, we have

$$I(a) = \int_0^a \frac{x^2 t \, \text{arctanh}(t)}{(1 - x^2 t^2)^{3/2}} dt = \int_0^a \left(\frac{1}{\sqrt{1 - x^2 t^2}} \right)' \text{arctanh}(t) dt$$

$$= \frac{\text{arctanh}(a)}{\sqrt{1 - x^2 a^2}} - \int_0^a \frac{1}{(1 - t^2)\sqrt{1 - x^2 t^2}} dt. \tag{6.27}$$

Letting the variable change $\tanh(u)/x = t$ in the last integral of (6.27), and rearranging, and returning to the variable in t, we have

$$\int_0^a \frac{1}{(1 - t^2)\sqrt{1 - x^2 t^2}} dt = x \int_0^{\text{arctanh}(xa)} \frac{\cosh(t)}{1 - (1 - x^2)\cosh^2(t)} dt$$

$$\left\{ \text{exploit the identity } \cosh^2(t) - \sinh^2(t) = 1 \right\}$$

$$= \frac{1}{x} \int_0^{\text{arctanh}(xa)} \frac{\cosh(t)}{1 - (\sqrt{1 - x^2}/x)^2 \sinh^2(t)} dt$$

$$= \frac{1}{x} \frac{\text{arctanh}(\sqrt{1 - x^2}/x \, \sinh(t))}{\sqrt{1 - x^2}/x} \Bigg|_{t=0}^{t=\text{arctanh}(xa)} = \frac{1}{\sqrt{1 - x^2}} \text{arctanh}\left(\frac{a\sqrt{1 - x^2}}{\sqrt{1 - a^2 x^2}} \right),$$
$$\tag{6.28}$$

where I used the simple facts that $\int_0^x \frac{1}{1 - a^2 t^2} dt = \frac{1}{a} \text{arctanh}(ax)$, $|ax| < 1$, and
$\sinh(\text{arctanh}(x)) = \frac{x}{\sqrt{1 - x^2}}$.

Combining (6.27) and (6.28), we get

$$I(a) = \int_0^a \frac{x^2 t \, \text{arctanh}(t)}{(1 - x^2 t^2)^{3/2}} dt = \frac{\text{arctanh}(a)}{\sqrt{1 - x^2 a^2}} - \frac{1}{\sqrt{1 - x^2}} \text{arctanh}\left(\frac{a\sqrt{1 - x^2}}{\sqrt{1 - a^2 x^2}} \right). \tag{6.29}$$

Observe that based on (6.29), we immediately have that

$$\int_0^1 \frac{x^2 t \, \text{arctanh}(t)}{(1 - x^2 t^2)^{3/2}} dt = \lim_{\substack{a \to 1 \\ a < 1}} I(a)$$

{use an artifice of calculation and cleverly distribute the limit}

$$= \lim_{\substack{a \to 1 \\ a < 1}} \underbrace{\left(\frac{1}{\sqrt{1 - x^2 a^2}} - \frac{1}{\sqrt{1 - x^2}} \right)}_{0} \text{arctanh}(a)$$

$$+ \frac{1}{\sqrt{1 - x^2}} \underbrace{\lim_{\substack{a \to 1 \\ a < 1}} \left(\operatorname{arctanh}(a) - \operatorname{arctanh}\left(\frac{a\sqrt{1 - x^2}}{\sqrt{1 - a^2 x^2}} \right) \right)}_{-1/2 \log(1 - x^2)} = -\frac{1}{2} \frac{\log(1 - x^2)}{\sqrt{1 - x^2}},$$

$$(6.30)$$

and both limits are straightforward by using a similar reasoning to the one given to the two limits found and calculated at the end of Sect. 3.44.

Finally, returning with (6.30) in (6.26), we conclude that

$$-\frac{\log(1 - x^2)}{\sqrt{1 - x^2}} = \sum_{n=0}^{\infty} x^{2n} \frac{1}{4^n} \binom{2n}{n} (2H_{2n} - H_n),$$

and the solution to the point (xvi) is complete. Note that the whole process of evaluating the integral in (6.30) is elementary!

At last, one nice task for the curious reader will be to discover how to exploit the power of such Cauchy products in various problems involving series like the ones given in the present book.

6.6 Good-to-Know Generating Functions: The First Part

Solution There are two major considerations that motivated the presence of this section in the book you are just reading: on one hand it is about the usefulness of these results in the work with integrals and series, and on the other hand, it is about the challenge of answering the following question: *How to proceed to be comfortable with the extraction of the series at the points (vi) and (vii)?* Even for those with experience in the work with series, these points might have a scary look.

In the following, I'll present a possible way to go, and it will be up to you to assess and decide how well this approach answers the question above.

The point (i) of the problem also appears in *(Almost) Impossible Integrals, Sums, and Series*, which you may find in [76, Chapter 4, p.284], and I'll take this opportunity for providing another solution. Also, the points (ii) and (iii) are spread in many calculations. For example, the points (ii) and (iv), with the first closed form, appear spread in [76, Chapter 6, Section 6.10, pp. 347–355]. But how about the second and third closed forms of the points (iv) and (v)?

So, let's begin with the point (i) of the problem! In the following, I'll use that $H_n^{(m)} = \sum_{k=1}^{n} \frac{1}{k^m}$, which then I'll combine with the change of summation order. Thus, we write

$$\sum_{n=1}^{\infty} x^n H_n^{(m)} = \sum_{n=1}^{\infty} x^n \sum_{k=1}^{n} \frac{1}{k^m} = \sum_{k=1}^{\infty} \frac{1}{k^m} \sum_{n=k}^{\infty} x^n = \frac{1}{1-x} \sum_{k=1}^{\infty} \frac{x^k}{k^m} = \frac{\mathrm{Li}_m(x)}{1-x},$$

where we understand that $\mathrm{Li}_1(x) = -\log(1-x)$, and the point (i) of the problem is complete. Such ways to go are known in mathematical literature and good to consider given the elegance of the approach. For a different solution, see [76, Chapter 6, pp. 348–349].

The result at the point (ii) is immediately derived by considering the case $m = 1$ of the result from the point (i), and we get

$$\sum_{n=1}^{\infty} x^n \frac{H_n}{n+1} = \sum_{n=1}^{\infty} \frac{1}{x} \int_0^x t^n \, dt \, H_n = \frac{1}{x} \int_0^x \sum_{n=1}^{\infty} t^n H_n \, dt = -\frac{1}{x} \int_0^x \frac{\log(1-t)}{1-t} \, dt$$

$$= \frac{1}{2} \frac{\log^2(1-x)}{x},$$

and the point (ii) of the problem is complete.

Next, the generating function at the point (iii) may be readily extracted based on the result from the point (ii), by rearranging and reindexing, and then we write

$$\frac{1}{2} \log^2(1-x) = \sum_{n=1}^{\infty} x^{n+1} \frac{H_n}{n+1} = \sum_{n=1}^{\infty} x^{n+1} \frac{H_{n+1} - 1/(n+1)}{n+1} = \sum_{n=1}^{\infty} x^n \frac{H_n}{n} - \underbrace{\sum_{n=1}^{\infty} \frac{x^n}{n^2}}_{\mathrm{Li}_2(x)},$$

whence we obtain that

$$\sum_{n=1}^{\infty} x^n \frac{H_n}{n} = \frac{1}{2} \log^2(1-x) + \mathrm{Li}_2(x),$$

and the first equality of the point (iii) is proved.

The second equality of the point (iii) comes straightforward from the Landen's dilogarithmic identity (see [21, Chapter 1, p.5], [70, Chapter 2, p.107]), $\mathrm{Li}_2(x) + \mathrm{Li}_2\left(\frac{x}{x-1}\right) = -\frac{1}{2} \log^2(1-x)$, and thus we have

$$\sum_{n=1}^{\infty} x^n \frac{H_n}{n} = \frac{1}{2} \log^2(1-x) + \mathrm{Li}_2(x) = -\mathrm{Li}_2\left(\frac{x}{x-1}\right),$$

and the point (iii) of the problem is complete. Another solution to the second equality above may be found in [76, Chapter 6, p.356]. The second equality is very useful in a subsequent derivation as you'll see!

Further, the first equality of the point (iv) is obtained at once by recognizing in the calculations the appearance of a logarithmic integral already treated in *(Almost)*

Impossible Integrals, Sums, and Series,

$$\int_0^x \frac{\log^2(1-t)}{t} dt$$

$$= \log(x)\log^2(1-x) + 2\log(1-x)\operatorname{Li}_2(1-x) - 2\operatorname{Li}_3(1-x) + 2\zeta(3)$$

$$= 2\zeta(2)\log(1-x) - \log(x)\log^2(1-x) - 2\log(1-x)\operatorname{Li}_2(x) - 2\operatorname{Li}_3(1-x) + 2\zeta(3),$$
$$(6.31)$$

where the first equality with a proof appears in [76, Chapter 3, p.65]. Essentially, for an evaluation leading to the two equalities, the integral in (6.31) requires integrations by parts and the dilogarithm function reflection formula $\operatorname{Li}_2(x) + \operatorname{Li}_2(1-x) = \zeta(2) - \log(x)\log(1-x)$ (see [21, Chapter 1, p.5], [70, Chapter 2, p.107]).

So, using (6.31), we write

$$\sum_{n=1}^{\infty} x^n \frac{H_n}{(n+1)^2} = \frac{1}{x}\sum_{n=1}^{\infty} \int_0^x t^n dt \frac{H_n}{n+1} = \frac{1}{x}\int_0^x \sum_{n=1}^{\infty} t^n \frac{H_n}{n+1} dt = \frac{1}{2}\frac{1}{x}\int_0^x \frac{\log^2(1-t)}{t} dt$$

$$= \frac{1}{2}\frac{\log(x)\log^2(1-x)}{x} + \frac{\log(1-x)\operatorname{Li}_2(1-x)}{x} - \frac{\operatorname{Li}_3(1-x)}{x} + \zeta(3)\frac{1}{x}$$

$$= \zeta(2)\frac{\log(1-x)}{x} - \frac{1}{2}\frac{\log(x)\log^2(1-x)}{x} - \frac{\log(1-x)\operatorname{Li}_2(x)}{x}$$

$$- \frac{\operatorname{Li}_3(1-x)}{x} + \zeta(3)\frac{1}{x},$$

and the first equality of the point (iv) is proved. In the calculations above, I also used the result from the point (ii).

Now, to get the second and third equalities, we will exploit the second and third equalities given at the point (v), and then we write

$$\sum_{n=1}^{\infty} x^n \frac{H_n}{(n+1)^2} = \sum_{n=1}^{\infty} x^n \frac{H_{n+1} - 1/(n+1)}{(n+1)^2}$$

{reindex the series and expand it}

$$= \sum_{n=1}^{\infty} x^{n-1}\frac{H_n}{n^2} - \sum_{n=1}^{\infty} \frac{x^{n-1}}{n^3} = \sum_{n=1}^{\infty} x^{n-1}\frac{H_n}{n^2} - \frac{\operatorname{Li}_3(x)}{x}$$

$$= 2\zeta(2)\frac{\log(1-x)}{x} + \frac{1}{6}\frac{\log^3(1-x)}{x} - \frac{\log(x)\log^2(1-x)}{x} - \frac{\log(1-x)\operatorname{Li}_2(x)}{x}$$

$$- \frac{\text{Li}_3(x)}{x} - 2\frac{\text{Li}_3(1-x)}{x} - \frac{1}{x}\text{Li}_3\left(\frac{x}{x-1}\right) + 2\zeta(3)\frac{1}{x}$$

$$= \frac{\text{Li}_3(x)}{x} + \frac{1}{x}\text{Li}_3\left(\frac{x}{x-1}\right) - \frac{\log(1-x)\text{Li}_2(x)}{x} - \frac{1}{6}\frac{\log^3(1-x)}{x},$$

and the point (iv) of the problem is complete.

As regards the first equality at the point (v), we can simply exploit the first equality from the previous point, already proved, and then we write

$$\sum_{n=1}^{\infty} x^n \frac{H_n}{n^2} = \sum_{n=1}^{\infty} x^n \frac{H_n - 1/n}{n^2} + \underbrace{\sum_{n=1}^{\infty} \frac{x^n}{n^3}}_{\text{Li}_3(x)}$$

$$\overset{\substack{\text{reindex} \\ \text{the series}}}{=} \sum_{n=1}^{\infty} x^{n+1} \frac{H_{n+1} - 1/(n+1)}{(n+1)^2} + \text{Li}_3(x) = \sum_{n=1}^{\infty} x^{n+1} \frac{H_n}{(n+1)^2} + \text{Li}_3(x)$$

$$= \zeta(2)\log(1-x) - \frac{1}{2}\log(x)\log^2(1-x) - \log(1-x)\text{Li}_2(x)$$

$$+ \text{Li}_3(x) - \text{Li}_3(1-x) + \zeta(3),$$

and the first equality of the point (v) is proved.

Before going further, we might take a break and make sure that everything that has been done so far is clear and easy to do.

How about the second and third equalities of the point (v) you might wonder! I would like to remind you that the second and third equalities from the previous point are extracted based on the second and third equalities from the present point.

In a joint work with mathematician Moti Levy (Rehovot, Israel) that led to the materialization of the paper *Euler Sum Involving Tail* (see [92, 2019–2020]), we use the dilogarithm function reflection formula,

$$\text{Li}_2(x) + \text{Li}_2(1-x) = \zeta(2) - \log(x)\log(1-x) \tag{6.32}$$

and the Landen's trilogarithm function identity, (see [21, Chapter 1, p.155], [70, Chapter 2, p.113])

$$\text{Li}_3(x) + \text{Li}_3(1-x) + \text{Li}_3\left(\frac{x}{x-1}\right)$$

$$= \zeta(3) + \zeta(2)\log(1-x) - \frac{1}{2}\log(x)\log^2(1-x) + \frac{1}{6}\log^3(1-x), \tag{6.33}$$

for obtaining the third closed form based on the first closed form.

What if we could prove the second and third equalities by keeping things simple, without a direct use of the Landen's trilogarithm function identity? That is a tempting question to answer! The third equality offers the possibility to construct an elegant way of deriving the results from the points (vi) and (vii).

Let's start the derivation of the second equality by considering the result from the point (iii) with the second closed form, $\displaystyle\sum_{n=1}^{\infty} t^n \frac{H_n}{n} = -\operatorname{Li}_2\left(\frac{t}{t-1}\right)$, where if we multiply both sides by $1/(t(1-t))$ and then integrate from $t = 0$ to $t = x$, we get

$$-\int_0^x \frac{1}{t(1-t)} \operatorname{Li}_2\left(\frac{t}{t-1}\right) dt = -\int_0^x \left(\operatorname{Li}_3\left(\frac{t}{t-1}\right)\right)' dt = -\operatorname{Li}_3\left(\frac{x}{x-1}\right)$$

$$= \int_0^x \sum_{n=1}^{\infty} \frac{t^n}{t(1-t)} \frac{H_n}{n} dt = \int_0^x \sum_{n=1}^{\infty} \left(t^{n-1}\frac{H_n}{n} + \frac{t^n}{1-t}\frac{H_n}{n}\right) dt$$

{expand and reverse the order of summation and integration}

$$= \sum_{n=1}^{\infty} \int_0^x t^{n-1} dt \frac{H_n}{n} + \sum_{n=1}^{\infty} \frac{H_n}{n} \int_0^x \frac{t^n}{1-t} dt = \sum_{n=1}^{\infty} x^n \frac{H_n}{n^2} - \sum_{n=1}^{\infty} \frac{H_n}{n} \int_0^x (\log(1-t))' t^n dt$$

{integrate by parts in the remaining integral}

$$= \sum_{n=1}^{\infty} x^n \frac{H_n}{n^2} - \log(1-x) \sum_{n=1}^{\infty} x^n \frac{H_n}{n} + \sum_{n=1}^{\infty} H_n \int_0^x t^{n-1} \log(1-t) dt$$

{for the second series use the first closed form at the point (iii)}

$$= \sum_{n=1}^{\infty} x^n \frac{H_n}{n^2} - \frac{1}{2}\log^3(1-x) - \log(1-x)\operatorname{Li}_2(x) + \int_0^x \log(1-t) \sum_{n=1}^{\infty} t^{n-1} H_n dt$$

$$= \sum_{n=1}^{\infty} x^n \frac{H_n}{n^2} - \frac{1}{2}\log^3(1-x) - \log(1-x)\operatorname{Li}_2(x) - \int_0^x \frac{\log^2(1-t)}{t(1-t)} dt$$

{use the partial fraction expansion and calculate the elementary integral}

$$= \sum_{n=1}^{\infty} x^n \frac{H_n}{n^2} - \frac{1}{6}\log^3(1-x) - \log(1-x)\operatorname{Li}_2(x) - \int_0^x \frac{\log^2(1-t)}{t} dt$$

{consider the result in (6.31), the second equality}

$$= \sum_{n=1}^{\infty} x^n \frac{H_n}{n^2} - 2\zeta(2)\log(1-x) - \frac{1}{6}\log^3(1-x) + \log(x)\log^2(1-x)$$

$$+ \log(1-x)\operatorname{Li}_2(x) + 2\operatorname{Li}_3(1-x) - 2\zeta(3),$$

whence we obtain

$$\sum_{n=1}^{\infty} x^n \frac{H_n}{n^2}$$

$$= 2\zeta(2)\log(1-x) + \frac{1}{6}\log^3(1-x) - \log(x)\log^2(1-x) - \log(1-x)\operatorname{Li}_2(x)$$

$$- 2\operatorname{Li}_3(1-x) - \operatorname{Li}_3\left(\frac{x}{x-1}\right) + 2\zeta(3),$$

and the second equality of the point (v) is proved.

All we need to do further is to consider the two representations of $\displaystyle\sum_{n=1}^{\infty} x^n \frac{H_n}{n^2}$ and then eliminate the two terms involving $\log(x)\log^2(1-x)$ that immediately leads to

$$\sum_{n=1}^{\infty} x^n \frac{H_n}{n^2} = 2\operatorname{Li}_3(x) + \operatorname{Li}_3\left(\frac{x}{x-1}\right) - \log(1-x)\operatorname{Li}_2(x) - \frac{1}{6}\log^3(1-x),$$

which is the third closed form needed, and the solution to the point (v) of the problem is complete.

A great bonus: It's nice to observe that by considering the equality between the first two given closed forms of $\displaystyle\sum_{n=1}^{\infty} x^n \frac{H_n}{n^2}$, we immediately get the Landen's trilogarithm function identity in (6.33).

Now, we have finally arrived at the last two points of the problem!

So, to get the result at the point (vi), we consider the result from the point (iv), with the third closed form, and then we write

$$\sum_{n=1}^{\infty} x^n \frac{H_n}{(n+1)^3} = \frac{1}{x}\sum_{n=1}^{\infty}\int_0^x t^n \, dt \frac{H_n}{(n+1)^2} = \frac{1}{x}\int_0^x \sum_{n=1}^{\infty} t^n \frac{H_n}{(n+1)^2} \, dt$$

$$= \frac{1}{x}\int_0^x \left(\frac{\operatorname{Li}_3(t)}{t} + \frac{1}{t}\operatorname{Li}_3\left(\frac{t}{t-1}\right) - \frac{\log(1-t)\operatorname{Li}_2(t)}{t} - \frac{1}{6}\frac{\log^3(1-t)}{t}\right) dt$$

{expand the integral and integrate the easy parts}

$$= \frac{\operatorname{Li}_4(x)}{x} + \frac{1}{2}\frac{(\operatorname{Li}_2(x))^2}{x} + \frac{1}{x}\int_0^x \frac{1}{t}\operatorname{Li}_3\left(\frac{t}{t-1}\right)dt - \frac{1}{6}\frac{1}{x}\int_0^x \frac{\log^3(1-t)}{t}dt.$$

(6.34)

For the first remaining integral in (6.34), we write

$$\int_0^x \frac{1}{t}\operatorname{Li}_3\left(\frac{t}{t-1}\right)dt = \int_0^x \frac{1-t}{t(1-t)}\operatorname{Li}_3\left(\frac{t}{t-1}\right)dt = \int_0^x \frac{1}{t(1-t)}\operatorname{Li}_3\left(\frac{t}{t-1}\right)dt$$

$$-\int_0^x \frac{1}{1-t}\operatorname{Li}_3\left(\frac{t}{t-1}\right)dt = \operatorname{Li}_4\left(\frac{x}{x-1}\right) + \int_0^x (\log(1-t))'\operatorname{Li}_3\left(\frac{t}{t-1}\right)dt$$

$$\overset{\substack{\text{integrate by}\\\text{parts}}}{=} \operatorname{Li}_4\left(\frac{x}{x-1}\right) + \log(1-x)\operatorname{Li}_3\left(\frac{x}{x-1}\right) - \int_0^x \frac{\log(1-t)}{t(1-t)}\operatorname{Li}_2\left(\frac{t}{t-1}\right)dt$$

$$= \operatorname{Li}_4\left(\frac{x}{x-1}\right) + \log(1-x)\operatorname{Li}_3\left(\frac{x}{x-1}\right) - \frac{1}{2}\left(\operatorname{Li}_2\left(\frac{x}{x-1}\right)\right)^2.$$

(6.35)

Then, for the other remaining integral in (6.34), we proceed with integration by parts as in the case of the integral in (6.31), and then we get

$$\int_0^x \frac{\log^3(1-t)}{t}dt = \log(t)\log^3(1-t)\Big|_{t=0}^{t=x} + 3\int_0^x \frac{\log(t)\log^2(1-t)}{1-t}dt$$

$$= \log(x)\log^3(1-x) + 3\int_0^x (\operatorname{Li}_2(1-t))'\log^2(1-t)dt$$

$$= \log(x)\log^3(1-x) + 3\operatorname{Li}_2(1-t)\log^2(1-t)\Big|_{t=0}^{t=x} + 6\int_0^x \frac{\operatorname{Li}_2(1-t)\log(1-t)}{1-t}dt$$

$$= \log(x)\log^3(1-x) + 3\operatorname{Li}_2(1-x)\log^2(1-x) - 6\int_0^x (\operatorname{Li}_3(1-t))'\log(1-t)dt$$

$$= \log(x)\log^3(1-x) + 3\operatorname{Li}_2(1-x)\log^2(1-x) - 6\operatorname{Li}_3(1-t)\log(1-t)\Big|_{t=0}^{t=x}$$

$$- 6\int_0^x \frac{\operatorname{Li}_3(1-t)}{1-t}dt$$

$$= \log(x)\log^3(1-x) + 3\log^2(1-x)\operatorname{Li}_2(1-x) - 6\log(1-x)\operatorname{Li}_3(1-x)$$

$$+ 6\operatorname{Li}_4(1-x) - 6\zeta(4).$$

(6.36)

Collecting the results from (6.35) and (6.36) in (6.34), where we also express $\text{Li}_2(1-x)$ by the identity in (6.32), then $\text{Li}_2\left(\dfrac{x}{x-1}\right)$ by the Landen's dilogarithmic identity, mentioned while proving the second equality from the point (iii) above, and finally $\text{Li}_3\left(\dfrac{x}{x-1}\right)$ by the Landen's trilogarithmic identity in (6.33), we arrive at

$$\sum_{n=1}^{\infty} x^n \frac{H_n}{(n+1)^3}$$

$$= \zeta(4)\frac{1}{x} + \zeta(3)\frac{\log(1-x)}{x} + \frac{1}{2}\zeta(2)\frac{\log^2(1-x)}{x} + \frac{1}{24}\frac{\log^4(1-x)}{x}$$

$$-\frac{1}{6}\frac{\log(x)\log^3(1-x)}{x} - \frac{\log(1-x)\text{Li}_3(x)}{x} + \frac{\text{Li}_4(x)}{x} - \frac{\text{Li}_4(1-x)}{x} + \frac{1}{x}\text{Li}_4\left(\frac{x}{x-1}\right),$$

and the solution to the point (vi) of the problem is complete.

Finally, to get the result from the last point, we'll want to exploit the result from the previous point, and then we write

$$\sum_{n=1}^{\infty} x^n \frac{H_n}{n^3} = \sum_{n=1}^{\infty} x^n \frac{H_n - 1/n}{n^3} + \underbrace{\sum_{n=1}^{\infty} \frac{x^n}{n^4}}_{\text{Li}_4(x)}$$

$$\overset{\substack{\text{reindex} \\ \text{the series}}}{=} \sum_{n=1}^{\infty} x^{n+1} \frac{H_{n+1} - 1/(n+1)}{(n+1)^3} + \text{Li}_4(x) = \sum_{n=1}^{\infty} x^{n+1} \frac{H_n}{(n+1)^3} + \text{Li}_4(x)$$

$$= \zeta(4) + \zeta(3)\log(1-x) + \frac{1}{2}\zeta(2)\log^2(1-x) + \frac{1}{24}\log^4(1-x)$$

$$-\frac{1}{6}\log(x)\log^3(1-x) - \log(1-x)\text{Li}_3(x) + 2\text{Li}_4(x) - \text{Li}_4(1-x) + \text{Li}_4\left(\frac{x}{x-1}\right),$$

and the solution to the point (vii) of the problem is complete.

During the calculations, also the *challenging question*, which involves a special derivation of the Landen's trilogarithmic identity, has been answered. In the next section, we'll meet more curious ordinary generating functions good to know!

6.7   Good-to-Know Generating Functions: The Second Part

Solution We continue the journey in the realm of *good-to-know* ordinary generating functions I started in the previous section. In the book *(Almost) Impossible Integrals, Sums, and Series*, one may find the result at the point (*i*) stated in [76, Chapter 6, Sect. 4.10, p.284], and a solution exploiting telescoping sums may be found in [76, Chapter 6, pp. 349–350]. Since I'll need the mentioned result in the subsequent calculations, I'll consider this as an opportunity for also presenting another solution to it based on the Cauchy product of two series.

If we consider that $\log^2(1-x) = 2\sum_{n=1}^{\infty} x^{n+1} \dfrac{H_n}{n+1} = 2\sum_{n=1}^{\infty} x^{n+1} \dfrac{H_{n+1} - 1/(n+1)}{n+1}$

$= 2\sum_{n=1}^{\infty} x^n \left(\dfrac{H_n}{n} - \dfrac{1}{n^2} \right)$, found at the point (*ii*) in the previous section, and $\dfrac{x}{1-x} =$

$\sum_{n=1}^{\infty} x^n$, and then apply the Cauchy product of two series, as seen in (6.15), Sect. 6.5, we get

$$x\frac{\log^2(1-x)}{1-x} = 2\left(\sum_{n=1}^{\infty} x^n \left(\frac{H_n}{n} - \frac{1}{n^2}\right)\right)\left(\sum_{n=1}^{\infty} x^n\right) = 2\sum_{n=1}^{\infty} x^{n+1} \sum_{k=1}^{n} \left(\frac{H_k}{k} - \frac{1}{k^2}\right)$$

$$\left\{ \text{use that } \sum_{k=1}^{n} \frac{H_k}{k} = \frac{1}{2}(H_n^2 + H_n^{(2)}) \text{ which is proved in [76, Chapter 3, p.60]} \right\},$$

$$= \sum_{n=1}^{\infty} x^{n+1}(H_n^2 - H_n^{(2)}),$$

where if we assume that $x \neq 0$, we arrive at

$$\frac{\log^2(1-x)}{1-x} = \sum_{n=1}^{\infty} x^n (H_n^2 - H_n^{(2)}), \tag{6.37}$$

and in the last result, we may also safely consider $x = 0$ since both sides equal 0 when $x = 0$, and thus $|x| < 1$.

Now, the result in (6.37), in the form of a generalization involving the elementary symmetric polynomials, also appears in [76, Chapter 6, pp. 354–355], where a solution is constructed by exploiting telescoping sums.

Since we have that $\sum_{n=1}^{\infty} x^n H_n^{(2)} = \dfrac{\operatorname{Li}_2(x)}{1-x}$, which appears in a generalized form in the previous section, then by combining this result with the one in (6.37), we immediately obtain the desired result

$$\sum_{n=1}^{\infty} x^n H_n^2 = \frac{1}{1-x}(\log^2(1-x) + \text{Li}_2(x)),$$

and the point (i) of the problem is finalized.

Passing to the point (ii) of the problem, we'll consider to exploit the result from the point (i), and then we write

$$\sum_{n=1}^{\infty} x^n \frac{H_n^2}{n+1} = \frac{1}{x} \sum_{n=1}^{\infty} \int_0^x t^n dt\, H_n^2 = \frac{1}{x} \int_0^x \sum_{n=1}^{\infty} t^n H_n^2 dt$$

$$= \frac{1}{x} \int_0^x \frac{\log^2(1-t)}{1-t} dt + \frac{1}{x} \int_0^x \frac{\text{Li}_2(t)}{1-t} dt$$

{integrate by parts in the second integral}

$$= -\frac{1}{3} \frac{\log^3(1-x)}{x} - \frac{\log(1-x)\text{Li}_2(x)}{x} - \frac{1}{x} \int_0^x \frac{\log^2(1-t)}{t} dt$$

{employ the first equality in (6.31), the previous section, and rearrange}

$$= 2\frac{\text{Li}_3(1-x)}{x} - \frac{\log(1-x)\text{Li}_2(1-x)}{x} - \frac{\log(1-x)}{x}(\text{Li}_2(x) + \text{Li}_2(1-x))$$

$$-\frac{1}{3} \frac{\log^3(1-x)}{x} - \frac{\log(x)\log^2(1-x)}{x} - 2\zeta(3)\frac{1}{x}$$

$$= 2\frac{\text{Li}_3(1-x)}{x} - \frac{\log(1-x)\text{Li}_2(1-x)}{x} - \frac{1}{3}\frac{\log^3(1-x)}{x} - \zeta(2)\frac{\log(1-x)}{x} - 2\zeta(3)\frac{1}{x},$$

where in the calculations above I also used the dilogarithm function reflection formula (see [21, Chapter 1, p.5], [70, Chapter 2, p.107]), $\text{Li}_2(x) + \text{Li}_2(1-x) = \zeta(2) - \log(x)\log(1-x)$, and the point (ii) of the problem is finalized.

Next, for the point (iii) of the problem, we exploit again the result from the point (i), and then we write

$$\sum_{n=1}^{\infty} x^n \frac{H_n^2}{n} = \sum_{n=1}^{\infty} \int_0^x t^{n-1} dt\, H_n^2 = \int_0^x \sum_{n=1}^{\infty} t^{n-1} H_n^2 dt$$

$$= \int_0^x \frac{1}{t(1-t)}(\log^2(1-t) + \text{Li}_2(t)) dt$$

{use the partial fraction decomposition and wisely split the integral}

$$= \int_0^x \frac{\text{Li}_2(t)}{t} dt + \int_0^x \left(\frac{\text{Li}_2(t)}{1-t} + \frac{\log^2(1-t)}{t} \right) dt + \int_0^x \frac{\log^2(1-t)}{1-t} dt$$

$$= \text{Li}_3(x) - \log(1-x)\text{Li}_2(x) - \frac{1}{3}\log^3(1-x),$$

where in the calculations I used that $\dfrac{d}{dx}(\log(1-x)\text{Li}_2(x)) = -\dfrac{\text{Li}_2(x)}{1-x} - \dfrac{\log^2(1-x)}{x}$, and the point (iii) of the problem is finalized.

Then, as regards the point (iv), we need to exploit the point (ii) of the problem, and therefore we have

$$\sum_{n=1}^{\infty} x^n \frac{H_n^2}{(n+1)^2} = \frac{1}{x} \sum_{n=1}^{\infty} \int_0^x t^n dt \frac{H_n^2}{n+1} = \frac{1}{x} \int_0^x \sum_{n=1}^{\infty} t^n \frac{H_n^2}{n+1} dt$$

$$= \frac{1}{x} \int_0^x \left(2\frac{\text{Li}_3(1-t)}{t} - \frac{\log(1-t)\text{Li}_2(1-t)}{t} - \frac{1}{3}\frac{\log^3(1-t)}{t} \right.$$

$$\left. - \zeta(2)\frac{\log(1-t)}{t} - 2\zeta(3)\frac{1}{t} \right) dt$$

{employ (6.32) and (6.33), the previous section, and then expand the integral}

$$= \frac{1}{x} \int_0^x \frac{\log(1-t)\text{Li}_2(t)}{t} dt - 2\frac{1}{x} \int_0^x \frac{\text{Li}_3(t)}{t} dt - 2\frac{1}{x} \int_0^x \frac{1}{t} \text{Li}_3\left(\frac{t}{t-1} \right) dt$$

$$= -2\zeta(3)\frac{\log(1-x)}{x} - \zeta(2)\frac{\log^2(1-x)}{x} - \frac{1}{12}\frac{\log^4(1-x)}{x} - \frac{\log^2(1-x)\text{Li}_2(1-x)}{x}$$

$$+ \frac{1}{2}\frac{(\text{Li}_2(x))^2}{x} + 2\frac{\log(1-x)\text{Li}_3(x)}{x} + 2\frac{\log(1-x)\text{Li}_3(1-x)}{x}$$

$$- 2\frac{\text{Li}_4(x)}{x} - 2\frac{1}{x}\text{Li}_4\left(\frac{x}{x-1} \right),$$

where in the calculations I used the integral result in (6.35), the preceding section, and then I expressed $\text{Li}_2\left(\dfrac{x}{x-1} \right)$ by the Landen's dilogarithmic identity (see [21, Chapter 1, p.5], [70, Chapter 2, p.107]), $\text{Li}_2(x) + \text{Li}_2\left(\dfrac{x}{x-1} \right) = -\dfrac{1}{2}\log^2(1-x)$; next $\text{Li}_2(x)$, appearing in $\dfrac{\log^2(1-x)\text{Li}_2(x)}{x}$, by the dilogarithm function reflection formula in (6.32), the previous section; and finally $\text{Li}_3\left(\dfrac{x}{x-1} \right)$ by the Landen's trilogarithmic identity in (6.33), the preceding section, and the point (iv) of the problem is finalized.

For the point (v) of the problem, we'll exploit the point (iii), and then we get

$$\sum_{n=1}^{\infty} x^n \frac{H_n^2}{n^2} = \sum_{n=1}^{\infty} \int_0^x t^{n-1} dt \frac{H_n^2}{n} = \int_0^x \sum_{n=1}^{\infty} t^{n-1} \frac{H_n^2}{n} dt$$

$$= \int_0^x \frac{\text{Li}_3(t)}{t} dt - \int_0^x \frac{\log(1-t)\text{Li}_2(t)}{t} dt - \frac{1}{3} \int_0^x \frac{\log^3(1-t)}{t} dt$$

$$= 2\zeta(4) - \frac{1}{3}\log(x)\log^3(1-x) - \log^2(1-x)\text{Li}_2(1-x) + \frac{1}{2}(\text{Li}_2(x))^2$$

$$+ 2\log(1-x)\text{Li}_3(1-x) + \text{Li}_4(x) - 2\text{Li}_4(1-x),$$

where the last integral is calculated in (6.36), the previous section, and the point (v) of the problem is finalized.

The variants with H_n^2 are now all calculated, and we'll want to pass to the variants involving $H_n^{(2)}$ in the summands.

So, for point (vi) of the problem, we'll make use of the ordinary generating function in (4.32), with $m = 2$, Sect. 4.6, that is, $\displaystyle\sum_{n=1}^{\infty} x^n H_n^{(2)} = \frac{\text{Li}_2(x)}{1-x}$, and then we obtain that

$$\sum_{n=1}^{\infty} x^n \frac{H_n^{(2)}}{n+1} = \frac{1}{x} \sum_{n=1}^{\infty} \int_0^x t^n dt\, H_n^{(2)} = \frac{1}{x} \int_0^x \sum_{n=1}^{\infty} t^n H_n^{(2)} dt = \frac{1}{x} \int_0^x \frac{\text{Li}_2(t)}{1-t} dt$$

$$= -\frac{1}{x} \int_0^x (\log(1-t))' \text{Li}_2(t) dt \overset{\underset{\text{integrate by}}{\text{parts}}}{=} -\frac{\log(1-x)\text{Li}_2(x)}{x} - \frac{1}{x} \int_0^x \frac{\log^2(1-t)}{t} dt$$

$$= 2\frac{\text{Li}_3(1-x)}{x} - \frac{\log(1-x)\text{Li}_2(1-x)}{x} - \frac{\log(1-x)}{x}(\text{Li}_2(x) + \text{Li}_2(1-x))$$

$$- \frac{\log(x)\log^2(1-x)}{x} - 2\zeta(3)\frac{1}{x}$$

$$= 2\frac{\text{Li}_3(1-x)}{x} - \frac{\log(1-x)\text{Li}_2(1-x)}{x} - \zeta(2)\frac{\log(1-x)}{x} - 2\zeta(3)\frac{1}{x}$$

$$= 2\frac{\text{Li}_3(1-x)}{x} + \frac{\log(1-x)\text{Li}_2(x)}{x} + \frac{\log(x)\log^2(1-x)}{x}$$

$$- 2\zeta(2)\frac{\log(1-x)}{x} - 2\zeta(3)\frac{1}{x},$$

where the last two equalities are derived with the help of the dilogarithm function reflection formula that also appeared during the calculations from the point (ii), and the point (vi) of the problem is finalized. In the calculations above I also used the result in (6.31), the first equality, the previous section.

The point (vii) is immediately derived based on the result from the point (vi), and then we have

$$\sum_{n=1}^{\infty} x^n \frac{H_n^{(2)}}{n} = \sum_{n=1}^{\infty} x^n \frac{H_n^{(2)} - 1/n^2}{n} + \underbrace{\sum_{n=1}^{\infty} \frac{x^n}{n^3}}_{\text{Li}_3(x)}$$

$$\underset{\text{the series}}{\overset{\text{reindex}}{=}} \sum_{n=1}^{\infty} x^{n+1} \frac{H_{n+1}^{(2)} - 1/(n+1)^2}{n+1} + \text{Li}_3(x) = \sum_{n=1}^{\infty} x^{n+1} \frac{H_n^{(2)}}{n+1} + \text{Li}_3(x)$$

$$= \text{Li}_3(x) + 2\,\text{Li}_3(1-x) - \log(1-x)\,\text{Li}_2(1-x) - \zeta(2)\log(1-x) - 2\zeta(3)$$

$$= \text{Li}_3(x) + 2\,\text{Li}_3(1-x) + \log(1-x)\,\text{Li}_2(x) + \log(x)\log^2(1-x)$$

$$- 2\zeta(2)\log(1-x) - 2\zeta(3),$$

and the point (vii) of the problem is finalized.

Regarding the point $(viii)$, we need to exploit the point (vi), and then we write

$$\sum_{n=1}^{\infty} x^n \frac{H_n^{(2)}}{(n+1)^2} = \frac{1}{x} \sum_{n=1}^{\infty} \int_0^x t^n dt \frac{H_n^{(2)}}{n+1} = \frac{1}{x} \int_0^x \sum_{n=1}^{\infty} t^n \frac{H_n^{(2)}}{n+1} dt$$

{combine the result from the point (vi), the second equality,}

{and the Landen's trilogarithmic identity in (6.33), Sect. 6.6}

{and then expand the resulting integral into four integrals}

$$= \frac{1}{3}\frac{1}{x} \int_0^x \frac{\log^3(1-t)}{t} dt + \frac{1}{x} \int_0^x \frac{\log(1-t)\,\text{Li}_2(t)}{t} dt - 2\frac{1}{x} \int_0^x \frac{\text{Li}_3(t)}{t} dt$$

$$- 2\frac{1}{x} \int_0^x \frac{1}{t} \text{Li}_3\left(\frac{t}{t-1}\right) dt$$

$$= -\frac{1}{2}\frac{(\text{Li}_2(x))^2}{x} - 2\frac{\text{Li}_4(x)}{x} + \frac{1}{3}\frac{1}{x} \int_0^x \frac{\log^3(1-t)}{t} dt - 2\frac{1}{x} \int_0^x \frac{1}{t} \text{Li}_3\left(\frac{t}{t-1}\right) dt$$

$$= -2\zeta(4)\frac{1}{x} - 2\zeta(3)\frac{\log(1-x)}{x} - \zeta(2)\frac{\log^2(1-x)}{x} + \frac{1}{3}\frac{\log(x)\log^3(1-x)}{x}$$

$$-\frac{1}{12}\frac{\log^4(1-x)}{x} + \frac{1}{2}\frac{(\operatorname{Li}_2(x))^2}{x} + 2\frac{\log(1-x)\operatorname{Li}_3(x)}{x} - 2\frac{\operatorname{Li}_4(x)}{x}$$

$$+ 2\frac{\operatorname{Li}_4(1-x)}{x} - 2\frac{1}{x}\operatorname{Li}_4\left(\frac{x}{x-1}\right),$$

where in the calculations I used the integral results in (6.35) and (6.36), Sect. 6.6, and then I expressed $\operatorname{Li}_2\left(\dfrac{x}{x-1}\right)$ by the Landen's dilogarithmic identity stated in the solution to the point (iv); next $\operatorname{Li}_2(x)$, appearing in $\dfrac{\log^2(1-x)\operatorname{Li}_2(x)}{x}$, by the dilogarithm function reflection formula stated at the end of the solution to the point (ii); and finally $\operatorname{Li}_3\left(\dfrac{x}{x-1}\right)$ by the Landen's trilogarithmic identity in (6.33), the preceding section, and the point $(viii)$ of the problem is finalized.

The point (ix) is extracted based on the result from the point $(viii)$, and then we write that

$$\sum_{n=1}^{\infty} x^n \frac{H_n^{(2)}}{n^2} = \sum_{n=1}^{\infty} x^n \frac{H_n^{(2)} - 1/n^2}{n^2} + \underbrace{\sum_{n=1}^{\infty} \frac{x^n}{n^4}}_{\operatorname{Li}_4(x)}$$

$$\overset{\text{reindex}}{\underset{\text{the series}}{=}} \sum_{n=1}^{\infty} x^{n+1} \frac{H_{n+1}^{(2)} - 1/(n+1)^2}{(n+1)^2} + \operatorname{Li}_4(x) = \sum_{n=1}^{\infty} x^{n+1} \frac{H_n^{(2)}}{(n+1)^2} + \operatorname{Li}_4(x)$$

$$= -2\zeta(4) - 2\zeta(3)\log(1-x) - \zeta(2)\log^2(1-x) - \frac{1}{12}\log^4(1-x)$$

$$+ \frac{1}{3}\log(x)\log^3(1-x) + \frac{1}{2}(\operatorname{Li}_2(x))^2 + 2\log(1-x)\operatorname{Li}_3(x) - \operatorname{Li}_4(x)$$

$$+ 2\operatorname{Li}_4(1-x) - 2\operatorname{Li}_4\left(\frac{x}{x-1}\right),$$

and the point (ix) of the problem is finalized.

For the point (x) of the problem, we start from the point (i), with $m = 3$, Sect. 4.6, that is, $\displaystyle\sum_{n=1}^{\infty} x^n H_n^{(3)} = \frac{\operatorname{Li}_3(x)}{1-x}$, and then we write

$$\sum_{n=1}^{\infty} x^n \frac{H_n^{(3)}}{n+1} = \frac{1}{x} \sum_{n=1}^{\infty} \int_0^x t^n \, dt \, H_n^{(3)} = \frac{1}{x} \int_0^x \sum_{n=1}^{\infty} t^n H_n^{(3)} dt = \frac{1}{x} \int_0^x \frac{\operatorname{Li}_3(t)}{1-t} dt$$

$$= -\frac{1}{x} \int_0^x (\log(1-t))' \operatorname{Li}_3(t) dt \overset{\substack{\text{integrate by} \\ \text{parts}}}{=} -\frac{\log(1-x)\operatorname{Li}_3(x)}{x}$$

$$+ \frac{1}{x} \int_0^x \frac{\log(1-t)\operatorname{Li}_2(t)}{t} dt = -\frac{\log(1-x)\operatorname{Li}_3(x)}{x} - \frac{1}{2}\frac{1}{x} \int_0^x ((\operatorname{Li}_2(t))^2)' dt$$

$$= -\frac{\log(1-x)\operatorname{Li}_3(x)}{x} - \frac{1}{2}\frac{(\operatorname{Li}_2(x))^2}{x},$$

and the point (x) of the problem is finalized.

Finally, for the last point, we follow a similar style to the one at the previous point, and then we write

$$\sum_{n=1}^{\infty} x^n \frac{H_n^{(3)}}{n} = \sum_{n=1}^{\infty} \int_0^x t^{n-1} dt \, H_n^{(3)} = \int_0^x \sum_{n=1}^{\infty} t^{n-1} H_n^{(3)} dt = \int_0^x \frac{\operatorname{Li}_3(t)}{t(1-t)} dt$$

$$= \int_0^x \frac{\operatorname{Li}_3(t)}{t} dt + \int_0^x \frac{\operatorname{Li}_3(t)}{1-t} dt = \operatorname{Li}_4(x) - \log(1-x)\operatorname{Li}_3(x) - \frac{1}{2}\operatorname{Li}_2(x))^2,$$

and the point (xi) of the problem is finalized. Observe the integral $\int_0^x \frac{\operatorname{Li}_3(t)}{1-t} dt$ has been calculated at the previous point.

It is worth noting that, like in the previous section, the most advanced integrals to calculate are $\int_0^x \frac{\log^3(1-t)}{t} dt$ and $\int_0^x \frac{1}{t}\operatorname{Li}_3\left(\frac{t}{t-1}\right) dt$.

Now, we are preparing to enter the next section where we'll continue to explore other interesting ordinary generating functions!

6.8 Good-to-Know Generating Functions: The Third Part

Solution The encounter of the generating functions in the previous two sections and the related results arising during their calculations will be found very useful during the extraction of the ones in the current section.

For the moment, we'll put aside the first four points of the problem and return to them later. A brief examination of the last four ordinary generating functions in the problem statement immediately makes us think of two generating functions presented in *(Almost) Impossible Integrals, Sums, and Series*, that is,

$$\sum_{n=1}^{\infty} x^n H_n^3 = \frac{1}{1-x}\left(\frac{3}{2}\log(x)\log^2(1-x) - 3\zeta(2)\log(1-x)\right.$$

$$\left. - \log^3(1-x) + \mathrm{Li}_3(x) + 3\,\mathrm{Li}_3(1-x) - 3\zeta(3)\right) \tag{6.38}$$

and

$$\sum_{n=1}^{\infty} x^n H_n H_n^{(2)} = \frac{1}{1-x}\left(\frac{1}{2}\log(x)\log^2(1-x) + \mathrm{Li}_3(x) + \mathrm{Li}_3(1-x)\right.$$

$$\left. - \zeta(2)\log(1-x) - \zeta(3)\right). \tag{6.39}$$

The results in (6.38) and (6.39) appear with solutions involving telescoping sums in [76, Chapter 6, pp. 350–354], and it is not hard to see that they might be a good starting point in the extraction of the generating functions at the points (v)–$(viii)$.

So, for the point (v) of the problem, we want to exploit the generating function in (6.38), and then we write

$$\sum_{n=1}^{\infty} x^n \frac{H_n^3}{n+1} = \frac{1}{x}\sum_{n=1}^{\infty}\int_0^x t^n dt\, H_n^3 = \frac{1}{x}\int_0^x \sum_{n=1}^{\infty} t^n H_n^3 dt$$

{exploit the result in (6.38) and then expand the integral}

$$= \frac{3}{2}\frac{1}{x}\int_0^x \frac{\log(t)\log^2(1-t)}{1-t}dt - 3\zeta(2)\frac{1}{x}\int_0^x \frac{\log(1-t)}{1-t}dt - \frac{1}{x}\int_0^x \frac{\log^3(1-t)}{1-t}dt$$

$$+ \frac{1}{x}\int_0^x \frac{\mathrm{Li}_3(t)}{1-t}dt + 3\frac{1}{x}\int_0^x \frac{\mathrm{Li}_3(1-t)}{1-t}dt - 3\zeta(3)\frac{1}{x}\int_0^x \frac{1}{1-t}dt$$

$$= 3\zeta(4)\frac{1}{x} + 3\zeta(3)\frac{\log(1-x)}{x} + \frac{3}{2}\zeta(2)\frac{\log^2(1-x)}{x} + \frac{1}{4}\frac{\log^4(1-x)}{x} - 3\frac{\mathrm{Li}_4(1-x)}{x}$$

$$+ \frac{3}{2}\frac{1}{x}\int_0^x \frac{\log(t)\log^2(1-t)}{1-t}dt + \frac{1}{x}\int_0^x \frac{\mathrm{Li}_3(t)}{1-t}dt. \tag{6.40}$$

For the first remaining integral in (6.40), we kick off with an integration by parts, and then we write

$$\int_0^x \frac{\log(t)\log^2(1-t)}{1-t}dt = -\frac{1}{3}\log(x)\log^3(1-x) + \frac{1}{3}\int_0^x \frac{\log^3(1-t)}{t}dt$$

{make use of the integral result in (6.36), Sect. 6.6}

$$= \log^2(1-x)\,\mathrm{Li}_2(1-x) - 2\log(1-x)\,\mathrm{Li}_3(1-x) + 2\,\mathrm{Li}_4(1-x) - 2\zeta(4). \qquad (6.41)$$

Then, for the second remaining integral in (6.40), we consider again integration by parts that gives

$$\int_0^x \frac{\mathrm{Li}_3(t)}{1-t}\,dt = -\int_0^x (\log(1-t))'\,\mathrm{Li}_3(t)\,dt = -\log(1-x)\,\mathrm{Li}_3(x)$$

$$+ \int_0^x \frac{\log(1-t)\,\mathrm{Li}_2(t)}{t}\,dt = -\log(1-x)\,\mathrm{Li}_3(x) - \frac{1}{2}\int_0^x \left((\mathrm{Li}_2(t))^2\right)'\,dt$$

$$= -\log(1-x)\,\mathrm{Li}_3(x) - \frac{1}{2}(\mathrm{Li}_2(x))^2. \qquad (6.42)$$

Collecting the results from (6.41) and (6.42) in (6.40), we conclude that

$$\sum_{n=1}^{\infty} x^n \frac{H_n^3}{n+1}$$

$$= 3\zeta(3)\frac{\log(1-x)}{x} + \frac{3}{2}\zeta(2)\frac{\log^2(1-x)}{x} + \frac{1}{4}\frac{\log^4(1-x)}{x} + \frac{3}{2}\frac{\log^2(1-x)\,\mathrm{Li}_2(1-x)}{x}$$

$$- \frac{1}{2}\frac{(\mathrm{Li}_2(x))^2}{x} - \frac{\log(1-x)\,\mathrm{Li}_3(x)}{x} - 3\frac{\log(1-x)\,\mathrm{Li}_3(1-x)}{x},$$

and the solution to the point (v) of the problem is complete.

Next, for the point (vi) of the problem, we'll want to exploit again the generating function employed at the previous point, and then write

$$\sum_{n=1}^{\infty} x^n \frac{H_n^3}{n} = \sum_{n=1}^{\infty} \int_0^x t^{n-1}\,dt\,H_n^3 = \int_0^x \sum_{n=1}^{\infty} t^{n-1} H_n^3\,dt$$

{use the generating function in (6.38) combined with the Landen's trilogarithmic}

{identity in (6.33), found in Sect. 6.6, and afterwards expand the resulting integral}

$$= -\frac{1}{2}\int_0^x \frac{\log^3(1-t)}{t(1-t)}\,dt - 2\int_0^x \frac{\mathrm{Li}_3(t)}{t(1-t)}\,dt - 3\int_0^x \frac{1}{t(1-t)}\mathrm{Li}_3\left(\frac{t}{t-1}\right)dt$$

$$= -\frac{1}{2}\int_0^x \frac{\log^3(1-t)}{1-t}\,dt - 2\int_0^x \frac{\mathrm{Li}_3(t)}{t}\,dt - \frac{1}{2}\int_0^x \frac{\log^3(1-t)}{t}\,dt - 2\int_0^x \frac{\mathrm{Li}_3(t)}{1-t}\,dt$$

$$-3 \int_0^x \frac{1}{t(1-t)} \operatorname{Li}_3\left(\frac{t}{t-1}\right) dt$$

$$= 3\zeta(4) - \frac{1}{2}\log(x)\log^3(1-x) + \frac{1}{8}\log^4(1-x) - \frac{3}{2}\log^2(1-x)\operatorname{Li}_2(1-x)$$

$$+ (\operatorname{Li}_2(x))^2 + 2\log(1-x)\operatorname{Li}_3(x) + 3\log(1-x)\operatorname{Li}_3(1-x) - 2\operatorname{Li}_4(x)$$

$$- 3\operatorname{Li}_4(1-x) - 3\operatorname{Li}_4\left(\frac{x}{x-1}\right),$$

where in the calculations, I used the integral results in (6.36), Sect. 6.6, and (6.42) found during the calculations at the previous point, and the solution to the point (vi) of the problem is complete.

Further, for the point (vii) of the problem, we consider the generating function in (6.39), and then we have

$$\sum_{n=1}^{\infty} x^n \frac{H_n H_n^{(2)}}{n+1} = \frac{1}{x}\sum_{n=1}^{\infty} \int_0^x t^n dt\, H_n H_n^{(2)} = \frac{1}{x}\int_0^x \sum_{n=1}^{\infty} t^n H_n H_n^{(2)} dt$$

{exploit the result in (6.39) and then expand the integral}

$$= \frac{1}{2}\frac{1}{x}\int_0^x \frac{\log(t)\log^2(1-t)}{1-t}dt + \frac{1}{x}\int_0^x \frac{\operatorname{Li}_3(t)}{1-t}dt + \frac{1}{x}\int_0^x \frac{\operatorname{Li}_3(1-t)}{1-t}dt$$

$$- \zeta(2)\frac{1}{x}\int_0^x \frac{\log(1-t)}{1-t}dt - \zeta(3)\frac{1}{x}\int_0^x \frac{1}{1-t}dt$$

$$= \zeta(3)\frac{\log(1-x)}{x} + \frac{1}{2}\zeta(2)\frac{\log^2(1-x)}{x} + \frac{1}{2}\frac{\log^2(1-x)\operatorname{Li}_2(1-x)}{x}$$

$$- \frac{1}{2}\frac{(\operatorname{Li}_2(x))^2}{x} - \frac{\log(1-x)\operatorname{Li}_3(x)}{x} - \frac{\log(1-x)\operatorname{Li}_3(1-x)}{x},$$

where the first two integrals already appeared during the calculations to the point (v) of the problem, in (6.41) and (6.42), and the solution to the point (vii) of the problem is complete.

As regards the point $(viii)$, we'll consider again the generating function used at the preceding point, and then we have

$$\sum_{n=1}^{\infty} x^n \frac{H_n H_n^{(2)}}{n} = \sum_{n=1}^{\infty} \int_0^x t^{n-1} dt\, H_n H_n^{(2)} = \int_0^x \sum_{n=1}^{\infty} t^{n-1} H_n H_n^{(2)} dt$$

{use the generating function in (6.39) combined with the Landen's trilogarithmic} {identity in (6.33), found in Sect. 6.6, and afterwards expand the resulting integral}

$$= \frac{1}{6} \int_0^x \frac{\log^3(1-t)}{t} dt + \frac{1}{6} \int_0^x \frac{\log^3(1-t)}{1-t} dt - \int_0^x \frac{1}{t(1-t)} \operatorname{Li}_3\left(\frac{t}{t-1}\right) dt$$

$$= -\zeta(4) + \frac{1}{6} \log(x) \log^3(1-x) - \frac{1}{24} \log^4(1-x) + \frac{1}{2} \log^2(1-x) \operatorname{Li}_2(1-x)$$

$$- \log(1-x) \operatorname{Li}_3(1-x) + \operatorname{Li}_4(1-x) - \operatorname{Li}_4\left(\frac{x}{x-1}\right),$$

where in the calculations, I used the integral result in (6.36), Sect. 6.6, and the solution to the point $(viii)$ of the problem is complete.

Now, it's time to return to the first four generating functions where I'll use a strategy involving telescoping sums as I proceeded for the ones in [76, Chapter 4, Section 4.10, p.284].

Let's begin with the generating function at the point (i), which we might denote by $G(x)$. Therefore, if we multiply its both sides by $1 - x$, we arrive at

$$\sum_{n=1}^{\infty} x^n (1-x) H_n^4 = (1-x) G(x). \tag{6.43}$$

By considering the left-hand side of (6.43) and rearranging to get a telescoping sum, we write

$$\sum_{n=1}^{\infty} x^n (1-x) H_n^4 = \sum_{n=1}^{\infty} x^n H_n^4 - \sum_{n=1}^{\infty} x^{n+1} \left(H_{n+1} - \frac{1}{n+1}\right)^4$$

$$\underset{\text{the second series}}{\overset{\text{reindex}}{=}} \sum_{n=1}^{\infty} x^n H_n^4 - \sum_{n=1}^{\infty} x^n \left(H_n - \frac{1}{n}\right)^4$$

$$= 4 \sum_{n=1}^{\infty} x^n \frac{H_n}{n^3} - 6 \sum_{n=1}^{\infty} x^n \frac{H_n^2}{n^2} + 4 \sum_{n=1}^{\infty} x^n \frac{H_n^3}{n} - \underbrace{\sum_{n=1}^{\infty} \frac{x^n}{n^4}}_{\operatorname{Li}_4(x)}$$

$$= 4\zeta(4) + 4\zeta(3) \log(1-x) + 2\zeta(2) \log^2(1-x) + \frac{2}{3} \log^4(1-x)$$

$$- \frac{2}{3} \log(x) \log^3(1-x) + (\operatorname{Li}_2(x))^2 + 4 \log(1-x) \operatorname{Li}_3(x) - 7 \operatorname{Li}_4(x)$$

$$- 4\operatorname{Li}_4(1-x) - 8\operatorname{Li}_4\left(\frac{x}{x-1}\right),$$

where in the calculations, I used the generating functions given in (4.38), Sect. 4.6; then (4.44), Sect. 4.7; and (vi), the current section, which together lead to the generating function in (6.43), and the solution to the point (i) of the problem is complete.

Next, for the generating function at the point (ii), which we denote by $P(x)$, we multiply both sides by $1 - x$, thus leading to

$$\sum_{n=1}^{\infty} x^n(1-x)(H_n^{(2)})^2 = (1-x)P(x). \tag{6.44}$$

So, we focus on the left-hand side of (6.44), which we rearrange to get a telescoping sum, and then we write

$$\sum_{n=1}^{\infty} x^n(1-x)(H_n^{(2)})^2 = \sum_{n=1}^{\infty} x^n(H_n^{(2)})^2 - \sum_{n=1}^{\infty} x^{n+1}\left(H_{n+1}^{(2)} - \frac{1}{(n+1)^2}\right)^2$$

$$\underset{\text{the second series}}{\overset{\text{reindex}}{=}} \sum_{n=1}^{\infty} x^n(H_n^{(2)})^2 - \sum_{n=1}^{\infty} x^n\left(H_n^{(2)} - \frac{1}{n^2}\right)^2 = 2\sum_{n=1}^{\infty} x^n\frac{H_n^{(2)}}{n^2} - \underbrace{\sum_{n=1}^{\infty} \frac{x^n}{n^4}}_{\operatorname{Li}_4(x)}$$

$$= -4\zeta(4) - 4\zeta(3)\log(1-x) - 2\zeta(2)\log^2(1-x) - \frac{1}{6}\log^4(1-x)$$

$$+ \frac{2}{3}\log(x)\log^3(1-x) + (\operatorname{Li}_2(x))^2 + 4\log(1-x)\operatorname{Li}_3(x) - 3\operatorname{Li}_4(x)$$

$$+ 4\operatorname{Li}_4(1-x) - 4\operatorname{Li}_4\left(\frac{x}{x-1}\right),$$

where in the calculations, I used the generating function in (4.48), Sect. 4.7, leading us to the generating function in (6.44), and the solution to the point (ii) of the problem is complete.

Further, regarding the generating function at the point (iii), which this time we denote by $Q(x)$, we consider again to multiply both sides by $1 - x$ that gives

$$\sum_{n=1}^{\infty} x^n(1-x)H_n H_n^{(3)} = (1-x)Q(x). \tag{6.45}$$

At this point, we rearrange the left-hand side of (6.45) to obtain a telescoping sum, and then we write

$$\sum_{n=1}^{\infty} x^n (1-x) H_n H_n^{(3)}$$

$$= \sum_{n=1}^{\infty} x^n H_n H_n^{(3)} - \sum_{n=1}^{\infty} x^{n+1} \left(H_{n+1} - \frac{1}{n+1}\right) \left(H_{n+1}^{(3)} - \frac{1}{(n+1)^3}\right)$$

$$\overset{\underset{\text{reindex}}{\text{the second series}}}{=} \sum_{n=1}^{\infty} x^n H_n H_n^{(3)} - \sum_{n=1}^{\infty} x^n \left(H_n - \frac{1}{n}\right) \left(H_n^{(3)} - \frac{1}{n^3}\right)$$

$$= \sum_{n=1}^{\infty} x^n \frac{H_n}{n^3} + \sum_{n=1}^{\infty} x^n \frac{H_n^{(3)}}{n} - \underbrace{\sum_{n=1}^{\infty} \frac{x^n}{n^4}}_{\text{Li}_4(x)}$$

$$= \zeta(4) + \zeta(3) \log(1-x) + \frac{1}{2}\zeta(2) \log^2(1-x) + \frac{1}{24} \log^4(1-x)$$

$$- \frac{1}{6} \log(x) \log^3(1-x) - \frac{1}{2}(\text{Li}_2(x))^2 - 2 \log(1-x) \text{Li}_3(x) + 2 \text{Li}_4(x)$$

$$- \text{Li}_4(1-x) + \text{Li}_4 \left(\frac{x}{x-1}\right),$$

where in the calculations, I used the generating functions given in (4.38), Sect. 4.6, and then (4.50), Sect. 4.7, which gives the generating function in (6.45), and the solution to the point (iii) of the problem is complete.

Finally, for the generating function at the point (iv), we may proceed in a similar style as I did for the ones at the points (i)–(iii), but I'll go differently this time.

I'll use two results already presented in *(Almost) Impossible Integrals, Sums, and Series*, that is

$$\sum_{n=1}^{\infty} x^n (H_n^4 - 6H_n^2 H_n^{(2)} + 8H_n H_n^{(3)} + 3(H_n^{(2)})^2 - 6H_n^{(4)}) = \frac{\log^4(1-x)}{1-x} \quad (6.46)$$

and

$$\sum_{n=1}^{\infty} \frac{x^n}{n} (H_n^4 + 6H_n^2 H_n^{(2)} + 8H_n H_n^{(3)} + 3(H_n^{(2)})^2 + 6H_n^{(4)}) = -24 \text{Li}_5 \left(\frac{x}{x-1}\right),$$

$$(6.47)$$

found in [76, Chapter 4, p. 285, Chapter 6, p. 355].

If we differentiate both sides of (6.47) and then rearrange, we get

$$\sum_{n=1}^{\infty} x^n (H_n^4 + 6H_n^2 H_n^{(2)} + 8H_n H_n^{(3)} + 3(H_n^{(2)})^2 + 6H_n^{(4)}) = -24\frac{1}{1-x} \operatorname{Li}_4\left(\frac{x}{x-1}\right).$$

(6.48)

At last, if we subtract (6.46) from (6.48) and then combine the result with the case $m = 4$ of (4.32), Sect. 4.6, that is, $\sum_{n=1}^{\infty} x^n H_n^{(4)} = \dfrac{\operatorname{Li}_4(x)}{1-x}$, we obtain that

$$\sum_{n=1}^{\infty} x^n H_n^2 H_n^{(2)} = \frac{1}{1-x}\left(-\frac{1}{12}\log^4(1-x) - \operatorname{Li}_4(x) - 2\operatorname{Li}_4\left(\frac{x}{x-1}\right)\right),$$

and the solution to the point (iv) of the problem is complete.

Since for the extraction of the result at the point (i) I also needed the result from the point (vi), it is easy to see why I preferred to let the proof of the result from the former point at a later moment.

In the next section, I will continue to treat generating functions, but with an interesting atypical summand structure! Are you ready to explore them?

6.9   Good-to-Know Generating Functions: The Fourth Part

Solution The first result of the current section is an atypical generating function found and derived in the paper *Euler Sum Involving Tail* (see [92, 2019–2020]). At the moment mathematician Moti Levy (Rehovot, Israel) invited me to join the work on the paper *Euler Sum Involving Tail*, he already derived the first generating function and calculated the first main atypical harmonic series of the paper. In fact, for the first solution to the point (i), I'll follow his lines of attack, but with some slight modifications.

Both the first and third atypical generating functions are constructed by also exploiting generating functions derived in the previous sections.

For a first solution to the point (i) of the problem, we start with proving the following auxiliary result:

$$\sum_{n=1}^{k-1} x^n H_n^{(p)} = \frac{1}{1-x}\left(-x^k H_k^{(p)} + \sum_{m=1}^{k} \frac{x^m}{m^p}\right).$$

(6.49)

Proof We start with the simple fact that $H_n^{(p)} = \sum\limits_{m=1}^{n} \dfrac{1}{m^p}$, and considering to change the summation order, we have

$$\sum_{n=1}^{k-1} x^n H_n^{(p)} = \sum_{n=1}^{k-1}\sum_{m=1}^{n} \frac{x^n}{m^p} = \sum_{m=1}^{k-1}\sum_{n=m}^{k-1} \frac{x^n}{m^p} = -\frac{1}{1-x}\sum_{m=1}^{k-1} \frac{x^k - x^m}{m^p}$$

$$= -\frac{x^k}{1-x}H_{k-1}^{(p)} + \frac{1}{1-x}\sum_{m=1}^{k-1}\frac{x^m}{m^p} = -\frac{x^k}{1-x}\left(H_k^{(p)} - \frac{1}{k^p}\right) + \frac{1}{1-x}\sum_{m=1}^{k-1}\frac{x^m}{m^p}$$

$$= \frac{1}{1-x}\left(-x^k H_k^{(p)} + \sum_{m=1}^{k}\frac{x^m}{m^p}\right),$$

which is the result to prove in (6.49). ∎

So, returning to the main result from the first point where we change the summation order and then consider the result from (6.49), the case $p = 1$, we have

$$\sum_{n=1}^{\infty} x^n H_n \sum_{k=n+1}^{\infty}\frac{H_k}{k^2} = \sum_{k=1}^{\infty}\frac{H_k}{k^2}\sum_{n=1}^{k-1}x^n H_n = -\frac{1}{1-x}\sum_{k=1}^{\infty}x^k\frac{H_k^2}{k^2} + \frac{1}{1-x}\sum_{k=1}^{\infty}\frac{H_k}{k^2}\sum_{m=1}^{k}\frac{x^m}{m}.$$
$$\tag{6.50}$$

Based on the power series $-\log(1-x) = \sum\limits_{n=1}^{\infty}\dfrac{x^n}{n}$, we have that

$$\sum_{m=1}^{k}\frac{x^m}{m} = -\log(1-x) - \sum_{m=k}^{\infty}\frac{x^{m+1}}{m+1}. \tag{6.51}$$

Plugging the result from (6.51) in the second left series of (6.50), we arrive at

$$\sum_{k=1}^{\infty}\frac{H_k}{k^2}\sum_{m=1}^{k}\frac{x^m}{m} = -\log(1-x)\sum_{k=1}^{\infty}\frac{H_k}{k^2} - \sum_{k=1}^{\infty}\frac{H_k}{k^2}\sum_{m=k}^{\infty}\frac{x^{m+1}}{m+1}$$

$$\overset{\substack{\text{reverse the order}\\\text{of summation}}}{=} -2\zeta(3)\log(1-x) - \sum_{m=1}^{\infty}\sum_{k=1}^{m}\frac{x^{m+1}}{m+1}\frac{H_k}{k^2}$$

$$= -2\zeta(3)\log(1-x) - \int_0^x \sum_{m=1}^{\infty}\sum_{k=1}^{m}t^m\frac{H_k}{k^2}dt$$

$$= -2\zeta(3)\log(1-x) - \int_0^x \frac{1}{1-t}\sum_{k=1}^{\infty} t^k \frac{H_k}{k^2} dt$$

$$= -2\zeta(3)\log(1-x) + \frac{1}{6}\int_0^x \frac{\log^3(1-t)}{1-t}dt + \int_0^x \frac{\log(1-t)\operatorname{Li}_2(t)}{1-t}dt$$

$$- 2\int_0^x \frac{\operatorname{Li}_3(t)}{1-t}dt - \int_0^x \frac{1}{1-t}\operatorname{Li}_3\left(\frac{t}{t-1}\right)dt, \qquad (6.52)$$

where in the calculations, I used that $\sum_{k=1}^{\infty}\dfrac{H_k}{k^2} = 2\zeta(3)$, which is the case $n = 2$ of the Euler sum generalization in (6.149), Sect. 6.19; next the Cauchy product of two series, in a reversed sense, since we start from the right-hand side of (6.15), Sect. 6.5, in order to obtain the fourth equality; and then the third form of the generating function in (4.36), Sect. 4.6.

Now, for the second integral in (6.52), we integrate by parts and then use the integral result in (6.36), Sect. 6.6, that gives

$$\int_0^x \frac{\log(1-t)\operatorname{Li}_2(t)}{1-t}dt = -\frac{1}{2}\int_0^x (\log^2(1-t))'\operatorname{Li}_2(t)dt$$

$$= -\frac{1}{2}\log^2(1-x)\operatorname{Li}_2(x) - \frac{1}{2}\int_0^x \frac{\log^3(1-t)}{t}dt$$

$$= 3\zeta(4) - \frac{1}{2}\log(x)\log^3(1-x) - \frac{1}{2}\log^2(1-x)\operatorname{Li}_2(x) - \frac{3}{2}\log^2(1-x)\operatorname{Li}_2(1-x)$$

$$+ 3\log(1-x)\operatorname{Li}_3(1-x) - 3\operatorname{Li}_4(1-x), \qquad (6.53)$$

where to obtain the last equality, I also applied the dilogarithm function reflection formula (see [21, Chapter 1, p.5], [70, Chapter 2, p.107]), $\operatorname{Li}_2(x) + \operatorname{Li}_2(1-x) = \pi^2/6 - \log(x)\log(1-x)$.

Next, for the last integral in (6.52), I'll prefer to exploit the calculation flow of the integral result in (6.35), Sect. 6.6, and then we write

$$\int_0^x \frac{1}{1-t}\operatorname{Li}_3\left(\frac{t}{t-1}\right)dt$$

$$= \frac{1}{2}\left(\operatorname{Li}_2\left(\frac{x}{x-1}\right)\right)^2 - \log(1-x)\operatorname{Li}_3\left(\frac{x}{x-1}\right)$$

$$= -\zeta(3)\log(1-x) - \frac{1}{2}\zeta(2)\log^2(1-x) - \frac{1}{24}\log^4(1-x) - \frac{1}{2}\log^2(1-x)\operatorname{Li}_2(1-x)$$

$$+ \frac{1}{2}(\text{Li}_2(x))^2 + \log(1-x)\,\text{Li}_3(x) + \log(1-x)\,\text{Li}_3(1-x), \qquad (6.54)$$

where in the calculations, I used both the Landen's dilogarithmic identity (see [21, Chapter 1, p.5], [70, Chapter 2, p.107]), $\text{Li}_2(x) + \text{Li}_2\left(\dfrac{x}{x-1}\right) = -\dfrac{1}{2}\log^2(1-x)$, and the Landen's trilogarithmic identity in (6.33), Sect. 6.6, and finally the dilogarithm function reflection formula, $\text{Li}_2(x) + \text{Li}_2(1-x) = \pi^2/6 - \log(x)\log(1-x)$ for the dilogarithm appearing in $\log^2(1-x)\,\text{Li}_2(x)$.

Collecting the results from (6.53) and (6.54) in (6.52), and at the same time using the integral result given in (6.42), the previous section, we arrive at

$$\sum_{k=1}^{\infty} \frac{H_k}{k^2} \sum_{m=1}^{k} \frac{x^m}{m}$$

$$= 3\zeta(4) - \zeta(3)\log(1-x) - \frac{1}{2}\log^2(1-x)\,\text{Li}_2(1-x) + \frac{1}{2}(\text{Li}_2(x))^2$$

$$+ \log(1-x)\,\text{Li}_3(x) + 2\log(1-x)\,\text{Li}_3(1-x) - 3\,\text{Li}_4(1-x). \qquad (6.55)$$

At last, if we consider the generating function from (4.44), Sect. 4.7, and (6.55) in (6.50), we obtain that

$$\sum_{n=1}^{\infty} x^n H_n \sum_{k=n+1}^{\infty} \frac{H_k}{k^2}$$

$$= \frac{1}{1-x}\left(\zeta(4) - \zeta(3)\log(1-x) + \frac{1}{3}\log(x)\log^3(1-x) + \frac{1}{2}\log^2(1-x)\,\text{Li}_2(1-x) \right.$$

$$\left. + \log(1-x)\,\text{Li}_3(x) - \text{Li}_4(1-x) - \text{Li}_4(x) \right),$$

and the first solution to the point (i) of the problem is complete.

For a second solution to the point (i) of the problem, we'll want to exploit telescoping sums as in the case of the generating functions at the points (i)–(iii) in the previous section.

So, if we denote the generating function by $G(x)$ and multiply its both sides by $1-x$, we get that

$$\sum_{n=1}^{\infty} x^n (1-x) H_n \sum_{k=n+1}^{\infty} \frac{H_k}{k^2} = (1-x)G(x). \qquad (6.56)$$

Now, by considering the left-hand side of (6.56) and rearranging to get a telescoping sum, we write

$$\sum_{n=1}^{\infty} x^n (1-x) H_n \sum_{k=n+1}^{\infty} \frac{H_k}{k^2}$$

$$= \sum_{n=1}^{\infty} x^n H_n \sum_{k=n+1}^{\infty} \frac{H_k}{k^2} - \sum_{n=1}^{\infty} x^{n+1} \left(H_{n+1} - \frac{1}{n+1} \right) \left(\frac{H_{n+1}}{(n+1)^2} + \sum_{k=n+2}^{\infty} \frac{H_k}{k^2} \right)$$

$$\underset{\substack{\text{reindex} \\ \text{the second} \\ \text{outer series}}}{=} \sum_{n=1}^{\infty} x^n H_n \sum_{k=n+1}^{\infty} \frac{H_k}{k^2} - \sum_{n=1}^{\infty} x^n \left(H_n - \frac{1}{n} \right) \left(\frac{H_n}{n^2} + \sum_{k=n+1}^{\infty} \frac{H_k}{k^2} \right)$$

$$= \sum_{n=1}^{\infty} x^n \frac{H_n}{n^3} - \sum_{n=1}^{\infty} x^n \frac{H_n^2}{n^2} + \sum_{n=1}^{\infty} \frac{x^n}{n} \sum_{k=n+1}^{\infty} \frac{H_k}{k^2} \underset{\substack{\text{reindex} \\ \text{the last series}}}{=} \sum_{n=1}^{\infty} x^n \frac{H_n}{n^3} - \sum_{n=1}^{\infty} x^n \frac{H_n^2}{n^2}$$

$$+ \sum_{k=1}^{\infty} \frac{H_k}{k^2} \left(-\frac{x^k}{k} + \sum_{n=1}^{k} \frac{x^n}{n} \right) = -\sum_{n=1}^{\infty} x^n \frac{H_n^2}{n^2} + \sum_{k=1}^{\infty} \frac{H_k}{k^2} \left(\sum_{n=1}^{k} \frac{x^n}{n} \right),$$

which leads to (6.50), the previous solution, and thus we conclude that

$$\sum_{n=1}^{\infty} x^n H_n \sum_{k=n+1}^{\infty} \frac{H_k}{k^2}$$

$$= -\frac{1}{1-x} \sum_{n=1}^{\infty} x^n \frac{H_n^2}{n^2} + \frac{1}{1-x} \sum_{k=1}^{\infty} \frac{H_k}{k^2} \left(\sum_{n=1}^{k} \frac{x^n}{n} \right)$$

$$= \frac{1}{1-x} \left(\zeta(4) - \zeta(3) \log(1-x) + \frac{1}{3} \log(x) \log^3(1-x) + \frac{1}{2} \log^2(1-x) \operatorname{Li}_2(1-x) \right.$$

$$\left. + \log(1-x) \operatorname{Li}_3(x) - \operatorname{Li}_4(1-x) - \operatorname{Li}_4(x) \right),$$

and the second solution to the point (i) of the problem is complete.

As regards the second point of the problem, we simply follow the steps in (6.50), (6.51), and (6.52), with $\sum_{k=1}^{\infty} \frac{H_k}{k^s} \sum_{m=1}^{k} \frac{x^m}{m}$ instead of $\sum_{k=1}^{\infty} \frac{H_k}{k^2} \sum_{m=1}^{k} \frac{x^m}{m}$, and thus we get

$$\sum_{n=1}^{\infty} x^n H_n \sum_{k=n+1}^{\infty} \frac{H_k}{k^s}$$

$$= -\frac{\log(1-x)}{1-x} \sum_{k=1}^{\infty} \frac{H_k}{k^s} - \frac{1}{1-x} \sum_{k=1}^{\infty} x^k \frac{H_k^2}{k^s} - \frac{1}{1-x} \int_0^x \frac{1}{1-t} \sum_{k=1}^{\infty} t^k \frac{H_k}{k^s} dt,$$

and the point (ii) of the problem is complete. The generalization is also found and derived in the paper *Euler Sum Involving Tail* (see [92, 2019–2020]), first obtained by Moti Levy.

What about the third point, one might hasten to ask! A key observation is that we might exploit the case $p = 1$ of the generalization in (6.49) such that we may relate the calculation to the generating function at the point (i),

$$\sum_{n=1}^{k-1} x^n H_n = \frac{1}{1-x} \left(-x^k H_k + \sum_{n=1}^{k} \frac{x^n}{n} \right). \tag{6.57}$$

So, if we multiply both sides of (6.57) by $\log(x)$ and then integrate from $x = 0$ to $x = 1$, and rearrange, we arrive at

$$\sum_{n=1}^{k} \frac{1}{n} (\zeta(2) - H_n^{(2)}) = \zeta(2) H_k - H_k H_k^{(2)} + \sum_{n=1}^{k-1} \frac{H_n}{(n+1)^2} = \zeta(2) H_k - H_k H_k^{(2)}$$

$$+ \sum_{n=1}^{k-1} \frac{H_{n+1} - 1/(n+1)}{(n+1)^2} = \zeta(2) H_k - H_k H_k^{(2)} - H_k^{(3)} + \sum_{n=1}^{k} \frac{H_n}{n^2}, \tag{6.58}$$

where to get the last equality, I reindexed and expanded the sum. Also, in the calculations based on (6.57), I used the fact that $\int_0^1 t^n \frac{\log(t)}{1-t} dt =$ $\int_0^1 t^n \log(t) \sum_{k=1}^{\infty} t^{k-1} dt = \sum_{k=1}^{\infty} \int_0^1 t^{k+n-1} \log(t) dt = -\sum_{k=1}^{\infty} \frac{1}{(k+n)^2} = -(\zeta(2) - H_n^{(2)})$.

To get (6.58) differently, we may simply apply Abel's summation (see [76, Chapter 2, pp.39–40]), or we may exploit the generalized result in (6.102), Sect. 6.13, with, say, $p = 2$ and $q = 1$, that gives

$$\sum_{n=1}^{k} \frac{H_n^{(2)}}{n} + \sum_{n=1}^{k} \frac{H_n}{n^2} = H_k H_k^{(2)} + H_k^{(3)}. \tag{6.59}$$

Now, we observe that if we let $k \to \infty$ in (6.58), we immediately get

$$\sum_{n=1}^{\infty} \frac{1}{n}(\zeta(2) - H_n^{(2)}) = \lim_{k\to\infty}\left(H_k(\zeta(2) - H_k^{(2)}) - H_k^{(3)} + \sum_{n=1}^{k}\frac{H_n}{n^2}\right) = \zeta(3),$$

(6.60)

which is another result we need.

In the calculations above, I used some simple facts. For example, I considered that $\lim_{k\to\infty} H_k(\zeta(2)-H_k^{(2)}) = 0$, because when k is large, $\zeta(2)-H_k^{(2)}$ behaves like $1/k$, which is straightforward to see by simple inequalities involving telescoping sums. Hence all reduces to proving that $\lim_{k\to\infty}\dfrac{H_k}{k} = 0$, which may be easily shown by the *Stolz–Cesàro theorem* (see [15, Appendix, pp.263–266]). Also, observe that since H_k behaves like $\log(k)$, when k is large, and here we may recall the limit definition of the Euler-Mascheroni constant, $\gamma = \lim_{n\to\infty}(H_n - \log(n))$, it can be enough to show that $\lim_{x\to\infty}\dfrac{\log(x)}{x} = 0$, which is straightforward by *L'Hospital's rule*.

Coming back to the main series from the third point and employing (6.60) and (6.58), we have

$$\sum_{n=1}^{\infty} x^n H_n \sum_{k=n+1}^{\infty} \frac{1}{k}(\zeta(2) - H_k^{(2)})$$

$$= \sum_{n=1}^{\infty} x^n H_n \underbrace{\sum_{k=1}^{\infty}\frac{1}{k}(\zeta(2) - H_k^{(2)})}_{\zeta(3)} - \sum_{n=1}^{\infty} x^n H_n \left(\sum_{k=1}^{n}\frac{1}{k}(\zeta(2) - H_k^{(2)})\right)$$

{after employing (6.58), expand the second series, and rearrange}

$$= \sum_{n=1}^{\infty} x^n H_n \left(2\zeta(3) - \sum_{k=1}^{n}\frac{H_k}{k^2}\right) - \zeta(2)\sum_{n=1}^{\infty} x^n H_n^2 + \sum_{n=1}^{\infty} x^n H_n^2 H_n^{(2)}$$

$$+ \sum_{n=1}^{\infty} x^n H_n H_n^{(3)} - \zeta(3)\sum_{n=1}^{\infty} x^n H_n$$

$$= \sum_{n=1}^{\infty} x^n H_n \left(\sum_{k=n+1}^{\infty}\frac{H_k}{k^2}\right) - \zeta(2)\sum_{n=1}^{\infty} x^n H_n^2 + \sum_{n=1}^{\infty} x^n H_n^2 H_n^{(2)}$$

$$+ \sum_{n=1}^{\infty} x^n H_n H_n^{(3)} - \zeta(3)\sum_{n=1}^{\infty} x^n H_n$$

$$= \frac{1}{1-x}\left(2\zeta(4) + \zeta(3)\log(1-x) - \frac{1}{24}\log^4(1-x) - \frac{1}{3}\log(x)\log^3(1-x)\right)$$

$$-\frac{1}{2}\log^2(1-x)\,\mathrm{Li}_2(x)-\zeta(2)\,\mathrm{Li}_2(x)-\frac{1}{2}(\mathrm{Li}_2(x))^2-\log(1-x)\,\mathrm{Li}_3(x)$$

$$-2\,\mathrm{Li}_4(1-x)-\mathrm{Li}_4\left(\frac{x}{x-1}\right)\bigg),$$

and the solution to the point (iii) of the problem is complete. Note that all the resulting generating functions have been previously calculated, either in the current section, or in the preceding sections treating generating functions.

Finding creative ways of extracting such generating functions is always an exciting experience. What other ways would you like to consider here? We prepare now to jump in the next section where we'll have to deal with generating functions involving skew-harmonic numbers!

6.10   Good-to-Know Generating Functions: The Fifth Part

Solution A first encounter of the results involving the skew-harmonic numbers already took place in the first chapter. Generating functions involving skew-harmonic numbers, like the ones in the present section, have already been treated in the literature, and a good example is the paper, *Power series with skew-harmonic numbers, dilogarithms, and double integrals* (see [8]), in which one may find a collection of such interesting results. The curious reader might also want to check it.

Now, as regards the point (i) of the problem, we go with the strategy provided in the paper above and consider the Cauchy product of two series, with $-\mathrm{Li}_m(-x) = \sum_{n=1}^{\infty}(-1)^{n-1}\dfrac{x^n}{n^m}$ and $\dfrac{1}{1-x}=\sum_{n=1}^{\infty}x^{n-1}$, which gives

$$-\frac{\mathrm{Li}_m(-x)}{1-x}=\left(\sum_{n=1}^{\infty}(-1)^{n-1}\frac{x^n}{n^m}\right)\left(\sum_{n=1}^{\infty}x^{n-1}\right)=\sum_{n=1}^{\infty}x^n\left(\sum_{k=1}^{n}(-1)^{k-1}\frac{1}{k^m}\right)$$

$$=\sum_{n=1}^{\infty}x^n\overline{H}_n^{(m)},$$

and the first solution to the point (i) of the problem is complete.

For a second elegant solution, we exploit the simple fact that $\dfrac{x^k}{1-x}=\sum_{n=k}^{\infty}x^n$, which we combine with the change of summation order, and then we write

$$-\frac{\mathrm{Li}_m(-x)}{1-x}=\sum_{k=1}^{\infty}(-1)^{k-1}\frac{1}{k^m}\frac{x^k}{1-x}=\sum_{k=1}^{\infty}(-1)^{k-1}\frac{1}{k^m}\left(\sum_{n=k}^{\infty}x^n\right)$$

$$= \sum_{n=1}^{\infty} x^n \left(\sum_{k=1}^{n} (-1)^{k-1} \frac{1}{k^m} \right) = \sum_{n=1}^{\infty} x^n \overline{H}_n^{(m)},$$

and the second solution to the point (i) of the problem is complete.

For a third solution, we may start by denoting $\sum_{n=1}^{\infty} x^n \overline{H}_n^{(m)} = G(x)$ and then try to exploit telescoping sums as, say, in the second solution to the point (i) from the previous section. In fact, upon multiplying both sides of the mentioned equality by $1 - x$, we have that

$$\sum_{n=1}^{\infty} x^n (1 - x) \overline{H}_n^{(m)} = (1 - x) G(x). \tag{6.61}$$

So, if we take the left-hand side of (6.61), we write

$$\sum_{n=1}^{\infty} x^n (1-x) \overline{H}_n^{(m)} = \sum_{n=1}^{\infty} x^n \overline{H}_n^{(m)} - \sum_{n=1}^{\infty} x^{n+1} \left(\overline{H}_{n+1}^{(m)} - (-1)^n \frac{1}{(n+1)^m} \right)$$

$$\overset{\text{reindex}}{\underset{\text{the 2nd series}}{=}} \sum_{n=1}^{\infty} x^n \overline{H}_n^{(m)} - \sum_{n=1}^{\infty} x^n \left(\overline{H}_n^{(m)} - (-1)^{n-1} \frac{1}{n^m} \right)$$

$$= -\sum_{n=1}^{\infty} \frac{(-x)^n}{n^m} = -\operatorname{Li}_m(-x),$$

whence we obtain that

$$\sum_{n=1}^{\infty} x^n \overline{H}_n^{(m)} = -\frac{\operatorname{Li}_m(-x)}{1-x},$$

and the third solution to the point (i) of the problem is complete.

It's easy to get the result at the point (ii) if we exploit the generating function from the previous point, the case $m = 1$, that is, $\sum_{n=1}^{\infty} x^n \overline{H}_n = \dfrac{\log(1+x)}{1-x}$ together with integration. In general, given a dilogarithmic structure of the form $\operatorname{Li}_2(f(x)) = -\displaystyle\int_0^{f(x)} \frac{\log(1-t)}{t} dt$, if we differentiate both sides, we get that

$$(\operatorname{Li}_2(f(x)))' = -\frac{d}{dx} \left(\int_0^{f(x)} \frac{\log(1-t)}{t} dt \right) = -\log(1-f(x)) \frac{f'(x)}{f(x)}, \tag{6.62}$$

which is helpful to recognize in the work with some integrals.

So, returning to (ii), rearranging, and using (6.62), we have that

$$\sum_{n=1}^{\infty} x^n \frac{\overline{H}_n}{n+1} = \sum_{n=1}^{\infty} \frac{1}{x} \int_0^x t^n dt\, \overline{H}_n = \frac{1}{x} \int_0^x \sum_{n=1}^{\infty} t^n \overline{H}_n dt = \frac{1}{x} \int_0^x \frac{\log(1+t)}{1-t} dt$$

$$= \frac{1}{x} \int_0^x \frac{\log((1+t)/2)}{1-t} dt + \log(2) \frac{1}{x} \int_0^x \frac{1}{1-t} dt$$

$$= \frac{1}{x} \int_0^x \log\left(1 - \frac{1-t}{2}\right) \frac{((1-t)/2)'}{(t-1)/2} dt + \log(2) \frac{1}{x} \int_0^x \frac{1}{1-t} dt = \frac{1}{x} \text{Li}_2\left(\frac{1-t}{2}\right) \Big|_{t=0}^{t=x}$$

$$- \log(2) \frac{\log(1-x)}{x} = \frac{1}{x} \text{Li}_2\left(\frac{1-x}{2}\right) - \frac{1}{x} \text{Li}_2\left(\frac{1}{2}\right) - \log(2) \frac{\log(1-x)}{x}$$

$$= \frac{1}{x} \left(\frac{1}{2} \log^2(2) - \frac{1}{2} \zeta(2) - \log(2) \log(1-x) + \text{Li}_2\left(\frac{1-x}{2}\right) \right),$$

and the solution to the point (ii) of the problem is complete. In the calculations, I used the special value of dilogarithm, $\text{Li}_2\left(\frac{1}{2}\right) = \frac{\pi^2}{12} - \frac{1}{2} \log^2(2) = \frac{1}{2} \zeta(2) - \frac{1}{2} \log^2(2)$.

Further, the generating function at the point (iii) may be readily extracted based on the result from the point (ii) by rearranging and reindexing, and then we write

$$\frac{1}{2} \log^2(2) - \frac{1}{2} \zeta(2) - \log(2) \log(1-x) + \text{Li}_2\left(\frac{1-x}{2}\right) = \sum_{n=1}^{\infty} x^{n+1} \frac{\overline{H}_n}{n+1}$$

$$= \sum_{n=1}^{\infty} x^{n+1} \frac{\overline{H}_{n+1} - (-1)^n/(n+1)}{n+1} = \sum_{n=1}^{\infty} x^n \frac{\overline{H}_n - (-1)^{n-1}/n}{n}$$

$$= \sum_{n=1}^{\infty} x^n \frac{\overline{H}_n}{n} + \sum_{n=1}^{\infty} (-1)^n \frac{x^n}{n^2} = \sum_{n=1}^{\infty} x^n \frac{\overline{H}_n}{n} + \text{Li}_2(-x),$$

whence we obtain that

$$\sum_{n=1}^{\infty} x^n \frac{\overline{H}_n}{n}$$

$$= \frac{1}{2} \log^2(2) - \frac{1}{2} \zeta(2) - \log(2) \log(1-x) - \text{Li}_2(-x) + \text{Li}_2\left(\frac{1-x}{2}\right), \quad (6.63)$$

and the solution to the point (iii) of the problem is complete.

Instead of deriving the generating function at the point (iv), I'll skip it for the moment and derive the one from the last point. So, in order to get the result we need, we start by using (1.99), Sect. 1.21, where if we replace the integration variable by t, further multiply both sides by $(-1)^{n-1}x^n/n$, then consider the summation from $n = 1$ to ∞, and rearrange, we obtain that

$$\sum_{n=1}^{\infty} x^n \frac{\overline{H}_n}{n^2}$$

$$= \log(2) \sum_{n=1}^{\infty} \frac{x^n}{n^2} - \log(2) \sum_{n=1}^{\infty} (-1)^n \frac{x^n}{n^2} - \int_0^1 \frac{\log(1+t)}{t} \sum_{n=1}^{\infty} (-1)^{n-1} \frac{(xt)^n}{n} dt$$

$$= \log(2) \operatorname{Li}_2(x) - \log(2) \operatorname{Li}_2(-x) - \int_0^1 \frac{\log(1+t)\log(1+xt)}{t} dt. \qquad (6.64)$$

For the remaining logarithmic integral in (6.64), we want to consider the fact that $\log(1+t)\log(1+xt) = \frac{1}{2}\left(\log^2(1+t) + \log^2(1+xt) - \log^2\left(\frac{1+t}{1+xt}\right)\right)$ that gives

$$\int_0^1 \frac{\log(1+t)\log(1+xt)}{t} dt$$

$$= \frac{1}{2}\int_0^1 \frac{\log^2(1+t)}{t} dt + \frac{1}{2}\int_0^1 \frac{\log^2(1+xt)}{t} dt - \frac{1}{2}\int_0^1 \frac{1}{t}\log^2\left(\frac{1+t}{1+xt}\right) dt. \qquad (6.65)$$

Now, the first integral in the right-hand side of (6.65) is already given and calculated in *(Almost) Impossible Integrals, Sums, and Series* (see [76, Chapter 1, p. 4, Chapter 3, pp. 72–73]), and we have that

$$\int_0^1 \frac{\log^2(1+t)}{t} dt = \frac{1}{4}\zeta(3). \qquad (6.66)$$

The result in (6.66) may also be extracted from (6.67) below upon setting $x = 1$.

Then, for the second integral in (6.65) we might employ again results presented in *(Almost) Impossible Integrals, Sums, and Series* (see [76, Chapter 1, p.3]). However, this time we need a different closed form and want to show that

$$\int_0^1 \frac{\log^2(1+xt)}{t} dt = \int_0^x \frac{\log^2(1+t)}{t} dt$$

$$= -\frac{1}{3}\log^3(1+x) - 2\log(1+x)\operatorname{Li}_2(-x) + 2\operatorname{Li}_3(-x) + 2\operatorname{Li}_3\left(\frac{x}{1+x}\right). \qquad (6.67)$$

Proof By letting the variable change $xt = u$, rearranging, and splitting, we get that

$$\int_0^1 \frac{\log^2(1+xt)}{t}\,dt = \int_0^x \frac{\log^2(1+u)}{u}\,du = \int_0^x \frac{\log^2(1+u)}{u(1+u)}\,du + \int_0^x \frac{\log^2(1+u)}{1+u}\,du$$

$$\left\{ \text{exploit that } \left(\operatorname{Li}_2\left(\frac{x}{1+x} \right) \right)' = \frac{\log(1+x)}{x(1+x)} \text{ to prepare an integration by parts} \right\}$$

$$= \int_0^x \left(\operatorname{Li}_2\left(\frac{u}{1+u} \right) \right)' \log(1+u)\,du + \frac{1}{3}\log^3(1+x) = \log(1+x)\operatorname{Li}_2\left(\frac{x}{1+x} \right)$$

$$+ \frac{1}{3}\log^3(1+x) + \int_0^x \frac{1}{u(1+u)}\operatorname{Li}_2\left(\frac{u}{1+u} \right)du - \int_0^x \frac{1}{u}\operatorname{Li}_2\left(\frac{u}{1+u} \right)du$$

$$\left\{ \text{for the first integral, use that } \left(\operatorname{Li}_3\left(\frac{x}{1+x} \right) \right)' = \frac{1}{x(1+x)}\operatorname{Li}_2\left(\frac{x}{1+x} \right), \right\}$$

{and for the first term and the integrand in the second resulting integral}

$$\left\{ \text{exploit the Landen's identity, } \operatorname{Li}_2(x) + \operatorname{Li}_2\left(\frac{x}{x-1} \right) = -\frac{1}{2}\log^2(1-x) \right\}$$

$$= -\frac{1}{6}\log^3(1+x) - \log(1+x)\operatorname{Li}_2(-x) + \operatorname{Li}_3\left(\frac{x}{1+x} \right)$$

$$+ \underbrace{\int_0^x \frac{\operatorname{Li}_2(-u)}{u}\,du}_{\operatorname{Li}_3(-x)} + \frac{1}{2}\int_0^x \frac{\log^2(1+u)}{u}\,du,$$

whence we extract the desired integral result,

$$\int_0^1 \frac{\log^2(1+xt)}{t}\,dt = \int_0^x \frac{\log^2(1+u)}{u}\,du$$

$$= -\frac{1}{3}\log^3(1+x) - 2\log(1+x)\operatorname{Li}_2(-x) + 2\operatorname{Li}_3(-x) + 2\operatorname{Li}_3\left(\frac{x}{1+x} \right),$$

which brings an end to the solution of the auxiliary result in (6.67). ∎

Next, for the last integral in (6.65), we let the variable change $\dfrac{1+t}{1+xt} = u$ that gives

$$\int_0^1 \frac{1}{t}\log^2\left(\frac{1+t}{1+xt} \right)dt$$

$$= \int_{2/(1+x)}^{1} \frac{(1-x)\log^2(u)}{(1-u)(1-xu)} du = \int_{2/(1+x)}^{1} \frac{\log^2(u)}{1-u} du - \int_{2/(1+x)}^{1} \frac{x\log^2(u)}{1-xu} du$$

{in the first integral make the variable change $1/u = t$ and then split the integral}

$$= \int_{(1+x)/2}^{1} \frac{\log^2(u)}{u} du + \int_{(1+x)/2}^{1} \frac{\log^2(u)}{1-u} du - \int_{2/(1+x)}^{1} \frac{x\log^2(u)}{1-xu} du$$

$$= -\frac{2}{3}\log^3(2) + 2\zeta(3) + 2\log^2(2)\log(1+x) - 2\log(2)\log^2(1+x) + \frac{2}{3}\log^3(1+x)$$

$$- 2(\log(2) - \log(1+x))\left(\operatorname{Li}_2\left(\frac{1+x}{2}\right) + \operatorname{Li}_2\left(\frac{2x}{1+x}\right)\right) - 2\operatorname{Li}_3(x)$$

$$- 2\operatorname{Li}_3\left(\frac{1+x}{2}\right) + 2\operatorname{Li}_3\left(\frac{2x}{1+x}\right), \tag{6.68}$$

where in the calculations I used that $\displaystyle\int \frac{x\log^2(u)}{1-xu} du = -\log^2(u)\log(1-xu) - 2\log(u)\operatorname{Li}_2(xu) + 2\operatorname{Li}_3(xu) + C$, readily obtained with simple integrations by parts.

To get a simpler form of (6.68), we want to prove and use the identity

$$\operatorname{Li}_2\left(\frac{1+x}{2}\right) + \operatorname{Li}_2\left(\frac{2x}{1+x}\right)$$

$$= \frac{1}{2}\zeta(2) - \frac{1}{2}\log^2(2) + \log(2)\log(1+x) - \frac{1}{2}\log^2(1+x) - \operatorname{Li}_2(-x) + \operatorname{Li}_2(x). \tag{6.69}$$

Proof The solution is straightforward, and it is based on differentiating and integrating back the left-hand side of (6.69). So, we write that

$$\operatorname{Li}_2\left(\frac{1+x}{2}\right) + \operatorname{Li}_2\left(\frac{2x}{1+x}\right) = \int_0^x \left(\operatorname{Li}_2\left(\frac{1+t}{2}\right) + \operatorname{Li}_2\left(\frac{2t}{1+t}\right)\right)' dt + \operatorname{Li}_2\left(\frac{1}{2}\right)$$

$$= \log(2)\int_0^x \frac{1}{1+t} dt - \int_0^x \frac{\log(1+t)}{1+t} dt - \int_0^x \frac{\log(1-t)}{t} dt + \int_0^x \frac{\log(1+t)}{t} dt$$

$$+ \frac{1}{2}\zeta(2) - \frac{1}{2}\log^2(2)$$

$$= \frac{1}{2}\zeta(2) - \frac{1}{2}\log^2(2) + \log(2)\log(1+x) - \frac{1}{2}\log^2(1+x) - \operatorname{Li}_2(-x) + \operatorname{Li}_2(x),$$

which finalizes the solution to the auxiliary result in (6.69). ∎

Upon plugging the result from (6.69) in (6.68), we arrive at

$$\int_0^1 \frac{1}{t} \log^2\left(\frac{1+t}{1+xt}\right) dt$$

$$= \frac{1}{3}\log^3(2) - \log(2)\zeta(2) + 2\zeta(3) + (\zeta(2) - \log^2(2))\log(1+x) + \log(2)\log^2(1+x)$$

$$- \frac{1}{3}\log^3(1+x) - 2\log(2)\operatorname{Li}_2(x) + 2\log(2)\operatorname{Li}_2(-x) + 2\log(1+x)\operatorname{Li}_2(x)$$

$$- 2\log(1+x)\operatorname{Li}_2(-x) - 2\operatorname{Li}_3(x) - 2\operatorname{Li}_3\left(\frac{1+x}{2}\right) + 2\operatorname{Li}_3\left(\frac{2x}{1+x}\right). \quad (6.70)$$

Collecting the results from (6.66), (6.67), and (6.70) in (6.65), we get that

$$\int_0^1 \frac{\log(1+t)\log(1+xt)}{t} dt$$

$$= \frac{1}{2}\log(2)\zeta(2) - \frac{7}{8}\zeta(3) - \frac{1}{6}\log^3(2) + \frac{1}{2}(\log^2(2) - \zeta(2))\log(1+x)$$

$$- \frac{1}{2}\log(2)\log^2(1+x) + \log(2)\operatorname{Li}_2(x) - \log(2)\operatorname{Li}_2(-x) - \log(1+x)\operatorname{Li}_2(x)$$

$$+ \operatorname{Li}_3(x) + \operatorname{Li}_3(-x) + \operatorname{Li}_3\left(\frac{1+x}{2}\right) + \operatorname{Li}_3\left(\frac{x}{1+x}\right) - \operatorname{Li}_3\left(\frac{2x}{1+x}\right). \quad (6.71)$$

At last, if we plug the result from (6.71) in (6.64), we obtain that

$$\sum_{n=1}^{\infty} x^n \frac{\overline{H}_n}{n^2}$$

$$= \frac{1}{6}\log^3(2) - \frac{1}{2}\log(2)\zeta(2) + \frac{7}{8}\zeta(3) - \frac{1}{2}(\log^2(2) - \zeta(2))\log(1+x)$$

$$+ \frac{1}{2}\log(2)\log^2(1+x) + \log(1+x)\operatorname{Li}_2(x) - \operatorname{Li}_3(x) - \operatorname{Li}_3(-x)$$

$$- \operatorname{Li}_3\left(\frac{1+x}{2}\right) - \operatorname{Li}_3\left(\frac{x}{1+x}\right) + \operatorname{Li}_3\left(\frac{2x}{1+x}\right),$$

and the solution to the point (v) of the problem is complete.

Returning to the point (iv) and exploiting the result previously derived, where we consider to rearrange and reindex, we write

$$\sum_{n=1}^{\infty} x^{n+1} \frac{\overline{H}_n}{(n+1)^2} = \sum_{n=1}^{\infty} x^{n+1} \frac{\overline{H}_{n+1} - (-1)^n/(n+1)}{(n+1)^2} = \sum_{n=1}^{\infty} x^n \frac{\overline{H}_n}{n^2} + \underbrace{\sum_{n=1}^{\infty} (-1)^n \frac{x^n}{n^3}}_{\text{Li}_3(-x)},$$

whence we obtain that

$$\sum_{n=1}^{\infty} x^n \frac{\overline{H}_n}{(n+1)^2}$$

$$= \frac{1}{x}\left(\frac{1}{6} \log^3(2) - \frac{1}{2} \log(2)\zeta(2) + \frac{7}{8}\zeta(3) - \frac{1}{2}(\log^2(2) - \zeta(2))\log(1+x) \right.$$

$$+ \frac{1}{2}\log(2)\log^2(1+x) + \log(1+x)\,\text{Li}_2(x) - \text{Li}_3(x)$$

$$\left. - \text{Li}_3\left(\frac{1+x}{2}\right) - \text{Li}_3\left(\frac{x}{1+x}\right) + \text{Li}_3\left(\frac{2x}{1+x}\right) \right),$$

and the solution to the point (iv) of the problem is complete.

It is clear that at some point, the calculations of either $\displaystyle\sum_{n=1}^{\infty} x^n \frac{\overline{H}_n}{n^2}$ or

$\displaystyle\sum_{n=1}^{\infty} x^n \frac{\overline{H}_n}{(n+1)^2}$ become (somewhat) challenging especially if we want to perform
them elegantly, and I mainly refer to the evaluation of the integral in (6.71).

How would we go differently? I would ask the curious reader looking forward to
such a challenge. A different reduction of such series to manageable integrals, and
their evaluations, may be found in [67, pp. 76–79].

6.11   Good-to-Know Generating Functions: The Sixth Part

Solution If you had a chance to take a look at the generating functions presented in
Sect. 4.7, you noticed that at the first point, we have $\displaystyle\sum_{n=1}^{\infty} x^n H_n^2 = \frac{1}{1-x}(\log^2(1-$
$x) + \text{Li}_2(x))$. Now, if we think of the skew-harmonic numbers, it might be natural to
ask ourselves how the generating function of the squared skew-harmonic numbers
looks like. An answer may already be found in the same paper mentioned in the
beginning part of the previous section (see [8]), which I'll also briefly give below.

Essentially, the author in [8] exploits the powerful identity in (1.120), the case $p = 1$, Sect. 1.23, which in his paper is proved by induction, that is,

$$\sum_{k=1}^{n}(-1)^{k-1}\frac{\overline{H}_k}{k} = \frac{1}{2}\left(\overline{H}_n^2 + H_n^{(2)}\right).$$

So, if we multiply both sides of the mentioned identity by x^n and then sum from $n = 1$ to ∞ and rearrange, we arrive at

$$\sum_{n=1}^{\infty}x^n\overline{H}_n^2 = 2\sum_{n=1}^{\infty}x^n\left(\sum_{k=1}^{n}(-1)^{k-1}\frac{\overline{H}_k}{k}\right) - \sum_{n=1}^{\infty}x^n H_n^{(2)}$$

{the second series in the right-hand side is the case $m = 2$ of (4.32), Sect. 4.6}

$$= 2\sum_{n=1}^{\infty}x^n\left(\sum_{k=1}^{n}(-1)^{k-1}\frac{\overline{H}_k}{k}\right) - \frac{\text{Li}_2(x)}{1-x}. \tag{6.72}$$

For the remaining series in (6.72), we first observe that based on (4.64), Sect. 4.10, where we replace x by $-x$, we get

$$\sum_{n=1}^{\infty}x^n(-1)^{n-1}\frac{\overline{H}_n}{n}$$

$$= \frac{1}{2}\zeta(2) - \frac{1}{2}\log^2(2) + \log(2)\log(1+x) + \text{Li}_2(x) - \text{Li}_2\left(\frac{1+x}{2}\right). \tag{6.73}$$

Now, if we consider the Cauchy product given in (6.15), Sect. 6.5, for the series in (6.73) and $\sum_{n=1}^{\infty}x^{n-1} = \frac{1}{1-x}$, we have that

$$\sum_{n=1}^{\infty}x^n\left(\sum_{k=1}^{n}(-1)^{k-1}\frac{\overline{H}_k}{k}\right) = \left(\sum_{n=1}^{\infty}x^n(-1)^{n-1}\frac{\overline{H}_n}{n}\right)\left(\sum_{n=1}^{\infty}x^{n-1}\right)$$

$$= \frac{1}{1-x}\left(\frac{1}{2}\zeta(2) - \frac{1}{2}\log^2(2) + \log(2)\log(1+x) + \text{Li}_2(x) - \text{Li}_2\left(\frac{1+x}{2}\right)\right). \tag{6.74}$$

At last, combining (6.72) and (6.74), we conclude that

$$\sum_{n=1}^{\infty}x^n\overline{H}_n^2$$

$$= \frac{1}{1-x}\left(\zeta(2) - \log^2(2) + 2\log(2)\log(1+x) + \text{Li}_2(x) - 2\text{Li}_2\left(\frac{1+x}{2}\right)\right),$$

and the first solution to the point (i) of the problem is complete. Good to observe that in (6.74), we could have also considered the change of summation order.

For a second solution, we may simply exploit a strategy with telescoping sums as in the case of the third solution given to the first point in the previous section. Then, if we consider the notation $\sum_{n=1}^{\infty} x^n \overline{H}_n^2 = G(x)$ and then multiply both sides by $1 - x$, we get that

$$\sum_{n=1}^{\infty} x^n (1 - x)\overline{H}_n^2 = (1 - x)G(x). \tag{6.75}$$

So, starting from the left-hand side of (6.75), we write that

$$\sum_{n=1}^{\infty} x^n (1 - x)\overline{H}_n^2 = \sum_{n=1}^{\infty} x^n \overline{H}_n^2 - \sum_{n=1}^{\infty} x^{n+1} \left(\overline{H}_{n+1} + \frac{(-1)^{n-1}}{n+1} \right)^2$$

$$\underset{\text{the second series}}{\overset{\text{reindex}}{=\!=\!=}} \sum_{n=1}^{\infty} x^n \overline{H}_n^2 - \sum_{n=1}^{\infty} x^n \left(\overline{H}_n - (-1)^{n-1}\frac{1}{n} \right)^2 = -2\sum_{n=1}^{\infty} (-x)^n \frac{\overline{H}_n}{n} - \underbrace{\sum_{n=1}^{\infty} \frac{x^n}{n^2}}_{\text{Li}_2(x)}$$

{make use of the generating function in (4.64), Sect. 4.10}

$$= \zeta(2) - \log^2(2) + 2\log(2)\log(1 + x) + \text{Li}_2(x) - 2\,\text{Li}_2\left(\frac{1+x}{2} \right),$$

which gives the generating function in (6.75), and the second solution to the point (i) of the problem is complete.

For the point (ii) of the problem, we'll exploit the result from the point (i), and then we write

$$\sum_{n=1}^{\infty} x^n \frac{\overline{H}_n^2}{n+1} = \frac{1}{x} \sum_{n=1}^{\infty} \int_0^x t^n dt\, \overline{H}_n^2 = \frac{1}{x} \int_0^x \sum_{n=1}^{\infty} t^n \overline{H}_n^2 dt$$

$$= (\zeta(2) - \log^2(2))\frac{1}{x} \underbrace{\int_0^x \frac{1}{1-t}dt}_{-\log(1-x)} + 2\log(2)\frac{1}{x}\int_0^x \frac{\log(1+t)}{1-t}dt + \frac{1}{x}\int_0^x \frac{\text{Li}_2(t)}{1-t}dt$$

$$- 2\frac{1}{x}\int_0^x \frac{1}{1-t}\text{Li}_2\left(\frac{1+t}{2} \right)dt. \tag{6.76}$$

Regarding the second integral in (6.76), we exploit the result in (6.62), the previous section, and then we get

$$\int_0^x \frac{\log(1+t)}{1-t}dt = -\int_0^x \log\left(1 - \frac{1-t}{2}\right)\frac{((1-t)/2)'}{(1-t)/2}dt + \log(2)\int_0^x \frac{1}{1-t}dt$$

$$= \text{Li}_2\left(\frac{1-x}{2}\right) - \text{Li}_2\left(\frac{1}{2}\right) - \log(2)\log(1-x)$$

$$= \frac{1}{2}\left(\log^2(2) - \zeta(2)\right) - \log(2)\log(1-x) + \text{Li}_2\left(\frac{1-x}{2}\right). \tag{6.77}$$

Then, for the third integral in (6.76), we integrate by parts, and then we have

$$\int_0^x \frac{\text{Li}_2(t)}{1-t}dt = -\int_0^x (\log(1-t))' \text{Li}_2(t)dt = -\log(1-x)\text{Li}_2(x) - \int_0^x \frac{\log^2(1-t)}{t}dt$$

$$= -\log(x)\log^2(1-x) - \log(1-x)\text{Li}_2(x) - 2\log(1-x)\text{Li}_2(1-x)$$

$$+ 2\text{Li}_3(1-x) - 2\zeta(3)$$

$$= -\zeta(2)\log(1-x) - \log(1-x)\text{Li}_2(1-x) + 2\text{Li}_3(1-x) - 2\zeta(3), \tag{6.78}$$

where in the calculations, I used the result in (6.31), Sect. 6.6, the first equality, and the dilogarithm function reflection formula (see [21, Chapter 1, p.5], [70, Chapter 2, p.107]), $\text{Li}_2(x) + \text{Li}_2(1-x) = \pi^2/6 - \log(x)\log(1-x)$.

For the last integral in (6.76), we use the dilogarithm function reflection formula, previously stated, then rearrange, and expand the integral that together lead to

$$\int_0^x \frac{1}{1-t}\text{Li}_2\left(\frac{1+t}{2}\right)dt$$

$$= \zeta(2)\int_0^x \frac{1}{1-t}dt - \int_0^x \frac{1}{1-t}\text{Li}_2\left(\frac{1-t}{2}\right)dt + \log(2)\int_0^x \frac{1}{1-t}\log\left(\frac{1+t}{2}\right)dt$$

$$- \int_0^x \frac{1}{1-t}\log\left(\frac{1+t}{2}\right)\log(1-t)dt$$

$$= \frac{1}{6}\log^3(2) + \frac{1}{2}\log(2)\zeta(2) - \frac{7}{4}\zeta(3) - \zeta(2)\log(1-x) - \log\left(\frac{1-x}{2}\right)\text{Li}_2\left(\frac{1-x}{2}\right)$$

$$+ 2\text{Li}_3\left(\frac{1-x}{2}\right), \tag{6.79}$$

where during the calculations I used the simple facts that $\int_0^x \frac{1}{1-t} dt =$

$-\log(1-x)$, then $\int_0^x \frac{1}{1-t} \mathrm{Li}_2\left(\frac{1-t}{2}\right) dt = -\mathrm{Li}_3\left(\frac{1-t}{2}\right)\Big|_{t=0}^{t=x} = \mathrm{Li}_3\left(\frac{1}{2}\right) -$

$\mathrm{Li}_3\left(\frac{1-x}{2}\right)$, next $\int_0^x \frac{1}{1-t} \log\left(\frac{1+t}{2}\right) dt = \mathrm{Li}_2\left(\frac{1-t}{2}\right)\Big|_{t=0}^{t=x} = \mathrm{Li}_2\left(\frac{1-x}{2}\right) -$

$\mathrm{Li}_2\left(\frac{1}{2}\right)$, further $\int_0^x \frac{1}{1-t} \log\left(\frac{1+t}{2}\right) \log(1-t) dt = \int_0^x \left(\mathrm{Li}_2\left(\frac{1-t}{2}\right)\right)'$

$\log(1-t) dt = \log(1-x) \mathrm{Li}_2\left(\frac{1-x}{2}\right) + \int_0^x \frac{1}{1-t} \mathrm{Li}_2\left(\frac{1-t}{2}\right) dt = \log(1-x)$

$\mathrm{Li}_2\left(\frac{1-x}{2}\right) - \mathrm{Li}_3\left(\frac{1-x}{2}\right) + \mathrm{Li}_3\left(\frac{1}{2}\right)$. I also considered the special values of

the dilogarithm and trilogarithm, that is, $\mathrm{Li}_2\left(\frac{1}{2}\right) = \frac{1}{2}\left(\frac{\pi^2}{6} - \log^2(2)\right)$ and

$\mathrm{Li}_3\left(\frac{1}{2}\right) = \frac{7}{8}\zeta(3) + \frac{1}{6}\log^3(2) - \frac{1}{2}\log(2)\zeta(2)$ (e.g., exploiting the Landen's
trilogarithmic identity in (4.39), Sect. 4.6).

Finally, collecting the results from (6.77), (6.78), and (6.79) in (6.76), we
conclude that

$$\sum_{n=1}^{\infty} x^n \frac{\overline{H}_n^2}{n+1}$$

$$= \frac{1}{x}\left(\frac{2}{3}\log^3(2) - 2\log(2)\zeta(2) + \frac{3}{2}\zeta(3) - \log^2(2)\log(1-x) - \log(1-x)\mathrm{Li}_2(1-x)\right.$$

$$\left. + 2\log(1-x)\mathrm{Li}_2\left(\frac{1-x}{2}\right) + 2\mathrm{Li}_3(1-x) - 4\mathrm{Li}_3\left(\frac{1-x}{2}\right)\right),$$

and the solution to the point (ii) of the problem is complete.

The generating function at the point (iii) will be extracted by exploiting the one
given at the previous point, and then we have

$$\sum_{n=1}^{\infty} x^n \frac{\overline{H}_n^2}{n} = \sum_{n=1}^{\infty} x^n \frac{(\overline{H}_n - (-1)^{n-1}/n)^2}{n} + 2\sum_{n=1}^{\infty} x^n (-1)^{n-1}\frac{\overline{H}_n}{n^2} - \underbrace{\sum_{n=1}^{\infty} \frac{x^n}{n^3}}_{\mathrm{Li}_3(x)}$$

$$\underset{\text{the series}}{\overset{\text{reindex}}{=}} \sum_{n=1}^{\infty} x^{n+1}\frac{(\overline{H}_{n+1} - (-1)^n/(n+1))^2}{n+1} - 2\sum_{n=1}^{\infty}(-x)^n\frac{\overline{H}_n}{n^2} - \mathrm{Li}_3(x)$$

$$= \sum_{n=1}^{\infty} x^{n+1} \frac{\overline{H}_n^2}{n+1} - 2 \sum_{n=1}^{\infty} (-x)^n \frac{\overline{H}_n}{n^2} - \operatorname{Li}_3(x)$$

$$= \frac{1}{3} \log^3(2) - \log(2)\zeta(2) + \frac{7}{4}\zeta(3) - \log(2)\log^2(1-x) + \frac{1}{3}\log^3(1-x)$$

$$+ \log(1-x)\operatorname{Li}_2(x) - 2\log(1-x)\operatorname{Li}_2(-x) + 2\log(1-x)\operatorname{Li}_2\left(\frac{1-x}{2}\right)$$

$$- \operatorname{Li}_3(x) + 2\operatorname{Li}_3(-x) - 2\operatorname{Li}_3\left(\frac{1-x}{2}\right) - 2\operatorname{Li}_3\left(\frac{2x}{x-1}\right),$$

where in the calculations, I also used the generating function in (4.66), Sect. 4.10; next the dilogarithm function reflection formula, stated above during the calculations in (6.78), for expressing $\operatorname{Li}_2(1-x)$; and then the Landen's trilogarithm function identity in (6.33), Sect. 6.6, for expressing $\operatorname{Li}_3\left(\frac{x}{x-1}\right)$, and the solution to the point (iii) of the problem is complete.

To get the result from the point (iv), we need the result from the point (4.62), the case $m = 2$, Sect. 4.10, that is, $\sum_{n=1}^{\infty} x^n \overline{H}_n^{(2)} = -\frac{\operatorname{Li}_2(-x)}{1-x}$, and then we write

$$\sum_{n=1}^{\infty} x^n \frac{\overline{H}_n^{(2)}}{n+1} = \frac{1}{x} \sum_{n=1}^{\infty} \int_0^x t^n dt\, \overline{H}_n^{(2)} = \frac{1}{x} \int_0^x \sum_{n=1}^{\infty} t^n \overline{H}_n^{(2)} dt = -\frac{1}{x} \int_0^x \frac{\operatorname{Li}_2(-t)}{1-t} dt$$

$$= \frac{1}{x} \int_0^x (\log(1-t))' \operatorname{Li}_2(-t) dt \overset{\substack{\text{integrate by} \\ \text{parts}}}{=} \frac{\log(1-x)\operatorname{Li}_2(-x)}{x}$$

$$+ \frac{1}{x} \int_0^x \frac{\log(1-t)\log(1+t)}{t} dt. \tag{6.80}$$

For the remaining integral in (6.80), it is convenient to exploit the algebraic identity, $\log(1-x)\log(1+x) = \frac{1}{2}\log^2(1-x^2) - \frac{1}{2}\log^2(1-x) - \frac{1}{2}\log^2(1+x)$, and then we have that

$$\int_0^x \frac{\log(1-t)\log(1+t)}{t} dt$$

$$= \frac{1}{2} \int_0^x \frac{\log^2(1-t^2)}{t} dt - \frac{1}{2} \int_0^x \frac{\log^2(1-t)}{t} dt - \frac{1}{2} \int_0^x \frac{\log^2(1+t)}{t} dt. \tag{6.81}$$

Now, for the first remaining integral in (6.81), we let the variable change $t^2 = u$, and then we write

$$\int_0^x \frac{\log^2(1-t^2)}{t} dt = \frac{1}{2} \int_0^{x^2} \frac{\log^2(1-u)}{u} du$$

$$= \zeta(2)\log(1-x^2) - \frac{1}{2}\log(x^2)\log^2(1-x^2) - 2\log(1-x)\operatorname{Li}_2(x) - 2\log(1+x)\operatorname{Li}_2(x)$$

$$- 2\log(1-x)\operatorname{Li}_2(-x) - 2\log(1+x)\operatorname{Li}_2(-x) - \operatorname{Li}_3(1-x^2) + \zeta(3), \quad (6.82)$$

where in the calculations, I combined the result in (6.31), the second equality, Sect. 6.6, and the dilogarithm function identity, $\operatorname{Li}_2(x) + \operatorname{Li}_2(-x) = \frac{1}{2}\operatorname{Li}_2(x^2)$. It is worth observing that we needed to be careful with the argument of the first logarithm found in the second term above since x may also take negative values (or we may count complex values).

By collecting the results from (6.82), (6.31), the second equality, Sect. 6.6, and (6.67), Sect. 6.10, in (6.81), we arrive at

$$\int_0^x \frac{\log(1-t)\log(1+t)}{t} dt$$

$$= \frac{1}{6}\log^3(1+x) - \zeta(2)\log(1-x) + \frac{1}{2}\zeta(2)\log(1-x^2) + \frac{1}{2}\log(x)\log^2(1-x)$$

$$- \frac{1}{4}\log(x^2)\log^2(1-x^2) - \log(1+x)\operatorname{Li}_2(x) - \log(1-x)\operatorname{Li}_2(-x) - \operatorname{Li}_3(-x)$$

$$+ \operatorname{Li}_3(1-x) - \frac{1}{2}\operatorname{Li}_3(1-x^2) - \operatorname{Li}_3\left(\frac{x}{1+x}\right) - \frac{1}{2}\zeta(3). \quad (6.83)$$

At last, by plugging the result from (6.83) in (6.80), we obtain that

$$\sum_{n=1}^{\infty} x^n \frac{\overline{H}_n^{(2)}}{n+1}$$

$$= \frac{1}{x}\left(\frac{1}{6}\log^3(1+x) - \zeta(2)\log(1-x) + \frac{1}{2}\zeta(2)\log(1-x^2) + \frac{1}{2}\log(x)\log^2(1-x)\right.$$

$$- \frac{1}{4}\log(x^2)\log^2(1-x^2) - \log(1+x)\operatorname{Li}_2(x) - \operatorname{Li}_3(-x) + \operatorname{Li}_3(1-x) - \frac{1}{2}\operatorname{Li}_3(1-x^2)$$

$$\left. - \operatorname{Li}_3\left(\frac{x}{1+x}\right) - \frac{1}{2}\zeta(3)\right),$$

and the solution to the point (iv) of the problem is complete.

Regarding the last point of the problem, I'll exploit the result from the previous point, and then we write

$$\sum_{n=1}^{\infty} x^n \frac{\overline{H}_n^{(2)}}{n} = \sum_{n=1}^{\infty} x^n \frac{\overline{H}_n^{(2)} - (-1)^{n-1}/n^2}{n} - \underbrace{\sum_{n=1}^{\infty}(-1)^n \frac{x^n}{n^3}}_{\text{Li}_3(-x)}$$

$$\stackrel{\substack{\text{reindex} \\ \text{the series}}}{=} \sum_{n=1}^{\infty} x^{n+1} \frac{\overline{H}_{n+1}^{(2)} - (-1)^n/(n+1)^2}{n+1} - \text{Li}_3(-x) = \sum_{n=1}^{\infty} x^{n+1} \frac{\overline{H}_n^{(2)}}{n+1} - \text{Li}_3(-x)$$

$$= \frac{1}{6}\log^3(1+x) - \zeta(2)\log(1-x) + \frac{1}{2}\zeta(2)\log(1-x^2) + \frac{1}{2}\log(x)\log^2(1-x)$$

$$-\frac{1}{4}\log(x^2)\log^2(1-x^2) - \log(1+x)\,\text{Li}_2(x) - 2\,\text{Li}_3(-x) + \text{Li}_3(1-x) - \frac{1}{2}\text{Li}_3(1-x^2)$$

$$-\text{Li}_3\left(\frac{x}{1+x}\right) - \frac{1}{2}\zeta(3),$$

and the solution to the point (v) of the problem is complete.

Having completed this section, we are preparing to pass to the last section with generating functions (involving skew-harmonic numbers, again)!

6.12 Good-to-Know Generating Functions: The Seventh Part

Solution If you've already been exposed to the previous two sections with solutions, or even better, to the previous six sections involving generating functions, it won't be difficult to figure out how to attack and prove the results in the present section.

These results are possibly less known or not known in the mathematical literature, but happily, they do not require more sophisticated approaches for their extractions than the previous generating functions.

For a first solution to the result at the point (i), I'll want to creatively exploit the result in (6.49), the case $p = 1$, Sect. 6.9, $\sum_{n=1}^{k-1} x^n H_n = \frac{1}{1-x}\left(-x^k H_k + \sum_{n=1}^{k} \frac{x^n}{n}\right)$.

Since we have that $\overline{H}_n = \log(2) - \sum_{k=n+1}^{\infty}(-1)^{k-1}\frac{1}{k}$ because $\lim_{n\to\infty} \overline{H}_n = \log(2)$, where a simple proof may be found in [76, Chapter 6, pp. 337–338], we write

$$\sum_{n=1}^{\infty} x^n H_n \overline{H}_n = \sum_{n=1}^{\infty} x^n H_n \left(\log(2) - \sum_{k=n+1}^{\infty} (-1)^{k-1} \frac{1}{k} \right)$$

$$= \log(2) \sum_{n=1}^{\infty} x^n H_n - \sum_{n=1}^{\infty} x^n H_n \sum_{k=n+1}^{\infty} (-1)^{k-1} \frac{1}{k}$$

{for the first series use the result in (4.32), the case $m = 1$, Sect. 4.6,}

{and regarding the second series, reverse the order of summation}

$$= -\log(2) \frac{\log(1-x)}{1-x} - \sum_{k=1}^{\infty} (-1)^{k-1} \frac{1}{k} \sum_{n=1}^{k-1} x^n H_n$$

$$= -\log(2) \frac{\log(1-x)}{1-x} - \frac{1}{1-x} \sum_{k=1}^{\infty} (-x)^k \frac{H_k}{k} - \frac{1}{1-x} \sum_{k=1}^{\infty} (-1)^{k-1} \frac{1}{k} \sum_{n=1}^{k} \frac{x^n}{n}.$$

(6.84)

For the last series in (6.84), we want to reverse the order of summation, and then we write

$$\sum_{k=1}^{\infty} (-1)^{k-1} \frac{1}{k} \sum_{n=1}^{k} \frac{x^n}{n} = \sum_{n=1}^{\infty} \frac{x^n}{n} \sum_{k=n}^{\infty} (-1)^{k-1} \frac{1}{k} = \sum_{n=1}^{\infty} \frac{x^n}{n} \left(\log(2) - \overline{H}_n + (-1)^{n-1} \frac{1}{n} \right)$$

$$= \log(2) \underbrace{\sum_{n=1}^{\infty} \frac{x^n}{n}}_{-\log(1-x)} - \underbrace{\sum_{n=1}^{\infty} \frac{(-x)^n}{n^2}}_{\text{Li}_2(-x)} - \sum_{n=1}^{\infty} x^n \frac{\overline{H}_n}{n} = \frac{1}{2} (\zeta(2) - \log^2(2)) - \text{Li}_2 \left(\frac{1-x}{2} \right),$$

(6.85)

where the third series is given in (4.64), Sect. 4.10.

Since the first series in (6.84) is immediately extracted based on (4.34), Sect. 4.6, by replacing x by $-x$, that is, $\sum_{k=1}^{\infty} (-x)^k \frac{H_k}{k} = \frac{1}{2} \log^2(1+x) + \text{Li}_2(-x)$, then by collecting this result and the one from (6.85) in (6.84), we conclude that

$$\sum_{n=1}^{\infty} x^n H_n \overline{H}_n$$

$$= \frac{1}{1-x} \left(\frac{1}{2} \log^2(2) - \frac{1}{2} \zeta(2) - \log(2) \log(1-x) - \frac{1}{2} \log^2(1+x) \right.$$

$$\left. - \text{Li}_2(-x) + \text{Li}_2 \left(\frac{1-x}{2} \right) \right),$$

and the first solution to the point (i) is finalized.

For a second solution, we may exploit the strategy involving telescoping sums as already met in the previous sections (e.g., see the second solution to the generating function given at the first point of the previous section). Thus, upon considering the notation $\sum_{n=1}^{\infty} x^n H_n \overline{H}_n = G(x)$ and multiplying both sides by $1 - x$, we obtain that

$$\sum_{n=1}^{\infty} (1 - x)x^n H_n \overline{H}_n = (1 - x)G(x). \tag{6.86}$$

Now, we focus on the left-hand side of (6.86), and then we write

$$\sum_{n=1}^{\infty} (1 - x)x^n H_n \overline{H}_n$$

$$= \sum_{n=1}^{\infty} x^n H_n \overline{H}_n - \sum_{n=1}^{\infty} x^{n+1} \left(H_{n+1} - \frac{1}{n+1} \right) \left(\overline{H}_{n+1} - (-1)^n \frac{1}{n+1} \right)$$

$$\overset{\substack{\text{reindex} \\ \text{the second series}}}{=} \sum_{n=1}^{\infty} x^n H_n \overline{H}_n - \sum_{n=1}^{\infty} x^n \left(H_n - \frac{1}{n} \right) \left(\overline{H}_n - (-1)^{n-1} \frac{1}{n} \right)$$

$$= \sum_{n=1}^{\infty} \frac{(-x)^n}{n^2} - \sum_{n=1}^{\infty} (-x)^n \frac{H_n}{n} + \sum_{n=1}^{\infty} x^n \frac{\overline{H}_n}{n}$$

$$= \frac{1}{2}\log^2(2) - \frac{1}{2}\zeta(2) - \log(2)\log(1-x) - \frac{1}{2}\log^2(1+x) - \text{Li}_2(-x) + \text{Li}_2\left(\frac{1-x}{2} \right),$$

which immediately leads to the generating function in (6.86), and the second solution to the point (i) is finalized. Observe that both $\sum_{n=1}^{\infty} (-x)^n \frac{H_n}{n}$ and $\sum_{n=1}^{\infty} x^n \frac{\overline{H}_n}{n}$ already appeared in the previous solution.

For the point (ii) of the problem, it is natural to think of using the result from the previous point, and then we have

$$\sum_{n=1}^{\infty} x^n \frac{H_n \overline{H}_n}{n+1} = \frac{1}{x} \sum_{n=1}^{\infty} \int_0^x t^n \, dt \, H_n \overline{H}_n = \frac{1}{x} \int_0^x \sum_{n=1}^{\infty} t^n H_n \overline{H}_n \, dt$$

$$= \left(\frac{1}{2}\log^2(2) - \frac{1}{2}\zeta(2)\right)\frac{1}{x}\underbrace{\int_0^x \frac{1}{1-t}dt}_{-\log(1-x)} - \log(2)\frac{1}{x}\underbrace{\int_0^x \frac{\log(1-t)}{1-t}dt}_{-1/2\log^2(1-x)}$$

$$-\frac{1}{2x}\int_0^x \frac{\log^2(1+t)}{1-t}dt - \frac{1}{x}\int_0^x \frac{\mathrm{Li}_2(-t)}{1-t}dt + \frac{1}{x}\underbrace{\int_0^x \frac{1}{1-t}\mathrm{Li}_2\left(\frac{1-t}{2}\right)dt}_{\mathrm{Li}_3(1/2)-\mathrm{Li}_3((1-x)/2)}.$$

$$(6.87)$$

From the solution to the point (ii) given in Sect. 6.10, we know that $\int_0^x \frac{\log((1+t)/2)}{1-t}$

$dt = \frac{1}{2}(\log^2(2) - \zeta(2)) + \mathrm{Li}_2\left(\frac{1-x}{2}\right)$, which we use in the subsequent operations, and then considering the third integral in (6.87) and rearranging, we write

$$\int_0^x \frac{\log^2(1+t)}{1-t}dt = \int_0^x \frac{\log((1+t)/2)\log(1+t)}{1-t}dt + \log(2)\int_0^x \frac{\log((1+t)/2)}{1-t}dt$$

$$+ \log^2(2)\int_0^x \frac{1}{1-t}dt$$

$$\left\{\text{for the first resulting integral use } f' = \frac{d}{dt}\int_0^t \frac{\log((1+u)/2)}{1-u}du\right\}$$

$$\{\text{and } g = \log(1+t), \text{ and afterwards apply integration by parts}\}$$

$$= \frac{1}{2}\log(2)(\log^2(2) - \zeta(2)) - \log^2(2)\log(1-x) + \log(2)\mathrm{Li}_2\left(\frac{1-x}{2}\right)$$

$$+ \log(1+x)\mathrm{Li}_2\left(\frac{1-x}{2}\right) - \int_0^x \frac{1}{1+t}\mathrm{Li}_2\left(\frac{1-t}{2}\right)dt$$

$$= \frac{2}{3}\log^3(2) - \frac{7}{4}\zeta(3) - \log^2(2)\log(1-x) - \zeta(2)\log(1+x) + \log(2)\mathrm{Li}_2\left(\frac{1-x}{2}\right)$$

$$+ \log(2)\mathrm{Li}_2\left(\frac{1+x}{2}\right) + \log(1+x)\mathrm{Li}_2\left(\frac{1-x}{2}\right) - \log(1+x)\mathrm{Li}_2\left(\frac{1+x}{2}\right)$$

$$+ 2\mathrm{Li}_3\left(\frac{1+x}{2}\right), \qquad\qquad (6.88)$$

where in the calculations, I also exploited the integral result in (6.79), Sect. 6.11.

Since we easily observe based on (4.62), the case $m = 2$, Sect. 4.10, that
$-\int_0^x \frac{\mathrm{Li}_2(-t)}{1-t}dt = \sum_{n=1}^{\infty} x^{n+1} \frac{\overline{H}_n^{(2)}}{n+1}$, then if we combine this result with the one
in (4.70), Sect. 4.11, and then plug all in (6.87), together with the result in (6.88), we conclude that

$$\sum_{n=1}^{\infty} x^n \frac{H_n \overline{H}_n}{n+1}$$

$$= \frac{1}{x}\left(\frac{1}{3}\log^3(2) - \log(2)\zeta(2) + \frac{5}{4}\zeta(3) - \frac{1}{2}\log^2(2)\log(1-x) + \frac{1}{2}\log(2)\log^2(1-x)\right.$$

$$+ \frac{1}{2}\zeta(2)\log(1+x) - \frac{1}{2}\log(2)\log^2(1+x) + \frac{1}{6}\log^3(1+x) + \frac{1}{2}\log(x)\log^2(1-x)$$

$$- \frac{1}{4}\log(x^2)\log^2(1-x^2) + \frac{1}{2}\log(1-x)\log^2(1+x) - \log(1+x)\mathrm{Li}_2(x)$$

$$+ \log(1+x)\mathrm{Li}_2\left(\frac{1+x}{2}\right) - \mathrm{Li}_3(-x) + \mathrm{Li}_3(1-x) - \mathrm{Li}_3\left(\frac{1-x}{2}\right)$$

$$\left. - \frac{1}{2}\mathrm{Li}_3(1-x^2) - \mathrm{Li}_3\left(\frac{1+x}{2}\right) - \mathrm{Li}_3\left(\frac{x}{1+x}\right)\right),$$

where in the calculations, I also used the dilogarithm function reflection formula (see [21, Chapter 1, p.5], [70, Chapter 2, p.107]), $\mathrm{Li}_2(x) + \mathrm{Li}_2(1-x) = \pi^2/6 - \log(x)\log(1-x)$, to express $\mathrm{Li}_2\left(\frac{1-x}{2}\right)$ differently, and the solution to the point (ii) is finalized.

Further, to obtain the result at the point (iii), I'll exploit the result from the previous point, and then we write

$$\sum_{n=1}^{\infty} x^{n+1} \frac{H_n \overline{H}_n}{n+1} = \sum_{n=1}^{\infty} x^{n+1} \frac{(H_{n+1} - 1/(n+1))(\overline{H}_{n+1} - (-1)^n/(n+1))}{n+1}$$

$$\stackrel{\substack{\text{reindex}\\\text{the series}}}{=} \sum_{n=1}^{\infty} x^n \frac{(H_n - 1/n)(\overline{H}_n - (-1)^{n-1}/n)}{n}$$

$$= \sum_{n=1}^{\infty} x^n \frac{H_n \overline{H}_n}{n} + \sum_{n=1}^{\infty}(-x)^n \frac{H_n}{n^2} - \sum_{n=1}^{\infty} x^n \frac{\overline{H}_n}{n^2} - \underbrace{\sum_{n=1}^{\infty} \frac{(-x)^n}{n^3}}_{\mathrm{Li}_3(-x)},$$

whence we obtain that

$$\sum_{n=1}^{\infty} x^n \frac{H_n \overline{H}_n}{n}$$

$$= \operatorname{Li}_3(-x) - \sum_{n=1}^{\infty}(-x)^n \frac{H_n}{n^2} + \sum_{n=1}^{\infty} x^n \frac{\overline{H}_n}{n^2} + \sum_{n=1}^{\infty} x^{n+1} \frac{H_n \overline{H}_n}{n+1}$$

$$= \frac{1}{2}\log^3(2) - \frac{3}{2}\log(2)\zeta(2) + \frac{17}{8}\zeta(3) - \frac{1}{2}\log^2(2)\log(1-x)$$

$$+ \frac{1}{2}\log(2)\log^2(1-x) - \frac{1}{2}(\log^2(2) - 2\zeta(2))\log(1+x) + \frac{1}{3}\log^3(1+x)$$

$$+ \frac{1}{2}\log(x)\log^2(1-x) - \frac{1}{4}\log(x^2)\log^2(1-x^2) + \frac{1}{2}\log(1-x)\log^2(1+x)$$

$$+\log(1+x)\operatorname{Li}_2(-x)+\log(1+x)\operatorname{Li}_2\left(\frac{1+x}{2}\right)-\operatorname{Li}_3(x)-3\operatorname{Li}_3(-x)+\operatorname{Li}_3(1-x)$$

$$-\operatorname{Li}_3\left(\frac{1-x}{2}\right)-\frac{1}{2}\operatorname{Li}_3(1-x^2)-2\operatorname{Li}_3\left(\frac{1+x}{2}\right)-3\operatorname{Li}_3\left(\frac{x}{1+x}\right)+\operatorname{Li}_3\left(\frac{2x}{1+x}\right),$$

where in the calculations, I employed the result from the previous point; then the series in (4.36), the third closed form, Sect. 4.6; next the result in (4.66), Sect. 4.10; and finally the Landen's trilogarithmic identity in (6.33), Sect. 6.6, to express $\operatorname{Li}_3(1+x)$ differently, and the solution to the point (iii) is finalized.

For extracting the generating function at the point (iv), I will use the result in (6.49), the case $p = 2$, Sect. 6.9, that is, $\displaystyle\sum_{n=1}^{k-1} x^n H_n^{(2)} = \frac{1}{1-x}\left(-x^k H_k^{(2)} + \sum_{n=1}^{k} \frac{x^n}{n^2}\right)$, where we follow a calculations flow as in the first solution to the first point of the current section, and using that $\overline{H}_n = \log(2) - \displaystyle\sum_{k=n+1}^{\infty}(-1)^{k-1}\frac{1}{k}$, we write

$$\sum_{n=1}^{\infty} x^n \overline{H}_n H_n^{(2)} = \sum_{n=1}^{\infty} x^n H_n^{(2)}\left(\log(2) - \sum_{k=n+1}^{\infty}(-1)^{k-1}\frac{1}{k}\right)$$

$$= \log(2)\sum_{n=1}^{\infty} x^n H_n^{(2)} - \sum_{n=1}^{\infty} x^n H_n^{(2)} \sum_{k=n+1}^{\infty}(-1)^{k-1}\frac{1}{k}$$

{for the first series use the result in (4.32), the case $m = 2$, Sect. 4.6,}

{and regarding the second series, reverse the order of summation}

$$= \log(2)\frac{\text{Li}_2(x)}{1-x} - \sum_{k=1}^{\infty}(-1)^{k-1}\frac{1}{k}\sum_{n=1}^{k-1}x^n H_n^{(2)}$$

$$= \log(2)\frac{\text{Li}_2(x)}{1-x} - \frac{1}{1-x}\sum_{k=1}^{\infty}(-x)^k\frac{H_k^{(2)}}{k} - \frac{1}{1-x}\sum_{k=1}^{\infty}(-1)^{k-1}\frac{1}{k}\sum_{n=1}^{k}\frac{x^n}{n^2}. \quad (6.89)$$

For the last series in (6.89), we want to reverse the order of summation, and then we write

$$\sum_{k=1}^{\infty}(-1)^{k-1}\frac{1}{k}\sum_{n=1}^{k}\frac{x^n}{n^2} = \sum_{n=1}^{\infty}\frac{x^n}{n^2}\sum_{k=n}^{\infty}(-1)^{k-1}\frac{1}{k} = \sum_{n=1}^{\infty}\frac{x^n}{n^2}\left(\log(2) - \overline{H}_n + (-1)^{n-1}\frac{1}{n}\right)$$

$$= \log(2)\sum_{n=1}^{\infty}\frac{x^n}{n^2} + \sum_{n=1}^{\infty}(-1)^{n-1}\frac{x^n}{n^3} - \sum_{n=1}^{\infty}x^n\frac{\overline{H}_n}{n^2} = \log(2)\text{Li}_2(x) - \text{Li}_3(-x) - \sum_{n=1}^{\infty}x^n\frac{\overline{H}_n}{n^2}.$$

$$(6.90)$$

If we plug the result from (6.90) in (6.89), we arrive at

$$\sum_{n=1}^{\infty}x^n\overline{H}_n H_n^{(2)}$$

$$= \frac{\text{Li}_3(-x)}{1-x} + \frac{1}{1-x}\sum_{n=1}^{\infty}x^n\frac{\overline{H}_n}{n^2} - \frac{1}{1-x}\sum_{n=1}^{\infty}(-x)^n\frac{H_n^{(2)}}{n}$$

$$= \frac{1}{1-x}\left(\frac{1}{6}\log^3(2) - \frac{1}{2}\log(2)\zeta(2) + \frac{7}{8}\zeta(3) - \frac{1}{2}(\log^2(2) - \zeta(2))\log(1+x)\right)$$

$$+ \frac{1}{2}\log(2)\log^2(1+x) - \frac{1}{3}\log^3(1+x) + \log(1+x)\text{Li}_2(x) - \log(1+x)\text{Li}_2(-x)$$

$$- \text{Li}_3(x) + \text{Li}_3(-x) - \text{Li}_3\left(\frac{1+x}{2}\right) + \text{Li}_3\left(\frac{x}{1+x}\right) + \text{Li}_3\left(\frac{2x}{1+x}\right)\Big),$$

where in the calculations, I used the results in (4.66), Sect. 4.10, (4.46), the second equality, Sect. 4.7, and the Landen's trilogarithmic identity in (6.33), Sect. 6.6, to express $\text{Li}_3(1+x)$ differently, and the solution to the point (iv) is finalized.

For a second solution, the curious reader might also want to exploit telescoping sums as in the second solution to the first point.

Before proceeding with the calculations to the point (v), where we want to employ a similar strategy, we observe that $\overline{H}_n^{(2)} = 1 - \dfrac{1}{2^2} + \cdots + (-1)^{n-1}\dfrac{1}{n^2} =$

$$\sum_{k=1}^{n}(-1)^{k-1}\frac{1}{k^2} = \underbrace{\sum_{k=1}^{\infty}(-1)^{k-1}\frac{1}{k^2}}_{1/2\zeta(2)} - \sum_{k=n+1}^{\infty}(-1)^{k-1}\frac{1}{k^2}, \text{ where in the calculations}$$

I used $\displaystyle\sum_{k=1}^{\infty}(-1)^{k-1}\frac{1}{k^2} = \sum_{k=1}^{\infty}\frac{1}{k^2} - 2\sum_{k=1}^{\infty}\frac{1}{(2k)^2} = \frac{1}{2}\sum_{k=1}^{\infty}\frac{1}{k^2} = \frac{1}{2}\zeta(2).$

Therefore, by also exploiting (6.49), the case $p = 1$, Sect. 6.9, $\displaystyle\sum_{n=1}^{k-1}x^n H_n =$

$\dfrac{1}{1-x}\left(-x^k H_k + \displaystyle\sum_{n=1}^{k}\frac{x^n}{n}\right)$, we obtain that

$$\sum_{n=1}^{\infty}x^n H_n \overline{H}_n^{(2)} = \sum_{n=1}^{\infty}x^n H_n\left(\frac{1}{2}\zeta(2) - \sum_{k=n+1}^{\infty}(-1)^{k-1}\frac{1}{k^2}\right)$$

$$= \frac{1}{2}\zeta(2)\sum_{n=1}^{\infty}x^n H_n - \sum_{n=1}^{\infty}x^n H_n\sum_{k=n+1}^{\infty}(-1)^{k-1}\frac{1}{k^2}$$

{for the first series use the result in (4.32), the case $m = 1$, Sect. 4.6,}

{and regarding the second series, reverse the order of summation}

$$= -\frac{1}{2}\zeta(2)\frac{\log(1-x)}{1-x} - \sum_{k=1}^{\infty}(-1)^{k-1}\frac{1}{k^2}\sum_{n=1}^{k-1}x^n H_n$$

$$= -\frac{1}{2}\zeta(2)\frac{\log(1-x)}{1-x} - \frac{1}{1-x}\sum_{k=1}^{\infty}(-x)^k\frac{H_k}{k^2} - \frac{1}{1-x}\sum_{k=1}^{\infty}(-1)^{k-1}\frac{1}{k^2}\sum_{n=1}^{k}\frac{x^n}{n}.$$

$$(6.91)$$

For the second remaining series in (6.91), we swap the summation order, and then we have

$$\sum_{k=1}^{\infty}(-1)^{k-1}\frac{1}{k^2}\sum_{n=1}^{k}\frac{x^n}{n} = \sum_{n=1}^{\infty}\frac{x^n}{n}\sum_{k=n}^{\infty}(-1)^{k-1}\frac{1}{k^2} = \sum_{n=1}^{\infty}\frac{x^n}{n}\left(\frac{1}{2}\zeta(2) - \overline{H}_n^{(2)} + \frac{(-1)^{n-1}}{n^2}\right)$$

$$= -\frac{1}{2}\zeta(2)\log(1-x) - \text{Li}_3(-x) - \sum_{n=1}^{\infty}x^n\frac{\overline{H}_n^{(2)}}{n}. \qquad (6.92)$$

Further, by plugging the result from (6.92) in (6.91), we obtain that

$$\sum_{n=1}^{\infty} x^n H_n \overline{H}_n^{(2)}$$

$$= \frac{\text{Li}_3(-x)}{1-x} - \frac{1}{1-x}\sum_{n=1}^{\infty}(-x)^n\frac{H_n}{n^2} + \frac{1}{1-x}\sum_{n=1}^{\infty}x^n\frac{\overline{H}_n^{(2)}}{n}$$

$$= \frac{1}{1-x}\left(-\frac{1}{2}\zeta(3) - \zeta(2)\log(1-x) + \frac{1}{2}\zeta(2)\log(1-x^2) + \frac{1}{3}\log^3(1+x)\right.$$

$$+ \frac{1}{2}\log(x)\log^2(1-x) - \frac{1}{4}\log(x^2)\log^2(1-x^2) - \log(1+x)\text{Li}_2(x)$$

$$+ \log(1+x)\text{Li}_2(-x) - 3\text{Li}_3(-x) + \text{Li}_3(1-x) - \frac{1}{2}\text{Li}_3(1-x^2) - 2\text{Li}_3\left(\frac{x}{1+x}\right)\right),$$

where in the calculations I employed the series results in (4.36), the third closed form, Sect. 4.6, and (4.71), Sect. 4.11, and the solution to the point (v) is finalized.

As mentioned at the end of the previous solution, for a getting a second solution, the curious reader might want to consider using telescoping sums as seen in the second solution to the first point.

Of course, it is natural to ask ourselves how we would apply a strategy alike for calculating the series at the point (vi). First, we want to prove and use a result similar to (6.49), Sect. 6.9, that is

$$\sum_{n=1}^{k-1} x^n \overline{H}_n^{(p)} = \frac{1}{1-x}\left(-x^k \overline{H}_k^{(p)} + \sum_{m=1}^{k}(-1)^{m-1}\frac{x^m}{m^p}\right), \qquad (6.93)$$

which is derived by following the same strategy[3] in the mentioned section.

---

[3] Simple manipulations with sums immediately show that $\sum_{n=1}^{k-1} x^n \overline{H}_n^{(p)} = \sum_{n=1}^{k-1}\sum_{m=1}^{n}(-1)^{m-1}\frac{x^n}{m^p} =$

$$\sum_{m=1}^{k-1}(-1)^{m-1}\sum_{n=m}^{k-1}\frac{x^n}{m^p} = -\frac{1}{1-x}\sum_{m=1}^{k-1}(-1)^{m-1}\frac{x^k-x^m}{m^p} = -\frac{x^k}{1-x}\overline{H}_{k-1}^{(p)} +$$

$$\frac{1}{1-x}\sum_{m=1}^{k-1}(-1)^{m-1}\frac{x^m}{m^p} = -\frac{x^k}{1-x}\left(\overline{H}_k^{(p)} - (-1)^{k-1}\frac{1}{k^p}\right) + \frac{1}{1-x}\sum_{m=1}^{k-1}(-1)^{m-1}\frac{x^m}{m^p} =$$

$$\frac{1}{1-x}\left(-x^k\overline{H}_k^{(p)} + \sum_{m=1}^{k}(-1)^{m-1}\frac{x^m}{m^p}\right).$$

In the strategy I prepare, we need (6.93), the case $p = 1$, that is, $\displaystyle\sum_{n=1}^{k-1} x^n \overline{H}_n =$

$\dfrac{1}{1-x}\left(-x^k \overline{H}_k + \displaystyle\sum_{n=1}^{k}(-1)^{n-1}\dfrac{x^n}{n}\right)$, and by using the same starting strategy as in

the previous solution, we write

$$\sum_{n=1}^{\infty} x^n \overline{H}_n \overline{H}_n^{(2)} = \sum_{n=1}^{\infty} x^n \overline{H}_n \left(\frac{1}{2}\zeta(2) - \sum_{k=n+1}^{\infty}(-1)^{k-1}\frac{1}{k^2}\right)$$

$$= \frac{1}{2}\zeta(2)\sum_{n=1}^{\infty} x^n \overline{H}_n - \sum_{n=1}^{\infty} x^n \overline{H}_n \sum_{k=n+1}^{\infty}(-1)^{k-1}\frac{1}{k^2}$$

{for the first series use the result in (4.62), the case $m = 1$, Sect. 4.10,}

{and regarding the second series, reverse the order of summation}

$$= \frac{1}{2}\zeta(2)\frac{\log(1+x)}{1-x} - \sum_{k=1}^{\infty}(-1)^{k-1}\frac{1}{k^2}\sum_{n=1}^{k-1} x^n \overline{H}_n$$

$$= \frac{1}{2}\zeta(2)\frac{\log(1+x)}{1-x} - \frac{1}{1-x}\sum_{k=1}^{\infty}(-x)^k\frac{\overline{H}_k}{k^2} - \frac{1}{1-x}\sum_{k=1}^{\infty}(-1)^{k-1}\frac{1}{k^2}\sum_{n=1}^{k}(-1)^{n-1}\frac{x^n}{n}.$$
$$\tag{6.94}$$

Then, for the second remaining series in (6.94), we change the summation order, and thus we write

$$\sum_{k=1}^{\infty}(-1)^{k-1}\frac{1}{k^2}\sum_{n=1}^{k}(-1)^{n-1}\frac{x^n}{n} = \sum_{n=1}^{\infty}(-1)^{n-1}\frac{x^n}{n}\sum_{k=n}^{\infty}(-1)^{k-1}\frac{1}{k^2}$$

$$= \sum_{n=1}^{\infty}(-1)^{n-1}\frac{x^n}{n}\left(\frac{1}{2}\zeta(2) - \overline{H}_n^{(2)} + \frac{(-1)^{n-1}}{n^2}\right)$$

$$= \frac{1}{2}\zeta(2)\log(1+x) + \operatorname{Li}_3(x) + \sum_{n=1}^{\infty}(-x)^n\frac{\overline{H}_n^{(2)}}{n}. \tag{6.95}$$

At last, by plugging (6.95) in (6.94), we arrive at

$$\sum_{n=1}^{\infty} x^n \overline{H}_n \overline{H}_n^{(2)}$$

$$= -\frac{\text{Li}_3(x)}{1-x} - \frac{1}{1-x} \sum_{n=1}^{\infty} (-x)^n \frac{\overline{H}_n}{n^2} - \frac{1}{1-x} \sum_{n=1}^{\infty} (-x)^n \frac{\overline{H}_n^{(2)}}{n}$$

$$= \frac{1}{1-x} \left(\frac{5}{8} \zeta(3) + \frac{1}{2} \log(2)\zeta(2) - \frac{1}{6} \log^3(2) + \frac{1}{2} (\log^2(2) + 2\zeta(2)) \log(1-x) \right.$$

$$- \frac{1}{2} \log(2) \log^2(1-x) + \frac{1}{6} \log^3(1-x) - \frac{1}{2} \zeta(2) \log(1+x) - \frac{1}{6} \log^3(1+x)$$

$$- \log(x) \log^2(1-x) + \frac{1}{4} \log(x^2) \log^2(1-x^2) + 2 \text{Li}_3(-x) - 2 \text{Li}_3(1-x)$$

$$\left. + \frac{1}{2} \text{Li}_3(1-x^2) + \text{Li}_3 \left(\frac{1-x}{2} \right) + \text{Li}_3 \left(\frac{x}{1+x} \right) - \text{Li}_3 \left(\frac{2x}{x-1} \right) \right),$$

where in the calculations I used the generating functions in (4.66), Sect. 4.10, and (4.71), Sect. 4.11, together with the Landen's trilogarithmic identity in (6.33), Sect. 6.6, to express both $\text{Li}_3(1+x)$ and $\text{Li}_3 \left(\dfrac{x}{x-1} \right)$ differently, and the solution to the point (vi) is finalized.

For the generating function at the point (vii), I'll consider presenting two solutions. So, for a first solution, I'll employ the result in (1.123), Sect. 1.23, where if we multiply both sides by x^n, consider the summation from $n=1$ to ∞, and then rearrange, we write

$$\sum_{n=1}^{\infty} x^n \overline{H}_n^3$$

$$= -2 \sum_{n=1}^{\infty} x^n \overline{H}_n^{(3)} - 3 \sum_{n=1}^{\infty} x^n \overline{H}_n H_n^{(2)} + 3 \sum_{n=1}^{\infty} x^n \sum_{k=1}^{n} (-1)^{k-1} \frac{\overline{H}_k^2 + H_k^{(2)}}{k}. \qquad (6.96)$$

Changing the summation order in the last series of (6.96), we have

$$\sum_{n=1}^{\infty} x^n \sum_{k=1}^{n} (-1)^{k-1} \frac{\overline{H}_k^2 + H_k^{(2)}}{k} = \sum_{k=1}^{\infty} (-1)^{k-1} \frac{\overline{H}_k^2 + H_k^{(2)}}{k} \sum_{n=k}^{\infty} x^n$$

$$= -\frac{1}{1-x} \sum_{k=1}^{\infty} (-x)^k \frac{\overline{H}_k^2 + H_k^{(2)}}{k}. \qquad (6.97)$$

Now, we combine the results in (6.96) and (6.97) that lead to

$$\sum_{n=1}^{\infty} x^n \overline{H}_n^3$$

$$= -2 \sum_{n=1}^{\infty} x^n \overline{H}_n^{(3)} - 3 \sum_{n=1}^{\infty} x^n \overline{H}_n H_n^{(2)} - 3 \frac{1}{1-x} \sum_{n=1}^{\infty} (-x)^n \frac{\overline{H}_n^2}{n} - 3 \frac{1}{1-x} \sum_{n=1}^{\infty} (-x)^n \frac{H_n^{(2)}}{n}$$

$$= \frac{1}{1-x} \left(-\frac{3}{2} \log^3(2) + \frac{9}{2} \log(2)\zeta(2) - \frac{63}{8}\zeta(3) + \frac{3}{2}(\log^2(2) - \zeta(2))\log(1+x) \right.$$

$$+ \frac{3}{2} \log(2) \log^2(1+x) - \log^3(1+x) + 3 \log(1+x) \operatorname{Li}_2(x) - 3 \log(1+x) \operatorname{Li}_2(-x)$$

$$- 6 \log(1+x) \operatorname{Li}_2 \left(\frac{1+x}{2} \right) - 3 \operatorname{Li}_3(x) + 5 \operatorname{Li}_3(-x) + 9 \operatorname{Li}_3 \left(\frac{1+x}{2} \right)$$

$$\left. + 3 \operatorname{Li}_3 \left(\frac{x}{1+x} \right) + 3 \operatorname{Li}_3 \left(\frac{2x}{1+x} \right) \right),$$

where in the calculations I used (4.62), the case $m = 3$, Sect. 4.10; then (4.75), Sect. 4.12; next (4.69), Sect. 4.11; and finally (4.46), the second equality, Sect. 4.7, together with the Landen's trilogarithmic identity in (6.33), Sect. 6.6, to express $\operatorname{Li}_3(1+x)$ differently, and the first solution to the point (vii) is finalized.

To obtain a second solution, we might like to consider the use of telescoping sums, with a flow similar to the one of the second solution found at the point (i).

By considering the notation $\sum_{n=1}^{\infty} x^n \overline{H}_n^3 = P(x)$ and multiplying both sides by $1-x$, we have that

$$\sum_{n=1}^{\infty} (1-x) x^n \overline{H}_n^3 = (1-x) P(x). \tag{6.98}$$

At this point, we want to address the left-hand side of (6.98), and then we write

$$\sum_{n=1}^{\infty} (1-x) x^n \overline{H}_n^3 = \sum_{n=1}^{\infty} x^n \overline{H}_n^3 - \sum_{n=1}^{\infty} x^{n+1} \left(\overline{H}_{n+1} - \frac{(-1)^n}{n+1} \right)^3$$

$$\overset{\underset{\mathrm{reindex}}{\underset{\mathrm{the\ second\ series}}{\underset{\mathrm{and\ then\ expand\ it}}{=}}}}{} -3\sum_{n=1}^{\infty}(-x)^n\frac{\overline{H}_n^2}{n} - 3\sum_{n=1}^{\infty}x^n\frac{\overline{H}_n}{n^2} - \underbrace{\sum_{n=1}^{\infty}\frac{(-x)^n}{n^3}}_{\text{Li}_3(-x)}$$

$$= -\frac{3}{2}\log^3(2) + \frac{9}{2}\log(2)\zeta(2) - \frac{63}{8}\zeta(3) + \frac{3}{2}(\log^2(2) - \zeta(2))\log(1+x)$$

$$+ \frac{3}{2}\log(2)\log^2(1+x) - \log^3(1+x) + 3\log(1+x)\text{Li}_2(x) - 3\log(1+x)\text{Li}_2(-x)$$

$$- 6\log(1+x)\text{Li}_2\left(\frac{1+x}{2}\right) - 3\text{Li}_3(x) + 5\text{Li}_3(-x) + 9\text{Li}_3\left(\frac{1+x}{2}\right)$$

$$+ 3\text{Li}_3\left(\frac{x}{1+x}\right) + 3\text{Li}_3\left(\frac{2x}{1+x}\right),$$

which leads to the generating function in (6.98), and the second solution to the point (vii) is finalized. Observe the resulting series also appeared in previous calculations.

For the last two generating functions, we'll want to use again telescoping sums as in the previous solution. So, let's make use of the notation,

$$\sum_{n=1}^{\infty}(1-x)x^n H_n \overline{H}_n^2 = (1-x)Q(x). \tag{6.99}$$

Now, let's develop the left-hand side of (6.99), and then we write

$$\sum_{n=1}^{\infty}(1-x)x^n H_n\overline{H}_n^2 = \sum_{n=1}^{\infty}x^n H_n\overline{H}_n^2 - \sum_{n=1}^{\infty}x^{n+1}\left(H_{n+1} - \frac{1}{n+1}\right)\left(\overline{H}_{n+1} - \frac{(-1)^n}{n+1}\right)^2$$

$$\overset{\underset{\mathrm{reindex}}{\underset{\mathrm{the\ second\ series}}{=}}}{} \sum_{n=1}^{\infty}x^n H_n\overline{H}_n^2 - \sum_{n=1}^{\infty}x^n\left(H_n - \frac{1}{n}\right)\left(\overline{H}_n - (-1)^{n-1}\frac{1}{n}\right)^2$$

$$= \underbrace{\sum_{n=1}^{\infty}\frac{x^n}{n^3}}_{\text{Li}_3(x)} - \sum_{n=1}^{\infty}x^n\frac{H_n}{n^2} - 2\sum_{n=1}^{\infty}(-x)^n\frac{H_n\overline{H}_n}{n} + 2\sum_{n=1}^{\infty}(-x)^n\frac{\overline{H}_n}{n^2} + \sum_{n=1}^{\infty}x^n\frac{\overline{H}_n^2}{n}$$

$$= \frac{1}{4}\zeta(3) + \log(2)\zeta(2) - \frac{1}{3}\log^3(2) + 2\zeta(2)\log(1-x) + \frac{1}{3}\log^3(1-x)$$

$$- (2\zeta(2) - \log^2(2))\log(1+x) - \log(2)\log^2(1+x) - \frac{1}{3}\log^3(1+x)$$

$$-\frac{3}{2}\log(x)\log^2(1-x)+\frac{1}{2}\log(x^2)\log^2(1-x^2)-\log^2(1-x)\log(1+x)$$

$$-\operatorname{Li}_3(x)+4\operatorname{Li}_3(-x)-3\operatorname{Li}_3(1-x)+\operatorname{Li}_3(1-x^2)+2\operatorname{Li}_3\left(\frac{1+x}{2}\right)$$

$$+2\operatorname{Li}_3\left(\frac{x}{1+x}\right)-2\operatorname{Li}_3\left(\frac{2x}{x-1}\right),$$

where in the calculations I used (4.36), the third closed form, Sect. 4.6; then (4.74), Sect. 4.12; next (4.66), Sect. 4.10; and further (4.69), Sect. 4.11, together with the Landen's trilogarithmic identity in (6.33), Sect. 6.6, to express both $\operatorname{Li}_3(1+x)$ and $\operatorname{Li}_3\left(\dfrac{x}{x-1}\right)$ differently, which gives the generating function in (6.99), and the solution to the point $(viii)$ is finalized.

Ultimately, for the last point of the problem, we'll want to use telescoping sums as in the preceding solution, and by using the notation $\displaystyle\sum_{n=1}^{\infty}x^n H_n^2 \overline{H}_n = R(x)$, where we multiply both sides by $1-x$, we get

$$\sum_{n=1}^{\infty}(1-x)x^n H_n^2 \overline{H}_n = (1-x)R(x). \tag{6.100}$$

At this step, we want to focus on the left-hand side of (6.100), and then we write

$$\sum_{n=1}^{\infty}(1-x)x^n H_n^2 \overline{H}_n$$

$$=\sum_{n=1}^{\infty}x^n H_n^2 \overline{H}_n - \sum_{n=1}^{\infty}x^{n+1}\left(H_{n+1}-\frac{1}{n+1}\right)^2\left(\overline{H}_{n+1}-(-1)^n\frac{1}{n+1}\right)$$

$$\overset{\substack{\text{reindex}\\\text{the second series}}}{=}\sum_{n=1}^{\infty}x^n H_n^2 \overline{H}_n - \sum_{n=1}^{\infty}x^n\left(H_n-\frac{1}{n}\right)^2\left(\overline{H}_n-(-1)^{n-1}\frac{1}{n}\right)$$

$$=2\sum_{n=1}^{\infty}(-x)^n\frac{H_n}{n^2}-\sum_{n=1}^{\infty}(-x)^n\frac{H_n^2}{n}+2\sum_{n=1}^{\infty}x^n\frac{H_n\overline{H}_n}{n}-\sum_{n=1}^{\infty}x^n\frac{\overline{H}_n}{n^2}-\underbrace{\sum_{n=1}^{\infty}\frac{(-x)^n}{n^3}}_{\operatorname{Li}_3(-x)}$$

$$=\frac{5}{6}\log^3(2)-\frac{5}{2}\log(2)\zeta(2)+\frac{27}{8}\zeta(3)-\log^2(2)\log(1-x)+\log(2)\log^2(1-x)$$

$$-\frac{1}{2}(\log^2(2) - 3\zeta(2))\log(1 + x) - \frac{1}{2}\log(2)\log^2(1 + x) + \frac{2}{3}\log^3(1 + x)$$

$$+ \log(x)\log^2(1 - x) - \frac{1}{2}\log(x^2)\log^2(1 - x^2) + \log(1 - x)\log^2(1 + x)$$

$$- \log(1 + x)\operatorname{Li}_2(x) + \log(1 + x)\operatorname{Li}_2(-x) + 2\log(1 + x)\operatorname{Li}_2\left(\frac{1 + x}{2}\right)$$

$$- \operatorname{Li}_3(x) - 3\operatorname{Li}_3(-x) + 2\operatorname{Li}_3(1 - x) - \operatorname{Li}_3(1 - x^2) - 2\operatorname{Li}_3\left(\frac{1 - x}{2}\right)$$

$$- 3\operatorname{Li}_3\left(\frac{1 + x}{2}\right) - 3\operatorname{Li}_3\left(\frac{x}{1 + x}\right) + \operatorname{Li}_3\left(\frac{2x}{1 + x}\right),$$

where in the calculations I used the results in (4.36), the third closed form, Sect. 4.6, (4.42), Sect. 4.7, (4.74), Sect. 4.12, (4.66), Sect. 4.10, and the solution to the point (ix) is finalized.

Here I put an end to the long journey involving a seven-section series with generating functions! The curious reader might also want to think over constructing other ways of proving such results! As I also emphasized in *(Almost) Impossible Integrals, Sums, and Series*, I highly recommend you not to get limited to the solutions you find in my book, and if possible, try to find other approaches, too.

6.13 Two Nice Sums Related to the Generalized Harmonic Numbers, an Asymptotic Expansion Extraction, a Neat Representation of $\log^2(2)$, and a Curious Power Series

Solution In this section, we prepare for a good time with two sums involving the generalized harmonic numbers, and then we will play with some other questions related to them. These are those kinds of attractive questions where even if you feel you are stumped at some point, it might be better to be patient and invest enough efforts to reach that moment of satisfaction when you possibly say loudly, *"Yes, I did it!"*.

Now, if you had a chance to read my first book, *(Almost) Impossible Integrals, Sums, and Series*, and particularly studied the sums in [76, Chapter 4, pp. 287–288], you noticed that a powerful strategy there is to exploit the difference of a sum when one of its variables takes different values, and this way also works perfectly here. I'll use this strategy to get the first solutions to the sums at the points (i) and (ii).

So, if we focus on the sum at the point (i), denote it by $S(n) = \sum_{k=1}^{n} \frac{H_{k+n}}{k}$, and

then consider the difference suggested above, we have

$$S(n) - S(n-1) = \sum_{k=1}^{n} \frac{H_{k+n}}{k} - \sum_{k=1}^{n-1} \frac{H_{k+n-1}}{k}$$

{set apart the term of the first sum for $k = n$, and at the same time for}

$$\left\{ \text{the second sum we use the simple fact that } H_{k+n-1} = H_{k+n} - \frac{1}{k+n} \right\}$$

$$= \frac{H_{2n}}{n} + \sum_{k=1}^{n-1} \frac{H_{k+n}}{k} - \sum_{k=1}^{n-1} \frac{H_{k+n} - 1/(k+n)}{k} = \frac{H_{2n}}{n} + \sum_{k=1}^{n-1} \frac{1}{k(k+n)}$$

$$= \frac{H_{2n}}{n} + \frac{1}{n} \sum_{k=1}^{n-1} \frac{1}{k} - \frac{1}{n} \sum_{k=1}^{n-1} \frac{1}{k+n} = \frac{H_{2n}}{n} + \frac{H_{n-1}}{n} - \frac{1}{n}(H_{2n-1} - H_n)$$

$$\left\{ \text{use that } H_{n-1} = H_n - \frac{1}{n} \text{ and } H_{2n-1} = H_{2n} - \frac{1}{2n}, \text{ and then expand} \right\}$$

$$= 2\frac{H_n}{n} - \frac{1}{2n^2},$$

or in a more concise form, to better see the picture of the result above, we write

$$S(n) - S(n-1) = 2\frac{H_n}{n} - \frac{1}{2n^2}. \tag{6.101}$$

Then, if we replace n by i in (6.101) and then sum up both sides from $i = 1$ to n, where $S(0) = 0$, we obtain

$$\sum_{i=1}^{n} (S(i) - S(i-1)) = S(n) - S(0)$$

$$= S(n) = \sum_{k=1}^{n} \frac{H_{k+n}}{k} = 2\sum_{i=1}^{n} \frac{H_i}{i} - \frac{1}{2} \sum_{i=1}^{n} \frac{1}{i^2} = H_n^2 + \frac{1}{2}H_n^{(2)} = \frac{1}{2}\left(2H_n^2 + H_n^{(2)}\right),$$

and the first solution to the point (i) of the problem is finalized.

In the calculations I also used the fact that $\sum_{i=1}^{n} \frac{H_i}{i} = \frac{1}{2}\left(H_n^2 + H_n^{(2)}\right)$, and this is a particular case $(p = q = 1)$ of a more general result,

$$\sum_{k=1}^{n} \frac{H_k^{(p)}}{k^q} + \sum_{k=1}^{n} \frac{H_k^{(q)}}{k^p} = H_n^{(p+q)} + H_n^{(p)} H_n^{(q)}. \tag{6.102}$$

Proof The result is straightforward, and we have

$$\sum_{k=1}^{n} \frac{H_k^{(p)}}{k^q} = \sum_{k=1}^{n}\sum_{i=1}^{k} \frac{1}{k^q i^p} \overset{\overset{\text{reverse the order}}{\text{of summation}}}{=} \sum_{i=1}^{n}\sum_{k=i}^{n} \frac{1}{k^q i^p} = \sum_{i=1}^{n} \frac{1}{i^p}\left(\frac{1}{i^q} + H_n^{(q)} - H_i^{(q)}\right)$$

$$= \sum_{i=1}^{n} \frac{1}{i^{p+q}} + \sum_{i=1}^{n} \frac{H_n^{(q)}}{i^p} - \sum_{i=1}^{n} \frac{H_i^{(q)}}{i^p} = H_n^{(p+q)} + H_n^{(p)} H_n^{(q)} - \sum_{k=1}^{n} \frac{H_k^{(q)}}{k^p},$$

where a key step was to reverse the order of summation after the second equality. ∎

To get a second solution, we might like to exploit the fact that $\displaystyle\sum_{k=1}^{n} \frac{H_k}{k} = \frac{1}{2}\left(H_n^2 + H_n^{(2)}\right)$ by observing that the numerator of the main sum may be written as

$$H_{k+n} = H_k + \frac{1}{k+1} + \cdots + \frac{1}{k+n} = H_k + \sum_{i=1}^{n} \frac{1}{k+i}. \text{ Then, we have}$$

$$\sum_{k=1}^{n} \frac{H_{k+n}}{k} = \sum_{k=1}^{n} \frac{H_k + \displaystyle\sum_{i=1}^{n} \dfrac{1}{k+i}}{k} = \sum_{k=1}^{n} \frac{H_k}{k} + \sum_{k=1}^{n}\sum_{i=1}^{n} \frac{1}{k(k+i)}$$

$$= \frac{1}{2}\left(2H_n^2 + H_n^{(2)}\right).$$

Note that for the double sum in the calculations above, I used that $s =$

$$\underbrace{\sum_{k=1}^{n}\sum_{i=1}^{n} \frac{1}{k(k+i)}}_{} = \underbrace{\sum_{k=1}^{n}\sum_{i=1}^{n} \frac{1}{ik}}_{H_n^2} - \underbrace{\sum_{i=1}^{n}\sum_{k=1}^{n} \frac{1}{i(i+k)}}_{s}, \text{ whence, due to the symmetry,}$$

we obtain the desired value, $s = \displaystyle\sum_{k=1}^{n}\sum_{i=1}^{n} \frac{1}{k(k+i)} = \frac{1}{2}H_n^2$, and the second solution to the point (i) of the problem is finalized. A similar solution has also been found by Ramya Dutta (India), who is a talented problem solver.

Moving on and passing to the sum at the point (ii), we want to use the same strategy as the one in the first solution to the previous sum, where we may consider the notation $T(n) = \displaystyle\sum_{k=1}^{n} \frac{H_k}{k+n}$. Then, assuming that $n \geq 2$, we write

$$T(n) - T(n-1) = \sum_{k=1}^{n} \frac{H_k}{k+n} - \sum_{k=1}^{n-1} \frac{H_k}{k+n-1}$$

{reindex the second sum}

$$= \sum_{k=1}^{n} \frac{H_k}{k+n} - \sum_{k=0}^{n-2} \frac{H_{k+1}}{k+n} = \sum_{k=1}^{n} \frac{H_k}{k+n} - \sum_{k=0}^{n-2} \frac{H_k}{k+n} - \sum_{k=0}^{n-2} \frac{1}{(k+1)(k+n)}$$

$$= \frac{H_{n-1}}{2n-1} + \frac{H_n}{2n} - \frac{1}{n-1} \sum_{k=0}^{n-2} \left(\frac{1}{k+1} - \frac{1}{k+n} \right) = \frac{H_{n-1}}{2n-1} + \frac{H_n}{2n} - 2\frac{H_{n-1}}{n-1} + \frac{H_{2n-2}}{n-1},$$

and to put the result above in a single line, we have

$$T(n) - T(n-1) = \frac{H_{n-1}}{2n-1} + \frac{H_n}{2n} - 2\frac{H_{n-1}}{n-1} + \frac{H_{2n-2}}{n-1}. \qquad (6.103)$$

Next, if we replace n by i in (6.103) and then sum up both sides from $i = 2$ to n, where $T(1) = 1/2$, we obtain

$$\sum_{i=2}^{n} (T(i) - T(i-1)) = T(n) - T(1) = T(n) - \frac{1}{2}$$

$$= \sum_{i=2}^{n} \frac{H_{i-1}}{2i-1} + \frac{1}{2} \sum_{i=2}^{n} \frac{H_i}{i} - 2\sum_{i=2}^{n} \frac{H_{i-1}}{i-1} + \sum_{i=2}^{n} \frac{H_{2i-2}}{i-1}$$

$$= \sum_{i=2}^{n} \frac{H_{i-1}}{2i-1} + \frac{1}{2} \left(\sum_{i=1}^{n} \frac{H_i}{i} - 1 \right) - 2\left(\sum_{i=1}^{n} \frac{H_i}{i} - \frac{H_n}{n} \right) + \sum_{i=2}^{n} \frac{H_{2i-2}}{i-1}$$

$$= \sum_{i=2}^{n} \frac{H_{i-1}}{2i-1} + \sum_{i=2}^{n} \frac{H_{2i-2}}{i-1} - \frac{3}{2} \sum_{i=1}^{n} \frac{H_i}{i} + 2\frac{H_n}{n} - \frac{1}{2},$$

whence we get that

$$T(n) = \sum_{k=1}^{n} \frac{H_k}{k+n} = \sum_{i=2}^{n} \frac{H_{i-1}}{2i-1} + \sum_{i=2}^{n} \frac{H_{2i-2}}{i-1} - \frac{3}{2} \sum_{i=1}^{n} \frac{H_i}{i} + 2\frac{H_n}{n}$$

$$\left\{ \text{employ the result } \sum_{i=1}^{n} \frac{H_i}{i} = \frac{1}{2} \left(H_n^2 + H_n^{(2)} \right) \right\}$$

$$= \sum_{i=2}^{n} \frac{H_{i-1}}{2i-1} + \sum_{i=2}^{n} \frac{H_{2i-2}}{i-1} + 2\frac{H_n}{n} - \frac{3}{4} \left(H_n^2 + H_n^{(2)} \right)$$

$$= H_n(H_{2n} - H_n) - \frac{2}{n}(H_{2n} - H_n) + \frac{H_{2n}}{n} - \frac{1}{2}H_n^{(2)}.$$

Note that the two remaining sums above may be calculated together by applying Abel's summation (see [76, Chapter 2, pp.39–40]). For example, if we consider $\sum_{i=2}^{n} \frac{H_{2i-2}}{i-1}$ and take $a_i = \frac{1}{i-1}$ and $b_i = H_{2i-2}$, we get

$$\sum_{i=2}^{n} \frac{H_{2i-2}}{i-1} = H_{n-1}H_{2n} - \sum_{i=2}^{n} \frac{H_{i-1}}{2i-1} - \frac{1}{2}\sum_{i=2}^{n} \frac{H_i - 1/i}{i}$$

{start in the last sum from $i = 1$ and then expand it}

$$= \left(H_n - \frac{1}{n}\right)H_{2n} - \sum_{i=2}^{n} \frac{H_{i-1}}{2i-1} - \frac{1}{2}\sum_{i=1}^{n} \frac{H_i}{i} + \frac{1}{2}\sum_{i=1}^{n} \frac{1}{i^2}$$

$$\left\{ \text{use that } \sum_{i=1}^{n} \frac{H_i}{i} = \frac{1}{2}\left(H_n^2 + H_n^{(2)}\right) \text{ and } \sum_{i=1}^{n} \frac{1}{i^2} = H_n^{(2)} \right\}$$

$$= H_n H_{2n} - \frac{H_{2n}}{n} - \sum_{i=2}^{n} \frac{H_{i-1}}{2i-1} - \frac{1}{4}H_n^2 + \frac{1}{4}H_n^{(2)},$$

whence we obtain that

$$\sum_{i=2}^{n} \frac{H_{2i-2}}{i-1} + \sum_{i=2}^{n} \frac{H_{i-1}}{2i-1} = H_n H_{2n} - \frac{H_{2n}}{n} - \frac{1}{4}H_n^2 + \frac{1}{4}H_n^{(2)},$$

and the first solution to the point (ii) of the problem is finalized.

In order to obtain a second solution to the point (ii) of the problem, we reduce the calculation of the present sum to the sum from the point (i), which we assume it is already evaluated and we can use it. So, we want to use Abel's summation (see [76, Chapter 2, pp.39–40]) where we set $a_k = \frac{1}{k+n}$ and $b_k = H_k$, and then we have

$$\sum_{k=1}^{n} \frac{H_k}{k+n} = H_{n+1}(H_{2n} - H_n) - \sum_{k=1}^{n} \frac{1}{k+1}(H_{k+n} - H_n)$$

{reindex the sum and start from $k = 2$ to $k = n + 1$}

$$= \left(H_n + \frac{1}{n+1}\right)(H_{2n} - H_n) - \sum_{k=2}^{n+1} \frac{1}{k}(H_{k+n-1} - H_n)$$

{in the remaining sum start from $k = 1$ and leave out the term for $k = n + 1$}

$$= H_n(H_{2n} - H_n) - \sum_{k=1}^{n} \frac{1}{k}(H_{k+n-1} - H_n) = H_n(H_{2n} - H_n) - \sum_{k=1}^{n} \frac{1}{k}\left(H_{k+n} - \frac{1}{k+n} - H_n\right)$$

$$= H_n(H_{2n} - H_n) - \sum_{k=1}^{n} \frac{H_{k+n}}{k} + \frac{1}{n}\sum_{k=1}^{n} \frac{1}{k} - \frac{1}{n}\sum_{k=1}^{n} \frac{1}{k+n} + H_n\sum_{k=1}^{n} \frac{1}{k}$$

{for the first sum make use of the result from the point (i)}

$$= H_n(H_{2n} - H_n) - \frac{1}{2}\left(2H_n^2 + H_n^{(2)}\right) + \frac{H_n}{n} - \frac{1}{n}(H_{2n} - H_n) + H_n^2$$

$$= H_n(H_{2n} - H_n) - \frac{2}{n}(H_{2n} - H_n) + \frac{H_{2n}}{n} - \frac{1}{2}H_n^{(2)},$$

and the second solution to the point (ii) of the problem is finalized.

The next point I consider (very) enjoyable, particularly if you like the (little) challenges with asymptotic behaviors.

To begin with, I will state the asymptotic behavior of the harmonic number H_n in a simple form, to have a clear picture of it, since we need it in our solution,

$$H_n \approx \gamma + \log(n), \quad \text{as } n \to \infty \qquad (6.104)$$

and this comes from the limit definition of the Euler-Mascheroni constant, $\gamma = \lim_{n\to\infty}(H_n - \log(n))$. We can also write a more developed asymptotic behavior including the big-O notation, $H_n = \gamma + \log(n) + O\left(\frac{1}{n}\right)$, but the form in (6.104) is enough for our needs.

Now, we need an artifice of calculation, which is the *magical* part of the solution, and then we write the initial sum as follows:

$$\sum_{k=1}^{n} \frac{H_k}{k+n} = \sum_{k=1}^{n} \frac{H_k - \log(k) + \log(k) - \log(n) + \log(n)}{k+n}$$

$$= \underbrace{\sum_{k=1}^{n} \frac{H_k - \log(k)}{k+n}}_{A_n} + \underbrace{\sum_{k=1}^{n} \frac{\log(k/n)}{k+n}}_{B_n} + \underbrace{\log(n)\sum_{k=1}^{n} \frac{1}{k+n}}_{C_n}, \qquad (6.105)$$

which already gives us an idea about how the asymptotic behavior looks like.

If we split the sum A_n in (6.105) at $k = \lfloor\sqrt{n}\rfloor + 1$, where $\lfloor x \rfloor$ is the integer part of x, we get

$$A_n = \sum_{k=1}^{n} \frac{H_k - \log(k)}{k+n} = \underbrace{\sum_{k=1}^{\lfloor\sqrt{n}\rfloor} \frac{H_k - \log(k)}{k+n}}_{D_n} + \underbrace{\sum_{k=\lfloor\sqrt{n}\rfloor+1}^{n} \frac{H_k - \log(k)}{k+n}}_{E_n}. \quad (6.106)$$

Now, for the sum D_n in (6.106), we have

$$|D_n| = \left| \sum_{k=1}^{\lfloor\sqrt{n}\rfloor} \frac{H_k - \log(k)}{k+n} \right| \le \sum_{k=1}^{\lfloor\sqrt{n}\rfloor} \left| \frac{H_k - \log(k)}{k+n} \right| \le |M| \frac{\lfloor\sqrt{n}\rfloor}{n} = \frac{\lfloor\sqrt{n}\rfloor}{n} \le \frac{1}{\sqrt{n}},$$

$$(6.107)$$

where letting $n \to \infty$, we get that $\lim_{n\to\infty} D_n = 0$. Note that if we consider the sequence $h_k = H_k - \log(k)$, $k \ge 1$, which is strictly decreasing[4] and positive, then the maximum of h_k, denoted by M in (6.107), equals 1.

For the sum E_n in (6.106), we use that h_k is strictly decreasing and positive (positivity comes from the fact that since $H_k > \log(k+1)$, which is a consequence of the right-hand side of the double inequality in the footnote, then we also have $H_k > \log(k)$), which further easily yields the double inequality

$$(H_n - \log(n)) \sum_{k=\lfloor\sqrt{n}\rfloor+1}^{n} \frac{1}{k+n} \le \underbrace{\sum_{k=\lfloor\sqrt{n}\rfloor+1}^{n} \frac{H_k - \log(k)}{k+n}}_{E_n}$$

$$\le (H_{\lfloor\sqrt{n}\rfloor+1} - \log(\lfloor\sqrt{n}\rfloor + 1)) \sum_{k=\lfloor\sqrt{n}\rfloor+1}^{n} \frac{1}{k+n},$$

and since $\lim_{n\to\infty} \sum_{k=\lfloor\sqrt{n}\rfloor+1}^{n} \frac{1}{k+n} = \lim_{n\to\infty} (H_{2n} - H_{n+\lfloor\sqrt{n}\rfloor}) = \lim_{n\to\infty} \log\left(\frac{2n}{n + \lfloor\sqrt{n}\rfloor}\right) =$

$\log(2)$, via the limit definition of the Euler-Mascheroni constant, which also shows that $\lim_{n\to\infty} (H_{\lfloor\sqrt{n}\rfloor+1} - \log(\lfloor\sqrt{n}\rfloor + 1)) = \gamma$ and $\lim_{n\to\infty} (H_n - \log(n)) = \gamma$, we get

---

[4] *Why is that?*, some of you might ask. Since $h_k = H_k - \log(k)$, then we show that $h_{k+1} - h_k = \frac{1}{k+1} - \log(k+1) + \log(k) < 0$. Essentially, we need to prove that $\frac{1}{k+1} < \log(k+1) - \log(k)$, and this is immediately obtained by applying the *mean value theorem* to the function $f(x) = \log(x)$ on the interval $[k, k+1]$, $\forall k \ge 1$, $k \in \mathbb{Z}$, that gives $\log(k+1) - \log(k) = \frac{1}{c_k}$, $c_k \in (k, k+1)$, where since $\frac{1}{k+1} < \frac{1}{c_k} < \frac{1}{k}$, we obtain the double inequality $\frac{1}{k+1} < \log(k+1) - \log(k) < \frac{1}{k}$ which shows the inequality we need holds.

$$\lim_{n \to \infty} E_n = \sum_{k=\lfloor \sqrt{n} \rfloor + 1}^{n} \frac{H_k - \log(k)}{k + n} = \gamma \log(2). \tag{6.108}$$

Next, we view the sum B_n in (6.105) as a Riemann sum, and letting $n \to \infty$, we turn it into an improper Riemann integral. Thus, we write

$$\lim_{n \to \infty} B_n = \lim_{n \to \infty} \sum_{k=1}^{n} \frac{\log(k/n)}{k + n} = \lim_{n \to \infty} \frac{1}{n} \sum_{k=1}^{n} \frac{\log(k/n)}{k/n + 1} = \int_0^1 \frac{\log(x)}{1 + x} dx$$

$$= \int_0^1 \log(x) \sum_{n=1}^{\infty} (-1)^{n-1} x^{n-1} dx = \sum_{n=1}^{\infty} (-1)^{n-1} \int_0^1 x^{n-1} \log(x) dx$$

$$= -\sum_{n=1}^{\infty} (-1)^{n-1} \frac{1}{n^2} = -\frac{\pi^2}{12}. \tag{6.109}$$

Lastly, for C_n in (6.105), when n is large, we have

$$C_n = \log(n) \sum_{k=1}^{n} \frac{1}{k + n} = \log(n)(H_{2n} - H_n) \approx \log(2) \log(n), \text{ as } n \to \infty, \tag{6.110}$$

where I used the asymptotic behavior of H_n in (6.104).

Combining (6.105), (6.106), (6.107), (6.108), (6.109), and (6.110), we obtain that

$$\sum_{k=1}^{n} \frac{H_k}{k + n} \approx \gamma \log(2) - \frac{\pi^2}{12} + \log(2) \log(n), \text{ as } n \to \infty,$$

and the solution to the point (iii) of the problem is finalized.

Surely, the asymptotic behavior is also easily extracted from the precise value of the sum given at the point (ii), but remember that the point we just finished required to avoid the use of it and asked us for finding a different way. For example, we can also extract the following form with the big-O notation:

$$\sum_{k=1}^{n} \frac{H_k}{k + n} = \gamma \log(2) - \frac{\pi^2}{12} + \log(2) \log(n) + O\left(\frac{\log(n)}{n}\right), \text{ as } n \to \infty.$$

What about the next point? The solution to the point (iv) will be a fast one since we can easily reduce it to the use of the result from the previous point.

So, we rearrange the expression under the limit and write

$$\lim_{n\to\infty}\sum_{k=1}^{n}\left(\frac{H_{k+n}}{k+3n}-\frac{nH_k}{(k+n)(k+2n)}\right)=\lim_{n\to\infty}\left(\sum_{k=1}^{n}\frac{H_{k+n}}{k+3n}+\sum_{k=1}^{n}\frac{H_k}{k+2n}-\sum_{k=1}^{n}\frac{H_k}{k+n}\right)$$

{reindex the first sum and start from $k=n+1$ to $k=2n$}

$$=\lim_{n\to\infty}\left(\sum_{k=1+n}^{2n}\frac{H_k}{k+2n}+\sum_{k=1}^{n}\frac{H_k}{k+2n}-\sum_{k=1}^{n}\frac{H_k}{k+n}\right)$$

{note we can merge the first two sums}

$$=\lim_{n\to\infty}\left(\underbrace{\sum_{k=1}^{2n}\frac{H_k}{k+2n}}_{S_{2n}}-\underbrace{\sum_{k=1}^{n}\frac{H_k}{k+n}}_{S_n}\right)=\lim_{n\to\infty}\left(S_{2n}-S_n\right)$$

$$=\lim_{n\to\infty}\left(\gamma\log(2)-\frac{\pi^2}{12}+\log(2)\log(2n)-\left(\gamma\log(2)-\frac{\pi^2}{12}+\log(2)\log(n)\right)\right)$$

$$=\log^2(2),$$

where the penultimate equality comes immediately from the asymptotic behavior of $\sum_{k=1}^{n}\dfrac{H_k}{k+n}$ stated at the previous point, and the solution to the point (iv) of the problem is finalized.

For the last point of the problem, if we know what to do, then everything goes smoothly. The generating function to recall and use is $\sum_{k=1}^{\infty}x^k H_k=-\dfrac{\log(1-x)}{1-x}$, $|x|<1$, which is the case $m=1$ in (4.32), Sect. 4.6, and then we write

$$\int_0^1\frac{\log\left(1-xy^2\right)}{y(xy-1)\left(1-xy^2\right)}dy=\int_0^1\frac{1}{y(1-xy)}\sum_{k=1}^{\infty}(xy^2)^k H_k dy$$

{let's rearrange a bit the integrand to see better the next step}

$$=\int_0^1\sum_{k=1}^{\infty}\frac{(xy)^k}{1-xy}y^{k-1}H_k dy$$

$$\left\{ \text{observe and use that } \frac{(xy)^k}{1-xy} = \sum_{n=k}^{\infty} (xy)^n \right\}$$

$$= \int_0^1 \sum_{k=1}^{\infty} \left(\sum_{n=k}^{\infty} x^n H_k y^{k+n-1} \right) \mathrm{d}y = \sum_{k=1}^{\infty} \left(\sum_{n=k}^{\infty} x^n H_k \int_0^1 y^{k+n-1} \mathrm{d}y \right)$$

$$= \sum_{k=1}^{\infty} \left(\sum_{n=k}^{\infty} x^n \frac{H_k}{k+n} \right) \overset{\substack{\text{reverse the order} \\ \text{of summation}}}{=} \sum_{n=1}^{\infty} \left(\sum_{k=1}^{n} \frac{H_k}{k+n} \right) x^n$$

{for the inner sum make use of the result from the point (ii)}

$$= \sum_{n=1}^{\infty} \left(H_n (H_{2n} - H_n) - \frac{2}{n} (H_{2n} - H_n) + \frac{H_{2n}}{n} - \frac{1}{2} H_n^{(2)} \right) x^n,$$

and thus the coefficient of the power series is

$$c_n = H_n (H_{2n} - H_n) - \frac{2}{n} (H_{2n} - H_n) + \frac{H_{2n}}{n} - \frac{1}{2} H_n^{(2)},$$

and the solution to the point (v) of the problem is finalized.

Besides the nice applications of the sum result at the point (ii) we met in this section, we will also find it (very) useful later during the extraction process of the atypical series involving the harmonic numbers of the type H_{2n}.

6.14   Opening the World of Harmonic Series with Beautiful Series That Require Athletic Movements During Their Calculations: The First (Enjoyable) Part

Solution One of the most beautiful and powerful ways to approach the world of harmonic series is by using series manipulations, often with no use of integrals. Sometimes, despite the power offered by the series manipulations, the way to go is not so easy to figure out.

On the other hand, sometimes one may feel more comfortable to simply use integrals to calculate certain harmonic series, like the series given in this section, since the picture with the way to go is easier to decipher.

In the following, I'll use a strategy with simple steps to calculate both series and make everything as clear as possible.

For the first point of the problem, we start the fact that $\sum\limits_{n=1}^{\infty} \dfrac{1}{n(n+k)} = \dfrac{H_k}{k}$, and

then we write

$$S = \sum_{k=1}^{\infty} \frac{H_k}{2k(2k+1)} = \sum_{k=1}^{\infty} \frac{1}{2k+1} \left(\sum_{n=1}^{\infty} \frac{1}{n(2n+2k)} \right)$$

{reverse the order of summation}

$$= \sum_{n=1}^{\infty} \frac{1}{n} \left(\sum_{k=1}^{\infty} \frac{1}{(2k+1)(2n+2k)} \right) = \sum_{n=1}^{\infty} \frac{1}{n(2n-1)} \left(\sum_{k=1}^{\infty} \left(\frac{1}{2k+1} - \frac{1}{2n+2k} \right) \right)$$

{add and subtract $1/(2k)$}

$$= \sum_{n=1}^{\infty} \frac{1}{n(2n-1)} \left(\sum_{k=1}^{\infty} \left(-\frac{1}{2k} + \frac{1}{2k+1} + \frac{1}{2k} - \frac{1}{2n+2k} \right) \right)$$

$$= \sum_{n=1}^{\infty} \frac{1}{n(2n-1)} \left(\sum_{k=1}^{\infty} \left(-\frac{1}{2k} + \frac{1}{2k+1} \right) \right) + \frac{1}{2} \sum_{n=1}^{\infty} \frac{1}{n(2n-1)} \left(\sum_{k=1}^{\infty} \left(\frac{1}{k} - \frac{1}{n+k} \right) \right)$$

$$\left\{ \text{use that } \sum_{k=1}^{\infty} \left(-\frac{1}{2k} + \frac{1}{2k+1} \right) = -1 + \sum_{k=1}^{\infty} (-1)^{k-1} \frac{1}{k} = \log(2) - 1, \right\}$$

$$\left\{ \text{together with the simple fact that } \sum_{k=1}^{\infty} \left(\frac{1}{k} - \frac{1}{n+k} \right) = H_n \right\}$$

$$= 2(\log(2) - 1) \sum_{n=1}^{\infty} \left(\frac{1}{2n-1} - \frac{1}{2n} \right) + \frac{1}{2} \sum_{n=1}^{\infty} \frac{H_n}{n(2n-1)}$$

$$\left\{ \text{use that } \sum_{n=1}^{\infty} \left(\frac{1}{2n-1} - \frac{1}{2n} \right) = \sum_{n=1}^{\infty} (-1)^{n-1} \frac{1}{n} = \log(2), \right\}$$

{and then reindex the second series and start from $n = 0$}

$$= 2\log(2)(\log(2) - 1) + \frac{1}{2} \sum_{n=0}^{\infty} \frac{H_n + 1/(n+1)}{(n+1)(2n+1)}$$

{leave out the term for $n = 0$, use the partial fraction expansion,}

$$\left\{ \text{and rearrange the summand using that } H_{n+1} = H_n + \frac{1}{n+1} \right\}$$

$$= 2\log(2)(\log(2) - 1) + \frac{1}{2} + \frac{1}{2}\sum_{n=1}^{\infty}\left(\frac{4}{2n+1} - \frac{4}{2n+2} + 2\frac{H_n}{2n+1} - \frac{H_{n+1}}{n+1}\right)$$

{inside the summand add and subtract H_n/n, and then wisely split the series}

$$= 2\log(2)(\log(2) - 1) + \frac{1}{2}$$

$$+ 2\sum_{n=1}^{\infty}\left(\frac{1}{2n+1} - \frac{1}{2n+2}\right) - \underbrace{\sum_{n=1}^{\infty}\frac{H_n}{2n(2n+1)}}_{S} + \frac{1}{2}\sum_{n=1}^{\infty}\left(\frac{H_n}{n} - \frac{H_{n+1}}{n+1}\right)$$

$$\left\{\text{use that } \sum_{n=1}^{\infty}\left(\frac{1}{2n+1} - \frac{1}{2n+2}\right) = -\frac{1}{2} + \sum_{k=1}^{\infty}(-1)^{k-1}\frac{1}{k} = -\frac{1}{2} + \log(2)\right\}$$

$$= 2\log^2(2) - \frac{1}{2} - S + \frac{1}{2}\lim_{N\to\infty}\sum_{n=1}^{N}\left(\frac{H_n}{n} - \frac{H_{n+1}}{n+1}\right)$$

$$= 2\log^2(2) - \frac{1}{2} - S + \frac{1}{2}\lim_{N\to\infty}\left(1 - \frac{H_{N+1}}{N+1}\right)$$

$$= 2\log^2(2) - S,$$

whence we obtain that

$$S = \log^2(2),$$

where in the last calculations I also used the *Stolz–Cesàro theorem* (see [15, Appendix, pp.263–266]) in order to the prove that $\lim_{N\to\infty}\dfrac{H_N}{N} = 0$, and the solution to the point (i) of the problem is complete.

To calculate the series at the point (ii), we may start with a different development of the series from the point (i), and using the fact that $\sum_{n=1}^{k}\dfrac{1}{n} = H_k$, we write

$$\log^2(2) = \sum_{k=1}^{\infty}\frac{H_k}{2k(2k+1)} = \sum_{k=1}^{\infty}\sum_{n=1}^{k}\frac{1}{2k(2k+1)n}$$

$$\underset{\substack{\text{reverse the order}\\\text{of summation}}}{=} \sum_{n=1}^{\infty}\frac{1}{n}\sum_{k=n}^{\infty}\frac{1}{2k(2k+1)} = \sum_{n=1}^{\infty}\frac{1}{n}\left(\left(\sum_{k=1}^{\infty} - \sum_{k=1}^{n-1}\right)\left(\frac{1}{2k} - \frac{1}{2k+1}\right)\right)$$

$$= \sum_{n=1}^{\infty} \frac{1}{n} \left(\sum_{k=1}^{\infty} \left(\frac{1}{2k} - \frac{1}{2k+1} \right) - \sum_{k=1}^{n-1} \left(\frac{1}{2k} - \frac{1}{2k+1} \right) \right)$$

$$\left\{ \text{note and use that } \sum_{k=1}^{\infty} \left(\frac{1}{2k} - \frac{1}{2k+1} \right) = 1 - \sum_{k=1}^{\infty} (-1)^{k-1} \frac{1}{k} = 1 - \log(2) \right\}$$

$$\left\{ \text{together with the fact that } \sum_{k=1}^{n-1} \left(\frac{1}{2k} - \frac{1}{2k+1} \right) = 1 - H_{2n} + H_n - 1/(2n) \right\}$$

$$= \sum_{n=1}^{\infty} \frac{1}{n} \left(H_{2n} - H_n - \log(2) + \frac{1}{2n} \right) = \sum_{n=1}^{\infty} \frac{1}{n} (H_{2n} - H_n - \log(2)) + \frac{1}{2} \sum_{n=1}^{\infty} \frac{1}{n^2}$$

$$\left\{ \text{the value of the last series comes from the Basel problem, } \sum_{n=1}^{\infty} \frac{1}{n^2} = \zeta(2) = \frac{\pi^2}{6}, \right\}$$

{which you might also find with a simple solution in [76, Chapter 3, pp. 55–57]}

$$= \frac{\pi^2}{12} + \lim_{N \to \infty} \sum_{n=1}^{N} \frac{1}{n} (H_{2n} - H_n - \log(2))$$

$$= \frac{\pi^2}{12} + \lim_{N \to \infty} \left(2 \sum_{n=1}^{N} \frac{H_{2n}}{2n} - \sum_{n=1}^{N} \frac{H_n}{n} - \log(2) \sum_{n=1}^{N} \frac{1}{n} \right)$$

$$= \frac{\pi^2}{12} + \lim_{N \to \infty} \left(\sum_{n=1}^{2N} (-1)^n \frac{H_n}{n} + \sum_{n=1}^{2N} \frac{H_n}{n} - \sum_{n=1}^{N} \frac{H_n}{n} - \log(2) \sum_{n=1}^{N} \frac{1}{n} \right)$$

$$\left\{ \text{rearrange and use that } \sum_{k=1}^{n} \frac{H_k}{k} = \frac{1}{2} \left(H_n^2 + H_n^{(2)} \right), \text{ based on (6.102), Sect. 6.13} \right\}$$

$$= \frac{\pi^2}{12} - \sum_{n=1}^{\infty} (-1)^{n-1} \frac{H_n}{n} + \frac{1}{2} \lim_{N \to \infty} \left(H_{2N}^2 - H_N^2 + H_{2N}^{(2)} - H_N^{(2)} - 2\log(2) H_N \right)$$

{use the asymptotic expansion, $H_n = \gamma + \log(n) + O(1/n)$, as $n \to \infty$, }

$$\left\{ \text{in order to prove that } \lim_{N \to \infty} \left(H_{2N}^2 - H_N^2 + H_{2N}^{(2)} - H_N^{(2)} - 2\log(2) H_N \right) = \log^2(2) \right\}$$

$$= \frac{\pi^2}{12} + \frac{1}{2}\log^2(2) - \sum_{n=1}^{\infty}(-1)^{n-1}\frac{H_n}{n},$$

whence we obtain that

$$\sum_{n=1}^{\infty}(-1)^{n-1}\frac{H_n}{n} = \frac{1}{2}\left(\zeta(2) - \log^2(2)\right),$$

and the solution to the point (ii) of the problem is complete. Alternatively, for another creative solution we may exploit the series result in (6.129), Sect. 6.16,
$$\sum_{n=1}^{\infty}\frac{H_{2n+1} - H_n}{2n(2n+1)}.$$

What a dynamite solution! All the simple mathematical connections involved in the calculations create such a beautiful picture of the solution!

The solution to the series at the point (ii) is built immediately on the use of the series from the point (i). Now, the series $\displaystyle\sum_{n=1}^{\infty}\frac{1}{n}(H_{2n} - H_n - \log(2))$ already appeared in *(Almost) Impossible Integrals, Sums, and Series*, in [76, Chapter 3, p.250], where it plays the part of an auxiliary result, and here I essentially used a part of the flow of its solution described there.

6.15   Opening the World of Harmonic Series with Beautiful Series That Require Athletic Movements During Their Calculations: The Second (Enjoyable) Part

Solution As you may guess, especially if you covered the past section, I'll attack the present problem by using a strategy similar to the one in the previous section. And remember, I want to accomplish the calculations by mainly using series manipulations, with no use of integrals!

Basically, upon using simple operations with series, I'll create a system of two different relations with the two series that will allow me to extract the desired values.

Afterward, we would also like to attack the enjoyable *challenging question*, found at the point (iii) of the problem, which involves a sum of two alternating harmonic series of weight 4. So, we have much fun ahead!

In the first instance, let's consider the simple fact that $H_k^{(2)} = \displaystyle\sum_{n=1}^{\infty}\left(\frac{1}{n^2} - \frac{1}{(k+n)^2}\right)$,

and then the series at the point (i) may be written as follows:

$$\sum_{k=1}^{\infty} \frac{H_k^{(2)}}{2k(2k+1)} = \sum_{k=1}^{\infty} \frac{1}{2k(2k+1)} \left(\sum_{n=1}^{\infty} \left(\frac{1}{n^2} - \frac{1}{(k+n)^2} \right) \right)$$

{reverse the order of summation and wisely expand the series}

$$= \sum_{n=1}^{\infty} \frac{1}{n^2} \sum_{k=1}^{\infty} \frac{1}{2k(2k+1)} - \sum_{n=1}^{\infty} \left(\sum_{k=1}^{\infty} \frac{1}{2k(k+n)^2} \right) + \sum_{n=1}^{\infty} \left(\sum_{k=1}^{\infty} \frac{1}{(2k+1)(k+n)^2} \right)$$

$$\left\{ \text{note and use that } \sum_{k=1}^{\infty} \left(\frac{1}{2k} - \frac{1}{2k+1} \right) = 1 - \sum_{k=1}^{\infty} (-1)^{k-1} \frac{1}{k} = 1 - \log(2) \right\}$$

$$= \zeta(2) - \log(2)\zeta(2) - \underbrace{\sum_{n=1}^{\infty} \left(\sum_{k=1}^{\infty} \frac{1}{2k(k+n)^2} \right)}_{T} + \underbrace{\sum_{n=1}^{\infty} \left(\sum_{k=1}^{\infty} \frac{1}{(2k+1)(k+n)^2} \right)}_{U}.$$

$$(6.111)$$

For the double series T in (6.111), we write

$$T = \sum_{n=1}^{\infty} \left(\sum_{k=1}^{\infty} \frac{1}{2k(k+n)^2} \right) = \frac{1}{2} \sum_{n=1}^{\infty} \frac{1}{n} \left(\sum_{k=1}^{\infty} \frac{1}{k(k+n)} \right) - \sum_{n=1}^{\infty} \left(\sum_{k=1}^{\infty} \frac{1}{2n(k+n)^2} \right)$$

$$\left\{ \text{for the inner series of the first double series use that } \sum_{k=1}^{\infty} \frac{1}{k(k+n)} = \frac{H_n}{n}, \right\}$$

{and then reverse the order of summation in the second double series}

$$= \frac{1}{2} \sum_{n=1}^{\infty} \frac{H_n}{n^2} - \underbrace{\sum_{k=1}^{\infty} \left(\sum_{n=1}^{\infty} \frac{1}{2n(k+n)^2} \right)}_{T} = \zeta(3) - T,$$

whence we obtain that

$$T = \frac{1}{2}\zeta(3), \qquad (6.112)$$

where $\sum_{n=1}^{\infty} \frac{H_n}{n^2} = 2\zeta(3)$ is the particular case $n = 2$ of the Euler sum generalization in (6.149), Sect. 6.19.

Next, for the double series U in (6.111), we have

$$U = \sum_{n=1}^{\infty} \left(\sum_{k=1}^{\infty} \frac{1}{(2k+1)(k+n)^2} \right)$$

$$= 4 \underbrace{\sum_{n=1}^{\infty} \frac{1}{(2n-1)^2} \left(\sum_{k=1}^{\infty} \left(\frac{1}{2k+1} - \frac{1}{2k+2n} \right) \right)}_{U_1} - \underbrace{\sum_{n=1}^{\infty} \left(\sum_{k=1}^{\infty} \frac{1}{(2n-1)(k+n)^2} \right)}_{U_2}.$$

$$(6.113)$$

As regards the series U_1 in (6.113), we need an artifice of calculation for the summand of the inner series, and then we write

$$U_1 = \sum_{n=1}^{\infty} \frac{1}{(2n-1)^2} \left(\sum_{k=1}^{\infty} \left(\frac{1}{2k+1} - \frac{1}{2k+2n} \right) \right)$$

$$= \sum_{n=1}^{\infty} \frac{1}{(2n-1)^2} \left(\sum_{k=1}^{\infty} \left(\frac{1}{2k+1} - \frac{1}{2k} + \frac{1}{2k} - \frac{1}{2k+2n} \right) \right)$$

$$= \sum_{n=1}^{\infty} \frac{1}{(2n-1)^2} \sum_{k=1}^{\infty} \left(\frac{1}{2k+1} - \frac{1}{2k} \right) + \frac{1}{2} \sum_{n=1}^{\infty} \frac{1}{(2n-1)^2} \left(\sum_{k=1}^{\infty} \left(\frac{1}{k} - \frac{1}{k+n} \right) \right)$$

$$\left\{ \text{consider that } \sum_{k=1}^{\infty} \left(\frac{1}{2k+1} - \frac{1}{2k} \right) = -1 + \sum_{k=1}^{\infty} (-1)^{k-1} \frac{1}{k} = \log(2) - 1, \right\}$$

$$\left\{ \text{then observe and use that } \sum_{n=1}^{\infty} \frac{1}{(2n-1)^2} = \sum_{n=1}^{\infty} \frac{1}{n^2} - \sum_{n=1}^{\infty} \frac{1}{(2n)^2} = \frac{3}{4} \zeta(2), \right\}$$

$$\left\{ \text{and employ the harmonic number representation, } \sum_{k=1}^{\infty} \left(\frac{1}{k} - \frac{1}{k+n} \right) = H_n \right\}$$

$$= \frac{3}{4} \log(2) \zeta(2) - \frac{3}{4} \zeta(2) + \frac{1}{2} \sum_{n=1}^{\infty} \frac{H_n}{(2n-1)^2}$$

{reindex the series and carefully expand it}

$$= \frac{3}{4} \log(2) \zeta(2) - \frac{3}{4} \zeta(2) + \frac{1}{2} + \frac{1}{2} \sum_{n=1}^{\infty} \frac{H_n}{(2n+1)^2} + \sum_{n=1}^{\infty} \left(\frac{1}{2n+2} - \frac{1}{2n+1} \right)$$

$$+\sum_{n=1}^{\infty}\frac{1}{(2n+1)^2} = \frac{7}{8}\zeta(3) - \log(2), \tag{6.114}$$

where to get the last equality above I also considered the case $m = 1$ of the result in (4.100), Sect. 4.19.

Further, regarding the series U_2 in (6.113), we start with reindexing the inner series and changing the summation order, and then we have

$$U_2 = \sum_{n=1}^{\infty}\left(\sum_{k=1}^{\infty}\frac{1}{(2n-1)(k+n)^2}\right) = \sum_{n=1}^{\infty}\left(\sum_{k=n+1}^{\infty}\frac{1}{(2n-1)k^2}\right)$$

$$\overset{\text{reverse the order}}{\underset{\text{of summation}}{=}} \sum_{k=1}^{\infty}\frac{1}{k^2}\left(\sum_{n=1}^{k-1}\frac{1}{2n-1}\right) = \sum_{k=1}^{\infty}\frac{1}{k^2}\left(-\frac{1}{2k-1}+\sum_{n=1}^{k}\frac{1}{2n-1}\right)$$

$$\left\{\text{note and use that } \sum_{n=1}^{k}\frac{1}{2n-1} = H_{2k} - \frac{1}{2}H_k\right\}$$

$$= 4\sum_{k=1}^{\infty}\frac{H_{2k}}{(2k)^2} - \frac{1}{2}\sum_{k=1}^{\infty}\frac{H_k}{k^2} - \sum_{k=1}^{\infty}\frac{1}{k^2(2k-1)}$$

$$\left\{\text{for the first sum use that } \sum_{n=1}^{\infty}a_{2n} = \frac{1}{2}\left(\sum_{n=1}^{\infty}a_n - \sum_{n=1}^{\infty}(-1)^{n-1}a_n\right)\right\}$$

$$\{\text{and afterwards carefully expand the third series}\}$$

$$= \frac{3}{2}\sum_{k=1}^{\infty}\frac{H_k}{k^2} - 2\sum_{k=1}^{\infty}(-1)^{k-1}\frac{H_k}{k^2} - 4\sum_{k=1}^{\infty}\left(\frac{1}{2k-1}-\frac{1}{2k}\right) + \sum_{k=1}^{\infty}\frac{1}{k^2}$$

$$= \zeta(2) + 3\zeta(3) - 4\log(2) - 2\sum_{k=1}^{\infty}(-1)^{k-1}\frac{H_k}{k^2}, \tag{6.115}$$

where in the calculations I also used the case $n = 2$ of the generalization in (6.149), Sect. 6.19.

If we plug the results from (6.114) and (6.115) in (6.113), we get

$$U = \sum_{n=1}^{\infty}\left(\sum_{k=1}^{\infty}\frac{1}{(2k+1)(k+n)^2}\right) = \frac{1}{2}\zeta(3) - \zeta(2) + 2\sum_{k=1}^{\infty}(-1)^{k-1}\frac{H_k}{k^2}. \tag{6.116}$$

Now, by plugging the results from (6.112) and (6.116) in (6.111), we obtain a first critical relation

$$2\sum_{k=1}^{\infty}(-1)^{k-1}\frac{H_k}{k^2} - \sum_{k=1}^{\infty}\frac{H_k^{(2)}}{2k(2k+1)} = \log(2)\zeta(2). \tag{6.117}$$

To get a second critical relation with the main series, we first start with proving a more general result

$$\sum_{k=1}^{\infty}\frac{H_k^{(m)}}{2k(2k+1)}$$

$$= \frac{1}{2}((m+2)(2^{m-1}-1)+1)\zeta(m+1) - \log(2)\zeta(m)$$

$$-\frac{1}{2}(2^{m-1}-1)\sum_{k=1}^{m-2}\zeta(k+1)\zeta(m-k) - 2^{m-1}\sum_{n=1}^{\infty}(-1)^{n-1}\frac{H_n}{n^m}, \quad m \geq 2. \tag{6.118}$$

Proof Essentially, we consider a flow similar to the one in the second part of the previous section, and then, using that $H_k^{(m)} = \sum_{n=1}^{k}\frac{1}{n^m}$, we write

$$\sum_{k=1}^{\infty}\frac{H_k^{(m)}}{2k(2k+1)} = \sum_{k=1}^{\infty}\left(\sum_{n=1}^{k}\frac{1}{2k(2k+1)n^m}\right)$$

$$\overset{\text{reverse the order}}{\underset{\text{of summation}}{=}} \sum_{n=1}^{\infty}\frac{1}{n^m}\left(\sum_{k=n}^{\infty}\frac{1}{2k(2k+1)}\right) = \sum_{n=1}^{\infty}\frac{1}{n^m}\left(\left(\sum_{k=1}^{\infty}-\sum_{k=1}^{n-1}\right)\left(\frac{1}{2k}-\frac{1}{2k+1}\right)\right)$$

$$= \sum_{n=1}^{\infty}\frac{1}{n^m}\left(\sum_{k=1}^{\infty}\left(\frac{1}{2k}-\frac{1}{2k+1}\right) - \sum_{k=1}^{n-1}\left(\frac{1}{2k}-\frac{1}{2k+1}\right)\right)$$

$$\left\{\text{note and use that } \sum_{k=1}^{\infty}\left(\frac{1}{2k}-\frac{1}{2k+1}\right) = 1 - \sum_{k=1}^{\infty}(-1)^{k-1}\frac{1}{k} = 1 - \log(2)\right\}$$

$$\left\{\text{together with the fact that } \sum_{k=1}^{n-1}\left(\frac{1}{2k}-\frac{1}{2k+1}\right) = 1 - H_{2n} + H_n - 1/(2n)\right\}$$

$$= \sum_{n=1}^{\infty}\frac{1}{n^m}\left(H_{2n} - H_n - \log(2) + \frac{1}{2n}\right)$$

$$= 2^m \sum_{n=1}^{\infty} \frac{H_{2n}}{(2n)^m} - \sum_{n=1}^{\infty} \frac{H_n}{n^m} - \log(2) \sum_{n=1}^{\infty} \frac{1}{n^m} + \frac{1}{2} \sum_{n=1}^{\infty} \frac{1}{n^{m+1}}$$

$$\left\{ \text{for the first sum use that } \sum_{n=1}^{\infty} a_{2n} = \frac{1}{2} \left(\sum_{n=1}^{\infty} a_n - \sum_{n=1}^{\infty} (-1)^{n-1} a_n \right) \right\}$$

$$= (2^{m-1} - 1) \sum_{n=1}^{\infty} \frac{H_n}{n^m} - 2^{m-1} \sum_{n=1}^{\infty} (-1)^{n-1} \frac{H_n}{n^m} - \log(2)\zeta(m) + \frac{1}{2}\zeta(m+1)$$

{the value of the first generalized series is given in (6.149), Sect. 6.19}

$$= \frac{1}{2}((m+2)(2^{m-1} - 1) + 1)\zeta(m+1) - \log(2)\zeta(m)$$

$$- \frac{1}{2}(2^{m-1} - 1) \sum_{k=1}^{m-2} \zeta(k+1)\zeta(m-k) - 2^{m-1} \sum_{n=1}^{\infty} (-1)^{n-1} \frac{H_n}{n^m},$$

and the proof is complete. ∎

Returning in (6.118) and setting $m = 2$, we get a second critical relation

$$2 \sum_{k=1}^{\infty} (-1)^{k-1} \frac{H_k}{k^2} + \sum_{k=1}^{\infty} \frac{H_k^{(2)}}{2k(2k+1)} = \frac{5}{2}\zeta(3) - \log(2)\zeta(2). \qquad (6.119)$$

At last, combining the relations in (6.117) and (6.119), we obtain that

$$\sum_{k=1}^{\infty} (-1)^{k-1} \frac{H_k}{k^2} = \frac{5}{8}\zeta(3)$$

and

$$\sum_{k=1}^{\infty} \frac{H_k^{(2)}}{2k(2k+1)} = \frac{5}{4}\zeta(3) - \log(2)\zeta(2),$$

and the solution to the points (i) and (ii) of the problem is complete.

An important note: The curious reader might like to observe that for extracting the value of the series $\sum_{k=1}^{\infty} (-1)^{k-1} \frac{H_k}{k^2}$, it is enough to wisely reduce the double series $\sum_{n=1}^{\infty} \left(\sum_{k=1}^{\infty} \frac{1}{(2k+1)(k+n)^2} \right)$ to two different results containing the mentioned series such that we can extract the desired value.

How about the challenging question? Well, admittedly, the way to go is not obvious, but one might be inspired by the previous work in this section. In fact, we need a similar strategy to the one above, but this time, we'll focus on approaching the double series $\sum_{n=1}^{\infty}\left(\sum_{k=1}^{\infty}\dfrac{1}{(2k+1)(k+n)^3}\right)$ in two different ways. Also, you can observe the similarity between this strategy I have just suggested and the one in the comment above under *An important note*.

Let's proceed with the partial fraction expansion, and then we write

$$V = \sum_{n=1}^{\infty}\left(\sum_{k=1}^{\infty}\frac{1}{(2k+1)(k+n)^3}\right)$$

$$= 8\underbrace{\sum_{n=1}^{\infty}\frac{1}{(2n-1)^3}\left(\sum_{k=1}^{\infty}\left(\frac{1}{2k+1}-\frac{1}{2k+2n}\right)\right)}_{V_1} - 2\underbrace{\sum_{n=1}^{\infty}\frac{1}{(2n-1)^2}\left(\sum_{k=1}^{\infty}\frac{1}{(k+n)^2}\right)}_{V_2}$$

$$-\underbrace{\sum_{n=1}^{\infty}\frac{1}{2n-1}\left(\sum_{k=1}^{\infty}\frac{1}{(k+n)^3}\right)}_{V_3}. \qquad (6.120)$$

Now, for the series V_1 in (6.120), we have a similar flow to the one of U_1 in (6.114), and then we write

$$V_1 = \sum_{n=1}^{\infty}\frac{1}{(2n-1)^3}\left(\sum_{k=1}^{\infty}\left(\frac{1}{2k+1}-\frac{1}{2k+2n}\right)\right)$$

$$= \sum_{n=1}^{\infty}\frac{1}{(2n-1)^3}\left(\sum_{k=1}^{\infty}\left(\frac{1}{2k+1}-\frac{1}{2k}+\frac{1}{2k}-\frac{1}{2k+2n}\right)\right)$$

$$= \sum_{n=1}^{\infty}\frac{1}{(2n-1)^3}\sum_{k=1}^{\infty}\left(\frac{1}{2k+1}-\frac{1}{2k}\right)+\frac{1}{2}\sum_{n=1}^{\infty}\frac{1}{(2n-1)^3}\left(\sum_{k=1}^{\infty}\left(\frac{1}{k}-\frac{1}{k+n}\right)\right)$$

$$\left\{\text{consider that }\sum_{k=1}^{\infty}\left(\frac{1}{2k+1}-\frac{1}{2k}\right)=-1+\sum_{k=1}^{\infty}(-1)^{k-1}\frac{1}{k}=\log(2)-1,\right\}$$

$$\left\{\text{then observe and use that }\sum_{n=1}^{\infty}\frac{1}{(2n-1)^3}=\sum_{n=1}^{\infty}\frac{1}{n^3}-\sum_{n=1}^{\infty}\frac{1}{(2n)^3}=\frac{7}{8}\zeta(3),\right\}$$

$$\left\{ \text{and employ the harmonic number representation, } \sum_{k=1}^{\infty} \left(\frac{1}{k} - \frac{1}{k+n} \right) = H_n \right\}$$

$$= \frac{7}{8} \log(2) \zeta(3) - \frac{7}{8} \zeta(3) + \frac{1}{2} \sum_{n=1}^{\infty} \frac{H_n}{(2n-1)^3}$$

{reindex the series and carefully expand it}

$$= \frac{7}{8} \log(2) \zeta(3) - \frac{7}{8} \zeta(3) + \frac{1}{2} + \frac{1}{2} \sum_{n=1}^{\infty} \frac{H_n}{(2n+1)^3} + \sum_{n=1}^{\infty} \left(\frac{1}{2n+1} - \frac{1}{2n+2} \right)$$

$$- \sum_{n=1}^{\infty} \frac{1}{(2n+1)^2} + \sum_{n=1}^{\infty} \frac{1}{(2n+1)^3} = \log(2) + \frac{45}{64} \zeta(4) - \frac{3}{4} \zeta(2), \qquad (6.121)$$

where in the calculations I also used the particular case $p = 3$ of the generalization in (4.102), Sect. 4.19.

Then, for the series V_2 in (6.120), we start with reindexing the inner series and then changing the summation order. Thus, we have

$$V_2 = \sum_{n=1}^{\infty} \frac{1}{(2n-1)^2} \left(\sum_{k=1}^{\infty} \frac{1}{(k+n)^2} \right) = \sum_{n=1}^{\infty} \frac{1}{(2n-1)^2} \left(\sum_{k=n+1}^{\infty} \frac{1}{k^2} \right)$$

$$= \sum_{k=1}^{\infty} \frac{1}{k^2} \left(\sum_{n=1}^{k-1} \frac{1}{(2n-1)^2} \right) = \sum_{k=1}^{\infty} \frac{1}{k^2} \left(\sum_{n=1}^{k} \frac{1}{(2n-1)^2} \right) - \sum_{k=1}^{\infty} \frac{1}{k^2(2k-1)^2}$$

$$\left\{ \text{employ the fact that } \sum_{n=1}^{k} \frac{1}{(2n-1)^2} = H_{2k}^{(2)} - \frac{1}{4} H_k^{(2)} \right\}$$

$$= \sum_{k=1}^{\infty} \frac{H_{2k}^{(2)} - 1/4 H_k^{(2)}}{k^2} + 8 \sum_{k=1}^{\infty} \left(\frac{1}{2k-1} - \frac{1}{2k} \right) - \sum_{k=1}^{\infty} \frac{1}{k^2} - 4 \sum_{k=1}^{\infty} \frac{1}{(2k-1)^2}$$

$$\left\{ \text{note and use that } \sum_{k=1}^{\infty} \left(\frac{1}{2k-1} - \frac{1}{2k} \right) = \sum_{k=1}^{\infty} (-1)^{k-1} \frac{1}{k} = \log(2) \right\}$$

$$= 4 \sum_{k=1}^{\infty} \frac{H_{2k}^{(2)}}{(2k)^2} - \frac{1}{4} \sum_{k=1}^{\infty} \frac{H_k^{(2)}}{k^2} + 8 \log(2) - 4 \zeta(2)$$

$$\left\{\text{for the first series use that } \sum_{n=1}^{\infty} a_{2n} = \frac{1}{2}\left(\sum_{n=1}^{\infty} a_n - \sum_{n=1}^{\infty}(-1)^{n-1}a_n\right)\right\}$$

$$= \frac{7}{4}\sum_{k=1}^{\infty}\frac{H_k^{(2)}}{k^2} - 2\sum_{k=1}^{\infty}(-1)^{k-1}\frac{H_k^{(2)}}{k^2} - 4\zeta(2) + 8\log(2)$$

$$\left\{\text{setting } p = q = 2, \text{ with } n \to \infty \text{ in (6.102), Sect. 6.13, we get } \sum_{k=1}^{\infty}\frac{H_k^{(2)}}{k^2} = \frac{7}{4}\zeta(4)\right\}$$

$$= \frac{49}{16}\zeta(4) - 4\zeta(2) + 8\log(2) - 2\sum_{k=1}^{\infty}(-1)^{k-1}\frac{H_k^{(2)}}{k^2}. \tag{6.122}$$

Further, for the series V_3 in (6.120), we proceed as in the previous case with the series V_2, and then we write

$$V_3 = \sum_{n=1}^{\infty}\frac{1}{2n-1}\left(\sum_{k=1}^{\infty}\frac{1}{(k+n)^3}\right) \overset{\substack{\text{reindex}\\\text{the inner series}}}{=} \sum_{n=1}^{\infty}\frac{1}{2n-1}\left(\sum_{k=n+1}^{\infty}\frac{1}{k^3}\right)$$

$$\overset{\substack{\text{reverse the order}\\\text{of summation}}}{=} \sum_{k=1}^{\infty}\frac{1}{k^3}\left(\sum_{n=1}^{k-1}\frac{1}{2n-1}\right) = \sum_{k=1}^{\infty}\frac{1}{k^3}\left(\sum_{n=1}^{k}\frac{1}{2n-1}\right) - \sum_{k=1}^{\infty}\frac{1}{k^3(2k-1)}$$

$$\left\{\text{employ the fact that } \sum_{n=1}^{k}\frac{1}{2n-1} = H_{2k} - \frac{1}{2}H_k\right\}$$

$$= \sum_{k=1}^{\infty}\frac{H_{2k} - 1/2H_k}{k^3} - 8\sum_{k=1}^{\infty}\left(\frac{1}{2k-1} - \frac{1}{2k}\right) + 2\sum_{k=1}^{\infty}\frac{1}{k^2} + \sum_{k=1}^{\infty}\frac{1}{k^3}$$

$$\left\{\text{note and use that } \sum_{k=1}^{\infty}\left(\frac{1}{2k-1} - \frac{1}{2k}\right) = \sum_{k=1}^{\infty}(-1)^{k-1}\frac{1}{k} = \log(2)\right\}$$

$$= 8\sum_{k=1}^{\infty}\frac{H_{2k}}{(2k)^3} - \frac{1}{2}\sum_{k=1}^{\infty}\frac{H_k}{k^3} - 8\log(2) + 2\zeta(2) + \zeta(3)$$

$$\left\{\text{for the first series use that } \sum_{n=1}^{\infty} a_{2n} = \frac{1}{2}\left(\sum_{n=1}^{\infty} a_n - \sum_{n=1}^{\infty}(-1)^{n-1}a_n\right)\right\}$$

$$= \frac{7}{2} \sum_{k=1}^{\infty} \frac{H_k}{k^3} - 4 \sum_{k=1}^{\infty} (-1)^{k-1} \frac{H_k}{k^3} - 8 \log(2) + 2\zeta(2) + \zeta(3)$$

{the first series is the case $n = 3$ of the generalization in (6.149), Sect. 6.19}

$$= \frac{35}{8} \zeta(4) + \zeta(3) + 2\zeta(2) - 8 \log(2) - 4 \sum_{k=1}^{\infty} (-1)^{k-1} \frac{H_k}{k^3}. \tag{6.123}$$

Collecting the values of the series V_1, V_2, and V_3 from (6.121), (6.122), and (6.123) in (6.120), we get the series V expressed as

$$V = 4 \sum_{k=1}^{\infty} (-1)^{k-1} \frac{H_k}{k^3} + 4 \sum_{k=1}^{\infty} (-1)^{k-1} \frac{H_k^{(2)}}{k^2} - \zeta(3) - \frac{39}{8} \zeta(4). \tag{6.124}$$

On the other hand, by a different approach of (6.120), we have

$$V = \sum_{n=1}^{\infty} \left(\sum_{k=1}^{\infty} \frac{1}{(2k+1)(k+n)^3} \right) = \sum_{k=1}^{\infty} \left(\sum_{n=1}^{\infty} \frac{1}{(2k+1)(k+n)^3} \right)$$

{reindex the inner series and start from $n = k + 1$}

$$= \sum_{k=1}^{\infty} \left(\sum_{n=k+1}^{\infty} \frac{1}{(2k+1)n^3} \right) \overset{\substack{\text{reverse the order} \\ \text{of summation}}}{=} \sum_{n=1}^{\infty} \frac{1}{n^3} \left(\sum_{k=1}^{n-1} \frac{1}{2k+1} \right)$$

$$\left\{ \text{use that } \sum_{k=1}^{n-1} \frac{1}{2k+1} = H_{2n} - \frac{1}{2} H_n - 1 \right\}$$

$$= \sum_{n=1}^{\infty} \frac{H_{2n} - 1/2 H_n - 1}{n^3} = 8 \sum_{n=1}^{\infty} \frac{H_{2n}}{(2n)^3} - \frac{1}{2} \sum_{n=1}^{\infty} \frac{H_n}{n^3} - \sum_{n=1}^{\infty} \frac{1}{n^3}$$

$$\left\{ \text{for the first sum use that } \sum_{n=1}^{\infty} a_{2n} = \frac{1}{2} \left(\sum_{n=1}^{\infty} a_n - \sum_{n=1}^{\infty} (-1)^{n-1} a_n \right) \right\}$$

$$= \frac{7}{2} \sum_{n=1}^{\infty} \frac{H_n}{n^3} - 4 \sum_{n=1}^{\infty} (-1)^{n-1} \frac{H_n}{n^3} - \zeta(3) = \frac{35}{8} \zeta(4) - \zeta(3) - 4 \sum_{n=1}^{\infty} (-1)^{n-1} \frac{H_n}{n^3}. \tag{6.125}$$

Finally, if we combine the results from (6.124) and (6.125), we conclude that

$$2\sum_{k=1}^{\infty}(-1)^{k-1}\frac{H_k}{k^3} + \sum_{k=1}^{\infty}(-1)^{k-1}\frac{H_k^{(2)}}{k^2} = \frac{37}{16}\zeta(4),$$

and the solution to the challenging question is complete.

In one of the next sections, you'll find a generalization of the series at the point (ii) built on a clever use of series manipulations, without employing integrals. Here, it is worth mentioning that this last relation also appears in *(Almost) Impossible Integrals, Sums, and Series*, but in that work I also used integrals (for details, see [76, Chapter 6, pp. 505–506]).

Also, good to know that similar ideas may be further exploited for succeeding with other extractions involving harmonic series of higher weights.

6.16   A Special Harmonic Series in Disguise Involving Nice Tricks

Solution As a maker of mathematical problems and solutions one may have various memories related to the *creations* obtained, and sometimes one is beautifully hit with nostalgia, in particular when one remembers some amazing moments related to this process of creating mathematics.

The present section is about such an amazing moment!

This is the first problem I submitted to *La Gaceta de la RSME*, and it was published in Vol. 18, No. 3 (2015). In the following, I'll present my original solution in the problem submission, which also answers the (fascinating) *challenging question*. Indeed, it is a really good challenge to prove the result without using integrals!

In the solution below, we'll see that after an application of Abel's summation, we get, in the right-hand side, two times the initial series, with the opposite sign, plus other manageable series. Overall, this simple effect simplifies everything a lot.

Before performing the calculations, let's observe that the summand of our series may be written in terms of harmonic numbers as follows:

$$\sum_{n=1}^{\infty}\frac{1}{n(2n+1)}\left(\frac{1}{n+1}+\frac{1}{n+2}+\cdots+\frac{1}{2n+1}\right)^2 = 2\underbrace{\sum_{n=1}^{\infty}\frac{1}{2n(2n+1)}\left(H_{2n+1}-H_n\right)^2}_{S}.$$

$$(6.126)$$

Now, if we consider Abel's summation, the series version in (6.7), Sect. 6.2, for the series S in (6.126), where we set $a_n = \dfrac{1}{2n(2n+1)}$ and $b_n = (H_{2n+1}-H_n)^2$, we get

$$S = 2\sum_{n=1}^{\infty} \frac{1}{2n(2n+1)} (H_{2n+1} - H_n)^2 = 2 \lim_{N\to\infty} \sum_{k=1}^{N} \frac{1}{2k(2k+1)} (H_{2N+3} - H_{N+1})^2$$

$$- 2\sum_{n=1}^{\infty} \frac{2(H_{2n+1} - H_n) - \dfrac{1}{(2n+2)(2n+3)}}{(2n+2)(2n+3)} (H_{2n+1} - H_n - 1)$$

$$\left\{ \text{observe that } \sum_{k=1}^{N} \frac{1}{2k(2k+1)} = \sum_{k=1}^{N} \left(\frac{1}{2k} - \frac{1}{2k+1} \right) = 1 + H_N - H_{2N+1} \right\}$$

$$\left\{ = \frac{2N}{2N+1} + H_N - H_{2N} \text{ and use the fact that } \lim_{N\to\infty} (H_{2N} - H_N) = \log(2), \right\}$$

{where the last limit with harmonic numbers is obtained by using the limit}

$$\left\{ \text{definition of the Euler-Mascheroni constant, } \gamma = \lim_{N\to\infty} (H_N - \log(N)), \text{ and} \right\}$$

{at the same time reindex the resulting series and start again from $n = 1$}

$$= 2\log^2(2) - 2\log^3(2)$$

$$- 2\sum_{n=1}^{\infty} \frac{2(H_{2n-1} - H_{n-1}) - \dfrac{1}{2n(2n+1)}}{2n(2n+1)} (H_{2n-1} - H_{n-1} - 1)$$

{rearrange the summand of the series}

$$= 2\log^2(2) - 2\log^3(2)$$

$$- 2\sum_{n=1}^{\infty} \left(\frac{2(H_{2n+1} - H_n)}{2n(2n+1)} + \frac{1}{(2n(2n+1))^2} \right) \left(H_{2n+1} - H_n + \frac{1}{2n(2n+1)} - 1 \right)$$

$$= 2\log^2(2) - 2\log^3(2)$$

$$- 2\cdot2\underbrace{\sum_{n=1}^{\infty} \frac{1}{2n(2n+1)} (H_{2n+1} - H_n)^2}_{S} - 6\underbrace{\sum_{n=1}^{\infty} \frac{H_{2n+1} - H_n}{(2n(2n+1))^2}}_{T} + 4\underbrace{\sum_{n=1}^{\infty} \frac{H_{2n+1} - H_n}{2n(2n+1)}}_{U}$$

$$-2\underbrace{\sum_{n=1}^{\infty}\frac{1}{(2n(2n+1))^3}}_{10-6\log(2)-3\zeta(2)-3/4\zeta(3)}+2\underbrace{\sum_{n=1}^{\infty}\frac{1}{(2n(2n+1))^2}}_{2\log(2)+\zeta(2)-3}$$

$$=16\log(2)+2\log^2(2)-2\log^3(2)+8\zeta(2)+\frac{3}{2}\zeta(3)-26-2S-6T+4U. \quad (6.127)$$

Based on (6.127), we obtain that

$$S=\frac{16}{3}\log(2)+\frac{2}{3}\log^2(2)-\frac{2}{3}\log^3(2)+\frac{8}{3}\zeta(2)+\frac{1}{2}\zeta(3)-\frac{26}{3}-2T+\frac{4}{3}U. \quad (6.128)$$

Now, to calculate the series U in (6.128), we follow the same reasoning as in the main calculations above, and applying Abel's summation, the series version in (6.7), Sect. 6.2, where we set $a_n=\dfrac{1}{2n(2n+1)}$ and $b_n=H_{2n+1}-H_n$, we get

$$U=\sum_{n=1}^{\infty}\frac{H_{2n+1}-H_n}{2n(2n+1)}=\lim_{N\to\infty}\sum_{k=1}^{N}\frac{1}{2k(2k+1)}(H_{2N+3}-H_{N+1})$$

$$-\sum_{n=1}^{\infty}\frac{1}{(2n+2)(2n+3)}(H_{2n+1}-H_n-1)$$

$$=\log(2)-\log^2(2)-\sum_{n=1}^{\infty}\frac{1}{(2n+2)(2n+3)}(H_{2n+1}-H_n-1)$$

{reindex the series and start again from $n=1$}

$$=\log(2)-\log^2(2)-\sum_{n=1}^{\infty}\frac{1}{2n(2n+1)}(H_{2n-1}-H_{n-1}-1)$$

{rearrange the summand of the series}

$$=\log(2)-\log^2(2)-\sum_{n=1}^{\infty}\frac{1}{2n(2n+1)}\left(H_{2n+1}-H_n+\frac{1}{2n(2n+1)}-1\right)$$

$$=\log(2)-\log^2(2)-\underbrace{\sum_{n=1}^{\infty}\frac{H_{2n+1}-H_n}{2n(2n+1)}}_{U}-\underbrace{\sum_{n=1}^{\infty}\frac{1}{(2n(2n+1))^2}}_{\zeta(2)+2\log(2)-3}+\underbrace{\sum_{n=1}^{\infty}\frac{1}{2n(2n+1)}}_{1-\log(2)},$$

whence we obtain that

$$U = \sum_{n=1}^{\infty} \frac{H_{2n+1} - H_n}{2n(2n+1)} = 2 - \frac{1}{2}\zeta(2) - \log(2) - \frac{1}{2}\log^2(2).$$ (6.129)

To calculate the series T in (6.128), we proceed as follows:

$$T = \sum_{n=1}^{\infty} \frac{H_{2n+1} - H_n}{(2n(2n+1))^2} = \sum_{n=1}^{\infty} (H_{2n+1} - H_n) \left(\frac{1}{(2n)^2} - \frac{2}{2n(2n+1)} + \frac{1}{(2n+1)^2} \right)$$

$$= \sum_{n=1}^{\infty} \left(\frac{H_{2n}}{(2n)^2} + \frac{H_{2n+1}}{(2n+1)^2} \right) - \frac{1}{4} \sum_{n=1}^{\infty} \frac{H_n}{n^2} - \sum_{n=1}^{\infty} \frac{H_n}{(2n+1)^2}$$

$$+ \underbrace{\sum_{n=1}^{\infty} \frac{1}{(2n+1)(2n)^2}}_{1/4\zeta(2) + \log(2) - 1} \underbrace{-2 \sum_{n=1}^{\infty} \frac{H_{2n+1} - H_n}{2n(2n+1)}}_{U}$$

$$\left\{ \text{note and use that } \sum_{n=1}^{\infty} \left(\frac{H_{2n}}{(2n)^2} + \frac{H_{2n+1}}{(2n+1)^2} \right) = -1 + \sum_{n=1}^{\infty} \frac{H_n}{n^2} \right\}$$

$$= \frac{1}{4}\zeta(2) + \log(2) - 2 + \frac{3}{4} \sum_{n=1}^{\infty} \frac{H_n}{n^2} - \sum_{n=1}^{\infty} \frac{H_n}{(2n+1)^2} - 2U$$

$$= \frac{5}{4}\zeta(2) + \frac{3}{2}\log(2)\zeta(2) - \frac{1}{4}\zeta(3) + 3\log(2) + \log^2(2) - 6,$$ (6.130)

where in the calculations I used that $\sum_{n=1}^{\infty} \frac{H_n}{n^2} = 2\zeta(3)$, which is the particular case $n = 2$ of the Euler sum generalization in (6.149), Sect. 6.19; then I employed the fact that $\sum_{n=1}^{\infty} \frac{H_n}{(2n+1)^2} = \frac{7}{4}\zeta(3) - \frac{3}{2}\log(2)\zeta(2)$, which is the case $p = 2$ of the generalization in (4.102), Sect. 4.19, and lastly the value of the series U is given in (6.129).

Collecting the values of the series U and T, which are given in (6.129) and (6.130), and plugging them in (6.128), we conclude that

$$S = \sum_{n=1}^{\infty} \frac{1}{n(2n+1)} \left(\frac{1}{n+1} + \frac{1}{n+2} + \cdots + \frac{1}{2n+1} \right)^2$$

$$= \zeta(3) - \frac{1}{2}\zeta(2) - 3\log(2)\zeta(2) - 2\log(2) - 2\log^2(2) - \frac{2}{3}\log^3(2) + 6,$$

and the solution is finalized.

It's definitely one of my favorite creations for which I wanted to dedicate a section in this book, a problem that always brings me a special form of nostalgia.

I hope you have also had a great time with this series problem and particularly enjoyed the *challenging question*!

6.17 A Few Nice Generalized Series: Most of Them May Be Seen as Applications of *The Master Theorem of Series*

Solution Generalized harmonic series with a similar structure have already been presented in the book *(Almost) Impossible Integrals, Sums, and Series*, beautifully derived by elementary means. The new thing is that now in our identities, we also have atypical harmonic numbers of the form H_{2n}, or the generalized skew-harmonic number, $\overline{H}_n^{(m)}$. They are interesting and, of course, useful as you'll find later!

As you may guess, for most of these generalized series, I'll use a strategy similar to the ones presented in the mentioned book and derive them by using *The Master Theorem of Series*:

(The first version) If k is a positive integer with $\mathscr{M}(k) = m(1) + m(2) + \cdots + m(k)$, and $m(k)$ are real numbers, where $\lim_{k \to \infty} m(k) = 0$, then the following double equality holds:

$$\sum_{k=1}^{\infty} \frac{\mathscr{M}(k)}{(k+1)(k+n+1)} = m(1)\left(\frac{H_n}{n} - \frac{1}{n+1}\right) + \frac{1}{n}\sum_{j=1}^{n}\sum_{k=1}^{\infty} \frac{m(k+1)}{j+k+1}$$

$$= \frac{1}{n}\sum_{j=1}^{n}\sum_{k=1}^{\infty} \frac{m(k)}{j+k},$$

where $H_n = 1 + \frac{1}{2} + \cdots + \frac{1}{n}$ is the nth harmonic number.

(The second version, the relaxed version) If k is a positive integer with $\mathscr{M}(k) = m(1) + m(2) + \cdots + m(k)$, and $m(k)$ are real numbers, where $\lim_{k \to \infty} \frac{\mathscr{M}(k)}{k} = 0$, then the stated double equality follows.

Proofs to *The Master Theorem of Series* may be found both in [73] and [76, Chapter 6, pp. 369–372].

If you read my first book, in the solution section dedicated to *The Master Theorem of Series*, I wrote that I developed this theorem for generating identities with the generalized harmonic numbers, which we may further exploit in order to derive harmonic series or sums of harmonic series of various weights.

This is exactly the *tool* I'll use to derive the desired results with series involving atypical harmonic numbers of the form H_{2n} and $\overline{H}_n^{(m)}$ to three points of the problem!

So, we use the opposite sides of *The Master Theorem of Series* stated above and set $m(k) = \dfrac{1}{2k-1}$ and $\mathcal{M}(k) = 1 + \dfrac{1}{3} + \cdots + \dfrac{1}{2k-1} = H_{2k} - \dfrac{1}{2}H_k$, which gives

$$\sum_{k=1}^{\infty} \frac{H_{2k} - 1/2\,H_k}{(k+1)(k+n+1)} = \frac{1}{n}\sum_{j=1}^{n}\sum_{k=1}^{\infty} \frac{1}{(2k-1)(j+k)}$$

{use partial fraction decomposition}

$$= 2\frac{1}{n}\sum_{j=1}^{n}\frac{1}{2j+1}\sum_{k=1}^{\infty}\left(\frac{1}{2k-1} - \frac{1}{2j+2k}\right)$$

{add and subtract $1/(2k)$ inside the series summand, and then split the series}

$$= 2\frac{1}{n}\sum_{j=1}^{n}\frac{1}{2j+1}\sum_{k=1}^{\infty}\left(\frac{1}{2k-1} - \frac{1}{2k}\right) + \frac{1}{n}\sum_{j=1}^{n}\frac{1}{2j+1}\sum_{k=1}^{\infty}\left(\frac{1}{k} - \frac{1}{j+k}\right)$$

$$\left\{\text{consider the series, } \sum_{k=1}^{\infty}\left(\frac{1}{2k-1} - \frac{1}{2k}\right) = \sum_{k=1}^{\infty}(-1)^{k-1}\frac{1}{k} = \log(2),\right\}$$

$$\left\{\text{together with the following simple result, } \sum_{k=1}^{\infty}\left(\frac{1}{k} - \frac{1}{n+k}\right) = H_n\right\}$$

$$= 2\log(2)\frac{1}{n}\sum_{j=1}^{n}\frac{1}{2j+1} + \frac{1}{n}\sum_{j=1}^{n}\frac{H_j}{2j+1}$$

$$\left\{\text{observe that } \sum_{k=1}^{n}\frac{1}{2k+1} = H_{2n} - \frac{1}{2}H_n - 1 + \frac{1}{2n+1}\right\}$$

$$= 2\log(2)\frac{1}{n}\left(H_{2n} - \frac{1}{2}H_n\right) - 4\log(2)\frac{1}{n}\frac{1}{2n+1} + \frac{1}{n}\sum_{i=1}^{n}\frac{H_i}{2i+1},$$

and the solution to the point (i) of the problem is complete.

As regards the point (ii) of the problem, we may simply combine the result at the point (i) and the fact that $\displaystyle\sum_{k=1}^{\infty} \frac{H_k}{(k+1)(k+n+1)} = \frac{H_n^2 + H_n^{(2)}}{2n}$, and then we immediately obtain the desired result

$$\sum_{k=1}^{\infty} \frac{H_{2k}}{(k+1)(k+n+1)}$$

$$= \frac{1}{4}\frac{1}{n}(H_n^2 + H_n^{(2)}) + 2\log(2)\frac{1}{n}\left(H_{2n} - \frac{1}{2}H_n\right) - 4\log(2)\frac{1}{2n+1} + \frac{1}{n}\sum_{i=1}^{n} \frac{H_i}{2i+1},$$

and the solution to the point (ii) of the problem is complete.

The identity $\displaystyle\sum_{k=1}^{\infty} \frac{H_k}{(k+1)(k+n+1)} = \frac{H_n^2 + H_n^{(2)}}{2n}$ is already given in a generalized form in [76, Chapter 4, Sect. 4.16, p.289] where I derived it by using *The Master Theorem of Series*. The curious reader might want to give it a try and finish it as suggested by using the settings, $m(k) = 1/k$ and $\mathscr{M}(k) = H_k$ (before taking a look at the given reference).

Of course, for this second point of the problem, we also have the possibility to employ directly the settings $m(k) = H_{2k} - H_{2k-2} = \dfrac{1}{2k} + \dfrac{1}{2k-1}$ and $\mathscr{M}(k) = H_{2k}$ in *The Master Theorem of Series*.

To get a solution to the third point, I'll exploit integrals, and we write that

$$\sum_{k=1}^{\infty} \frac{(-1)^{k-1} H_k}{(k+1)(k+n+1)} = \sum_{k=1}^{\infty}(-1)^{k-1} H_k \int_0^1 y^{k+n} \left(\int_0^1 x^k dx\right) dy$$

{reverse the order of integration and summation}

$$= -\int_0^1 y^n \left(\int_0^1 \sum_{k=1}^{\infty}(-1)^k (xy)^k H_k dx\right) dy$$

{exploit the generating function in (4.32), Sect. 4.6}

$$= \int_0^1 y^n \left(\int_0^1 \frac{\log(1+xy)}{1+xy}dx\right) dy = \frac{1}{2}\int_0^1 y^{n-1} \log^2(1+y)dy$$

{make use of the integral result in (1.102), Sect. 1.21}

$$= \frac{1}{2}\log^2(2)\frac{1}{n} + \frac{1}{2}\log^2(2)(-1)^{n-1}\frac{1}{n} - \log(2)(-1)^{n-1}\frac{H_n}{n} - \log(2)(-1)^{n-1}\frac{\overline{H}_n}{n}$$

$$+ (-1)^{n-1} \frac{H_n \overline{H}_n}{n} + (-1)^{n-1} \frac{\overline{H}_n^{(2)}}{n} - (-1)^{n-1} \frac{1}{n} \sum_{k=1}^{n} (-1)^{k-1} \frac{H_k}{k}$$

$$= \frac{1}{2} \log^2(2) \frac{1}{n} + \frac{1}{2} \log^2(2)(-1)^{n-1} \frac{1}{n} - \log(2)(-1)^{n-1} \frac{H_n}{n} - \log(2)(-1)^{n-1} \frac{\overline{H}_n}{n}$$

$$+ (-1)^{n-1} \frac{1}{n} \sum_{k=1}^{n} \frac{\overline{H}_k}{k},$$

and the solution to the point (iii) of the problem is complete.

The generalization at the point (iv) will probably make us think about the analogous series presented in *(Almost) Impossible Integrals, Sums, and Series* (see [76, Chapter 4, Sect. 4.16, p.289]), where we employ a similar strategy. This time we play with a curious generalization involving $\overline{H}_n^{(m)}$, which is the nth generalized skew-harmonic number of order m.

In order to prove the result, we consider the use of *The Master Theorem of Series*, the second equality, where if we set $\mathcal{M}(k) = \overline{H}_k^{(m)} = 1 - \frac{1}{2^m} + \cdots + (-1)^{k-1} \frac{1}{k^m}$, $m(k) = (-1)^{k-1} \frac{1}{k^m}$, we have that

$$\sum_{k=1}^{\infty} \frac{\overline{H}_k^{(m)}}{(k+1)(k+n+1)} = \frac{1}{n} \sum_{j=1}^{n} \underbrace{\sum_{k=1}^{\infty} \frac{(-1)^{k-1}}{k^m (j+k)}}_{S_{m,j}}. \tag{6.131}$$

Now, the series in the right-hand side of (6.131) can be written as

$$S_{m,j} = \sum_{k=1}^{\infty} \frac{(-1)^{k-1}}{k^m (j+k)} = \sum_{k=1}^{\infty} (-1)^{k-1} \frac{(j+k)-j}{k^{m+1}(j+k)} = \sum_{k=1}^{\infty} \frac{(-1)^{k-1}}{k^{m+1}} - j \underbrace{\sum_{k=1}^{\infty} \frac{(-1)^{k-1}}{k^{m+1}(j+k)}}_{S_{m+1,j}}$$

$$= \eta(m+1) - j S_{m+1,j},$$

or to put it in a simpler way,

$$S_{m,j} + j S_{m+1,j} = \eta(m+1). \tag{6.132}$$

Multiplying both sides of (6.132) by $(-1)^{m-1} j^m$ gives

$$(-1)^{m-1} j^m S_{m,j} - (-1)^m j^{m+1} S_{m+1,j} = (-1)^{m-1} j^m \eta(m+1). \tag{6.133}$$

Further, upon replacing m by i in (6.133) and considering the summation from $i = 1$ to $m - 1$, we obtain

$$\sum_{i=1}^{m-1} \left((-1)^{i-1} j^i S_{i,j} - (-1)^i j^{i+1} S_{i+1,j} \right) = j S_{1,j} - (-1)^{m-1} j^m S_{m,j}$$

$$= \sum_{i=1}^{m-1} (-1)^{i-1} j^i \eta(i+1),$$

and since $S_{1,j} = \sum_{k=1}^{\infty} \dfrac{(-1)^{k-1}}{k(j+k)} = \log(2)\dfrac{1}{j} + \log(2)\dfrac{(-1)^{j-1}}{j} - (-1)^{j-1}\dfrac{\overline{H}_j}{j}$ we get

$$S_{m,j} = \log(2)\frac{(-1)^{m-1}}{j^m} + \log(2)\frac{(-1)^{j+m}}{j^m} - (-1)^{j+m}\frac{\overline{H}_j}{j^m}$$

$$- (-1)^{m-1} \sum_{i=1}^{m-1} (-1)^{i-1} j^{i-m} \eta(i+1),$$

or if we reindex the last sum,

$$S_{m,j} = \underbrace{\log(2)\frac{(-1)^{j+m}}{j^m} - (-1)^{j+m}\frac{\overline{H}_j}{j^m}}_{a_{m,j}}$$

$$+ (-1)^{m-1} \sum_{i=1}^{m} (-1)^{i-1} j^{i-m-1} \eta(i). \tag{6.134}$$

Plugging the result from (6.134) in (6.131), and using the notation $a_{m,j}$ in the penultimate line of (6.134), we obtain that

$$\sum_{k=1}^{\infty} \frac{\overline{H}_k^{(m)}}{(k+1)(k+n+1)}$$

$$= \frac{1}{n} \sum_{j=1}^{n} \underbrace{\sum_{k=1}^{\infty} \frac{(-1)^{k-1}}{k^m(j+k)}}_{S_{m,j}} = \frac{1}{n} \sum_{j=1}^{n} \left(a_{m,j} + (-1)^{m-1} \sum_{i=1}^{m} (-1)^{i-1} j^{i-m-1} \eta(i) \right)$$

$$= \frac{1}{n} \sum_{j=1}^{n} a_{m,j} + \frac{(-1)^{m-1}}{n} \sum_{j=1}^{n} \sum_{i=1}^{m} (-1)^{i-1} j^{i-m-1} \eta(i)$$

{reverse the summation order in the double sum}

$$= \frac{1}{n} \sum_{j=1}^{n} a_{m,j} + \frac{(-1)^{m-1}}{n} \sum_{i=1}^{m} (-1)^{i-1} \eta(i) \sum_{j=1}^{n} \frac{1}{j^{m-i+1}}$$

$$= \frac{1}{n} \sum_{j=1}^{n} a_{m,j} + \frac{(-1)^{m-1}}{n} \sum_{i=1}^{m} (-1)^{i-1} \eta(i) H_n^{(m-i+1)}$$

{use the extended form of $a_{m,j}$ and replace j by i in the first sum}

$$= \log(2)(-1)^{m-1} \frac{\overline{H}_n^{(m)}}{n} - \frac{(-1)^{m-1}}{n} \sum_{i=1}^{n} (-1)^{i-1} \frac{\overline{H}_i}{i^m}$$

$$+ \frac{(-1)^{m-1}}{n} \sum_{i=1}^{m} (-1)^{i-1} \eta(i) H_n^{(m-i+1)}.$$

Observing that the case $m = 1$ is straightforward if we use that $\sum_{k=1}^{n} (-1)^{k-1} \frac{H_k}{k} = \frac{1}{2} \left(H_n^2 + H_n^{(2)} \right)$, which is the case $p = 1$ in (1.120), Sect. 1.23, we conclude that

$$\sum_{k=1}^{\infty} \frac{\overline{H}_k^{(m)}}{(k+1)(k+n+1)} = \begin{cases} \log(2)\dfrac{H_n}{n} - \dfrac{1}{2}\dfrac{H_n^{(2)}}{n} + \log(2)\dfrac{\overline{H}_n}{n} - \dfrac{1}{2}\dfrac{\overline{H}_n^2}{n}, & m = 1; \\[3mm] \log(2)(-1)^{m-1}\dfrac{\overline{H}_n^{(m)}}{n} \\[2mm] + \dfrac{(-1)^{m-1}}{n}\displaystyle\sum_{i=1}^{m}(-1)^{i-1}\eta(i)H_n^{(m-i+1)} \\[3mm] - \dfrac{(-1)^{m-1}}{n}\displaystyle\sum_{i=1}^{n}(-1)^{i-1}\dfrac{\overline{H}_i}{i^m}, & m \geq 2, \end{cases}$$

and the solution to the point (iv) of the problem is complete.

It is nice to see again and again the beauty and power of the series manipulations, which I considered at the points (i), (ii), and (iv) of the problem!

6.18   Useful Relations Involving Polygamma with the Argument $n/2$ and the Generalized Skew-Harmonic Numbers

Solution It was a rather twisted story that finally led to the creation of this section. In order to derive a certain Fourier series and bring it to a specific form, I needed a special logarithmic integral (rather stubborn to bring it to a satisfactory closed form) where finding alternative ways of expressing the polylogarithmic values of the type treated in this section represented a key!

It will be perhaps your pleasure to discover in this book the results and sections alluded above in case you didn't do it yet.

The nucleus of the present section will be built by starting with deriving the following result:

$$\sum_{k=1}^{\infty}(-1)^{k-1}\frac{1}{k+x} = \log(2) + 2\frac{1}{x} - \psi(1+x) + \psi\left(\frac{x}{2}\right). \qquad (6.135)$$

Proof We have from *(Almost) Impossible Integrals, Sums, and Series* (see [76, Chapter 1, p.3]) that $\displaystyle\int_0^1 \frac{t^{x-1}}{1+t}dt = \psi(x) - \psi\left(\frac{x}{2}\right) - \log(2)$, and if we combine it with the recurrence relation of Digamma function, $\psi(1+x) = \psi(x) + \frac{1}{x}$, we get

$$\psi(1+x) - \psi\left(\frac{x}{2}\right) - \frac{1}{x} - \log(2) = \int_0^1 \frac{t^{x-1}}{1+t}dt = \int_0^1 t^{x-1}\sum_{k=1}^{\infty}(-1)^{k-1}t^{k-1}dt$$

$$= \sum_{k=1}^{\infty}(-1)^{k-1}\int_0^1 t^{k+x-2}dt = \sum_{k=1}^{\infty}(-1)^{k-1}\frac{1}{k+x-1} \stackrel{\substack{\text{reindex}\\\text{the series}}}{=} \sum_{k=0}^{\infty}(-1)^k\frac{1}{k+x}$$

{leave out the first term of the series}

$$= \frac{1}{x} - \sum_{k=1}^{\infty}(-1)^{k-1}\frac{1}{k+x},$$

whence the desired result follows. The result, in a slightly different form, also appears in [15, 3.32., p.144], with a different solution strategy. ∎

Now, we are ready to attack and prove the result at the point (i)! Replacing x by n, a positive integer, in (6.135), and using that $H_n = \psi(n+1) + \gamma$, we have

$$\log(2) + 2\frac{1}{n} - \psi(1+n) + \psi\left(\frac{n}{2}\right) = \log(2) + \gamma + 2\frac{1}{n} - H_n + \psi\left(\frac{n}{2}\right)$$

$$= \sum_{k=1}^{\infty}(-1)^{k-1}\frac{1}{k+n} \overset{\text{reindex}}{\underset{\text{the series}}{=}} (-1)^n \sum_{k=n+1}^{\infty}(-1)^{k-1}\frac{1}{k} = -(-1)^{n-1}\underbrace{\sum_{k=1}^{\infty}(-1)^{k-1}\frac{1}{k}}_{\log(2)}$$

$$+ (-1)^{n-1}\sum_{k=1}^{n}(-1)^{k-1}\frac{1}{k} = -\log(2)(-1)^{n-1} + (-1)^{n-1}\overline{H}_n,$$

whence we get that

$$\psi\left(\frac{n}{2}\right) = -\log(2) - \gamma - \log(2)(-1)^{n-1} - 2\frac{1}{n} + H_n + (-1)^{n-1}\overline{H}_n,$$

and the solution to the point (i) is complete.

The result at the point (ii) is straightforward if we use the recurrence relation of Digamma function, $\psi(1+x) = \psi(x) + \dfrac{1}{x}$, where if we replace x by $n/2$ and combine all with the result at the previous point, we arrive at

$$\psi\left(\frac{n}{2}+1\right) = -\log(2) - \gamma - \log(2)(-1)^{n-1} + H_n + (-1)^{n-1}\overline{H}_n,$$

and the solution to the point (ii) is complete.

So, we want now to progress to the generalization given at the point (iii). Some simple results are needed, and we use that $\dfrac{d^m}{dx^m}\left(\dfrac{1}{x+k}\right)\bigg|_{x=n} = (-1)^m\dfrac{m!}{(n+k)^{m+1}}$, then $\dfrac{d^m}{dx^m}(\psi(x+1))\bigg|_{x=n} = \psi^{(m)}(n+1) = (-1)^{m-1}m!\left(\zeta(m+1) - H_n^{(m+1)}\right)$, and the fact that $\dfrac{d^m}{dx^m}\left(\psi\left(\dfrac{x}{2}\right)\right)\bigg|_{x=n} = \dfrac{1}{2^m}\psi^{(m)}\left(\dfrac{n}{2}\right)$.

With these results in mind, we differentiate m times both sides of (6.135) and then replace x by n, a positive integer. Consequently, we get

$$-2(-1)^{m-1}m!\frac{1}{n^{m+1}} - (-1)^{m-1}m!\left(\zeta(m+1) - H_n^{(m+1)}\right) + \frac{1}{2^m}\psi^{(m)}\left(\frac{n}{2}\right)$$

$$= -(-1)^{m-1}m!\sum_{k=1}^{\infty}(-1)^{k-1}\frac{1}{(k+n)^{m+1}} \overset{\text{reindex}}{\underset{\text{the series}}{=}} (-1)^{m+n}m!\sum_{k=n+1}^{\infty}(-1)^{k-1}\frac{1}{k^{m+1}}$$

$$= (-1)^{m+n}m!\sum_{k=1}^{\infty}(-1)^{k-1}\frac{1}{k^{m+1}} - (-1)^{m+n}m!\sum_{k=1}^{n}(-1)^{k-1}\frac{1}{k^{m+1}}$$

$$= (-1)^{m+n}m!(1-2^{-m})\zeta(m+1) - (-1)^{m+n}m!\overline{H}_n^{(m+1)},$$

whence we obtain that

$$\psi^{(m)}\left(\frac{n}{2}\right)$$

$$= (-1)^{m-1}2^{m+1}m!\frac{1}{n^{m+1}} + (-1)^{m+n}(2^m - 1)m!\zeta(m+1)$$

$$+ (-1)^{m-1}2^m m!\left(\zeta(m+1) - H_n^{(m+1)}\right) - (-1)^{m+n}2^m m!\overline{H}_n^{(m+1)},$$

and the solution to the point (iii) is complete.

We have now a way of expressing the Polygamma function with an argument of the type $n/2$, where n is a positive integer, by using the generalized harmonic number and the generalized skew-harmonic number!

At last, using the recurrence relation of the Polygamma function, $\psi^{(m)}(x+1) = \psi^{(m)}(x) + (-1)^m\frac{m!}{x^{m+1}}$, which is derived by differentiation from the recurrence relation of Digamma function, $\psi(x+1) = \psi(x) + \frac{1}{x}$, then replacing x by $n/2$, and combining all with the previous generalized result, we get

$$\psi^{(m)}\left(\frac{n}{2} + 1\right)$$

$$= (-1)^{m+n}(2^m - 1)m!\zeta(m+1) + (-1)^{m-1}2^m m!\left(\zeta(m+1) - H_n^{(m+1)}\right)$$

$$- (-1)^{m+n}2^m m!\overline{H}_n^{(m+1)},$$

and the solution to the point (iv) is complete.

What else? Well, based on the recurrence relation of the Polygamma function above, $\psi^{(m)}(x+1) = \psi^{(m)}(x) + (-1)^m\frac{m!}{x^{m+1}}$, if we replace x by k, and then make the sum from $k = 1$ to $n - 1$, we immediately have by telescoping sums that

$$\sum_{k=1}^{n-1}\left(\psi^{(m)}(k+1) - \psi^{(m)}(k)\right) = \psi^{(m)}(n) - \psi^{(m)}(1) = (-1)^m m!\sum_{k=1}^{n-1}\frac{1}{k^{m+1}}.$$
$$\tag{6.136}$$

At the same time, if we let $n \to \infty$ in (6.136) and count that when n is large $\psi^{(m)}(n)$ behaves like $(-1)^{m-1}(m-1)!/n^m$, which is easy to see if we combine differentiation with the asymptotic expansion of Digamma function, $\psi(x) = \log(x) + O(1/x)$, we get

$$\psi^{(m)}(1) = (-1)^{m-1}m!\zeta(m+1). \tag{6.137}$$

Combining (6.136) and (6.137), we arrive at

$$\psi^{(m)}(n) = (-1)^{m-1}m!\zeta(m+1) + (-1)^{m}m!\sum_{k=1}^{n-1}\frac{1}{k^{m+1}}$$

$$= (-1)^{m-1}m!\zeta(m+1) + (-1)^{m}m!H_{n-1}^{(m+1)}. \tag{6.138}$$

For example, the result in (6.138) promptly explains the fact that $\psi^{(m)}(n+1) = (-1)^{m-1}m!\left(\zeta(m+1) - H_n^{(m+1)}\right)$, which I used in the solution to the third point.

6.19   A Key Classical Generalized Harmonic Series

Solution Let's prepare for some interesting and enjoyable generalizations with harmonic series! One of these generalizations will represent a part of the *means* needed to establish another very useful generalization with harmonic series. In fact, to make things explicit, the result at the point (i) will help us in the calculation process of the important and challenging generalization, $\sum_{n=1}^{\infty}(-1)^{n-1}\dfrac{H_n}{n^{2m}}$, $m \geq 1$, $m \in \mathbb{N}$, often arising in the calculation of many other (advanced) harmonic series.

The main idea in the present solutions follows the strategy I presented in my paper, *A new powerful strategy of calculating a class of alternating Euler sums* (see [77]) which I uploaded on *ResearchGate* on June 25, 2019, soon after the publishing of my first book, *(Almost) Impossible Integrals, Sums, and Series*, where I also calculated the harmonic series generalization, $\sum_{k=1}^{\infty}\dfrac{H_k}{(2k+1)^{2m}}$, which is the series result I present at the point (i).

Before proceeding with the main calculation to the first point of the problem, let's recall the relation between the harmonic number and the Digamma function, that is, $H_n = \psi(n+1) + \gamma$, which allows an extension of the harmonic number H_n to the non-integer values of n. Then, we also recollect the following relation between the Polygamma function and the generalized harmonic number,

$$\psi^{(m)}(n+1) = (-1)^{m-1}m!\sum_{k=n+1}^{\infty}\frac{1}{k^{m+1}} = (-1)^{m-1}m!\left(\zeta(m+1) - H_n^{(m+1)}\right), \; m \geq 1,$$

$$\tag{6.139}$$

which may be obtained from the series representation of the Polygamma function (also, see the end of the previous section),

$$\psi^{(m)}(x) = (-1)^{m-1}m!\sum_{k=0}^{\infty}\frac{1}{(x+k)^{m+1}}. \tag{6.140}$$

The first key result we need in our calculations is

$$\sum_{k=1}^{\infty} \frac{H_k}{(k+1)(k+n+1)} = \frac{H_n^2 + H_n^{(2)}}{2n} = \frac{(\gamma + \psi(n+1))^2 + \zeta(2) - \psi^{(1)}(n+1)}{2n}.$$

$$(6.141)$$

The identity in (6.141), stated in the form with the first equality, is part of a generalization I presented in my first book (see [76, Chapter 4, p.289]). Alternatively, one can write $\displaystyle\sum_{k=1}^{\infty} \frac{H_k}{(k+1)(k+n+1)} = \sum_{k=1}^{\infty} \int_0^1 x^{k+n} \frac{H_k}{k+1} dx =$ $\displaystyle\int_0^1 \sum_{k=1}^{\infty} x^{k+n} \frac{H_k}{k+1} dx = \frac{1}{2} \int_0^1 x^{n-1} \log^2(1-x)dx$, which is a derivative form of the Beta function, and that immediately leads to the desired result (e.g., with the simple way presented in [12]). The second equality in (6.141) points out the possibility of an extension of the identity to real numbers by the use of the Polygamma function, a form we need since we'll use differentiation.

So, if we multiply the opposite sides of the result in (6.141) by n and then differentiate m times with respect to n, we obtain

$$\frac{d^m}{dn^m}\left\{\sum_{k=1}^{\infty} \frac{n H_k}{(k+1)(k+n+1)}\right\} = (-1)^{m-1} m! \sum_{k=1}^{\infty} \frac{H_k}{(k+n+1)^{m+1}}$$

$$= \frac{1}{2}\frac{d^m}{dn^m}\left\{(\gamma + \psi(n+1))^2 + \zeta(2) - \psi^{(1)}(n+1)\right\},$$

and considering the second equality and rearraging it, we have

$$\sum_{k=1}^{\infty} \frac{H_k}{(k+n+1)^{m+1}} = \frac{(-1)^{m-1}}{2 \cdot m!}\frac{d^m}{dn^m}\left\{(\gamma + \psi(n+1))^2 + \zeta(2) - \psi^{(1)}(n+1)\right\}$$

$$= \frac{(-1)^{m-1}}{2 \cdot m!}\left(\frac{d^m}{dn^m}\left\{(\gamma + \psi(n+1))^2\right\} + \frac{d^m}{dn^m}\left\{\zeta(2) - \psi^{(1)}(n+1)\right\}\right).$$

$$(6.142)$$

Now, if we focus on the right-hand side of (6.142) and use the *general Leibniz rule*, $(f \cdot g)^n = \displaystyle\sum_{k=0}^{n} \binom{n}{k} f^{(n-k)} \cdot g^{(k)}$, we get that

$$\frac{d^m}{dn^m}\left\{(\gamma + \psi(n+1))^2\right\} = \left(\underbrace{(\gamma + \psi(n+1))}_{f} \cdot \underbrace{(\gamma + \psi(n+1))}_{g}\right)^{(m)}$$

$$= \sum_{k=0}^{m} \binom{m}{k} (\gamma + \psi(n+1))^{(m-k)} \cdot (\gamma + \psi(n+1))^{(k)}$$

{leave out the terms of the sum for $k = 0$ and $k = m$}

$$= 2(\gamma+\psi(n+1))^{(m)} \cdot (\gamma+\psi(n+1)) + \sum_{k=1}^{m-1} \binom{m}{k} (\gamma+\psi(n+1))^{(m-k)} \cdot (\gamma+\psi(n+1))^{(k)}$$

$$\left\{ \text{observe and use that } (\gamma + \psi(n+1))^{(m)} = \psi^{(m)}(n+1), \ m \geq 1 \right\}$$

$$= 2\psi^{(m)}(n+1) \cdot (\gamma+\psi(n+1)) + \sum_{k=1}^{m-1} \binom{m}{k} \psi^{(m-k)}(n+1) \cdot \psi^{(k)}(n+1). \qquad (6.143)$$

At the same time, we have

$$\frac{d^m}{dn^m} \left(\zeta(2) - \psi^{(1)}(n+1) \right) = -\psi^{(m+1)}(n+1), \ m \geq 1. \qquad (6.144)$$

Then, we return with the results from (6.143) and (6.144) in (6.142) that gives

$$\sum_{k=1}^{\infty} \frac{H_k}{(k+n+1)^{m+1}}$$

$$= \frac{(-1)^{m-1}}{2 \cdot m!} \left(2\psi^{(m)}(n+1) \cdot (\gamma+\psi(n+1)) - \psi^{(m+1)}(n+1) \right.$$

$$\left. + \sum_{k=1}^{m-1} \binom{m}{k} \psi^{(m-k)}(n+1) \cdot \psi^{(k)}(n+1) \right). \qquad (6.145)$$

Before starting the main calculations for each point of the problem, we need to prepare two more key results. We require the fact that

$$\psi\left(\frac{1}{2}\right) = -\gamma - 2\log(2), \qquad (6.146)$$

where we need to recall and use the series representation of Digamma function,
$\psi(x) = -\gamma + \sum_{n=0}^{\infty} \left(\frac{1}{n+1} - \frac{1}{n+x} \right)$. Now, if we plug $x = 1/2$ in the mentioned series form of the Digamma function, we have

$$\psi\left(\frac{1}{2}\right) = -\gamma + \sum_{n=0}^{\infty}\left(\frac{1}{n+1} - \frac{2}{2n+1}\right) = -\gamma + \sum_{n=0}^{\infty}\left(\frac{2}{2n+2} - \frac{2}{2n+1}\right)$$

$$= -\gamma - 2\sum_{n=0}^{\infty}\left(\frac{1}{2n+1} - \frac{1}{2n+2}\right) = -\gamma - 2\sum_{n=1}^{\infty}\frac{(-1)^{n-1}}{n} = -\gamma - 2\log(2),$$

where the last alternating series is a well-known one.

Then, we also require another result,

$$\psi^{(k)}\left(\frac{1}{2}\right) = (-1)^{k-1}k!(2^{k+1} - 1)\zeta(k+1), \ k \geq 1, \tag{6.147}$$

where we need the series representation of the Polygamma function in (6.140) with $x = 1/2$. Then, the extraction process flows smoothly

$$\psi^{(k)}\left(\frac{1}{2}\right) = (-1)^{k-1}k!2^{k+1}\sum_{n=0}^{\infty}\frac{1}{(2n+1)^{k+1}}$$

$$= (-1)^{k-1}k!2^{k+1}\left(\sum_{n=0}^{\infty}\frac{1}{(n+1)^{k+1}} - \sum_{n=0}^{\infty}\frac{1}{(2n+2)^{k+1}}\right)$$

$$= (-1)^{k-1}k!2^{k+1}\left(\sum_{n=0}^{\infty}\frac{1}{(n+1)^{k+1}} - \frac{1}{2^{k+1}}\sum_{n=0}^{\infty}\frac{1}{(n+1)^{k+1}}\right)$$

$$= (-1)^{k-1}k!(2^{k+1} - 1)\zeta(k+1).$$

We are ready now to prove the main results!

For the first main point, we replace m by $2m - 1$ in (6.145), set $n = -1/2$, and multiply both sides by $1/2^{2m}$, and then we have

$$\sum_{k=1}^{\infty}\frac{H_k}{(2k+1)^{2m}}$$

$$= \frac{1}{(2m-1)!2^{2m+1}}\left(2\psi^{(2m-1)}\left(\frac{1}{2}\right)\cdot\left(\gamma + \psi\left(\frac{1}{2}\right)\right) - \psi^{(2m)}\left(\frac{1}{2}\right)\right.$$

$$\left. + \underbrace{\sum_{k=1}^{2m-2}\binom{2m-1}{k}\psi^{(2m-k-1)}\left(\frac{1}{2}\right)\cdot\psi^{(k)}\left(\frac{1}{2}\right)}_{f(k,m)}\right)$$

$$\left\{ \text{use the symmetry to write } \sum_{k=1}^{2m-2} f(k,m) = 2\sum_{k=1}^{m-1} f(k,m) \text{ and} \right\}$$

{at the same time make use of the results in (6.146) and (6.147)}

$$= 2m\left(1 - \frac{1}{2^{2m+1}}\right)\zeta(2m+1) - 2\log(2)\left(1 - \frac{1}{2^{2m}}\right)\zeta(2m)$$

$$- \frac{1}{2^{2m}}\sum_{k=1}^{m-1}(1 - 2^{k+1})(1 - 2^{2m-k})\zeta(k+1)\zeta(2m-k),$$

and the solution to the point (i) of the problem is complete.

To make the step above clearer, if necessary, where I invoked the symmetry, just note that we can write

$$\sum_{k=1}^{2m-2}\binom{2m-1}{k}\psi^{(2m-k-1)}\left(\frac{1}{2}\right)\cdot\psi^{(k)}\left(\frac{1}{2}\right)$$

$$= \left(\sum_{k=1}^{m-1} + \sum_{k=m}^{2m-2}\right)\binom{2m-1}{k}\psi^{(2m-k-1)}\left(\frac{1}{2}\right)\cdot\psi^{(k)}\left(\frac{1}{2}\right),$$

where if we expand the previous line and make the variable change $2m - k - 1 = i$ in the second sum and then return to the notation in k, we arrive at

$$\sum_{k=m}^{2m-2}\binom{2m-1}{k}\psi^{(2m-k-1)}\left(\frac{1}{2}\right)\cdot\psi^{(k)}\left(\frac{1}{2}\right)$$

$$= \sum_{k=1}^{m-1}\binom{2m-1}{k}\psi^{(2m-k-1)}\left(\frac{1}{2}\right)\cdot\psi^{(k)}\left(\frac{1}{2}\right),$$

which clearly explains the result invoked above.

For the second main point, we replace m by $2m$ in (6.145), set $n = -1/2$, and multiply both sides by $1/2^{2m+1}$, and then we get

$$\sum_{k=1}^{\infty}\frac{H_k}{(2k+1)^{2m+1}}$$

$$= \frac{1}{(2m)!2^{2m+2}}\left(\psi^{(2m+1)}\left(\frac{1}{2}\right) - 2\psi^{(2m)}\left(\frac{1}{2}\right)\cdot\left(\gamma + \psi\left(\frac{1}{2}\right)\right)\right)$$

$$- \sum_{k=1}^{2m-1} \binom{2m}{k} \underbrace{\psi^{(2m-k)}\left(\frac{1}{2}\right) \cdot \psi^{(k)}\left(\frac{1}{2}\right)}_{f(k,m)}$$

$$\left\{ \text{use the symmetry to write } \sum_{k=1}^{2m-1} f(k,m) = 2\sum_{k=1}^{m-1} f(k,m) + f(m,m) \right\}$$

{and at the same time make use of the results in (6.146) and (6.147)}

$$= \left(1 - \frac{1}{2^{2m+2}}\right)(2m+1)\zeta(2m+2)$$

$$- \left(1 - \frac{1}{2^{m+1}}\right)^2 \zeta^2(m+1) - 2\log(2)\left(1 - \frac{1}{2^{2m+1}}\right)\zeta(2m+1)$$

$$- \frac{1}{2^{2m+1}} \sum_{k=1}^{m-1}(1 - 2^{k+1})(1 - 2^{2m-k+1})\zeta(k+1)\zeta(2m-k+1),$$

and the solution to the point (ii) of the problem is complete.

Finally, I will present a general form of the series, and upon considering $m+1 = p$, $p \geq 2$, in (6.145), setting $n = -1/2$, and multiplying both sides by $1/2^p$, we obtain

$$\sum_{k=1}^{\infty} \frac{H_k}{(2k+1)^p}$$

$$= \frac{(-1)^{p-1}}{(p-1)!2^{p+1}}\left(\psi^{(p)}\left(\frac{1}{2}\right) - 2\psi^{(p-1)}\left(\frac{1}{2}\right) \cdot \left(\gamma + \psi\left(\frac{1}{2}\right)\right)\right)$$

$$- \sum_{k=1}^{p-2}\binom{p-1}{k}\psi^{(p-k-1)}\left(\frac{1}{2}\right) \cdot \psi^{(k)}\left(\frac{1}{2}\right)\right)$$

{make use of the results in (6.146) and (6.147)}

$$= p\left(1 - \frac{1}{2^{p+1}}\right)\zeta(p+1) - 2\log(2)\left(1 - \frac{1}{2^p}\right)\zeta(p)$$

$$- \frac{1}{2^{p+1}} \sum_{k=1}^{p-2}(1 - 2^{k+1})(1 - 2^{p-k})\zeta(k+1)\zeta(p-k),$$

and the solution to the point (iii) of the problem is complete.

Before closing the section, there is a *Bonus* moment we might like to enjoy! If we go back in (6.145) and let $n \to 0$, we obtain

$$\sum_{k=1}^{\infty} \frac{H_k}{(k+1)^{m+1}}$$

$$= \frac{(-1)^{m-1}}{2 \cdot m!} \left(2\psi^{(m)}(1) \cdot (\gamma + \psi(1)) - \psi^{(m+1)}(1) + \sum_{k=1}^{m-1} \binom{m}{k} \psi^{(m-k)}(1) \cdot \psi^{(k)}(1) \right)$$

{use the result in (6.139), where we set $n = 0$, that immediately leads to}

$$\left\{ \text{the fact that } \psi^{(m)}(1) = (-1)^{m-1} m! \zeta(m+1), \text{ and at the same time} \right\}$$

$$\left\{ \text{we note that since } \psi(1) = -\gamma, \text{ we get } 2\psi^{(m)}(1) \cdot (\gamma + \psi(1)) = 0 \right\}$$

$$= \frac{1}{2}(m+1)\zeta(m+2) - \frac{1}{2}\sum_{k=1}^{m-1} \zeta(k+1)\zeta(m-k+1),$$

or if we replace $m + 1$ by n, $n \geq 2$, $n \in \mathbb{N}$, we get

$$\sum_{k=1}^{\infty} \frac{H_k}{(k+1)^n} = \frac{1}{2}n\zeta(n+1) - \frac{1}{2}\sum_{k=1}^{n-2} \zeta(k+1)\zeta(n-k). \tag{6.148}$$

Further, since $\displaystyle\sum_{k=1}^{\infty} \frac{H_k}{(k+1)^n} = \sum_{k=1}^{\infty} \frac{H_{k+1} - 1/(k+1)}{(k+1)^n} = \sum_{k=1}^{\infty} \frac{H_k}{k^n} - \sum_{k=1}^{\infty} \frac{1}{k^{n+1}} =$

$\displaystyle\sum_{k=1}^{\infty} \frac{H_k}{k^n} - \zeta(n+1)$, if we combine this result with the one in (6.148), we obtain that

$$\sum_{k=1}^{\infty} \frac{H_k}{k^n} = \frac{1}{2}(n+2)\zeta(n+1) - \frac{1}{2}\sum_{k=1}^{n-2} \zeta(k+1)\zeta(n-k), \tag{6.149}$$

that is one of the most famous and oldest results with harmonic series in the mathematical literature, first discovered by Leonhard Euler (1707–1783), which is an essential result we (very) often return to in the work with harmonic series.

The identity in (6.141) is one of the many identities with such a summand structure where one can exploit differentiation to obtain useful results with harmonic series. We'll meet again such strategies in some of the subsequent sections.

6.20 Revisiting Two Classical Challenging Alternating Harmonic Series, Calculated by Exploiting a Beta Function Form

Solution The words *pretty hard nuts to crack* could be the way many of you might like to describe the first experience with the two alternating harmonic series I present in this section. Without using some clever approaches, both series may be perceived as pretty challenging, and they often arise in many calculations with integrals involving logarithms, polylogarithms. So, we want to know how to calculate them! Moreover, it would be magnificent if we had simple means to derive them!

Both series are already treated in the book *(Almost) Impossible Integrals, Sums, and Series*, as you may see in [76, Chapter 6, pp. 503–505] and [76, Chapter 6, pp. 508–513]. Considering their resistance, every new solution is a moment of joy, which is doubled by the fact that in this section, we aim to calculate them in a simple, elegant way.

The two points of the problem were also considered in my paper, *An easy approach to two classical Euler sums, $\sum_{n=1}^{\infty}(-1)^{n-1}\frac{H_n}{n^3}$ and $\sum_{n=1}^{\infty}(-1)^{n-1}\frac{H_n}{n^4}$*, that took shape in [88, February 15, 2020].

Let's begin with the part of the title, *a Beta function form*,[5] and then write it,

$$\int_0^1 \frac{x^{a-1} + x^{b-1}}{(1+x)^{a+b}}dx = B(a,b), \qquad (6.150)$$

which is also stated in **3.216.1** from [17].

Now, we prepare to perform the calculations to the harmonic series at the point (i). *How exactly would we like to begin?*

If we use the Beta function defined by $B(x,y) = \int_0^1 t^{x-1}(1-t)^{y-1}dt$, $\Re(x)$, $\Re(y) > 0$, and then exploit its derivatives, we may immediately show that

---

[5] We have the following special result, $I = \int_0^1 \frac{x^{a-1} + x^{b-1}}{(1+x)^{a+b}}dx = B(a,b)$. To see this is true, making the variable change $x = 1/y$, and then returning to the variable in x, we get that $I = \int_1^{\infty} \frac{x^{a-1} + x^{b-1}}{(1+x)^{a+b}}dx$. Upon adding up the two integrals, we arrive at the desired result, $I = \frac{1}{2}\int_0^{\infty} \frac{x^{a-1} + x^{b-1}}{(1+x)^{a+b}}dx = \frac{1}{2}\left(\int_0^{\infty} \frac{x^{a-1}}{(1+x)^{a+b}}dx + \int_0^{\infty} \frac{x^{b-1}}{(1+x)^{a+b}}dx\right) = \frac{1}{2}(B(a,b) + B(b,a)) = B(a,b)$, where I also used the integral representation of the Beta function, $\int_0^{\infty} \frac{x^{a-1}}{(1+x)^{a+b}}dx = B(a,b)$, and the fact that $B(a,b) = B(b,a)$ due to symmetry. By the variable change $\frac{x}{1+x} = t$ in the last integral, we arrive at the definition of the Beta function, $B(a,b) = \int_0^1 t^{a-1}(1-t)^{b-1}dt$.

$$\lim_{\substack{x\to 0 \\ y\to 0}} \frac{\partial^3}{\partial x^2 \partial y} B(x, y) = \int_0^1 \frac{\log^2(t)\log(1-t)}{t(1-t)}dt = -\int_0^1 \log^2(t) \sum_{n=1}^{\infty} t^{n-1} H_n dt$$

$$= -\sum_{n=1}^{\infty} H_n \int_0^1 t^{n-1}\log^2(t)dt = -2\sum_{n=1}^{\infty} \frac{H_n}{n^3} = -\frac{5}{2}\zeta(4), \qquad (6.151)$$

where the limit with the derivatives of the Beta function has been reduced to the case $n = 3$ of the Euler sum generalization in (6.149), Sect. 6.19. After the second equality in (6.151), I also used the well-known fact that $\sum_{n=1}^{\infty} x^n H_n = -\dfrac{\log(1-x)}{1-x}$, which is the case $m = 1$ in (4.32), Sect. 4.6.

From a different perspective, if we consider the result $\lim_{\substack{x\to 0 \\ y\to 0}} \dfrac{\partial^3}{\partial x^2 \partial y} B(x, y) = -\dfrac{5}{2}\zeta(4)$ in (6.151) and combine it with the special form of the Beta function in (6.150), we get that

$$-\frac{5}{2}\zeta(4) = \lim_{\substack{x\to 0 \\ y\to 0}} \frac{\partial^3}{\partial x^2 \partial y} B(x, y)$$

$$= 3\underbrace{\int_0^1 \frac{\log(t)\log^2(1+t)}{t}dt}_{X} - \underbrace{\int_0^1 \frac{\log^2(t)\log(1+t)}{t}dt}_{Y} - 2\underbrace{\int_0^1 \frac{\log^3(1+t)}{t}dt}_{Z}.$$

$$(6.152)$$

For the integral X in (6.152), we write

$$X = \int_0^1 \frac{\log(t)\log^2(1+t)}{t}dt = 2\int_0^1 \log(t) \sum_{n=1}^{\infty} (-1)^{n-1} t^n \frac{H_n}{n+1}dt$$

$$= 2\sum_{n=1}^{\infty} (-1)^{n-1}\frac{H_n}{n+1} \int_0^1 t^n \log(t)dt = -2\sum_{n=1}^{\infty} (-1)^{n-1} \frac{H_{n+1} - 1/(n+1)}{(n+1)^3}$$

{reindex the series and expand it}

$$= 2\sum_{n=1}^{\infty} (-1)^{n-1}\frac{H_n}{n^3} - 2\sum_{n=1}^{\infty} (-1)^{n-1}\frac{1}{n^4} = 2\sum_{n=1}^{\infty} (-1)^{n-1}\frac{H_n}{n^3} - \frac{7}{4}\zeta(4), \qquad (6.153)$$

where I used that $\sum_{n=1}^{\infty} (-1)^{n-1} t^n \dfrac{H_n}{n+1} = \dfrac{1}{2}\dfrac{\log^2(1+t)}{t}$, based on (4.33), Sect. 4.6.

• Then, for the integral Y in (6.152), we have

$$Y = \int_0^1 \frac{\log^2(t)\log(1+t)}{t}dt = \int_0^1 \log^2(t)\sum_{n=1}^\infty (-1)^{n-1}\frac{t^{n-1}}{n}dt$$

$$= \sum_{n=1}^\infty (-1)^{n-1}\frac{1}{n}\int_0^1 t^{n-1}\log^2(t)dt = 2\sum_{n=1}^\infty (-1)^{n-1}\frac{1}{n^4} = \frac{7}{4}\zeta(4). \qquad (6.154)$$

Next, for the integral Z in (6.152), we get

$$Z = \int_0^1 \frac{\log^3(1+t)}{t}dt \overset{1/(1+t)=u}{=} -\int_{1/2}^1 \frac{\log^3(u)}{u(1-u)}du$$

$$= -\int_{1/2}^1 \frac{\log^3(u)}{1-u}du - \int_{1/2}^1 \frac{\log^3(u)}{u}du$$

$$= \int_0^{1/2} \frac{\log^3(u)}{1-u}du - \int_0^1 \frac{\log^3(u)}{1-u}du - \int_{1/2}^1 \frac{\log^3(u)}{u}du$$

$$= \int_0^{1/2} \sum_{n=1}^\infty u^{n-1}\log^3(u)du - \int_0^1 \sum_{n=1}^\infty u^{n-1}\log^3(u)du + \frac{1}{4}\log^4(2)$$

{reverse the order of summation and integration}

$$= \sum_{n=1}^\infty \int_0^{1/2} u^{n-1}\log^3(u)du - \sum_{n=1}^\infty \int_0^1 u^{n-1}\log^3(u)du + \frac{1}{4}\log^4(2)$$

$$= -6\sum_{n=1}^\infty \frac{1}{n^4 2^n} - 6\log(2)\sum_{n=1}^\infty \frac{1}{n^3 2^n} - 3\log^2(2)\sum_{n=1}^\infty \frac{1}{n^2 2^n} - \log^3(2)\sum_{n=1}^\infty \frac{1}{n 2^n}$$

$$+ 6\sum_{n=1}^\infty \frac{1}{n^4} + \frac{1}{4}\log^4(2)$$

$$= 6\zeta(4) - \frac{21}{4}\log(2)\zeta(3) + \frac{3}{2}\log^2(2)\zeta(2) - \frac{1}{4}\log^4(2) - 6\operatorname{Li}_4\left(\frac{1}{2}\right), \qquad (6.155)$$

where I also used the special value of dilogarithm function,

$$\operatorname{Li}_2\left(\frac{1}{2}\right) = \frac{1}{2}(\zeta(2) - \log^2(2)), \qquad (6.156)$$

and the special value of trilogarithm function (e.g., exploiting the Landen's trilogarithmic identity in (4.39), Sect. 4.6),

$$\text{Li}_3\left(\frac{1}{2}\right) = \frac{7}{8}\zeta(3) - \frac{1}{2}\log(2)\zeta(2) + \frac{1}{6}\log^3(2). \tag{6.157}$$

Collecting the values of the integrals X, Y, and Z from (6.153), (6.154), and (6.155) in (6.152), we arrive the desired result

$$\sum_{n=1}^{\infty}(-1)^{n-1}\frac{H_n}{n^3}$$

$$= \frac{11}{4}\zeta(4) - \frac{7}{4}\log(2)\zeta(3) + \frac{1}{2}\log^2(2)\zeta(2) - \frac{1}{12}\log^4(2) - 2\text{Li}_4\left(\frac{1}{2}\right),$$

and the solution to the point (i) of the problem is finalized.

Passing to the next point of the problem, we first observe we can write that

$$\int_0^1 \frac{\log^2(t)\log^2(1+t)}{t}dt = 2\int_0^1 \log^2(t)\sum_{n=1}^{\infty}(-1)^{n-1}t^n\frac{H_n}{n+1}dt$$

$$\left\{\text{make use of the fact that } \sum_{n=1}^{\infty}(-1)^{n-1}t^n\frac{H_n}{n+1} = \frac{1}{2}\frac{\log^2(1+t)}{t}\right\}$$

$$= 2\sum_{n=1}^{\infty}(-1)^{n-1}\frac{H_n}{n+1}\int_0^1 t^n\log^2(t)dt = 4\sum_{n=1}^{\infty}(-1)^{n-1}\frac{H_{n+1}-1/(n+1)}{(n+1)^4}$$

{reindex the series and expand it}

$$= \frac{15}{4}\zeta(5) - 4\sum_{n=1}^{\infty}(-1)^{n-1}\frac{H_n}{n^4}. \tag{6.158}$$

The *magical* part comes at this point when we want to exploit the special form of the Beta function in (6.150) to deal with the integral in (6.158). The victorious step is to observe that we can actually relate the mentioned integral *in two relevant ways* to the special form of the Beta function in (6.150), more accurately to limits involving its derivatives. It sounds interesting, doesn't it?

Considering to differentiate four times the Beta function form in (6.150), *in two different, relevant ways*, we get

$$\lim_{\substack{a\to 0 \\ b\to 0}} \frac{\partial^4}{\partial a^3 \partial b} B(a,b)$$

$$= 2 \int_0^1 \frac{\log^4(1+t)}{t} dt - 4 \int_0^1 \frac{\log(t) \log^3(1+t)}{t} dt + 3 \int_0^1 \frac{\log^2(t) \log^2(1+t)}{t} dt$$

$$- \int_0^1 \frac{\log^3(t) \log(1+t)}{t} dt, \tag{6.159}$$

and

$$\lim_{\substack{a \to 0 \\ b \to 0}} \frac{\partial^4}{\partial a^2 \partial b^2} B(a,b)$$

$$= 2 \int_0^1 \frac{\log^4(1+t)}{t} dt - 4 \int_0^1 \frac{\log(t) \log^3(1+t)}{t} dt + 2 \int_0^1 \frac{\log^2(t) \log^2(1+t)}{t} dt. \tag{6.160}$$

A careful look at (6.159) and (6.160) reveals a beautiful fact, that is, if we combine them, then we may express the integral in (6.158) in a very convenient way,

$$\int_0^1 \frac{\log^2(t) \log^2(1+t)}{t} dt$$

$$= \lim_{\substack{a \to 0 \\ b \to 0}} \frac{\partial^4}{\partial a^3 \partial b} B(a,b) - \lim_{\substack{a \to 0 \\ b \to 0}} \frac{\partial^4}{\partial a^2 \partial b^2} B(a,b) + \int_0^1 \frac{\log^3(t) \log(1+t)}{t} dt. \tag{6.161}$$

For the two limits with the derivatives of the Beta function in (6.161), we use the Beta function definition, $B(x,y) = \int_0^1 t^{x-1}(1-t)^{y-1} dt$, $\Re(x), \Re(y) > 0$, and then reduce them to particular cases of the Euler sum generalization in (6.149), Sect. 6.19. Then, we write that

$$\lim_{\substack{a \to 0 \\ b \to 0}} \frac{\partial^4}{\partial a^3 \partial b} B(a,b) = \int_0^1 \frac{\log^3(t) \log(1-t)}{t(1-t)} dt = - \int_0^1 \log^3(t) \sum_{n=1}^{\infty} t^{n-1} H_n dt$$

$$= - \sum_{n=1}^{\infty} H_n \int_0^1 t^{n-1} \log^3(t) dt = 6 \sum_{n=1}^{\infty} \frac{H_n}{n^4} = 18\zeta(5) - 6\zeta(2)\zeta(3), \tag{6.162}$$

where the last Euler sum is the particular case $n = 4$ in (6.149), Sect. 6.19. Also, observe that I used $\sum_{n=1}^{\infty} x^n H_n = -\frac{\log(1-x)}{1-x}$.

Further, for the other limit, we write

$$\lim_{\substack{a \to 0 \\ b \to 0}} \frac{\partial^4}{\partial a^2 \partial b^2} B(a, b) = \int_0^1 \frac{\log^2(t) \log^2(1 - t)}{t(1 - t)} dt$$

$$= \int_0^1 \left(\frac{\log^2(t) \log^2(1 - t)}{t} + \frac{\log^2(t) \log^2(1 - t)}{1 - t} \right) dt$$

{expand and let $1 - t = u$ in one of the integrals, and}

{then return to the variable in t, and add up the two integrals}

$$= 2 \int_0^1 \frac{\log^2(t) \log^2(1 - t)}{t} dt$$

$$\left\{ \text{make use of } \sum_{n=1}^{\infty} t^n \frac{H_n}{n + 1} = \frac{1}{2} \frac{\log^2(1 - t)}{t} \right\}$$

$$= 4 \int_0^1 \log^2(t) \sum_{n=1}^{\infty} t^n \frac{H_n}{n + 1} dt = 4 \sum_{n=1}^{\infty} \frac{H_n}{n + 1} \int_0^1 t^n \log^2(t) dt$$

$$= 8 \sum_{n=1}^{\infty} \frac{H_{n+1} - 1/(n + 1)}{(n + 1)^4} \overset{\substack{\text{reindex the series} \\ \text{and expand it}}}{=} 8 \sum_{n=1}^{\infty} \frac{H_n}{n^4} - 8 \sum_{n=1}^{\infty} \frac{1}{n^5}$$

$$= 16\zeta(5) - 8\zeta(2)\zeta(3), \tag{6.163}$$

and the last Euler sum is the particular case $n = 4$ in (6.149), Sect. 6.19.

At last, the remaining integral in (6.161) is straightforward, and then we write

$$\int_0^1 \frac{\log^3(t) \log(1 + t)}{t} dt = \int_0^1 \log^3(t) \sum_{n=1}^{\infty} (-1)^{n-1} \frac{t^{n-1}}{n} dt$$

$$= \sum_{n=1}^{\infty} (-1)^{n-1} \frac{1}{n} \int_0^1 t^{n-1} \log^3(t) dt = -6 \sum_{n=1}^{\infty} (-1)^{n-1} \frac{1}{n^5} = -\frac{45}{8} \zeta(5). \tag{6.164}$$

Collecting the results from (6.162), (6.163), and (6.164) in (6.161), we obtain

$$\int_0^1 \frac{\log^2(t) \log^2(1 + t)}{t} dt = 2\zeta(2)\zeta(3) - \frac{29}{8} \zeta(5). \tag{6.165}$$

Finally, combining the results in (6.165) and (6.158), we arrive at the desired result

$$\sum_{n=1}^{\infty}(-1)^{n-1}\frac{H_n}{n^4} = \frac{59}{32}\zeta(5) - \frac{1}{2}\zeta(2)\zeta(3),$$

and the solution to the point (ii) of the problem is finalized.

This section has been such a nice adventure with the Beta function forms!

What's next? The next section has been waiting for us with a splendid generalization of the series from the point (ii), mainly based on series manipulations!

6.21 A Famous Classical Generalization with Alternating Harmonic Series, Derived by a New Special Way

Solution If you read my first book, *(Almost) Impossible Integrals, Sums, and Series,* you probably remember that I treated the series $\sum_{k=1}^{\infty}(-1)^{k-1}\frac{H_k}{k^2}$ and $\sum_{k=1}^{\infty}(-1)^{k-1}\frac{H_k}{k^4}$ in a dedicated section where, for example, the second series is obtained by exploiting integral results involving polylogarithms (see [76, Chapter 6, pp. 508–513]). At the end of the section from the foregoing reference I wrote, *It would be interesting finding more ways of deriving such series since they seem to be pretty resistant.*

Moreover, a long-standing personal challenge was to find a solution to the present generalization that also avoids the use of integrals, one that is focused mainly on the series manipulations. This happy moment eventually happened and led to the materialization of my paper *A new powerful strategy of calculating a class of alternating Euler sums* in [77, June 25, 2019], but it was a bit too late for having a chance to include it in my first book.

In the first part of the section, I'll prepare the auxiliary results we need for the main calculations. Surely, the image of the usefulness of all these will become very clear in the second part of the section where they come into play.

So, one of the auxiliary results we need is

$$\sum_{k=1}^{\infty}\frac{1}{2k(2k+2n-1)} = \frac{1}{2n-1}\left(H_{2n} - \frac{1}{2}H_n - \log(2)\right), \tag{6.166}$$

where $n \geq 1$ is a natural number and $H_n = \sum_{k=1}^{n}\frac{1}{k}$ denotes the nth harmonic number.

Proof We already met this result, in a slightly different form, in [76, Chapter 6, p. 531], and in the following, I'll present a proof based on the same steps presented in the foregoing reference. By a simple partial fraction decomposition, we have

$$\sum_{k=1}^{\infty}\frac{1}{2k(2k+2n-1)} = \frac{1}{2n-1}\sum_{k=1}^{\infty}\left(\frac{1}{2k} - \frac{1}{2n+2k-1}\right)$$

$$= \frac{1}{2n-1} \sum_{k=1}^{\infty} \left(\frac{1}{2k} - \frac{1}{2k-1} + \frac{1}{2k-1} - \frac{1}{2n+2k-1} \right)$$

$$= \frac{1}{2n-1} \sum_{k=1}^{\infty} \left(\frac{1}{2k} - \frac{1}{2k-1} \right) + \frac{1}{2n-1} \sum_{k=1}^{\infty} \left(\frac{1}{2k-1} - \frac{1}{2n+2k-1} \right)$$

$$= -\frac{\log(2)}{2n-1} + \frac{1}{2n-1} \sum_{k=1}^{n} \frac{1}{2k-1} = \frac{1}{2n-1} \left(H_{2n} - \frac{1}{2} H_n - \log(2) \right),$$

and the proof of the first auxiliary result is complete. Observe that in the calculations above I used that $\sum_{k=1}^{\infty} \left(\frac{1}{2k-1} - \frac{1}{2k} \right) = \sum_{k=1}^{\infty} \frac{(-1)^{k-1}}{k} = \log(2)$, where the last series[6] is well-known.   ∎

Next, another auxiliary result we need is

$$\sum_{n=1}^{\infty} \frac{1}{(2k+2n-1)(2n-1)^{2m-1}}$$

$$= \frac{1}{2^{2m-1}k^{2m-1}} \left(H_{2k} - \frac{1}{2} H_k \right) - \frac{1}{2^{2m}} \sum_{i=1}^{m-1} \left(\frac{2^{2i}-1}{k^{2m-2i}} \zeta(2i) - \frac{2^{2i+1}-1}{k^{2m-2i-1}} \zeta(2i+1) \right),$$

$$(6.167)$$

where $k, m \geq 1$ are positive integers, $H_n = \sum_{k=1}^{n} \frac{1}{k}$ denotes the nth harmonic number, and ζ represents the Riemann zeta function.

Proof We may start with rearranging the summand, and then we write

$$\sum_{n=1}^{\infty} \frac{1}{(2k+2n-1)(2n-1)^{2m-1}} = \sum_{n=1}^{\infty} \frac{(2k+2n-1)-2k}{(2k+2n-1)(2n-1)^{2m}}$$

$$= \sum_{n=1}^{\infty} \frac{1}{(2n-1)^{2m}} - 2k \sum_{n=1}^{\infty} \frac{(2k+2n-1)-2k}{(2k+2n-1)(2n-1)^{2m+1}}$$

---

[6] A fast and elegant solution can be constructed based on the limit definition of the Euler-Mascheroni constant, $\lim_{n\to\infty} (H_n - \log(n)) = \gamma$, and then our alternating series may be written as $\sum_{k=1}^{\infty} (-1)^{k-1} \frac{1}{k} = \lim_{n\to\infty} \sum_{k=1}^{2n} (-1)^{k-1} \frac{1}{k} = \lim_{n\to\infty} (H_{2n} - H_n) = \lim_{n\to\infty} (H_{2n} - \log(2n) - (H_n - \log(n)) + \log(2)) = \lim_{n\to\infty} (H_{2n} - \log(2n)) - \lim_{n\to\infty} (H_n - \log(n)) + \log(2) = \log(2)$, and the calculations are finalized.

$$= \sum_{n=1}^{\infty} \frac{1}{(2n-1)^{2m}} - 2k \sum_{n=1}^{\infty} \frac{1}{(2n-1)^{2m+1}} + 4k^2 \sum_{n=1}^{\infty} \frac{1}{(2k+2n-1)(2n-1)^{2m+1}},$$

where if we consider the multiplication by $2^{2m-1}k^{2m-1}$ of the opposite sides of the result above, we have

$$\underbrace{\sum_{n=1}^{\infty} \frac{2^{2m-1}k^{2m-1}}{(2k+2n-1)(2n-1)^{2m-1}}}_{S_{k,m}} - \underbrace{\sum_{n=1}^{\infty} \frac{2^{2m+1}k^{2m+1}}{(2k+2n-1)(2n-1)^{2m+1}}}_{S_{k,m+1}}$$

$$= 2^{2m-1}k^{2m-1} \left(\sum_{n=1}^{\infty} \frac{1}{(2n-1)^{2m}} - 2k \sum_{n=1}^{\infty} \frac{1}{(2n-1)^{2m+1}} \right)$$

$$\left\{ \text{for both absolutely convergent series use that } \sum_{n=1}^{\infty} a_n = \sum_{n=1}^{\infty} a_{2n-1} + \sum_{n=1}^{\infty} a_{2n} \right\}$$

$$= 2^{2m-1}k^{2m-1} \left(\sum_{n=1}^{\infty} \frac{1}{n^{2m}} - \sum_{n=1}^{\infty} \frac{1}{(2n)^{2m}} - 2k \sum_{n=1}^{\infty} \frac{1}{n^{2m+1}} + 2k \sum_{n=1}^{\infty} \frac{1}{(2n)^{2m+1}} \right)$$

$$= \frac{1}{2} k^{2m-1}(2^{2m} - 1)\zeta(2m) - \frac{1}{2} k^{2m}(2^{2m+1} - 1)\zeta(2m+1). \qquad (6.168)$$

Now, if we consider the opposite sides of (6.168) where we replace m by i and then make the sum from $i = 1$ to $m - 1$, we get

$$\sum_{i=1}^{m-1} (S_{k,i} - S_{k,i+1}) = S_{k,1} - S_{k,m}$$

$$= 2k \sum_{n=1}^{\infty} \frac{1}{(2k+2n-1)(2n-1)} - 2^{2m-1}k^{2m-1} \sum_{n=1}^{\infty} \frac{1}{(2k+2n-1)(2n-1)^{2m-1}}$$

$$= \frac{1}{2} \sum_{i=1}^{m-1} \left(k^{2i-1}(2^{2i} - 1)\zeta(2i) - k^{2i}(2^{2i+1} - 1)\zeta(2i+1) \right),$$

where if we divide both sides of the last equality by $2^{2m-1}k^{2m-1}$, then we are able to extract the desired result

$$\sum_{n=1}^{\infty} \frac{1}{(2k+2n-1)(2n-1)^{2m-1}}$$

$$= \frac{1}{2^{2m-2}k^{2m-2}} \sum_{n=1}^{\infty} \frac{1}{(2k+2n-1)(2n-1)}$$

$$- \frac{1}{2^{2m}} \sum_{i=1}^{m-1} \left(\frac{2^{2i}-1}{k^{2m-2i}} \zeta(2i) - \frac{2^{2i+1}-1}{k^{2m-2i-1}} \zeta(2i+1) \right)$$

$$= \frac{1}{2^{2m-1}k^{2m-1}} \left(H_{2k} - \frac{1}{2}H_k \right) - \frac{1}{2^{2m}} \sum_{i=1}^{m-1} \left(\frac{2^{2i}-1}{k^{2m-2i}} \zeta(2i) - \frac{2^{2i+1}-1}{k^{2m-2i-1}} \zeta(2i+1) \right),$$

where to get the last equality I also used the simple fact that $\displaystyle\sum_{n=1}^{\infty} \frac{1}{(2k+2n-1)(2n-1)}$

$$= \frac{1}{2k} \sum_{n=1}^{\infty} \left(\frac{1}{2n-1} - \frac{1}{2k+2n-1} \right) = \frac{1}{2k} \sum_{n=1}^{k} \frac{1}{2n-1} = \frac{1}{2k} \left(H_{2k} - \frac{1}{2}H_k \right), \text{ and}$$

the proof of the second auxiliary result is complete. ∎

Let's pick up now a function, in general $f : \mathbb{N} \to \mathbb{R}$, but also with some possible restrictions on this domain (e.g., $f(s) = \zeta(s)$, where $\Re(s) > 1$, which we'll need in one of the applications of the result below). Then, we want to prove that

$$\sum_{k=1}^{m-1} f(k+1)f(2m-k) = \sum_{k=1}^{m-1} f(2k)f(2m-2k+1), \ m \geq 1. \qquad (6.169)$$

Proof Let's observe first that if we let the variable change $2m - k - 1 = i$ for the sum $\displaystyle\sum_{k=1}^{m-1} f(k+1)f(2m-k)$, we have

$$\sum_{k=1}^{m-1} f(k+1)f(2m-k) = \sum_{i=m}^{2m-2} f(i+1)f(2m-i) = \sum_{k=m}^{2m-2} f(k+1)f(2m-k). \qquad (6.170)$$

So, based on the result in (6.170), we obtain that

$$\sum_{k=1}^{m-1} f(k+1)f(2m-k) = \frac{1}{2} \sum_{k=1}^{2m-2} f(k+1)f(2m-k)$$

{split the sum according to the parity}

$$= \frac{1}{2} \sum_{k=1}^{m-1} f(2k)f(2m-2k+1) + \frac{1}{2} \sum_{k=1}^{m-1} f(2k+1)f(2m-2k)$$

{in the second sum make the variable change $m - k = i$}

$$= \frac{1}{2} \sum_{k=1}^{m-1} f(2k) f(2m - 2k + 1) + \frac{1}{2} \sum_{i=1}^{m-1} f(2i) f(2m - 2i + 1)$$

$$= \sum_{k=1}^{m-1} f(2k) f(2m - 2k + 1),$$

which brings an end to our proof.   ∎

Now, we are ready to extract another two auxiliary results we need in our main calculations. So, applying the result in (6.169) for $f(s) = \zeta(s)$ and $f(s) = (1 - 2^s)\zeta(s)$, we obtain

$$\sum_{k=1}^{m-1} \zeta(k + 1)\zeta(2m - k) = \sum_{k=1}^{m-1} \zeta(2k)\zeta(2m - 2k + 1) \qquad (6.171)$$

and

$$\sum_{k=1}^{m-1}(1 - 2^{k+1})(1 - 2^{2m-k})\zeta(k + 1)\zeta(2m - k)$$

$$= \sum_{k=1}^{m-1}(1 - 2^{2k})(1 - 2^{2m-2k+1})\zeta(2k)\zeta(2m - 2k + 1), \qquad (6.172)$$

and the calculations to these two auxiliary results are complete.

We are finally ready to start the main calculations!

Returning to the result in (6.166) where we multiply both sides by $1/(2n - 1)^{2m-1}$, we get that

$$\sum_{n=1}^{\infty} \left(\sum_{k=1}^{\infty} \frac{1}{2k(2k + 2n - 1)(2n - 1)^{2m-1}} \right)$$

$$= \sum_{n=1}^{\infty} \frac{H_{2n}}{(2n - 1)^{2m}} - \frac{1}{2} \sum_{n=1}^{\infty} \frac{H_n}{(2n - 1)^{2m}} - \log(2) \sum_{n=1}^{\infty} \frac{1}{(2n - 1)^{2m}}$$

{reindex the second series and start from $n = 0$ to ∞}

{and at the same time leave out the term for $n = 0$}

$$= \sum_{n=1}^{\infty} \frac{H_{2n-1} + 1/(2n)}{(2n-1)^{2m}} - \frac{1}{2} \sum_{n=1}^{\infty} \frac{H_n + 1/(n+1)}{(2n+1)^{2m}} - \frac{1}{2} - \log(2) \sum_{n=1}^{\infty} \frac{1}{(2n-1)^{2m}}$$

$$= \sum_{n=1}^{\infty} \frac{H_{2n-1}}{(2n-1)^{2m}} - \frac{1}{2} \sum_{n=1}^{\infty} \frac{H_n}{(2n+1)^{2m}} - \log(2) \sum_{n=1}^{\infty} \frac{1}{(2n-1)^{2m}}$$

$$+ \underbrace{\sum_{n=1}^{\infty} \left(\frac{1}{2n(2n-1)^{2m}} - \frac{1}{(2n+2)(2n+1)^{2m}} \right)}_{\text{A telescoping sum leading to } 1/2} - \frac{1}{2}$$

$$\left\{ \text{for the first series use that } \sum_{n=1}^{\infty} a_{2n-1} = \frac{1}{2} \left(\sum_{n=1}^{\infty} a_n + \sum_{n=1}^{\infty} (-1)^{n-1} a_n \right) \right\}$$

$$= \frac{1}{2} \sum_{n=1}^{\infty} \frac{H_n}{n^{2m}} + \frac{1}{2} \sum_{n=1}^{\infty} (-1)^{n-1} \frac{H_n}{n^{2m}} - \frac{1}{2} \sum_{n=1}^{\infty} \frac{H_n}{(2n+1)^{2m}} - \log(2) \left(1 - \frac{1}{2^{2m}} \right) \zeta(2m)$$

{make use of the results in (6.149), Sect. 6.19, and (4.100), Sect. 4.19}

$$= \frac{1}{2^{2m+1}} (m - 4^m(m-1)) \zeta(2m+1) - \frac{1}{2} \sum_{k=1}^{m-1} \zeta(k+1) \zeta(2m-k)$$

$$+ \frac{1}{2^{2m+1}} \sum_{k=1}^{m-1} (1 - 2^{k+1})(1 - 2^{2m-k}) \zeta(k+1) \zeta(2m-k) + \frac{1}{2} \sum_{n=1}^{\infty} (-1)^{n-1} \frac{H_n}{n^{2m}}$$

{employ the sum transformations in (6.171) and (6.172)}

$$= \frac{1}{2^{2m+1}} (m - 4^m(m-1)) \zeta(2m+1) - \frac{1}{2} \sum_{k=1}^{m-1} \zeta(2k) \zeta(2m-2k+1)$$

$$+ \frac{1}{2^{2m+1}} \sum_{k=1}^{m-1} (1 - 2^{2k})(1 - 2^{2m-2k+1}) \zeta(2k) \zeta(2m-2k+1) + \frac{1}{2} \sum_{n=1}^{\infty} (-1)^{n-1} \frac{H_n}{n^{2m}}.$$

$$(6.173)$$

On the other hand, reversing the order of the double series in (6.173), we have that

$$\sum_{n=1}^{\infty}\left(\sum_{k=1}^{\infty}\frac{1}{2k(2k+2n-1)(2n-1)^{2m-1}}\right)=\sum_{k=1}^{\infty}\left(\sum_{n=1}^{\infty}\frac{1}{2k(2k+2n-1)(2n-1)^{2m-1}}\right)$$

{make use of the result in (6.167)}

$$=\sum_{k=1}^{\infty}\frac{1}{2k}\left(\frac{2H_{2k}-H_k}{2^{2m}k^{2m-1}}-\frac{1}{2^{2m}}\sum_{i=1}^{m-1}\left(\frac{2^{2i}-1}{k^{2m-2i}}\zeta(2i)-\frac{2^{2i+1}-1}{k^{2m-2i-1}}\zeta(2i+1)\right)\right)$$

$$=\sum_{k=1}^{\infty}\frac{H_{2k}}{(2k)^{2m}}-\frac{1}{2^{2m+1}}\sum_{k=1}^{\infty}\frac{H_k}{k^{2m}}-\frac{1}{2^{2m+1}}\sum_{i=1}^{m-1}(2^{2i}-1)\zeta(2i)\sum_{k=1}^{\infty}\frac{1}{k^{2m-2i+1}}$$

$$+\frac{1}{2^{2m+1}}\sum_{i=1}^{m-1}(2^{2i+1}-1)\zeta(2i+1)\sum_{k=1}^{\infty}\frac{1}{k^{2m-2i}}$$

$$\left\{\text{for the first series use that }\sum_{n=1}^{\infty}a_{2n}=\frac{1}{2}\left(\sum_{n=1}^{\infty}a_n-\sum_{n=1}^{\infty}(-1)^{n-1}a_n\right)\right\}$$

$$=\frac{1}{2}\left(1-\frac{1}{2^{2m}}\right)\sum_{k=1}^{\infty}\frac{H_k}{k^{2m}}-\frac{1}{2}\sum_{k=1}^{\infty}(-1)^{k-1}\frac{H_k}{k^{2m}}$$

$$-\frac{1}{2^{2m+1}}\sum_{i=1}^{m-1}(2^{2i}-1)\zeta(2i)\zeta(2m-2i+1)$$

$$+\frac{1}{2^{2m+1}}\sum_{i=1}^{m-1}(2^{2i+1}-1)\zeta(2i+1)\zeta(2m-2i)$$

{consider the Euler sum in (6.149), Sect. 6.19}

$$=\frac{1}{2}(m+1)\left(1-\frac{1}{2^{2m}}\right)\zeta(2m+1)-\frac{1}{2}\left(1-\frac{1}{2^{2m}}\right)\sum_{k=1}^{m-1}\zeta(k+1)\zeta(2m-k)$$

$$-\frac{1}{2^{2m+1}}\sum_{k=1}^{m-1}(2^{2k}-1)\zeta(2k)\zeta(2m-2k+1)$$

$$+ \frac{1}{2^{2m+1}} \sum_{k=1}^{m-1} (2^{2k+1} - 1)\zeta(2k+1)\zeta(2m-2k) - \frac{1}{2} \sum_{k=1}^{\infty} (-1)^{k-1} \frac{H_k}{k^{2m}}$$

{for the first sum use the transformation in (6.171), and for the third sum}

{make the variable change $m - k = i$ and then return to the variable in k}

$$= \frac{1}{2}(m+1)\left(1 - \frac{1}{2^{2m}}\right)\zeta(2m+1) - \frac{1}{2}\left(1 - \frac{1}{2^{2m}}\right) \sum_{k=1}^{m-1} \zeta(2k)\zeta(2m-2k+1)$$

$$- \frac{1}{2^{2m+1}} \sum_{k=1}^{m-1} (2^{2k} - 1)\zeta(2k)\zeta(2m-2k+1)$$

$$+ \frac{1}{2^{2m+1}} \sum_{k=1}^{m-1} (2^{2m-2k+1} - 1)\zeta(2k)\zeta(2m-2k+1) - \frac{1}{2} \sum_{n=1}^{\infty} (-1)^{n-1} \frac{H_n}{n^{2m}}. \qquad (6.174)$$

At last, combining the results in (6.173) and (6.174), which we further simplify and rearrange, we arrive at

$$\sum_{n=1}^{\infty} (-1)^{n-1} \frac{H_n}{n^{2m}}$$

$$= m\left(1 - \frac{1}{2^{2m}}\right)\zeta(2m+1) - \frac{1}{2^{2m+1}}\zeta(2m+1)$$

$$- \sum_{k=1}^{m-1} \left(1 - 2^{1-2k}\right)\zeta(2k)\zeta(2m-2k+1)$$

{use the relation between the Dirichlet eta function}

{and the Riemann zeta function, $\eta(s) = (1 - 2^{1-s})\zeta(s)$}

$$= \left(m + \frac{1}{2}\right)\eta(2m+1) - \frac{1}{2}\zeta(2m+1) - \sum_{k=1}^{m-1} \eta(2k)\zeta(2m-2k+1),$$

and the solution to the main result is finalized.

To easily catch the idea behind the generalization, one may work out first some simple particular cases like $m = 1, 2$ and then follow the solution above and see what happens in every step of the solution.

6.22   Seven Useful Generalized Harmonic Series

Solution The calculations in this section will flow pretty fast and smoothly since the key integrals needed have already been evaluated in the third chapter! For example, the first, the second, the fifth, and the sixth generalized harmonic series are immediately extracted by exploiting some of their corresponding integrals. My solutions to the generalized alternating harmonic series at the points (i) and (ii) had also been presented in [46, May 23, 2019].

Let's see how exactly to do it! One simple fact we want to use for some evaluations is that $\dfrac{1}{k^m} = (-1)^{m-1}\dfrac{1}{(m-1)!}\displaystyle\int_0^1 x^{k-1}\log^{m-1}(x)\mathrm{d}x$ (found in [76, Chapter 1, p.1]), and then returning to the first point of the problem, we write

$$\sum_{n=1}^{\infty}(-1)^{n-1}\frac{H_n^{(m)}}{n} = \sum_{n=1}^{\infty}(-1)^{n-1}\frac{1}{n}\sum_{k=1}^{n}\frac{1}{k^m}$$

$$= (-1)^{m-1}\frac{1}{(m-1)!}\sum_{n=1}^{\infty}(-1)^{n-1}\frac{1}{n}\sum_{k=1}^{n}\int_0^1 x^{k-1}\log^{m-1}(x)\mathrm{d}x$$

{reverse the order of integration and summation}

$$= (-1)^{m-1}\frac{1}{(m-1)!}\int_0^1 \log^{m-1}(x)\sum_{n=1}^{\infty}(-1)^{n-1}\frac{1}{n}\sum_{k=1}^{n}x^{k-1}\mathrm{d}x$$

$$= (-1)^{m-1}\frac{1}{(m-1)!}\int_0^1 \log^{m-1}(x)\sum_{n=1}^{\infty}(-1)^{n-1}\frac{1}{n}\cdot\frac{1-x^n}{1-x}\mathrm{d}x$$

$$= (-1)^{m}\frac{1}{(m-1)!}\int_0^1 \frac{\log^{m-1}(x)\log\left(\dfrac{1+x}{2}\right)}{1-x}\mathrm{d}x$$

{exploit the integral result in (1.131), Sect. 1.27}

$$= \frac{1}{2}\left(m\zeta(m+1) - \sum_{k=1}^{m}\eta(k)\eta(m-k+1)\right),$$

and the solution to the point (i) is finalized. A different route of solving the problem may be found in [67, p.214].

In a similar way as before, for the second point of the problem, we write

$$\sum_{n=1}^{\infty}(-1)^{n-1}\frac{H_{2n}^{(m)}}{n} = \sum_{n=1}^{\infty}(-1)^{n-1}\frac{1}{n}\sum_{k=1}^{2n}\frac{1}{k^m}$$

$$= (-1)^{m-1}\frac{1}{(m-1)!}\sum_{n=1}^{\infty}(-1)^{n-1}\frac{1}{n}\sum_{k=1}^{2n}\int_0^1 x^{k-1}\log^{m-1}(x)dx$$

{reverse the order of integration and summation}

$$= (-1)^{m-1}\frac{1}{(m-1)!}\int_0^1 \log^{m-1}(x)\sum_{n=1}^{\infty}(-1)^{n-1}\frac{1}{n}\sum_{k=1}^{2n}x^{k-1}dx$$

$$= (-1)^{m-1}\frac{1}{(m-1)!}\int_0^1 \log^{m-1}(x)\sum_{n=1}^{\infty}(-1)^{n-1}\frac{1}{n}\cdot\frac{1-x^{2n}}{1-x}dx$$

$$= (-1)^m\frac{1}{(m-1)!}\int_0^1 \frac{\log^{m-1}(x)\log\left(\frac{1+x^2}{2}\right)}{1-x}dx$$

{exploit the integral result in (1.132), Sect. 1.27}

$$= m\zeta(m+1) - \frac{1}{2^{m+1}}\sum_{k=1}^{m}\eta(k)\eta(m-k+1) - \sum_{k=1}^{m}\beta(k)\beta(m-k+1),$$

and the solution to the point (ii) is finalized.

The extraction of the harmonic series at the third point is straightforward if we exploit the series from the point (i), and then we write

$$\sum_{n=1}^{\infty}(-1)^{n-1}\frac{H_n^{(m)}}{n} = \sum_{n=1}^{\infty}(-1)^{n-1}\frac{1}{n}\sum_{k=1}^{n}\frac{1}{k^m} \overset{\substack{\text{reverse the order}\\\text{of summation}}}{=} \sum_{k=1}^{\infty}\frac{1}{k^m}\sum_{n=k}^{\infty}(-1)^{n-1}\frac{1}{n}$$

$$= \sum_{k=1}^{\infty}\frac{1}{k^m}\left((-1)^{k-1}\frac{1}{k} + \log(2) - \overline{H}_k\right) = \sum_{k=1}^{\infty}(-1)^{k-1}\frac{1}{k^{m+1}} + \log(2)\sum_{k=1}^{\infty}\frac{1}{k^m}$$

$$- \sum_{k=1}^{\infty}\frac{\overline{H}_n}{k^m} = \log(2)\zeta(m) + \eta(m+1) - \sum_{k=1}^{\infty}\frac{\overline{H}_n}{k^m},$$

whence we obtain that

$$\sum_{n=1}^{\infty} \frac{\overline{H}_n}{n^m} s$$

$$= \log(2)\zeta(m) - \frac{1}{2}m\zeta(m+1) + \eta(m+1) + \frac{1}{2}\sum_{k=1}^{m} \eta(k)\eta(m-k+1),$$

and the solution to the point (iii) is finalized. A different approach may be found in [67, pp. 196–200].

To get the value of the series at the fourth point, we want to use Abel's summation, the series version in (6.7), Sect. 6.2 which we may apply directly to the series from the point (ii), where if we set $a_n = (-1)^{n-1}1/n$ and $b_n = H_{2n}^{(m)}$, we get

$$\sum_{n=1}^{\infty} (-1)^{n-1} \frac{H_{2n}^{(m)}}{n} = \log(2)\zeta(m) - \frac{1}{2^m}\sum_{n=1}^{\infty} \frac{\overline{H}_{n+1} - (-1)^n/(n+1)}{(n+1)^m} - \sum_{n=1}^{\infty} \frac{\overline{H}_n}{(2n+1)^m}$$

$$= \log(2)\zeta(m) + \frac{1}{2^m}\eta(m+1) - \frac{1}{2^m}\sum_{n=1}^{\infty} \frac{\overline{H}_n}{n^m} - \sum_{n=1}^{\infty} \frac{\overline{H}_n}{(2n+1)^m},$$

whence upon considering the values of the series at the points (ii) and (iii), we conclude that

$$\sum_{n=1}^{\infty} \frac{\overline{H}_n}{(2n+1)^m}$$

$$= \log(2)(1 - 2^{-m})\zeta(m) - m(1 - 2^{-m-1})\zeta(m+1) + \sum_{k=1}^{m} \beta(k)\beta(m-k+1),$$

and the solution to the point (iv) is finalized.

What about the harmonic series at the fifth point? We may proceed as I did for the first two points of the problem, and then we write

$$\sum_{n=1}^{\infty} (-1)^{n-1} \frac{\overline{H}_{2n}^{(m)}}{n} = \sum_{n=1}^{\infty} (-1)^{n-1} \frac{1}{n} \sum_{k=1}^{2n} (-1)^{k-1} \frac{1}{k^m}$$

$$= (-1)^{m-1} \frac{1}{(m-1)!} \sum_{n=1}^{\infty} (-1)^{n-1} \frac{1}{n} \sum_{k=1}^{2n} (-1)^{k-1} \int_0^1 x^{k-1} \log^{m-1}(x) dx$$

{reverse the order of integration and summation}

$$= (-1)^{m-1} \frac{1}{(m-1)!} \int_0^1 \log^{m-1}(x) \sum_{n=1}^{\infty} (-1)^{n-1} \frac{1}{n} \sum_{k=1}^{2n} (-1)^{k-1} x^{k-1} dx$$

$$= (-1)^{m-1} \frac{1}{(m-1)!} \int_0^1 \log^{m-1}(x) \sum_{n=1}^{\infty} (-1)^{n-1} \frac{1}{n} \cdot \frac{1-x^{2n}}{1+x} dx$$

$$= (-1)^m \frac{1}{(m-1)!} \int_0^1 \frac{\log^{m-1}(x) \log\left(\frac{1+x^2}{2}\right)}{1+x} dx$$

{exploit the integral result in (1.133), Sect. 1.27}

$$= m\eta(m+1) + \frac{1}{2^{m+1}} \sum_{k=1}^{m} \eta(k)\eta(m-k+1) - \sum_{k=1}^{m} \beta(k)\beta(m-k+1),$$

and the solution to the point (v) is finalized.

Next, to calculate the series at the point (vi), we proceed as follows:

$$\sum_{n=1}^{\infty} (-1)^{n-1} \frac{\overline{H}_n}{n^{2m}} = \sum_{n=1}^{\infty} (-1)^{n-1} \frac{\overline{H}_{n-1} + (-1)^{n-1}/n}{n^{2m}} = \sum_{n=1}^{\infty} (-1)^{n-1} \frac{\overline{H}_{n-1}}{n^{2m}}$$

$$+ \sum_{n=1}^{\infty} \frac{1}{n^{2m+1}} = \zeta(2m+1) + \sum_{n=1}^{\infty} (-1)^{n-1} \frac{\overline{H}_{n-1}}{n^{2m}}$$

{reindex the resulting series and start from $n = 1$}

$$= \zeta(2m+1) - \sum_{n=1}^{\infty} (-1)^{n-1} \frac{\overline{H}_n}{(n+1)^{2m}}$$

$$= \zeta(2m+1) + \frac{1}{(2m-1)!} \sum_{n=1}^{\infty} (-1)^{n-1} \overline{H}_n \int_0^1 x^n \log^{2m-1}(x) dx$$

$$= \zeta(2m+1) - \frac{1}{(2m-1)!} \int_0^1 \log^{2m-1}(x) \sum_{n=1}^{\infty} (-x)^n \overline{H}_n dx$$

{employ the generating function in (4.62), $m = 1$, Sect. 4.10}

$$= \zeta(2m+1) - \frac{1}{(2m-1)!} \int_0^1 \frac{\log^{2m-1}(x) \log(1-x)}{1+x} dx$$

{make use of the generalized integral in (1.130), Sect. 1.26}

$$= \log(2)\zeta(2m) - \frac{1}{2^{2m+1}}(m2^{2m+1} - 2m - 1)\zeta(2m + 1)$$

$$+ \left(1 - \frac{1}{2^{2m-1}}\right) \sum_{k=1}^{m-1} \zeta(2k)\zeta(2m - 2k + 1)$$

$$- \sum_{k=1}^{m-1} \eta(2k)\zeta(2m - 2k + 1) + \sum_{k=0}^{m-1} \eta(2k + 1)\eta(2m - 2k),$$

and the solution to the point (vi) is finalized. Another solution may be found in [67, p.203].

Further, the series result at the seventh point is straightforward if we exploit the previous result and combine it we the change of summation order. So, we write

$$\sum_{n=1}^{\infty}(-1)^{n-1}\frac{\overline{H}_n^{(2m)}}{n} = \sum_{n=1}^{\infty}(-1)^{n-1}\frac{1}{n}\sum_{k=1}^{n}(-1)^{k-1}\frac{1}{k^{2m}}$$

$$= \sum_{k=1}^{\infty}(-1)^{k-1}\frac{1}{k^{2m}}\sum_{n=k}^{\infty}(-1)^{n-1}\frac{1}{n} = \sum_{k=1}^{\infty}(-1)^{k-1}\frac{1}{k^{2m}}\left((-1)^{k-1}\frac{1}{k} + \log(2) - \overline{H}_k\right)$$

$$= \sum_{k=1}^{\infty}\frac{1}{k^{2m+1}} + \log(2)\sum_{k=1}^{\infty}(-1)^{k-1}\frac{1}{k^{2m}} - \sum_{k=1}^{\infty}(-1)^{k-1}\frac{\overline{H}_k}{k^{2m}}$$

$$= \zeta(2m + 1) + \log(2)\eta(2m) - \sum_{k=1}^{\infty}(-1)^{k-1}\frac{\overline{H}_k}{k^{2m}}$$

$$= \log(2)\eta(2m) - \log(2)\zeta(2m) + \frac{1}{2^{2m+1}}((m + 1)2^{2m+1} - 2m - 1)\zeta(2m + 1)$$

$$- \left(1 - \frac{1}{2^{2m-1}}\right) \sum_{k=1}^{m-1} \zeta(2k)\zeta(2m - 2k + 1)$$

$$+ \sum_{k=1}^{m-1} \eta(2k)\zeta(2m - 2k + 1) - \sum_{k=0}^{m-1} \eta(2k + 1)\eta(2m - 2k),$$

and the solution to the point (vii) is finalized.

We meet particular cases of such series in other sections during the extraction of some results. Good to know and use them when needed!

6.23   A Special Challenging Harmonic Series of Weight 4, Involving Harmonic Numbers of the Type H_{2n}

Solution The unusual thing about this series is the product $H_n H_{2n}$ in the numerator, which makes the series troublesome. In general, we want to reduce the calculations to simpler harmonic series, the classical ones.

How do we want to proceed? Well, we could exploit identities that naturally involve the appearance of H_{2n}. One of them, a powerful one, also met in my first book *(Almost) Impossible Integrals, Sums, and Series*, is

$$- \log(1+x)\log(1-x) = \sum_{k=1}^{\infty} x^{2k} \frac{H_{2k} - H_k}{k} + \frac{1}{2} \sum_{k=1}^{\infty} \frac{x^{2k}}{k^2}, \quad |x| < 1, \qquad (6.175)$$

which is immediately yielded by the Cauchy product of two series as seen in [76, Chapter 6, p.344].

Recall the version of the Cauchy product of two series which states that

$$\left(\sum_{n=1}^{\infty} a_n \right) \left(\sum_{n=1}^{\infty} b_n \right) = \sum_{n=1}^{\infty} \left(\sum_{k=1}^{n} a_k b_{n-k+1} \right),$$

where both series $\sum_{n=1}^{\infty} a_n$ and $\sum_{n=1}^{\infty} b_n$ are absolutely convergent.

Now, if we are inspired to return to Sect. 1.21 and exploit the result in (1.100), then we can extract a solution.

So, for a first solution, we multiply both sides of the result in (6.175) by $\log(1 + x)/x$ and integrate from $x = 0$ to $x = 1$ that gives

$$- \int_0^1 \frac{\log(1-x)\log^2(1+x)}{x} dx$$

$$= \int_0^1 \log(1+x) \sum_{k=1}^{\infty} x^{2k-1} \frac{H_{2k} - H_k}{k} dx + \frac{1}{2} \int_0^1 \log(1+x) \sum_{k=1}^{\infty} \frac{x^{2k-1}}{k^2} dx$$

{reverse the order of summation and integration}

$$= \sum_{k=1}^{\infty} \frac{H_{2k} - H_k}{k} \int_0^1 x^{2k-1} \log(1+x) dx + \frac{1}{2} \sum_{k=1}^{\infty} \frac{1}{k^2} \int_0^1 x^{2k-1} \log(1+x) dx$$

{make use of the result in (1.100), Sect. 1.21, and then expand the series}

$$= \frac{1}{2}\sum_{k=1}^{\infty}\frac{H_k^2}{k^2} - \frac{1}{4}\sum_{k=1}^{\infty}\frac{H_k}{k^3} + 2\sum_{k=1}^{\infty}\frac{H_{2k}}{(2k)^3} + 2\sum_{k=1}^{\infty}\frac{H_{2k}^2}{(2k)^2} - \sum_{k=1}^{\infty}\frac{H_k H_{2k}}{k^2}$$

$$\left\{ \text{for the third and fourth series use that } \sum_{k=1}^{\infty} a_{2k} = \frac{1}{2}\left(\sum_{k=1}^{\infty} a_k - \sum_{k=1}^{\infty}(-1)^{k-1} a_k\right)\right\}$$

{and then we replace the letter of the summation variable and use n instead of k}

$$= \frac{3}{2}\sum_{n=1}^{\infty}\frac{H_n^2}{n^2} + \frac{3}{4}\sum_{n=1}^{\infty}\frac{H_n}{n^3} - \sum_{n=1}^{\infty}(-1)^{n-1}\frac{H_n}{n^3} - \sum_{n=1}^{\infty}(-1)^{n-1}\frac{H_n^2}{n^2} - \sum_{n=1}^{\infty}\frac{H_n H_{2n}}{n^2}.$$

$$(6.176)$$

At the same time, by exploiting the simple identity, $\displaystyle\sum_{n=1}^{\infty}(-1)^{n-1} x^{n+1}\frac{H_n}{n+1} =$

$\frac{1}{2}\log^2(1+x)$, based on (4.33), Sect. 4.6, we have that

$$-\int_0^1 \frac{\log(1-x)\log^2(1+x)}{x}dx = -2\int_0^1 \log(1-x)\sum_{n=1}^{\infty}(-1)^{n-1} x^n\frac{H_n}{n+1}dx$$

{reverse the order of summation and integration}

$$= -2\sum_{n=1}^{\infty}(-1)^{n-1}\frac{H_n}{n+1}\int_0^1 x^n\log(1-x)dx$$

$$\left\{\text{use the integral } \int_0^1 x^{n-1}\log(1-x)dx = -\frac{H_n}{n}, \text{ as seen in } (3.10), \text{Sect. 3.3}\right\}$$

$$= 2\sum_{n=1}^{\infty}(-1)^{n-1}\frac{H_n H_{n+1}}{(n+1)^2} = 2\sum_{n=1}^{\infty}(-1)^{n-1}\frac{(H_{n+1}-1/(n+1))H_{n+1}}{(n+1)^2}$$

{reindex the series and then expand it}

$$= 2\sum_{n=1}^{\infty}(-1)^{n-1}\frac{H_n}{n^3} - 2\sum_{n=1}^{\infty}(-1)^{n-1}\frac{H_n^2}{n^2}. \qquad (6.177)$$

Next, by combining the results in (6.176) and (6.177), we get

$$\sum_{n=1}^{\infty}\frac{H_n H_{2n}}{n^2} = \frac{3}{2}\sum_{n=1}^{\infty}\frac{H_n^2}{n^2} + \frac{3}{4}\sum_{n=1}^{\infty}\frac{H_n}{n^3} - 3\sum_{n=1}^{\infty}(-1)^{n-1}\frac{H_n}{n^3} + \sum_{n=1}^{\infty}(-1)^{n-1}\frac{H_n^2}{n^2}.$$

$$(6.178)$$

Before going further with the calculations in (6.178), I'll present and extract a result very good to know which essentially presents a relation between the alternating series, $\sum_{n=1}^{\infty}(-1)^{n-1}\dfrac{H_n}{n^3}$ and $\sum_{n=1}^{\infty}(-1)^{n-1}\dfrac{H_n^2}{n^2}$, which as you can see are also found in our last relation in (6.178). The main point here is that once we know the value of any of the two series, then we'll be able to extract the value of the other series by using such a relation.

Let's return to the integral in (6.177) where if we use the algebraic identity $a^2b = 1/6((a+b)^3 - (a-b)^3 - 2b^3)$, with $b = \log(1-x)$ and $a = \log(1+x)$, we have

$$\int_0^1 \frac{\log(1-x)\log^2(1+x)}{x}dx$$

$$= \underbrace{\frac{1}{6}\int_0^1 \frac{\log^3(1-x^2)}{x}dx}_{X} - \underbrace{\frac{1}{6}\int_0^1 \frac{1}{x}\log^3\left(\frac{1+x}{1-x}\right)dx}_{Y} - \underbrace{\frac{1}{3}\int_0^1 \frac{\log^3(1-x)}{x}dx}_{Z}.$$

$$(6.179)$$

For the integral X in (6.179), we obtain

$$X = \int_0^1 \frac{\log^3(1-x^2)}{x}dx \overset{1-x^2=y}{=} \frac{1}{2}\int_0^1 \frac{\log^3(y)}{1-y}dy = \frac{1}{2}\int_0^1 \sum_{n=1}^{\infty}y^{n-1}\log^3(y)dy$$

$$= \frac{1}{2}\sum_{n=1}^{\infty}\int_0^1 y^{n-1}\log^3(y)dy = -3\sum_{n=1}^{\infty}\frac{1}{n^4} = -3\zeta(4). \qquad (6.180)$$

Then, for the integral Y in (6.179), we get

$$Y = \int_0^1 \frac{1}{x}\log^3\left(\frac{1+x}{1-x}\right)dx \overset{\frac{1-x}{1+x}=y}{=} -2\int_0^1 \frac{\log^3(y)}{1-y^2}dy = -2\int_0^1 \sum_{n=1}^{\infty}y^{2n-2}\log^3(y)dy$$

$$= -2\sum_{n=1}^{\infty}\int_0^1 y^{2n-2}\log^3(y)dy = 12\sum_{n=1}^{\infty}\frac{1}{(2n-1)^4} = \frac{\pi^4}{8} = \frac{45}{4}\zeta(4). \qquad (6.181)$$

Lastly, for the integral Z in (6.179), we have

$$Z = \int_0^1 \frac{\log^3(1-x)}{x}dx \overset{x=y^2}{=} 2\int_0^1 \frac{\log^3(1-y^2)}{y}dy = -6\zeta(4), \qquad (6.182)$$

since $Z = 2X$ and X is the integral in (6.180).

Collecting the values of the integrals X, Y, and Z from (6.180), (6.181), and (6.182) in (6.179), we obtain that

$$\int_0^1 \frac{\log(1-x)\log^2(1+x)}{x}dx = -\frac{3}{8}\zeta(4). \tag{6.183}$$

With the result in (6.183), we return to (6.177) that gives

$$\sum_{n=1}^\infty (-1)^{n-1}\frac{H_n}{n^3} - \sum_{n=1}^\infty (-1)^{n-1}\frac{H_n^2}{n^2} = \frac{3}{16}\zeta(4), \tag{6.184}$$

which is the auxiliary result we wanted to obtain.

Now, let's take a closer look at the right-hand side in (6.178). So, the first series

is the famous Au-Yeung series, $\displaystyle\sum_{n=1}^\infty \frac{H_n^2}{n^2} = \frac{17}{4}\zeta(4)$, which is already calculated

by series manipulations in [76, Chapter 6, pp. 392–393], or if you want an elegant
strategy with integrals, then see [91]. The second series is the particular case $n = 3$
of the classical Euler sum in (6.149), Sect. 6.19. The third series is found calculated
in (4.103), Sect. 4.20, and [76, Chapter 6, pp. 502–505], and the fourth series

is $\displaystyle\sum_{n=1}^\infty (-1)^{n-1}\frac{H_n^2}{n^2} = \frac{41}{16}\zeta(4) - \frac{7}{4}\log(2)\zeta(3) + \frac{1}{2}\log^2(2)\zeta(2) - \frac{1}{12}\log^4(2) -$

$2\operatorname{Li}_4\left(\dfrac{1}{2}\right)$, which is calculated in [76, Chapter 6, pp. 506–508], but based on the

result in (6.184), we may express it by using $\displaystyle\sum_{n=1}^\infty (-1)^{n-1}\frac{H_n}{n^3}$ as I'll do it below.

Finally, continuing the calculations in (6.178) by first using (6.184), we write

$$\sum_{n=1}^\infty \frac{H_n H_{2n}}{n^2} = \frac{3}{2}\sum_{n=1}^\infty \frac{H_n^2}{n^2} + \frac{3}{4}\sum_{n=1}^\infty \frac{H_n}{n^3} - 2\sum_{n=1}^\infty (-1)^{n-1}\frac{H_n}{n^3} - \frac{3}{16}\zeta(4)$$

$$= \frac{13}{8}\zeta(4) + \frac{7}{2}\log(2)\zeta(3) - \log^2(2)\zeta(2) + \frac{1}{6}\log^4(2) + 4\operatorname{Li}_4\left(\frac{1}{2}\right),$$

and the first solution to the main series is finalized. My first strategy has also been
presented in [56].

Next, to get a second solution to the main series, we use the result in (4.82),
Sect. 4.13, where we multiply both sides by $1/n^2$ and then sum from $n = 1$ to ∞
that gives

$$\sum_{n=1}^\infty \frac{1}{n^2}\left(\sum_{k=1}^n \frac{H_k}{k+n}\right) \overset{\substack{\text{reverse the order}\\\text{of summation}}}{=} \sum_{k=1}^\infty \left(\sum_{n=k}^\infty \frac{H_k}{n^2(k+n)}\right)$$

$$= \sum_{k=1}^\infty \left(\sum_{n=k}^\infty \left(\frac{H_k}{kn^2} - \frac{H_k}{kn(k+n)}\right)\right) = \sum_{k=1}^\infty \left(\sum_{n=k}^\infty \frac{H_k}{kn^2}\right) - \sum_{k=1}^\infty \frac{H_k}{k}\left(\sum_{n=k}^\infty \frac{1}{n(k+n)}\right)$$

{reverse the order of summation in the first double series and in}

$$\left\{ \text{the second one use the simple fact, } \sum_{n=k}^{\infty} \frac{1}{n(n+k)} = \frac{H_{2k-1} - H_{k-1}}{k} \right\}$$

$$= \sum_{n=1}^{\infty} \frac{1}{n^2} \left(\sum_{k=1}^{n} \frac{H_k}{k} \right) - \sum_{k=1}^{\infty} \frac{H_k(H_{2k} - H_k + 1/(2k))}{k^2}$$

$$\left\{ \text{use } \sum_{k=1}^{n} \frac{H_k}{k} = \frac{1}{2}(H_n^2 + H_n^{(2)}), \text{ which is the case } p = q = 1 \text{ in (6.102), Sect. 6.13} \right\}$$

{and then expand the second series and replace the summation variable k by n}

$$= \frac{3}{2} \sum_{n=1}^{\infty} \frac{H_n^2}{n^2} + \frac{1}{2} \sum_{n=1}^{\infty} \frac{H_n^{(2)}}{n^2} - \frac{1}{2} \sum_{n=1}^{\infty} \frac{H_n}{n^3} - \sum_{n=1}^{\infty} \frac{H_n H_{2n}}{n^2}. \tag{6.185}$$

The other side yielded by the multiplication of (4.82), Sect. 4.13, by $1/n^2$ is

$$\sum_{n=1}^{\infty} \frac{H_n H_{2n}}{n^2} - \sum_{n=1}^{\infty} \frac{H_n^2}{n^2} - 8 \sum_{n=1}^{\infty} \frac{H_{2n}}{(2n)^3} + 2 \sum_{n=1}^{\infty} \frac{H_n}{n^3} - \frac{1}{2} \sum_{n=1}^{\infty} \frac{H_n^{(2)}}{n^2}$$

$$\left\{ \text{for the third series use that } \sum_{n=1}^{\infty} a_{2n} = \frac{1}{2} \left(\sum_{n=1}^{\infty} a_n - \sum_{n=1}^{\infty} (-1)^{n-1} a_n \right) \right\}$$

$$= \sum_{n=1}^{\infty} \frac{H_n H_{2n}}{n^2} - \sum_{n=1}^{\infty} \frac{H_n^2}{n^2} - \frac{1}{2} \sum_{n=1}^{\infty} \frac{H_n^{(2)}}{n^2} - 2 \sum_{n=1}^{\infty} \frac{H_n}{n^3} + 4 \sum_{n=1}^{\infty} (-1)^{n-1} \frac{H_n}{n^3}. \tag{6.186}$$

At last, by combining (6.185) and (6.186), we arrive at the desired result

$$\sum_{n=1}^{\infty} \frac{H_n H_{2n}}{n^2} = \frac{5}{4} \sum_{n=1}^{\infty} \frac{H_n^2}{n^2} + \frac{1}{2} \sum_{n=1}^{\infty} \frac{H_n^{(2)}}{n^2} + \frac{3}{4} \sum_{n=1}^{\infty} \frac{H_n}{n^3} - 2 \sum_{n=1}^{\infty} (-1)^{n-1} \frac{H_n}{n^3}$$

$$= \frac{13}{8}\zeta(4) + \frac{7}{2}\log(2)\zeta(3) - \log^2(2)\zeta(2) + \frac{1}{6}\log^4(2) + 4\operatorname{Li}_4\left(\frac{1}{2}\right),$$

and the second solution to the main series is finalized. Compared to the previous solution, observe that this time we also needed the fact that $\sum_{n=1}^{\infty} \frac{H_n^{(2)}}{n^2} = \frac{7}{4}\zeta(4)$, which is obtained by setting $p = q = 2$ in (6.102), Sect. 6.13, and then letting $n \to \infty$.

Note that above I also stated and used that

$$\sum_{n=k}^{\infty} \frac{1}{n(n+k)} = \frac{H_{2k-1} - H_{k-1}}{k} = \frac{H_{2k} - H_k + 1/(2k)}{k}. \tag{6.187}$$

Proof For a nice derivation of the result in (6.187), we may observe that we can write the summand as $\dfrac{1}{n(n+k)} = \dfrac{1}{k}\sum_{i=1}^{k} \dfrac{1}{(i+n-1)(i+n)}$, which further gives

$$\sum_{n=k}^{\infty} \frac{1}{n(n+k)} = \frac{1}{k}\sum_{n=k}^{\infty}\left(\sum_{i=1}^{k}\frac{1}{(i+n-1)(i+n)}\right) = \frac{1}{k}\sum_{i=1}^{k}\left(\sum_{n=k}^{\infty}\left(\frac{1}{i+n-1} - \frac{1}{i+n}\right)\right)$$

$$= \frac{1}{k}\sum_{i=1}^{k}\left(\lim_{N\to\infty}\sum_{n=k}^{N}\left(\frac{1}{i+n-1} - \frac{1}{i+n}\right)\right) = \frac{1}{k}\sum_{i=1}^{k}\lim_{N\to\infty}\left(\frac{1}{i+k-1} - \frac{1}{i+N}\right)$$

$$= \frac{1}{k}\sum_{i=1}^{k}\frac{1}{i+k-1} = \frac{H_{2k-1} - H_{k-1}}{k} = \frac{H_{2k} - H_k + 1/(2k)}{k},$$

which brings an end to the solution of the identity in (6.187).   ∎

How about a third solution? I'll cleverly exploit $\dfrac{1}{2}\displaystyle\sum_{n=1}^{\infty}\dfrac{(2x)^{2n}}{n^2\binom{2n}{n}} = \arcsin^2(x)$

(which also appears in [76, Chapter 4, p.279]). If we replace x by $\sin(x)$ in the mentioned formula, we write

$$\frac{1}{2}\sum_{n=1}^{\infty}\sin^{2n}(x)\frac{2^{2n}}{n^2\binom{2n}{n}} = x^2. \tag{6.188}$$

Further, we also need to show that

$$\int_0^{\pi/2} \sin^{2n}(x)\log^2(\sin(x))\,\mathrm{d}x$$

$$= \frac{\pi}{4}\frac{1}{2^{2n}}\binom{2n}{n}\left(2\log^2(2) + \frac{\pi^2}{6} + 4\log(2)H_n + 2H_n^2 - 4\log(2)H_{2n} + 2H_{2n}^2\right.$$

$$\left. + H_n^{(2)} - 2H_{2n}^{(2)} - 4H_n H_{2n}\right). \tag{6.189}$$

Proof It is straightforward to get the desired result if we start from Wallis' integral case, $\int_0^{\pi/2} \sin^{2n}(x)\,dx = \dfrac{\pi}{2^{2n+1}} \binom{2n}{n} = \dfrac{\pi}{2^{2n+1}} \dfrac{\Gamma(2n+1)}{\Gamma^2(n+1)}$, differentiate the opposite sides two times with respect to n, and then divide them by 4.   ∎

Multiplying the opposite sides of (6.188) by $\log^2(\sin(x))$, then integrating from $x = 0$ to $x = \pi/2$, and using the value of the integral in (1.236), Sect. 1.54, we have

$$\int_0^{\pi/2} x^2 \log^2(\sin(x))\,dx = \frac{1}{24}\log^4(2)\pi + \frac{1}{2}\log(2)\pi\zeta(3) - \frac{3}{320}\pi^5 + \pi \operatorname{Li}_4\left(\frac{1}{2}\right)$$

$$= \frac{1}{2}\int_0^{\pi/2} \sum_{n=1}^{\infty} \sin^{2n}(x)\log^2(\sin(x)) \frac{2^{2n}}{n^2\binom{2n}{n}}\,dx$$

{reverse the order of summation and integration}

$$= \frac{1}{2}\sum_{n=1}^{\infty} \frac{2^{2n}}{n^2\binom{2n}{n}} \int_0^{\pi/2} \sin^{2n}(x)\log^2(\sin(x))\,dx$$

{consider the result in (6.189) and then expand the series}

$$= \frac{\pi}{8}\left(2\log^2(2) + \frac{\pi^2}{6}\right)\sum_{n=1}^{\infty}\frac{1}{n^2} + \log(2)\frac{\pi}{2}\sum_{n=1}^{\infty}\frac{H_n}{n^2} + \frac{\pi}{4}\sum_{n=1}^{\infty}\frac{H_n^2}{n^2} - 2\log(2)\pi\sum_{n=1}^{\infty}\frac{H_{2n}}{(2n)^2}$$

$$+ \pi\sum_{n=1}^{\infty}\frac{H_{2n}^2}{(2n)^2} + \frac{\pi}{8}\sum_{n=1}^{\infty}\frac{H_n^{(2)}}{n^2} - \pi\sum_{n=1}^{\infty}\frac{H_{2n}^{(2)}}{(2n)^2} - \frac{\pi}{2}\sum_{n=1}^{\infty}\frac{H_n H_{2n}}{n^2}$$

$$\left\{\text{for the 4th, 5th, and 7th series use that } \sum_{n=1}^{\infty}a_{2n} = \frac{1}{2}\left(\sum_{n=1}^{\infty}a_n - \sum_{n=1}^{\infty}(-1)^{n-1}a_n\right)\right\}$$

$$= \log^2(2)\frac{\pi^3}{24} + \frac{\pi^5}{288} - \log(2)\frac{\pi}{2}\sum_{n=1}^{\infty}\frac{H_n}{n^2} + \frac{3}{4}\pi\sum_{n=1}^{\infty}\frac{H_n^2}{n^2} - \frac{3}{8}\pi\sum_{n=1}^{\infty}\frac{H_n^{(2)}}{n^2}$$

$$+ \log(2)\pi\sum_{n=1}^{\infty}(-1)^{n-1}\frac{H_n}{n^2} - \frac{\pi}{2}\sum_{n=1}^{\infty}(-1)^{n-1}\frac{H_n^2}{n^2} + \frac{\pi}{2}\sum_{n=1}^{\infty}(-1)^{n-1}\frac{H_n^{(2)}}{n^2}$$

$$- \frac{\pi}{2}\sum_{n=1}^{\infty}\frac{H_n H_{2n}}{n^2},$$

and if we count that a part of the series already appeared at the end of the first and second solutions, where their references are given, and additionally consider that the value of the first series is the particular case $n = 2$ of the Euler sum generalization in (6.149), Sect. 6.19, then the fourth series is the case $m = 1$ of the alternating Euler sum in (4.105), Sect. 4.21, and the sixth series appears in (6.190), Sect. 6.24, we arrive at the desired closed form

$$\sum_{n=1}^{\infty} \frac{H_n H_{2n}}{n^2} = \frac{13}{8}\zeta(4) + \frac{7}{2}\log(2)\zeta(3) - \log^2(2)\zeta(2) + \frac{1}{6}\log^4(2) + 4\operatorname{Li}_4\left(\frac{1}{2}\right),$$

and the third solution to the main series is finalized.

The curious reader eager to explore more ways and get a fourth solution might exploit Parseval's theorem combined with the Fourier-like series in (6.376), Sect. 6.52.

Which of these solutions do you prefer? All of them have their own *magic*, and it is so nice to see how elegantly things can be done (although some ways are clearly non-obvious—in particular, the last two).

We prepare now to jump in the next section where I will continue treating series of weight 4, preparing the ground for the calculation of a series similar to the one in this section, but with a slightly different denominator!

6.24 Two Useful Atypical Harmonic Series of Weight 4 with Denominators of the Type $(2n + 1)^2$

Solution If you already followed the previous section, you probably noticed that at the end of it, I was referring to using this section in order to calculate (in the next section) a series similar to the one in the previous section; more exactly it is about the harmonic series $\sum_{n=1}^{\infty} \dfrac{H_n H_{2n}}{(2n + 1)^2}$. In fact, to get a solution to the series in the next section we need the series result at the point (ii) in the current section. Also, in the present section, we need the series from the point (i) in order to derive the series at the point (ii).

For the series at the point (i), we make use of Abel's summation, the series version in (6.7), Sect. 6.2, where we set $a_n = 1/(2n + 1)^2$ and $b_n = H_n^{(2)}$, and if we also take into account that $\sum_{k=1}^{n} \dfrac{1}{(2k + 1)^2} = H_{2n}^{(2)} - \dfrac{1}{4}H_n^{(2)} + \dfrac{1}{(2n + 1)^2} - 1$, we write

$$\sum_{n=1}^{\infty} \frac{H_n^{(2)}}{(2n+1)^2} = \lim_{N\to\infty} \left(H_{2N}^{(2)} - \frac{1}{4} H_N^{(2)} + \frac{1}{(2N+1)^2} - 1 \right) H_{N+1}^{(2)}$$

$$- \sum_{n=1}^{\infty} \frac{1}{(n+1)^2} \left(H_{2n+1}^{(2)} - \frac{1}{4} H_n^{(2)} - 1 \right)$$

{reindex the series and start from $n = 1$}

$$= \frac{15}{8}\zeta(4) - \zeta(2) - \sum_{n=1}^{\infty} \frac{1}{n^2} \left(H_{2n-1}^{(2)} - \frac{1}{4} H_{n-1}^{(2)} - 1 \right)$$

$$= \frac{15}{8}\zeta(4) - \zeta(2) + \sum_{n=1}^{\infty} \frac{1}{n^2} + \frac{1}{4}\sum_{n=1}^{\infty} \frac{H_n^{(2)}}{n^2} - 4\sum_{n=1}^{\infty} \frac{H_{2n}^{(2)}}{(2n)^2}$$

$$\left\{ \text{and for the third series use that } \sum_{k=1}^{\infty} a_{2k} = \frac{1}{2} \left(\sum_{k=1}^{\infty} a_k - \sum_{k=1}^{\infty} (-1)^{k-1} a_k \right) \right\}$$

$$= \frac{15}{8}\zeta(4) - \frac{7}{4}\sum_{n=1}^{\infty} \frac{H_n^{(2)}}{n^2} + 2\sum_{n=1}^{\infty} (-1)^{n-1} \frac{H_n^{(2)}}{n^2}$$

{for the first series use (6.102), Sect. 6.13, with $p = q = 2$, and $n \to \infty$}

$$= \frac{1}{3}\log^4(2) - 2\log^2(2)\zeta(2) + 7\log(2)\zeta(3) - \frac{121}{16}\zeta(4) + 8\operatorname{Li}_4\left(\frac{1}{2}\right),$$

where in the calculations I also used that

$$\sum_{n=1}^{\infty} (-1)^{n-1} \frac{H_n^{(2)}}{n^2}$$

$$= \frac{1}{6}\log^4(2) - \log^2(2)\zeta(2) + \frac{7}{2}\log(2)\zeta(3) - \frac{51}{16}\zeta(4) + 4\operatorname{Li}_4\left(\frac{1}{2}\right), \qquad (6.190)$$

which is found calculated in [76, Chapter 6, pp. 505–506], by nicely relating the series $\sum_{n=1}^{\infty} (-1)^{n-1} \frac{H_n^{(2)}}{n^2}$ to the series $\sum_{n=1}^{\infty} (-1)^{n-1} \frac{H_n}{n^3}$, and the point (*i*) of the problem is complete. The suggested harmonic series relation is also given as a challenging question with special requirements in (4.90), Sect. 4.15. Another way to obtain (6.190) is achieved by exploiting the generating function in (4.48), Sect. 4.7.

To deal with the second point of the problem, we want to explore a result from the previously mentioned book,

$$\sum_{k=1}^{\infty} \frac{H_k^2 - H_k^{(2)}}{(k+1)(k+n+1)} = \frac{H_n^3 + 3H_n H_n^{(2)} + 2H_n^{(3)}}{3n}$$

$$= \frac{(\psi(n+1)+\gamma)^3 + 3(\psi(n+1)+\gamma)(\zeta(2) - \psi^{(1)}(n+1))}{3n}$$

$$+ \frac{2(\zeta(3) + 1/2\psi^{(2)}(n+1))}{3n}, \tag{6.191}$$

where the first equality is given with a proof in [76, Chapter 6, p.381] and [73]. The second equality is obtained by the use of the Polygamma function and may be viewed as a way to achieve the extension to the real numbers.

Multiplying the opposite sides of (6.191) by n, differentiating once with respect to n, setting $n = -1/2$, and rearranging everything, we get

$$\sum_{k=1}^{\infty} \frac{H_k^2 - H_k^{(2)}}{(2k+1)^2} = \sum_{k=1}^{\infty} \frac{H_k^2}{(2k+1)^2} - \sum_{k=1}^{\infty} \frac{H_k^{(2)}}{(2k+1)^2}$$

$$= \frac{1}{4}\bigg((\psi(n+1)+\gamma)^2 \psi^{(1)}(n+1) + \psi^{(1)}(n+1)(\zeta(2) - \psi^{(1)}(n+1))$$

$$- \psi^{(2)}(n+1)(\psi(n+1)+\gamma) + 1/3\psi^{(3)}(n+1) \bigg)\bigg|_{n=-1/2}$$

$$= \frac{15}{4}\zeta(4) - 7\log(2)\zeta(3) + 3\log^2(2)\zeta(2), \tag{6.192}$$

where in the calculations I used that $\psi\left(\frac{1}{2}\right) = -\gamma - 2\log(2)$, $\psi^{(1)}\left(\frac{1}{2}\right) = 3\zeta(2)$, $\psi^{(2)}\left(\frac{1}{2}\right) = -14\zeta(3)$, and $\psi^{(3)}\left(\frac{1}{2}\right) = 90\zeta(4)$, which are given in (6.146) and (6.147), Sect. 6.19.

Finally, if we combine the results in (6.192) and the one at the point (i), we immediately conclude that

$$\sum_{n=1}^{\infty} \frac{H_n^2}{(2n+1)^2}$$

$$= \frac{1}{3}\log^4(2) + \log^2(2)\zeta(2) - \frac{61}{16}\zeta(4) + 8\operatorname{Li}_4\left(\frac{1}{2}\right),$$

and the point (ii) of the problem is complete.

In the solution to the point (i), Abel's summation displays again its amazing power, and then the calculations get reduced to a known alternating harmonic series of weight 4.

Also, the powerful strategy in the solution to the point (ii), which combines the special identity in (6.191) and differentiation in a clever way, has been previously presented in *(Almost) Impossible Integrals, Sums, and Series* in a slightly different form. For example, in the footnote from [76, Chapter 6, p.454], I explained how to show that

$$\sum_{k=1}^{\infty} \frac{H_k^3 - 3H_k H_k^{(2)} + 2H_k^{(3)}}{(k+1)^2} = 6\zeta(5),$$

by using such a strategy, but by employing a more advanced identity than the one in (6.191). Also, in [76, Chapter 6, p.528], I suggested the use of such a strategy to calculate two integrals. It is also worth mentioning my paper *A new powerful strategy of calculating a class of alternating Euler sums* (see [77]) where I used such a strategy that you may also find exploited in Sect. 6.19.

The harmonic series at the point (i) together with my solution were also presented in [30, June 12, 2019].

With these results in hand, we are ready to attack the series in the next section!

6.25 Another Special Challenging Harmonic Series of Weight 4, Involving Harmonic Numbers of the Type H_{2n}

Solution If you covered the final part of the fourth chapter in my first book, *(Almost) Impossible Integrals, Sums, and Series* (see [76, Chapter 4, p.313]), you observed a fascinating sum of seven series, which leads to $\zeta(4)$,

$$\zeta(4)$$

$$= \frac{8}{5}\sum_{n=1}^{\infty} \frac{H_n H_{2n}}{n^2} + \frac{64}{5}\sum_{n=1}^{\infty} \frac{H_{2n}^2}{(2n+1)^2} + \frac{64}{5}\sum_{n=1}^{\infty} \frac{H_{2n}}{(2n+1)^3}$$

$$- \frac{8}{5}\sum_{n=1}^{\infty} \frac{H_{2n}^2}{n^2} - \frac{32}{5}\sum_{n=1}^{\infty} \frac{H_n H_{2n}}{(2n+1)^2} - \frac{64}{5}\log(2)\sum_{n=1}^{\infty} \frac{H_{2n}}{(2n+1)^2} - \frac{8}{5}\sum_{n=1}^{\infty} \frac{H_{2n}^{(2)}}{n^2}.$$

$$(6.193)$$

Let's observe that the result in (6.193) is more than just a beautiful representation of $\zeta(4)$, since it allows us to express the series we want to calculate in terms of series that are already known to us, and to see this, we only have to rearrange some of the series above and express them in terms of simpler series.

To get a solution, we focus on the right-hand side of (6.193), and first observe the series $\sum_{n=1}^{\infty} \dfrac{H_n H_{2n}}{n^2}$ is calculated in three different ways in Sect. 6.23. Further, for the second series, we write

$$\sum_{n=1}^{\infty} \frac{H_{2n}^2}{(2n+1)^2} = \sum_{n=1}^{\infty} \frac{(H_{2n+1} - 1/(2n+1))^2}{(2n+1)^2}$$

$$= \sum_{n=1}^{\infty} \frac{H_{2n+1}^2}{(2n+1)^2} - 2\sum_{n=1}^{\infty} \frac{H_{2n+1}}{(2n+1)^3} + \sum_{n=1}^{\infty} \frac{1}{(2n+1)^4}$$

{reindex all three series}

$$= \sum_{n=1}^{\infty} \frac{H_{2n-1}^2}{(2n-1)^2} - 2\sum_{n=1}^{\infty} \frac{H_{2n-1}}{(2n-1)^3} + \sum_{n=1}^{\infty} \frac{1}{(2n-1)^4}$$

$$\left\{ \text{for the first two series use that } \sum_{n=1}^{\infty} a_{2n-1} = \frac{1}{2}\left(\sum_{n=1}^{\infty} a_n + \sum_{n=1}^{\infty} (-1)^{n-1} a_n \right) \right\}$$

$$= \frac{1}{2}\sum_{n=1}^{\infty} \frac{H_n^2}{n^2} + \frac{1}{2}\sum_{n=1}^{\infty} (-1)^{n-1}\frac{H_n^2}{n^2} - \sum_{n=1}^{\infty} \frac{H_n}{n^3} - \sum_{n=1}^{\infty} (-1)^{n-1}\frac{H_n}{n^3} + \frac{15}{16}\zeta(4)$$

$$= \frac{11}{32}\zeta(4) + \frac{7}{8}\log(2)\zeta(3) - \frac{1}{4}\log^2(2)\zeta(2) + \frac{1}{24}\log^4(2) + \operatorname{Li}_4\left(\frac{1}{2}\right), \qquad (6.194)$$

where the first series is the Au-Yeung series, $\sum_{n=1}^{\infty} \left(\dfrac{H_n}{n} \right)^2 = \dfrac{17}{4}\zeta(4)$, which you may find calculated in [76, Chapter 6, pp. 392–393], [91]; the second series is calculated in [76, Chapter 6, pp. 506–508], or it can be extracted by combining the identity in (6.184), Sect. 6.23, and the series in (4.103), Sect. 4.20; the third series is the particular case $n = 3$ of the Euler sum generalization in (6.149), Sect. 6.19; and the fourth series is elegantly calculated in Sect. 6.20.

Next, for the third series in (6.193), we have

$$\sum_{n=1}^{\infty} \frac{H_{2n}}{(2n+1)^3} = \sum_{n=1}^{\infty} \frac{H_{2n+1} - 1/(2n+1)}{(2n+1)^3}$$

{reindex the series and expand it}

$$= \sum_{n=1}^{\infty} \frac{H_{2n-1}}{(2n-1)^3} - \sum_{n=1}^{\infty} \frac{1}{(2n-1)^4} = \sum_{n=1}^{\infty} \frac{H_{2n-1}}{(2n-1)^3} - \frac{15}{16}\zeta(4)$$

$$\left\{\text{exploit the fact that } \sum_{n=1}^{\infty} a_{2n-1} = \frac{1}{2}\left(\sum_{n=1}^{\infty} a_n + \sum_{n=1}^{\infty}(-1)^{n-1}a_n\right)\right\}$$

$$= \frac{1}{2}\sum_{n=1}^{\infty}\frac{H_n}{n^3} + \frac{1}{2}\sum_{n=1}^{\infty}(-1)^{n-1}\frac{H_n}{n^3} - \frac{15}{16}\zeta(4)$$

$$= \frac{17}{16}\zeta(4) - \frac{7}{8}\log(2)\zeta(3) + \frac{1}{4}\log^2(2)\zeta(2) - \frac{1}{24}\log^4(2) - \operatorname{Li}_4\left(\frac{1}{2}\right), \quad (6.195)$$

where the first series is the particular case $n = 3$ of the Euler sum generalization in (6.149), Sect. 6.19, and the second series is calculated both in Sect. 6.20 and [76, Chapter 6, pp. 503–505].

Thereafter, for the fourth harmonic series in (6.193), we use the fact that

$$\sum_{n=1}^{\infty} a_{2n} = \frac{1}{2}\left(\sum_{n=1}^{\infty} a_n - \sum_{n=1}^{\infty}(-1)^{n-1}a_n\right), \text{ and then we write}$$

$$\sum_{n=1}^{\infty}\frac{H_{2n}^2}{n^2} = 4\sum_{n=1}^{\infty}\frac{H_{2n}^2}{(2n)^2} = 2\sum_{n=1}^{\infty}\frac{H_n^2}{n^2} - 2\sum_{n=1}^{\infty}(-1)^{n-1}\frac{H_n^2}{n^2}$$

$$= \frac{27}{8}\zeta(4) + \frac{7}{2}\log(2)\zeta(3) - \log^2(2)\zeta(2) + \frac{1}{6}\log^4(2) + 4\operatorname{Li}_4\left(\frac{1}{2}\right), \quad (6.196)$$

where we meet again the Au-Yeung series which is calculated in [76, Chapter 6, pp. 392–393], [91], and the second series is extracted immediately if we combine the identity in (6.184), Sect. 6.23, and the series in (4.103), Sect. 4.20.

Then, with respect to the sixth series in (6.193), we rearrange it and write that

$$\sum_{n=1}^{\infty}\frac{H_{2n}}{(2n+1)^2} = \sum_{n=1}^{\infty}\frac{H_{2n+1} - 1/(2n+1)}{(2n+1)^2}$$

$$\{\text{reindex the series and expand it}\}$$

$$= \sum_{n=1}^{\infty}\frac{H_{2n-1}}{(2n-1)^2} - \sum_{n=1}^{\infty}\frac{1}{(2n-1)^3} = \sum_{n=1}^{\infty}\frac{H_{2n-1}}{(2n-1)^2} - \frac{7}{8}\zeta(3)$$

$$\left\{\text{consider using that } \sum_{n=1}^{\infty} a_{2n-1} = \frac{1}{2}\left(\sum_{n=1}^{\infty} a_n + \sum_{n=1}^{\infty}(-1)^{n-1}a_n\right)\right\}$$

$$= \frac{1}{2}\sum_{n=1}^{\infty}\frac{H_n}{n^2} + \frac{1}{2}\sum_{n=1}^{\infty}(-1)^{n-1}\frac{H_n}{n^2} - \frac{7}{8}\zeta(3) = \frac{7}{16}\zeta(3), \quad (6.197)$$

where the first series is the particular case $n = 2$ of the Euler sum generalization in (6.149), Sect. 6.19, and the second series is the particular case $m = 1$ of the alternating Euler sum generalization in (4.105), Sect. 4.21. The second series may also be found calculated in a different style in [76, Chapter 6, p.509].

Finally, for the last harmonic series in (6.193), we may use the fact that $\sum\limits_{n=1}^{\infty} a_{2n} = \frac{1}{2}\left(\sum\limits_{n=1}^{\infty} a_n - \sum\limits_{n=1}^{\infty} (-1)^{n-1} a_n\right)$ that gives

$$\sum_{n=1}^{\infty} \frac{H_{2n}^{(2)}}{n^2} = 4 \sum_{n=1}^{\infty} \frac{H_{2n}^{(2)}}{(2n)^2} = 2 \sum_{n=1}^{\infty} \frac{H_n^{(2)}}{n^2} - 2 \sum_{n=1}^{\infty} (-1)^{n-1} \frac{H_n^{(2)}}{n^2}$$

$$= \frac{79}{8}\zeta(4) + 2\log^2(2)\zeta(2) - 7\log(2)\zeta(3) - \frac{1}{3}\log^4(2) - 8\operatorname{Li}_4\left(\frac{1}{2}\right), \quad (6.198)$$

where for the first series, I used (6.102), Sect. 6.13, with $p = q = 2$, and let $n \to \infty$ that gives $\sum\limits_{n=1}^{\infty} \frac{H_n^{(2)}}{n^2} = \frac{7}{4}\zeta(4)$, and the second series is already stated in Sect. 6.24, in (6.190).

The final step that will give us the desired result is to collect all needed results that appear in (4.113), Sect. 4.23, (6.194), (6.195), (6.196), (6.197), and (6.198), and then plug them in the identity from (6.193),

$$\sum_{n=1}^{\infty} \frac{H_n H_{2n}}{(2n+1)^2}$$

$$= \frac{1}{12}\log^4(2) - \frac{1}{2}\log^2(2)\zeta(2) + \frac{7}{8}\log(2)\zeta(3) - \frac{1}{4}\zeta(4) + 2\operatorname{Li}_4\left(\frac{1}{2}\right),$$

and the solution to the main series is finalized.

Clearly not an obvious solution! Surely, a key step to sort out things here is to establish a relation among the series we want to calculate and other known series such that we can make possible an extraction of the value of the desired series. The strategy found in [76, Chapter 6, pp. 530–532] also works great for more advanced harmonic series, and such an example we'll see in one of the next sections.

6.26   A First Uncommon Series with the Tail of the Riemann Zeta Function $\zeta(2) - H_{2n}^{(2)}$, Related to Weight 4 Harmonic Series

Solution One of the natural things to do from the very beginning is to imagine our series in a simpler form, with $H_n^{(2)}$ instead of $H_{2n}^{(2)}$,

$$\sum_{n=1}^{\infty} \frac{H_n}{n}\left(\zeta(2) - H_n^{(2)}\right) = \frac{7}{4}\zeta(4), \tag{6.199}$$

and this version is known and easy to calculate by using Abel's summation as presented in [76, Chapter 6, pp. 479–480] for a slightly different series. Also, it is possible to nicely calculate the series in (6.199) by exploiting the strategy used for getting the generalization from [76, Chapter 6, pp. 483–484]. The version in (6.199) previously appeared in [15, p.149] where the author performed the calculations by using and combining logarithmic and dilogarithmic integrals.

Now, Abel's summation also works great here as a first step in the calculation of the main series.

So, let's consider $a_n = H_n/n$ and $b_n = \zeta(2) - H_{2n}^{(2)}$ and then apply Abel's summation, the series version in (6.7), Sect. 6.2, where we also consider the fact that $\sum_{k=1}^{n} \frac{H_k}{k} = \frac{1}{2}(H_n^2 + H_n^{(2)})$, which is also explained in ([76, Chapter 3, p.60]), and then we write

$$\sum_{n=1}^{\infty} \frac{H_n}{n}\left(\zeta(2) - H_{2n}^{(2)}\right)$$

$$= \frac{1}{2}\sum_{n=1}^{\infty} \frac{H_n^2 + H_n^{(2)}}{(2n+1)^2} + \frac{1}{8}\sum_{n=1}^{\infty} \frac{(H_{n+1} - 1/(n+1))^2 + (H_{n+1}^{(2)} - 1/(n+1)^2)}{(n+1)^2}$$

{reindex the second series and wisely expand it}

$$= \frac{1}{2}\sum_{n=1}^{\infty} \frac{H_n^2}{(2n+1)^2} + \frac{1}{2}\sum_{n=1}^{\infty} \frac{H_n^{(2)}}{(2n+1)^2} + \frac{1}{8}\sum_{n=1}^{\infty} \frac{H_n^2 + H_n^{(2)}}{n^2} - \frac{1}{4}\sum_{n=1}^{\infty} \frac{H_n}{n^3}$$

{the first two series are found in (4.114) and (4.115), Sect. 4.24, and the last}

{harmonic series is the particular case $n = 3$ of the Euler sum in (6.149), Sect. 6.19}

$$= \frac{1}{3}\log^4(2) - \frac{1}{2}\log^2(2)\zeta(2) + \frac{7}{2}\log(2)\zeta(3) - 6\zeta(4) + 8\operatorname{Li}_4\left(\frac{1}{2}\right)$$

$$+ \frac{1}{8} \sum_{n=1}^{\infty} \frac{H_n^2 + H_n^{(2)}}{n^2}$$

$$= \frac{1}{3} \log^4(2) - \frac{1}{2} \log^2(2) \zeta(2) + \frac{7}{2} \log(2) \zeta(3) - \frac{21}{4} \zeta(4) + 8 \operatorname{Li}_4 \left(\frac{1}{2} \right),$$

where the last series, $\sum_{n=1}^{\infty} \dfrac{H_n^2 + H_n^{(2)}}{n^2}$, can be calculated elegantly by using the

powerful elementary integral $\displaystyle\int_0^1 x^{n-1} \log^2(1-x) dx = \dfrac{H_n^2 + H_n^{(2)}}{n}$ in [76, Chapter 1, p.2], where if we multiply both sides by $1/n$ and then consider the summation from $n = 1$ to ∞, we obtain

$$\sum_{n=1}^{\infty} \frac{H_n^2 + H_n^{(2)}}{n^2} = \sum_{n=1}^{\infty} \frac{1}{n} \int_0^1 x^{n-1} \log^2(1-x) dx$$

{change the order of integration and summation}

$$= \int_0^1 \sum_{n=1}^{\infty} \frac{x^{n-1}}{n} \log^2(1-x) dx = - \int_0^1 \frac{\log^3(1-x)}{x} dx \overset{1-x=y}{=} - \int_0^1 \frac{\log^3(y)}{1-y} dy$$

$$= - \int_0^1 \sum_{n=1}^{\infty} y^{n-1} \log^3(y) dy = - \sum_{n=1}^{\infty} \int_0^1 y^{n-1} \log^3(y) dy = 6 \sum_{n=1}^{\infty} \frac{1}{n^4} = 6\zeta(4),$$

$$(6.200)$$

and the solution is complete.

After the use of Abel's summation, everything works fast since the values of the challenging resulting harmonic series are given in Sect. 4.24. Also, observe that I preferred to keep the series form, $\sum_{n=1}^{\infty} \dfrac{H_n^2 + H_n^{(2)}}{n^2}$, without splitting it, which is a good idea as you can see in the last part part of the solution where I used the logarithmic integral, $\displaystyle\int_0^1 x^{n-1} \log^2(1-x) dx = \dfrac{H_n^2 + H_n^{(2)}}{n}$.

Do you like the atypical harmonic series with the Riemann zeta tails? Well, the next section is waiting for us with another beautiful harmonic series alike.

6.27   A Second Uncommon Series with the Tail of the Riemann Zeta Function $\zeta(2) - H_n^{(2)}$, Related to Weight 4 Harmonic Series

Solution If we compare the series in this section with the one in the previous section, we notice that in this case the simple argument n and the double argument $2n$ of the two harmonic numbers are switched, and we have H_{2n} and $H_n^{(2)}$.

We may obtain a solution by starting with the use of an integral representation of the tail of the Riemann zeta function. For example, in the book *(Almost) Impossible Integrals, Sums, and Series*, I successfully used the representation

$$\zeta(p) - 1 - \frac{1}{2^p} - \cdots - \frac{1}{n^p} = \sum_{k=1}^{\infty} \frac{1}{(n+k)^p} = \frac{(-1)^{p-1}}{(p-1)!} \sum_{k=1}^{\infty} \int_0^1 x^{k+n-1} \log^{p-1}(x) dx$$

$$= \frac{(-1)^{p-1}}{(p-1)!} \int_0^1 \frac{x^n \log^{p-1}(x)}{1-x} dx, \tag{6.201}$$

for calculating the generalizations in [76, Chapter 4, Section 4.46, p.304]. Then, for our specific problem, we need the case $p = 2$ in (6.201),

$$\zeta(2) - H_n^{(2)} = \zeta(2) - 1 - \frac{1}{2^2} - \cdots - \frac{1}{n^2} = -\sum_{k=1}^{\infty} \int_0^1 x^{k+n-1} \log(x) dx$$

$$= -\int_0^1 \frac{x^n \log(x)}{1-x} dx \overset{x=y^2}{=} -4 \int_0^1 \frac{y^{2n+1} \log(y)}{1-y^2} dy. \tag{6.202}$$

Why is making the variable change in (6.202) a good step to take? Things become clear if we take a close look at the harmonic number in front of the tail.

So, by using the result in (6.202), we write our series as follows:

$$\sum_{n=1}^{\infty} \frac{H_{2n}}{n} \left(\zeta(2) - H_n^{(2)} \right) = -8 \sum_{n=1}^{\infty} H_{2n} \int_0^1 x^{2n-1} dx \int_0^1 \frac{y^{2n+1} \log(y)}{1-y^2} dy$$

{reverse the order of integration and summation}

$$= -8 \int_0^1 \left(\int_0^1 \frac{y \log(y)}{x(1-y^2)} \sum_{n=1}^{\infty} (xy)^{2n} H_{2n} dx \right) dy$$

$$\left\{ \text{use that } \sum_{n=1}^{\infty} x^{2n} H_{2n} = \frac{2x \operatorname{arctanh}(x) - \log(1-x^2)}{2(1-x^2)} \right\}$$

$$= -4 \int_0^1 \frac{y \log(y)}{1 - y^2} \left(\int_0^1 \frac{2xy \arctanh(xy) - \log(1 - (xy)^2)}{x(1 - (xy)^2)} dx \right) dy$$

$$= -4 \int_0^1 \frac{y \log(y)}{1 - y^2} \left(\arctanh^2(xy) + \frac{1}{4} \log^2(1 - (xy)^2) + \frac{1}{2} \operatorname{Li}_2 \left((xy)^2 \right) \Big|_{x=0}^{x=1} \right) dy$$

{expand the integral after taking the limits of integration}

$$= -4 \int_0^1 \frac{y \log(y) \arctanh^2(y)}{1 - y^2} dy - \int_0^1 \frac{y \log(y) \log^2(1 - y^2)}{1 - y^2} dy$$

$$- 2 \int_0^1 \frac{y \log(y) \operatorname{Li}_2(y^2)}{1 - y^2} dy$$

$\left\{ \text{in the last two integrals let } y^2 = z \text{ and then return to the notation in } y \right\}$

$$= \underbrace{-4 \int_0^1 \frac{y \log(y) \arctanh^2(y)}{1 - y^2} dy}_{I_1} - \frac{1}{4} \underbrace{\int_0^1 \frac{\log(y) \log^2(1 - y)}{1 - y} dy}_{I_2}$$

$$- \frac{1}{2} \underbrace{\int_0^1 \frac{\log(y) \operatorname{Li}_2(y)}{1 - y} dy}_{I_3} . \tag{6.203}$$

For the integral I_1 in (6.203), we have

$$I_1 = \int_0^1 \frac{y \log(y) \arctanh^2(y)}{1 - y^2} dy \stackrel{(1-y)/(1+y)=t}{=} \frac{1}{8} \int_0^1 \frac{(1 - t) \log \left(\frac{1 - t}{1 + t} \right) \log^2(t)}{t(1 + t)} dt$$

$$= \frac{1}{8} \underbrace{\int_0^1 \frac{\log(1 - t) \log^2(t)}{t} dt}_{J_1} - \frac{1}{8} \underbrace{\int_0^1 \frac{\log(1 + t) \log^2(t)}{t} dt}_{J_2} + \frac{1}{4} \underbrace{\int_0^1 \frac{\log(1 + t) \log^2(t)}{1 + t} dt}_{J_3}$$

$$- \frac{1}{4} \underbrace{\int_0^1 \frac{\log(1 - t) \log^2(t)}{1 + t} dt}_{J_4} . \tag{6.204}$$

Now, for the integral J_1 in (6.204), we write

$$J_1 = \int_0^1 \frac{\log(1-t)\log^2(t)}{t}\,dt = -\int_0^1 \sum_{n=1}^{\infty} \frac{t^{n-1}}{n}\log^2(t)\,dt = -\sum_{n=1}^{\infty} \frac{1}{n}\int_0^1 t^{n-1}\log^2(t)\,dt$$

$$= -2\sum_{n=1}^{\infty}\frac{1}{n^4} = -2\zeta(4). \tag{6.205}$$

Then, for the integral J_2 in (6.204), we have

$$J_2 = \int_0^1 \frac{\log(1+t)\log^2(t)}{t}\,dt = \int_0^1 \sum_{n=1}^{\infty}(-1)^{n-1}\frac{t^{n-1}}{n}\log^2(t)\,dt$$

$$= \sum_{n=1}^{\infty}\frac{(-1)^{n-1}}{n}\int_0^1 t^{n-1}\log^2(t)\,dt = 2\sum_{n=1}^{\infty}(-1)^{n-1}\frac{1}{n^4} = \frac{7}{4}\zeta(4). \tag{6.206}$$

Next, for the integral J_3 in (6.204), we get

$$J_3 = \int_0^1 \frac{\log(1+t)\log^2(t)}{1+t}\,dt = \int_0^1 \log^2(t)\sum_{n=1}^{\infty}(-1)^{n-1}t^n H_n\,dt$$

{reverse the order of summation and integration}

$$= \sum_{n=1}^{\infty}(-1)^{n-1}H_n\int_0^1 t^n\log^2(t)\,dt = 2\sum_{n=1}^{\infty}(-1)^{n-1}\frac{H_{n+1} - 1/(n+1)}{(n+1)^3}$$

{reindex the series and expand it}

$$= 2\sum_{n=1}^{\infty}(-1)^{n-1}\frac{1}{n^4} - 2\sum_{n=1}^{\infty}(-1)^{n-1}\frac{H_n}{n^3}$$

{employ the series result in (4.103), Sect. 4.20}

$$= \frac{1}{6}\log^4(2) - \log^2(2)\zeta(2) + \frac{7}{2}\log(2)\zeta(3) - \frac{15}{4}\zeta(4) + 4\operatorname{Li}_4\left(\frac{1}{2}\right), \tag{6.207}$$

where I used $\sum_{n=1}^{\infty}x^n H_n = -\dfrac{\log(1-x)}{1-x}$, which is the case $m = 1$ in (4.32), Sect.

4.6, in order to obtain and employ the fact that $\sum_{n=1}^{\infty}(-1)^{n-1}x^n H_n = \dfrac{\log(1+x)}{1+x}$.

Further, for the integral J_4 in (6.204), we use that $\displaystyle\int_0^1 x^{n-1}\log(1-x)dx =$
$-\dfrac{H_n}{n} = -\dfrac{\psi(n+1)+\gamma}{n}$, found in (3.10), Sect. 3.3, which gives

$$J_4 = \int_0^1 \frac{\log(1-t)\log^2(t)}{1+t}dt = \int_0^1 \log(1-t)\log^2(t)\sum_{n=1}^{\infty}(-1)^{n-1}t^{n-1}dt$$

$$= \sum_{n=1}^{\infty}(-1)^{n-1}\int_0^1 t^{n-1}\log^2(t)\log(1-t)dt$$

$$= \sum_{n=1}^{\infty}(-1)^{n-1}\frac{d^2}{dn^2}\left(\int_0^1 t^{n-1}\log(1-t)dt\right) = \sum_{n=1}^{\infty}(-1)^{n-1}\frac{d^2}{dn^2}\left(-\frac{\psi(n+1)+\gamma}{n}\right)$$

$$= 2\zeta(3)\sum_{n=1}^{\infty}(-1)^{n-1}\frac{1}{n} + 2\zeta(2)\sum_{n=1}^{\infty}(-1)^{n-1}\frac{1}{n^2} - 2\sum_{n=1}^{\infty}(-1)^{n-1}\frac{H_n}{n^3}$$

$$- 2\sum_{n=1}^{\infty}(-1)^{n-1}\frac{H_n^{(2)}}{n^2} - 2\sum_{n=1}^{\infty}(-1)^{n-1}\frac{H_n^{(3)}}{n}$$

$$= \zeta(4) + \log^2(2)\zeta(2) - \frac{1}{6}\log^4(2) - 4\operatorname{Li}_4\left(\frac{1}{2}\right), \tag{6.208}$$

where the series $\displaystyle\sum_{n=1}^{\infty}(-1)^{n-1}\frac{H_n}{n^3}$ is given in (4.103), Sect. 4.20; then the series
$\displaystyle\sum_{n=1}^{\infty}(-1)^{n-1}\frac{H_n^{(2)}}{n^2}$ is stated in (6.190), Sect. 6.24; and the last series is given
in (4.106), the case $m = 3$, Sect. 4.22. Note that above I could have exploited
directly the first generalization in Sect. 1.24.

Gathering the results from (6.205), (6.206), (6.207), and (6.208) in (6.204), we obtain

$$I_1 = \int_0^1 \frac{y\log(y)\operatorname{arctanh}^2(y)}{1-y^2}dy$$

$$= \frac{1}{12}\log^4(2) - \frac{1}{2}\log^2(2)\zeta(2) + \frac{7}{8}\log(2)\zeta(3) - \frac{53}{32}\zeta(4) + 2\operatorname{Li}_4\left(\frac{1}{2}\right). \tag{6.209}$$

Let's go further with the next integral in (6.203), I_2, and then we write

$$I_2 = \int_0^1 \frac{\log(y)\log^2(1-y)}{1-y}dy \overset{1-y=z}{=} \int_0^1 \frac{\log(1-z)\log^2(z)}{z}dz$$

$$= J_1 = -2\zeta(4). \tag{6.210}$$

At last, for the integral I_3, we have

$$I_3 = \int_0^1 \frac{\log(y)\operatorname{Li}_2(y)}{1-y}dy = \int_0^1 \log(y)\sum_{n=1}^\infty y^n H_n^{(2)}dy = \sum_{n=1}^\infty H_n^{(2)}\int_0^1 y^n \log(y)dy$$

$$= -\sum_{n=1}^\infty \frac{H_{n+1}^{(2)}-1/(n+1)^2}{(n+1)^2} = \sum_{n=1}^\infty \frac{1}{n^4} - \sum_{n=1}^\infty \frac{H_n^{(2)}}{n^2} = -\frac{3}{4}\zeta(4), \tag{6.211}$$

where in the calculations I used the case $m = 2$ of (4.32), Sect. 4.6, which also appears in a generalized form in [76, Chapter 4, p.284], together with the series result, $\sum_{n=1}^\infty \frac{H_n^{(2)}}{n^2} = \frac{7}{4}\zeta(4)$, which is immediately extracted by setting $p = q = 2$ in (6.102) from Sect. 6.13 and then lettting $n \to \infty$.

Collecting the values of the integrals I_1, I_2, and I_3 from (6.209), (6.210), and (6.211) in (6.203), we conclude that

$$\sum_{n=1}^\infty \frac{H_{2n}}{n}\left(\zeta(2) - H_n^{(2)}\right)$$

$$= \frac{15}{2}\zeta(4) - \frac{7}{2}\log(2)\zeta(3) + 2\log^2(2)\zeta(2) - \frac{1}{3}\log^4(2) - 8\operatorname{Li}_4\left(\frac{1}{2}\right),$$

and the solution to the main series is finalized.

Using the integral representation of the tail of the Riemann zeta function leads to an interesting solution which flows pretty smoothly, and here I also take into account some of the necessary subresults which happily are already known to us.

In the calculation process, more exactly in (6.203), I also used that $\sum_{n=1}^\infty x^{2n} H_{2n} = \frac{2x\operatorname{arctanh}(x) - \log(1-x^2)}{2(1-x^2)}$. To make this point clearer, we may recall that $\sum_{n=1}^\infty x^n H_n = -\frac{\log(1-x)}{1-x}$, and also consider the variant with x replaced by $-x$, and then observe that $\sum_{n=1}^\infty x^{2n} H_{2n} = \frac{1}{2}\left(\sum_{n=1}^\infty x^n H_n + \sum_{n=1}^\infty (-1)^n x^n H_n\right)$, which immediately gives us the result stated above.

6.28 A Third Uncommon Series with the Tail of the Riemann Zeta Function $\zeta(2) - H_{2n}^{(2)}$, Related to Weight 4 Harmonic Series

Solution We continue the calculations of the harmonic series with the tail of the Riemann zeta function, and in this section, we'll find a series similar to the previous two ones, but this time the double argument $2n$ will be found in both harmonic numbers inside the summand of the main series.

As in the preceding section, the starting point lays around the identity in (6.202) that gives us the following auxiliary result:

$$\zeta(2) - H_{2n}^{(2)} = -\int_0^1 \frac{x^{2n} \log(x)}{1-x} dx. \tag{6.212}$$

So, using the result in (6.212), the main series may be written as

$$\sum_{n=1}^{\infty} \frac{H_{2n}}{n} \left(\zeta(2) - H_{2n}^{(2)}\right) = -2 \sum_{n=1}^{\infty} H_{2n} \int_0^1 y^{2n-1} dy \int_0^1 \frac{x^{2n} \log(x)}{1-x} dx$$

{reverse the order of integration and summation}

$$= -2 \int_0^1 \left(\int_0^1 \frac{\log(x)}{y(1-x)} \sum_{n=1}^{\infty} (xy)^{2n} H_{2n} dy\right) dx$$

$$\left\{\text{use that } \sum_{n=1}^{\infty} x^{2n} H_{2n} = \frac{2x \operatorname{arctanh}(x) - \log(1-x^2)}{2(1-x^2)}\right\}$$

$$= -\int_0^1 \frac{\log(x)}{1-x} \left(\int_0^1 \frac{2xy \operatorname{arctanh}(xy) - \log(1-(xy)^2)}{y(1-(xy)^2)} dy\right) dx$$

$$= -\int_0^1 \frac{\log(x)}{1-x} \left(\operatorname{arctanh}^2(xy) + \frac{1}{4} \log^2(1-(xy)^2) + \frac{1}{2} \operatorname{Li}_2\left((xy)^2\right) \Big|_{y=0}^{y=1}\right) dx$$

{expand the integral after taking the limits of integration}

$$= -\int_0^1 \frac{\log(x) \operatorname{arctanh}^2(x)}{1-x} dx - \frac{1}{4} \int_0^1 \frac{\log(x) \log^2(1-x^2)}{1-x} dx$$

$$- \frac{1}{2} \int_0^1 \frac{\log(x) \operatorname{Li}_2(x^2)}{1-x} dx$$

$$\left\{ \text{use that } \text{arctanh}^2(x) = 1/4(\log^2(1+x) - 2\log(1+x)\log(1-x) + \log^2(1-x)) \right\}$$

$$\left\{ \text{and then } \log^2(1-x^2) = \log^2(1-x) + 2\log(1-x)\log(1+x) + \log^2(1+x) \right\}$$

{and afterwards expand the first two integrals and simplify}

$$= -\frac{1}{2}\int_0^1 \frac{\log(x)\log^2(1-x)}{1-x}dx - \frac{1}{2}\int_0^1 \frac{\log(x)\log^2(1+x)}{1-x}dx$$

$$-\frac{1}{2}\int_0^1 \frac{\log(x)\text{Li}_2(x^2)}{1-x}dx$$

{for the last integral use the identity $1/2\,\text{Li}_2(x^2) = \text{Li}_2(x) + \text{Li}_2(-x)$ and rearrange}

$$= -\frac{1}{2}\int_0^1 \frac{\log(x)\log^2(1-x)}{1-x}dx - \int_0^1 \frac{\log(x)\text{Li}_2(x)}{1-x}dx$$

$$-\int_0^1 \frac{\log(x)(1/2\log^2(1+x) + \text{Li}_2(-x))}{1-x}dx$$

{the first two integrals are calculated in the previous section in (6.210) and (6.211)}

$$= \frac{7}{4}\zeta(4) - \int_0^1 \frac{\log(x)(1/2\log^2(1+x) + \text{Li}_2(-x))}{1-x}dx. \tag{6.213}$$

From the additional information to the work in (6.207), the previous section, we have that $\sum_{n=1}^{\infty}(-1)^{n-1}y^n H_n = \dfrac{\log(1+y)}{1+y}$, where if we divide both sides by y and integrate from $y = 0$ to $y = x$, we get

$$\sum_{n=1}^{\infty}(-1)^{n-1}x^n\frac{H_n}{n} = \int_0^x \frac{\log(1+y)}{y(1+y)}dy = \int_0^x \frac{\log(1+y)}{y}dy - \int_0^x \frac{\log(1+y)}{1+y}dy$$

$$= -\frac{1}{2}\log^2(1+x) - \text{Li}_2(-x). \tag{6.214}$$

At this point, to approach the remaining integral in (6.213), we consider the result in (6.214), and then we write

$$\int_0^1 \frac{\log(x)(1/2\log^2(1+x) + \text{Li}_2(-x))}{1-x}dx = -\int_0^1 \frac{\log(x)}{1-x}\sum_{n=1}^{\infty}(-1)^{n-1}x^n\frac{H_n}{n}dx$$

{reverse the order of summation and integration}

$$= \sum_{n=1}^{\infty} (-1)^{n-1} \frac{H_n}{n} \int_0^1 -\frac{x^n \log(x)}{1-x} dx$$

$$\left\{ \text{exploit the result in (6.202), the previous section, } -\int_0^1 \frac{x^n \log(x)}{1-x} dx = \zeta(2) - H_n^{(2)} \right\}$$

$$= \sum_{n=1}^{\infty} (-1)^{n-1} \frac{H_n}{n} (\zeta(2) - H_n^{(2)}) = \zeta(2) \sum_{n=1}^{\infty} (-1)^{n-1} \frac{H_n}{n} - \sum_{n=1}^{\infty} (-1)^{n-1} \frac{H_n H_n^{(2)}}{n}$$

{the value of the first series is immediately extracted by setting $x = 1$ in (6.214)}

$$= \frac{5}{4}\zeta(4) - \frac{1}{2}\log^2(2)\zeta(2) - \sum_{n=1}^{\infty} (-1)^{n-1} \frac{H_n H_n^{(2)}}{n}. \tag{6.215}$$

And we have just arrived at the tricky part of the solution! How would you like to continue at this point?

To avoid the troublesome calculations, we may cleverly combine two powerful identities we already met in *(Almost) Impossible Integrals, Sums, and Series*. One of them is $\int_0^1 x^{n-1} \log^3(1-x) dx = -\dfrac{H_n^3 + 3H_n H_n^{(2)} + 2H_n^{(3)}}{n}$ (see [76, Chapter 1, p.2]), where if we multiply both sides by $(-1)^{n-1}$ and then consider the summation from $n = 1$ to ∞, we get

$$-\sum_{n=1}^{\infty} (-1)^{n-1} \frac{H_n^3}{n} - 3\sum_{n=1}^{\infty} (-1)^{n-1} \frac{H_n H_n^{(2)}}{n} - 2\sum_{n=1}^{\infty} (-1)^{n-1} \frac{H_n^{(3)}}{n}$$

$$= \sum_{n=1}^{\infty} (-1)^{n-1} \int_0^1 x^{n-1} \log^3(1-x) dx = \int_0^1 \log^3(1-x) \sum_{n=1}^{\infty} (-1)^{n-1} x^{n-1} dx$$

$$= \int_0^1 \frac{\log^3(1-x)}{1+x} dx \overset{1-x=y}{=} \int_0^1 \frac{\log^3(y)}{2-y} dy = \frac{1}{2} \int_0^1 \frac{\log^3(y)}{1-y/2} dy$$

$$= \frac{1}{2} \int_0^1 \sum_{n=1}^{\infty} \left(\frac{y}{2}\right)^{n-1} \log^3(y) dy$$

{reverse the order of summation and integration}

$$= \sum_{n=1}^{\infty} \frac{1}{2^n} \int_0^1 y^{n-1} \log^3(y) dy = -6\sum_{n=1}^{\infty} \frac{1}{n^4 2^n} = -6\operatorname{Li}_4\left(\frac{1}{2}\right). \tag{6.216}$$

The other identity we need is $\displaystyle\sum_{n=1}^{\infty} x^n(H_n^3 - 3H_n H_n^{(2)} + 2H_n^{(3)}) = -\frac{\log^3(1-x)}{1-x}$

(see [76, Chapter 6, p.355]). To check and convince yourself the identity holds for $|x| < 1$, you may consider a strategy similar to the one provided for the generalization of this identity which may be found in [76, Chapter 6, pp. 354–355], where the use of the elementary symmetric polynomials can be avoided if desired. Essentially, we multiply both sides by $1-x$ and then try to build and use telescoping sums for getting a simpler form. In the calculation process, one might also like to exploit that $\displaystyle\sum_{n=1}^{\infty} x^n(H_n^2 - H_n^{(2)}) = \frac{\log^2(1-x)}{1-x}$, and for this last identity, we use the same way involving the multiplication of both sides by $1-x$ and then try to exploit telescoping sums in the left-hand side to establish the result.

Now, if we replace x by $-x$ in $\displaystyle\sum_{n=1}^{\infty} x^n(H_n^3 - 3H_n H_n^{(2)} + 2H_n^{(3)}) = -\frac{\log^3(1-x)}{1-x}$,

then multiply both sides by $-1/x$, and integrate from $x = 0$ to $x = 1$, we get

$$\int_0^1 \sum_{n=1}^{\infty}(-1)^{n-1}x^{n-1}(H_n^3 - 3H_n H_n^{(2)} + 2H_n^{(3)})\mathrm{d}x$$

{reverse the order of summation and integration}

$$\sum_{n=1}^{\infty}(-1)^{n-1}(H_n^3 - 3H_n H_n^{(2)} + 2H_n^{(3)})\int_0^1 x^{n-1}\mathrm{d}x$$

$$= \sum_{n=1}^{\infty}(-1)^{n-1}\frac{H_n^3}{n} - 3\sum_{n=1}^{\infty}(-1)^{n-1}\frac{H_n H_n^{(2)}}{n} + 2\sum_{n=1}^{\infty}(-1)^{n-1}\frac{H_n^{(3)}}{n}$$

$$= \int_0^1 \frac{\log^3(1+x)}{x(1+x)}\mathrm{d}x = \int_0^1 \frac{\log^3(1+x)}{x}\mathrm{d}x - \underbrace{\int_0^1 \frac{\log^3(1+x)}{1+x}\mathrm{d}x}_{1/4\log^4(2)}$$

{the first resulting integral is calculated in (6.155), Sect. 4.20}

$$= 6\zeta(4) - \frac{21}{4}\log(2)\zeta(3) + \frac{3}{2}\log^2(2)\zeta(2) - \frac{1}{2}\log^4(2) - 6\operatorname{Li}_4\left(\frac{1}{2}\right). \qquad (6.217)$$

So, if we combine the results in (6.216) and (6.217), we arrive at

$$\sum_{n=1}^{\infty}(-1)^{n-1}\frac{H_n H_n^{(2)}}{n}$$

$$= \frac{1}{12} \log^4(2) - \frac{1}{4} \log^2(2)\zeta(2) + \frac{7}{8} \log(2)\zeta(3) - \zeta(4) + 2\operatorname{Li}_4\left(\frac{1}{2}\right). \quad (6.218)$$

The solution to the series in (6.218) presented above follows my way in [55].

Then, if we plug the value of the series from (6.218) in (6.215), we get

$$\int_0^1 \frac{\log(x)(1/2 \log^2(1+x) + \operatorname{Li}_2(-x))}{1-x} dx$$

$$= \frac{9}{4}\zeta(4) - \frac{7}{8} \log(2)\zeta(3) - \frac{1}{4} \log^2(2)\zeta(2) - \frac{1}{12} \log^4(2) - 2\operatorname{Li}_4\left(\frac{1}{2}\right). \quad (6.219)$$

Finally, upon plugging the result from (6.219) in (6.213), we arrive at the desired result

$$\sum_{n=1}^{\infty} \frac{H_{2n}}{n}\left(\zeta(2) - H_{2n}^{(2)}\right)$$

$$= \frac{1}{12} \log^4(2) + \frac{1}{4} \log^2(2)\zeta(2) - \frac{1}{2}\zeta(4) + \frac{7}{8} \log(2)\zeta(3) + 2\operatorname{Li}_4\left(\frac{1}{2}\right), \quad (6.220)$$

and the solution is complete.

As regards *the tricky part of the solution*, in the book *(Almost) Impossible Integrals, Sums, and Series* (see [76, Chapter 4, p.284]), the curious reader may find another useful identity for the calculation of the series in (6.218),

$$\sum_{n=1}^{\infty} x^n H_n H_n^{(2)}$$

$$= \frac{1}{1-x}\left(\frac{1}{2} \log(x) \log^2(1-x) + \operatorname{Li}_3(x) + \operatorname{Li}_3(1-x) - \zeta(2) \log(1-x) - \zeta(3)\right).$$

So good to have multiple tools in hand for dealing with the problems!

6.29   A Fourth Uncommon Series with the Tail of the Riemann Zeta Function $\zeta(2) - H_n^{(2)}$, Related to Weight 4 Harmonic Series

Solution In this section, we keep working on another atypical harmonic series with the tail of the Riemann zeta function. Observe that now the fraction in front of the tail is $\dfrac{H_n}{2n+1}$ compared, for example, to $\dfrac{H_{2n}}{n}$ in the previous section.

So, if we multiply, say, both sides of (4.93), Sect. 4.17, by $1/n$ and consider the summation from $n = 1$ to ∞, then for the left-hand side, we get

$$\sum_{n=1}^{\infty} \frac{1}{n}\left(\sum_{k=1}^{\infty}\frac{H_{2k}}{(k+1)(k+n+1)}\right) = \sum_{k=1}^{\infty}\frac{H_{2k}}{k+1}\left(\sum_{n=1}^{\infty}\frac{1}{n(n+k+1)}\right)$$

$$\left\{\text{employ the fact that } \sum_{n=1}^{\infty}\frac{1}{n(n+k)} = \frac{H_k}{k}\right\}$$

$$= \sum_{k=1}^{\infty}\frac{H_{2k}H_{k+1}}{(k+1)^2} = \sum_{k=1}^{\infty}\frac{(H_{2k+2}-1/(2k+1)-1/(2k+2))H_{k+1}}{(k+1)^2}$$

{reindex the series and then properly expand it}

$$= \sum_{k=1}^{\infty}\frac{H_k H_{2k}}{k^2} - \frac{1}{2}\sum_{k=1}^{\infty}\frac{H_k}{k^3} + \sum_{k=1}^{\infty}\frac{H_k}{k^2} - 2\sum_{k=1}^{\infty}\frac{H_k}{k(2k-1)}$$

$$= \zeta(4) + 2\zeta(3) + \frac{7}{2}\log(2)\zeta(3) - \log^2(2)\zeta(2) - 8\log(2) + 4\log^2(2) + \frac{1}{6}\log^4(2)$$

$$+ 4\operatorname{Li}_4\left(\frac{1}{2}\right), \tag{6.221}$$

where the value of first series is given in (4.113), Sect. 4.23; the second and third series are the cases $n = 2, 3$ of the Euler sum generalization in (6.149), Sect. 6.19; and the value of the last harmonic series is extracted by using (4.86), Sect. 4.14, as follows:

$$\log^2(2) = \sum_{k=1}^{\infty}\frac{H_k}{2k(2k+1)} = \sum_{k=1}^{\infty}\left(\frac{H_k}{2k} - \frac{H_k}{2k+1}\right)$$

{add and subtract $H_k/(2k-1)$, rearrange, and wisely expand}

$$= -\sum_{k=1}^{\infty}\frac{H_k}{2k(2k-1)} + \sum_{k=1}^{\infty}\left(\frac{H_k}{2k-1} - \frac{H_{k+1}}{2k+1}\right) + 2\sum_{k=1}^{\infty}\left(\frac{1}{2k+1} - \frac{1}{2k+2}\right)$$

{note the second series may be reduced to a telescoping sum}

$$= -\sum_{k=1}^{\infty}\frac{H_k}{2k(2k-1)} + 2\log(2),$$

whence we get that $\displaystyle\sum_{k=1}^{\infty}\frac{H_k}{2k(2k-1)} = 2\log(2) - \log^2(2)$.

Further, considering the summation from $n = 1$ to ∞ for the right-hand side of (4.93), Sect. 4.17, multiplied by $1/n$, we have

$$\frac{1}{4}\sum_{n=1}^{\infty}\frac{H_n^2 + H_n^{(2)}}{n^2} + 8\log(2)\sum_{n=1}^{\infty}\frac{H_{2n}}{(2n)^2} - \log(2)\sum_{n=1}^{\infty}\frac{H_n}{n^2} - 8\log(2)\sum_{n=1}^{\infty}\frac{1}{2n(2n+1)}$$

$$+ \sum_{n=1}^{\infty}\frac{1}{n^2}\sum_{i=1}^{n}\frac{H_i}{2i+1}$$

{the first series is given in (6.200), Sect. 6.26, then for the second one we may}

$$\left\{\text{use a transformation of the type } \sum_{n=1}^{\infty}a_{2n} = \frac{1}{2}\left(\sum_{n=1}^{\infty}a_n - \sum_{n=1}^{\infty}(-1)^{n-1}a_n\right),\right\}$$

$$\left\{\text{next for the fourth series use that } \sum_{n=1}^{\infty}\frac{1}{2n(2n+1)} = \sum_{n=1}^{\infty}\left(\frac{1}{2n} - \frac{1}{2n+1}\right)\right\}$$

$$\left\{= 1 - \sum_{n=1}^{\infty}(-1)^{n-1}\frac{1}{n} = 1 - \log(2), \text{ where the last series is well-known, }\right\}$$

{and finally, for the last harmonic series change the summation order}

$$= \frac{3}{2}\zeta(4) + 3\log(2)\sum_{n=1}^{\infty}\frac{H_n}{n^2} - 4\log(2)\sum_{n=1}^{\infty}(-1)^{n-1}\frac{H_n}{n^2}$$

$$- 8\log(2)(1 - \log(2)) + \sum_{i=1}^{\infty}\frac{H_i}{2i+1}\sum_{n=i}^{\infty}\frac{1}{n^2}$$

{the first series is the case $n = 2$ of generalization in (6.149), Sect. 6.19, and}

{the second series is the case $m = 1$ of the generalization in (4.105), Sect. 4.21}

$$= \frac{3}{2}\zeta(4) + \frac{7}{2}\log(2)\zeta(3) - 8\log(2)(1 - \log(2)) + \sum_{i=1}^{\infty}\frac{H_i}{2i+1}\sum_{n=i+1}^{\infty}\frac{1}{n^2}$$

$$+ \sum_{i=1}^{\infty}\frac{H_i}{(2i+1)i^2} = \frac{3}{2}\zeta(4) + \frac{7}{2}\log(2)\zeta(3) - 8\log(2)(1 - \log(2))$$

$$+ \sum_{i=1}^{\infty} \frac{H_i}{2i+1}\left(\zeta(2) - H_i^{(2)}\right) + \sum_{i=1}^{\infty} \frac{H_i}{i^2} - 4 \sum_{i=1}^{\infty} \frac{H_i}{2i(2i+1)}$$

{the penultimate harmonic series is the case $n = 2$ of the generalization in}

{(6.149), Sect. 6.19, and the last harmonic series is given in (4.86), Sect. 4.14}

$$= \frac{3}{2}\zeta(4) + 2\zeta(3) + \frac{7}{2}\log(2)\zeta(3) - 8\log(2)(1 - \log(2)) - 4\log^2(2)$$

$$+ \sum_{i=1}^{\infty} \frac{H_i}{2i+1}\left(\zeta(2) - H_i^{(2)}\right). \tag{6.222}$$

Essentially, at this point, we are done by simply combining the results in (6.221) and (6.222) and considering to replace i by n. Therefore, we obtain

$$\sum_{n=1}^{\infty} \frac{H_n}{2n+1}\left(\zeta(2) - H_n^{(2)}\right)$$

$$= \frac{1}{6}\log^4(2) - \log^2(2)\zeta(2) - \frac{1}{2}\zeta(4) + 4\operatorname{Li}_4\left(\frac{1}{2}\right),$$

and the first solution to the main series is finalized.

For a second solution, I'll make use of the result in (1.103), Sect. 1.21, where we multiply both sides by $1/n$ and then consider the summation from $n = 1$ to ∞ that gives

$$\sum_{n=1}^{\infty} \frac{1}{n} \int_0^1 x^{2n-1} \log^2(1+x)dx = \int_0^1 \sum_{n=1}^{\infty} \frac{x^{2n-1}}{n} \log^2(1+x)dx$$

$$= -\int_0^1 \frac{\log(1-x^2)\log^2(1+x)}{x}dx$$

$$= 8\log(2) \sum_{n=1}^{\infty} \frac{H_{2n}}{(2n)^2} - \log(2) \sum_{n=1}^{\infty} \frac{H_n}{n^2} - \frac{1}{2} \sum_{n=1}^{\infty} \frac{H_{2n}^2 + H_{2n}^{(2)}}{n^2} + \frac{1}{4} \sum_{n=1}^{\infty} \frac{H_n^2 + H_n^{(2)}}{n^2}$$

$$+ \sum_{n=1}^{\infty} \frac{1}{n^2} \sum_{k=1}^{n-1} \frac{H_k}{2k+1}$$

$$\left\{ \text{for the first series use that } \sum_{n=1}^{\infty} a_{2n} = \frac{1}{2}\left(\sum_{n=1}^{\infty} a_n - \sum_{n=1}^{\infty} (-1)^{n-1} a_n\right), \text{ then for} \right\}$$

$$\left\{ \text{the third series use the logarithmic integral } \int_0^1 x^{n-1} \log^2(1-x)dx = \frac{H_n^2 + H_n^{(2)}}{n} \right\}$$

{in [76, Chapter 1, p.2], next the fourth series is given in (6.200), Sect. 6.26}

{and for the last series we might like to reverse the order of summation}

$$= 3\log(2) \sum_{n=1}^{\infty} \frac{H_n}{n^2} - 4\log(2) \sum_{n=1}^{\infty} (-1)^{n-1} \frac{H_n}{n^2} + \int_0^1 \frac{\log^2(1-x)\log(1-x^2)}{x} dx$$

$$+ \frac{3}{2}\zeta(4) + \sum_{k=1}^{\infty} \frac{H_k}{2k+1} \sum_{n=k+1}^{\infty} \frac{1}{n^2}$$

{the first series is the case $n = 2$ of the Euler sum in (6.149), Sect. 6.19, and}

{the second series is the case $m = 1$ of the Euler sum in (4.105), Sect. 4.21}

$$= \frac{3}{2}\zeta(4) + \frac{7}{2}\log(2)\zeta(3) + \int_0^1 \frac{\log^2(1-x)\log(1-x^2)}{x} dx$$

$$+ \sum_{k=1}^{\infty} \frac{H_k}{2k+1} \left(\zeta(2) - H_k^{(2)} \right),$$

whence, upon rearranging, we obtain that

$$\sum_{k=1}^{\infty} \frac{H_k}{2k+1} \left(\zeta(2) - H_k^{(2)} \right)$$

$$= -\frac{7}{2}\log(2)\zeta(3) - \frac{3}{2}\zeta(4) - \int_0^1 \frac{\log^3(1-x)}{x} dx - \int_0^1 \frac{\log^3(1+x)}{x} dx$$

$$- \int_0^1 \frac{\log(1-x)\log^2(1+x) + \log^2(1-x)\log(1+x)}{x} dx$$

$$\left\{ \text{exploit the algebraic identity } ab^2 + a^2 b = 1/3((a+b)^3 - a^3 - b^3) \right\}$$

{for the last integral, where we set $a = \log(1-x)$ and $b = \log(1+x)$,}

{and then expand the logarithmic integral into simpler integrals}

$$= -\frac{7}{2}\log(2)\zeta(3) - \frac{3}{2}\zeta(4) - \frac{2}{3}\int_0^1 \frac{\log^3(1+x)}{x}dx - \frac{2}{3}\int_0^1 \frac{\log^3(1-x)}{x}dx$$

$$-\frac{1}{3}\int_0^1 \frac{\log^3(1-x^2)}{x}dx$$

$$\left\{ \text{for the last integral make the variable change } x^2 = y \right\}$$

$$\{\text{and then add the integral to the penultimate one}\}$$

$$= -\frac{7}{2}\log(2)\zeta(3) - \frac{3}{2}\zeta(4) - \frac{2}{3}\int_0^1 \frac{\log^3(1+x)}{x}dx - \frac{5}{6}\int_0^1 \frac{\log^3(1-x)}{x}dx$$

$$= \frac{1}{6}\log^4(2) - \log^2(2)\zeta(2) - \frac{1}{2}\zeta(4) + 4\operatorname{Li}_4\left(\frac{1}{2}\right),$$

where the two remaining integrals are calculated and given in (6.155), Sect. 6.20, and (6.182), Sect. 6.23, and the second solution to the main series is finalized, which also answers the *challenging question*.

Both solutions are non-obvious, particularly the first one that is built on *The Master Theorem of Series*. Surely, in the work with harmonic series, one wants to get used to a panel of *tools* that usually helps you create solutions (e.g., the logarithmic integrals from *(Almost) Impossible Integrals, Sums, and Series*; see [76, Chapter 1, p.2]). Sometimes these tools we need, like the key identities in these solutions, are not at our disposal immediately, and we have to obtain them by research efforts.

We prepare now to enter another section with enjoyable harmonic series involving the tail of the Riemann zeta function!

6.30   A Fifth Uncommon Series with the Tail of the Riemann Zeta Function $\zeta(2) - H_{2n}^{(2)}$, Related to Weight 4 Harmonic Series

Solution *Wizardry* is that one word that would best describe this section, and I here also take into account the fact that in the present solution, I'll answer the *challenging question*. I wonder if you agree with me after a thorough read!

The beauty of the present solution lies in its subtleties, the non-obvious mathematical connections we make. It's always exciting to choose a path that is not well-trodden, and this is how I'll proceed in the following.

Now, if we take a look at the series in the previous section and compare it with this one, we observe that this time we have $H_{2n}^{(2)}$ instead of $H_n^{(2)}$ in the Riemann zeta tail. And remember, the *challenging question* asks us to also exploit the series from the previous section.

Let's prepare the first *magical* step, and we start by observing the simple fact that

$$\sum_{k=n}^{\infty} \frac{1}{(2k+1)^2} = \sum_{k=n+1}^{\infty} \frac{1}{(2k-1)^2} = \underbrace{\sum_{k=1}^{\infty} \frac{1}{(2k-1)^2}}_{3/4\zeta(2)} - \sum_{k=1}^{n} \frac{1}{(2k-1)^2}$$

$$\left\{ \text{use that } \sum_{k=1}^{n} \frac{1}{(2k-1)^2} = \sum_{k=1}^{2n} \frac{1}{k^2} - \sum_{k=1}^{n} \frac{1}{(2k)^2} = H_{2n}^{(2)} - \frac{1}{4} H_n^{(2)} \right\}$$

$$= \frac{3}{4}\zeta(2) - H_{2n}^{(2)} + \frac{1}{4} H_n^{(2)}$$

{observe the last result may be put in a form with two Riemann zeta tails}

$$= \zeta(2) - H_{2n}^{(2)} - \frac{1}{4}\left(\zeta(2) - H_n^{(2)}\right). \tag{6.223}$$

If we multiply the opposite sides of (6.223) by $\dfrac{H_n}{2n+1}$, consider the summation from $n=1$ to ∞, and rearrange, we obtain that

$$\sum_{n=1}^{\infty} \frac{H_n}{2n+1}\left(\zeta(2) - H_{2n}^{(2)}\right)$$

$$= \frac{1}{4} \sum_{n=1}^{\infty} \frac{H_n}{2n+1}\left(\zeta(2) - H_n^{(2)}\right) + \sum_{n=1}^{\infty} \frac{H_n}{2n+1}\left(\sum_{k=n}^{\infty} \frac{1}{(2k+1)^2}\right)$$

{the first series is calculated in the previous section}

$$= \frac{1}{24}\log^4(2) - \frac{1}{4}\log^2(2)\zeta(2) - \frac{1}{8}\zeta(4) + \operatorname{Li}_4\left(\frac{1}{2}\right)$$

$$+ \sum_{n=1}^{\infty} \frac{H_n}{2n+1}\left(\sum_{k=n}^{\infty} \frac{1}{(2k+1)^2}\right). \tag{6.224}$$

Next, we prepare the second *magical* step. We begin with reversing the order of summation for the remaining series in (6.224) that further gives

$$\sum_{n=1}^{\infty} \frac{H_n}{2n+1}\left(\sum_{k=n}^{\infty} \frac{1}{(2k+1)^2}\right) = \sum_{k=1}^{\infty} \frac{1}{(2k+1)^2}\left(\sum_{n=1}^{k} \frac{H_n}{2n+1}\right)$$

{we may change the variable letters to more convenient ones}

$$= \sum_{n=1}^{\infty} \frac{1}{(2n+1)^2} \underbrace{\left(\sum_{i=1}^{n} \frac{H_i}{2i+1} \right)}_{S}. \tag{6.225}$$

Then, we may observe a very nice fact! We may cleverly exploit the identity in (4.92), Sect. 4.17, to reduce the series in (6.225) to more manageable ones! Doing that, we also meet the requirement of the *challenging question* that asks us to also make use of *The Master Theorem of Series*. Since the result in (4.92), Sect. 4.17, I obtain by the use of *The Master Theorem of Series*, I have also checked this last requirement point!

Upon multiplying both sides of (4.92), Sect. 4.17, by $n/(2n+1)^2$, considering the summation from $n = 1$ to ∞, and then rearranging, we arrive at

$$S = \sum_{n=1}^{\infty} \frac{1}{(2n+1)^2} \left(\sum_{i=1}^{n} \frac{H_i}{2i+1} \right)$$

$$= \sum_{n=1}^{\infty} \frac{n}{(2n+1)^2} \left(\sum_{k=1}^{\infty} \frac{H_{2k} - 1/2H_k}{(k+1)(k+n+1)} \right) + 4\log(2) \sum_{n=1}^{\infty} \frac{n}{(2n+1)^3}$$

$$- 2\log(2) \sum_{n=1}^{\infty} \frac{H_{2n}}{(2n+1)^2} + \log(2) \sum_{n=1}^{\infty} \frac{H_n}{(2n+1)^2}$$

{reverse the order of summation in the first series and then}

{consider to bring the second and third series to simplified forms}

$$= \sum_{k=1}^{\infty} \frac{H_{2k} - 1/2H_k}{k+1} \underbrace{\left(\sum_{n=1}^{\infty} \frac{n}{(2n+1)^2(n+k+1)} \right)}_{S_1} - 2\log(2) \underbrace{\sum_{n=1}^{\infty} \frac{H_{2n+1}}{(2n+1)^2}}_{S_2}$$

$$+ \log(2) \underbrace{\sum_{n=1}^{\infty} \frac{H_n}{(2n+1)^2}}_{S_3} - 2\log(2) + \frac{3}{2}\log(2)\zeta(2). \tag{6.226}$$

For now, we focus on the inner series of S_1 in (6.226), and then, using partial fraction expansion, we write

$$\sum_{n=1}^{\infty} \frac{n}{(2n+1)^2(n+k+1)}$$

$$= \frac{2(k+1)}{(2k+1)^2} \sum_{n=1}^{\infty} \left(\frac{1}{2n+1} - \frac{1}{2n+2k+2} \right) - \frac{1}{2k+1} \sum_{n=1}^{\infty} \frac{1}{(2n+1)^2}$$

{for the first series, add and subtract $1/(2n)$, and then split it carefully}

$$= \frac{2(k+1)}{(2k+1)^2} \sum_{n=1}^{\infty} \left(\frac{1}{2n+1} - \frac{1}{2n} \right) + \frac{k+1}{(2k+1)^2} \sum_{n=1}^{\infty} \left(\frac{1}{n} - \frac{1}{n+k+1} \right)$$

$$+ \left(1 - \frac{3}{4}\zeta(2) \right) \frac{1}{2k+1}$$

$$\left\{ \text{consider that } \sum_{n=1}^{\infty} \left(\frac{1}{2n+1} - \frac{1}{2n} \right) = -1 + \sum_{n=1}^{\infty}(-1)^{n-1}\frac{1}{n} = \log(2) - 1, \right\}$$

$$\left\{ \text{and for the second series we have the fact that } \sum_{k=1}^{\infty} \left(\frac{1}{k} - \frac{1}{k+n} \right) = H_n \right\}$$

$$= \left(\log(2) - \frac{3}{4}\zeta(2) \right) \frac{1}{2k+1} + \log(2)\frac{1}{(2k+1)^2} + \frac{1}{2}\frac{H_k}{2k+1} + \frac{1}{2}\frac{H_k}{(2k+1)^2}.$$
$$\tag{6.227}$$

Returning with the result from (6.227) in the series S_1 in (6.226), which then we also expand, we have

$$S_1 = \sum_{k=1}^{\infty} \frac{H_{2k} - 1/2H_k}{k+1} \left(\sum_{n=1}^{\infty} \frac{n}{(2n+1)^2(n+k+1)} \right)$$

$$= \frac{3}{2}\zeta(2) \sum_{k=1}^{\infty} \left(\frac{1}{2k+2} - \frac{1}{2k+1} \right) - \frac{3}{8}\zeta(2) \sum_{k=1}^{\infty} \frac{1}{(k+1)^2} + \frac{3}{2}\zeta(2) \sum_{k=1}^{\infty} \frac{1}{(2k+1)^2}$$

$$- 2\log(2) \sum_{k=1}^{\infty} \frac{1}{(2k+1)^3} + \frac{3}{4}\zeta(2) \sum_{k=1}^{\infty} \frac{H_k}{(2k+1)(2k+2)}$$

$$- \frac{3}{2}\zeta(2) \sum_{k=1}^{\infty} \left(\frac{H_{2k+1}}{2k+1} - \frac{H_{2k+2}}{2k+2} \right) - \log(2) \sum_{k=1}^{\infty} \frac{H_k}{(2k+1)^2} + 2\log(2) \sum_{k=1}^{\infty} \frac{H_{2k+1}}{(2k+1)^2}$$

$$-\frac{1}{2}\sum_{k=1}^{\infty}\frac{H_k^2}{(2k+1)^2}+\sum_{k=1}^{\infty}\frac{H_k H_{2k}}{(2k+1)^2}. \tag{6.228}$$

Combining (6.226) and (6.228), the series S_2 and S_3 vanish, and we get

$$S=\sum_{n=1}^{\infty}\frac{1}{(2n+1)^2}\left(\sum_{i=1}^{n}\frac{H_i}{2i+1}\right)$$

$$=\frac{3}{2}\zeta(2)\sum_{k=1}^{\infty}\left(\frac{1}{2k+2}-\frac{1}{2k+1}\right)+\frac{3}{4}\zeta(2)\sum_{k=1}^{\infty}\frac{H_k}{(2k+1)(2k+2)}$$

$$-\frac{3}{2}\zeta(2)\sum_{k=1}^{\infty}\left(\frac{H_{2k+1}}{2k+1}-\frac{H_{2k+2}}{2k+2}\right)-\frac{1}{2}\sum_{k=1}^{\infty}\frac{H_k^2}{(2k+1)^2}+\sum_{k=1}^{\infty}\frac{H_k H_{2k}}{(2k+1)^2}$$

$$-\frac{9}{8}\zeta(2)+\frac{15}{8}\zeta(4)+\frac{3}{2}\log(2)\zeta(2)-\frac{7}{4}\log(2)\zeta(3)$$

$$=\frac{83}{32}\zeta(4)-\frac{7}{8}\log(2)\zeta(3)-\log^2(2)\zeta(2)-\frac{1}{12}\log^4(2)-2\operatorname{Li}_4\left(\frac{1}{2}\right), \tag{6.229}$$

where for the first series I used that $\displaystyle\sum_{k=1}^{\infty}\left(\frac{1}{2k+2}-\frac{1}{2k+1}\right)=\frac{1}{2}-$

$\displaystyle\sum_{k=1}^{\infty}(-1)^{k-1}\frac{1}{k}=\frac{1}{2}-\log(2)$. Then, for the value of the second series in (6.229), I cleverly exploited the series in (4.86), Sect. 4.14,

$$\underbrace{\sum_{k=1}^{\infty}\frac{H_k}{2k(2k+1)}}_{\log^2(2)}+\sum_{k=1}^{\infty}\frac{H_k}{(2k+1)(2k+2)}=\sum_{k=1}^{\infty}\left(\frac{H_k}{2k(2k+1)}+\frac{H_k}{(2k+1)(2k+2)}\right)$$

{use the partial fraction expansion and then simplify}

$$=\frac{1}{2}\sum_{k=1}^{\infty}\left(\frac{H_k}{k}-\frac{H_{k+1}}{k+1}+\frac{1}{(k+1)^2}\right)=\frac{1}{2}\lim_{N\to\infty}\sum_{k=1}^{N}\left(\frac{H_k}{k}-\frac{H_{k+1}}{k+1}\right)+\frac{1}{2}\sum_{k=1}^{\infty}\frac{1}{(k+1)^2}$$

$$=\frac{1}{2}\lim_{N\to\infty}\left(1-\frac{H_{N+1}}{N+1}\right)+\frac{1}{2}(\zeta(2)-1)=\frac{1}{2}\zeta(2),$$

whence we obtain that

$$\sum_{k=1}^{\infty} \frac{H_k}{(2k+1)(2k+2)} = \frac{1}{2}\zeta(2) - \log^2(2). \tag{6.230}$$

For the third series in (6.229), we note that $\displaystyle\sum_{k=1}^{\infty}\left(\frac{H_{2k+1}}{2k+1} - \frac{H_{2k+2}}{2k+2}\right) = -\frac{1}{4} +$

$\displaystyle\sum_{k=1}^{\infty}(-1)^{k-1}\frac{H_k}{k} = \frac{1}{2}\zeta(2) - \frac{1}{2}\log^2(2) - \frac{1}{4}$, where the last series is found in (4.106),
the case $m = 1$, Sect. 4.22.

Finally, the values of the last two series in (6.229) are given in (4.115), Sect. 4.24, and (4.116), Sect. 4.25, respectively.

Good to know: Alternatively, the curious reader might also exploit the identities in (1.103) and (1.104), Sect. 1.21, to get a different approach to the series S in (6.229).

Combining the result in (6.229) and the ones in (6.224) and (6.225), we conclude that

$$\sum_{n=1}^{\infty} \frac{H_n}{2n+1}\left(\zeta(2) - H_{2n}^{(2)}\right)$$

$$= \frac{79}{32}\zeta(4) - \frac{7}{8}\log(2)\zeta(3) - \frac{5}{4}\log^2(2)\zeta(2) - \frac{1}{24}\log^4(2) - \operatorname{Li}_4\left(\frac{1}{2}\right),$$

and the solution is complete.

The very tricky part in this solution appears from the very beginning when one wants to imagine an identity like the one in (6.223) that may be further combined with an application of *The Master Theorem of Series*, in this case the identity in (4.92), Sect. 4.17, in order to extract the value of the desired series.

The curious reader might also like to consider an approach similar to the one in Sect. 6.28 for getting another solution and then use the integral representation of the Riemann zeta tail, $\zeta(2) - H_{2n}^{(2)} = -\displaystyle\int_0^1 \frac{x^{2n}\log(x)}{1-x}\mathrm{d}x$, for reducing the series, in a first phase, to an integral. So, just another possible way one might like to explore for more fun!

6.31   A Sixth Uncommon Series with the Tail of the Riemann Zeta Function $\zeta(2) - H_n^{(2)}$, Related to Weight 4 Harmonic Series

Solution It may be surprising to see that in this section the *challenging question* acts more like a helping hand, and this will be well understood from the solution below.

One of the winning choices in this section is again the use of Abel's summation. However, before reaching this point, let's rearrange a bit the main series we want to calculate.

First, we might like to consider the main series with the fraction $\dfrac{H_{2n+1}}{2n+1}$ instead of $\dfrac{H_{2n}}{2n+1}$, and for doing that, we write

$$\sum_{n=1}^{\infty} \frac{H_{2n}}{2n+1}\left(\zeta(2) - H_n^{(2)}\right) = \sum_{n=1}^{\infty} \frac{H_{2n+1} - 1/(2n+1)}{2n+1}\left(\zeta(2) - H_n^{(2)}\right)$$

$$= \sum_{n=1}^{\infty} \frac{H_{2n+1}}{2n+1}\left(\zeta(2) - H_n^{(2)}\right) - \zeta(2)\underbrace{\sum_{n=1}^{\infty} \frac{1}{(2n+1)^2}}_{3/4\zeta(2) - 1} + \sum_{n=1}^{\infty} \frac{H_n^{(2)}}{(2n+1)^2}$$

{the value of the last series may be found in (4.114), Sect. 4.24}

$$= \frac{1}{3}\log^4(2) - 2\log^2(2)\zeta(2) + 7\log(2)\zeta(3) + \zeta(2) - \frac{151}{16}\zeta(4) + 8\operatorname{Li}_4\left(\frac{1}{2}\right)$$

$$+ \underbrace{\sum_{n=1}^{\infty} \frac{H_{2n+1}}{2n+1}\left(\zeta(2) - H_n^{(2)}\right)}_{S}. \tag{6.231}$$

Now, if we consider the series that appears in the *challenging question*, multiply it by $1/2$, then denote it by T, that is, $T = \sum_{n=1}^{\infty} \frac{H_{2n}}{2n}\left(\zeta(2) - H_n^{(2)}\right)$, and add it to the series S in (6.231), we arrive at

$$S + T = \sum_{n=1}^{\infty}\left(\frac{H_{2n}}{2n} + \frac{H_{2n+1}}{2n+1}\right)\left(\zeta(2) - H_n^{(2)}\right), \tag{6.232}$$

which at first sight might look even more complicated than the initial series. However, there is some captivating *magic* here that is highlighted by Abel's summation.

So, applying Abel's summation in (6.232), the series version in (6.7), Sect. 6.2, where we consider $a_n = \dfrac{H_{2n}}{2n} + \dfrac{H_{2n+1}}{2n+1}$ and $b_n = \zeta(2) - H_n^{(2)}$, we have

$$S + T = \sum_{n=1}^{\infty} \left(\frac{H_{2n}}{2n} + \frac{H_{2n+1}}{2n+1} \right) \left(\zeta(2) - H_n^{(2)} \right)$$

$$= \underbrace{\lim_{N \to \infty} \sum_{n=1}^{N} \left(\frac{H_{2n}}{2n} + \frac{H_{2n+1}}{2n+1} \right) \left(\zeta(2) - H_{N+1}^{(2)} \right)}_{0}$$

$$+ \sum_{n=1}^{\infty} \frac{1}{(n+1)^2} \left(\sum_{k=1}^{n} \left(\frac{H_{2k}}{2k} + \frac{H_{2k+1}}{2k+1} \right) \right)$$

$$= \sum_{n=1}^{\infty} \frac{1}{(n+1)^2} \left(\sum_{k=1}^{n} \left(\frac{H_{2k}}{2k} + \frac{H_{2k+1}}{2k+1} \right) \right)$$

$$\left\{ \text{observe and use that } \sum_{k=1}^{n} \left(\frac{H_{2k}}{2k} + \frac{H_{2k+1}}{2k+1} \right) = -\frac{H_{2n+2}}{2n+2} + \sum_{k=2}^{2n+2} \frac{H_k}{k} \right\}$$

$$= \sum_{n=1}^{\infty} \frac{1}{(n+1)^2} \left(\sum_{k=2}^{2n+2} \frac{H_k}{k} \right) - \frac{1}{2} \sum_{n=1}^{\infty} \frac{H_{2n+2}}{(n+1)^3}$$

{reindex both series and start from $n = 1$}

$$= \sum_{n=1}^{\infty} \frac{1}{n^2} \left(\sum_{k=2}^{2n} \frac{H_k}{k} \right) - 4 \sum_{n=1}^{\infty} \frac{H_{2n}}{(2n)^3}$$

$$\left\{ \text{use } \sum_{k=1}^{n} \frac{H_k}{k} = \frac{1}{2} \left(H_n^2 + H_n^{(2)} \right), \text{ based on } p = q = 1 \text{ in (6.102), Sect. 6.13} \right\}$$

$$= 2 \sum_{n=1}^{\infty} \frac{H_{2n}^2}{(2n)^2} + 2 \sum_{n=1}^{\infty} \frac{H_{2n}^{(2)}}{(2n)^2} - 4 \sum_{n=1}^{\infty} \frac{H_{2n}}{(2n)^3} - \sum_{n=1}^{\infty} \frac{1}{n^2}$$

$$\left\{ \text{for the first three series employ that } \sum_{n=1}^{\infty} a_{2n} = \frac{1}{2} \left(\sum_{n=1}^{\infty} a_n - \sum_{n=1}^{\infty} (-1)^{n-1} a_n \right) \right\}$$

$$= \sum_{n=1}^{\infty} \frac{H_n^2 + H_n^{(2)}}{n^2} - 2 \sum_{n=1}^{\infty} \frac{H_n}{n^3} - \sum_{n=1}^{\infty} (-1)^{n-1} \frac{H_n^2}{n^2} + 2 \sum_{n=1}^{\infty} (-1)^{n-1} \frac{H_n}{n^3}$$

$$- \sum_{n=1}^{\infty} (-1)^{n-1} \frac{H_n^{(2)}}{n^2} - \zeta(2)$$

{the value of the first series is given in (6.200), Sect. 6.26, then the second series}

{is the case $n = 3$ of the Euler sum generalization in (6.149), Sect. 6.19, next,}

$$\left\{ \text{using (6.184), Sect. 6.23 we express } \sum_{n=1}^{\infty} (-1)^{n-1} \frac{H_n^2}{n^2} \text{ via } \sum_{n=1}^{\infty} (-1)^{n-1} \frac{H_n}{n^3} \text{ and} \right\}$$

$$\left\{ \text{using (4.90), Sect. 4.15, we express } \sum_{n=1}^{\infty} (-1)^{n-1} \frac{H_n^{(2)}}{n^2} \text{ via } \sum_{n=1}^{\infty} (-1)^{n-1} \frac{H_n}{n^3} \right\}$$

$$= \frac{11}{8} \zeta(4) - \zeta(2) + 3 \sum_{n=1}^{\infty} (-1)^{n-1} \frac{H_n}{n^3}$$

{the value of the remaining series is given in (4.103), Sect. 4.20}

$$= \frac{77}{8} \zeta(4) - \zeta(2) - \frac{21}{4} \log(2)\zeta(3) + \frac{3}{2} \log^2(2)\zeta(2) - \frac{1}{4} \log^4(2) - 6 \operatorname{Li_4}\left(\frac{1}{2}\right).$$
$$(6.233)$$

Since the value of T may be extracted from (4.118), Sect. 4.27, then based on (6.233) we obtain that

$$S = \sum_{n=1}^{\infty} \frac{H_{2n+1}}{2n+1} \left(\zeta(2) - H_n^{(2)} \right)$$

$$= \frac{77}{8} \zeta(4) - \zeta(2) - \frac{21}{4} \log(2)\zeta(3) + \frac{3}{2} \log^2(2)\zeta(2) - \frac{1}{4} \log^4(2) - 6 \operatorname{Li_4}\left(\frac{1}{2}\right) - T$$

$$= \frac{47}{8} \zeta(4) - \zeta(2) - \frac{7}{2} \log(2)\zeta(3) + \frac{1}{2} \log^2(2)\zeta(2) - \frac{1}{12} \log^4(2) - 2 \operatorname{Li_4}\left(\frac{1}{2}\right).$$
$$(6.234)$$

Returning with the value from (6.234) in (6.231), we obtain the desired result

$$\sum_{n=1}^{\infty} \frac{H_{2n}}{2n+1}\left(\zeta(2) - H_n^{(2)}\right)$$

$$= \frac{1}{4}\log^4(2) - \frac{57}{16}\zeta(4) + \frac{7}{2}\log(2)\zeta(3) - \frac{3}{2}\log^2(2)\zeta(2) + 6\operatorname{Li}_4\left(\frac{1}{2}\right),$$

and the solution is complete.

I guess some of you might like a little explanation on the limit tending to 0 that appears in the first part of the calculations in (6.233)! *How would we explain that elegantly?*, you might wonder. It is about the following limit I want to prove

$$\lim_{N\to\infty} \left(\zeta(2) - H_{N+1}^{(2)}\right) \sum_{n=1}^{N}\left(\frac{H_{2n}}{2n} + \frac{H_{2n+1}}{2n+1}\right) = 0. \qquad (6.235)$$

We note that $\zeta(2) - H_{N+1}^{(2)} = \displaystyle\sum_{n=N+2}^{\infty} \frac{1}{n^2}$, and since we have

$$\underbrace{\sum_{n=N+2}^{\infty}\left(\frac{1}{n} - \frac{1}{n+1}\right)}_{1/(N+2)} \leq \sum_{n=N+2}^{\infty}\frac{1}{n^2} \leq \underbrace{\sum_{n=N+2}^{\infty}\left(\frac{1}{n-1} - \frac{1}{n}\right)}_{1/(N+1)},$$

we observe that when N is large, we arrive at the asymptotic behavior

$$\left(\zeta(2) - H_{N+1}^{(2)}\right) \sim \frac{1}{N}, \text{ as } N \to \infty. \qquad (6.236)$$

Based on the result in (6.236), the limit in (6.235) gets reduced to calculating

$$\lim_{N\to\infty} \frac{1}{N}\sum_{n=1}^{N}\left(\frac{H_{2n}}{2n} + \frac{H_{2n+1}}{2n+1}\right),$$

where if we apply the *Stolz–Cesàro theorem* (see [15, Appendix, pp.263–266]), we arrive at

$$\lim_{N\to\infty}\left(\frac{H_{2N+2}}{2N+2} + \frac{H_{2N+3}}{2N+3}\right),$$

but since $\lim\limits_{N\to\infty} H_N/N = 0$ by *Stolz-Cesàro theorem*, we obtain the value of the limit,

$$\lim_{N\to\infty}\left(\frac{H_{2N+2}}{2N+2} + \frac{H_{2N+3}}{2N+3}\right) = 0,$$

which brings an end to the proof of the limit result in (6.235).

Finally, note that all three key alternating harmonic series of weight 4 appearing during the extraction result in (6.233), that is, $\sum\limits_{n=1}^{\infty}(-1)^{n-1}\dfrac{H_n^2}{n^2}$, $\sum\limits_{n=1}^{\infty}(-1)^{n-1}\dfrac{H_n}{n^3}$, and

$\sum\limits_{n=1}^{\infty}(-1)^{n-1}\dfrac{H_n^{(2)}}{n^2}$ are also found and calculated in *(Almost) Impossible Integrals, Sums, and Series* (see [76, Chapter 4, pp. 309–310]).

The curious reader interested in another way to go might also easily observe that by using the integral representation of the Riemann zeta tail, $\zeta(2) - 1 - \dfrac{1}{2^2} - \cdots - \dfrac{1}{n^2} = -\int_0^1 \dfrac{x^n \log(x)}{1-x}dx$, the series reduces to calculating

$$\sum_{n=1}^{\infty}\frac{H_{2n}}{2n+1}\left(\zeta(2) - H_n^{(2)}\right) = \int_0^1 \frac{\left(\log^2(1+x) - \log^2(1-x)\right)\log(x)}{1-x^2}dx,$$

which is an integral representation that looks like another good point from which one might like to continue the calculations.

6.32   A Seventh Uncommon Series with the Tail of the Riemann Zeta Function $\zeta(2) - H_{2n}^{(2)}$, Related to Weight 4 Harmonic Series

Solution In the whole row of uncommon series with the tail of the Riemann zeta function, related to the harmonic series of weight 4, I have presented so far, this one doesn't have a polylogarithmic component! No appearance of $\mathrm{Li}_4\left(\dfrac{1}{2}\right)$! And, indeed, the closed form looks more appealing this way!

The first solution will be *designed* by taking into account the part (*i*) of the *challenging question* that asks us to make use of one of the six previous results involving the Riemann zeta tail.

Surely, one might also think of considering a solution similar to the one in the previous section, but this time, I'll act a bit differently, and for doing that, I'll start from a classical series form involving the Riemann zeta function,

$$\sum_{n=1}^{\infty} \frac{H_n}{n} \left(\zeta(2) - H_n^{(2)} \right) = \frac{7}{4} \zeta(4), \tag{6.237}$$

which also appears in [76, Chapter 6, p.490].

To briefly see the result in (6.237) holds, we set $a_n = H_n/n$ and $b_n = \zeta(2) - H_n^{(2)}$ and then apply Abel's summation, the series version in (6.7), Sect. 6.2, and using the fact that $\sum_{k=1}^{n} \frac{H_k}{k} = \frac{1}{2} \left(H_n^2 + H_n^{(2)} \right)$, which is the case $p = q = 1$ in (6.102), Sect. 6.13, we write

$$\sum_{n=1}^{\infty} \frac{H_n}{n} \left(\zeta(2) - H_n^{(2)} \right) = \frac{1}{2} \sum_{n=1}^{\infty} \frac{(H_{n+1} - 1/(n+1))^2 + H_{n+1}^{(2)} - 1/(n+1)^2}{(n+1)^2}$$

{reindex the series and expand it}

$$= \frac{1}{2} \sum_{n=1}^{\infty} \frac{H_n^2 + H_n^{(2)}}{n^2} - \sum_{n=1}^{\infty} \frac{H_n}{n^3} = \frac{7}{4} \zeta(4),$$

where the value of the first series is given in (6.200), Sect. 6.26, and the second series is the case $n = 3$ of the Euler sum generalization in (6.149), Sect. 6.19.

A *magical* moment will come in place now! All we have to do is to exploit the parity in (6.237), and then we write

$$\frac{7}{4} \zeta(4) = \sum_{n=1}^{\infty} \frac{H_n}{n} \left(\zeta(2) - H_n^{(2)} \right)$$

$$= \zeta(2) - 1 + \frac{1}{2} \sum_{n=1}^{\infty} \frac{H_{2n}}{n} \left(\zeta(2) - H_{2n}^{(2)} \right) + \sum_{n=1}^{\infty} \frac{H_{2n+1}}{2n+1} \left(\zeta(2) - H_{2n+1}^{(2)} \right)$$

{rearrange the last series and then expand it}

$$= \zeta(2) - 1 + \underbrace{\frac{1}{2} \sum_{n=1}^{\infty} \frac{H_{2n}}{n} \left(\zeta(2) - H_{2n}^{(2)} \right)}_{T} + \underbrace{\sum_{n=1}^{\infty} \frac{H_{2n}}{2n+1} \left(\zeta(2) - H_{2n}^{(2)} \right)}_{S}$$

$$- \sum_{n=1}^{\infty} \frac{H_{2n+1}}{(2n+1)^3} - \sum_{n=1}^{\infty} \frac{H_{2n+1}^{(2)}}{(2n+1)^2} + \zeta(2) \sum_{n=1}^{\infty} \frac{1}{(2n+1)^2} + \sum_{n=1}^{\infty} \frac{1}{(2n+1)^4}$$

$$\left\{\text{for the last four series use that } \sum_{n=1}^{\infty} a_n = a_1 + \sum_{n=1}^{\infty} a_{2n} + \sum_{n=1}^{\infty} a_{2n+1}\right\}$$

$$= \frac{1}{2}T + S + \frac{45}{16}\zeta(4) - \sum_{n=1}^{\infty}\frac{H_n}{n^3} - \sum_{n=1}^{\infty}\frac{H_n^{(2)}}{n^2} + \sum_{n=1}^{\infty}\frac{H_{2n}}{(2n)^3} + \sum_{n=1}^{\infty}\frac{H_{2n}^{(2)}}{(2n)^2}$$

$$\left\{\text{for the last two series use that } \sum_{n=1}^{\infty} a_{2n} = \frac{1}{2}\left(\sum_{n=1}^{\infty} a_n - \sum_{n=1}^{\infty}(-1)^{n-1} a_n\right)\right\}$$

$$= \frac{1}{2}T + S + \frac{45}{16}\zeta(4) - \frac{1}{2}\sum_{n=1}^{\infty}\frac{H_n}{n^3} - \frac{1}{2}\sum_{n=1}^{\infty}\frac{H_n^{(2)}}{n^2}$$

$$- \frac{1}{2}\sum_{n=1}^{\infty}(-1)^{n-1}\frac{H_n}{n^3} - \frac{1}{2}\sum_{n=1}^{\infty}(-1)^{n-1}\frac{H_n^{(2)}}{n^2}$$

{the third series is the case $n = 3$ of the Euler sum in (6.149), Sect. 6.19, next,}

{the fourth series is given by $p = q = 2$ with $n \to \infty$ in (6.102), Sect. 6.19,}

$$\left\{\text{and we state } \sum_{n=1}^{\infty}(-1)^{n-1}\frac{H_n^{(2)}}{n^2} \text{ via } \sum_{n=1}^{\infty}(-1)^{n-1}\frac{H_n}{n^3} \text{ using (4.90), Sect. 4.15}\right\}$$

$$= \frac{1}{2}T + S + \frac{5}{32}\zeta(4) + \frac{1}{2}\sum_{n=1}^{\infty}(-1)^{n-1}\frac{H_n}{n^3}$$

{the value of the last series is given in (4.103), Sect. 4.20}

$$= \frac{1}{2}T + S + \frac{49}{32}\zeta(4) - \frac{7}{8}\log(2)\zeta(3) + \frac{1}{4}\log^2(2)\zeta(2) - \frac{1}{24}\log^4(2) - \text{Li}_4\left(\frac{1}{2}\right),$$

and since the value of T is known from (4.119), Sect. 4.28, we obtain the desired result

$$S = \sum_{n=1}^{\infty}\frac{H_{2n}}{2n+1}\left(\zeta(2) - H_{2n}^{(2)}\right)$$

$$= \frac{15}{32}\zeta(4) + \frac{7}{16}\log(2)\zeta(3) - \frac{3}{8}\log^2(2)\zeta(2),$$

and the solution is finalized.

In order to get a second solution and answer the point (ii) of the *challenging question*, we might like to build first an integral representation of the series by exploiting the integral representation of the Riemann zeta tail as it appears in Sect. 6.28, the result in (6.212), $\zeta(2) - H_{2n}^{(2)} = -\int_0^1 \frac{x^{2n}\log(x)}{1-x}dx$, and then we write

$$\sum_{n=1}^{\infty} \frac{H_{2n}}{2n+1}\left(\zeta(2) - H_{2n}^{(2)}\right) = -\sum_{n=1}^{\infty} H_{2n}\int_0^1 y^{2n}dy \int_0^1 \frac{x^{2n}\log(x)}{1-x}dx$$

{reverse the order of integration and summation}

$$= -\int_0^1 \left(\int_0^1 \frac{\log(x)}{1-x}\sum_{n=1}^{\infty}(xy)^{2n} H_{2n}dy\right)dx$$

$$\left\{\text{use that } \sum_{n=1}^{\infty}x^{2n} H_{2n} = \frac{2x\,\text{arctanh}(x) - \log(1-x^2)}{2(1-x^2)}\right\}$$

$$= -\frac{1}{2}\int_0^1 \frac{\log(x)}{1-x}\left(\int_0^1 \frac{2xy\,\text{arctanh}(xy) - \log(1-(xy)^2)}{1-(xy)^2}dy\right)dx$$

$$= -\frac{1}{2}\int_0^1 \frac{\log(x)}{1-x}\left(-\frac{\text{arctanh}(xy)\log(1-(xy)^2)}{x}\Big|_{y=0}^{y=1}\right)dx$$

{take the limits of integration and then expand the integral}

$$= \frac{1}{2}\int_0^1 \frac{\text{arctanh}(x)\log(x)\log(1-x^2)}{x(1-x)}dx$$

$$= \frac{1}{4}\int_0^1 \frac{\log(x)(\log^2(1+x) - \log^2(1-x))}{x(1-x)}dx$$

$$= \frac{1}{4}\int_0^1 (\log(x) - \log(1-x))'\log(x)(\log^2(1+x) - \log^2(1-x))dx$$

{integrate by parts and then expand}

$$= \frac{1}{4}\int_0^1 \frac{\log(1-x)\log^2(1+x)}{x}dx + \frac{1}{2}\int_0^1 \frac{\log(1-x)\log(x)\log(1+x)}{1+x}dx$$

$$-\frac{1}{4}\int_0^1 \frac{\log^3(1-x)}{x}dx + \frac{1}{2}\int_0^1 \frac{\log^2(1-x)\log(x)}{1-x}dx + \frac{1}{4}\int_0^1 \frac{\log^2(1-x)\log(x)}{x}dx$$

$$-\frac{1}{2}\int_0^1 \frac{\log(1-x)\log^2(x)}{1-x}dx - \frac{1}{2}\int_0^1 \frac{\log(1+x)\log^2(x)}{1+x}dx$$

$$-\frac{1}{4}\int_0^1 \frac{\log^2(1+x)\log(x)}{x}dx$$

{the first integral is calculated in Sect. 6.23 by exploiting algebraic identities;}

{for the fourth integral integrate by parts and add it to the third integral; then in}

{the fifth integral let the variable change $1-x = y$ and then add it to the sixth one;}

{finally, integrate by parts the last integral and add it to the penultimate integral}

$$= -\frac{3}{32}\zeta(4) + \frac{1}{2}\int_0^1 \frac{\log(1-x)\log(x)\log(1+x)}{1+x}dx - \frac{1}{12}\int_0^1 \frac{\log^3(1-x)}{x}dx$$

$$-\frac{1}{4}\int_0^1 \frac{\log(1-x)\log^2(x)}{1-x}dx - \frac{1}{4}\int_0^1 \frac{\log(1+x)\log^2(x)}{1+x}dx. \qquad (6.238)$$

For the first remaining integral in (6.238), we exploit the algebraic identity $(a + b)^2 - (a - b)^2 = 4ab$, where we set $a = \log(1 - x)$ and $b = \log(1 + x)$, and then we have

$$\int_0^1 \frac{\log(1-x)\log(x)\log(1+x)}{1+x}dx$$

$$= \frac{1}{4}\int_0^1 \frac{\log^2\left(1-x^2\right)\log(x)}{1-x^2}(1-x)dx - \frac{1}{4}\int_0^1 \frac{\log^2\left(\dfrac{1-x}{1+x}\right)\log(x)}{1+x}dx$$

$\left\{\text{in the first integral we want to let the variable change } x^2 = y \text{ and in the second}\right\}$

$\left\{\text{integral we want to make the variable change } \dfrac{1-x}{1+x} = y, \text{ and then expand them}\right\}$

$$= \frac{1}{2}\int_0^1 \frac{\log(1+y)\log^2(y)}{1+y}dy - \frac{1}{16}\int_0^1 \frac{\log^2(1-y)\log(y)}{1-y}dy$$

$$+ \frac{1}{16}\int_0^1 \frac{\log^2(1-y)\log(y)}{(1-y)\sqrt{y}}dy - \frac{1}{4}\int_0^1 \frac{\log(1-y^2)\log^2(y)}{1-y^2}(1-y)dy$$

{for the second resulting integral we want to apply integration by parts and}

$\left\{\text{in the last integral we make the variable change } y^2 = x \text{ and then expand it}\right\}$

$$= \frac{1}{32} \int_0^1 \frac{\log(1-x)\log^2(x)}{1-x}dx - \frac{1}{48} \int_0^1 \frac{\log^3(1-x)}{x}dx$$

$$+ \frac{1}{2} \int_0^1 \frac{\log(1+x)\log^2(x)}{1+x}dx - \frac{1}{32} \int_0^1 \frac{\log(1-x)\log^2(x)}{(1-x)\sqrt{x}}dx$$

$$+ \frac{1}{16} \int_0^1 \frac{\log^2(1-x)\log(x)}{(1-x)\sqrt{x}}dx. \tag{6.239}$$

If we plug the result from (6.239) in (6.238), where we also con-
sider that $\int_0^1 \frac{\log^3(1-x)}{x}dx = -6\zeta(4)$, which may be found calculated
in (6.182), Sect. 6.23, and then the fact that $\int_0^1 \frac{\log(1-x)\log^2(x)}{1-x}dx =$

$$-\int_0^1 \sum_{n=1}^{\infty} x^n \log^2(x) H_n dx = -\sum_{n=1}^{\infty} H_n \int_0^1 x^n \log^2(x)dx = -2\sum_{n=1}^{\infty} \frac{H_n}{(n+1)^3} =$$

$-\frac{1}{2}\zeta(4)$, based on (6.148), the case $n = 3$, Sect. 6.19, the main series turns into

$$\sum_{n=1}^{\infty} \frac{H_{2n}}{2n+1}\left(\zeta(2) - H_{2n}^{(2)}\right)$$

$$= \frac{75}{128}\zeta(4) - \frac{1}{64} \underbrace{\int_0^1 \frac{\log(1-x)\log^2(x)}{(1-x)\sqrt{x}}dx}_{X} + \frac{1}{32} \underbrace{\int_0^1 \frac{\log^2(1-x)\log(x)}{(1-x)\sqrt{x}}dx}_{Y}.$$

$$\tag{6.240}$$

One clear point to note in (6.240): both remaining integrals are forms of the
derivatives of the Beta function. The calculation of both integrals is related to the
use of applications of *The Master Theorem of Series*.

So, for the integral X in (6.240), we start from the identity $\sum_{n=1}^{\infty} x^n H_n =$

$-\frac{\log(1-x)}{1-x}$, which is the case $m = 1$ in (4.32), Sect. 4.6, and then we write

$$X = \int_0^1 \frac{\log(1-x)\log^2(x)}{(1-x)\sqrt{x}}dx = -\int_0^1 \sum_{n=1}^{\infty} x^{n-1/2} H_n \log^2(x)dx$$

{reverse the order of summation and integration}

$$= -\sum_{n=1}^{\infty} H_n \int_0^1 x^{n-1/2} \log^2(x) dx = -16 \sum_{n=1}^{\infty} \frac{H_n}{(2n+1)^3}$$

$$= 28 \log(2)\zeta(3) - \frac{45}{2}\zeta(4), \qquad (6.241)$$

where the value of the last series is the case $p = 3$ of the generalization in (4.102), Sect. 4.19, and thus I have obtained a fast evaluation of the integral X!

Then, for the integral Y in (6.240), we proceed in a similar style as before, and happily things will be done fast again as you'll see in the following.

By using the identity $\sum_{n=1}^{\infty} x^n (H_n^2 - H_n^{(2)}) = \dfrac{\log^2(1-x)}{1-x}$ (see [76, Chapter 6, p.355]), we write that

$$Y = \int_0^1 \frac{\log^2(1-x)\log(x)}{(1-x)\sqrt{x}} dx = \int_0^1 \sum_{n=1}^{\infty} x^{n-1/2} \log^2(x)(H_n^2 - H_n^{(2)}) dx$$

{reverse the order of summation and integration}

$$= \sum_{n=1}^{\infty}(H_n^2 - H_n^{(2)}) \int_0^1 x^{n-1/2} \log(x) dx = -4 \sum_{n=1}^{\infty} \frac{H_n^2 - H_n^{(2)}}{(2n+1)^2}$$

$$= 28 \log(2)\zeta(3) - 12 \log^2(2)\zeta(2) - 15\zeta(4), \qquad (6.242)$$

where happily the resulting series is calculated in Sect. 6.24, (6.192).

Finally, by plugging the results from (6.241) and (6.242) in (6.240), we arrive at the desired result

$$\sum_{n=1}^{\infty} \frac{H_{2n}}{2n+1}\left(\zeta(2) - H_{2n}^{(2)}\right)$$

$$= \frac{15}{32}\zeta(4) + \frac{7}{16}\log(2)\zeta(3) - \frac{3}{8}\log^2(2)\zeta(2).$$

Good to know: From the calculations in (6.238), we also get the curious integral

$$\int_0^1 \frac{\log^2(1+x)\log(x)}{x(1-x)} dx = \frac{7}{4}\log(2)\zeta(3) - \frac{3}{2}\log^2(2)\zeta(2) - \frac{5}{8}\zeta(4). \qquad (6.243)$$

An interesting fact to observe with respect to the integral in (6.243) is that if we use the partial fraction decomposition, split it, and try to calculate the integrals separately, we'll get closed forms that also contain a polylogarithmic component. Also, one may observe that by splitting the integral in (6.243) and the one in (6.219), Sect. 6.28, we get a common integral.

This section ends the myriad of sections dedicated to the calculation of the uncommon harmonic series with the tail of the Riemann zeta function related to harmonic series of weight 4.

It's time to jump in the next section where we are supposed to work with unusual harmonic series of weight 5!

6.33   On the Calculation of an Essential Harmonic Series of Weight 5, Involving Harmonic Numbers of the Type H_{2n}

Solution With the opening of this section, we step into a more advanced world of harmonic series, involving harmonic numbers of the type H_{2n}, the ones with a weight 5 structure, and the series I'll treat now may be seen as a cornerstone since it plays an important part in the derivation of many other harmonic series alike.

My evaluation of the current series also appears in the article *On the calculation of two essential harmonic series with a weight 5 structure, involving harmonic numbers of the type H_{2n}* that was published in the *Journal of Classical Analysis*, Vol. 16, No. 1, 2020, and the solution presented will follow the way from this work.

To get a solution, we start from the series representation of $\log(1+y)\log(1-y)$,

$$- \log(1 + y) \log(1 - y) = \sum_{n=1}^{\infty} y^{2n} \left(\frac{H_{2n} - H_n}{n} + \frac{1}{2n^2} \right), \quad |y| < 1, \qquad (6.244)$$

which is straightforward if we use the Cauchy product of two series as presented in [76, Chapter 6, p.344], and this is also the key identity I used in the first solution to its *little brother* of weight four, that is, $\displaystyle\sum_{n=1}^{\infty} \frac{H_n H_{2n}}{n^2}$, which is calculated in Sect. 6.23.

If we divide both sides of (6.244) by y and then integrate from $y = 0$ to $y = x$, we obtain that

$$- \int_0^x \frac{\log(1 + y) \log(1 - y)}{y} dy = \sum_{n=1}^{\infty} x^{2n} \left(\frac{H_{2n} - H_n}{2n^2} + \frac{1}{4n^3} \right). \qquad (6.245)$$

Further, upon multiplying both sides of (6.245) by $\log(1 + x)/x$ and then integrating from $x = 0$ to $x = 1$, we get

$$- \int_0^1 \frac{\log(1 + x)}{x} \left(\int_0^x \frac{\log(1 + y) \log(1 - y)}{y} dy \right) dx$$

$$= \int_0^1 \log(1 + x) \sum_{n=1}^{\infty} x^{2n-1} \left(\frac{H_{2n} - H_n}{2n^2} + \frac{1}{4n^3} \right) dx$$

{reverse the order of summation and integration}

$$= \sum_{n=1}^{\infty} \left(\frac{H_{2n} - H_n}{2n^2} + \frac{1}{4n^3} \right) \int_0^1 x^{2n-1} \log(1+x) dx$$

{use the integral result in (1.100), Sect. 1.21}

$$= \sum_{n=1}^{\infty} \frac{H_{2n} - H_n}{2n} \left(\frac{H_{2n} - H_n}{2n^2} + \frac{1}{4n^3} \right)$$

$$= 2\sum_{n=1}^{\infty} \frac{H_{2n}^2}{(2n)^3} + 2\sum_{n=1}^{\infty} \frac{H_{2n}}{(2n)^4} + \frac{1}{4}\sum_{n=1}^{\infty} \frac{H_n^2}{n^3} - \frac{1}{8}\sum_{n=1}^{\infty} \frac{H_n}{n^4} - \frac{1}{2}\sum_{n=1}^{\infty} \frac{H_n H_{2n}}{n^3}$$

$$\left\{ \text{for the first two series use that } \sum_{n=1}^{\infty} a_{2n} = \frac{1}{2} \left(\sum_{n=1}^{\infty} a_n - \sum_{n=1}^{\infty} (-1)^{n-1} a_n \right) \right\}$$

$$= \frac{5}{4}\sum_{n=1}^{\infty} \frac{H_n^2}{n^3} + \frac{7}{8}\sum_{n=1}^{\infty} \frac{H_n}{n^4} - \sum_{n=1}^{\infty} (-1)^{n-1} \frac{H_n^2}{n^3} - \sum_{n=1}^{\infty} (-1)^{n-1} \frac{H_n}{n^4} - \frac{1}{2}\sum_{n=1}^{\infty} \frac{H_n H_{2n}}{n^3}.$$

$$(6.246)$$

At the same time, if we integrate by parts for the integral in (6.246), we have

$$\int_0^1 \frac{\log(1+x)}{x} \left(\int_0^x \frac{\log(1+y)\log(1-y)}{y} dy \right) dx$$

$$= -\int_0^1 (\text{Li}_2(-x))' \left(\int_0^x \frac{\log(1+y)\log(1-y)}{y} dy \right) dx$$

$$= -\text{Li}_2(-x) \int_0^x \frac{\log(1+y)\log(1-y)}{y} dy \Big|_{x=0}^{x=1} + \int_0^1 \frac{\log(1-x)\log(1+x)\text{Li}_2(-x)}{x} dx$$

$$= -\frac{5}{16}\zeta(2)\zeta(3) + \int_0^1 \frac{\log(1-x)\log(1+x)\text{Li}_2(-x)}{x} dx, \qquad (6.247)$$

where in the calculations I used that $\int_0^1 \frac{\log(1-x)\log(1+x)}{x} dx = -\frac{5}{8}\zeta(3)$, which is a result already given in [76, Chapter 1, p.4]. Also, note that based on (6.245) we get that $\int_0^1 \frac{\log(1-x)\log(1+x)}{x} dx = -\frac{1}{2}\sum_{n=1}^{\infty} \frac{H_n}{n^2} +$

$$\sum_{n=1}^{\infty} (-1)^{n-1} \frac{H_n}{n^2} - \frac{1}{4}\sum_{n=1}^{\infty} \frac{1}{n^3} = -\frac{5}{8}\zeta(3), \text{ where the first harmonic series is the}$$

particular case $n = 2$ of the Euler sum generalization in (6.149), Sect. 6.19, and the second harmonic series is the particular case $m = 1$ of the alternating Euler sum generalization in (4.105), Sect. 4.21.

Finally, combining the results in (6.246) and (6.247), we are able to extract the value of the desired series, and we obtain that

$$\sum_{n=1}^{\infty} \frac{H_n H_{2n}}{n^3}$$

$$= -\frac{5}{8}\zeta(2)\zeta(3) + \frac{7}{4}\sum_{n=1}^{\infty}\frac{H_n}{n^4} + \frac{5}{2}\sum_{n=1}^{\infty}\frac{H_n^2}{n^3} - 2\sum_{n=1}^{\infty}(-1)^{n-1}\frac{H_n}{n^4} - 2\sum_{n=1}^{\infty}(-1)^{n-1}\frac{H_n^2}{n^3}$$

$$+ 2\int_0^1 \frac{\log(1-x)\log(1+x)\operatorname{Li}_2(-x)}{x}dx$$

$$= \frac{307}{16}\zeta(5) - \frac{1}{2}\zeta(2)\zeta(3) - 7\log^2(2)\zeta(3) + \frac{8}{3}\log^3(2)\zeta(2) - \frac{8}{15}\log^5(2)$$

$$- 16\log(2)\operatorname{Li}_4\left(\frac{1}{2}\right) - 16\operatorname{Li}_5\left(\frac{1}{2}\right),$$

where in the calculations I used that the value of the first series is the particular case $n = 4$ of the Euler sum generalization in (6.149), Sect. 6.19; then the second series, $\sum_{n=1}^{\infty}\frac{H_n^2}{n^3} = \frac{7}{2}\zeta(5) - \zeta(2)\zeta(3)$, may be found both in [76, Chapter 4, p.293] and in my article *A new proof for a classical quadratic harmonic series* that was published in the *Journal of Classical Analysis*, Vol. 8, No. 2, 2016 (see [72]); next the third series is given in (4.104) Sect. 4.20; further the value of the last series is given in (3.331) Sect. 3.48; and finally, the value of the resulting integral is found in (1.216) Sect. 1.48, and the solution is complete.

Maybe we could use the strategy in the second part of Sect. 6.23 to get a second solution? Some curious readers might like to ask! Indeed, it's a good point to observe this possibility! However, there is a little issue! If in the case of the series $\sum_{n=1}^{\infty}\frac{H_n H_{2n}}{n^2}$ things worked perfectly with the use of the identity (4.82) in Sect. 4.13, this time the cancelation of the series $\sum_{n=1}^{\infty}\frac{H_n H_{2n}}{n^3}$ has happened and we cannot extract its value in a similar way. But we get *something else. What exactly?*, you might wonder. More details you'll find in one of the upcoming sections!

6.34 More Helpful Atypical Harmonic Series of Weight 5 with Denominators of the Type $(2n+1)^2$ and $(2n+1)^3$

Solution Did you manage to pass through Sect. 6.24? If *Yes*, then there is a good chance you can immediately sketch a way to go for the first three points of the problem. In that section, I calculated two similar series from the weight 4 series class.

So, for the series at the point (i), we employ Abel's summation, the series version in (6.7), Sect. 6.2, where we set $a_n = 1/(2n+1)^3$ and $b_n = H_n^{(2)}$, and if we also consider that $\displaystyle\sum_{k=1}^{n} \frac{1}{(2k+1)^3} = H_{2n}^{(3)} - \frac{1}{8} H_n^{(3)} + \frac{1}{(2n+1)^3} - 1$, we have

$$\sum_{n=1}^{\infty} \frac{H_n^{(2)}}{(2n+1)^3} = \lim_{N\to\infty} \left(H_{2N}^{(3)} - \frac{1}{8} H_N^{(3)} + \frac{1}{(2N+1)^3} - 1 \right) H_{N+1}^{(2)}$$

$$- \sum_{n=1}^{\infty} \frac{1}{(n+1)^2} \left(H_{2n+1}^{(3)} - \frac{1}{8} H_n^{(3)} - 1 \right)$$

$$\overset{\substack{\text{reindex} \\ \text{the series}}}{=} \frac{7}{8}\zeta(2)\zeta(3) - \zeta(2) - \sum_{n=1}^{\infty} \frac{1}{n^2} \left(H_{2n-1}^{(3)} - \frac{1}{8} H_{n-1}^{(3)} - 1 \right)$$

$$= \frac{7}{8}\zeta(2)\zeta(3) - \zeta(2) + \sum_{n=1}^{\infty} \frac{1}{n^2} + \frac{1}{8}\sum_{n=1}^{\infty} \frac{H_n^{(3)}}{n^2} - 4\sum_{n=1}^{\infty} \frac{H_{2n}^{(3)}}{(2n)^2}$$

$$\left\{ \text{for the third series use that } \sum_{k=1}^{\infty} a_{2k} = \frac{1}{2}\left(\sum_{k=1}^{\infty} a_k - \sum_{k=1}^{\infty}(-1)^{k-1}a_k \right) \right\}$$

$$= \frac{7}{8}\zeta(2)\zeta(3) - \frac{15}{8}\sum_{n=1}^{\infty} \frac{H_n^{(3)}}{n^2} + 2\sum_{n=1}^{\infty}(-1)^{n-1}\frac{H_n^{(3)}}{n^2}$$

$$= \frac{49}{8}\zeta(2)\zeta(3) - \frac{93}{8}\zeta(5),$$

where in the calculations I used that $\displaystyle\sum_{n=1}^{\infty} \frac{H_n^{(3)}}{n^2} = \frac{11}{2}\zeta(5) - 2\zeta(2)\zeta(3)$ (see [76, Chapter 6, p.386]) and $\displaystyle\sum_{n=1}^{\infty}(-1)^{n-1}\frac{H_n^{(3)}}{n^2} = \frac{3}{4}\zeta(2)\zeta(3) - \frac{21}{32}\zeta(5)$ ([76, Chapter 4, p.311]), and the solution to the point (i) of the problem is complete.

Further, for the series at the point (ii), we employ again Abel's summation, the
series version in (6.7), Sect. 6.2, where we set $a_n = 1/(2n+1)^2$ and $b_n = H_n^{(3)}$,
and if we also consider that $\displaystyle\sum_{k=1}^{n} \frac{1}{(2k+1)^2} = H_{2n}^{(2)} - \frac{1}{4}H_n^{(2)} + \frac{1}{(2n+1)^2} - 1$, we
have

$$\sum_{n=1}^{\infty} \frac{H_n^{(3)}}{(2n+1)^2} = \lim_{N\to\infty} \left(H_{2N}^{(2)} - \frac{1}{4}H_N^{(2)} + \frac{1}{(2N+1)^2} - 1 \right) H_{N+1}^{(3)}$$

$$- \sum_{n=1}^{\infty} \frac{1}{(n+1)^3} \left(H_{2n+1}^{(2)} - \frac{1}{4}H_n^{(2)} - 1 \right)$$

$$\underset{\substack{\text{reindex} \\ \text{the series}}}{=} \frac{3}{4}\zeta(2)\zeta(3) - \zeta(3) - \sum_{n=1}^{\infty} \frac{1}{n^3} \left(H_{2n-1}^{(2)} - \frac{1}{4}H_{n-1}^{(2)} - 1 \right)$$

$$= \frac{3}{4}\zeta(2)\zeta(3) - \zeta(3) + \sum_{n=1}^{\infty} \frac{1}{n^3} + \frac{1}{4}\sum_{n=1}^{\infty} \frac{H_n^{(2)}}{n^3} - 8\sum_{n=1}^{\infty} \frac{H_{2n}^{(2)}}{(2n)^3}$$

$$\left\{ \text{for the third series use that } \sum_{k=1}^{\infty} a_{2k} = \frac{1}{2}\left(\sum_{k=1}^{\infty} a_k - \sum_{k=1}^{\infty} (-1)^{k-1} a_k \right) \right\}$$

$$= \frac{3}{4}\zeta(2)\zeta(3) - \frac{15}{4}\sum_{n=1}^{\infty} \frac{H_n^{(2)}}{n^3} + 4\sum_{n=1}^{\infty} (-1)^{n-1} \frac{H_n^{(2)}}{n^3}$$

$$= \frac{31}{2}\zeta(5) - 8\zeta(2)\zeta(3),$$

where in the calculations I used that $\displaystyle\sum_{n=1}^{\infty} \frac{H_n^{(2)}}{n^3} = 3\zeta(2)\zeta(3) - \frac{9}{2}\zeta(5)$ (see [76,
Chapter 6, p.386]) and $\displaystyle\sum_{n=1}^{\infty} (-1)^{n-1} \frac{H_n^{(2)}}{n^3} = \frac{5}{8}\zeta(2)\zeta(3) - \frac{11}{32}\zeta(5)$ ([76, Chapter 4,
p.311]), and the solution to the point (ii) of the problem is complete.

We return to (6.191), Sect. 6.24, where if we multiply both sides by n, then
differentiate twice with respect to n, set $n = -1/2$, and rearrange, we get

$$\sum_{k=1}^{\infty} \frac{H_k^2 - H_k^{(2)}}{(2k+1)^3} = \sum_{k=1}^{\infty} \frac{H_k^2}{(2k+1)^3} - \sum_{k=1}^{\infty} \frac{H_k^{(2)}}{(2k+1)^3}$$

$$= -\frac{1}{8}(\psi(n+1) + \gamma)\left(\psi^{(1)}(n+1)\right)^2 + \frac{1}{16}(\psi(n+1) + \gamma)\psi^{(3)}(n+1)$$

$$-\frac{1}{16}(\psi(n+1)+\gamma)^2\psi^{(2)}(n+1)+\frac{3}{16}\psi^{(1)}(n+1)\psi^{(2)}(n+1)$$

$$-\frac{1}{16}\zeta(2)\psi^{(2)}(n+1)-\frac{1}{48}\psi^{(4)}(n+1)\Big|_{n=-1/2}$$

$$=\frac{31}{2}\zeta(5)-\frac{45}{8}\log(2)\zeta(4)+\frac{7}{2}\log^2(2)\zeta(3)-7\zeta(2)\zeta(3),\qquad(6.248)$$

where in the calculations I used that $\psi\left(\frac{1}{2}\right)=-\gamma-2\log(2)$, $\psi^{(1)}\left(\frac{1}{2}\right)=3\zeta(2)$, $\psi^{(2)}\left(\frac{1}{2}\right)=-14\zeta(3)$, $\psi^{(3)}\left(\frac{1}{2}\right)=90\zeta(4)$, and $\psi^{(4)}\left(\frac{1}{2}\right)=-744\zeta(5)$, which are given in (6.146) and (6.147), Sect. 6.19.

Finally, if we combine the results in (6.248) and the one from the point (i), we conclude that

$$\sum_{n=1}^{\infty}\frac{H_n^2}{(2n+1)^3}$$

$$=\frac{7}{2}\log^2(2)\zeta(3)-\frac{45}{8}\log(2)\zeta(4)+\frac{31}{8}\zeta(5)-\frac{7}{8}\zeta(2)\zeta(3),$$

and the solution to the point (iii) of the problem is complete.

To attack the series at the fourth point, I'll consider the powerful Cauchy product in (4.22), Sect. 4.5, where if we multiply its both sides by $\operatorname{arctanh}(x)/x$ and then integrate from $x=0$ to $x=1$, we get

$$\int_0^1\frac{\operatorname{arctanh}^2(x)\operatorname{Li}_2(x^2)}{x}dx=\frac{1}{4}\int_0^1\frac{\log^2(1-x)\operatorname{Li}_2(x^2)}{x}dx$$

$$+\frac{1}{4}\int_0^1\frac{\log^2(1+x)\operatorname{Li}_2(x^2)}{x}dx-\frac{1}{2}\int_0^1\frac{\log(1-x)\log(1+x)\operatorname{Li}_2(x^2)}{x}dx$$

$$=\int_0^1\sum_{n=1}^{\infty}x^{2n}\left(4\frac{H_{2n}}{(2n+1)^2}+\frac{H_n^{(2)}}{2n+1}\right)\operatorname{arctanh}(x)dx$$

{reverse the order of summation and integration}

$$=\sum_{n=1}^{\infty}\int_0^1 x^{2n}\operatorname{arctanh}(x)dx\left(4\frac{H_{2n}}{(2n+1)^2}+\frac{H_n^{(2)}}{2n+1}\right)$$

{employ the integral in (1.111), Sect. 1.22, and expand the series}

$$= -4\log(2) \sum_{n=1}^{\infty} \frac{1}{(2n-1)^4} + 4\log(2) \sum_{n=1}^{\infty} \frac{H_{2n-1}}{(2n-1)^3} + \log(2) \sum_{n=1}^{\infty} \frac{H_n^{(2)}}{(2n+1)^2}$$

$$+ 2\sum_{n=1}^{\infty} \frac{H_n H_{2n}}{(2n+1)^3} + \frac{1}{2} \sum_{n=1}^{\infty} \frac{H_n H_n^{(2)}}{(2n+1)^2},$$

where the first two series have been reindexed, and using that $\displaystyle\sum_{n=1}^{\infty} \frac{1}{(2n-1)^4} =$

$\displaystyle\sum_{n=1}^{\infty} \frac{1}{n^4} - \sum_{n=1}^{\infty} \frac{1}{(2n)^4} = \frac{15}{16}\zeta(4)$, next $\displaystyle\sum_{n=1}^{\infty} \frac{H_{2n-1}}{(2n-1)^3} = \frac{1}{2}\left(\sum_{n=1}^{\infty} \frac{H_n}{n^3} + \sum_{n=1}^{\infty} (-1)^{n-1}\frac{H_n}{n^3}\right)$

$= 2\zeta(4) - \frac{7}{8}\log(2)\zeta(3) + \frac{1}{4}\log^2(2)\zeta(2) - \frac{1}{24}\log^4(2) - \text{Li}_4\left(\frac{1}{2}\right)$, where I used

that the first resulting series is the particular case $n = 3$ of the Euler sum in (6.149), Sect. 6.19, and the second one is given in (4.103), Sect. 4.20, further the values of the series in (4.114), Sect. 4.24, (4.130), Sect. 4.35, together with the values of the integrals in (1.211), (1.214), Sect. 1.47, and (1.217), Sect. 1.48, we have that

$$\sum_{n=1}^{\infty} \frac{H_n H_n^{(2)}}{(2n+1)^2}$$

$$= \frac{7}{4}\zeta(2)\zeta(3) - \frac{589}{32}\zeta(5) + \frac{121}{8}\log(2)\zeta(4) - 7\log^2(2)\zeta(3) + \frac{4}{3}\log^3(2)\zeta(2)$$

$$- \frac{2}{15}\log^5(2) + 16\,\text{Li}_5\left(\frac{1}{2}\right),$$

and the solution to the point (iv) of the problem is complete.

For the last series, I'll use a flow similar to the one in (6.191), Sect. 6.24, except that now we consider another identity from [76, Chapter 4, p.291], that is

$$\sum_{k=1}^{\infty} \frac{H_k^3 - 3H_k H_k^{(2)} + 2H_k^{(3)}}{(k+1)(k+n+1)}$$

$$= \frac{(\psi(n+1)+\gamma)^4 + 6(\psi(n+1)+\gamma)^2(\zeta(2)-\psi^{(1)}(n+1))}{4n}$$

$$+ \frac{8(\psi(n+1)+\gamma)(\zeta(3)+1/2\psi^{(2)}(n+1)) + 3(\zeta(2)-\psi^{(1)}(n+1))^2}{4n}$$

$$+ \frac{3}{2}\frac{(\zeta(4)-1/6\psi^{(3)}(n+1))}{n}, \tag{6.249}$$

and if we multiply both sides of (6.249) by n, then differentiate once with respect to n, set $n = -1/2$, and rearrange, we have

$$\sum_{k=1}^{\infty} \frac{H_k^3 - 3H_k H_k^{(2)} + 2H_k^{(3)}}{(2k+1)^2} = \sum_{k=1}^{\infty} \frac{H_k^3}{(2k+1)^2} - 3\sum_{k=1}^{\infty} \frac{H_k H_k^{(2)}}{(2k+1)^2} + 2\sum_{k=1}^{\infty} \frac{H_k^{(3)}}{(2k+1)^2}$$

$$= \frac{1}{16} \frac{d}{dn} \bigg((\psi(n+1) + \gamma)^4 + 6(\psi(n+1) + \gamma)^2 (\zeta(2) - \psi^{(1)}(n+1))$$

$$+ 8(\psi(n+1) + \gamma)(\zeta(3) + 1/2\psi^{(2)}(n+1)) + 3(\zeta(2) - \psi^{(1)}(n+1))^2$$

$$+ 6(\zeta(4) - 1/6\psi^{(3)}(n+1)) \bigg) \bigg|_{n=-1/2}$$

$$= \frac{93}{2}\zeta(5) - \frac{39}{2}\zeta(2)\zeta(3) - 6\log^3(2)\zeta(2) + 21\log^2(2)\zeta(3) - \frac{45}{2}\log(2)\zeta(4),$$

$$(6.250)$$

where the limit can be calculated either by using *Mathematica* or manually, exploiting the Polygamma values in (6.146) and (6.147), Sect. 6.19.

Finally, plugging the values of the series from the points (ii) and (iv) in (6.250), we conclude that

$$\sum_{n=1}^{\infty} \frac{H_n^3}{(2n+1)^2}$$

$$= \frac{7}{4}\zeta(2)\zeta(3) - \frac{1271}{32}\zeta(5) + \frac{183}{8}\log(2)\zeta(4) - 2\log^3(2)\zeta(2) - \frac{2}{5}\log^5(2)$$

$$+ 48\operatorname{Li}_5\left(\frac{1}{2}\right),$$

and the solution to the point (v) of the problem is complete.

Finding simple solutions to derive series like the last two ones remains an appealing ongoing project the curious reader might like to consider!

6.35   On the Calculation of Another Essential Harmonic Series of Weight 5, Involving Harmonic Numbers of the Type H_{2n}

Solution The harmonic series in this section also appears as a main series in the article *On the calculation of two essential harmonic series with a weight 5 structure, involving harmonic numbers of the type H_{2n}*, which was published in the *Journal of Classical Analysis*, Vol. 16, No. 1, 2020, besides the one given in Sect. 4.33.

The solution will follow the calculations given in the mentioned paper. Similar ideas and *tools* are the basis of the solution to the result in Sect. 4.25, where one may find the corresponding series from the weight 4 series class.

Now, recall that $\displaystyle\sum_{k=1}^{\infty} \frac{H_k}{(k+1)(k+n+1)} = \frac{H_n^2 + H_n^{(2)}}{2n}$, which is a particular case of a generalization in [76, Chapter 4, p.289], where if we replace n by $2n$, then multiply both sides by $1/n^2$, and consider the summation from $n = 1$ to ∞, we have

$$2\sum_{n=1}^{\infty} \frac{H_{2n}^2}{(2n)^3} + 2\sum_{n=1}^{\infty} \frac{H_{2n}^{(2)}}{(2n)^3} = \sum_{n=1}^{\infty}\left(\sum_{k=1}^{\infty} \frac{H_k}{(k+1)(k+2n+1)n^2}\right)$$

{split the series according to the parity of the variable k}

$$= \frac{1}{4}\sum_{n=1}^{\infty}\left(\sum_{k=1}^{\infty} \frac{H_{2k-1}}{k(k+n)n^2}\right) + \sum_{n=1}^{\infty}\left(\sum_{k=1}^{\infty} \frac{H_{2k}}{(2k+1)(2k+2n+1)n^2}\right)$$

$$\overset{\substack{\text{reverse the order}\\\text{of summations}}}{=} \frac{1}{4}\sum_{k=1}^{\infty}\left(\sum_{n=1}^{\infty} \frac{H_{2k-1}}{k(k+n)n^2}\right) + \sum_{k=1}^{\infty} \frac{H_{2k}}{2k+1}\left(\sum_{n=1}^{\infty} \frac{1}{(2k+2n+1)n^2}\right)$$

$$= \frac{1}{4}\sum_{k=1}^{\infty}\left(\sum_{n=1}^{\infty} \frac{H_{2k-1}}{k^2 n^2}\right) - \frac{1}{4}\sum_{k=1}^{\infty} \frac{H_{2k-1}}{k^2}\left(\sum_{n=1}^{\infty} \frac{1}{n(k+n)}\right) + \sum_{k=1}^{\infty} \frac{H_{2k}}{(2k+1)^2}\left(\sum_{n=1}^{\infty} \frac{1}{n^2}\right)$$

$$- 4\sum_{k=1}^{\infty} \frac{H_{2k}}{(2k+1)^2}\left(\sum_{n=1}^{\infty} \frac{1}{(2k+2n+1)2n}\right)$$

$$= \frac{\zeta(2)}{4}\sum_{k=1}^{\infty} \frac{H_{2k}-1/(2k)}{k^2} - \frac{1}{4}\sum_{k=1}^{\infty} \frac{H_k(H_{2k}-1/(2k))}{k^3} + \zeta(2)\sum_{k=1}^{\infty} \frac{H_{2k+1}-1/(2k+1)}{(2k+1)^2}$$

$$- 4\sum_{k=1}^{\infty} \frac{H_{2k}}{(2k+1)^2}\left(\frac{1}{(2k+1)^2} + \frac{H_{2k}}{2k+1} - \frac{H_k}{2(2k+1)} - \log(2)\frac{1}{2k+1}\right)$$

$$\overset{k=n}{=} \frac{1}{8}\sum_{n=1}^{\infty} \frac{H_n}{n^4} + \zeta(2)\sum_{n=1}^{\infty} \frac{H_{2n}}{(2n)^2} + \zeta(2)\sum_{n=1}^{\infty} \frac{H_{2n+1}}{(2n+1)^2} + 4\log(2)\sum_{n=1}^{\infty} \frac{H_{2n+1}}{(2n+1)^3}$$

$$+ 4\sum_{n=1}^{\infty} \frac{H_{2n+1}}{(2n+1)^4} - 4\sum_{n=1}^{\infty} \frac{H_{2n+1}^2}{(2n+1)^3} + 2\sum_{n=1}^{\infty} \frac{H_n H_{2n}}{(2n+1)^3} - \frac{1}{4}\sum_{n=1}^{\infty} \frac{H_n H_{2n}}{n^3}$$

$$+ \zeta(2) - \zeta(2)\zeta(3) - \frac{15}{4}\log(2)\zeta(4) + 4\log(2),$$

and exploiting the simple facts that $\sum_{n=1}^{\infty} a_n = a_1 + \sum_{n=1}^{\infty} a_{2n} + \sum_{n=1}^{\infty} a_{2n+1}$ and

$\sum_{n=1}^{\infty} a_{2n+1} = -a_1 + \frac{1}{2}\left(\sum_{n=1}^{\infty} a_n + \sum_{n=1}^{\infty}(-1)^{n-1}a_n\right)$, which are applicable here since the series are absolutely convergent, we arrive at

$$\sum_{n=1}^{\infty} \frac{H_n H_{2n}}{(2n+1)^3}$$

$$= \frac{1}{2}\sum_{n=1}^{\infty} \frac{H_n^{(2)}}{n^3} + \frac{3}{2}\sum_{n=1}^{\infty} \frac{H_n^2}{n^3} - \frac{1}{2}\zeta(2)\sum_{n=1}^{\infty} \frac{H_n}{n^2} - \log(2)\sum_{n=1}^{\infty} \frac{H_n}{n^3} - \frac{17}{16}\sum_{n=1}^{\infty} \frac{H_n}{n^4} + \frac{1}{8}\sum_{n=1}^{\infty} \frac{H_n H_{2n}}{n^3}$$

$$- \log(2)\sum_{n=1}^{\infty} (-1)^{n-1}\frac{H_n}{n^3} - \sum_{n=1}^{\infty}(-1)^{n-1}\frac{H_n}{n^4} - \frac{1}{2}\sum_{n=1}^{\infty}(-1)^{n-1}\frac{H_n^{(2)}}{n^3} + \frac{1}{2}\sum_{n=1}^{\infty}(-1)^{n-1}\frac{H_n^2}{n^3}$$

$$+ \frac{15}{8}\log(2)\zeta(4) + \frac{1}{2}\zeta(2)\zeta(3)$$

$$= \frac{1}{12}\log^5(2) - \frac{1}{2}\log^3(2)\zeta(2) + \frac{7}{4}\log^2(2)\zeta(3) - \frac{17}{8}\log(2)\zeta(4) + \frac{31}{128}\zeta(5)$$

$$+ 2\log(2)\,\mathrm{Li}_4\left(\frac{1}{2}\right),$$

where during the calculations I used $\sum_{n=1}^{\infty} \frac{1}{(2k+2n+1)2n} = \frac{1}{(2k+1)^2} + \frac{H_{2k}}{2k+1} -$
$\frac{H_k}{2(2k+1)} - \frac{\log(2)}{2k+1}$, which is straightforward to obtain if we employ the partial
fraction expansion, add and subtract $1/(2n+1)$ inside the summand, and then split
the series (the details of such an approach may be found in [76, Chapter 6, p.531]);
then the Euler sum generalization in (6.149), the cases $n = 2, 3, 4$, Sect. 6.19; next
the results in (4.103) and (4.104), Sect. 4.20, or for the latter result, we may use the
alternating Euler sum generalization in (4.105), the case $m = 2$, Sect. 4.21; next
$\sum_{n=1}^{\infty} \frac{H_n^{(2)}}{n^3} = 3\zeta(2)\zeta(3) - \frac{9}{2}\zeta(5)$ (see [76, Chapter 6, p.386]), $\sum_{n=1}^{\infty} \frac{H_n^2}{n^3} = \frac{7}{2}\zeta(5) -$
$\zeta(2)\zeta(3)$ (see [76, Chapter 4, p.293]); further $\sum_{n=1}^{\infty}(-1)^{n-1}\frac{H_n^{(2)}}{n^3} = \frac{5}{8}\zeta(2)\zeta(3) -$
$\frac{11}{32}\zeta(5)$, $\sum_{n=1}^{\infty}(-1)^{n-1}\frac{H_n^2}{n^3} = \frac{2}{15}\log^5(2) - \frac{11}{8}\zeta(2)\zeta(3) - \frac{19}{32}\zeta(5) + \frac{7}{4}\log^2(2)\zeta(3) -$

$$\frac{2}{3} \log^3(2)\zeta(2) + 4 \log(2) \operatorname{Li}_4\left(\frac{1}{2}\right) + 4 \operatorname{Li}_5\left(\frac{1}{2}\right) \text{ (see [76, Chapter 4, p.311]); and}$$

last but not least, the *essential* harmonic series in (4.124), Sect. 4.33.

The series in the present section together with a part of the ones in the previous two sections may also serve, as you'll see, as powerful auxiliary results during the derivations of the unusual series involving the tail of the Riemann zeta function, related to the weight 5 harmonic series, in the next sections to come.

6.36   A First Unusual Series with the Tail of the Riemann Zeta Function $\zeta(3) - H_{2n}^{(3)}$, Related to Weight 5 Harmonic Series

Solution This section opens a series of seven sections that treat unusual (and spectacular) series involving the tail of the Riemann zeta function, related to weight 5 harmonic series. For their calculations, I'll get inspiration from the sections where their corresponding series of weight 4 may be found (e.g., for the present section, we may consider the strategy given in Sect. 6.26).

So, if we employ Abel's summation, the series version in (6.7), Sect. 6.2, where we set $a_n = H_n/n$ and $b_n = \zeta(3) - H_{2n}^{(3)}$, and we also consider the fact that $\sum_{k=1}^{n} \frac{H_k}{k} = \frac{1}{2}(H_n^2 + H_n^{(2)})$, which is the case $p = q = 1$ in (6.102), Sect. 6.13, then we get that

$$\sum_{n=1}^{\infty} \frac{H_n}{n}\left(\zeta(3) - H_{2n}^{(3)}\right)$$

$$= \frac{1}{2} \sum_{n=1}^{\infty} \frac{H_n^2 + H_n^{(2)}}{(2n+1)^3} + \frac{1}{16} \sum_{n=1}^{\infty} \frac{(H_{n+1} - 1/(n+1))^2 + (H_{n+1}^{(2)} - 1/(n+1)^2)}{(n+1)^3}$$

{reindex the second series and wisely expand it}

$$= \frac{1}{2} \sum_{n=1}^{\infty} \frac{H_n^2}{(2n+1)^3} + \frac{1}{2} \sum_{n=1}^{\infty} \frac{H_n^{(2)}}{(2n+1)^3} + \frac{1}{16} \sum_{n=1}^{\infty} \frac{H_n^2}{n^3} + \frac{1}{16} \sum_{n=1}^{\infty} \frac{H_n^{(2)}}{n^3} - \frac{1}{8} \sum_{n=1}^{\infty} \frac{H_n}{n^4}$$

$$= \frac{23}{8}\zeta(2)\zeta(3) - \frac{69}{16}\zeta(5) + \frac{7}{4}\log^2(2)\zeta(3) - \frac{45}{16}\log(2)\zeta(4),$$

where the first two harmonic series are given in (4.125), (4.127), Sect. 4.34; next

$$\sum_{n=1}^{\infty} \frac{H_n^{(2)}}{n^3} = 3\zeta(2)\zeta(3) - \frac{9}{2}\zeta(5) \text{ (see [76, Chapter 6, p.386]),} \sum_{n=1}^{\infty} \frac{H_n^2}{n^3} = \frac{7}{2}\zeta(5) -$$

$\zeta(2)\zeta(3)$ (see [76, Chapter 4, p.293]); then the Euler sum generalization in (6.149), the case $n = 4$, Sect. 6.19, and the solution is complete.

Thanks to the preparation in advance of the results in Sect. 4.34, the whole flow of the solution is short, and we immediately arrive at the desired value!

6.37 A Second Unusual Series with the Tail of the Riemann Zeta Function $\zeta(3) - H_n^{(3)}$, Related to Weight 5 Harmonic Series

Solution I would like to emphasize from the very beginning that the calculations in this section will bring to the surface some wonderful advanced logarithmic integrals, which happily are given in the first chapter and calculated in the third chapter, thus reducing the efforts needed in this section.

So, we'll want to follow the ideas for the corresponding series from the weight 4 series class found in Sect. 6.27, and upon considering the result in (6.201), Sect. 6.27, with $p = 3$, we get

$$\zeta(3) - H_n^{(3)} = \zeta(3) - 1 - \frac{1}{2^3} - \cdots - \frac{1}{n^3} = \frac{1}{2}\sum_{k=1}^{\infty}\int_0^1 x^{k+n-1}\log^2(x)dx$$

$$= \frac{1}{2}\int_0^1 \frac{x^n\log^2(x)}{1-x}dx \stackrel{x=y^2}{=} 4\int_0^1 \frac{y^{2n+1}\log^2(y)}{1-y^2}dy. \tag{6.251}$$

Then, by using (6.251), we write

$$\sum_{n=1}^{\infty} \frac{H_{2n}}{n}\left(\zeta(3) - H_n^{(3)}\right) = 8\sum_{n=1}^{\infty} H_{2n}\int_0^1 x^{2n-1}dx\int_0^1 \frac{y^{2n+1}\log^2(y)}{1-y^2}dy$$

{reverse the order of integration and summation}

$$= 8\int_0^1\left(\int_0^1 \frac{y\log^2(y)}{x(1-y^2)}\sum_{n=1}^{\infty}(xy)^{2n}H_{2n}dx\right)dy$$

$$\left\{\text{use }\sum_{n=1}^{\infty}x^{2n}H_{2n} = \frac{2x\operatorname{arctanh}(x) - \log(1-x^2)}{2(1-x^2)}, \text{based on (4.32)}, m = 1, \text{Sect. 4.6}\right\}$$

$$= 4 \int_0^1 \frac{y \log^2(y)}{1 - y^2} \left(\int_0^1 \frac{2xy \operatorname{arctanh}(xy) - \log(1 - (xy)^2)}{x(1 - (xy)^2)} dx \right) dy$$

$$= 4 \int_0^1 \frac{y \log^2(y)}{1 - y^2} \left(\operatorname{arctanh}^2(xy) + \frac{1}{4} \log^2(1 - (xy)^2) + \frac{1}{2} \operatorname{Li}_2\left((xy)^2\right) \Big|_{x=0}^{x=1} \right) dy$$

{expand the integral after taking the limits of integration}

$$= 4 \int_0^1 \frac{y \log^2(y) \operatorname{arctanh}^2(y)}{1 - y^2} dy + \int_0^1 \frac{y \log^2(y) \log^2(1 - y^2)}{1 - y^2} dy$$

$$+ 2 \int_0^1 \frac{y \log^2(y) \operatorname{Li}_2(y^2)}{1 - y^2} dy$$

$\left\{ \text{in the last two integrals let } y^2 = z \text{ and then return to the notation in } y \right\}$

$$= 4 \underbrace{\int_0^1 \frac{y \log^2(y) \operatorname{arctanh}^2(y)}{1 - y^2} dy}_{I_1} + \frac{1}{8} \underbrace{\int_0^1 \frac{\log^2(y) \log^2(1 - y)}{1 - y} dy}_{I_2}$$

$$+ \frac{1}{4} \underbrace{\int_0^1 \frac{\log^2(y) \operatorname{Li}_2(y)}{1 - y} dy}_{I_3}. \tag{6.252}$$

For the integral I_1 in (6.252), we have

$$I_1 = \int_0^1 \frac{y \log^2(y) \operatorname{arctanh}^2(y)}{1 - y^2} dy \overset{(1-y)/(1+y)=t}{=} \frac{1}{8} \int_0^1 \frac{(1 - t) \log^2\left(\frac{1 - t}{1 + t}\right) \log^2(t)}{t(1 + t)} dt$$

$$= \frac{1}{8} \int_0^1 \frac{\log^2(1 - t) \log^2(t)}{t} dt + \frac{1}{8} \int_0^1 \frac{\log^2(1 + t) \log^2(t)}{t} dt$$

$$- \frac{1}{4} \int_0^1 \frac{\log^2(1 + t) \log^2(t)}{1 + t} dt - \frac{1}{4} \int_0^1 \frac{\log^2(1 - t) \log^2(t)}{1 + t} dt$$

$$- \frac{1}{4} \int_0^1 \frac{\log(1 - t) \log^2(t) \log(1 + t)}{t} dt + \frac{1}{2} \int_0^1 \frac{\log(1 - t) \log^2(t) \log(1 + t)}{1 + t} dt. \tag{6.253}$$

Next, for the first resulting integral in (6.253), we use (4.33), Sect. 4.6, that gives

$$\int_0^1 \frac{\log^2(1-t)\log^2(t)}{t}dt = 2\int_0^1 \log^2(t)\sum_{n=1}^\infty t^n \frac{H_n}{n+1}dt$$

$$= 2\sum_{n=1}^\infty \frac{H_n}{n+1}\int_0^1 t^n \log^2(t)dt = 4\sum_{n=1}^\infty \frac{H_n}{(n+1)^4} = 8\zeta(5) - 4\zeta(2)\zeta(3),$$

(6.254)

where the resulting harmonic series is the case $n = 4$ of the generalization in (6.148), Sect. 6.19.

Then, for the second resulting integral in (6.253), we use again (4.33), Sect. 4.6, which leads to

$$\int_0^1 \frac{\log^2(1+t)\log^2(t)}{t}dt = 2\int_0^1 \log^2(t)\sum_{n=1}^\infty (-1)^{n-1}t^n \frac{H_n}{n+1}dt$$

$$= 2\sum_{n=1}^\infty (-1)^{n-1}\frac{H_n}{n+1}\int_0^1 t^n \log^2(t)dt = 4\sum_{n=1}^\infty (-1)^{n-1}\frac{H_{n+1}-1/(n+1)}{(n+1)^4}$$

$$\overset{\substack{\text{reindex and}\\\text{expand}}}{=} 4\sum_{n=1}^\infty (-1)^{n-1}\frac{1}{n^5} - 4\sum_{n=1}^\infty (-1)^{n-1}\frac{H_n}{n^4} = 2\zeta(2)\zeta(3) - \frac{29}{8}\zeta(5), \quad (6.255)$$

where in the calculations I used the result in (4.105), the case $m = 2$, Sect. 4.21.

As regards the third resulting integral in (6.253), we rearrange it by applying once integration by parts, and then we arrive at

$$\int_0^1 \frac{\log^2(1+t)\log^2(t)}{1+t}dt = \frac{1}{3}\int_0^1 (\log^3(1+t))'\log^2(t)dt$$

$$= -\frac{2}{3}\int_0^1 \frac{\log^3(1+t)\log(t)}{t}dt. \quad (6.256)$$

If we plug in (6.253) the results from (6.254), (6.255), and (6.256), afterwards the ones in (1.198), (1.200), and (1.199), Sect. 1.45, and finally
$\int_0^1 \frac{\log(1-t)\log^2(t)\log(1+t)}{t}dt = \frac{3}{4}\zeta(2)\zeta(3) - \frac{27}{16}\zeta(5)$ (see [76, Chapter 1, p.6]), we obtain that

$$I_1 = \int_0^1 \frac{y\log^2(y)\operatorname{arctanh}^2(y)}{1-y^2}dy$$

$$= \frac{217}{32}\zeta(5) + \frac{7}{16}\zeta(2)\zeta(3) - \frac{53}{16}\log(2)\zeta(4) + \frac{1}{3}\log^3(2)\zeta(2) - \frac{1}{10}\log^5(2)$$

$$- 4\log(2)\operatorname{Li}_4\left(\frac{1}{2}\right) - 8\operatorname{Li}_5\left(\frac{1}{2}\right). \tag{6.257}$$

Let's proceed with the next integral in (6.252), I_2, and then we write

$$I_2 = \int_0^1 \frac{\log^2(y)\log^2(1-y)}{1-y}dy \overset{1-y=t}{=} \int_0^1 \frac{\log^2(1-t)\log^2(t)}{t}dt$$

$$= 8\zeta(5) - 4\zeta(2)\zeta(3) \tag{6.258}$$

where I used the integral result in (6.254).

Finally, for the integral I_3 in (6.252), we use the generating function in (4.32), the case $m = 2$, Sect. 4.6,

$$I_3 = \int_0^1 \frac{\log^2(y)\operatorname{Li}_2(y)}{1-y}dy = \int_0^1 \log^2(y)\sum_{n=1}^{\infty} y^n H_n^{(2)}dy = \sum_{n=1}^{\infty} H_n^{(2)}\int_0^1 y^n \log^2(y)dy$$

$$= 2\sum_{n=1}^{\infty} \frac{H_{n+1}^{(2)} - 1/(n+1)^2}{(n+1)^3} = 2\sum_{n=1}^{\infty} \frac{H_n^{(2)}}{n^3} - 2\sum_{n=1}^{\infty} \frac{1}{n^5} = 6\zeta(2)\zeta(3) - 11\zeta(5),$$

$$\tag{6.259}$$

where in the calculations I used $\displaystyle\sum_{n=1}^{\infty} \frac{H_n^{(2)}}{n^3} = 3\zeta(2)\zeta(3) - \frac{9}{2}\zeta(5)$ (see [76, Chapter 6, p.386]).

Collecting the values of the integrals I_1, I_2, and I_3 from (6.257), (6.258), and (6.259) in (6.252), we conclude that

$$\sum_{n=1}^{\infty} \frac{H_{2n}}{n}\left(\zeta(3) - H_n^{(3)}\right)$$

$$= \frac{203}{8}\zeta(5) + \frac{11}{4}\zeta(2)\zeta(3) - \frac{53}{4}\log(2)\zeta(4) + \frac{4}{3}\log^3(2)\zeta(2) - \frac{2}{5}\log^5(2)$$

$$- 16\log(2)\operatorname{Li}_4\left(\frac{1}{2}\right) - 32\operatorname{Li}_5\left(\frac{1}{2}\right),$$

and the solution to the main series is finalized.

Good to keep in mind the starting strategy in this section while preparing to attack the series from the next section!

6.38 A Third Unusual Series with the Tail of the Riemann Zeta Function $\zeta(3) - H_{2n}^{(3)}$, Related to Weight 5 Harmonic Series

Solution As suggested at the end of the previous section, we'll want to consider a starting strategy similar to the one there. Observe that compared to the previous two sections, this time, the double argument $2n$ will be found in both harmonic numbers of the summand. More exactly, now we have H_{2n} and $H_{2n}^{(3)}$.

The solution to the corresponding series from the weight 4 series class found in Sect. 6.28 may be seen as a guiding light for building the present solution.

First, let's see that if we consider (6.251), Sect. 6.37, where we replace n by $2n$, we have that

$$\zeta(3) - H_{2n}^{(3)} = \frac{1}{2} \int_0^1 \frac{x^{2n} \log^2(x)}{1-x} dx. \qquad (6.260)$$

Then, by employing the result in (6.260), the main series becomes

$$\sum_{n=1}^{\infty} \frac{H_{2n}}{n} \left(\zeta(3) - H_{2n}^{(3)} \right) = \sum_{n=1}^{\infty} H_{2n} \int_0^1 y^{2n-1} dy \int_0^1 \frac{x^{2n} \log^2(x)}{1-x} dx$$

{reverse the order of integration and summation}

$$= \int_0^1 \left(\int_0^1 \frac{\log^2(x)}{y(1-x)} \sum_{n=1}^{\infty} (xy)^{2n} H_{2n} dy \right) dx$$

$$\left\{ \text{use that } \sum_{n=1}^{\infty} x^{2n} H_{2n} = \frac{2x \operatorname{arctanh}(x) - \log(1-x^2)}{2(1-x^2)} \right\}$$

$$= \frac{1}{2} \int_0^1 \frac{\log^2(x)}{1-x} \left(\int_0^1 \frac{2xy \operatorname{arctanh}(xy) - \log(1-(xy)^2)}{y(1-(xy)^2)} dy \right) dx$$

$$= \frac{1}{2} \int_0^1 \frac{\log^2(x)}{1-x} \left(\operatorname{arctanh}^2(xy) + \frac{1}{4} \log^2(1-(xy)^2) + \frac{1}{2} \operatorname{Li}_2\left((xy)^2\right) \Big|_{y=0}^{y=1} \right) dx$$

{expand the integral after taking the limits of integration}

$$= \frac{1}{2} \int_0^1 \frac{\log^2(x) \operatorname{arctanh}^2(x)}{1-x} dx + \frac{1}{8} \int_0^1 \frac{\log^2(x) \log^2(1-x^2)}{1-x} dx$$

$$+\frac{1}{4}\int_0^1 \frac{\log^2(x)\,\mathrm{Li}_2(x^2)}{1-x}dx$$

{for the last integral use the identity $1/2\,\mathrm{Li}_2(x^2) = \mathrm{Li}_2(x)+\mathrm{Li}_2(-x)$ and rearrange}

$$=\frac{1}{4}\int_0^1 \frac{\log^2(x)\log^2(1-x)}{1-x}dx + \frac{1}{2}\int_0^1 \frac{\log^2(x)\,\mathrm{Li}_2(x)}{1-x}dx$$

$$+\frac{1}{2}\int_0^1 \frac{\log^2(x)(1/2\log^2(1+x)+\mathrm{Li}_2(-x))}{1-x}dx$$

{the first two integrals are calculated in the previous section in (6.258) and (6.259)}

$$= 2\zeta(2)\zeta(3)-\frac{7}{2}\zeta(5)+\frac{1}{2}\int_0^1 \frac{\log^2(x)(1/2\log^2(1+x)+\mathrm{Li}_2(-x))}{1-x}dx. \qquad (6.261)$$

Since $\displaystyle\sum_{n=1}^{\infty}(-1)^{n-1}x^n\frac{H_n}{n} = -\frac{1}{2}\log^2(1+x)-\mathrm{Li}_2(-x)$, as you can see in (6.214), Sect. 6.28, for the remaining integral in (6.261), we write

$$\int_0^1 \frac{\log^2(x)(1/2\log^2(1+x)+\mathrm{Li}_2(-x))}{1-x}dx = -\int_0^1 \frac{\log^2(x)}{1-x}\sum_{n=1}^{\infty}(-1)^{n-1}x^n\frac{H_n}{n}dx$$

{reverse the order of summation and integration}

$$= -\sum_{n=1}^{\infty}(-1)^{n-1}\frac{H_n}{n}\int_0^1 \frac{x^n\log^2(x)}{1-x}dx$$

$$\left\{\text{exploit the result in (6.251), Sect. 6.37,}\ \int_0^1 \frac{x^n\log^2(x)}{1-x}dx = 2(\zeta(3)-H_n^{(3)})\right\}$$

$$= -2\sum_{n=1}^{\infty}(-1)^{n-1}\frac{H_n}{n}(\zeta(3)-H_n^{(3)})$$

$$= -2\zeta(3)\ \underbrace{\sum_{n=1}^{\infty}(-1)^{n-1}\frac{H_n}{n}}_{1/2(\zeta(2)-\log^2(2))}\ +2\sum_{n=1}^{\infty}(-1)^{n-1}\frac{H_n H_n^{(3)}}{n}$$

$$= \frac{1}{3}\log^3(2)\zeta(2) + \frac{7}{4}\log^2(2)\zeta(3) - \frac{49}{8}\log(2)\zeta(4) + \frac{167}{16}\zeta(5) - \frac{9}{8}\zeta(2)\zeta(3)$$

$$- \frac{1}{10} \log^5(2) - 4 \log(2) \operatorname{Li}_4\left(\frac{1}{2}\right) - 8 \operatorname{Li}_5\left(\frac{1}{2}\right), \qquad (6.262)$$

where the last equality is obtained by using the alternating harmonic series of weight 5, $\displaystyle\sum_{n=1}^{\infty} (-1)^{n-1} \frac{H_n H_n^{(3)}}{n} = \frac{1}{6} \log^3(2)\zeta(2) + \frac{3}{8} \log^2(2)\zeta(3) - \frac{49}{16} \log(2)\zeta(4) +$

$\dfrac{167}{32}\zeta(5) - \dfrac{1}{16}\zeta(2)\zeta(3) - \dfrac{1}{20}\log^5(2) - 2\log(2)\operatorname{Li}_4\left(\dfrac{1}{2}\right) - 4\operatorname{Li}_5\left(\dfrac{1}{2}\right)$, which happily
is found and calculated in [76, Chapter 6, Section 6.58, pp. 523–529].

Finally, by combining (6.261) and (6.262), we conclude that

$$\sum_{n=1}^{\infty} \frac{H_{2n}}{n}\left(\zeta(3) - H_{2n}^{(3)}\right)$$

$$= \frac{1}{6}\log^3(2)\zeta(2) + \frac{7}{8}\log^2(2)\zeta(3) - \frac{49}{16}\log(2)\zeta(4) + \frac{55}{32}\zeta(5) + \frac{23}{16}\zeta(2)\zeta(3)$$

$$- \frac{1}{20}\log^5(2) - 2\log(2)\operatorname{Li}_4\left(\frac{1}{2}\right) - 4\operatorname{Li}_5\left(\frac{1}{2}\right),$$

and the solution to the main series is finalized.

If you have reacted by saying something like *This was a short solution!*, I would like to point out that this was possible due to the fact that the present solution naturally and fast leads to an advanced alternating harmonic series of weight 5 which is calculated in *(Almost) Impossible Integrals, Sums, and Series*, as you can see above.

If you enjoy these kinds of problems and are ready for more, let's pass to the next section where another similar series is waiting for us!

6.39 A Fourth Unusual Series with the Tail of the Riemann Zeta Function $\zeta(3) - H_n^{(3)}$, Related to Weight 5 Harmonic Series

Solution We want to *reload* the ideas we met during the calculation of the corresponding series from the weight 4 class found in Sect. 6.29. Compared to the previous three sections involving series with tails of the Riemann zeta function, we observe that now the denominator of the fraction in front of the tail is $2n + 1$.

A powerful way of approaching the series is to use the identity in (4.93), Sect. 4.17, where if we multiply its both sides by $1/n^2$ and consider the summation from $n = 1$ to ∞, then for the left-hand side, we get

$$\sum_{n=1}^{\infty} \frac{1}{n^2} \left(\sum_{k=1}^{\infty} \frac{H_{2k}}{(k+1)(k+n+1)} \right) = \sum_{k=1}^{\infty} \frac{H_{2k}}{k+1} \left(\sum_{n=1}^{\infty} \frac{1}{n^2(n+k+1)} \right)$$

$$= \sum_{k=1}^{\infty} \frac{H_{2k}}{(k+1)^2} \left(\underbrace{\sum_{n=1}^{\infty} \frac{1}{n^2}}_{\zeta(2)} - \sum_{n=1}^{\infty} \frac{1}{n(n+k+1)} \right)$$

$$\left\{ \text{employ the fact that } \sum_{n=1}^{\infty} \frac{1}{n(n+k)} = \frac{H_k}{k} \text{ and then expand the outer series} \right\}$$

$$= \zeta(2) \sum_{k=1}^{\infty} \frac{H_{2k}}{(k+1)^2} - \sum_{k=1}^{\infty} \frac{H_{2k} H_{k+1}}{(k+1)^3}$$

$$= \zeta(2) \sum_{k=1}^{\infty} \frac{H_{2k+2} - 1/(2k+1) - 1/(2k+2)}{(k+1)^2}$$

$$- \sum_{k=1}^{\infty} \frac{(H_{2k+2} - 1/(2k+1) - 1/(2k+2)) H_{k+1}}{(k+1)^3}$$

$$\{\text{reindex the series and then expand them}\}$$

$$= \frac{5}{2}\zeta(4) - 4\log(2)\zeta(2) - \frac{1}{2}\zeta(2)\zeta(3) + 8\sum_{k=1}^{\infty} \frac{H_k}{2k(2k-1)} - 2\sum_{k=1}^{\infty} \frac{H_k}{k^2}$$

$$- \sum_{k=1}^{\infty} \frac{H_k}{k^3} + \frac{1}{2} \sum_{k=1}^{\infty} \frac{H_k}{k^4} + 4\zeta(2) \sum_{k=1}^{\infty} \frac{H_{2k}}{(2k)^2} - \sum_{k=1}^{\infty} \frac{H_k H_{2k}}{k^3}$$

$$\left\{ \text{for the fifth series use that } \sum_{n=1}^{\infty} a_{2n} = \frac{1}{2} \left(\sum_{n=1}^{\infty} a_n - \sum_{n=1}^{\infty} (-1)^{n-1} a_n \right) \right\}$$

$$= \frac{5}{2}\zeta(4) - 4\log(2)\zeta(2) - \frac{1}{2}\zeta(2)\zeta(3) + 8\sum_{k=1}^{\infty} \frac{H_k}{2k(2k-1)} + 2(\zeta(2)-1) \sum_{k=1}^{\infty} \frac{H_k}{k^2}$$

$$- \sum_{k=1}^{\infty} \frac{H_k}{k^3} + \frac{1}{2} \sum_{k=1}^{\infty} \frac{H_k}{k^4} - 2\zeta(2) \sum_{k=1}^{\infty} (-1)^{k-1} \frac{H_k}{k^2} - \sum_{k=1}^{\infty} \frac{H_k H_{2k}}{k^3}$$

$$= \frac{8}{15} \log^5(2) - 8\log^2(2) + 16\log(2) - \left(4\log(2) + \frac{8}{3}\log^3(2)\right)\zeta(2)$$

$$+ 7\log^2(2)\zeta(3) - 4\zeta(3) + \frac{5}{4}\zeta(4) - \frac{283}{16}\zeta(5) + \frac{9}{4}\zeta(2)\zeta(3)$$

$$+ 16\log(2)\operatorname{Li}_4\left(\frac{1}{2}\right) + 16\operatorname{Li}_5\left(\frac{1}{2}\right), \tag{6.263}$$

where in the calculations above I used that $\displaystyle\sum_{k=1}^{\infty} \frac{H_k}{2k(2k-1)} = 2\log(2) - \log^2(2)$,

which is given with a solution under the result in (6.221), Sect. 6.29; then the Euler sum generalization in (6.149), Sect. 6.19, the cases $n = 2, 3, 4$; next the alternating Euler sum in (4.105), the case $m = 1$, Sect. 4.21; and finally the harmonic series in (4.124), Sect. 4.33.

Next, considering the summation from $n = 1$ to ∞ for the right-hand side of (4.93), Sect. 4.17, multiplied by $1/n^2$, we have

$$16\log(2)\sum_{n=1}^{\infty} \frac{H_{2n}}{(2n)^3} - \log(2)\sum_{n=1}^{\infty} \frac{H_n}{n^3} + \frac{1}{4}\sum_{n=1}^{\infty} \frac{H_n^2}{n^3} + \frac{1}{4}\sum_{n=1}^{\infty} \frac{H_n^{(2)}}{n^3}$$

$$- 4\log(2)\underbrace{\sum_{n=1}^{\infty} \frac{1}{n^2(2n+1)}}_{\zeta(2) + 4\log(2) - 4} + \sum_{n=1}^{\infty} \frac{1}{n^3}\sum_{i=1}^{n} \frac{H_i}{2i+1}$$

$$\left\{ \text{for the first series use that } \sum_{n=1}^{\infty} a_{2n} = \frac{1}{2}\left(\sum_{n=1}^{\infty} a_n - \sum_{n=1}^{\infty} (-1)^{n-1}a_n\right), \right\}$$

{and in the last series use the change of summation order of the type}

$$\left\{ \sum_{n=1}^{\infty}\sum_{i=1}^{n} f(i,n) = \sum_{i=1}^{\infty}\sum_{n=i}^{\infty} f(i,n), \text{ and then leave out the term for } n = i \right\}$$

$$= 16\log(2) - 16\log^2(2) - 4\log(2)\zeta(2) + 8\sum_{n=1}^{\infty} \frac{H_n}{2n(2n+1)} - 2\sum_{n=1}^{\infty} \frac{H_n}{n^2}$$

$$+ (7\log(2) + 1)\sum_{n=1}^{\infty} \frac{H_n}{n^3} - 8\log(2)\sum_{n=1}^{\infty}(-1)^{n-1}\frac{H_n}{n^3} + \frac{1}{4}\sum_{n=1}^{\infty} \frac{H_n^2}{n^3} + \frac{1}{4}\sum_{n=1}^{\infty} \frac{H_n^{(2)}}{n^3}$$

$$+ \sum_{i=1}^{\infty} \frac{H_i}{2i+1} \sum_{n=i+1}^{\infty} \frac{1}{n^3}$$

$$= 16 \log(2) - 8 \log^2(2) + \frac{2}{3} \log^5(2) - 4 \log(2)(1 + \log^2(2))\zeta(2) + 14 \log^2(2)\zeta(3)$$

$$- \frac{53}{4} \log(2)\zeta(4) - 4\zeta(3) + \frac{5}{4}\zeta(4) - \frac{1}{4}\zeta(5) + \frac{1}{2}\zeta(2)\zeta(3) + 16 \log(2) \operatorname{Li}_4\left(\frac{1}{2}\right)$$

$$+ \sum_{n=1}^{\infty} \frac{H_n}{2n+1} \left(\zeta(3) - H_n^{(3)}\right), \qquad (6.264)$$

where in the calculations I used (4.86), Sect. 4.14; then the Euler sum generalization in (6.149), Sect. 6.19, the cases $n = 2, 3$; next the famous alternating Euler sum in (4.103), Sect. 4.20; and finally $\sum_{n=1}^{\infty} \frac{H_n^2}{n^3} = \frac{7}{2}\zeta(5) - \zeta(2)\zeta(3)$ (see [76, Chapter 4, p.293]) and $\sum_{n=1}^{\infty} \frac{H_n^{(2)}}{n^3} = 3\zeta(2)\zeta(3) - \frac{9}{2}\zeta(5)$ (see [76, Chapter 6, p.386]).

Combining (6.263) and (6.264), we arrive at

$$\sum_{n=1}^{\infty} \frac{H_n}{2n+1} \left(\zeta(3) - H_n^{(3)}\right)$$

$$= \frac{7}{4}\zeta(2)\zeta(3) - \frac{279}{16}\zeta(5) + \frac{4}{3} \log^3(2)\zeta(2) - 7 \log^2(2)\zeta(3) + \frac{53}{4} \log(2)\zeta(4)$$

$$- \frac{2}{15} \log^5(2) + 16 \operatorname{Li}_5\left(\frac{1}{2}\right),$$

and the first solution is finalized.

The framework of the first solution is built by exploiting a key identity in a series form as seen above. Now, in the following, I'll construct a solution by exploiting a key identity in an integral form. So, we go back to Sect. 1.21, the identity in (1.103), where if we multiply both sides by $1/n^2$ and consider the summation from $n = 1$ to ∞, then for the left-hand side, we have

$$\sum_{n=1}^{\infty} \frac{1}{n^2} \int_0^1 x^{2n-1} \log^2(1+x)\mathrm{d}x = \int_0^1 \frac{\log^2(1+x)}{x} \sum_{n=1}^{\infty} \frac{x^{2n}}{n^2}\mathrm{d}x$$

$$= \int_0^1 \frac{\log^2(1+x)\operatorname{Li}_2(x^2)}{x}\mathrm{d}x$$

$$= \frac{4}{5}\log^5(2) - \frac{13}{4}\zeta(2)\zeta(3) - \frac{281}{16}\zeta(5) - 4\log^3(2)\zeta(2) + \frac{21}{2}\log^2(2)\zeta(3)$$

$$+ 24\log(2)\operatorname{Li}_4\left(\frac{1}{2}\right) + 24\operatorname{Li}_5\left(\frac{1}{2}\right), \tag{6.265}$$

where luckily the resulting challenging integral is found in (1.214), Sect. 1.47.

Further, taking into account the summation from $n = 1$ to ∞ for the right-hand side of (1.103), Sect. 1.21, multiplied by $1/n^2$, we get

$$16\log(2)\sum_{n=1}^{\infty}\frac{H_{2n}}{(2n)^3} - 4\sum_{n=1}^{\infty}\frac{H_{2n}^2}{(2n)^3} - 4\sum_{n=1}^{\infty}\frac{H_{2n}^{(2)}}{(2n)^3} - \log(2)\sum_{n=1}^{\infty}\frac{H_n}{n^3} + \frac{1}{4}\sum_{n=1}^{\infty}\frac{H_n^2}{n^3}$$

$$+ \frac{1}{4}\sum_{n=1}^{\infty}\frac{H_n^{(2)}}{n^3} + \sum_{n=1}^{\infty}\frac{1}{n^3}\sum_{k=1}^{n-1}\frac{H_k}{2k+1}$$

$$\left\{\text{for the first three series use that } \sum_{n=1}^{\infty}a_{2n} = \frac{1}{2}\left(\sum_{n=1}^{\infty}a_n - \sum_{n=1}^{\infty}(-1)^{n-1}a_n\right),\right\}$$

{and at the same time, in the last sum change the summation order}

$$= 7\log(2)\sum_{n=1}^{\infty}\frac{H_n}{n^3} - \frac{7}{4}\sum_{n=1}^{\infty}\frac{H_n^2}{n^3} - \frac{7}{4}\sum_{n=1}^{\infty}\frac{H_n^{(2)}}{n^3} - 8\log(2)\sum_{n=1}^{\infty}(-1)^{n-1}\frac{H_n}{n^3}$$

$$+ 2\sum_{n=1}^{\infty}(-1)^{n-1}\frac{H_n^2}{n^3} + 2\sum_{n=1}^{\infty}(-1)^{n-1}\frac{H_n^{(2)}}{n^3} + \sum_{k=1}^{\infty}\frac{H_k}{2k+1}\sum_{n=k+1}^{\infty}\frac{1}{n^3}$$

$$= \frac{14}{15}\log^5(2) - \frac{16}{3}\log^3(2)\zeta(2) + \frac{35}{2}\log^2(2)\zeta(3) - \frac{53}{4}\log(2)\zeta(4) - \frac{1}{8}\zeta(5)$$

$$- 5\zeta(2)\zeta(3) + 24\log(2)\operatorname{Li}_4\left(\frac{1}{2}\right) + 8\operatorname{Li}_5\left(\frac{1}{2}\right) + \sum_{n=1}^{\infty}\frac{H_n}{2n+1}\left(\zeta(3) - H_n^{(3)}\right), \tag{6.266}$$

where most of the resulting series appeared during the first solution, and beside them we also need that $\displaystyle\sum_{n=1}^{\infty}(-1)^{n-1}\frac{H_n^2}{n^3} = \frac{2}{15}\log^5(2) - \frac{11}{8}\zeta(2)\zeta(3) - \frac{19}{32}\zeta(5) +$

$\frac{7}{4}\log^2(2)\zeta(3) - \frac{2}{3}\log^3(2)\zeta(2) + 4\log(2)\operatorname{Li}_4\left(\frac{1}{2}\right) + 4\operatorname{Li}_5\left(\frac{1}{2}\right)$, which is given in

(3.331), Sect. 3.48, and then $\sum_{n=1}^{\infty} (-1)^{n-1} \dfrac{H_n^{(2)}}{n^3} = \dfrac{5}{8}\zeta(2)\zeta(3) - \dfrac{11}{32}\zeta(5)$ (see [76, Chapter 4, p.311]).

At last, when (6.265) and (6.266) come together, we immediately obtain that

$$\sum_{n=1}^{\infty} \frac{H_n}{2n+1}\left(\zeta(3) - H_n^{(3)}\right)$$

$$= \frac{7}{4}\zeta(2)\zeta(3) - \frac{279}{16}\zeta(5) + \frac{4}{3}\log^3(2)\zeta(2) - 7\log^2(2)\zeta(3) + \frac{53}{4}\log(2)\zeta(4)$$

$$- \frac{2}{15}\log^5(2) + 16\operatorname{Li}_5\left(\frac{1}{2}\right),$$

and the second solution is finalized.

It is always exciting to have more perspectives of approaching a problem! So far, we have two ways to go! And here is an *irresistible* question to address further: *What other ways would you like to try in order to solve the problem differently?*

6.40   A Fifth Unusual Series with the Tail of the Riemann Zeta Function $\zeta(3) - H_{2n}^{(3)}$, Related to Weight 5 Harmonic Series

Solution Recall that the solution to the corresponding series from the weight 4 series class in Sect. 6.30 was built by taking into account the given *challenging question*, which is similar to the one expected to be answered in the present section. So, we might want to proceed with similar steps to the ones considered in the solution of the section previously mentioned.

In order to perform the calculations, we first need to prepare a starting key result, that is, we want to observe that

$$\sum_{k=n}^{\infty} \frac{1}{(2k+1)^3} \overset{\substack{\text{reindex} \\ \text{the series}}}{=} \sum_{k=n+1}^{\infty} \frac{1}{(2k-1)^3} = \underbrace{\sum_{k=1}^{\infty} \frac{1}{(2k-1)^3}}_{7/8\zeta(3)} - \sum_{k=1}^{n} \frac{1}{(2k-1)^3}$$

$$\left\{ \text{use that } \sum_{k=1}^{n} \frac{1}{(2k-1)^3} = \sum_{k=1}^{2n} \frac{1}{k^3} - \sum_{k=1}^{n} \frac{1}{(2k)^3} = H_{2n}^{(3)} - \frac{1}{8}H_n^{(3)} \right\}$$

$$= \frac{7}{8}\zeta(3) - H_{2n}^{(3)} + \frac{1}{8}H_n^{(3)}$$

{next, the last result may be put in a form with two Riemann zeta tails}

$$= \zeta(3) - H_{2n}^{(3)} - \frac{1}{8}\left(\zeta(3) - H_n^{(3)}\right). \tag{6.267}$$

If we multiply the opposite sides of (6.267) by $\dfrac{H_n}{2n+1}$, consider the summation from $n = 1$ to ∞, and rearrange, we get that

$$\sum_{n=1}^{\infty} \frac{H_n}{2n+1}\left(\zeta(3) - H_{2n}^{(3)}\right)$$

$$= \frac{1}{8}\sum_{n=1}^{\infty} \frac{H_n}{2n+1}\left(\zeta(3) - H_n^{(3)}\right) + \sum_{n=1}^{\infty} \frac{H_n}{2n+1}\left(\sum_{k=n}^{\infty} \frac{1}{(2k+1)^3}\right)$$

{note the first series is calculated in the previous section}

$$= \frac{7}{32}\zeta(2)\zeta(3) - \frac{279}{128}\zeta(5) + \frac{1}{6}\log^3(2)\zeta(2) - \frac{7}{8}\log^2(2)\zeta(3) + \frac{53}{32}\log(2)\zeta(4)$$

$$- \frac{1}{60}\log^5(2) + 2\operatorname{Li}_5\left(\frac{1}{2}\right) + \sum_{n=1}^{\infty} \frac{H_n}{2n+1}\left(\sum_{k=n}^{\infty} \frac{1}{(2k+1)^3}\right). \tag{6.268}$$

Reversing the order of summation for the remaining series in (6.268) gives

$$\sum_{n=1}^{\infty} \frac{H_n}{2n+1}\left(\sum_{k=n}^{\infty} \frac{1}{(2k+1)^3}\right) = \sum_{k=1}^{\infty} \frac{1}{(2k+1)^3}\left(\sum_{n=1}^{k} \frac{H_n}{2n+1}\right)$$

{we may replace the variable letters by more convenient ones}

$$= \underbrace{\sum_{n=1}^{\infty} \frac{1}{(2n+1)^3}\left(\sum_{i=1}^{n} \frac{H_i}{2i+1}\right)}_{S}. \tag{6.269}$$

Further, we want to exploit the identity in (4.93), Sect. 4.17, to reduce the series in (6.269) to more doable ones, which is a second key result in our calculations.

So, upon multiplying both sides of (4.93), Sect. 4.17, by $n/(2n+1)^3$, considering the summation from $n = 1$ to ∞, and then rearranging, we arrive at

$$S = \sum_{n=1}^{\infty} \frac{1}{(2n+1)^3}\left(\sum_{i=1}^{n} \frac{H_i}{2i+1}\right)$$

$$= \sum_{n=1}^{\infty} \frac{n}{(2n+1)^3} \left(\sum_{k=1}^{\infty} \frac{H_{2k}}{(k+1)(k+n+1)} \right) + 4\log(2) \sum_{n=1}^{\infty} \frac{n}{(2n+1)^4}$$

$$- 2\log(2) \sum_{n=1}^{\infty} \frac{H_{2n}}{(2n+1)^3} + \log(2) \sum_{n=1}^{\infty} \frac{H_n}{(2n+1)^3} - \frac{1}{4} \sum_{n=1}^{\infty} \frac{H_n^2}{(2n+1)^3}$$

$$- \frac{1}{4} \sum_{n=1}^{\infty} \frac{H_n^{(2)}}{(2n+1)^3}$$

{swap the order of summation in the first series}

$$= \underbrace{\sum_{k=1}^{\infty} \frac{H_{2k}}{k+1} \left(\sum_{n=1}^{\infty} \frac{n}{(2n+1)^3(n+k+1)} \right)}_{S_1} - 2\log(2) \sum_{n=1}^{\infty} \frac{H_{2n+1}}{(2n+1)^3}$$

$$+ \log(2) \sum_{n=1}^{\infty} \frac{H_n}{(2n+1)^3} - \frac{1}{4} \sum_{n=1}^{\infty} \frac{H_n^2}{(2n+1)^3} - \frac{1}{4} \sum_{n=1}^{\infty} \frac{H_n^{(2)}}{(2n+1)^3}$$

$$- 2\log(2) + \frac{7}{4}\log(2)\zeta(3)$$

$$= S_1 + \frac{1}{12}\log^5(2) - \frac{1}{2}\log^3(2)\zeta(2) + \frac{7}{8}(2\log(2) - \log^2(2))\zeta(3) - \frac{19}{16}\log(2)\zeta(4)$$

$$+ \frac{31}{16}\zeta(5) - \frac{21}{16}\zeta(2)\zeta(3) + 2\log(2)\operatorname{Li}_4\left(\frac{1}{2}\right), \qquad (6.270)$$

where in the calculations above, I used that $\sum_{n=1}^{\infty} \frac{H_{2n+1}}{(2n+1)^3} = -1 + \frac{1}{2}\sum_{n=1}^{\infty} \frac{H_n}{n^3} +$

$\frac{1}{2} \sum_{n=1}^{\infty} (-1)^{n-1} \frac{H_n}{n^3} = -1 + 2\zeta(4) - \frac{7}{8}\log(2)\zeta(3) + \frac{1}{4}\log^2(2)\zeta(2) - \frac{1}{24}\log^4(2) -$

$\operatorname{Li}_4\left(\frac{1}{2}\right)$, and then noticed that the third series is the case $m = 1$ of (4.101),
Sect. 4.19, and the last two series are found in Sect. 4.34.

Further, we concentrate on the inner series of S_1 in (6.270), and then, using partial fraction expansion, we write

$$\sum_{n=1}^{\infty} \frac{n}{(2n+1)^3(n+k+1)}$$

$$= -\frac{2(k+1)}{(2k+1)^3} \sum_{n=1}^{\infty} \left(\frac{1}{2n+1} - \frac{1}{2n+2k+2} \right) + \frac{1}{2k+1} \sum_{n=1}^{\infty} \frac{1}{(2n+1)^2}$$

$$+ \frac{1}{(2k+1)^2} \sum_{n=1}^{\infty} \frac{1}{(2n+1)^2} - \frac{1}{2k+1} \sum_{n=1}^{\infty} \frac{1}{(2n+1)^3}$$

{for the first series, add and subtract $1/(2n)$, and then split it carefully}

$$= -\frac{2(k+1)}{(2k+1)^3} \sum_{n=1}^{\infty} \left(\frac{1}{2n+1} - \frac{1}{2n} \right) - \frac{k+1}{(2k+1)^3} \sum_{n=1}^{\infty} \left(\frac{1}{n} - \frac{1}{1+k+n} \right)$$

$$+ \frac{1}{8} (6\zeta(2) - 7\zeta(3)) \frac{1}{2k+1} - \left(1 - \frac{3}{4}\zeta(2) \right) \frac{1}{(2k+1)^2}$$

$$\left\{ \text{consider that } \sum_{n=1}^{\infty} \left(\frac{1}{2n+1} - \frac{1}{2n} \right) = -1 + \sum_{n=1}^{\infty} (-1)^{n-1} \frac{1}{n} = \log(2) - 1, \right\}$$

$$\left\{ \text{and for the second series we have the fact that } \sum_{n=1}^{\infty} \left(\frac{1}{n} - \frac{1}{k+n} \right) = H_k \right\}$$

$$= \frac{1}{8} (6\zeta(2) - 7\zeta(3)) \frac{1}{2k+1} - \left(\log(2) - \frac{3}{4}\zeta(2) \right) \frac{1}{(2k+1)^2} - \log(2) \frac{1}{(2k+1)^3}$$

$$- \frac{1}{2} \frac{H_k}{(2k+1)^2} - \frac{1}{2} \frac{H_k}{(2k+1)^3}. \tag{6.271}$$

Returning with the result from (6.271) to the series S_1 in (6.270), which then we also expand and rearrange, we get

$$S_1 = \sum_{k=1}^{\infty} \frac{H_{2k}}{k+1} \left(\sum_{n=1}^{\infty} \frac{n}{(2n+1)^3(n+k+1)} \right)$$

$$= \frac{7}{4}\zeta(3) \sum_{k=1}^{\infty} \left(\frac{1}{2k+2} - \frac{1}{2k+1} \right) + \frac{7}{4}\zeta(3) \sum_{k=1}^{\infty} \left(\frac{H_{2k+2}}{2k+2} - \frac{H_{2k+1}}{2k+1} \right)$$

$$+ \frac{3}{2}\zeta(2) \sum_{k=1}^{\infty} \frac{H_{2k+1}}{(2k+1)^2} - 2\log(2) \sum_{k=1}^{\infty} \frac{H_{2k+1}}{(2k+1)^3} - \sum_{k=1}^{\infty} \frac{H_k H_{2k}}{(2k+1)^3}$$

$$- 2\log(2) + \frac{3}{2}\zeta(2) - \frac{21}{16}\zeta(3) + \frac{15}{8}\log(2)\zeta(4) - \frac{7}{16}\zeta(2)\zeta(3)$$

$$= \frac{7}{8}(\log^2(2) - 2\log(2))\zeta(3) + \frac{21}{32}\zeta(2)\zeta(3) - \frac{31}{128}\zeta(5), \tag{6.272}$$

where the first two series (the second one with a changed sign) are given after the calculations in (6.229), Sect. 6.30, then we have $\displaystyle\sum_{k=1}^{\infty} \frac{H_{2k+1}}{(2k+1)^2} = -1 + \frac{1}{2}\sum_{k=1}^{\infty} \frac{H_k}{k^2} +$

$\displaystyle\frac{1}{2}\sum_{k=1}^{\infty}(-1)^{k-1}\frac{H_k}{k^2} = -1 + \frac{21}{16}\zeta(3)$, where I used the Euler sum generalization in (6.149), the case $n = 2$, Sect. 6.19, and (4.105), the case $m = 1$, Sect. 4.21; next the fourth series is also stated after the calculations in (6.270); and finally, the last series is given in (4.130), Sect. 4.35.

Combining (6.272) and (6.270), we obtain that

$$S = \sum_{n=1}^{\infty} \frac{1}{(2n+1)^3}\left(\sum_{i=1}^{n} \frac{H_i}{2i+1}\right)$$

$$= \frac{1}{12}\log^5(2) - \frac{1}{2}\log^3(2)\zeta(2) - \frac{19}{16}\log(2)\zeta(4) + \frac{217}{128}\zeta(5) - \frac{21}{32}\zeta(2)\zeta(3)$$

$$+ 2\log(2)\operatorname{Li}_4\left(\frac{1}{2}\right). \tag{6.273}$$

At last, if we combine (6.273), (6.269), and (6.268), we conclude that

$$\sum_{n=1}^{\infty} \frac{H_n}{2n+1}\left(\zeta(3) - H_{2n}^{(3)}\right)$$

$$= \frac{1}{15}\log^5(2) - \frac{1}{3}\log^3(2)\zeta(2) - \frac{7}{8}\log^2(2)\zeta(3) + \frac{15}{32}\log(2)\zeta(4) - \frac{31}{64}\zeta(5)$$

$$- \frac{7}{16}\zeta(2)\zeta(3) + 2\log(2)\operatorname{Li}_4\left(\frac{1}{2}\right) + 2\operatorname{Li}_5\left(\frac{1}{2}\right),$$

and the solution is complete.

Again, as mentioned at the end of the previous section, it is always exciting to have more ways to go for such problems. *What other ways would we like to design?* Of course, we might try to exploit the fact that $\zeta(3) - H_{2n}^{(3)} = \dfrac{1}{2}\displaystyle\int_0^1 \frac{x^{2n}\log^2(x)}{1-x}dx.$

6.41 A Sixth Unusual Series with the Tail of the Riemann Zeta Function $\zeta(3) - H_n^{(3)}$, Related to Weight 5 Harmonic Series

Solution After following the previous sections with series of this type, it's pretty easy to figure out that it's a *routine* to get inspiration from its corresponding series from the weight 4 series class, which this time may be found in Sect. 6.31.

To begin with, we want to rearrange the series and get the form of it with $\dfrac{H_{2n+1}}{2n+1}$, instead of $\dfrac{H_{2n}}{2n+1}$, and then we write that

$$\sum_{n=1}^{\infty} \frac{H_{2n}}{2n+1}\left(\zeta(3) - H_n^{(3)}\right) = \sum_{n=1}^{\infty} \frac{H_{2n+1} - 1/(2n+1)}{2n+1}\left(\zeta(3) - H_n^{(3)}\right)$$

$$= \sum_{n=1}^{\infty} \frac{H_{2n+1}}{2n+1}\left(\zeta(3) - H_n^{(3)}\right) - \zeta(3)\underbrace{\sum_{n=1}^{\infty} \frac{1}{(2n+1)^2}}_{3/4\zeta(2) - 1} + \sum_{n=1}^{\infty} \frac{H_n^{(3)}}{(2n+1)^2}$$

{the value of the last series may be found in (4.126), Sect. 4.34}

$$= \zeta(3) + \frac{31}{2}\zeta(5) - \frac{35}{4}\zeta(2)\zeta(3) + \underbrace{\sum_{n=1}^{\infty} \frac{H_{2n+1}}{2n+1}\left(\zeta(3) - H_n^{(3)}\right)}_{S}. \tag{6.274}$$

Next, if we consider the series that appears in the *challenging question*, multiply it by $1/2$, and denote it by T, that is, $T = \sum_{n=1}^{\infty} \frac{H_{2n}}{2n}\left(\zeta(3) - H_n^{(3)}\right)$, and add it to the series S in (6.274), we obtain

$$S + T = \sum_{n=1}^{\infty} \left(\frac{H_{2n}}{2n} + \frac{H_{2n+1}}{2n+1}\right)\left(\zeta(3) - H_n^{(3)}\right), \tag{6.275}$$

which you might think doesn't look *friendlier* than the initial series, but this form is a more manageable one.

So, applying Abel's summation in (6.275), the series version in (6.7), Sect. 6.2, where we consider $a_n = \dfrac{H_{2n}}{2n} + \dfrac{H_{2n+1}}{2n+1}$ and $b_n = \zeta(3) - H_n^{(3)}$, we have

$$S + T = \sum_{n=1}^{\infty} \left(\frac{H_{2n}}{2n} + \frac{H_{2n+1}}{2n+1} \right) \left(\zeta(3) - H_n^{(3)} \right)$$

$$= \sum_{n=1}^{\infty} \frac{1}{(n+1)^3} \left(\sum_{k=1}^{n} \left(\frac{H_{2k}}{2k} + \frac{H_{2k+1}}{2k+1} \right) \right)$$

$$\left\{ \text{observe and use that } \sum_{k=1}^{n} \left(\frac{H_{2k}}{2k} + \frac{H_{2k+1}}{2k+1} \right) = -\frac{H_{2n+2}}{2n+2} + \sum_{k=2}^{2n+2} \frac{H_k}{k} \right\}$$

$$= \sum_{n=1}^{\infty} \frac{1}{(n+1)^3} \left(\sum_{k=2}^{2n+2} \frac{H_k}{k} \right) - \frac{1}{2} \sum_{n=1}^{\infty} \frac{H_{2n+2}}{(n+1)^4}$$

$$\{ \text{reindex both series and start from } n = 1 \}$$

$$= \sum_{n=1}^{\infty} \frac{1}{n^3} \left(\sum_{k=2}^{2n} \frac{H_k}{k} \right) - 8 \sum_{n=1}^{\infty} \frac{H_{2n}}{(2n)^4}$$

$$\left\{ \text{use that } \sum_{k=1}^{n} \frac{H_k}{k} = \frac{1}{2} \left(H_n^2 + H_n^{(2)} \right) \text{ via } p = q = 1 \text{ in (6.102), Sect. 6.13} \right\}$$

$$= 4 \sum_{n=1}^{\infty} \frac{H_{2n}^2}{(2n)^3} + 4 \sum_{n=1}^{\infty} \frac{H_{2n}^{(2)}}{(2n)^3} - 8 \sum_{n=1}^{\infty} \frac{H_{2n}}{(2n)^4} - \sum_{n=1}^{\infty} \frac{1}{n^3}$$

$$\left\{ \text{for the first three series employ that } \sum_{n=1}^{\infty} a_{2n} = \frac{1}{2} \left(\sum_{n=1}^{\infty} a_n - \sum_{n=1}^{\infty} (-1)^{n-1} a_n \right) \right\}$$

$$= 2 \sum_{n=1}^{\infty} \frac{H_n^2}{n^3} + 2 \sum_{n=1}^{\infty} \frac{H_n^{(2)}}{n^3} - 4 \sum_{n=1}^{\infty} \frac{H_n}{n^4} + 4 \sum_{n=1}^{\infty} (-1)^{n-1} \frac{H_n}{n^4} - 2 \sum_{n=1}^{\infty} (-1)^{n-1} \frac{H_n^2}{n^3}$$

$$- 2 \sum_{n=1}^{\infty} (-1)^{n-1} \frac{H_n^{(2)}}{n^3} - \sum_{n=1}^{\infty} \frac{1}{n^3}$$

$$= \frac{4}{3} \log^3(2) \zeta(2) - \frac{7}{2} \log^2(2) \zeta(3) - \frac{4}{15} \log^5(2) - \zeta(3) - \frac{19}{4} \zeta(5) + \frac{15}{2} \zeta(2) \zeta(3)$$

$$- 8 \log(2) \operatorname{Li}_4 \left(\frac{1}{2} \right) - 8 \operatorname{Li}_5 \left(\frac{1}{2} \right), \tag{6.276}$$

where during the calculations, I considered the following series results, $\sum_{n=1}^{\infty} \dfrac{H_n^2}{n^3} =$

$\dfrac{7}{2}\zeta(5) - \zeta(2)\zeta(3)$ (see [76, Chapter 4, p.293]); then $\sum_{n=1}^{\infty} \dfrac{H_n^{(2)}}{n^3} = 3\zeta(2)\zeta(3) - \dfrac{9}{2}\zeta(5)$

(see [76, Chapter 6, p.386]); further the third series is the case $n = 4$ of the Euler sum generalization in (6.149), Sect. 6.19; next the fourth series is the case $m = 2$ of the alternating Euler sum generalization in (4.105), Sect. 4.21; and

finally $\sum_{n=1}^{\infty}(-1)^{n-1}\dfrac{H_n^2}{n^3} = \dfrac{2}{15}\log^5(2) - \dfrac{11}{8}\zeta(2)\zeta(3) - \dfrac{19}{32}\zeta(5) + \dfrac{7}{4}\log^2(2)\zeta(3) -$

$\dfrac{2}{3}\log^3(2)\zeta(2) + 4\log(2)\operatorname{Li}_4\left(\dfrac{1}{2}\right) + 4\operatorname{Li}_5\left(\dfrac{1}{2}\right)$, which is given in (3.331), Sect.

3.48, and $\sum_{n=1}^{\infty}(-1)^{n-1}\dfrac{H_n^{(2)}}{n^3} = \dfrac{5}{8}\zeta(2)\zeta(3) - \dfrac{11}{32}\zeta(5)$ (see [76, Chapter 4, p.311]).

Since the value of T may be extracted from (4.132), Sect. 4.37, then based on (6.276), we get that

$$S = \sum_{n=1}^{\infty} \frac{H_{2n+1}}{2n+1}\left(\zeta(3) - H_n^{(3)}\right)$$

$$= \frac{2}{3}\log^3(2)\zeta(2) - \frac{7}{2}\log^2(2)\zeta(3) + \frac{53}{8}\log(2)\zeta(4) - \frac{1}{15}\log^5(2) - \zeta(3) - \frac{279}{16}\zeta(5)$$

$$+ \frac{49}{8}\zeta(2)\zeta(3) + 8\operatorname{Li}_5\left(\frac{1}{2}\right). \tag{6.277}$$

Finally, returning with the value from (6.277) in (6.274), we arrive at the desired result

$$\sum_{n=1}^{\infty} \frac{H_{2n}}{2n+1}\left(\zeta(3) - H_n^{(3)}\right)$$

$$= \frac{2}{3}\log^3(2)\zeta(2) - \frac{7}{2}\log^2(2)\zeta(3) + \frac{53}{8}\log(2)\zeta(4) - \frac{1}{15}\log^5(2) - \frac{31}{16}\zeta(5)$$

$$- \frac{21}{8}\zeta(2)\zeta(3) + 8\operatorname{Li}_5\left(\frac{1}{2}\right),$$

and the solution is complete.

We observe the solution above also answers the *challenging question*. Now, the curious reader might also explore other ways to go. For example, one might start from exploiting the integral representation of the Riemann zeta tail, $\zeta(3) - H_n^{(3)} =$

$4\displaystyle\int_0^1 \frac{x^{2n+1}\log^2(x)}{1-x^2}dx$, which shows the series reduces to evaluating

$$\sum_{n=1}^{\infty} \frac{H_{2n}}{2n+1}\left(\zeta(3) - H_n^{(3)}\right) = \int_0^1 \frac{\left(\log^2(1-x) - \log^2(1+x)\right)\log^2(x)}{1-x^2}dx,$$

and this could be viewed as another point from which one might want to proceed with the calculations.

6.42   A Seventh Unusual Series with the Tail of the Riemann Zeta Function $\zeta(3) - H_{2n}^{(3)}$, Related to Weight 5 Harmonic Series

Solution If we take a look at the first solution to the corresponding series from the weight 4 series class, which may be found in Sect. 6.32, then it is not hard to realize this time we want to start from the harmonic series with the Riemann zeta tail,

$$\sum_{n=1}^{\infty} \frac{H_n}{n}\left(\zeta(3) - H_n^{(3)}\right) = 2\zeta(2)\zeta(3) - \frac{7}{2}\zeta(5), \tag{6.278}$$

which also appears in [76, Chapter 4, p.303]. A solution may be immediately set up if we consider applying Abel's summation with $a_n = H_n/n$ and $b_n = \zeta(3) - H_n^{(3)}$, the series version in (6.7), Sect. 6.2 (for details, see [76, Chapter 6, pp. 479–480]).

The key step we need to make now is to return to (6.278) and exploit the parity that leads to

$$2\zeta(2)\zeta(3) - \frac{7}{2}\zeta(5) = \sum_{n=1}^{\infty} \frac{H_n}{n}\left(\zeta(3) - H_n^{(3)}\right)$$

$$= \zeta(3) - 1 + \sum_{n=1}^{\infty} \frac{H_{2n}}{2n}\left(\zeta(3) - H_{2n}^{(3)}\right) + \sum_{n=1}^{\infty} \frac{H_{2n+1}}{2n+1}\left(\zeta(3) - H_{2n+1}^{(3)}\right)$$

{rearrange the last series and then expand it}

$$= \zeta(3) - 1 + \underbrace{\frac{1}{2}\sum_{n=1}^{\infty} \frac{H_{2n}}{n}\left(\zeta(3) - H_{2n}^{(3)}\right)}_{T} + \underbrace{\sum_{n=1}^{\infty} \frac{H_{2n}}{2n+1}\left(\zeta(3) - H_{2n}^{(3)}\right)}_{S}$$

$$- \sum_{n=1}^{\infty} \frac{H_{2n+1}}{(2n+1)^4} - \sum_{n=1}^{\infty} \frac{H_{2n+1}^{(3)}}{(2n+1)^2} + \zeta(3)\sum_{n=1}^{\infty} \frac{1}{(2n+1)^2} + \sum_{n=1}^{\infty} \frac{1}{(2n+1)^5}$$

$$\left\{\text{for the 3rd and 4th series use } \sum_{n=1}^{\infty} a_{2n+1} = -a_1 + \frac{1}{2}\left(\sum_{n=1}^{\infty} a_n + \sum_{n=1}^{\infty} (-1)^{n-1} a_n\right)\right\}$$

$$= \frac{1}{2}T + S - \frac{1}{2}\sum_{n=1}^{\infty} \frac{H_n}{n^4} - \frac{1}{2}\sum_{n=1}^{\infty} \frac{H_n^{(3)}}{n^2} - \frac{1}{2}\sum_{n=1}^{\infty} (-1)^{n-1}\frac{H_n}{n^4} - \frac{1}{2}\sum_{n=1}^{\infty} (-1)^{n-1}\frac{H_n^{(3)}}{n^2}$$

$$+ \frac{3}{4}\zeta(2)\zeta(3) + \frac{31}{32}\zeta(5)$$

$$= \frac{1}{2}T + S + \frac{17}{8}\zeta(2)\zeta(3) - \frac{31}{8}\zeta(5), \tag{6.279}$$

where in the calculations I used (6.149), the case $n = 4$, Sect. 6.19; then $\displaystyle\sum_{n=1}^{\infty} \frac{H_n^{(3)}}{n^2} =$

$\dfrac{11}{2}\zeta(5) - 2\zeta(2)\zeta(3)$ (see [76, Chapter 6, p.386]); next the alternating Euler sum

in (4.105), the case $m = 2$, Sect. 4.21; and finally $\displaystyle\sum_{n=1}^{\infty} (-1)^{n-1}\frac{H_n^{(3)}}{n^2} = \frac{3}{4}\zeta(2)\zeta(3) -$

$\dfrac{21}{32}\zeta(5)$ ([76, Chapter 4, p.311]).

   Now, since the value of the series T is given in (4.133), Sect. 4.38, if we plug it in (6.279), we can extract the desired result

$$S = \sum_{n=1}^{\infty} \frac{H_{2n}}{2n+1}\left(\zeta(3) - H_{2n}^{(3)}\right)$$

$$= \frac{1}{40}\log^5(2) - \frac{1}{12}\log^3(2)\zeta(2) - \frac{7}{16}\log^2(2)\zeta(3) + \frac{49}{32}\log(2)\zeta(4) - \frac{31}{64}\zeta(5)$$

$$- \frac{27}{32}\zeta(2)\zeta(3) + \log(2)\operatorname{Li}_4\left(\frac{1}{2}\right) + 2\operatorname{Li}_5\left(\frac{1}{2}\right),$$

and the solution is finalized.

   The curious reader might also want to try a more direct way by exploiting that
$\zeta(3) - H_{2n}^{(3)} = \dfrac{1}{2}\displaystyle\int_0^1 \frac{x^{2n}\log^2(x)}{1-x}dx$, which leads to the manageable integral,

$$\frac{1}{8}\int_0^1 \frac{(\log^2(1-x) - \log^2(1+x))\log^2(x)}{x(1-x)}dx,$$

where one may use integration by parts and/or simple algebraic identities.

6.43 Three More Spectacular Harmonic Series of Weight 5, Involving Harmonic Numbers of the Type H_{2n} and $H_{2n}^{(2)}$

Solution *Three More Spectacular Harmonic Series* ... that also happen to be (very) special and challenging in some sense. The first two ones I also treated separately in two different papers. The third harmonic series will be a delightful extraction based on a special Cauchy product presented in the fourth chapter.

Now, among all the non-trivial, non-alternating harmonic series of weight 5 involving harmonic numbers of the type H_{2n}, only the series at the first point has a simple closed form, expressed in terms of values of the Riemann zeta function, with no polylogarithmic value! That's curious and intriguing! For its evaluation, I'll exploit the strategy given in my paper, *The evaluation of a special harmonic series with a weight 5 structure, involving harmonic numbers of the type H_{2n}* (see [84, October 1, 2019]).

So, considering the Cauchy product in (4.23), Sect. 4.5, where we replace x by x^2, and then multiplying both sides by $\log(1-x)/x$ and using that $\int_0^1 x^{n-1} \log(1-x)dx = -\dfrac{H_n}{n}$, which is also given in (3.10), Sect. 3.3, we write

$$\int_0^1 \sum_{n=1}^\infty x^{2n-1} \log(1-x) \left(3\frac{1}{n^3} - 2\frac{H_n}{n^2} - \frac{H_n^{(2)}}{n}\right) dx$$

{reverse the order of summation and integration}

$$= \sum_{n=1}^\infty \left(3\frac{1}{n^3} - 2\frac{H_n}{n^2} - \frac{H_n^{(2)}}{n}\right) \int_0^1 x^{2n-1} \log(1-x)dx$$

{after integrating, expand the series}

$$= \frac{1}{2} \sum_{n=1}^\infty \frac{H_{2n} H_n^{(2)}}{n^2} + \sum_{n=1}^\infty \frac{H_n H_{2n}}{n^3} - 24 \sum_{n=1}^\infty \frac{H_{2n}}{(2n)^4}$$

$$\left\{\text{for the last series use that } \sum_{n=1}^\infty a_{2n} = \frac{1}{2}\left(\sum_{n=1}^\infty a_n - \sum_{n=1}^\infty (-1)^{n-1} a_n\right)\right\}$$

$$= \frac{1}{2} \sum_{n=1}^\infty \frac{H_{2n} H_n^{(2)}}{n^2} + \sum_{n=1}^\infty \frac{H_n H_{2n}}{n^3} - 12 \sum_{n=1}^\infty \frac{H_n}{n^4} + 12 \sum_{n=1}^\infty (-1)^{n-1} \frac{H_n}{n^4}$$

$$= \int_0^1 \frac{\log(1-x) \log(1-x^2) \operatorname{Li}_2(x^2)}{x} dx$$

$$= \int_0^1 \frac{\log(1-x)\log(1+x)\operatorname{Li}_2(x^2)}{x}dx + \int_0^1 \frac{\log^2(1-x)\operatorname{Li}_2(x^2)}{x}dx$$

$$= \frac{271}{32}\zeta(5) + \frac{39}{8}\zeta(2)\zeta(3) + \frac{8}{3}\log^3(2)\zeta(2) - 7\log^2(2)\zeta(3) - \frac{8}{15}\log^5(2)$$

$$- 16\log(2)\operatorname{Li}_4\left(\frac{1}{2}\right) - 16\operatorname{Li}_5\left(\frac{1}{2}\right), \qquad (6.280)$$

where to get the last equality, I also used the values of the integrals given in (1.217), Sect. 1.48, and then (1.211), Sect. 1.47.

If we also consider in (6.280) the values of the series in (4.124), Sect. 4.33, together with the particular cases $n = 4$ of the Euler sum generalization in (6.149), Sect. 6.19, and $m = 2$ of the alternating Euler generalization in (4.105), Sect. 4.21, we arrive at the desired result

$$\sum_{n=1}^{\infty} \frac{H_{2n}H_n^{(2)}}{n^2} = \frac{101}{16}\zeta(5) - \frac{5}{4}\zeta(2)\zeta(3),$$

and the solution to the point (i) is complete.

A different, interesting approach may be found in [47], but at some point, they arrive at $\sum_{n=1}^{\infty} \dfrac{H_n H_{2n}}{n^3}$, as in my solution above, which involves a closed form with polylogarithmic values.

This is a mysterious series! How would we calculate it by avoiding entirely deriving intermediate results involving polylogarithmic values? That may be one possible natural question you may have! Did you see the second solution in Sect. 3.52? Perhaps a clever grouping of integrals/series and then calculating them together might help circumvent the appearance of the polylogarithmic values.

The next series is part of my article, *The calculation of a harmonic series with a weight 5 structure, involving the product of harmonic numbers, $H_n H_{2n}^{(2)}$* (see [82, October 10, 2019]), and in the solution below, I'll follow the strategy described there.

Let's start from the identity $\sum_{n=1}^{\infty} x^n H_n^{(2)} = \dfrac{\operatorname{Li}_2(x)}{1-x}$, which is the particular case $m = 2$ of (4.32), Sect. 4.6, immediately leading to

$$\sum_{n=1}^{\infty} x^{2n} H_{2n}^{(2)} = \frac{1}{2}\left(\frac{\operatorname{Li}_2(x)}{1-x} + \frac{\operatorname{Li}_2(-x)}{1+x}\right), \quad |x| < 1. \qquad (6.281)$$

If we take the result in (6.281), in variable y, then multiply both sides by $1/y$, and integrate from $y = 0$ to $y = x$, we have

$$\int_0^x \sum_{n=1}^{\infty} y^{2n-1} H_{2n}^{(2)} dy = \sum_{n=1}^{\infty} H_{2n}^{(2)} \int_0^x y^{2n-1} dy = \frac{1}{2} \sum_{n=1}^{\infty} x^{2n} \frac{H_{2n}^{(2)}}{n}$$

$$= \frac{1}{2} \int_0^x \left(\frac{\text{Li}_2(y)}{y(1-y)} + \frac{\text{Li}_2(-y)}{y(1+y)} \right) dy. \tag{6.282}$$

Upon multiplying both sides of the last equality in (6.282) by $\log(1 - x^2)/x$, then integrating from $x = 0$ to $x = 1$, using that $\int_0^1 x^{2n-1} \log(1 - x^2) dx = -\dfrac{H_n}{2n}$, which is straightforward to extract based on (3.10), Sect. 3.3, and rearranging, we obtain

$$\sum_{n=1}^{\infty} \frac{H_n H_{2n}^{(2)}}{n^2} = -2 \int_0^1 \frac{\log(1-x^2)}{x} \left(\int_0^x \left(\frac{\text{Li}_2(y)}{y(1-y)} + \frac{\text{Li}_2(-y)}{y(1+y)} \right) dy \right) dx$$

$$\overset{\substack{\text{reverse the order} \\ \text{of integration}}}{=} -2 \int_0^1 \left(\frac{\text{Li}_2(y)}{y(1-y)} + \frac{\text{Li}_2(-y)}{y(1+y)} \right) \left(\int_y^1 \frac{\log(1-x^2)}{x} dx \right) dy$$

$$= \int_0^1 \left(\frac{\text{Li}_2(y)}{y(1-y)} + \frac{\text{Li}_2(-y)}{y(1+y)} \right) \left(\zeta(2) - \text{Li}_2(y^2) \right) dy$$

$$\overset{y=x}{=} \underbrace{\frac{1}{2} \zeta(2) \int_0^1 \frac{\text{Li}_2(x^2)}{x} dx}_{1/2\zeta(3)} - \zeta(2) \int_0^1 \frac{\text{Li}_2(-x)}{1+x} dx - \frac{1}{2} \int_0^1 \frac{(\text{Li}_2(x^2))^2}{x} dx$$

$$+ \int_0^1 \frac{\text{Li}_2(-x) \text{Li}_2(x^2)}{1+x} dx + \int_0^1 \frac{\text{Li}_2(x) \left(\zeta(2) - \text{Li}_2(x^2) \right)}{1-x} dx, \tag{6.283}$$

where to get the last equality, I also used the dilogarithmic identity, $\text{Li}_2(x) + \text{Li}_2(-x) = 1/2 \text{Li}_2(x^2)$.

As regards the third remaining integral in (6.283), we have

$$\int_0^1 \frac{(\text{Li}_2(x^2))^2}{x} dx \overset{x^2=y}{=} \frac{1}{2} \int_0^1 \frac{(\text{Li}_2(y))^2}{y} dy = \frac{1}{2} \int_0^1 \sum_{n=1}^{\infty} \frac{y^{n-1}}{n^2} \text{Li}_2(y) dy$$

$$= \frac{1}{2} \sum_{n=1}^{\infty} \frac{1}{n^2} \int_0^1 y^{n-1} \text{Li}_2(y) dy = \frac{1}{2} \zeta(2) \sum_{n=1}^{\infty} \frac{1}{n^3} - \frac{1}{2} \sum_{n=1}^{\infty} \frac{H_n}{n^4} = \zeta(2)\zeta(3) - \frac{3}{2} \zeta(5),$$

$$\tag{6.284}$$

where in the calculations I used that $\int_0^1 x^{n-1} \text{Li}_2(x) dx = \zeta(2) \dfrac{1}{n} - \dfrac{H_n}{n^2}$, as seen in (3.327), Sect. 3.46, and the particular case $n = 4$ of the Euler sum generalization in (6.149), Sect. 6.19.

Next, for the fourth remaining integral in (6.283), we combine integration by parts and the dilogarithmic identity, $\text{Li}_2(x) + \text{Li}_2(-x) = 1/2 \text{Li}_2(x^2)$, and then we get

$$\int_0^1 \frac{\text{Li}_2(-x) \text{Li}_2(x^2)}{1+x} dx = \int_0^1 (\log(1+x))' \text{Li}_2(-x) \text{Li}_2(x^2) dx = -\frac{5}{4} \log(2) \zeta(4)$$

$$+ 2 \int_0^1 \frac{\log^2(1+x) \text{Li}_2(-x)}{x} dx + \int_0^1 \frac{\log^2(1+x) \text{Li}_2(x^2)}{x} dx$$

$$+ 2 \int_0^1 \frac{\log(1-x) \log(1+x) \text{Li}_2(-x)}{x} dx$$

$$= \frac{4}{5} \log^5(2) - 4 \log^3(2) \zeta(2) + \frac{21}{2} \log^2(2) \zeta(3) - \frac{5}{4} \log(2) \zeta(4) - \frac{23}{8} \zeta(2) \zeta(3)$$

$$- \frac{283}{16} \zeta(5) + 24 \log(2) \text{Li}_4 \left(\frac{1}{2} \right) + 24 \text{Li}_5 \left(\frac{1}{2} \right), \tag{6.285}$$

and the closed forms of the resulting integrals are given in (1.213) and (1.214), Sect. 1.47, and then in (1.216), Sect. 1.48.

Further, for the last remaining integral in (6.283), we integrate by parts, and then we write

$$\int_0^1 \frac{\text{Li}_2(x) \left(\zeta(2) - \text{Li}_2 \left(x^2 \right) \right)}{1-x} dx = - \int_0^1 (\log(1-x))' \text{Li}_2(x) \left(\zeta(2) - \text{Li}_2 \left(x^2 \right) \right) dx$$

$$= -\zeta(2) \int_0^1 \frac{\log^2(1-x)}{x} dx + 2 \int_0^1 \frac{\log^2(1-x) \text{Li}_2(x)}{x} dx$$

$$+ \int_0^1 \frac{\log^2(1-x) \text{Li}_2(x^2)}{x} dx + 2 \int_0^1 \frac{\log(1-x) \log(1+x) \text{Li}_2(x)}{x} dx$$

$$= \frac{25}{4} \zeta(2) \zeta(3) - \frac{39}{32} \zeta(5) + \frac{4}{3} \log^3(2) \zeta(2) - \frac{7}{2} \log^2(2) \zeta(3) - \frac{4}{15} \log^5(2)$$

$$- 8 \log(2) \text{Li}_4 \left(\frac{1}{2} \right) - 8 \text{Li}_5 \left(\frac{1}{2} \right), \tag{6.286}$$

where the first integral is straightforward, $\displaystyle\int_0^1 \frac{\log^2(1-x)}{x}dx \overset{1-x=y}{=} \int_0^1 \frac{\log^2(y)}{1-y}dy$

$= \displaystyle\int_0^1 \sum_{n=1}^{\infty} y^{n-1}\log^2(y)dy = 2\sum_{n=1}^{\infty}\frac{1}{n^3} = 2\zeta(3)$, and the other three integrals are

found in (1.209) and (1.211), Sect. 1.47, and (1.215), Sect. 1.48.

Counting that $\displaystyle\int_0^1 \frac{\operatorname{Li}_2(-x)}{1+x}dx = \frac{1}{4}\zeta(3) - \frac{1}{2}\log(2)\zeta(2)$, which is found at the end
of the second solution to the point (i) in Sect. 3.20, and then collecting the values of
the integrals from (6.284), (6.285), and (6.286) in (6.283), we arrive at

$$\sum_{n=1}^{\infty}\frac{H_n H_{2n}^{(2)}}{n^2}$$

$$= \frac{23}{8}\zeta(2)\zeta(3) - \frac{581}{32}\zeta(5) - \frac{8}{3}\log^3(2)\zeta(2) + 7\log^2(2)\zeta(3) + \frac{8}{15}\log^5(2)$$

$$+ 16\log(2)\operatorname{Li}_4\left(\frac{1}{2}\right) + 16\operatorname{Li}_5\left(\frac{1}{2}\right),$$

and the solution to the point (ii) is complete. Another solution is given in [49].

The last point of the problem is *almost magically* straightforward if we exploit
the special Cauchy product in (4.22), Sect. 4.5, and then observe that the resulting
integrals and series have already been calculated in other sections!

Multiplying both sides of (4.22), Sect. 4.5, by $\log(1-x)/x$ and then considering
$\displaystyle\int_0^1 x^{n-1}\log(1-x)dx = -\frac{H_n}{n}$, as it appears in (3.10), Sect. 3.3, we have

$$\int_0^1 \frac{\operatorname{arctanh}(x)\log(1-x)\operatorname{Li}_2(x^2)}{x}dx$$

$$= \frac{1}{2}\int_0^1 \frac{\log(1-x)\log(1+x)\operatorname{Li}_2(x^2)}{x}dx - \frac{1}{2}\int_0^1 \frac{\log^2(1-x)\operatorname{Li}_2(x^2)}{x}dx$$

$$= \int_0^1 \sum_{n=1}^{\infty} x^{2n}\left(4\frac{H_{2n}}{(2n+1)^2} + \frac{H_n^{(2)}}{2n+1}\right)\log(1-x)dx$$

{reverse the order of summation and integration}

$$= \sum_{n=1}^{\infty}\left(4\frac{H_{2n}}{(2n+1)^2} + \frac{H_n^{(2)}}{2n+1}\right)\int_0^1 x^{2n}\log(1-x)dx$$

$$= -4 \sum_{n=1}^{\infty} \frac{H_{2n+1}\left(H_{2n+1} - \frac{1}{2n+1}\right)}{(2n+1)^3} - \sum_{n=1}^{\infty} \frac{\left(H_{2n} + \frac{1}{2n+1}\right) H_n^{(2)}}{(2n+1)^2}$$

{reindex the first series and then expand both series}

$$= 4 \sum_{n=1}^{\infty} \frac{H_{2n-1}}{(2n-1)^4} - 4 \sum_{n=1}^{\infty} \frac{H_{2n-1}^2}{(2n-1)^3} - \sum_{n=1}^{\infty} \frac{H_n^{(2)}}{(2n+1)^3} - \sum_{n=1}^{\infty} \frac{H_{2n} H_n^{(2)}}{(2n+1)^2}$$

$$\left\{ \text{exploit the simple fact that } \sum_{n=1}^{\infty} a_{2n-1} = \frac{1}{2}\left(\sum_{n=1}^{\infty} a_n + \sum_{n=1}^{\infty} (-1)^{n-1} a_n\right) \right\}$$

$$= 2 \sum_{n=1}^{\infty} \frac{H_n}{n^4} - 2 \sum_{n=1}^{\infty} \frac{H_n^2}{n^3} + 2 \sum_{n=1}^{\infty} (-1)^{n-1} \frac{H_n}{n^4} - 2 \sum_{n=1}^{\infty} (-1)^{n-1} \frac{H_n^2}{n^3}$$

$$- \sum_{n=1}^{\infty} \frac{H_n^{(2)}}{(2n+1)^3} - \sum_{n=1}^{\infty} \frac{H_{2n} H_n^{(2)}}{(2n+1)^2},$$

where if we use the Euler sum generalization in (6.149), the case $n = 4$, Sect. 6.19;
then $\sum_{n=1}^{\infty} \frac{H_n^2}{n^3} = \frac{7}{2}\zeta(5) - \zeta(2)\zeta(3)$ (see [76, Chapter 4, p.293]); next the alternating
Euler sum generalization in (4.105), the case $m = 2$, Sect. 4.21; further the harmonic
series in (3.331), Sect. 3.48, (4.125), Sect. 4.34, all together with the values of the
integrals in (1.217), Sect. 1.48 and (1.211), Sect. 1.47, we arrive at

$$\sum_{n=1}^{\infty} \frac{H_{2n} H_n^{(2)}}{(2n+1)^2}$$

$$= \frac{713}{64}\zeta(5) - \frac{21}{16}\zeta(2)\zeta(3) - \frac{7}{2}\log^2(2)\zeta(3) + \frac{4}{3}\log^3(2)\zeta(2) - \frac{4}{15}\log^5(2)$$

$$- 8\log(2)\operatorname{Li}_4\left(\frac{1}{2}\right) - 8\operatorname{Li}_5\left(\frac{1}{2}\right),$$

and the solution to the point (iii) is complete.

Another solution to this last series problem may be found in [48].

As a final thought, before closing the current section, I would emphasize that the
series problems alike may often turn to be very difficult to calculate without some
key results in hand. Also, it remains an ongoing project to find simple, fast ways to
calculate such series. The curious reader might not want to miss such challenges!

6.44   Two Atypical Sums of Series, One of Them Involving the Product of the Generalized Harmonic Numbers $H_n^{(3)} H_n^{(6)}$

Solution The sum of series at the point (ii) represents a problem I submitted years ago to *MathProblems journal*, Vol. 6, No. 2 (see [59, Problem **157**, p.560]). Then, it is clear from the problem statement that the sum of series at the point (i) is an auxiliary result we need during the calculations to the sum of series at the point (ii). Additionally, it is good to know that the sum of series at the first point is a classical one that may also be found in [14].

Ready? Let's embark on an exciting little journey and start from a *somewhat* unexpected double series, that is

$$\sum_{k=1}^{\infty}\left(\sum_{n=1}^{\infty}\frac{1}{k^4(k+n)^4}\right)=\sum_{k=1}^{\infty}\left(\sum_{n=1}^{\infty}\frac{1}{n^4}\left(\frac{1}{k}-\frac{1}{k+n}\right)^4\right)$$

$$=\underbrace{\sum_{k=1}^{\infty}\frac{1}{k^4}\left(\sum_{n=1}^{\infty}\frac{1}{n^4}\right)}_{\zeta^2(4)=7/6\zeta(8)}-4\sum_{k=1}^{\infty}\left(\sum_{n=1}^{\infty}\frac{1}{n^4k^3(k+n)}\right)+6\sum_{k=1}^{\infty}\left(\sum_{n=1}^{\infty}\frac{1}{n^4k^2(k+n)^2}\right)$$

$$-4\sum_{k=1}^{\infty}\left(\sum_{n=1}^{\infty}\frac{1}{n^4k(k+n)^3}\right)+\sum_{k=1}^{\infty}\left(\sum_{n=1}^{\infty}\frac{1}{n^4(k+n)^4}\right),$$

whence, since the opposite double series vanish, we get

$$4\sum_{k=1}^{\infty}\left(\sum_{n=1}^{\infty}\frac{1}{n^4k^3(k+n)}\right)-6\sum_{k=1}^{\infty}\left(\sum_{n=1}^{\infty}\frac{1}{n^4k^2(k+n)^2}\right)+4\sum_{k=1}^{\infty}\left(\sum_{n=1}^{\infty}\frac{1}{n^4k(k+n)^3}\right)$$

$$=\frac{7}{6}\zeta(8), \tag{6.287}$$

where I used the fact that we have $\displaystyle\sum_{k=1}^{\infty}\left(\sum_{n=1}^{\infty}\frac{1}{k^4(k+n)^4}\right)=\sum_{n=1}^{\infty}\left(\sum_{k=1}^{\infty}\frac{1}{k^4(k+n)^4}\right)$

$=\displaystyle\sum_{k=1}^{\infty}\left(\sum_{n=1}^{\infty}\frac{1}{n^4(k+n)^4}\right)<\infty$, which allowed the cancelation of the two double series above. Also, observe that during the calculations, I exploited that $\zeta(4)=\dfrac{\pi^4}{90}$ and $\zeta(8)=\dfrac{\pi^8}{9450}$, which may be extracted based on (6.315), Sect. 6.47.

Alternatively, another excellent way to go for the extraction of the zeta values needed, which uses Fourier series, is provided by Paul in [65, Chapter 3, Epilogue, pp. 360–363].

Continuing in (6.287) with the partial fraction expansion, we write

$$\frac{7}{6}\zeta(8)$$

$$= 4\underbrace{\sum_{n=1}^{\infty}\frac{1}{n^5}\left(\sum_{k=1}^{\infty}\frac{1}{k^3}\right)}_{\zeta(3)\zeta(5)} - 10\underbrace{\sum_{n=1}^{\infty}\frac{1}{n^6}\left(\sum_{k=1}^{\infty}\frac{1}{k^2}\right)}_{\zeta(2)\zeta(6)} + 20\sum_{n=1}^{\infty}\left(\sum_{k=1}^{\infty}\frac{1}{n^7}\left(\frac{1}{k} - \frac{1}{k+n}\right)\right)$$

$$- 10\sum_{n=1}^{\infty}\left(\sum_{k=1}^{\infty}\frac{1}{n^6(k+n)^2}\right) - 4\sum_{n=1}^{\infty}\left(\sum_{k=1}^{\infty}\frac{1}{n^5(k+n)^3}\right)$$

$$\left\{\text{use the simple fact that } H_n = \sum_{k=1}^{\infty}\left(\frac{1}{k} - \frac{1}{k+n}\right)\right\}$$

$$= 4\zeta(3)\zeta(5) - 10\zeta(2)\zeta(6) + 20\sum_{n=1}^{\infty}\frac{H_n}{n^7} - 10\sum_{n=1}^{\infty}\frac{1}{n^6}\left(\zeta(2) - H_n^{(2)}\right)$$

$$- 4\sum_{n=1}^{\infty}\frac{1}{n^5}\left(\zeta(3) - H_n^{(3)}\right)$$

{expand the last two series with Riemann zeta tail and at the same time}

{exploit the Euler sum generalization in (6.149), the case $n = 7$, Sect. 6.19}

$$= \frac{35}{3}\zeta(8) - 20\zeta(3)\zeta(5) + 10\sum_{n=1}^{\infty}\frac{H_n^{(2)}}{n^6} + 4\sum_{n=1}^{\infty}\frac{H_n^{(3)}}{n^5},$$

whence we obtain that

$$5\sum_{n=1}^{\infty}\frac{H_n^{(2)}}{n^6} + 2\sum_{n=1}^{\infty}\frac{H_n^{(3)}}{n^5} = 10\zeta(3)\zeta(5) - \frac{21}{4}\zeta(8),$$

where in the calculations above I also used that $\zeta(6) = \dfrac{\pi^6}{945}$ besides the zeta values given immediately after the result in (6.287), and the solution to the point (i) of the problem is finalized.

Returning to the first series of the main sum of series found at the point (ii) and applying Abel's summation, the series version in (6.7), Sect. 6.2, where we set $a_n = 1$ and $b_n = \zeta(3)\zeta(6) - H_n^{(3)} H_n^{(6)}$, we write that

$$\sum_{n=1}^{\infty} \left(\zeta(3)\zeta(6) - H_n^{(3)} H_n^{(6)} \right)$$

$$= \underbrace{\lim_{N \to \infty} N \left(\zeta(3)\zeta(6) - H_{N+1}^{(3)} H_{N+1}^{(6)} \right)}_{0} + \sum_{n=1}^{\infty} n \left(H_{n+1}^{(3)} H_{n+1}^{(6)} - H_n^{(3)} H_n^{(6)} \right)$$

$$\overset{\substack{\text{reindex} \\ \text{the series}}}{=} \sum_{n=2}^{\infty} (n-1) \left(H_n^{(3)} H_n^{(6)} - H_{n-1}^{(3)} H_{n-1}^{(6)} \right) = \sum_{n=1}^{\infty} (n-1) \left(H_n^{(3)} H_n^{(6)} - H_{n-1}^{(3)} H_{n-1}^{(6)} \right)$$

$$= \sum_{n=1}^{\infty} (n-1) \left(H_n^{(3)} H_n^{(6)} - \left(H_n^{(3)} - \frac{1}{n^3} \right) \left(H_n^{(6)} - \frac{1}{n^6} \right) \right)$$

$$= \sum_{n=1}^{\infty} \frac{H_n^{(3)}}{n^5} + \sum_{n=1}^{\infty} \frac{H_n^{(6)}}{n^2} - \sum_{n=1}^{\infty} \frac{H_n^{(3)}}{n^6} - \sum_{n=1}^{\infty} \frac{H_n^{(6)}}{n^3} - \sum_{n=1}^{\infty} \frac{1}{n^8} + \sum_{n=1}^{\infty} \frac{1}{n^9}$$

$$= \sum_{n=1}^{\infty} \frac{H_n^{(3)}}{n^5} - \sum_{n=1}^{\infty} \frac{H_n^{(2)}}{n^6} - \left(\sum_{n=1}^{\infty} \frac{H_n^{(3)}}{n^6} + \sum_{n=1}^{\infty} \frac{H_n^{(6)}}{n^3} \right) + \frac{5}{3}\zeta(8) + \zeta(9)$$

$$= \frac{5}{3}\zeta(8) - \zeta(3)\zeta(6) + \sum_{n=1}^{\infty} \frac{H_n^{(3)}}{n^5} - \sum_{n=1}^{\infty} \frac{H_n^{(2)}}{n^6}, \tag{6.288}$$

where in the calculations I used the transformation $\displaystyle\sum_{n=1}^{\infty} \frac{H_n^{(6)}}{n^2} = \frac{8}{3}\zeta(8) - \sum_{n=1}^{\infty} \frac{H_n^{(2)}}{n^6}$, which is obtained by using (6.102), with $p = 2$, $q = 6$ (or $p = 6$, $q = 2$), and letting $n \to \infty$, Sect. 6.13, and then the fact that by using the same result, with $p = 3$ and $q = 6$ (or $p = 6$, $q = 3$), and letting $n \to \infty$, we get $\displaystyle\sum_{n=1}^{\infty} \frac{H_n^{(3)}}{n^6} +$

$\displaystyle\sum_{n=1}^{\infty} \frac{H_n^{(6)}}{n^3} = \zeta(9) + \zeta(3)\zeta(6)$. Moreover, it is easy to see that the residual limit after applying Abel's summation is 0, and this is because we have the following asymptotic expansion behaviors, $H_N^{(3)} = \zeta(3) + O(1/N^2)$ and $H_N^{(6)} = \zeta(6) + O(1/N^5)$. A simple reasoning behind such estimations is given during the extraction of (6.236), Sect. 6.31.

At this point, we are ready to finalize the calculations, and then we write

$$2 \sum_{n=1}^{\infty} \left(\zeta(3)\zeta(6) - H_n^{(3)} H_n^{(6)} \right) + 7 \sum_{n=1}^{\infty} \frac{H_n^{(2)}}{n^6}$$

{make use of the result in (6.288)}

$$= \frac{10}{3}\zeta(8) - 2\zeta(3)\zeta(6) + 2 \sum_{n=1}^{\infty} \frac{H_n^{(3)}}{n^5} + 5 \sum_{n=1}^{\infty} \frac{H_n^{(2)}}{n^6}$$

{exploit result result from the point (i)}

$$= 10\zeta(3)\zeta(5) - 2\zeta(3)\zeta(6) - \frac{23}{12}\zeta(8),$$

and the solution to the point (ii) of the problem is finalized.

The present problem definitely poses some difficulties, especially for those that are not aware of the advancement and challenges in the world of harmonic series. For instance, one might be tempted to assume that if $\sum_{n=1}^{\infty} \frac{H_n^{(2)}}{n^2} = \frac{7}{4}\zeta(4)$ and $\sum_{n=1}^{\infty} \frac{H_n^{(2)}}{n^4} = \zeta^2(3) - \frac{1}{3}\zeta(6)$, then also the case $\sum_{n=1}^{\infty} \frac{H_n^{(2)}}{n^6}$ might have such a nice closed form, however, in this case no closed form alike is known in the mathematical literature at the moment of writing the present book. *That sounds strange and unexpected*, you might think. Some details on this matter I also provided in *(Almost) Impossible Integrals, Sums, and Series*, as seen in [76, Chapter 6, Sect. 6.21, p.384] (the calculable harmonic series given above as examples may be found evaluated in the same book). Then, another difficult part is to know how to group the series in order to calculate them together, not taken separately. Assuming we know these steps and properly group the series, then we are ready to start the evaluation of the sum of series (which might be found challenging, as well). Good to know that in the mathematical literature, more similar examples with sums of harmonic series are known, which we want to evaluate together, not separately (see [14]).

6.45   Amazing, Unexpected Relations with Alternating and Non-alternating Harmonic Series of Weight 5 and 7

Solution For a thorough understanding of the *story* in this section, you might also like to cover Sect. 6.33 (if you didn't do it yet!). In the last part of the mentioned section, I said that trying to use the strategy in the second solution to the series

$\sum_{n=1}^{\infty} \dfrac{H_n H_{2n}}{n^2}$, which may be found in Sect. 6.23, leads to an undesired cancelation

of the series to evaluate there, that is, $\sum_{n=1}^{\infty} \dfrac{H_n H_{2n}}{n^3}$, and then I specified that we get

something else. It's time to find out more about it!

The famous scientist and inventor Alexander Graham Bell (1847–1922), best known for inventing the telephone, said once, *When one door closes, another opens.* That cancelation mentioned above actually opens the door to *a wonderful result*! By using the strategy in the second solution from Sect. 6.23, where this time both sides of the identity in (4.82) from Sect. 4.13 are multiplied by $1/n^3$, we get that

$$\sum_{n=1}^{\infty} \frac{1}{n^3} \left(\sum_{k=1}^{n} \frac{H_k}{k+n} \right) \overset{\substack{\text{reverse the order} \\ \text{of summation}}}{=\!=\!=} \sum_{k=1}^{\infty} \left(\sum_{n=k}^{\infty} \frac{H_k}{n^3(k+n)} \right)$$

$$= \sum_{k=1}^{\infty} \left(\sum_{n=k}^{\infty} \left(\frac{H_k}{kn^3} - \frac{H_k}{k^2 n^2} + \frac{H_k}{k^2 n(k+n)} \right) \right)$$

$$= \sum_{k=1}^{\infty} \frac{H_k}{k} \left(\sum_{n=k}^{\infty} \frac{1}{n^3} \right) - \sum_{k=1}^{\infty} \frac{H_k}{k^2} \left(\sum_{n=k}^{\infty} \frac{1}{n^2} \right) + \sum_{k=1}^{\infty} \frac{H_k}{k^2} \left(\sum_{n=k}^{\infty} \frac{1}{n(k+n)} \right)$$

{reverse the order of summation in the first double series and for the inner}

{sum of the third double series use the result found in (6.187), Sect. 6.23}

$$= \sum_{n=1}^{\infty} \frac{1}{n^3} \left(\sum_{k=1}^{n} \frac{H_k}{k} \right) - \sum_{k=1}^{\infty} \frac{H_k}{k^2} \left(\zeta(2) - H_k^{(2)} + \frac{1}{k^2} \right)$$

$$+ \sum_{k=1}^{\infty} \frac{H_k(H_{2k} - H_k + 1/(2k))}{k^3}$$

$$\left\{ \text{use } \sum_{k=1}^{n} \frac{H_k}{k} = \frac{1}{2}(H_n^2 + H_n^{(2)}) \text{ based on } p = q = 1 \text{ in (6.102), Sect. 6.13, and} \right\}$$

{then expand the last two series and replace the summation variable k by n}

$$= -\zeta(2) \sum_{n=1}^{\infty} \frac{H_n}{n^2} - \frac{1}{2} \sum_{n=1}^{\infty} \frac{H_n}{n^4} + \frac{1}{2} \sum_{n=1}^{\infty} \frac{H_n^{(2)}}{n^3} - \frac{1}{2} \sum_{n=1}^{\infty} \frac{H_n^2}{n^3} + \sum_{n=1}^{\infty} \frac{H_n H_n^{(2)}}{n^2}$$

$$+ \sum_{n=1}^{\infty} \frac{H_n H_{2n}}{n^3}. \tag{6.289}$$

The other side given by the multiplication of (4.82), Sect. 4.13, by $1/n^3$ is

$$\sum_{n=1}^{\infty} \frac{H_n H_{2n}}{n^3} - \sum_{n=1}^{\infty} \frac{H_n^2}{n^3} - 16 \sum_{n=1}^{\infty} \frac{H_{2n}}{(2n)^4} + 2 \sum_{n=1}^{\infty} \frac{H_n}{n^4} - \frac{1}{2} \sum_{n=1}^{\infty} \frac{H_n^{(2)}}{n^3}$$

$$\left\{ \text{for the third sum use that } \sum_{n=1}^{\infty} a_{2n} = \frac{1}{2} \left(\sum_{n=1}^{\infty} a_n - \sum_{n=1}^{\infty} (-1)^{n-1} a_n \right) \right\}$$

$$= \sum_{n=1}^{\infty} \frac{H_n H_{2n}}{n^3} - \sum_{n=1}^{\infty} \frac{H_n^2}{n^3} - 6 \sum_{n=1}^{\infty} \frac{H_n}{n^4} + 8 \sum_{n=1}^{\infty} (-1)^{n-1} \frac{H_n}{n^4} - \frac{1}{2} \sum_{n=1}^{\infty} \frac{H_n^{(2)}}{n^3}. \qquad (6.290)$$

Combining the results in (6.289) and (6.290), we obtain

$$8 \sum_{n=1}^{\infty} (-1)^{n-1} \frac{H_n}{n^4} - \sum_{n=1}^{\infty} \frac{H_n H_n^{(2)}}{n^2}$$

$$= \frac{11}{2} \sum_{n=1}^{\infty} \frac{H_n}{n^4} - \zeta(2) \sum_{n=1}^{\infty} \frac{H_n}{n^2} + \frac{1}{2} \sum_{n=1}^{\infty} \frac{H_n^2}{n^3} + \sum_{n=1}^{\infty} \frac{H_n^{(2)}}{n^3}$$

$$= \frac{55}{4} \zeta(5) - 5\zeta(2)\zeta(3),$$

where, after the first equality sign, the first two series are particular cases of the Euler sum generalization in (6.149), Sect. 6.19; then $\sum_{n=1}^{\infty} \frac{H_n^2}{n^3} = \frac{7}{2}\zeta(5) - \zeta(2)\zeta(3)$ is given in [76, Chapter 4, p.293] and in my article *A new proof for a classical quadratic harmonic series* that was published in the *Journal of Classical Analysis*, Vol. 8, No. 2, 2016 (see [72]); and the last series, $\sum_{n=1}^{\infty} \frac{H_n^{(2)}}{n^3} = 3\zeta(2)\zeta(3) - \frac{9}{2}\zeta(5)$, is given in [76, Chapter 6, p.386] and in the previously mentioned article, and the solution to the point (i) is finalized.

Both $\sum_{n=1}^{\infty} (-1)^{n-1} \frac{H_n}{n^4}$ and $\sum_{n=1}^{\infty} \frac{H_n H_n^{(2)}}{n^2}$ may be viewed to some extent as hard nuts to crack, especially when we want to perform the calculations by series manipulations! Knowing the value of either of them and using the identity above, we extract the value of the other series!

Good to know that the series of weight 5, $\sum_{n=1}^{\infty} \frac{H_n H_n^{(2)}}{n^2} = \zeta(2)\zeta(3) + \zeta(5)$, is calculated by series manipulations in *(Almost) Impossible Integrals, Sums, and Series* (for details, see [76, Chapter 6, pp. 398–401]).

Passing to the point (ii) of the problem, we proceed in a similar style as before and multiply both sides of the identity in (4.82), Sect. 4.13, by $1/n^5$ that gives

$$\sum_{n=1}^{\infty} \frac{1}{n^5} \left(\sum_{k=1}^{n} \frac{H_k}{k+n} \right) \overset{\text{reverse the order}}{\underset{\text{of summation}}{=}} \sum_{k=1}^{\infty} \left(\sum_{n=k}^{\infty} \frac{H_k}{n^5(k+n)} \right)$$

$$= \sum_{k=1}^{\infty} \left(\sum_{n=k}^{\infty} \left(\frac{H_k}{kn^5} - \frac{H_k}{k^2 n^4} + \frac{H_k}{k^3 n^3} - \frac{H_k}{k^4 n^2} + \frac{H_k}{k^4 n(k+n)} \right) \right)$$

$$= \sum_{k=1}^{\infty} \frac{H_k}{k} \sum_{n=k}^{\infty} \frac{1}{n^5} - \sum_{k=1}^{\infty} \frac{H_k}{k^2} \sum_{n=k}^{\infty} \frac{1}{n^4} + \sum_{k=1}^{\infty} \frac{H_k}{k^3} \sum_{n=k}^{\infty} \frac{1}{n^3} - \sum_{k=1}^{\infty} \frac{H_k}{k^4} \sum_{n=k}^{\infty} \frac{1}{n^2}$$

$$+ \sum_{k=1}^{\infty} \frac{H_k}{k^4} \sum_{n=k}^{\infty} \frac{1}{n(n+k)}$$

{reverse the order of summation in the first double series, and then, for}

{the inner sum of the fifth double series use the result in (6.187), Sect. 6.23}

$$= \sum_{n=1}^{\infty} \frac{1}{n^5} \sum_{k=1}^{n} \frac{H_k}{k} - \sum_{k=1}^{\infty} \frac{H_k}{k^2} \left(\zeta(4) - H_k^{(4)} + \frac{1}{k^4} \right)$$

$$+ \sum_{k=1}^{\infty} \frac{H_k}{k^3} \left(\zeta(3) - H_k^{(3)} + \frac{1}{k^3} \right) - \sum_{k=1}^{\infty} \frac{H_k}{k^4} \left(\zeta(2) - H_k^{(2)} + \frac{1}{k^2} \right)$$

$$+ \sum_{k=1}^{\infty} \frac{H_k(H_{2k} - H_k + 1/(2k))}{k^5}$$

$$\left\{ \text{use } \sum_{k=1}^{n} \frac{H_k}{k} = \frac{1}{2}(H_n^2 + H_n^{(2)}), \text{ based on } p = q = 1 \text{ in (6.102), Sect. 6.13, and} \right\}$$

{then expand the last four series and replace the summation variable k by n}

$$= -\zeta(4) \sum_{n=1}^{\infty} \frac{H_n}{n^2} + \zeta(3) \sum_{n=1}^{\infty} \frac{H_n}{n^3} - \zeta(2) \sum_{n=1}^{\infty} \frac{H_n}{n^4} - \frac{1}{2} \sum_{n=1}^{\infty} \frac{H_n}{n^6} - \frac{1}{2} \sum_{n=1}^{\infty} \frac{H_n^2}{n^5}$$

$$+ \frac{1}{2} \sum_{n=1}^{\infty} \frac{H_n^{(2)}}{n^5} + \sum_{n=1}^{\infty} \frac{H_n H_n^{(4)}}{n^2} + \sum_{n=1}^{\infty} \frac{H_n H_n^{(2)}}{n^4} - \sum_{n=1}^{\infty} \frac{H_n H_n^{(3)}}{n^3} + \sum_{n=1}^{\infty} \frac{H_n H_{2n}}{n^5}.$$

$$(6.291)$$

Then, the other side given by the multiplication of (4.82), Sect. 4.13, by $1/n^5$ is

$$\sum_{n=1}^{\infty} \frac{H_n H_{2n}}{n^5} - \sum_{n=1}^{\infty} \frac{H_n^2}{n^5} - 64 \sum_{n=1}^{\infty} \frac{H_{2n}}{(2n)^6} + 2 \sum_{n=1}^{\infty} \frac{H_n}{n^6} - \frac{1}{2} \sum_{n=1}^{\infty} \frac{H_n^{(2)}}{n^5}$$

$$\left\{ \text{for the third sum use that } \sum_{n=1}^{\infty} a_{2n} = \frac{1}{2} \left(\sum_{n=1}^{\infty} a_n - \sum_{n=1}^{\infty} (-1)^{n-1} a_n \right) \right\}$$

$$= \sum_{n=1}^{\infty} \frac{H_n H_{2n}}{n^5} - \sum_{n=1}^{\infty} \frac{H_n^2}{n^5} - 30 \sum_{n=1}^{\infty} \frac{H_n}{n^6} + 32 \sum_{n=1}^{\infty} (-1)^{n-1} \frac{H_n}{n^6} - \frac{1}{2} \sum_{n=1}^{\infty} \frac{H_n^{(2)}}{n^5}.$$
$$(6.292)$$

Combining the results in (6.291) and (6.292), we obtain

$$32 \sum_{n=1}^{\infty} (-1)^{n-1} \frac{H_n}{n^6} + \sum_{n=1}^{\infty} \frac{H_n H_n^{(3)}}{n^3} - \sum_{n=1}^{\infty} \frac{H_n H_n^{(4)}}{n^2} - \sum_{n=1}^{\infty} \frac{H_n H_n^{(2)}}{n^4}$$

$$= -\zeta(4) \sum_{n=1}^{\infty} \frac{H_n}{n^2} + \zeta(3) \sum_{n=1}^{\infty} \frac{H_n}{n^3} - \zeta(2) \sum_{n=1}^{\infty} \frac{H_n}{n^4} + \frac{59}{2} \sum_{n=1}^{\infty} \frac{H_n}{n^6} + \frac{1}{2} \sum_{n=1}^{\infty} \frac{H_n^2}{n^5} + \sum_{n=1}^{\infty} \frac{H_n^{(2)}}{n^5}$$

$$= 111\zeta(7) - 28\zeta(2)\zeta(5) - 27\zeta(3)\zeta(4), \qquad (6.293)$$

where, after the first equality sign, the first four series are particular cases of the Euler sum generalization in (6.149), Sect. 6.19, then $\sum_{n=1}^{\infty} \frac{H_n^2}{n^5} = 6\zeta(7) - \zeta(2)\zeta(5) - \frac{5}{2}\zeta(3)\zeta(4)$ is given in [76, Chapter 4, p.293], and the series $\sum_{n=1}^{\infty} \frac{H_n^{(2)}}{n^5} = 5\zeta(2)\zeta(5) + 2\zeta(3)\zeta(4) - 10\zeta(7)$ is calculated in [76, Chapter 6, p.389].

Fine! But how would we turn this identity into a useful tool, you might hurry to ask! In fact, the identity is amazingly useful when combined with some other results.

For the beginning, let's observe that the series $\sum_{n=1}^{\infty} (-1)^{n-1} \frac{H_n}{n^6} = \frac{377}{128}\zeta(7) - \frac{7}{8}\zeta(3)\zeta(4) - \frac{1}{2}\zeta(2)\zeta(5)$ is a particular case of (4.105), Sect. 4.21, and then the result in (6.293) can be written as

$$\sum_{n=1}^{\infty} \frac{H_n H_n^{(3)}}{n^3} - \sum_{n=1}^{\infty} \frac{H_n H_n^{(4)}}{n^2} - \sum_{n=1}^{\infty} \frac{H_n H_n^{(2)}}{n^4}$$

$$= \frac{67}{4}\zeta(7) - 12\zeta(2)\zeta(5) + \zeta(3)\zeta(4). \qquad (6.294)$$

Now, if you carefully went through the weight 7-related identities I presented, derived, and exploited in *(Almost) Impossible Integrals, Sums, and Series*, and spent some time with them, you might remember the key identity

$$\sum_{n=1}^{\infty} \frac{H_n H_n^{(4)}}{n^2} - \sum_{n=1}^{\infty} \frac{H_n H_n^{(2)}}{n^4} = \frac{1}{2} \left(5\zeta(2)\zeta(5) - \frac{9}{2}\zeta(3)\zeta(4) \right), \qquad (6.295)$$

which may be found in [76, Chapter 4, p.297].

If we use (6.295) in (6.294) to express $\sum_{n=1}^{\infty} \frac{H_n H_n^{(4)}}{n^2}$ in terms of $\sum_{n=1}^{\infty} \frac{H_n H_n^{(2)}}{n^4}$, then we arrive at

$$\sum_{n=1}^{\infty} \frac{H_n H_n^{(3)}}{n^3} - 2\sum_{n=1}^{\infty} \frac{H_n H_n^{(2)}}{n^4} = \frac{67}{4}\zeta(7) - \frac{5}{4}\zeta(3)\zeta(4) - \frac{19}{2}\zeta(2)\zeta(5). \qquad (6.296)$$

This is a marvelous moment!, the first sentence that naturally comes to mind. We already have a similar identity to the one in (6.296), which may be found in [76, Chapter 4, p.297] (see the second identity below),

$$\begin{cases} \sum_{n=1}^{\infty} \dfrac{H_n H_n^{(3)}}{n^3} - 2\sum_{n=1}^{\infty} \dfrac{H_n H_n^{(2)}}{n^4} = \dfrac{67}{4}\zeta(7) - \dfrac{5}{4}\zeta(3)\zeta(4) - \dfrac{19}{2}\zeta(2)\zeta(5) \\ \sum_{n=1}^{\infty} \dfrac{H_n H_n^{(3)}}{n^3} + 2\sum_{n=1}^{\infty} \dfrac{H_n H_n^{(2)}}{n^4} = 4\zeta(7) + \dfrac{7}{4}\zeta(3)\zeta(4) - \dfrac{3}{2}\zeta(2)\zeta(5), \end{cases}$$

$$(6.297)$$

that immediately leads to the extraction of both series. So, we have

$$\sum_{n=1}^{\infty} \frac{H_n H_n^{(2)}}{n^4} = 2\zeta(2)\zeta(5) + \frac{3}{4}\zeta(3)\zeta(4) - \frac{51}{16}\zeta(7) \qquad (6.298)$$

and

$$\sum_{n=1}^{\infty} \frac{H_n H_n^{(3)}}{n^3} = \frac{83}{8}\zeta(7) - \frac{11}{2}\zeta(2)\zeta(5) + \frac{1}{4}\zeta(3)\zeta(4). \qquad (6.299)$$

Furthermore, by combining (6.295) and (6.298), we get that

$$\sum_{n=1}^{\infty} \frac{H_n H_n^{(4)}}{n^2} = \frac{9}{2}\zeta(2)\zeta(5) - \frac{3}{2}\zeta(3)\zeta(4) - \frac{51}{16}\zeta(7). \qquad (6.300)$$

Extracting harmonic series of weight 7 like the ones in (6.298), (6.299), and (6.300) is always a big challenge as seen in *(Almost) Impossible Integrals, Sums, and Series*, particularly if we want to do it by elementary series manipulations. So good to have more options in hand!

Finally, the curious reader may go further and exploit the strategy above to get relations with more advanced harmonic series. Here is a relation with harmonic series of weight 9 obtained by similar means,

$$
\sum_{n=1}^{\infty} \frac{H_n H_n^{(6)}}{n^2} - \sum_{n=1}^{\infty} \frac{H_n H_n^{(5)}}{n^3} + \sum_{n=1}^{\infty} \frac{H_n H_n^{(4)}}{n^4} - \sum_{n=1}^{\infty} \frac{H_n H_n^{(3)}}{n^5} + \sum_{n=1}^{\infty} \frac{H_n H_n^{(2)}}{n^6}
$$

$$
= \frac{1}{3}\zeta^3(3) - 2\zeta(3)\zeta(6) + 59\zeta(2)\zeta(7) + 10\zeta(4)\zeta(5) - \frac{629}{6}\zeta(9). \qquad (6.301)
$$

Surely, one may continue and try to combine the identity in (6.301) and other identities in order to get simpler sums of series (e.g., see [76, Chapter 4, p.297]).

I'll put an end here to the present section, and, of course, the curious reader may go further and explore other possibilities, too.

6.46 A Quintet of Advanced Harmonic Series of Weight 5 Involving Skew-Harmonic Numbers

Solution The first two harmonic series are also part of my paper, *Two advanced harmonic series of weight 5 involving skew-harmonic numbers* (see [75, December 15, 2019]), and I'll consider the ideas presented in the mentioned paper and show two solutions for each of the first two points of the problem.

The harmonic series from the last three points will turn to be beautiful extractions where we'll also use special logarithmic integrals involving harmonic numbers, and the series given at the point (i), when needed.

To get a first solution, I'll use a less known Cauchy product, that is, the one in (4.29), Sect. 4.5, where if we replace x by $-x$ and then integrate both sides from $x = 0$ to $x = 1$, we get

$$
\int_0^1 \frac{(\operatorname{Li}_2(-x))^2}{1+x}\,dx = -\int_0^1 \sum_{n=1}^{\infty} x^n(-1)^{n-1}\left((H_n^{(2)})^2 - 5H_n^{(4)} + 4\sum_{k=1}^{n} \frac{H_k}{k^3}\right) dx
$$

{reverse the order of summation and integration}

$$
= -\sum_{n=1}^{\infty} \int_0^1 x^n(-1)^{n-1}\left((H_n^{(2)})^2 - 5H_n^{(4)} + 4\sum_{k=1}^{n} \frac{H_k}{k^3}\right) dx
$$

$$
= -\sum_{n=1}^{\infty} (-1)^{n-1}\frac{1}{n+1}\left(\left(H_{n+1}^{(2)} - \frac{1}{(n+1)^2}\right)^2 - 5\left(H_{n+1}^{(4)} - \frac{1}{(n+1)^4}\right)\right)
$$

$$+4\sum_{k=1}^{n+1}\frac{H_k}{k^3}-4\frac{H_{n+1}}{(n+1)^3}\bigg)$$

{reindex the series and expand it}

$$=\frac{45}{8}\zeta(5)-5\sum_{n=1}^{\infty}(-1)^{n-1}\frac{H_n^{(4)}}{n}+\sum_{n=1}^{\infty}(-1)^{n-1}\frac{(H_n^{(2)})^2}{n}-2\sum_{n=1}^{\infty}(-1)^{n-1}\frac{H_n^{(2)}}{n^3}$$

$$+4\sum_{n=1}^{\infty}\frac{(-1)^{n-1}}{n}\sum_{k=1}^{n-1}\frac{H_k}{k^3}. \tag{6.302}$$

Reversing the order of summation in the last series of (6.302), we have

$$\sum_{n=1}^{\infty}\frac{(-1)^{n-1}}{n}\sum_{k=1}^{n-1}\frac{H_k}{k^3}=\sum_{k=1}^{\infty}\frac{H_k}{k^3}\sum_{n=k+1}^{\infty}\frac{(-1)^{n-1}}{n}=\sum_{k=1}^{\infty}\frac{H_k}{k^3}\left(\log(2)-\overline{H}_k\right)$$

$$=\log(2)\sum_{k=1}^{\infty}\frac{H_k}{k^3}-\sum_{k=1}^{\infty}\frac{H_k\overline{H}_k}{k^3}=\frac{5}{4}\log(2)\zeta(4)-\sum_{k=1}^{\infty}\frac{H_k\overline{H}_k}{k^3}, \tag{6.303}$$

where in the calculations I used the particular case $n = 3$ of the Euler sum generalization in (6.149), Sect. 6.19.

For the integral in (6.302), we integrate by parts once and use the integral result in (1.213), Sect. 1.47,

$$\int_0^1\frac{(\text{Li}_2(-x))^2}{1+x}dx=\int_0^1(\log(1+x))'(\text{Li}_2(-x))^2dx=\log(1+x)(\text{Li}_2(-x))^2\bigg|_{x=0}^{x=1}$$

$$+2\int_0^1\frac{\log^2(1+x)\,\text{Li}_2(-x)}{x}dx=\frac{5}{8}\log(2)\zeta(4)+2\int_0^1\frac{\log^2(1+x)\,\text{Li}_2(-x)}{x}dx$$

$$=\frac{4}{15}\log^5(2)-\frac{4}{3}\log^3(2)\zeta(2)+\frac{7}{2}\log^2(2)\zeta(3)+\frac{5}{8}\log(2)\zeta(4)-\frac{1}{4}\zeta(2)\zeta(3)$$

$$-\frac{125}{16}\zeta(5)+8\log(2)\,\text{Li}_4\left(\frac{1}{2}\right)+8\,\text{Li}_5\left(\frac{1}{2}\right). \tag{6.304}$$

Collecting the results from (6.303) and (6.304) in (6.302), where we also consider the particular case $m = 4$ of the harmonic series generalization in (4.106), Sect. 4.22, further $\sum_{n=1}^{\infty}(-1)^{n-1}\frac{(H_n^{(2)})^2}{n}=\frac{1}{5}\log^5(2)-\frac{2}{3}\log^3(2)\zeta(2)+$

$\frac{29}{4}\log(2)\zeta(4) - \frac{259}{16}\zeta(5) + \frac{5}{8}\zeta(2)\zeta(3) + 8\log(2)\operatorname{Li}_4\left(\frac{1}{2}\right) + 16\operatorname{Li}_5\left(\frac{1}{2}\right)$, which

happily is found and calculated in [76, Chapter 6, Section 6.58, pp. 523–529], and

finally employ, $\displaystyle\sum_{n=1}^{\infty}(-1)^{n-1}\frac{H_n^{(2)}}{n^3} = \frac{5}{8}\zeta(2)\zeta(3) - \frac{11}{32}\zeta(5)$ (see [76, Chapter 4,

p.311]), we have that

$$\sum_{n=1}^{\infty}\frac{H_n\overline{H}_n}{n^3}$$

$$= \frac{1}{6}\log^3(2)\zeta(2) - \frac{7}{8}\log^2(2)\zeta(3) + 4\log(2)\zeta(4) - \frac{193}{64}\zeta(5) + \frac{3}{8}\zeta(2)\zeta(3)$$

$$- \frac{1}{60}\log^5(2) + 2\operatorname{Li}_5\left(\frac{1}{2}\right),$$

and the first solution to the point (i) is complete.

For a second solution, recall the relations between the skew-harmonic numbers and harmonic numbers based on parity, $\overline{H}_{2n} = H_{2n} - H_n$ and $\overline{H}_{2n+1} = H_{2n+1} - H_n = H_{2n} - H_n + 1/(2n+1)$, since I'll make a reduction to the atypical harmonic series of weight 5, involving harmonic numbers of the type H_{2n}.

Splitting the main series according to the parity, we write

$$\sum_{n=1}^{\infty}\frac{H_n\overline{H}_n}{n^3} = 1 + \sum_{n=1}^{\infty}\frac{H_{2n}\overline{H}_{2n}}{(2n)^3} + \sum_{n=1}^{\infty}\frac{H_{2n+1}\overline{H}_{2n+1}}{(2n+1)^3}$$

$$= 1 + \sum_{n=1}^{\infty}\frac{H_{2n}^2}{(2n)^3} + \sum_{n=1}^{\infty}\frac{H_{2n+1}^2}{(2n+1)^3} - \sum_{n=1}^{\infty}\frac{H_n}{(2n+1)^4} - \frac{1}{8}\sum_{n=1}^{\infty}\frac{H_n H_{2n}}{n^3} - \sum_{n=1}^{\infty}\frac{H_n H_{2n}}{(2n+1)^3}$$

$$= \sum_{n=1}^{\infty}\frac{H_n^2}{n^3} - \sum_{n=1}^{\infty}\frac{H_n}{(2n+1)^4} - \frac{1}{8}\sum_{n=1}^{\infty}\frac{H_n H_{2n}}{n^3} - \sum_{n=1}^{\infty}\frac{H_n H_{2n}}{(2n+1)^3}$$

$$= \frac{1}{6}\log^3(2)\zeta(2) - \frac{7}{8}\log^2(2)\zeta(3) + 4\log(2)\zeta(4) - \frac{193}{64}\zeta(5) + \frac{3}{8}\zeta(2)\zeta(3)$$

$$- \frac{1}{60}\log^5(2) + 2\operatorname{Li}_5\left(\frac{1}{2}\right),$$

where during the calculations, I also used that $\displaystyle\sum_{n=1}^{\infty}\frac{H_n^2}{n^3} = \frac{7}{2}\zeta(5) - \zeta(2)\zeta(3)$ (see [76, Chapter 4, p.293]); then the second resulting series is the case $m = 2$ of the result in (4.100), Sect. 4.19; and the last two series are given in (4.124), Sect. 4.33, and (4.130), Sect. 4.35, and the second solution to the point (i) is complete.

We are on the move to the next point of the problem where for a first solution, I would like to exploit the point (i) of the problem. If we consider that $\dfrac{(H_n - \overline{H}_n)^2}{n^3} = \dfrac{H_n^2}{n^3} - 2\dfrac{H_n \overline{H}_n}{n^3} + \dfrac{(\overline{H}_n)^2}{n^3}$ and then split the series $\displaystyle\sum_{n=1}^{\infty} \dfrac{(H_n - \overline{H}_n)^2}{n^3}$ based on parity, using that for $m = 2n$ and $m = 2n + 1$ we have $H_m - \overline{H}_m = H_n$, we write

$$\sum_{n=1}^{\infty} \frac{(\overline{H}_n)^2}{n^3}$$

$$= \sum_{n=1}^{\infty} \frac{(H_n - \overline{H}_n)^2}{n^3} - \sum_{n=1}^{\infty} \frac{H_n^2}{n^3} + 2 \sum_{n=1}^{\infty} \frac{H_n \overline{H}_n}{n^3} = \sum_{n=1}^{\infty} \frac{H_n^2}{(2n + 1)^3} - \frac{7}{8} \sum_{n=1}^{\infty} \frac{H_n^2}{n^3} + 2 \sum_{n=1}^{\infty} \frac{H_n \overline{H}_n}{n^3}$$

$$= \frac{1}{3} \log^3(2)\zeta(2) + \frac{7}{4} \log^2(2)\zeta(3) + \frac{19}{8} \log(2)\zeta(4) - \frac{167}{32}\zeta(5) - \frac{1}{30} \log^5(2)$$

$$+ \frac{3}{4}\zeta(2)\zeta(3) + 4 \operatorname{Li}_5\left(\frac{1}{2}\right),$$

where in the calculations I used the series result in (4.127), Sect. 4.34, then the fact that $\displaystyle\sum_{n=1}^{\infty} \frac{H_n^2}{n^3} = \frac{7}{2}\zeta(5) - \zeta(2)\zeta(3)$ (see [76, Chapter 4, p.293]), and last but not least, the value of the series at the previous point, and the first solution to the point (ii) is complete.

Following the same reduction style presented in the second solution to the point (i) of the problem, we have

$$\sum_{n=1}^{\infty} \frac{(\overline{H}_n)^2}{n^3} = 1 + \frac{1}{8} \sum_{n=1}^{\infty} \frac{(\overline{H}_{2n})^2}{n^3} + \sum_{n=1}^{\infty} \frac{(\overline{H}_{2n+1})^2}{(2n + 1)^3}$$

$$= 1 + \frac{1}{8} \sum_{n=1}^{\infty} \frac{H_n^2}{n^3} + \frac{1}{8} \sum_{n=1}^{\infty} \frac{H_{2n}^2}{n^3} + \sum_{n=1}^{\infty} \frac{H_{2n+1}^2}{(2n + 1)^3} + \sum_{n=1}^{\infty} \frac{H_n^2}{(2n + 1)^3} - 2 \sum_{n=1}^{\infty} \frac{H_n}{(2n + 1)^4}$$

$$- \frac{1}{4} \sum_{n=1}^{\infty} \frac{H_n H_{2n}}{n^3} - 2 \sum_{n=1}^{\infty} \frac{H_n H_{2n}}{(2n + 1)^3}$$

$$= \frac{9}{8} \sum_{n=1}^{\infty} \frac{H_n^2}{n^3} + \sum_{n=1}^{\infty} \frac{H_n^2}{(2n + 1)^3} - 2 \sum_{n=1}^{\infty} \frac{H_n}{(2n + 1)^4} - \frac{1}{4} \sum_{n=1}^{\infty} \frac{H_n H_{2n}}{n^3} - 2 \sum_{n=1}^{\infty} \frac{H_n H_{2n}}{(2n + 1)^3}$$

$$= \frac{1}{3}\log^3(2)\zeta(2) + \frac{7}{4}\log^2(2)\zeta(3) + \frac{19}{8}\log(2)\zeta(4) - \frac{167}{32}\zeta(5) - \frac{1}{30}\log^5(2)$$

$$+ \frac{3}{4}\zeta(2)\zeta(3) + 4\operatorname{Li}_5\left(\frac{1}{2}\right),$$

where all five resulting series have already appeared during the previous solutions of the section, and the last two series arose during the second solution to the first point of the problem, and the second solution to the point (ii) is complete.

Let's pass now to the last three points of the problem! As for the third point of the problem, we start from the case $m = 1$ of the identity in (1.127), Sect. 1.25, where if we multiply both sides by $(-1)^{n-1}H_n/n$ and take the sum from $n = 1$ to ∞, we get

$$-\frac{1}{2}\zeta(2)\sum_{n=1}^{\infty}\frac{H_n}{n^2} - \log(2)\sum_{n=1}^{\infty}\frac{H_n}{n^3} - \log(2)\sum_{n=1}^{\infty}(-1)^{n-1}\frac{H_n}{n^3} + \sum_{n=1}^{\infty}\frac{H_n\overline{H}_n}{n^3} + \sum_{n=1}^{\infty}\frac{H_n\overline{H}_n^{(2)}}{n^2}$$

$$= \sum_{n=1}^{\infty}(-1)^{n-1}\frac{H_n}{n}\int_0^1 x^{n-1}\log(x)\log(1+x)dx$$

{reverse the order of integration and summation}

$$= -\int_0^1 \frac{\log(x)\log(1+x)}{x}\sum_{n=1}^{\infty}(-x)^n\frac{H_n}{n}dx$$

{use the generating function in (4.34), the first equality, Sect. 4.6, and expand}

$$= -\frac{1}{2}\int_0^1 \frac{\log(x)\log^3(1+x)}{x}dx - \int_0^1 \frac{\log(1+x)\log(x)\operatorname{Li}_2(-x)}{x}dx$$

{the values of the integrals are given in (1.198), Sect. 1.45, and (6.458), Sect. 6.59}

$$= \frac{1}{5}\log^5(2) - \log^3(2)\zeta(2) + \frac{21}{8}\log^2(2)\zeta(3) - \frac{41}{16}\zeta(5) - \frac{15}{8}\zeta(2)\zeta(3)$$

$$+ 6\log(2)\operatorname{Li}_4\left(\frac{1}{2}\right) + 6\operatorname{Li}_5\left(\frac{1}{2}\right),$$

where if we focus on the opposite sides and also consider the cases $n = 2, 3$ of the Euler sum generalization in (6.149), Sect. 6.19, then the alternating Euler sum in (4.103), Sect. 4.20, and at last the value of the series at the point (i), we manage to obtain the desired result,

$$\sum_{n=1}^{\infty} \frac{H_n \overline{H}_n^{(2)}}{n^2}$$

$$= \frac{29}{64}\zeta(5) - \frac{5}{4}\zeta(2)\zeta(3) + \frac{7}{4}\log^2(2)\zeta(3) - \frac{2}{3}\log^3(2)\zeta(2) + \frac{2}{15}\log^5(2)$$

$$+ 4\log(2)\operatorname{Li}_4\left(\frac{1}{2}\right) + 4\operatorname{Li}_5\left(\frac{1}{2}\right),$$

and the solution to the point (iii) is complete.

Next, jumping to the fourth point of the problem, we want to start from the fact that $\sum_{k=1}^{n} \frac{H_k^{(2)}}{k^2} = \frac{1}{2}\left(H_n^{(4)} + (H_n^{(2)})^2\right)$, which is obtained if we set $p = q = 2$ in (6.102), Sect. 6.13, and combining this fact together with Abel's summation, the series version in (6.7), Sect. 6.2, where we set $a_n = \frac{H_n^{(2)}}{n^2}$ and $b_n = \overline{H}_n$, we obtain

$$\sum_{n=1}^{\infty} \frac{\overline{H}_n H_n^{(2)}}{n^2}$$

$$= \frac{7}{4}\log(2)\zeta(4) - \frac{1}{2}\sum_{n=1}^{\infty}(-1)^n \frac{H_{n+1}^{(4)} - 1/(n+1)^4 + (H_{n+1}^{(2)} - 1/(n+1)^2)^2}{(n+1)}$$

{reindex the series and then expand it}

$$= \frac{7}{4}\log(2)\zeta(4) - \frac{1}{2}\sum_{n=1}^{\infty}(-1)^{n-1}\frac{H_n^{(4)}}{n} + \sum_{n=1}^{\infty}(-1)^{n-1}\frac{H_n^{(2)}}{n^3} - \frac{1}{2}\sum_{n=1}^{\infty}(-1)^{n-1}\frac{(H_n^{(2)})^2}{n}$$

$$= \frac{27}{4}\zeta(5) + \frac{1}{2}\zeta(2)\zeta(3) + \frac{1}{3}\log^3(2)\zeta(2) - \frac{23}{16}\log(2)\zeta(4) - \frac{1}{10}\log^5(2)$$

$$- 4\log(2)\operatorname{Li}_4\left(\frac{1}{2}\right) - 8\operatorname{Li}_5\left(\frac{1}{2}\right),$$

where the first series is the particular case $m = 4$ of the generalization in (4.106), Sect. 4.22, then the second series is $\sum_{n=1}^{\infty}(-1)^{n-1}\frac{H_n^{(2)}}{n^3} = \frac{5}{8}\zeta(2)\zeta(3) - \frac{11}{32}\zeta(5)$ (see [76, Chapter 4, p.311]), and the third alternating harmonic series needed is

$$\sum_{n=1}^{\infty}(-1)^{n-1}\frac{(H_n^{(2)})^2}{n} = \frac{1}{5}\log^5(2) - \frac{2}{3}\log^3(2)\zeta(2) + \frac{29}{4}\log(2)\zeta(4) - \frac{259}{16}\zeta(5) +$$

$\frac{5}{8}\zeta(2)\zeta(3)+8\log(2)\operatorname{Li}_4\left(\frac{1}{2}\right)+16\operatorname{Li}_5\left(\frac{1}{2}\right)$, found and calculated in [76, Chapter 6, Section 6.58, pp. 523–529], and the solution to the point (iv) is complete.

To calculate the series from the last point, we might like to recall the result in (1.128), the case $m = 1$, Sect. 1.25, where if we multiply both sides by $1/n^2$, and consider the summation from $n = 1$ to ∞, we get

$$\zeta(2)\zeta(3) - \frac{15}{4}\log(2)\zeta(4) + \log(2)\sum_{n=1}^{\infty}\frac{H_n^{(2)}}{n^2} - \sum_{n=1}^{\infty}\frac{H_n^{(3)}}{n^2} + \frac{1}{2}\zeta(2)\sum_{n=1}^{\infty}\frac{\overline{H}_n}{n^2}$$

$$+ \log(2)\sum_{n=1}^{\infty}\frac{\overline{H}_n^{(2)}}{n^2} - \sum_{n=1}^{\infty}\frac{\overline{H}_n\overline{H}_n^{(2)}}{n^2} = \int_0^1 \frac{\operatorname{Li}_2(x)\log(x)\log(1+x)}{1-x}dx$$

{make use of the generating function in (4.62), the case $m = 1$, Sect. 4.10}

$$= \int_0^1 \operatorname{Li}_2(x)\log(x)\sum_{n=1}^{\infty}x^n\overline{H}_n dx$$

$$= \sum_{n=1}^{\infty}\overline{H}_n\int_0^1 x^n\log(x)\operatorname{Li}_2(x)dx = \sum_{n=1}^{\infty}\overline{H}_n\frac{d}{dn}\left(\int_0^1 x^n\operatorname{Li}_2(x)dx\right)$$

{make use of the result in (3.327), Sect. 3.46}

$$= \sum_{n=1}^{\infty}\overline{H}_n\frac{d}{dn}\left(\zeta(2)\frac{1}{n+1} - \frac{\psi(n+2)+\gamma}{(n+1)^2}\right)$$

$$= \sum_{n=1}^{\infty}\left(\overline{H}_{n+1} - (-1)^n\frac{1}{n+1}\right)\left(\frac{H_{n+1}^{(2)}}{(n+1)^2} + 2\frac{H_{n+1}}{(n+1)^3} - 2\zeta(2)\frac{1}{(n+1)^2}\right)$$

{reindex the series and expand it}

$$= \frac{3}{2}\zeta(2)\zeta(3) - 2\sum_{n=1}^{\infty}(-1)^{n-1}\frac{H_n}{n^4} - \sum_{n=1}^{\infty}(-1)^{n-1}\frac{H_n^{(2)}}{n^3} - 2\zeta(2)\sum_{n=1}^{\infty}\frac{\overline{H}_n}{n^2}$$

$$+ 2\sum_{n=1}^{\infty}\frac{H_n\overline{H}_n}{n^3} + \sum_{n=1}^{\infty}\frac{\overline{H}_n H_n^{(2)}}{n^2}. \tag{6.305}$$

Using the strategy of reversing the order of summation, we note that

$$\sum_{n=1}^{\infty} \frac{\overline{H}_n^{(2)}}{n^2} = \sum_{n=1}^{\infty} \frac{1}{n^2} \sum_{k=1}^{n} (-1)^{k-1} \frac{1}{k^2} = \sum_{k=1}^{\infty} (-1)^{k-1} \frac{1}{k^2} \sum_{n=k}^{\infty} \frac{1}{n^2}$$

$$= \sum_{k=1}^{\infty} (-1)^{k-1} \frac{1}{k^2} \left(\zeta(2) - H_k^{(2)} + \frac{1}{k^2} \right) = \zeta(2) \sum_{k=1}^{\infty} (-1)^{k-1} \frac{1}{k^2}$$

$$+ \sum_{k=1}^{\infty} (-1)^{k-1} \frac{1}{k^4} - \sum_{k=1}^{\infty} (-1)^{k-1} \frac{H_k^{(2)}}{k^2}$$

$$= \frac{85}{16} \zeta(4) - \frac{7}{2} \log(2)\zeta(3) + \log^2(2)\zeta(2) - \frac{1}{6} \log^4(2) - 4\operatorname{Li}_4\left(\frac{1}{2}\right), \qquad (6.306)$$

where the last series is given in (6.190), Sect. 6.24.

Returning to the result in (6.305), having in mind the opposite sides of it, and considering that $\sum_{n=1}^{\infty} \frac{H_n^{(2)}}{n^2} = \frac{7}{4}\zeta(4)$, which is obtained by setting $p = q = 2$ and letting $n \to \infty$ in (6.102), Sect. 6.13, then using that $\sum_{n=1}^{\infty} \frac{H_n^{(3)}}{n^2} = \frac{11}{2}\zeta(5) - 2\zeta(2)\zeta(3)$ (see [76, Chapter 6, p.386]), next counting the particular case $m = 2$ of the generalization in (4.108), Sect. 4.22, the value of the series in (6.306), then the value of the series in (4.104), Sect. 4.20, or the case $m = 2$ of the alternating Euler sum generalization in (4.105), Sect. 4.21, further $\sum_{n=1}^{\infty} (-1)^{n-1} \frac{H_n^{(2)}}{n^3} = \frac{5}{8}\zeta(2)\zeta(3) - \frac{11}{32}\zeta(5)$ (see [76, Chapter 4, p.311]), and finally the harmonic series at the points (i) and (iv), we arrive at

$$\sum_{n=1}^{\infty} \frac{\overline{H}_n \overline{H}_n^{(2)}}{n^2}$$

$$= \frac{1}{3} \log^3(2)\zeta(2) - \frac{7}{4} \log^2(2)\zeta(3) + \frac{49}{8} \log(2)\zeta(4) - \frac{23}{8}\zeta(5) - \frac{3}{4}\zeta(2)\zeta(3)$$

$$- \frac{1}{30} \log^5(2) + 4\operatorname{Li}_5\left(\frac{1}{2}\right),$$

and the solution to the point (v) is complete.

The curious reader looking for thrilling adventures may consider evaluating many other series alike using similar ideas and strategies as the ones presented above!

6.47 Fourier Series Expansions of the Bernoulli Polynomials

Solution Taking a look at the first two points of the problem, we immediately recognize that the given results are *classical* (see [17, **1.443.1**, **1.443.2**, p.46]). The ones at the last two points are immediately derived based on the previous ones. The usefulness of such results is well understood during the extractions of the results in Sect. 1.53.

Since in this section we deal with Bernoulli numbers[7] and Bernoulli polynomials, called after the renowned Swiss mathematician Jacob Bernoulli (1655–1705), some basic information about them, in terms of definitions and properties, will be given below.

By conventions we may have two variants of the Bernoulli numbers, one where $B_1 = -1/2$ (also possibly denoted by B_1^-) and one with $B_1 = 1/2$ (also found denoted by B_1^+), a point the reader might want to pay attention to. Moreover, a good fact to keep in mind is that one can easily switch from one convention to another by using the simple fact that $B_n^+ = (-1)^n B_n^-$, and this is possible due to (6.309) presented below (a careful read of the results beneath will make this fact clear).

Now, a way of defining Bernoulli numbers is achieved by the generating function

$$\frac{x}{e^x - 1} = \sum_{n=0}^{\infty} x^n \frac{B_n}{n!}. \tag{6.307}$$

Reindexing the series in (6.307), multiplying both sides by $e^x - 1 = \sum_{n=1}^{\infty} \frac{x^n}{n!}$, and using the Cauchy product of two series, we get that

$$x = (e^x - 1) \sum_{n=1}^{\infty} x^{n-1} \frac{B_{n-1}}{(n-1)!} = \left(\sum_{n=1}^{\infty} x^{n-1} \frac{B_{n-1}}{(n-1)!} \right) \left(\sum_{n=1}^{\infty} \frac{x^n}{n!} \right)$$

---

[7] Bernoulli numbers are introduced in the mathematical literature by Jacob Bernoulli during the study of the sums of powers of consecutive integers, $\sum_{k=1}^{n} k^p = 1^p + 2^p + \cdots + n^p$. Historical details and mathematical facts about Bernoulli numbers and Bernoulli polynomials may be found beautifully presented in the title, *Bernoulli Numbers and Zeta Functions* by Tsuneo Arakawa, Tomoyoshi Ibukiyama, Masanobu Kaneko, Springer (2014).

$$= \sum_{n=1}^{\infty} \frac{x^n}{n!} \left(\sum_{k=1}^{n} \binom{n}{k-1} B_{k-1} \right) = \sum_{n=1}^{\infty} \frac{x^n}{n!} \left(\sum_{k=0}^{n-1} \binom{n}{k} B_k \right),$$

and comparing the opposite sides, we observe that for $n \geq 2$ we have

$$\sum_{k=0}^{n-1} \binom{n}{k} B_k = 0. \tag{6.308}$$

A similar way to go is presented in [62, Chapter 13, pp. 357–358]. In view of (6.308) we have: $B_0 = 1$, $B_1 = -1/2$, $B_2 = 1/6$, $B_3 = 0$, $B_4 = -1/30$, and so on.

Except for B_1, all Bernoulli numbers of odd index are equal to 0. This can be easily seen if we consider $f(x) = \dfrac{x}{e^x - 1} + \dfrac{x}{2} - 1 = \dfrac{x}{2} \left(\dfrac{e^{x/2} + e^{-x/2}}{e^{x/2} - e^{-x/2}} \right) - 1 = \displaystyle\sum_{n=2}^{\infty} x^n \dfrac{B_n}{n!}$ and observe that $f(x)$ is even, that is, $f(x) = f(-x)$, which means that we expect to have $B_n = (-1)^n B_n$, $n \geq 2$. Hence, for odd values of n, with $n \geq 3$, we arrive at $B_n = 0$, or to put it differently

$$B_{2n+1} = 0, \; n \geq 1. \tag{6.309}$$

As regards (6.309), one may also consult [3, Chapter 1, pp. 10–11], [62, Chapter 13, p.358].

On the other hand, we have the following definition of the Bernoulli polynomials by the generating function:

$$\frac{t e^{xt}}{e^t - 1} = \sum_{n=0}^{\infty} t^n \frac{B_n(x)}{n!}, \; n \geq 0. \tag{6.310}$$

Considering the generating functions in (6.307) and (6.310), and using the Cauchy product of two series, we have

$$\sum_{n=0}^{\infty} t^n \frac{B_n(x)}{n!} = \frac{t}{e^t - 1} \cdot e^{xt} = \left(\sum_{n=0}^{\infty} t^n \frac{B_n}{n!} \right) \left(\sum_{n=0}^{\infty} \frac{(xt)^n}{n!} \right) = \sum_{n=0}^{\infty} \frac{t^n}{n!} \left(\sum_{k=0}^{n} \binom{n}{k} B_k x^{n-k} \right),$$

whence by comparing the coefficients of the series at the opposite sides, we get

$$B_n(x) = \sum_{k=0}^{n} \binom{n}{k} B_k x^{n-k}, \tag{6.311}$$

which is a result we'll need during the main calculations.

Then, in view of (6.311), we have $B_0(x) = 1$, $B_1(x) = x - 1/2$, $B_2(x) = x^2 - x + 1/6$, $B_3(x) = x^3 - 3/2x^2 + 1/2x$, $B_4(x) = x^4 - 2x^3 + x^2 - 1/30$, and so on. Also, we want to observe that if we set $x = 0$ in (6.310) and compare it to (6.307), we have

$$B_n(0) = B_n, \tag{6.312}$$

and this is another result we need during the main calculations.

Based on (6.310), it is easy to see that

$$B'_n(x) = n B_{n-1}(x), \quad n \geq 1, \tag{6.313}$$

since $\displaystyle\sum_{n=1}^{\infty} t^n \frac{B'_n(x)}{n!} = \sum_{n=0}^{\infty} t^n \frac{B'_n(x)}{n!} = t \cdot \frac{te^{xt}}{e^t - 1} = \sum_{n=0}^{\infty} t^{n+1} \frac{B_n(x)}{n!} =$

$\displaystyle\sum_{n=1}^{\infty} t^n \frac{B_{n-1}(x)}{(n-1)!}.$

At this point, by exploiting (6.313), we arrive at the integral

$$\int_a^x B_n(t)dt = \frac{B_{n+1}(x) - B_{n+1}(a)}{n+1}, \tag{6.314}$$

which is another important result we need during the main calculations.

Next, I'll prepare one of the central auxiliary results needed in the main calculations, which is again a classical result,

$$B_{2n} = 2(-1)^{n-1} \frac{(2n)!}{(2\pi)^{2n}} \zeta(2n). \tag{6.315}$$

Proof Preparing to go the famous Euler way! First, we take Euler's infinite product for the sine, that is, $\sin(x) = x \displaystyle\prod_{n=1}^{\infty} \left(1 - \frac{x^2}{n^2\pi^2}\right)$ (see [10, pp. 251–252]), where if we replace x by πx, take log of both sides, differentiate, and then rearrange, we obtain that

$$\pi x \cot(\pi x) = 1 - 2x^2 \sum_{n=1}^{\infty} \frac{1}{n^2} \frac{1}{1 - x^2/n^2} = 1 - 2 \sum_{n=1}^{\infty} \left(\sum_{k=1}^{\infty} \left(\frac{x^2}{n^2}\right)^k\right)$$

$$\overset{\substack{\text{reverse the order} \\ \text{of summation}}}{=} 1 - 2 \sum_{k=1}^{\infty} x^{2k} \left(\sum_{n=1}^{\infty} \frac{1}{n^{2k}}\right) = 1 - 2 \sum_{k=1}^{\infty} x^{2k} \zeta(2k), \quad |x| < 1. \tag{6.316}$$

At the same time, since based on the generating function in (6.307), we can write that $\dfrac{x}{e^x - 1} + \dfrac{x}{2} = 1 + \sum\limits_{n=2}^{\infty} x^n \dfrac{B_n}{n!}$, if we consider (6.309), we get that

$$\frac{x}{e^x - 1} + \frac{x}{2} = \frac{1}{2} x \frac{e^x + 1}{e^x - 1} = 1 + \sum_{n=1}^{\infty} x^{2n} \frac{B_{2n}}{(2n)!}. \tag{6.317}$$

If we replace x by $i2\pi x$ in the second equality of (6.317) and use that $\sin(z) = \dfrac{e^{iz} - e^{-iz}}{2i}$ and $\cos(z) = \dfrac{e^{iz} + e^{-iz}}{2}$, we write that

$$i\pi x \frac{e^{i2\pi x} + 1}{e^{i2\pi x} - 1} = \pi x \frac{(e^{i\pi x} + e^{-i\pi x})/2}{(e^{i\pi x} - e^{-i\pi x})/(2i)} = \pi x \cot(\pi x) = 1 + \sum_{n=1}^{\infty} x^{2n}(-1)^n(2\pi)^{2n} \frac{B_{2n}}{(2n)!},$$

or if we focus on the last equality,

$$\pi x \cot(\pi x) = 1 - \sum_{n=1}^{\infty} x^{2n}(-1)^{n-1}(2\pi)^{2n} \frac{B_{2n}}{(2n)!}. \tag{6.318}$$

At last, comparing the coefficients of the series in (6.316) and (6.318), the result in (6.315) follows, and we bring an end to the solution of the auxiliary result. ∎

Recall that the first section in *(Almost) Impossible Integrals, Sums, and Series* also asks as a challenging question to solve the Basel problem by using the integral result presented in that section. At this point, we observe that (6.315) offers us another way of extracting the value of $\zeta(2)$, but at the same time also the values of all $\zeta(2n)$, $n \geq 1$. *Simply wonderful, isn't it?*

We are ready now to get a jump on things! To make it very clear, the second equality for each of the results found at all four main points of the problem is immediately obtained if we employ the identity in (6.311).

Now, one of the most popular Fourier series that appear in many calculations is $\sum\limits_{n=1}^{\infty} \dfrac{\sin(nt)}{n} = \dfrac{\pi - t}{2}$, $0 < t < 2\pi$ (see **1.441.1** in [17]), and if we make the variable change $2\pi x = t$, we get that

$$\sum_{n=1}^{\infty} \frac{\sin(2\pi nx)}{n} = \pi \left(\frac{1}{2} - x \right) = (-1)^1 \pi B_1(x), \ 0 < x < 1, \tag{6.319}$$

where I also considered the use of the Bernoulli polynomials. At the end of Sect. 3.12, I showed that $\sum\limits_{n=1}^{\infty} (-1)^{n-1} \dfrac{\sin(2n\theta)}{n} = \theta$, $\theta \in \left(-\dfrac{\pi}{2}, \dfrac{\pi}{2} \right)$, and if we let here the variable change $2\theta = \pi - t$, we arrive at the Fourier series considered for getting (6.319). Of course, one can do it more directly by considering

$\Im\left\{ \log\left(1 - e^{i\theta}\right) \right\}$. At last, we can derive it by using the usual way[8] described in [71, pp. 28–29].

Keeping in mind (6.314), (6.312), and (6.308), we integrate the opposite sides of (6.319) from $x = 0$ to $x = t$, then replace t by x, and rearrange,

$$\sum_{n=1}^{\infty} \frac{\cos(2\pi nx)}{n^2} - \sum_{n=1}^{\infty} \frac{1}{n^2} = (-1)^2 \pi^2 \left(B_2(x) - B_2(0)\right) = (-1)^2 \pi^2 B_2(x) - \frac{\pi^2}{6},$$

and if we use the well-known value, $\displaystyle\sum_{n=1}^{\infty} \frac{1}{n^2} = \zeta(2) = \frac{\pi^2}{6}$, {see the comments after (6.318)}, we get

$$\sum_{n=1}^{\infty} \frac{\cos(2\pi nx)}{n^{2 \cdot 1}} = (-1)^2 \pi^2 B_2(x) = (-1)^{1-1} \frac{1}{2} \frac{(2\pi)^{2 \cdot 1}}{1 \cdot 2} B_2(x). \qquad (6.320)$$

Proceeding similarly as in the previous case, integrating both sides of the first equality in (6.320) from $x = 0$ to $x = t$, then replacing t by x, and rearranging, we have

$$\sum_{n=1}^{\infty} \frac{\sin(2\pi nx)}{n^{2 \cdot 2 - 1}} = (-1)^2 \frac{2}{3} \pi^3 B_3(x) = (-1)^2 \frac{1}{2} \frac{(2\pi)^{2 \cdot 2 - 1}}{1 \cdot 2 \cdot 3} B_3(x), \qquad (6.321)$$

which is immediately shown by the integral formula in (6.314) and the fact that $B_3(0) = B_3 = 0$, based on (6.312) and (6.309).

So, continuing in (6.321) with integration as before, we get that $\displaystyle\sum_{n=1}^{\infty} \frac{\cos(2\pi nx)}{n^{2 \cdot 2}} =$

$(-1)^{2-1} \frac{1}{2} \frac{(2\pi)^{2 \cdot 2}}{1 \cdot 2 \cdot 3 \cdot 4} B_4(x),$ $\displaystyle\sum_{n=1}^{\infty} \frac{\sin(2\pi nx)}{n^{2 \cdot 3 - 1}} = (-1)^3 \frac{1}{2} \frac{(2\pi)^{2 \cdot 3 - 1}}{1 \cdot 2 \cdot 3 \cdot 4 \cdot 5} B_5(x),$ and so on.

Note that in the previous work, I acted as if the values of the main series were unknown, not given in the problem statement, thus showing that it is possible to easily see the pattern of the generalizations from the first two main points.

Let's prepare now for induction! We assume, based on the previous investigations on some particular cases, that the following result holds:

---

[8] It is worth mentioning a great classical book on Fourier series, called exactly like that, *Fourier Series*, by Georgi P. Tolstov, Dover Publications, New York, 1976. One of the first examples in the book, with a worked-out solution, is about expanding in Fourier series $f(x) = x$, where $0 < x < 2\pi$, and this leads exactly and promptly to the stated Fourier series.

$$\sum_{k=1}^{\infty} \frac{\sin(2k\pi t)}{k^{2n-1}} = (-1)^n \frac{1}{2} \frac{(2\pi)^{2n-1}}{(2n-1)!} B_{2n-1}(t).$$

(6.322)

Then, we want to show that if (6.322) is true, this implies that $\displaystyle\sum_{k=1}^{\infty} \frac{\sin(2k\pi t)}{k^{2n+1}} =$

$(-1)^{n+1} \dfrac{1}{2} \dfrac{(2\pi)^{2n+1}}{(2n+1)!} B_{2n+1}(t)$ is true, as well.

Integrating both sides of (6.322) from $t = 0$ to $t = x$, using (6.314), (6.312), and (6.315), and rearranging, we have

$$\sum_{k=1}^{\infty} \frac{\cos(2k\pi x)}{k^{2n}} - \underbrace{\sum_{k=1}^{\infty} \frac{1}{k^{2n}}}_{\zeta(2n)} = (-1)^{n+1} \frac{1}{2} \frac{(2\pi)^{2n}}{(2n)!} \left(B_{2n}(x) - B_{2n} \right)$$

$$= (-1)^{n+1} \frac{1}{2} \frac{(2\pi)^{2n}}{(2n)!} \left(B_{2n}(x) - 2(-1)^{n-1} \frac{(2n)!}{(2\pi)^{2n}} \zeta(2n) \right),$$

or

$$\sum_{k=1}^{\infty} \frac{\cos(2k\pi x)}{k^{2n}} = (-1)^{n+1} \frac{1}{2} \frac{(2\pi)^{2n}}{(2n)!} B_{2n}(x).$$

(6.323)

After another *round* of integration, this time involving (6.323), from $x = 0$ to $x = t$, we obtain that

$$\sum_{k=1}^{\infty} \frac{\sin(2k\pi t)}{k^{2n+1}} = (-1)^{n+1} \frac{1}{2} \frac{(2\pi)^{2n+1}}{(2n+1)!} \left(B_{2n+1}(t) - B_{2n+1} \right)$$

$$= (-1)^{n+1} \frac{1}{2} \frac{(2\pi)^{2n+1}}{(2n+1)!} B_{2n+1}(t),$$

and we have arrived at the desired result aimed by induction, and thus we have finalized the induction procedure, and the solution to the point (i) of the problem is complete. During the last calculations, I used (6.314), (6.312), and (6.309).

Since now we have the point (i) proved, based on (6.323), we arrive at the second main result, and the solution to the point (ii) of the problem is complete. Of course, if I had started from the point (ii) first, then I could have used induction again.

The last two points of the problem are straightforward. For example, noting the simple facts that $\displaystyle\sum_{k=1}^{\infty} (-1)^{k-1} \frac{\sin(2k\pi x)}{k^{2n-1}} = \sum_{k=1}^{\infty} \frac{\sin(2k\pi x)}{k^{2n-1}} - 2 \sum_{k=1}^{\infty} \frac{\sin(2(2k)\pi x)}{(2k)^{2n-1}}$

and $\displaystyle\sum_{k=1}^{\infty}(-1)^{k-1}\frac{\cos(2k\pi x)}{k^{2n}} = \sum_{k=1}^{\infty}\frac{\cos(2k\pi x)}{k^{2n}} - 2\sum_{k=1}^{\infty}\frac{\cos(2(2k)\pi x)}{(2k)^{2n}}$, where x
follows the restrictions in the problem statement, and using the series results from the first two main points, we conclude that

$$\sum_{k=1}^{\infty}(-1)^{k-1}\frac{\sin(2k\pi x)}{k^{2n-1}} = (-1)^{n-1}\frac{\pi^{2n-1}}{(2n-1)!}\left(B_{2n-1}(2x) - 2^{2(n-1)}B_{2n-1}(x)\right)$$

and

$$\sum_{k=1}^{\infty}(-1)^{k-1}\frac{\cos(2k\pi x)}{k^{2n}} = (-1)^{n-1}\frac{\pi^{2n}}{(2n)!}\left(2^{2n-1}B_{2n}(x) - B_{2n}(2x)\right),$$

and the solutions to the points (iii) and (iv) of the problem are complete.

The Fourier series from the first two points of the problem are also found derived in [2, Chapter 12, p.267], [3, Chapter 4, pp. 59–62].

6.48 Stunning Fourier Series with log(sin(x)) and log(cos(x)) Raised to Positive Integer Powers, Related to Harmonic Numbers

Solution I'll start this section by reminding you that in *(Almost) Impossible Integrals, Sums, and Series* I presented the derivations of two powerful Fourier series, that is, for $\tan(x)\log(\sin(x))$, $\cot(x)\log(\cos(x))$, and a Fourier-like series for $\log(\sin(x))\log(\cos(x))$, together with applications of them (see [76, Chapter 3, pp. 242–252]). Well, at this point, one might be at least tempted to think about the way the Fourier series of $\log^2(\sin(x))$ and/or $\log^2(\cos(x))$ would look like. *Did you think about them?*

In the work below, for deriving the given Fourier series, I'll mainly propose solutions constructed based on the use of some special integrals.

In order to prove the result at the point (i), I'll state first the following integral result we need:

$$\int_0^{\infty}\log(\sinh(y))e^{-2ny}\mathrm{d}y = -\frac{1}{2}\log(2)\frac{1}{n} + \frac{1}{4}\frac{1}{n^2} - \frac{1}{2}\frac{H_n}{n}. \tag{6.324}$$

Proof A good idea is to start with the variable change $y = -\log(\sqrt{t})$ in an attempt to turn the integral into known logarithmic integrals, and then we get

$$\int_0^\infty \log(\sinh(y))e^{-2ny}\mathrm{d}y = \frac{1}{2}\int_0^1 t^{n-1}\log\left(\frac{1-t}{2\sqrt{t}}\right)\mathrm{d}t$$

$$= -\frac{1}{2}\log(2)\underbrace{\int_0^1 t^{n-1}\mathrm{d}t}_{1/n} -\frac{1}{4}\underbrace{\int_0^1 t^{n-1}\log(t)\mathrm{d}t}_{-1/n^2} +\frac{1}{2}\int_0^1 t^{n-1}\log(1-t)\mathrm{d}t$$

$$= -\frac{1}{2}\log(2)\frac{1}{n} + \frac{1}{4}\frac{1}{n^2} - \frac{1}{2}\frac{H_n}{n},$$

where in the calculations I used that $\int_0^1 x^{n-1}\log(1-x)\mathrm{d}x = -\frac{H_n}{n}$, which you may also find stated and proved in (3.10), Sect. 3.3. ∎

Another result we want to prove and use is

$$\sum_{n=1}^\infty \cos(2nx)e^{-2ny} = -\frac{1}{2} - \frac{1}{2}\frac{\sinh(2y)}{\cos(2x) - \cosh(2y)}. \tag{6.325}$$

Proof Before proceeding with the derivation of the result above, we observe that for $|p| < 1$, we have

$$\sum_{n=1}^\infty (pe^{ix})^n = \lim_{N\to\infty}\sum_{n=1}^N (pe^{ix})^n = \lim_{N\to\infty} p\frac{1 - (pe^{ix})^N}{e^{-ix} - p}$$

$$= \frac{p}{e^{-ix} - p} = \frac{p(e^{ix} - p)}{(e^{-ix} - p)(e^{ix} - p)}$$

$$= \frac{p(\cos(x) - p)}{1 - 2p\cos(x) + p^2} + i\frac{p\sin(x)}{1 - 2p\cos(x) + p^2},$$

where by equating the real and imaginary parts, we arrive at

$$\sum_{n=1}^\infty p^n \sin(nx) = \frac{p\sin(x)}{1 - 2p\cos(x) + p^2} \tag{6.326}$$

and

$$\sum_{n=1}^\infty p^n \cos(nx) = \frac{p(\cos(x) - p)}{1 - 2p\cos(x) + p^2}. \tag{6.327}$$

Essentially, the derivation flow of (6.326) and (6.327) is the same as the one in [76, Chapter 3, p.244], based on the Euler's formula, $e^{ix} = \cos(x) + i\sin(x)$. When the values of $\sum_{n=1}^{\infty} p^n \sin(nx)$ and $\sum_{n=1}^{\infty} p^n \cos(nx)$ are given, it is easier to figure out that alternatively one might want to multiply both sides of (6.326) and (6.327) by $1 - 2p\cos(x) + p^2$ and then reduce all to telescoping sums.

For example, exploiting the well-known trigonometric identity, $\sin(a)\cos(b) = \frac{1}{2}(\sin(a+b) + \sin(a-b))$, we immediately obtain that

$$\sum_{n=1}^{\infty} p^n \sin(nx)(1 - 2p\cos(x) + p^2)$$

$$= \sum_{n=1}^{\infty} \left(p^{n+2} \sin(nx) - p^{n+1} \sin((n-1)x) + p^n \sin(nx) - p^{n+1} \sin((n+1)x) \right)$$

$$= \underbrace{\lim_{N \to \infty} \sum_{n=1}^{N} \left(p^{n+2} \sin(nx) - p^{n+1} \sin((n-1)x) \right)}_{\lim_{N \to \infty} p^{N+2} \sin(Nx) = 0}$$

$$+ \underbrace{\lim_{N \to \infty} \sum_{n=1}^{N} \left(p^n \sin(nx) - p^{n+1} \sin((n+1)x) \right)}_{\lim_{N \to \infty} \left(p \sin(x) - p^{N+1} \sin((N+1)x) \right) = p\sin(x)} = p\sin(x).$$

Further, by using the trigonometric identity, $\cos(a)\cos(b) = \frac{1}{2}(\cos(a+b) + \cos(a-b))$, we get

$$\sum_{n=1}^{\infty} p^n \cos(nx)(1 - 2p\cos(x) + p^2)$$

$$= \sum_{n=1}^{\infty} \left(p^{n+2} \cos(nx) - p^{n+1} \cos((n-1)x) + p^n \cos(nx) - p^{n+1} \cos((n+1)x) \right)$$

$$= \lim_{N \to \infty} \sum_{n=1}^{N} \left(p^{n+2} \cos(nx) - p^{n+1} \cos((n-1)x) \right)$$

$$\underbrace{\lim_{N \to \infty} \left(p^{N+2} \cos(Nx) - p^2 \right) = -p^2}$$

$$+ \underbrace{\lim_{N \to \infty} \sum_{n=1}^{N} \left(p^n \cos(nx) - p^{n+1} \cos((n+1)x) \right)}_{} = p(\cos(x) - p).$$

$$\underbrace{\lim_{N \to \infty} \left(p \cos(x) - p^{N+1} \cos((N+1)x) \right) = p \cos(x)}$$

At last, the result from (6.325) is extracted by replacing x by $2x$ and setting $p = e^{-2y}$ in (6.327). ∎

Now, we multiply both sides of (6.324) by $\cos(2nx)$, then consider the summation from $n = 1$ to ∞, next change the order of integration and summation, and employ the result in (6.325). That means we have the following lines of calculations:

$$-\frac{1}{2} \sum_{n=1}^{\infty} \left(\log(2)\frac{1}{n} - \frac{1}{2n^2} + \frac{H_n}{n} \right) \cos(2nx) = \sum_{n=1}^{\infty} \cos(2nx) \int_0^{\infty} \log(\sinh(y)) e^{-2ny} dy$$

$$= \int_0^{\infty} \log(\sinh(y)) \sum_{n=1}^{\infty} \cos(2nx) e^{-2ny} dy$$

$$= -\frac{1}{2} \int_0^{\infty} \log(\sinh(y)) \left(1 + \frac{\sinh(2y)}{\cos(2x) - \cosh(2y)} \right) dy$$

$$\overset{\sinh(y)=t}{=\!=\!=} \underbrace{-\frac{1}{2} \int_0^{\infty} \left(\frac{\log(t)}{\sqrt{1+t^2}} - \frac{t \log(t)}{\sin^2(x) + t^2} \right) dt}_{\text{Also check the end of the section!}}$$

$$= -\frac{1}{2} \lim_{s \to 0} \frac{d}{ds} \int_0^{\infty} \left(\frac{t^s}{\sqrt{1+t^2}} - \frac{t^{s+1}}{\sin^2(x) + t^2} \right) dt$$

$$= -\frac{1}{2} \lim_{s \to 0} \frac{d}{ds} \left(\int_0^{\infty} \frac{t^s}{\sqrt{1+t^2}} dt - \int_0^{\infty} \frac{t^{s+1}}{\sin^2(x) + t^2} dt \right). \qquad (6.328)$$

Above I also employed the hyperbolic identities, $\sinh(2\operatorname{arcsinh}(x)) = 2x\sqrt{1+x^2}$ and $\cosh(2\operatorname{arcsinh}(x)) = 2x^2 + 1$. Both remaining integrals can be reduced to the Beta function form[9] $B(a,b) = \displaystyle\int_0^\infty \frac{x^{a-1}}{(1+x)^{a+b}}dx$. So, for the first integral in (6.328), we make the variable change $t^2 = u$ and then consider the previous Beta function form that leads to

$$\int_0^\infty \frac{t^s}{\sqrt{1+t^2}}dt = \frac{1}{2}\int_0^\infty \frac{u^{(1+s)/2-1}}{\sqrt{1+u}}du = \frac{1}{2}B\left(\frac{1+s}{2}, -\frac{s}{2}\right)$$

$$\left\{\text{use the Beta–Gamma identity, } B(x,y) = \frac{\Gamma(x)\Gamma(y)}{\Gamma(x+y)}, \text{ and } \Gamma\left(\frac{1}{2}\right) = \sqrt{\pi}\right\}$$

$$= \frac{1}{2\sqrt{\pi}}\Gamma\left(\frac{1+s}{2}\right)\Gamma\left(-\frac{s}{2}\right). \tag{6.329}$$

Observe that $\Gamma\left(\dfrac{1}{2}\right) = \sqrt{\pi}$ used above is well-known and also discussed in [76, Chapter 3, p.196]. In short, setting $s = 1/2$ in Euler's reflection formula, $\Gamma(a)\Gamma(1-a) = \dfrac{\pi}{\sin(a\pi)}$, we arrive at the announced result. Or we may see that $\Gamma\left(\dfrac{1}{2}\right) = \displaystyle\int_0^\infty t^{-1/2}e^{-t}dt \overset{t=u^2}{=} 2\int_0^\infty e^{-u^2}du$, where the last integral is the Gaussian integral, which one may also attack by exploiting polar coordinates.[10]

For the other integral in (6.328), we make the variable change $t = \sin(x)\sqrt{u}$. Thus, we have that

$$\int_0^\infty \frac{t^{s+1}}{\sin^2(x)+t^2}dt = \frac{1}{2}\sin^s(x)\int_0^\infty \frac{u^{(s/2+1)-1}}{1+u}du = \frac{1}{2}\sin^s(x)B\left(\frac{s}{2}+1, -\frac{s}{2}\right)$$

$$\left\{\text{use the Beta–Gamma identity, } B(x,y) = \frac{\Gamma(x)\Gamma(y)}{\Gamma(x+y)}, \text{ and } \Gamma(1) = 1\right\}$$

---

[9] Starting from the Beta function definition $B(a,b) = \displaystyle\int_0^1 x^{a-1}(1-x)^{b-1}dx$, $\Re(a) > 0$, $\Re(b) > 0$, and letting the variable change $\dfrac{x}{1-x} = t$, we arrive at the Beta function form we need, $B(a,b) = \displaystyle\int_0^\infty \frac{t^{a-1}}{(1+t)^{a+b}}dt$. Professor Victor H. Moll dedicated two chapters to the Beta function in his book, *Special Integrals of Gradshteyn and Ryzhik: the Proofs - Volume I (2014)*, where also the Beta function form discussed above may be found, together with many other (interesting) forms and ways to derive them.

[10] The technique of evaluating the famous Gaussian integral by using polar coordinates is wonderful and well-known in the mathematical literature. Did you try it? It is also nicely described by Paul in his recent title, *In Pursuit of Zeta-3: The World's Most Mysterious Unsolved Math Problem* (2021).

$$= \frac{1}{2} \sin^s(x) \Gamma\left(-\frac{s}{2}\right) \Gamma\left(1 + \frac{s}{2}\right) = -\frac{\pi}{2} \sin^s(x) \csc\left(\frac{\pi}{2}s\right), \qquad (6.330)$$

where the last equality comes from Euler's reflection formula, $\Gamma(x)\Gamma(1 - x) = \frac{\pi}{\sin(\pi x)}$.

Combining (6.328), (6.329), and (6.330), where we consider the restriction $-1 < s < 0$, we arrive at

$$\sum_{n=1}^{\infty} \left(\log(2)\frac{1}{n} - \frac{1}{2n^2} + \frac{H_n}{n}\right) \cos(2nx)$$

$$= \lim_{s \to 0^-} \frac{d}{ds}\left(\frac{1}{2\sqrt{\pi}} \Gamma\left(\frac{1+s}{2}\right) \Gamma\left(-\frac{s}{2}\right) + \frac{\pi}{2} \csc\left(\frac{\pi}{2}s\right) \sin^s(x)\right)$$

$$= -\frac{1}{2}\log^2(2) - \frac{\pi^2}{24} + \frac{1}{2}\log^2(\sin(x)),$$

whence we obtain that

$$\log^2(\sin(x))$$

$$= \log^2(2) + \frac{\pi^2}{12} + 2\sum_{n=1}^{\infty} \left(\log(2)\frac{1}{n} - \frac{1}{2n^2} + \frac{H_n}{n}\right) \cos(2nx), \ 0 < x < \pi,$$

and the solution to the point (i) of the problem is complete. For the limit above, we needed to collect some facts that allowed us to easily calculate it.

First, we recall the Laurent series for $\Gamma(x)$ about $x = 0$, $\Gamma(x) = \frac{1}{x} - \gamma + \frac{1}{2}\left(\gamma^2 + \frac{\pi^2}{6}\right)x - \frac{1}{6}\left(\gamma^3 + \gamma\frac{\pi^2}{2} + 2\varsigma(3)\right)x^2 + \cdots$ (see [76, Chapter 3, p.199], [60, Chapter 0, pp.3–6]) that gives the useful limited expansion $\Gamma(x) = \frac{1}{x} - \gamma + \frac{1}{2}\left(\gamma^2 + \frac{\pi^2}{6}\right)x + O(x^2)$.

Then, by collecting the following expansions, $\frac{1}{2^{x-1}} = 2 - 2\log(2)x + \log^2(2)x^2 + O(x^3)$ and

$$\frac{\Gamma(x)}{\Gamma(x/2)} = \frac{1/x - \gamma + ((\gamma^2 + \pi^2/6)/2)x + O(x^2)}{2/x - \gamma + ((\gamma^2 + \pi^2/6)/4)x + O(x^2)}$$

$$= \frac{1/2 - (\gamma/2)x + ((\gamma^2 + \pi^2/6)/4)x^2 + O(x^3)}{1 - (\gamma/2)x + ((\gamma^2 + \pi^2/6)/8)x^2 + O(x^3)}$$

$$= \left(\frac{1}{2} - \frac{1}{2}\gamma x + \frac{1}{4}\left(\gamma^2 + \frac{\pi^2}{6}\right)x^2 + O(x^3)\right)\left(1 + \frac{1}{2}\gamma x\right.$$

$$\left. - \frac{1}{8}\left(\gamma^2 + \frac{\pi^2}{6}\right)x^2 + \left(\frac{1}{2}\gamma x - \frac{1}{8}\left(\gamma^2 + \frac{\pi^2}{6}\right)x^2\right)^2 + O(x^5)\right)$$

$$= \frac{1}{2} - \frac{1}{4}\gamma x + \frac{1}{32}(2\gamma^2 + \pi^2)x^2 + O(x^3), \qquad (6.331)$$

in the Legendre duplication formula, $\Gamma\left(x + \frac{1}{2}\right) = \sqrt{\pi}\,\frac{1}{2^{2x-1}}\frac{\Gamma(2x)}{\Gamma(x)}$, with $x = s/2$, we get $\Gamma\left(\frac{1+s}{2}\right) = \sqrt{\pi}\,\frac{1}{2^{s-1}}\frac{\Gamma(s)}{\Gamma(s/2)} = \sqrt{\pi} - \frac{\sqrt{\pi}}{2}(2\log(2) + \gamma)s + \frac{\sqrt{\pi}}{16}(8\log^2(2) + 8\log(2)\gamma + 2\gamma^2 + \pi^2)s^2 + O(s^3)$. A solution to the Legendre duplication formula may be found in [96], but also other ways to go are possible.[11]

We also needed that $\sin^s(x) = e^{\log(\sin(x))s} = 1 + \log(\sin(x))s + \frac{1}{2}\log^2(\sin(x))s^2 + O(s^3)$ together with $\csc\left(\frac{\pi}{2}s\right) = \frac{1}{\sin((\pi/2)s)} = \frac{2}{\pi}\frac{1}{s\left(1 - (\pi^2/24)s^2 + O(s^4)\right)} = \frac{2}{\pi}\frac{1}{s}\left(1 + \frac{\pi^2}{24}s^2 + O(s^4)\right) = \frac{2}{\pi}\frac{1}{s} + \frac{\pi}{12}s + O(s^3)$, where both easily follow from exploiting known power series.[12]

---

[11] In *(Almost) Impossible Integrals, Sums, and Series*, page 68, I mentioned that a way of proving the Digamma identity $\psi(2x) = \frac{1}{2}\psi(x) + \frac{1}{2}\psi\left(x + \frac{1}{2}\right) + \log(2)$ is achieved by using the Legendre duplication formula, $\Gamma\left(x + \frac{1}{2}\right) = \sqrt{\pi}\,\frac{1}{2^{2x-1}}\frac{\Gamma(2x)}{\Gamma(x)}$. However, we can also go the other way, that is, by proving the Digamma identity differently, say, as presented in the mentioned book on pages 68 − 69, and then use it to derive the Legendre duplication formula by exploiting integration.

[12] Since the power series of the exponential function is $e^x = \sum_{n=0}^{\infty}\frac{x^n}{n!}$, we immediately obtain $\sin^s(x) = e^{\log(\sin(x))s} = 1 + \log(\sin(x))s + \frac{1}{2}\log^2(\sin(x))s^2 + O(s^3)$. Further, for $\csc\left(\frac{\pi}{2}s\right)$, I used limited forms of the power expansions $\sin(x) = \sum_{n=1}^{\infty}(-1)^{n-1}\frac{x^{2n-1}}{(2n-1)!}$ and $\frac{1}{1-x} = \sum_{n=1}^{\infty}x^{n-1}$, that is, $\sin(x) = x - \frac{x^3}{6} + O(x^5)$ and $\frac{1}{1-x} = 1 + x + O(x^2)$. So, I used the first limited expansion for showing that $\frac{1}{\sin((\pi/2)s)} = \frac{2}{\pi}\frac{1}{s\left(1 - (\pi^2/24)s^2 + O(s^4)\right)}$ and the second limited expansion to indicate that $\frac{1}{1 - (\pi^2/24)s^2 + O(s^4)} = 1 + \frac{\pi^2}{24}s^2 + O(s^4)$. Observe that during the extraction of the limited expansion of $\frac{\Gamma(x)}{\Gamma(x/2)}$, I also exploited a useful limited expansion of $\frac{1}{1-x} = \sum_{n=1}^{\infty}x^{n-1}$, posed in the form $\frac{1}{1-x} = 1 + x + x^2 + O(x^3)$.

I guess at this point one might react saying, *It's way easier to calculate it by using Mathematica!* And, of course, I agree, but at least we know how to deal with such limits, step by step, with pen and paper, and no other computational aid. Furthermore, the other limits appearing at the next points can be treated in a similar manner for obtaining the desired values.

Another natural way to go for obtaining a solution is achieved by exploiting (4.33), Sect. 4.6, in the form $2 \sum_{n=1}^{\infty} x^{n+1} \frac{H_n}{n+1} = \log^2(1-x)$, where we set $x = e^{i2x}$, as similarly presented in [67, pp. 127–128]. However, one might want to note that the form obtained by the last approach is $\log^2(2\sin(x)) = \left(\frac{\pi}{2} - x\right)^2 + 2 \sum_{n=1}^{\infty} \frac{H_{n-1}}{n} \cos(2nx)$, which is different, and we need some additional work to get the form given at the point (i).

As regards the Fourier series at the point (ii), I'll use the result from the previous point where it is enough to replace x by $\pi/2 - x$, and then we get

$$\log^2(\cos(x))$$

$$= \log^2(2) + \frac{\pi^2}{12} - 2 \sum_{n=1}^{\infty} (-1)^{n-1} \left(\log(2)\frac{1}{n} - \frac{1}{2n^2} + \frac{H_n}{n} \right) \cos(2nx), \quad -\frac{\pi}{2} < x < \frac{\pi}{2},$$

and the solution to the point (ii) of the problem is complete.

For the Fourier series at the point (iii), we need the squared log version of the integral in (6.324), that is

$$\int_0^{\infty} \log^2(\sinh(y)) e^{-2ny} dy$$

$$= \left(\frac{1}{2} \log^2(2) + \frac{\pi^2}{12} \right) \frac{1}{n} - \frac{1}{2} \log(2)\frac{1}{n^2} + \frac{1}{4}\frac{1}{n^3} + \log(2)\frac{H_n}{n} - \frac{1}{2}\frac{H_n}{n^2} + \frac{1}{2}\frac{H_n^2}{n}.$$
$$\tag{6.332}$$

Proof By letting the variable change $y = -\log(\sqrt{t})$, we get

$$\int_0^{\infty} \log^2(\sinh(y)) e^{-2ny} dy = \frac{1}{2} \int_0^1 t^{n-1} \log^2\left(\frac{1-t}{2\sqrt{t}} \right) dt$$

$$= \frac{1}{2} \log^2(2) \underbrace{\int_0^1 t^{n-1} dt}_{1/n} + \frac{1}{2} \log(2) \underbrace{\int_0^1 t^{n-1} \log(t) dt}_{-1/n^2} + \frac{1}{8} \underbrace{\int_0^1 t^{n-1} \log^2(t) dt}_{2/n^3}$$

$$- \log(2) \int_0^1 t^{n-1} \log(1-t) dt + \frac{1}{2} \int_0^1 t^{n-1} \log^2(1-t) dt$$

$$-\frac{1}{2}\int_0^1 t^{n-1}\log(t)\log(1-t)dt$$

$$= \left(\frac{1}{2}\log^2(2)+\frac{\pi^2}{12}\right)\frac{1}{n}-\frac{1}{2}\log(2)\frac{1}{n^2}+\frac{1}{4}\frac{1}{n^3}+\log(2)\frac{H_n}{n}-\frac{1}{2}\frac{H_n}{n^2}+\frac{1}{2}\frac{H_n^2}{n},$$

where in the calculations I used $\displaystyle\int_0^1 x^{n-1}\log(1-x)dx = -\frac{H_n}{n}$ that may be found

stated and proved in (3.10), Sect. 3.3, then $\displaystyle\int_0^1 x^{n-1}\log^2(1-x)dx = \frac{H_n^2+H_n^{(2)}}{n}$
(see [76, Chapter 1, p.2]), and the last integral is the particular case $m = 1$ found
in (1.125), Sect. 1.24. ∎

By multiplying both sides of (6.332) by $\cos(2nx)$ and then considering similar
calculation lines as in (6.328), (6.329), and (6.330) (see this time we have a squared
log), we arrive at

$$\sum_{n=1}^{\infty}\left(\left(\frac{1}{2}\log^2(2)+\frac{\pi^2}{12}\right)\frac{1}{n}-\frac{1}{2}\log(2)\frac{1}{n^2}+\frac{1}{4}\frac{1}{n^3}+\log(2)\frac{H_n}{n}\right.$$

$$\left.-\frac{1}{2}\frac{H_n}{n^2}+\frac{1}{2}\frac{H_n^2}{n}\right)\cos(2nx)$$

$$= -\frac{1}{2}\lim_{s\to 0^-}\frac{d^2}{ds^2}\left(\int_0^{\infty}\frac{t^s}{\sqrt{1+t^2}}dt-\int_0^{\infty}\frac{t^{s+1}}{\sin^2(x)+t^2}dt\right)$$

$$= -\frac{1}{2}\lim_{s\to 0^-}\frac{d^2}{ds^2}\left(\frac{1}{2\sqrt{\pi}}\Gamma\left(\frac{1+s}{2}\right)\Gamma\left(-\frac{s}{2}\right)+\frac{\pi}{2}\csc\left(\frac{\pi}{2}s\right)\sin^s(x)\right)$$

$$= -\frac{1}{12}\log(2)\pi^2-\frac{1}{6}\log^3(2)-\frac{1}{4}\zeta(3)-\frac{\pi^2}{24}\log(\sin(x))-\frac{1}{6}\log^3(\sin(x)),$$

where if we exploit the Fourier series,[13] $\displaystyle\log(\sin(x)) = -\log(2)-\sum_{n=1}^{\infty}\frac{\cos(2nx)}{n}$, 0
$< x < \pi$ (see **1.441.2** in [17]), and rearrange, we arrive at

---

[13] The curious reader might want to observe that we could also successfully use the main strategy
described in current section for deriving the basic, known Fourier series of $\log(\sin(x))$, $\log(\cos(x))$.
This would mean to use that $\displaystyle\int_0^{\infty}\log^k(\sinh(y))e^{-2ny}dy\Big|_{k=0} = \int_0^{\infty}e^{-2ny}dy = \frac{1}{2n}$.

$$\sum_{n=1}^{\infty}\left(-\left(3\log^2(2)+\frac{\pi^2}{4}\right)\frac{1}{n}+3\log(2)\frac{1}{n^2}-\frac{3}{2}\frac{1}{n^3}-6\log(2)\frac{H_n}{n}\right.$$

$$\left.+3\frac{H_n}{n^2}-3\frac{H_n^2}{n}\right)\cos(2nx)$$

$$=\frac{1}{4}\log(2)\pi^2+\log^3(2)+\frac{3}{2}\zeta(3)+\log^3(\sin(x)),$$

whence we obtain that

$$\log^3(\sin(x))$$

$$=-\log^3(2)-\frac{1}{4}\log(2)\pi^2-\frac{3}{2}\zeta(3)+\sum_{n=1}^{\infty}\left(\left(-3\log^2(2)-\frac{\pi^2}{4}\right)\frac{1}{n}+3\log(2)\frac{1}{n^2}\right.$$

$$\left.-\frac{3}{2}\frac{1}{n^3}-6\log(2)\frac{H_n}{n}+3\frac{H_n}{n^2}-3\frac{H_n^2}{n}\right)\cos(2nx),\ 0<x<\pi,$$

and the solution to the point (*iii*) of the problem is complete. For calculating the critical limit above, I just employed *Mathematica* in order to avoid the type of calculations you found in the solution to the point (*i*).

An alternative solution would be to exploit $\log^3(1-x)=-3\sum_{n=1}^{\infty}\frac{x^{n+1}}{n+1}(H_n^2-H_n^{(2)})$, which is immediately obtained by integrating the identity $\frac{\log^2(1-x)}{1-x}=\sum_{n=1}^{\infty}x^n(H_n^2-H_n^{(2)})$, easily derivable by using telescoping sums, as presented in [76, Chapter 6, pp. 354–355], and then set $x=e^{i2x}$.

The result at the point (*iv*) is straightforward if we replace x by $\pi/2-x$ in the previous Fourier series, and then we have

$$\log^3(\cos(x))$$

$$=-\log^3(2)-\frac{1}{4}\log(2)\pi^2-\frac{3}{2}\zeta(3)+\sum_{n=1}^{\infty}(-1)^{n-1}\left(\left(3\log^2(2)+\frac{\pi^2}{4}\right)\frac{1}{n}\right.$$

$$\left.-3\log(2)\frac{1}{n^2}+\frac{3}{2}\frac{1}{n^3}+6\log(2)\frac{H_n}{n}-3\frac{H_n}{n^2}+3\frac{H_n^2}{n}\right)\cos(2nx),\ -\frac{\pi}{2}<x<\frac{\pi}{2},$$

and the solution to the point (*iv*) of the problem is complete.

Next, for the Fourier series at the point (v), we need the cubed log version of the integral in (6.324), that is

$$\int_0^\infty \log^3(\sinh(y))e^{-2ny}\,dy$$

$$= -\frac{1}{4}(2\log^3(2)+\log(2)\pi^2+3\zeta(3))\frac{1}{n}+\frac{1}{8}(6\log^2(2)+\pi^2)\frac{1}{n^2}-\frac{3}{4}\log(2)\frac{1}{n^3}+\frac{3}{8}\frac{1}{n^4}$$

$$-\frac{1}{4}(6\log^2(2)+\pi^2)\frac{H_n}{n}+\frac{3}{2}\log(2)\frac{H_n}{n^2}-\frac{3}{4}\frac{H_n}{n^3}-\frac{3}{2}\log(2)\frac{H_n^2}{n}$$

$$+\frac{3}{4}\frac{H_n^2}{n^2}-\frac{1}{2}\frac{H_n^3}{n}-\frac{1}{4}\frac{H_n^{(3)}}{n}. \tag{6.333}$$

Proof Like in the solutions to the points (i) and (iii), we want to let the variable change $y = -\log(\sqrt{t})$ that gives

$$\int_0^\infty \log^3(\sinh(y))e^{-2ny}\,dy = \frac{1}{2}\int_0^1 t^{n-1}\log^3\left(\frac{1-t}{2\sqrt{t}}\right)dt$$

$$= -\frac{1}{2}\log^3(2)\underbrace{\int_0^1 t^{n-1}\,dt}_{1/n} -\frac{3}{4}\log^2(2)\underbrace{\int_0^1 t^{n-1}\log(t)\,dt}_{-1/n^2} -\frac{3}{8}\log(2)\underbrace{\int_0^1 t^{n-1}\log^2(t)\,dt}_{2/n^3}$$

$$-\frac{1}{16}\underbrace{\int_0^1 t^{n-1}\log^3(t)\,dt}_{-6/n^4} +\frac{3}{2}\log^2(2)\int_0^1 t^{n-1}\log(1-t)\,dt$$

$$-\frac{3}{2}\log(2)\int_0^1 t^{n-1}\log^2(1-t)\,dt +\frac{1}{2}\int_0^1 t^{n-1}\log^3(1-t)\,dt$$

$$+\frac{3}{2}\log(2)\int_0^1 t^{n-1}\log(t)\log(1-t)\,dt +\frac{3}{8}\int_0^1 t^{n-1}\log^2(t)\log(1-t)\,dt$$

$$-\frac{3}{4}\int_0^1 t^{n-1}\log(t)\log^2(1-t)\,dt$$

$$= -\frac{1}{4}(2\log^3(2)+\log(2)\pi^2+3\zeta(3))\frac{1}{n}+\frac{1}{8}(6\log^2(2)+\pi^2)\frac{1}{n^2}-\frac{3}{4}\log(2)\frac{1}{n^3}+\frac{3}{8}\frac{1}{n^4}$$

$$- \frac{1}{4}(6\log^2(2) + \pi^2)\frac{H_n}{n} + \frac{3}{2}\log(2)\frac{H_n}{n^2} - \frac{3}{4}\frac{H_n}{n^3} - \frac{3}{2}\log(2)\frac{H_n^2}{n}$$

$$+ \frac{3}{4}\frac{H_n^2}{n^2} - \frac{1}{2}\frac{H_n^3}{n} - \frac{1}{4}\frac{H_n^{(3)}}{n},$$

where in the calculations I used $\displaystyle\int_0^1 x^{n-1}\log(1-x)\mathrm{d}x = -\frac{H_n}{n}$ that may be found

stated and proved in (3.10), Sect. 3.3, then $\displaystyle\int_0^1 x^{n-1}\log^2(1-x)\mathrm{d}x = \frac{H_n^2 + H_n^{(2)}}{n}$

and $\displaystyle\int_0^1 x^{n-1}\log^3(1-x)\mathrm{d}x = -\frac{H_n^3 + 3H_n H_n^{(2)} + 2H_n^{(3)}}{n}$ (see [76, Chapter 1, p.2]),

next the last three integrals are the particular cases $m = 1, 2$ in (1.125), and the case $m = 1$ in (1.126), Sect. 1.24, and the proof to the auxiliary result is done. ∎

If we multiply both sides of (6.333) by $\cos(2nx)$ and then consider similar calculation lines as in (6.328), (6.329), and (6.330) (see this time we have a cubed log), we obtain that

$$\sum_{n=1}^{\infty}\left(-\frac{1}{4}(2\log^3(2) + \log(2)\pi^2 + 3\zeta(3))\frac{1}{n} + \frac{1}{8}(6\log^2(2) + \pi^2)\frac{1}{n^2} - \frac{3}{4}\log(2)\frac{1}{n^3}\right.$$

$$+ \frac{3}{8}\frac{1}{n^4} - \frac{1}{4}(6\log^2(2) + \pi^2)\frac{H_n}{n} + \frac{3}{2}\log(2)\frac{H_n}{n^2} - \frac{3}{4}\frac{H_n}{n^3} - \frac{3}{2}\log(2)\frac{H_n^2}{n}$$

$$\left. + \frac{3}{4}\frac{H_n^2}{n^2} - \frac{1}{2}\frac{H_n^3}{n} - \frac{1}{4}\frac{H_n^{(3)}}{n}\right)\cos(2nx)$$

$$= -\frac{1}{2}\lim_{s\to 0}\frac{\mathrm{d}^3}{\mathrm{d}s^3}\left(\int_0^{\infty}\frac{t^s}{\sqrt{1+t^2}}\mathrm{d}t - \int_0^{\infty}\frac{t^{s+1}}{\sin^2(x)+t^2}\mathrm{d}t\right)$$

$$= -\frac{1}{2}\lim_{s\to 0^-}\frac{\mathrm{d}^3}{\mathrm{d}s^3}\left(\frac{1}{2\sqrt{\pi}}\Gamma\left(\frac{1+s}{2}\right)\Gamma\left(-\frac{s}{2}\right) + \frac{\pi}{2}\csc\left(\frac{\pi}{2}s\right)\sin^s(x)\right)$$

$$= \frac{1}{8}\log^4(2) + \log^2(2)\frac{\pi^2}{8} + \frac{29}{1920}\pi^4 + \frac{3}{4}\log(2)\zeta(3)$$

$$- \frac{\pi^2}{16}\log^2(\sin(x)) - \frac{1}{8}\log^4(\sin(x)),$$

where if we plug in the Fourier series of $\log^2(\sin(x))$, calculated in the beginning part of the section, we obtain that

$$\log^4(\sin(x))$$

$$= \log^4(2) + \frac{1}{2}\log^2(2)\pi^2 + 6\log(2)\zeta(3) + \frac{19}{240}\pi^4$$

$$+ \sum_{n=1}^{\infty}\left((4\log^3(2) + \log(2)\pi^2 + 6\zeta(3))\frac{1}{n} - \left(6\log^2(2) + \frac{\pi^2}{2}\right)\frac{1}{n^2} + 6\log(2)\frac{1}{n^3}\right.$$

$$- 3\frac{1}{n^4} + (12\log^2(2) + \pi^2)\frac{H_n}{n} - 12\log(2)\frac{H_n}{n^2} + 6\frac{H_n}{n^3} + 12\log(2)\frac{H_n^2}{n}$$

$$\left. - 6\frac{H_n^2}{n^2} + 4\frac{H_n^3}{n} + 2\frac{H_n^{(3)}}{n}\right)\cos(2nx), \quad 0 < x < \pi; \qquad (6.334)$$

and the solution to the point (v) of the problem is complete. Again, for calculating the critical limit above, I just employed *Mathematica*, in order to avoid the type of calculations you found in the solution to the point (i). Surely, the curious reader may take the path of manually calculating the limit by following the same ideas presented in the solution to the first point.

Already a routine to note that the cosine version given at the last point is extracted from the previous Fourier series, by replacing x by $\pi/2 - x$, and thus we have

$$\log^4(\cos(x))$$

$$= \log^4(2) + \frac{1}{2}\log^2(2)\pi^2 + 6\log(2)\zeta(3) + \frac{19}{240}\pi^4$$

$$- \sum_{n=1}^{\infty}(-1)^{n-1}\left((4\log^3(2) + \log(2)\pi^2 + 6\zeta(3))\frac{1}{n} - \left(6\log^2(2) + \frac{\pi^2}{2}\right)\frac{1}{n^2}\right.$$

$$+ 6\log(2)\frac{1}{n^3} - 3\frac{1}{n^4} + (12\log^2(2) + \pi^2)\frac{H_n}{n} - 12\log(2)\frac{H_n}{n^2} + 6\frac{H_n}{n^3} + 12\log(2)\frac{H_n^2}{n}$$

$$\left. - 6\frac{H_n^2}{n^2} + 4\frac{H_n^3}{n} + 2\frac{H_n^{(3)}}{n}\right)\cos(2nx), \quad -\frac{\pi}{2} < x < \frac{\pi}{2},$$

and the solution to the point (vi) of the problem is complete.

Now, I would like to emphasize an important point for the curious reader and make it clear that with the integral after the antepenultimate equality of (6.328), we may also proceed differently! And, essentially, we could also try using such a way at other points.

For example, by splitting the integral at $t = 1$, making the variable change $1/t = \sqrt{u}$ in the second integral, and then expanding, and rearranging, we have

$$\int_0^\infty \left(\frac{\log(t)}{\sqrt{1+t^2}} - \frac{t \log(t)}{\sin^2(x) + t^2} \right) dt$$

$$= \int_0^1 \frac{\log(u)}{\sqrt{1+u^2}} du + \frac{1}{4} \int_0^1 \frac{\log(u)}{u} \left(1 - \frac{1}{\sqrt{1+u}} \right) du - \int_0^1 \frac{\csc^2(x) u \log(u)}{1 + \csc^2(x) u^2} du$$

$$- \frac{1}{4} \int_0^1 \frac{\sin^2(x) \log(u)}{1 + \sin^2(x) u} du,$$

from where the curious reader might enjoy to continue the calculations. For instance, the first integral is met and calculated in (3.304), Sect. 3.42, and the second integral is found calculated in Sect. 3.43. Also, as regards the last two integrals, one might like to show, by exploiting simple, well-known results, that $\int_0^1 \frac{\csc^2(x) \log(u)}{1 + \csc^2(x) u} du +$

$\int_0^1 \frac{\sin^2(x) \log(u)}{1 + \sin^2(x) u} du = -\frac{\pi^2}{6} - 2 \log^2(\sin(x)), \ 0 < x < \pi.$

Experimental calculations on small cases suggest that virtually we could extract the Fourier series for any $\log^n(\sin(x))$ and $\log^n(\cos(x))$, where n si a positive integer. While the cases for $n = 2$ are known, the more advanced cases like $n = 3, 4, \ldots$ seem to be less known or new in the mathematical literature. We'll find them very useful in the work with some special integrals and series.

Further, it is interesting to note that compared to the alternative solutions suggested at the end of the solutions to the points (i) and (iii) (and similar solutions can be found at all points), the primary solutions involving $\int_0^\infty \log^p(\sinh(x)) e^{-2nx} dx, \ n, p \in \mathbb{N}$, lead us directly to forms of $\log^m(\sin(x))$ and $\log^m(\cos(x))$ without containing terms with x^n. Of course, these forms containing x^n can be brought to the forms presented in the statement section.

6.49   More Stunning Fourier Series, Related to Atypical Harmonic Numbers (Skew-Harmonic Numbers)

Solution As suggested in the title of the section, this time, we want to derive beautiful Fourier series with coefficients involving the skew-harmonic numbers.

So, I'll start the present section in a style simiar to the one found in the previous section, but now I'll first state explicitly the two powerful Fourier series of $\tan(x) \log(\sin(x))$ and $\cot(x) \log(\cos(x))$ in [76, Chapter 3, pp. 242–252],

$$\sum_{n=1}^\infty \left(\psi \left(\frac{n+1}{2} \right) - \psi \left(\frac{n}{2} \right) - \frac{1}{n} \right) \sin(2nx) = -\tan(x) \log(\sin(x)) \qquad (6.335)$$

and

$$\sum_{n=1}^{\infty}(-1)^{n-1}\left(\psi\left(\frac{n+1}{2}\right)-\psi\left(\frac{n}{2}\right)-\frac{1}{n}\right)\sin(2nx)=-\cot(x)\log(\cos(x)),$$

$$(6.336)$$

where $0 < x < \dfrac{\pi}{2}$. However, for (6.335), the validity of the Fourier series result may be extended to $x \in \left(0,\dfrac{\pi}{2}\right)\cup\left(\dfrac{\pi}{2},\pi\right)$, and for (6.336), we may have an extension to $x \in \left(-\dfrac{\pi}{2},0\right)\cup\left(0,\dfrac{\pi}{2}\right)$.

At this point, I'll show you the beautiful fact that the coefficient $\psi\left(\dfrac{n+1}{2}\right)-\psi\left(\dfrac{n}{2}\right)-\dfrac{1}{n}$ can be expressed differently, by using skew-harmonic numbers. So, we want to prove that

$$\psi\left(\frac{n+1}{2}\right)-\psi\left(\frac{n}{2}\right)-\frac{1}{n}=2\log(2)(-1)^{n-1}+\frac{1}{n}-2(-1)^{n-1}\overline{H}_n.\quad(6.337)$$

Proof We know from [76, Chapter 1, p.3] the following Digamma function related integral, $\displaystyle\int_0^1\frac{x^{s-1}}{1+x}dx=\frac{1}{2}\left(\psi\left(\frac{1+s}{2}\right)-\psi\left(\frac{s}{2}\right)\right)$, where if we set $s=n+1$ and then consider the Digamma recurrence relation, $\psi(1+x)=\psi(x)+\dfrac{1}{x}$, we get

$$\int_0^1\frac{x^n}{1+x}dx=\frac{1}{n}+\frac{1}{2}\psi\left(\frac{n}{2}\right)-\frac{1}{2}\psi\left(\frac{n+1}{2}\right).\quad(6.338)$$

If we integrate by parts once in (6.338) and then rearrange, we have

$$\int_0^1 x^{n-1}\log(1+x)dx=\log(2)\frac{1}{n}-\frac{1}{n^2}+\frac{1}{2}\frac{1}{n}\left(\psi\left(\frac{n+1}{2}\right)-\psi\left(\frac{n}{2}\right)\right).$$

$$(6.339)$$

Finally, combining (6.339) and the result in (1.99), Sect. 1.21, we arrive at the announced auxiliary result in (6.337). Alternatively, we can directly employ the result in (4.96), Sect. 4.18.   ∎

In view of (6.337), the two Fourier series in (6.335) and (6.336) get the form

$$\sum_{n=1}^{\infty}\left(2\log(2)+(-1)^{n-1}\frac{1}{n}-2\overline{H}_n\right)\sin(2nx)=-\cot(x)\log(\cos(x))\quad(6.340)$$

and

$$\sum_{n=1}^{\infty}(-1)^{n-1}\left(2\log(2)+(-1)^{n-1}\frac{1}{n}-2\overline{H}_n\right)\sin(2nx)=-\tan(x)\log(\sin(x)).$$

(6.341)

Now, we are able to provide a first solution to the point (i) of the problem. Observing that $\left(\text{Li}_2\left(\sin^2(x)\right)\right)'=-4\cot(x)\log(\cos(x))$, then considering (6.340) in variable t, integrating from $t=0$ to $t=x$, and rearranging, we get

$$\text{Li}_2\left(\sin^2(x)\right)$$

$$=2\sum_{n=1}^{\infty}\left(2\log(2)+(-1)^{n-1}\frac{1}{n}-2\overline{H}_n\right)\left(\frac{1}{n}-\frac{\cos(2nx)}{n}\right)$$

$$=\frac{\pi^2}{6}-2\log^2(2)-2\sum_{n=1}^{\infty}\left(2\log(2)\frac{1}{n}+(-1)^{n-1}\frac{1}{n^2}-2\frac{\overline{H}_n}{n}\right)\cos(2nx),$$

and the first solution to the point (i) is finalized.

Observe that above I also used the following series result that can be derived by exploiting (6.337) and (6.338):

$$\sum_{n=1}^{\infty}(-1)^{n-1}\frac{1}{n}\left(2\log(2)(-1)^{n-1}+\frac{1}{n}-2(-1)^{n-1}\overline{H}_n\right)$$

$$=\sum_{n=1}^{\infty}(-1)^{n-1}\frac{1}{n}\left(\frac{1}{n}-2\int_0^1\frac{x^n}{1+x}dx\right)=\sum_{n=1}^{\infty}(-1)^{n-1}\frac{1}{n^2}-2\sum_{n=1}^{\infty}\frac{(-1)^{n-1}}{n}\int_0^1\frac{x^n}{1+x}dx$$

$$=\frac{\pi^2}{12}-2\int_0^1\frac{1}{1+x}\sum_{n=1}^{\infty}(-1)^{n-1}\frac{x^n}{n}dx=\frac{\pi^2}{12}-2\int_0^1\frac{\log(1+x)}{1+x}dx=\frac{\pi^2}{12}-\log^2(2).$$

For a second solution to the point (i), we might think of the key integral in (6.324) I used in the previous section, but this time, I'll consider a slightly modified version of it, that is, I'll use $\cosh(y)$ instead of $\sinh(y)$. In fact, I'll prove that

$$\int_0^{\infty}\log(\cosh(y))e^{-2ny}dy=\frac{1}{2}\log(2)(-1)^{n-1}\frac{1}{n}+\frac{1}{4}\frac{1}{n^2}-\frac{1}{2}(-1)^{n-1}\frac{\overline{H}_n}{n}.$$

(6.342)

Proof Upon letting the variable charge $y = -\log(\sqrt{t})$, we obtain that

$$\int_0^\infty \log(\cosh(y)) e^{-2ny} dy = \frac{1}{2} \int_0^1 t^{n-1} \log\left(\frac{1+t}{2\sqrt{t}}\right) dt$$

$$= -\frac{1}{2} \log(2) \underbrace{\int_0^1 t^{n-1} dt}_{1/n} - \frac{1}{4} \underbrace{\int_0^1 t^{n-1} \log(t) dt}_{-1/n^2} + \frac{1}{2} \int_0^1 t^{n-1} \log(1+t) dt$$

$$= \frac{1}{2} \log(2) (-1)^{n-1} \frac{1}{n} + \frac{1}{4} \frac{1}{n^2} - \frac{1}{2} (-1)^{n-1} \frac{\overline{H}_n}{n},$$

where to get the last equality, I used the integral result in (1.99), Sect. 1.21, which brings an end to the auxiliary result in (6.342). ∎

Upon multiplying both sides of (6.342) by $(-1)^{n-1} \cos(2nx)$ and considering the summation from $n = 1$ to ∞, we write

$$\sum_{n=1}^\infty \left(\frac{1}{2} \log(2) \frac{1}{n} + \frac{1}{4} (-1)^{n-1} \frac{1}{n^2} - \frac{1}{2} \frac{\overline{H}_n}{n} \right) \cos(2nx)$$

$$= \sum_{n=1}^\infty (-1)^{n-1} \cos(2nx) \int_0^\infty \log(\cosh(y)) e^{-2ny} dy$$

$$= \int_0^\infty \log(\cosh(y)) \sum_{n=1}^\infty (-1)^{n-1} \cos(2nx) e^{-2ny} dy$$

$$= \frac{1}{2} \int_0^\infty \log(\cosh(y)) \left(1 - \frac{\sinh(2y)}{\cos(2x) + \cosh(2y)} \right) dy$$

$$\overset{1/\cosh^2(y)=t}{=} \frac{1}{8} \int_0^1 \left(\frac{\log(t)}{t} \left(1 - \frac{1}{\sqrt{1-t}} \right) + \frac{\sin^2(x) \log(t)}{1 - \sin^2(x)t} \right) dt$$

$$= \frac{1}{8} \int_0^1 \frac{\log(t)}{t} \left(1 - \frac{1}{\sqrt{1-t}} \right) dt + \frac{1}{8} \int_0^1 \frac{\sin^2(x) \log(t)}{1 - \sin^2(x)t} dt, \qquad (6.343)$$

and for the cosine series above, I used the result in (6.327), Sect. 6.48, where we replace x by $2x$ and set $p = -e^{-2y}$. Also, one needs the hyperbolic identities, $\sinh(2 \operatorname{arccosh}(x)) = 2x\sqrt{x^2 - 1}$, $x \geq 1$, and $\cosh(2 \operatorname{arccosh}(x)) = 2x^2 - 1$, $x \geq 1$.

Instead of integrating by parts, as I'll do for a similar integral in the solution at the point (iii), we reduce the first integral in (6.343) to a Beta function limit,

$$\int_0^1 \frac{\log(t)}{t}\left(1 - \frac{1}{\sqrt{1-t}}\right) dt = \lim_{s\to 0^+} \frac{d}{ds}\int_0^1 t^{s-1}\left(1 - \frac{1}{\sqrt{1-t}}\right) dt$$

$$= \lim_{s\to 0^+} \frac{d}{ds}\left(\int_0^1 t^{s-1}dt - \underbrace{\int_0^1 t^{s-1}(1-t)^{1/2-1}dt}_{\text{A Beta function form, } B(s, 1/2)}\right)$$

$$\left\{\text{exploit the Beta–Gamma identity, } B(x, y) = \frac{\Gamma(x)\Gamma(y)}{\Gamma(x+y)}\right\}$$

$$= \lim_{s\to 0^+} \frac{d}{ds}\left(\frac{1}{s} - \frac{\Gamma(1/2)\,\Gamma(s)}{\Gamma(1/2+s)}\right) = \lim_{s\to 0^+} \frac{d}{ds}\left(\frac{1}{s} - \frac{\sqrt{\pi}\,\Gamma(s)}{\Gamma(1/2+s)}\right)$$

$$= \frac{\pi^2}{6} - 2\log^2(2), \tag{6.344}$$

where in the calculations I used the limited expansion $\dfrac{\Gamma(s)}{\Gamma(1/2+s)} = \dfrac{1}{\sqrt{\pi}}\dfrac{1}{s} +$ $2\log(2)\dfrac{1}{\sqrt{\pi}} + \dfrac{1}{6}\dfrac{1}{\sqrt{\pi}}(12\log^2(2) - \pi^2)s + O(s^2)$ immediately derived with the help of the limited expansions extracted from the previous section, that is, $\Gamma(s) = \dfrac{1}{s} - \gamma + \dfrac{1}{2}\left(\gamma^2 + \dfrac{\pi^2}{6}\right)s + O(s^2)$ and $\Gamma\left(\dfrac{1}{2} + s\right) = \sqrt{\pi} - \sqrt{\pi}(2\log(2) + \gamma)s + \dfrac{\sqrt{\pi}}{4}(8\log^2(2) + 8\log(2)\gamma + 2\gamma^2 + \pi^2)s^2 + O(s^3)$. After using these two mentioned limited expansions in $\dfrac{\sqrt{\pi}\,\Gamma(s)}{\Gamma(1/2+s)}$ and simplifying $\sqrt{\pi}$, we also need to use the limited expansion $\dfrac{1}{1-x} = 1 + x + x^2 + O(x^3)$.

Yeah, a Beta function game, but what different strategies might we consider here? a question you might ponder over! Let's try a way with dilogarithms and put the first integral of (6.343) in a different form in order to facilitate the upcoming operations,

$$\int_0^1 \frac{\log(t)}{t}\left(1 - \frac{1}{\sqrt{1-t}}\right) dt = \lim_{a\to 0^+}\int_a^1 \frac{\log(t)}{t}\left(1 - \frac{1}{\sqrt{1-t}}\right) dt$$

$$= \lim_{a\to 0^+}\left(\int_a^1 \frac{\log(t)}{t}dt - \int_a^1 \frac{\log(t)}{t\sqrt{1-t}}dt\right) = \lim_{a\to 0^+}\left(-\frac{1}{2}\log^2(a) - \int_a^1 \frac{\log(t)}{t\sqrt{1-t}}dt\right).$$

$$\tag{6.345}$$

Letting the variable change $\dfrac{(1 - \sqrt{1 - t})^2}{t} = u$, or $t = \dfrac{4u}{(1 + u)^2}$, we write

$$\int_a^1 \frac{\log(t)}{t\sqrt{1 - t}} dt = \int_{(1-\sqrt{1-a})^2/a}^1 \frac{1}{u} \log\left(\frac{4u}{(1 + u)^2}\right) du = 2\log(2) \int_{(1-\sqrt{1-a})^2/a}^1 \frac{1}{u} du$$

$$+ \int_{(1-\sqrt{1-a})^2/a}^1 \frac{\log(u)}{u} du - 2 \int_{(1-\sqrt{1-a})^2/a}^1 \frac{\log(1 + u)}{u} du$$

$$= -\frac{\pi^2}{6} - 2\log(2) \log\left(\frac{(1 - \sqrt{1 - a})^2}{a}\right) - \frac{1}{2} \log^2\left(\frac{(1 - \sqrt{1 - a})^2}{a}\right)$$

$$- 2\operatorname{Li}_2\left(-\frac{(1 - \sqrt{1 - a})^2}{a}\right). \tag{6.346}$$

By plugging the result from (6.346) in (6.345), and rearranging, we get that

$$\int_0^1 \frac{\log(t)}{t} \left(1 - \frac{1}{\sqrt{1 - t}}\right) dt = \lim_{a \to 0^+} \int_a^1 \frac{\log(t)}{t} \left(1 - \frac{1}{\sqrt{1 - t}}\right) dt$$

$$= \frac{\pi^2}{6} + 2 \lim_{a \to 0^+} \log(a) \log\left(\frac{2(1 - \sqrt{1 - a})}{a}\right) + 4\log(2) \lim_{a \to 0^+} \log\left(\frac{1 - \sqrt{1 - a}}{a}\right)$$

$$+ 2 \lim_{a \to 0^+} \log^2\left(\frac{1 - \sqrt{1 - a}}{a}\right) + 2 \lim_{a \to 0^+} \operatorname{Li}_2\left(-\frac{(1 - \sqrt{1 - a})^2}{a}\right) = \frac{\pi^2}{6} - 2\log^2(2),$$

$$\tag{6.347}$$

where all limits are straightforward[14] with the help of *L'Hospital's rule*, when needed.

---

[14] The limit of the form $\displaystyle\lim_{a \to 0^+} \frac{1 - \sqrt{1 - a}}{a}$ is immediately obtained by *L'Hospital's rule*,
and we get $\dfrac{1}{2}$. As regards the first limit, we can write it as a product of two limits,

$$\underbrace{\lim_{a \to 0^+} \log(a) \log\left(\frac{2(1 - \sqrt{1 - a})}{a}\right)}_{0} = \underbrace{\lim_{a \to 0^+} a \log(a)}_{} \cdot \underbrace{\lim_{a \to 0^+} \frac{1}{a} \log\left(\frac{2(1 - \sqrt{1 - a})}{a}\right)}_{1/4} = 0, \text{ where}$$

each limit is calculated by using *L'Hospital's rule*.

For the second integral in (6.343), we have

$$\int_0^1 \frac{\sin^2(x)\log(t)}{1 - \sin^2(x)t}\, dt = -\,\text{Li}_2(\sin^2(x)), \tag{6.348}$$

where I used that $\int_0^1 \dfrac{y\log^n(x)}{1 - yx}\, dx = (-1)^n n!\,\text{Li}_{n+1}(y)$, $y \in (-\infty, 1]$, which may be found in [76, Chapter 1, p.4]

Combining (6.343), (6.344) or (6.347), and (6.348), and rearranging, we obtain that

$$\text{Li}_2\left(\sin^2(x)\right)$$

$$= \frac{\pi^2}{6} - 2\log^2(2) - 2\sum_{n=1}^\infty \left(2\log(2)\frac{1}{n} + (-1)^{n-1}\frac{1}{n^2} - 2\frac{\overline{H}_n}{n}\right)\cos(2nx),$$

and the second solution to the point (i) is finalized.

As regards the Fourier series at the second point, I'll use the result from the previous point where we replace x by $\pi/2 - x$, and then we get

$$\text{Li}_2\left(\cos^2(x)\right)$$

$$= \frac{\pi^2}{6} - 2\log^2(2) + 2\sum_{n=1}^\infty (-1)^{n-1}\left(2\log(2)\frac{1}{n} + (-1)^{n-1}\frac{1}{n^2} - 2\frac{\overline{H}_n}{n}\right)\cos(2nx),$$

and the solution to the point (ii) is finalized.

To prove the Fourier series result at the point (iii) I'll want to use a way similar to the one found in the second solution to the first point of the problem, but this time, I'll consider the squared log version of the integral in (6.342),

$$\int_0^\infty \log^2(\cosh(y))e^{-2ny}\, dy$$

$$= \frac{1}{24}(\pi^2 - 12\log^2(2))(-1)^{n-1}\frac{1}{n} + \frac{1}{2}\log(2)(-1)^{n-1}\frac{1}{n^2} + \frac{1}{4}\frac{1}{n^3}$$

$$- \log(2)(-1)^{n-1}\frac{H_n}{n} - \frac{1}{2}(-1)^{n-1}\frac{\overline{H}_n}{n^2} - \frac{1}{2}(-1)^{n-1}\frac{\overline{H}_n^{(2)}}{n} + (-1)^{n-1}\frac{1}{n}\sum_{k=1}^n \frac{\overline{H}_k}{k}.$$

$$\tag{6.349}$$

Proof By letting the variable change $y = -\log(\sqrt{t})$, we have that

$$\int_0^\infty \log^2(\cosh(y))e^{-2ny}dy = \frac{1}{2}\int_0^1 t^{n-1}\log^2\left(\frac{1+t}{2\sqrt{t}}\right)dt$$

$$= \frac{1}{2}\log^2(2)\underbrace{\int_0^1 t^{n-1}dt}_{1/n} + \frac{1}{2}\log(2)\underbrace{\int_0^1 t^{n-1}\log(t)dt}_{-1/n^2} + \frac{1}{8}\underbrace{\int_0^1 t^{n-1}\log^2(t)dt}_{2/n^3}$$

$$- \log(2)\int_0^1 t^{n-1}\log(1+t)dt + \frac{1}{2}\int_0^1 t^{n-1}\log^2(1+t)dt$$

$$- \frac{1}{2}\int_0^1 t^{n-1}\log(t)\log(1+t)dt$$

$$= \frac{1}{24}(\pi^2 - 12\log^2(2))(-1)^{n-1}\frac{1}{n} + \frac{1}{2}\log(2)(-1)^{n-1}\frac{1}{n^2} + \frac{1}{4}\frac{1}{n^3}$$

$$- \log(2)(-1)^{n-1}\frac{H_n}{n} - \frac{1}{2}(-1)^{n-1}\frac{\overline{H}_n}{n^2} - \frac{1}{2}(-1)^{n-1}\frac{\overline{H}_n^{(2)}}{n} + (-1)^{n-1}\frac{1}{n}\sum_{k=1}^n \frac{\overline{H}_k}{k},$$

where to get the last equality, I used the results in (1.99), (1.102), Sect. 1.21, and the case $m = 1$ of the generalization in (1.127), Sect. 1.25. ∎

Next, if we multiply both sides of (6.349) by $(-1)^{n-1}\cos(2nx)$, then consider the summation from $n = 1$ to ∞, and follow almost the same steps shown in (6.343), where this time we have $\log^2(\cosh(y))$ instead of $\log(\cosh(y))$, we arrive at

$$\sum_{n=1}^\infty \left(\frac{1}{24}(\pi^2 - 12\log^2(2))\frac{1}{n} + \frac{1}{2}\log(2)\frac{1}{n^2} + \frac{1}{4}(-1)^{n-1}\frac{1}{n^3}\right.$$

$$\left. - \log(2)\frac{H_n}{n} - \frac{1}{2}\frac{\overline{H}_n}{n^2} - \frac{1}{2}\frac{\overline{H}_n^{(2)}}{n} + \frac{1}{n}\sum_{k=1}^n \frac{\overline{H}_k}{k}\right)\cos(2nx)$$

$$= -\frac{1}{16}\int_0^1 \frac{\log^2(t)}{t}\left(1 - \frac{1}{\sqrt{1-t}}\right)dt - \frac{1}{16}\int_0^1 \frac{\sin^2(x)\log^2(t)}{1-\sin^2(x)t}dt$$

{for the first integral integrate by parts, and for the second one consider}

$$\left\{\int_0^1 \frac{y\log^n(x)}{1-yx}dx = (-1)^n n!\operatorname{Li}_{n+1}(y), \ y \in (-\infty, 1] \ (\text{see [76, Chapter 1, p.4]})\right\}$$

$$= -\frac{1}{96}\int_0^1 \log^3(t)(1-t)^{-1/2-1}dt - \frac{1}{8}\mathrm{Li}_3(\sin^2(x))$$

$$= -\frac{1}{96}\lim_{\substack{x\to 1\\ y\to -1/2}} \frac{\partial^3}{\partial x^3} B(x,y) - \frac{1}{8}\mathrm{Li}_3(\sin^2(x))$$

$$= \frac{1}{6}\log^3(2) - \log(2)\frac{\pi^2}{24} + \frac{1}{4}\zeta(3) - \frac{1}{8}\mathrm{Li}_3(\sin^2(x)),$$

from where we obtain that

$$\mathrm{Li}_3(\sin^2(x))$$

$$= \frac{4}{3}\log^3(2) - \log(2)\frac{\pi^2}{3} + 2\zeta(3) + \sum_{n=1}^{\infty}\left(\left(4\log^2(2) - \frac{\pi^2}{3}\right)\frac{1}{n}\right.$$

$$\left. -4\log(2)\frac{1}{n^2} - 2(-1)^{n-1}\frac{1}{n^3} + 8\log(2)\frac{H_n}{n} + 4\frac{\overline{H}_n}{n^2} + 4\frac{\overline{H}_n^{(2)}}{n} - 8\frac{1}{n}\sum_{k=1}^{n}\frac{\overline{H}_k}{k}\right)\cos(2nx)$$

$$= \frac{4}{3}\log^3(2) - \log(2)\frac{\pi^2}{3} + 2\zeta(3) + \sum_{n=1}^{\infty}\left(\left(4\log^2(2) - \frac{\pi^2}{3}\right)\frac{1}{n} - 4\log(2)\frac{1}{n^2}\right.$$

$$\left. -2\frac{(-1)^{n-1}}{n^3} + 8\log(2)\frac{H_n}{n} + 4\frac{\overline{H}_n}{n^2} - 4\frac{\overline{H}_n^{(2)}}{n} - 8\frac{H_n\overline{H}_n}{n} + 8\frac{1}{n}\sum_{k=1}^{n}(-1)^{k-1}\frac{H_k}{k}\right)\cos(2nx),$$

where the second equality follows based on the other closed form given in (1.102), Sect. 1.21, and the solution to the point (iii) is finalized. The calculations of the Beta function limit can be done either by using *Mathematica* or manually. The reader might be concerned because we need $y \to -1/2$, but according to the Beta function definition, we known that $B(x,y) = \int_0^1 t^{x-1}(1-t)^{y-1}dt$, $\Re(x), \Re(y) > 0$. However, since we have a third derivative with respect to x, we note that $\int_0^1 t^{x-1}\log^3(t)(1-t)^{y-1}dt$ is defined for $y > -3$. Also, observe that $\int_0^1 \frac{\log^2(t)}{t}\left(1 - \frac{1}{\sqrt{1-t}}\right)dt$ might be treated in a style similar to the one I treated the integral $\int_0^1 \frac{\log(t)}{t}\left(1 - \frac{1}{\sqrt{1-t}}\right)dt$ in the second solution to the point (i).

As for the Fourier series at the fourth point, I'll use the result from the previous point where we replace x by $\pi/2 - x$ and then we obtain that

$$\mathrm{Li}_3(\cos^2(x))$$

$$= \frac{4}{3}\log^3(2) - \log(2)\frac{\pi^2}{3} + 2\zeta(3) - \sum_{n=1}^{\infty}(-1)^{n-1}\left(\left(4\log^2(2) - \frac{\pi^2}{3}\right)\frac{1}{n}\right.$$

$$\left. -4\log(2)\frac{1}{n^2} - 2(-1)^{n-1}\frac{1}{n^3} + 8\log(2)\frac{H_n}{n} + 4\frac{\overline{H}_n}{n^2} + 4\frac{\overline{H}_n^{(2)}}{n} - 8\frac{1}{n}\sum_{k=1}^{n}\frac{\overline{H}_k}{k}\right)\cos(2nx)$$

$$= \frac{4}{3}\log^3(2) - \log(2)\frac{\pi^2}{3} + 2\zeta(3) - \sum_{n=1}^{\infty}(-1)^{n-1}\left(\left(4\log^2(2) - \frac{\pi^2}{3}\right)\frac{1}{n} - 4\log(2)\frac{1}{n^2}\right.$$

$$\left. -2\frac{(-1)^{n-1}}{n^3} + 8\log(2)\frac{H_n}{n} + 4\frac{\overline{H}_n}{n^2} - 4\frac{\overline{H}_n^{(2)}}{n} - 8\frac{H_n\overline{H}_n}{n} + 8\frac{1}{n}\sum_{k=1}^{n}(-1)^{k-1}\frac{H_k}{k}\right)\cos(2nx),$$

and the solution to the point (iv) is finalized.

The curious reader might also go further and extract the following two Fourier series where the coefficients are kept in an integral form,

$$\operatorname{Li}_4(\sin^2(x))$$

$$= \frac{\pi^4}{40} - 4\log(2)\zeta(3) + \log^2(2)\frac{\pi^2}{3} - \frac{2}{3}\log^4(2)$$

$$- \frac{8}{3}\sum_{n=1}^{\infty}(-1)^{n-1}\left(\int_0^1 t^{n-1}\log^3\left(\frac{1+t}{2\sqrt{t}}\right)dt\right)\cos(2nx) \qquad (6.350)$$

and

$$\operatorname{Li}_4(\cos^2(x))$$

$$= \frac{\pi^4}{40} - 4\log(2)\zeta(3) + \log^2(2)\frac{\pi^2}{3} - \frac{2}{3}\log^4(2)$$

$$+ \frac{8}{3}\sum_{n=1}^{\infty}\left(\int_0^1 t^{n-1}\log^3\left(\frac{1+t}{2\sqrt{t}}\right)dt\right)\cos(2nx). \qquad (6.351)$$

As in the previous section, experimental calculations on small cases show that we could extract the Fourier series for any $\operatorname{Li}_n(\sin^2(x))$ and $\operatorname{Li}_n(\cos^2(x))$, at least in forms with coefficients expressed in terms of integrals of the type mentioned in (6.350) and (6.351). In contrast to the previous section, the *superstar* integral of the current section is $\int_0^{\infty}\log^p(\cosh(x))e^{-2nx}dx$, $n, p \in \mathbb{N}$ (recall that in the preceding section, we have the integral variant with $\sinh(x)$ instead of $\cosh(x)$).

And to share a last thought with the curious reader, recall that in the beginning of the section I mentioned the reference [76, Chapter 3, pp.242–252] while referring to two powerful Fourier series in my first book. Now, the curious reader might like to know that following a very similar procedure described there, where this time we use $\int_0^\infty \tanh(x)e^{-nx}\,dx$ instead of $\int_0^\infty \tanh(x)e^{-2nx}\,dx$, we arrive at

$$\tan(x)\log(1-\cos(x)) = -\sum_{n=1}^\infty \left(\int_0^1 t^{n/2-1}\frac{1-t}{1+t}dt\right)\sin(nx), \qquad (6.352)$$

and if we integrate (6.352) once, we obtain the spectacular and useful Fourier series,

$$\operatorname{Li}_2(\cos(x)) = \frac{\pi^2}{24} - \frac{1}{2}\log^2(2) + \sum_{n=1}^\infty \frac{1}{n}\left(\int_0^1 t^{n/2-1}\frac{1-t}{1+t}dt\right)\cos(nx). \qquad (6.353)$$

6.50 And More Stunning Fourier Series, Related to Atypical Harmonic Numbers (Skew-Harmonic Numbers)

Solution In this section, we go further and extract more interesting Fourier series, which are powerful tools to have in hand in various calculations. During the derivations of the desired main results, we'll also find useful Fourier series in the previous sections as we'll note right from the first point.

One way to attack the point (i) of the problem is to use the Landen's dilogarithmic identity, $\operatorname{Li}_2(x) + \operatorname{Li}_2\left(\frac{x}{x-1}\right) = -\frac{1}{2}\log^2(1-x)$ (see [21, Chapter 1, p.5], [70, Chapter 2, p.107]), where if we replace x by $\sin^2(x)$ and rearrange, we get

$$\operatorname{Li}_2(-\tan^2(x)) = -\frac{1}{2}\log^2(\cos^2(x)) - \operatorname{Li}_2(\sin^2(x)). \qquad (6.354)$$

Now, since the Fourier series of the functions in the right-hand side of (6.354) are obtained from (4.155), Sect. 4.48, and (4.160), Sect. 4.49, we immediately arrive at

$$\operatorname{Li}_2\left(-\tan^2(x)\right)$$

$$= -\frac{\pi^2}{3} + 4\sum_{n=1}^\infty \left(\log(2)\frac{1}{n} + \log(2)\frac{(-1)^{n-1}}{n} + (-1)^{n-1}\frac{H_n}{n} - \frac{\overline{H}_n}{n}\right)\cos(2nx),$$

and the first solution to the point (i) is finalized.

For a second solution to the point (i), we start from the integral

$$\int_0^\infty \log(\tanh(y))e^{-2ny}dy = -\frac{1}{2}\log(2)\frac{1}{n} - \frac{1}{2}\log(2)\frac{(-1)^{n-1}}{n} - \frac{1}{2}\frac{H_n}{n} + \frac{1}{2}(-1)^{n-1}\frac{\overline{H}_n}{n}.$$

$$(6.355)$$

Proof We make the variable change $y = -\log(\sqrt{t})$ that gives

$$\int_0^\infty \log(\tanh(y))e^{-2ny}dy = \frac{1}{2}\int_0^1 t^{n-1}\log\left(\frac{1-t}{1+t}\right)dt$$

$$= \frac{1}{2}\int_0^1 t^{n-1}\log(1-t)dt - \frac{1}{2}\int_0^1 t^{n-1}\log(1+t)dt$$

$$= -\frac{1}{2}\log(2)\frac{1}{n} - \frac{1}{2}\log(2)\frac{(-1)^{n-1}}{n} - \frac{1}{2}\frac{H_n}{n} + \frac{1}{2}(-1)^{n-1}\frac{\overline{H}_n}{n},$$

where in the calculations I used the integral results in (3.10), Sect. 3.3, and (1.99), Sect. 1.21. ∎

Upon multiplying both sides of (6.355) by $(-1)^{n-1}\cos(2nx)$ and considering the summation from $n = 1$ to ∞, we write

$$\sum_{n=1}^\infty \left(-\frac{1}{2}\log(2)\frac{1}{n} - \frac{1}{2}\log(2)\frac{(-1)^{n-1}}{n} - \frac{1}{2}(-1)^{n-1}\frac{H_n}{n} + \frac{1}{2}\frac{\overline{H}_n}{n}\right)\cos(2nx)$$

$$= \sum_{n=1}^\infty (-1)^{n-1}\cos(2nx)\int_0^\infty \log(\tanh(y))e^{-2ny}dy$$

$$= \int_0^\infty \log(\tanh(y))\sum_{n=1}^\infty (-1)^{n-1}\cos(2nx)e^{-2ny}dy$$

{employ (6.327), Sect. 6.48, where we replace x by $2x$ and set $p = -e^{-2y}$}

$$= \frac{1}{2}\int_0^\infty \log(\tanh(y))\left(1 - \frac{\sinh(2y)}{\cos(2x)+\cosh(2y)}\right)dy$$

$$\underset{\tanh(y)=\sqrt{t}}{=} \frac{1}{8}\int_0^1 \frac{\log(t)}{\sqrt{t}(1-t)}dt - \frac{1}{8}\int_0^1 \frac{\log(t)}{1-t}dt - \frac{1}{8}\int_0^1 \frac{\tan^2(x)\log(t)}{1+\tan^2(x)t}dt$$

$$= -\frac{\pi^2}{24} - \frac{1}{8}\text{Li}_2\left(-\tan^2(x)\right),$$

and if we rearrange, we get that

$$\mathrm{Li}_2\left(-\tan^2(x)\right)$$

$$= -\frac{\pi^2}{3} + 4\sum_{n=1}^{\infty}\left(\log(2)\frac{1}{n} + \log(2)\frac{(-1)^{n-1}}{n} + (-1)^{n-1}\frac{H_n}{n} - \frac{\overline{H}_n}{n}\right)\cos(2nx),$$

where in the calculations I used that $\displaystyle\int_0^1 \frac{\log(t)}{\sqrt{t(1-t)}}dt = \int_0^1 \log(t)\sum_{n=1}^{\infty}t^{n-3/2}dt =$

$$\sum_{n=1}^{\infty}\int_0^1 t^{n-3/2}\log(t)dt = -4\sum_{n=1}^{\infty}\frac{1}{(2n-1)^2} = -4\left(\sum_{n=1}^{\infty}\frac{1}{n^2} - \sum_{n=1}^{\infty}\frac{1}{(2n)^2}\right) =$$

$$-3\sum_{n=1}^{\infty}\frac{1}{n^2} = -\frac{\pi^2}{2}, \text{ then } \int_0^1\frac{\log(t)}{1-t}dt = \int_0^1\sum_{n=1}^{\infty}t^{n-1}\log(t)dt = \sum_{n=1}^{\infty}\int_0^1$$

$$t^{n-1}\log(t)dt = -\sum_{n=1}^{\infty}\frac{1}{n^2} = -\frac{\pi^2}{6}, \text{ and finally } \int_0^1\frac{\tan^2(x)\log(t)}{1+\tan^2(x)t}dt =$$

$\mathrm{Li}_2\left(-\tan^2(x)\right)$, by following the reasoning in (6.348), Sect. 6.49, and the second solution to the point (i) is finalized. Also, after performing the variable changes in the main calculation above I used the hyperbolic identities, $\sinh(2\arctan(x)) = 2x/(1-x^2)$ and $\cosh(2\arctan(x)) = (1+x^2)/(1-x^2)$, $x \in (-1, 1)$.

The result at the point (ii) is straightforward if we employ the Fourier series from the previous point and replace x by $\pi/2 - x$ that gives

$$\mathrm{Li}_2\left(-\cot^2(x)\right)$$

$$= -\frac{\pi^2}{3} - 4\sum_{n=1}^{\infty}(-1)^{n-1}\left(\log(2)\frac{1}{n} + \log(2)\frac{(-1)^{n-1}}{n} + (-1)^{n-1}\frac{H_n}{n} - \frac{\overline{H}_n}{n}\right)\cos(2nx),$$

and the solution to the point (ii) is finalized.

We step further and prepare to extract the Fourier series at the point (iii), but before doing that, we need to prove the following auxiliary result:

$$\int_0^{\infty}\log^2(\tanh(y))e^{-2ny}dy$$

$$= \frac{\pi^2}{12}\frac{1}{n} + \frac{\pi^2}{12}(-1)^{n-1}\frac{1}{n} + \log(2)\frac{H_n}{n} - \log(2)(-1)^{n-1}\frac{H_n}{n} + \frac{1}{2}\frac{H_n^2}{n} + \log(2)\frac{\overline{H}_n}{n}$$

$$- \log(2)(-1)^{n-1}\frac{\overline{H}_n}{n} - \frac{1}{2}\frac{\overline{H}_n^2}{n} + (-1)^{n-1}\frac{\overline{H}_n^{(2)}}{n} + (-1)^{n-1}\frac{H_n\overline{H}_n}{n}$$

$$- 2(-1)^{n-1} \frac{1}{n} \sum_{k=1}^{n} (-1)^{k-1} \frac{H_k}{k}$$

$$= \frac{\pi^2}{12} \frac{1}{n} + \frac{\pi^2}{12} (-1)^{n-1} \frac{1}{n} + \log(2) \frac{H_n}{n} - \log(2)(-1)^{n-1} \frac{H_n}{n} + \frac{1}{2} \frac{H_n^2}{n} + \log(2) \frac{\overline{H}_n}{n}$$

$$- \log(2)(-1)^{n-1} \frac{\overline{H}_n}{n} - \frac{1}{2} \frac{\overline{H}_n^2}{n} - (-1)^{n-1} \frac{\overline{H}_n^{(2)}}{n} - (-1)^{n-1} \frac{H_n \overline{H}_n}{n}$$

$$+ 2(-1)^{n-1} \frac{1}{n} \sum_{k=1}^{n} \frac{\overline{H}_k}{k}. \tag{6.356}$$

Proof As in the second solution to the point (i), we let $y = -\log(\sqrt{t})$ that gives

$$\int_0^\infty \log^2(\tanh(y)) e^{-2ny} dy = \frac{1}{2} \int_0^1 t^{n-1} \log^2 \left(\frac{1-t}{1+t} \right) dt$$

$$= \frac{1}{2} \int_0^1 t^{n-1} \log^2(1-t) dt - \int_0^1 t^{n-1} \log(1-t) \log(1+t) dt + \frac{1}{2} \int_0^1 t^{n-1} \log^2(1+t) dt$$

$$= \frac{\pi^2}{12} \frac{1}{n} + \frac{\pi^2}{12} (-1)^{n-1} \frac{1}{n} + \log(2) \frac{H_n}{n} - \log(2)(-1)^{n-1} \frac{H_n}{n} + \frac{1}{2} \frac{H_n^2}{n} + \log(2) \frac{\overline{H}_n}{n}$$

$$- \log(2)(-1)^{n-1} \frac{\overline{H}_n}{n} - \frac{1}{2} \frac{\overline{H}_n^2}{n} + (-1)^{n-1} \frac{\overline{H}_n^{(2)}}{n} + (-1)^{n-1} \frac{H_n \overline{H}_n}{n}$$

$$- 2 \frac{(-1)^{n-1}}{n} \sum_{k=1}^{n} (-1)^{k-1} \frac{H_k}{k}$$

$$= \frac{\pi^2}{12} \frac{1}{n} + \frac{\pi^2}{12} (-1)^{n-1} \frac{1}{n} + \log(2) \frac{H_n}{n} - \log(2)(-1)^{n-1} \frac{H_n}{n} + \frac{1}{2} \frac{H_n^2}{n} + \log(2) \frac{\overline{H}_n}{n}$$

$$- \log(2)(-1)^{n-1} \frac{\overline{H}_n}{n} - \frac{1}{2} \frac{\overline{H}_n^2}{n} - (-1)^{n-1} \frac{\overline{H}_n^{(2)}}{n} - (-1)^{n-1} \frac{H_n \overline{H}_n}{n}$$

$$+ 2(-1)^{n-1} \frac{1}{n} \sum_{k=1}^{n} \frac{\overline{H}_k}{k},$$

where in the calculations I used that $\int_0^1 x^{n-1} \log^2(1-x) dx = \dfrac{H_n^2 + H_n^{(2)}}{n}$, which appears in [76, Chapter 1, Section 1.3, p.2], and then the results in (1.102), Sect. 1.21, and (1.117), Sect. 1.22. ∎

Following the steps in the second solution to the point (i), after multiplying the double equality in (6.356) by $(-1)^{n-1}\cos(2nx)$ and considering the summation from $n = 1$ to ∞, in the leftmost-hand side, we have

$$\sum_{n=1}^{\infty}(-1)^{n-1}\cos(2nx)\int_0^{\infty}\log^2(\tanh(y))e^{-2ny}dy$$

$$= \int_0^{\infty}\log^2(\tanh(y))\sum_{n=1}^{\infty}(-1)^{n-1}\cos(2nx)e^{-2ny}dy$$

$$= \frac{1}{2}\int_0^{\infty}\log^2(\tanh(y))\left(1 - \frac{\sinh(2y)}{\cos(2x)+\cosh(2y)}\right)dy$$

$$\overset{\tanh(y)=\sqrt{t}}{=} \frac{1}{16}\int_0^1\frac{\log^2(t)}{\sqrt{t}(1-t)}dt - \frac{1}{16}\int_0^1\frac{\log^2(t)}{1-t}dt - \frac{1}{16}\int_0^1\frac{\tan^2(x)\log^2(t)}{1+\tan^2(x)t}dt$$

$$= \frac{3}{4}\zeta(3) + \frac{1}{8}\operatorname{Li}_3\left(-\tan^2(x)\right), \tag{6.357}$$

where in the calculations I used that $\displaystyle\int_0^1\frac{\log^2(t)}{\sqrt{t}(1-t)}dt = \int_0^1\sum_{n=1}^{\infty}t^{n-3/2}\log^2(t) =$

$\displaystyle\sum_{n=1}^{\infty}\int_0^1 t^{n-3/2}\log^2(t)dt = 16\sum_{n=1}^{\infty}\frac{1}{(2n-1)^3} = 16\left(\sum_{n=1}^{\infty}\frac{1}{n^3} - \sum_{n=1}^{\infty}\frac{1}{(2n)^3}\right) =$

$\displaystyle 14\sum_{n=1}^{\infty}\frac{1}{n^3} = 14\zeta(3), \int_0^1\frac{\log^2(t)}{1-t}dt = \int_0^1\sum_{n=1}^{\infty}t^{n-1}\log^2(t)dt = \sum_{n=1}^{\infty}\int_0^1$

$\displaystyle t^{n-1}\log^2(t)dt = 2\sum_{n=1}^{\infty}\frac{1}{n^3} = 2\zeta(3),$ and finally $\displaystyle\int_0^1\frac{\tan^2(x)\log^2(t)}{1+\tan^2(x)t}dt =$

$-2\operatorname{Li}_3\left(-\tan^2(x)\right),$ by considering the reasoning in (6.348), Sect. 6.49.

Since above I assumed that all sides of (6.356) are multiplied by $(-1)^{n-1}\cos(2nx)$ and then I considered making the summation from $n = 1$ to ∞, and using (6.357), we arrive at

$$\operatorname{Li}_3\left(-\tan^2(x)\right)$$

$$= -6\zeta(3) + \sum_{n=1}^{\infty}\left(\frac{2}{3}\pi^2\frac{1}{n} + \frac{2}{3}\pi^2(-1)^{n-1}\frac{1}{n} - 8\log(2)\frac{H_n}{n} + 8\log(2)(-1)^{n-1}\frac{H_n}{n}\right.$$

$$\left. + 4(-1)^{n-1}\frac{H_n^2}{n} - 8\log(2)\frac{\overline{H}_n}{n} + 8\log(2)(-1)^{n-1}\frac{\overline{H}_n}{n} - 4(-1)^{n-1}\frac{\overline{H}_n^2}{n}\right.$$

$$+8\frac{\overline{H}_n^{(2)}}{n}+8\frac{H_n\overline{H}_n}{n}-16\frac{1}{n}\sum_{k=1}^{n}(-1)^{k-1}\frac{H_k}{k}\Bigg)\cos(2nx)$$

$$=-6\zeta(3)+\sum_{n=1}^{\infty}\Bigg(\frac{2}{3}\pi^2\frac{1}{n}+\frac{2}{3}\pi^2(-1)^{n-1}\frac{1}{n}-8\log(2)\frac{H_n}{n}+8\log(2)(-1)^{n-1}\frac{H_n}{n}$$

$$+4(-1)^{n-1}\frac{H_n^2}{n}-8\log(2)\frac{\overline{H}_n}{n}+8\log(2)(-1)^{n-1}\frac{\overline{H}_n}{n}-4(-1)^{n-1}\frac{\overline{H}_n^2}{n}$$

$$-8\frac{\overline{H}_n^{(2)}}{n}-8\frac{H_n\overline{H}_n}{n}+16\frac{1}{n}\sum_{k=1}^{n}\frac{\overline{H}_k}{k}\Bigg)\cos(2nx),$$

and the solution to the point (iii) is finalized.

The result at the point (iv) is immediately obtained if we use the Fourier series from the previous point where we replace x by $\pi/2-x$ that leads to

$$\text{Li}_3\left(-\cot^2(x)\right)$$

$$=-6\zeta(3)-\sum_{n=1}^{\infty}(-1)^{n-1}\Bigg(\frac{2}{3}\pi^2\frac{1}{n}+\frac{2}{3}\pi^2(-1)^{n-1}\frac{1}{n}-8\log(2)\frac{H_n}{n}$$

$$+8\log(2)(-1)^{n-1}\frac{H_n}{n}+4(-1)^{n-1}\frac{H_n^2}{n}-8\log(2)\frac{\overline{H}_n}{n}+8\log(2)(-1)^{n-1}\frac{\overline{H}_n}{n}$$

$$-4(-1)^{n-1}\frac{\overline{H}_n^2}{n}+8\frac{\overline{H}_n^{(2)}}{n}+8\frac{H_n\overline{H}_n}{n}-16\frac{1}{n}\sum_{k=1}^{n}(-1)^{k-1}\frac{H_k}{k}\Bigg)\cos(2nx)$$

$$=-6\zeta(3)-\sum_{n=1}^{\infty}(-1)^{n-1}\Bigg(\frac{2}{3}\pi^2\frac{1}{n}+\frac{2}{3}\pi^2(-1)^{n-1}\frac{1}{n}-8\log(2)\frac{H_n}{n}$$

$$+8\log(2)(-1)^{n-1}\frac{H_n}{n}+4(-1)^{n-1}\frac{H_n^2}{n}-8\log(2)\frac{\overline{H}_n}{n}+8\log(2)(-1)^{n-1}\frac{\overline{H}_n}{n}$$

$$-4(-1)^{n-1}\frac{\overline{H}_n^2}{n}-8\frac{\overline{H}_n^{(2)}}{n}-8\frac{H_n\overline{H}_n}{n}+16\frac{1}{n}\sum_{k=1}^{n}\frac{\overline{H}_k}{k}\Bigg)\cos(2nx)$$

and the solution to the point (iv) is finalized.

In this final part of the section, I have a similar comment to the one in the previous section, and it may be observed that we could extract the Fourier series for any

$\text{Li}_n(-\tan^2(x))$ and $\text{Li}_n(-\cot^2(x))$, at least in forms with coefficients expressed in terms of integrals. For performing such extractions, we may build solutions by exploiting an integral of the form $\int_0^\infty \log^p(\tanh(x))e^{-2nx}dx$, $n, p \in \mathbb{N}$, as seen in the work presented above.

6.51   Yet Other Stunning Fourier Series, This Time with the Coefficients Mainly Kept in an Integral Form

Solution In some calculations, you might find (very) useful to consider the Fourier series with the coefficients in the form of integrals. In this section, I keep the Fourier series in such forms, except that for the point (i) I also provide the form of the Fourier coefficient in terms of (skew-)harmonic numbers. As seen in the solutions from the previous sections related to derivations of Fourier series, we can get the present Fourier series either in the form with coefficients expressed as integrals or in the form with coefficients posed in terms of the generalized (skew-)harmonic numbers.

In the following, I'll bring together the results we need taking into account that $\log^2(\tan(x)) = \log^2(\sin(x)) - 2\log(\sin(x))\log(\cos(x)) + \log^2(\cos(x))$, $0 < x < \dfrac{\pi}{2}$.

We know from *(Almost) Impossible Integrals, Sums, and Series* (see [76, Chapter 3, p.248]) that

$$\sum_{n=1}^\infty \left(\psi\left(n + \frac{1}{2}\right) - \psi(n) - \frac{1}{2n}\right)\frac{\sin^2(2nx)}{n} = \log(\sin(x))\log(\cos(x)), \ 0 < x < \frac{\pi}{2},$$

(6.358)

which may be put in the form

$$\sum_{n=1}^\infty (1-(-1)^{n-1})\left(\psi\left(\frac{n+1}{2}\right) - \psi\left(\frac{n}{2}\right) - \frac{1}{n}\right)\frac{\sin^2(nx)}{n} = \log(\sin(x))\log(\cos(x)).$$

(6.359)

Also, from [76, Chapter 3, p.244], we know that

$$\int_0^1 t^{n-1}\frac{1-t}{1+t}dt = \psi\left(\frac{n+1}{2}\right) - \psi\left(\frac{n}{2}\right) - \frac{1}{n}.$$

(6.360)

Thus, combining (6.359) and (6.360), we get

$$\log(\sin(x))\log(\cos(x)) = \sum_{n=1}^\infty (1-(-1)^{n-1})\left(\int_0^1 t^{n-1}\frac{1-t}{1+t}dt\right)\frac{\sin^2(nx)}{n}$$

$$\left\{ \text{use the trigonometric identity, } 1 - \cos(2x) = 2\sin^2(x) \right\}$$

$$= \frac{1}{2} \sum_{n=1}^{\infty} (1 - (-1)^{n-1}) \underbrace{\int_0^1 t^{n-1} \frac{1}{n} \frac{1-t}{1+t} dt}$$

$$2\log^2(2) - \pi^2/12$$

$$- \frac{1}{2} \sum_{n=1}^{\infty} \left((1 - (-1)^{n-1}) \int_0^1 t^{n-1} \frac{1}{n} \frac{1-t}{1+t} dt \right) \cos(2nx)$$

$$= \log^2(2) - \frac{\pi^2}{24} - \frac{1}{2} \sum_{n=1}^{\infty} \left((1 - (-1)^{n-1}) \int_0^1 t^{n-1} \frac{1}{n} \frac{1-t}{1+t} dt \right) \cos(2nx),$$

$$(6.361)$$

where in the calculations I used that

$$\sum_{n=1}^{\infty} (1 - (-1)^{n-1}) \left(\int_0^1 t^{n-1} \frac{1}{n} \frac{1-t}{1+t} dt \right) = \int_0^1 \frac{1-t}{1+t} \sum_{n=1}^{\infty} (1 - (-1)^{n-1}) \frac{t^{n-1}}{n} dt$$

$$= - \underbrace{\int_0^1 \frac{\log(1-t)}{t} dt}_{-\pi^2/6} - \underbrace{\int_0^1 \frac{\log(1+t)}{t} dt}_{\pi^2/12} + 2 \underbrace{\int_0^1 \frac{\log(1+t)}{1+t} dt}_{\log^2(2)/2} + 2 \int_0^1 \frac{\log(1-t)}{1+t} dt$$

$$\{\text{the last integral is calculated in (3.163), Sect. 3.21}\}$$

$$= 2\log^2(2) - \frac{\pi^2}{12}.$$

On the other hand, from the solutions to the points (*i*) and (*ii*) of Sect. 6.48, we have that

$$\log^2(\sin(x))$$

$$= \log^2(2) + \frac{\pi^2}{12} - 2 \sum_{n=1}^{\infty} \left(\int_0^1 t^{n-1} \log \left(\frac{1-t}{2\sqrt{t}} \right) dt \right) \cos(2nx) \qquad (6.362)$$

and

$$\log^2(\cos(x))$$

$$= \log^2(2) + \frac{\pi^2}{12} + 2 \sum_{n=1}^{\infty} (-1)^{n-1} \left(\int_0^1 t^{n-1} \log \left(\frac{1-t}{2\sqrt{t}} \right) dt \right) \cos(2nx). \qquad (6.363)$$

So, at this point, if we combine (6.361), (6.362), and (6.363), considering the trigonometric identity mentioned at the beginning of the present solution, we get

$$\log^2(\tan(x)) = \log^2(\cot(x))$$

$$= \frac{\pi^2}{4} + \sum_{n=1}^{\infty}\left((1 - (-1)^{n-1})\int_0^1 t^{n-1}\left(\frac{1}{n}\frac{1-t}{1+t} - 2\log\left(\frac{1-t}{2\sqrt{t}}\right)\right)dt\right)\cos(2nx),$$

which is the first desired form of the Fourier series given at the point (i).

To get the second equality, we might want to observe that if we combine (6.361), (6.360), and (4.96), Sect. 4.18, we get

$$\log(\sin(x))\log(\cos(x)) = \log^2(2) - \frac{\pi^2}{24}$$

$$+ \frac{1}{2}\sum_{n=1}^{\infty}\left((1 - (-1)^{n-1})\left(2\log(2)\frac{1}{n} + (-1)^{n-1}\frac{1}{n^2} - 2\frac{\overline{H}_n}{n}\right)\right)\cos(2nx).$$

$$\tag{6.364}$$

Returning to the simple trigonometric identity, $\log^2(\tan(x)) = \log^2(\sin(x)) - 2\log(\sin(x))\log(\cos(x)) + \log^2(\cos(x))$, $0 < x < \frac{\pi}{2}$, and then considering (6.364) and (4.154), (4.155), Sect. 4.48, we obtain that

$$\log^2(\tan(x)) = \log^2(\cot(x))$$

$$= \frac{\pi^2}{4} + 2\sum_{n=1}^{\infty}(1 - (-1)^{n-1})\left(\frac{H_n}{n} + \frac{\overline{H}_n}{n}\right)\cos(2nx),$$

which is the second form of the Fourier series found at the first point of the problem, and the solution to the point (i) is finalized.

What other ways to go? you might wonder. For example, we may start from the fact[15] that $\log(\tan(x)) = -2\arctanh(e^{i2x}) + i\frac{\pi}{2}$, $0 < x < \frac{\pi}{2}$, which shows that

---

[15] Since $\arctanh(x) = \frac{1}{2}\log\left(\frac{1+x}{1-x}\right)$, and then $\sin(x) = \frac{e^{ix} - e^{-ix}}{2i} = \frac{i}{2}e^{-ix}(1 - e^{i2x})$ and $\cos(x) = \frac{e^{ix} + e^{-ix}}{2} = \frac{1}{2}e^{-ix}(1 + e^{i2x})$, we easily see that $\arctanh(e^{i2x}) = \frac{1}{2}\log\left(\frac{1+e^{i2x}}{1-e^{i2x}}\right) = \frac{1}{2}\log(i\cot(x)) = -\frac{1}{2}\log(\tan(x)) + i\frac{\pi}{4}$, where for the last equality, I used that $\log(a + ib) = \log(b) + i\frac{\pi}{2}$, $a = 0, b > 0$.

$$\log^2(\tan(x)) = -\frac{\pi^2}{4} - i2\pi \arctanh(e^{i2x}) + 4 \arctanh^2(e^{i2x})$$

$$= \frac{\pi^2}{4} + 4 \arctanh^2(e^{i2x}) + i\pi \log(\tan(x)), \tag{6.365}$$

On the other hand, since we have that $\displaystyle\sum_{n=1}^{\infty} \frac{(e^{i2x})^{2n-1}}{2n-1} = \arctanh(e^{i2x})$, if we consider the Cauchy product of two series, we get

$$\arctanh^2(e^{i2x}) = \left(\sum_{n=1}^{\infty} \frac{(e^{i2x})^{2n-1}}{2n-1}\right)^2 = \sum_{n=1}^{\infty}\left(\sum_{k=1}^{n} \frac{e^{i4nx}}{(2k-1)(2n-2k+1)}\right)$$

$$= \frac{1}{2}\sum_{n=1}^{\infty} \frac{e^{i4nx}}{n}\left(\sum_{k=1}^{n}\left(\frac{1}{2k-1} + \frac{1}{2n-2k+1}\right)\right) = \sum_{n=1}^{\infty}\left(\frac{H_{2n}}{n} - \frac{1}{2}\frac{H_n}{n}\right)e^{i4nx}. \tag{6.366}$$

At last, combining (6.365) and (6.366), and then taking the real part, we obtain

$$\log^2(\tan(x)) = \log^2(\cot(x)) = \frac{\pi^2}{4} + 4\sum_{n=1}^{\infty}\left(\frac{H_{2n}}{n} - \frac{1}{2}\frac{H_n}{n}\right)\cos(4nx)$$

$$\overset{\overline{H}_{2n} \underset{=}{=} H_{2n} - H_n}{=} \frac{\pi^2}{4} + 2\sum_{n=1}^{\infty}(1-(-1)^{n-1})\left(\frac{H_n}{n} + \frac{\overline{H}_n}{n}\right)\cos(2nx),$$

and the second solution to the point (i), which involves the second form of the Fourier series, is finalized.

For the point (ii) of the problem, let's recollect the trilogarithm identity,

$$\text{Li}_3(-x) - \text{Li}_3\left(-\frac{1}{x}\right) = -\frac{\pi^2}{6}\log(x) - \frac{1}{6}\log^3(x) \text{ (see [21, Appendix, p.296])},$$

where if we replace x by $\tan^2(x)$ and rearrange, we obtain that

$$\log^3(\tan(x)) = -\frac{\pi^2}{4}\log(\tan(x)) - \frac{3}{4}\text{Li}_3(-\tan^2(x)) + \frac{3}{4}\text{Li}_3(-\cot^2(x)), \ 0 < x < \frac{\pi}{2}. \tag{6.367}$$

Further, if we consider that $\displaystyle\int_0^1 \frac{y\log^n(x)}{1-yx}dx = (-1)^n n!\, \text{Li}_{n+1}(y)$, $y \in (-\infty, 1]$, which may be found in [76, Chapter 1, p.4], and then take the trilogarithmic parts in the right-hand side of (6.367), we write

$$\text{Li}_3(-\tan^2(x)) - \text{Li}_3(-\cot^2(x)) = \frac{1}{2}\int_0^1 \frac{\cot^2(x)\log^2(t)}{1+\cot^2(x)t}dt - \frac{1}{2}\int_0^1 \frac{\tan^2(x)\log^2(t)}{1+\tan^2(x)t}dt$$

$$= \frac{1}{2} \int_0^1 \frac{(\cot^4(x) - 1)\log^2(t)}{(\cot^2(x) + t)(1 + \cot^2(x)t)} dt \overset{t=\left(\frac{1-u}{1+u}\right)^2}{=} 32\cos(2x) \int_0^1 \frac{(1 - u^2)\operatorname{arctanh}^2(u)}{1 - 2\cos(4x)u^2 + u^4} du.$$

$$(6.368)$$

Then, on one hand, it is easy to note that $\frac{1}{2} \int_0^1 \frac{\cot^2(x)}{1 + \cot^2(x)t} dt = -\log(\sin(x))$, and on the other hand, we have $\frac{1}{2} \int_0^1 \frac{\tan^2(x)}{1 + \tan^2(x)t} dt = -\log(\cos(x))$. Now, if we combine these two preceding results, we get

$$\log(\tan(x)) = \frac{1}{2} \int_0^1 \frac{\tan^2(x)}{1 + \tan^2(x)t} dt - \frac{1}{2} \int_0^1 \frac{\cot^2(x)}{1 + \cot^2(x)t} dt$$

$$= \frac{1}{2} \int_0^1 \frac{1 - \cot^4(x)}{(\cot^2(x) + t)(1 + \cot^2(x)t)} dt \qquad (6.369)$$

$$\overset{t=\left(\frac{1-u}{1+u}\right)^2}{=} -2\cos(2x) \int_0^1 \frac{1 - u^2}{1 - 2\cos(4x)u^2 + u^4} du.$$

Collecting the results from (6.368) and (6.369) in (6.367), we obtain that

$$\log^3(\tan(x))$$

$$= 12 \int_0^1 \frac{2\cos(2x)(1 - u^2)}{1 - 2\cos(4x)u^2 + u^4} \left(\frac{\pi^2}{48} - \operatorname{arctanh}^2(u)\right) du$$

$$= 12 \int_0^1 \left(\frac{\pi^2}{48} - \operatorname{arctanh}^2(u)\right) \sum_{n=1}^{\infty}(1 + (-1)^{n-1})u^{n-1}\cos(2nx)du$$

{reverse the order of summation and integration}

$$= 12 \sum_{n=1}^{\infty} \left((1 + (-1)^{n-1}) \int_0^1 u^{n-1}\left(\frac{\pi^2}{48} - \operatorname{arctanh}^2(u)\right) du\right) \cos(2nx),$$

and the solution to the point (ii) is finalized. In order to get the penultimate equality above, I exploited (6.327), Sect. 6.48, which gives $\sum_{n=1}^{\infty} \left((1+(-1)^{n-1})u^{n-1}\cos(2nx)\right)$

$$= \frac{2\cos(2x)(1 - u^2)}{1 - 2\cos(4x)u^2 + u^4}.$$

The result at the point (iii) is straightforward if we consider the Fourier series at the preceding point and the simple fact that $\log^3(\cot(x)) = -\log^3(\tan(x))$, and then we arrive at

$$\log^3(\cot(x))$$

$$= -12 \sum_{n=1}^{\infty} \left((1 + (-1)^{n-1}) \int_0^1 t^{n-1} \left(\frac{\pi^2}{48} - \operatorname{arctanh}^2(t) \right) dt \right) \cos(2nx),$$

and the solution to the point (iii) is finalized.

Passing to the point (iv), we have at least two possibilities to act that involve the use of some Fourier series previously met (counting both the current section and the preceding ones).

So, one way to go is based on exploiting the Landen's trilogarithmic identity in (4.39), Sect. 4.6, where if we replace x by $\cos^2(x)$ we get

$$\log(\cos(x)) \log^2(\sin(x))$$

$$= \frac{1}{4}\zeta(3) + \frac{\pi^2}{12} \log(\sin(x)) + \frac{1}{3} \log^3(\sin(x)) - \frac{1}{4} \operatorname{Li}_3\left(-\cot^2(x) \right)$$

$$- \frac{1}{4} \operatorname{Li}_3(\sin^2(x)) - \frac{1}{4} \operatorname{Li}_3(\cos^2(x)), \ 0 < x < \frac{\pi}{2}. \tag{6.370}$$

Now, from the solution to the point (iii), Sect. 6.48, we have

$$\log^3(\sin(x))$$

$$= -\log^3(2) - \frac{1}{2} \log(2)\pi^2 - \frac{3}{2}\zeta(3) - \frac{\pi^2}{4} \log(\sin(x))$$

$$- 3 \sum_{n=1}^{\infty} \left(\int_0^1 t^{n-1} \log^2\left(\frac{1-t}{2\sqrt{t}} \right) dt \right) \cos(2nx). \tag{6.371}$$

Further, based on the solutions to the points (iii) and (iv), Sect. 6.50, we get

$$\operatorname{Li}_3\left(-\cot^2(x) \right)$$

$$= -6\zeta(3) - 16 \sum_{n=1}^{\infty} \left(\int_0^1 t^{n-1} \operatorname{arctanh}^2(t) dt \right) \cos(2nx). \tag{6.372}$$

Next, from the solution to the point (iii) in Sect. 6.49, we get the following Fourier series form with the coefficient expressed in terms of an integral,

$$\operatorname{Li}_3(\sin^2(x)) = \frac{4}{3} \log^3(2) - \frac{1}{3} \log(2)\pi^2 + 2\zeta(3)$$

$$- 4 \sum_{n=1}^{\infty} (-1)^{n-1} \left(\int_0^1 t^{n-1} \log^2\left(\frac{1+t}{2\sqrt{t}} \right) dt \right) \cos(2nx). \tag{6.373}$$

By simply replacing x by $\pi/2 - x$ in (6.373), we get

$$\mathrm{Li}_3(\cos^2(x)) = \frac{4}{3}\log^3(2) - \frac{1}{3}\log(2)\pi^2 + 2\zeta(3)$$

$$+ 4\sum_{n=1}^{\infty}\left(\int_0^1 t^{n-1}\log^2\left(\frac{1+t}{2\sqrt{t}}\right)dt\right)\cos(2nx). \tag{6.374}$$

Collecting (6.371), (6.372), (6.373), and (6.374) in (6.370), we arrive at

$$\log(\cos(x))\log^2(\sin(x))$$

$$= \frac{1}{4}\zeta(3) - \log^3(2) + \sum_{n=1}^{\infty}\left(\int_0^1 t^{n-1}\left(4\operatorname{arctanh}^2(t) - \log^2\left(\frac{1-t}{2\sqrt{t}}\right)\right.\right.$$

$$\left.\left. - \left(1 - (-1)^{n-1}\right)\log^2\left(\frac{1+t}{2\sqrt{t}}\right)\right)dt\right)\cos(2nx),$$

and the solution to the point (iv) is finalized.

For a different approach, the curious reader might think of exploiting the algebraic identity, $ab^2 = \frac{1}{6}((a-b)^3 + (a+b)^3 - 2a^3)$, where if we set $a = \log(\cos(x))$ and $b = \log(\sin(x))$, we arrive at

$$\log(\cos(x))\log^2(\sin(x))$$

$$= \frac{1}{6}\left(\log^3(\cot(x)) + \log^3\left(\frac{1}{2}\sin(2x)\right) - 2\log^3(\cos(x))\right), \ 0 < x < \frac{\pi}{2}. \tag{6.375}$$

All the work left now is to reduce the right-hand side of (6.375) to other Fourier series already calculated.

Finally, for the Fourier series given at the last point, we replace x by $\pi/2 - x$ in the Fourier series from the previous point, which leads to

$$\log(\sin(x))\log^2(\cos(x))$$

$$= \frac{1}{4}\zeta(3) - \log^3(2) - \sum_{n=1}^{\infty}(-1)^{n-1}\left(\int_0^1 t^{n-1}\left(4\operatorname{arctanh}^2(t) - \log^2\left(\frac{1-t}{2\sqrt{t}}\right)\right.\right.$$

$$\left.\left. - \left(1 - (-1)^{n-1}\right)\log^2\left(\frac{1+t}{2\sqrt{t}}\right)\right)dt\right)\cos(2nx),$$

and the solution to the point (v) is finalized.

At last, I would like to emphasize that also the coefficients of the Fourier series from the points (ii)–(v) can be brought to forms involving (skew-)harmonic numbers as in the previous related sections, which the curious reader can quickly extract by using the necessary results that are all found in the present book.

6.52 A Pair of (Very) Challenging Alternating Harmonic Series with a Weight 4 Structure, Involving Harmonic Numbers of the Type H_{2n}

Solution *From Agony to Ecstasy* ... could have been a good part to consider in the title of the section, too. And it would have been well deserved! People with experience in the realm of the harmonic series will immediately realize that we have in front of our eyes two (very) challenging atypical alternating harmonic series.

To make it clear, calculating these series might be one of the most wonderful mathematical experiences for the lovers of harmonic series, *but there is a trick!* To get solutions, you will have to face the power of these two very resistant harmonic series, especially if you want to get solutions by real methods, and the whole process may take time, which means that one wants to be stubborn, ready for a long journey before reaching that desired moment when one will want to jump for joy.

Let's kick off with the point (i) of the problem! The *magical* result we need in our calculations is obtained by combining the simple fact that $\log(\sin(x)) \log(\cos(x)) = \frac{1}{4} \log^2 \left(\frac{1}{2} \sin(2x) \right) - \frac{1}{4} \log^2(\tan(x))$ and the powerful Fourier-like series,

$$\sum_{n=1}^{\infty} \left(2H_{2n} - 2H_n + \frac{1}{2n} - 2\log(2) \right) \frac{\sin^2(2nx)}{n}$$

$$= \log(\sin(x)) \log(\cos(x)), \ 0 < x < \frac{\pi}{2}, \tag{6.376}$$

presented and derived in *(Almost) Impossible Integrals, Sums, and Series* (see [76, Chapter 3, p.248]), that together give

$$\sum_{n=1}^{\infty} \left(2H_{2n} - 2H_n + \frac{1}{2n} - 2\log(2) \right) \frac{\sin^2(2nx)}{n}$$

$$= \frac{1}{4} \log^2 \left(\frac{1}{2} \sin(2x) \right) - \frac{1}{4} \log^2(\tan(x)), \ 0 < x < \frac{\pi}{2}. \tag{6.377}$$

A second auxiliary result we need, which is the transformation of a useful integral into a sum of series by the Cauchy product of two series in (4.16), Sect. 4.5, is

$$\int_0^{\pi/4} x \log^2(\tan(x)) dx = \int_0^1 \frac{\arctan(x) \log^2(x)}{1 + x^2} dx$$

$$= \frac{1}{4}\sum_{n=1}^{\infty}(-1)^{n-1}\frac{H_{2n}}{n^3} + \frac{1}{96}\log^4(2) - \frac{1}{16}\log^2(2)\zeta(2) + \frac{7}{32}\log(2)\zeta(3)$$

$$-\frac{11}{32}\zeta(4) + \frac{1}{4}\operatorname{Li}_4\left(\frac{1}{2}\right). \tag{6.378}$$

Proof The first equality is straightforward if we let the variable change, $\tan(x) = y$. Further, to prove the second equality in (6.378), we employ the Cauchy product in (4.16), Sect. 4.5, and then we have

$$\int_0^{\pi/4} x\log^2(\tan(x))\mathrm{d}x = \int_0^1 \frac{\arctan(x)\log^2(x)}{1+x^2}\mathrm{d}x$$

$$= \frac{1}{2}\int_0^1 \log^2(x)\sum_{n=1}^{\infty}(-1)^{n-1}x^{2n-1}(2H_{2n} - H_n)\mathrm{d}x$$

{reverse the order of summation and integration}

$$= \frac{1}{2}\sum_{n=1}^{\infty}(-1)^{n-1}(2H_{2n} - H_n)\int_0^1 x^{2n-1}\log^2(x)\mathrm{d}x$$

$$\left\{\text{use that } \int_0^1 x^{2n-1}\log^2(x)\mathrm{d}x = \frac{1}{4n^3} \text{ and then expand the resulting series}\right\}$$

$$= \frac{1}{4}\sum_{n=1}^{\infty}(-1)^{n-1}\frac{H_{2n}}{n^3} - \frac{1}{8}\sum_{n=1}^{\infty}(-1)^{n-1}\frac{H_n}{n^3}$$

{for the second series use the result in (4.103), Sect. 4.20}

$$= \frac{1}{4}\sum_{n=1}^{\infty}(-1)^{n-1}\frac{H_{2n}}{n^3} + \frac{1}{96}\log^4(2) - \frac{1}{16}\log^2(2)\zeta(2) + \frac{7}{32}\log(2)\zeta(3)$$

$$-\frac{11}{32}\zeta(4) + \frac{1}{4}\operatorname{Li}_4\left(\frac{1}{2}\right).$$

and the proof to the second auxiliary result is finalized.   ∎

A third auxiliary result we might like to consider in our calculations is

$$\int_0^{\pi/2} x\log(\sin(x))\mathrm{d}x = \frac{7}{16}\zeta(3) - \frac{3}{4}\log(2)\zeta(2), \tag{6.379}$$

which appears in Sect. 3.29, during the calculations to the point (v) of the problem.

And a last auxiliary result to prepare, before starting the main calculations, is

$$\sum_{n=1}^{\infty} \frac{1}{n}(H_{2n} - H_n - \log(2)) = \log^2(2) - \frac{1}{2}\zeta(2). \tag{6.380}$$

Proof A first solution is already found in *(Almost) Impossible Integrals, Sums, and Series* (see [76, Chapter 3, p.250]).

For a second solution, we consider the result in (6.376), where if we integrate both sides from $x = 0$ to $x = \pi/2$ and then change the order of summation and integration, we obtain that

$$\sum_{n=1}^{\infty} \frac{1}{n}\left(2H_{2n} - 2H_n + \frac{1}{2n} - 2\log(2)\right) \int_0^{\pi/2} \sin^2(2nx)dx$$

$$\left\{\text{use that } \int_0^{\pi/2} \sin^2(2nx)dx = \frac{\pi}{4}, n \geq 1, \text{ and then carefully split the series}\right\}$$

$$= \frac{\pi}{2}\sum_{n=1}^{\infty}\frac{1}{n}(H_{2n} - H_n - \log(2)) + \frac{\pi}{8}\underbrace{\sum_{n=1}^{\infty}\frac{1}{n^2}}_{\pi^2/6} = \int_0^{\pi/2} \log(\sin(x))\log(\cos(x))dx,$$

from which we obtain that

$$\sum_{n=1}^{\infty}\frac{1}{n}(H_{2n} - H_n - \log(2))$$

$$= \frac{2}{\pi}\underbrace{\int_0^{\pi/2}\log(\sin(x))\log(\cos(x))dx}_{\log^2(2)\pi/2 - \pi^3/48} - \frac{\pi^2}{24}$$

$$= \log^2(2) - \frac{1}{2}\zeta(2),$$

where the resulting integral is calculated in (3.133), Sect. 3.19. ∎

It's time to pass to the main calculations to the point (i) of the problem! So, if we multiply both sides of (6.377) by x, integrate from $x = 0$ to $x = \pi/4$, and reverse the order of summation and integration, we obtain in the left-hand side that

$$\sum_{n=1}^{\infty}\frac{1}{n}\left(2H_{2n} - 2H_n + \frac{1}{2n} - 2\log(2)\right)\int_0^{\pi/4} x\sin^2(2nx)dx$$

$$\left\{\text{use that } \int_0^{\pi/4} x \sin^2(2nx)\,dx = \frac{1}{32n^2} + \frac{(-1)^{n-1}}{32n^2} + \frac{\pi^2}{64} \text{ and carefully split the series}\right\}$$

$$= \frac{1}{2}\sum_{n=1}^{\infty}\frac{H_{2n}}{(2n)^3} - \frac{1}{16}\sum_{n=1}^{\infty}\frac{H_n}{n^3} + \frac{\pi^2}{32}\sum_{n=1}^{\infty}\frac{1}{n}(H_{2n} - H_n - \log(2)) + \frac{1}{16}\sum_{n=1}^{\infty}(-1)^{n-1}\frac{H_{2n}}{n^3}$$

$$- \frac{1}{16}\sum_{n=1}^{\infty}(-1)^{n-1}\frac{H_n}{n^3} + \frac{\pi^2}{128}\sum_{n=1}^{\infty}\frac{1}{n^2} - \frac{1}{16}\log(2)\sum_{n=1}^{\infty}\frac{1}{n^3} + \frac{1}{64}\sum_{n=1}^{\infty}\frac{1}{n^4}$$

$$- \frac{1}{16}\log(2)\sum_{n=1}^{\infty}(-1)^{n-1}\frac{1}{n^3} + \frac{1}{64}\sum_{n=1}^{\infty}(-1)^{n-1}\frac{1}{n^4}$$

$$= \frac{5}{192}\log^4(2) + \frac{1}{32}\log^2(2)\zeta(2) + \frac{7}{16}\log(2)\zeta(3) - \frac{365}{512}\zeta(4)$$

$$+ \frac{5}{8}\operatorname{Li}_4\left(\frac{1}{2}\right) + \frac{1}{16}\sum_{n=1}^{\infty}(-1)^{n-1}\frac{H_{2n}}{n^3}, \tag{6.381}$$

where I used that $\displaystyle\sum_{n=1}^{\infty}\frac{H_{2n}}{(2n)^3} = \frac{1}{2}\left(\sum_{n=1}^{\infty}\frac{H_n}{n^3} - \sum_{n=1}^{\infty}(-1)^{n-1}\frac{H_n}{n^3}\right)$, then $\displaystyle\sum_{n=1}^{\infty}\frac{H_n}{n^3} = \frac{5}{4}\zeta(4)$ is the case $n = 3$ of the Euler sum generalization in (6.149), Sect. 6.19, the third series is given in (6.380), and the fifth series is given in (4.103), Sect. 4.20.

As regards the right-hand side of (6.377), which is multiplied by x and then integrated from $x = 0$ to $x = \pi/4$, we have

$$\frac{1}{4}\int_0^{\pi/4} x \log^2\left(\frac{1}{2}\sin(2x)\right)dx - \frac{1}{4}\int_0^{\pi/4} x \log^2(\tan(x))dx$$

{in the first integral make the variable change $2x = y$ and then expand}

$$= \frac{1}{16}\log^2(2)\int_0^{\pi/2} x\,dx - \frac{1}{8}\log(2)\int_0^{\pi/2} x\log(\sin(x))dx + \frac{1}{16}\int_0^{\pi/2} x\log^2(\sin(x))dx$$

$$- \frac{1}{4}\int_0^{\pi/4} x\log^2(\tan(x))dx$$

$$= \frac{3}{16}\log^2(2)\zeta(2) - \frac{7}{64}\log(2)\zeta(3) + \frac{25}{512}\zeta(4) - \frac{1}{16}\sum_{n=1}^{\infty}(-1)^{n-1}\frac{H_{2n}}{n^3}, \tag{6.382}$$

where the second integral is given in (6.379), the third integral is found in (1.234), Sect. 1.54, and the last integral is reduced to a useful series as seen in (6.378).

Finally, if we combine (6.381) and (6.382), we conclude that

$$\sum_{n=1}^{\infty}(-1)^{n-1}\frac{H_{2n}}{n^3}$$

$$= \frac{195}{32}\zeta(4) - \frac{35}{8}\log(2)\zeta(3) + \frac{5}{4}\log^2(2)\zeta(2) - \frac{5}{24}\log^4(2) - 5\operatorname{Li}_4\left(\frac{1}{2}\right),$$

and the solution to the point (i) of the problem is complete.

My solution above also appeared, in large steps, in [27, August 26, 2020]. It is definitely one of the best solutions I obtained in the work with harmonic series (but after much struggle since this harmonic series is so resistant to various ways of attacking it).

To get a different way to the point (i) of the problem, one may employ contour integration and proceed as presented in [26].

What about the next point? The happy fact here is that in *(Almost) Impossible Integrals, Sums, and Series*, I already have a relation established between the two harmonic series presented in this section,

$$\sum_{n=1}^{\infty}(-1)^{n-1}\frac{H_{2n}}{n^3} + \sum_{n=1}^{\infty}(-1)^{n-1}\frac{H_{2n}^{(2)}}{n^2} = 2G^2 + \frac{37}{64}\zeta(4), \qquad (6.383)$$

which is not hard to prove with the right *tools* in hand as you may see in [76, Chapter 6, pp. 529–530]

Therefore, if we combine (6.383) with the result from the point (i), we obtain that

$$\sum_{n=1}^{\infty}(-1)^{n-1}\frac{H_{2n}^{(2)}}{n^2}$$

$$= 2G^2 - \frac{353}{64}\zeta(4) + \frac{35}{8}\log(2)\zeta(3) - \frac{5}{4}\log^2(2)\zeta(2) + \frac{5}{24}\log^4(2) + 5\operatorname{Li}_4\left(\frac{1}{2}\right),$$

and the solution to the point (ii) of the problem is complete.

Looking back over the whole section, we note that the critical part is represented by the attack of the series $\sum_{n=1}^{\infty}(-1)^{n-1}\frac{H_{2n}}{n^3}$ by means of the powerful form of the Fourier-like series in (6.377). Of course, even after this step, things might be pretty complicated without knowing how to proceed further $\left(\text{e.g., finding a way to}\right.$ calculate $\left.\int_0^{\pi/2} x\log^2(\sin(x))dx\right)$. So, much practice and experience are needed.

To conclude, one thing remains clear (from a personal perspective): fighting such harmonic series represents one of the most memorable (mathematical) moments!

6.53   Important Tetralogarithmic Values and More (Curious) Challenging Alternating Harmonic Series with a Weight 4 Structure, Involving Harmonic Numbers H_{2n}

Solution The extraction of the tetralogarithmic values in this section is possible due to the calculation of the key series in the previous section, but before proceeding with those calculations, I'll make a short review of the dilogarithmic and trilogarithmic values involving arguments of the type $1 \pm i$ and $(1 \pm i)/2$.

Using that $\log(\pm i) = \pm i\pi/2$, $\log(1 \pm i) = \dfrac{1}{2}\log(2) \pm i\dfrac{\pi}{4}$, since $\log(a + ib) =$ $\log(\sqrt{a^2 + b^2}) + i\arctan(b/a)$, $a > 0$, then the fact that

$$\mathrm{Li}_2(\pm i) = -\frac{1}{4}\sum_{n=1}^{\infty}\frac{(-1)^{n-1}}{n^2} \pm i\sum_{n=1}^{\infty}\frac{(-1)^{n-1}}{(2n-1)^2} = -\frac{1}{8}\zeta(2) \pm iG, \qquad (6.384)$$

which also shows that for the real and imaginary parts we have,

$$\Re\{\mathrm{Li}_2(\pm i)\} = -\frac{1}{8}\zeta(2) \qquad (6.385)$$

and

$$\Im\{\mathrm{Li}_2(\pm i)\} = \pm G, \qquad (6.386)$$

and returning to the dilogarithm function reflection formula, $\mathrm{Li}_2(z) + \mathrm{Li}_2(1 - z) = \zeta(2) - \log(z)\log(1 - z)$ (see [21, Chapter 1, p.5], [70, Chapter 2, p.107]), where we set $z = \pm i$, we get

$$\mathrm{Li}_2(1 \pm i) = \frac{3}{8}\zeta(2) \pm i\left(G + \log(2)\frac{\pi}{4}\right). \qquad (6.387)$$

Taking the real and imaginary parts of (6.387), we obtain

$$\Re\{\mathrm{Li}_2(1 \pm i)\} = \frac{3}{8}\zeta(2) \qquad (6.388)$$

and

$$\Im\{\mathrm{Li}_2(1 \pm i)\} = \pm G \pm \log(2)\frac{\pi}{4}. \qquad (6.389)$$

Based on (6.384) and Landen's identity, $\mathrm{Li}_2(z) + \mathrm{Li}_2(z/(z - 1)) = -\dfrac{1}{2}\log^2(1 - z)$, where we set $z = \pm i$, we arrive at

$$\mathrm{Li}_2\left(\frac{1 \pm i}{2}\right) = \frac{5}{16}\zeta(2) - \frac{1}{8}\log^2(2) \pm i\left(G - \log(2)\frac{\pi}{8}\right). \qquad (6.390)$$

Taking the real and imaginary parts of (6.390), we get

$$\Re\left\{\mathrm{Li}_2\left(\frac{1\pm i}{2}\right)\right\} = \frac{5}{16}\zeta(2) - \frac{1}{8}\log^2(2) \qquad (6.391)$$

and

$$\Im\left\{\mathrm{Li}_2\left(\frac{1\pm i}{2}\right)\right\} = \pm G \mp \log(2)\frac{\pi}{8} \qquad (6.392)$$

These ways to go with respect to the dilogarithmic values presented above are well-known in the mathematical literature.

Recall that the case $\Im\left\{\mathrm{Li}_2\left(\frac{1+i}{2}\right)\right\} = G - \log(2)\frac{\pi}{8}$ is already found in (1.21), Sect. 1.6, and at the end of Sect. 3.6, we may find a way of extracting this value by means involving integrals.

Sometimes, it is enough for us to know a relation between $\mathrm{Li}_2((1\pm i)/2)$ and $\mathrm{Li}_2(1\mp i)$, and for doing this, we might use the well-known dilogarithm function identity (see [21, Chapter 1, p.4]) $\mathrm{Li}_2(-x) + \mathrm{Li}_2(-1/x) = -\zeta(2) - 1/2\log^2(x)$, which also appears in (3.52), Sect. 3.10. Setting $x = \pm i - 1$ in the latter identity and combining all with the results $\log(-1 + i) = 1/2\log(2) + i3/4\pi$, $\log(-1 - i) = 1/2\log(2) - i3/4\pi$, based upon the facts that $\log(a + ib) = 1/2\log(a^2 + b^2) + i(\arctan(b/a) + \pi)$, $a < 0$, $b \geq 0$, and $\log(a + ib) = 1/2\log(a^2 + b^2) + i(\arctan(b/a) - \pi)$, $a < 0$, $b < 0$, we get

$$\mathrm{Li}_2\left(\frac{1\pm i}{2}\right) + \mathrm{Li}_2(1\mp i) = \frac{11}{16}\zeta(2) - \frac{1}{8}\log^2(2) \mp i\frac{3}{8}\log(2)\pi. \qquad (6.393)$$

Upon considering the real and imaginary parts of (6.393), we have

$$\Re\left\{\mathrm{Li}_2\left(\frac{1\pm i}{2}\right)\right\} + \Re\{\mathrm{Li}_2(1\mp i)\} = \frac{11}{16}\zeta(2) - \frac{1}{8}\log^2(2) \qquad (6.394)$$

and

$$\Im\left\{\mathrm{Li}_2\left(\frac{1\pm i}{2}\right)\right\} + \Im\{\mathrm{Li}_2(1\mp i)\} = \mp\frac{3}{8}\log(2)\pi. \qquad (6.395)$$

For (6.394) and (6.395), we may also exploit the facts that $\Re\{\mathrm{Li}_2((1+i)/2)\} = \Re\{\mathrm{Li}_2((1-i)/2)\}$, $\Im\{\mathrm{Li}_2((1+i)/2)\} = -\Im\{\mathrm{Li}_2((1-i)/2)\}$, $\Re\{\mathrm{Li}_2(1+i)\} = \Re\{\mathrm{Li}_2(1-i)\}$, $\Im\{\mathrm{Li}_2(1+i)\} = -\Im\{\mathrm{Li}_2(1-i)\}$, which are explained in the part of this section where I treat the tetralogarithmic values.

Now, we make a jump to the trilogarithmic area, and for some calculations, I'll use the strategy in my paper, *A special way of extracting the real part of the Trilogarithm*, $\mathrm{Li}_3\left(\frac{1\pm i}{2}\right)$ (see [81, December 10, 2019]).

The first step is to prove the following auxiliary logarithmic integral:

$$\int_0^\infty \frac{\log^2(x)}{(1+x)(1+(1-a)x)}dx = -2\zeta(2)\frac{\log(1-a)}{a} - \frac{1}{3}\frac{\log^3(1-a)}{a}, \quad a < 1, a \in \mathbb{R}.$$
(6.396)

Proof Based on (3.233), Sect. 3.31, the version with $c = 1$, we have $\int_0^\infty \frac{x^{s-1}}{1+x}dx = \pi\csc(\pi s)$, which we'll use in the main calculations below where we start by using the partial fraction decomposition and expanding the integral

$$\int_0^\infty \frac{\log^2(x)}{(1+x)(1+(1-a)x)}dx$$

$$= \frac{1}{a}\lim_{s\to 1^-}\frac{\partial^2}{\partial s^2}\left(\int_0^\infty \frac{x^{s-1}}{1+x}dx - (1-a)\underbrace{\int_0^\infty \frac{x^{s-1}}{1+(1-a)x}dx}_{\text{let }(1-a)x = t}\right)$$

$$= \frac{1}{a}\lim_{s\to 1^-}\frac{\partial^2}{\partial s^2}\left((1-(1-a)^{1-s})\int_0^\infty \frac{x^{s-1}}{1+x}dx\right)$$

$$= \frac{\pi}{a}\lim_{s\to 1^-}\frac{\partial^2}{\partial s^2}\left((1-(1-a)^{1-s})\csc(\pi s)\right)$$

$$= \frac{\pi}{a}\lim_{s\to 0^+}\frac{\partial^2}{\partial s^2}\left((1-(1-a)^s)\csc(\pi s)\right) = -2\zeta(2)\frac{\log(1-a)}{a} - \frac{1}{3}\frac{\log^3(1-a)}{a},$$

where the calculation of the limit can be done either with *Mathematica* or manually (one can get the proper inspiration from the first part of Sect. 6.48 where I used certain useful limited expansions in the work with limits).   ∎

Next, we pass to the second step where we want to prove that the following special integral identity, with $a < 1, a \in \mathbb{R}$, holds

$$\int_0^1 \frac{\log(x)\log(1-x)}{1-ax}dx$$

$$= \zeta(2)\frac{\log(1-a)}{a} + \frac{1}{6}\frac{\log^3(1-a)}{a} + \frac{1}{a}\operatorname{Li}_3(a) - \frac{1}{a}\operatorname{Li}_3\left(\frac{a}{a-1}\right).$$
(6.397)

Proof If we use the simple fact that $\log^2\left(\dfrac{x}{1-x}\right) = \log^2(x) - 2\log(x)\log(1-x) + \log^2(1-x)$, then we have

$$\int_0^1 \frac{\log(x)\log(1-x)}{1-ax}dx$$

$$= \frac{1}{2}\int_0^1 \frac{\log^2(x)}{1-ax}dx + \frac{1}{2}\int_0^1 \frac{\log^2(1-x)}{1-ax}dx - \frac{1}{2}\int_0^1 \frac{\log^2\left(\dfrac{x}{1-x}\right)}{1-ax}dx$$

$$= \frac{1}{2}\int_0^1 \frac{\log^2(x)}{1-ax}dx + \frac{1}{2(1-a)}\int_0^1 \frac{\log^2(x)}{1-\dfrac{a}{a-1}x}dx - \frac{1}{2}\int_0^\infty \frac{\log^2(x)}{(1+x)(1+(1-a)x)}dx$$

$$= \zeta(2)\frac{\log(1-a)}{a} + \frac{1}{6}\frac{\log^3(1-a)}{a} + \frac{1}{a}\operatorname{Li}_3(a) - \frac{1}{a}\operatorname{Li}_3\left(\frac{a}{a-1}\right),$$

where above I used the integral result in (3.45), Sect. 3.10, the case $n = 2$, and (6.396). Besides, during the calculations above I considered the variable changes $1 - x = y$ and $x/(1-x) = y$. ∎

Upon multiplying both sides of (6.397) by $-a$, then plugging $a = \mp i$ in (6.397), and taking the real part, we get

$$\Re\left\{\int_0^1 \frac{\pm i\log(x)\log(1-x)}{1\pm ix}dx\right\} = \Re\left\{\int_0^1 \frac{(x\pm i)\log(x)\log(1-x)}{1+x^2}dx\right\}$$

$$= \int_0^1 \frac{x\log(x)\log(1-x)}{1+x^2}dx = \frac{41}{64}\zeta(3) - \frac{9}{16}\log(2)\zeta(2) = -\zeta(2)\Re\{\log(1\pm i)\}$$

$$- \frac{1}{6}\Re\{\log^3(1\pm i)\} - \Re\{\operatorname{Li}_3(\mp i)\} + \Re\left\{\operatorname{Li}_3\left(\frac{\mp i}{\mp i - 1}\right)\right\},$$

whence we obtain that

$$\Re\left\{\operatorname{Li}_3\left(\frac{1\pm i}{2}\right)\right\} = \frac{1}{48}\log^3(2) - \frac{5}{32}\log(2)\zeta(2) + \frac{35}{64}\zeta(3), \tag{6.398}$$

where in the calculations above, I also used that $\displaystyle\int_0^1 \frac{x\log(x)\log(1-x)}{1+x^2}dx = \frac{1}{16}\left(\frac{41}{4}\zeta(3) - 9\log(2)\zeta(2)\right)$, found and calculated in [76, Chapter 1, p.8], and

$$\text{Li}_3(\pm i) = \sum_{n=1}^{\infty} \frac{(\pm i)^n}{n^3} = -\sum_{n=1}^{\infty}(-1)^{n-1}\frac{1}{(2n)^3} \pm i \sum_{n=1}^{\infty}(-1)^{n-1}\frac{1}{(2n-1)^3}$$

$$= -\frac{3}{32}\zeta(3) \pm i\frac{\pi^3}{32}, \tag{6.399}$$

which gives

$$\Re\{\text{Li}_3(\pm i)\} = -\frac{3}{32}\zeta(3) \tag{6.400}$$

and

$$\Im\{\text{Li}_3(\pm i)\} = \pm\frac{\pi^3}{32}. \tag{6.401}$$

The series of the imaginary part in (6.399) is obtained by considering (4.150), Sect. 4.47, with $n = 2$ and $x = 1/4$.

Now, using the well-known dilogarithm function identity (see [21, Chapter 1, p.4]) $\text{Li}_2(-x) + \text{Li}_2(-1/x) = -\zeta(2) - 1/2\log^2(x)$, dividing its both sides by x, and integrating once and considering the integration constant, we obtain $\text{Li}_3(-x) - \text{Li}_3(-1/x) = -\zeta(2)\log(x) - 1/6\log^3(x)$. Setting $x = \pm i - 1$ and proceeding as with the dilogarithmic cases and rearranging, we have

$$\text{Li}_3\left(\frac{1\pm i}{2}\right) - \text{Li}_3(1 \mp i) = \frac{1}{48}\log^3(2) - \frac{11}{192}\log(2)\pi^2 \pm i\left(\frac{7}{128}\pi^3 + \frac{3}{32}\log^2(2)\pi\right). \tag{6.402}$$

Based on (6.402), the real and imaginary parts are immediately revealed,

$$\Re\left\{\text{Li}_3\left(\frac{1\pm i}{2}\right)\right\} - \Re\left\{\text{Li}_3(1 \mp i)\right\} = \frac{1}{48}\log^3(2) - \frac{11}{32}\log(2)\zeta(2) \tag{6.403}$$

and

$$\Im\left\{\text{Li}_3\left(\frac{1\pm i}{2}\right)\right\} - \Im\left\{\text{Li}_3(1 \mp i)\right\} = \pm\frac{7}{128}\pi^3 \pm \frac{3}{32}\log^2(2)\pi. \tag{6.404}$$

Regarding (6.403) and (6.404), we may also exploit that $\Re\left\{\text{Li}_3\left((1 + i)/2\right)\right\} = \Re\left\{\text{Li}_3\left((1 - i)/2\right)\right\}$, $\Im\left\{\text{Li}_3\left((1 + i)/2\right)\right\} = -\Im\left\{\text{Li}_3\left((1 - i)/2\right)\right\}$, $\Re\left\{\text{Li}_3(1 + i)\right\} = \Re\left\{\text{Li}_3(1 - i)\right\}$, $\Im\left\{\text{Li}_3(1 + i)\right\} = -\Im\left\{\text{Li}_3(1 - i)\right\}$, and all these facts are explained below where I consider the tetralogarithmic values.

By also considering (6.398) in (6.403), we arrive at

$$\Re\{\text{Li}_3(1 \pm i)\} = \frac{3}{16}\log(2)\zeta(2) + \frac{35}{64}\zeta(3). \tag{6.405}$$

Ready now for starting with the first point of the problem? A very simple observation to start with is that given $f(x) = \sum\limits_{n=0}^{\infty} a_n x^n$, if we set $x = \pm i$, assume the convergence, and take the real part of both sides, we observe that $\Re\{f(\pm i)\} = \sum\limits_{n=0}^{\infty}(-1)^n a_{2n}$, which means that $\Re\{f(i)\} = \Re\{f(-i)\}$, and analogously, when considering the imaginary part, we have $\Im\{f(\pm i)\} = \pm\sum\limits_{n=0}^{\infty}(-1)^n a_{2n+1}$. It is exactly what happens if we consider $f(x) = \mathrm{Li}_4\left(x/(x-1)\right)$, set $x = \pm i$, and take the real part! Recall from *(Almost) Impossible Integrals, Sums, and Series* the extensions of the Landen's dilogarithmic identity in terms of series, which may be found in [76, Chapter 4, p.285]. There I have stated that $-\dfrac{1}{6}\sum\limits_{n=1}^{\infty}\dfrac{x^n}{n}(H_n^3 + 3H_n H_n^{(2)} + 2H_n^{(3)}) = \mathrm{Li}_4\left(\dfrac{x}{x-1}\right)$, and now the picture of things gets very clear! So, we may safely conclude that

$$\Re\left\{\mathrm{Li}_4\left(\frac{1+i}{2}\right)\right\} = \Re\left\{\mathrm{Li}_4\left(\frac{1-i}{2}\right)\right\},$$

and the solution to the point (i) of the problem is complete. Of course, we may proceed similarly for the Polylogarithm, as a generalization of the tetralogarithmic case treated above, but also for many other situations.

Could we act differently above? Yes, we can! For a different solution, we may consider the use of the extended version to the complex plane of the result in (3.13), Sect. 3.3. The details of such an approach will be found at the next point of the problem.

Essentially, for getting the result at the point (ii), assuming that we considered the first suggested solution at the previous point, the one involving power series, we need similar *tools* to the ones needed for getting the trilogarithmic ones in (6.405). Starting now from $\mathrm{Li}_3(-x) - \mathrm{Li}_3(-1/x) = -\zeta(2)\log(x) - 1/6\log^3(x)$, previously derived, and dividing its both sides by x, integrating once, and considering the integration constant, we get $\mathrm{Li}_4(-x) + \mathrm{Li}_4(-1/x) = -7/4\zeta(4) - 1/2\zeta(2)\log^2(x) - 1/24\log^4(x)$. Setting $x = \pm i - 1$ in the latter identity and rearranging, we get

$$\mathrm{Li}_4\left(\frac{1\pm i}{2}\right) + \mathrm{Li}_4(1\mp i)$$

$$= \frac{1313}{92160}\pi^4 + \frac{11}{768}\log^2(2)\pi^2 - \frac{1}{384}\log^4(2) \mp i\left(\log^3(2)\frac{\pi}{64} + \frac{7}{256}\log(2)\pi^3\right),$$

$$(6.406)$$

and if we consider the real and imaginary parts of (6.406), we have

$$\Re\left\{\text{Li}_4\left(\frac{1\pm i}{2}\right)\right\}+\Re\left\{\text{Li}_4(1\mp i)\right\}=\frac{1313}{1024}\zeta(4)+\frac{11}{128}\log^2(2)\zeta(2)-\frac{1}{384}\log^4(2)$$

$$(6.407)$$

and

$$\Im\left\{\text{Li}_4\left(\frac{1\pm i}{2}\right)\right\}+\Im\left\{\text{Li}_4(1\mp i)\right\}=\mp\log^3(2)\frac{\pi}{64}\mp\frac{7}{256}\log(2)\pi^3. \quad (6.408)$$

Based on the point (i) of the problem, the first term in the left-hand side of (6.407) has the same value for both variants of the argument, and that also means $\Re\left\{\text{Li}_4(1\mp i)\right\}$ leads to one single value for both variants of the argument. So, we obtain that

$$\Re\left\{\text{Li}_4\left(1+i\right)\right\}=\Re\left\{\text{Li}_4\left(1-i\right)\right\},$$

and the first solution to the point (ii) of the problem is complete.

For a second solution, as suggested at the end of the previous point, we may exploit, say, the result in (3.45), Sect. 3.10, where if we set $z=1\pm i$, $n=3$, and rearrange, we get

$$\text{Li}_4(1\pm i)=\frac{1}{6}\int_0^1\frac{(2t-1)\log^3(t)}{2t^2-2t+1}dt\mp i\frac{1}{6}\int_0^1\frac{\log^3(t)}{2t^2-2t+1}dt. \quad (6.409)$$

Taking the real part of (6.409), we immediately get the desired result, and the solution to the point (ii) is complete. Moreover, if we take the imaginary part of both sides of (6.409), we find that

$$\Im\left\{\text{Li}_4(1+i)\right\}=-\Im\left\{\text{Li}_4\left(1-i\right)\right\}. \quad (6.410)$$

Of course, the approach considered in (6.409) can be further extended in a similar style to the Polylogarithm.

Before performing the extraction of the desired tetralogarithmic values, we want to observe that

$$\sum_{n=1}^{\infty}(-1)^{n-1}\frac{1}{(2n-1)^4}\overset{\substack{\text{exploit}\\\text{the parity}}}{=}\frac{1}{256}\left(\sum_{n=1}^{\infty}\frac{1}{(n-3/4)^4}-\sum_{n=1}^{\infty}\frac{1}{(n-1/4)^4}\right)$$

$$\left\{\text{use Polygamma series representation, } \psi^{(m)}(z)=(-1)^{m-1}m!\sum_{k=0}^{\infty}\frac{1}{(z+k)^{m+1}}\right\}$$

$$= \frac{1}{1536}\left(\psi^{(3)}\left(\frac{1}{4}\right) - \psi^{(3)}\left(\frac{3}{4}\right)\right) = \frac{1}{768}\psi^{(3)}\left(\frac{1}{4}\right) - \frac{15}{16}\zeta(4), \qquad (6.411)$$

where to get the last equality in (6.411), I used the Polygamma function reflection formula, $(-1)^m \psi^{(m)}(1-x) - \psi^{(m)}(x) = \pi \dfrac{d^m}{dx^m}\cot(\pi x)$, with $m = 3$ and $x = \dfrac{1}{4}$ to derive and use the identity, $\psi^{(3)}\left(\dfrac{1}{4}\right) + \psi^{(3)}\left(\dfrac{3}{4}\right) = 16\pi^4$.

Thus, based on (6.411), we extract other key results

$$\mathrm{Li}_4(\pm i) = \sum_{n=1}^{\infty}\frac{(\pm i)^n}{n^4} = -\sum_{n=1}^{\infty}(-1)^{n-1}\frac{1}{(2n)^4} \pm i\sum_{n=1}^{\infty}(-1)^{n-1}\frac{1}{(2n-1)^4}$$

$$= -\frac{7}{128}\zeta(4) \pm i\left(\frac{1}{768}\psi^{(3)}\left(\frac{1}{4}\right) - \frac{15}{16}\zeta(4)\right), \qquad (6.412)$$

which leads to

$$\Re\{\mathrm{Li}_4(\pm i)\} = -\frac{7}{128}\zeta(4) \qquad (6.413)$$

and

$$\Im\{\mathrm{Li}_4(\pm i)\} = \pm\frac{1}{768}\psi^{(3)}\left(\frac{1}{4}\right) \mp \frac{15}{16}\zeta(4). \qquad (6.414)$$

At this point, we are ready to extract the tetralogarithmic values given in the problem! Setting $x = \pm i$ in (4.38), Sect. 4.6, and taking the real part, we have

$$\Re\left\{\sum_{n=1}^{\infty}(\pm i)^n\frac{H_n}{n^3}\right\} = -\frac{1}{8}\sum_{n=1}^{\infty}(-1)^{n-1}\frac{H_{2n}}{n^3}$$

$$= \zeta(4) + \zeta(3)\Re\{\log(1 \mp i)\} + \frac{1}{2}\zeta(2)\Re\{\log^2(1 \mp i)\} + \frac{1}{24}\Re\{\log^4(1 \mp i)\}$$

$$- \frac{1}{6}\Re\{\log(\pm i)\log^3(1 \mp i)\} - \Re\{\log(1 \mp i)\,\mathrm{Li}_3(\pm i)\} + 2\Re\{\mathrm{Li}_4(\pm i)\}$$

$$- \Re\{\mathrm{Li}_4(1 \mp i)\} + \Re\left\{\mathrm{Li}_4\left(\frac{1 \mp i}{2}\right)\right\},$$

where if we use the key series in (4.173), Sect. 4.52, together with the simple facts that $\log(\pm i) = \pm i\pi/2$, $\log(1 \pm i) = 1/2\log(2) \pm i\pi/4$ (see the beginning of the section), then (6.399), (6.413), and at last combining the result from the point (i) and (6.407) and rearranging, we obtain

$$\Re\left\{\mathrm{Li}_4\left(\frac{1\pm i}{2}\right)\right\} = \frac{343}{1024}\zeta(4) - \frac{5}{128}\log^2(2)\zeta(2) + \frac{1}{96}\log^4(2) + \frac{5}{16}\mathrm{Li}_4\left(\frac{1}{2}\right),$$

and the solution to the point (iii) of the problem is complete.

The fourth point of the problem is straightforward if we combine the result from the previous point and (6.407), thus giving

$$\Re\{\mathrm{Li}_4(1\pm i)\} = \frac{485}{512}\zeta(4) + \frac{1}{8}\log^2(2)\zeta(2) - \frac{5}{384}\log^4(2) - \frac{5}{16}\mathrm{Li}_4\left(\frac{1}{2}\right),$$

and the solution to the point (iv) of the problem is complete.

The series from the last three points are a *routine* if we consider the strategy applied at the point (iii)! In short, considering the generating functions in (4.44), Sect. 4.7, (4.56) and (4.58), Sect. 4.8, plugging in $x = \pm i$, and also counting the resulting values needed (which are all given within the present section) for getting the desired closed forms, we conclude that

$$\sum_{n=1}^{\infty}(-1)^{n-1}\frac{H_{2n}^2}{n^2}$$

$$= 2G^2 - \log(2)\pi G + \frac{231}{32}\zeta(4) - \frac{35}{16}\log(2)\zeta(3) + \log^2(2)\zeta(2) - \frac{5}{48}\log^4(2)$$

$$- 2\pi\Im\left\{\mathrm{Li}_3\left(\frac{1+i}{2}\right)\right\} - \frac{5}{2}\mathrm{Li}_4\left(\frac{1}{2}\right),$$

then

$$\sum_{n=1}^{\infty}(-1)^{n-1}\frac{H_{2n}^3}{n}$$

$$= 2G^2 - \frac{3}{4}\log(2)\pi G + \frac{1055}{256}\zeta(4) - \frac{93}{64}\log(2)\zeta(3) + \frac{21}{32}\log^2(2)\zeta(2) - \frac{1}{32}\log^4(2)$$

$$- \frac{3}{2}\pi\Im\left\{\mathrm{Li}_3\left(\frac{1+i}{2}\right)\right\},$$

and finally,

$$\sum_{n=1}^{\infty}(-1)^{n-1}\frac{H_{2n}H_{2n}^{(2)}}{n}$$

$$= \frac{1}{4}\log(2)\pi G - \frac{137}{128}\zeta(4) + \frac{35}{64}\log(2)\zeta(3) - \frac{3}{8}\log^2(2)\zeta(2) + \frac{5}{96}\log^4(2)$$

$$+ \frac{\pi}{2} \Im \left\{ \operatorname{Li}_3 \left(\frac{1+i}{2} \right) \right\} + \frac{5}{4} \operatorname{Li}_4 \left(\frac{1}{2} \right),$$

and the solutions to the points (v), (vi), and (vii) of the problem are complete.

With these results in hand, we'll be able to attack a larger panel of series problems, and some of them we'll meet right in the upcoming sections!

6.54 Two Alternating Euler Sums Involving Special Tails, a Joint Work with Moti Levy, Plus Two Newer Ones

Solution The (unusual) harmonic series at the points (i) and (ii) represent the core of the paper *Euler Sum Involving Tail* (see [92, 2019–2020]), a pleasant joint work with mathematician Moti Levy (Rehovot, Israel), which aims to evaluate two series proposed in Gazeta Matematică, Seria A, 37 (116), (1–2), 2019, counted as an *open problem*.

The last two points of the problem involve ideas met at the point (i), but we also need advanced results involving the tetralogarithm with a complex argument.

The first part of the problem is straightforward since the key generating function I want to employ is already found in (4.59), Sect. 4.9, where if we set $x = -1$ and rearrange, we get

$$\sum_{n=1}^{\infty} (-1)^{n-1} H_n \sum_{k=n+1}^{\infty} \frac{H_k}{k^2}$$

$$= -\frac{1}{2} \left(\zeta(4) - \log(2)\zeta(3) + \frac{1}{3} \log(-1) \log^3(2) + \frac{1}{2} \log^2(2) \operatorname{Li}_2(2) \right.$$

$$\left. + \log(2) \operatorname{Li}_3(-1) - \operatorname{Li}_4(2) - \operatorname{Li}_4(-1) \right)$$

$$= \frac{1}{16} \zeta(4) + \frac{7}{8} \log(2)\zeta(3) + \frac{1}{8} \log^2(2)\zeta(2) - \frac{1}{48} \log^4(2) - \frac{1}{2} \operatorname{Li}_4 \left(\frac{1}{2} \right),$$

where in the calculations I also combined the dilogarithm function identity (see [21, Chapter 1, p.4]) $\operatorname{Li}_2(-x) + \operatorname{Li}_2(-1/x) = -\zeta(2) - 1/2 \log^2(x)$, where I set $x = -2$ and considered that, since $\log(a + ib) = \frac{1}{2} \log(a^2 + b^2) + i \left(\arctan \left(\frac{b}{a} \right) + \pi \right)$, $a < 0$, $b \geq 0$, we have $\log(-a) = \log(a) + i\pi$, $0 < a$, and the special value $\operatorname{Li}_2(1/2) = 1/2\zeta(2) - 1/2 \log^2(2)$ to get that $\operatorname{Li}_2(2) = 3/2\zeta(2) - i \log(2)\pi$. Also, I used the Tetralogarithm function identity, derived in the previous section, $\operatorname{Li}_4(-x) + \operatorname{Li}_4(-1/x) = -7/4\zeta(4) - 1/2\zeta(2) \log^2(x) - 1/24 \log^4(x)$, where I set $x = 2$, to obtain that $\text{Li}_4(2) = 2\zeta(4) + \log^2(2)\zeta(2) - 1/24\log^4(2) - \text{Li}_4(1/2) - i\log^3(2)\pi/6$, and then the special values, $\text{Li}_3(-1) = -3/4\zeta(3)$ and $\text{Li}_4(-1) = -7/8\zeta(4)$.

For a second solution to the point (i), check the second solution below to the point (ii) of the problem.

How about the second point of the problem? For a first solution, we may exploit the result in (4.60), Sect. 4.9, the case $s = 3$, $x = -1$, where if we multiply both sides by -1, we obtain that

$$\sum_{n=1}^{\infty}(-1)^{n-1}H_n\sum_{k=n+1}^{\infty}\frac{H_k}{k^3}$$

$$= \frac{1}{2}\log(2)\sum_{k=1}^{\infty}\frac{H_k}{k^3} - \frac{1}{2}\sum_{k=1}^{\infty}(-1)^{k-1}\frac{H_k^2}{k^3} + \frac{1}{2}\int_0^{-1}\frac{1}{1-t}\sum_{k=1}^{\infty}t^k\frac{H_k}{k^3}dt. \qquad (6.415)$$

In the last integral in (6.415), we make the variable change $-t = u$, which yields

$$\int_0^{-1}\frac{1}{1-t}\sum_{k=1}^{\infty}t^k\frac{H_k}{k^3}dt = \int_0^1\frac{1}{1+t}\sum_{k=1}^{\infty}(-1)^{k-1}t^k\frac{H_k}{k^3}dt$$

$$= \int_0^1(\log(1+t))'\sum_{k=1}^{\infty}(-1)^{k-1}t^k\frac{H_k}{k^3}dt$$

{integrate by parts and reverse the order of summation and integration}

$$= \log(2)\sum_{k=1}^{\infty}(-1)^{k-1}\frac{H_k}{k^3} - \sum_{k=1}^{\infty}(-1)^{k-1}\frac{H_k}{k^2}\int_0^1 t^{k-1}\log(1+t)dt$$

{exploit the result in (1.99), Sect. 1.21}

$$= \sum_{k=1}^{\infty}\frac{H_k\overline{H}_k}{k^3} - \log(2)\sum_{k=1}^{\infty}\frac{H_k}{k^3}. \qquad (6.416)$$

Plugging (6.416) in (6.415), we arrive at

$$\sum_{n=1}^{\infty}(-1)^{n-1}H_n\sum_{k=n+1}^{\infty}\frac{H_k}{k^3} = \frac{1}{2}\sum_{k=1}^{\infty}\frac{H_k\overline{H}_k}{k^3} - \frac{1}{2}\sum_{k=1}^{\infty}(-1)^{k-1}\frac{H_k^2}{k^3},$$

where if we count that the first resulting series is found in (4.145), Sect. 4.46, and the second one is given in (3.331), Sect. 3.48, we conclude that

$$\sum_{n=1}^{\infty}(-1)^{n-1}H_n\sum_{k=n+1}^{\infty}\frac{H_k}{k^3}$$

$$=\frac{5}{12}\log^3(2)\zeta(2)+2\log(2)\zeta(4)+\frac{7}{8}\zeta(2)\zeta(3)-\frac{21}{16}\log^2(2)\zeta(3)$$

$$-\frac{155}{128}\zeta(5)-\frac{3}{40}\log^5(2)-2\log(2)\operatorname{Li}_4\left(\frac{1}{2}\right)-\operatorname{Li}_5\left(\frac{1}{2}\right),$$

and the first solution to the point (ii) is finalized.

In the mentioned paper, the first solution led to a reduction to manageable integrals, but as you see above, my aim was to get a reduction to manageble series previously calculated, where, of course, one of them is found in *(Almost) Impossible Integrals, Sums, and Series*.

Regarding a second solution to the point (ii), I'll consider the *magic* of the simple series manipulations in order to get the desired series reduction.

Let me first prove the simple fact that

$$\sum_{k=1}^{n}(-1)^{k-1}H_k=\frac{1}{2}\left((-1)^{n-1}H_n+\overline{H}_n\right). \qquad (6.417)$$

Proof Keeping in mind that $\sum_{k=1}^{n}(-1)^{k-1}=\frac{1}{2}(1+(-1)^{n-1})$, which immediately

gives $\sum_{k=i}^{n}(-1)^{k-1}=\frac{1}{2}((-1)^{i-1}+(-1)^{n-1})$, we return to the main sum where we

consider that $H_k=\sum_{i=1}^{k}\frac{1}{i}$, and then we write

$$\sum_{k=1}^{n}(-1)^{k-1}H_k=\sum_{k=1}^{n}\sum_{i=1}^{k}(-1)^{k-1}\frac{1}{i}\overset{\substack{\text{reverse the order}\\\text{of summation}}}{=}\sum_{i=1}^{n}\sum_{k=i}^{n}(-1)^{k-1}\frac{1}{i}$$

$$=\frac{1}{2}\sum_{i=1}^{n}\frac{(-1)^{i-1}+(-1)^{n-1}}{i}=\frac{1}{2}\left((-1)^{n-1}H_n+\overline{H}_n\right),$$

which brings an end to the auxiliary result. ∎

So, returning to the main series, reversing the order of summation, and using (6.417), we write

$$\sum_{n=1}^{\infty}(-1)^{n-1}H_n\sum_{k=n+1}^{\infty}\frac{H_k}{k^3}=\sum_{k=1}^{\infty}\frac{H_k}{k^3}\sum_{n=1}^{k-1}(-1)^{n-1}H_n=\frac{1}{2}\sum_{k=1}^{\infty}\frac{H_k}{k^3}\left(\overline{H}_k-(-1)^{k-1}H_k\right)$$

$$= \frac{1}{2} \sum_{k=1}^{\infty} \frac{H_k \overline{H}_k}{k^3} - \frac{1}{2} \sum_{k=1}^{\infty} (-1)^{k-1} \frac{H_k^2}{k^3}$$

$$= \frac{5}{12} \log^3(2)\zeta(2) + 2\log(2)\zeta(4) + \frac{7}{8}\zeta(2)\zeta(3) - \frac{21}{16}\log^2(2)\zeta(3)$$

$$- \frac{155}{128}\zeta(5) - \frac{3}{40}\log^5(2) - 2\log(2)\operatorname{Li}_4\left(\frac{1}{2}\right) - \operatorname{Li}_5\left(\frac{1}{2}\right),$$

and the second solution to the point (ii) is finalized. Observe the same harmonic series appeared in the final part of the solution, exactly as in the first solution.

In the paper *Euler Sum Involving Tail* (see [92, 2019–2020]), the second solution to the point (ii) was based on exploiting the straightforward identities
$$\sum_{k=1}^{2n-1}(-1)^k H_k = \frac{1}{2}H_n - H_{2n} \text{ and } \sum_{k=1}^{2n}(-1)^k H_k = \frac{1}{2}H_n, \text{ further combined with}$$
Abel's summation that led to a reduction to atypical harmonic series of weight 5, involving harmonic numbers of the type H_{2n}, like the ones in (4.124), Sect. 4.33, and (4.130), Sect. 4.35.

Finally, the last two results are immediately obtained if we set $x = i$ in the generating function in (4.59), Sect. 4.9, then take the real and imaginary parts, and plug in the needed values which are presented in the previous section, and thus we obtain

$$\sum_{n=1}^{\infty}(-1)^{n-1} H_{2n} \sum_{k=2n+1}^{\infty} \frac{H_k}{k^2}$$

$$= \frac{\pi^2}{64}G + \frac{1}{16}\log(2)\pi G - \frac{1}{16}\log^2(2)G + \frac{55}{18432}\pi^4 + \frac{7}{512}\log(2)\pi^3 + \frac{5}{768}\log^2(2)\pi^2$$

$$+ \log^3(2)\frac{\pi}{384} + \frac{35}{256}\pi\zeta(3) + \frac{35}{128}\log(2)\zeta(3) - \frac{5}{768}\log^4(2) - \frac{1}{1536}\psi^{(3)}\left(\frac{1}{4}\right)$$

$$+ \frac{1}{2}\Im\left\{\operatorname{Li}_4\left(\frac{1+i}{2}\right)\right\} - \frac{5}{32}\operatorname{Li}_4\left(\frac{1}{2}\right)$$

and

$$\sum_{n=1}^{\infty}(-1)^{n-1} H_{2n-1} \sum_{k=2n}^{\infty} \frac{H_k}{k^2}$$

$$= \frac{\pi^2}{64}G - \frac{1}{16}\log(2)\pi G - \frac{1}{16}\log^2(2)G + \frac{137}{18432}\pi^4 + \frac{7}{512}\log(2)\pi^3 - \frac{5}{768}\log^2(2)\pi^2$$

$$+ \log^3(2)\frac{\pi}{384} + \frac{35}{256}\pi\zeta(3) - \frac{35}{128}\log(2)\zeta(3) + \frac{5}{768}\log^4(2) - \frac{1}{1536}\psi^{(3)}\left(\frac{1}{4}\right)$$

$$+ \frac{1}{2}\Im\left\{\mathrm{Li}_4\left(\frac{1+i}{2}\right)\right\} + \frac{5}{32}\mathrm{Li}_4\left(\frac{1}{2}\right),$$

and the solutions to the points (iii) and (iv) of the problem are finalized.

Both last harmonic series have exotic closed forms, you might immediately remark! Besides, these last two harmonic series also involve advanced results with the tetralogarithm, which in this case it is about $\Re\{\mathrm{Li}_4(1 - i)\}$, and here we need its simpler form to get the desired closed form!

6.55 A (Very) Hard Nut to Crack (An Alternating Harmonic Series with a Weight 4 Structure, Involving Harmonic Numbers of the Type H_{2n})

Solution The appearance of that (key) integral in the statement section may already be considered to some extent a helpful hand in the process of finding a solution. However, to put it bluntly, it is still challenging to obtain full solutions. As indicated by the title, it is *a (very) hard nut to crack*, or simply *a beast*.

The first solution is (highly) *non-obvious*, and it involves the integral in (1.251), Sect. 1.57, in a modified form, obtained after applying the variable change $(1 - t)/(1 + t) = u$, as it appears in (3.381), Sect. 3.57, and then, upon rearranging, we get

$$\int_0^1 \frac{\arctan^2(t)\arctanh(t)}{t}dt$$

$$= \frac{\pi^4}{64} - \frac{\pi^2}{4}\underbrace{\int_0^1 \frac{\arctanh(t)}{t}dt}_{\pi^2/8} + \pi\int_0^1 \frac{\arctan(t)\arctanh(t)}{t}dt, \qquad (6.418)$$

where the first resulting integral in the right-hand side is given in (3.34), Sect. 3.9.

Next, for the second integral in the right-hand side of (6.418), we write

$$\int_0^1 \frac{\arctan(t)\arctanh(t)}{t}dt = \int_0^1 \frac{(\pi/4 - \arctan((1 - t)/(1 + t)))\arctanh(t)}{t}dt$$

$$= \frac{\pi}{4}\underbrace{\int_0^1 \frac{\arctanh(t)}{t}dt}_{\pi^2/8} - \underbrace{\int_0^1 \frac{\arctan((1 - t)/(1 + t))\arctanh(t)}{t}dt}_{(1 - t)/(1 + t) = u}$$

{after making the variable change, expand the integral}

$$\overset{u=t}{=} \frac{\pi^3}{32} + \frac{1}{2} \underbrace{\int_0^1 \frac{\arctan(t)\log(t)}{1+t} dt}_{\log(2)G/2 - \pi^3/64} + \frac{1}{2} \int_0^1 \frac{\arctan(t)\log(t)}{1-t} dt. \qquad (6.419)$$

The first integral in (6.419) is already found in [76, Chapter 1, p.14] and [43], and for the second integral in (6.419), we consider the following parameterized integral,
$$I(a) = \int_0^1 \frac{\log(1+at)\log(t)}{1-t} dt, \text{ where if we differentiate with respect to } a, \text{ split,}$$
and use (3.13), Sect. 3.3, we get that

$$I'(a) = \int_0^1 \frac{t\log(t)}{(1-t)(1+at)} dt = \frac{1}{1+a} \underbrace{\int_0^1 \frac{\log(t)}{1-t} dt}_{-\zeta(2)} - \frac{1}{1+a} \int_0^1 \frac{\log(t)}{1+at} dt$$

$$= -\zeta(2)\frac{1}{1+a} - \frac{\text{Li}_2(-a)}{a} + \frac{\text{Li}_2(-a)}{1+a}. \qquad (6.420)$$

Replacing a by x in (6.420) and integrating from $x = 0$ to $x = a$, we have

$$\int_0^a I'(x)dx = I(a) - I(0) = I(a) = \int_0^1 \frac{\log(1+at)\log(t)}{1-t} dt = -\zeta(2) \int_0^a \frac{1}{1+x} dx$$

$$- \int_0^a \frac{\text{Li}_2(-x)}{x} dx + \int_0^a \frac{\text{Li}_2(-x)}{1+x} dx = -\zeta(2)\log(1+a) - \text{Li}_3(-a)$$

$$+ \int_0^a (\log(1+x))' \text{Li}_2(-x)dx = -\zeta(2)\log(1+a) - \text{Li}_3(-a)$$

$$+ \log(1+a)\text{Li}_2(-a) + \int_0^a \frac{\log^2(1+x)}{x} dx$$

{exploit the generalized integral result in (6.67), Sect. 6.10}

$$= -\zeta(2)\log(1+a) - \frac{1}{3}\log^3(1+a) - \log(1+a)\text{Li}_2(-a) + \text{Li}_3(-a) + 2\text{Li}_3\left(\frac{a}{1+a}\right). \qquad (6.421)$$

Now, if we set $a = i$ in (6.421), take the imaginary part of the needed sides, and consider the logarithm with a complex argument, $\log(a + ib) = \log(\sqrt{a^2 + b^2}) + i\arctan(b/a)$, $a > 0$, together with the simple results, $\text{Li}_2(-i) =$

$$-\frac{1}{4}\sum_{n=1}^{\infty}\frac{(-1)^{n-1}}{n^2} - i\sum_{n=1}^{\infty}\frac{(-1)^{n-1}}{(2n-1)^2} = -\frac{1}{8}\zeta(2) - iG \text{ and then } \Im\{\text{Li}_3(-i)\} =$$

$$\Im\left\{-i\sum_{n=1}^{\infty}\frac{(-1)^{n-1}}{(2n-1)^3}\right\} = \Im\left\{-i\frac{\pi^3}{32}\right\} = -\frac{\pi^3}{32}, \text{ where the last series is obtained by}$$

considering (4.150), Sect. 4.47, with $n = 2$ and $x = 1/4$, we arrive at

$$\Im\left\{\int_0^1 \frac{\log(1+it)\log(t)}{1-t}dt\right\} = \int_0^1 \frac{\arctan(t)\log(t)}{1-t}dt$$

$$= \frac{1}{2}\log(2)G - \frac{\pi}{16}\log^2(2) - \frac{\pi^3}{16} + 2\Im\left\{\text{Li}_3\left(\frac{1+i}{2}\right)\right\}. \tag{6.422}$$

Upon plugging (6.422) in (6.419), we obtain that

$$\int_0^1 \frac{\arctan(t)\text{arctanh}(t)}{t}dt = \frac{1}{2}\log(2)G - \frac{\pi}{32}\log^2(2) - \frac{\pi^3}{128} + \Im\left\{\text{Li}_3\left(\frac{1+i}{2}\right)\right\}. \tag{6.423}$$

By combining (6.423) and (6.418), we get

$$\int_0^1 \frac{\arctan^2(t)\text{arctanh}(t)}{t}dt$$

$$= \frac{1}{2}\log(2)\pi G - \frac{3}{16}\log^2(2)\zeta(2) - \frac{135}{64}\zeta(4) + \pi\Im\left\{\text{Li}_3\left(\frac{1+i}{2}\right)\right\}. \tag{6.424}$$

So, a first crucial part of the proof is already done! Further, I'll want to turn the integral in (6.424) into a sum of series, and to do this, I'll combine the Cauchy product in (4.17), Sect. 4.5, and (1.110), Sect. 1.22, and then we have

$$\int_0^1 \frac{\arctan^2(t)\text{arctanh}(t)}{t}dt = \int_0^1 \left(\frac{1}{2}\sum_{n=1}^{\infty}(-1)^{n-1}t^{2n-1}\frac{2H_{2n}-H_n}{n}\right)\text{arctanh}(t)dt$$

$$= \frac{1}{2}\sum_{n=1}^{\infty}(-1)^{n-1}\frac{2H_{2n}-H_n}{n}\int_0^1 t^{2n-1}\text{arctanh}(t)dt$$

$$= \frac{1}{4}\sum_{n=1}^{\infty}(-1)^{n-1}\frac{2H_{2n}-H_n}{n}\left(\frac{H_{2n}}{n} - \frac{1}{2}\frac{H_n}{n}\right)$$

$$= \frac{1}{8}\sum_{n=1}^{\infty}(-1)^{n-1}\frac{H_n^2}{n^2} + \frac{1}{2}\sum_{n=1}^{\infty}(-1)^{n-1}\frac{H_{2n}^2}{n^2} - \frac{1}{2}\sum_{n=1}^{\infty}(-1)^{n-1}\frac{H_n H_{2n}}{n^2}$$

$$= G^2 - \frac{1}{2}\log(2)\pi G + \frac{503}{128}\zeta(4) - \frac{21}{16}\log(2)\zeta(3) + \frac{9}{16}\log^2(2)\zeta(2) - \frac{1}{16}\log^4(2)$$

$$- \pi\Im\left\{\operatorname{Li}_3\left(\frac{1+i}{2}\right)\right\} - \frac{3}{2}\operatorname{Li}_4\left(\frac{1}{2}\right) - \frac{1}{2}\sum_{n=1}^{\infty}(-1)^{n-1}\frac{H_n H_{2n}}{n^2}, \qquad (6.425)$$

where the first series is $\displaystyle\sum_{n=1}^{\infty}(-1)^{n-1}\frac{H_n^2}{n^2} = \frac{41}{16}\zeta(4) - \frac{7}{4}\log(2)\zeta(3) + \frac{1}{2}\log^2(2)\zeta(2) -$

$\dfrac{1}{12}\log^4(2) - 2\operatorname{Li}_4\left(\dfrac{1}{2}\right)$, and it is extracted if we combine (6.184), Sect. 6.23, and (4.103), Sect. 4.20, and then the second (advanced) series is found in (4.179), Sect. 4.53.

Combining the results in (6.424) and (6.425), we conclude that

$$\sum_{n=1}^{\infty}(-1)^{n-1}\frac{H_n H_{2n}}{n^2}$$

$$= 2G^2 - 2\log(2)\pi G - \frac{1}{8}\log^4(2) + \frac{3}{2}\log^2(2)\zeta(2) - \frac{21}{8}\log(2)\zeta(3) + \frac{773}{64}\zeta(4)$$

$$- 4\pi\Im\left\{\operatorname{Li}_3\left(\frac{1+i}{2}\right)\right\} - 3\operatorname{Li}_4\left(\frac{1}{2}\right),$$

and the first solution is finalized.

For a second solution, we might want to start with the following integral, $\Re\left\{\displaystyle\int_0^{\infty}\frac{(\pi/2 - \arctan(t))^2\,\operatorname{arctanh}(t)}{t}\,dt\right\}$. *Of course, but why?* Do you remember the integral in the statement section (which also appears in the previous solution)? It is an artifice of calculations that will finally lead to a useful result involving that integral (and it won't be necessary to evaluate it, although the extraction of its value will be straightforward with the key result in hand)!

Before going further, we need to prove and use that

$$\frac{(\pi/2 - \arctan(t))^2}{t} = 2\int_0^1 \frac{x\,\operatorname{arctanh}(x)}{t(t^2 + x^2)}\,dx, \quad t > 0. \qquad (6.426)$$

Proof It is straightforward to prove the result if we replace a by $1/t$ in (1.31), Sect. 1.9, use that $\arctan(x) + \arctan(1/x) = \pi/2\operatorname{sgn}(x)$, $x \in \mathbb{R} \setminus \{0\}$, and then multiply both sides by $1/t^3$, and the result follows. ∎

So, using (6.426) we write

$$\Re\left\{\int_0^{\infty}\frac{(\pi/2 - \arctan(t))^2\,\operatorname{arctanh}(t)}{t}\,dt\right\}$$

$$= \Re\left\{2\int_0^\infty \operatorname{arctanh}(t)\left(\int_0^1 \frac{x\operatorname{arctanh}(x)}{t(t^2+x^2)}dx\right)dt\right\}$$

$$\overset{\substack{\text{reverse the order} \\ \text{of integration}}}{=} 2\int_0^1 x\operatorname{arctanh}(x)\left(\int_0^\infty \frac{\Re\{\operatorname{arctanh}(t)\}}{t(x^2+t^2)}dt\right)dx$$

$$\left\{\text{employ the fact that P. V.}\int_0^1 \frac{t}{1-t^2u^2}du = \Re\{\operatorname{arctanh}(t)\},\ x\in\mathbb{R}\setminus\{\pm 1\}\right\}$$

$$= 2\int_0^1 x\operatorname{arctanh}(x)\left(\int_0^\infty \left(\text{P. V.}\int_0^1 \frac{1}{(x^2+t^2)(1-t^2u^2)}du\right)dt\right)dx$$

$$\overset{\substack{\text{reverse the order} \\ \text{of integration}}}{=} 2\int_0^1 x\operatorname{arctanh}(x)\left(\int_0^1 \left(\text{P. V.}\int_0^\infty \frac{1}{(x^2+t^2)(1-u^2t^2)}dt\right)du\right)dx$$

$$= \pi\int_0^1 \operatorname{arctanh}(x)\left(\int_0^1 \frac{1}{1+x^2u^2}du\right)dx = \pi\int_0^1 \frac{\arctan(x)\operatorname{arctanh}(x)}{x}dx.$$
$$(6.427)$$

During the calculations in (6.427), I employed

$$\text{P. V.}\int_0^\infty \frac{1}{(x^2+t^2)(1-u^2t^2)}dt$$

$$= \frac{u^2}{1+u^2x^2}\underbrace{\text{P. V.}\int_0^\infty \frac{1}{1-u^2t^2}dt}_{0} + \frac{1}{1+x^2u^2}\underbrace{\int_0^\infty \frac{1}{x^2+t^2}dt}_{\pi/(2x)} = \frac{\pi}{2}\frac{1}{x(1+x^2u^2)},$$
$$(6.428)$$

where I used the simple fact that P. V. $\displaystyle\int_0^\infty \frac{1}{1-u^2t^2}dt \overset{ut=v}{=} \frac{1}{u}\text{P. V.}\int_0^\infty \frac{1}{1-v^2}dv$

$$= \frac{1}{u}\lim_{\epsilon\to 0^+}\left(\underbrace{\int_0^{1-\epsilon}\frac{1}{1-v^2}dv}_{1/2\log(2/\epsilon-1)} + \underbrace{\int_{1+\epsilon}^\infty \frac{1}{1-v^2}dv}_{-1/2\log(2/\epsilon+1)}\right) = \frac{1}{2u}\lim_{\epsilon\to 0^+}\log\left(\frac{2-\epsilon}{2+\epsilon}\right) = 0.$$

Returning to the key integral I started the second solution with and using the identities, $\arctan(x)+\arctan(1/x) = \pi/2$, $x>0$, and $\operatorname{arctanh}(1/x)-\operatorname{arctanh}(x) = i\pi/2$, $x>1$, where the latter may follow immediately by exploiting the fact that $\log(a+ib) = \dfrac{1}{2}\log(a^2+b^2) + i\left(\arctan\left(\dfrac{b}{a}\right)+\pi\right)$, $a<0$, $b\geq 0$, which shows that $\log(-a) = \log(a)+i\pi$, $0<a$, we write

$$\Re\left\{\int_0^\infty \frac{(\pi/2 - \arctan(t))^2 \arctanh(t)}{t} dt\right\} = \int_0^1 \frac{(\pi/2 - \arctan(t))^2 \arctanh(t)}{t} dt$$

$$+ \Re\left\{\underbrace{\int_1^\infty \frac{\arctan^2(1/t)(\arctanh(1/t) - i\pi/2)}{t} dt}_{\text{let } 1/t = u}\right\}$$

{after taking the real part, expand both integrals}

$$= \frac{\pi^2}{4}\underbrace{\int_0^1 \frac{\arctanh(t)}{t} dt}_{\pi^2/8} - \pi\int_0^1 \frac{\arctan(t)\arctanh(t)}{t} dt + 2\int_0^1 \frac{\arctan^2(t)\arctanh(t)}{t} dt,$$

$$(6.429)$$

where the first resulting integral in the right-hand side is given in (3.34), Sect. 3.9.

Combining (6.429) and (6.427), and then rearranging, we may get the exotic result

$$\zeta(4) = \frac{32}{45}\pi\int_0^1 \frac{\arctan(t)\arctanh(t)}{t} dt - \frac{32}{45}\int_0^1 \frac{\arctan^2(t)\arctanh(t)}{t} dt.$$

$$(6.430)$$

At this point, there is some overlap in terms of the final steps of the two solutions. To get a useful form of the result in the right-hand side of (6.430), we need the value of the integral in (6.423) combined with the result in (6.425), and then we arrive at the desired result

$$\sum_{n=1}^\infty (-1)^{n-1} \frac{H_n H_{2n}}{n^2}$$

$$= 2G^2 - 2\log(2)\pi G - \frac{1}{8}\log^4(2) + \frac{3}{2}\log^2(2)\zeta(2) - \frac{21}{8}\log(2)\zeta(3) + \frac{773}{64}\zeta(4)$$

$$- 4\pi\Im\left\{\text{Li}_3\left(\frac{1+i}{2}\right)\right\} - 3\text{Li}_4\left(\frac{1}{2}\right),$$

and the second solution is finalized.

Despite its (high) difficulty, undoubtedly, it allows us to obtain wonderful solutions when hard work, perseverance, and a reasonable amount of creativity are put together. The series also appeared in [42].

6.56 Another (Very) Hard Nut to Crack (An Alternating Harmonic Series with a Weight 4 Structure, Involving Harmonic Numbers of the Type H_{2n})

Solution Like in the previous section, we prepare to face another *beast*, an alternating harmonic series with a weight 4 structure, involving numbers of the type H_{2n}. For this result, I would also like to emphasize that to my best knowledge, I'm not aware if the value of the series has ever been presented before in the mathematical literature.

I'll build a strategy such that I'll be able to exploit the integrals given in the statement section, and then we start with the integral $\Re\left\{\displaystyle\int_0^\infty \frac{\operatorname{arctanh}(t)\operatorname{Li}_2(-t^2)}{t(1+t^2)}dt\right\}$.

Before continuing I would like to remind you, as already seen in the second solution of the previous section, that P. V. $\displaystyle\int_0^1 \frac{t}{1-t^2u^2}du = \Re\{\operatorname{arctanh}(t)\}$, $x \in \mathbb{R}\setminus\{\pm 1\}$, and at the same time, based on (3.13), the case $n = 1$, Sect. 3.3, we have $\displaystyle\int_0^1 \frac{t^2\log(x)}{1+t^2x}dx = \operatorname{Li}_2(-t^2)$, which are two results we need to use in the following calculations.

So, rearranging the starting integral mentioned above, we write

$$\Re\left\{\int_0^\infty \frac{\operatorname{arctanh}(t)\operatorname{Li}_2(-t^2)}{t(1+t^2)}dt\right\} = \int_0^\infty \frac{\Re\{\operatorname{arctanh}(t)\}\operatorname{Li}_2(-t^2)}{t(1+t^2)}dt$$

$$= \int_0^\infty \left(\int_0^1 \left(\text{P. V.}\int_0^1 \frac{t^2\log(x)}{(1-t^2u^2)(1+t^2x)(1+t^2)}du\right)dx\right)dt$$

$$\underset{\substack{\text{reverse the order}\\\text{of integration}}}{=} \int_0^1 \left(\int_0^1 \left(\text{P. V.}\int_0^\infty \frac{t^2\log(x)}{(1-u^2t^2)(1+xt^2)(1+t^2)}dt\right)du\right)dx$$

$$= \int_0^1 \log(x)\left(\int_0^1 \frac{u^2}{(x+u^2)(1+u^2)}\underbrace{\left(\text{P. V.}\int_0^\infty \frac{1}{1-u^2t^2}dt\right)}_{0}du\right)dx$$

$$\underbrace{\phantom{\int_0^1 \log(x)\left(\int_0^1 \frac{u^2}{(x+u^2)(1+u^2)}\right)dx}}_{0}$$

$$-\underbrace{\int_0^1 \frac{\log(x)}{1-x}\left(\int_0^1 \frac{1}{1+u^2}\left(\int_0^\infty \frac{1}{1+t^2}dt\right)du\right)dx}_{-15/8\zeta(4)}$$

$$+ \int_0^1 \frac{x \log(x)}{1-x} \left(\int_0^1 \frac{1}{x+u^2} \left(\underbrace{\int_0^\infty \frac{1}{1+xt^2} dt}_{\pi/(2\sqrt{x})} \right) du \right) dx$$

$$= \frac{15}{8} \zeta(4) + \frac{\pi}{2} \int_0^1 \frac{\sqrt{x} \log(x)}{1-x} \left(\int_0^1 \frac{1}{x+u^2} du \right) dx$$

$$= \frac{15}{8} \zeta(4) + \frac{\pi}{2} \int_0^1 \frac{\arctan(1/\sqrt{x}) \log(x)}{1-x} dx$$

$$\overset{x=y^2}{=} \frac{15}{8} \zeta(4) + 2\pi \int_0^1 \frac{y(\pi/2 - \arctan(y)) \log(y)}{1-y^2} dy = \frac{15}{8} \zeta(4)$$

$$+ \pi^2 \int_0^1 \frac{y \log(y)}{1-y^2} dy - \pi \int_0^1 \frac{\arctan(y) \log(y)}{1-y} dy + \pi \int_0^1 \frac{\arctan(y) \log(y)}{1+y} dy$$

$$= \frac{3}{8} \log^2(2) \zeta(2) + \frac{75}{32} \zeta(4) - 2\pi \Im \left\{ \mathrm{Li}_3 \left(\frac{1+i}{2} \right) \right\}, \tag{6.431}$$

where for the first integral above, we have that $\int_0^1 \frac{y \log(y)}{1-y^2} dy \overset{y^2=v}{=} \frac{1}{4} \int_0^1 \frac{\log(v)}{1-v} dv$

$$= \frac{1}{4} \int_0^1 \log(v) \sum_{n=1}^\infty v^{n-1} dv = \frac{1}{4} \sum_{n=1}^\infty \int_0^1 v^{n-1} \log(v) dv = -\frac{1}{4} \sum_{n=1}^\infty \frac{1}{n^2} = -\frac{1}{4} \zeta(2);$$

then the value of the second integral is given in (6.422), the previous section; and finally, the last integral is $\int_0^1 \frac{\arctan(y) \log(y)}{1+y} dy = \frac{1}{2} \log(2) G - \frac{\pi^3}{64}$, already met in *(Almost) Impossible Integrals, Sums, and Series* (see [76, Chapter 1, p.14]) and [43].

Returning to the main integral I started with, then splitting it, and using the well-known dilogarithm function identity, $\mathrm{Li}_2(-x) + \mathrm{Li}_2(-1/x) = -\frac{\pi^2}{6} - \frac{1}{2} \log^2(x)$, $x > 0$, and the identity $\operatorname{arctanh}(1/x) - \operatorname{arctanh}(x) = i\pi/2$, $x > 1$, which appeared in the previous section, we write

$$\Re \left\{ \int_0^\infty \frac{\operatorname{arctanh}(t) \mathrm{Li}_2(-t^2)}{t(1+t^2)} dt \right\} = \int_0^1 \frac{\operatorname{arctanh}(t) \mathrm{Li}_2(-t^2)}{t(1+t^2)} dt$$

$$+ \Re \left\{ \underbrace{\int_1^\infty \frac{(\operatorname{arctanh}(1/t) - i\pi/2)(-\pi^2/6 - 2\log^2(t) - \operatorname{Li}_2(-1/t^2))}{t(1+t^2)} dt}_{\text{let } 1/t = u} \right\}$$

{after taking the real part, expand both integrals}

$$= -\frac{\pi^2}{6} \int_0^1 \frac{t \operatorname{arctanh}(t)}{1+t^2} dt - 2 \int_0^1 \frac{t \operatorname{arctanh}(t) \log^2(t)}{1+t^2} dt - \int_0^1 \frac{\operatorname{arctanh}(t) \operatorname{Li}_2(-t^2)}{t} dt$$

$$+ 2 \int_0^1 \frac{\operatorname{arctanh}(t) \operatorname{Li}_2(-t^2)}{t(1+t^2)} dt. \tag{6.432}$$

Letting the variable change $(1 - t)/(1 + t) = u$ in the first integral of (6.432), and returning to the variable in t, we have that

$$\int_0^1 \frac{t \operatorname{arctanh}(t)}{1+t^2} dt = \frac{1}{2} \underbrace{\int_0^1 \frac{t \log(t)}{1+t^2} dt}_{\text{let } t^2 = u} - \frac{1}{2} \int_0^1 \frac{\log(t)}{1+t} dt = -\frac{3}{8} \int_0^1 \frac{\log(t)}{1+t} dt$$

$$= -\frac{3}{8} \sum_{n=1}^\infty (-1)^{n-1} \int_0^1 t^{n-1} \log(t) dt = \frac{3}{8} \sum_{n=1}^\infty (-1)^{n-1} \frac{1}{n^2} = \frac{3}{16} \zeta(2). \tag{6.433}$$

For the second and third integrals in (6.432), in a first phase, we'll want to make a transformation into series, and then, by exploiting (1.110), Sect. 1.22, we write

$$\int_0^1 \frac{t \operatorname{arctanh}(t) \log^2(t)}{1+t^2} dt = \int_0^1 \sum_{n=1}^\infty (-1)^{n-1} t^{2n-1} \operatorname{arctanh}(t) \log^2(t) dt$$

{reverse the order of summation and integration}

$$= \sum_{n=1}^\infty (-1)^{n-1} \int_0^1 t^{2n-1} \log^2(t) \operatorname{arctanh}(t) dt$$

$$= \frac{1}{4} \sum_{n=1}^\infty (-1)^{n-1} \frac{d^2}{dn^2} \left(\int_0^1 t^{2n-1} \operatorname{arctanh}(t) dt \right)$$

$$= \frac{1}{16} \sum_{n=1}^\infty (-1)^{n-1} \frac{d^2}{dn^2} \left(\frac{2\psi(2n+1) - \psi(n+1) + \gamma}{n} \right)$$

$$
= -\frac{7}{8}\zeta(3)\underbrace{\sum_{n=1}^{\infty}(-1)^{n-1}\frac{1}{n}}_{\log(2)} - \frac{\pi^2}{16}\underbrace{\sum_{n=1}^{\infty}(-1)^{n-1}\frac{1}{n^2}}_{\pi^2/12} - \frac{1}{8}\sum_{n=1}^{\infty}(-1)^{n-1}\frac{H_n}{n^3}
$$

$$
+ \frac{1}{4}\sum_{n=1}^{\infty}(-1)^{n-1}\frac{H_{2n}}{n^3} - \frac{1}{8}\sum_{n=1}^{\infty}(-1)^{n-1}\frac{H_n^{(2)}}{n^2} + \frac{1}{2}\sum_{n=1}^{\infty}(-1)^{n-1}\frac{H_{2n}^{(2)}}{n^2}
$$

$$
- \frac{1}{8}\sum_{n=1}^{\infty}(-1)^{n-1}\frac{H_n^{(3)}}{n} + \sum_{n=1}^{\infty}(-1)^{n-1}\frac{H_{2n}^{(3)}}{n}. \tag{6.434}
$$

Next, we have

$$
\int_0^1 \frac{\operatorname{arctanh}(t)\,\operatorname{Li}_2(-t^2)}{t}\,dt
$$

$$
= \int_0^1 \operatorname{arctanh}(t)\sum_{n=1}^{\infty}(-1)^n\frac{t^{2n-1}}{n^2}\,dt = \sum_{n=1}^{\infty}(-1)^n\frac{1}{n^2}\int_0^1 t^{2n-1}\operatorname{arctanh}(t)\,dt
$$

{employ the result in (1.110), Sect. 1.22}

$$
= \frac{1}{4}\sum_{n=1}^{\infty}(-1)^{n-1}\frac{H_n}{n^3} - \frac{1}{2}\sum_{n=1}^{\infty}(-1)^{n-1}\frac{H_{2n}}{n^3}. \tag{6.435}
$$

Multiplying both sides of (6.434) by 2 and adding it to the result in (6.435), we get that

$$
2\int_0^1 \frac{t\,\operatorname{arctanh}(t)\log^2(t)}{1+t^2}\,dt + \int_0^1 \frac{\operatorname{arctanh}(t)\,\operatorname{Li}_2(-t^2)}{t}\,dt
$$

$$
= -\frac{7}{4}\log(2)\zeta(3) - \frac{15}{16}\zeta(4) - \frac{1}{4}\sum_{n=1}^{\infty}(-1)^{n-1}\frac{H_n^{(2)}}{n^2} - \frac{1}{4}\sum_{n=1}^{\infty}(-1)^{n-1}\frac{H_n^{(3)}}{n}
$$

$$
+ 2\sum_{n=1}^{\infty}(-1)^{n-1}\frac{H_{2n}^{(3)}}{n} + \sum_{n=1}^{\infty}(-1)^{n-1}\frac{H_{2n}^{(2)}}{n^2}
$$

$$
= \frac{1}{6}\log^4(2) - \frac{91}{32}\zeta(4) + \frac{7}{4}\log(2)\zeta(3) - \log^2(2)\zeta(2) + 4\operatorname{Li}_4\left(\frac{1}{2}\right), \tag{6.436}
$$

where the value of the first series is stated in (6.190), Sect. 6.24; then the second and third series are the particular cases $m = 3$ of the generalizations in (4.106) and (4.107), Sect. 4.22; and the value of the last series is given in (4.174), Sect. 4.52.

Further, for the last integral in (6.432), we combine the case $m = 2$ of (4.32), Sect. 4.6, and the integral result in (1.110), Sect. 1.22, that lead to

$$\int_0^1 \frac{\operatorname{arctanh}(t) \operatorname{Li}_2(-t^2)}{t(1+t^2)} dt = \int_0^1 \operatorname{arctanh}(t) \sum_{n=1}^{\infty} (-1)^n t^{2n-1} H_n^{(2)} dt$$

{reverse the order of summation and integration}

$$= \sum_{n=1}^{\infty} (-1)^n H_n^{(2)} \int_0^1 t^{2n-1} \operatorname{arctanh}(t) dt$$

$$= \frac{1}{4} \sum_{n=1}^{\infty} (-1)^{n-1} \frac{H_n H_n^{(2)}}{n} - \frac{1}{2} \sum_{n=1}^{\infty} (-1)^{n-1} \frac{H_{2n} H_n^{(2)}}{n}$$

$$= \frac{1}{48} \log^4(2) - \frac{1}{16} \log^2(2)\zeta(2) + \frac{7}{32} \log(2)\zeta(3) - \frac{1}{4}\zeta(4) + \frac{1}{2}\operatorname{Li}_4\left(\frac{1}{2}\right)$$

$$- \frac{1}{2} \sum_{n=1}^{\infty} (-1)^{n-1} \frac{H_{2n} H_n^{(2)}}{n}, \qquad (6.437)$$

where I used the series found and calculated in (6.218), Sect. 6.28.

Plugging (6.433), (6.436), and (6.437) in (6.432), we arrive at

$$\Re\left\{ \int_0^{\infty} \frac{\operatorname{arctanh}(t) \operatorname{Li}_2(-t^2)}{t(1+t^2)} dt \right\} = \frac{15}{8}\zeta(4) - \frac{21}{16} \log(2)\zeta(3) + \frac{7}{8} \log^2(2)\zeta(2)$$

$$- \frac{1}{8} \log^4(2) - 3\operatorname{Li}_4\left(\frac{1}{2}\right) - \sum_{n=1}^{\infty} (-1)^{n-1} \frac{H_{2n} H_n^{(2)}}{n}. \qquad (6.438)$$

Finally, combining (6.431) and (6.438), we conclude that

$$\sum_{n=1}^{\infty} (-1)^{n-1} \frac{H_{2n} H_n^{(2)}}{n}$$

$$= \frac{1}{2} \log^2(2)\zeta(2) - \frac{21}{16} \log(2)\zeta(3) - \frac{15}{32}\zeta(4) - \frac{1}{8} \log^4(2)$$

$$+ 2\pi \Im\left\{ \operatorname{Li}_3\left(\frac{1+i}{2}\right) \right\} - 3\operatorname{Li}_4\left(\frac{1}{2}\right),$$

and the solution is complete.

As a final point, the curious reader might also take and calculate separately the integrals from (6.434) and (6.435), since all the resulting series are given in the book. Therefore, we immediately obtain that

$$\int_0^1 \frac{t \arctanh(t) \log^2(t)}{1+t^2} dt = \frac{1}{24} \log^4(2) - \frac{1}{4} \log^2(2)\zeta(2) - \frac{31}{128}\zeta(4) + \text{Li}_4\left(\frac{1}{2}\right)$$
(6.439)

and

$$\int_0^1 \frac{\arctanh(t) \text{Li}_2(-t^2)}{t} dt$$

$$= \frac{1}{12} \log^4(2) - \frac{1}{2} \log^2(2)\zeta(2) + \frac{7}{4} \log(2)\zeta(3) - \frac{151}{64}\zeta(4) + 2\text{Li}_4\left(\frac{1}{2}\right).$$
(6.440)

What other options would we have for such a series problem? This is a good question to consider for ending the current section!

6.57 Two Harmonic Series with a Wicked Look, Involving Skew-Harmonic Numbers and Harmonic Numbers H_{2n}

Solution Like in the previous section, when we look at each of the summands, we identify a weight 4 structure, but this time with both skew-harmonic numbers and harmonic numbers of the type H_{2n}! And, indeed, they have a wicked look!

We'll see that, in fact, they may be reduced to results previously derived, including special polylogarithmic values with a complex argument (derived in Sect. 6.53).

For the series at the point (i), we consider the integral result in (1.99), Sect. 1.21, where if we multiply both sides by $(-1)^{n-1}H_{2n}/n$ and consider the summation from $n = 1$ to ∞, we get

$$\log(2) \sum_{n=1}^{\infty} \frac{H_{2n}}{n^2} + \log(2) \sum_{n=1}^{\infty} (-1)^{n-1} \frac{H_{2n}}{n^2} - \sum_{n=1}^{\infty} \frac{\overline{H}_n H_{2n}}{n^2} = 2\log(2) \sum_{n=1}^{\infty} \frac{H_n}{n^2}$$

$$- 2\log(2) \sum_{n=1}^{\infty} (-1)^{n-1} \frac{H_n}{n^2} - 4\log(2)\Re\left\{\sum_{n=1}^{\infty} i^n \frac{H_n}{n^2}\right\} - \sum_{n=1}^{\infty} \frac{\overline{H}_n H_{2n}}{n^2}$$

{exploit the Euler sum generalizations in (6.149), Sect. 6.19, (4.105), Sec. 4.21,}

{and consider the generating function in (4.36), Sect. 4.6, the third closed form}

$$= \log(2)\pi G + \frac{21}{16}\log(2)\zeta(3) - \sum_{n=1}^{\infty}\frac{\overline{H}_n H_{2n}}{n^2} = \sum_{n=1}^{\infty}(-1)^{n-1}\frac{H_{2n}}{n}\int_0^1 x^{n-1}\log(1+x)dx$$

{reverse the order of integration and summation, and rearrange}

$$\stackrel{x=t^2}{=} \int_0^1 \frac{\log(1+t^2)}{t}\sum_{n=1}^{\infty}(-1)^{n-1}t^{2n}\frac{2H_{2n}}{n}dt$$

$$= \int_0^1 \frac{\log(1+t^2)}{t}\sum_{n=1}^{\infty}(-1)^{n-1}t^{2n}\frac{2H_{2n}-H_n}{n}dt$$

$$- \int_0^1 \frac{\log(1+t^2)}{t}\sum_{n=1}^{\infty}(-1)^n t^{2n}\frac{H_n}{n}dt$$

{exploit the Cauchy product in (4.17), Sect. 4.5, and (4.34), Sect. 4.6}

$$= 2\int_0^1 \frac{\arctan^2(t)\log(1+t^2)}{t}dt - \frac{1}{2}\int_0^1 \frac{\log^3(1+t^2)}{t}dt - \int_0^1 \frac{\log(1+t^2)\operatorname{Li}_2(-t^2)}{t}dt. \tag{6.441}$$

For the first resulting integral in (6.441), since we have $\log(a+ib) = \log(\sqrt{a^2+b^2}) + i\arctan(b/a)$, $a > 0$, we may write

$$\int_0^1 \frac{\arctan^2(t)\log(1+t^2)}{t}dt = \frac{1}{12}\int_0^1 \frac{\log^3(1+t^2)}{t}dt - \frac{2}{3}\Re\left\{\int_0^1 \frac{\log^3(1-it)}{t}dt\right\}, \tag{6.442}$$

Regarding the first integral in the right-hand side of (6.442), we let the variable change $t^2 = u$ that gives

$$\int_0^1 \frac{\log^3(1+t^2)}{t}dt = \frac{1}{2}\int_0^1 \frac{\log^3(1+u)}{u}du$$

{the resulting integral is already found in (6.155), Sect. 6.20}

$$= 3\zeta(4) - \frac{21}{8}\log(2)\zeta(3) + \frac{3}{4}\log^2(2)\zeta(2) - \frac{1}{8}\log^4(2) - 3\operatorname{Li}_4\left(\frac{1}{2}\right). \tag{6.443}$$

Next, letting the variable change $it = u$ in the second integral from the right-hand side of (6.442) and using (6.36), Sect. 6.6, we obtain that

$$\Re\left\{\int_0^1 \frac{\log^3(1-it)}{t}dt\right\} = \Re\left\{\int_0^i \frac{\log^3(1-u)}{u}du\right\}$$

$$= \Re\{\log(i)\log^3(1-i)\} + 3\Re\{\log^2(1-i)\operatorname{Li}_2(1-i)\} - 6\Re\{\log(1-i)\operatorname{Li}_3(1-i)\}$$

$$+ 6\Re\{\operatorname{Li}_4(1-i)\} - 6\zeta(4)$$

$$= \frac{1359}{256}\zeta(4) - \frac{105}{64}\log(2)\zeta(3) + \frac{3}{4}\log^2(2)\zeta(2) - \frac{3}{4}\log(2)\pi G - \frac{5}{64}\log^4(2)$$

$$- \frac{3}{2}\pi\Im\left\{\operatorname{Li}_3\left(\frac{1+i}{2}\right)\right\} - \frac{15}{8}\operatorname{Li}_4\left(\frac{1}{2}\right), \tag{6.444}$$

where during the calculations I used that $\log(i) = i\pi/2$, $\log(1-i) = 1/2\log(2) - i\pi/4$, since $\log(a+ib) = \log(\sqrt{a^2+b^2}) + i\arctan(b/a)$, $a > 0$, then (6.387), (6.405), (6.404), Sect. 6.53, to express $\Im\{\operatorname{Li}_3(1-i)\}$ in terms of $\Im\{\operatorname{Li}_3((1+i)/2)\}$, and (4.178), Sect. 4.53.

Plugging (6.443) and (6.444) in (6.442), we get

$$\int_0^1 \frac{\arctan^2(t)\log(1+t^2)}{t}dt$$

$$= \frac{1}{24}\log^4(2) + \frac{1}{2}\log(2)\pi G - \frac{7}{16}\log^2(2)\zeta(2) + \frac{7}{8}\log(2)\zeta(3) - \frac{421}{128}\zeta(4)$$

$$+ \pi\Im\left\{\operatorname{Li}_3\left(\frac{1+i}{2}\right)\right\} + \operatorname{Li}_4\left(\frac{1}{2}\right). \tag{6.445}$$

The same strategy of reducing the integral to known polylogarithmic values is also used by M. H. Zhao.

Collecting (6.443), (6.445), and the following integral, $\displaystyle\int_0^1 \frac{\log(1+t^2)\operatorname{Li}_2(-t^2)}{t}$

$$dt = -\frac{1}{4}(\operatorname{Li}_2(-t^2))^2\Big|_{t=0}^{t=1} = -\frac{5}{32}\zeta(4),$$ in (6.441), we are able to extract the desired series,

$$\sum_{n=1}^{\infty} \frac{\overline{H}_n H_{2n}}{n^2}$$

$$= \frac{507}{64}\zeta(4) - \frac{7}{4}\log(2)\zeta(3) + \frac{5}{4}\log^2(2)\zeta(2) - \frac{7}{48}\log^4(2)$$

$$- 2\pi\Im\left\{\operatorname{Li}_3\left(\frac{1+i}{2}\right)\right\} - \frac{7}{2}\operatorname{Li}_4\left(\frac{1}{2}\right),$$

and the solution to the point (i) of the problem is complete.

Alternatively, for getting a slightly different solution, the curious reader might start by combining the result in (6.445) and the Cauchy product in (4.17), Sect. 4.5. The series version with H_n instead of H_{2n}, that is, $\displaystyle\sum_{n=1}^{\infty} \frac{\overline{H}_n H_n}{n^2}$, was already presented and calculated in [44]. That *minor change* makes a world of difference in terms of difficulty between the two series.

Passing to the second point of the problem, we may proceed as with the previous point and then we start with multiplying both sides of (1.99), Sect. 1.21, by H_{2n}/n, and considering the summation from $n = 1$ to ∞, which gives

$$\log(2)\sum_{n=1}^{\infty}\frac{H_{2n}}{n^2} + \log(2)\sum_{n=1}^{\infty}(-1)^{n-1}\frac{H_{2n}}{n^2} - \sum_{n=1}^{\infty}(-1)^{n-1}\frac{\overline{H}_n H_{2n}}{n^2}$$

{the sum of the first two series is calculated at the previous point}

$$= \log(2)\pi G + \frac{21}{16}\log(2)\zeta(3) - \sum_{n=1}^{\infty}(-1)^{n-1}\frac{\overline{H}_n H_{2n}}{n^2}$$

$$= \sum_{n=1}^{\infty}\frac{H_{2n}}{n}\int_0^1 x^{n-1}\log(1+x)dx \overset{x=t^2}{=} \int_0^1 \frac{\log(1+t^2)}{t}\sum_{n=1}^{\infty}t^{2n}\frac{2H_{2n}}{n}dt$$

$$= \int_0^1 \frac{\log(1+t^2)}{t}\sum_{n=1}^{\infty}t^{2n}\frac{2H_{2n}-H_n}{n}dt + \int_0^1 \frac{\log(1+t^2)}{t}\sum_{n=1}^{\infty}t^{2n}\frac{H_n}{n}dt$$

{exploit the Cauchy product in (4.19), Sect. 4.5, and (4.34), Sect. 4.6}

$$= 2\int_0^1 \frac{\operatorname{arctanh}^2(t)\log(1+t^2)}{t}dt + \frac{1}{2}\int_0^1 \frac{\log^2(1-t^2)\log(1+t^2)}{t}dt$$

$$+ \int_0^1 \frac{\log(1+t^2)\operatorname{Li}_2(t^2)}{t}dt. \tag{6.446}$$

Making the variable change $t^2 = u$ in the second integral in (6.446), we get

$$\int_0^1 \frac{\log^2(1-t^2)\log(1+t^2)}{t}dt = \frac{1}{2}\int_0^1 \frac{\log^2(1-u)\log(1+u)}{u}du$$

$$= \frac{1}{24}\log^4(2) - \frac{1}{4}\log^2(2)\zeta(2) + \frac{7}{8}\log(2)\zeta(3) - \frac{5}{16}\zeta(4) + \operatorname{Li}_4\left(\frac{1}{2}\right), \tag{6.447}$$

where the resulting integral is found in (3.325), Sect. 3.45.

Then, letting the variable change $t^2 = u$ in the third integral of (6.446), and using $\int_0^1 x^{n-1} \operatorname{Li}_2(x)dx = \zeta(2)\dfrac{1}{n} - \dfrac{H_n}{n^2}$, as seen in (3.327), Sect. 3.46, we write

$$\int_0^1 \frac{\log(1+t^2)\operatorname{Li}_2(t^2)}{t}dt = \frac{1}{2}\int_0^1 \frac{\log(1+u)\operatorname{Li}_2(u)}{u}du$$

$$= \frac{1}{2}\int_0^1 \sum_{n=1}^{\infty}(-1)^{n-1}\frac{u^{n-1}}{n}\operatorname{Li}_2(u)du = \frac{1}{2}\sum_{n=1}^{\infty}(-1)^{n-1}\frac{1}{n}\int_0^1 u^{n-1}\operatorname{Li}_2(u)du$$

$$= \frac{1}{2}\zeta(2)\sum_{n=1}^{\infty}(-1)^{n-1}\frac{1}{n^2} - \frac{1}{2}\sum_{n=1}^{\infty}(-1)^{n-1}\frac{H_n}{n^3}$$

$$= \frac{1}{24}\log^4(2) - \frac{1}{4}\log^2(2)\zeta(2) + \frac{7}{8}\log(2)\zeta(3) - \frac{3}{4}\zeta(4) + \operatorname{Li}_4\left(\frac{1}{2}\right). \qquad (6.448)$$

At last, collecting the results from (1.202), Sect. 1.45, (6.447), and (6.448) in (6.446), we arrive at

$$\sum_{n=1}^{\infty}(-1)^{n-1}\frac{\overline{H}_n H_{2n}}{n^2}$$

$$= \frac{5}{48}\log^4(2) - \frac{5}{8}\log^2(2)\zeta(2) + \frac{7}{2}\log(2)\zeta(3) - \frac{77}{32}\zeta(4) + \log(2)\pi G$$

$$- 2G^2 + \frac{5}{2}\operatorname{Li}_4\left(\frac{1}{2}\right),$$

and the solution to the point (ii) of the problem is complete.

In order to get a somewhat different solution, the curious reader might exploit the Cauchy product in (4.19), Sect. 4.5.

As a final thought, one might be stumped to see that between the two given harmonic series, the alternating version is actually way easier. Usually, the simpler harmonic series to evaluate are the non-alternating ones, but when skew-harmonic numbers appear in the summand, *things might be different in terms of difficulty*!

6.58   Nice Series with the Reciprocal of the Central Binomial Coefficient and the Generalized Harmonic Number

Solution If you are used to the work involving the well-known power series of $\dfrac{\arcsin(x)}{\sqrt{1-x^2}}$, it might not be hard to see how to start out at the first two points of the problem involving the reciprocal of the central binomial coefficient.

In fact, if you had a chance to see my solution to a similar series presented in [76, Chapter 6, p.334], which I attack by a similar strategy, then it is easy to figure out how to start out here. Although known in the mathematical literature, evaluating elegantly their corresponding integral representations to get a solution may be a challenging *round* as you have *probably* seen in the third chapter.

Any useful starting idea for the last two points of the problem? This might be just a natural, immediate question to expect from you. This time, besides the reciprocal of the central binomial coefficient, the summands also contain particular cases of the generalized harmonic number, in one case $H_n^{(3)}$, and in the other one, we have $H_n^{(4)}$. And later we'll see that the series at the last two points are strongly related to the series from the first two points.

Let's begin with the point (i) of the problem! As suggested above, we need the fact that $\dfrac{\arcsin(x)}{\sqrt{1-x^2}} = \displaystyle\sum_{n=1}^{\infty} \dfrac{(2x)^{2n-1}}{n\dbinom{2n}{n}}$ (proofs of this result may also be found in [76, Chapter 6, pp. 331–333]), and then we write

$$\sum_{n=1}^{\infty} \frac{2^{2n}}{n^4\dbinom{2n}{n}} = 4\sum_{n=1}^{\infty} \frac{2^{2n}}{n\dbinom{2n}{n}} \int_0^1 x^{2n-1}\log^2(x)dx$$

{reverse the order of integration and summation}

$$= 8\int_0^1 \log^2(x) \sum_{n=1}^{\infty} \frac{(2x)^{2n-1}}{n\dbinom{2n}{n}}dx = 8\int_0^1 \frac{\arcsin(x)}{\sqrt{1-x^2}}\log^2(x)dx$$

$$\overset{x=\sin(y)}{=} 8\int_0^{\pi/2} y\log^2(\sin(y))dy$$

$$= \frac{1}{3}\log^4(2) + 4\log^2(2)\zeta(2) - \frac{19}{4}\zeta(4) + 8\operatorname{Li}_4\left(\frac{1}{2}\right),$$

where the last integral is calculated in (1.234), Sect. 1.54, and the point (i) of the problem is finalized. The series also appears in [67, p.308].

We attack the point (ii) in a similar style as before, and then we have

$$\sum_{n=1}^{\infty} \frac{2^{2n}}{n^5 \binom{2n}{n}} = -\frac{8}{3} \sum_{n=1}^{\infty} \frac{2^{2n}}{n \binom{2n}{n}} \int_0^1 x^{2n-1} \log^3(x) dx$$

{reverse the order of integration and summation}

$$= -\frac{16}{3} \int_0^1 \log^3(x) \sum_{n=1}^{\infty} \frac{(2x)^{2n-1}}{n \binom{2n}{n}} dx = -\frac{16}{3} \int_0^1 \frac{\arcsin(x)}{\sqrt{1-x^2}} \log^3(x) dx$$

$$\overset{x=\sin(y)}{=\!=\!=} -\frac{16}{3} \int_0^{\pi/2} y \log^3(\sin(y)) dy$$

$$= \frac{2}{15} \log^5(2) + \frac{8}{3} \log^3(2) \zeta(2) - \frac{19}{2} \log(2) \zeta(4) + 6\zeta(2)\zeta(3)$$

$$+ \frac{31}{8} \zeta(5) - 16 \operatorname{Li}_5\left(\frac{1}{2}\right),$$

where the last integral is calculated in (1.238), Sect. 1.54, and the point (ii) of the problem is finalized.

Before starting the main solutions to the last two points of the problem, we'll want to return to Sect. 4.3 and use the result (4.8), where if we replace n by $n-1$ and consider that $\binom{2n-2}{n-1} = \frac{n}{2(2n-1)} \binom{2n}{n}$, we arrive at

$$\sum_{k=1}^{n-1} \frac{2^{2k}}{k(2k+1)\binom{2k}{k}} = 2 - \frac{2^{2n}}{n\binom{2n}{n}}. \qquad (6.449)$$

So, we begin the main calculations with an artifice of calculation, and we write

$$\sum_{n=1}^{\infty} \frac{2^{2n} H_n^{(3)}}{n(2n+1)\binom{2n}{n}} = \sum_{n=1}^{\infty} \frac{2^{2n}\left(\zeta(3) - \left(\zeta(3) - H_n^{(3)}\right)\right)}{n(2n+1)\binom{2n}{n}}$$

$$= \zeta(3) \sum_{n=1}^{\infty} \frac{2^{2n}}{n(2n+1)\binom{2n}{n}} - \sum_{n=1}^{\infty} \frac{2^{2n}}{n(2n+1)\binom{2n}{n}}\left(\zeta(3) - H_n^{(3)}\right)$$

{for the first series combine the result in (6.449) and the asymptotic}

$$\left\{ \text{expansion behavior of the central binomial coefficient,} \ \binom{2n}{n} \approx \frac{4^n}{\sqrt{\pi n}}, \right\}$$

$$\left\{ \text{and for the second series use that } \zeta(3) - H_n^{(3)} = \sum_{k=1}^{\infty} \frac{1}{(n+k)^3} = \sum_{k=n+1}^{\infty} \frac{1}{k^3} \right\}$$

$$= 2\zeta(3) - \sum_{n=1}^{\infty} \frac{2^{2n}}{n(2n+1)\binom{2n}{n}} \left(\sum_{k=n+1}^{\infty} \frac{1}{k^3} \right)$$

{reverse the order of summation and then use the result in (6.449)}

$$= 2\zeta(3) - \sum_{k=1}^{\infty} \frac{1}{k^3} \left(\sum_{n=1}^{k-1} \frac{2^{2n}}{n(2n+1)\binom{2n}{n}} \right) = 2\zeta(3) - \sum_{k=1}^{\infty} \frac{1}{k^3} \left(2 - \frac{2^{2k}}{k\binom{2k}{k}} \right)$$

$$= \sum_{k=1}^{\infty} \frac{2^{2k}}{k^4 \binom{2k}{k}}$$

$$= \frac{1}{3}\log^4(2) + 4\log^2(2)\zeta(2) - \frac{19}{4}\zeta(4) + 8\operatorname{Li}_4\left(\frac{1}{2}\right), \qquad (6.450)$$

where the last equality follows based on the point (i) of the section, and the point (iii) of the problem is finalized.

Finally, for the last point of the problem, we proceed in a similar style as before, and then we have

$$\sum_{n=1}^{\infty} \frac{2^{2n} H_n^{(4)}}{n(2n+1)\binom{2n}{n}} = \sum_{n=1}^{\infty} \frac{2^{2n}\left(\zeta(4) - \left(\zeta(4) - H_n^{(4)}\right)\right)}{n(2n+1)\binom{2n}{n}}$$

$$= \zeta(4) \sum_{n=1}^{\infty} \frac{2^{2n}}{n(2n+1)\binom{2n}{n}} - \sum_{n=1}^{\infty} \frac{2^{2n}}{n(2n+1)\binom{2n}{n}} \left(\zeta(4) - H_n^{(4)}\right)$$

{for the first series combine the result in (6.449) and the asymptotic}

$$\left\{ \text{expansion behavior of the central binomial coefficient, } \binom{2n}{n} \approx \frac{4^n}{\sqrt{\pi n}}, \right\}$$

$$\left\{ \text{and for the second series use that } \zeta(4) - H_n^{(4)} = \sum_{k=1}^{\infty} \frac{1}{(n+k)^4} = \sum_{k=n+1}^{\infty} \frac{1}{k^4} \right\}$$

$$= 2\zeta(4) - \sum_{n=1}^{\infty} \frac{2^{2n}}{n(2n+1)\binom{2n}{n}} \left(\sum_{k=n+1}^{\infty} \frac{1}{k^4} \right)$$

{reverse the order of summation and then use the result in (6.449)}

$$= 2\zeta(4) - \sum_{k=1}^{\infty} \frac{1}{k^4} \left(\sum_{n=1}^{k-1} \frac{2^{2n}}{n(2n+1)\binom{2n}{n}} \right) = 2\zeta(4) - \sum_{k=1}^{\infty} \frac{1}{k^4} \left(2 - \frac{2^{2k}}{k\binom{2k}{k}} \right)$$

$$= \sum_{k=1}^{\infty} \frac{2^{2k}}{k^5 \binom{2k}{k}}$$

$$= \frac{2}{15} \log^5(2) + \frac{8}{3} \log^3(2)\zeta(2) - \frac{19}{2} \log(2)\zeta(4) + 6\zeta(2)\zeta(3)$$

$$+ \frac{31}{8}\zeta(5) - 16 \operatorname{Li}_5\left(\frac{1}{2}\right), \tag{6.451}$$

where the last equality follows based on the point (ii) of the section, and the point (iv) of the problem is finalized.

It is nice, and to some extent surprising, to see how beautifully the series from the last two points are reduced to the series from the first two points. Surely, we may view the whole process as a special transformation, where a telescoping sum presented in Sect. 4.3 plays a crucial part.

6.59 Marvellous Binoharmonic Series Forged with Nice Ideas

Solution Behind the beautiful results in this section, there is also a key result with a spectacular story. If you read *Irresistible Integrals*, you probably came across at some point the section called, *Several Evaluations for* $\zeta(2)$, which is one of the nicest parts of the book (see [6, pp. 225–227]). Now, one of the proofs there, called Matsuoka's Proof, after the name of the discoverer of the proof (1961), is built based on the use of Wallis' integral $I_n = \int_0^{\pi/2} \cos^{2n}(x)\mathrm{d}x$ and its *relative* $J_n = \int_0^{\pi/2} x^2 \cos^{2n}(x)\mathrm{d}x$, leading to one of the most beautiful solutions to the Basel problem, showing that $\sum_{n=1}^{\infty} \dfrac{1}{n^2} = \zeta(2) = \dfrac{\pi^2}{6}$. The integral J_n is also the star of the present section, which I'll exploit for both points of the problem!

Following the use of the recurrence relation idea given in the mentioned book, if we consider for I_n two integrations by parts, $n \geq 1$, $n \in \mathbb{N}$, we get

$$I_n = \int_0^{\pi/2} \cos^{2n}(x)\mathrm{d}x = \int_0^{\pi/2} x' \cos^{2n}(x)\mathrm{d}x = 2n \int_0^{\pi/2} x \sin(x) \cos^{2n-1}(x)\mathrm{d}x$$

$$= n \int_0^{\pi/2} (x^2)' \sin(x) \cos^{2n-1}(x)\mathrm{d}x$$

$$= -n \int_0^{\pi/2} x^2 \left(\cos^{2n}(x) - (2n-1)(1 - \cos^2(x)) \cos^{2n-2}(x) \right) \mathrm{d}x$$

$$= n(2n-1) \underbrace{\int_0^{\pi/2} x^2 \cos^{2n-2}(x)\mathrm{d}x}_{J_{n-1}} - 2n^2 \underbrace{\int_0^{\pi/2} x^2 \cos^{2n}(x)\mathrm{d}x}_{J_n},$$

or to put it simply,

$$I_n = n(2n-1)J_{n-1} - 2n^2 J_n. \tag{6.452}$$

Upon using in (6.452) the well-known closed form of Wallis' integral, the even case, $\int_0^{\pi/2} \sin^{2n}(x)\mathrm{d}x = \int_0^{\pi/2} \cos^{2n}(x)\mathrm{d}x = \dfrac{\pi}{2} \cdot \dfrac{(2n-1)!!}{(2n)!!} = \dfrac{\pi}{2^{2n+1}} \binom{2n}{n}$, we get

$$\frac{\pi}{2} \cdot \frac{(2n-1)!!}{(2n)!!} = n(2n-1)J_{n-1} - 2n^2 J_n,$$

and if we multiply both sides of the last result by $(2n-2)!!/(n \cdot (2n-1)!!)$, we arrive at

$$\frac{\pi}{4} \frac{1}{n^2} = \frac{(2n-2)!!}{(2n-3)!!} J_{n-1} - \frac{(2n)!!}{(2n-1)!!} J_n,$$

where if we replace n by k, and then make the summation from $k = 1$ to $k = n$, we have

$$\frac{\pi}{4}\sum_{k=1}^{n}\frac{1}{k^2} = \frac{\pi}{4}H_n^{(2)} = \sum_{k=1}^{n}\left(\frac{(2k-2)!!}{(2k-3)!!}J_{k-1} - \frac{(2k)!!}{(2k-1)!!}J_k\right) = J_0 - \frac{(2n)!!}{(2n-1)!!}J_n$$

$$= \frac{\pi^3}{24} - \frac{(2n)!!}{(2n-1)!!}J_n,$$

whence we obtain that

$$J_n = \frac{(2n-1)!!}{(2n)!!}\frac{\pi}{4}\left(\frac{\pi^2}{6} - H_n^{(2)}\right),$$

or we may put it in the form,

$$J_n = \int_0^{\pi/2} x^2 \cos^{2n}(x)\mathrm{d}x = \frac{\pi}{4}\frac{1}{2^{2n}}\binom{2n}{n}\left(\zeta(2) - H_n^{(2)}\right), \qquad (6.453)$$

which is exactly what we need in the subsequent calculations!

The result in (6.453) may also be seen as a result of the manipulation of another famous parameterized integral, that is

$$\int_0^{\pi/2} \cos^{v-1}(x)\cos(ax)\mathrm{d}x = \frac{\pi}{v2^v\,\mathrm{B}\left(\dfrac{v+a+1}{2}, \dfrac{v-a+1}{2}\right)}, \ \Re\{v\} > 0,$$

$$(6.454)$$

appearing as the result **3.631.9** in [17]. A short proof of (6.454), based on contour integration, may be found in [45]. For example, replacing v by $2n + 1$ in (6.454), then differentiating two times with respect to a, and finally letting $a \to 0$, we arrive at the result in (6.453).

Returning to the series at the point (i) of the problem, when we look at the summand, it is not hard to guess what we could use there! It is about the result in (6.453)! Then, we write that

$$\sum_{n=1}^{\infty}\frac{1}{n^3 2^{2n}}\binom{2n}{n}\left(\zeta(2) - H_n^{(2)}\right) = \frac{4}{\pi}\sum_{n=1}^{\infty}\frac{1}{n^3}\int_0^{\pi/2} x^2 \cos^{2n}(x)\mathrm{d}x$$

{reverse the order of integration and summation}

$$= \frac{4}{\pi}\int_0^{\pi/2} x^2 \sum_{n=1}^{\infty}\frac{\cos^{2n}(x)}{n^3}\mathrm{d}x = \frac{4}{\pi}\int_0^{\pi/2} x^2 \mathrm{Li}_3(\cos^2(x))\mathrm{d}x. \qquad (6.455)$$

Now, we might use the Fourier series in (4.163), Sect. 4.49, but instead of using the coefficient of the series expressed in terms of harmonic numbers, I'll use the integral form as it appears in its derivation in Sect. 6.49, that is

$$\text{Li}_3(\cos^2(x))$$

$$= \frac{4}{3}\log^3(2) - \log(2)\frac{\pi^2}{3} + 2\zeta(3) + 4\sum_{n=1}^{\infty}\left(\int_0^1 t^{n-1}\log^2\left(\frac{1+t}{2\sqrt{t}}\right)dt\right)\cos(2nx).$$
                                                                                            (6.456)

Multiplying both sides of (6.456) by x^2, integrating from $x = 0$ to $x = \pi/2$, reversing the order of integration and summation, using that $\int_0^{\pi/2} x^2\cos(2nx)dx = \frac{\pi}{4}\frac{(-1)^n}{n^2}$, and then taking the summation, we arrive at

$$\int_0^{\pi/2} x^2\,\text{Li}_3(\cos^2(x))dx = \frac{\pi^3}{12}\zeta(3) + \log^3(2)\frac{\pi^3}{18} - \log(2)\frac{\pi^5}{72}$$

$$+ \pi\int_0^1 \frac{\text{Li}_2(-t)}{t}\log^2\left(\frac{1+t}{2\sqrt{t}}\right)dt = \frac{\pi^3}{12}\zeta(3) + \log^3(2)\frac{\pi^3}{18} - \log(2)\frac{\pi^5}{72}$$

$$+ \log^2(2)\pi\underbrace{\int_0^1 \frac{\text{Li}_2(-t)}{t}dt}_{-3/4\zeta(3)} + \log(2)\pi\underbrace{\int_0^1 \frac{\log(t)\,\text{Li}_2(-t)}{t}dt}_{7/720\pi^4} + \frac{\pi}{4}\underbrace{\int_0^1 \frac{\log^2(t)\,\text{Li}_2(-t)}{t}dt}_{-15/8\zeta(5)}$$

$$- 2\log(2)\pi\underbrace{\int_0^1 \frac{\log(1+t)\,\text{Li}_2(-t)}{t}dt}_{-\pi^4/288} - \pi\int_0^1 \frac{\log(1+t)\log(t)\,\text{Li}_2(-t)}{t}dt$$

$$+ \pi\int_0^1 \frac{\log^2(1+t)\,\text{Li}_2(-t)}{t}dt.$$ (6.457)

Let's observe the fifth integral in (6.59) is the case $m = 1$ of the generalized integral in (1.207), Sect. 1.46, that gives

$$\int_0^1 \frac{\log(1+t)\log(t)\,\text{Li}_2(-t)}{t}dt = \frac{45}{16}\zeta(5) - 2\sum_{n=1}^{\infty}(-1)^{n-1}\frac{H_n}{n^4} - \sum_{n=1}^{\infty}(-1)^{n-1}\frac{H_n^{(2)}}{n^3}$$

$$= \frac{3}{8}\zeta(2)\zeta(3) - \frac{17}{32}\zeta(5),$$ (6.458)

where in the calculations I used that the first alternating harmonic series is given in (4.104), Sect. 4.20, or it may be seen as the case $m = 2$ of the alternating Euler sum generalization in (4.105), Sect. 4.21, and the second one is $\sum_{n=1}^{\infty}(-1)^{n-1}\frac{H_n^{(2)}}{n^3} =$

$\frac{5}{8}\zeta(2)\zeta(3) - \frac{11}{32}\zeta(5)$ (see [76, Chapter 4, p.311]).

Returning with the result from (6.458) in (6.59), where we note the first four integrals are straightforward either by direct integration or by integrations by parts $\left(\text{e.g., note that } \int_0^1 \frac{\log(1+t)\,\text{Li}_2(-t)}{t}dt = -\frac{1}{2}(\text{Li}_2(-t))^2 + C\right)$, and the last integral is already given in (1.213), Sect. 1.47, we get that

$$\int_0^{\pi/2} x^2\,\text{Li}_3(\cos^2(x))dx$$

$$= \log(2)\frac{\pi^5}{360} - \log^3(2)\frac{\pi^3}{18} + \frac{2}{15}\log^5(2)\pi + \log^2(2)\pi\zeta(3) - \frac{123}{32}\pi\zeta(5)$$

$$+ 4\log(2)\pi\,\text{Li}_4\left(\frac{1}{2}\right) + 4\pi\,\text{Li}_5\left(\frac{1}{2}\right). \tag{6.459}$$

At last, by plugging (6.459) in (6.455), we conclude that

$$\sum_{n=1}^{\infty} \frac{1}{n^3 2^{2n}}\binom{2n}{n}\left(\zeta(2) - H_n^{(2)}\right)$$

$$= \frac{8}{15}\log^5(2) - \frac{4}{3}\log^3(2)\zeta(2) + 4\log^2(2)\zeta(3) + \log(2)\zeta(4) - \frac{123}{8}\zeta(5)$$

$$+ 16\log(2)\,\text{Li}_4\left(\frac{1}{2}\right) + 16\,\text{Li}_5\left(\frac{1}{2}\right),$$

and the solution to the point (i) of the problem is finalized.

Next, to attack the second point of the problem, we first assume the extension of n to real numbers in (6.453),

$$\int_0^{\pi/2} x^2\cos^{2n}(x)dx = \frac{\pi}{4}\frac{1}{2^{2n}}\frac{\Gamma(2n+1)}{\Gamma^2(n+1)}\psi^{(1)}(n+1), \tag{6.460}$$

where I have also considered the passing to Gamma function and Trigamma function before differentiating.

Upon differentiating once both sides of (6.460) with respect to n, and rearranging, we immediately arrive at

$$\int_0^{\pi/2} x^2\log(\cos(x))\cos^{2n}(x)dx$$

$$= \frac{\pi}{4}\frac{1}{2^{2n}}\binom{2n}{n}\left((H_{2n} - H_n - \log(2))\left(\zeta(2) - H_n^{(2)}\right) - \left(\zeta(3) - H_n^{(3)}\right)\right), \tag{6.461}$$

and we immediately see the right-hand side is very similar to the summand of our series! Yes, we'll use it in our calculations!

Multiplying both sides of (6.461) by $-(4/\pi)H_{n-1}/n$ and considering the summation from $n = 1$ to ∞, we obtain that

$$\sum_{n=1}^{\infty} \frac{H_{n-1}}{n2^{2n}} \binom{2n}{n} \left(\left(\zeta(3) - H_n^{(3)} \right) - (H_{2n} - H_n - \log(2)) \left(\zeta(2) - H_n^{(2)} \right) \right)$$

$$= -\frac{4}{\pi} \sum_{n=1}^{\infty} \frac{H_{n-1}}{n} \int_0^{\pi/2} x^2 \log(\cos(x)) \cos^{2n}(x) dx$$

$$= -\frac{4}{\pi} \int_0^{\pi/2} x^2 \log(\cos(x)) \sum_{n=1}^{\infty} \cos^{2n}(x) \frac{H_{n-1}}{n} dx$$

{the *magical* part is coming now when I employ (4.33), Sect. 4.6}

$$= -\frac{8}{\pi} \int_0^{\pi/2} x^2 \log^2(\sin(x)) \log(\cos(x)) dx$$

$$= \frac{17}{8} \zeta(5) - \frac{1}{2} \zeta(2)\zeta(3) + 2\log^3(2)\zeta(2) - \log^2(2)\zeta(3) - \frac{9}{4} \log(2)\zeta(4),$$

and the solution to the point (ii) of the problem is finalized. Note that happily the last integral appears in (1.245), Sect. 1.55! Such a *magical* reduction to a manageable integral! Observe that above I exploited that when multiplying both sides of (4.33), Sect. 4.6, by x and then reindexing, we obtain that $\frac{1}{2} \log^2(1-x) =$

$$\sum_{n=1}^{\infty} x^{n+1} \frac{H_n}{n+1} = \sum_{n=2}^{\infty} x^n \frac{H_{n-1}}{n} = \sum_{n=1}^{\infty} x^n \frac{H_{n-1}}{n}, \text{ and the last equality is true since}$$

$$H_0 = \sum_{k=1}^{0} \frac{1}{k} = 0.$$

You might agree that all auxiliary results presented above and the ways they have been used for deriving the main results form together a spectacular firework!

6.60 Presenting an Appealing Triple Infinite Series Together with an Esoteric-Looking Functional Equation

Solution The ending section of the present book has just showed up, and I want to announce that it has much in common with the last section in *(Almost) Impossible Integrals, Sums, and Series*! If you agree with me, the most intriguing, counterintuitive point of the problem seems to be represented by the functional equation! *Really, how does it happen that the series satisfies such a beautiful relation?* This is one of the possible thoughts after seeing it!

The points of this section could have been included in my first book, but at that moment, I felt I needed more time to make investigations around these results.

So, let's start first with a key observation, involving Gamma and Beta functions,

$$\frac{\Gamma(a_1)\Gamma(a_2)\Gamma(a_3)}{\Gamma(a_1 + a_2 + a_3)} = \frac{\Gamma(a_1)\Gamma(a_2)}{\Gamma(a_1 + a_2)} \cdot \frac{\Gamma(a_1 + a_2)\Gamma(a_3)}{\Gamma(a_1 + a_2 + a_3)} = B(a_1, a_2) \cdot B(a_1 + a_2, a_3),$$

(6.462)

which is the particular case $n = 3$ of the more general case in [76, Chapter 6, p.532]

So, in view of (6.462), we observe we can write

$$\frac{i!j!k!}{(i + j + k + 2)!} = \frac{\Gamma(i + 1)\Gamma(j + 1)\Gamma(k + 1)}{\Gamma(i + j + k + 3)} = B(i+1, j+1) \cdot B(i+j+2, k+1).$$

(6.463)

Returning to the series at the point (i), and using (6.463), we have

$$\Lambda(x) = \sum_{i=0}^{\infty} \left(\sum_{j=0}^{\infty} \left(\sum_{k=0}^{\infty} x^{i+j+k} \frac{i!j!k!}{(i + j + k + 2)!} \right) \right)$$

$$= \sum_{i=0}^{\infty} \left(\sum_{j=0}^{\infty} \left(\sum_{k=0}^{\infty} x^{i+j+k} B(i + 1, j + 1) \cdot B(i + j + 2, k + 1) \right) \right)$$

$$= \sum_{i=0}^{\infty} \left(\sum_{j=0}^{\infty} \left(\sum_{k=0}^{\infty} x^{i+j+k} \int_0^1 t^i (1 - t)^j dt \cdot \int_0^1 u^{i+j+1}(1 - u)^k du \right) \right)$$

{reverse the order of integration and summation}

$$= \int_0^1 \left(\int_0^1 \sum_{i=0}^{\infty} \left(\sum_{j=0}^{\infty} \left(\sum_{k=0}^{\infty} u(xtu)^i (xu(1 - t))^j (x(1 - u))^k \right) \right) dt \right) du$$

$$= \int_0^1 \left(\int_0^1 \frac{u}{(1 - x + xu)(1 - xut)(1 - xu + xut)} dt \right) du$$

$$= \int_0^1 \left(\int_0^1 \frac{u}{(2 - xu)(1 - x + xu)(1 - xut)} dt \right) du$$

$$+ \int_0^1 \left(\int_0^1 \frac{u}{(2 - xu)(1 - x + xu)(1 - xu + xut)} dt \right) du$$

$$= -2\frac{1}{x} \int_0^1 \frac{\log(1 - xu)}{(2 - xu)(1 - x + xu)} du$$

$$= 2\frac{1}{x(x - 3)} \int_0^1 \frac{\log(1 - xu)}{2 - xu} du + 2\frac{1}{x(x - 3)} \int_0^1 \frac{\log(1 - xu)}{1 - x + xu} du. \qquad (6.464)$$

At this point, I could use the result in (3.23), Sect. 3.6, but this time, I'll choose a different route. So, considering the first remaining integral in (6.464), and integrating by parts, we get

$$\int_0^1 \frac{\log(1 - xu)}{2 - xu} du = -\frac{1}{x} \int_0^1 \log(1 - xu)(\log(2 - xu))' du$$

$$= -\frac{\log(1 - xu)\log(2 - xu)}{x}\Big|_{u=0}^{u=1} - \int_0^1 \frac{\log(1 + (1 - xu))}{1 - ux} du = -\frac{\log(1 - x)\log(2 - x)}{x}$$

$$- \frac{1}{x} \operatorname{Li}_2(-1 + xu)\Big|_{u=0}^{u=1} = -\frac{1}{2}\zeta(2)\frac{1}{x} - \frac{\log(1 - x)\log(2 - x)}{x} - \frac{\operatorname{Li}_2(-(1 - x))}{x}.$$
$$(6.465)$$

Further, in order to evaluate the second integral in (6.464), we first cleverly integrate by parts, and then we write

$$\int_0^1 \frac{\log(1 - xu)}{1 - x + xu} du = \int_0^1 \frac{\log(1 - xu)}{x} \left(\log\left(\frac{1 - x + xu}{2 - x}\right)\right)' du$$

$$= \frac{\log(1 - xu)}{x} \log\left(\frac{1 - x + xu}{2 - x}\right)\Big|_{u=0}^{u=1} - \frac{1}{x} \int_0^1 \log\left(1 - \frac{1 - xu}{2 - x}\right) \frac{2 - x}{1 - xu} \left(\frac{1 - xu}{2 - x}\right)' du$$

$$= -\log(1 - x)\log(2 - x) + \frac{1}{x}\operatorname{Li}_2\left(\frac{1 - xu}{2 - x}\right)\Big|_{u=0}^{u=1}$$

$$= -\frac{\log(1 - x)\log(2 - x)}{x} - \frac{1}{x}\operatorname{Li}_2\left(\frac{1}{2 - x}\right) + \frac{1}{x}\operatorname{Li}_2\left(\frac{1 - x}{2 - x}\right). \qquad (6.466)$$

The form in (6.466) can be simplified and reduced to a form involving a single dilogarithmic value. In order to do that, we might want to recall the identity

$$\operatorname{Li}_2\left(\frac{1}{1 + x}\right) - \operatorname{Li}_2(-x) = \zeta(2) - \frac{1}{2}\log(1 + x)\log\left(\frac{1 + x}{x^2}\right), \quad x > 0, \qquad (6.467)$$

which is immediately proved by differentiating and integrating back the left-hand side and deriving the integration constant.

Based on (6.467), where we replace x by $1 - x$, we get

$$\text{Li}_2\left(\frac{1}{2-x}\right) = \zeta(2) - \frac{1}{2}\log(2-x)\log\left(\frac{2-x}{(1-x)^2}\right) + \text{Li}_2(-(1-x)), \ x < 1.$$

$$(6.468)$$

Then, recalling the dilogarithm function reflection formula (see [21, Chapter 1, p.5], [70, Chapter 2, p.107]), $\text{Li}_2(x) + \text{Li}_2(1-x) = \zeta(2) - \log(x)\log(1-x)$, and upon replacing x by $(1-x)/(2-x)$, and rearranging, we immediately obtain that

$$\text{Li}_2\left(\frac{1-x}{2-x}\right) = \zeta(2) + \log\left(\frac{1-x}{2-x}\right)\log(2-x) - \text{Li}_2\left(\frac{1}{2-x}\right)$$

{exploit the result in (6.468)}

$$= -\frac{1}{2}\log^2(2-x) - \text{Li}_2(-(1-x)).\tag{6.469}$$

Returning with the results from (6.468) and (6.469) in (6.466), we get

$$\int_0^1 \frac{\log(1-xu)}{1-x+xu}du = -\zeta(2)\frac{1}{x} - 2\frac{\log(1-x)\log(2-x)}{x} - 2\frac{\text{Li}_2(-(1-x))}{x}.$$

$$(6.470)$$

Collecting the results from (6.465) and (6.470) in (6.464), we arrive at

$$\Lambda(x) = \sum_{i=0}^{\infty}\sum_{j=0}^{\infty}\sum_{k=0}^{\infty} x^{i+j+k}\frac{i!j!k!}{(i+j+k+2)!}$$

$$= 6\frac{1}{(3-x)x^2}\left(\frac{1}{2}\zeta(2) + \log(1-x)\log(2-x) + \text{Li}_2(-(1-x))\right),$$

and the solution to the point (i) is complete. If at the sight of the problem things looked scary, now we are pleasantly surprised by the simple and beautiful form of the triple series! In the problem statement I assumed that $|x| < 1$, but we may go further and also extract the values of the series for $x = \pm 1$. So, by letting $x \to 1^-$ in the result above, we get that $\Lambda(1) = \sum_{i=0}^{\infty}\sum_{j=0}^{\infty}\sum_{k=0}^{\infty} x^{i+j+k}\frac{i!j!k!}{(i+j+k+2)!} = \frac{3}{2}\zeta(2)$, and by setting $x = -1$, we obtain the following closed form with a polylogarithmic value, $\Lambda(-1) = \sum_{i=0}^{\infty}\sum_{j=0}^{\infty}\sum_{k=0}^{\infty}(-1)^{i+j+k}\frac{i!j!k!}{(i+j+k+2)!} = \frac{3}{2}\log(2)\log(3) +$

$\frac{3}{4}\zeta(2) + \frac{3}{2}\text{Li}_2(-2) = \frac{3}{2}\log(2)\log(3) - \frac{3}{4}\log^2(2) - \frac{3}{4}\zeta(2) - \frac{3}{2}\text{Li}_2\left(-\frac{1}{2}\right)$, and the last form is obtained with the dilogarithm identity used at the next point.

Now, proving the functional equation at the second point is straightforward if we use the main result above and combine it with the dilogarithm function identity, $\text{Li}_2(-x) + \text{Li}_2(-1/x) = -\zeta(2) - 1/2\log^2(x)$, $x > 0$, leading immediately to

$$(3+\theta)\theta^2\Lambda(-\theta) + \frac{(3+2\theta)\theta^2}{(1+\theta)^3}\Lambda\left(\frac{\theta}{\theta+1}\right) = 3\log^2(1+\theta),$$

where if we consider the initial condition, $|x| < 1$, we have that $-1/2 < \theta < 1$, and the solution to the point (ii) is complete.

From one perspective, we have reached the end of our second odyssey, but from another perspective, for the curious readers eager to explore further the power of the techniques and results presented in the book, this odyssey might only mean the beginning, a long road with many other challenges, and therefore let me wish all of you much success! I strongly hope my second title has brought you more enjoyable problems and solutions and has offered useful ideas that could also be tweaked for solving more appealing problems you might be interested in.

If you have had a lot of fun, a great time with my mathematics, then I feel my mission is fully accomplished, and I'm very happy for you, dear reader!

Adieu!

Cornel Ioan Vălean

References

1. Ahmed, Z.: Ahmed's Integral: the maiden solution (2014). https://arxiv.org/abs/1411.5169, v2
2. Apostol, T.M.: Introduction to Analytic Number Theory. Springer, New York (1976)
3. Arakawa, T., Ibukiyama, T., Kaneko, M.: Bernoulli Numbers and Zeta Functions. Springer Monographs in Mathematics. Springer, Tokyo (2014)
4. Berndt, B.: Ramanujan's Notebooks, Part I. Springer, New York (1985)
5. Berndt, B.C., Straub, A.: Certain Integrals Arising from Ramanujan's Notebooks. https://arxiv.org/pdf/1509.00886.pdf (2015)
6. Boros, G., Moll, V.H.: Irresistible Integrals, Symbolics, Analysis and Experiments in the Evaluation of Integrals. Cambridge University Press, Cambridge (2004)
7. Boros, G., Moll, V.H.: Sums of arctangents and some formulas of Ramanujan. Scientia **11**, 13–24 (2005)
8. Boyadzhiev, K.N.: Power series with skew-harmonic numbers, dilogarithms, and double integrals. Tatra Mt. Math. Publ. **56**, 93–108 (2013)
9. Cantarini, M., D'Aurizio, J.: On the interplay between hypergeometric series, Fourier-Legendre expansions and Eulers sums. https://arxiv.org/abs/1806.08411
10. Choudary, A.D.R., Niculescu, C.P.: Real Analysis on Intervals. Springer, New Delhi (2014)
11. Duren, P.L.: Invitation to Classical Analysis. American Mathematical Society, Providence (2012)
12. Dutta, R.: Evaluation of a cubic Euler sum. J. Class. Anal. **9**(2), 151–159 (2016)
13. Fichtenholz, G.M.: Differential und Integralrechnung. Band 2, zweite Auflage. VEB Deutscher Verlag der Wissenschaften, Berlin (1966)
14. Flajolet, P., Salvy, B.: Euler sums and contour integral representations. Exp. Math. **7**, 15–35 (1998)
15. Furdui, O.: Limits, Series and Fractional Part Integrals. Problems in Mathematical Analysis. Springer, New York (2013)
16. Gleason, A.M., Greenwood, R.E., Kelly, L.M.: The William Lowell Putnam Mathematical Competition. Problems and Solutions: 1938–1964. Mathematical Association of America, Washington (1980)
17. Gradshteyn, I.S., Ryzhik, I.M.: In: Zwillinger, D., Moll, V. (eds.) Table of Integrals, Series, and Products, 8th edn. Academic, New York (2015)
18. Graham, R., Knuth, D., Patashnik, O.: Concrete Mathematics, 2nd edn. Addison Wesley, Boston (1994)
19. Johnson, W.P.: Down with Weierstrass! Am. Math. Mon. **127**(7), 649–653 (2020). https://tandfonline.com/doi/abs/10.1080/00029890.2020.1763122

© The Author(s), under exclusive license to Springer Nature Switzerland AG 2023   813
C. I. Vălean, *More (Almost) Impossible Integrals, Sums, and Series*, Problem
Books in Mathematics, https://doi.org/10.1007/978-3-031-21262-8

20. La Gaceta de la RSME (Spain): A solution to the problem 398. http://gaceta.rsme.es/abrir.php?id=1635 (2021)
21. Lewin, L.: Polylogarithms and Associated Functions. North-Hollan, New York (1981)
22. Mathematics Stack Exchange: https://math.stackexchange.com/q/3339892
23. Mathematics Stack Exchange: https://math.stackexchange.com/q/3425231
24. Mathematics Stack Exchange: https://math.stackexchange.com/q/3426424
25. Mathematics Stack Exchange: https://math.stackexchange.com/q/1640940
26. Mathematics Stack Exchange: https://math.stackexchange.com/q/3302793
27. Mathematics Stack Exchange: https://math.stackexchange.com/q/3803762
28. Mathematics Stack Exchange: https://math.stackexchange.com/q/542741
29. Mathematics Stack Exchange: https://math.stackexchange.com/q/3006106
30. Mathematics Stack Exchange: https://math.stackexchange.com/q/3259984
31. Mathematics Stack Exchange: https://math.stackexchange.com/q/816253
32. Mathematics Stack Exchange: https://math.stackexchange.com/q/966471
33. Mathematics Stack Exchange: https://math.stackexchange.com/q/128515
34. Mathematics Stack Exchange: https://math.stackexchange.com/q/3905908
35. Mathematics Stack Exchange: https://math.stackexchange.com/q/472994
36. Mathematics Stack Exchange: https://math.stackexchange.com/q/407420
37. Mathematics Stack Exchange: https://math.stackexchange.com/q/4384783
38. Mathematics Stack Exchange: https://math.stackexchange.com/q/3325928
39. Mathematics Stack Exchange: https://math.stackexchange.com/q/979460
40. Mathematics Stack Exchange: https://math.stackexchange.com/q/4188260
41. Mathematics Stack Exchange: https://math.stackexchange.com/q/4310602
42. Mathematics Stack Exchange: https://math.stackexchange.com/q/3552194
43. Mathematics Stack Exchange: https://math.stackexchange.com/q/1842284
44. Mathematics Stack Exchange: https://math.stackexchange.com/q/3522967
45. Mathematics Stack Exchange: https://math.stackexchange.com/q/936418
46. Mathematics Stack Exchange: https://math.stackexchange.com/q/3236584
47. Mathematics Stack Exchange: https://math.stackexchange.com/q/3350339
48. Mathematics Stack Exchange: https://math.stackexchange.com/q/3372879
49. Mathematics Stack Exchange: https://math.stackexchange.com/q/3353705
50. Mathematics Stack Exchange: https://math.stackexchange.com/q/908108
51. Mathematics Stack Exchange: https://math.stackexchange.com/q/805298
52. Mathematics Stack Exchange: https://math.stackexchange.com/q/1289593
53. Mathematics Stack Exchange: https://math.stackexchange.com/q/771277
54. Mathematics Stack Exchange: https://math.stackexchange.com/q/2394836
55. Mathematics Stack Exchange: https://math.stackexchange.com/q/3528838
56. Mathematics Stack Exchange: https://math.stackexchange.com/q/3261717
57. Mathematics Stack Exchange: https://math.stackexchange.com/q/2591269
58. Mathematics Stack Exchange: https://math.stackexchange.com/q/4374105
59. MathProblems Journal: Problems and Solutions. Problem 157. **6**(2) (2016). www.mathproblems-ks.org
60. Miller, P.D.: Applied Asymptotic Analysis. Graduate Studies in Mathematics, vol. 75. American Mathematical Society, Providence (2006)
61. Mladenović, P.: Combinatorics. A Problem-Based Approach. Springer, Cham (2019)
62. Moll, V: Numbers and Functions: From a Classical-Experimental Mathematician's Point of View. American Mathematical Society, Providence (2012)
63. Moll, V.: Special Integrals of Gradshteyn and Ryzhik. The Proofs, vol. I. CRC Press, Taylor and Francis Group/Chapman and Hall, Boca Raton/London (2014)
64. Moll, V.: Special Integrals of Gradshteyn and Ryzhik. The Proofs, vol. II. CRC Press, Taylor and Francis Group/Chapman and Hall, Boca Raton/London (2015)
65. Nahin, P.J.: Inside Interesting Integrals, 1st edn. Springer, New York (2014)
66. Nahin, P.J.: Inside Interesting Integrals, 2nd edn. Springer, New York (2020)

67. Olaikhan, A.S.: An Introduction to the Harmonic Series and Logarithmic Integrals: For High School Students Up to Researchers, 1st edn. Independent Publishing, p. 338 (2021). Paperback version

68. Romanian Mathematical Magazine, vol. 22 (2018). The Problem U.13. https://www.ssmrmh.ro/2019/01/24/old-rmm-22/

69. Sprugnoli, R.: Sums of reciprocals of the central binomial coefficients. Integers **6**, A27 (2006)

70. Srivastava, H.M., Choi, J.: Series Associated with the Zeta and Related Functions. Springer (originally published by Kluwer), Dordrecht (2001)

71. Tolstov, G.P.: Fourier Series. Dover Publications, New York (1976)

72. Vălean, C.I.: A new proof for a classical quadratic harmonic series. J. Class. Anal. **8**(2), 155–161 (2016)

73. Vălean, C.I.: A master theorem of series and an evaluation of a cubic harmonic series. J. Class. Anal. **10**(2), 97–107 (2017)

74. Vălean, C. I.: Problem 12054, problems and solutions. Am. Math. Mon. **125**(6), 562–570 (2018). https://tandfonline.com/doi/abs/10.1080/00029890.2018.1460990

75. Vălean, C.I.: Two advanced harmonic series of weight 5 involving skew-harmonic numbers. https://www.researchgate.net/publication/337937502 (2019)

76. Vălean, C.I.: (Almost) Impossible Integrals, Sums, and Series, 1st edn. Springer, Cham (2019)

77. Vălean, C.I.: A new powerful strategy of calculating a class of alternating Euler sums. https://www.researchgate.net/publication/333999069 (2019)

78. Vălean, C.I.: A note presenting the generalization of a special logarithmic integral. https://www.researchgate.net/publication/335149209 (2019)

79. Vălean, C.I.: A simple idea to calculate a class of polylogarithmic integrals by using the Cauchy product of squared Polylogarithm function. https://www.researchgate.net/publication/337739055 (2019)

80. Vălean, C.I.: A simple strategy of calculating two alternating harmonic series generalizations. https://www.researchgate.net/publication/333339284 (2019)

81. Vălean, C.I.: A special way of extracting the real part of the Trilogarithm, $\mathrm{Li}_3\left(\dfrac{1 \pm i}{2}\right)$. https://www.researchgate.net/publication/337868999 (2019)

82. Vălean, C.I.: The calculation of a harmonic series with a weight 5 structure, involving the product of harmonic numbers, $H_n H_{2n}^{(2)}$. https://www.researchgate.net/publication/336378340 (2019)

83. Vălean, C.I.: The derivation of eighteen special challenging logarithmic integrals. https://www.researchgate.net/publication/334598773 (2019)

84. Vălean, C.I.: The evaluation of a special harmonic series with a weight 5 structure, involving harmonic numbers of the type H_{2n}. https://www.researchgate.net/publication/336148969 (2019)

85. Vălean, C.I.: A new perspective on the evaluation of the logarithmic integral, $\int_0^1 \frac{\log(x)\log^3(1+x)}{x}\,\mathrm{d}x$. https://www.researchgate.net/publication/339024876 (2020)

86. Vălean, C.I.: A note on two elementary logarithmic integrals, $\int_0^1 x^{2n-1}\log(1+x)\mathrm{d}x$ and $\int_0^1 x^{2n}\log(1+x)\mathrm{d}x$. https://www.researchgate.net/publication/342703290 (2020)

87. Vălean, C.I.: A symmetry-related treatment of two fascinating sums of integrals. https://www.researchgate.net/publication/340953717 (2020)

88. Vălean, C.I.: An easy approach to two classical Euler sums, $\sum_{n=1}^{\infty}(-1)^{n-1}\frac{H_n}{n^3}$ and $\sum_{n=1}^{\infty}(-1)^{n-1}\frac{H_n}{n^4}$. https://www.researchgate.net/publication/339290253 (2020)

89. Vălean, C.I.: A short presentation of a parameterized logarithmic integral with a Cauchy principal value meaning. https://www.researchgate.net/publication/348910585 (2021)

90. Vălean, C.I.: Two identities with special dilogarithmic values. https://www.researchgate.net/publication/348522432 (2021)

91. Vălean, C.I., Furdui, O.: Reviving the quadratic series of Au-Yeung. J. Class. Anal. **6**(2), 113–118 (2015)

92. Vălean, C.I., Levy, M.: Euler sum involving tail. https://www.researchgate.net/publication/339552441 (2020)

93. Weisstein, E.W.: Dirichlet Eta Function. http://mathworld.wolfram.com/DirichletEtaFunction.html
94. Weisstein, E.W.: Double Factorial. http://mathworld.wolfram.com/DoubleFactorial.html
95. Weisstein, E.W.: Gaussian Integral. http://mathworld.wolfram.com/GaussianIntegral.html
96. Weisstein, E.W.: Legendre Duplication Formula. http://mathworld.wolfram.com/LegendreDuplicationFormula.html
97. Weisstein, E. W.: Stirling's Approximation. http://mathworld.wolfram.com/StirlingsApproximation.html
98. Weisstein, E. W.: Chebyshev's sum inequality. http://mathworld.wolfram.com/ChebyshevSumInequality.html
99. Wilf, H.S.: generatingfunctionology, 3rd edn. A K Peters Ltd., Wellesley (2006)
100. Zhao, M.H.: On logarithmic integrals, harmonic sums and variations. https://arxiv.org/abs/1911.12155, v13 (2020)

Printed in the United States
by Baker & Taylor Publisher Services

Printed in the United States
by Baker & Taylor Publisher Services